USEFUL CONSTANTS

speed of light	$c = 2.998 \times 10^{8}$ m/s
Planck's constant	$h = 6.626 \times 10^{-27}$ erg·s
Wien law constant	$\lambda_{max}T = 0.29$ cm/deg
angstrom unit	$\text{Å} = 10^{-1}$ nm $= 10^{-4}$ μm $= 10^{-10}$ m
Celsius degree	°C = 9/5 °F
ton TNT (energy)	$= 4.2 \times 10^{16}$ erg
astronomical unit	AU $= 1.496 \times 10^{8}$ km
parsec	pc = 206265 AU = 3.262 LY
light year	LY $= 6.324 \times 10^{4}$ AU = 0.3065 pc
mass of Earth	$M_{\oplus} = 5.977 \times 10^{21}$ ton
radius of Earth	$R_{\oplus} = 6378$ km
mass of sun	$M_{\odot} = 1.989 \times 10^{27}$ ton
radius of sun	$R_{\odot} = 6.96 \times 10^{5}$ km
luminosity of sun	$L_{\odot} = 3.83 \times 10^{33}$ erg/s

C2 all
C3 all
C4 omit 4.2 d p56
omit simplified print

C5 include 5.1 b, i
omit 5.1 e
5.3 b

EXPLORATION OF THE UNIVERSE

fifth edition

George O. Abell
University of California, Los Angeles

David Morrison
University of Hawaii

Sidney C. Wolff
Kitt Peak National Observatory

Saunders Golden Sunburst Series
SAUNDERS COLLEGE PUBLISHING
Philadelphia New York Chicago
San Francisco Montreal Toronto
London Sydney Tokyo Mexico City
Rio de Janeiro Madrid

Address orders to:
383 Madison Avenue
New York, NY 10017

Address editorial correspondence to:
210 West Washington Square
Philadelphia, PA 19105

Text Typeface: Plantin
Compositor: The Clarinda Company
Acquisitions Editor: John J. Vondeling
Project Editor: Margaret Mary Kerrigan
Copy Editor: Joanne Fraser
Art Director: Carol C. Bleistine
Text and Cover Designer: Lawrence R. Didona
Layout Artist: Edward A. Butler
Text Artwork: Larry Ward
Production Manager: Tim Frelick
Assistant Production Manager: JoAnn Melody

Cover credit: Image of Halley's comet taken on January 20, 1986 with a CCD camera mounted on a 61-inch (154-cm) telescope in the Catalina mountains near Tucson, Arizona. The false color image was created by assigning color to the various intensity levels in the black and white image. The image was taken through a 35-Å wide filter centered on an emission band of H_2O^+. Most of the structure seen is due to the interaction of the solar wind with the H_2O^+ ions which are being swept into the tail. The observers were Uwe Fink, Al Schultz, Mike DiSanti, and Reiner Fink. (Lunar and Planetary Laboratory, University of Arizona)

Library of Congress Cataloging-in-Publication Data

Abell, George Ogden, 1927–1983
Exploration of the Universe.

(Saunders golden sunburst series)
Bibliography: p.
Includes index.
1. Astronomy. I. Morrison, David, 1940–
II. Wolff, Sidney C. III. Title.
QB45.A16 1987 520 86-26045
ISBN 0-03-005143-6

EXPLORATION OF THE UNIVERSE, FIFTH EDITION 0-03-005143-6

Library of Congress catalog card number 86-26045.

7890 042 98765432

CBS COLLEGE PUBLISHING
Saunders College Publishing
Holt, Rinehart and Winston
The Dryden Press

PREFACE

These are exciting times for science in general and for astronomy in particular. We have sent spacecraft to explore seven of the nine planets and to land on the surfaces of Venus, the Moon, and Mars. Three probes have even plunged through the head of Comet Halley. We have discovered the dying glow of the primeval fireball that began the expansion of the universe and have used theory to probe to within a fraction of a second of this cosmic beginning. We have identified gravitational lenses in space, and we believe we have found black holes. New telescopes on Earth and in space constantly reveal a universe richer, more varied, and more violent than had been suspected by previous generations.

At present, however, much research is in a period of retrenchment, when budget austerity has curtailed expenditures and has severely limited the U.S. programs of planetary exploration and the launching of scientific Earth satellites. Recently there has been more interest in space weapons than space science. Yet the situation is not as bleak as it may appear. The U.S. will soon launch the Hubble Space Telescope, an instrument that will be the first true space observatory. Private funds are providing for the construction of a new telescope in Hawaii that is four times the size of the Palomar 200-inch. And international interest has grown in space science, as shown by the highly successful Soviet planetary exploration program and by a number of new astronomical spacecraft being launched by Europe and Japan.

We are also pleased that public interest in the new frontiers of astronomy continues to grow. A variety of high-quality publications interpret astronomy and other sciences for the layperson, and in the colleges our classes are more crowded than ever. Students and the public alike are becoming more sophisticated and more fervent in their desire to understand as much as they can of what we have learned about the cosmos. People seem to be aware of what astronomy can offer to the human perspective. What they are often not aware of, however, is the method of science—the exacting procedure and rigid rules that have permitted science its steady progress. They look for quick answers, not realizing that there are no final authorities, [illegible] process that leads tow[illegible]better but always inadequate understanding of nature.

That communications gap becomes especially obvious when we note that many people, thirsty for knowledge about new frontiers, have turned to all manner of unreliable sources for their information. Even in this era when science and technology are crucial to our lives, the public is bombarded by misinformation and pseudoscience. We scientists have an obligation to the public to increase our efforts in presenting an honest view of science. We should show not only how the universe is, but also how, by simple rational processes, we can probe its mysteries.

This is what *Exploration of the Universe* is all about. It is, of course, a textbook in astronomy—specifically for a rather comprehensive one-year course in astronomy for the liberal arts student or an intensive one-term course. But from the time George Abell introduced this text in 1964, *Exploration* has aspired to a broader goal than instruction in the discipline of astronomy. His objective, which we also embrace, was to provide an introduction to the rational exploration of nature and a glimpse of the history and character of science and scientific thinking.

George Abell's untimely death in 1983 deprived us of a leading teacher and popularizer of science. His legacy is still with us, however, in the tangible form of his extensive writing as well as in the personal influence he had on tens of thousands of his students and colleagues. Perhaps his most important contribution was *Exploration of the Universe,* the first modern text in astronomy and the standard against which all other texts were judged. This book has served for 20 years as the definitive introduction to astronomy, suitable for the beginning undergraduate but at the same time an indispensable reference for the shelves of graduate students and professional astronomers. In this and subsequent editions of *Exploration,* we intend to maintain the high and uncompromising standards which George Abell set.

We realize that *Exploration* can be a rather large dose for a first-year course for the nonscience student, and especially so for one taking less than a full-year course. Nevertheless, we felt it was important, as in the previous editions of *Exploration,* to try to be as comprehensive as we could—not just for the science student who may find the book a useful reference, but also for the interested lay nonspecialist [illegible] questions often arise amid the answers to old ones.

There are several ways in which portions of this text could be used effectively for a one-semester course in astronomy. Chapters 1–8, 12–21, and 38 constitute a one-semester introduction to the solar system. Chapters 1, 8–11, and 22–37 could be used for an introductory course on stars, the Milky Way Galaxy, and extragalactic systems. For a one-semester introduction to modern astronomy, with emphasis on the evolution of astronomical objects, one could omit much of the historical material and some of the details of how observations are made and make use of Chapters 1, 4, 8, 12–21, 22.1–22.2, 23–24, 26, 28–30, 32, 33.6, 34–35, and 37. Some references to other chapters will be required, but are explicitly noted in the text. If this last approach is adopted, all amplifying and mathematical material in special type (set off by double horizontal lines) should be omitted. Individual preferences about the specific material to be included will vary, of course. We urge, however, against too thorough a coverage of the earlier material at the expense of not reaching the grand messages of the final chapters.

This fifth edition of *Exploration of the Universe* is a complete revision. Much out-of-date material has been deleted, and much new material has been added. Chapters 12 through 21 on the planetary system have been completely rewritten to reflect current thinking about the planets, which are truly worlds with their own geological and chemical histories. In addition to adopting the viewpoint of comparative planetology, we have incorporated recent results from the Venera 15 and 16 radar maps of Venus, the Venera 17 and 18 balloon experiments in the atmosphere of Venus, the Voyager 2 encounter with Uranus, and the VEGA and Giotto probes to Comet Halley. Reorganization has permitted separate chapters on Venus and Mars, and also the separation of the Jovian planets from discussion of their satellite and ring systems.

In the Fifth Edition, the material on star formation and evolution has been substantially reorganized and expanded. Star clusters (Chapter 31) are now found following stellar evolution, so that their properties can be interpreted in evolutionary terms. The three chapters (35 through 37) on galaxies and cosmology have been expanded and rewritten to reflect our increasing knowledge of this aspect of astronomy, including the role of dark matter, the discovery of voids in the spatial distribution of galaxies, and the new concept of the inflationary universe.

Throughout the text the results from new instruments on the ground and in space are incorporated. A special effort has been made to integrate the data from different spectral regions rather than to emphasize the techniques by which the observations were obtained. We thus concentrate our attention on the objects themselves and on the physical processes taking place, rather than on the source of the data. However, the new observational tools of the astronomer are described in more detail than previously in Chapters 9 and 10, which are also largely new in this edition. Finally, the material on extraterrestrial life has been moved to the last chapter (38) and combined with a discussion of recent empirical evidence for the existence of planets around other stars.

As with previous editions, we have avoided mathematics beyond the simplest algebra. Sections containing mathematics or other slightly more technical material are printed in small type and can be omitted without interrupting the sequence of topics or the logic of the presentation. Also, the more challenging exercises are marked with a star. As before, we try to stress that astronomy is a very human endeavor and have related it to those men and women who created our science. Inevitably we have had to be selective about the individuals mentioned by name, and we ask our colleagues' understanding if the choices often seem arbitrary. Finally, because of our concern over the widespread confusion between science and pseudoscience, we have briefly addressed astrology and other popular fads.

Many people have reviewed parts of this book and have offered valuable suggestions. For this edition, these include L. Blitz (U. Maryland), D. P. Cruikshank (U. Hawaii), M. J. Drake (U. Arizona), Harland Epps (U. of California, Los Angeles), L. Esposito (U. Colorado), A. Fraknoi (Astronomical Society of the Pacific), J. Gallagher (Lowell Observatory), M. Jura (UCLA), D. M. Hunten (U. Arizona), J. Lewis (U. Arizona), J. Liebert (U. Arizona), T. Owen (SUNY), E. Rosenthal (U. Hawaii), T. Simon (U. Hawaii), M. V. Sykes (U. Arizona), D. J. Tholen (U. Hawaii), and J. Veverka (Cornell U.) For their special help in providing illustrations we thank M. J. Carr (USGS), D. P. Cruikshank (U. Hawaii), U. Fink (U. Arizona), M. Gentry (NASA/JSC), H. Hammel (U. Hawaii), J. Head (Brown U.), K. Kiel (U. New Mexico), M. Longair (Royal Observatory Edinburgh), A. McEwen (Arizona State U.), H. Masursky (USGS), and J. Van der Woude (NASA/JPL). David Morrison wishes to thank Director Eugene Levy and his colleagues at the Lunar and Planetary Laboratory of the University of Arizona for their hospitality during the time when most of the work in the revision was carried out, and Sidney C. Wolff thanks the staff of Kitt Peak National Observatory for their help in bringing us up to date in various areas of astronomy. Obviously none of the reviewers has seen the final version of the manuscript, and responsibility for errors and omissions falls entirely on us.

We also want to express appreciation for the great assistance we have received from our editors at Saunders, especially John Vondeling and Margaret Mary Kerrigan. It has been a pleasure to undertake this revision of a classic text, and we hope you, the readers, enjoy the product of these efforts.

David Morrison
Sidney C. Wolff
Honolulu and Tucson
August 1986

CONTENTS

EXPLORATION OF THE UNIVERSE

fifth edition

John Muir (1828–1914), Scots-American naturalist, was largely responsible for the establishment of Yosemite and other national parks. He wrote, "When we try to pick out anything by itself, we find it hitched to everything else in the universe." (Sierra Club)

1 THE UNITY OF THE UNIVERSE

The Ancients believed in a sort of unity between the heavens and the Earth. To be sure, in classical Greece the Earth was thought to be composed of base stuff—the four "elements," earth, water, air, and fire—and the heavens of crystalline material, but the heavens were the realms of the gods; the planets *were* gods in some early cultures. And the gods, presumably, controlled or influenced human affairs. Earthly events seemed chaotic and unpredictable, but the Ancients recognized a regularity in the motions in the heavens and quite naturally hoped that by understanding the motions of their planet gods, they would better understand the individual lots of men and women. They thus sought a unity between the Earth and heavens through the primitive religion of astrology. Ironically, today, in our 20th-century enlightenment, a large fraction of all people still believe in that ancient religion.

But there *is* a real unity, and one far grander and more beautiful than our ancestors could possibly have imagined. That real unity is in the basic structure of matter everywhere in the universe, and in the laws of nature—the rules that govern how everything works.

1.1 THE UNITY OF STRUCTURE

We have learned that all matter is made of the same stuff—the matter of the Earth and Moon (by direct analyses of their rocks), of the other planets (through analysis by space flybys and landers) (Figure 1.1), and of the stars and even the remotest galaxies (from studying their spectra). This stuff is not the elements earth, water, air, and fire of antiquity, but approximately a hundred different kinds of atoms that make up the hundred or so naturally occurring elements and, in various combinations, the molecules of the billions of kinds of chemical compounds.

Each atom has a *nucleus* with a positive electrical charge, and surrounding it negatively charged *electrons,* carrying a combined charge equal, but opposite in sign, to that of the nucleus. These atoms are held together, and different atoms are bound in molecules, by the *electromagnetic* forces that act between electrically charged particles. The electromagnetic force is the second strongest of the four forces known in nature. In its binding of molecules together in solids, that force accounts for the rigidity of steel!

The sizes of atoms are exceedingly tiny, typically about one hundred-millionth of a centimeter in radius. Yet atoms are almost empty space, far more so than is the solar system. The Earth's distance from the sun is about a hundred times the sun's diameter, but on the average the electron in a hydrogen atom is distant from the atomic nucleus by about 100,000 times the size of that nucleus. By volume, atoms are emptier than the inner part of the solar system by a thousand million times.

But even the tiny nucleus is not a single entity; it is composed of still smaller particles called *protons* and *neutrons.* They are similar in size, and each has a mass of almost 2000 times that of an electron. But

Figure 1.1 Surface of Mars photographed by the Viking 1 lander in 1976. (NASA/JPL)

while the neutron is, as its name implies, electrically neutral, the proton carries a positive charge equal in magnitude to the negative charge carried by the electron. The number of protons in the nucleus of an atom (equal to the number of external electrons) determines what kind of atom it is and what kind of element that kind of atom makes up. An atom of hydrogen, for example, has a single proton in its nucleus, usually no neutrons (that is, its nucleus is a proton), and one external electron. The helium atom has a nucleus with 2 protons and usually 2 neutrons. Lithium has 3 protons, oxygen has 8, and uranium has 92. Except for hydrogen, the number of neutrons in a nucleus is usually roughly comparable to the number of protons.

The positively charged protons in an atomic nucleus all repel each other (similar electrical charges repel, and opposite charges attract). Then why doesn't the nucleus fly apart? Because its particles are bound together by the *strong nuclear force;* it is about 100 times as strong as the electromagnetic force of repulsion between two protons. The strong force is the strongest of the forces of nature.

Neutrons themselves are stable in an atomic nucleus, but outside the nucleus after about 11 minutes, on the average, a neutron spontaneously breaks up into a proton, an electron, and another particle called an *antineutrino* (more about these later). The force that is involved in the decay of the neutron is the *weak force;* this third of the known forces is about a thousand times weaker than the electromagnetic force.

The deeply significant point is that everything, everywhere, is basically the same. So far as we know, from the time when the very first stable matter formed in the universe, it was made up of the same things: mainly protons, electrons, and neutrons. All matter since then has been subjected to the same kinds of forces and has obeyed the same laws.

1.2 THE UNITY OF BEHAVIOR

The seeds of modern science began to sprout following the Reformation. The 17th-century German Johannes Kepler discovered for the first time certain simple mathematical rules that describe accurately the motions of the planets (for example, that they move in precise elliptical paths about the sun). His contemporary, the Italian Galileo Galilei, discovered some other precise rules that describe the behavior of bodies on Earth (for example, that all freely falling bodies near the Earth's surface pick up speed—*accelerate*—at the same rate) (Figure 1.2). Later in the same century, Isaac Newton showed that Kepler's celestial rules and Galileo's terrestrial ones are united by the same underlying laws. Newton had the insight to recognize that the force that makes planets fall in ellipses about the sun and the force that makes apples fall with uniform acceleration near the Earth's surface are different manifestations of the same thing: gravitation.

Gravitation is the fourth and weakest force of nature, yet it was the first to be discovered. Gravitation is an incredibly weak force; the gravitational attraction between the proton and electron in a hydrogen atom is weaker than the electromagnetic attraction between their opposite (plus and minus) charges by 10^{39} (one followed by 39 zeros) times! Yet Newton's discovery of gravitation (Figure 1.3) preceded by nearly two centuries a comparable understanding of any of the other forces. Why?

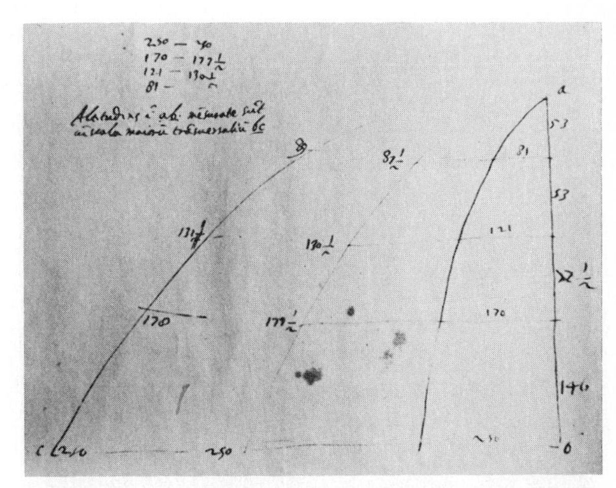

Figure 1.2 A reproduction of a page of Galileo's notes concerning his investigation of the paths of falling bodies.

Nuclear forces act only over the very tiny dimensions of the atomic nucleus; the electromagnetic force acts only between electrically charged particles, and most of bulk matter is electrically neutral. Nature seems to have provided equal quantities of positive and negative charge in the universe, and the very electromagnetic force that acts between charges has assured that plus and minus charges are distributed more or less uniformly. Thus over large distances matter appears neutral, and only gravitational forces are important. Very locally, over the dimensions of atoms and molecules, the positive and negative charges of nearby protons and elec-

Figure 1.3 Newton's home, Woolsthorpe Manor, where he claimed to have conceived his ideas on gravitation. The legend is that he was inspired by the fall of an apple from the tree in the foreground. (G. O. Abell)

Figure 1.4 Jupiter and its four largest satellites, in a composite picture made of Voyager images. (NASA/JPL)

trons *do not* completely cancel, so they pull on each other and bind atoms and molecules together and to adjacent molecules. But bulk matter is electrically neutral, so it exerts no net electrical attraction or repulsion on its surroundings.

On the other hand, we are all aware of the gravitational force, despite the great weakness of its attraction between ordinary masses, because we are very close to a body of astronomical mass—the Earth. Gravitation depends only on the total amount (mass) of matter, not on charge, so if enough mass is collected together, as in the Earth, its gravitational pull can be quite significant, as everyone knows who has tumbled on an ice rink.

1.3 THE UNIVERSE

The Earth is a small planet in the solar system. Four of the system's nine planets are enormous in comparison; Jupiter has more than 300 times the amount of matter that the Earth does (Figure 1.4). But the solar system is dominated by a far larger body, with 1000 times the mass of Jupiter: the sun.

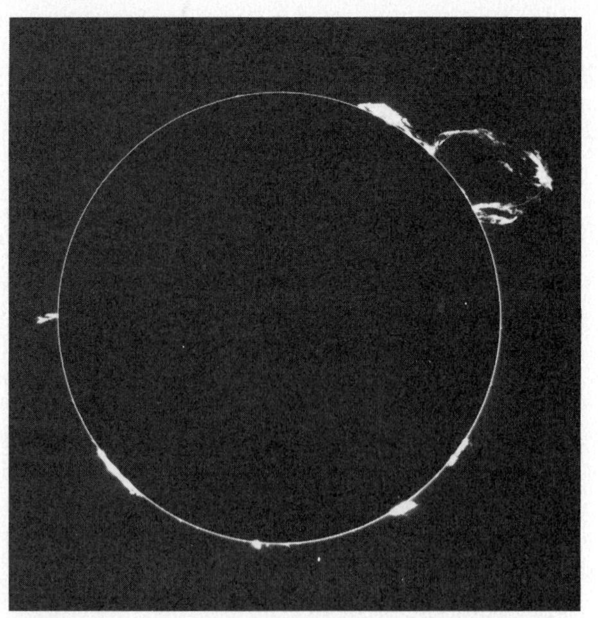

Figure 1.5 The sun, showing active features in its atmosphere. (University of Hawaii Mees Solar Observatory)

The sun's gravitation is great enough to keep all nine planets, a hundred thousand or more asteroids (or minor planets), thousands of millions of comets, and a far, far greater number yet of tiny chunks of ice and/or rock and metal—the meteoroids—all revolving about it.

The highest speed possible (as we shall see in Chapter 11) is the speed of light—3×10^{10} cm/s (300,000 km/s, or 186,000 miles/s). It takes light just over a second to reach Earth from our nearest celestial neighbor, the Moon; the Moon is just over one *light second* away. The sun is 400 times the Moon's distance, and it takes light 8 minutes to come from the sun; it is 8 *light minutes* away. The other planets range from light minutes to light hours away.

The sun is an enormous ball of tremendously hot gas, so hot that all of its chemical elements, including those like iron and tungsten that are solid on Earth, are vaporized to the gaseous state (Figure 1.5). At its surface, the sun's temperature is about 6000 K, but at its center the solar temperature ranges up to many millions of degrees.* Deep in the sun's interior, the thermonuclear conversion of hy-

* In astronomy, we almost always express temperature in Kelvins (K), which are Celsius degrees but measured from absolute zero: −273°C; see Appendix 5.

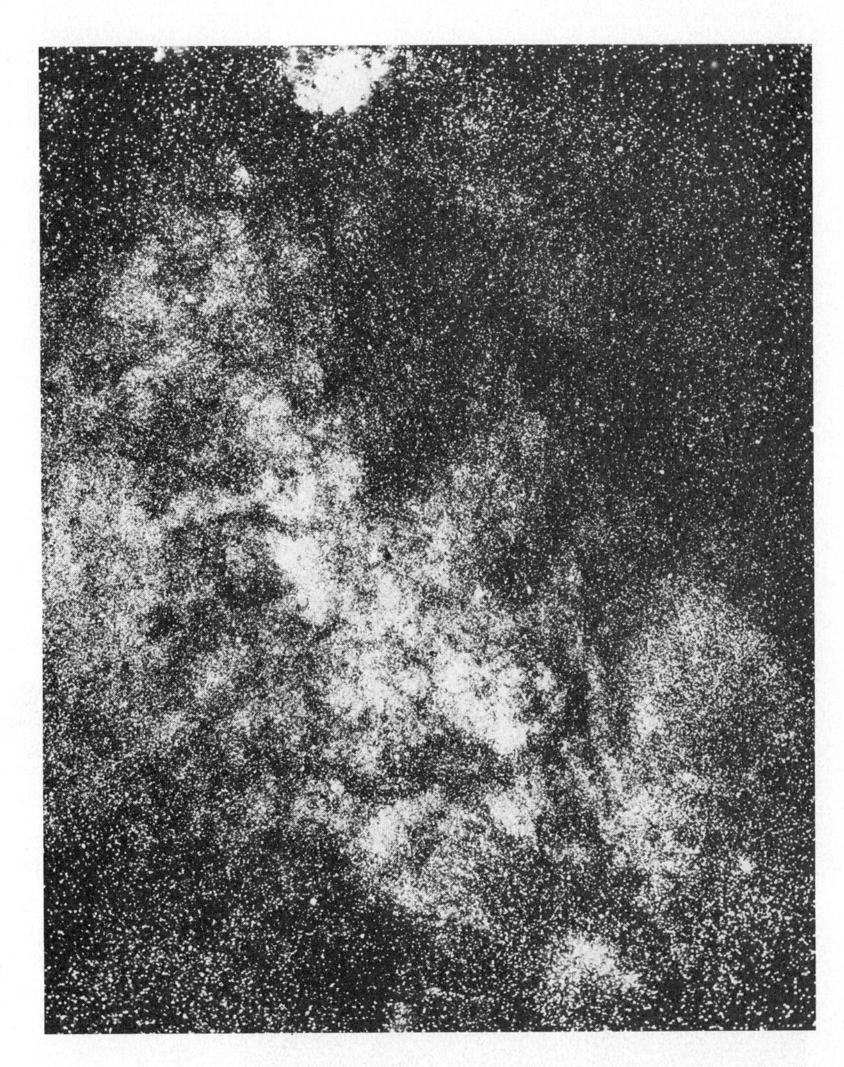

Figure 1.6 A portion of the Milky Way—the galaxy in which we live. (Yerkes Observatory)

drogen to helium gives rise to the sun's enormous outpouring of energy. The sun has a third of a million times the mass of the Earth and more than a million times the Earth's volume. Yet the sun is an ordinary star.

The thousands of stars we see around us in the sky, and the many millions revealed by telescopes, are among the few hundred thousand million stars that make up our *Galaxy*. Those stars are suns, more or less like our own sun, but at distances of *light years*. (A *light year*, abbreviated LY, is the distance light travels in one year: 9.46 million million kilometers, or about 6 million million miles.) The entire Galaxy is a wheel-shaped system about 100,000 LY across, with the sun far from its center, perhaps two-thirds to three-quarters of the way to an edge of the wheel. As we look into the sky in directions that take our line of sight edge-on through the Galaxy, the very many remote stars in those directions produce a faint glow of light—the *Milky Way*, an irregular luminous band completely circling the sky (Figure 1.6). Our Galaxy is therefore often called the Milky Way Galaxy, or simply the Milky Way.

The Galaxy rotates, just as the planets revolve about the sun, but the sun, carrying our solar system with it, revolves about the center of the Milky Way Galaxy in its galactic orbit, at a speed of 200 to 300 km/s, in about 200 million years. Probably many of the other stars revolving about the nucleus of our Galaxy also have planets revolving about them, but those other stars are so remote that we have not yet been able to detect any planets associated with them.

Our Galaxy is not the end of the story. There are millions, probably thousands of millions, of other galaxies in the observable universe (Figure 1.7). These galaxies, like remote islands, each with its thousands of millions of suns, are separated from each other by distances many times their own diameters. Still, even galaxies tend to group in *clusters of galaxies,* and those clusters of galaxies, perhaps also with individual galaxies not members of clusters, are parts of still larger systems called *superclusters.*

Figure 1.7 The neighboring large spiral galaxy in Andromeda. (Mount Wilson and Las Campañas Observatories)

We have been tossing off words describing objects of incredible mass and dimensions, and distances of unimaginable extent, as though they were apples and inches. There is no possibility that someone not already familiar with these concepts can grasp them on a first reading. That, of course, is one purpose of studying a whole course in astronomy. Still, it is worthwhile to try to give a feeling for the scale of the universe, no matter how hard it may be to grasp it thoroughly in one run-through.

1.4 THE SCALE OF THE UNIVERSE

The Earth is a nearly spherical body about 13,000 km (8000 miles) in diameter. The Moon, about one-fourth the Earth's diameter, is 30 of those earth diameters away. We can actually present a scale drawing of the Earth and Moon on a page of this book (Figure 1.8).

The average distance of the Earth from the sun is called an *astronomical unit* (AU). That radius of the Earth's nearly circular (actually elliptical) orbit is about 100 times the diameter of the sun, but the whole orbit of the Moon would fit easily inside the sun itself (Figure 1.9). We cannot, therefore, show the Earth, sun, and Moon to scale on the same page. But if we show the sun as a small dot (Figure 1.10a), we can see the relative sizes of the orbits of the inner planets (Mercury, Venus, Earth, and Mars) to the correct size. Even then, we must change the scale (Figure 1.10b) to show the orbits of the outer planets (Jupiter, Saturn, Uranus, Neptune, and Pluto) with the right sizes relative to the size of the orbit of Mars. Pluto's average distance from the sun is 40 times that of the Earth; that is, Pluto is 40 AU from the sun.

The nearest star beyond the sun is, in contrast, 300,000 AU away. On the scale of Figure 1.10b, its distance would be more than a kilometer. That star is 4 LY away; most visible stars are hundreds or even thousands of light years away.

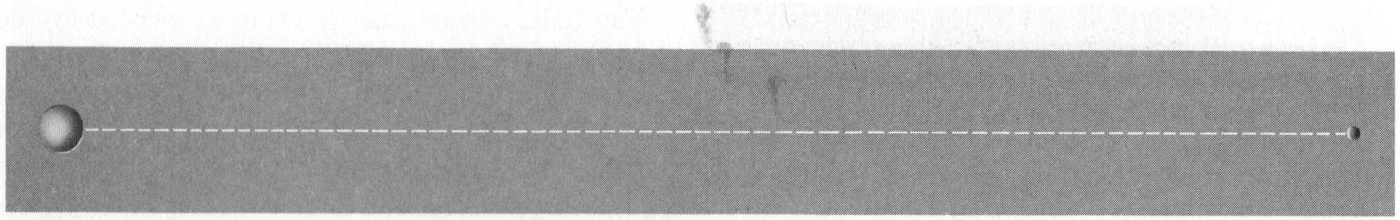

Figure 1.8 The Earth (left) and the Moon drawn to scale.

Suppose we make a rough scale drawing, showing the stars within 10 LY of the sun. In Figure 1.11a, the circle represents a sphere of 10 LY radius centered on the sun. Roughly ten stars are included. Now we change scale. In Figure 1.11b, the

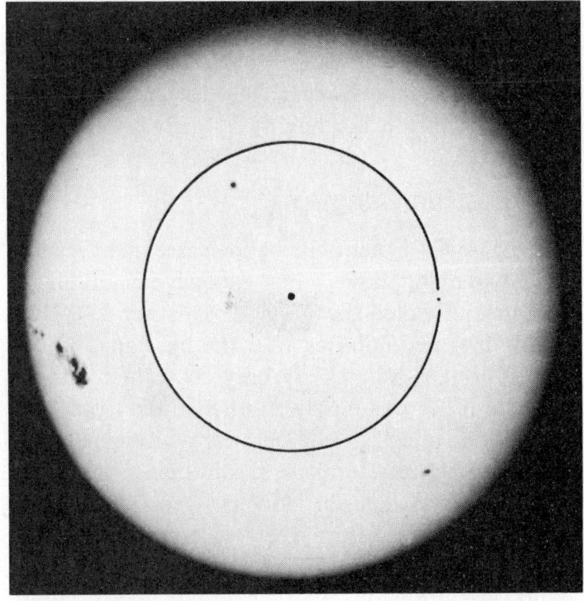

Figure 1.9 The Moon's orbit, to scale, superimposed on a photograph of the sun. (Griffith Observatory)

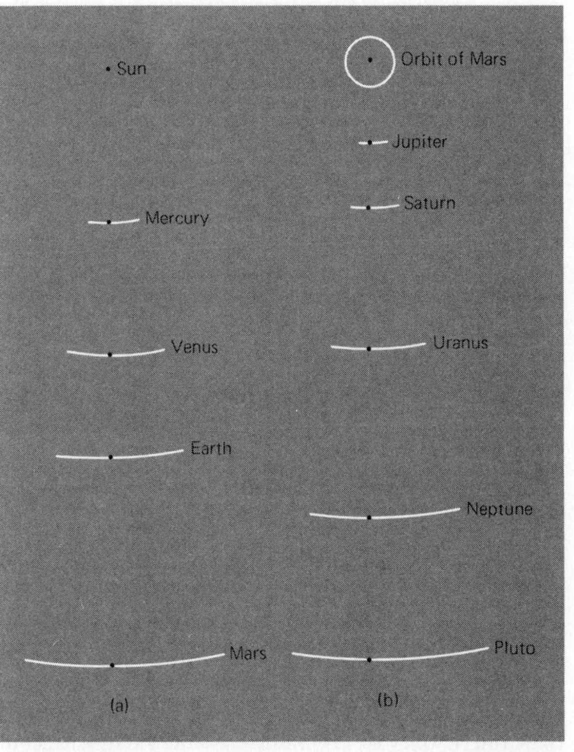

Figure 1.10 The distances of the planets from the sun. (a) The inner planets. (b) The outer planets to the scale of the orbit of Mars.

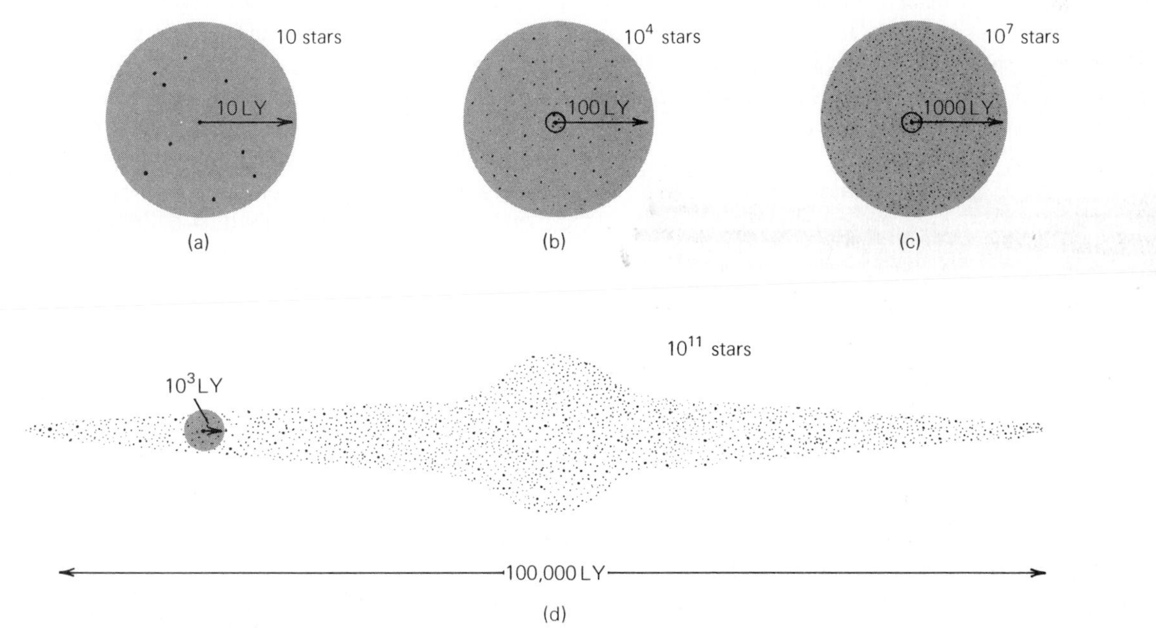

Figure 1.11 The distribution of stars around the sun within (a) 10 LY; (b) 100 LY; (c) 1000 LY; (d) the Galaxy.

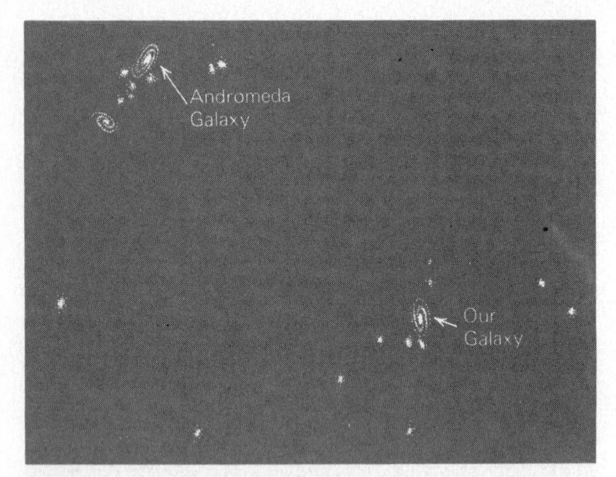

Figure 1.12 Schematic diagram of the Local Group of galaxies, approximately to scale.

sphere of 10 LY radius is the small circle in the center, and the larger circle represents a sphere of 100 LY radius—ten times as great in radius, but 1000 times as great in volume. In that sphere we would find approximately 10,000 stars. Similarly, the sphere of 100 LY radius is the small circle in Figure 1.11c, while the larger circle represents a sphere 1000 LY in radius within which there are ten million (10^7) stars. In our next change of scale, Figure 1.11d, the stars do thin out, but in some dimensions before others. We have begun to reach the boundaries of our Galaxy, but at first only in its thin dimension. The entire wheel-shaped Galaxy, in an edge-on cross section, is shown in Figure 1.11d, and the sphere of 1000 LY radius is the small circle on the left side, with the invisible sun deep in its center.

As are most (if not all) galaxies, our own Milky Way Galaxy is part of a cluster, actually a collection of about two dozen galaxies called the *Local Group*. The Local Group is about 3 million light years across and is shown, roughly to correct scale, in Figure 1.12. Far beyond its boundaries are other such groups, and rare in space, typically tens to hundreds of millions of light years apart, are the great clusters of hundreds or even thousands of member galaxies each. One such is the Hercules cluster (Figure 1.13).

The most remote clusters of galaxies yet identified are one to two thousand times as far away from us, on the scale of Figure 1.12, as the size of a page of this book, perhaps half a kilometer or so. Even farther off are the remote *quasars* (Chapter 35), and yet beyond the quasars is the remote glow from the past—a glow of radio radiation that has been traveling through space to reach us from all directions for at least 10 thousand million years. When that radiation began its journey through space it was light, not radio waves. It has been transformed by the expansion of the universe and is our best evidence today that the universe has evolved from a hot, dense state that existed far in the past.

(a) The Big Bang

The radio waves mentioned above are, in a very real sense, the dying glow of that explosive beginning of our universe called the *big bang* (Chapter 37). During the first few minutes after the big bang, atomic nuclei formed, but mainly only those of hydrogen and helium. It was nearly another million years before the universe cooled (because of its expansion) enough to permit electrons to join nuclei and make atoms. At this stage the universe became transparent, so light could flow freely through it; it is just this radiation that we now observe as radio waves.

It was probably another billion years before matter collected together to form galaxies and stars. In the centers of stars atoms were reheated, and thermonuclear reactions fused hydrogen into helium, providing the energy by which the stars shine. Later, in the interiors of certain stars, atoms of helium fused into those of heavier elements—carbon, oxygen, nitrogen, and silicon, for example—those elements that can form rocky planets.

Subsequently, these stars ejected some of their matter, so enriched in heavier elements, into interstellar space, to eventually condense into new stars. One such star that formed about 4.6 billion years ago, with its system of planets, is our sun. Much of the material of the sun and, we think, virtually all of that of the Earth, consists of atoms synthesized by nuclear reactions in earlier-generation stars. We discuss all these goings-on in later chapters.

1.5 THE NATURE OF SCIENCE

There are large gaps in our full understanding of the origin of the universe through the big bang, as well as of the life cycles and evolution of stars, just as there are gaps in our knowledge of the evolution of life on Earth. But the general picture is fairly

Figure 1.13 A cluster of galaxies in Hercules. (Mount Wilson and Las Campañas Observatories)

clear and becomes clearer as research continues to advance the scientific frontier.

That is not to say, however, that science ever provides, or even attempts to provide, the absolute truth or the ultimate answers to everything. For example, we can apply known physical laws—our best theory—to the conditions of the early universe, but that does not tell us where its matter and energy came from in the first place, or how the matter and radiation got into that hot dense state. Perhaps our ideas about the big bang are even completely wrong; we are, after all, extrapolating known theory to quite a limit in space and time. Very many hypotheses at the scientific frontier (and the big bang is certainly at that frontier) will turn out to be wrong. It is far less likely that our ideas of electricity and magnetism, of gravitation (as formulated in the general theory of relativity), or of the behavior of planets in the solar system will be wrong, for these ideas have been extraordinarily well tested.

It is the testing that is the part and parcel of science. Science is not just a collection of knowledge, of figures and test tubes. It is an organized *method* for exploring nature. Science involves three steps: (1) An observation or an experimental result is noted (say, the falling of a body with uniform acceleration). (2) A hypothesis is advanced to describe that result in terms of a more general model (a theory of gravitation). (3) The model is then used to predict new observations or the results of new experiments (that the Moon's orbit must be an ellipse, or that an unknown planet, Neptune, is causing irregularities in the motion of Uranus).

Then we must check whether these predictions hold up. Since most predictions in science are *quantitative,* involving numerical values that can be calculated from mathematical equations, these checks usually require accurate measurements of the phenomena being studied. Often it is only through increasing the precision of the observations that a distinction between two competing theories can be made. For example, ordinary gravitational theory and Einstein's theory of general relativity make identical predictions about everyday experience, but only diverge under extreme conditions, such as velocities near the speed of light.

If the predictions of a theory are not confirmed by observation or experiment, the hypothesis must be discarded or modified. If the new data confirm the predictions, it is still necessary to devise additional tests, with ever broader applications, until the hypothesis has been subjected to the most ruthless scrutiny. Even then, it may later prove wrong or at least to have limited usefulness, but the longer it survives and the more documented it becomes, the better are its chances to become a part of established theory. As we shall see in Chapter 37, the big bang has survived three crucial tests and is certainly worthy of being taken very seriously, but it is by no means a *fact*—not yet. And the fabric of even the best of scientific theories may begin to tear when it is pushed to limits beyond the regime in which it is well tested, as did Newtonian gravitational and mechanical theory in the 20th century, when it was challenged with speeds near that of light or with gravitational fields of enormous strength. Even so, when applied with proper respect for its limitations, Newtonian theory is of incalculable value.

So it is that third step in the *scientific method*, the test of the hypothesis, that is crucial. Without such tests and checks, a model is mere speculation and not part of science. For example, an assertion that a civilization of people exactly like us exists on a planet revolving about a star in a remote galaxy, or the contention that far beyond the limits of our observations are other universes, are not scientific hypotheses or theories—at least not today—for there is no way to test them. Perhaps the assertions are true, but in the absence of any possibility of verifying them, their truth or error is irrelevant and certainly has nothing to do with science.

To be sure, science has limitations. It must operate by strict rules, and very many questions cannot be asked, let alone answered, in science. Many other valid and highly valuable areas of human activity, art and music for example, have great importance even though they lie outside of science. Still, science has proved itself to be a remarkably successful way of studying nature. It has provided models by which we can understand a great deal, and which have enabled us to develop a highly advanced technology.

And if it has not yet uncovered the whole truth of the universe (and it never can), science has revealed a marvelous unity in the universe; as in a Greek tragedy, it has a unity of time, place, and action. Time and place (time and space), as we shall see, are inextricably interrelated, for even at this moment we observe the remote past of the universe as we look to its remote parts. And everywhere—then, there, here, and now—we find the same kinds of stuff: atoms, electrons, and so on. Unity of action is even more remarkable. The laws of nature, so far as we can tell, are truly universal.

Nor can they be suspended, even for a moment. It was when people stopped believing in magic that science began to advance. Magic is the antithesis of science. The fake psychics and fortune-tellers notwithstanding, if we could really suspend nature's laws we would have utter chaos. Far from everything being possible, *nothing* would be possible!

As a final reminder of the great unity of the universe, recall that, according to the results of our current research, the very atoms that comprise our own bodies were formed in the centers of past-generation stars. We are, ourselves, quite literally, made of stardust!

EXERCISES

1. If a gas is *ionized*—that is, if the electrons are stripped from the nuclei of its atoms, so that each freed electron moves about in the gas in the same way an individual molecule would—the gas can be compressed to a far higher density than ordinary solids can. Explain why. (Assume that high temperatures or other effects prevent ions from reuniting with electrons, thereby becoming neutral atoms.)
2. If we were to try to communicate, say by radio waves (which travel with the speed of light), with a hypothetical inhabitant of a planet revolving about a star 100 LY away, how long would we have to wait after transmitting a question before we could expect to receive an answer?
3. From the data given in this chapter, calculate how many circles the size of the Earth's orbit would be required to reach across the length of a diameter of the Galaxy, if they were laid out barely touching each other, as in a chain. How many such circles would it take to reach a remote cluster of galaxies, say, 300 million LY distant?
4. Review how many changes of scale are required to prepare a series of scale drawings on standard pieces of paper (8½ by 11 inches), beginning with a diagram that shows the Earth as a circle and ending with one that shows the correct relative distance for a remote cluster of galaxies. By what factor must the linear size of the largest distance on one diagram be reduced to show it as the smallest one on the next diagram in the series?
5. Give at least three examples of questions that are improper in the realm of science.

Claudius Ptolemaeus (Ptolemy) (second century) was one of the great astronomers of antiquity. He devised a system of cosmology that described the motions of the planets so satisfactorily that there was no substantial change until the time of Copernicus 13 centuries later. (Burdy Library, photograph by Owen Gingerich)

2 EARLY ASTRONOMY: MYTH AND SCIENCE

2.1 EARLIEST ASTRONOMERS

Speculations on the nature of the universe must date from prehistoric times. It is difficult to state definitely when the earliest observations of a more or less quantitative sort were made or when astronomy as a science began. Certainly, in many of the ancient civilizations the regularity of the motions of celestial bodies was recognized, and attempts were made to keep track of and predict celestial phenomena. In particular, the invention of and keeping of a calendar requires at least some knowledge of astronomy—the basic units of the calendar being the day, the month (originally, the 29- and 30-day cycling of the Moon's phases), and the year of seasons.

The Chinese had a working calendar and had determined the length of the year several centuries before Christ (B.C.). About 350 B.C., the astronomer Shih Shen prepared what was probably the earliest star catalogue, containing about 800 entries. The Chinese also kept rather accurate records of comets, meteors, and fallen meteorites from 700 B.C. Records were made of sunspots visible to the naked eye and of what the Chinese called "guest stars," stars that are normally too faint to be seen but suddenly flare up to become visible for a few weeks or months (such a star is now called a *nova*). The most significant of the Chinese observations of nova outbursts was of the great supernova of 1054 A.D. in the constellation of Taurus. Today's remnant of that cosmic explosion is the Crab nebula, a chaotic, expanding mass of gas (see Chapter 32).

The Babylonians, Assyrians, and pre-Christian Egyptian astronomers also knew the approximate length of the year from early times. By a few centuries B.C., the Egyptians had adopted a calendar based on a 365-day year. Of particular significance to them was the date when the bright star Sirius could first be seen in the dawn sky, rising just before the sun. This predawn rising of Sirius coincided fairly well with the average time of the annual flooding of the Nile, which gave the astronomer-priests the ability to predict very roughly when this economically important event could be expected to occur.

There is evidence of ancient astronomical knowledge in other parts of the world as well. The Maya in Central America developed a sophisticated calendar and made astronomical observations in a period contemporary with early European civilizations (Chapter 6), and the Polynesians learned to navigate by means of celestial observations, over hundreds of kilometers of ocean separating their islands.

Figure 2.1 Stonehenge, a megalithic monument, possibly an observatory or calendar-keeping device, located in the Salisbury plain of England. (Courtesy E. C. Krupp, Griffith Observatory)

Particularly interesting are monuments left by Bronze Age people in northwestern Europe, especially in the British Isles. The best-preserved of the monuments is Stonehenge (Figure 2.1), about 13 km from Salisbury in southwest England. It is a complex array of stones, ditches, and holes arranged in concentric circles. Carbon dating and other studies show that Stonehenge was built during three periods ranging from about 2500 B.C. to about 1700 B.C. Some of the stones are aligned with the directions of the sun during its rising and setting at critical times of the year (such as the beginnings of summer and winter), and it is widely thought that at least one function of the monument was connected with the noting of these occasions. However, only some of the many hundreds of other monuments have alignments that can be interpreted as astronomical.

2.2 EARLY GREEK ASTRONOMY

The high point in ancient science was in the Greek culture from 600 B.C. to 400 A.D. The earliest Greeks were not scientists in the modern sense. They were often more interested in solving abstract geometrical problems, reasoning from given axioms, than in making original observations. Yet, in that Greek reservoir of ideas and inspiration, many observations were carried out, with the result that science in general and astronomy in particular were raised to a level unsurpassed until the 16th century.

(a) Early Concepts of the Sky

THE CELESTIAL SPHERE

If we gaze upward at the sky on a clear night, we cannot avoid the impression that the sky is a great hollow spherical shell with the Earth at the center. The early Greeks regarded the sky as just such a *celestial sphere;* some apparently thought of it as an actual sphere of a crystalline material, with the stars embedded in it like tiny jewels. The sphere, they reasoned, must be of very great size, for if its surface were close to the Earth, as one moved from place to place he would see an apparent angular displacement in the directions of the stars.

Of course, at any one time we see only a hemisphere overhead, but with the smallest effort of imagination we can envision the remaining hemisphere, that part of the sky that lies below the horizon. If we watch the sky for several hours, we see that the celestial sphere is gradually and continually changing its orientation. The effect is caused simply by the rotation of the Earth, which carries us under successively different portions of the sphere. Following along with us must be our *horizon,* that line in the distance at which the ground seems to dip out of sight, providing a demarcation between Earth and sky. (The horizon may, of course, be hidden from view by mountains, trees, buildings, or, in large cities, smog.) As our horizon tips down in the direction that the Earth's rotation carries us, stars hitherto hidden beyond it appear to rise above it. In the opposite direction the horizon tips up, and stars hitherto visible appear to set behind it. Analogously, as we round a curve in a mountain road, new scenery comes into view while old scenery disappears behind us.

The direction around the sky toward which the Earth's rotation carries us is *east;* the opposite direction is *west.* The Greeks, unaware of the Earth's rotation, imagined that the celestial sphere rotated about an axis that passed through the Earth. As it turned, it carried the stars up in the east, across the sky, and down in the west.

CELESTIAL POLES

A careful observer will notice that some stars do not rise or set. As seen from the Northern Hemisphere, there is a point in the sky some distance above the northern horizon about which the whole celestial sphere appears to turn. As stars circle about that point, those close enough to it can pass beneath it without dipping below the northern horizon. A star exactly at the point would appear motionless in the sky. Today the star *Polaris* (the North Star) is within 1° of this pivot point of the heavens.

The Greeks regarded that pivot point as one end of the axis about which the celestial sphere rotates. We know today that it is the Earth that spins about an axis through its North and South Poles. An extension of the axis would appear to intersect the sky at points in line with the North and South Poles of the Earth but, because of the virtually infinite size of the celestial sphere, immensely far away. As the Earth rotates about its polar axis, the sky appears to turn in the opposite direction about those *north* and *south celestial poles.*

An observer at the North Pole of the Earth would see the north celestial pole directly overhead (at the *zenith*). The stars would all appear to circle about the sky parallel to the horizon, none rising or setting. An observer at the Earth's equator, on the other hand, would see the celestial poles at the north and south points on the horizon. As the sky apparently turned about these points, all the stars would appear to rise straight up in the east and set straight down in the west. For an observer at an arbitrary place in the Northern Hemisphere (for example, in Greece), the north celestial pole would appear at a point between the zenith and the north point on the horizon, its location depending on the relative distances from the equator and North Pole

Figure 2.2 Time exposure showing trails left by stars as a consequence of the apparent rotation of the celestial sphere. (Lick Observatory)

of the Earth. The stars that were not always above the horizon would rise at an oblique angle in the east, arc across the sky, and set obliquely in the west.

RISING AND SETTING OF THE SUN

The sun is always present at some position on the celestial sphere. When the apparent rotation of the sphere carries the sun above the horizon, the brilliant sunlight scattered about by the molecules of the Earth's atmosphere produces the blue sky that hides the stars that are also above the horizon. The early Greeks were aware that the stars were there during the day as well as at night.

ANNUAL MOTION OF THE SUN

The Greeks were also aware, as were the Chinese, Babylonians, and Egyptians before them, that the sun gradually changes its position on the celestial sphere, moving each day about 1° to the east among the stars. Of course, the daily westward rotation of the celestial sphere (or eastward rotation of the Earth) carries the sun, like everything else in the heavens, to the west across the sky. Each day, however, the sun rises, on the average, about four minutes later with respect to the stars; the celestial

Figure 2.3 Time exposure showing star trails in the region of the north celestial pole. The bright trail below the center was made by Polaris (the North Star), which is about one degree away from the true pole. (Yerkes Observatory)

sphere (or Earth) must make just a bit more than one complete rotation to bring the sun up again. The sun, in other words, has an independent motion of its own in the sky, quite apart from the daily apparent rotation of the celestial sphere.

In the course of one year, the sun completes a circuit of the celestial sphere. The time for this circuit, which is also the period of the seasonal cycle, can be taken as the definition of the *year*. The early peoples mapped the sun's eastward journey among the stars. This apparent path of the sun is called the *ecliptic* (because eclipses can occur only when the Moon is on or near it). The sun's motion on the ecliptic is in fact merely an illusion produced by another motion of the Earth—its annual revolution about the sun. As we look at the sun from different places in our orbit, we see it projected against different stars in the background, or we would, at least, if we could see the stars in the daytime; in practice, we must deduce what stars lie behind and beyond the sun by observing the stars visible in the opposite direction at night. After a year, when we have completed one trip around the sun, it has apparently completed one circuit of the sky along the ecliptic. We have an analogous experience if we walk around a campfire at night; we see the flames appear successively in front of each of the people seated about the fire.

It was also noted by the ancients that the ecliptic does not lie in a plane perpendicular to the line between the celestial poles, but is inclined at an angle of about 23½° to that plane. This angle is called the *obliquity* of the ecliptic and was measured surprisingly accurately by several ancient observers. The obliquity of the ecliptic, as we shall see, is responsible for the seasons and also for the invariable tilt in the axes of terrestrial globe maps.

FIXED AND WANDERING STARS

The sun is not the only moving object among the stars. The Moon and each of the five planets visible to the unaided eye—Mercury, Venus, Mars, Jupiter, and Saturn—change their positions in the sky from day to day. The Moon, being the Earth's nearest celestial neighbor, has the fastest apparent motion; it completes a trip around the sky in about 1 month. During a single day, of course, these objects rise and set, as do the sun and the stars. We are referring here to their independent motions among the stars, superimposed on the daily rotation of the celestial sphere. The Greeks distinguished between what they called the *fixed* stars, the real stars that appeared to maintain fixed patterns among themselves throughout many generations, and the *wandering stars* or *planets*. The Greek word *planet* means "wanderer." Today, we do not regard the sun and Moon as planets, but the Greeks applied the term to all seven of the moving objects in the sky. Much of ancient astronomy was devoted to observing and predicting their motions. In fact, they give us the names for the seven days of our week; Sunday is the sun's day, Monday the Moon's day, and Saturday is Saturn's day. We have only to look at the names of the other days of the week in the Romance languages to see that they are named for the remaining planets.

THE ZODIAC

The individual paths of the Moon and planets in the sky all lie close to the ecliptic, although not exactly on it. The reason is that the paths of the planets about the sun, and of the Moon about the Earth, are all in nearly the same plane, as if they were marbles rolling about on the top of a table. The planets and Moon are always found in the sky within a narrow belt 18° wide centered on the ecliptic, called the *zodiac*. The apparent motions of the planets in the sky result from a combination of their actual motions and the motion of the Earth about the sun, and consequently they are somewhat complex.

CONSTELLATIONS

The backdrop for the motions of the "wanderers" in the sky is the canopy of stars themselves. Like the Chinese and the Egyptians, the Greeks had divided the sky into *constellations,* apparent configurations of stars. Modern astronomers still make use of these constellations to denote approximate locations in the sky, much as geographers use political areas to denote the locations of places on the Earth. The boundaries between the modern constellations are imaginary lines in the sky running north-south and east-west, so that every point in the sky falls in one constellation or another.

Many of the 88 recognized constellations are of Greek origin and bear names that are Latin translations of those given them by the Greeks. Today, the lay person is often puzzled because the constellations seldom resemble the people or animals for which they were named. In all likelihood, the Greeks themselves did not name groupings of stars because they resembled actual people or objects, but rather named sections of the sky in *honor* of the characters in their mythology, and then fitted the configurations of stars to the animals and people as best they could.

(b) The First Greek Astronomers

THE IONIAN SCHOOL

The earliest Greek scientists were the Ionians, who lived in Turkey. Pythagoras (who died ca. 497 B.C.) was originally an Ionian, but he later founded a school of his own in southern Italy. He pictured a series of concentric spheres, in which each of the seven moving objects—the planets, the sun, and the Moon—was carried by a separate sphere from the one that carried the stars, so that the motions of the planets resulted from independent rotations of the different spheres about the Earth. These motions gave rise to harmonious sounds, the *music of the spheres,* which only the most gifted ear could hear.

Pythagoras also believed that the Earth, Moon, and other heavenly bodies were spherical. It is doubtful that he had a sound reason for this belief, but it may have stemmed from the realization that the Moon shines only by reflected sunlight, and that the Moon's sphericity is indicated by the curved shape of the *terminator,* the demarcation line between its illuminated and dark portions. If he had so reasoned that the Moon is round, the sphericity of the Earth might have seemed to follow by analogy.

Another member of the Pythagorian school was Philolaus, who lived in the following century. He may have been the first person to introduce the concept that the Earth is in motion, although he did not think this motion was around the sun, but rather around an invisible "fire" that marked the center of the universe. This bold idea may have had some influence on later Greek thought. Other Greek philosophers of the sixth to fourth centuries B.C. who are said to have believed in a moving Earth are Hicetas, Heracleides, and Ecphantus. Centuries later Copernicus, in his *De Revolutionibus,* quoted the Pythagoreans as authorities for his own doctrines.

In their invention of cosmological schemes, the Greeks did not always necessarily attempt to describe what they regarded as reality. Rather, they were often trying to find a scheme—a model—that would *describe the phenomena* and would predict events (eclipses, configurations of the planets, and so on). The epicycles of Ptolemy, developed later, may similarly be regarded as mathematical representations of the motions of planets in the sky.

(c) The Moon's Phases

Aristotle (384–322 B.C.), most famous of the Greek philosophers, wrote encyclopedic treatises on nearly every field. Aristotle's writings tell us that such phenomena as phases of the Moon and eclipses were understood at least in the fourth century B.C. The basic concepts are so important to the development of astronomy that we shall consider them here rather than in the later chapters that deal more directly with these topics.

The Moon's changing shape during the month results from the fact that it is not itself luminous but is illuminated by sunlight. Because of its sphericity, only half of the Moon is illuminated, that is, having daylight, at one time—the half turned toward the sun. The apparent shape of the Moon in the sky depends simply on how much of its daylight hemisphere is turned to our view.

Even in Aristotle's time it was known that the sun is more distant than the Moon. This was sur-

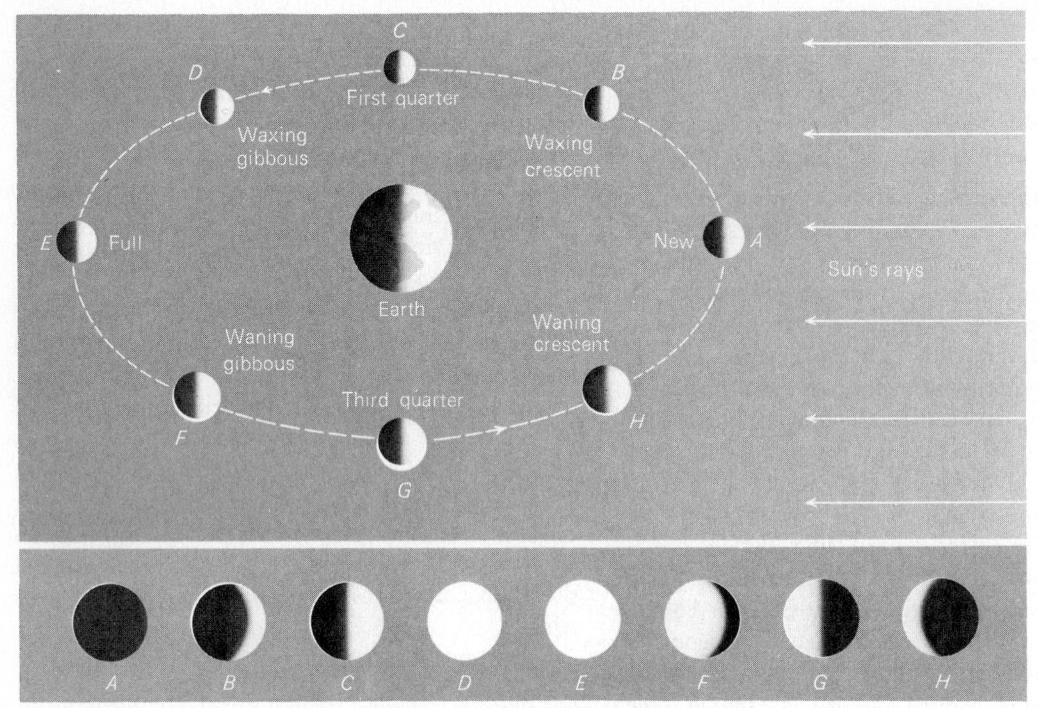

Figure 2.4 Phases of the Moon. In the upper part, the Moon's orbit is viewed obliquely. Below is shown the appearance of the Moon from the Earth.

mised from the sun's slower apparent motion among the stars on the celestial sphere and also from the fact that the Moon occasionally passes exactly between the Earth and sun and temporarily hides the sun from view (*solar eclipse*). Thus when the Moon is in the same general direction from Earth as the sun (position *A* in Figure 2.4), its daylight side is turned away from the Earth. Because its night side—the side turned toward us—is dark and invisible, we do not see the Moon in that position. The phase of the Moon is then *new*. (Perhaps it would seem more reasonable to call it "no moon" instead of "new moon," for we do not see any Moon at all.) To appear silhouetted in front of the sun, producing a solar eclipse, the Moon must be at the new phase and must also lie on the line joining the Earth and sun (see Chapter 7). A solar eclipse does not occur at every new moon because the plane of the Moon's orbit is inclined slightly (about 5°) to the plane of the ecliptic; hence the new moon usually lies above or below the Earth-sun line.

A few days after new moon, the Moon reaches position *B*, and from the Earth we see a small part of its daylight hemisphere. The illuminated crescent increases in size on successive days as the Moon moves farther and farther around the sky away from the direction of the sun. During these days the Moon is in the *waxing crescent* phase. About a week after new moon, the Moon is one quarter of the way around the sky from the sun (position *C*) and is at the *first quarter* phase. Here the line from the Earth to the Moon is at right angles to the line from the Earth to the sun and half of the Moon's daylight side is visible—it appears as a half moon.

During the week after the first quarter phase we see more and more of the Moon's illuminated hemisphere, and the Moon is in the *waxing gibbous* phase (position *D*). Finally, about two weeks after new moon, the Moon (at *E*) and the sun are opposite each other in the sky; the side of the Moon turned toward the sun is also turned toward the Earth; we have *full moon*. During the next two weeks the Moon goes through the same phases in reverse order—through *waning gibbous*, *third* (or *last*) *quarter*, and *waning crescent*. Occasionally the full moon passes through the Earth's shadow, which of course extends outward in space in the direction opposite the sun. This is a *lunar eclipse*.

Figure 2.5 Partially eclipsed Moon moving out of the Earth's shadow. (Yerkes Observatory)

If you find difficulty in picturing the phases of the Moon from this verbal account, try a simple experiment: Stand about six feet in front of a bright electric light outdoors at night and hold in your hand a small round object such as a tennis ball or an orange. If the object is then viewed from various sides, the portions of its illuminated hemisphere that are visible will represent the analogous phases of the Moon.

(d) The Spherical Shape of the Earth

Another important topic discussed by Aristotle was the shape of the Earth. He cited two convincing arguments for the Earth's sphericity. First is the fact that during a lunar eclipse, as the Moon enters or emerges from the Earth's shadow, the shape of the shadow seen on the Moon is always round (Figure 2.5). Only a spherical object always produces a round shadow. If the Earth were a disk, for example, there would be some occasions when the sunlight would be striking the disk edge on, and the shadow on the Moon would be a line.

As a second argument, Aristotle explained that northbound travelers observe the stars near the north celestial pole to be higher in the sky than is observed at home, and different stars pass through the zenith. Conversely, when one travels to more southern latitudes the stars near the north celestial pole are seen lower in the sky, and some stars that are never above the horizon at home are seen to rise and move across the southern sky. The only possible explanation is that the travelers' horizons had tipped to the north or south, respectively, which indicates that they must have moved over a curved surface of the Earth. As a third piece of evidence that the Earth is round, Aristotle mentioned that elephants had been observed to the east in India and also to the west in Morocco; evidently, those two places must not be far apart! But he also advanced a theoretical argument that material falling to a center would take on a spherical shape—an idea consistent with the gravitational theory of Newton two millennia later.

(e) The Motion of the Earth

It is interesting that Aristotle pointed out that the apparent daily motion of the sky can be explained by a hypothesis of the rotation of either the celestial sphere or the Earth. He rejected the latter explanation. He also considered the possibility that the Earth revolves about the sun rather than the sun about the Earth. He discarded this *heliocentric* hypothesis in the light of an argument that has been used many times since. Aristotle explained that if the Earth moved about the sun we would be observing the stars from successively different places along our orbit, and their apparent directions in the sky would then change continually during the year.

Any apparent shift in the direction of an object as a result of motion of the observer is called *parallax*. An annual shifting in the apparent directions of the stars that results from the Earth's orbital motion is called *stellar parallax*. For the nearer stars it is observable with modern telescopes, but it is impossible to measure with the naked eye because of the great distances of even the nearest stars. Indeed, Tycho Brahe in the 16th century was unable to detect stellar parallax and concluded that the Earth is stationary.

2.3 LATER GREEK ASTRONOMY

The early Greeks, as we have seen, were aware of, and to some extent understood, the phenomena of the sky. Remarkable progress, however, was made in the centuries following Aristotle, especially by the school of astronomers centered in Alexandria, where Greek science attained its greatest heights.

(a) Aristarchus of Samos

Especially interesting is Aristarchus of Samos (ca. 310–230 B.C.), who is reported to have believed that the Earth revolves about the sun. We know of this, however, only from the writings of others, for only one manuscript of Aristarchus survives: "On the Sizes and Distances of the Sun and Moon." But this document alone is remarkable and deserves some discussion.

Aristarchus opens his treatise with several postulates, the "givens" that are needed to proceed with a geometrical proof. The most essential of these are (1) that the Moon receives its light from the sun, (2) that it appears half full when the angle in the sky between the Moon and sun is 3° less than a right angle (that is, 87°), and (3) that the diameter of the Earth's shadow at the Moon's distance is twice the size of the Moon. (He also implicitly assumed that the Moon's orbit about the Earth is a perfect circle.) From these assumptions, Aristarchus, using the rules of Euclidean geometry, derives that (1) the distance of the sun is more than 18 but less than 20 times the distance of the Moon and (2) the ratio of the sun's diameter to that of the Earth is more than 19 to 3 but less than 43 to 6.

In other words, Aristarchus found that the sun is about 19 times as far away as the Moon is (the correct figure is about 400) and that the sun's diameter is about 7 times the Earth's (the correct value is 109). A reading of his account suggests that to Aristarchus the entire exercise is no more than an interesting geometry problem. There is no mention at all of where the numbers used in his postulates came from. Perhaps he made crude estimates; perhaps they were someone else's estimates; we do not know. But there is no suggestion that Aristarchus himself actually made any careful observations or measurements. To him, it was, we repeat, an exercise in geometry.

On the other hand, the basic ideas were ingenious and beautiful in their simplicity. Moreover these ideas were applied later by other astronomers—especially by Hipparchus—to attempt an accurate determination of the size and distance of the Moon. It is interesting, therefore, to see how the method can work. The following, we emphasize, is *not* the procedure or reasoning of Aristarchus (whose geometry is actually rather tedious) but is a description of how *we* would be able to derive these astronomical dimensions, given Aristarchus' assumptions.

The Moon appears exactly *half full* (first and last quarters) when the terminator—the line dividing the light and dark halves—is a perfectly straight line as viewed from the Earth. But the Moon is spherical, and the terminator, being a line upon its surface, must be curved. Thus, the only way it can appear straight is for us to view it exactly edge on. That is, the plane of the terminator must contain the line of sight from the Earth to the center of the Moon. When that is true, the line from the Moon to the Earth must be at right angles to the line from the Moon to the sun. In Figure 2.6 these right angles are *EMS* and *EM'S*. Now we see that because the sun is not infinitely far away, by assumption, the points *M'*, *E*, and *M* do not lie along a straight line. Hence the Moon, moving at a uniform rate, should require a shorter time to go from *M'* to *M* than from *M* to *M'*. We could use the difference between these intervals from third quarter to first quarter Moon and from first quarter to third quarter to determine the angle *M'EM*. For example, if the period from *M* to *M'* were, say, twice that from *M'* to *M*, the angle *M'EM* would be a third of a circle, or 120°, and the angle *SEM* would be 60°. We have no idea how Aristarchus arrived at the figure 87°. Even with our modern equipment of the late 20th century, we could not observe the instants of quarter moon with sufficient accuracy to determine the *ES*/*EM* ratio meaningfully, because of the sun's great distance.

However determined, the angle *M'EM* can be constructed inside a circle representing the Moon's orbit; the lines *MS* and *M'S*, drawn tangent to the circle at *M* and *M'*, intersect at *S*, thus determining the position of the sun and hence its distance in terms of the size of the Moon's orbit.

To find the relative sizes of the sun and Moon, we use the information that the Earth's shadow at the Moon's distance is twice the size of the Moon (the correct ratio is about 8 to 3). Now it is well known that the sun and Moon appear to be the same *angular size* in the sky. By angular size we mean the angle subtended by the diameter of an object, that is, the angle of intersection between two lines drawn from a point on the Earth (for example, the observer's eye) to opposite ends of a diameter of the object. The sun and the Moon each has an angular size of about ½°. If, as Aristarchus had determined, the sun is 19 times as distant as the Moon, it must also be 19 times as big to appear the same size. Aristarchus grossly overesti-

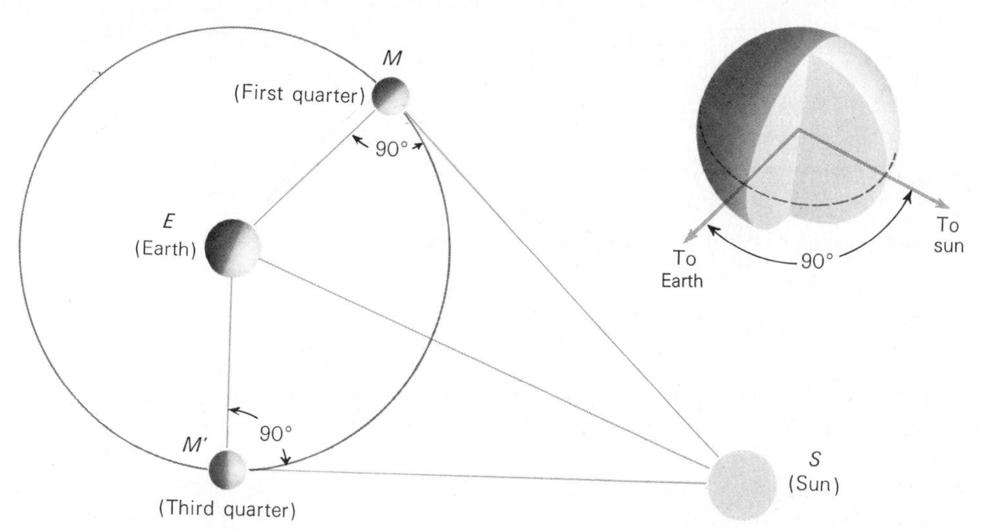

Figure 2.6 Aristarchus' method of determining the relative distances of the sun and Moon from the Earth.

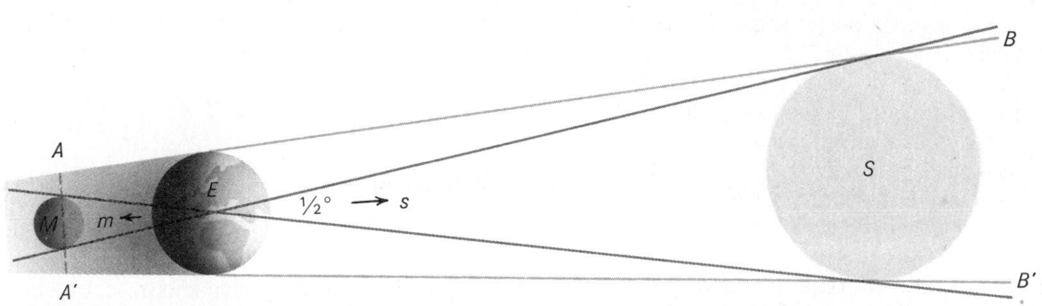

Figure 2.7 The principle by which Aristarchus could determine the relative sizes of the sun, Moon, and Earth.

mated the angular sizes of the sun and Moon to be about 2° each (perhaps the error was intentional to emphasize the geometry rather than reality). With such data we could find the relative sizes of the Earth, the Moon, and the sun by geometrical construction.

We illustrate the geometrical principles of the construction in Figure 2.7. First, at *E*, which represents the center of the Earth, we draw two lines that intersect at an angle of ½°. During a lunar eclipse the sun and Moon are opposite in the sky; thus in direction *s* the ½° angle can be considered as representing the angular diameter of the sun, and in direction *m* the angular diameter of the Moon. The sun, *S*, and Moon, *M*, can now be drawn in, and at arbitrary distances from *E* as long as the distance *ES* is 19 times the distance *EM*. Now at *M*, the diameter of the Earth's shadow, *AA′*, can be constructed at twice the size of the Moon. Because the rays of sunlight, in which the Earth casts its shadow, travel in straight lines, the lines *AB* and *A′B′*, drawn tangent to the sun at *B* and *B′*, must also be tangent to the Earth. Thus, finally, the sphere of the Earth can be drawn in to proper scale at *E*. We have now constructed a scale drawing of the Earth, Moon, and sun. We need only measure with a ruler to obtain their relative sizes.

Perhaps it was his finding that the sun was seven times the Earth's diameter that led Aristarchus to the conclusion that the sun, not the Earth, was at the center of the universe. At any rate, he is the first person of whom we have knowledge who professed a belief in the heliocentric hypothesis—that the Earth goes about the sun. He also postulated that the stars must be extremely distant to account for the fact that their parallaxes could not be observed.

(b) Measurement of the Earth by Eratosthenes

Aristarchus had derived the dimensions of the sun and Moon, but only in terms of the size of the Earth. The

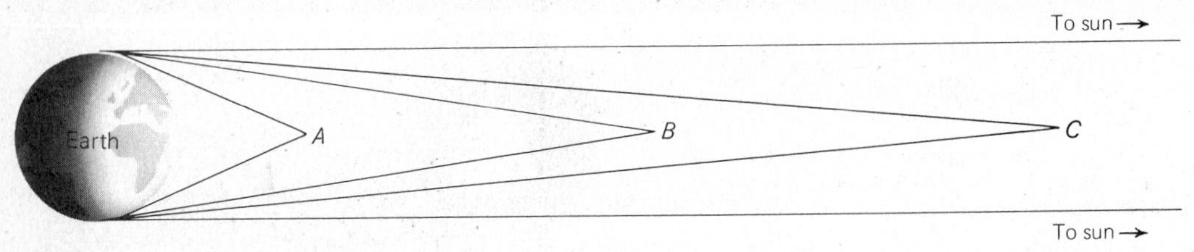

Figure 2.8 The more distant an object, the more nearly parallel are the rays of light coming from it.

latter was not accurately known to him. The first fairly accurate determination of the Earth's diameter was made by Eratosthenes (276–195 or 196 B.C.), an astronomer of the Alexandrian school.

To appreciate Eratosthenes' technique for measuring the Earth, which is in principle the same as many modern methods, we must understand that the sun is so distant from the Earth compared with its size, even by Aristarchus' value, that the sun's rays intercepted by all parts of the Earth approach it along sensibly parallel lines. Imagine a light source near the Earth, say at position *A* in Figure 2.8. Its rays strike different parts of the Earth along diverging paths. From a light source at *B*, or at *C*, still farther away, the angle between rays that strike extreme parts of the Earth is smaller. The more distant the source, the smaller the angle between the rays. For a source *infinitely* distant, the rays travel along parallel paths. The sun is not, of course, infinitely far away, but light rays striking the Earth from a point on the sun diverge from each other by at most an angle of less than one third of a minute of arc (⅓′), far too small to be observed with the unaided eye. As a consequence, if people all over the Earth who could see the sun were to point at it, their fingers would all be pointing in the same direction—they would all be parallel to each other. The concept that rays of light from the sun, planets, and stars approach the Earth along parallel lines is vital to the art of celestial navigation—the determination of position at sea.

Eratosthenes noticed that at Syene, Egypt, now modern Aswân, on the first day of summer, sunlight struck the bottom of a vertical well at noon, which indicated that Syene was on a direct line from the center of the Earth to the sun. At the corresponding time and date in Alexandria, 5000 stadia north of Syene (the *stadium* was a Greek unit of length), he observed that the sun was not directly overhead but slightly south of the zenith, so that its rays made an angle with the vertical equal to 1/50 of a circle (about 7°). Yet the sun's rays striking the two cities are parallel to each other. Therefore (see Figure 2.9), Alexandria must be one-fiftieth of the Earth's circumference north of Syene, and the Earth's circumference must be 50 × 5000, or 250,000, stadia. The figure was later revised to 252,000, so that each degree on the Earth's surface would have exactly 700 stadia.

It is not possible to evaluate precisely the accuracy of Eratosthenes' solution because there is doubt as to which of the various kinds of Greek stadia he used. If it was the common Olympic stadium, his result was about 20 percent too large. According to another interpretation, he used a stadium equal to about ⅙ km, in which case his figure was within 1 percent of the correct value of 40,000 km. The diameter of the Earth is found from the circumference, of course, by dividing the latter by π.

(c) Hipparchus

The greatest astronomer of pre-Christian antiquity was Hipparchus, who was born in Nicaea in Bithynia. The dates of his life are not accurately known, but he carried out his work at Rhodes, and possibly also at Alexandria, in the period from 160 to 127 B.C. Many of the phenomena Hipparchus detected are quite subtle, and the measurements he made—all without optical aid—were remarkable for the time.

HIPPARCHUS' STAR CATALOGUE

Hipparchus erected an observatory on the island of Rhodes and built instruments with which he measured as accurately as possible the directions of objects in the sky. He compiled a star catalogue of about 850 entries. He designated for each star its celestial coordinates, that is, quantities analogous to latitude and longitude that specify its position (direction) in the sky. He also divided the stars according to their apparent brightnesses into six categories, or *magnitudes,* and specified the magnitude of each star. In the course of his observations of the stars, and in comparing his data with older observations, he made one of his most remarkable discoveries: the position in the sky of the north celestial pole had altered over the previous century and a half. Hipparchus correctly deduced that the direction of the axis about which the celestial sphere

COLOR PLATE 1 Armillary sphere of Antonio Santucci delle Pomerance, made for the Grand Duke Fernando I Medici in 1593. (Istituto e Museo di Storia della Scienza di Firenze)

COLOR PLATE 2a Telescopes used by Galileo. The longest has a wooden tube covered with paper, a focal length of 1.33 m, and an aperture of 26 mm. (Istituto e Museo di Storia della Scienza di Firenza)

COLOR PLATE 2b The Jantar Mantar in Delhi, India, is one of the last and best-preserved astronomical observatories from the pre-telescopic period. Its huge masonry instruments for measuring celestial positions were constructed in 1724 by Maharaja Jai Singh II. (Photo by D. Morrison)

COLOR PLATE 2c First editions of some books of great historical interest. (Crawford Library; Courtesy Astronomer Royal for Scotland, Royal Observatory Edinburgh)

COLOR PLATE 3a The Space Shuttle in orbit. (NASA)

COLOR PLATE 3b The Shuttle with its cargo bay doors open. (NASA)

COLOR PLATE 4a The Earth from space. Views like this help us appreciate that our Earth is indeed a planet. (NASA)

COLOR PLATE 4b The eclipsed Moon

COLOR PLATE 4c The solar eclipse of March 7, 1970. (NASA)

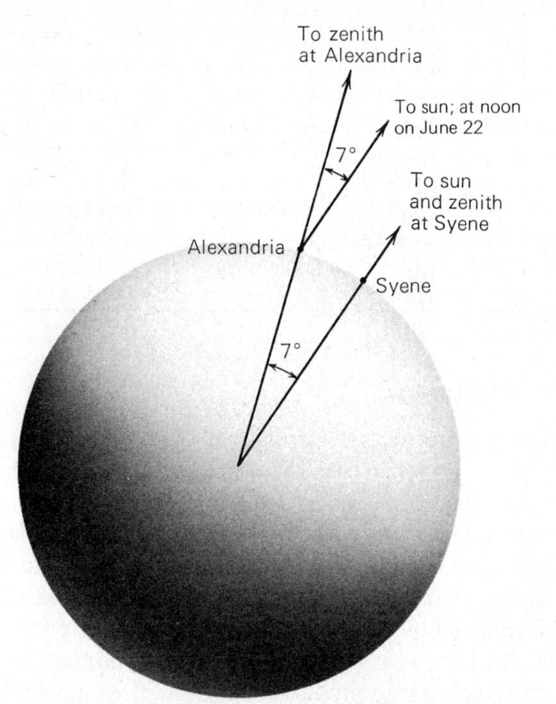

Figure 2.9 Eratosthenes' method of determining the size of the Earth.

appears to rotate continually changes. The real explanation for the phenomenon is that the direction of the Earth's rotational axis changes slowly because of the gravitational influence of the Moon and the sun, much as a top's axis describes a conical path as the Earth's gravitation tries to tumble the top over. This variation in orientation of the Earth's axis, called *precession,* requires about 26,000 years for one cycle.

OTHER MEASUREMENTS

Hipparchus, refining the technique first applied by Aristarchus, also obtained a good estimate of the Moon's size and distance. He used the correct value of ½° for the angular diameters of the sun and Moon and also the correct value 8⁄3 for the ratio of the diameter of the Earth's shadow to the diameter of the Moon. He tried several values for the relative distances of the sun and the Moon, including the value found by Aristarchus, but found that the exact distance assumed for the sun, provided it was large, did not have much effect on the figures he derived for the Moon. He found the Moon's distance to be 59 times the Earth's radius; the correct number is 60.

He determined the length of the year to within 6 minutes and even analyzed his possible errors, estimating that he could not be farther off than about 15 minutes. He also carefully observed the motions of the sun, Moon, and planets and found a method by which he could predict the position of the sun on any date of the year with an accuracy equal to the best observations and the position of the Moon with somewhat less accuracy. His work made possible the reliable prediction of eclipses, and with the information he left, astronomers thereafter could predict a lunar eclipse to within an hour or so.

THE MOTIONS OF THE SUN AND THE MOON

Hipparchus' study of the motion of the sun deserves special mention. The Earth's true orbit around the sun is not a circle but an ellipse; the Earth's distance from the sun and its orbital speed both vary slightly. Now we can account for the apparent motion of the sun by imagining it to move around the Earth in an elliptical path of exactly the same shape as the Earth's orbit. This apparent path of the sun, as we have seen, is the ecliptic. Because we see the sun's apparent orbit edge on (from the inside), the ecliptic is a circle around the sky. Moreover, the sun's eastward rate of motion on the ecliptic varies, exactly as the Earth's orbital speed varies. The variation in speed is slight but is observable.

Eudoxus of Cnidas (ca. 408–355 B.C.) had accounted for the sun's motion approximately by representing it with a series of rotating spheres pivoted one on the other. Later the mathematician Apollonius of Perga (latter half of the third century B.C.) suggested that the motions of all the heavenly bodies could be represented equally well by a combination of uniform circular motions. By uniform circular motion is meant a motion at a uniform speed about the circumference of a circle. Because the circle is the simplest geometrical figure, and because uniform motion seemed the most natural kind, Hipparchus, following the suggestion of Apollonius, attempted to find a combination of uniform circular motions that would account for the sun's apparently irregular behavior.

The plan he adopted was to represent the sun's orbit by an *eccentric,* a circle, but with the Earth slightly off center (Figure 2.10). The scheme was highly successful because the true orbit of the Earth is very close to a circle with the sun just off center. Now, one effect of the sun's variable speed on the ecliptic is to produce an inequality in the lengths of the seasons. Although the inequality had been known before, Hipparchus remeasured the small differences between the seasons' durations and from them deduced that the Earth's distance from the center of the sun's orbit must be 1/24 of the sun's distance. He found further that the Earth and sun were nearest each other in early December, which was correct at that time. (The date has changed over the thousands of years because of precession and, to a lesser extent, because of a slow

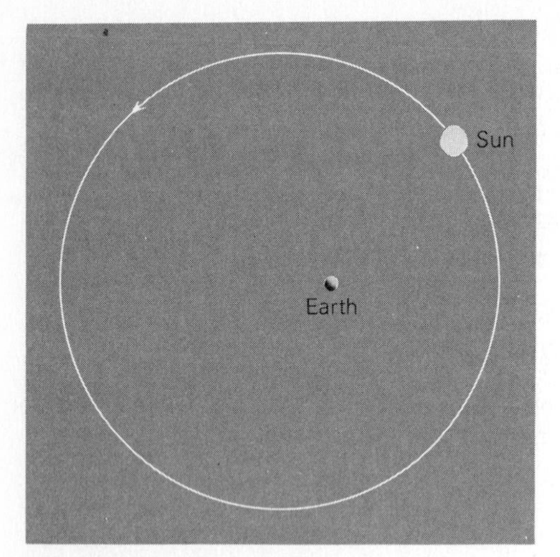

Figure 2.10 The eccentric.

motion of the long axis of the Earth's elliptical orbit; the closest approach now occurs in early January.)

Hipparchus pointed out that he could also have represented the sun's apparent motion by presuming it to move on the circumference of a portable circle called an *epicycle,* whose center, in turn, revolves about the Earth in a circle called a *deferent* (Figure 2.11). He considered the eccentric a simpler and thus preferable system.

The Moon's motion is more complicated, and Hipparchus was not quite so successful in finding a geometrical scheme to describe it. According to the model he adopted, the Moon went in a circle about a point near the Earth (an eccentric), but the center of the eccentric also revolved slowly about the Earth. Hipparchus measured the 9-year period of this revolution, as well as a 19-year period during which the intersections of the Moon's orbit with the ecliptic slide completely around the ecliptic, and the 5° inclination between the Moon's orbit and the ecliptic. The apparent motions of the planets are even more complicated than that of the Moon. Hipparchus thus declined to fit the planets into a cosmological scheme but rather made careful observations of their positions for use by later investigators.

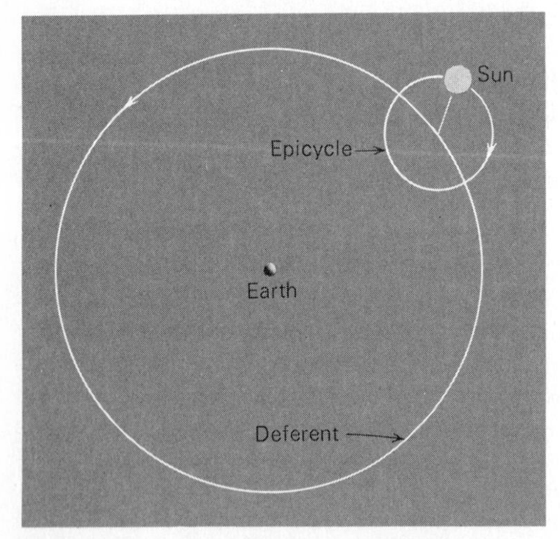

Figure 2.11 The deferent and epicycle.

(d) Ptolemy

The last great Greek astronomer of antiquity was Claudius Ptolemy (or Ptolemaeus), who lived around 140 A.D. He compiled a series of 13 volumes on astronomy known as the *Almagest*. All of the *Almagest* does not deal with Ptolemy's own work, for it includes a compilation of the astronomical achievements of the past, principally of Hipparchus. In fact, it is our main source of information about Greek astronomy. The *Almagest* also contains the contributions of Ptolemy himself.

THE DISTANCE TO THE MOON

One of Ptolemy's accomplishments was a new measurement of the distance to the Moon. The method he used, the principle of which is illustrated in Figure 2.12, makes use of the Moon's parallax, discussed by Hipparchus in connection with solar eclipses. Suppose we could observe the Moon directly overhead. We would have to be, then, at position *A* on the Earth, on a line between the center of the Earth *E,* and the center of the Moon *M.* Suppose that at the same time someone else at position *B* were to observe the angle *ZBM* between the Moon's direction and the point directly over his head, *Z.* The angle *MBE* would then be determined in the triangle *MBE* (it is 180° minus angle *ZBM*). The distance from *A* to *B* determines the angle *BEM.* For example, if *A* is one-twelfth of the way around the Earth from *B,* the angle *BEM* is 30°. The side *BE* is of course the radius of the Earth. We therefore know two angles and an included side of the triangle *MBE.* It is now possible to determine, either by trigonometry or geometrical construction, the distance *EM* between the centers of the Earth and Moon.

In practice, we do not need another observer at *B,* for the rotation of the Earth will carry us over there in a few hours anyway, and we can observe the angle *ZBM* then. We shall have to correct, however, for the

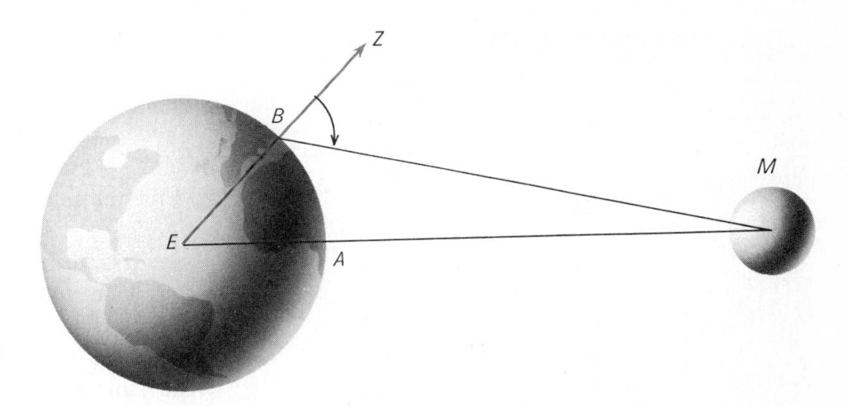

Figure 2.12 Ptolemy's method of finding the distance to the Moon.

motion of the Moon in its orbit during the interval between our two observations; the Moon's motion being known, the correction is a detail easily accomplished. Using the principle described, Ptolemy determined the Moon's distance to be 59 times the radius of the Earth—very nearly the correct value.

PTOLEMY'S SCHEME OF COSMOLOGY

Ptolemy's most important original contribution was a geometrical representation of the solar system that predicted the motions of the planets with considerable accuracy. Hipparchus, having determined by observation that earlier theories of the motions of the planets did not fit their actual behavior, and not having enough data on hand to solve the problem himself, instead amassed observational material for posterity to use. Ptolemy supplemented the material with observations of his own and with it produced a cosmological hypothesis that endured until the time of Copernicus.

The complicating factor in the analysis of the planetary motions is that their apparent wanderings in the sky result from the combination of their own motions and the Earth's orbital revolution. Notice, in Figure 2.13, the orbit of the Earth and the orbit

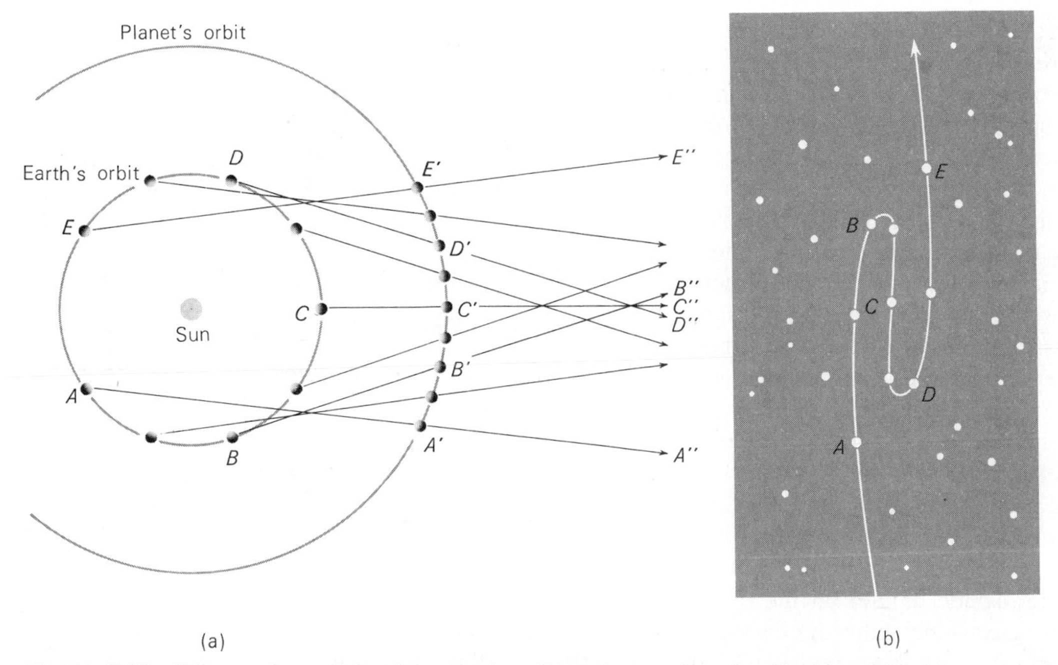

Figure 2.13 Retrograde motion of a superior planet (external to the Earth's orbit). (a) Actual positions of the planet and the Earth. (b) The apparent path of the planet as seen from the moving Earth, against the background of stars.

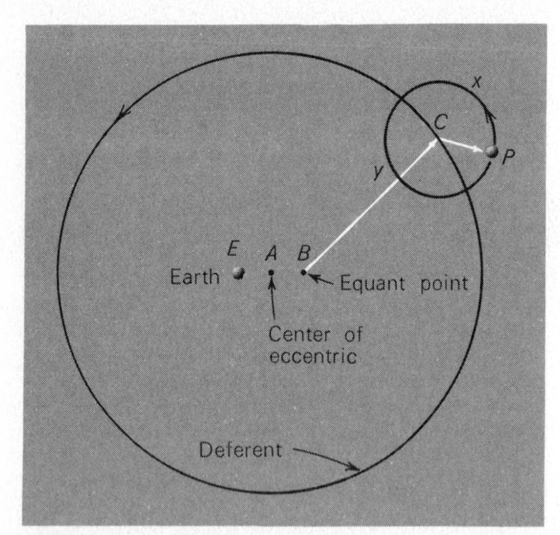

Figure 2.14 Ptolemy's cosmological system based on geocentric motion; the diagram shows the deferent, epicycle, eccentric, and equant.

of a hypothetical planet farther from the sun than the Earth. The Earth travels around the sun in the same direction as the planet and in nearly the same plane, but has a higher orbital speed. Consequently it periodically overtakes the planet, like a faster race car on the inside track. The apparent directions of the planet, seen from the Earth, are shown at successive intervals of time along lines $AA'A''$, $BB'B''$, and so on. In the right side of the figure we see the resultant apparent path of the planet among the stars. From positions B to D, as the Earth passes the planet, it appears to drift backward, to the *west* in the sky, even though it is actually moving to the *east*. Similarly, a slowly moving car appears to drift backward with respect to the distant scenery when we pass it in a faster-moving car. As the Earth rounds its orbit toward position E, the planet again takes up its usual eastward motion in the sky. The temporary westward motion of a planet as the Earth swings between it and the sun is called *retrograde* motion. (During and after its retrograde motion, the planet's apparent path in the sky does not trace exactly over itself because of the slight inclinations between the orbits of the Earth and other planets. Thus, the retrograde path is shown as an open loop in Figure 2.13.) Obviously, we need a different explanation for retrograde motion on the hypothesis that the planet is revolving about the Earth.

Ptolemy solved the problem by having a planet P (Figure 2.14) revolve in an epicyclic orbit about C. The center of the epicycle C in turn revolved in the deferent about the Earth. When the planet is at position x, it is moving in its epicyclic orbit in the same direction as the point C moves about the Earth, and the planet appears to be moving eastward. When the planet is at y, however, its epicyclic motion is in the opposite direction to the motion of C. By choosing the right combination of speeds and distances, Ptolemy succeeded in having the planet moving westward at the right speed at y and for the correct interval of time. However, because the planets, as does the Earth, travel about the sun in elliptical orbits, their actual behavior cannot be represented accurately by so simple a scheme of uniform circular motions. Consequently, Ptolemy made the deferent an eccentric, centered not on the Earth, but slightly away from the Earth at A. Furthermore, he had the center of the epicycle, C, move at a uniform angular rate, not around A, or E, but at point B, called the *equant*, on the opposite side of A from the Earth.

It is a tribute to the genius of Ptolemy as a mathematician that he was able to conceive such a complex system to account successfully for the observations. His hypothesis, with some modifications, was accepted as absolute authority throughout the Middle Ages, until it finally gave way to the heliocentric theory in the 17th century. In the *Almagest,* however, Ptolemy made no claim that his cosmological model described reality. He intended his scheme rather as a mathematical representation to predict the positions of the planets at any time. Modern astronomers do the same thing with algebraic formulas. Our modern mathematical methods were not available to Ptolemy; he had to use geometry.

2.4 ASTROLOGY

Modern research has shown that all matter in the universe is composed of atoms—and the same kinds of atoms. Thus our Viking space probes of Mars and our telescopic spectra of the light from the most remote quasars indicate that, whatever we do not yet understand about Mars and the quasars, at least they are made of the same stuff that makes up our own bodies.

Still, we cannot fault the ancients for assuming that the luminous orbs in the sky, the stars and planets, are made of "heavenly" substances and not of the "earthly" elements we find at home. In fact, the realization that celestial worlds are actually worlds and not ethereal substance is relatively recent in the history of science. Small wonder, then, that the ancients regarded the planets (including the sun and Moon), which alone moved about among the stars on the celestial sphere, as having special significance. Thus the planets came to be associated with the gods of ancient mythologies; in some cases, they were themselves thought of as gods. Even in the comparatively sophisticated Greece of antiquity, the planets had the names of gods and were credited with having the same powers and influences as the gods whose names they bore. From such ideas grew the religion of astrology.

Astrology began, we think, in the valley of the Euphrates and Tigris Rivers a millennium or so before Christ. The Mesopotamians and the Babylonians, believing that the planets and their motions influenced the fortunes of kings and nations, practiced what we call *mundane* astrology. When the Babylonian culture was absorbed by the Greeks, their astrology gradually influenced the entire western world and eventually spread to the Orient as well. By the third or second century B.C. the Greeks democratized astrology by developing the tradition that the planets influenced the life of every individual. In particular, they believed that the configuration of the planets at the moment of a person's birth affected his personality and fortune. This form of astrology, known as *natal* astrology, reached its acme with Ptolemy in the second century A.D. Ptolemy, as famous for his astrology as for his astronomy, compiled the *Tetrabiblos,* a treatise on astrology that remains the "bible" of the subject even today.

(a) The Horoscope

The key to natal astrology is the *horoscope,* a chart that shows the positions of the planets in the sky at the moment of an individual's birth. The charting of a horoscope, as of any map, requires the use of coordinates. The celestial coordinates used by astrology, in antiquity as well as today, are analogous to, and share a common origin with, those used by astronomers.

First, the planets (including the sun and the Moon—classed as planets by the ancients) are located in the sky with respect to the fixed stars on the celestial sphere by specifying their positions in the zodiac—the belt centered on the ecliptic that contains the planets. For the purposes of astrology, the zodiac is divided into twelve sectors called *signs,* each 30° long. Second, the constantly turning celestial sphere, with its stars and the planets, must have its orientation specified with respect to the Earth at the time and place of the subject's birth. For this purpose the sky is divided into twelve regions, called *houses,* that are fixed with respect to the horizon. Each day the turning sky carries the planets and signs through all of the houses. An example of a horoscope (George Abell's) is shown in Figure 2.15.

(b) Interpretation of the Horoscope

There are more or less standardized rules for the interpretation of the horoscope, many or most of which (at least in Western schools of astrology) are derived from the *Tetrabiblos* of Ptolemy. Each sign, each house, and each planet, the latter supposedly acting as a center of force, is associated with particular matters.

The interpretation of a horoscope is a very complicated business, and whereas the rules may be standardized, how each rule is to be weighed and applied is a matter of judgment—and "art." It also means that it is very difficult to tie astrology down to specific predictions.

The interpretation of an individual's horoscope, charted for the time and place of his birth, is *natal* astrology; his characteristics and fortunes, presumably, depend on his natal horoscope. Another branch of the subject is *horary* astrology, which purports to answer direct specific questions by casting a horoscope for the time and place at which the question was first posed. Horary astrology might be used, for example, to find whether the coming Monday would be a good time for a particular business deal.

A modern variant of natal astrology is *sun-sign astrology,* which uses only one element of the horoscope, the sign occupied by the sun at the time of a person's birth. Although even professional astrologers do not place much trust in such a limited scheme, which tries to fit everyone into just 12 groups, sun-sign astrology is the mainstay of newspaper astrology columns and party games, and apparently many people take it quite seriously. A recent poll showed that more than half of the teenagers, in the U.S. said they "believed in astrology."

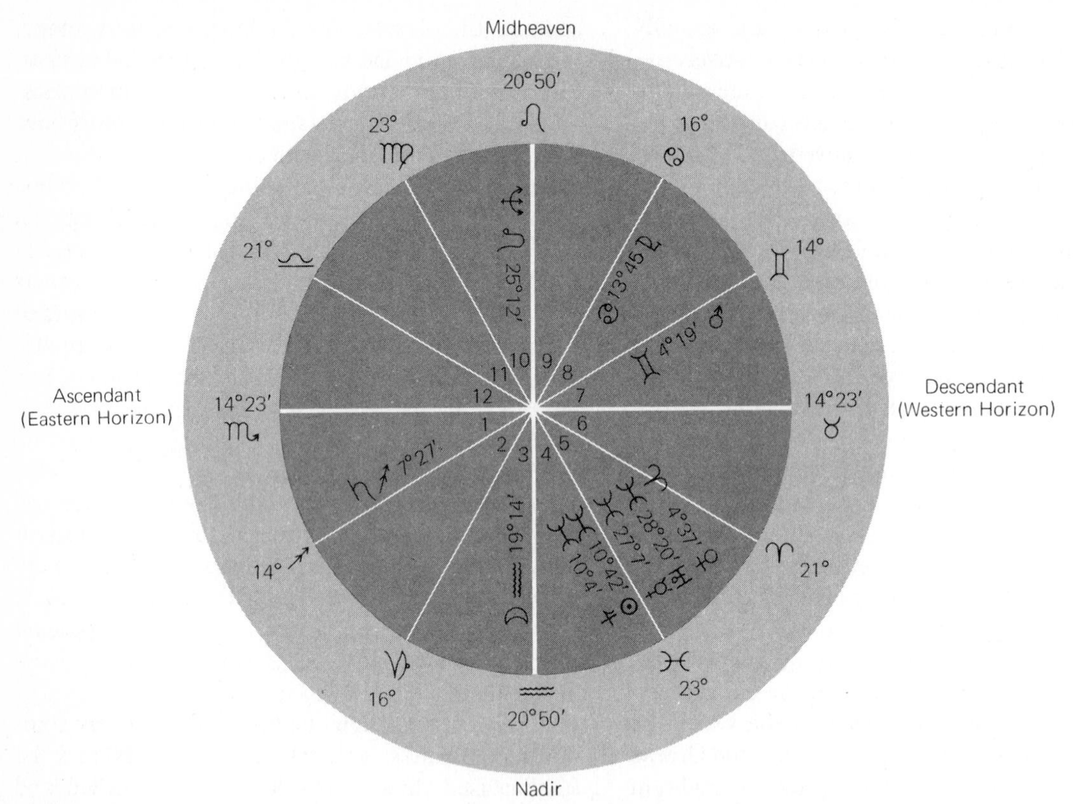

Figure 2.15 Natal horoscope of G. O. Abell, who was born in Los Angeles, California, on March 1, 1927, at 10:50 P.M., PST. The 12 pie-shaped sectors represent the 12 houses, and the outer circular zone represents the zodiac. The definition of houses used in preparing this horoscope is that of Placedus, in which, as the rotating celestial sphere carries the planets around the sky, each place in the zodiac spends equal time in each of the six houses above the horizon (diurnal houses) and also in the six houses below the horizon (nocturnal houses); however, the time required for an object to pass through a diurnal house is not the same as that required for it to pass through a nocturnal house except for objects on the celestial equator. The boundaries between the houses (cusps) intersect the ecliptic in the zodiacal signs indicated by their symbols in the outer circular zone. The number beside each sign symbol is the angular distance of the cusp from the beginning of that sign. The position of each planet is shown in the house it occupied at the instant of birth. Beside the symbol for the planet is the symbol of the zodiacal sign it was also in at that time, and the angular distance of the planet from the beginning of that sign. The places where the horizon intersects the zodiac are shown, and also the highest point of the ecliptic in the sky (midheaven) and its lowest point below the horizon, or nadir. (This astrological definition of the nadir is different from the astronomical one, in which the nadir is directly opposite the zenith.)

(c) Value of Astrology

Today, with our knowledge of the nature of the planets as physical bodies, composed as they are of rocks and fluids, it is hard to imagine that the directions of these planets in the sky at the moment of one's birth could have anything to do with his personality or future. The gravitational influence of the Moon and sun on tides is unquestionable, but tides produced on a person by a book in his hand are millions of times as strong as those produced by all the planets combined. The sun's light and heat are obviously of great importance to us, but even minute variations in the sun's irradiation are millions of times as great as the combined light of the planets. Jupiter (and to a lesser extent, the other planets) has a strong magnetic field and emits radio waves, but their detection requires magnetometers carried on space probes and large radio telescopes. The feeble radio signals from a small 1000-watt transmitter 100 miles away reach us with a strength millions of times as great as the radio waves from Jupiter, and can be picked up by a pocket radio. Even the magnet in the loudspeaker of that radio produces around the listeners a magnetic field enormously stronger than does Jupiter. Moreover, the distances of the planets from the Earth vary greatly, and any gravitational and radiation effects would vary as the inverse square of their distances—factors ignored by astrology.

Astrology would have to argue that there are unknown forces exerted by the planets that depend on their configurations with respect to each other and with respect to arbitrary coordinate systems invented by man—forces for which there is not a whit of solid evidence. Are astronauts on the Moon similarly affected by the same kind of force exerted by the Earth? Or is the Earth, alone, subject to these unknown laws of nature?

In the most orthodox astrology, one's entire life (and death) is predetermined by his natal horoscope. If a man dies in an auto accident at the age of 63 because someone else ran a stoplight, are we supposed to assume that all of the complicated chain of events that led to the circumstances of his being in that accident were blueprinted by the planets at the instant of his birth, but that all would have been different if he had been born two hours later? Most of us would find this assumption so incredible that we would need the most overwhelming evidence of its validity before taking it seriously. In the tens of centuries of astrology, no such evidence has been presented.

One could argue, on the other hand, that astrology only works statistically; that other influences—heredity and environment, for example—are important too, and that astrological influences are only important as tendencies, everything else being equal. In that case the reality of astrological effects could only be tested statistically. From time to time astrologers have presented statistical "proofs" of astrology, but not one survives objective scientific scrutiny.

During the past few years, a number of statistical tests of the predictive power of natal astrology have been carried out. The simplest of these examine sun-sign astrology to determine whether some signs are more likely than others to be associated with such objective measures of success as winning Olympic medals, earning high corporate salaries, or achieving elective office or high military rank. You could make such a test yourself using, for example, the birthdates of all members of Congress or of all members of the U.S. Olympic Team. But more sophisticated studies have also been done, involving horoscopes calculated for thousands of individuals. (With modern computers, the once laborious process of calculating a horoscope is practically instantaneous.) The results of all of these studies are the same: there is no evidence that natal astrology has any predictive power, even in a statistical sense.*

In retrospect, we can understand the belief in astrology on the part of ancient peoples who thought the heavenly bodies to be made of celestial material different from the elements that compose the Earth, and to be placed in the sky by their gods for the benefit of mankind. In the light of modern knowledge, the astrological claims seem so far-fetched as to be ludicrous. Because we would not expect the supposed influences, even in a statistical sense, we would want solid evidence and demonstrable predictions. Physical scientists and others who have investigated the subject with the hope of finding some grain of validity in it have found negative results. Virtually all scientists reject astrology as an unfounded superstition. Yet it continues to appeal to the popular fancy. The hope of predicting the future by magical or mystical means, and perhaps of transferring one's responsibilities and the blame for one's failures and misfortunes to an omnipotent power, continues to be a strong attraction. Moreover, it may simply be "fun" to speculate about the unknown and unprovable no matter how little basis there may be for it. Many astrologers today acknowledge that astrology cannot be proven by statistics or by experiment, but assert that it must be "known" or "realized" as knowledge or truth. In this context, it is outside the realm of science, and no rational argument based on the rules of science is relevant. To many astrology is still a religion, and hence is outside the scope of our consideration here.

---

* Most of these results have appeared or been summarized in the quarterly journal *Skeptical Enquirer,* which is devoted to the scientific investigation of paranormal phenomena. Authors Abell and Morrison have both served as consultants on astronomy and as members of the editorial board for this journal.

One fact remains: The practice of astrology in ancient times required the knowledge of the motions of the planets in order to construct horoscopes for past or future events. The quest to find a mechanism for charting the planets, joined with a natural curiosity about nature, stimulated centuries of observations and calculations, leading—as we shall see—to our modern technology.

EXERCISES

1. Where on Earth are all stars above the horizon at one time or another?
2. Where on Earth is only half the sky ever above the horizon?
3. Look up the names of the days of the week in French, Italian, and Spanish, and compare them with the names of the planets.
4. About what time of day or night does the Moon rise when it is full? When it is new?
5. Why can an eclipse of the Moon never occur on the day following a solar eclipse?
6. As seen by a terrestrial observer, which (if any) of the following can never appear in the opposite direction in the sky from the sun? in the same direction? at an angle of 90° from the sun? (a) Mars; (b) a star; (c) the sun; (d) Earth; (e) Jupiter; (f) the Moon; (g) Venus; (h) Mercury.
7. The Earth's diameter is about three and two-thirds times the diameter of the Moon. What is the angular diameter of the Earth as seen by an observer on the Moon?
8. Suppose Eratosthenes had found that at Alexandria at noon on the first day of summer the line to the sun makes an angle of 30° with the vertical. What then would he have found for the Earth's circumference?
9. Suppose Eratosthenes' results for the Earth's circumference were quite accurate. If the diameter of the Earth is 12,740 km, evaluate the length of his stadium in kilometers.
10. Why would Eratosthenes's method not have worked if the Earth were flat, like a pancake?
11. You are on a strange planet. You note that the stars do not rise or set, but that they circle around parallel to the horizon. Then you travel over the surface of the planet in one direction for 10,000 km, and at that new place you find that the stars rise straight up from the horizon in the east and set straight down in the west. What is the circumference of the planet? ***Answer:*** 40,000 km
12. Is retrograde motion observed for an inferior planet? Explain.
13. Many people try to use pseudostatistical arguments to justify their beliefs in a pseudoscience. Try the experiment of flipping a coin ten times and then recording the number of heads that turn up. Do this experiment 100 or more times (several people can flip coins at the same time and then pool results, thereby saving labor). Prepare a table showing how many times no head was obtained (ten tails in a row), how many times one head was obtained, how many times two heads, and so on. Make a graph showing the same data. What was the most frequent number of heads? What fraction of the time were less than three or more than seven heads obtained? If an event occurring only 1 percent of the time is enough to arouse your suspicions, how many heads would you have to obtain in a single experiment to question the honesty of the coin?

Nicolaus Copernicus (Mikolaj Kopernik) (1473–1543), Polish cleric and scientist, played a leading role in the emergence of modern science. While he could not prove that the Earth revolves about the sun, he presented such compelling arguments for this idea that he turned the tide of cosmological thought, laying the foundations upon which Galileo and Kepler so effectively built in the following century.

3

COPERNICUS AND THE HELIOCENTRIC HYPOTHESIS

In the 13 centuries following Ptolemy the most significant astronomical investigations were made by the Hindus and Arabs. The Hindus invented our system of numbers with place counting by tens. The Arabs brought the Hindu system of numbers to Europe and developed trigonometry. They also had access to some of the records of the Greek astronomers. Their greatest contribution was to provide continuity between ancient astronomy and the development of modern astronomy in the Renaissance.

Astronomy made no major advances in medieval Europe, where the prevailing philosophy was acceptance of the dogma of authority. Medieval cosmology combined the crystalline spheres of Pythagoras (as perpetuated by Aristotle) with the epicycles of Ptolemy. Astrology was widely practiced, however, and an interest in the motions of the planets was thus kept alive. Then came the Renaissance; in science the rebirth was clearly embodied in Nicholas Copernicus.

3.1 COPERNICUS

Nicholas Copernicus (in Polish, Mikolaj Kopernik, 1473–1543) was born in Torun on the Vistula in Poland. His training was in law and medicine, but Copernicus' main interest was astronomy and mathematics. By the time he had reached middle age, he was well known as an authority on astronomy.

Copernicus' great contribution to science was a critical reappraisal of the existing theories of cosmology and the development of a new model of the solar system. His unorthodox idea that the sun, not the Earth, is the center of the solar system had become known by 1515, chiefly through an early manuscript circulated by him and his friends.

His ideas were set forth in detail in his *De Revolutionibus*, published in the year of his death (Figure 3.1). Supervision over the publication of the book fell into the hands of a Lutheran preacher, Andrew Osiander, who was probably responsible for the augmented title of the work—*De Revolution-*

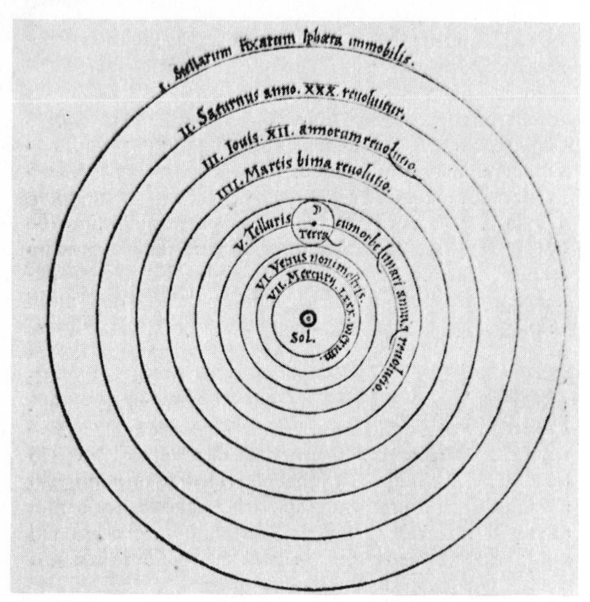

Figure 3.1 Geocentric plan of the solar system in the first edition of Copernicus' *De Revolutionibus.* (Crawford Collection, Royal Observatory Edinburgh)

ibus Orbium Celestium (On the Revolutions of the Celestial Spheres). Osiander wrote a preface, which he neglected to sign, expressing the view that science presented only an abstract mathematical hypothesis and implying that the theory set forth in the book was only a convenient calculating scheme. The preface was almost certainly in contradiction to Copernicus' own feelings.

In *De Revolutionibus,* Copernicus sets forth certain postulates from which he derives his system of planetary motions. His postulates include the assumptions that the universe is spherical and that the motions of the heavenly bodies must be made up of combinations of uniform circular motions; thus Copernicus was not free of all traditional prejudices. Yet, he evidently found something orderly and pleasing in the heliocentric system, and his defense of it was elegant and persuasive. His ideas, although not widely accepted until more than a century after his death, never disappeared and were ultimately of immense influence.

(a) Planetary Motions According to Copernicus

A person moving uniformly is not necessarily aware of his motion. We have all experienced the phenomenon of seeing an adjacent train, car, or ship appear to change position, only to discover that it is we who are moving (or vice versa). Copernicus argued that the apparent annual motion of the sun about the Earth could be equally well represented by a motion of the Earth about the sun, and that the rotation of the celestial sphere could be accounted for by assuming that the Earth rotates about a fixed axis while the celestial sphere is stationary. To the objection that if the Earth rotated about an axis it would fly into pieces, Copernicus answered that if such motion would tear the Earth apart, the even faster motion (because of its greater size) of the celestial sphere required by the alternative hypothesis would be even more devastating to it.

The important point that Copernicus made in *De Revolutionibus* is that the Earth is but one of six (then known) planets that revolve about the sun. Given this, he was able to work out the correct general picture of the solar system. He placed the planets, starting nearest the sun, in the order Mercury, Venus, Earth, Mars, Jupiter, and Saturn. Further, he deduced that the nearer a planet is to the sun, the greater is its orbital speed. Thus the retrograde motions of the planets (Section 2.3d) were easily understood without the necessity for epicycles. Also, Copernicus worked out the correct approximate scale of the solar system. To understand how, it will be helpful to define a few terms that describe the positions of planets in their orbits. These are illustrated in Figure 3.2.

A *superior planet* is any planet whose orbit is larger than that of the Earth, that is, a planet that is farther from the sun than the Earth is (Mars, Jupiter, and Saturn). An *inferior planet* is a planet closer to the sun than the Earth is (Venus and Mercury).

Every now and then, the Earth passes between a superior planet and the sun. Then that planet appears in exactly the opposite direction in the sky from the sun—or at least as nearly opposite as is allowed by the slight differences of inclination among the planes of the orbits of the planets. At such time, the planet rises at sunset, is above the horizon all night long, and sets at sunrise. We look one way to see the sun, and in the opposite direction to see the planet. The planet is then said to be in *opposition.*

On other occasions, a superior planet is on the other side of the sun from the Earth. It is then in

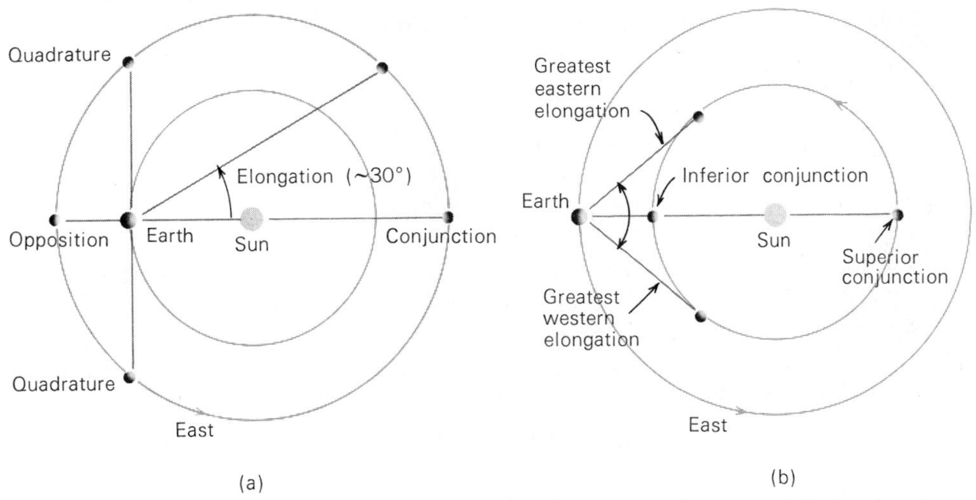

Figure 3.2 (a) Configurations of a superior planet; (b) configurations of an inferior planet.

the same direction from the Earth as the sun is, and of course is not visible. At such time, the planet is said to be in *conjunction*.

In between these extremes (but not halfway between), a superior planet may appear 90° away from the sun in the sky, so that a line from the Earth to the sun makes a right angle with the line from the Earth to the planet. Then the planet is said to be at *quadrature*. At quadrature, a planet rises or sets at either noon or midnight.

The angle formed at the Earth between the Earth-planet direction and the Earth-sun direction is called the planet's *elongation*. In other words, the elongation of a planet is its angular distance from the sun as seen from the Earth. At conjunction, a planet has an elongation of 0°, at opposition 180°, and at quadrature 90°.

An inferior planet can never be at opposition, for its orbit lies entirely within that of the Earth. The greatest angular distance from the sun, on either the east or west side, that the inferior planet can attain is called its *greatest eastern elongation* or *greatest western elongation*.

When an inferior planet passes between the Earth and sun, it is in the same direction from Earth as the sun and is said to be in *inferior conjunction*. When it passes on the far side of the sun from the Earth, and is again in the same direction as the sun, it is said to be at *superior conjunction*.

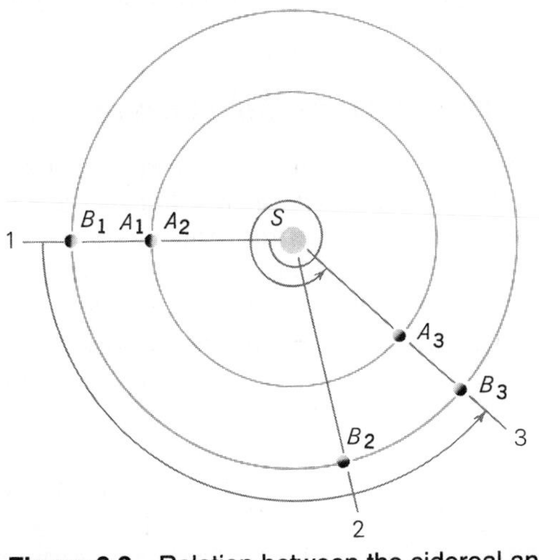

Figure 3.3 Relation between the sidereal and synodic periods of a planet.

SIDEREAL AND SYNODIC PERIODS OF A PLANET

Copernicus recognized the distinction between the *sidereal period* of a planet—that is, its actual period of revolution about the sun with respect to the fixed stars—and its *synodic period*, its apparent period of revolution about the sky with respect to the sun. The synodic period is also the time required for it to return to the same configuration, such as the time from opposition to opposition or from conjunction to conjunction.

Consider two planets, A and B, A moving faster in a smaller orbit (Figure 3.3). At position

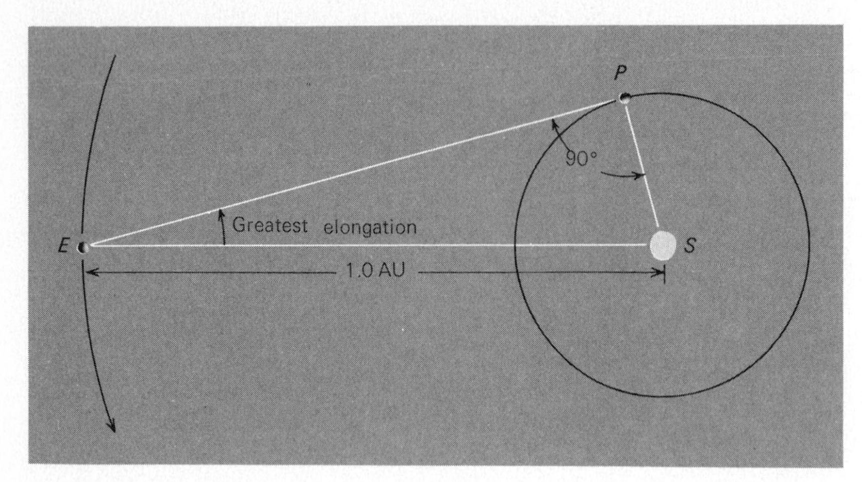

Figure 3.4 Determination of the distance of an inferior (inner) planet from the sun, relative to the Earth's distance.

(1), planet A passes between B and the sun S. Planet B is at opposition as seen from A, and A is in inferior conjunction as seen from B. When A has made one revolution about the sun and has returned to position (1), B has, in the meantime, moved on to position (2). In fact, A does not catch up with B until both planets reach position (3). Now planet A has gained one full lap on B. Planet A has revolved in its orbit through 360° *plus* the angle that B has described in traveling from position (1) to position (3) in its orbit. The time required for the faster-moving planet to gain a lap on the slower-moving one is the synodic period of one with respect to the other. If B is the Earth and A an inferior planet, the synodic period of A is the time required for the inferior planet to gain a lap on the Earth; if A is the Earth and B a superior planet, the synodic period of B is the time for the Earth to gain a lap on the superior planet.

What is observed directly from the Earth is the synodic, not the sidereal period of a planet. By reasoning along the lines outlined in the last paragraph, however, we can deduce the sidereal periods of the planets from their synodic periods. Let a planet's sidereal period be P years and its synodic period S years. In S years, the Earth, completing one revolution per year, must make S trips around the sun. (The quantity S, of course, can be less than 1, in which case the Earth would complete less than one circuit.) The other planet, completing one revolution in P years, would make, in S years, S/P trips around the sun. Consider first an inferior planet. It has made one more trip around the sun during its synodic period than has the Earth, so $S + 1 = S/P$, which, by rearrangement of terms, can be written

$$\frac{1}{P} = 1 + \frac{1}{S} \qquad \text{for an } \textit{inferior planet.}$$

For a superior planet, it is the Earth that gains the extra lap, and $S = S/P + 1$, which can be written

$$\frac{1}{P} = 1 - \frac{1}{S} \qquad \text{for a } \textit{superior planet.}$$

As an example, consider Jupiter, whose synodic period is 1.09211 years. Since Jupiter is a superior planet, $1/P = 1 - 1/1.09211 = 1 - 0.91566$, or $1/P = 0.08434$. Thus, $P = 1/0.08434 = 11.86$ years.

RELATIVE DISTANCES OF THE PLANETS

Copernicus was able to find the planets' distances from the sun relative to the Earth's. For the sake of illustration, let us assume that the orbits of the planets are precisely circular, even though that assumption is an oversimplification. The problem is particularly simple for the inferior planets. When an inferior planet is at greatest elongation (Figure 3.4), the line of sight from the Earth to the planet, EP, must be tangent to the orbit of the planet, and hence perpendicular to the line from the planet to the sun, PS. We have, therefore, a right triangle, EPS. The angle PES is observed (it is the greatest elongation), and the side ES is the Earth's distance from the sun. The planet's distance from the sun

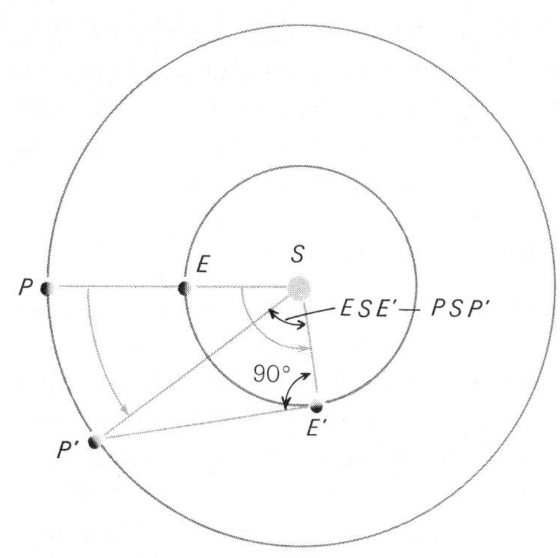

Figure 3.5 Determination of the distance of a superior (outer) planet from the sun, relative to the Earth's distance.

can then be found, in terms of the Earth's distance, by geometrical construction or by trigonometric calculation.

As a simple illustration of the procedure by which the distance of a superior planet can be found, suppose (Figure 3.5) the planet P is at opposition. We can now time the interval until the planet is next at quadrature; the planet is then at P' and the Earth at E'. With a knowledge of the sidereal periods of the planet and the Earth, we can calculate the fractions of their respective orbits that have been traversed by the two bodies. Thus the angles PSP' and ESE' can be determined, and subtraction gives the angle $P'SE'$ in the right triangle $P'SE'$. The side SE' is the Earth's distance from the sun, so enough data are available to solve the triangle and find the planet's distance from the sun, $P'S$ (again in terms of the Earth's distance), by construction or calculation.

The values obtained by Copernicus for the distances of the various planets from the sun, in units of Earth's distance, are summarized in Table 3.1. Also given are the values determined by modern measurement.

So far, we have discussed the Copernican theory as though Copernicus regarded the planets as having circular orbits centered on the sun. However, we recall that centuries earlier Ptolemy had introduced epicycles, eccentrics, and equants to account for those minor irregularities that arise because of deviations from uniform circular motion (actually because the true orbits of planets are ellipses). Copernicus rejected the equants of Ptolemy as unworthy of the perfection of heavenly bodies and instead introduced a system of eccentrics and small epicycles to take care of the irregularities.

TABLE 3.1 Distances of Planets from the Sun

PLANET	COPERNICUS	MODERN
Mercury	0.38	0.387
Venus	0.72	0.723
Earth	1.00	1.00
Mars	1.52	1.52
Jupiter	5.22	5.20
Saturn	9.18	9.54

Harvard astronomer Owen Gingerich, who has a special interest in the history of astronomy, has searched out in various libraries, public and private, more than 500 copies of early editions of *De Revolutionibus* to inspect the handwritten marginal annotations of the original owners of the books. Many of these readers turned out to be famous astronomers themselves. Gingerich has noted that the early part of *De Revolutionibus,* in which Copernicus sets forth his general plan of the heliocentric hypothesis, usually have rather few annotations; evidently this part, with its radical new cosmology, was not very attractive to astronomers of the 16th century. Well-read and marked-up portions of the books were the later ones, dealing with the rather dull details by which Copernicus was able to account for the motions of the planets without equants. It is an interesting comment on the insights of scholars of the generation following Copernicus. But of course not *all* astronomers took Copernicus' cosmological ideas lightly.

Philosophically, the main point of Copernicus' idea is that the Earth is not something special, but merely one of the several planets in revolution about the sun. The idea that we are at a *typical,* rather than a *special* place in the universe, is sometimes referred to as the *Copernican cosmological principle.* It will be brought up again and again in our investigation of the universe as a whole.

Contrary to popular belief, Copernicus did not *prove* that the Earth revolves about the sun. In fact, with some adjustments the old Ptolemaic system could have accounted as well for the motions of the planets in the sky. But the Ptolemaic cosmology was clumsy and lacked the beauty and coherence of its successor. Copernicus made the Earth an astronomical body, which brought a kind of unity to the universe. It was, to borrow from Neil Armstrong, a "giant step for mankind."

3.2 TYCHO BRAHE

Three years after the publication of *De Revolutionibus,* Tycho Brahe (1546–1601) was born of a family of Danish nobility. Tycho (as he is generally known) developed an early interest in astronomy and as a young man made significant astronomical observations.

In 1572 he observed a nova or "new star" (now believed to be a supernova—see Chapter 32) that rivaled the planet Venus in brilliance. Tycho observed the star for 16 months until it disappeared from naked-eye visibility. Now, we have seen (Section 2.3c) that the Moon exhibits a diurnal parallax, or apparent displacement in direction, because of the rotation of the Earth, which constantly shifts our position of observation. The effect is the same whether we regard it as being caused by the Earth's rotation or a rotation of the celestial sphere carrying the Moon about us. Tycho, despite the most careful observations, was unable to detect any parallax of his nova and accordingly concluded that it must be more distant than the Moon. This conclusion was of the utmost importance, for it showed that changes can occur in the celestial sphere, generally regarded as perfect and unchanging, apart from the regular motions of the planets.

The reputation of the young Tycho Brahe as an astronomer gained him the patronage of Frederick II, and in 1576 Tycho was able to establish a fine astronomical observatory on the Danish island of Hveen. The chief building of the observatory was named *Uraniborg*. The facilities at Hveen included a library, laboratory, living quarters, workshops, a printing press, and even a jail. There, for 20 years, Tycho and his assistants carried out the most complete and accurate astronomical observations yet made.

Unfortunately, Tycho was both arrogant and extravagant, and after Frederick II died, the new king, Christian IV, lost patience with the astronomer and eventually discontinued his support. Thus, in 1597 Tycho was forced to leave Denmark. He took up residence near Prague, taking with him some of his instruments and most of his records. There, as court astronomer for Emperor Rudolph II of Bohemia, Tycho Brahe spent the remaining years of his life analyzing the data accumulated over 20 years of observation. In 1600, the year before his death, he secured the assistance of a most able young mathematician, Johannes Kepler, who, like Tycho, was in exile from his native land.

(a) Tycho's Observations

Tycho, like others of his time and before him, believed that comets were luminous vapors in the Earth's atmosphere. In 1577, however, a bright comet appeared for which he could observe no parallax. Tycho concluded that the comet was at least three times as distant as the Moon and guessed that it probably revolved around the sun, in contradiction to earlier beliefs. Other comets were observed by him or his students in 1580, 1582, 1585, 1590, 1593, and 1596.

Tycho is most famous for his very accurate observations of the positions of the stars and planets. With instruments of his own design, he was able to make observations accurate to the limit of vision with the naked eye. The positions of the nine fundamental stars in his excellent star catalogue were accurate in most cases to within 1 arcminute. Only in one case was he off by as much as 2 arcminutes, and this was a star whose position was distorted by atmospheric refraction (see Chapter 8).

Tycho's observations included a continuous record of the positions of the sun, Moon, and planets. His daily observations of the sun, extending over years and comprising thousands of individual sightings, led to solar tables that were good to within 1 arcminute. He re-evaluated nearly every astronomical constant and determined the length of the year to within one second. His extensive and precise observations of planetary positions enabled him to note that there were variations in the positions of the planets from those given in published tables, and he even noted regularities in the variations.

(b) Tycho's Cosmology

Tycho rejected the Copernican heliocentric hypothesis on what seemed at the time to be very sound grounds. First, he found it difficult to reconcile a moving Earth with certain Biblical statements, nor could he even imagine an object as heavy and "sluggish" as the Earth to be in motion. The fact that he could not detect a parallax for even a single star, moreover, meant that the stars would have to be enormously distant if the Earth revolved around the

Figure 3.6 Tycho Brahe's observatory, Uraniborg. (Yerkes Observatory)

sun. The great void that would be required between the orbit of Saturn and the stars would alone have been enough to make him doubt the motion of the Earth; even more convincing to Tycho was the fact that he believed that he could measure the angular sizes of stars. The brightest of them he thought to be 2 arcminutes across. Now, the farther away an object is, the larger must be its true size in order that it have a given angular diameter. Tycho could not detect as much as 1 arcminute of parallax for any star, so it followed that the stars were so distant that, to have angular diameters of 2 arcminutes, their actual sizes would have to be twice the size of the entire orbit of the Earth. If they were still farther away, their diameters would have to be proportionally greater. (Later telescopic observations showed that the stars, unlike the planets, appear as luminous points; their disklike appearance to the naked eye is illusory.)

Tycho did, however, suggest an original system of cosmology, although it was not worked out in full detail. He envisioned the Earth in the center, with the sun revolving about the Earth each year, and with the other planets revolving about the sun in the order Mercury, Venus, Mars, Jupiter, and Saturn (Figure 3.7).

3.3 KEPLER

Johannes Kepler (1571–1630) was born in Weil-der-Stadt, Württemberg (southwestern Germany). He attended college at Tübingen and studied for a theological career. There he learned the principles of the Copernican system. He became an early con-

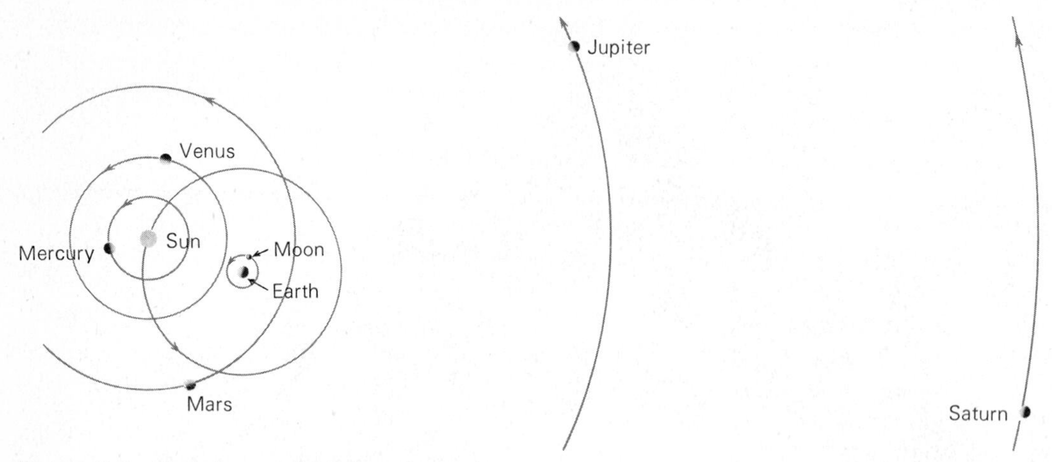

Figure 3.7 Tycho's model for the solar system.

vert to the heliocentric hypothesis and defended it in arguments with his fellow scholars.

In 1594, because of his facility as a mathematician, he was offered a position teaching mathematics and astronomy at the secondary school at Graz. As part of his duties at Graz, he prepared almanacs that gave astronomical and astrological data. Eventually, however, the power of the Catholic church in Graz grew to the point where Kepler, a Protestant, was forced to quit his post. Accordingly, he went to Prague to serve as an assistant to Tycho Brahe.

Tycho set Kepler to work trying to find a satisfactory theory of planetary motion—one that was compatible with the long series of observations made at Hveen. Brahe, however, was reluctant to supply Kepler with enough data to enable him to make substantial progress; perhaps Brahe was afraid of being "scooped" by the young mathematician. After Tycho's death, though, Kepler succeeded him as mathematician to the emperor Rudolph and obtained possession of the majority of Tycho's records. Their study occupied most of Kepler's time for more than 20 years.

(a) The Investigation of Mars

Kepler's most detailed study was of Mars, for which the observational data were the most extensive. He published the first results of his work in 1609 in *The New Astronomy*, or *Commentaries on the Motions of Mars*. He had spent several years trying to fit various combinations of circular motion, including eccentrics and equants, to the observed motion of Mars, but without success. At one point he found a hypothesis that agreed with observations to within 8′ (about one quarter the diameter of the full moon), but he believed that Tycho's observations could not have been in error by even this small amount, and so, with characteristic integrity, he discarded the hypothesis. Finally, Kepler tried to represent the orbits of Mars with an oval, and soon discovered that the orbit could be fitted very well by a curve known as an *ellipse*.

PROPERTIES OF THE ELLIPSE

Next to the circle, the ellipse is the simplest kind of closed curve. It belongs to a family of curves known as *conic sections* (Figure 3.8). A conic section is simply the curve of intersection between a hollow cone (whose base is presumed to extend downward indefinitely) and a plane that cuts through it. If the plane is perpendicular to the axis of the cone (or parallel to its base), the intersection is a circle. If the plane is inclined at an arbitrary angle, but still cuts completely through the surface of the cone, the resulting curve is an ellipse. If the plane is parallel to a line in the surface of the cone, it never quite cuts all the way through the cone, and the curve of intersection is open at one end. Such a curve is called a *parabola*. If the plane is inclined at an even smaller angle to the axis of the cone, an open curve results that is called a *hyperbola*. The ellipse, then, ranges from a circle at one extreme to a parabola at

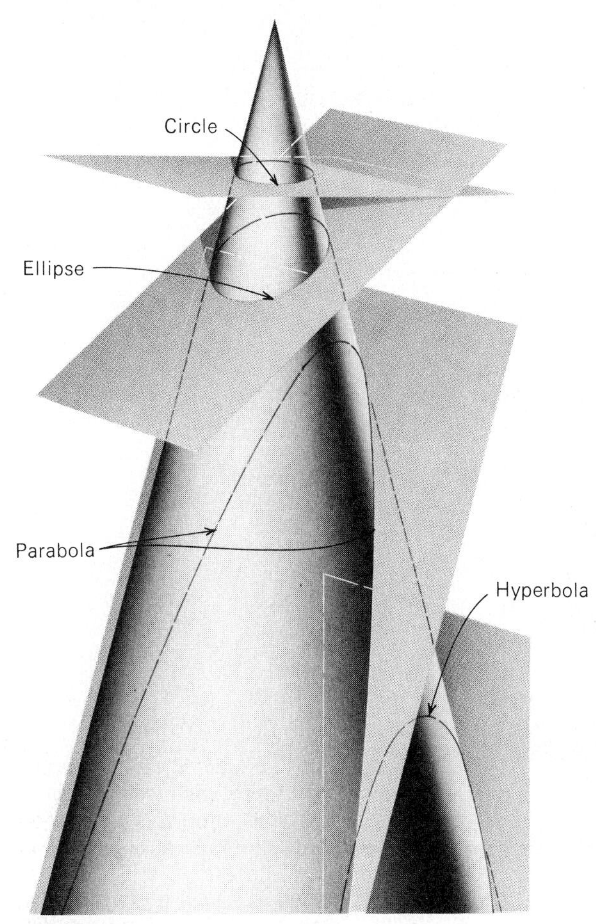

Figure 3.8 Conic sections.

the other. The parabola separates the family of ellipses from the family of hyperbolas.

An interesting and important property of an ellipse is that from *any point* on the curve the sum of the distances to two points inside the ellipse, called the *foci* of the ellipse, is the same. This property suggests a simple way to draw an ellipse. The ends of a length of string are tied to two tacks pushed through a sheet of paper into a drawing board, so that the string is slack. If a pencil is then pushed against the string, so that the string is held taut, and then slid against the string around the tacks (Figure 3.9), the curve that results is an ellipse; at any point where the pencil may be, the sum of the distances from the pencil to the two tacks is a constant length—the length of the string. The tacks, of course, are at the two foci of the ellipse.

The maximum diameter of the ellipse is called its *major axis*. Half the distance, that is, the distance from the center of the ellipse to one end, is the *semimajor axis*. The *size* of an ellipse depends on the length of the major axis. The *shape* of an ellipse depends on how close together the two foci are compared to the major axis. The ratio of the distance between the foci to the major axis is the length of the string, and the eccentricity is the distance between the tacks divided by the length of the string. If the foci (or tacks) coincide, the ellipse is a circle; a circle is, then, an ellipse of eccentricity zero. Ellipses of various shapes are obtained by varying the spacing of the tacks (as long as they are not farther apart than the length of the string). If one tack is removed to an infinite distance, and if enough

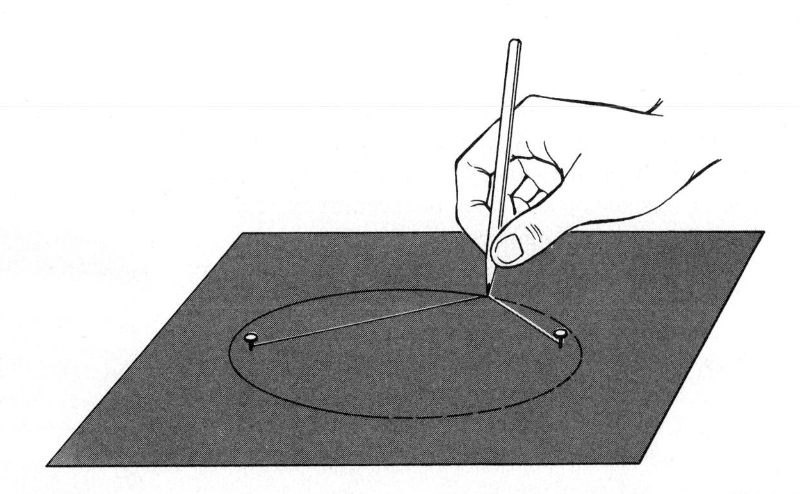

Figure 3.9 Drawing an ellipse.

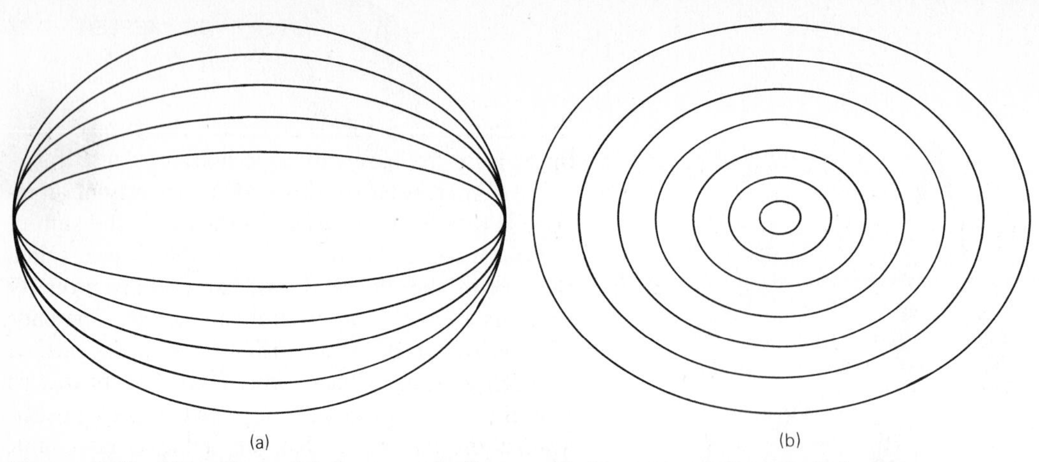

Figure 3.10 The ellipse. (a) A group of ellipses with the same major axis but various eccentricities. (b) Ellipses with the same eccentricity but various major axes.

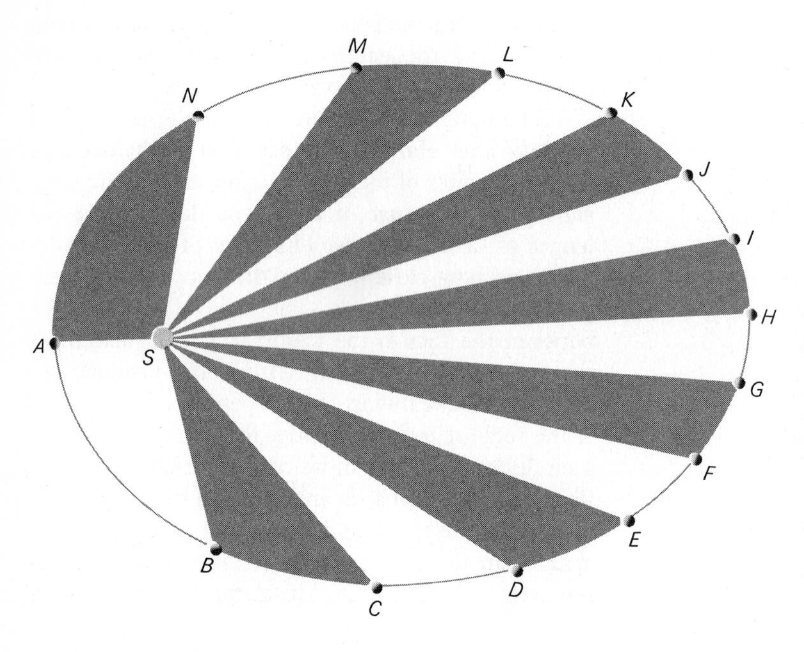

Figure 3.11 Kepler's law of equal areas. A planet moves most rapidly on its elliptical orbit when it is at position *A*, nearest the sun *(S)* at one focus of the ellipse. The orbital speed of the planet varies is such a way that in equal intervals of time it moves distances *AB, BC, CD,* and so on, so that the regions swept out by the line connecting the planet to the sun (alternating shaded and clear zones) are always the same area.

string is available, "our end" of the resulting, infinitely long ellipse is a parabola. A parabola has an eccentricity of one. An ellipse is completely specified by its major axis and its eccentricity. Figure 3.10 shows several ellipses.

Kepler found that Mars has an orbit that is an ellipse and that the sun is at one focus (the other focus is empty). The eccentricity of the orbit of Mars is only about 0.1; the orbit, drawn to scale, would be practically indistinguishable from a circle. It is a tribute to Tycho's observations and to Kepler's perseverance that he was able to determine that the orbit was an ellipse at all.

THE VARYING SPEED OF MARS

Before he saw that the orbit of Mars could be represented accurately by an ellipse, Kepler had already investigated the manner in which the planet's orbital speed varied. After some calculation, he found that Mars speeds up as it comes closer to the sun and slows down as it pulls away from the sun. Kepler expressed this relation by imagining that the sun and Mars are connected by a straight, elastic line. As Mars travels in its elliptical orbit around the sun, in equal intervals of time the areas swept out in space by this imaginary line are always equal (Figure 3.11). This relation is commonly called the *law of equal areas*.

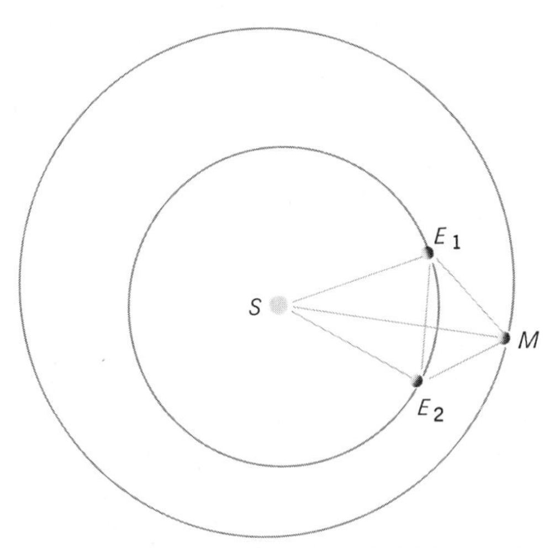

Figure 3.12 Kepler's method of triangulating the distance to Mars using observations made at different points along the Earth's orbit.

HOW KEPLER DETERMINED THE ORBIT OF MARS

Kepler determined the distance between Mars and the sun at various positions of the planet in its orbit by the process of *triangulation*. In Figure 3.12, S represents the sun and M represents Mars at some point in its path around the sun. Suppose we observe Mars when the Earth is at E_1. The angle SE_1M at the Earth between Mars and the sun is observable. Since the sidereal period of Mars is 687 days, after 687 days Mars will return to point M. The Earth, meanwhile, will have completed nearly two full revolutions around the sun and will be at E_2. Angle SE_2M can now be observed. In exactly 2 years, or 730½ days, the Earth will have returned to E_1. The Earth is short by $730\frac{1}{2} - 687 = 43\frac{1}{2}$ days of completing two revolutions about the sun. Thus the angle E_1SE_2 is known—it is the angle through which the Earth moves in 43½ days. Lines SE_1 and SE_2 are each the Earth's distance from the sun. Thus two sides and an included angle of the triangle E_1SE_2 are known, and the triangle can be solved for the side E_1E_2, in terms of the distance from the Earth to the sun, and for the angles SE_1E_2 and SE_2E_1.

Subtraction of angles SE_1E_2 and SE_2E_1 from SE_1M and SE_2M, respectively, gives the angles E_2E_1M and E_1E_2M, both in the triangle E_1ME_2. In that latter triangle since two angles and an included side are now known, sides E_1M and E_2M and the third angle can be found. Finally, the distance of Mars from the sun (but in terms of the Earth's distance) can be found from either triangle SE_1M or SE_2M.

Kepler found the distance of Mars from the sun at five points along its orbit by choosing from Tycho's records the elongations of Mars on each of five pairs of dates separated from each other by intervals of 687 days.

KEPLER'S FIRST TWO LAWS OF PLANETARY MOTION SUMMARIZED

We may summarize the most important contributions in *The New Astronomy*, or *Commentaries on the Motions of Mars*, by stating what are now known as Kepler's first two *laws of planetary motion:*

KEPLER'S FIRST LAW: Each planet moves about the sun in an orbit that is an ellipse, with the sun at one focus of the ellipse.

KEPLER'S SECOND LAW (THE LAW OF AREAS): The straight line joining a planet and the sun sweeps out equal areas in the orbital plane in equal intervals of time.

At the time of publication of the *New Astronomy* (1609), Kepler appears to have demonstrated the validity of these two laws only for the case of Mars. However, he expressed the opinion that they held also for the other planets.

(b) The Harmony of the Worlds

Kepler believed in an underlying harmony in nature, and he constantly searched for numerological relations in the celestial realm. It was a great personal triumph, therefore, that he found a simple algebraic relation between the lengths of the semimajor axes of the planets' orbits and their sidereal periods. Because planetary orbits are elliptical, the distance between a given planet and the sun varies. Now, the major axis of a planet's orbit is the sum of its maximum and minimum distances from the sun. Therefore, half of this sum, the semimajor axis, can be thought of as the *average* distance of a planet from the sun. In a circular orbit, the semimajor axis is simply the radius of the circle.

Kepler published his discovery in 1619 in *The Harmony of the Worlds*. The relation is now known as his *third*, or *harmonic*, *law*.

KEPLER'S THIRD LAW: The squares of the sidereal periods of the planets are in direct proportion to the cubes of the semimajor axes of their orbits.

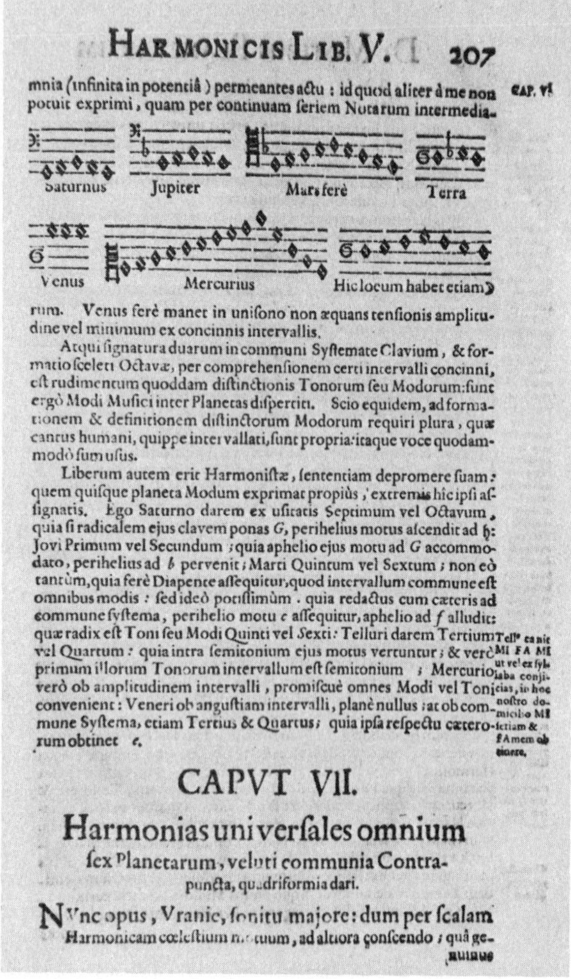

HARMONICIS LIB. V. 207

mnia (infinita in potentiâ) permeantes actu: id quod aliter à me non potuit exprimi, quam per continuam ſeriem Notarum intermedia-

CAP. VI

rum. Venus ferè manet in uniſono non æquans tenſionis amplitudine vel minimum ex concinnis intervallis.

Atqui ſignatura duarum in communi Syſtemate Clavium, & formatio ſceleti Octavæ, per comprehenſionem certi intervalli concinni, eſt rudimentum quoddam diſtinctionis Tonorum ſeu Modorum: ſunt ergò Modi Muſici inter Planetas diſpertiti. Scio equidem, ad formationem & definitionem diſtinctorum Modorum requiri plura, quæ cantus humani, quippe intervallati, ſunt propria: itaque voce quodammodò ſum uſus.

Liberum autem erit Harmoniſtæ, ſententiam depromere ſuam: quem quiſque planeta Modum exprimat propiùs, extremis hîc ipſi aſſignatis. Ego Saturno darem ex uſitatis Septimum vel Octavum, quia ſi radicalem ejus clavem ponas *G*, perihelius motus aſcendit ad ♮: Jovi Primum vel Secundum; quia aphelio ejus motu ad *G* accommodato, perihelius ad *b* pervenit; Marti Quintum vel Sextum; non eò tantùm, quia ferè Diapente aſſequitur, quod intervallum commune eſt omnibus modis: ſed ideò potiſſimùm. quia redactus cum cæteris ad commune ſyſtema, perihelio motu *c* aſſequitur, aphelio ad *f* alludit: quæ radix eſt Toni ſeu Modi Quinti vel Sexti: Telluri darem Tertium vel Quartum: quia intra ſemitonium ejus motus vertuntur; & verò primum illorum Tonorum intervallum eſt ſemitonium; Mercurio verò ob amplitudinem intervalli, promiſcuè omnes Modi vel Toni convenient: Veneri ob anguſtiam intervalli, planè nullus; at ob commune Syſtema, etiam Tertius & Quartus; quia ipſa reſpectu cæterorum obtinet *e*.

Tell⁹ canit MI FA MI ut vel ex ſyllaba conjicias, in hoc noſtro domicilio MIſeriam & FAmem obtinere.

CAPVT VII.

Harmonias univerſales omnium ſex Planetarum, veluti communia Contrapuncta, quadriformia dari.

NVnc opus, Vranie, ſonitu majore: dum per ſcalam Harmonicam cœleſtium motuum, ad altiora conſcendo; quâ genuinue

Figure 3.13 Detail from Kepler's *Harmony of the Worlds*. (Crawford Collection, Royal Observatory Edinburgh)

It is simplest to express Kepler's third law with the algebraic equation:

$$P^2 = Ka^3,$$

where P represents the sidereal period of the planet, a is the semimajor axis of its orbit, and K is a numerical constant whose value depends on the kinds of units chosen to measure time and distance. It is convenient to choose for the unit of time the Earth's period—the year—and for the unit of distance the semimajor axis of the Earth's orbit—the *astronomical unit* (AU). Note that the values found for the distances of the planets from the sun, both as determined by Copernicus and by Kepler, are in astronomical units. With this choice of units, $K = 1$, and Kepler's third law can be written

$$P^2 = a^3.$$

We see that to arrive at his third law it was not necessary for Kepler to know the *actual* distances of the planets from the sun (say, in kilometers), only the distance in units of the Earth's distance, the astronomical unit. The length of the astronomical unit in kilometers was not determined accurately until later.

As an example of Kepler's third law, consider Mars. The semimajor axis, a, of Mars' orbit is 1.524 AU. The cube of 1.524 is 3.54. According to the above formula, the period of Mars, in years, should be the square root of 3.54, or 1.88 years, a result that is in agreement with observations. Table 3.2 gives for each of the six planets known to Kepler the modern values of a, P, a^3, and P^2. To the limit of accuracy of the data given, we see that Kepler's law holds exactly, except for Jupiter and Saturn, for which there are very slight discrepancies. Decades later, Newton gave an explanation for the discrepancies, but within the limit of accuracy of the observational data available in 1619, Kepler was justified in considering his formula to be exact.

Much of the rest of *Harmony of the Worlds* deals with Kepler's attempts to associate numerical relations in the solar system with music; indeed, he tried to derive notes of music played by the planets as they move harmoniously in their orbits. The Earth, for example, plays the notes *mi, fa, mi,* which he took to symbolize the "*miseria* (misery), *fames* (famine), *miseria*" of our planet.

(c) The *Epitome*

In 1618, 1620, and 1621, Kepler published (in installments) his text on astronomy, the *Epitome of the Copernican Astronomy*. The book includes accounts of discoveries, both by himself and by Galileo, and firmly supports the Copernican view. Here, for the first time, Kepler implies that his first two laws had been tested and found valid for the other planets besides Mars (including the Earth) and for the Moon. Also, he states that the harmonic (third) law applies to the motions of the four newly discovered satellites of Jupiter as well as to the motions of the planets about the sun.

TABLE 3.2 Observational Test of Kepler's Third Law

PLANET	SEMIMAJOR AXIS OF ORBIT, a (AU)	SIDEREAL PERIOD, P (Years)	a^3	P^2
Mercury	0.387	0.241	0.058	0.058
Venus	0.723	0.615	0.378	0.378
Earth	1.000	1.000	1.000	1.000
Mars	1.524	1.881	3.537	3.537
Jupiter	5.203	11.862	140.8	140.7
Saturn	9.534	29.456	867.9	867.7

KEPLER'S DISCUSSION OF THE DISTANCE TO THE SUN

We recall (Section 2.3) that both Hipparchus and Ptolemy had rather accurately measured the Moon's distance to be about 60 times the radius of the Earth. Earlier, Aristarchus (Section 2.3a) had found that the sun was 18 to 20 times as far away as the Moon, which placed the sun at a distance of at most 1200 Earth radii. This figure survived until the 17th century, when Kepler revised it upward. Kepler was not able to detect any *diurnal parallax* of Mars, that is, any apparent shift in direction caused by the rotation of the Earth carrying the observer from one side of the Earth to the other. Now the distance to Mars was known only in terms of the Earth's distance from the sun—the astronomical unit; the actual assumed distance to Mars would be proportional to whatever value was assumed for the astronomical unit. If the latter were only 1200 Earth radii, Mars should be near enough to allow observation of such a daily parallax. Kepler concluded that the astronomical unit must be at least three times the accepted figure—still a value seven times too small but nevertheless an improvement.

3.4 GALILEO

Galileo Galilei (1564–1642), the great Italian contemporary of Kepler, was born in Pisa. Galileo, like Copernicus, began training for a medical career, but he had little interest in the subject and later switched to mathematics. In school he incurred the wrath of his professors by refusing to accept on faith dogmatic statements based solely on the authority of great writers of the past. From his classmates he gained the nickname "Wrangler."

For financial reasons, Galileo was never able to complete his formal university training. Nevertheless, his exceptional ability as a mathematician gained him the post, in 1589, of professor of mathematics and astronomy at the university at Pisa. In 1592 he obtained a far better position at the university at Padua, where he remained until 1610, when he left to become mathematician to the Grand Duke of Tuscany. While at Padua he became famous throughout Europe as a brilliant lecturer and as a foremost scientific investigator.

(a) Galileo's Experiments in Mechanics

Galileo's greatest contributions were in the field of *mechanics* (the study of motion and the actions of forces on bodies). The principles of mechanics outlined by Aristotle had still not been completely discarded. Although the seeds of experimental science had been sown by certain of the later Greek scholars, notably Archimedes, the practice of performing experiments to learn physical laws was not standard procedure even in Galileo's time.

Galileo experimented with pendulums, with balls rolling down inclined planes, with light and mirrors, with falling bodies, and many other objects. Aristotle had said that heavy objects fall faster than lighter ones. Galileo argued that if a heavy and light object were dropped together, even from a great height, both would hit the ground at practically the same time. What little difference there was could easily be accounted for by the resistance of the air. Galileo then tested these conflicting ideas by performing the experiment—still a novel approach in his day to understanding natural phenomena.

LAWS OF MOTION

In the course of his experiments, Galileo discovered laws that invariably described the behavior of physical objects. The most far-reaching of these is the *law of inertia* (now known universally as Newton's first law). The inertia of a body is that property of the body that resists any change of motion. It was familiar to all persons then as it is to us now that if a body is at rest it tends to remain at rest, and requires some outside influence to start it in motion. Rest was thus generally regarded as the *natural state of matter*. Galileo showed, however, that rest was no more natural than motion. If an object is slid along a rough horizontal floor, it soon comes to rest, because friction between it and the floor acts as a retarding force. However, if the floor and object are both highly polished, the body, given the same initial speed, will slide farther before coming to rest. On a smooth layer of ice, it will slide farther still. Galileo noted that the less the retarding force, the less the body's tendency to slow down, and he reasoned that if all resisting effects could be removed (for example, the friction of the floor or ground, and of the air) the body would continue in a steady state of motion indefinitely. In fact, he argued, not only is a force required to start an object moving from rest, but a force is also required to slow down, stop, speed up, or change the direction of a moving object.

Galileo also studied the way bodies accelerated, that is, changed their speed, as they fell freely, or rolled down inclined planes. He found that such bodies accelerate uniformly, that is, in equal intervals of time they gain equal increments in speed. Galileo formulated these newly found laws in precise mathematical terms that enabled one to predict, in future experiments, how far and how fast bodies would move in various lengths of time. It remained for Newton to incorporate and generalize Galileo's principles into a few simple laws so fundamental that they have become the basis of a great part of our modern technology (Chapter 4).

(b) Galileo's Astronomical Contributions

Sometime in the 1590s Galileo accepted the Copernican hypothesis of the solar system. In Roman Catholic Italy, this was not a popular philosophy, for the Church authorities still upheld the ideas of Aristotle and Ptolemy. It was primarily because of Galileo that in 1616 the Church issued a prohibition decree which stated that the Copernican doctrine was "false and absurd" and was not to be held or defended.

The prevailing notion of the time was that the celestial bodies belonged to the realm of the heavens where all is perfect, unchanging, and incorruptible. Perpetual circular motion, being the "perfect" kind of motion, was regarded as the natural state of affairs for those heavenly bodies. Once Galileo had established the principle of inertia—that on the Earth bodies in undisturbed motion remain in motion—it was no longer necessary to ascribe any special status to the fact that the planets remain perpetually in orbit. By the same token, even the Earth could continue to move, once started. What *does* need to be explained is why the planets move in curved paths around the sun rather than in straight lines. Evidently, Galileo was sufficiently imbued with Aristotelian concepts that he accepted uniform circular celestial motion without subjecting the planets to the same objective scrutiny that he applied in his terrestrial experiments.

In answer to the common objection that objects could not remain on the Earth if it were in motion, Galileo noted that if a stone is dropped from the masthead of a moving ship it does not fall behind the ship and land in the water beyond its stern, but rather lands at the foot of the mast, for the stone already has a forward inertia gained from its common motion with the ship before it is dropped. In an analogous way, objects on the Earth would not be swept off and left behind if the Earth were moving, for they share the Earth's forward motion.

GALILEO'S TELESCOPES

It is not certain when the principle was first conceived of combining two or more pieces of glass to produce an instrument that enlarged distant objects, making them appear nearer. Claims for the discovery exist as early as the time of Roger Bacon (13th century). At any rate, the first telescopes that attracted much notice were made by the Dutch spectacle-maker Hans Lippershey in 1608. Galileo heard of the discovery in 1609, and without ever having seen an assembled telescope, he constructed one of his own with a three-power magnification, that is, it made distant objects appear three times nearer and larger. He quickly built other instruments, his best with a magnification of about 30.

SIDEREAL MESSENGER

The usefulness of the telescope for terrestrial observations (including its military applications) was apparent to many people. But the idea that the telescope could also reveal new insights about the heavens was less obvious. There was a long tradition that the human eye was the best possible measure of truth, while lenses and mirrors distorted. This idea is preserved today in the expression "it was done with mirrors," meaning that someone has tricked us.

Galileo himself was slow to turn his new optical toy toward the sky, and he first seems to have con-

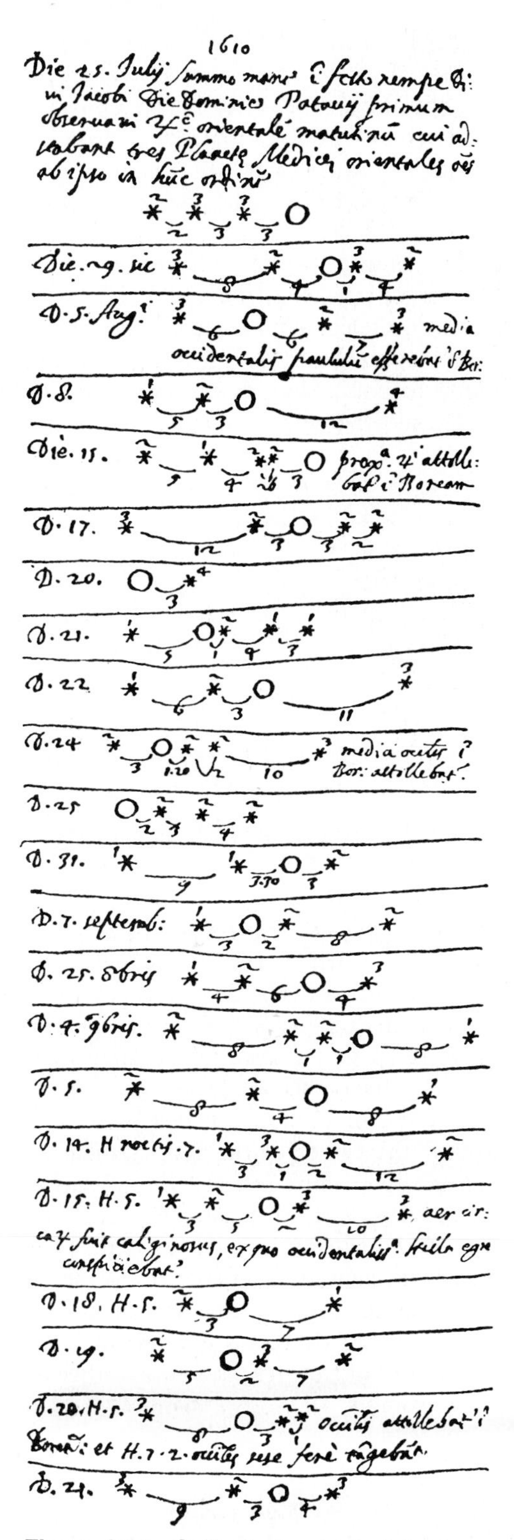

Figure 3.14 Galileo's drawings of Jupiter and its satellites.

ducted extensive tests to convince himself that the magnified image was an accurate representation of a distant scene. Late in 1609 he began his astronomical work. While he was not the only person to try using optical aids to study the sky, Galileo applied himself to this task with his characteristic care and persistence. In 1610 he startled the world by publishing a list of his remarkable discoveries in a small book, *Sidereal Messenger (Sidereus Nuncius)*.

Galileo found that many stars too faint to be seen with the naked eye became visible with his telescope. In particular, he found that some nebulous blurs resolved into many stars (for example, the Praesepe in Cancer) and that the Milky Way was made up of multitudes of individual stars. He found that Jupiter had four satellites or moons revolving about it with periods ranging from just under 2 days to about 17 days (12 other satellites of Jupiter have been found since). This discovery was particularly important because it showed that there could be centers of motion that in turn are in motion. It had been argued that if the Earth were in motion the Moon would be left behind, because it could hardly keep up with the rapidly moving planet. Yet here were Jupiter's satellites doing exactly that!

PHASES OF VENUS

Another important telescopic discovery that strongly supported the Copernican view was the fact that Venus goes through phases like the Moon. In the Ptolemaic system, Venus is always closer to the Earth than is the sun, and thus, because Venus never has more than about 45° elongation, it would never be able to turn its fully illuminated surface to our view—it would always appear as a crescent. Galileo, however, saw that Venus went through both crescent and gibbous phases, and concluded that it must travel around the sun, passing at times behind and beyond it, rather than revolving directly around the Earth (Figure 3.15). Mercury also goes through all phases.

IRREGULARITIES IN THE HEAVENS

Galileo's observations revealed much about our nearest neighbor, the Moon. He saw craters, mountain ranges, valleys, and flat dark areas that he guessed might be water (the dark *maria*, or "seas," on the Moon were thought to be water until long after Galileo's time). Not only did these discoveries show that the heavenly bodies, regarded as perfect,

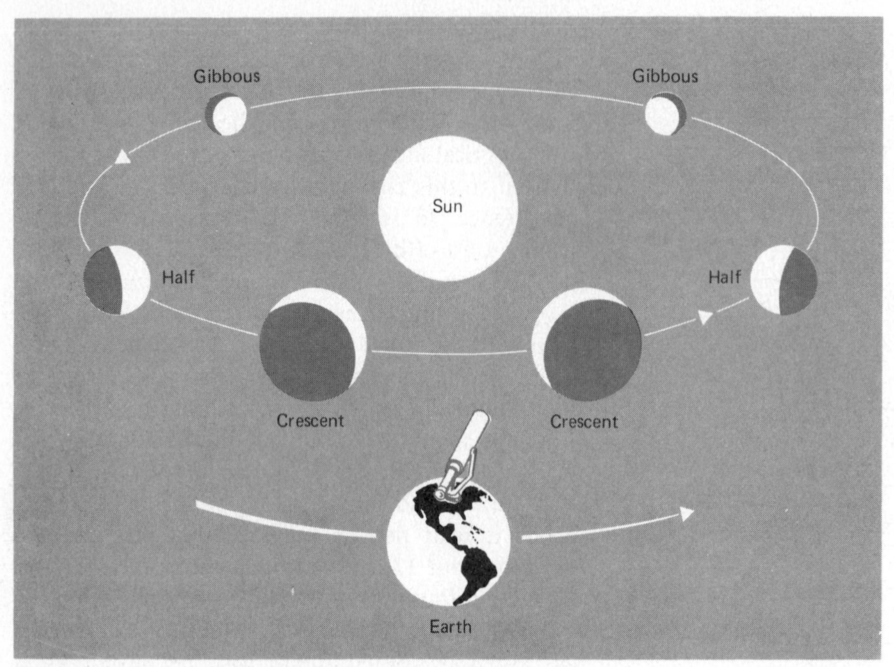

Figure 3.15 Phases of an inferior (inner) planet.

smooth, and incorruptible, do indeed have irregularities, as does the Earth, but they showed the Moon to be not so dissimilar to the Earth, which suggested that the Earth, too, could belong to the realm of celestial bodies.

Galileo also found that Saturn seemed strange, with what looked like appendages on either side of the planet. The best explanation he could suggest was that Saturn had two very large satellites, making it a triple planet. It was not until 1655 that Huygens, using improved optics, recognized the magnificent system of rings around Saturn (Chapter 18).

One of Galileo's most disturbing observations, to his contemporaries, was of spots on the sun, showing that this body also had "blemishes." Sunspots are now known to be large, comparatively cool areas on the sun that appear dark because of their contrast with the brighter and hotter solar surface. Sunspots are temporary, lasting usually only a few weeks to a few months. Large sunspots actually had been observed before, with the unaided eye, but were generally regarded either as something in the Earth's atmosphere or as planets between the Earth and the sun silhouetting themselves against the sun's disk in the sky. In fact, some of Galileo's critics attempted to explain the spots as satellites revolving about the sun.

Galileo observed the spots to move, day by day, across the disk of the sun. He also noted that they moved most rapidly when near the center of the sun's disk and increasingly slowly as they approached the limb (the sun's limb is its apparent "edge" as we see it in the sky). Often, after about two weeks, the same spots would reappear on the opposite limb, and move slowly at first, then more rapidly toward the center of the disk. Galileo explained that the spots must be either on the surface of the sun or very close to it and that they were carried around the sun by its own rotation. Their variable speed, he showed, is an effect of foreshortening; when near the center of the sun's disk they are being carried directly across our line of sight, but near the limb most of their motion is either toward or away from us. He determined the sun's period of rotation to be a little under a month.

(c) *Dialogue on the Two Great World Systems*

As we have seen, Galileo had accumulated a great deal of evidence to support the Copernican system.

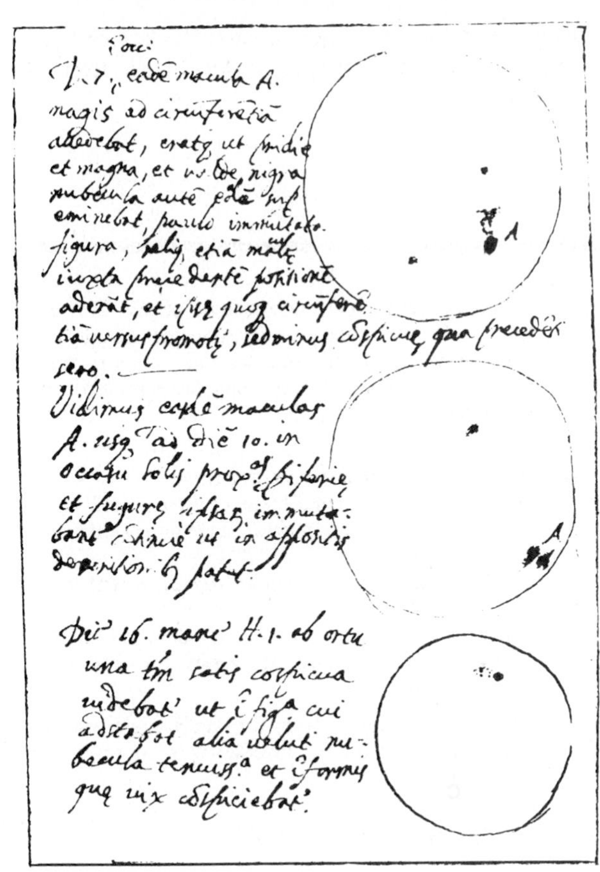

Figure 3.16 Galileo's drawings of sunspots. (Yerkes Observatory)

Figure 3.17 Villa Galileo, in Arcetri (near Florence), where Galileo spent the last years of his life in home imprisonment. The plaque between the front windows invites passersby to stop and contemplate, with respect, the great observer of the skies. (Courtesy Professor G. Godoli, Arcetri Observatory)

By the decree of 1616 he was forbidden to "hold or defend" the odious hypothesis, but he still hoped to convert his countrymen to the heliocentric view. He finally prevailed upon his long-time acquaintance, Pope Urban VIII, to allow him to publish a book that explained fully all arguments for and against the Copernican system, not for the purpose of extolling it, but merely to examine it, and to show those of other nationalities that Italians were not ignorant of new theories.

The book appeared in 1632 under the title *Dialogue on the Two Great World Systems (Dialogo dei Due Massimi Sistemi)*. The *Dialogue* is written in Italian (not Latin) to reach a large audience and is a magnificent and unanswerable argument for Copernican astronomy. It is in the form of a conversation, lasting four days, among three philosophers: Salviati, the most brilliant and the one through whom Galileo generally expresses his own views; Sagredo, who is usually quick to see the truth of Salviati's arguments; and Simplicio, an Aristotelian philosopher who brings up all the usual objections to the Copernican system, which Salviati promptly shows to be absurd.

It is pointed out in the preface to the *Dialogue* that the arguments to follow are merely a mathematical fantasy, and that divine knowledge assures us of the immobility of the Earth. This was thinly cloaked irony, however, and Galileo's enemies acted quickly to build a case against him. He was called before the Roman Inquisition on the charge of believing and holding doctrines that are false and contrary to Divine Scriptures. Galileo was forced to plead guilty and deny his own doctrines. His life sentence was commuted to confinement in his own home at Arcetri, near Florence, for the last ten years of his life. At the time of his inquisition, he was nearly 70.

The *Dialogue* joined Copernicus' *De Revolutionibus* and Kepler's *Epitome* on the *Index of Prohibited*

Books. It was removed from the *Index*, however, in 1835. In 1980, Pope John Paul II ordered a re-examination of the evidence against Galileo, which led to his exoneration and eliminated the last vestige of resistance to the Copernican revolution from the Roman Catholic Church.

EXERCISES

1. A friend tells you that the Earth cannot be rotating, as the scientists claim, because if a point on the equator were being carried eastward at about 1000 mi/hr, as claimed, a baseball pitcher could throw a ball straight up in the air, and the Earth and pitcher would move to the east out from under the ball, which would land some distance behind (or to the west) of the pitcher. How might you straighten out this friend of yours?
2. Which if any of the following can never appear at opposition? at conjunction? at quadrature?
   (a) Jupiter (d) Venus (g) Mercury
   (b) Earth (e) Saturn (h) Moon
   (c) Sun (f) Mars
3. Does full moon occur at intervals of the sidereal or synodic revolution of the Moon about the Earth? Why?
4. If a superior planet had a synodic period equal to its sidereal period, what would this period be? Which planet most closely approximates this condition?
5. What would be the sidereal period of an inferior planet that appeared at greatest western elongation exactly once a year?
6. The synodic period of Saturn is 1.03513 sidereal years. What is its sidereal period?
   Answer: 29.5 years
7. What would be the distance from the sun, in astronomical units, of an inferior planet that had a greatest elongation of 30°? Assume circular orbits for the planet and the Earth.
8. Draw an ellipse by the procedure described, using a string and two tacks. Arrange the tacks so that they are separated by one-tenth the length of the string. Comment on the appearance of your ellipse. This (if you have been careful in your construction) is approximately the shape of the orbit of Mars.
9. At best the positional accuracy attainable with the naked eye approaches one minute of arc. Nevertheless Tycho was able to determine the length of the year to within about one second of time, during which time the sun moves along the ecliptic through an angle of only 0.04 seconds of arc. How could he have done this?
10. (a) What is the major axis of a circle?
    (b) Where is the second focus of a parabola?
    (c) Which conic section could have an eccentricity of 0.3?
    (d) Which conic section could have an eccentricity of 3.2?
11. What is the eccentricity of the orbit of a planet whose distance from the sun varies from 180 million to 220 million km?
12. The Earth's distance from the sun varies from 147.2 million to 152.1 million km. What is the eccentricity of its orbit?
    Answer: 0.016
13. Consider Kepler's third law. Carefully explain why $K = 1$ when a is measured in astronomical units and P in years.
14. What would be the period of a planet whose orbit has a semimajor axis of 4 AU?
15. What would be the distance from the sun of a planet whose period is 45.66 days?
    Answer: 0.25 AU
16. Suppose Kepler's laws apply to the motion of Jupiter's satellites around that planet, and that one of the satellites has a period 5.196 times as long as another one. What would be the ratio of a semimajor axes of their orbits?
    Answer: 3:1
17. Galileo's observations of the phases of Venus ruled out Ptolemy's system of cosmology. Did they also rule out Tycho Brahe's system? Why, or why not?

Johannes Kepler (1571–1630), German mathematician and astronomer, was a contemporary of Galileo, living during the tumultuous period of the Counter-Reformation and the Thirty Years' War. His discovery of the basic quantitative laws that describe planetary motion placed the heliocentric cosmology of Copernicus on a firm mathematical basis and made possible Newton's later formulation of the laws of motion and of universal gravitation.

4

GRAVITATION: ACTION AT A DISTANCE

Kepler deduced that a force from the sun pulled on the planets. He did not determine the mathematical nature of this force, but as we have seen, he did discover some of the rules of planetary motion that result from it. Meanwhile, Galileo discovered some of the laws that describe the behavior of falling bodies, for example, that a freely falling body near the surface of the Earth accelerates uniformly. Although he was not able to measure the precise value of that acceleration, it is now known that every second that a body falls (in a vacuum) its speed increases by 980 cm/s. Newton unified these insights by showing that the force of gravitation that accelerates falling bodies near the Earth is the same force that keeps the Moon in its orbit around the Earth and the planets in their orbits about the sun.

Indeed, Newton's principles of mechanics and law of gravitation are so general and powerful that in the century following his death they strongly influenced the prevailing philosophy. Many thinkers held that the basic rules of nature were finally known, that all that remained was to fill in minor details. The philosophy of determinism reigned: namely, that every action in the universe follows by mechanistic laws from conditions immediately preceding the action. We shall see in the coming chapters, however, that this view of the ultimate success of science was overoptimistic. Newton's laws are not absolute; they fail to describe the motions of electrons in atoms, for example, for which the quantum theory is needed, and they fail when speeds are involved that are not small compared to that of light, in which case relativity provides a better model.

But the fact that Newton's laws are limited does not invalidate them; within the realm in which they work, they provide a magnificent description of the behavior of material objects. The success of our space program and of our other technology attests to the validity of Newtonian mechanics. Moreover, the Newtonian description of nature is so beautifully simple that, at least in essence, it can be understood and appreciated by those not specializing in science.

4.1 NEWTON'S PRINCIPLES OF MECHANICS

Isaac Newton (1643–1727) was born at Woolsthorpe, in Lincolnshire, England, only four days less than one year after the death of Galileo. (Newton was born on Christmas Day, 1642, according to the calendar in use at his time, but by the modern Gregorian calendar his birth date was January 4, 1643.)

Newton entered Trinity College at Cambridge in 1661 and eight years later was appointed Lucasian Professor of Mathematics, a post that he held during most of his productive career. As a young man in college, he became interested in natural philosophy, as science was called then. The university was closed during the plague years of 1665 and 1666, during which Newton returned to Woolsthorpe. He wrote later that it was in those years that he worked out the main outline of his ideas on mechanics and gravitation. But on his later return to Cambridge, his research was mainly in mathematics and optics, and it was to be nearly two decades before he turned his attention again to gravitation.

Newton's return to gravitation was almost fortuitous. Physicist Robert Hooke, architect Christopher Wren, and astronomer Edmund Halley had all come independently to some notion of the law of gravitation and had realized that a force of attraction toward the sun must become weaker in proportion to the square of the distance from the sun. None, however, was able to solve the problem of how a planet should move under the influence of such a force. In 1684, Halley chanced to consult Newton on the matter. He was astonished to hear that Newton had solved the problem years previously and had found that the orbit of a planet should be an ellipse. Although Newton was unable to find his original notes containing the mathematical proof, he was able to re-solve the problem, and a short time later sent the demonstration of the proof to Halley. Early the following year Newton submitted a formal paper on the subject to the Royal Society. This treatise, consisting of four theorems and seven problems, was to become the nucleus of his great work on mechanics and gravitation, The *Mathematical Principles of Natural Philosophy*, usually known by the abbreviated form of its Latin title, *Principia*. Newton worked on the *Principia* for a year and a half during 1685 and 1686. It was published under the imprimatur of the

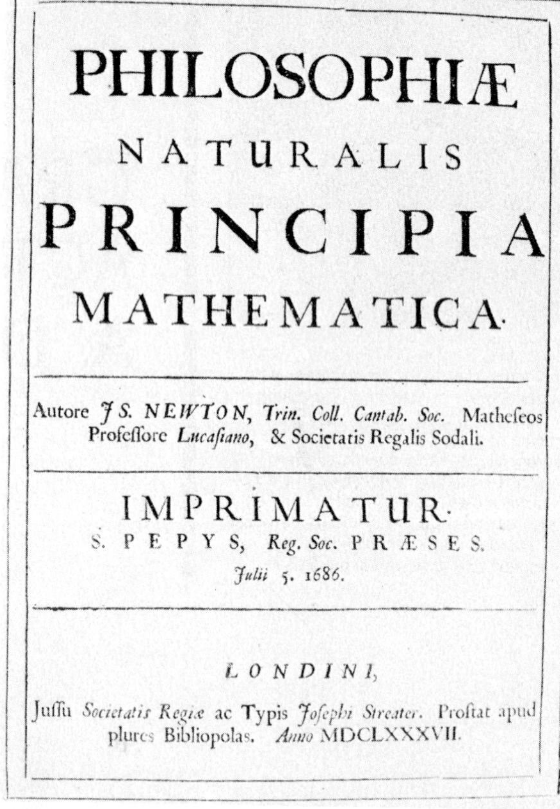

PHILOSOPHIÆ
NATURALIS
PRINCIPIA
MATHEMATICA.

Autore *J S. NEWTON*, *Trin. Coll. Cantab. Soc.* Matheseos Professore *Lucasiano*, & Societatis Regalis Sodali.

IMPRIMATUR.
S. PEPYS, *Reg. Soc.* PRÆSES.
Julii 5. 1686.

LONDINI,
Jussu *Societatis Regiæ* ac Typis *Josephi Streater*. Prostat apud plures Bibliopolas. *Anno* MDCLXXXVII.

Figure 4.1 Title page of the first edition of Newton's *Principia*. (History of Science Collections, University of Oklahoma Libraries)

Royal Society of London, and supervision of its publication was in the hands of Halley. As it turned out, the Society at that time was in financial difficulties, and Halley himself covered the cost of publication from his own personal funds.

In the *Principia* Newton gives his three laws of motion:

> I. Every body continues in a state of rest, or of uniform motion in a straight line, unless it is compelled to change that state by forces impressed upon it.
>
> II. The change of motion is proportional to the force impressed; and is made in the direction of the straight line in which that force is impressed.
>
> III. To every action there is always an equal and opposite reaction: or, the mutual actions of two bodies upon each other are always equal, and act in opposite directions.

Galileo had arrived at the first two laws, although he did not state them as precisely as Newton did. They are, however, deeper than Galileo could have realized. To appreciate Newtonian mechanics, we must understand thoroughly the meanings of certain terms.

(a) Some Basic Concepts: Length

In science, terms must be defined *operationally;* that is, we must supply a recipe or procedure for defining a quantity, so that someone else repeating our observation or experiment will obtain the same result. Consider, for example, *length.* It obviously will not do to define length as "how long something is." Suppose we wished to measure the length of a desk top. We might find a stick and define it as a unit of length; we could call that unit, say, a "thinga." Then by laying out the stick along the desk top we might learn that it has a length of 3.5 thingas.

The international standard unit of length is the *meter (m).* The meter was originally intended to be 10^{-7} times the distance from the equator of the Earth to the North Pole. For years, however, the actual standard was a metal bar stored in a Paris vault. The modern standard was defined in 1960 at a General Conference sponsored by the International Bureau of Weights and Measures; it is equal in length to 1,650,763.73 times the wavelength (in a vacuum) of the orange light emitted by a certain isotope of krypton (^{86}Kr) under specified conditions of pressure and temperature. One hundredth of a meter is a *centimeter (cm).*

Areas of surfaces are defined as *square* measures. The area of our desk top, for example, might be the number of squares each one thinga on a side (and fraction thereof) it would take to cover its surface. The area of a rectangular surface is its length times its width.

Volumes of solids are *cubic* measures; for example, the volume of the desk might be the number of cubic containers of water, each one thinga on a side, it would take to fill a box the exact size and shape of the desk. The volume of a rectangular solid is its length times its width times its height.

Suppose we have two solids of identical shape but of different size—say, spheres 1 cm and 10 cm in radius, or a large man 2 m tall and an identical small one only 1 m tall. The surface area of one of the spheres (for example, the amount of paint needed to cover it) is proportional to the *square* of any of its linear dimensions, such as its radius. Thus the area of the larger sphere is 100 times that of the smaller (for 100 is the square of 10). Similarly, since 4 is the square of 2, it takes four times as much skin to cover the large man as it does the small man with half the large one's linear dimensions.

The volumes of solids are proportional to the *cubes* of their linear dimensions. Thus the sphere 10 cm in radius can hold 1000 times as much water as the one with a radius of 1 cm. Similarly, the large man has eight times as much flesh and bones as the small one. It is convenient to remember that areas are proportional to the squares, and volumes to the cubes of the linear dimensions of similarly shaped objects, whatever their shape may be.

(b) Time

Another basic concept is *time.* It is said that Galileo once used the beat of his pulse as a unit of time to measure the swing of a chandelier in church. He found that the time for one swing stayed the same, even though the length of the swing died down. The story of the swinging chandelier may be apocryphal, but Galileo did discover the law that determines the period of a pendulum—that the period of its oscillation depends only on the pendulum's length and not on the amount of arc of swing. Galileo later suggested that the pendulum would be a good device to regulate a clock, thus inventing the principle of the pendulum clock.

We could, of course, choose any convenient period for a unit of time. We could hold one end of that stick we used to define the thinga, and start the other end vibrating, like a tuning fork. The time of a single vibration might be called a "majig." We could count how many majigs it took for water to leak out of a can with a hole in the bottom.

A more universal unit of time is the *second (s).* It was originally meant to be 1/60 of a minute, which is 1/60 of a hour, which is 1/24 of a day. But the Earth does not rotate quite regularly enough to serve as an accurate enough standard for modern measurements. The 1967 General Conference turned to the *atomic clock* and defined the second as the duration of 9,192,631,770 periods of one of the radiations from a certain isotope of the cesium atom (^{133}Cs).

(c) Speed, Velocity, and Acceleration

With ways to measure length and time we can define *speed,* for example, how many thingas a hare

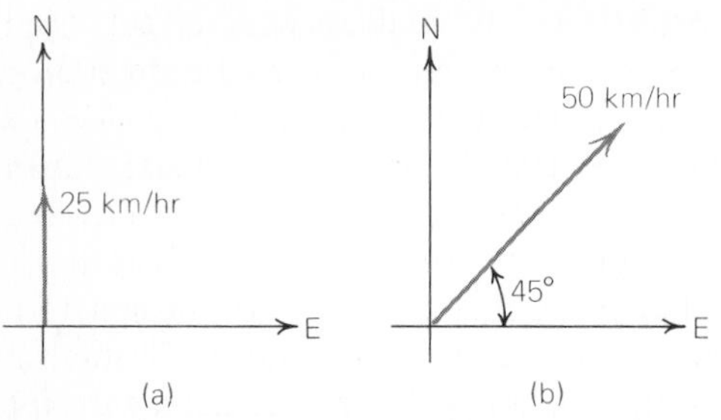

Figure 4.2 Two vectors, representing velocities at 25 km/hr to the north and 50 km/hr to the northeast.

can run in one majig. Now if we measured the distance the hare ran in 1000 majigs, that distance divided by 1000 is the *average* or *mean speed* of the hare. But if the hare had sped up part of the time, or had stopped to eat some grass along the way, his speed would be changing. At any point along his journey, the *instantaneous speed* of the hare is the number of thingas he would run in a majig if he kept that identical speed for one full majig. An automobile speedometer measures instantaneous speed, although we might have some trouble finding one calibrated in thingas per majig.

Speed and *velocity* are often confused. Velocity conveys more information than speed; it is a description of both the instantaneous speed and the *direction* of motion. Velocity is an example of a *vector* (Figure 4.2), a quantity that has both size (speed in this case) and direction.

Any change in velocity requires *acceleration*. Acceleration, therefore, involves a change of speed, or of direction, or both. Starting, stopping, speeding up, slowing down, or changing direction are all accelerations. Acceleration is also a vector. The magnitude of acceleration is the rate at which the velocity changes, and its direction is the direction of that change. The acceleration produced by gravity at the surface of the Earth, for example, is 980 cm/s per second (often written 980 cm/s^2), in a direction toward the center of the Earth.

(d) Newton's First Law—Momentum

A moving body tends to keep moving and a stationary body tends to remain at rest. *Momentum* is a measure of this state of motion. Momentum depends on speed, for clearly a body moving at 50 km/hr certainly has more "motion" than one moving 10 km/hr. But momentum also depends on the amount of matter in a moving object; an automobile going 30 km/hr certainly has more "motion," and is harder to speed up, stop, or turn than, say, a bicycle moving with the same speed.

Thus Newton defined momentum as proportional to velocity, and defined the constant of proportionality as *mass*. Mass is a quantity that characterizes the total amount of material in the body and is the property that gives the body its *inertia*, that is, that makes it resist acceleration. We have not yet defined mass operationally, but we shall see that Newton's third law provides a means of doing so. For now it is enough to say that our intuitive idea of what is usually meant by *weight* is actually *mass*. Weight, technically, is a measure of the gravitational pull upon an object. An astronaut on the Moon weighed only one-sixth of what he weighed on the Earth. In the Space Shuttle, an astronaut weighs essentially nothing, since the spacecraft is falling freely around the Earth. In both cases, however, the bulk or *mass* of the astronaut is unchanged from its value on the surface of the Earth.

Thus Newton's first law says that the product of a body's mass and velocity is constant if no outside force is applied to it. This means that motion is as natural a state as is rest. With no force on it, a moving body would go in a straight line at a constant speed forever.

(e) Newton's Second Law—Force

The second law of motion deals with changes in momentum. It states that if a force acts on a body, it produces a *change* in the momentum of the body that is in the direction of the applied force. The second law, then, defines *force*. The magnitude or strength of a force is defined as the *rate at which it produces a change* in the momentum of the body on which it acts.

Some familiar examples of forces are the pull of the Earth, the friction of air slowing down objects moving through it, the friction of the ground or a floor similarly slowing bodies, the impact of a bat on a baseball, the pressure exerted by air, and the thrust of a rocket engine.

Note that Newton's first law of motion is consistent with his second; when there is no force, the change in momentum is zero.

There are three ways in which the momentum of a body can change. Its velocity can change, or its mass, or both. Most often the mass of a body does not change when a force acts upon it; a change in momentum usually results from a change in velocity. Thus, in the vast majority of examples, the second law can be written as the simple formula

$$\text{force} = \text{mass} \times \text{acceleration},$$

because acceleration is the rate at which velocity changes. If the acceleration occurs in the same direction as the velocity, the body simply speeds up; if the acceleration occurs in the opposite direction to the velocity, the body slows down. If acceleration occurs exactly at *right angles* to the velocity, only the direction of motion of the body, and not its speed, changes.

The acceleration of falling bodies is downward (in the direction toward which the gravitational pull of the Earth is acting). Gravity accelerates a body in the direction it is already moving, and so simply speeds it up.

If a body is slid along a rough horizontal surface, it slows down uniformly in time. It is therefore accelerated in a direction opposite to its velocity. The acceleration is produced by the force of friction between the moving body and the rough surface.

A body accelerates, that is, its momentum changes, only when a force acts upon it. The application of a force can change both speed and direction of motion, and either of these changes is an acceleration. The mathematics needed to calculate acceleration and motion under external forces is *differential calculus,* a topic typically taught to first-year college students. Newton *invented* the differential calculus while an undergraduate at Cambridge University.

Two equal accelerations may correspond to entirely different forces. Consider the forces required to accelerate an automobile and a bicycle each to a speed of 30 km/hr in 20 seconds. Clearly, because of the car's greater mass, a proportionately greater force will be required to produce the necessary acceleration. Similarly, once the bodies are both moving at that speed, a far greater force is needed to *stop* the automobile as quickly as the bicycle.

(f) The Third Law—Reaction

Newton's third law of motion was a new idea. It states that all forces occur as *pairs* of forces that are mutually equal to and opposite each other. If a force is exerted on an object, it must be exerted by something else, and the object will exert an equal and opposite force back upon that something. All forces, in other words, must be mutual forces acting *between* two objects or things.

If a man pushes against his car, the car pushes back against him with an equal and opposite force, but if the man has his feet firmly implanted on the ground, the reaction force is transmitted through him to the Earth. Because of its enormously greater mass, the Earth accelerates far less than the car. Suppose a boy jumps off a table down to the ground. The force pulling him down is a mutual gravitational force between him and the Earth. Both he and the Earth suffer the same total change of momentum because of the influence of this mutual force. Of course, the boy does most of the moving; because of the greater mass of the Earth, it can experience the same change of momentum by accelerating only a negligible amount.

A more obvious manifestation of the mutual nature of forces between objects is familiar to all who have played baseball. The recoil of the bat shows that the ball exerts a force on the bat during the impact, just as the bat does on the ball. The momentum imparted to the bat by the ball is transmitted through the batter to the Earth, so the acceleration produced is far less than that suffered by the ball. Similarly, when a rifle is discharged, the force pushing the bullet out the muzzle is equal to that pushing backward upon the gun and marksman.

Here, in fact, is the principle of rockets—the force that discharges the exhaust gases from the rear of the rocket is accompanied by a force that shoves the rocket forward. The exhaust gases need not push against air or the Earth; a rocket operates best in a vacuum.

In all the cases considered above, a mutual force acts upon the two objects concerned; each object always experiences the same total change of momentum, but in opposite directions. Because momentum is the product of velocity and mass, the object of lesser mass will end up with proportionately greater velocity.

MEASURING MASS

We are now in a position to see how we can measure mass. Initially, some object must be adopted as a standard and said to have *unit mass,* for example, one cubic centimeter of water, defined as one *gram (g)*. In the metric standard system, the unit of mass is the *kilogram (kg)*, the mass of 1000 cubic centimeters of water (see Appendix 4). Then the mass of any other object can be found by measuring the relative acceleration produced when the same force acts on it and the standard.

Having found a way to measure mass, we can now express the value of a force numerically. The standard metric unit of force is the *newton*, which is the force necessary to give a mass of 1 kg an acceleration of 1 m/s^2. Also frequently used is the smaller unit the *dyne*, the force needed to accelerate 1 g by 1 cm/s^2. A small mosquito sitting on your arm exerts a force of about 1 dyne.

(g) Density

It is important not to confuse mass, volume, and *density*. Volume is simply a measure of the physical space occupied by a body, say, in cubic centimeters or liters. In short, the volume is the "size" of an object—it has nothing to do with its mass. A lady's wristwatch and an inflated balloon may both have the same mass, but they have very different volumes.

The watch and balloon are also very different in *density*, which is a measure of how much mass is contained within a given volume. Specifically, density is the ratio of mass to volume:

$$\text{density} = \frac{\text{mass}}{\text{volume}}$$

Sometimes in everyday language we use "heavy" and "light" as indications of density rather than weight, as, for instance, when we say that iron is heavy or that a well-baked pastry is light.

The units of density are grams per cubic centimeter (g/cm^3) or, equivalently, tons per cubic meter. Familiar materials span a considerable range in density, from gold (19 g/cm^3) through ordinary rock (2–3 g/cm^3) to water (1.0 g/cm^3) to artificial materials such as plastic insulating foam (less than 0.1 g/cm^3). In the astronomical universe, much more remarkable densities can be found, all the way from a comet's tail (10^{-15} g/cm^3) to a neutron star (10^{15} g/cm^3).

To sum up then, *mass* is "how much," *volume* is "how big," and *density* is "how tightly packed."

(h) Angular Momentum

Another useful concept is *angular momentum*, which measures the momentum of an object as it rotates or revolves about some fixed point. Just as ordinary momentum was defined as the product of two quantities—mass and speed—the angular momentum of an object is defined as the product of three quantities: mass, speed, and the distance from the fixed point around which the object turns.

If these three quantities remain constant, that is, if the motion takes place at a constant speed and at a fixed distance from the point of origin, then the angular momentum is also a constant. More generally, angular momentum is constant, or is *conserved*, in any rotating system in which no external forces act, or in which the only forces are directed toward or away from the point of origin. An example of such a system is a planet orbiting the sun, since the mutual gravitational forces act directly along the line joining the two objects, and there is no external force such as friction to slow the motion. Kepler's second law is an example of the conservation of angular momentum.

Conservation of angular momentum applies also to a solid body spinning around its own axis, like the Earth, or to a spinning gas cloud, such as those that are the birthplaces of stars and planets. However, such objects can certainly change their spin rate, if their size or configuration is altered. Indeed, the conservation of angular momentum dictates that the rotation will change if, for example, a dust cloud shrinks.

Altering the spin rate by rearranging the mass in a rotating system is exactly what is accomplished by figure skaters. Suppose a skater is spinning on the tip of her skate with arms outstretched. If she holds her body rigid, she rotates at a constant rate. However, if she pulls her arms in to her body, some parts of her mass are closer to their rotation axes, and hence decrease in angular momentum unless she compensates by spinning faster. Thus, a figure skater can start a spin with her arms out, and then pull them in, thereby spinning faster so that her angular momentum is conserved. She can slow down again by pushing her arms (or a free leg) out from her body.

You can perform the same experiment with an old-fashioned rotating piano stool. Purchase another copy of this book and sit on the stool while holding both books out from your body at arm's length.

Have a friend start you spinning—but not too fast, or you may not have the opportunity to finish reading even your first copy. Now pull your arms in so that the books are next to your body and you will experience a dramatic demonstration of the conservation of angular momentum.

The conservation of angular momentum is an important concept to an understanding of the formation of the solar system with its planets and their satellites, including the rings of Saturn, and even the formation of galaxies. We shall refer to it again in coming chapters.

(i) Acceleration in a Circular Orbit

It might be assumed that some force or power is required to keep the planets in motion. However, Galileo argued from the principle of inertia (Newton's first law of motion) that once started, the planets would remain in motion—that the state of motion for planets was as natural as for terrestrial objects. What does require explanation, however, is why the planets move in nearly circular orbits rather than in straight lines (the latter motion would eventually carry them away from the vicinity of the solar system). Galileo had not considered this problem.

By Newton's time, a number of investigators had considered the problem of circular motion. The correct solution to the problem was first published (in 1673) by the Dutch physicist Christian Huygens (1629–1695). However, Newton had found the solution independently in 1666.

For a body to move in a circular path rather than in a straight line, it must continually experience an acceleration toward the center of the circle. Such an acceleration is called *centripetal acceleration.* The central force that produces the centripetal acceleration *(centripetal force)* is, for a planet, an attraction between the planet and the sun. For a stone whirled about at the end of a string, the centripetal force is the tension in the string. With the help of Newton's laws of motion and some elementary mathematics we can calculate how great that central force has to be. We find that if a particle of mass m moves with a speed v on the circumference of a circle of radius r, the centripetal force is given by the formula

$$F = \frac{mv^2}{r}.$$

4.2 UNIVERSAL GRAVITATION

It is obvious that the Earth exerts a force of attraction upon all objects at its surface. This is a mutual force; a falling apple and the Earth are pulling on each other. Newton reasoned that this force of attraction between the Earth and objects on or near its surface might extend as far as the Moon and produce the centripetal acceleration required to keep the Moon in its orbit. He further speculated that there is a general force of attraction between *all* material bodies. If so, the attractive force between the sun and each of the planets could provide the centripetal acceleration necessary to keep each in its respective orbit.

Thus Newton hypothesized that there is a universal attraction between all bodies everywhere in space. Next he had to determine the mathematical nature of the attraction and test the hypothesis by using it to predict *new* phenomena. We shall now see how Newton formulated his law of universal gravitation.

(a) The Mathematical Description of Gravitation

For mathematical simplification we make the assumption that planets revolve around the sun in perfectly circular orbits. A more complicated analysis can be made to apply to the actual elliptical orbits. Using the results of Section 4.1i we find that the centripetal force that the sun must exert upon a planet of mass m_p, moving with speed v in a circular orbit of radius r, is

$$\text{force} = \frac{m_p v^2}{r}.$$

Now the period P of the planet, that is, the time required for the planet to go completely around the sun, is the circumference of its orbit ($2\pi r$) divided by its speed, or

$$P = \frac{2\pi r}{v}.$$

Solving the above equation for v, we find

$$v = \frac{2\pi r}{P}.$$

On the other hand, from Kepler's third law we know

that the squares of the periods of planets are in proportion to the cubes of their distances from the sun. Because the sun is observed to be almost at the center of a planet's orbit, that distance is very nearly the radius of the orbit, r, and we have

$$P^2 = Ar^3,$$

where A is a constant of proportionality whose value depends on the units used to measure time and distance. Combining the last two equations, we find

$$v^2 = \frac{4\pi^2 r^2}{P^2} = \frac{4\pi^2 r^2}{Ar^3} = \frac{4\pi^2}{Ar},$$

that is,

$$v^2 \propto \frac{1}{r},$$

where the symbol $\propto$ means "proportional to."

If we substitute the above formula for v^2 into the one expressing the sun's centripetal force on the planet, we obtain

$$\text{force} \propto \frac{m_p}{r^2}.$$

The centripetal force exerted on the planet by the sun must therefore be in proportion to the planet's mass and in inverse proportion to the square of the planet's distance from the sun. According to Newton's third law, however, the planet must exert an equal and opposite attractive force on the sun. If the gravitational attraction of the planet on the sun is to be given by the same mathematical formula as that for the attraction of the sun on the planet, the planet's force on the sun must be

$$\text{force} \propto \frac{m_s}{r^2},$$

where m_s is the sun's mass. Since this is a mutual force of attraction between the sun and planet, it must be proportional to both the mass of the sun and the mass of the planet; therefore, the attractive force between the two has the mathematical form

$$\text{force} \propto \frac{m_s m_p}{r^2}.$$

Both the sun and a planet revolving around it experience the same change of momentum as a result of this mutual force between them. The fact that the sun is observed to remain more or less at the center of the solar system while the planets revolve around it is evidence that the sun's mass must be enormously greater than that of the planets. Therefore, its acceleration is relatively small, but it is actually observable.

(b) The Law of Universal Gravitation

For Newton's hypothesis of universal attraction to be correct, there must be an attractive force between all pairs of objects everywhere whose value is given by the same mathematical formula as that above for the force between the sun and a planet. Thus the force F between two bodies of masses m_1 and m_2, and separated by a distance d, is

$$F = G\frac{m_1 m_2}{d^2}.$$

Here G, the constant of proportionality in the equation, is a number called the *constant of gravitation*, whose value depends on the units of mass, distance, and force used. The actual value of G has to be determined by laboratory measurements of the attractive force between two material bodies. If metric units are used (grams for mass, centimeters for distance, and dynes for force), G has the numerical value 6.67×10^{-8}.*

The above equation expresses Newton's law of universal gravitation, which is stated as follows:

> Between any two objects anywhere in space there exists a force of attraction that is in proportion to the product of the masses of the objects and in inverse proportion to the square of the distance between them.

Not only is there a force between the sun and each planet, but also between any two planets. Because of the sun's far greater mass, the dominant force felt by any planet is that between it and the sun. The attractive forces between the planets have relatively little influence. Similarly, there is a gravitational attraction between any two objects on Earth (for example, between two flying airplanes, or between the kitchen sink and a tree outside the

* The notation 6.67×10^{-8} means 0.0000000667; see Appendix 3.

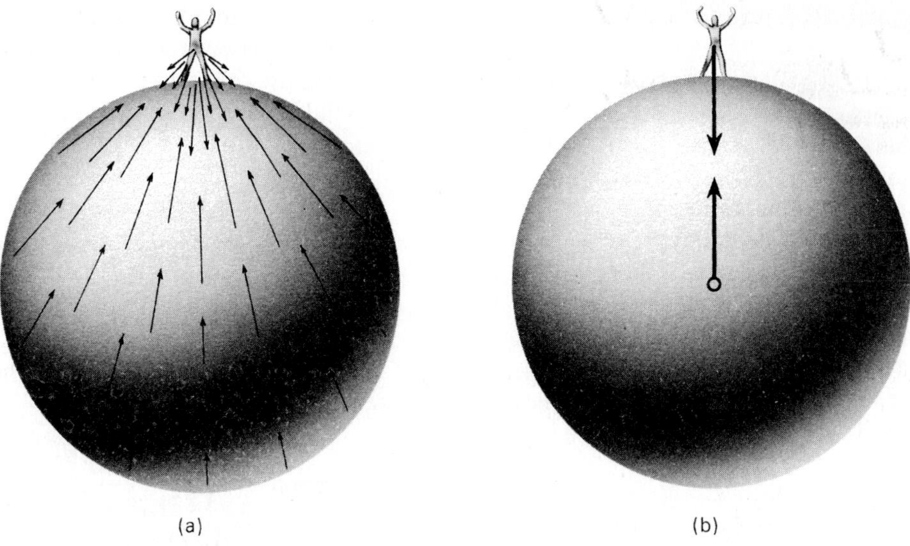

(a) (b)

Figure 4.3 The gravitational attraction of a sphere is as though all its mass were concentrated at its center, called the center of mass or center of gravity.

house), but this force is insignificant compared to the force between each of them and the very massive Earth.

Before we see how Newton tested his law of gravitation, let us investigate some of its other consequences.

(c) Weight

Newton hypothesized a force of attraction between all pairs of bodies. The Earth is a large spherical mass, however, that can be thought of as being composed of a large number of component parts. An object, say a person, on the surface of the Earth feels the simultaneous attractions of the many parts of the Earth pulling from many different directions. Exactly what is the resultant gravitational effect of the many parts of a sphere, each pulling independently upon a mass outside the surface of the sphere? Here was a difficult problem to which Newton had to find a solution before he could test his law of gravitation.

To solve the problem, Newton had to calculate the force between an object on the surface of the Earth and each infinitesimal piece of the Earth, and then calculate how all of these forces combined. It was necessary for him to invent and use a new method of mathematics which he called *inverse fluxions* (today we call it *integral calculus*). Fortunately, the solution to the problem gives a beautifully simple result. A spherical body acts gravitationally as though all its mass were concentrated at a point at its center (Figure 4.3).* This means that we can consider the Earth, the Moon, the sun, and the planets as geometrical points as far as their gravitational influences are concerned.

The gravitational force, then, between an object of mass m on the Earth and the Earth itself, of mass M, is equal to the constant of gravitation times the product of the masses m and M, divided by the square of the distance from the object to the center of the Earth. The latter distance is just the radius of the Earth, R. This gravitational force between the Earth and a body on its surface is the body's *weight*. Algebraically, the weight W of a body is given by

$$W = G\frac{mM}{R^2}.$$

We see that the weight of a body is proportional to its mass. This circumstance gives us another method of measuring mass, namely, by measuring the weight of a body. Whenever the mass of an ob-

---

* Strictly, the statement is correct only if the density distribution within the body is spherically symmetrical.

ject is determined by its gravitational influence, as by measuring its weight, the mass so determined is defined as a *gravitational mass*. So far, the results of all experiments indicate that gravitational mass is exactly equivalent to *inertial mass*.

Because the gravitational attraction between two bodies decreases as the square of their separation, an object weighs less if it is lifted above the surface of the Earth. A person actually weighs less at the top of a step ladder than on the ground. Careful laboratory experiments are able to detect such subtle changes in gravitational force. At 6400 km above the Earth's surface, a body is twice as far from the Earth's center as it would be at the surface, so an object there would weigh $\frac{1}{2}^2$, or $\frac{1}{4}$ times as much as at the surface of the Earth.

If an object is dropped from a height, the downward acceleration is equal to the force acting on it, that is, its weight, divided by its mass:

$$\text{acceleration} = \frac{W}{m} = G\frac{M}{R^2}.$$

Thus, although the gravitational force between the Earth and a massive object is greater than between the Earth and a less massive one, all objects experience the same accelerations and fall at the same rate, as Galileo had found.

(d) Test of Gravitation: The Apple and the Moon

Suppose an apple is dropped from a height above the surface of the Earth. We have seen that it accelerates 980 cm/s^2. If the hypothesis of gravitation is correct, the accelerations toward the Earth of the apple (at the Earth's surface) and of the Moon should both be given by the equation

$$\text{acceleration} = G\frac{M}{D^2},$$

where M is the mass of the Earth and D is the distance of the object in question from the center of the Earth. The acceleration, in other words, should be inversely proportional to the square of the distance from the Earth's center. The apple's distance is about 6400 km, and the Moon's distance is 384,403 km, about 60 times as far. Thus the acceleration of the Moon should be $\frac{1}{60}^2$, or $\frac{1}{3600}$, as much as the apple's.

If we assume that the Moon's orbit about the Earth is a perfect circle (its orbit is, indeed, very nearly circular), we can use the formula for centripetal acceleration found in Section 4.1i to calculate the Moon's acceleration:

$$\text{acceleration} = \frac{v^2}{D}.$$

The Moon's distance (the radius of its orbit) is 3.844×10^{10} cm. Its average orbital speed is 1.023×10^5 cm/s. If we substitute these numbers in the above formula we find

$$\text{Moon's acceleration} = \frac{(1.023 \times 10^5)^2}{3.844 \times 10^{10}} = 0.272 \text{ cm/s}^2.$$

The acceleration predicted by the law of gravitation is

$$\text{Moon's acceleration} = \frac{980}{(60)^2} = 0.272 \text{ cm/s}^2.$$

The law of gravitation predicts that because the Moon is 60 times as far from the center of the Earth as an apple is, its acceleration should be $\frac{1}{60}^2$ as much; that is exactly the acceleration that we observe for the Moon. The test gives results that are consistent with Newton's law of universal gravitation. It works! We can imagine the thrill that Newton must have felt in discovering that the same simple algebraic formula describes the law of gravitation that operates on the Earth, on the Moon and planets, and so far as we know, throughout the entire universe!

(e) The Mass of the Earth and the Determination of G

Note that in the formula for the weight of an object at the surface of the Earth, or for its acceleration toward the Earth, the constant of gravitation, G, always occurs multiplied by the mass of the Earth. If the latter were known, G could be evaluated at once from the known acceleration of gravity:

$$G = \frac{R^2 g}{M_E},$$

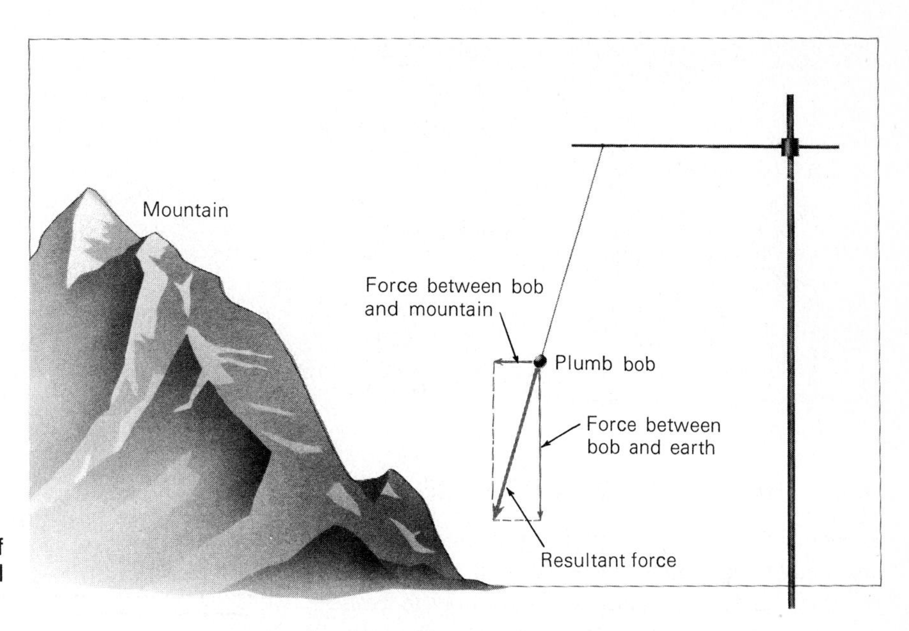

Figure 4.4 Deflection of plumb bob from the vertical by a nearby mountain.

where R and M_E are the radius and mass of the Earth and g is the known acceleration of gravity at the Earth's surface—980 cm/s^2. Conversely, if G were known, the mass of the Earth could be found:

$$M_E = \frac{R^2 g}{G}.$$

Hence the determination of G is equivalent to the determination of the mass of the Earth.

There are various methods of finding the mass of the Earth. All depend on a comparison of the gravitational attraction between some object of known mass and the Earth, with the attraction between two objects of known mass.

The first attempts to evaluate G (or, equivalently, the mass of the Earth) were made by comparing the pull of the Earth to that of large natural features, such as mountains. The most famous such experiment was carried out in 1774 by Nevil Maskelyne, Astronomer Royal and Director of the Royal Greenwich Observatory. Overlooking Scotland's Loch Rannoch is Mount Schiehallion, a shallow conical peak so symmetrical that it is feasible to estimate its total mass. Now such irregularities in the Earth's crust affect the "down" direction; a plumb bob suspended near a large mountain is pulled on dominantly by the massive Earth itself, but the mass of the mountain is enough to attract the bob slightly, and hence to draw it a bit off vertical (Figure 4.4).

Maskelyne measured the angle by which Mount Schiehallion deflected a plumb bob suspended at two different places, one on the north slope, and the other on the south slope, by sighting the direction of its supporting cord against distant stars. He found that the mountain deflected the bob by 5.63 arcseconds from the vertical, and thus determined the pull of Schiehallion on the bob compared with the pull of the Earth.

The strength of the pull on the bob by the mountain depends on the mountain's distance and mass, and that of the Earth depends on its mass and the distance from the surface to the center of the Earth (that is, the Earth's radius). The distances involved are known, so by estimating the mass of the mountain one obtains enough information to calculate the mass of the Earth (or G). Surveys of Schiehallion and samplings of its surface rocks were carried out and refined at various times, yielding fair estimates of the Earth's mass, but the difficulty of knowing the precise density of the mountain's interior makes the method necessarily crude.

Far more accurate were later techniques that could be carried out entirely within the laboratory. The results of the best determinations of the mass of the Earth give 5.98×10^{27}g, or about 6×10^{21} tons.

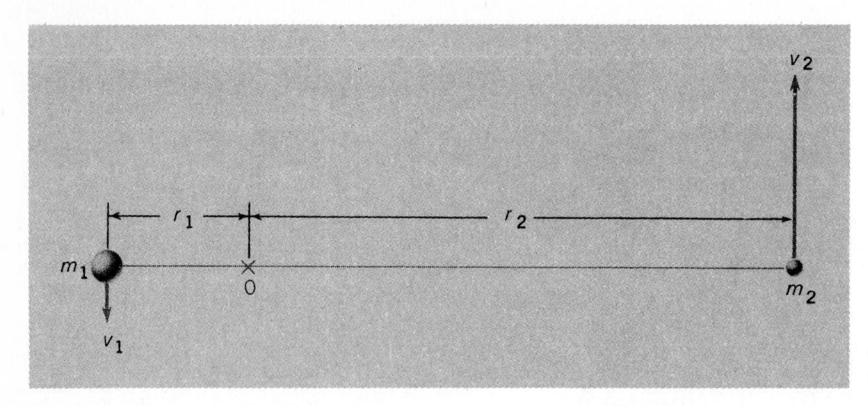

Figure 4.5 Center of mass; the two bodies, 1 and 2, mutually revolve about point O.

4.3 THE TWO-BODY PROBLEM

Newton's laws of motion and gravitation enable us to predict the motions of bodies under the influence of their mutual gravitation. They apply to planets, to spacecraft, and even to the fall of a sparrow. In the next chapter, we will look at some applications of these laws to relatively complex systems, in which the mutual interactions of a number of bodies must be considered. First, however, we describe the much simpler motions of two bodies, both of which are either point masses or spherically symmetrical, so that they act like point masses. This subject is called the *two-body problem.*

(a) Center of Mass

According to Newton's third law, the total momentum of an isolated system is conserved; that is, all changes of momentum within it are balanced. We can, therefore, define a point within the system that remains fixed (or moves uniformly) as if the entire mass of the system were concentrated at that point, called *the center of mass.* It can be shown that the center of mass of a complex body (which is a collection of point masses joined rigidly together) is that point at which the body balances when placed near a gravitating body; thus it is often also called the *center of gravity.*

The center of mass (or gravity) for two bodies is given a special name: the *barycenter*. It must lie on a line connecting the centers of the bodies. We now derive the location of the barycenter relative to the bodies. For simplicity we shall assume that each body revolves about it in a circular orbit. With somewhat more advanced mathematics, we would find that the result we derive is correct for any kind of motion of the two bodies—as long as they are acted on only by mutual forces between them.

Let the two bodies, of masses m_1 and m_2, revolve about and on opposite sides of the point O on the line between their centers (Figure 4.5). The distances of the two bodies from O are r_1 and r_2. To accelerate body 1 into a circular orbit, the force upon it must be $m_1v_1^2/r_1$. The force on body 2 is $m_2v_2^2/r_2$. As in Section 4.2, we can write the orbital speed of each body in terms of its period of revolution and its distance from the center of its orbit, that is,

$$v_1 = \frac{2\pi r_1}{P} \quad \text{and} \quad v_2 = \frac{2\pi r_2}{P}.$$

Of course, both bodies have the same period—the period of their mutual revolution.

The same force—the gravitational attraction between the bodies—produces the centripetal acceleration of each. Therefore, the centripetal force acting on each of them must be the same. Thus we can write

$$\frac{m_1v_1^2}{r_1} = \frac{m_2v_2^2}{r_2}.$$

or

$$\frac{m_1 4\pi^2 r_1^2}{P^2 r_1} = \frac{m_2 4\pi^2 r_2^2}{P^2 r_2}.$$

After cancellation of common factors, the above equation becomes

$$\frac{r_1}{r_2} = \frac{m_2}{m_1}.$$

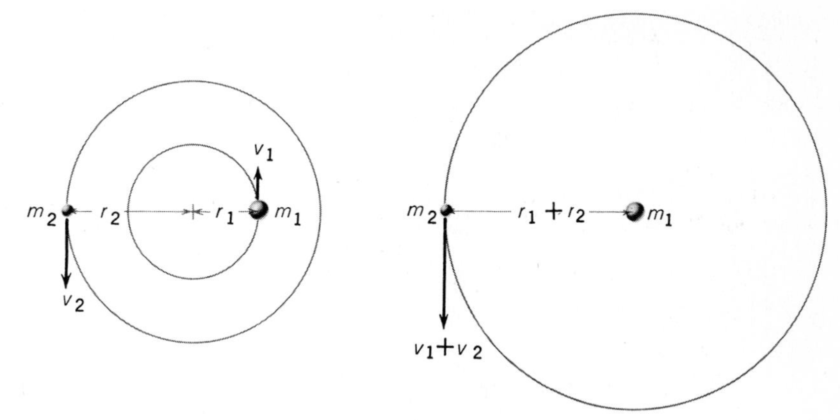

Figure 4.6 Relative orbit of two bodies.

We see that the point of mutual revolution of two bodies on a line between them is located so that the distance of each body from that point is in inverse proportion to its mass.

The concept of center of mass reminds us of two children on a seesaw. For the seesaw to balance properly, the fulcrum (support point) must be located proportionately closer to the child of greater mass. Similarly, the barycenter of two revolving bodies lies proportionately closer to the body of greater mass.

Because of the sun's far greater mass, the center of mass of a system of revolving bodies consisting of the sun and a planet lies very close to the center of the sun—in most cases, within the sun's surface.

(b) Relative Orbits

As two bodies revolve mutually about their barycenter, they must always maintain the same relative distances from it; if one body doubles its distance so must the other, and if one comes closer, the other must lessen its distance by a proportional amount. We see, then, that the orbits of the two bodies are similar to each other (that is, are the same shape), the sizes of the orbits being in inverse proportion to the masses of the bodies.

The similarity of the two orbits makes it easy to refer to the orbit of one object relative to the other. It is conventional to choose the center of the more massive body as reference. Of course neither body is actually fixed—both move about the barycenter—but it is often convenient to *regard* one (the more massive) as origin, and consider the motion of the other with respect to it. This is called the *relative orbit*. The Earth and Moon, for example, revolve about their barycenter on a line between their centers. The Earth is *not* at the center of the Moon's orbit, but we can and often do speak of the relative orbit of the Moon about the Earth's center.

The size of the relative orbit is the sum of the sizes of the individual orbits about the barycenter. If the bodies move in circular paths (Figure 4.6), the relative orbit is a circle with a radius equal to the sum of the distances of the two bodies from the barycenter (it is also equal to the distance between the bodies). If the orbits are ellipses, the relative orbit is an ellipse of major axis equal to the sum of the major axes of the elliptical orbits of the two bodies; all three ellipses have the same eccentricity. The speed of the less massive object in the relative orbit (its speed with respect to the more massive body) is the sum of the speeds of the two individual bodies about their barycenter.

We now turn to the motions of the planets, satellites, and other celestial objects.

(c) Orbital Motion Explained

If an object near the Earth's surface is dropped toward the ground, at the end of one second it has accelerated to a speed of 980 cm/s. Its average speed during that second is 490 cm/s. Thus it drops, in one second, through a distance of 490 cm. The Moon, which accelerates only 1/3600 as much, drops toward the Earth only about 490/3600 cm in one second, or about 0.14 cm. In other words, as the Moon moves forward in its orbit for one second and travels about 1 km it falls 0.14 cm toward the Earth. However, because of the Earth's curvature,

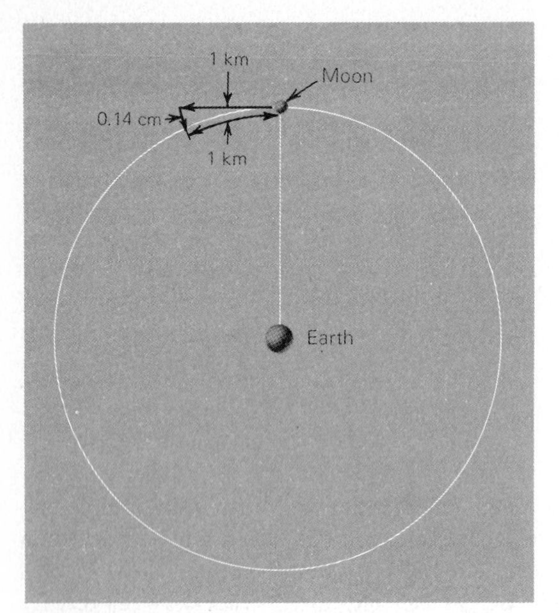

Figure 4.7 Velocity and acceleration of the Moon in its orbit about the Earth.

the ground has fallen away under the Moon by that same distance of 0.14 cm, so the Moon is still the same distance from the Earth. In this way the Moon literally "falls around the Earth." In the period of about one month, it has "fallen" through one complete circuit of the Earth and is back to its starting point (Figure 4.7).

Orbital motion is thus easily understood in terms of Newton's laws of motion and gravitation. At any given instant, the orbital speed of the Moon would tend to carry it off in space in a straight line tangent to its orbit. The gravitational attraction between the Earth and the Moon provides the proper centripetal force to accelerate the Moon into its nearly circular path. Consequently, the Moon continually falls toward the Earth without getting any closer to it. The orbital motions of the planets are similarly explained.

4.4 NEWTON'S DERIVATION OF KEPLER'S LAWS

Kepler's laws of planetary motion are *empirical* laws, that is, they describe the way the planets are *observed* to behave. Kepler himself did not know of the more fundamental laws or relationships from which his three laws of planetary motion follow. On the other hand, Newton's laws of motion and gravitation were proposed by him as the basis of all mechanics. Thus it should be possible to derive Kepler's laws from them. Newton did so. Today, the principal diagram Newton used in the proof (from the *Principia*) is reproduced on the British one-pound note (Figure 4.8).

Figure 4.8 Newton illustrated on the back side of a British one-pound note.

(a) Kepler's First Law

Consider a planet of mass m_p at a distance r from the sun moving with a speed v in a direction at right angles to the line from the planet to the sun. The centripetal force needed to keep the planet in a *circular* orbit, that is, at constant distance from the sun, is

$$\text{force} = \frac{m_p v^2}{r}.$$

Now suppose the gravitational force between the planet and the sun happens to be greater than the force given by the above equation. Then the planet will receive more acceleration than is necessary to keep it in a circular orbit, and it will move in somewhat closer to the sun. As it does so, its speed will increase, just as the speed of a falling stone increases as it approaches the ground.

Because of the planet's increased speed and decreased distance from the sun, a greater centripetal force is required to keep it at a constant distance from the sun. Eventually, as the planet continues to sweep in closer to the sun at higher and higher speed, a point will be reached at which the gravitational force between the two is no longer sufficient to produce enough centripetal acceleration to keep the planet from moving out away from the sun. Thus the planet will move outward as it rounds the sun until it has

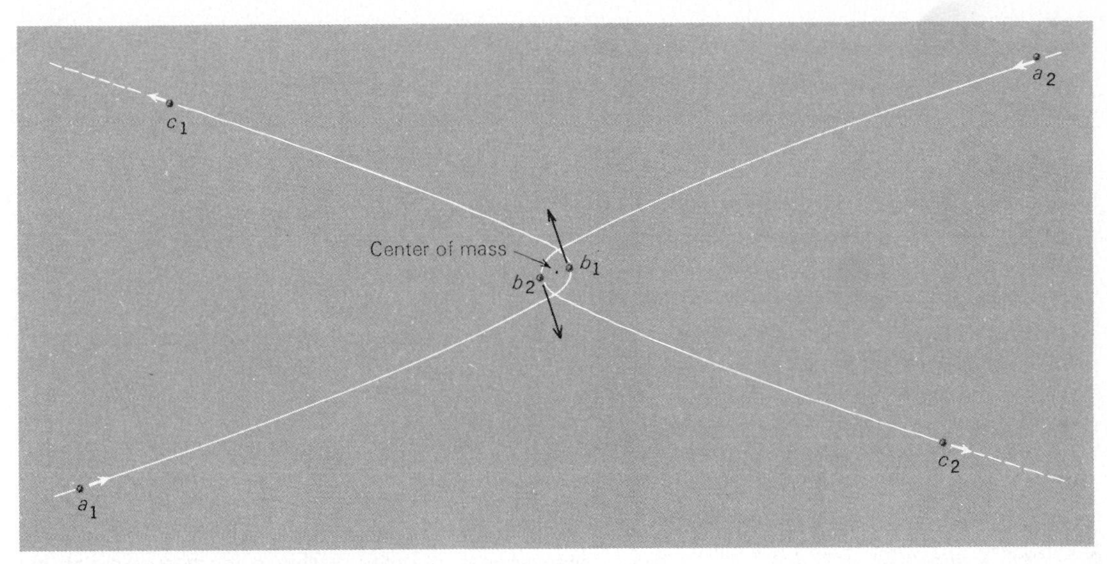

Figure 4.9 Relative hyperbolic orbits.

reached a position where the gravitational acceleration is again greater than the circular centripetal acceleration, and the process is repeated.

If the situation were reversed, and the planet were moving fast enough for the centripetal force required for circular motion to be greater than the gravitational attraction, the planet would move outward and consequently slow down until the gravitational force could pull it back again.

Thus we see, qualitatively, how a planet may follow an elliptical orbit. If, however, a planet had a high enough speed, the gravitational force between it and the sun might never be enough to provide sufficient centripetal force to hold the planet in the solar system, and the planet would move off into space. Its orbit would then be a *hyperbola* rather than a closed, elliptical path (Figure 4.9). There is a certain critical speed, which depends on the planet's distance from the sun, at which the planet can just barely escape the solar system along a *parabolic* orbit. This critical speed is called the *parabolic velocity,* or the *velocity of escape.*

To prove rigorously that the gravitational force between the sun and a planet must result in an orbit for planet that is either a circle, ellipse, parabola, or hyperbola is beyond the power of elementary algebra. Newton, in solving the problem, made use of his new *fluxions,* which we now know as differential calculus. He showed, in fact, that the gravitational interaction between *any* two bodies would result in an orbital motion of each body about the other that is some form of a *conic section* (see Section 3.3).

Circular and parabolic orbits require theoretically precise speeds that would not be expected to occur in nature; thus we would not expect to find a planet (or other object) with *exactly* a circular or parabolic orbit. The latter divides the family of elliptical (closed) from the family of hyperbolic (open) orbits that actually do occur. The planets, of course, do not have hyperbolic orbits or they would long since have receded into interstellar space; their orbits, then, must be elliptical, as found by Kepler.

(b) Kepler's Second Law

In the preceding paragraphs we saw how a planet speeds up as it approaches the sun and slows down as it pulls away, in qualitative agreement with Kepler's second law. Newton derived the second law rigorously with a simple geometrical proof of this law of equal areas, which we repeat here.

Consider a planet at *A* revolving about the sun at *S* (Figure 4.10). In a short interval of time, the planet's forward velocity would ordinarily carry it to *B*. However, the gravitational pull between it and the sun accelerates it to *C*. Since we are considering a very brief interval of time, we can regard the acceleration of the planet as being along a direction *BC,* parallel to *AS,* the direction from the planet to the sun at the beginning of the instant. The planet now has a velocity along the direction *AC*. In the next brief interval of time, equal in length to the first interval, the planet would ordinarily continue moving in a straight line at a constant speed and would end up at *D,* along the extension of *AC,* so that the distance *CD* was equal to the distance *AC*. However, again the sun accelerates the planet toward

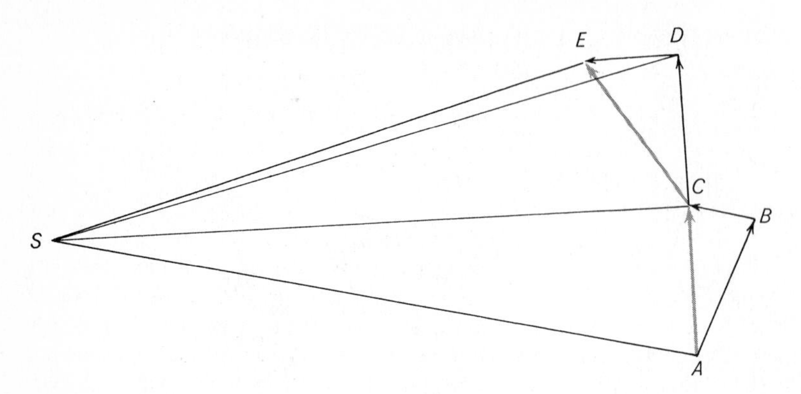

Figure 4.10 Geometrical proof of the law of areas.

it (now in direction *CS*), so the planet actually moves along *CE*.

Consider *AC* and *CD* to be the bases of the triangles *ASC* and *CSD*, respectively. Since *AC* = *CD*, the two triangles have equal bases. They also have the same altitude—the perpendicular distance of *S* from *AD* or its extension. Thus triangles *ASC* and *CSD*, having equal bases and altitudes, have equal areas.

Now *SC* is a common base of triangles *SEC* and *SDC*. Those triangles also have equal altitudes, the distance between the parallel lines *SC* and *ED*. Thus triangles *SEC* and *SDC* are equal in area.

Because triangles *ASC* and *CSE* are both equal in area to triangle *SCD*, they are equal in area to each other. These are the areas swept out by a line from the planet to the sun in two successive equal intervals of time. Many such brief intervals of time can be combined to show that the areas swept out in *any* two equal intervals of time are equal. Thus Kepler's second law is verified.

The argument given is rigorous only if the brief time intervals are infinitesimal. But we can imagine triangles *ASC* and *CSE* to be as small in area as we like, and the proof is still valid. Here is another example in which Newton employed differential calculus, letting time intervals approach zero.

In fact, we did not need to restrict ourselves to a planet moving around the sun. If any two objects revolve about each other under the influence of a central force, the law of equal areas will apply. The law of areas, as we have seen (Section 4.1h), is the geometrical manifestation of the conservation of angular momentum.

(c) Kepler's Third Law

We may now use the ideas we have developed to demonstrate a simple derivation of Kepler's third law, but again only for circular orbits. Newton derived the same form of the law for the more general elliptical orbits.

For each of two mutually revolving bodies, the gravitational attraction between the two provides the centripetal acceleration to keep them in circular orbits. If the bodies have masses m_1 and m_2 and distances r_1 and r_2 from their common center of mass, they are separated by a distance $r_1 + r_2$ and we can equate the gravitational force to the centripetal force for each body. Equating the formulas derived in Section 4.3 for centripetal force, we obtain for body 1,

$$\frac{Gm_1m_2}{(r_1 + r_2)^2} = \frac{m_1 4\pi^2 r_1}{P^2},$$

and for body 2,

$$\frac{Gm_1m_2}{(r_1 + r_2)^2} = \frac{m_2 4\pi^2 r_2}{P^2}.$$

If we cancel out the masses common to each side of each equation and add the two equations, we obtain

$$\frac{G(m_1 + m_2)}{(r_1 + r_2)^2} = \frac{4\pi^2}{P_2}(r_1 + r_2),$$

or

$$(m_1 + m_2)P^2 = \frac{4\pi^2}{G}(r_1 + r_2)^3.$$

Since $r_1 + r_2$ is the distance between the two bodies, we recognize it, in the case of a planet going around the sun in a circular orbit, as the semimajor axis, *a*, of the relative orbit. Then the above equation looks the same as the formula we gave in Chapter 3 for Kepler's third law, except for the factor $m_1 + m_2$ and the factor $4\pi^2/G$. The latter is simply a constant of proportionality. If the proper units are chosen for distance and time, *G* will take on such a value that $4\pi^2/G$ will equal unity.

We shall see in a moment why Kepler was unaware of the factor $(m_1 + m_2)$.

First, let us illustrate two systems of units such that the constant of proportionality $4\pi^2/G = 1$. One such system is to measure the sum of the masses of the revolving bodies, $m_1 + m_2$, in units of the combined mass of the sun and Earth, the period in years, and the separation of the bodies in astronomical units. Then if the equation is applied to the mutual revolution of the Earth and sun, everything in the equation other than the factor $4\pi^2/G$ is equal to unity, so it must be unity also. It is easy to see that another suitable choice of units is the combined mass of the Earth and the Moon for the sum of the masses, the sidereal month (the Moon's period of revolution about the Earth) for the period, and the Moon's mean distance from Earth for the separation of the bodies.

Newton derived his equation not only for the planets moving about the sun but also for any pair of mutually revolving bodies—two stars, a planet and a satellite, or even a plate and a spoon revolving about each other in space.

Newton's version of Kepler's third law differs from the original in that it contains as a factor the sum of the masses of the two revolving bodies. It will become clear why Kepler was not aware of that term if we note that we can consider the sun and Earth to be a pair of mutually revolving bodies. The sun has a mass of about 330,000 times that of the Earth. Thus the combined mass of the sun and the Earth is, to all intents and purposes, the mass of the sun itself, the Earth's mass being negligible in comparison. Suppose we choose the mass of the sun for our unit of mass. Then, in the Earth-sun system, $m_1 + m_2 = 1$. Furthermore, the sum of the masses of the sun and any other planet is also very nearly unity. Even Jupiter, the most massive planet, has only 1/1000 of the mass of the sun; for the sun and Jupiter, $m_1 + m_2 = 1.001$, a number so nearly equal to 1.000 that Kepler was unable to detect the difference from Tycho's observations. (However, the fact that the masses of Jupiter and Saturn are not completely negligible compared to the sun accounts, in part, for the slight discrepancies in Kepler's version of his third law as applied to Jupiter and Saturn; see Table 3.2). Thus, if we apply the equation Newton derived to the mutual revolution of the sun and a planet, and choose years and astronomical units for the units of time and distance, and the solar mass for the unit of mass, Newton's equation reduces to

$$(m_1 + m_2)P^2 = (1)P^2 = P^2 = a^3,$$

in agreement with Kepler's formulation of the law.

(d) Kepler's Laws Restated—Summary

We now restate Kepler's three laws of planetary motion in their more general form, as they were derived by Newton:

KEPLER'S FIRST LAW: If two bodies interact gravitationally, each will describe an orbit that is a conic section about the common center of mass of the pair. In particular, if the bodies are permanently associated, their orbits will be ellipses. If they are not permanently associated, their orbits will be hyperbolas.

KEPLER'S SECOND LAW: If two bodies revolve about each other under the influence of a central force (whether or not in a closed elliptical orbit), a line joining them sweeps out equal areas in the orbital plane in equal intervals of time.

KEPLER'S THIRD LAW: If two bodies revolve mutually about each other, the sum of their masses times the square of their period of mutual revolution is in proportion to the cube of the semimajor axis of the relative orbit of one about the other.

In metric units, the algebraic formulation of Newton's version of Kepler's third law is

$$(m_1 + m_2)P^2 = \frac{4\pi^2}{G}a^3,$$

where G is the constant of gravitation. (If the units of length, mass, and time are centimeters, grams, and seconds, respectively, G has the value 6.67×10^{-8}.) If either of the sets of units shown in Table 4.1 is used, the law becomes

$$(m_1 + m_2)P^2 = a^3.$$

TABLE 4.1 Examples of Systems of Units for Which $4\pi^2/G = 1$

	I	II
Units of $(m_1 + m_2)$	Sun's mass + Earth's mass	Earth's mass + Moon's mass
Units of P	Sidereal year	Sidereal month
Units of a	Astronomical unit	Mean distance of Moon from Earth

4.5 APPLICATION OF NEWTON'S LAWS

(a) Masses of Planets and Stars

Our only means of measuring the masses of astronomical bodies is to study the way in which they react gravitationally with other bodies. Newton's derivation of Kepler's third law, which includes a term involving the sum of the masses of the revolving bodies, is most useful for this purpose.

Consider a planet, such as Jupiter, that has one or more satellites revolving about it. We can select one of those satellites and regard it and its parent planet as a pair of mutually revolving bodies. We measure the period of revolution of the satellite (say, in sidereal months) and the distance of the satellite from the planet (in terms of the distance of the Moon from the Earth), and insert those values into the equation

$$m_1 + m_2 = \frac{a^3}{P^2}.$$

Since both a and P are observed, we can immediately calculate the combined mass of the planet and its satellite. Obviously most of this mass belongs to the planet, its satellites all being very small compared to it. Thus $m_1 + m_2$ is, essentially, the mass of the planet in terms of the mass of the Earth.

As a numerical example, Deimos, the outermost satellite of Mars, has a sidereal period of 1.262 days and a mean distance from the center of Mars of 23,500 km. In sidereal months, the period of the satellite is 1.262/27.3 = 0.0462. In units of the distance of the Moon from the Earth, Deimos has a distance from the center of Mars of 23,500/384,404 = 0.0611. Thus the mass of Mars plus the mass of Deimos is given by

$$m_{\text{Mars}} + m_{\text{Deimos}} = \frac{(0.0611)^3}{(0.0462)^2} = \frac{2.28 \times 10^{-4}}{2.13 \times 10^{-3}} = 0.11 \text{ Earth mass.}$$

Since Deimos is a very tiny satellite (only about 13 km across), its mass can be neglected compared with that of Mars, and we find that Mars has a mass of just over one-tenth that of the Earth.

In Chapter 25 we shall see that we use the same mathematical technique to determine the masses of stars that are members of binary-star systems (a binary star is a pair of stars that revolve around each other). In fact, we can use Newton's version of Kepler's third law to estimate the mass of our entire Galaxy (Chapter 34) or even of other galaxies (Chapter 35).

(b) Conservation of Energy

A system of mutually revolving bodies possesses a certain energy that does not change unless external forces are applied. The total energy of the system in turn determines the kinds of orbits that are possible. In the particular case of a two-body system, there is a simple equation, called the *energy equation,* that expresses this conservation of energy.

Consider two mutually revolving bodies of masses m_1 and m_2 with a relative orbit of semimajor axis a, the magnitude of the velocity, v, of one body with respect to the other at an instant when the bodies are at a distance r apart is given by the equation

$$v^2 = G(m_1 + m_2)\left(\frac{2}{r} - \frac{1}{a}\right).$$

This equation is called the *energy equation.* If the relative orbit of one body around the other is a circle, $r = a$, and the equation gives for the *circular velocity*

$$v^2 = G(m_1 + m_2)\frac{1}{r}.$$

If the relative orbit is a parabola, the bodies escape from each other, and the energy equation gives for the *parabolic* or *escape velocity,*

$$v^2 = G(m^1 + m_2)\frac{2}{r},$$

because a parabolic orbit can be considered an ellipse of eccentricity 1, with $a = \infty$ (infinity). Note that the velocity of escape is equal to the circular velocity times the square root of 2.

If two bodies have the most minute sideways motion with respect to each other, they cannot fall straight toward each other but will move in elliptical orbits about each other. If that sideways velocity is just great enough so that the centripetal force required for a circular orbit is exactly equal to the bodies' mutual gravitational attractive force, they will move about each other in circular orbits. This critical speed is the *circular velocity* given by the second to last equation above. A still higher sideways motion will produce elliptical orbits of larger major axes than the diameters of the circular orbits. A sideways velocity equal to the velocity of escape of one body with respect to the other will result in parabolic orbits, and still higher velocities give hyperbolic orbits.

The energy equation is quite general. Similar equations can also be written for large numbers of bodies gravitationally bound together, such as clusters of stars. In each case, there is a definite relationship among the mutual masses, speeds, and distances of the objects. One useful application in astronomy is to use the measured separations and speeds of stars to calculate the total mass of a cluster.

In the two-body case, the motion of one body with respect to the other can be in any direction, and the energy equation will still hold: if v is the speed of one body with respect to the other, the equation gives the corresponding value of a, the semimajor axis of the relative orbit. For any closed orbit (circle or ellipse), a must be positive and finite. A value of v greater than the parabolic velocity results in open or hyperbolic orbits. The semimajor axis of a hyperbolic orbit is taken as negative.

It is important to note that if two objects approach each other from a great distance in space they can never "capture" each other into elliptical orbits. Their mutual attraction will speed them up so that they pass each other with a relative speed greater than their mutual velocity of escape, and they will swing out away from each other again, moving in orbits that are hyperbolas (Figure 4.9). As an example, it is impossible to send a rocket to the Moon, and to cause it to move on an elliptical orbit about the Moon, without slowing it down when it is in the lunar vicinity. That is why a rocket intended for lunar orbit carries a *retrorocket* designed to reduce its speed at an appropriate time so that it can enter an elliptical orbit about the Moon. Otherwise, it would bypass the Moon on a hyperbolic orbit.

EXERCISES

1. What is the momentum of a body whose velocity is zero? Does the first law of motion (in the absence of a force the momentum of a body is constant) include the case of a body at rest?
2. Explain how a quantity of lead could be less massive than a quantity of feathers. How could a certain mass of lead be less dense than an equal mass of feathers?
3. If 24 g of material fills a cube 2 cm on a side, what is the density of the material?
4. Suppose two billiard balls collide. Ignoring frictional forces, describe their velocities and accelerations.
5. A body moves in a perfectly circular path at constant speed. What can be said about the presence or absence of forces in such a system?
6. Suppose a 10-g ball and a 1000-g ball are simultaneously dropped from the top of a building. (a) What is the ratio of the weights of the two objects? (b) Explain why they would strike the ground at the same time. (Neglect air resistance.)
7. ★ Did Newton need to know the size of the Earth in kilometers to test his theory of gravitation with the apple and Moon? Why or why not?
8. ★ Calculate the gravitational attraction between a man weighing 100 kg and a woman weighing 50 kg 10 m away from him. Compare this to the attraction between the woman and the Earth. (For the purpose of this calculation, assume the man and woman to be perfect spheres.)
9. How much would a 100 kg man weigh at 6400 km (about one earth radius) above the surface of the Earth? How much would he weigh at 25,600 km above the surface of the Earth?
   Answer to second part: about 4 kg
10. By what factor would a person's weight at the surface of the Earth be reduced if (a) the Earth had its present mass, but eight times its present volume? (b) the Earth had its present size but only one-third its present mass?
11. ★ At some point along a line connecting the Earth and the Moon a mass would experience equal but opposite forces due to the gravitational attractions of those two bodies. Verify that the distance of this point from the earth is about 346,000 km by calculating the acceleration at this point due to the Earth and showing that it is the same as that due to the moon. Consult Appendices 10 and 11.
12. How far is the barycenter from a star of three times the mass of the sun in a double-star system in which the other star has a mass equal to the sun's and a distance of 4 AU from the first star?
13. What is the period of mutual revolution of the two stars described in Exercise 12?
14. ★ Why does Newton's version of Kepler's third law have the form

$$(m_1 + m_2)P^2 = a^3,$$

with the constant of proportionality equal to unity, if $m_1 + m_2$ is in units of the combined mass of the Earth and Moon, P is sidereal months, and a in units of the Moon's distance? Find another such set of units, other than those given in the text, for which the constant of proportionality is unity.

15. A cow attempted to jump over the Moon but ended in an orbit around the Moon instead. Describe how the cow could be used to determine the mass of the Moon.

Sir Isaac Newton (1643–1727) had the insight to realize that the force that makes planets fall around the sun and the force that makes apples fall to the ground are different manifestations of the same thing: gravitation. Newton's work on the laws of motion, gravitation, optics, and mathematics laid the foundation for almost all physical science up to the 20th century.

5

GRAVITATION IN THE PLANETARY SYSTEM

Newton's laws of motion and the law of gravitation provide a framework within which to explore many aspects of the planetary system. To an excellent first approximation, the orbits of planets (Kepler's laws) can be understood in terms of the two-body problem discussed in Chapter 4. However, to calculate the exact paths of the planets, or to navigate a spacecraft from one planet to another, the attractions of other bodies must also be taken into account. The familiar phenomenon of the tides (and some less familiar applications of the same ideas) also require us to deal with gravitational theory involving more than two bodies. These calculations are considerably more complex than the two-body problem, but they can be carried out today with high precision using large computers.

5.1 ORBITS OF PLANETS AND SPACECRAFT

(a) Description of an Orbit

Celestial mechanics is the study of the motions of astronomical objects using gravitational theory. A classic problem in celestial mechanics is to determine the orbit of an object (such as a newly discovered comet) from observations only of its directions at various times as seen from the Earth. An orbit, once determined, then allows the future paths of the object to be calculated.

For the general characterization of an orbit, three quantities are required. These are the *size* (the semimajor axis), the *shape* (the eccentricity), and the *period of revolution.* These three items provide enough information, for example, to calculate the sum of the masses of the objects, an important objective in studying binary stars (stars in orbit around each other). However, to predict the path of a planet or comet or spacecraft, we must specify six quantities to determine the orbit fully. These six items of information are called the *elements* of the orbit. Some of the elements of the orbits of the planets are given in Appendix 9.

(b) Elements of Orbits

Two elements are needed to describe the size and shape of an orbit. We have already seen, for example (Section 3.3a), that the size and shape of an elliptical orbit can be specified by the semimajor axis and eccentricity of the ellipse. In fact, the same two quantities serve to specify the size and shape of *any* conic section, and hence of any orbit. Three other numbers are required to specify the orientation of the orbit with respect to some reference system, say, one defined by the Earth's orbit. A final element is needed to specify where the object is in its orbit at some particular time, so that its location at other times can be computed. A total of six such orbital elements are sufficient if the object is in orbit around the sun and if it has a mass that can be neglected in comparison with the sun's mass. If, however, the sum of the masses of two mutually revolving bodies is not known, a seventh datum is needed to specify their orbit completely. If the relative orbit is an ellipse, the period of mutual revolution of the bodies suffices for this seventh element. If it is a hyperbola, the *areal velocity* replaces the period. The areal velocity is the rate at which an imaginary line between the two bodies sweeps out an area in space with respect to one of the bodies.

TABLE 5.1 Elements of an Orbit

NAME	SYMBOL	DEFINITION
Semimajor axis	a	Half of the distance between the points nearest the foci on the conic that represents the orbit (usually measured in astronomical units).
Eccentricity	e	Distance between the foci of the conic divided by the major axis.
Inclination	i	Angle of intersection between the orbital planes of the object and of the Earth.
Longitude of the ascending node	Ω	Angle from the vernal equinox (where the ecliptic and celestial equator intersect with the sun crossing the equator from south to north), measured to the east along the ecliptic plane, to the point where the object crosses the ecliptic traveling from south to north (the ascending node).
Argument of perihelion	ω	Angle from the ascending node, measured in the plane of the object's orbit and in the direction of its motion, to the perihelion point (its closest approach to the sun).
Time of perihelion passage	T	One of the precise times that the object passed the perihelion point.
Period	P	The sidereal period of revolution of the object about the sun.

The six (or seven) orbital elements can be specified in a multitude of ways. In Table 5.1 is summarized the set of elements that is most conventional for describing the orbit of an object revolving about the sun. It must be emphasized, however, that other sets of data can be used for the elements of an orbit and indeed, often are used in modern practice. The elements described in Table 5.1 are illustrated in Figure 5.1.

If the set of elements given in Table 5.1 is used, the inclination and longitude of the ascending node, i and Ω, describe the orientation of the orbital plane. The argument of perihelion, ω, gives the orientation of the orbit in its plane. The semimajor axis a and eccentricity e give the size and form of the orbit. The time of perihelion passage T and period P are the data required to calculate the position of the object in its orbit. If the object is one of small mass circling the sun, the period is superfluous, for it can be obtained from the semimajor axis with the use of Kepler's third law. If the orbit is not an ellipse, the areal velocity rather than the period can be used.

(c) Satellite Orbits

The first artificial Earth satellite, Sputnik 1, was launched on October 4, 1957. Since that time, thousands of satellites have been placed into orbit around the Earth, and spacecraft have also orbited the Moon, Venus, and Mars.

Once an artificial satellite is in orbit, its behavior is no different from that of a natural satellite, such as our Moon or Phobos and Deimos at Mars. If the satellite is high enough to be free of atmospheric friction, it will remain in orbit forever, following Kepler's laws in a perfectly respectable way. However, although there is no difficulty in maintaining a satellite once it is in orbit, a great deal of energy is required to lift the spacecraft off the Earth and accelerate it to orbital speed.

To illustrate how a satellite is launched, imagine a gun on top of a high mountain, firing a bullet horizontally (Figure 5.2—adapted from a similar diagram by Newton—Figure 5.3). Imagine, further, that the friction of the air could be removed, and that all hindering objects, such as other mountains, buildings, and so on, are absent. Then the only force that acts on the bullet after it leaves the muzzle is the gravitational force between the bullet and Earth. If the bullet is fired with velocity v_a, it will continue to have that forward speed, but meanwhile the gravitational force acting upon it will ac-

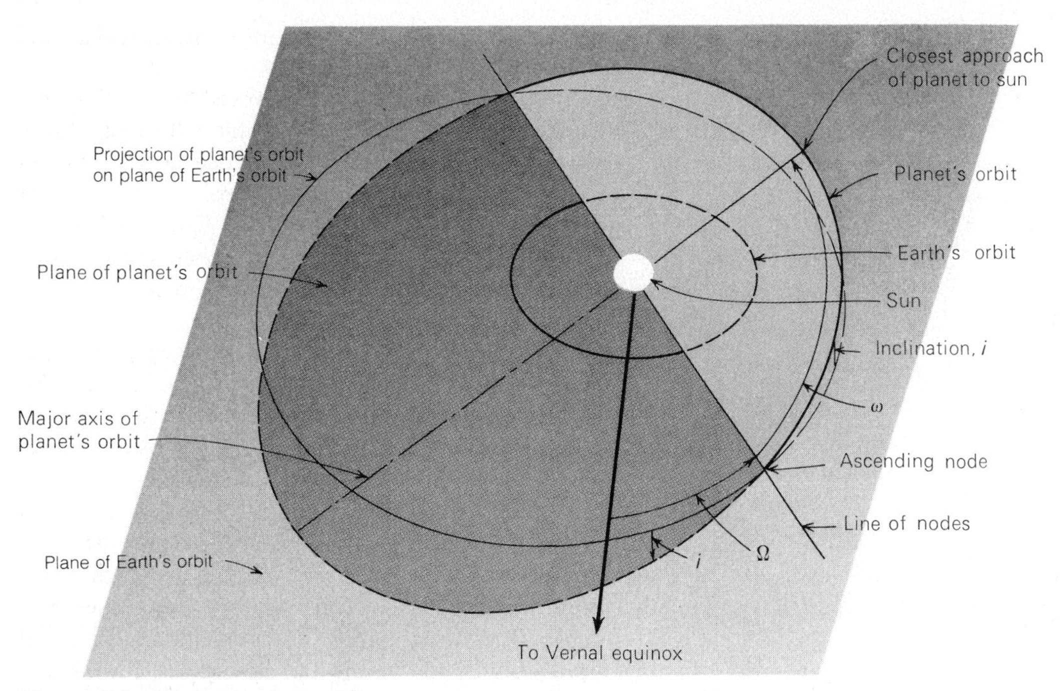

Figure 5.1 Elements of an orbit.

celerate it downward so that it strikes the ground at a. However, if it is given a higher muzzle velocity v_b, its higher forward speed will carry it farther before it hits the ground, for, regardless of its forward speed, its downward gravitational acceleration is the same. Thus this faster-moving bullet will strike the ground at b. If the bullet is given a high enough muzzle velocity, v_c, as it accelerates toward the

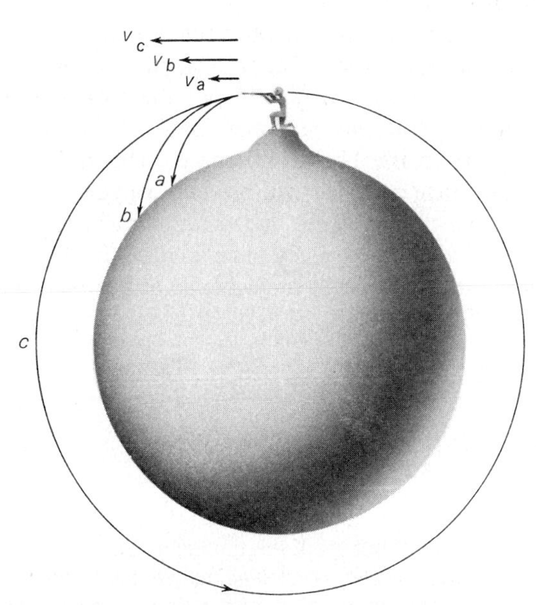

Figure 5.2 Firing a bullet into a satellite orbit.

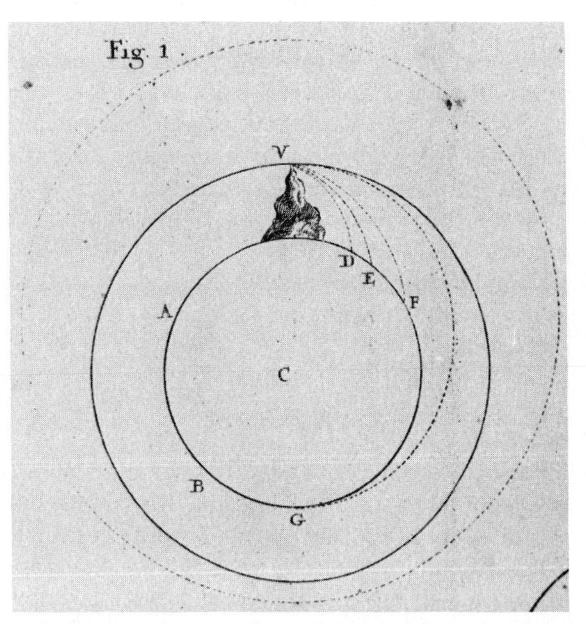

Figure 5.3 A diagram by Newton in his *De mundi systematic,* 1731 edition, illustrating the same concept shown in Figure 5.2. (Crawford Collection, Royal Observatory Edinburgh)

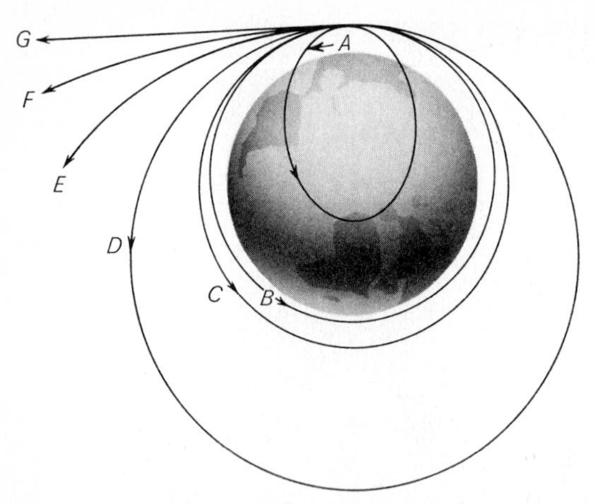

Figure 5.4 Various satellite orbits that result from different initial velocities parallel to the Earth's surface. *A* is an orbit that is intercepted by the solid Earth (like that of a military ballistic missile); *C* is a circular orbit; and *D, E, F,* etc., are orbits of increasing energy but all with the same perigee at the point of injection.

ground, the curved surface of the Earth will cause the ground to tip out from under it so that it remains the same distance above the ground, and "falls around" the Earth in a complete circle. This is another way of saying that at a critical speed v_c the gravitational force between the bullet and Earth is just sufficient to produce the centripetal acceleration needed for a circular orbit about the Earth. The speed v_c, the *circular satellite velocity* at the surface of the Earth, is about 8 km/s.

Novelist Jules Verne anticipated Earth satellites long ago. In one of his stories an enemy force was planning to bomb a city with a gigantic cannon ball. However, the cannon ball was propelled with too great a speed—in fact, the circular satellite velocity—so it passed harmlessly over the city and on into a circular orbit around the Earth.

(d) Possible Orbits

Suppose that a rocket is sent up to an altitude of a few hundred miles, then turned so that it is moving horizontally, and finally given a forward horizontal thrust. It will proceed in an orbit the size and shape of which depend critically on the exact direction and speed of the rocket at the instant of its "burnout," that is, the instant when the thrust supplied by its fuel is shut off. First, suppose that it is moving exactly horizontally, or parallel to the ground, at burnout. The possible kinds of orbits it can enter are shown in Figure 5.4.

If the rocket's burnout speed is less than the circular satellite velocity, its orbit will be an ellipse, with the center of the Earth at one focus of the ellipse. The *apogee* point of the orbit, that point that is *farthest* from the center of the Earth, will be the point of burnout; the *perigee* point (closest approach to the center of the Earth) will be halfway around the orbit from burnout.

If the burnout speed is substantially below the circular satellite velocity, most of its elliptical orbit will lie beneath the surface of the Earth (orbit *A*), where, of course, the satellite cannot travel; consequently, it will traverse only a small section of its orbit before colliding with the surface of the Earth (or more likely, burning up in the dense lower atmosphere). If the burnout speed is just slightly below the circular satellite velocity, the rocket may clear the surface (orbit *B*), although its orbit will probably lie too low in the atmosphere for the satellite to be longlived.

If the burnout speed were exactly the circular satellite velocity, a circular orbit centered on the center of the Earth would result (orbit *C*). It is extraordinarily unlikely that a missile could be given so accurate a direction and speed that a perfectly circular orbit could be achieved. A slightly greater burnout speed will produce an elliptical orbit with *perigee* at burnout point and apogee halfway around the orbit (orbit *D*).

A burnout speed equal to the velocity of escape from the Earth's surface, that is, the parabolic velocity (about 11 km/s), will put the rocket into a parabolic orbit that will just enable the vehicle to escape from the Earth into space (orbit *E*). A still higher burnout speed will produce a hyperbolic orbit in which the vehicle escapes the Earth with energy to spare (orbit *F*). The higher the burnout speed, the nearer will the orbit be to a straight line (orbit *G*).

(e) Application of the Energy Equation

We can apply the energy equation to the orbit of a satellite moving about the Earth. Let us measure speed in terms of the circular-satellite velocity at the Earth's surface, the masses in terms of the Earth's mass, and r and a in units of the radius of the Earth. In these units, the constant G is equal to unity, and the equation simplifies to

$$v^2 = \left(\frac{2}{r} - \frac{1}{a}\right).$$

Suppose a satellite is launched from a point near the Earth's surface (say, at an altitude of 300 km); r is 1.047 and v at that point is the burnout speed. Then the semimajor axis of the orbit, a (a measure of the size of the orbit), can easily be calculated if the burnout speed is known:

$$\frac{1}{a} = \frac{2}{r} - v^2.$$

Negative values of a correspond to hyperbolic orbits.

As an example, suppose the burnout speed is 10 km/s, or about 1.263 in units of the circular-satellite velocity. Then, we find for a

$$\frac{1}{a} = \frac{2}{1.047} - (1.263)^2 = 0.315,$$

or

$$a = 3.17 \text{ Earth radii.}$$

Such a satellite would have an apogee distance of about 33,760 km from the center of the Earth, or about 27,381 km above the surface.

The energy equation holds regardless of the direction the two bodies are moving with respect to each other. Note that there is no term in the equation that involves the direction in which a rocket is moving at burnout. Thus, even if the rocket were not moving parallel to the ground at burnout, the major axis of its orbit would depend only on its burnout speed (Figure 5.5). However, the *eccentricity,* or shape, of the orbit does depend on the direction of motion of the rocket. We see in Figure 5.5 that for a rocket launched into a satellite orbit near the surface of the Earth, unless the burnout direction is nearly parallel to the ground, the resulting orbit will be too eccentric to clear the surface of the Earth.

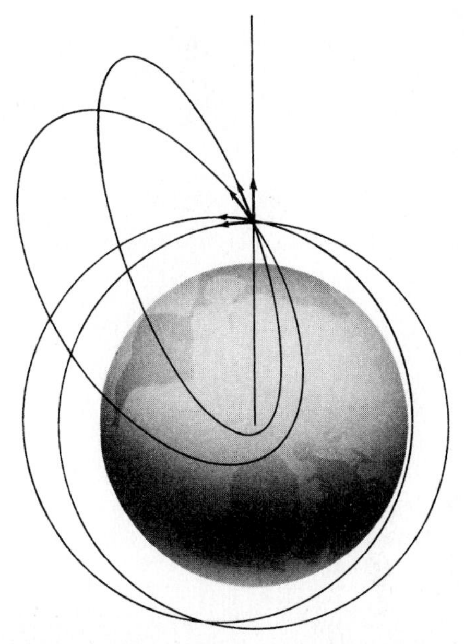

Figure 5.5 Various satellite orbits that result from the same initial speed but in different directions. All these orbits have the same major axis but different eccentricities.

(f) Launch Vehicles

Launch of an Earth satellite or an interplanetary spacecraft requires a rocket-powered launch vehicle. Only rocket engines can operate in the near-vacuum of the upper atmosphere or of near-Earth space, and only rockets have the high efficiency required to reach Earth orbit.

The simplest rocket engines are *solid fuel rockets,* which can be stored for long periods and require relatively little care. The 19th-century British rockets that were used in the attack on Baltimore ("the rockets' red glare"), most of the ICBMs of today (Minuteman, Polaris, etc.), and the Space Shuttle booster engines are all examples of solid fuel rockets. Their main disadvantages are that they cannot be restarted in space and are less efficient than liquid fuel engines.

Liquid fuel rockets were first developed in the U.S. and Germany in the 1930s and saw action in World War II in the form of the V-2 missile. The most efficient possible chemical rocket burns liquid hydrogen and liquid oxygen (to produce an exhaust of water vapor). Hydrogen/oxygen rockets powered the upper stages of the Saturn V Apollo vehicle used to land astronauts on the Moon, and they are used for the Centaur family of deep-space launch vehicles and for the Space Shuttle main engines (Figures 5.6 and 5.7). Since the hydrogen and oxygen must be stored and handled in liquid form at extremely low temperatures, other fuels are required for long-duration flights. The Shuttle's small

Figure 5.6 Launch of the Space Shuttle, the primary U.S. rocket vehicle of the 1980s.

Figure 5.7 The Shuttle in low Earth orbit.

maneuvering engines, and the rockets used for interplanetary flight, usually are powered by liquid fuels that do not require special cooling and that ignite spontaneously as they are mixed in the rocket combustion chamber.

Space-flight technology, so exotic a few decades ago, has spread around the world. Today there are hundreds of military and civilian launch vehicles manufactured by the governments of about a dozen nations as well as by private industry. The most powerful launch vehicle ever developed was the Saturn V (1968), with a thrust of 7.5 million pounds. Unfortunately, the Saturn V was discontinued in the early 1970s, and the last of these vehicles now lie rusting on the lawns of NASA centers at Cape Canaveral, Florida, and Houston, Texas. The U.S.S.R. is developing a still larger vehicle, rumored to have a thrust of 10 to 12 million pounds, but at this writing it has not been successfully launched.

Today the majority of the large civilian payloads are launched by the U.S.S.R., the U.S., or the European Space Agency (ESA), a consortium of European nations. The centerpiece of the U.S. space fleet is the Space Shuttle (or STS, Space Transportation System, in NASA jargon). The Shuttle can carry up about 30 tons to low Earth orbit and is unique in its ability to provide service to spacecraft in orbit and even return payloads to Earth. The U.S.S.R. is still using modifications of reliable expendable rockets that were first developed in the 1960s, although a new generation of vehicles, including a reusable shuttle, are now being tested. The European entry is the Ariane series of rockets, which are generating tough competition for the Shuttle in launching moderate-sized commercial payloads into Earth orbit.

(g) Earth Satellites

Sputnik 1, the first artificial Earth satellite, had an overall weight of about 4 tons, and a scientific instrumentation package of about 80 kg. The first American satellite, Explorer 1, launched January

Figure 5.8 The Voyager spacecraft, an example of an interplanetary robot craft. Voyager 2, launched in 1977, is exploring four planets and their ring and satellite systems.

31, 1958, was much smaller. Since then, the U.S.S.R. and the U.S. have each launched dozens of satellites each year, many of which remain in orbit. Additional nations that have launched their own Earth satellites include China, Japan, India, and ESA (the European consortium), while others have taken advantage of the launch services provided by the U.S., U.S.S.R., or ESA.

Most satellites are launched into low Earth orbit, since this requires the minimum launch energy. At the orbital speed of about 8 km/s, they circle the planet in about 90 minutes. These orbits are not stable indefinitely, since the drag generated by friction with the thin upper atmosphere eventually leads to a loss of energy and "decay" of the orbit. Upon re-entering the denser parts of the atmosphere, most satellites are burned up by atmospheric friction, although some solid parts may reach the surface. And of course, the Shuttle and other recoverable payloads are designed to survive re-entry intact.

If a satellite is aimed for a higher orbit, it is initially injected into a somewhat eccentric orbit with apogee (highest point) near the target altitude. A second firing of the rocket engine near apogee then imparts additional energy to raise the perigee and circularize the orbit at the desired position.

For many purposes, the most desirable altitude is that at which the orbital period exactly equals the rotation period of the Earth, 24 hours. Kepler's third law tells us that this altitude above the center of the Earth must be about 40,000 km. Here, a satellite seems stationary with respect to the rotating Earth beneath, which is ideal for most meteorological and communications applications. The orbit is called *geostationary*. As we will see later, Pluto's natural satellite Charon is in a stationary orbit with respect to its planet.

(h) Interplanetary Spacecraft

The exploration of the solar system has been carried out largely by robot spacecraft sent to the other planets (Figure 5.8). To escape Earth, these craft must achieve a velocity of more than 11 km/s, after which they coast to their targets, subject only to minor trajectory adjustments provided by small thruster rockets on board. In interplanetary flight these spacecraft follow Keplerian orbits around the sun, modified only when they pass near one of the planets (see Section 5.1i).

Most interplanetary probes have been *flybys*, which means that they have made the relevant observations of the planets in the brief periods during which they passed near their targets. Closeup observations by flybys are generally limited to a few days or less. An extreme example is provided by the 1986 Soviet and European flybys of Comet Halley,

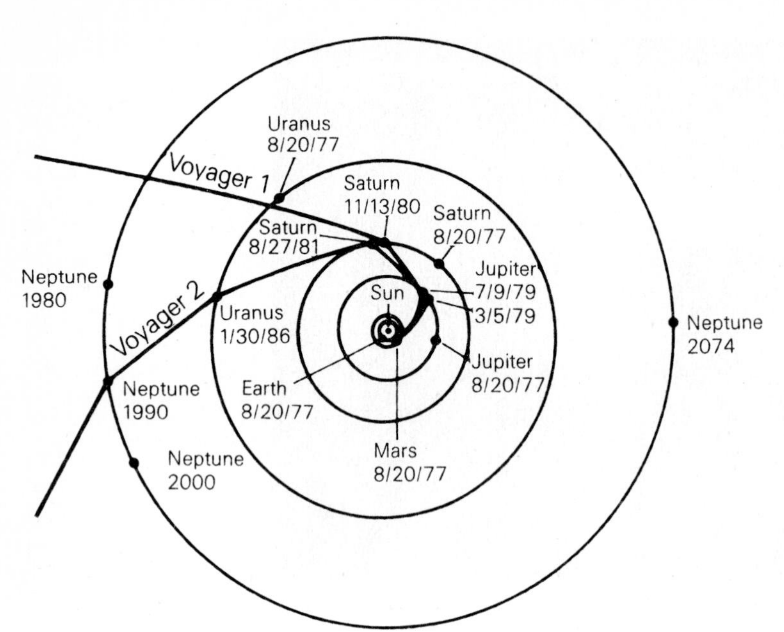

Figure 5.9 The flight paths of the Voyager spacecraft through the outer solar system, taking advantage of the gravitation of each planet to adjust the trajectory toward the next target.

which flashed past the small nucleus in less than a second.

While close to its target, a spacecraft is deflected into a modified orbit, either gaining or losing energy in the process. By carefully choosing the aim point in a planetary encounter, controllers have been able to redirect a flyby spacecraft to a second target. Such *gravity assist trajectories* were first used in 1974 to direct Mariner 10 from Venus to Mercury. Voyager 2 has used a series of gravity assisted encounters to yield successive flybys of Jupiter (1979), Saturn (1980), Uranus (1986), and Neptune (1989) (Figure 5.9).

If we wish to orbit a planet, we must slow the spacecraft with a rocket firing near its target, allowing it to be captured into an elliptical orbit. Mariner 9, in 1971, was the first spacecraft to go into orbit around another planet (Mars).

The next steps beyond a planetary orbiter are atmospheric entry probes and landers. The U.S.S.R. achieved the first successful probes of the atmosphere of Venus (in 1970), the first landers on the surface (1975), and the first instrumented balloons deployed in the Venerian atmosphere (1985). The Soviets also made the first landing on Mars (in 1971), but their results were superceded by the highly successful U.S. Viking entry probes and landers of 1976. We will describe some of the results of these missions in Chapters 15 through 18.

(i) Interplanetary Trajectories

We have now learned the principles of space travel. Spacecraft, once they have left the Earth, are astronomical bodies. They obey the same laws of celestial mechanics as the planets and other natural bodies in the solar system. In other words, spacecraft travel in orbits. If the space vehicles carry auxiliary rocket engines and extra fuel, it may be possible to alter their orbits at will, but the principles remain the same.

We shall illustrate one particular kind of space trajectory by showing one of the many possible ways to reach each of the planets Mars and Venus. The orbits to Mars and Venus we show are those that require the expenditure of the least energy as the rocket leaves the Earth and are thus the most economical of fuel. The orbits of the successful United States Mariner and Pioneer Venus probes, of the Mariner and Viking Mars probes, and of the similar Soviet probes, were all nearly of this type.

Suppose, for simplicity, that the orbits of Venus, Earth, and Mars are circles centered on the sun (when the slight ellipticity of planetary orbits is taken into account, the problem is similar but slightly more complicated). The least-energy orbit that will take us to Mars is an ellipse tangent to the Earth's orbit at the space vehicle's *perihelion* (closest approach to the sun) and tangent to the orbit of Mars at the vehicle's *aphelion* (farthest point from the sun) (Figure 5.10).

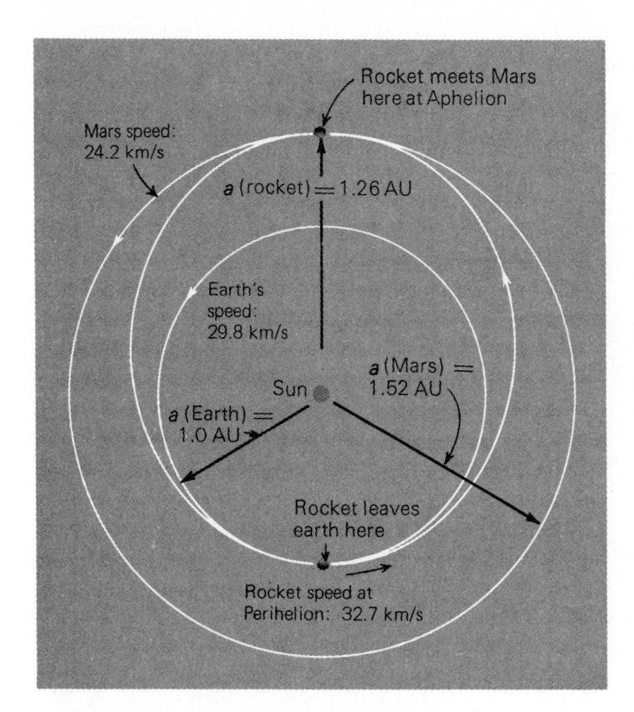

Figure 5.10 Least-energy orbit to Mars.

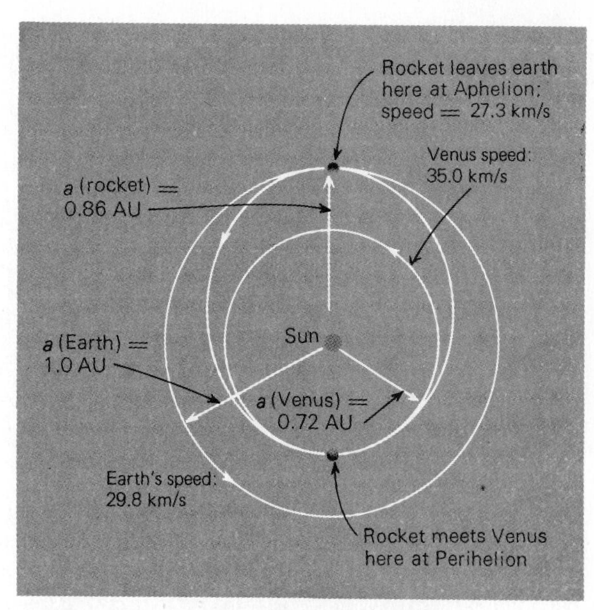

Figure 5.11 Least-energy orbit to Venus.

The Earth is traveling around the sun at the right speed for a circular orbit. For us on the Earth to enter the elliptical orbit to Mars, we must achieve a speed, in the same direction as the Earth is moving, that is slightly greater than the Earth's circular velocity (which is about 29.8 km/s). To calculate this speed, we employ the energy equation. The major axis of the elliptical orbit we want to achieve is the sum of the radii of the orbits of the Earth and Mars. Half of this major axis is the value a. The appropriate value of a is 1.26 AU. The value r is, of course, the Earth's distance from the sun, and $m_1 + m_2$ is the combined mass of the sun and the spaceship (the latter is negligible). The required speed turns out to be slightly under 33 km/s. Since the Earth is already moving 29.8 km/s, we need to leave the Earth with the proper speed and direction so that when we are far enough from it that its gravitational influence on us is negligible compared to the sun's, we are still moving in the same direction as the Earth with a speed relative to it of about 3 km/s.

We have now entered an orbit that will carry us out to the orbit of Mars. The time required for the trip can be found from Kepler's third law, because our spaceship is a planet. The period required to traverse the entire orbit is $a^{3/2}$ years if a is measured in astronomical units. The entire period of the orbit is thus $(1.26)^{3/2} = 1.41$ years. The time required to reach the aphelion point (Mars' orbit) is half of this, or about 8½ months. The trip will have to be planned very carefully so that when we reach the aphelion point of the least-energy orbit, Mars will be there at the same time.

The return trip from Mars to the Earth is of the same type as the trip from the Earth to Venus, to which we now turn our attention. The orbit to Venus is very similar to the orbit to Mars, except now it is at the *aphelion* point that the trajectory ellipse is tangent to the Earth's orbit, and at the *perihelion* point that it is tangent to the orbit of Venus (Figure 5.11). The semimajor axis of this orbit is half the sum of the radii of the orbits of the Earth and Venus, which is 0.86 AU. From the energy equation we find that the speed at the aphelion point in the orbit is 27.3 km/s, about 2.5 km/s *less* (rather than more) than the Earth's speed. The space vehicle would have to leave the Earth, as before, with enough speed so that when it has left the Earth's vicinity it has a speed with respect to the Earth of 2.5 km/s but in a direction *opposite* that of the Earth's motion. Then, relative to the sun, the vehicle is moving at the required 27.3 km/s and will reach the orbit of Venus along the desired elliptical orbit. The travel time to Venus, found as before from Kepler's third law, is about five months. Returning from Venus to Earth is similar to traveling from Earth to Mars.

From the foregoing discussion it is obvious that the Earth and the planet to be visited must be at a critical configuration at the time of launch, in order that the space vehicle meet the planet at the other end of the vehicle's heliocentric orbit. These critical configurations occur at intervals equal to the synodic period of the planet. In practice it is not necessary, and is seldom feasible, to launch the rocket at exactly the

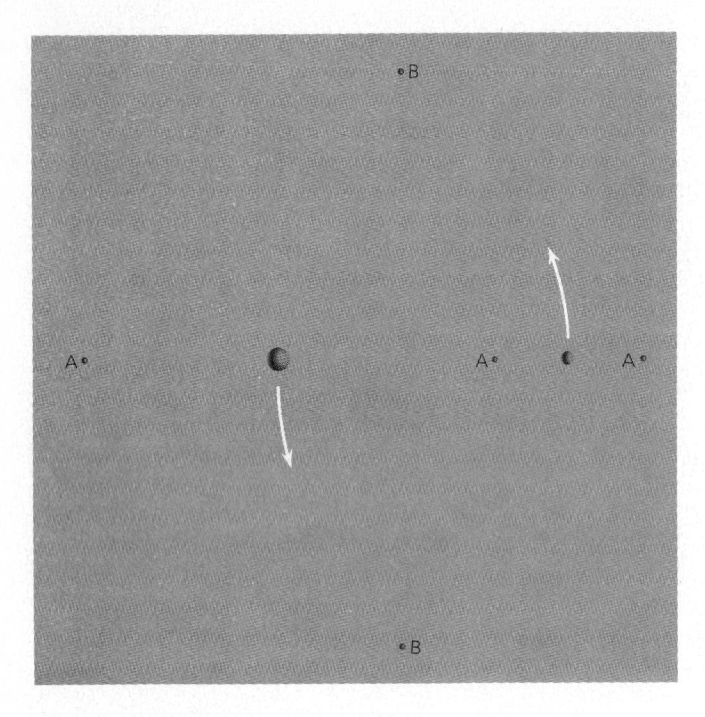

Figure 5.12 Lagrangian points, at each of which a body of small mass moves in a circular orbit, maintaining a fixed orientation with respect to two larger bodies mutually revolving in circular orbits.

proper instant to achieve the least-energy orbit. However, there is a short range of time (typically a few weeks) during which a *nearly* least-energy orbit can be achieved. The length of this time period, called a "window" in space jargon, depends on the thrust capabilities of the available rockets (that is, on how much energy, above the least possible needed, can be supplied by the rocket). "Windows" for Mars journeys occur at intervals of about 780 days; those for Venus trips at intervals of about 584 days.

5.2 THE MANY-BODY PROBLEM

(a) Perturbation Theory

The many-body problem is the problem of describing the motion of any body in a collection of many objects interacting under the influence of their mutual gravitation. Unlike the two-body problem, there are no simple ways to describe this motion. The trajectories are not closed curves like the ellipse or the circle, and the positions cannot be calculated from simple equations like those expressing Kepler's laws. Only a large computer can handle such problems with the precision necessary, for example, to compute the path of an interplanetary probe.

One of the ways to carry out such calculations involves *perturbation theory*. Fortunately, the problem is greatly simplified if one mass dominates in a system. For example, the sun dominates the solar system, and the orbits of planets are very close to their Keplerian ideal. In comparison to the gravitational force between any planet and the sun, the forces exerted by other planets are quite small. Thus the influences of the other planets can be regarded as small corrections or *perturbations* to be applied to the two-body solution. The analysis of such perturbations is especially well suited to modern computers.

(b) Lagrangian Points

There are some special cases of many-body problems that have interested mathematicians for centuries. One of these, involving a small object moving in the gravitational field of two large masses in circular orbits, was studied by the French mathematician Louis Lagrange (1736–1813).

Lagrange found that there are five positions in such a system where the small object, once placed, will move in a circular orbit, always maintaining a fixed orientation with respect to the two greater masses. These are known as the *Lagrangian points* associated with the two large masses (Figure 5.12).

A set of Lagrangian points can be identified near each of the planets, where the two large masses are the sun and the planet.

The Lagrangian points marked B in Figure 5.12 are stable, in that a small object brought near them will remain there and will not be forced away by the perturbations of other planets. (The three points, marked A, are not stable against perturbations.) Note that these two points each mark the tip of an equilateral triangle, with the sun and the planet at the other tips. Objects at these points follow the same orbit as the planet, but lead or trail it by 60 degrees. There are several examples in the solar system of Lagrangian orbits, including the so-called Trojan asteroids that lead and trail Jupiter (Chapter 19) and several of the small satellites of Saturn (Chapter 18). The two stable Lagrangian points of the Earth-Moon system, called L-4 and L-5, have also been suggested as suitable locations for future large space habitats.

(c) Nonspherical Bodies

Bodies with spherical symmetry act, gravitationally, as point masses, for which gravitational influences are easily calculated. In nature, however, most bodies are not exactly spherical, and the simple two-body theory does not give precise results. If the shape of a body deviates only slightly from a sphere, we usually approximate its gravitational influence by that produced by a point mass and treat the small effects of its asphericity as perturbations.

A common cause of the deformation of a star or planet from a perfect sphere is its rotation. In isolation, a nonrotating object will tend toward a perfectly spherical shape under the influence of its own gravitation, but a rotating body tends toward an *oblate spheroid,* which is flattened at the poles and bulging at the equator. Jupiter, for example, is noticeably flattened when seen in a telescope.

The rotational flattening of the Earth is slight but important for calculating the orbits of Earth satellites (including the Moon). The diameter of the Earth measured from pole to pole is 43 km less than the equatorial diameter. This amounts to 1 part in 298, which is referred to as the *oblateness* of the planet. The extra matter at the equator exerts additional gravitational attraction on satellites, beyond what would be expected for a spherical Earth. Except for the special cases of satellites in orbits either parallel or perpendicular to the equator, this force produces easily measurable perturbations. Indeed, it is by accurately tracking satellites that we have mapped the gravity field of the Earth and used this information to probe the interior structure of our planet.

5.3 DIFFERENTIAL GRAVITATIONAL FORCES

For most orbital calculations, we have seen that the gravitational effects of one body on another can be approximated by point masses. There are additional effects, however, that can distort the shape of one body in the gravitational field of another, giving rise to such phenomena as tides and precession. Tidal effects are also of critical importance for understanding planetary rings (Chapter 18) and the exchange of mass between members of binary star systems (Chapter 32).

(a) One Body's Attraction on Two Others

A *differential gravitational force* is the *difference* between the gravitational forces exerted on two neighboring particles by a third more distant body. Both are, of course, attracted by the third body, around which they may be in orbit. But relative to each other, the small differential force can be quite significant. The differential force will tend to pull the two particles away from each other, and if they are part of the same object, the force will distort it or perhaps even tear it apart.

Figure 5.13 Attraction of a large mass and two smaller ones.

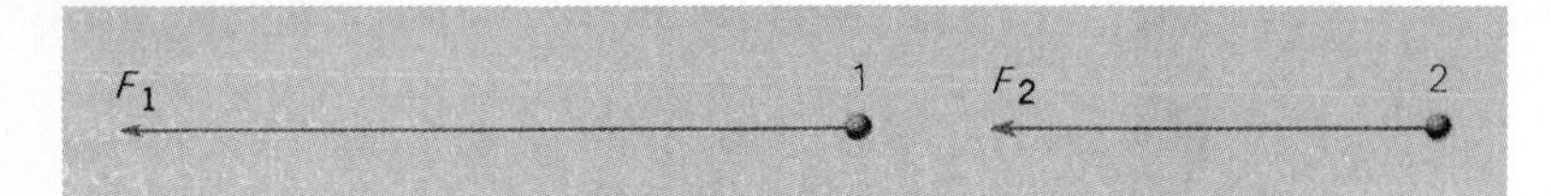

Figure 5.14 Forces on the smaller masses (Figure 5.13) shown as vectors.

As is derived below, the size of the differential gravitational force depends on the inverse *cube* of the distance from the third body, unlike the basic gravitational force itself, which varies with the inverse square power. Thus these forces, which give rise to tidal effects, are strongly concentrated near large bodies. If the distance decreases by half, the differential force rises by a factor of eight. We will look further at the consequences of this dependence of force on distance in Section 5.4e.

(b) Calculation of Differential Gravitational Force

The differential gravitational force can be calculated fairly easily. As an example, consider Figure 5.13, in which three bodies are shown in a line. These are either point masses or perfectly spherical objects whose gravitational effect on external objects is the same as that produced by point masses. To the left is a large body of mass M. To the right are two bodies, each of whose masses we shall assume, for ease of calculation, to be unity—say, each has a mass of 1 g. The first of the small bodies, body 1, is at a distance R from the large one; the other, body 2, is at a distance $R + d$.

The force of attraction between the large mass and body 1 is

$$F_1 = \frac{GM}{R^2},$$

and that between the large mass and body 2 is

$$F_2 = \frac{GM}{(R + d)^2}.$$

Note that F_2 is slightly *smaller* than F_1 because of the greater distance between the large mass and body 2. The difference $F_1 - F_2$ is the differential gravitational force of the large mass on the two smaller masses.

In Figure 5.14 the forces F_1 and F_2 are shown as vectors pointing toward the large mass to the left. Because the force on body 1 is greater than on body 2, the *differential* force tends to separate the two bodies.

Now the center of mass of two small bodies is halfway between them. If either of the two unit masses were at that point, the attraction it would feel toward the large body, M, would be

$$F_{CM} = \frac{GM}{(R + \frac{1}{2}d)^2}.$$

This force is intermediate between the force on body 1 and that on body 2. *With respect to the center of mass,* therefore, both body 1 and body 2 feel themselves pulled *outward.* If the bodies are free to move, they will separate unless their mutual gravitational attraction (not shown in Figure 5.14) is great enough to hold them together. In the example described in the preceding paragraphs, the differential gravitational force ΔF was found to be

$$\Delta F = F_1 - F_2 = \frac{GM}{R^2} - \frac{GM}{(R + d)^2}.$$

Combining the two terms of ΔF, we find, with simple algebra,

$$\Delta F = GM\frac{d(2R + d)}{R^2(R + d)^2}.$$

Now let us suppose that the distance R is very much greater than the distance d. In this case, $R + d$ is so nearly equal to R that we can write

$$R + d \approx R.$$

Similarly,

$$2R + d \approx 2R.$$

With this approximation, our equation for ΔF becomes

$$\Delta F = 2GM\frac{Rd}{R^4} = 2GM\frac{d}{R^3}.$$

Now let us denote by δF the differential force corresponding to a unit separation of the two small bodies, that is, for the case where $d = 1$. Then

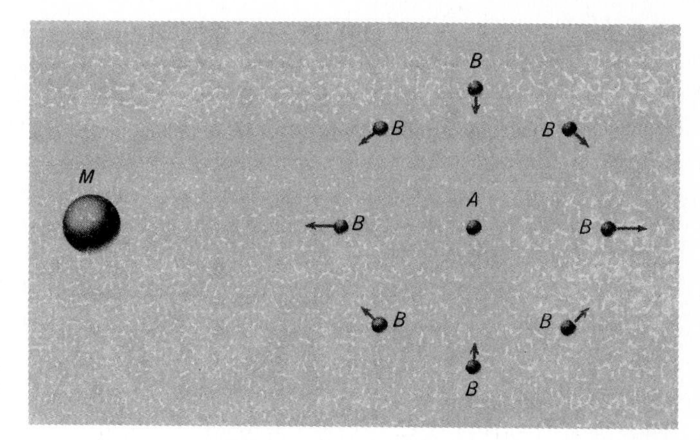

Figure 5.15 Vector differences between forces of attraction of mass M on each of masses A and B.

$$\delta F = \frac{2GM}{R^3},$$

and the total differential force is

$$\Delta F = d \times \delta F.$$

In the foregoing calculations it was assumed that the three bodies are in a line. In general, the bodies are not lined up, and the differential gravitational force between two of them is not simply the arithmetic difference between the forces exerted on each by the third body. Since a force is a *vector* (for it has both magnitude and direction), the difference must be calculated according to the rules of vector subtraction. In Figure 5.15 a mass M (to the left) is attracting each of two masses, A and B. Mass B is shown in various orientations with respect to the line between A and M. In each case the *vector difference* between the force of attraction of M on B and on A is shown. Usually, this differential gravitational force acts in such a way as to tend to separate A and B. However, when B is nearly at right angles to the line joining M and A, M's force on the two is in slightly different directions and tends to pull them closer together; then the differential force on B is directed more or less toward A.

5.4 TIDES

The effects of the differential force of the Moon's gravity on the Earth are noticed by anyone who lives near the sea, for it is this force that causes the ocean tides. While the association of high tides with the position of the Moon was noticed early in history, a satisfactory explanation of the tides awaited the development of the theory of gravitation.

(a) Earth Tides

First, we shall consider the effects of the Moon's attraction on the solid Earth. For the moment, we ignore the flattening of the Earth due to its rotation. Our planet can be regarded as being composed of a large number of particles, each of unit mass, all bound together by their mutual gravitational attraction and cohesive forces. The gravitational forces exerted by the Moon at several arbitrarily selected places in the Earth are illustrated in Figure 5.16. These forces differ slightly from each other because of the Earth's finite size; all parts are not equally distant from the Moon, nor are they all in exactly the same direction from the Moon. If the Earth re-

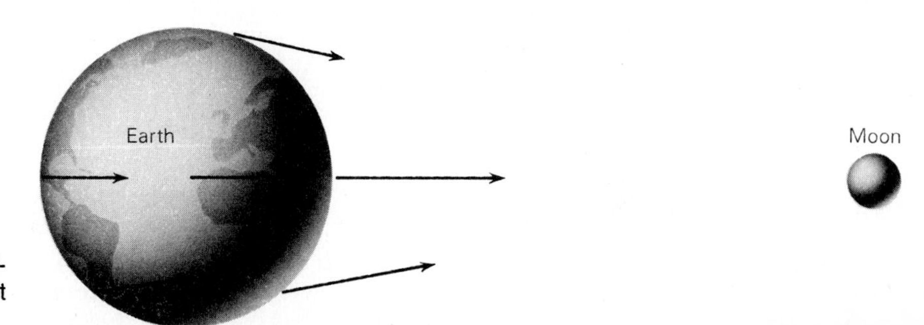

Figure 5.16 The Moon's differential attraction of different parts of the Earth.

tained a perfectly spherical shape, the resultant of all these forces would be that of the force on a point mass, equal to the mass of the Earth, and located at the Earth's center. Such is approximately true, because the Earth is nearly spherical, and it is this resultant force on the Earth that causes it to accelerate each month in an elliptical orbit about the barycenter of the Earth-Moon system.

The Earth, however, is not *perfectly* rigid. Consequently, the differential force of the Moon's attraction on different parts of the Earth causes the Earth to distort slightly. The side of the Earth nearest the Moon is attracted toward the Moon more strongly than is the center of the Earth, which, in turn, is attracted more strongly than is the side of the Earth opposite the Moon. Thus, the differential force tends to "stretch" the Earth slightly into a *prolate spheroid* with its major axis pointed toward the Moon. That is, the Earth takes on a shape such that a cross section whose plane contains the line between the centers of the Earth and Moon is an ellipse with its major axis in the Earth-Moon direction.

Figure 5.17 shows the forces (as vectors) that are acting at several points on the surface of the Earth. In each case, the forces are shown with respect to the Earth's center. The dashed vectors represent the forces due to the Earth's gravity, that is, the weights of various parts of the Earth. The solid vectors (much exaggerated in length) represent the differential gravitational forces due to the varying attraction of the Moon on different parts of the Earth. They are called the *tidal forces*. Those parts of the Earth closer to the Moon than the Earth's

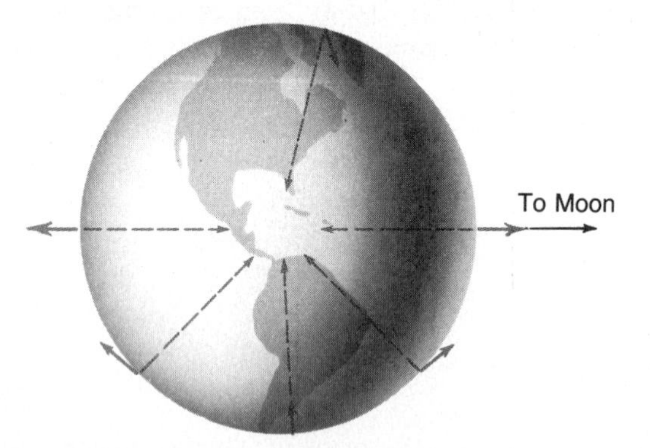

Figure 5.17 Gravitational and tidal forces at various places on the Earth's surface.

center are attracted more strongly toward the Moon than parts of the Earth near its center. Thus the tidal forces are directed *toward* the Moon. Those parts on the opposite side of the Earth are attracted less strongly than are parts at the Earth's center. The tidal forces there are directed *away* from the Moon.

In each case, the vector representing the force can be broken into two components, one in the *vertical* direction, that is, toward or away from the direction of the Earth's gravity, and one in the *horizontal* direction, along the surface of the Earth. The effect of the vertical component of the tidal force is to change slightly the weight of the surface rocks of the Earth. The effect of the horizontal component is to attempt to cause the surface regions of the Earth to flow horizontally.

If the Earth were perfectly spherical, its gravitational attraction for objects on its surface would be in a *vertical* direction, toward the center of the Earth. The actual Earth, however, distorts under the influence of the tidal forces and is not quite spherical (we are still ignoring distortion due to rotation). Consequently, the Earth's gravitational pull upon objects on its surface is not exactly in a direction perpendicular to the surface; there is a slight *horizontal* component in the gravitational pull of the Earth upon its surface regions (Figure 5.18).

If the Earth were fluid, like water, it would distort until all the horizontal components of the tidal forces were exactly balanced by the horizontal pull of the Earth at all points throughout it. Then the inward force on an object at the Earth's surface would be in a vertical direction. It would depend on two factors, those components of the Earth's gravitational attraction and of the tidal force that are normal to the surface of the Earth at that point.

Measures have been made to investigate the actual deformation of the Earth. It is found that the solid Earth does distort, as would a liquid, but only about one-third as much, because of the high rigidity of the Earth's interior. In fact, the rigidity of the Earth must exceed that of steel to account for the small degree of its tidal distortion, a result in agreement with seismic studies (Chapter 13). The maximum tidal distortion of the solid Earth amounts at its greatest to only about 20 cm.

As the Earth rotates, different parts of it are continually being carried under the Moon, so the direction and magnitude of the tidal force acting at any given place on the Earth's surface are con-

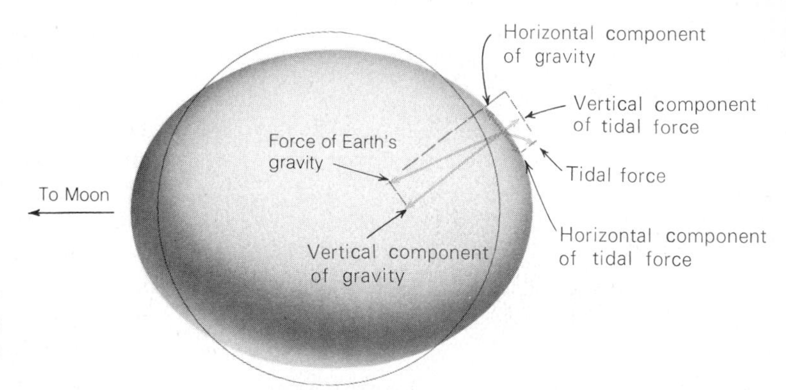

Figure 5.18 Deformation of the solid Earth under the influence of tidal forces (much exaggerated).

stantly changing. If the Earth were viscous, its distortional adjustment would lag somewhat as the tidal forces on it change. However, direct observation shows that the Earth readjusts its shape under the influence of the changing tidal force almost instantaneously. This circumstance implies that the Earth is not only almost perfectly rigid, but also highly elastic on the time scale of tides.

In summary, tidal forces on the Earth, that is, the differential gravitational forces of the Moon's attraction on different parts of the Earth with respect to its center, cause the solid Earth to distort continually from a spherical shape, rising up and down and tilting as a fluid surface would do, but by only about one third the amount. Furthermore, these deformations are nearly instantaneous, changing just as quickly as the tidal forces change due to the Earth's rotation. These facts show the Earth to be more rigid than steel and to be highly elastic.

Rotation, of course, also distorts the Earth's shape. The slight elongation of the Earth that results from the tidal distortions described above is superimposed on the equatorial bulge due to its rotation. The latter is a very much greater distortion than the distortion due to tides.

(b) Ideal Ocean Tides

In Figure 5.19 the vectors represent tidal forces (relative to the Earth's center) at various points on the Earth's surface. These forces are directed generally toward the Moon on the side of the Earth facing the Moon and away from the Moon on the opposite side. Within a zone around the Earth that is roughly the same distance from the Moon as the Earth's center, the tidal forces are directed more or less toward the center of the Earth. At these points, the attraction toward the Moon is the same in magnitude as it is at the center of the Earth, but because of the relatively small distance to the Moon, it is in a direction that tends to pull those points closer to the Earth's center. Each of the vectors (solid arrows) representing a tidal force in Figure 5.19 is resolved into components perpendicular to and parallel to the Earth's surface (dashed arrows). We have seen that if the Earth were fluid it would take on a shape such that points on its surface would feel no horizontal component of force. However, the Earth is sufficiently rigid to be distorted from a sphere by only about one third of the amount required to remove these horizontal forces. Consequently, objects at the surface of the Earth that are not restrained from horizontal motion, for

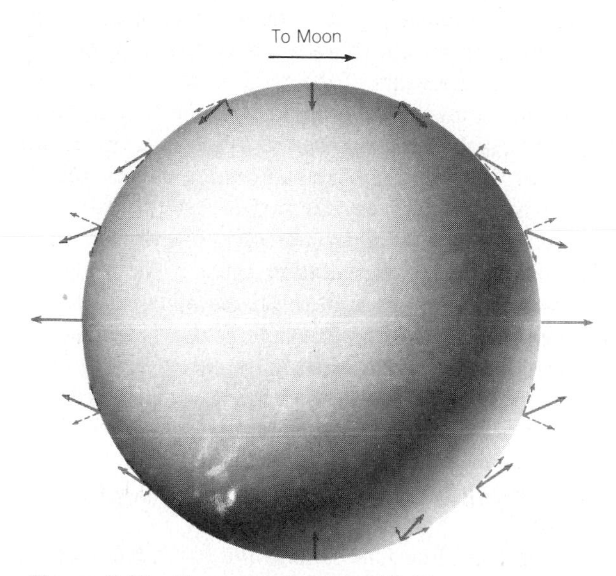

Figure 5.19 Components of the tidal forces.

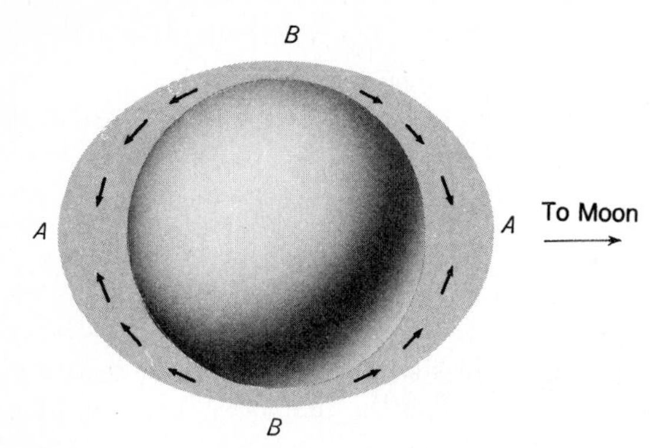

Figure 5.20 Tidal bulges in the "ideal" oceans.

example, the waters in the oceans, are free to flow in the direction of the horizontal components of the tidal forces. We shall assume, first, that the Earth is covered uniformly by a deep ocean, and investigate the nature of the tides produced in it.

The actual accelerations of the ocean waters caused by the horizontal components of the tidal force are very small. These forces, acting over a number of hours, however, produce motions of the water that result in measurable tidal bulges in the oceans. Water on the lunar side of the Earth is drawn toward the sublunar point (the point on the Earth where the Moon appears in the zenith), piling up water to greater depths on that side of the Earth, with the greatest depths at the sublunar point. On the opposite side of the Earth, water moves in the *opposite* direction, producing a tidal bulge on the side of the Earth opposite the Moon (Figure 5.20).

It is important to understand that it is the horizontal components of the tidal forces that produce the tidal bulges in the oceans. At the two opposite points on the Earth where the Moon is at the zenith and at the nadir, the tidal forces are exactly radial, that is, directed away from the Earth's center. At those points the horizontal components are zero, and there is no acceleration causing the water to flow along the surface of the Earth. The tidal forces serve only to reduce very slightly the weight of the water, but because of the low compressibility of water, its physical expansion (because of its reduced weight) is completely negligible. Thus, at these points, the tidal forces play virtually no role at all, even though that is where the water is piled up the most.

In a beltlike zone around the Earth from which the Moon appears on the horizon, the direction of the tidal force is *inward,* toward the center of the Earth, and again there are no horizontal components. The weight of the water is increased very slightly, but it is not appreciably compressed. Here, also, the tidal forces have no effect, although it is in this belt that the ocean level is lowest.

The tidal bulges in the oceans, then, do not result from the Moon compressing or expanding the water, nor from the Moon lifting the water "away from the Earth." Rather, the tidal bulges result from an actual flow of water over the Earth's surface, toward the regions below and opposite the Moon, causing the water to pile up to greater depths at those places. It is the horizontal components of the tidal forces that produce this flow; those components, or "tide-raising forces," are greatest in regions of the Earth intermediate between those from which the Moon appears at the zenith or the nadir (points A in Figure 5.20) and on the horizon (points B).

The tidal bulge on the side of the Earth *opposite* the Moon often seems mysterious to students who picture the tides as being formed by the Moon "lifting the water away from the Earth." What actually happens, as we have seen, is that the differential gravitational force of the Moon on the Earth tends to stretch the Earth, elongating it slightly toward the Moon. The solid Earth distorts slightly, but, because of its high rigidity, not enough to reach complete equilibrium with the tidal forces. Consequently, the ocean, moving freely over the Earth's surface, flows in such a way as to increase the elongation and piles up at points under and opposite the Moon.

In this section we have regarded the Earth as though its ocean waters were distributed uniformly over its surface. In this idealized picture, not actually realized even in the largest oceans, the tides would cause the depths of the ocean to range through only a few feet. The rotation of the Earth would carry an observer at any given place alternately into regions of deeper and shallower water. As he was being carried toward the regions under or opposite the Moon where the water was deepest, he would say, "the tide is coming in"; when carried away from those regions, he would say, "the tide is going out." During a day, he would be carried through two tidal bulges (one on each side of the Earth) and so would experience two "high tides"

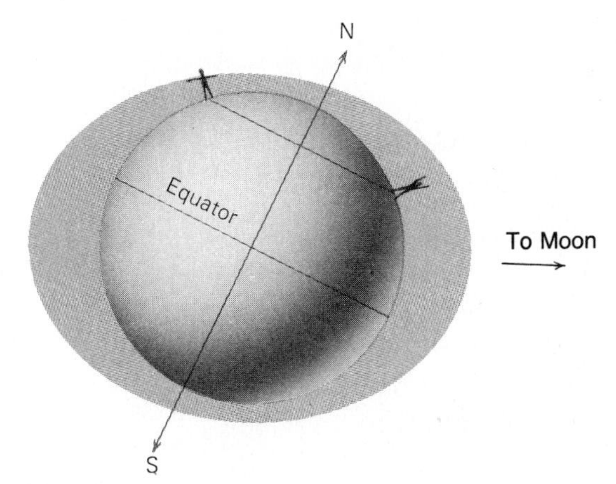

Figure 5.21 Inequality of the two "high tides" during a day.

and two "low tides."

The two high tides during a day need not be equally "high," however. For example, in northern or southern temperate latitudes, the axis of the tidal bulges is periodically inclined to the equator. The observer in the Northern Hemisphere (shown in Figure 5.21) would find the high tide on the side of the Earth facing the Moon much higher than the high tide half a day later. An observer in the Southern Hemisphere would find the opposite effect. In extreme cases there may appear to be only one "high tide" a day.

(c) Tides Produced by the Sun

The sun also produces tides on the Earth, although the sun is less than half as effective a tide-raising agent as the Moon. Actually, the gravitational attraction between the sun and the Earth is about 180 times as great as that between the Earth and the Moon. We recall, however, that the tidal force is the differential gravitational force of a body on the Earth. The sun's attraction for the Earth is much greater than the Moon's, but the sun is so distant that it attracts all parts of the Earth with almost equal strength. The Moon, on the other hand, is close enough for its attraction on the near side of the Earth to be substantially greater than its attraction on the far side. In other words, its *differential gravitational pull* on the Earth is greater than the sun's, even though its total gravitational attraction is less.

If there were no Moon, the tides produced by the sun would be all we would experience, and the tides would be less than half as great as those we now have. The Moon's tides, therefore, dominate. On the other hand, when the sun and Moon are lined up, that is, at new moon or full moon, the tides produced by the sun and Moon reinforce each other and are greater than normal. These are called *spring tides*. Spring tides (which have nothing to do with spring) are approximately the same, whether at new moon or full moon, because tidal bulges occur on both sides of the Earth—the side *toward* the Moon (or sun) and the side away from the Moon (or sun).

In contrast, when the Moon is at first quarter or last quarter, the tides produced by the sun partially cancel out the tides of the Moon, and the tides are lower than usual. These are *neap tides*. Spring and neap tides are illustrated in Figure 5.22.

Although spring tides are the highest type of tides, they are not all equally high, because the distances between the Earth and sun and the Earth and Moon (and hence the tide-raising effectiveness of these bodies) both vary. The Moon's distance varies by about 10 percent, and its tide-raising effectiveness varies by about 30 percent. The highest spring tides occur at those times when the Moon is also at perigee.

(d) The Complicated Nature of Actual Tides

The "simple" theory of tides, described in the preceding paragraphs, would be sufficient if the Earth were completely surrounded by very deep oceans, and if it rotated very slowly. However, the presence of land masses stopping the flow of water, the friction in the oceans and between oceans and the ocean floors, the rotation of the Earth, the variable depth of the ocean, winds, and so on, all complicate the picture.

Both the times and the heights of high tide vary considerably from place to place on the Earth. The Earth's rapid rotation causes the tide-raising forces within a given mass of water to vary too rapidly for the water to adjust completely to them. These forces, however, recurring periodically, set up forced oscillations in the ocean surfaces, so that the water over a large area rises and lowers in step. Consequently, the highest water does not necessarily occur when the Moon is highest in the sky (or lowest below the horizon), but rather when the os-

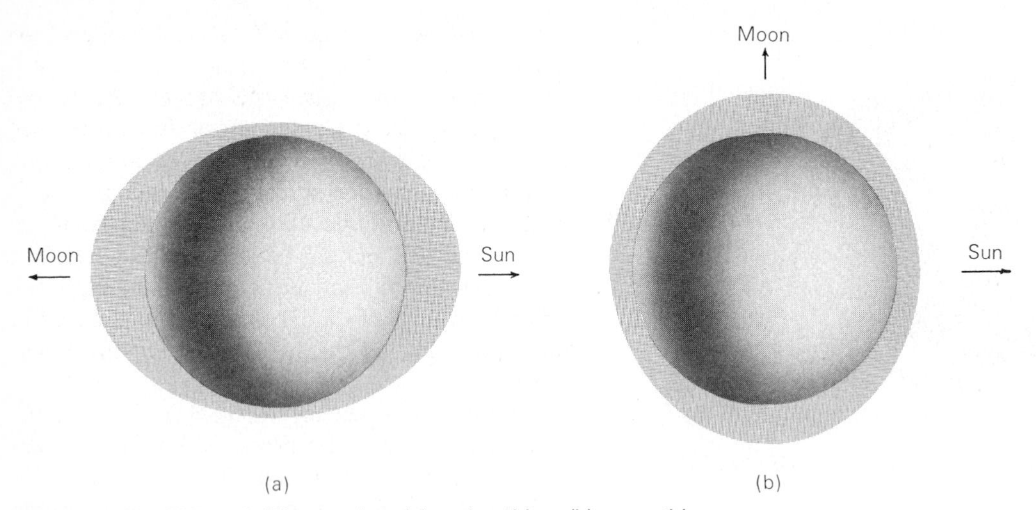

Figure 5.22 Tides of different size: (a) spring tides; (b) neap tides.

cillations of the ocean, produced by the tidal forces acting upon it, pile up the water to its greatest depth at that location. The latter depends critically upon the shape and depth of the adjacent ocean basin. The United States Coast and Geodetic Survey prepares and publishes each year the *Tide Tables,* which give the times and heights of tides at principal ports throughout the world.

Sometimes shallow coastal seas have such shapes and sizes that the natural frequency of oscil-

Figure 5.23 A 16-meter (52-foot) tidal range at Hantsport, Nova Scotia, at the head of the Bay of Fundy. These photographs were taken on January 29, 1979, a date when several factors were favorable for an exceptionally high tide: The Moon was just past new and was near perigee (nearest the Earth), the Earth was near perihelion (nearest the sun), and the barometric pressure was low. The low-tide photograph *(left)* was taken at 8:30 AM; the high-tide photo *(right)* was taken five hours later. The objects floating on the water are pieces of ice. (Courtesy Roy L. Bishop)

lation of water sloshing back and forth in the sea basins is very nearly the same as that of the tidal rise and fall of the water in the adjacent ocean. Then the ocean tides can set the water in these seas into strong resonance, like wind blowing into an organ pipe. The most famous such place is the Bay of Fundy between New Brunswick and Nova Scotia. The Bay of Fundy and Gulf of Maine act as a single oscillating system. The outer boundary of this system is the edge of the continental shelf with its approximately 50-fold increase in depth. The highest tides on Earth occur at the head of the Bay of Fundy, in Minas Basin. Under favorable circumstances the tidal range here can exceed 50 feet (Figure 5.23).

(e) Tides Elsewhere in the Solar System

Every planet produces tidal effects on its satellites, and every satellite raises tides upon its planet. One of the most obvious consequences of these forces is that nearly all of the satellites keep the same face turned toward their primaries, just as the Moon does toward the Earth. Only in this configuration does the tidal bulge remain fixed with respect to the solid body. For any other rotational rate, the satellite must turn with respect to the bulge, just as the Earth turns with respect to the tides raised by the Moon. The resulting friction has gradually slowed down the satellites until their rotation period is the same as their period of revolution.

The sun also raises tides on the planets. Since tidal effects decline as the inverse cube of the distance, they are much stronger for the inner planet, Mercury, than for any other. Mercury has, as expected, been slowed in its rotation by these solar tides, but the state it has reached is not one in which it always keeps the same face toward the sun. Rather, Mercury has a rotation period that is just ⅔ of its period of revolution, as we will see in Chapter 14.

One of the most dramatic examples of tidal forces at work in the solar system is provided by Jupiter's volcanically active satellite Io. Io is the innermost large satellite of Jupiter, and the tidal stresses upon it by the planet are immense. As described in Chapter 18, these stresses heat Io, ultimately resulting in the remarkable volcanic eruptions discovered by the Voyager spacecraft in 1979.

(f) Tides and Rings: The Tidal Stability Limit

Another important application of tidal theory results when we consider what happens to a solid object such as a small satellite in the proximity of a large planet. As noted in Section 5.3a, differential gravitational forces will tend to separate two particles, and could ultimately result in the disruption of such a small satellite, possibly giving rise to a ring system. It is interesting to calculate the location of the *tidal stability limit* within which such disruption might take place.

This calculation was first carried out about 1850 by the French mathematician Edouard Roche (1820–1883), and in his honor the tidal stability limit is often called the Roche limit. Roche considered the case of a satellite that is held together only by its self-gravitation, such as a liquid body or a loose collection of solid fragments. He found that if the satellite has the same density as the planet, the critical distance is 2.44 times the planet's radius. Somewhat different values result if other assumptions are made about the shape and density of the satellite, but in general the tidal stability limit is found to be at about 2.5 planet radii.

Only a satellite with no intrinsic strength would actually be disrupted at 2.5 planetary radii; real objects could come considerably closer before they were torn apart. After all, the Space Shuttle works regularly in Earth's orbit far inside our stability limit, and no one fears that it will be broken up by tidal forces. On the other hand, an astronaut working near the Shuttle would, if disabled, slowly drift away into space rather than being gravitationally attracted toward the Shuttle. This is what it means to be inside the stability limit for the Earth.

Perhaps more pertinent to the origin of planetary rings is the fact that inside the tidal stability limit two particles will not merge under their own gravitational attraction. Instead, they will separate, like the Shuttle and the astronaut. Thus fine material inside the limit is unable to join together to form a larger satellite. This is a problem we will return to in Chapter 18, when we discuss the rings of Jupiter, Saturn, Uranus, and Neptune.

5.5 PRECESSION

The Earth, because of its rapid rotation, is not perfectly spherical but has taken on the approximate

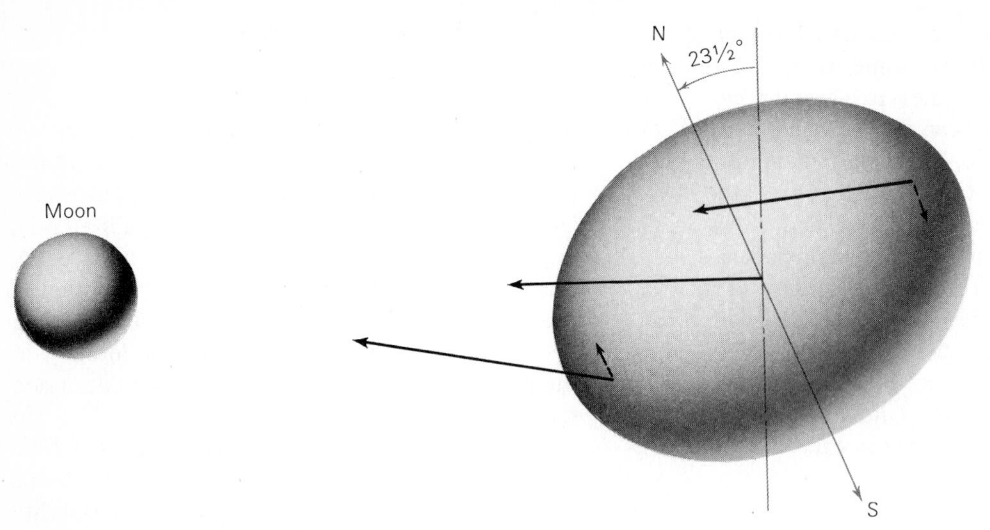

Figure 5.24 Differential force of the Moon on the oblate Earth tends to "erect" its axis, leading to precessional motion.

shape of an oblate spheroid; its equatorial diameter is 43 km greater than its polar diameter. As we have seen, the plane of the Earth's equator, and thus of its equatorial bulge, is inclined at about 23½° to the plane of the ecliptic, which, in turn, is inclined at 5° to the plane of the Moon's orbit. The differential gravitational forces of the sun and Moon upon the Earth not only cause the tides but also attempt to pull the equatorial bulge of the Earth into coincidence with the ecliptic.

The latter pull is illustrated in Figure 5.24. The solid arrows are vectors that represent the attraction of the Moon on representative parts of the Earth. The part of the Earth's equatorial bulge nearest the Moon is pulled more strongly than the part farthest from the Moon, and the Earth's center is pulled with an intermediate force. The dashed arrows show the differential forces with respect to the Earth's center. Note how they tend not only to "stretch" the Earth toward the Moon, but also to pull the equatorial bulge into the plane of the ecliptic. The differential force of the sun, although less than half as effective, does the same thing. Thus, the gravitational attractions of the sun and the Moon on the Earth act in such a way as to attempt to *change the direction of the Earth's axis of rotation,* so that it would stand perpendicular to the orbital plane of the Earth. To understand what actually takes place, we must digress for a moment to consider what happens when a similar force acts upon a top or gyroscope.

(a) Precession of a Gyroscope

Consider the top (a simple form of gyroscope) pictured in Figure 5.25. If the top's axis is not perfectly vertical, its weight (the force of gravity between it and the Earth) tends to topple it over. The actual force that acts to change the orientation of the axis of rotation of the top is that component of the top's weight that is perpendicular to its axis. We know from watching a top spin that the axis of the top does not fall toward the horizontal, but rather moves off in a direction *perpendicular to the plane defined by the axis and the force tending to change its orientation.* Until the spin of the top is slowed down by friction, the axis does not change its angle of inclination to the vertical (or to the floor), but rather describes a conical motion (a cone about the vertical line passing through the pivot point of the top). This conical motion of the top's axis is called *precession.*

(b) Precession of the Earth

The differential gravitational force of the sun on the Earth tends to pull the Earth's equatorial bulge into the plane of the ecliptic, and that of the Moon tends to pull the bulge into the plane of the Moon's orbit,

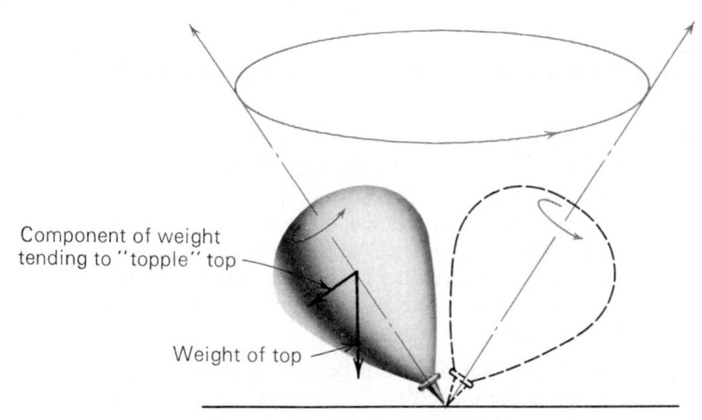

Figure 5.25 Precession of a top.

which is nearly in the ecliptic. These forces, in other words, tend to pull the Earth's axis into a direction approximately perpendicular to the ecliptic plane. Like a top, however, the Earth's axis does not yield in the direction of these forces, but precesses. The obliquity of the ecliptic remains approximately 23½°. The Earth's axis slides along the surface of an imaginary cone, perpendicular to the ecliptic, and with a half-angle at its apex of 23½° (see Figure 5.26). The precessional motion is exceedingly slow; one complete cycle of the axis about the cone requires about 26,000 years.

Precession is this motion of the axis of the Earth. Precession does not affect the cardinal directions on the Earth nor the positions of geographical places that are measured with respect to the Earth's rotational axis, but only the orientation of the axis with respect to the celestial sphere.

Precession does, however, affect the positions among the stars of the celestial poles, those points where extensions of the Earth's axis intersect the celestial sphere. In the 20th century, for example, the north celestial pole is very near Polaris. This was not always so. In the course of 26,000 years, the north celestial pole will move on the celestial sphere along an approximate circle of about 23½° radius, centered on the pole of the ecliptic (where the perpendicular to the Earth's orbit intersects the celestial sphere). This motion of the pole is shown in Figure 5.27. In about 12,000 years, the celestial pole will be fairly close to the bright star Vega.

As the positions of the poles change on the celestial sphere, so do the regions of the sky that are circumpolar, that is, that are perpetually above (or below) the horizon for an observer at any particular

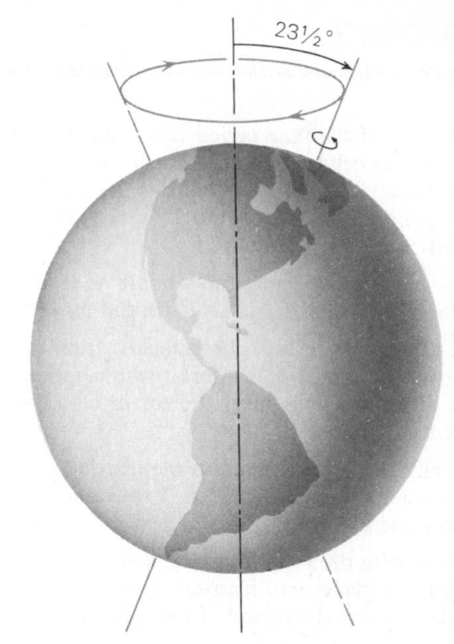

Figure 5.26 Precession of the Earth.

place on Earth. The Little Dipper, for example, will not always be circumpolar as seen from north temperate latitudes. Moreover, 2000 years ago, the Southern Cross was sometimes visible from parts of the continental United States. It was by noting the very gradual changes in the positions of stars with respect to the celestial poles that Hipparchus discovered precession in the second century B.C. (Section 2.3c).

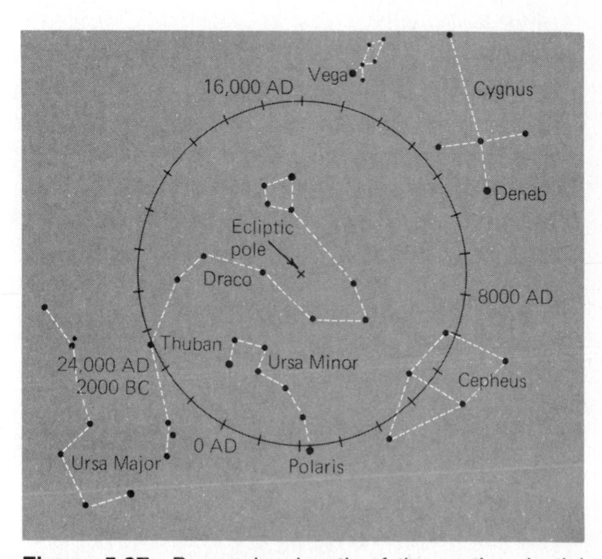

Figure 5.27 Precessional path of the north celestial pole among the northern stars.

EXERCISES

1. What would be the period of an artificial satellite in a circular orbit around the Earth with a radius equal to 96,000 km? (Assume that the Moon's distance and period are 384,000 km and 27⅓ days, respectively.)
Answer: About ½ week
2. As air friction causes a satellite to spiral inward closer to the Earth, its orbital speed *increases*. Why?
3. If a lunar probe is to be launched from the Earth's surface into an elliptical orbit whose apogee point is at the Moon, why must the eccentricity of the orbit be nearly one?
4. ★ Verify the periods given in the text for the times required to reach the planets Venus and Mars along least-energy orbits.
5. ★ Show why the times at which a space vehicle can be sent to a planet on a least-energy orbit occur at intervals of a synodic period of the planet.
6. ★ Describe how a space vehicle must be launched if it is to fall into the sun.
7. If a satellite has a nearly circular orbit at a critical distance from the Earth's center, it will have a period of revolution equal to one day and thus can appear stationary in the sky above a particular place on Earth. Calculate the radius of the orbit of such a *synchronous* satellite.
Answer: About 42,400 km
8. If the sun had eight times its present mass and the Earth's orbit were twice its present size, what would then be its orbital period?
9. ★ Find the separation d between two small bodies, each of unit mass, lined up with a large body of mass M, at a distance R from the nearest of the small bodies, such that the gravitational attraction between the small bodies is just equal to the differential gravitational force between them caused by their attraction to the large body. The answer should be in terms of G, M, and R.
10. Strictly speaking, should it be a 24-hour period during which there are two "high tides"? If not, what should the interval be?
11. Compute the relative tide-raising effectiveness of the sun and the Moon. For this approximate calculation, assume that the Earth is 80 times as massive as the Moon, that the sun is 300,000 times as massive as the Earth, and that the sun is 400 times as distant as the Moon.
Answer: Moon is ⁵⁄₃ times as effective
12. What will be the principal north circumpolar constellations as seen from Los Angeles (latitude 34° north) in the year 18,000?
13. What would be the annual motion of the equinoxes along the ecliptic if the entire precessional cycle required only 360 years?
14. If the precessional rate is about 50 arcseconds per year, show that the complete cycle is about 26,000 years.

Tycho Brahe (1546–1601), Danish astronomer whose extensive observations of the planets led to Kepler's discovery of their laws of motion. Brahe, the last major astronomer from the era before invention of the telescope, made meticulous measurements of the lengths of the seasons, precession of the equinoxes, length of the year, and nearly every other astronomical constant known at the time. (Yerkes Observatory)

THE CELESTIAL CLOCKWORK

In the preceding chapters we were concerned with the mechanics that govern the motions of celestial bodies. Now we turn our attention to the motions of the Earth, and to the relation between the Earth and the sky. It is these motions, of course, that define our measures of date and time, that give rise to the seasons, and that produce the constantly changing appearance of the sky. We begin with the rotation of the Earth about its axis.

6.1 EARTH AND SKY

(a) The Foucault Pendulum

We have seen that the apparent rotation of the celestial sphere could be accounted for either by a daily rotation of the sky around the Earth or by the rotation of the Earth itself. Since the time of Copernicus, it has been generally accepted that it was the Earth that turned, but not until the 19th century did the French physicist Jean Foucault provide a direct and unambiguous demonstration of this rotation. In 1851, he suspended a 60-m pendulum weighing about 25 kg from the domed ceiling of the Pantheon in Paris. He started the pendulum swinging evenly by drawing it to one side with a cord and then burning the cord. The direction of swing of the pendulum was recorded on a ring of sand placed on a table beneath its point of suspension. At the end of each swing a pointed stylus attached to the bottom of the bob cut a notch in the sand. Foucault had taken great care to avoid air currents and other influences that would disturb the direction of swing of the pendulum. Yet, after a few moments it became apparent that the plane of oscillation of the pendulum was slowly changing with respect to the ring of sand, and hence with respect to the Earth.

The only force acting upon the pendulum was that of gravity between it and the Earth, and, of course, this force was in a downward direction. If the Earth was stationary, there would be no force that could cause the plane of oscillation of the pendulum to alter, and, in accord with Newton's first law, the pendulum should continue to swing in the same direction. The fact that the pendulum slowly changed its direction of swing with respect to the Earth is proof that the Earth rotates.

It is comparatively easy to visualize a Foucault pendulum experiment at the North Pole. Here we can imagine the plane of swing of the pendulum maintaining a fixed direction in space with respect to the stars, while the Earth turns under it every day. Thus, at the North (or South) Pole, a pendu-

Figure 6.1 A Foucault pendulum. (Griffith Observatory)

lum would *appear* to rotate its plane of oscillation once completely in 24 hours (actually, 23 hours 56 minutes). At other places than the poles, the problem is complicated because the pendulum must always swing in a vertical plane that passes through the center of the Earth. That plane of oscillation obviously must change with respect to the stars.

We must think of the pendulum as measuring the rate at which the Earth turns around directly *beneath* it—that is, the rate of rotation of the Earth about an imaginary line from the center of the Earth out through the point of the pendulum's suspension. About this line the plane of swing of the pendulum does not rotate. If we imagine ourselves looking down upon the Earth's North Pole, we can "see" the Earth spinning beneath us like a phonograph record. On the other hand, if we imagine ourselves looking down on the Earth's equator, we do not see a rotation of the Earth beneath us, only a west-east translational motion. At intermediate latitudes we see beneath us a combination of west-east motion and a certain degree of rotation. At the equator, therefore, a pendulum would not appear to change direction of swing, while at latitudes intermediate between the equator and the poles its period—the time required for it to change its apparent plane of oscillation through 360°—would have a value somewhere between 24 hours and infinity, depending on the exact latitude. For example, at a latitude of 34° (the latitude of Los Angeles), the Foucault pendulum has a period of just under 43 hours. This is the time required for the Earth to turn around a line from its center through Los Angeles, or the time required for a spectator to be carried completely around the pendulum by the turning Earth.

It should be noted that the turning Earth also turns the support system for the pendulum, and consequently the wire and bob of the pendulum itself. However, the rotation of the wire and bob of the pendulum does not alter the direction of swing. Try the following simple experiment. Improvise a small pendulum, say a watch and watch chain. Swing the watch to and fro, holding the end of the

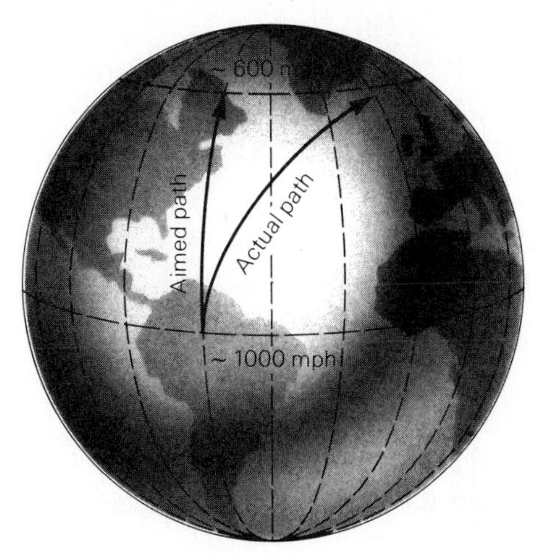

Figure 6.2 The Coriolis effect on a projectile fired northward from near the equator of the rotating Earth.

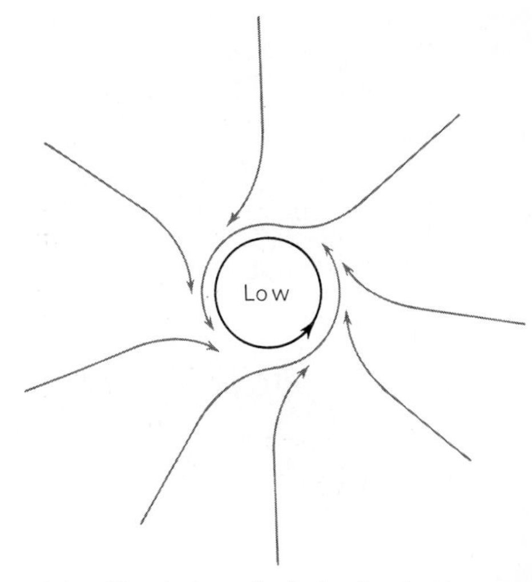

Figure 6.3 Circulation of winds about a low-pressure area in the Northern Hemisphere, induced by Coriolis forces.

chain in your fingers. Now twist the chain in your fingers; the watch will twist with the chain, but will *not* change its direction of swing.

(b) The Coriolis Effect

The apparent rotation of the plane of oscillation of the Foucault pendulum is a demonstration of the rotation of the Earth underneath a freely moving body. Any such apparent deflection in the motion of a body, resulting from the Earth's rotation, is called the *coriolis effect.* The moving body need not be the bob of a pendulum. Any object moving freely over the surface of the Earth appears to be deflected to the right in the Northern Hemisphere (to the left in the Southern Hemisphere) because of the rotation of the Earth beneath it. As an example of the effect, consider a projectile fired to the north from the equator.

The projectile starts its northward trip with an *eastward* velocity that it shares with the turning Earth just before it is fired (Figure 6.2); at the equator this eastward velocity is about 1700 km/hr.

There is no westward force on the projectile to slow it down, so it continues to move eastward after being fired. Proceeding northward over the curved surface of the Earth, however, it comes closer to the axis of the Earth's rotation. To conserve its angular momentum, the projectile's linear speed to the east must increase if its distance from the axis of rotation decreases. Meanwhile the ground beneath the northbound projectile moves eastward progressively slower, because that ground, closer to the Earth's axis, has less far to move in its daily rotation. We see, then, that the eastward speed of the projectile increases and that of the ground beneath it decreases. Thus, relative to the ground, the missile veers off to the east, that is, to the right for one looking in the direction of its motion.

A similar analysis would show that no matter in what direction a projectile moves, in the Northern Hemisphere it veers off to the *right,* and in the Southern Hemisphere to the *left* of its target. This effect must be corrected for in the firing of long-range artillery and of course, in the launching of missiles.

Winds blowing toward a low-pressure area similarly veer off to the right of this area (left in the Southern Hemisphere). However, the force continually trying to equalize the pressure of the air accelerates the wind toward the low-pressure area. The wind, rather than "falling" directly into the low center, is caused to circle *around* the low center by the inertia of the forward moving air (Figure 6.3). If it were not for the Earth's rotation, winds would blow directly into low-pressure regions, but because the winds veer off and miss the lows, they end up with a *cyclonic* motion. In the Northern Hemisphere, the winds blow around low-pressure centers in a *counterclockwise direction,* whether they be gentle cyclonic features or violent storms such as hurricanes and tornados. In the Southern Hemisphere

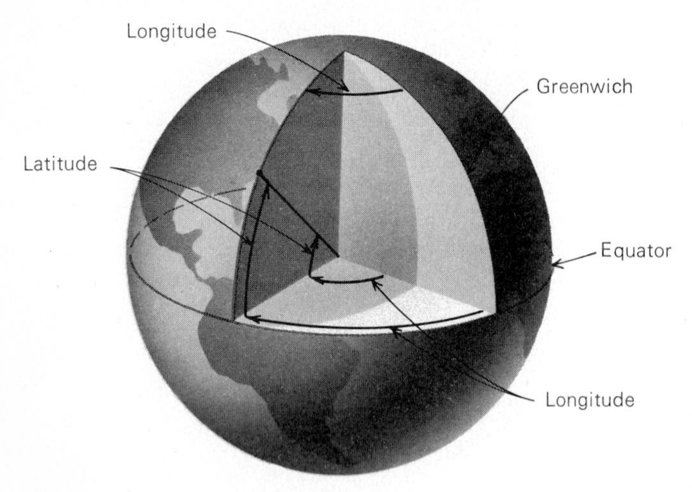

Figure 6.4 Latitude and longitude of Washington, D.C.

the winds are *reversed,* that is, storms there (such as the South Pacific and Indian Ocean typhoons) circulate in a *clockwise* direction.

(c) Positions on the Earth and Sky

To denote positions of places on the Earth, we must set up a system of coordinates on the Earth's surface. The Earth's axis of rotation (that is, the locations of its North and South Poles) is the basis for such a system.

A *great circle* is any circle on the surface of a sphere whose center is at the center of the sphere. The Earth's *equator* is a great circle on the Earth's surface halfway between the North and South Poles. We can also imagine a series of great circles that pass *through* the North and South Poles. These circles are called *meridians;* they intersect the equator at right angles.

A meridian can be imagined passing through an arbitrary point on the surface of the Earth (Figure 6.4). This meridian specifies the east-west location of that place. The *longitude* of the place is the number of degrees, minutes, and seconds of arc along the equator between the meridian passing through the place and the one passing through Greenwich, England, the site of the old Royal Observatory. Longitudes are measured either to the east or west of the Greenwich meridian from 0 to 180°. The convention of referring longitudes to the Greenwich meridian is of course completely arbitrary. As an example, the longitude of the bench mark in the clock house of the Naval Observatory in Washington, D.C., is 77°03′56″.7 W.* Note in Figure 6.4 that the number of degrees along the equator between the meridians of Greenwich and Washington is also the angle at which the planes of those two meridians intersect at the Earth's axis.

The *latitude* of a place is the number of degress, minutes, and seconds of arc measured along its meridian to the place from the equator. Latitudes are measured either to the north or south of the equator from 0 to 90°. As an example, the latitude of the above-mentioned bench mark is 38°55′14″.0 N. Note that the latitude of Washington is also the angular distance between it and the equator as seen from the center of the Earth.

(d) Astronomical Coordinate Systems

In denoting positions of objects in the sky, it is often convenient to make use of the fictitious *celestial sphere,* a concept, we recall, that many early peoples accepted literally (see Chapter 2). We can think of the celestial sphere as being a hollow shell of extremely large radius, centered on the observer. The celestial objects appear to be set in the inner surface of this sphere, so we can speak of their positions *on* the celestial sphere. We have devised coordinate systems, analogous to latitude and longitude, to

---

* Standard notation for writing angles uses ° for degrees, ′ for arcminutes (1′ = 1/60°), and ″ for arcseconds (1″ = 1/60′).

designate these positions. Of course we are really only denoting their *directions* in the sky.

The point on the celestial sphere directly above an observer (defined as opposite to the direction of a plumb bob) is the *zenith*. Straight down, 180° from the zenith, is the observer's *nadir*. Halfway between, and 90° from each, is the *horizon*. (This is the *celestial* horizon and will not necessarily coincide with the apparent horizon, which may be interrupted with such things as mountains, buildings, and trees.) Note that observers at different places have different zeniths, nadirs, and horizons.

CELESTIAL EQUATOR AND POLES

The apparent rotation of the sky takes place about an extension of the Earth's axis of rotation. That is, the sky appears to rotate about points directly in line with the North and South Poles of the Earth—the *north celestial pole* and the *south celestial pole*. Halfway between the celestial poles, and thus 90° from each, is the *celestial equator*, a great circle on the celestial sphere that is in the same plane as the Earth's equator; it would appear to pass directly through the zenith of a person on the equator of the Earth. Great circles passing through the celestial poles and intersecting the celestial equator at right angles (analogous to meridians on the Earth) are called *hour circles*.

THE CELESTIAL MERIDIAN

The great circle passing through the celestial poles and the zenith (and also through the nadir) is called the observer's *celestial meridian*. It coincides with the projection of his terrestrial meridian, as seen from the Earth's center, onto the celestial sphere. The celestial meridian intercepts the horizon at the *north* and *south* points. Halfway between these north and south points on the horizon are the *east* and *west* points.

As the Earth turns, the observer's terrestrial meridian moves under the celestial sphere, sweeping continually eastward around the sky. An equivalent way of putting it is to say that as the sky turns around the Earth, the stars pass by the observer's stationary celestial meridian.

It helps to visualize these circles in the sky if we imagine that the Earth is a hollow transparent spherical shell with the terrestrial coordinates (latitude and longitude) painted on it. Then if we imagine ourselves at the center of the Earth, looking out through its transparent surface to the sky, the terrestrial poles, equator, and meridians will be superimposed upon the celestial ones.

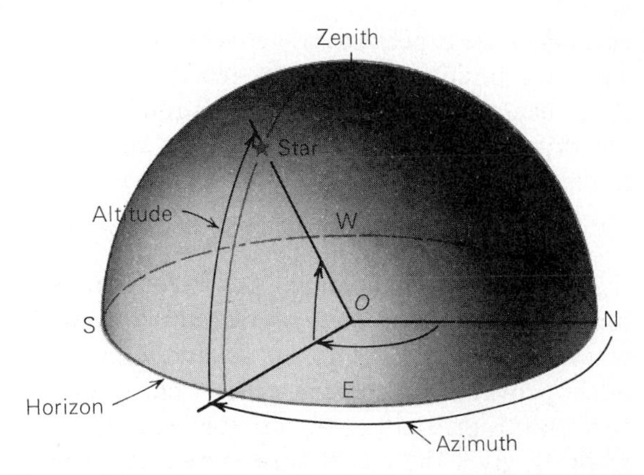

Figure 6.5 Altitude and azimuth, one set of coordinates used to define a position on the celestial sphere.

ALTITUDE AND AZIMUTH

The most obvious coordinate system is based on the horizon and zenith of the observer. Great circles passing through the zenith (*vertical circles*) intersect the horizon at right angles. Imagine a vertical circle through a particular star (Figure 6.5). The *altitude* of that star is the number of degrees along this circle from the horizon up to the star. It is also the angular "height" of the star as seen by the observer.

The *azimuth* is the number of degrees along the horizon to the vertical circle of the star from some reference point on the horizon. In astronomical tradition, azimuth formerly was measured from the south point on the observer's horizon, but in modern practice azimuth is measured from the north point, in conformity with the convention of navigators and engineers. In any case, azimuth is measured to the east (clockwise to one looking down from the sky) along the horizon from 0 to 360°.

RIGHT ASCENSION AND DECLINATION

The principal disadvantage of the altitude and azimuth system (the *horizon* system) is that as the Earth turns the coordinates of the celestial objects are constantly changing. It is desirable, therefore, to devise a coordinate system that is attached to the celestial sphere itself, just as the system of latitude and longitude is permanently attached to the Earth. Then the positions of the stars remain fixed rather than changing rapidly as the Earth's rotation makes the sky seem to rotate. A system that comes close to meeting these requirements is *right ascension* and *declination*, or the *equator system*.

Right ascension and *declination* bear the same relation to the celestial equator and poles that longitude and latitude do to the terrestrial equator and poles. *Declination* gives the arc distance of a star (or other point on the celestial sphere) along an hour circle north or south of the celestial equator. *Right ascension* gives the arc distance measured eastward along the celestial equator to the hour circle of the star from a reference point on the celestial equator. That reference point is the *vernal equinox*, one of the two points on the celestial sphere where the celestial equator and the ecliptic intersect. Because of precession, both the celestial equator and the vernal equinox slowly move with respect to the stars; thus the right ascension and declination of a star continually change, but the changes are so gradual as not to be important, for most purposes, over a period of one year or so. The lack of constancy of right ascension and declination makes the system less than ideal, but it is still the most convenient one available, for it is based on the celestial equator and is thus symmetrical with respect to the Earth's axis of rotation. Right ascension and declination are therefore very useful for pointing telescopes and moving them to follow the daily motions of the stars (Chapter 9).

Several celestial coordinate systems are in common use. Each has its advantages for special purposes and is important to astronomers. These systems are defined in Appendix 7.

(e) The Orientation of the Celestial Sphere

The next step is to determine the orientation of the celestial sphere with respect to the zenith and horizon of a particular observer on the Earth. At the North (or South) Pole, the problem is very simple indeed. The north celestial pole, directly over the earth's North Pole, appears at the zenith. The celestial equator, 90° from the celestial poles, lies along the horizon. An observer at one of the terrestrial poles would never see more than half the sky.

At the equator the problem is almost as simple. The celestial equator, in the same plane as the

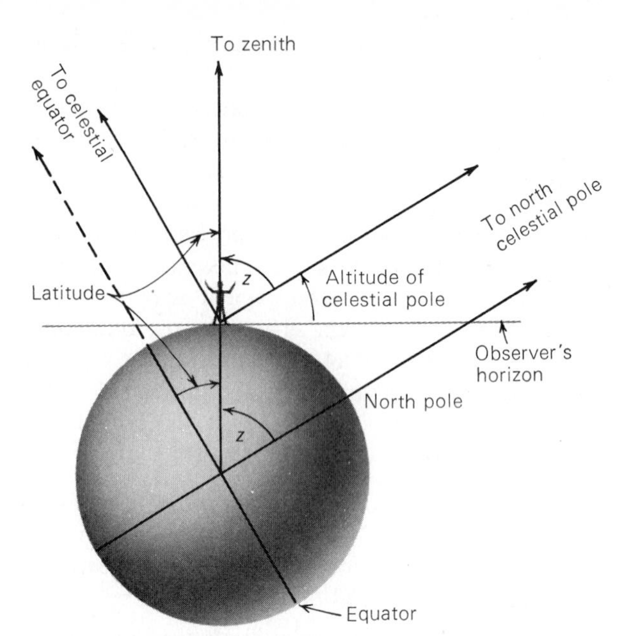

Figure 6.6 The altitude of the celestial pole equals the observer's latitude.

Earth's equator, passes through the zenith, and since it runs east and west, it intersects the horizon at the east and west points. The celestial poles, being 90° from the celestial equator, must be at the north and south points on the horizon. Evidently at points on the Earth between the equator and poles, one of the celestial poles must be a certain distance above the horizon.

Figure 6.6 shows an observer at an arbitrary latitude north of the equator; for a southern latitude the case would be exactly analogous. Since the terrestrial North Pole is on the observer's terrestrial meridian, the north celestial pole will have to be on his celestial meridian, at some altitude above the north point on the horizon. Suppose the angle from the observer's zenith down to the north celestial pole is z. Being on the celestial sphere, the north celestial pole is so distant that it is seen in the same direction as by an observer at the center of the Earth. Thus the angle at the center of the Earth between the observer and the north celestial pole is also z. (See Figure 6.6; the angles of intersection between each of two parallel lines and a third line are equal.)

We recognize that z is just 90° *minus* the observer's latitude. But also z is 90° (the altitude of the zenith) *minus* the altitude of the north celestial pole. Thus we see that *the altitude of the north (or south)*

celestial pole is equal to the observer's north (or south) latitude.

Finally, since the celestial equator is 90° from the celestial poles, it must cut through the east and west points on the horizon, tilt southward as it extends up above the horizon, and cross the celestial meridian a distance south of the zenith that is also equal to the observer's latitude.

(f) Nomenclature of Stars

While we are discussing the celestial sphere, it is in order to describe briefly the system for naming the stars and constellations. As has been stated (Section 2.2a), the ancients designated certain apparent groupings of stars in honor of characters or animals in their mythology. We retain most of these *constellations* today, although their number has been augmented to 88 (they are listed in Appendix 20). By action of the International Astronomical Union in 1928, the boundaries between constellations were established as east-west lines of constant declination and north-south segments of hour circles. Because of precession, over the years the constellation boundaries have gradually tilted slightly from precisely north-south and east-west. Although all constellation boundaries ran north-south and east-west, they nevertheless jogged about considerably, so that the modern constellations still contain most or all of the brighter stars assigned to them by the ancients. Consequently, the boundaries often delineate highly irregular regions of the sky, reminding one of the boundaries between Congressional districts that result from the gerrymandering practices of many state legislatures. At any rate, because every position on the celestial sphere (or every direction in the sky) lies in one or another constellation, we commonly use constellations today to designate the places of stars or other celestial objects.

Many of the brighter stars have proper names. Often these are Arabic names that describe the positions of the stars in the imagined figures that the Greek constellations represented. For example, *Deneb* is Arabic for "tail" and is the star that marks the tail of *Cygnus* (the Swan), and *Denebola* is the star at the tail of *Leo* (the Lion). Some star names, however, are of modern origin, for example, *Polaris,* the "pole star."

A superior designation of stars was introduced by the Bavarian, J. Bayer, in his *Atlas* of the constellations, published in 1603. He assigned successive letters of the Greek alphabet to the more conspicuous stars in each constellation, in approximate order of decreasing brightness. The full star designation is the Greek letter, followed by the genitive form of the constellation name. Thus Deneb, the brightest star in Cygnus, is α *Cygni,* and Denebola, the second brightest in Leo, is β *Leonis.* Bayer's ordering of stars by brightness was not always correct, and on occasion he deviated from the scheme altogether and assigned letters to stars of comparable brightness according to their geometrical arrangement in the constellation figure, for example, in the Big Dipper.

Fainter stars were subsequently given number designations, with the numbers increasing in order of the stars' right ascensions. The majority of stars, however, too faint to see without a telescope, are designated, if at all, only by their numbers in various catalogues. The most famous and extensive catalogue, which contains one third of a million stars, was complied in the years following 1837 by F. W. Argelander at the Bonn Observatory. Stars in this Bonn Catalogue, or *Bonner Durchmusterung,* are known by their *BD numbers.* The Bonner Durchmusterung was later extended to the part of the sky too far south to observe at Bonn and was eventually supplemented with a catalogue of the southernmost stars, made at Cordoba in Argentina.

Many other catalogues, with ever-increasing accuracy of the star positions they record, have been and are being compiled. Many star catalogues are prepared for special purposes, or list only certain types of stars, or stars in certain regions of the sky. A commonly used catalogue produced by the Harvard College Observatory gives the spectral types of the stars (Chapter 24); stars in this *Henry Draper Catalogue* are denoted by their *HD numbers.* Data for the nearest stars, listed in Appendix 13, are seen, from the variety of nomenclature, to have been selected from several different catalogues. Many stars, of course, are listed in more than one catalogue and thus bear various names—a circumstance that has sometimes confused even astronomers. The vast majority of stars, however, too faint and numerous to measure and catalogue, remain nameless.

(g) The Motion of the Sky as Seen from Different Places on the Earth

Imagine an observer at the Earth's North Pole. The celestial north pole is at the zenith and the celestial equator at the horizon. As the Earth rotates, the sky turns about a point directly overhead. The stars neither rise nor set; they circle the sky parallel to the horizon. Only that half of the sky that is north of the celestial equator is ever visible to this observer.

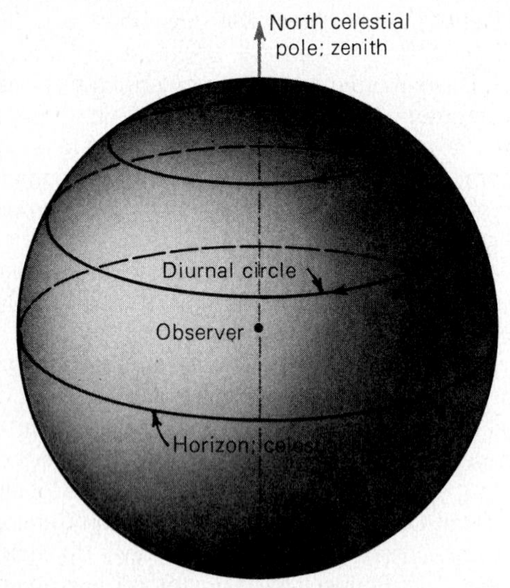

Figure 6.7 Sky from the North Pole; the apparent paths of all the stars are parallel to the horizon.

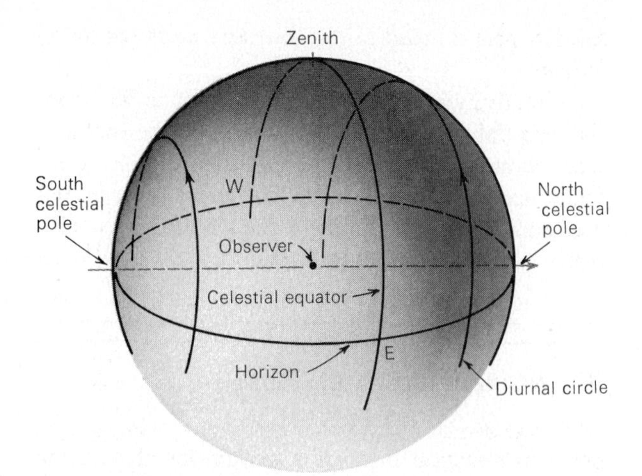

Figure 6.8 Sky from the equator; here, stars rise and set perpendicular to the horizon.

Similarly, an observer at the South Pole would only see the southern half of the sky (Figure 6.7).

The situation is very different for an observer at the equator (Figure 6.8). There the celestial poles, the points about which the sky turns, lie at the north and south points on the horizon. All stars rise and set; they move straight up from the east side of the horizon and set straight down on the west side. During a 24-hour period, all stars are above the horizon exactly half the time.

For an observer between the equator and North Pole, say at 34° north latitude, the situation is as depicted in Figure 6.9. Here the north celestial pole is 34° above the observer's northern horizon. The south celestial pole is 34° *below* the southern horizon. As the Earth turns, the whole sky seems to pivot about the north celestial pole, and the stars appear to circle around parallel to the celestial equator. For this observer, stars within 34° of the north celestial pole can never set. They are always above the horizon, day and night. This part of the sky is called the *north circumpolar zone* for the latitude 34° N. To observers in the United States, the Big and

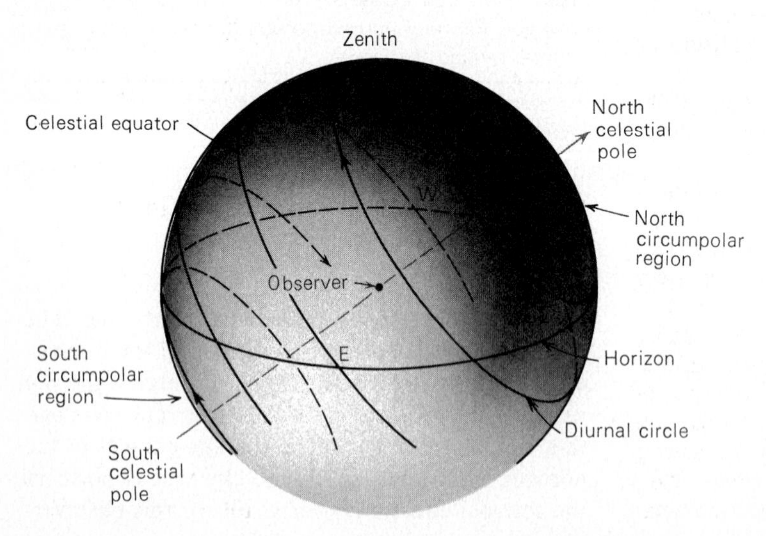

Figure 6.9 Sky from latitude 34° north. Stars in the north circumpolar region never set, while those in the south circumpolar region never rise at this latitude.

Little Dippers and Cassiopeia are examples of star groups that are in the north circumpolar zone. On the other hand, stars within 34° of the south celestial pole never rise. That part of the sky is the *south circumpolar zone*. To most U.S. observers, the Southern Cross is in that zone. At the North or South Pole the entire sky is circumpolar, half of it above the horizon and half below.

Stars north of the celestial equator, but outside the north circumpolar zone, in the greater parts of their daily paths, or *diurnal circles*, lie above the horizon; hence they are up more than half the time. Stars *on* the celestial equator are up exactly half the time, for their diurnal circle is the celestial equator; because it is a great circle, exactly half of it must be above the horizon. Stars south of the celestial equator, but outside the south circumpolar zone, are up less than half the time.

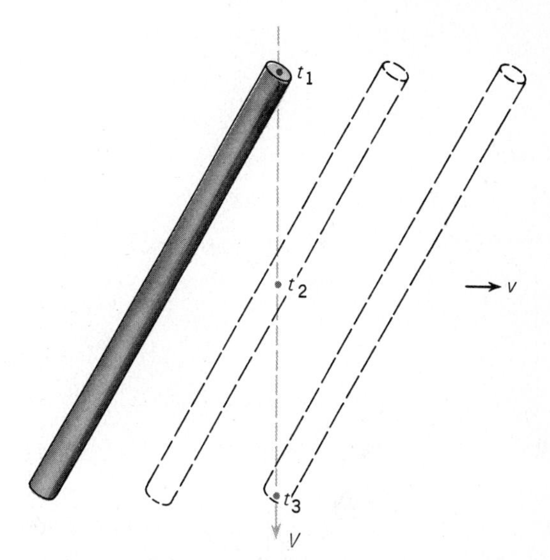

Figure 6.10 Raindrops falling through a moving drainpipe—an example of aberration.

6.2 THE SEASONS

(a) Proofs of the Earth's Revolution

We have seen (Chapter 2) that the Earth's apparent revolution about the sun produces an apparent annual motion of the sun along the ecliptic, thereby giving rise to our seasons and defining the length of the year. What evidence is there, however, that this apparent motion is real, and that the Earth actually orbits the sun?

If we adopt Newton's laws of motion and gravitation, it follows simply and directly that the Earth must revolve about the sun and not vice versa. It is obvious that *either* the Earth goes around the sun or the sun goes around the Earth. Thus we have a system of *two mutually revolving bodies*. The problem is simply to determine where the common center of revolution is. In Chapter 12 we shall see that the sun is about 330,000 times as massive as the Earth. Thus the common center of mass (Section 5.1) of the Earth-sun system must be less than 1/300,000 of the distance from the center of the sun to the center of the Earth. This puts it well inside the surface of the sun. Essentially, then, the Earth revolves around the sun.

There are also some geometrical consequences of the Earth's revolution that would be very difficult to explain if the Earth were assumed to be stationary. These are *stellar parallax* and *aberration of starlight*. We shall discuss stellar parallax in Chapter 22.

To understand aberration, consider the analogy of walking in the rain, holding a straight drainpipe (Figure 6.10). If the drainpipe is held vertically, and if the raindrops are assumed to fall vertically, they will fall through the length of the pipe only if you are standing still. If you walk forward, you must tilt the pipe slightly forward, so that drops entering the top will fall out the bottom without being swept up by the approaching inside wall of the pipe. If the raindrops fall with a speed V, and if you walk with a speed v, the distance by which the top of the pipe precedes the bottom, divided by the vertical distance between the top and bottom of the pipe, must be in the ratio v/V.

Similarly, because of the Earth's orbital motion, if starlight is to pass through the length of a telescope, the telescope must be tilted slightly forward in the direction of the Earth's motion. In other words, the apparent direction of a star is displaced slightly from its geometrical direction, and the displacement is in the direction of the Earth's orbital motion. Analogous to the tilt of the drainpipe, this forward tilt of the telescope is in the ratio of the speed of the Earth to the speed of light. The speed of light is about 10,000 times that of the Earth in its orbit, so the angle through which a telescope must be tilted forward is 1 part in 10,000 or about 20.5 arcseconds. The effect is greatest when

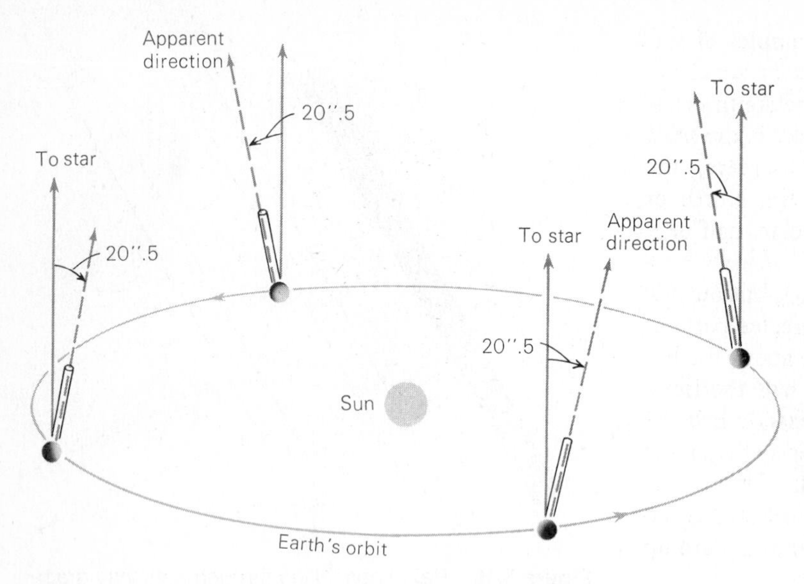

Figure 6.11 Aberration of starlight, resulting from the orbital motion of the Earth.

the Earth is moving at right angles to the direction of the star, and disappears when the Earth moves directly toward or away from the star. A star that is on the ecliptic appears to shift back and forth in a straight line during the year, for through part of the year the Earth is moving in one direction compared to the star and during the rest of the year the Earth is moving in the opposite direction. A star in a direction perpendicular to the Earth's orbit appears to describe a small circle in the sky, for its apparent direction is constantly displaced in the direction of the Earth's orbital motion from the direction it would have as seen from the sun. Stars between these extremes appear to shift their apparent directions along tiny elliptical paths of semimajor axis 20.5 arcseconds (Figure 6.11).

(b) The Seasons and Sunshine

The Earth's orbit around the sun is an ellipse rather than a circle, and our distance varies by about 3 percent from perihelion to aphelion. However, we all know that the changing distance of the Earth from the sun is not the cause of the seasons. The seasons result because the axis of rotation is tilted

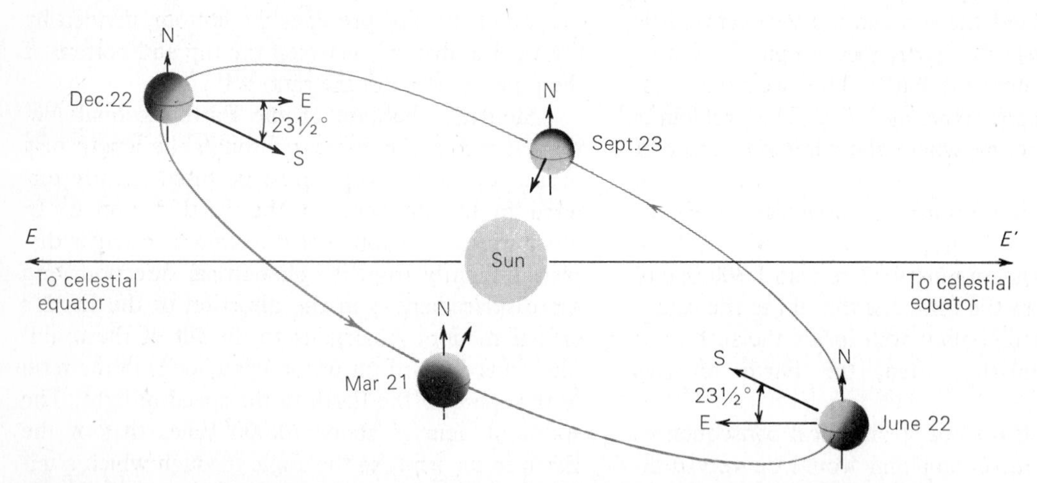

Figure 6.12 The seasons are caused by the inclination of the plane of the Earth's orbit to the plane of the equator.

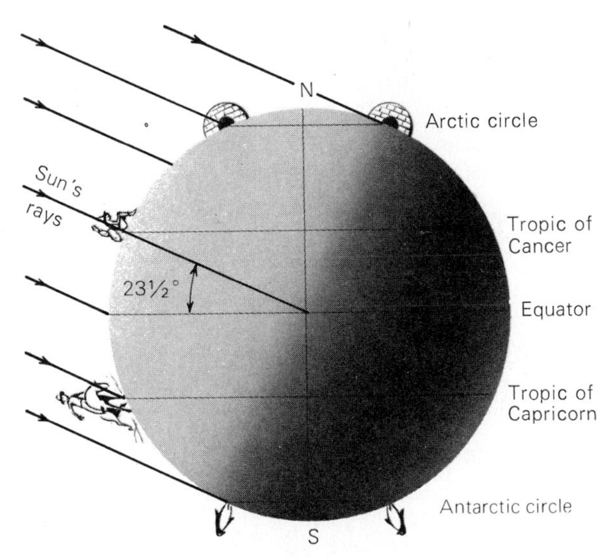

Figure 6.13 The Earth on June 22, the summer solstice in the Northern Hemisphere.

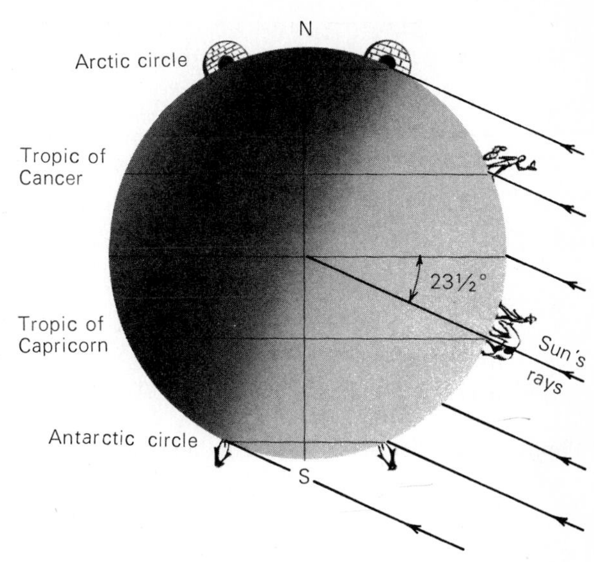

Figure 6.14 The Earth on December 22, the winter solstice in the Northern Hemisphere.

with respect to the orbit, that is, the plane in which the Earth revolves is not coincident with the Earth's equator. This angle of tilt, called by astronomers the *obliquity*, is about 23½° for the Earth.

Figure 6.12 shows the Earth's path around the sun. The line *EE*′ is in the plane of the celestial equator. In the figure the Earth appears to pass alternately above and below this plane, but the celestial sphere is so large, and the celestial equator so far away, that a line from the center of the Earth through the Earth's equator always points to the celestial equator.

We see in the figure that on about June 22 (the date of the *summer solstice*), the sun shines down most directly upon the Northern Hemisphere of the Earth. It appears 23½° *north* of the equator and thus on that date passes through the zenith of places on the Earth that are at 23½° north latitude. The situation is shown in detail in Figure 6.13. To an observer on the equator, the sun appears 23½° north of the zenith at noon. To a person at a latitude 23½° N, the sun is overhead at noon. This latitude on the Earth, at which the sun can appear at the zenith at noon on the first day of summer, is called the *Tropic of Cancer*. We see also in Figure 6.13 that the sun's rays shine down past the North Pole; in fact, all places within 23½° of the pole, that is, at a latitude greater than 66½° N, have sunshine for 24 hours on the first day of summer. The sun is as far north on this date as it can get; thus, 66½° is the southernmost latitude where sun can ever be seen for a full 24-hour period (the *midnight sun*); that circle of latitude is called the *Arctic Circle*.

During this time, the sun's rays shine very obliquely on the Southern Hemisphere. In fact, all places within 23½° of the South Pole—that is, south of latitude 66½° S (the *Antarctic Circle*)—have no sight of the sun for the entire 24-hour period.

The situation is reversed six months later, about December 22 (the date of the *winter solstice*), as is shown in Figure 6.14. Now it is the Arctic Circle that has a 24-hour night and the Antarctic Circle that has the midnight sun. At latitude 23½° S, the *Tropic of Capricorn*, the sun passes through the zenith at noon. It is winter in the northern hemisphere, summer in the southern.

Finally, we see in Figure 6.12 that on about March 21 and September 23 the sun appears to be in the direction of the celestial equator, and, on these dates, the equator itself is the diurnal circle for the sun. Every place on the Earth then receives exactly 12 hours of sunshine and 12 hours of night. These points, where the sun crosses the celestial equator, are called the *vernal* (spring) *equinox* and *autumnal* (fall) *equinox*. *Equinox* means "equal night."

Figure 6.15 is a map in which the sky is shown flattened out, as in a Mercator projection of the Earth. The equator runs along the middle of the map, and the ecliptic is shown as a wavy line cross-

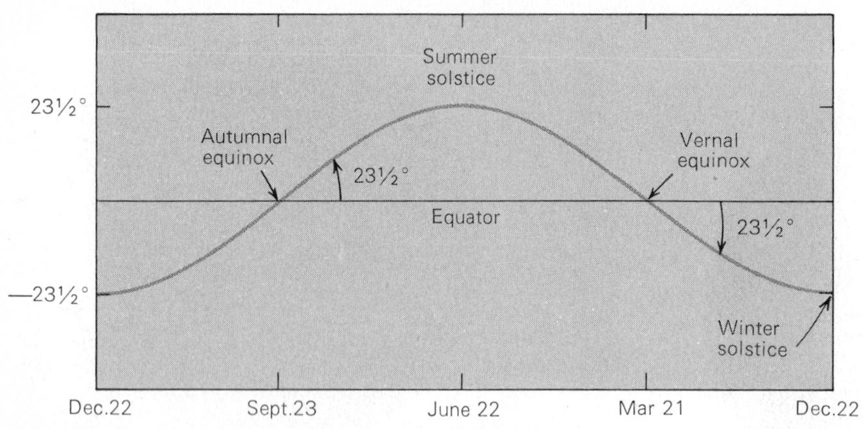

Figure 6.15 Plot of the ecliptic around the celestial equator, showing the apparent positions of the sun at the equinoxes and the solstices.

ing the equator at the two equinoxes. Both equator and ecliptic are, of course, great circles, but they cannot both be shown as straight lines on a flat surface. Notice that the ecliptic intersects the equator at an angle of 23½°, and that its northernmost extent is 23½° north of the equator (the summer solstice) and its southernmost extent is 23½° south of the equator (the winter solstice).

Figure 6.16 shows the aspect of the sky at a typical latitude in the United States. During the spring and summer, the sun is north of the equator and is thus up more than half the time. A typical spot in the United States, on the first day of summer (about June 22), receives about 14 or 15 hours of sunshine. Also, notice that the sun appears *high* in the sky, and so in these seasons the sunlight is more direct, and thus more effective in heating than in the fall and winter when the sun appears at a lower altitude in the sky.

In the fall and winter the sun is south of the equator, where most of its diurnal circle is below the horizon, and so it is up less than half the time. On about December 22, a typical city at, say 30° to 40° north latitude receives only nine or ten hours of sunshine. Also, the sun is low in the sky; a bundle of its rays is spread out over a larger area on the ground (Figure 6.17) than in summer; because the energy is spread out over a larger area, there is less for each square meter, and so the sun at low altitudes is less effective in heating the ground.

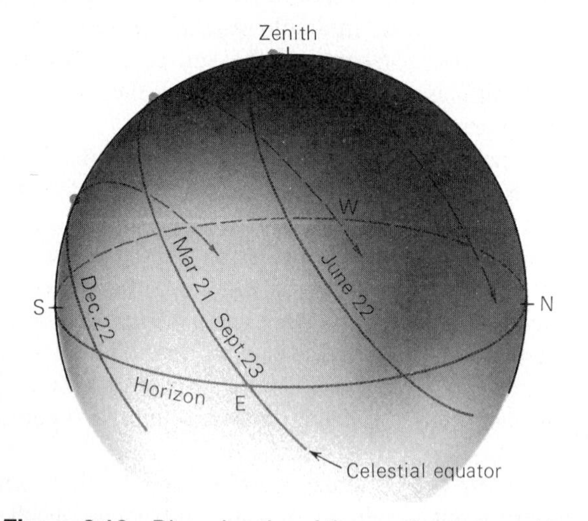

Figure 6.16 Diurnal paths of the sun for various dates at a typical place in the temperate Northern Hemisphere.

(c) The Seasons at Different Latitudes

At the equator all seasons are much the same. Every day of the year, the sun is up half the time, so there are always 12 hours of sunshine at the equator. About June 22, the sun crosses the meridian 23½° north of the zenith, and about December 22, 23½° south of the zenith.

The seasons become more pronounced as one travels north or south of the equator. At the Tropic of Cancer, on the date of the summer solstice, the sun is at the zenith at noon. On the date of the winter solstice, the sun crosses the meridian 47° south of the zenith. At the Arctic Circle, on the first day of summer the sun never sets, but at midnight can be seen just skimming the north point on the

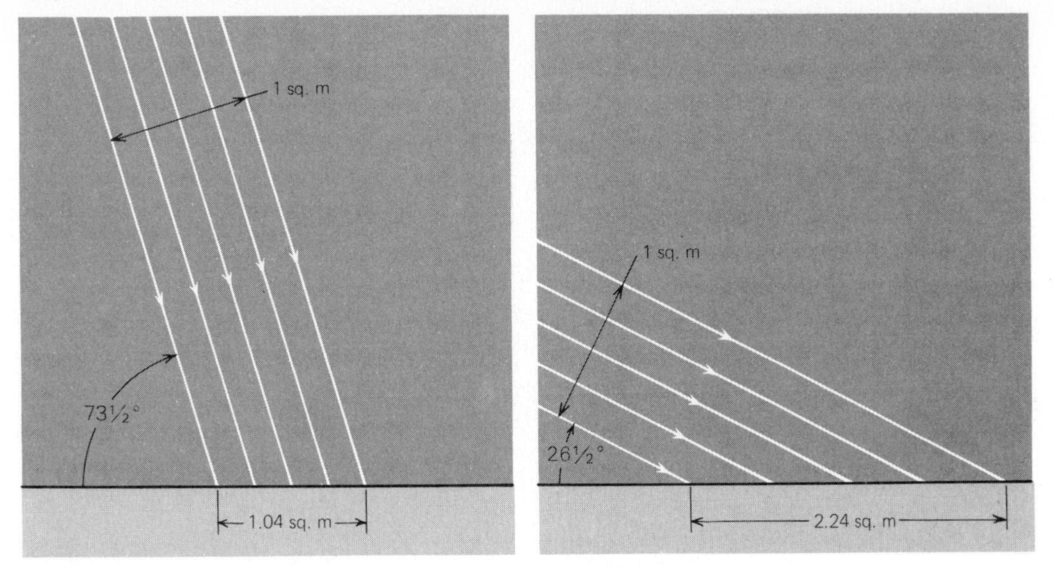

Figure 6.17 Effect of the sun's altitude. When the sun is low in the sky its rays are more oblique to the ground and are spread over a larger area than when the sun is high. This is just one of the reasons that winter is colder than summer.

horizon. About December 22, the sun does not quite rise at the Arctic Circle, but just gets up to the south point on the horizon at noon. Between the Tropic of Cancer and the Arctic Circle, the number of hours of sunshine and the noon altitude of the sun range between these two extremes.

We recall that at the North Pole, all celestial objects that are north of the celestial equator are always above the horizon and, as the Earth turns, circle around parallel to it. The sun is north of the celestial equator from about March 21 to September 23, and so at the North Pole the sun rises when it reaches the vernal equinox and sets when it reaches the autumnal equinox. There are six months of sunshine at the Pole. The sun reaches its maximum altitude of 23½° about June 22; before that date it climbs gradually higher each day, and after that it drops gradually lower. A navigator can easily tell when he is at the North Pole, for there the sun circles around the sky parallel to the horizon, getting no higher or lower (except gradually as the days go by.)*

In the Southern Hemisphere, the seasons are reversed from those in the north. While we are having summer in the United States, in Australia it is winter. Furthermore, in the Southern Hemisphere, the sun crosses the meridian generally to the *north* of the zenith. In Buenos Aires, you would want a house with a good *northern* exposure.

The Earth, in its elliptical orbit, reaches its closest approach to the sun about January 4. It is then said to be at *perihelion*. It is farthest from the sun, at *aphelion*, about July 5. We see, then, that the Earth is closest to the sun when it is winter in the north. However, it is summer in the Southern Hemisphere when the Earth is at perihelion, and the Earth is farthest from the sun during the Southern Hemisphere's winter. Therefore, we might expect the seasons to be somewhat more severe in the Southern Hemisphere than in the Northern. However, there is more ocean area in the Southern Hemisphere; this and other topographical factors are more important in their influence on climate than is the Earth's changing distance from the sun. For Mars, whose orbit is considerably more eccentric than the Earth's, the same kind of situation does have a pronounced effect upon the seasons.

* It is said that one botanist considered the North Pole to be an excellent place to raise sunflowers, because there were so many hours of sunshine during the summer months. He accordingly planted some there, and they did quite well for a while. However, sunflowers like to *face* the sun, and as they followed the sun around and around the sky, they ended by wringing their own necks!

(d) Precession of the Equinoxes

As the Earth's axis precesses in its conical motion, the equatorial plane retains (approximately) its 23½° inclination to the ecliptic plane; the obliquity of the ecliptic

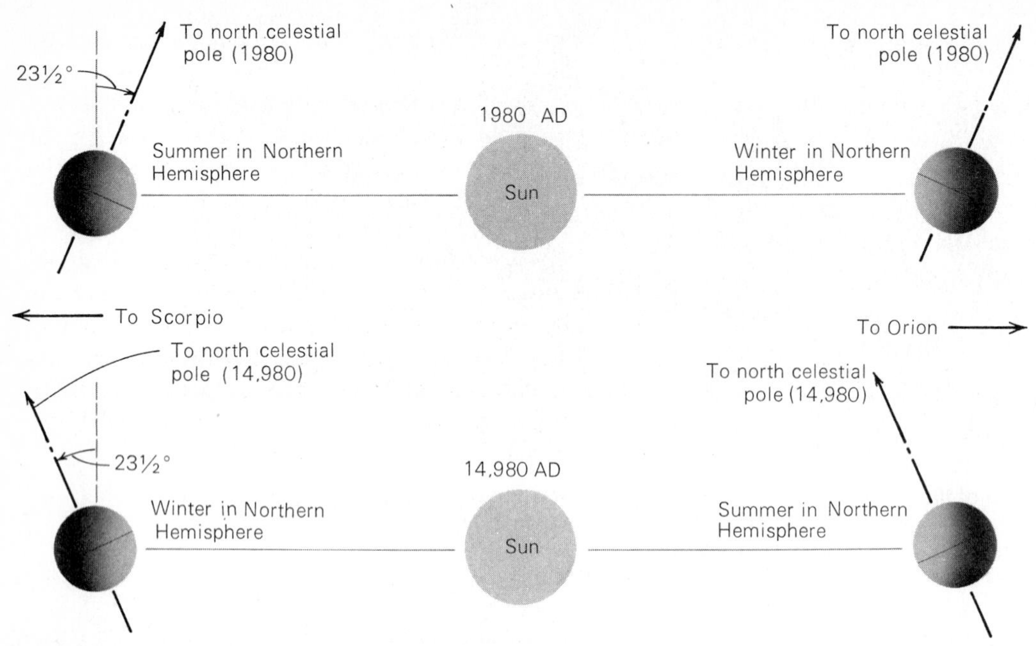

Figure 6.18 Summer and winter constellations change because of precession.

remains constant. However, the intersections of the celestial equator and the ecliptic (the equinoxes) must always be 90° from the celestial poles (because all points on the celestial equator are 90° from the celestial poles). Thus, as the poles move because of precession, the equinoxes slide around the sky, moving westward along the ecliptic. This motion is called the *precession of the equinoxes.* The angle through which the equinoxes move each year, the *annual precession,* is 1/26,000 of 360°, or about 50 arcseconds. Each year as the sun completes its *eastward* revolution about the sky with respect to, say, the vernal equinox, that equinox has moved *westward*, to meet the sun, about 50 arcseconds. Since it takes the sun about 20 minutes to move 50 arcseconds along the ecliptic (or, more accurately, because it takes the Earth that long to move through an angle of 50 arcseconds in its orbit about the sun), a *tropical year,* measured with respect to the equinoxes, is 20 minutes shorter than a *sidereal* year, measured with respect to the stars.

Precession has no important effect on the seasons. The Earth's axis retains its inclination to the ecliptic, so the Northern Hemisphere is still tipped toward the sun during one part of the year and away from it during the other. Our calendar year is based on the beginnings of the seasons (the times when the sun reaches the equinoxes and solstices), so spring in the Northern Hemisphere still begins in March, summer in June, and so on. The only effect is that as the precessional cycle goes on, a given season will occur when the Earth is in gradually different places in its orbit with respect to the stars. In the 20th century, for example, Orion is a *winter constellation;* we look out at night, away from the sun, and see Orion in the sky during the winter months. In 13,000 years, half a precessional cycle later, it will be summertime when we look out in the same direction, away from the sun, and see Orion. Similarly, Scorpius is a summer constellation now, whereas in the year 15,000 it will be a winter constellation (see Figure 6.18).

The vernal equinox is sometimes called the *first point of Aries,* because about 2000 years ago, when it received that name, it lay in the constellation of Aries. Now, because of precession, the vernal equinox has slid westward into the constellation of Pisces.

(e) The Many Motions of the Earth

In this chapter we have discussed in detail two of the Earth's motions: rotation and revolution. However, there are many other motions of the Earth. Here, for completeness, these motions are summarized:

1. The Earth *rotates* daily on its axis.
2. The Earth periodically shifts slightly with respect to its axis of rotation *(variation in latitude).*
3. The Earth *revolves* about the sun.

4. The gravitational pull of the sun and Moon on the Earth's equatorial bulge causes a very slow change in orientation of the axis of the Earth called *precession*.
5. Because the Moon's orbit is not quite in the plane of the ecliptic, and because of a slow change in orientation of the Moon's orbit, there is a small periodic motion of the Earth's axis superimposed upon precession, called *nutation*.
6. Actually, it is the center of mass of the Earth-Moon system, or *barycenter,* that revolves about the sun in an elliptical orbit. Each month the center of the Earth revolves about the barycenter.
7. The Earth shares the motion of the sun and the entire solar system among its neighboring stars. This *solar motion* is about 20 km/s.
8. The sun, with its neighboring stars, shares in the general *rotation of the Galaxy*. Our motion about the center of the Galaxy is about 250 km/s.
9. All other galaxies are observed to be in motion. Therefore, our *Galaxy is in motion* with respect to other galaxies in the universe.
10. In view of the many motions of the Earth, one might wonder what the absolute speed of the Earth is in the universe. To determine the absolute velocity of the earth was the object of the Michelson-Morley experiment, which we shall see (Chapter 11) gave null results. A fundamental postulate of Einstein's special theory of relativity is that it is not possible to define an absolute coordinate system with respect to which the absolute speed of an object in space can be determined. We shall see in Chapter 37, however, that we now believe we have measured the speed of our Galaxy with respect to the average distribution of the matter in the universe around us. That speed is about 500 km/s.

6.3 TIME

One of the most ancient uses of astronomy was the keeping of time and the calendar. From the earliest history in virtually every center of civilization—China, India, Mesopotamia, Egypt, Greece, the Mayan and Aztec civilizations in the western hemisphere, and evidently even in Bronze-Age England when Stonehenge was built—people followed the motions in the heavens to calculate the time and the date.

The measurement of time is based on the rotation of the Earth. As the Earth turns, objects in the sky appear to move around us, crossing the meridian each day. Time is determined by the position in the sky, with respect to the local meridian, of some reference object on the celestial sphere. The interval between successive meridian crossings or *transits* of that object is defined as a *day*. The actual length of a day depends on the reference object chosen; several different kinds of days, corresponding to different reference objects, are defined. Each kind of day is divided into 24 equal parts, called *hours*.

(a) The Passage of Time; Hour Angle

Time is reckoned by the angular distance around the sky that the reference object has moved since it last crossed the meridian. The motion of that point around the sky is like the motion of the hour hand on a 24-hour clock. The angle measured to the west along the celestial equator from the local meridian to the hour circle passing through any object (for example, a star) is that object's *hour angle*. (An hour circle is a great circle on the celestial sphere running north and south through the celestial poles. *Time* can be defined as the *hour angle of the reference object*.

As an example, suppose that the star *Rigel* is chosen as the reference for time. Then when Rigel is on the meridian it is $0^h0^m0^s$, "Rigel time." Twelve *Rigel hours* later, Rigel is halfway around the sky, at an hour angle of 180°, and the Rigel time is $12^h0^m0^s$. When Rigel is only 1° east of the meridian, and one *Rigel day* is nearly gone, the star is at an hour angle of 359°, and the Rigel time is $23^h56^m0^s$.

Time can be represented graphically by means of a *time diagram*, as in Figure 6.19. Here we imagine ourselves looking straight down on the north celestial pole from *outside* the celestial sphere. The pole appears as a point in the middle of the diagram, and the celestial equator appears as a circle centered on the pole. As the Earth turns to the east, the local meridian of an observer sweeps around the sky, so that its intersection with the celestial equator would move *counterclockwise* around the circle in the time diagram. However, it is customary to represent the observer's meridian as fixed, intersecting the equator, say, at the top of the diagram. Then

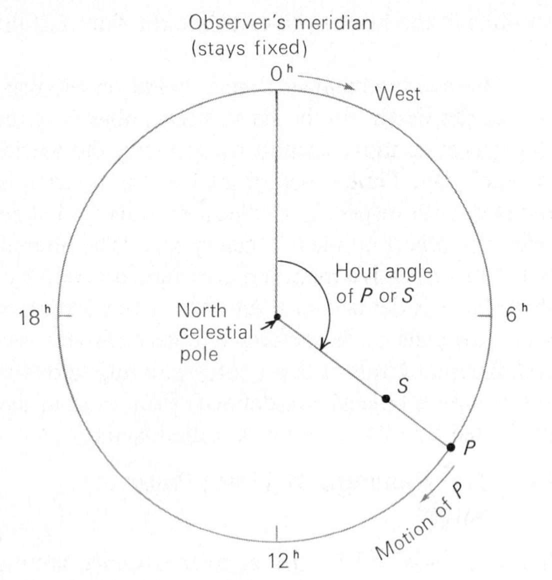

Figure 6.19 Time diagram.

the celestial sphere must be regarded as rotating *clockwise* with respect to the meridian. Let the reference object be denoted by S. Its hour circle intersects the equator at P, and as the celestial sphere rotates, the point P moves clockwise around the circle, like the hour hand of a clock. The hour angle of S (or P) increases uniformly with the rotation of the celestial sphere. In the diagram it is shown as about 120° (the time would then be about 8^h). Since the celestial equator intersects the horizon at the east and west points, P is below the observer's horizon in the time interval from 6 to 18 hours; the same is not true of S unless it happens to lie on the equator.

Because of the relation between hour angle and time, it is often convenient to measure angles in time units. In this notation, 24 hours corresponds to a full circle of 360°, 12 hours to 180°, 6 hours to 90°, and so on. One hour equals 15°, and 1° is four minutes of time.

Here we must distinguish between minutes and seconds of *time* (subdivisions of an hour), denoted by ^{m} and ^{s}, respectively, and minutes and seconds of *arc* (subdivisions of a degree), denoted by ′ and ″, respectively. The conversion between units of time and arc is given in Table 6.1.

Because we base time on the rotation of the Earth, it might seem that the Earth's rotation with respect to the stars would be the nearest we could come to defining the "true rotation rate" of the Earth. In practical terms, however, we have usually been more interested in defining a day by the apparent motion of the sun. After all, for most people the time of sunrise is more important than the time Arcturus or some other star rises, so with the exception of astronomers, we set our clocks to some version of sun time.

TABLE 6.1 Conversion Between Units of Time and Arc

TIME UNITS	ARC UNITS
24^h	360°
1^h	15°
4^m	1°
1^m	15′
4^s	1′
1^s	15″

(b) The Solar and Sidereal Days

The solar day is the period of the Earth's rotation with respect to the sun. The sidereal day is, instead, the time required for the Earth to make a complete rotation with respect to a point in space. The point chosen is the *vernal equinox*, defined as the point on the celestial sphere where the sun in its apparent path around the sky (the ecliptic) crosses the celestial equator from south to north, marking the first day of spring. Technically, the term "sidereal day" is a misnomer, because the vernal equinox slowly shifts its position in the sky as a result of precession. This movement is so slow, however, that a sidereal day is within 0.01 s of the true period of rotation of the Earth with respect to the stars.

A solar day is slightly longer than a sidereal day, as a study of Figure 6.20 will show. Suppose we start a day when the Earth is at A, with the sun on the meridian of an observer at point O on the Earth. The direction from the Earth to the sun, AS, if extended, points in the direction C among stars on the celestial sphere. After the Earth has made one rotation with respect to the stars, the same stars in direction C will again be on the local meridian to the observer at O. However, because the Earth has moved from A to B in its orbit about the sun during its rotation, the sun has not yet returned to the meridian of the observer but is still slightly to the east. The vernal equinox is so nearly fixed among the stars that the Earth has completed, essentially, one *sidereal* day, but to complete a *solar* day it must turn a little more to bring the sun back to the meridian.

In other words, a solar day is slightly *longer* than a sidereal day, or one complete rotation of the Earth. There are about 365 days in a year and 360° in a circle; thus the daily motion of the Earth in its orbit is about 1°. This 1° angle, *ASB,* is nearly the same as the additional angle over and above 360° through which the Earth must turn to complete a solar day. It takes the Earth about 4 minutes to turn through 1°. A solar day, therefore, is about 4 minutes longer than a sidereal day.

Each kind of day is subdivided into hours, minutes, and seconds. A unit of solar time (hour, minute, or second) is longer than the corresponding unit of sidereal time by about 1 part in 365. In units of solar time, one sideral day is $23^h56^m4^s.091$.

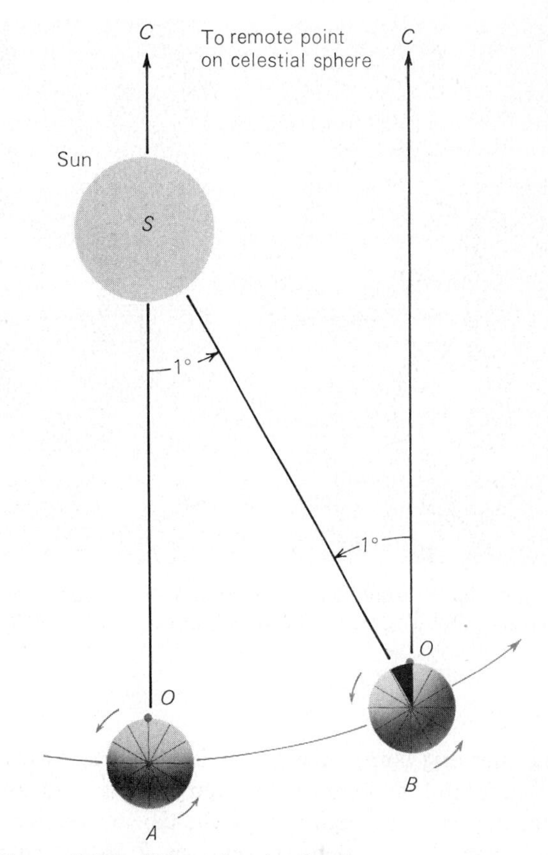

Figure 6.20 Sidereal and solar days.

(c) Sidereal Time

Sidereal time is based on the sidereal day with its subdivisions of sidereal hours, minutes, and seconds. It is defined as the *hour angle of the vernal equinox*. The sidereal day begins ($0^h0^m0^s$) when the vernal equinox is on the meridian.

Sidereal time is useful in astronomy and navigation. The common coordinate system used to denote positions of stars and planets on the celestial sphere (right ascension and declination) is referred to the celestial equator and the vernal equinox, much as latitude and longitude on the Earth are referred to the Earth's equator and the meridian of Greenwich, England (see Appendix 7). Therefore, the position of a star in the sky with respect to the observer's meridian is directly related to the sidereal time. Every observatory maintains clocks that keep accurate sidereal time.

One of us recalls a rather vivid demonstration from his undergraduate days of the slight difference between solar and sidereal time. In the 1950s, before electronic clocks, observatories used pendulum clocks, one keeping standard solar time and the other set to the sidereal rate. At his university, these two clocks were in the same room, which was also used by one unfortunate graduate student as his office. Their rates differed by one part in 365, so that every 365 seconds they would tick together, then slowly get out of phase until they were again coincident about 6 minutes later. This constant interplay between the two loudly ticking clocks should have been enough to drive anyone slowly mad, but actually the students using this peculiar office managed to survive. Perhaps your professor is a product of this conditioning.

(d) Apparent Solar Time

Just as sidereal time is reckoned by the hour angle of the vernal equinox, so *apparent solar time* is determined by the hour angle of the sun. At midday, apparent solar time, the sun is on the meridian. The hour angle of the sun is the time *past midday (post meridiem,* or P.M.). It is convenient to start the day not at noon, but at midnight. Therefore the elapsed apparent solar time since the beginning of a day is the hour angle of the sun *plus* 12 hours. During the first half of the day, the sun has not yet reached the meridian. We designate those hours as *before midday (ante meridiem,* or A.M.). We customarily start numbering the hours after noon over again, and designate them by P.M. to distinguish them from

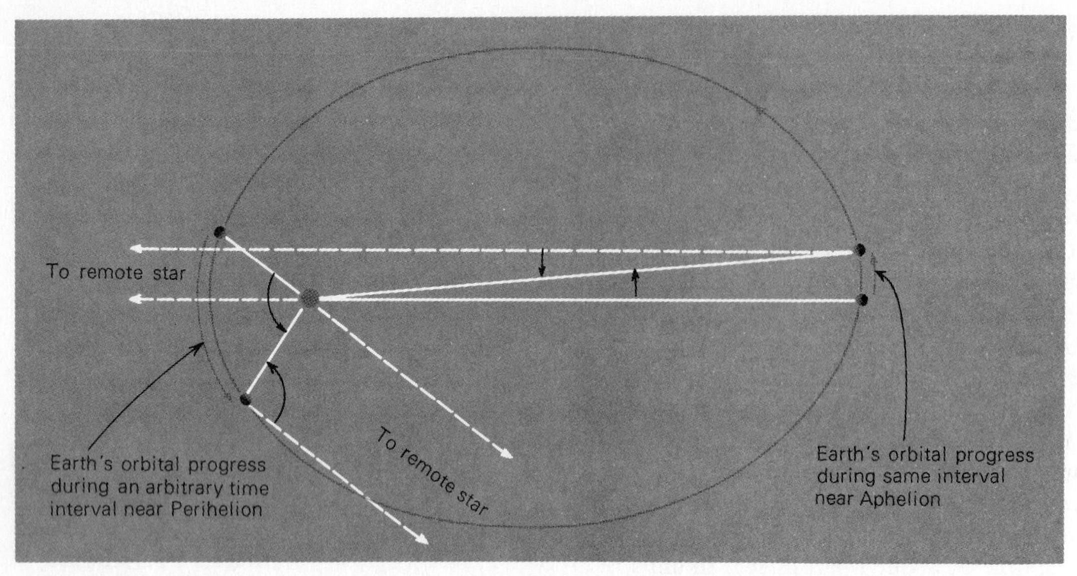

Figure 6.21 Variation in the length of an apparent solar day because of the Earth's variable orbital speed. (The effect, of course, is greatly exaggerated.)

the morning hours (A.M.). On the other hand, it is often useful to number the hours from 0 to 24, starting from the beginning of the day at midnight. For example, in various conventions, 7:46 P.M. may be written as 19^h46^m, 19:46, or simply 1946.

On about September 23, the sun passes through the autumnal equinox, halfway around the sky from the vernal equinox. On that date, at midnight, when the day begins, the vernal equinox is on the meridian, and so solar time and sidereal time are in agreement. With each succeeding day, however, sidereal time gains 3^m56^s on solar time, and the two kinds of time do not agree again until the daily difference between them accumulates to a full 24 hours—one year later.

Apparent solar time, defined as the hour angle of the sun plus 12 hours, is the most obvious and direct kind of solar time. It is the time that is kept by a sundial. In a sundial, a raised marker, or *gnomon,* casts a shadow whose direction indicates the hour angle of the sun. Apparent solar time was the time kept by people through many centuries.

The exact length of an apparent solar day, however, varies slightly during the year. Recall that the difference between an apparent solar day and a sidereal day, if time is counted from noon on one day, is the extra time required, after one rotation of the Earth with respect to the vernal equinox, to bring the sun back to the meridian. The length of this extra time depends on how far *east* of the meridian the sun is after the completion of one sidereal day. The Earth rotates to the east at a nearly constant rate of 1° every 4 sidereal minutes. Thus, if the sun were exactly 1° east of the meridian, about 4 sidereal minutes would be needed to bring it the rest of the way to the meridian. (Actually, it is just over four sidereal minutes, because the Earth is still advancing in its orbit during that period, which moves the sun another 10 arcseconds to the east along the ecliptic.) If the sun were more or less than 1° east of the meridian, the extra time required would be a little more or less than 4 sidereal minutes.

The length of the apparent solar day would be constant if the eastward progress of the sun, in its apparent annual journey around the sky, were precisely constant. However, there are two reasons why the amount by which the sun shifts to the east is not the same every day of the year.

The first reason is that the Earth's orbital speed varies. In accord with Kepler's second law—the law of areas—the Earth moves fastest when its is nearest the sun (perihelion) in early January and slowest when it is farthest from the sun (aphelion) in July. However, its rate of rotation is nearly constant. Consequently, it moves *farther* in its orbit during a sidereal day in January than in July (Figure 6.21).

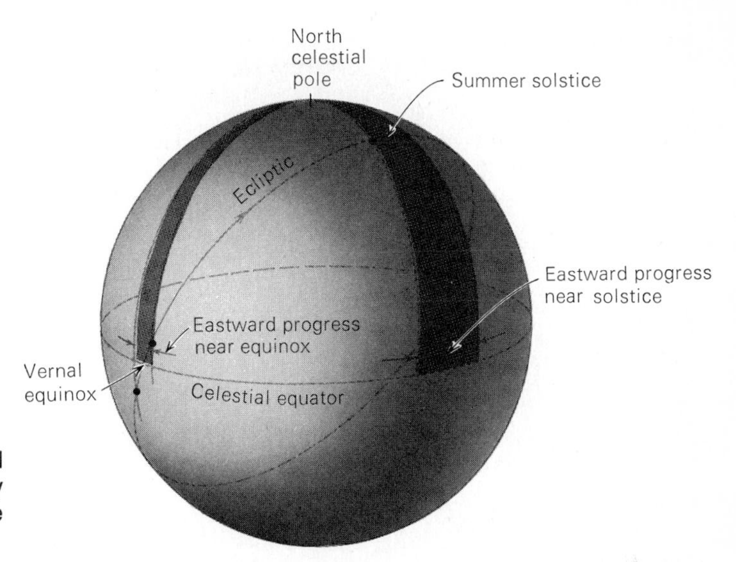

Figure 6.22 The sun's apparent eastward daily progress varies because of the obliquity of the ecliptic. (The obliquity, and hence the effect, have been greatly exaggerated.)

The sun's apparent motion along the ecliptic is just the result of the Earth's revolution, so the sun's daily progress to the east reflects the inequalities of the Earth's daily progress in its orbit. We see, then, that the extra amount by which the Earth must turn after a sidereal day to complete a rotation with respect to the sun is not always exactly the same.

The second reason for the variation in the rate of the sun's eastward progress, and the consequent nonuniformity in the length of the apparent solar day, is that the sun's path—the ecliptic—does not run exactly east and west in the sky, along the celestial equator, but is inclined to the equator by 23½°. Even if the Earth's orbit were circular, so that the sun moved uniformly along the *ecliptic*, the amount by which it moved to the east would vary slightly throughout the year. The situation is illustrated in Figure 6.22, which shows the celestial sphere, the celestial equator, and the ecliptic. To make the effect more obvious, the obliquity of the ecliptic is grossly exaggerated. Now suppose the sun moved equal distances along the ecliptic near March 21 and June 22; such equal distances are marked off on the ecliptic in the figure. Near the equinox, part of the sun's motion is northward, and it progresses less far to the *east* than it does along the ecliptic. At the solstice, on the other hand, not only is the sun moving due east but it is also north of the equator where the hour circles converge, so that a 1° advance on the ecliptic is *more* than a 1° advance to the east. A similar analysis shows the sun would also make more eastward progress near the winter solstice than near the autumnal equinox. With the actual 23½° obliquity, it turns out that a 1° advance on the ecliptic corresponds to 0.°92 advance to the east at the equinoxes and 1.°08 advance to the east at the solstices. Thus, even if the sun did move uniformly on the ecliptic, its eastward progress would be variable.

The apparent solar day is always *about* 4 minutes longer than a sidereal day, but because of the sun's variable progress to the east, the precise interval varies by up to one-half minute one way or the other. The variation can accumulate after a number of days to several minutes. After the invention of clocks that could run at a uniform rate, it became necessary to abandon the apparent solar day as the fundamental unit of time. Otherwise, all clocks would have to be adjusted to run at a different rate each day.

(e) Mean Solar Time

Mean solar time is based on the *mean solar day,* which has a duration equal to the *average* length of an apparent solar day. Mean solar time is defined as the hour angle plus 12 hours, of a fictitious point in the sky that moves uniformly to the east along the *celestial equator,* with approximately the same average eastern rate as the true sun. In other words, mean solar time is just apparent solar time averaged uniformly. Originally that fictitious point was called the *mean sun,* but in the modern definition of time standards, it is no longer related to the sun.

Although mean solar time has the advantage of progressing at a uniform rate, it is still inconvenient

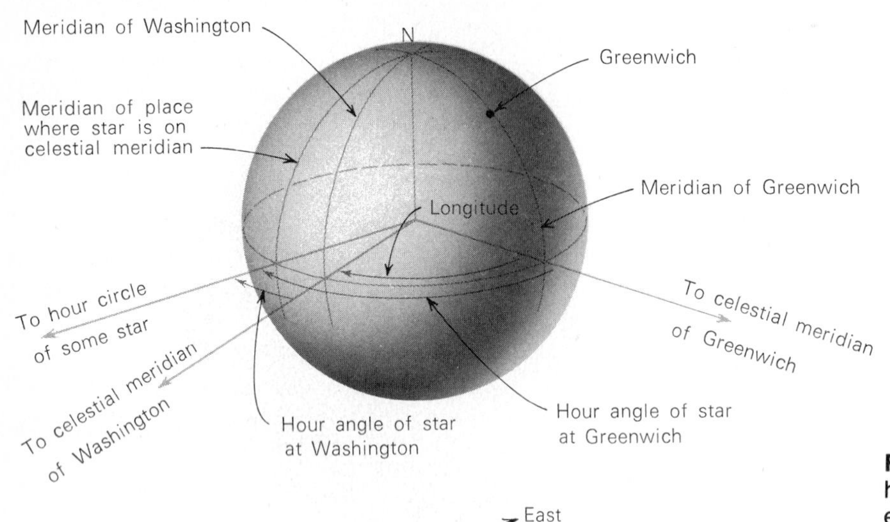

Figure 6.23 A difference in hour angle equals the difference in longitude.

for practical use. Recall that it is defined by the hour angle of the fictitious point. But hour angle refers to the local celestial meridian, which is different for every longitude on Earth (Figure 6.23). Thus, observers on different north-south lines on the Earth have a different hour angle of the point and hence a different mean solar time. If mean solar time were strictly observed, a person traveling east or west would have to reset his watch continually as his longitude changed, if it were always to read the local mean time correctly. For instance, a commuter traveling from Oyster Bay to New York City would have to adjust his watch as he rode through the East River tunnel, because Oyster Bay time is actually about 1.6 min more advanced than that of Manhattan.

(f) Standard and Daylight Saving Times

Until near the end of the last century, every city and town in the United States kept its own local mean time. With the development of railroads and the telegraph, however, the need for some kind of standardization became evident. In 1883 the nation was divided into four time zones. Within each zone, all places keep the same time, the local mean solar time of a standard meridian running more or less through the middle of each zone. Now a traveler resets his watch only when the time change has amounted to a full hour. For local convenience, the boundaries between the four time zones are chosen to correspond, as much as possible, to divisions between states. Mean solar time, so standardized, is called *standard time*. The standard time zones in the United States (not including Alaska and Hawaii) are Eastern Standard Time (EST), Central Standard Time (CST), Mountain Standard Time (MST), and Pacific Standard Time (PST), which respectively keep the mean times of the meridians of 75°, 90°, 105°, and 120° west longitude. Hawaii and Alaska both keep the time of the meridian 150° west longitude, two hours less advanced than Pacific Standard Time. Since 1884, standard time has been in use around the world by international agreement. Almost all countries have adopted one or more standard time zones, although the second largest nation, India, has settled on a half zone, being 5½ hours from Greenwich standard.

Daylight saving time is simply the local standard time of the place plus one hour. It has been adopted for spring and summer use in most states in the U.S. and in many countries to prolong the sunlight into evening hours, on the apparent theory that it is easier to change the time by government action than it would be for individuals or businesses to adjust their own schedules to produce the same effect.

(g) The International Date Line

The direction of the Earth's rotation is to the east. Therefore, places to the east of us must always have a time *more advanced* than ours. The celestial meridian of New York sweeps under the sun on the celestial sphere about three hours earlier than does the celestial meridian of San Francisco. Thus in New York the hour angle of the sun, the local time there, is three hours ahead of San Francisco time. The times of places halfway around the world from each other differ by 12 hours.

The fact that time is always more advanced to the east presents a problem. Suppose you travel eastward around the world. You pass into a new time zone, on the average, for about every 15° of longitude you travel, and each time you dutifully set your watch ahead an hour. By the time you have completed your trip, you have set your watch ahead through a full 24 hours, and thus gained a day over those who stayed at home.

The solution to this dilemma is the *international date line*, set by international agreement to run approximately along the 180° meridian of longitude. The date line runs about down the middle of the Pacific Ocean, although it jogs a bit in a few places to avoid cutting through groups of islands and through Alaska. By convention, at the date line, the date of the calendar is changed by one day. Crossing the date line from west to east, thus advancing your time, you compensate by decreasing the date; crossing from east to west, you increase the date by one day. We simply must accept that the date will differ in different cities at the same time. A well-known example is the "day of infamy" when the Imperial Japanese Navy bombed Pearl Harbor in Hawaii, known in the U.S. as Sunday, December 7, but taught to Japanese students as taking place on December 8, 1941.

If an ocean liner crosses the international date line from west to east just after its passengers have been served a Christmas dinner, the crew can begin the preparation of a second feast, for the next day will be Christmas again. On the other hand, one could skip Christmas altogether by crossing the date line from east to west.

(h) Summary of Time

We may briefly summarize the story of time as follows:

1. The ordinary day is based on a rotation of the Earth with respect to the sun (not the stars).
2. The length of the apparent solar day is slightly variable because of the Earth's variable orbital speed and the obliquity of the ecliptic.
3. Therefore, the average length of an apparent solar day is defined as the mean solar day.
4. Time based on the mean solar day is mean solar time. It is defined as the local hour angle of a fictitious point revolving annually on the celestial equator at a perfectly uniform eastward rate.
5. The mean solar time at any one place, called *local mean time*, varies continuously with longitude, so that two places a few kilometers east and west of each other have slightly different times.
6. Therefore time is standardized, so that each place in a certain region or zone keeps the same time—the local mean time of the standard meridian in that zone.
7. In many localities, standard time is advanced by one hour during part or all of the year to take advantage of the maximum number of hours of sunshine during the waking hours. This is daylight saving times.

6.4 THE CALENDAR

(a) The Length of the Year

The natural units of the calendar are the day, based on the period of rotation of the Earth; the month, based on the period of revolution of the Moon about the Earth, and the year, based on the period of revolution of the Earth about the sun. The difficulties in the the calendar have resulted from the fact that these three periods are not commensurable, that is, one does not divide evenly into any of the others.

The period of revolution of the Moon with respect to the stars, the *sidereal month*, is about 27⅓ days. However, the interval between corresponding phases of the Moon, the more obvious kind of month, is the Moon's period of revolution with respect to the sun, the *synodic month*, which has about 29½ days (Chapter 7).

There are at least three kinds of year. The period of revolution of the Earth about the sun with respect to the stars is called the *sidereal year*. Its length is 365.2564 mean solar days, or $365^d6^h9^m10^s$.

The period of revolution of the Earth with respect to the vernal equinox, that is, with respect to the beginnings of the various seasons, is the *tropical year*. Its length is 365.242199 mean solar days, or $365^d5^h48^m46^s$. Our calendar, to keep in step with the seasons, is based on the tropical year. Because of precession (Section 6.2d), the tropical year is slightly shorter than the sidereal year.

The third kind of year is the *anomalistic year*, the interval between two successive perihelion passages of the Earth. Its length is 365.2596 mean solar days, or $365^d6^h13^m53^s$. It differs from the sidereal year, because the major axis of the Earth's orbit slowly shifts in the plane of the Earth's orbital revolution. Perturbations by the other planets cause this shift.

(b) The Days of the Week

The week is an independent unit arbitrarily invented by man, although its length may have been based on the interval between the quarter phases of the Moon. The seven days of the week are named for the seven planets (including the sun and Moon) recognized by the ancients. In order of supposed decreasing distance from the Earth, these seven objects are Saturn, Jupiter, Mars, the sun, Venus, Mercury, and the Moon. (It was believed that the faster-moving objects were the nearest. The assumption is not necessarily correct—Mercury, the fastest-moving planet, does not come as close to the Earth as Venus.)

Each hour of the day was believed to be ruled by one of the planets in the order named. A particular day of the week was named for the planet that ruled it during the first hour of that day. The first day, Saturday, was ruled by Saturn during the first hour, Jupiter during the second, Mars during the third, and so on. Saturn thus ruled, in addition to the first hour, the eighth, fifteenth, and twenty-second. Jupiter and Mars were allotted the twenty-third and twenty-fourth hours of Saturday, leaving the first hour of the second day, Sunday, for the sun. If the scheme is continued, it is found that the moon rules the third day (Monday), and Mars, Mercury, Jupiter, and Venus the fourth, fifth, sixth, and seventh days. Our Anglo-Saxon names for the fourth through the seventh days come from the Teutonic equivalents of the Roman gods for which the planets Mars, Mercury, Jupiter, and Venus were named. The connection between those planets and the days of the week is even more obvious if we look, for example, at the Italian names for Tuesday, Wednesday, Thursday, and Friday: *Martedi*, *Mercoledi*, *Giovedi*, and *Venerdi*.

(c) Roman Calendars

The roots of our modern calendar go back to the Roman republican calendar, which derives from earlier Roman and Greek calendars dating from at least the eighth century B.C. The earliest Roman calendar probably had ten months, the last four of which have given us the names of our months, September, October, November, and December. But by the first century B.C., two additional months had been added.

The original Roman calendar was lunar. The months were based on the Moon's synodic period; each month began with a new moon. To give the months an average length of 29½ days (the lunar synodic period) the months had 29 and 30 days alternately. The difficulty with the lunar calendar is that 12 lunar months add up to only 354 days, whereas the tropical year has about 365¼ days. After about three years, the difference accumulates to a whole month. To keep their year in step with the year of the seasons, the ancients adopted the policy of intercalation, that is, they simply inserted a 13th month every third year or so. The normal 12-month years were "empty years," and the 13-month years were "full years."

The Roman republican calendar, in use by about 70 B.C., had 12 months. These months, and their duration in days, were Martius (31), Aprilis (29), Maius (31), Iunius (29), Quintilis (31), Sextilis (29), September (29), October (31), November (29), December (29), Ianuarius (29), and Februarius (28). The year thus had 355 days. From the middle of the second century B.C. January (Ianuarius) 1 officially marked the beginning of the year, although in the popular view the year ended with February 23. When an extra month had to be intercalated every two to four years to bring the average length of the year to 365¼ days, it was added immediately after February 23. Then followed the last five days of February, and March.

Unfortunately, the management of the calendar was left to the discretion of the priests, who greatly abused their authority by declaring as full years those in which their friends were in public office. The intercalation process became such a political football that by the time of the reign of Julius Caesar, a Roman traveler going from town to town could find himself going from year to year! Thus, in 46 B.C., Julius Caesar instigated a calendar reform.

At the advice of the Alexandrian astronomer Sosigenes, Caesar adopted a new calendar, which had 12 more or less equal months averaging about 30½ days in length rather than 29½. The features of the Julian calendar reform of 46 B.C. were as follows.

1. The lunar synodic month was abandoned as a basic unit in the calendar. Instead, each year contained 12 months, which contained a total of 365 days. Caesar distributed the ten extra days among the 12 months, but there appears to be a difference of opinion among historians as to which months originally had how many days.

2. The calendar was to be based on the tropical year, whose length had at that time been determined to be 365¼ days. Of course, one-fourth of a day could not be "tacked on" to the end of the calendar year. Therefore, *common years* were to contain only 365 days. However, after four years, this quarter-day per year adds up to one full day. Thus every fourth year was to have 366 days (a *leap year*), the extra day being added to February. The *average* length of the year would then be 365¼ days. Note that the process of leap year is analogous to the intercalation of extra months in "empty years" to make them "full years" of 13 months.

3. To bring the date of the vernal equinox, which had fallen badly out of place in the Roman republican calendar, back to its traditional date of March 25, Caesar intercalated three extra months in the year 46 B.C, bringing its length to 445 days. Forty-six B.C. was known as the "year of confusion." The Julian calendar was introduced, then, on January 1, 45 B.C.

After Caesar's death in 44 B.C., the month Quintilis (the fifth month in the original Roman calendar) was renamed in his honor (thus our name, July). Later, the Roman senate did some further juggling with the Julian calendar. Sextilis (originally the sixth month) was renamed in honor of Augustus Caesar, successor to Julius, and the present number of days for each month resulted.

The dates of observance of Easter and certain other religious holidays were fixed by order of the Council of Nicaea in 325 A.D. Easter, according to the rule adopted, falls on the first Sunday after the 14th day of the Moon (almost full moon) that occurs on or after March 21. At that time March 21 was the date of the vernal equinox. (The Sunday *after* full moon was specified intentionally to avoid the possibility of an occasional coincidence with the Jewish Passover.)

Note that between 45 B.C. and 325 A.D., the date of the vernal equinox had slipped back from March 25 to March 21. This was because the Julian year, with an average length of 365¼ days, is 11^m14^s longer than the tropical year of $365^d5^h48^m46^s$. The slight discrepancy had accumulated to just over three days in those four centuries.

(d) The Gregorian Calendar

By 1582, that 11 minutes and 14 seconds per year had added up to another ten days, so that the first day of spring was occurring on March 11. If the trend were allowed to continue, eventually Easter and the related days of observance would be occurring in early winter. Therefore, Pope Gregory XIII instituted a further calendar reform.

The Gregorian calendar reform consisted of two steps. First, ten days had to be dropped out of the calendar to bring the vernal equinox back to March 21, where it was at the time of the Council of Nicaea. This step was expediently accomplished. By proclamation the day following October 4, 1582, became October 15.

The second feature of the new Gregorian calendar was that the rule for leap year was changed so that the average length of the year would more closely approximate the tropical year. In the Julian calendar, every year divisible by four was a leap year, so that the average year was 365.250000 mean solar days in length. The error between this and the tropical year of 365.242199 mean solar days accumulates to a full day every 128 years. Ideally, therefore, one leap year should be made a common year, thus dropping one day, every 128 years. Such a rule, however, is cumbersome.

Instead, Gregory decreed that three out of every four century years, all leap years under the Julian calendar, would be common years henceforth. The rule was that only century years divisible by 400 should be leap years. Thus, 1700, 1800, and 1900, all divisible by four, and thus leap years in the old Julian calendar, were *not* leap years in the Gregorian calendar. On the other hand, the years 1600, and 2000, both divisible by 400, are leap years under both systems. The average length of this Gregorian year was 365.2425 mean solar days, and was correct to about one day in 3300 years.

The Catholic countries immediately put the Gregorian reform into effect, but countries under control of the Eastern Church and most Protestant countries did not adopt it until much later. It was 1752 when England and the American colonies finally made the change. The year 1700 had been a leap year in the Julian calendar but not the Gregorian; thus the discrepancy between the two systems had become 11 days. By parliamentary decree, September 2, 1752, was followed by September 14. Although special laws were passed to prevent such breaches of justice as landlords collecting a full month's rent for September, there were still riots, and people demanded their 11 days back. To make matters worse, in England it had been customary to follow the ancient practice of starting the year on March 25, originally the date of the vernal equinox. In 1752, however, the start of the year was moved back to January 1, so in England and the colonies 1751 had no months of January and February and had lost 24 days of March! We mark George Washington's Birthday on February 22, 1732, but at the time of his birth, a calendar would have read February 11, 1731. Russia did not abandon the Julian calendar until the time of the Bolshevik revolution. The Russians then had to omit 13 days to come into step with the rest of the world.

The Gregorian calendar has now been modified slightly to come into better conformity with the tropical year: the years 4000, 8000, 12,000, and so on, all leap years in the original Gregorian calendar, are now common years. The calendar is thus accurate to one day in about 20,000 years.

At a meeting of the Congress of the Orthodox Oriental Churches at Constantinople in 1923, a slightly improved version of the Gregorian calendar was adopted for the Eastern churches. This Eastern calendar is shorter than the Julian by seven days every 900 years, rather than three days every 400 years. The rule for leap year is that century years, when divided by 900, will be leap years only if the remainder is either 200 or 600. The years 2000 and 2400 will be leap years in both the Gregorian and Eastern Orthodox calendars. The years 2100, 2200, 2300, 2500, 2600, and 2700 will not be leap years in either. The two calendars will not diverge until 2800, which will be a leap year in the Gregorian calendar but not in the Eastern Orthodox. The Eastern Orthodox calendar year has an average length of 365.2422 mean solar days, very nearly that of the tropical year. Its error is only one day in 44,000 years.

(e) The Mayan Calendar

Of the various calendar systems of other ancient civilizations, one of the most interesting was that of the Maya, whose civilization, flourishing in the Yucatan area in Central America, was contemporary with the early European civilizations. The Mayan calendar was later adopted, at least in part, by the conquering Aztecs.

The Mayan calendar was more sophisticated and complicated than either the Roman or Julian calendar. Apparently, the Maya did not attempt to correlate their calendar accurately with the length of the year or lunar month. Rather, their calendar was a system for keeping track of the passage of days and for counting time far into the past or future. Among other purposes, their calendar was useful for predicting astronomical events, for example, the positions of Venus in the sky.

The Mayan calendar consisted of three simultaneous systems for counting days. The first was the sacred almanac, called the Tzolkin by modern archaeologists, which was somewhat analogous to our week of seven named days that recur in specified order perpetually. However, the Tzolkin had 20 named days; moreover, each day's name was accompanied by a number. The numbers ran from 1 to 13 and were then repeated. The first day of the sacred almanac was 1 *Imix,* the second 2 *Ik,* and so forth, up to the 13th day, which was 13 *Ben.* The next day, *Ix,* was accompanied by the number 1 again, that is, 1 *Ix.* The twentieth day was 7 *Ahau.* Then the day names started over with *Imix* again, but this time *Imix* appeared with the number 8: 8 *Imix.* In other words, the numbers were always out of phase with the day names. After 13 × 20, or 260 days. 1 *Imix* appeared again, after which the whole series was repeated, and so on, indefinitely. Thus, the Tzolkin was a counting system containing 260 combinations of numbers and names.

The second counting system of the Mayan calendar was a 365-day period that is approximately equal to the year. However, there was no intercalation of extra days, or "leap-year" scheme, so the 365-day period did not remain fixed with respect to the seasons.

Figure 6.24 The famous Aztec calendar (based on the Mayan calendar) on display in the Anthropological Museum in Mexico City. (G.O. Abell)

This 365-day period was divided into 18 *uinals* (analogous to months, but not equal to the Moon's period) of 20 days each, with five "unlucky" days tacked on as a 19th uinal. Each day was numbered according to its position in its uinal; thus the Maya would speak of 17 *Yaxkin,* much as we would say July 23. (Of course, there is no simple correspondence between the dates in our calendar and theirs.) To give both the Tzolkin day name and uinal date, the Maya might say, for example, 7 *Ik* 15 *Yaxkin;* analogously, we might say Thursday, January 11. In our calendar, January 11 can fall on a Thursday every several years (it would be every seven years if it were not for leap year), but a date in the Mayan calendar, specified like 7 *Ik* 15 *Yaxkin,* occurred exactly once every 18,980 days, or about every 52 years.

Finally, to specify completely a particular date, the Maya made use of what is called the *long count,* a perpetual tally of the days that had elapsed since a particular date about 3000 years in the past. The starting date was not meant to be that of "the beginning"—it was merely an arbitrary starting point, from which days could be counted. The significant feature of the long count is that it employed a vigesimal number system, that is, one based on 20 (rather than 10, as is our decimal system). The useful property of our decimal system is not, particularly, that it is based on the number 10 but that it employs the *zero,* without which arithmetic would be extremely tedious. (Any reader who questions this statement, should try multiplying 53,498 by 627 in Roman numerals.) The Maya made use of the zero in counting and arithmetic and employed a method of place value, analogous to our own method of writing numbers, many centuries before the Arabs introduced the concept to Europe.

EXERCISES

1. Show that the apparent deflection in the direction of swing of a Foucault pendulum in the Southern Hemisphere is to the *left,* rather than to the *right* as in the Northern Hemisphere.
2. What is the latitude of (a) the North Pole? (b) the South Pole?
3. Why has longitude no meaning at the North or South Pole?
4. Draw a diagram to show that for an observer south of the equator the altitude of the south celestial pole is equal to the latitude.
5. Prove that the celestial equator must pass through the east and west points on the horizon.
6. If a star rises in the northeast, in what direction does it set?
7. ★ Prove that if vertically falling raindrops dropping with a speed of V are to fall through a drainpipe, the pipe must be tilted forward so that its top precedes its bottom by a distance which, when divided by the vertical extent of the pipe, is in the ratio v/V, where v is the speed of the drainpipe.
8. Explain why New York has more hours of daylight on the first day of summer than does Los Angeles.
9. Suppose the obliquity of the ecliptic were only 16½°. What then would be the difference in latitude between the Arctic Circle and the Tropic of Cancer?
10. If the obliquity were only 16½°, what would be the effect on the seasons as compared to the actual obliquity of 23½°?
11. What are the approximate dates of sunrise and sunset at the South Pole? Would a lunar eclipse occurring in January be visible from there? Why or why not?

12. If there are 365¼ solar days per year, what is the daily motion of the Earth in its orbit in degrees per solar day?
13. If the sidereal month is 27⅓ days, show that the rotation period of the Earth with respect to the Moon is about 53 minutes longer than with respect to the stars.
14. If it is 3:00 P.M. local apparent time, what is the hour angle of the sun?
15. If a star rises at 8:30 P.M. tonight, at approximately what time will it rise two months from now?
16. If it is 1:00 A.M., July 17, at longitude 165° W, what are the time and date at longitude 165° E?
17. What is the greatest number of Sundays possible in February for the crew of a vessel making weekly sailings from Siberia to Alaska?
Answer: 10
18. Show that the Gregorian calendar will be in error by one day in about 3300 years.
19. If the Earth were to speed up in its orbit slightly, so that a tropical year were completed in exactly 365^d3^h, how often would we need to have a leap year? What rule, if any, would we need for century years?

Galileo Galilei (1564–1642) advocated that we perform experiments or make observations to ask Nature her ways, rather than deciding how things must be on the basis of preconceived notions. When Galileo turned the telescope to the sky, he found that things are not as philosophers had supposed, discovering sunspots, the mountains of the Moon, the phases of Venus, and the four large satellites of Jupiter—worlds that are still called the Galilean satellites.

7 THE MOON AND PLANETS IN THE SKY

Having examined in Chapter 6 the motions of the Earth and how those motions affect the appearance of the sky, we now turn to the apparent and real motions of the Moon and planets. The physical nature of the Moon is discussed in Chapter 14 and that of the other planets in Chapters 15 through 17.

7.1 ASPECTS OF THE MOON

The Moon, because of its proximity to the Earth, appears to move more rapidly in the sky than any other celestial object. The Moon looks to be the same size as the sun, subtending ½° of arc. As it travels around the Earth each month, it progresses through its cycle of phases, while always keeping the same hemisphere turned toward the Earth.

(a) Moonlight

The amount of light we receive from the Moon varies with its phase. When the Moon is full, its light is nearly bright enough to read by; we receive only about 10 percent as much light from the Moon at first and last quarter, and only one thousandth of the light of the full moon when the Moon appears as a thin crescent 20° from the sun in the sky.

Despite the brilliance of the full moon, it shines with less than 1/400,000 the light of the sun. Even if the entire visible hemisphere of the sky were packed with full moons, the illumination would be only about one-fifth or less of that in bright sunlight.

Because the Moon shines by reflected sunlight, we can calculate the Moon's reflecting power from its apparent brightness. The Moon and Earth are at about the same distance from the sun; consequently, the Moon receives as much sunlight per unit area of its surface as does the Earth. Calculation shows that only 7 percent of the incident light is reflected. The Moon thus absorbs most of the sunlight that falls upon it; its surface is quite dull. You can judge this for yourself by comparing the Moon in the daytime sky with a nearby sunlit building or mountain; although it receives about the same illumination, the Moon is clearly not as bright as most of your surroundings. The 93 percent of incident sunlight that is absorbed heats the surface of the Moon and is radiated into space as infrared radiation.

Astronomers use the term *albedo* (more technically, *Bond albedo*) for the fraction of incident light that is reflected from a body in the solar system. A more useful concept generally, however, is the *reflectivity* of the surface, which is a measure of how bright it is relative to a perfectly reflecting diffuse white material. Reflectivity, defined in this way, is not the same as albedo. The reflectivity of the Moon is 12 percent, meaning that it appears 12 percent as bright as if it were painted with a high-quality white paint. In describing other planets, we will usually note their reflectivities, not their albedos. When the bright part of the Moon is only a thin crescent, the "night" side of the Moon often appears faintly illuminated. Leonardo da Vinci (1452–1519) first explained this illumination as *earthshine,* light reflected by the Earth back to the Moon, just as moonlight often illuminates the night side of the Earth. Earthshine is, however, much brighter than moonshine, since the Earth has a reflectivity about 4 times that of the Moon, and an area that is about 16 times greater.

Figure 7.1 The Moon, the celestial body that we can see best through a telescope. When the resolution of a planetary image is high enough (typically a few kilometers), it is possible to identify the major landforms there and begin to trace its geological history. (NASA)

(b) The Moon Through a Telescope

Viewed with the naked eye, the Moon shows a faint pattern of light and dark regions that have been interpreted throughout the ages as outlining an animal, a human figure, or a human face (the "man in the Moon"). As we will see in Chapter 14, these light and dark patches represent two very different kinds of surface material, with the darker regions having been produced by extensive lunar volcanism. Without optical aid, however, no *topographic* detail can be discerned on the Moon. Only through Galileo's telescope could even the largest lunar mountains and craters be discovered.

The *resolution* of the human eye is about 200 km on the Moon, so that any feature less than 200 km in size cannot be made out. A telescope (Chapter 9) produces higher resolution images, up to the limit set by the shimmer in the Earth's atmosphere. Galileo's telescopes had a resolution of a few tens of kilometers, and a modern amateur astronomer's telescope can easily show objects on the Moon as small as a few kilometers across. Since topographic features such as mountains and valleys are typically a few kilometers to a few tens of kilometers in size, they are readily detected with such instruments (Figure 7.1).

As seen through a good pair of binoculars or a small telescope, the appearance of the Moon's surface changes dramatically with its phase. At full phase it shows almost no topographic detail, and you must look closely to see more than a few craters. This is because the sunlight illuminates the full moon straight on, and in this flat lighting no shadows are cast. Much more revealing is the view near first or last quarter, when the sunlight streams in from the side and topographic features cast sharp shadows. At a given resolution, it is almost always more rewarding to study a planetary surface under such *oblique lighting,* when the maximum information about the surface topography can be obtained.

The best resolution of the Moon from ground-based telescopes is about 1 km, ample to reveal many features of interest to the geologist. In contrast, the nearest planets (Mars and Venus) are more than 150 times farther away, with correspondingly lower telescopic resolution. At a resolution of 150 km or less, no topographic information can be obtained; indeed, the best views of Mars or Venus possible before the space age were equivalent to the *naked-eye* resolution of the Moon. No wonder astronomers had such a difficult time trying to understand the nature of our planetary neighbors!

(c) The Moon's Apparent Path in the Sky

If the Moon's position among the stars on the celestial sphere is carefully noted night after night, it is seen that the Moon changes its position rather rapidly, moving, on the average, about 13° to the east per day. In fact, during a single evening the Moon creeps visibly eastward among the stars. The Moon's apparent path around the celestial sphere is a great circle (or very nearly so) that intersects the sun's path, the ecliptic, at an angle of about 5°. The Moon's path intersects the ecliptic at two points on opposite sides of the celestial sphere. These points are called the *nodes* of the Moon's orbit.

The Moon's orbit is constantly and gradually changing because of perturbations, just as the orbits of artificial satellites change. The most important perturbations are caused by the gravitational attraction of the sun. One of the effects of the perturbations on the Moon's orbit is that the nodes slide westward along the ecliptic, completing one trip around the celestial sphere in about 18.6 years. This motion is called the Moon's *regression of the nodes.* Perturbations also cause the inclination of the Moon's orbit to the ecliptic to vary from 4°57′ to 5°20′; the average inclination is 5°9′.

If it were not for the regression of the nodes, the Moon's orbit would maintain a nearly fixed angle to the celestial equator. It maintains a nearly fixed angle of about 5° to the ecliptic, but because the nodes constantly shift, its angle of inclination to the equator varies from 23½ + 5°, or about 28½°, to 23½ − 5°, or about 18½° (23½° is the inclination of the ecliptic to the celestial equator).

(d) Delay in Moonrise from Day to Day

We have seen that the Moon's average eastward motion with respect to the stars is about 13° per day. The sun, on the other hand, appears to move to the east about 1° per day. With respect to the sun, therefore, the Moon moves eastward about 12° per day. As the Earth turns on its axis, the Moon, like other celestial objects, appears to rise in the east, move across the sky, and set in the west. But because of its daily eastward motion on the celestial sphere, it crosses the local meridian each day about 50 minutes later, on the average, than on the previous day. We could define this interval of 24^h50^m as the average length of an apparent lunar day.

Moonrise (and moonset) is similarly retarded from day to day. If the Moon did not move with respect to the sun, it would rise at nearly the same time from one day to the next. At moonrise, the Moon occupies some particular place on the celestial sphere. Approximately 24 hours later, the same place on the celestial sphere rises again, but the Moon in the meantime has moved off to the east, so moonrise does not occur until a little later. At the equator, the daily delay is the same as the Moon's delay in crossing the meridian. However, at other latitudes the Moon and stars rise obliquely to the horizon, rather than in a direction perpendicular to it. Consequently, the time required for the Earth to turn the sky westward through the angle representing the Moon's eastward motion is not necessarily the same as the daily delay in moonrise. The phenomenon of the harvest moon, discussed below, provides an excellent example. In the northern parts of the United States, the daily delay in moonrise can vary from a few minutes to well over an hour.

(e) The "Harvest Moon"

The *harvest moon* is the full moon that occurs nearest the date of the autumnal equinox. Because the Moon, when full, is opposite the sun in the sky, it must rise as the sun sets. When the sun is near the autumnal equinox, the full moon is near the vernal equinox, so at the time of the harvest moon the vernal equinox is rising with the full moon. When the vernal equinox is rising, the ecliptic makes its minimum angle with the horizon for an observer in intermediate northern latitudes.

Since the Moon's orbit lies within 5° of the ecliptic, it is evident that at the same hour on successive nights the Moon's apparent motion, being nearly parallel to the horizon, will not change the Moon's relation to the eastern horizon appreciably. In Figure 7.2 the full moon (position 1) is shown rising at sunset. At the same time on the next night it has moved about 12° along its orbit; but it is not very far below the horizon and will rise by moving along the line *AB*, parallel to the celestial equator. The Earth will not have to turn far to bring up the Moon on this second night. Thus, for several nights near full moon in late September or early October there will be bright moonlight in the early evening—a traditional aid to harvesters. The phenomenon of the harvest moon is most striking in northern latitudes.

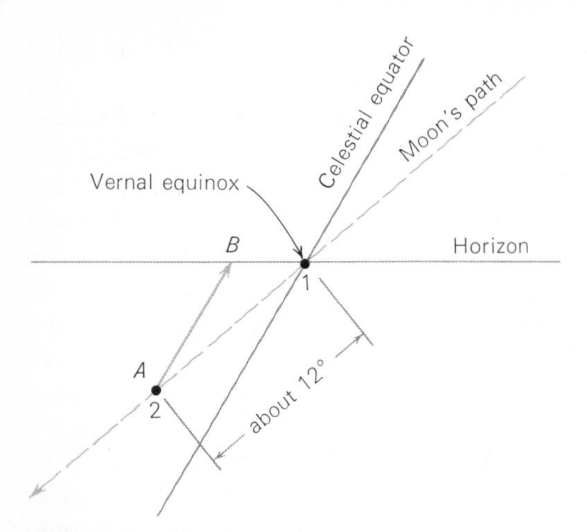

Figure 7.2 The harvest moon.

(f) The Progression of the Moon's Phases

The phases of the Moon (explained in Section 2.2d) were thoroughly understood by the ancients. The relation between any phase of the Moon and the Moon's corresponding position in the sky at any time of day should now be clear. Except at extreme northern or southern latitudes, one can easily tell where to look for the Moon in the sky from a knowledge of its phase.

In Figure 7.3 we imagine ourselves looking down upon the Earth and the Moon's orbit from the north. The Moon is shown in eight positions in its monthly circuit of the Earth. The sun is off to the right of the figure at a distance so great that its rays approach the Earth and all parts of the Moon's orbit along essentially parallel paths. The daylight sides of the Earth and Moon—the sides of those bodies turned toward the sun—are indicated. For each position of the Moon, its phase, that is, its appearance *as viewed from the Earth,* is shown just outside its orbit. Several observers are at various places on the Earth, *A*, *B*, *C*, and so on. The time of day, indicated for each observer, depends on the position of the sun in the sky with respect to his local meridian or, equivalently, on his position on the Earth with respect to the meridian where it is noon.

For person *A* it is 3:00 P.M. If he sees the Moon on the meridian, it must be in the waxing crescent phase (the Moon can be seen easily at noon when in this phase). If the Moon is in the waning crescent phase it is setting, for it lies on his western horizon. West is the direction away from which the

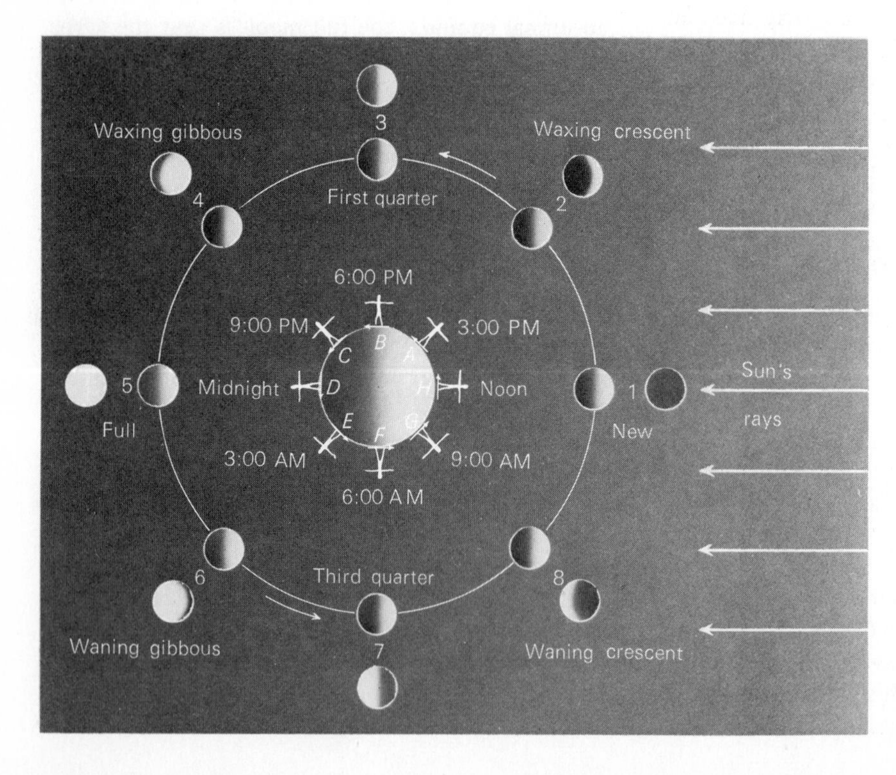

Figure 7.3 Phases of the Moon and the time of day. (The outer series of figures shows the Moon at various phases as seen in the sky from the Earth's surface.)

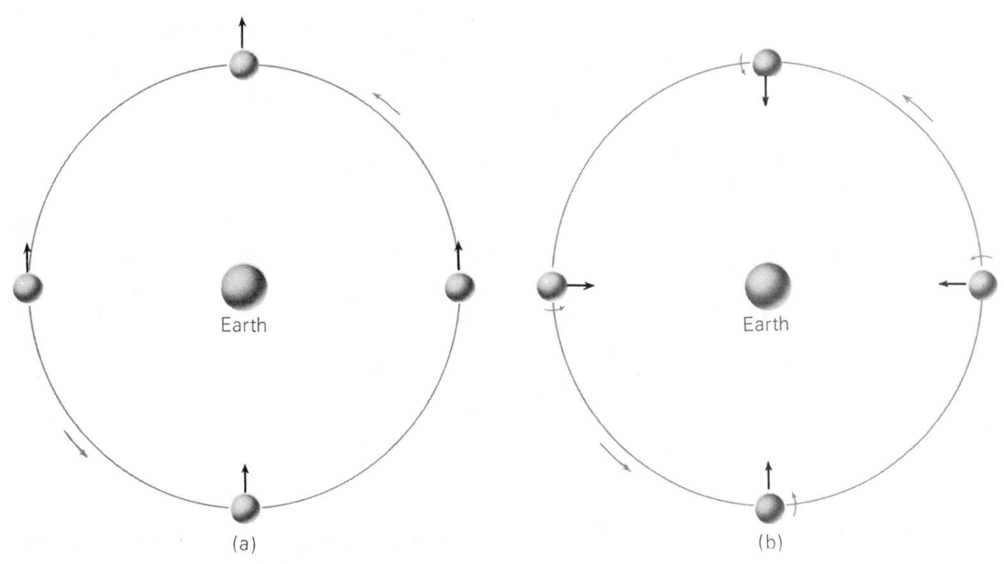

Figure 7.4 (a) If the Moon did not rotate, it would turn all its sides to our view. (b) Actually, it does rotate in the same period as it revolves, so we always see the same side. (Thus there is a "back side" or "far side," but certainly not a "dark side".)

turning Earth carries the observer, and his horizon lies in a plane tangent to the surface of the Earth at the point where he is standing. If the Moon is new it is in about the same direction as the sun in the western sky, and if it is at first quarter it is in the eastern sky. If the Moon is rising it must be in the waxing gibbous phase.

For person *B* it is 6:00 P.M. (Person *B* could be person *A* three hours later. During a period of even a full day, the Moon does not move enough for its phase to change appreciably.) For *B* the Moon is setting if new, and rising if full. If it is a waxing crescent or gibbous it is in the western or eastern sky, respectively. If it is in the first quarter phase, it appears on the meridian.

For person *D* it is midnight. If the Moon is full, it rose at sunset, and is now on the meridian. The first or third quarter Moon is just setting or rising, respectively. The waxing or waning gibbous Moons must appear in the western or eastern sky.

By studying Figure 7.3 we can tell where the Moon is in the sky at any time of day or night if we know its phase. For example, the full moon rises at sunset and sets at sunrise. The first quarter moon rises at noon and sets at midnight, and so on. We have ignored here the dependence of the Moon's position on the latitude of the observer. However, for most places on Earth the figure is a good enough approximation.

(g) The Rotation of the Moon

Even naked-eye observation is sufficient to determine that as the Moon goes about the Earth it keeps the same side toward the Earth. The same facial characteristics of the "man in the Moon" are always turned to our view. Although the Moon always presents the same side to us, it must not mislead us into thinking that it does not rotate on its axis. In Figure 7.4 the arrow on the Moon represents some lunar feature. If the Moon did not rotate, as in (a), we would see that feature part of the time, and part of the time we would see the other side of the Moon. Actually the Moon rotates on its axis with respect to the stars in the same period as it revolves about the Earth, and so always turns the same side toward us (b).

The coincidence of the periods of the Moon's rotation and revolution can hardly be accidental. It is believed to have come about as a result of the Earth's tidal forces on the Moon.

We often hear the back side of the Moon (the side we do not see) called the "dark side." Of course, the back side is dark no more frequently than the front side. Since the Moon rotates, the sun rises and sets on all sides of the Moon. The back side of the Moon is receiving full daylight at new moon; the dark side is then turned toward the Earth.

(h) The Size and Distance of the Moon

The Moon's distance from the Earth is only about 30 times the diameter of the Earth; consequently, the direction of the Moon differs slightly as seen from various places on the Earth. The astronomers of ancient Greece and Rome used this fact to determine geometrically the distance to the Moon, and modern astronomers with precision optical observations can do the same thing much better. However, all of these geometrical methods have recently been superceded by new techniques.

One of the most accurate ways of finding the distance to the Moon is by *radar*. The first successful radar contact with the Moon was achieved in 1946. In this technique, radio waves, focused into a beam by a powerful broadcasting antenna, are transmitted to the Moon; some of this energy is reflected back to the Earth. Since radio waves are a form of electromagnetic energy, they travel with the speed of light (Chapters 8 and 9); hence half the time between when the waves are transmitted and the echo is received, multiplied by the speed of light, gives the distance to the Moon. The time intervals involved can now be measured electronically to better than one-millionth of a second. By 1957 the Naval Research Laboratory had determined the Moon's distance by radar to within about 1 km.

More precise still is ranging by *laser* (Chapter 8). A laser is a device in which atoms are induced to emit light pulses that leave the device in a highly focused beam, in a very narrow range of frequency, and with the waves in each pulse accurately aligned with each other. One advantage of the laser beam is that it can be transmitted over large distances with high efficiency, and if it is reflected from an object, the round-trip travel time can be measured to one nanosecond (10^{-9}s).

Since late 1969 lasers have been attached to telescopes to send such light pulses to the Moon. Mirrors left on the Moon by Apollo astronauts and Soviet lunar probes reflect an extremely small fraction of this light back to Earth, where it is observed with the telescopes. In principle the laser technique can determine the distance to the Moon to within a few centimeters. At present this accuracy is not realized because the speed of light is known to a precision of only a few parts in 10 million (Chapter 11). Laser observations, however, can detect relative changes in the Moon's distance as it moves in its elliptical orbit, and can also check the effects of perturbations on the Moon.

The best determination to date gives for the distance from the center of the Earth to the center of the Moon the value 384,404 km, with an uncertainty of about 0.5 km.

The mean angular diameter of the Moon is 31′5″. From its angular size and its distance, the linear size of the Moon can easily be found. Because the method is the same as that applied to measure the diameters of planets and other astronomical objects that subtend a measurable angle, we shall explain the procedure in detail.

Notice in Figure 7.5 that because the Moon's angular size is relatively small its linear diameter is essentially a small arc of a circle, with the observer as center and with a radius equal to the Moon's distance. Obviously, the Moon's diameter is the same fraction of a complete circle as the angle subtended by the Moon is of 360°. A complete circle contains 1,296,000″ (there are 60″ per minute, 60′ per degree, and 360° in a circle). As seen from the center of the Earth, the Moon's mean angular diameter of

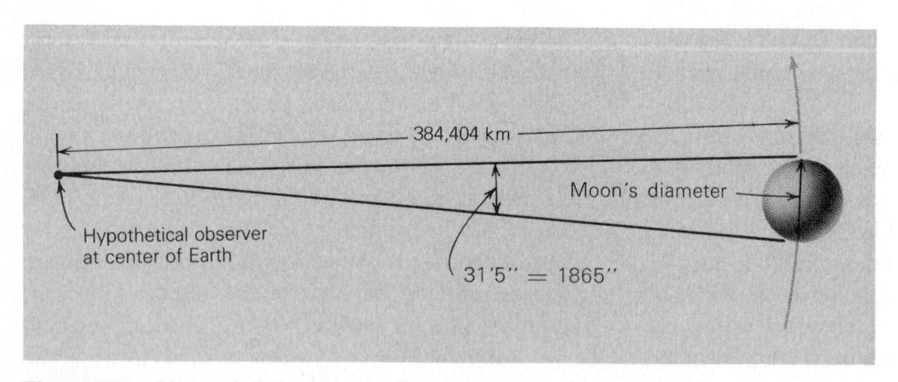

Figure 7.5 Measuring the Moon's diameter.

31′5″, or 1865″, is thus 1⁄695 of a circle. The Moon's diameter, therefore, is 1⁄695 of the circumference of a circle of radius 384,404 km. Since the circumference of a circle is 2π times its radius, we have

$$\text{diameter of Moon} = \frac{2\pi(384404)}{695} = 3475 \text{ km}.$$

This type of calculation can be generalized if we note that the distance along the arc of a circle of radius R subtended by 1″ is $2\pi R/1,296,000 = R/206,265$. Thus, if an object, say a planet, subtends an angle of α seconds, and has a distance D, its linear diameter d is given by

$$d = \frac{\alpha D}{206,265}.$$

Accurate calculations of the Moon's diameter give as a result 3475.9 km, with an uncertainty of a few hundredths of a kilometer. The Moon's diameter is a little over one quarter of the Earth's, and its volume is about 1⁄49 of the Earth's.

(i) The Mass of the Moon

The mass of the Moon is determined from an application of Kepler's laws. We saw in Section 4.3 that one body does not strictly revolve about another, but that the two bodies mutually revolve about their center of mass, the *barycenter*. The barycenter of the Earth-Moon system revolves annually in an elliptical orbit about the sun, while the Earth and Moon simultaneously revolve about the barycenter in a shorter period—the sidereal month. Careful observations of other planets (carried out today primarily by radar) show that the mean distance of the center of the Earth from the barycenter is 4672 km. Thus, the Earth and Moon jointly revolve about a point approximately 1707 km below the surface of the Earth.

The distances of two bodies from their barycenter are inversely proportional to their masses. The 4672 km distance of the Earth from the barycenter is 1⁄82.3 of the distance from the Earth to the Moon; hence the Moon is 81.3 times as far from the barycenter as the Earth and so is only 1⁄81.3 as massive. Since the Earth's mass is known to be 6.0×10^{21} tons, it follows that the mass of the Moon is 7.4×10^{19} tons.

(j) Tidal Friction and Evolution

One force that is believed to affect, over long periods, the mutual revolution of the Earth-Moon system is friction in the tides. As the Earth rotates, the tidal waters continually flow back and forth over each other, across ocean floors, and in and out over coastal shallows. The resulting friction within the water and between the water and the solid Earth draws a considerable amount of energy from the kinetic energy of rotation of the Earth and expends this energy in the form of heat. Even friction in the tidal distortions of the solid Earth and in atmospheric tides may play a role. The Earth, consequently, is slowing in its rate of rotation, and the day is gradually lengthening at the rate of about 0.002 s per day each century. The continuous dissipation of the rotational energy of the Earth is calculated to be approximately 2000 million horsepower.

Whereas the Earth slows down in its axial spin as it loses kinetic energy, the angular momentum of the Earth-Moon system must be conserved. According to a theory of tidal evolution worked out by Sir George Darwin (son of the naturalist), the Moon must slowly spiral outward away from the Earth, thus maintaining the total angular momentum of the Earth and the Moon. (But this is superimposed on a temporary *decrease* of the Moon's distance caused by a lowering of the eccentricity of the Earth's orbit.) As the Moon's distance from the Earth increases, its period of revolution must also increase and its orbital speed must decrease, in accord with Kepler's third law. The day and the month, in other words, will both lengthen. Eventually the day will catch up with the month in length, when both require about 47 of our present days.

7.2 SHADOWS AND ECLIPSES

Eclipses occur whenever any part of either the Earth or the Moon enters the shadow of the other. When the Moon's shadow strikes the Earth, people on Earth within that shadow see the sun covered at least partially by the Moon, that is, they witness a *solar eclipse*. When the Moon passes into the shadow of the Earth, people on the night side of the Earth see the Moon darken—a *lunar eclipse*.

A shadow is a region of space within which rays from a source of light are obstructed by an opaque body. Ordinarily, a shadow is not visible. Only when some opaque material, which will show the contrast between lighted and unlighted areas, inter-

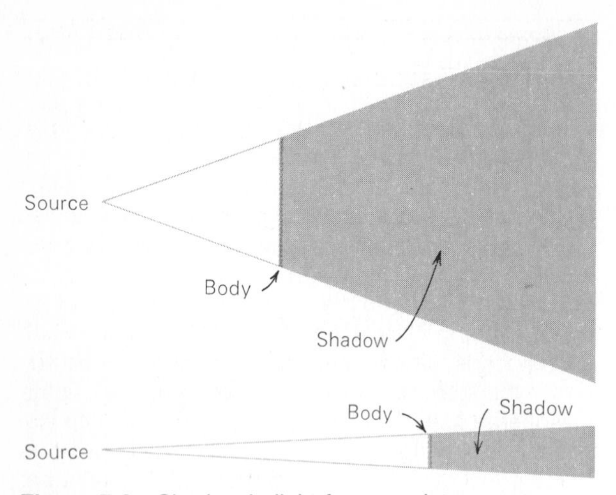

Figure 7.6 Shadow in light from a point source.

sects the shadow and the surrounding area does the shadow become visible.

(a) Shadow from a Point Source

If a source of illumination is a point source (Figure 7.6), the shadow cast by an opaque body has sharp boundaries. From a point inside the shadow, the light source is not visible; outside the shadow it is. The boundaries of the shadow diverge radially from the source. The more distant the source, the smaller is the angle of divergence. If the source is infinitely distant, the boundaries of the shadow are parallel.

(b) Shadow from an Extended Source

If a light source is not a point source but presents a finite angular size as seen from the opaque body, as is almost always the case, the shadow cast by the body is not limited to the inner part of the shadow, the *umbra*, where complete darkness prevails. From within the umbra, of course, no part of the light source is visible. At any point completely outside the shadow, there is no obscuration of light, and from such places the entire light source is visible. Between the umbra and the region of full light lies a space of partial illumination, within which the illumination ranges from complete darkness at the boundary of the umbra to full illumination at the outer boundary of the entire shadow. This transition region of the shadow is called the *penumbra*. From any point within it, a part, but not all, of the light source is visible.

As an illustration, consider the shadow cast by a spherical body, such as the Earth, the Moon, or a planet, in sunlight (Figure 7.7). It is obvious from the figure that because the umbra includes that region of space from which no part of the sun is visible, it must have the shape of a cone pointing away from the sun. The umbra of the Moon or a planet is sometimes called the *shadow cone* of that body. Everything on the night side of the Earth is within the umbra, or shadow cone, of the Earth.

The penumbra, on the other hand, is that region from any point within which only part of the sun is covered by the eclipsing body. It is clear from the figure that the penumbra has the shape of a truncated cone pointed *toward* the sun, and that it includes the umbra, as a reversed cone symmetrical about the same axis. The appearances of the sun from points A, B, C, and D in the shadow are shown in Figure 7.8.

The size of the umbra cast by an opaque sphere in sunlight depends on the size of the sphere and its

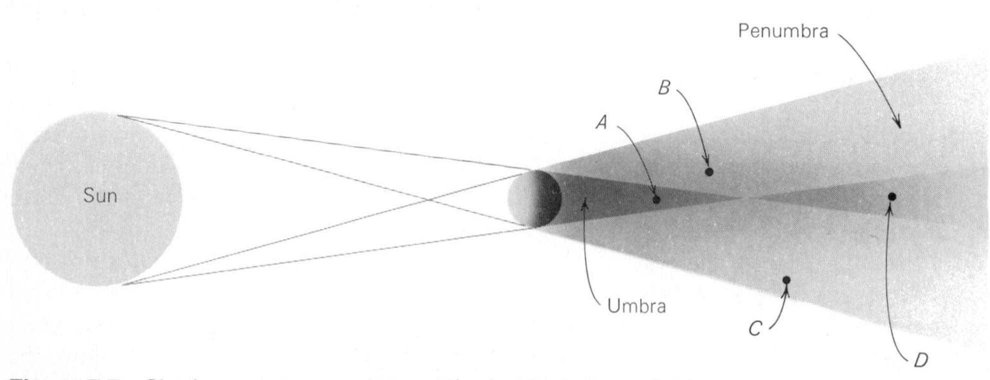

Figure 7.7 Shadow cast by an opaque spherical body in sunlight.

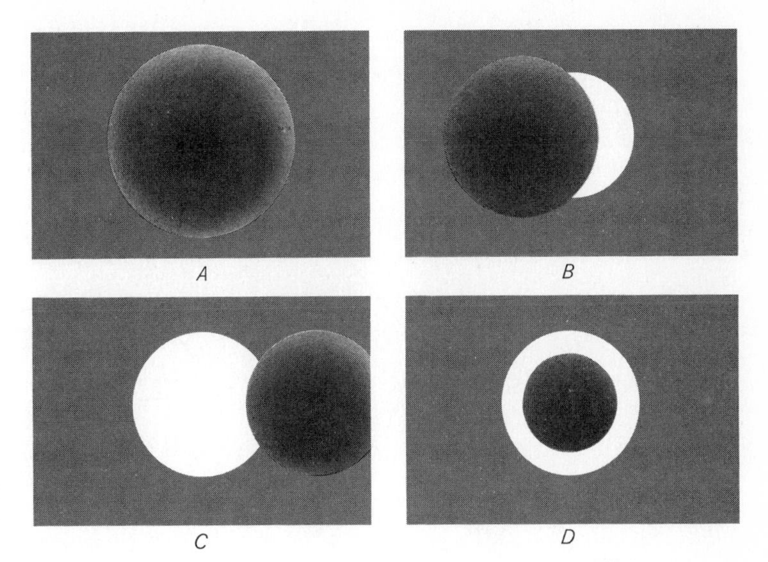

Figure 7.8 Appearance of the sun from points *A* to *D* in the shadow shown in Figure 7.7.

distance from the sun. The length of the umbra is directly proportional to the distance of the sphere from the sun. Figure 7.9 illustrates this relation.

(c) Eclipse Seasons

For the Moon to appear to cover the sun and thus to produce a solar eclipse, it must be in the same direction as the sun in the sky, that is, it must be at the *new* phase. For the Moon to enter the Earth's shadow and produce a lunar eclipse, it must be opposite the sun; that is, it must be at the *full* phase. Eclipses occur, therefore, only at new moon and at full moon. If the orbit of the Moon about the Earth lay exactly in the plane of the Earth's orbit about the sun—in the ecliptic—an eclipse of the sun would occur at every new moon and a lunar eclipse at every full moon. However, because the Moon's orbit is inclined at about 5° to the ecliptic, the new moon, in most cases, is not *exactly* in line with the sun, but is a little to the north or to the south of the sun in the sky. Similarly, the full moon usually passes a little south or north of the Earth's shadow.

However, if full or new moon occurs when the Moon is at or near one of the *nodes* of its orbit (where its orbit intercepts the ecliptic), an eclipse can occur. The line through the center of the Earth that connects the nodes of the Moon's orbit is called the *line of nodes*. If the direction of the sun lies along, or nearly along, the line of nodes, new or full moon occurs when the Moon is near a node, and an eclipse results. The situation is illustrated in Figure 7.10. The orientation of the Moon's orbit, and the line of nodes, nn', remains relatively fixed during a revolution of the Earth about the sun. There are, therefore, just two places in the Earth's orbit, points A and B, where the sun's direction lies along the line of nodes. It is only during the times in the year, roughly six months apart, when the Earth-sun line is approximately along the line of nodes that eclipses can occur. These times are called *eclipse seasons*.

Because of perturbations of the Moon's orbit, the line of nodes is gradually moving westward on the ecliptic, making one complete circuit in 18.6 years. Therefore, the eclipse seasons occur earlier each year by about 20 days; in 1986 the eclipse seasons were April and October.

Figure 7.9 Shadow lengths at various distances from the sun.

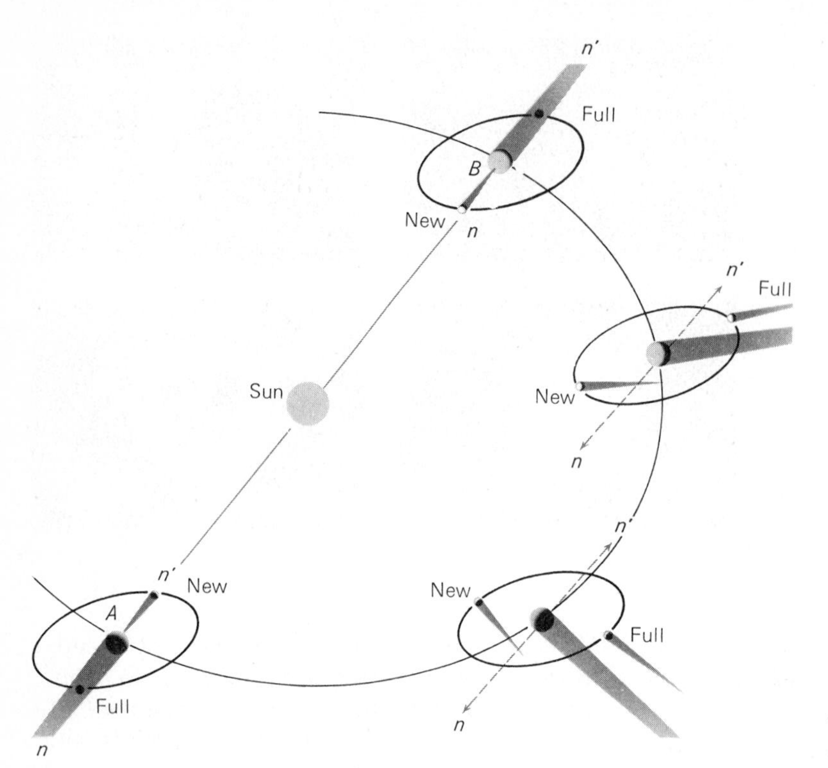

Figure 7.10 Eclipses occur only when the sun is along, or nearly along, the line of nodes.

(d) Geometry of a Lunar Eclipse

The geometry of lunar eclipses is shown in Figures 7.11 and 7.12. Unlike a solar eclipse, which is visible only in certain local areas on the Earth, a lunar eclipse is visible to everyone who can see the Moon, because its source of illumination is blocked off. Weather permitting, a lunar eclipse can be seen from the entire night side of the Earth, including those sections of the Earth that are carried into the

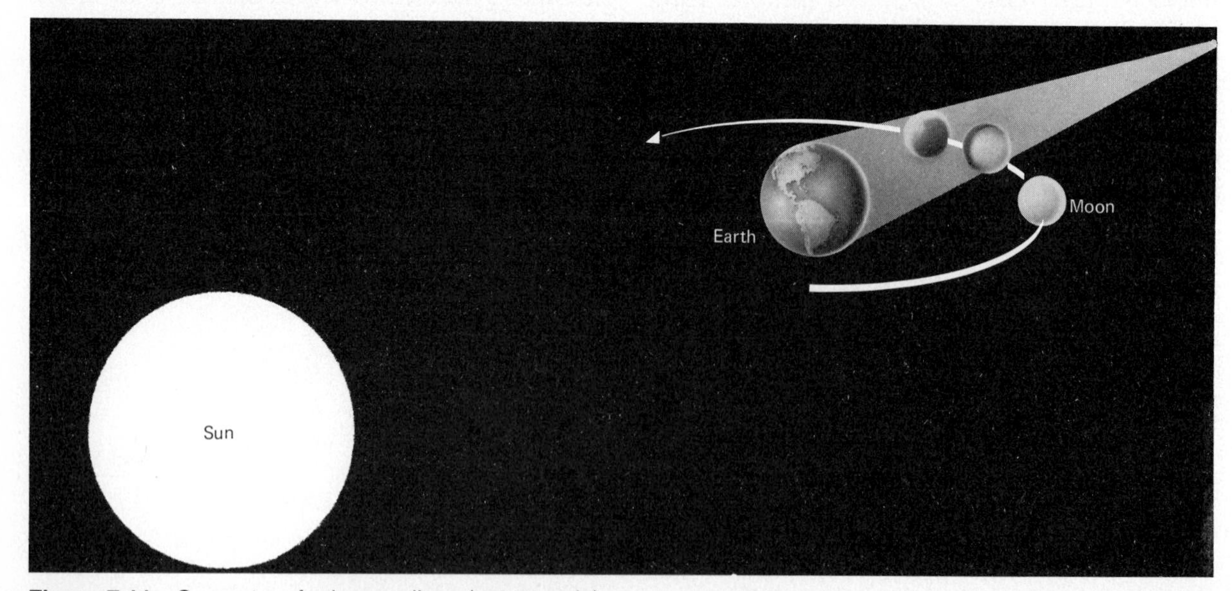

Figure 7.11 Geometry of a lunar eclipse (not to scale).

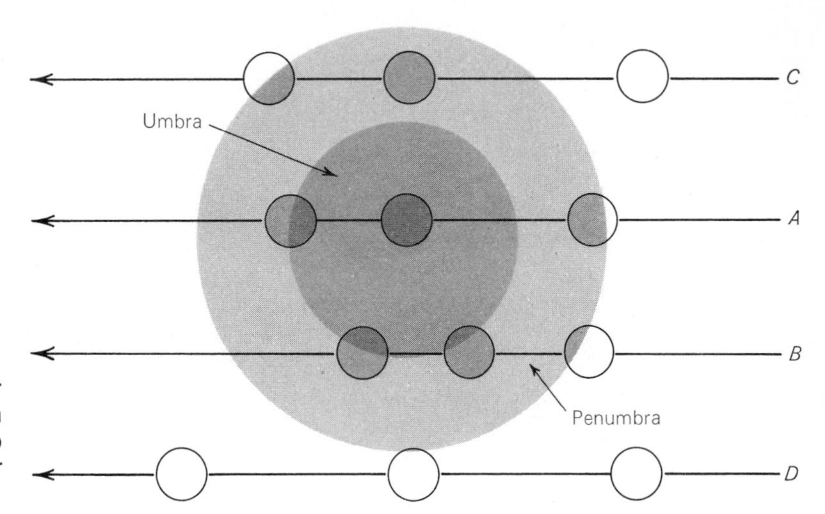

Figure 7.12 Different kinds of lunar eclipses. The penumbra pales from the full light of the outer boundary to the full dark of the umbra at the inner boundary.

Earth's umbra while the eclipse is in progress. Lunar eclipses, therefore, are observed far more frequently from a given place on Earth than are solar eclipses.

Figure 7.12 shows the cross section of the Earth's shadow at the Moon's distance from Earth. Since the cross-sectional diameter of a cone is proportional to the distance from the apex of the cone, the cross section of the Earth's shadow cone at the Moon's distance is in the same proportion to the size of the Earth as the Moon's distance from the end of the shadow is to the total length of the shadow. The umbra is thus found to be 9200 km in diameter at the Moon's distance. The value varies slightly from one eclipse to another because the Earth-Moon distance varies, and because the diameter of the umbra at the place where the Moon enters the shadow depends on the Moon's and the sun's distances at the time of the eclipse. The penumbra of the Earth's shadow averages about 16,000 km across at the Moon's distance.

In Figure 7.12 four of the many possible paths of the Moon through the Earth's shadow are shown. A total lunar eclipse occurs when the Moon passes completely into the umbra (path *A*). A *partial eclipse* occurs if only part of the Moon skims through the umbra (path *B*), and a *penumbral eclipse* occurs if the Moon passes through the penumbra, or partially through the penumbra, but does not come into contact with the umbra (paths *C* and *D*).

(e) Appearance of Lunar Eclipses

Penumbral eclipses usually go unnoticed even by astronomers. Only within about 1100 km of the umbra is the penumbra dark enough to produce a noticeable darkening on the Moon. However, the dimished illumination on the Moon's surface can be detected photometrically.

Every total or partial lunar eclipse must begin with a penumbral phase. About 20 minutes or so before the Moon reaches the shadow cone of the Earth, the side nearest the umbra begins to darken somewhat. At the moment called *first contact,* the limb of the Moon (the "edge" of its apparent disk in the sky) begins to dip into the umbra of the Earth. As the Moon moves farther and farther into the umbra, the curved shape of the Earth's shadow upon it is very apparent. In fact, Aristotle listed the round shape of the Earth's shadow as one of the earliest proofs of the fact that the Earth is spherical (Section 2.2).

If the eclipse is a partial one, the Moon never gets completely into the umbra of the Earth's shadow but passes on by, part of it remaining in the penumbra, where it still receives some sunlight. At *last contact* the Moon emerges from the umbra.

On the other hand, if the eclipse is a total one, at the instant of *second contact* the Moon is completely inside the umbra, and the total phase of the eclipse begins. Even when totally eclipsed, the Moon is still faintly visible, usually appearing a dull coppery red. Kepler explained this phenomenon in his treatise *Epitome.* The illumination on the eclipsed Moon is sunlight that has passed through the Earth's atmosphere and has been refracted by the air into the Earth's shadow (Figure 7.13).

The eclipse is darkest if the center of the lunar disk passes near the center of the umbra. The darkness of the lunar eclipse depends also upon weather

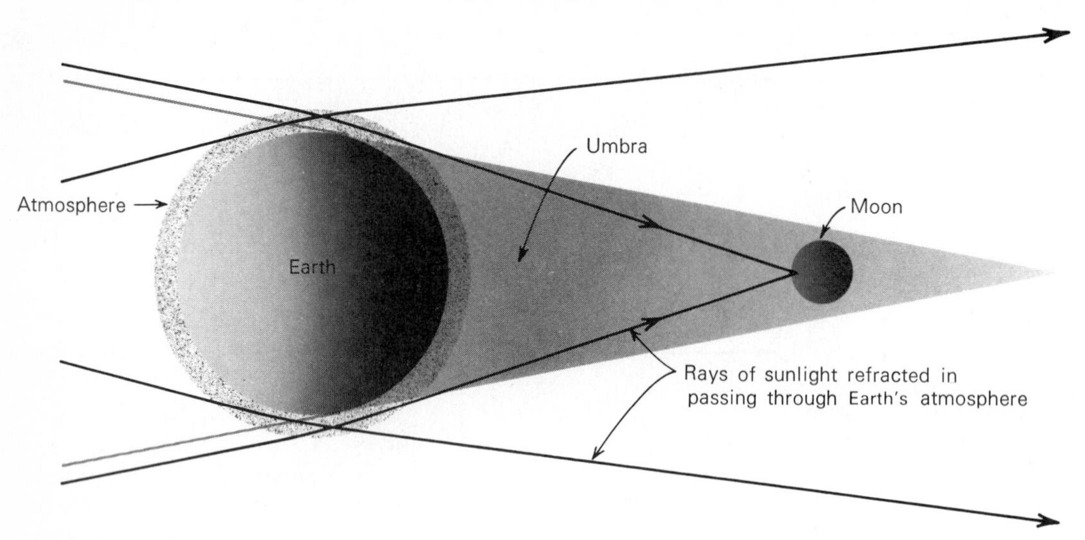

Figure 7.13 Illumination of the Moon during a total eclipse by sunlight refracted by the Earth's atmosphere into the Earth's shadow.

conditions around the terminator of the Earth. It is only here, on the line between day and night on the Earth, that sunlight passing through the atmosphere can be refracted into the shadow. Heavy cloudiness in that critical region will allow less light to pass through. The light striking the eclipsed Moon inside the umbra is reddish because red light of longer wavelengths penetrates through the long path of the Earth's atmosphere most easily.

Totality ends at *third contact,* when the Moon begins to leave the umbra. It passes through its partial phases to *last contact,* and finally emerges completely from the penumbra. The total duration of the eclipse depends on how closely the Moon's path approaches the axis of the shadow during the eclipse. The Moon's velocity with respect to the shadow is about 3400 km/hr; if it passes through the center of the shadow, therefore, about six hours will elapse from the time the Moon starts to enter the penumbra until it finally leaves it. The penumbral phases at the beginning and end of the eclipse last about one hour each, and each partial phase consumes at least one hour. The total phase can last as long as 1 hour 40 minutes if the eclipse is central.

7.3 ECLIPSES OF THE SUN

One of the coincidences of nature is that the two most prominent astronomical objects, the sun and the Moon, have so nearly the same apparent size in the sky. Although the sun is about 400 times as large in diameter as the Moon, it is also about 400 times farther away, so both objects subtend the same apparent angle of 0.5°.

The apparent or angular sizes of both the sun and the Moon vary slightly from time to time, as their respective distances from the Earth vary. The average angular diameter of the sun (as seen from the center of the Earth) is 31′59″, and the average angular diameter of the Moon is slightly less, 31′5″. However, the sun's apparent size can vary from the mean by about 1.7 percent and the Moon's by 7 percent. The maximum apparent size of the Moon is 33′16″, which is larger than the sun's apparent size, even at its largest. Therefore, if an eclipse of the sun occurs when the Moon is somewhat nearer than its average distance, the Moon can completely hide the sun, producing a *total solar eclipse.* In other words, a total eclipse of the sun occurs whenever the umbra of the Moon's shadow reaches the surface of the Earth.

(a) Geometry of a Total Solar Eclipse

The geometry of a total solar eclipse is illustrated in Figure 7.14. The Earth must be at a position in its orbit such that the direction of the sun is nearly along the line of nodes of the Moon's orbit. Fur-

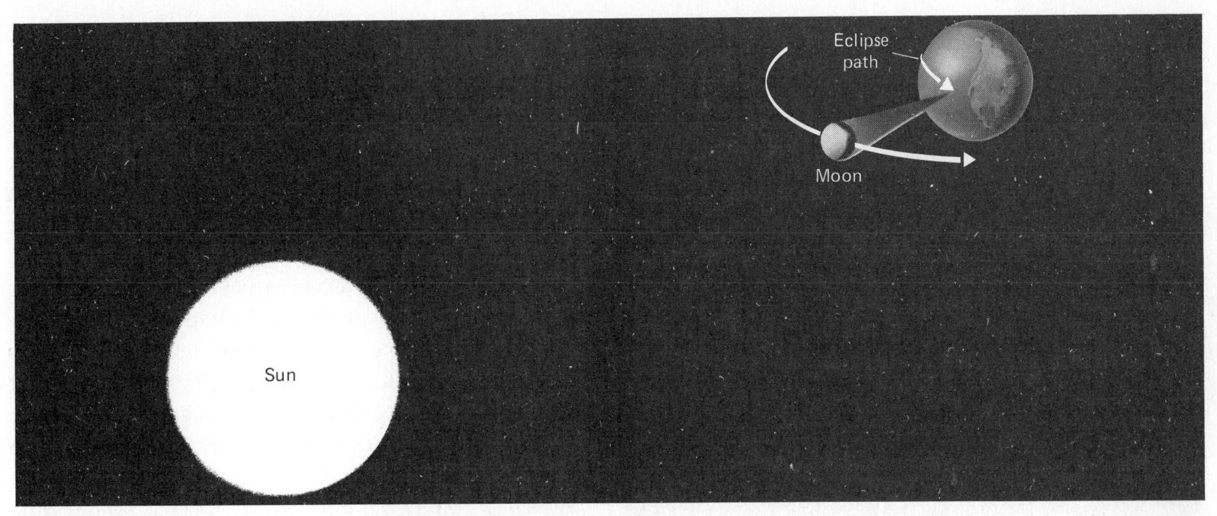

Figure 7.14 Geometry of a total solar eclipse (not to scale).

thermore, the Moon must be at a distance from the surface of the Earth that is less than the length of the umbra of the Moon's shadow. Then, at new moon, the Moon's umbra intersects the ground at a point on the Earth's surface. Anyone on the Earth within the small area covered by the Moon's umbra will not see the sun and will witness a total eclipse. The Moon's penumbra, on the other hand, covers a larger area of the Earth's surface. Any person within the penumbra will see part but not all of the sun eclipsed by the Moon—a partial solar eclipse. The regions of total and partial eclipse correspond to points *A* and *B* in Figures 7.7 and 7.8.

As the Moon moves eastward in its orbit with respect to the sun at about 3400 km/hr, its shadow sweeps eastward across the Earth at the same speed. The Earth, however, is rotating eastward at the same time, so the speed of the shadow with respect to a particular place on Earth is less than 3400 km/hr. At the equator, where the rotation of the Earth carries places eastward at about 1670 km/hr, the shadow moves relative to the Earth with a speed of about 1730 km/hr. In higher latitudes the speed is greater. In any case, the tip of the truncated cone of the umbra of the Moon's shadow sweeps along a thin band across the surface of the Earth, and the total solar eclipse is observed successively along this band (refer to Figure 7.14). This path across the Earth within which a total solar eclipse is visible (weather permitting) is called the *path of totality*. Within a zone about 3000 km on either side of the eclipse path, a partial solar eclipse is visible—the observer, inside this limit, being located in the penumbra of the shadow.

Because the Moon's umbra just barely reaches the Earth, the width of the eclipse path, within which a total eclipse can be seen, is very small. Under the most favorable conditions, the diameter of the shadow cone is 269 km wide at the Earth's surface. This would be the width of the path if the shadow struck the Earth perpendicularly near the Earth's equator. At far northern or southern latitudes, because the Moon's shadow falls obliquely on the ground, it can cover a path somewhat more than 269 km wide.

It does not take long for the Moon's umbra to sweep past a given point on Earth. The duration of totality may be only a brief instant. It can never exceed about 7½ minutes.

(b) Appearance of a Total Eclipse

A total solar eclipse is one of the most spectacular of natural phenomena. If you are anywhere near the path of totality of solar eclipse, it is well worth your while to travel into the eclipse path so that you may witness this rare and impressive event.

The very beginning of a solar eclipse is the *first contact*, when the Moon just begins to silhouette itself against the edge of the sun's disk. The *partial phase* follows during which more and more of the sun is covered by the Moon. *Second contact* usually occurs more than an hour after first contact, at the instant when the sun becomes completely hidden

Figure 7.15 Time-lapse photograph showing the Moon passing in front of the sun during a total solar eclipse. (American Museum of Natural History)

behind the Moon. In the few minutes immediately before second contact (the beginning of totality) the sky noticeably darkens; some flowers close up, and wildlife, especially birds, exhibit nocturnal behavior. One of us saw bats appear during totality of the eclipse of March 7, 1970, as he watched it from a Mexican desert. Because the diminished light that reaches the Earth must come solely from the edge of the sun's disk, and consequently from the higher layers in its atmosphere (see Chapter 28), the sky and landscape take on strange colors. In the last instant before totality, the only parts of the sun that are visible are those that shine through the lower valleys in the Moon's irregular profile and line up along the periphery of the advancing edge of the Moon—a phenomenon called *Baily's beads.* The final flash of sunlight through a lunar valley produces a brilliant flare on the disappearing crescent of the sun—the *diamond ring* effect. During totality, the sky is quite dark (like late twilight). The brighter planets are visible, and sometimes the brighter stars as well.

As Baily's beads disappear and the bright disk of the sun becomes entirely hidden behind the Moon, the *corona* flashes into view. The corona is the sun's outer tenuous atmosphere, consisting of

Figure 7.16 The solar corona is easily seen in this photo of the total solar eclipse of 7 March 1970, near a time of maximum sunspot activity. A radially symmetric, neutral-density filter in the focal plane of the camera was used to compensate for the increase in brightness of the corona near the limb of the sun. Photograph by Gordon Newkirk, Jr. (High Altitude Observatory, Boulder, Colorado, a Division of the National Center for Atmospheric Research. NCAR is operated by the University Corporation for Atmospheric Research under sponsorship of the National Science Foundation.)

Figure 7.17 Total solar eclipse of 30 June 1973, near a time of minimum sunspot activity. A radially graded filter was used, as in the photograph in Figure 7.16, with a technique developed by Gordon Newkirk, Jr. (High Altitude Observatory, a Division of the National Center for Atmospheric Research. NCAR is operated by the University Corporation for Atmospheric Research under sponsorship of the National Science Foundation.)

sparse gases that extend for millions of miles in all directions from the apparent surface of the sun. It is ordinarily not visible because the light of the corona is feeble compared to that from the underlying layers of the sun that radiate most of the solar energy into space. Only when the brilliant glare from the sun's visible disk is blotted out by the Moon during the total eclipse is the pearly white corona, the sun's outer extension, visible. (It is, however, possible to photograph the inner, brighter, part of the corona with an instrument called a *coronagraph,* a telescope in which a black disk in the telescope's focal plane produces an artificial eclipse, enabling the brightest parts of the corona to be studied at any time.)

Also, during a total solar eclipse, the chromosphere can be observed—the layer of gases just above the sun's visible surface. Prominences, great jets of gas extending above the sun's surface, are often seen. These outer parts of the sun's atmosphere are discussed more completely in Chapter 28.

The total phase of the eclipse ends, as abruptly as it began, with *third contact,* when the Moon begins to uncover the sun. Gradually the partial phases of the eclipse repeat themselves, in reverse order. At *last contact* the Moon has completely uncovered the sun.

In addition to being inspiring to watch, total eclipses of the sun have considerable astronomical value. Many data are obtained during eclipses that are otherwise not accessible. For example, during an eclipse we can determine the exact relative positions of the sun and Moon by timing the instants of the four contacts. We can take direct photographs and make spectrographic observations of the sun's outer atmosphere and prominences. We can measure the light and heat emitted by the corona. We can determine how meteorological conditions are affected by solar eclipses and can learn something about the light-scattering properties of the Earth's atmosphere.

(c) Annular and Partial Eclipses

More than half the time the Moon does not appear large enough in the sky to cover the sun completely, which means that the umbra of its shadow does not reach all the way to the surface of the Earth. When the Moon's shadow cone does not reach the Earth, a total eclipse is not possible. If an eclipse occurs under these circumstances, the Moon appears silhouetted against the sun, with a bright ring of sunlight surrounding it (Figure 7.18). Such eclipses are called *annular.* Annular eclipses are of relatively low interest to astronomer or layperson alike, since the

Figure 7.18 An annular eclipse of the sun. Bright spots are sunlight streaming through lunar valleys. (Lick Observatory)

Moon does not block enough of the light to reveal the sun's spectacular chromosphere and corona.

(d) How to Observe an Eclipse

The progress of an eclipse can be observed safely by holding a card with a small (1 mm) hole punched in it several feet above a white surface, such as a concrete sidewalk. The hole in the cardboard produces a pinhole camera image of the sun (Figure 7.19).

Although there are safe filters through which one can safely look at the sun directly, people are reported to have suffered permanent eye damage by looking at the sun through improper filters (or no filter at all!). In particular, neutral-density photographic filters are not safe, for they transmit infrared radiation that can cause severe damage to the retina.

Common sense (and pain) prevents most of us from looking at the sun directly on an ordinary day for more than a brief glance. Of course there is nothing about an eclipse that makes the radiation from the sun more dangerous than it is any other time; on the contrary, we receive less radiation from the sun when it is partly hidden by the Moon. It is *never* safe, however, to look steadily at the sun when it is still in *partial* eclipse; even the thin crescent of sunlight visible a few minutes before totality has a surface brightness great enough to burn and permanently destroy part of the retina. Unless you have a filter prepared especially for viewing the sun, it is best to watch the partial phases with a pinhole camera device, as described above.

Figure 7.19 How to watch a solar eclipse safely.

It is *perfectly safe*, however, to look at the sun directly when it is *totally eclipsed*, even through binoculars or telescopes. Unfortunately, unnecessary panic has often been created by well-meaning but uninformed public officials acting with the best intentions. One of us has witnessed two marvelous total eclipses in Australia, during which townspeople held newspapers over their heads for protection, and schoolchildren cowered indoors, with their heads under their desks. What a cheat to those people to have missed what would have been one of the most memorable experiences of their lifetimes! During totality, by all means look at the sun.

Nor should you be terrified of accidently catching a glimpse of the sun outside totality. How many times have you glanced at the sun on ordinary days

Figure 7.20 The shadow of Jupiter's satellite Io falling on the planet. Within the shadow a Jovian observer would witness a total solar eclipse. (NASA/JPL)

while driving a car or playing ball or tennis? Common sense made you look away at once. Do the same if you inadvertently glimpse the sun directly while it is partially eclipsed.

7.4 PHENOMENA RELATED TO ECLIPSES

Eclipses are not restricted to the Earth-Moon system; they take place on other planets in the solar system. Also common are situations in which one body passes in front of another, a circumstance that has been used extensively by astronomers to investigate planets, stars, and even distant quasars. These configurations are variously called *eclipses*, *occultations*, and *transits*.

(a) Eclipses in the Planetary System

Eclipses of the sun are visible from any planet with satellites. In the Jovian and Saturnian systems, with their many satellites, such eclipses are much more common than on Earth. Most of these are annular eclipses, but the larger Jovian and Saturnian satellites produce total solar eclipses. Figure 7.20 shows the umbra of the shadow cast on Jupiter by its innermost large satellite, Io; within this shadow a total solar eclipse is in progress. The Viking 1 lander photographed a solar eclipse from the Martian surface, produced when the shadow of Phobos fell on the spacecraft.

The equivalent of eclipses of the Moon also take place elsewhere in the solar system, whenever a satellite passes into the shadow cast by its planet. In the Jupiter system, the three inner Galilean satellites (as well as the four small satellites still closer to the planet) undergo an eclipse once in each revolution around the planet. As we will see in Chapter 18, studies of Io carried out during eclipse, when the sunlight is blocked and the surface cools, clearly reveal the presence of hot volcanic sources on Io. Callisto, the outer Galilean satellite, experiences eclipses when its orbit is properly aligned, but much of the time it passes above or below the shadow of Jupiter, just as the Moon usually misses the shadow of the Earth.

A third phenomenon that can occur in systems with multiple satellites is the eclipse of one satellite by another. While very rare, such mutual eclipses do take place in the Jovian system, when one Galilean satellite passes through the shadow of another.

(b) Lunar Occultations

The Moon often passes between the Earth and a star; the phenomenon is called a lunar *occultation*. The stars are so remote that the shadow of the Moon cast in the light of a star is extremely long and sensibly cylindrical. Because a star is virtually a point source, there is no penumbra. During an occultation, a star suddenly disappears as the eastern limb of the Moon crosses the line between the star and observer. If the Moon is at a phase between new and full, the eastern limb will not be illuminated and the star may appear to vanish mysteriously as the dark edge of the Moon covers it. Because the Moon moves through an angle about equal to its own diameter every hour, the longest time that a lunar occultation can last is about one hour. It can have a much shorter duration if the occultation is not central. Geometrically, occultations are equivalent to total solar eclipses, except that they are total eclipses of stars other than the sun.

The sudden disappearance of a star behind the limb of the Moon during an occultation is evidence that the Moon has no appreciable atmosphere. If there were one, the star would fade gradually as the Moon's limb approached it, because the starlight would traverse a long path of the lunar atmosphere. Occultations also demonstrate the extremely small angular sizes of the stars (owing to their great distances). If a star had an appreciable angular size, it would require a perceptible time to disappear behind the Moon, as is true during the partial phases of a total solar eclipse. Actually, the partial phases of occultations have been measured photoelectrically, but they are extremely brief, less than a few hundredths of a second. The angular sizes of stars cannot be observed directly in a telescope, but they can often be determined by various techniques (see Chapter 25). It has been possible to observe that stars of large computed angular size require longer to disappear behind the Moon than those of small angular size. Also observations of occultations of celestial radio sources have been useful in detecting the accurate positions and angular sizes of those objects.

Occultations of the brighter stars are listed in advance in various astronomical publications. The times and durations of the occultations and the places on Earth from which they are visible are given. Also listed are the comparatively rare occultations of planets by the Moon, and of stars by planets.

(c) Other Occultations

The term *occultation* is used generally to describe a situation in which one object passes in front of another, while an *eclipse* takes place when one object moves into the shadow of another. Thus, technically, a solar eclipse is an *occultation* of the sun by the Moon. Occultations of stars by solar-system objects have become a powerful tool for investigating planets, satellites, and even small asteroids.

Occultations of stars by planets are much rarer than their occultation by the Moon, because the angular size of planets is so much smaller. Typically, an occultation of a bright star by a planet takes place only once every few decades; for example, the occultation of Regulus, the brightest star in Leo, by Venus on July 7, 1959. However, with large telescopes, occultations of fainter stars can be used to good purpose, and today several occultations per year are observed for Uranus and Neptune, primarily to investigate their ring systems.

Occultations of faint rings are especially interesting phenomena because the technique is able to reveal fine detail that would otherwise require a spacecraft visit. As the ring passes in front of the star, the apparent brightness of the star fluctuates, tracing out the profile of absorbing ring material. As we will see in Chapter 18, effective resolutions of as small as a few kilometers are possible. The rings of both Uranus and Neptune were discovered by occultation observations.

Occultations of stars by planets are also useful for probing planetary atmospheres. Before spacecraft investigated them directly, much of what we knew about the vertical profiles of the atmospheres of Jupiter, Uranus, and Neptune were derived from detailed mathematical analysis of the brightness fluctuations of stars as their light passed through the successive layers of the planetary atmosphere. Atmospheric studies require brighter stars than are needed to trace out planetary rings, however, and suitable occultations occur only once every few years.

Asteroids, or minor planets, also occasionally occult stars. Careful timing of the duration of such occultations provides our most accurate measures of the diameters of asteroids, as we will see in Chapter 19.

Even more powerful than these naturally occurring occultations are the targeted occultations of spacecraft. Since the transmission of the radio signal from an interplanetary spacecraft is precisely controlled, the interaction of that radio signal with a planet or ring system provides more information than the occultation of light from a distant star. The occultation determination of the composition and surface pressure of the atmosphere of Mars carried out by the Mariner 4 spacecraft was an essential step toward the design of the Viking Mars landers. Much of what we know of the structure of the atmospheres of the outer planets comes from occultations of the Pioneer and Voyager spacecraft. We will learn more about the results of these measurements in Chapters 17 and 18.

(d) Transits

In an *occultation* the more distant object is smaller (in apparent size) than the object that passes in front of it. If the smaller object is in the foreground, the event is called a *transit*. The most familiar transits are those of the planets Mercury and Venus across the sun. On the rare occasions when either of these planets lines up directly between Earth and sun, we can see its slow passage across the solar disk, looking rather like a black sunspot. On the average, there are 13 transits of Mercury per century. Transits of Venus are more rare; the last took place in 1874 and 1888, and the next two will be on June 8, 2004, and June 6, 2012.

On almost any night at least one transit can be observed in the Jupiter system, where the Galilean satellites regularly pass in front of the planetary disk. And perhaps one should call an annular eclipse of the sun a transit of the Moon across the sun.

(e) Stellar "Eclipses"

Historically, some of the most important eclipse phenomena to astronomers have been those of *eclipsing binary stars*. A binary system consists of two stars that revolve about each other. Such stars are very common. When the plane of mutual revolution of such a double star happens to be oriented so that we see it nearly edge on, each star periodically passes partially or entirely behind the other. Thus we have an alternating sequence of stellar occultations and transits. Although the terminology is confusing, astronomers have traditionally called these occultations and transits *eclipses*.

While one star partially blocks the light from the other, the combined light received from the system is diminished. The most famous eclipsing binary star is the bright star Algol in the constellation of Perseus. Algol consists of two stars revolving around each other, a small bright one and a large faint one. About every 2 days and 21 hours, the bright star is occulted by the large dark companion, and the apparent brightness of Algol drops down to less than half normal—a change readily observable by the naked eye. Half a cycle later, when the bright star transits the faint one, the light hardly changes at all. A study of the light variation of such binary systems gives much information about the sizes and masses of their member stars, as we will see in Chapter 25.

EXERCISES

1. Consider a telescope with a resolution of 1 arcsecond. Determine the linear resolution (in kilometers) that is obtained with this telescope for each of the planets when it is closest to the Earth. Make the same calculation for the farthest distance of each planet from the Earth. (You can ignore the eccentricities of the Earth and planets for this calculation.)
2. The Hubble Space Telescope has a resolution of approximately 0.1 arcsecond. What is its linear resolution on Mars, Jupiter, and Saturn?
3. The Hubble Space Telescope is to be used to make a map of the asteroid Ceres, which has a diameter of about 950 km and a minimum distance from the Earth of about 1.7 AU. What is the number of resolution elements across the diameter of Ceres that can be obtained? What is the number of separate points (pixels) that can be distinguished on the disk of Ceres at any one time with this telescope?
4. When earthshine is brightest on the Moon, what

must be the phase of the Earth, as seen from the Moon?

5. If the Moon revolved from east to west rather than from west to east, would a synodic month be longer or shorter than a sidereal month? Why?

6. What is the phase of the Moon if (a) it rises at 3:00 P.M.? (b) it is on the meridian at 7:00 A.M.? (c) it sets at 10:00 A.M.?

7. What time does (a) the first quarter moon cross the meridian? (b) the third quarter moon set? (c) the new moon rise?

8. Describe the phases of the Earth as seen from the Moon. At what phase of the Moon (as seen from the Earth) would the Earth (as seen from the Moon) be a waning gibbous?

9. Suppose you lived in the crater Copernicus on the Moon. (a) How often would the sun rise? (b) How often would the Earth set? (c) Over what fraction of the time would you be able to see the stars?

10. The mean distance from the Earth to the Moon is 384,404 km, and light travels at 299,793 km/s. How long do radar waves take to make a round trip to the Moon?
Answer: 2.56 seconds

11. Suppose the Moon were exactly 386,000 km from the Earth and that the Earth's distance from the barycenter turned out to be 96,500 km. How much more massive than the Moon would the Earth be?

12. Sketch the shadow of a pencil cast by an electric lamp. Indicate the umbra and penumbra.

13. What is the angle at the apex of the shadow cone of the Earth? (*Hint:* What is the angular size of the sun?)

14. What are the relative lengths of the shadow cones of the Earth and the Moon? Assume that both objects are exactly the same distance from the sun. (Illustrate and show the method of your calculation, rather than looking up the figures in the chapter.)

15. Describe what an observer at the crater Copernicus on the Moon would observe during what would be a total solar eclipse as viewed from the Earth.

16. Does the Moon enter the shadow of the Earth from the east or west side? Explain why.

17. Describe the phenomenon observed by a spectator on the Moon while the Moon is eclipsed.

18. If the penumbra of the Earth's shadow is 16,000 km across, and if the Moon moves 3400 km/hr with respect to the shadow, why does it take six hours instead of only five hours to get completely through the penumbra?

19. Occultations of stars are used to measure the sizes of asteroids. Consider the asteroid Ceres passing in front of a star at an angular speed of 0.01 arcseconds per second of time. Suppose the maximum time the star is occulted is 65 seconds. What is the angular diameter of Ceres?

20. At the time of the occultation described in the previous exercise, Ceres is at a distance of 2.0 AU from the Earth. What is its linear diameter, in kilometers?

21. Suppose that an occultation of a star by an asteroid is seen by only one observer, who carefully measures the duration of the event. Using the procedure described in the previous two exercises, the observer calculates a diameter for the asteroid. Could the true diameter be substantially larger than this estimate? Substantially smaller? What are the main sources of uncertainty in such a determination of diameter from a single occultation observation?

22. Occultations of stars by the rings of Uranus have yielded resolutions of 10 km in determining ring structure. What would be the angular resolution (in arcseconds) that a space telescope would have to achieve to obtain equal resolution from Earth orbit? How close to Uranus would a spacecraft have to come to obtain equal resolution with a camera having angular resolution of 2 arcseconds?

Max Planck (1858–1947), German physicist, established the quantized nature of light and other electromagnetic radiation. The multicentered Max-Planck Institut of West Germany is named in tribute to his contributions. (American Institute of Physics, Niels Bohr Library)

8

ENERGY FROM SPACE: THE ELECTROMAGNETIC SPECTRUM

The Earth is constantly exposed to energy of various forms from space. In this chapter we discuss the nature and properties of some of this cosmic energy and how it interacts with matter; in the next two chapters we shall describe the techniques and instruments we use to observe it. Among these various forms of energy are:

1. *Electromagnetic radiation,* if we include that which we receive from the sun, is by far the most important kind of energy reaching the Earth; most of this chapter and the next two are concerned with it.
2. *Cosmic rays* are charged *particles,* mostly the nuclei of atoms, that strike the molecules of the Earth's upper atmosphere, producing tremendous numbers of secondary subatomic particles that rain down to the Earth's surface. The total energy we receive in the form of cosmic rays exceeds that which we get from starlight. We discuss these particles near the end of this chapter.
3. *Neutrinos* are produced by the decay of certain subatomic particles and also in many of the nuclear reactions that keep stars shining. Until 1980 they were generally thought to have no mass, in which case they must travel with the speed of light (Chapter 11). Several recent experiments, however, have yielded some evidence that suggests neutrinos may have very small masses, and if so they cannot have speeds quite as great as that of light. In any case, they pass through great amounts of matter without influencing it in the slightest and so are very difficult to detect. Nevertheless, the universe must be completely bathed in the radiation of neutrinos. We discuss them further in Chapter 29.
4. Einstein's general relativity theory predicts that motions of matter must generate *gravitational waves.* They must be everywhere around us, but gravitational waves are extraordinarily weak and enormously difficult to detect. We shall describe them, and our efforts to detect them, in Chapter 33.

8.1 ELECTROMAGNETIC RADIATION

We expect that astronomers will continue to investigate astronomical objects by means of electromagnetic radiation for many decades to come. However, the overwhelming majority of all the radiation from space that will be observed throughout our lifetimes is already in space on its way to us. What is the nature of this energy approaching us with the speed of light from all directions, waiting to be sampled by our telescopes for the century to come—and beyond? What are the secrets it holds, and what are

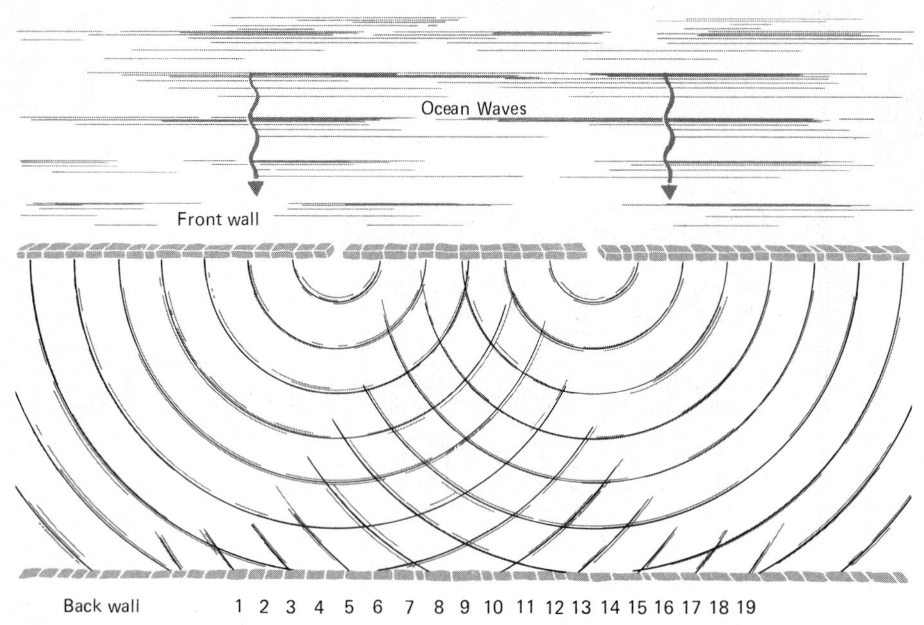

Figure 8.1 Interference of water waves passing through two openings.

the revelations it will give us about those objects it left years, centuries, even thousands of millions of years ago?

Light is the most familiar form of electromagnetic energy. Actually, visible light constitutes only a small part of a wide range of different kinds of electromagnetic radiation, which differ from each other only in *wavelength* but are called by different names: gamma rays, X-rays, ultraviolet, light, infrared, and radio waves. By its omnipresence, light is familiar to us all. Yet to describe its nature, or to represent it with pictorial conceptions, is extremely difficult. The best we can do is to describe its properties with mathematical models. Most of the characteristics of electromagnetic radiation can be described adequately if it is represented as energy propagated in waves, although no medium is required to transmit the waves. Electromagnetic radiation is so named because it is the propagation of a disturbance in an electromagnetic field, which could be caused, say, by the oscillatory motion of a charge. All forms of electromagnetic radiation must travel (in a vacuum) with the same speed, denoted c. We shall have more to say of this in Chapter 11.

Any wave motion can be characterized by a *wavelength*. Ocean waves provide an analogy. The wavelength is simply the distance separating successive wave crests. Various forms of electromagnetic energy differ from each other only in their wavelengths. Those with the longest waves, ranging up to many kilometers in length, are called *radio waves*. Forms of electromagnetic energy of successively shorter wavelengths are called, respectively, infrared radiation, light, ultraviolet radiation, X-rays, and finally the very short-wave gamma rays. Long infrared waves are sometimes referred to as radiant heat. All these forms of radiation are the same basic kind of energy and could be thought of as different kinds of light. Here, however, we shall reserve the term "light" to describe those wavelengths of electromagnetic radiation that, by their action upon the organs of vision, stimulate sight.

(a) Wave Motion: Interference and Diffraction

If you stand in the water at the beach, just out beyond the breakers, you can experience the waves passing by you, with the water getting deeper as each crest, and shallower as each trough, moves in. Similarly, as light passes a point, it varies in intensity (as the ocean water does in depth); the waves of light pass by so rapidly, however—some thousand million million per second—that you are unaware of the wavelike nature of light.

To explain one of the many experiments that demonstrate the wave properties of light, it will be useful to begin with an analogy. Imagine ocean

Figure 8.2 Interference fringes of light.

waves approaching a wall, labeled "front wall" in Figure 8.1. Let us suppose there are two openings, or gates, in that front wall. The waves of water are stopped, or reflected back, at every point along that front wall, except where it can flow through the open gates. Now, as the waves of water flow through the gates, they spread out into semicircular waves, radiating from each gate just as the ripples in a pond radiate from the place where a pebble is dropped. But the waves flowing through the two different openings intermingle; where two crests happen to come together, the water is especially deep, and where two troughs happen to combine, it is especially shallow.

Now consider the heights of the water on a second wall, behind the first one, the "back wall" in Figure 8.1. At points 1,3,5,7,9, . . . , along the wall, wave crests from the two gates always arrive at the same time. Between the successive arrivals of two crests, wave troughs from the two gates arrive together. Thus at those odd-numbered points the water alternately is very high and very low, as the successive crests and troughs of the waves from the two gates reinforce each other. At intermediate points, however, 2,4,6,8, . . . , the wave crest from one gate always meets a trough from the other, so the two sets of waves cancel each other.

If you have followed the preceding discussion, you will realize that if you were standing at the odd-numbered points (in the figure) you would experience large-amplitude waves striking you (*amplitude* simply means the height of the waves compared to the midpoint between crest and trough). In between, at the even-numbered points, the waves would disappear. Well, exactly the same experiment can be done with light and with the same result.

It works best if we select light of a very pure color, and let that light, from a distant source, approach an opaque surface with two narrow slits, exactly analogous to the gates in the front wall in Figure 8.1. Now suppose you put a white card or screen some distance behind the slits. Light passing through the two slits spreads out, just as the ocean waves do, and strikes the screen behind the slits. However, you will not see the screen uniformly illuminated by the light, but rather crossed by a series of bright and dark patterns or *fringes*, of the same color as that of the light source (as in Figure 8.2). The reason is that light consists of waves, and where the wave crests and troughs of the light passing through the two slits reinforce each other at the screen, you see a bright fringe of light, just as you see large-amplitude waves at the odd-numbered points on the back wall in Figure 8.1. In between, though, the troughs and crests of the light from the two slits cancel each other, and you see no light; those places are the dark fringes in Figure 8.2 and correspond to the even-numbered places on the back wall in Figure 8.1.

The exact spacing of the fringes on the screen, or the distance between the odd-numbered points on the back wall in Figure 8.1, depends on how far apart the crests of the wave are; that distance between wave crests is called the *wavelength* of the waves. Typically ocean waves have wavelengths of tens of meters, but visible light typically has wavelengths of 0.00005 cm. This phenomenon, whereby light waves reinforce each other or cancel out each other and produce fringes, is called *interference*. The fact that light can display such interference is evidence that it propagates as waves.

It is easy to see interference in light with a very simple experiment. Place in front of one of your eyes two straight-edged cards, with the edges close together, so that they form a narrow gap or slit between them. Now look at a bright source of light through the slit. As you narrow the gap between the cards, you will see a succession of bright and dark lines running along the gap. These are interference fringes, evidence of interference of light waves alternately canceling and reinforcing each other to produce dark and bright bands.

The interference bands are formed in this experiment because of *diffraction*. The phenomenon of diffraction is a deflection of light when it passes the edge of an opaque object; light spreads itself out,

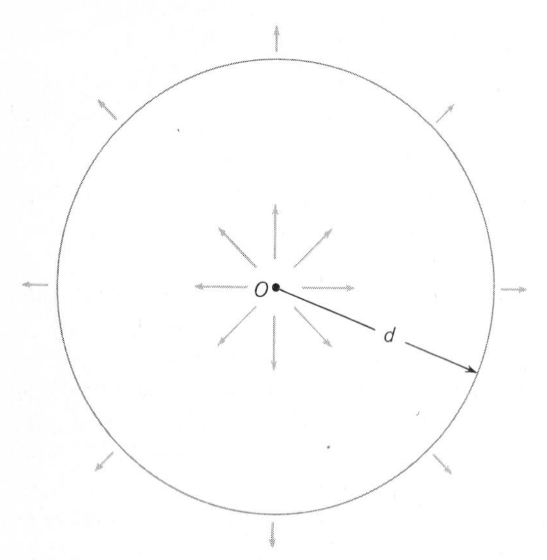

Figure 8.3 Inverse-square law of light.

slightly, into the shadow of the object it passes. The phenomenon is familiar also in sound: out of sight is *not* out of earshot! In the above experiment, the light passing the edges of the cards on either side of the slit fans out. The image of fringes you see as you look through the slit is produced by waves from different places across the width of the slit, which must travel different distances, resulting in alternate reinforcement and cancellation of the waves.

Diffraction is important in many astronomical applications; among other things it provides a theoretical limitation to the resolving power of a telescope. It is also central to the operation of many spectrographs (Chapter 9). Other phenomena whereby the direction of light is changed are *reflection* and *refraction*, both discussed later in this chapter.

(b) Propagation of Light: The Inverse-Square Law

Light propagates through empty space in a straight line (but in Chapter 33 we consider more deeply the meaning of "straight line"). An important property of the propagation of electromagnetic energy is the *inverse-square law*, a property that also belongs to the propagation of other kinds of energy, for example, sound. The amount of energy that would be picked up by a telescope, or an eye, or any other detecting device of fixed area, located at any given distance d from a light source O, is proportional to the amount of energy crossing each square centimeter of the surface of an imaginary sphere of radius d (Figure 8.3). A certain finite amount of energy is emitted by the source in a given time interval. That energy spreads out over a larger and larger area as it moves away from the source. When it has moved out a distance d, it is spread over a sphere of area $4\pi d^2$ ($4\pi d^2$ is the surface area of a sphere of radius d). If E is the amount of energy in question, an amount $E/4\pi d^2$ passes through each square centimeter (if d is measured in centimeters).

We see, then, that the amount of energy passing through a unit area *decreases* with the *square of the distance from the source*. At distances d_1 and d_2 from a light source, the amounts of energy received by a telescope (or other detecting device), ℓ_1 and ℓ_2, are in the proportion:

$$\frac{\ell_1}{\ell_2} = \frac{4\pi d_2^2}{4\pi d_1^2} = \left(\frac{d_2}{d_1}\right)^2.$$

The above relation is known as the inverse-square law of light propagation. In this respect, the propagation of radiation is similar to the effectiveness of gravity, because the force of gravitation between two attracting masses is inversely proportional to the square of their separation. Figure 8.4 also illustrates the inverse-square law of propagation.

(c) The Electromagnetic Spectrum

We have seen that all forms of electromagnetic energy have certain characteristics in common: their wavelike method of propagation and their speed. Nevertheless, the various kinds of radiant energy, differing from each other in wavelength, are detected by very different means.

Radio waves have the longest wavelength, ranging up to several kilometers. Those used in shortwave communication and in television have wavelengths ranging from centimeters to meters. When these waves pass a conductor, such as a radio antenna, they induce in it a feeble current of electricity, which can be amplified and processed so that it can be recorded on magnetic disk or tape, or can drive a loudspeaker.

The shortest wavelengths of radio radiation, between a few millimeters and a fraction of a millimeter (hundreds of micrometers), merge into the *submillimeter* and *infrared* parts of the spectrum. Ra-

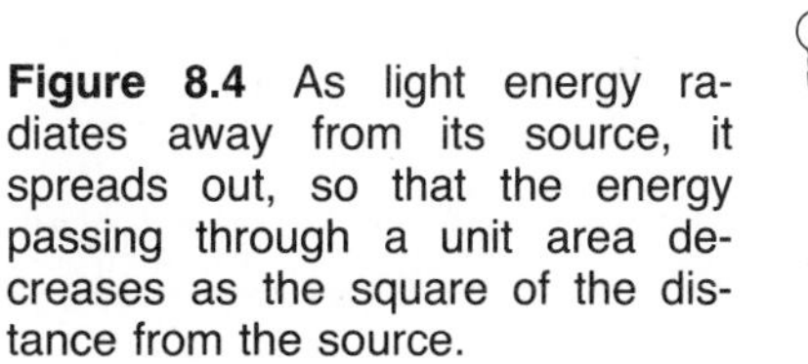

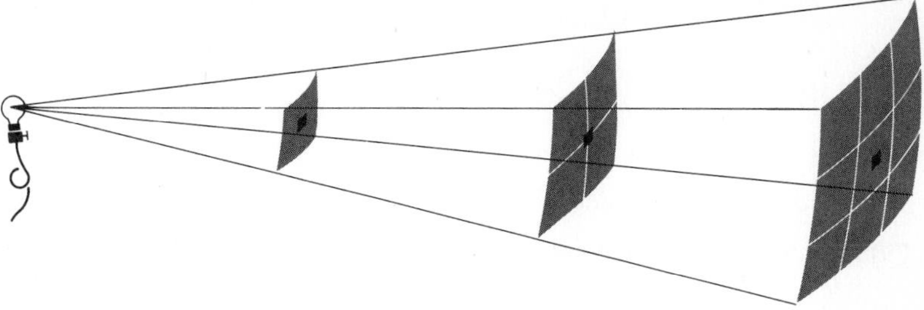

Figure 8.4 As light energy radiates away from its source, it spreads out, so that the energy passing through a unit area decreases as the square of the distance from the source.

diation all the way down to visible wavelengths (short of one micrometer) is called infrared.

Over most of the infrared part of the spectrum the most sensitive detectors are bolometers, bits of germanium or silicon semiconductor material that change their electrical properties as their temperature changes. From about 10 micrometers shortward, a variety of additional semiconductor detectors have been developed, each able to generate a tiny electrical signal when infrared radiation strikes them. Finally, at the shortest infrared wavelengths, detectors developed for visible light, such as photographic plates and television-type cameras, can be used.

Electromagnetic radiation with a wavelength in the range 0.0004 to 0.0007 mm comprises *visible light*. It is more convenient to express the wavelength in smaller units, such as the micrometer (1 μm = 0.001 mm) or the nanometer (1 nm = 0.001 μm). In these units, the wavelengths of visible light are 0.7 to 0.4 μm or 700 to 400 nm. Astronomers and spectroscopists also have a long tradition of using a different unit, the *angstrom*. One angstrom (abbreviated Å) is 0.0001 μm or 0.1 nm in length. Visible light, then, has wavelengths that range from about 4000 to 7000 Å.

The exact wavelength of visible light determines its color. Radiation with a wavelength in the range of 4000 to 4500 Å gives the visual impression of the color violet. Radiations of successively longer wavelengths give the impression of the colors blue, green, yellow, orange, and red, respectively. The array of colors of visible light is called the *spectrum*. A mixture of light of all wavelengths, in about the same relative proportions as are found in the light emitted into space by the sun, gives the impression of white light.

Light not only affects the organs of vision but can also be detected photographically and photoelectrically. A photographic film or plate detects and records chemical changes that result when radiation strikes it. Photoelectric and television-type detectors (which are discussed in Chapter 9) utilize physical changes, such as the ejection of an electron, that occur when some substances are exposed to light radiation.

Radiation of wavelengths too short to be visible to the eye is called *ultraviolet*. Radiation with wavelengths less than about 200 Å comprise X-rays. Both ultraviolet radiation and X-rays, like visible light, can be detected photographically and photoelectrically.

Electromagnetic radiation of the shortest wavelength, to less than 0.1 Å (10^{-8} mm), is called *gamma radiation*. Gamma rays are often emitted in the course of nuclear reactions and by radioactive elements. Gamma radiation is generated in the deep interiors of stars; it is gradually degraded into visible light by repeated absorption and reemission by the gases that comprise stars.

The array of radiation of all wavelengths, from radio waves to gamma rays, is called the *electromagnetic spectrum*.

(d) Relation Between Wavelength, Frequency, and Speed

Let us once again compare the propagation of light to the propagation of ocean waves. While an ocean wave travels forward, the water itself is displaced only in a vertical direction. A stick of wood floating in the water merely bobs up and down as the waves move along the surface of the water. Waves that propagate with this kind of motion are called *transverse waves*.

Light also propagates with a transverse wave motion, and travels with its highest possible speed through a perfect vacuum. In this respect light differs markedly from *sound*, which is a physical vibration of matter. Sound does not travel at all through a vacuum. The displacements of the matter

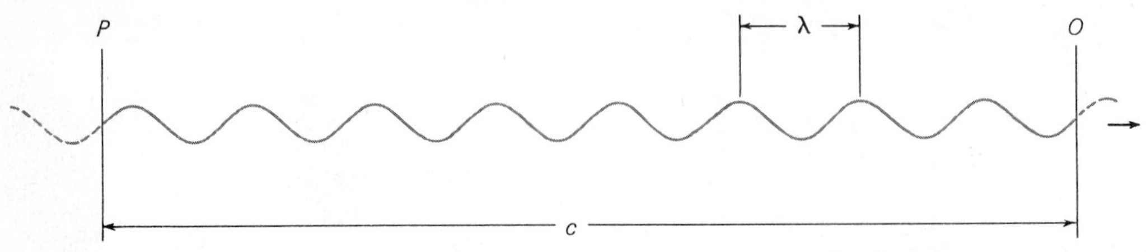

Figure 8.5 Relation between wavelength, frequency, and the speed of radiation.

that carry a sound impulse are in a *longitudinal* direction, that is, in the direction of the propagation, rather than at right angles to it. Sound is actually a traveling wave of alternate compressions and rarefactions of the matter through which it moves. Of course, sound also travels far more slowly than electromagnetic radiation—only about ⅓ km/s through air at sea level.

For any kind of wave motion, sound or light, we can derive a simple relation between wavelength and *frequency*. The frequency of light (or sound) is the rate at which wave crests pass a given point, that is, the number of wave crests that pass per second. Imagine a long train of waves moving to the right, past point O (Figure 8.5), at a speed c. If c is measured, say, in centimeters per second, we can measure back to a distance of c centimeters to the left of O and find the point P along the wave train that will just reach the point O after a period of one second. The frequency f of the wave train—the number of waves that pass O during that second—is obviously the number of waves between P and O. That number of waves, times the length of each, λ, is equal to the distance c. Thus we see that for any wave motion, the speed of propagation equals the frequency times the wavelength, or symbolically,

$$c = f \times \lambda.$$

(It is traditional to denote the length of a wave of electromagnetic radiation by λ, the Greek letter for "ℓ".)

(e) Heat and Radiation

A warm or hot solid is composed of molecules and atoms that are in continuous vibration. A gas consists of molecules that are flying about freely at high speed, continually bumping into each other, and bombarding against the surrounding matter. That energy of motion is called *heat*. The hotter the solid or gas, the more rapid is the motion of those molecules, and its *temperature* is just a measure of the average energy of those particles. One of the principles of *thermodynamics* (the *second law*) is that heat always tends to transfer from a hot object to a cooler one. Thus a solid or gas that is at a higher temperature than its surroundings radiates some of its heat energy into those surroundings, thereby cooling.

A century ago physicists were interested in the properties of this emitted radiation—that is, how much of which kind of radiation is radiated by warm and hot bodies? The situation is quite complicated, because bodies absorb some radiation, reflect some, and transmit some. A blue sweater, for example, reflects more of the relatively short-wave blue light than it does other colors; that is why it looks blue. A piece of black coal reflects relatively little of any visible light, and a window pane transmits most light through without either absorbing or reflecting it.

All these bodies, though, absorb some radiation, especially when the entire electromagnetic spectrum is considered; moreover, all bodies will eventually come into equilibrium with their surroundings, until they re-emit energy at a rate which, averaged over time, is exactly the rate at which they absorb it.

When confronted with a complex problem, as this one is, the scientist usually tries to get a start in her analysis by finding a circumstance in which the problem is simplified. To this end, physicists invented the *ideal* or *perfect radiator*, a hypothetical body that completely absorbs every kind of electromagnetic radiation incident on it, reaches some equilibrium temperature, and then reradiates that energy as rapidly as it absorbs it. Because the perfect radiator absorbs everything and reflects and transmits nothing, it is also called a *black body*.

A black body, however, is not so called because it necessarily *looks* black; since it is radiating energy, it might be very bright indeed. It happens

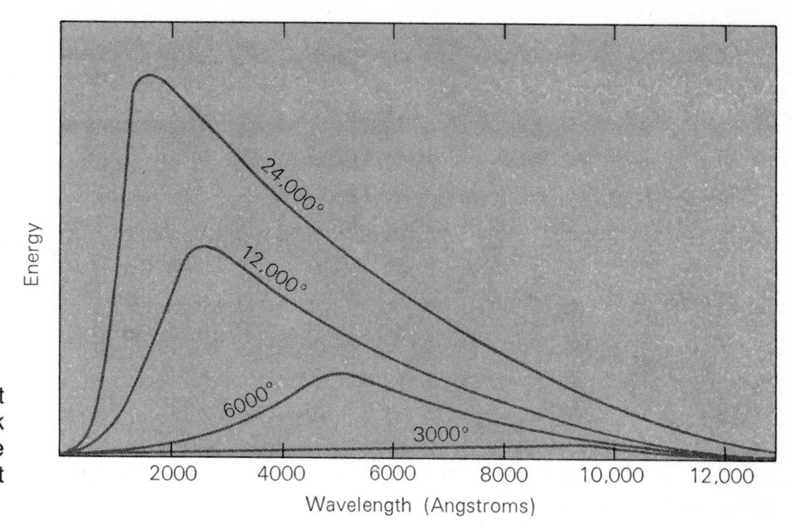

Figure 8.6 Energy emitted at different wavelengths for perfect radiators (black bodies) at several temperatures. (The curves are schematic only and are not plotted accurately to scale.)

that a piece of coal, which is a crude approximation to a black body, does look black, but that is because (at least at room temperature) its reradiated energy is mostly in the invisible infrared; if we could see infrared radiation, we would find that coal to be glowing brightly. Stars happen to be good black bodies, because the gases they are made of are very opaque to virtually all electromagnetic radiation, and only from a star's outermost layers can energy escape into space. A star like the sun, of course, hardly looks black.

(f) The Perfect Radiator

Laboratory devices that are close approximations to perfect radiators were invented long ago. One, for example, is an enclosed chamber with well-insulated walls that are painted black on the inside. The internal energy, nearly all absorbed and re-emitted by the interior walls, can be observed through a tiny hole in the chamber. Such a device was used to learn experimentally the nature of black-body radiation and to test the derived radiation laws. It was found that the distribution of energy at different wavelengths *(spectral energy distribution)* emitted by a given unit area of such a black body depends only on the body's temperature; two different black bodies of the same temperature always radiate in exactly the same way. Figure 8.6 shows, schematically, how much electromagnetic energy at various wavelengths black bodies of several different temperatures radiate from each square centimeter of their surfaces.

We note three interesting things about the graph in the figure. First, every black body emits some radiation at *every* wavelength. Second, at *every* wavelength, a hotter black body emits *more* energy than a cooler one. Third, for a black body at each temperature there is a certain wavelength at which it radiates a maximum amount of energy, and the *higher* the temperature, the shorter that wavelength of peak radiation.

The third point is expressed precisely by *Wien's law,* which states that the wavelength of peak emission is inversely proportional to the temperature (or, equivalently, the product of its temperature and that wavelength of maximum radiation is always the same number); if we denote by λ_{max} that wavelength of maximum radiation, we have

$$\lambda_{max} = \frac{\text{constant}}{T}.$$

When the wavelength is expressed in centimeters and the temperature in Kelvins, the constant has the value 0.2897 cm · K. The sun, for example, has a temperature of about 6000 K and emits its maximum light at about 5000 Å. Wien's law means that a star with twice the sun's temperature emits its maximum light at 2500 Å. Such a star appears blue to us, because most of its radiation is at short wavelengths. Some stars are so hot that most of their energy is emitted in the invisible ultraviolet. On the other hand, a star with a temperature of only 3000 K, half the sun's temperature, emits its maximum light at 10,000 Å—in the infrared. Note that since all black bodies emit some energy at all wavelengths, even the cool stars emit a little ultraviolet light. Moreover, not only do the very hot stars emit some infrared and visible light, but they emit *more* of it (per unit area) than the cool stars do (remember the second point in the foregoing paragraph).

So a consequence of Wien's law is that we can calculate the temperature of a star from observing the wavelength at which it emits the most radiation. Since that also is what determines the color of a star, its color tells us its temperature. In practice we do not even have to observe the wavelength of maximum light; whenever we observe the color index of a star (Chapter 24), we are comparing the intensity of its radiation in two different wavelength regions. We can determine at what temperature a black body would have the same ratio of intensity in the same two wavelength regions and thereby find the star's temperature. A more accurate comparison with a black body can be obtained if more than two spectral regions are observed. A stellar temperature determined by comparison of the spectral distribution of the star's radiation with that of a black body is called a *color temperature*.

(g) The Color of the Sun

Now the sun's wavelength of maximum radiation is that of green light; yet the sun appears *white*, not green. The reason is that white light is the particular admixture of different wavelengths we receive from the sun (or, approximately, from a black body with a temperature of 6000 K). A piece of white paper looks white because it reflects all wavelengths about equally to our eye, so the reflected light from a white paper has the same distribution of wavelengths as sunlight—or any other source of white light. Stars that have higher temperatures than the sun look blue or violet to the eye, and those of lower temperature look yellow, orange, or red, because the peak of their light is at those shorter or longer wavelengths, respectively; a star you might expect to look green has about the same temperature as the sun and is white, more or less by definition. We would have to filter out most of the other colors to make sunlight look green.

Most people (even some astronomers, but *not* solar astronomers) describe the sun as *yellow*. It is not. Snow is a pretty good reflector of all visible wavelengths; the sun is roughly the color of snow. Then why do people think it is yellow? Perhaps because it is hard (indeed, dangerous!) to look at the sun when it is high in the sky. We can and often do look at the sun comfortably when it is low in the sky, and we see it through a long path of air, which scatters the short wavelengths out of the beam, allowing only the long waves to pass through. Consequently the sun, when seen low or setting, generally appears yellow, orange, or red. Also, in grade school when we drew and colored the sun we used yellow crayons, because white crayon doesn't show up well on white paper. Maybe it's just traditional to regard the sun as yellow. Still, it's white.

(h) The Stefan-Boltzmann Law

Since a hot black body emits more energy at all wavelengths than a cool one, the combined amount of electromagnetic radiation over all wavelengths emitted by a black body is extremely sensitive to its temperature. If we sum up the contributions from all parts of the spectrum, we obtain the total energy emitted by a black body over all wavelengths. The total energy emitted per second per square centimeter by a black body at a temperature T is given by the equation known as the *Stefan-Boltzmann law:*

$$E(T) = \sigma T^4.$$

The constant σ, called the *Stefan-Boltzmann constant,* has the value 5.672×10^{-5}, if T is in K and $E(T)$ is in ergs/cm^2/s.

Now imagine two stars of the same size and distance, one of which is blue and the other red. The blue star, being hotter, radiates very much more efficiently and so appears far, far brighter than the red one. But what if a red and a blue star, both the same distance, appeared *equally* bright? Knowing that the red star is an inefficient radiator (per unit area), you realize that it must be very much larger to appear as bright as the blue one.

So at once you see how we can use the Stefan-Boltzmann law to figure out approximately how big a star is. First, from its color, we find the wavelength of the star's maximum radiation; that tells us the temperature. The temperature, with the Stefan-Boltzmann law, tells us how much energy the star radiates from every square centimeter of its surface. Now if we know the star's distance, we can observe how much energy it emits altogether from its entire surface. Knowing how much it radiates from each square centimeter and knowing how much it radiates altogether tell us how many square centimeters there are in its surface, that is, its size.

The Earth, Moon, and planets are not very good black bodies. Typically, they reflect about half of the sunlight incident on them, although it's not the same for all of them; Venus, for example, reflects more than half, and the Moon less than 10 percent. But they all absorb some of the sunlight hitting them, and they must reradiate this energy. We see the planets in the sky by the sunlight they reflect. It's easy to calculate how much sunlight strikes them, so from their brightnesses we find out how much solar radiation each planet must absorb, which is, of course, how much it also emits. Now the Stefan-Boltzmann law enables us to calculate the temperature of a black body that emits the same amount of energy a particular planet does. Although a planet is not really a perfect radiator, its surface temperature will be in the general neighborhood of that black body's temperature.

Thus we find that the Earth's temperature should average somewhere around 300 K—about room temperature. (Actually, 293 K is 20°C, or 68°F.) Mars, being farther from the sun and thus receiving less energy from it than we do, is a little cooler. Mercury, being closer than we, is quite a bit hotter.

Wien's law tells us the wavelength at which a black body of the same temperature as a planet would emit its maximum radiation; roughly, this is the wavelength of the peak of the energy radiated by the planet, too. The Earth's temperature of 300 K is about 1/20 the sun's 6000 K; if the sun radiates most strongly in green light of wavelength 5000 Å, the Earth must radiate most strongly at 100,000 Å (or about 10 μm)—way out in the infrared. Everything about us, therefore, including ourselves, is radiating infrared energy. If we had infrared vision, we would easily see each other in the dark; in fact, we would find that our world had continuous illumination. "Night" is night only to visible sunlight; there is no night on Earth to its own radiation.

Similarly, the other planets radiate in the infrared. We can measure the radiation they emit with infrared detectors attached to our telescopes. By the way, by now you can appreciate one of the problems of the astronomer who specializes in infrared observations. The telescope dome and everything in it, including the telescope itself and the astronomer, are all shining brightly in the very wavelengths we are trying to detect from celestial objects. It's rather like trying to watch a motion picture show with the theater lights on. The astronomer must take very special pains to shield the detector from all radiation except that reflected from the telescope mirror.

(i) The Quantization of Light; Photons

We have seen that quite a bit was known a century ago about the way things radiate. At that time, however, there was not yet a physical theory that accounted for the properties of radiation from black bodies, that is, those particular graphs shown in Figure 8.6. We did not understand radiation in the same sense that we understand the motions of the planets in terms of the law of gravitation. Now it was known that currents moving in a wire emit electromagnetic waves (for example, radio waves), so it was thought that charged particles (electrons) within the atoms of the radiating substances were oscillating back and forth to emit all kinds of electromagnetic energy. But if those electrons could oscillate in every way, there was no reason why a plot of the energy emitted at various wavelengths should look like the graphs in Figure 8.6.

The answer to that riddle was part of the start of the quantum theory—one of the great new branches of physics of the 20th century. In 1900 the German physicist Max Planck succeeded in deriving a theoretical formula for those graphs in Figure 8.6.* To do so, he had only to assume that the tiny atomic oscillators in the radiating bodies can emit energy at any given wavelength or frequency only in certain discrete amounts. In other words, at each frequency the energy leaving a radiating particle is always some multiple of a minute unit of energy. That unit of energy is the frequency of the radiation multiplied by a tiny number—always the *same* number, symbolized h. That number, h, is now called

---

* If T is the temperature in Kelvins, λ the wavelength in centimeters, and k Boltzmann's constant (1.37×10^{-16}), $E(\lambda,T)$, the energy in ergs emitted per unit wavelength interval per second per square centimeter and into unit solid angle, is

$$E(\lambda,T) = \frac{2hc^2}{\lambda^5} \frac{1}{e^{hc/\lambda kT} - 1}.$$

Wien's law can be derived from Planck's formula by finding the wavelength at which the derivative of Planck's formula equals zero, and Stefan's law by integrating Planck's formula.

Planck's constant. If energy is measured in ergs and frequency in cycles or waves per second, Planck's constant has the value $h = 6.62 \times 10^{-27}$ erg · s.

Since 1887 (Section 8.1c) it has been known that any metal will emit electrons when struck by electromagnetic radiation if that radiation is of short enough wavelength (or high enough frequency). In other words, the radiation must have a certain critical or threshold frequency. Radiation of *just* the threshold frequency barely dislodges electrons from the metal, but that of *higher* frequency dislodges them and gives them some energy of motion (kinetic energy) also.

The subject was clarified by Albert Einstein in 1905. Einstein proposed that electromagnetic radiation itself travels in units of energy, each unit having energy equal to h (Planck's constant) times the frequency. Moreover, each electron emitted results from the metal absorbing just *one* of those units of radiant energy. Now it takes a certain minimum energy to remove an electron from the metal. Thus, of the units of electromagnetic energy that are absorbed, the only ones that can cause the ejection of electrons are those which have high enough frequencies to possess that requisite energy. On the other hand, any radiation unit of *higher* energy (that is, higher frequency) can be absorbed, and the excess energy absorbed goes into kinetic energy (energy of motion) of the dislodged electron.

In short, Einstein showed that the phenomenon of radiation causing the ejection of electrons—called the *photoelectric effect*—is evidence that electromagnetic energy is itself *quantized,* consisting of little packets of energy. These packets of radiant energy are now called *photons.* Einstein received the Nobel prize for this contribution to our understanding of the nature of radiation—not for his better known relativity theory!

(j) Particles and Waves

The photoelectric effect and many other experiments show that electromagnetic energy acts as though it were made up of little energetic particles—photons. But we have already seen that electromagnetic energy also propagates as *waves.* So these photons themselves must be waves, but always containing discrete amounts of energy. Thus photons of the shortest wavelength and highest frequency—gamma rays—have the highest energy, and those of radio waves have the lowest energy; visible light is intermediate.

How can we reconcile our concept of photons of light and other electromagnetic radiation as having the properties of both particles and waves? There are two ways: First, we can think of each photon itself as waves propagating in all directions through space, like ripples in a pond. The photon can do all the things waves can do—it can pass through two slits at once and interfere with itself, for example. But as soon as we *observe* it, say by its absorption in the retina of the eye, in a photographic emulsion, or by means of the photoelectric effect, those waves immediately disappear, and the photon itself is known to have been at the place where it was observed.

Second, we can think of those waves as just a mathematical formula, which tells us the probability of the photon—itself a localized particle—being in any particular place. Thus the wave crests are the places where the photon is more likely to be found, and the wave troughs places where it is less likely to be. We never, of course, know where it *really* is until we observe it.

Perhaps the second way of thinking of photons makes physical sense—one that is conceptually credible. But actually, both ways are exactly equivalent as far as predicting results of experiments are concerned, and science, after all, doesn't concern itself with what a photon "really" is, but only *how it behaves.* Unless we can devise an experiment to distinguish between these viewpoints, it does not even make sense—within the realm of science, anyway—to ask which is *correct.*

In fact, it turns out that material particles—like protons and electrons, and yes, even billiard balls—can similarly be thought of as waves. We shall see that we cannot say just where an electron (for example) is located at any given instant. The best we can do is write down an equation that gives us the probability of its having various locations. The equation turns out to be the equation for a system of waves; we are, in other words, exactly where we are with the photon and may as well think of the electron itself as a wave.

These concepts may seem less satisfying than the simple, mechanical, predictable system of Newtonian theory, but it is just as successful in accounting for the behavior of matter on the atomic level as Newton's laws are on the scale of the solar system. And the two theories are not contradictory, either. If you flip an honest coin in a random way one million times, you can safely predict that very close to 50 percent of the flips will come up heads. The gentlemen who run the casinos at Las Vegas make their fortunes by knowing exactly how statistics of large numbers work out. Similarly, planets and billiard balls have absolutely enormous numbers of atomic particles. Just as we cannot predict whether an individual gambler will win a particular hand of blackjack, so the behavior of each particle in a billiard ball is uncertain. But the behavior of the total ensemble is highly predictable. As the casino operator knows that on the average the customers will lose (and for that matter, so will each individual if he plays long enough), so the physicist knows how the billiard ball will carom, behaving according to the average of the motions of its tremendous numbers of atoms.

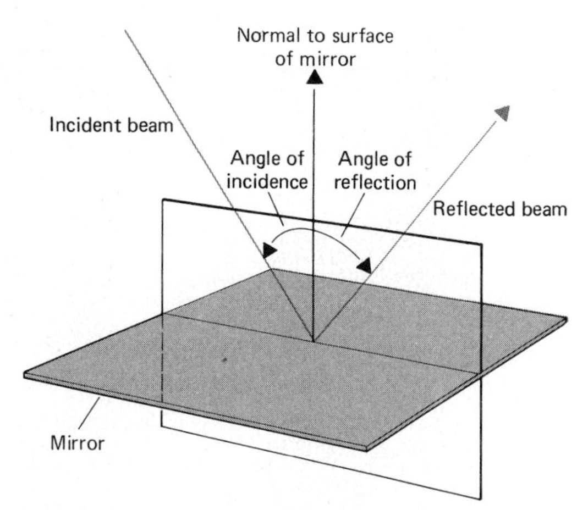

Figure 8.7 Law of reflection.

8.2 OPTICS

We now turn our attention to some properties of visible light. Those that are most important to the design and construction of telescopes and other astronomical instruments can be summarized simply in three laws of *geometrical optics*—the laws of *reflection, refraction,* and *dispersion.*

(a) Reflection

The law of reflection describes the manner in which light is reflected from a smooth, shiny surface. The *normal* to a surface at some point is simply a line or direction perpendicular to that surface at that point. If light strikes a shiny surface, its direction must make a certain angle with the normal to the surface at the point where it strikes. That angle is the *angle of incidence.* The angle that the reflected beam of light makes with the normal is called the *angle of reflection.* The law of reflection states that the angle of reflection is equal to the angle of incidence and that the reflected beam lies in the plane formed by the normal and the incident beam (Figure 8.7).

(b) Refraction

The law of refraction deals with the deflection of light when it passes from one kind of transparent medium into another. The *index of refraction* of a transparent substance is a measure of the degree to which the speed of light is diminished while passing through it; specifically, it is the ratio of the speed of light in a vacuum to that in the substance. Usually, media of higher densities have higher indices of refraction. The index of refraction of a vacuum is, by definition, exactly 1.0. That of air at sea level is about 1.00029. Water has an index of about 1.3; crown glass and flint glass have indices of about 1.5 and 1.6, respectively. Diamond has the high index of refraction of 2.4.

The law of refraction states that if light passes from one medium into a second one of a different index of refraction, the angle the light beam makes with the normal to the interface between the two substances is always *less* in the medium of higher index. Thus, if light goes from air into glass or water, it is bent *toward* the normal to the interface (that is, the direction perpendicular to the surface), while if it goes from water or glass into air, it is bent *away* from the normal.

It is the refraction of light when it passes from the water into the air that makes the handle of a spoon appear bent if a spoon is immersed in a glass of water. Similarly, light entering the Earth's atmosphere from space is slightly bent. The light from stars, planets, the sun, and the Moon is bent, upon entering the Earth's atmosphere, in such a way as to make the object appear to be at a greater altitude above the horizon than it actually is. Atmospheric refraction is greatest for objects near the horizon. It raises the apparent altitude of objects on the horizon by about ½°. The refraction of light passing through glass and the Earth's atmosphere is illustrated in Figures 8.8 and 8.9, respectively.

The illusion that the Moon (or sun) looks larger near the horizon than when it is high in the sky is *not* due to refraction. Actually, refraction raises the lower limb of the Moon more than the upper, so that the Moon really looks smaller and oval near the

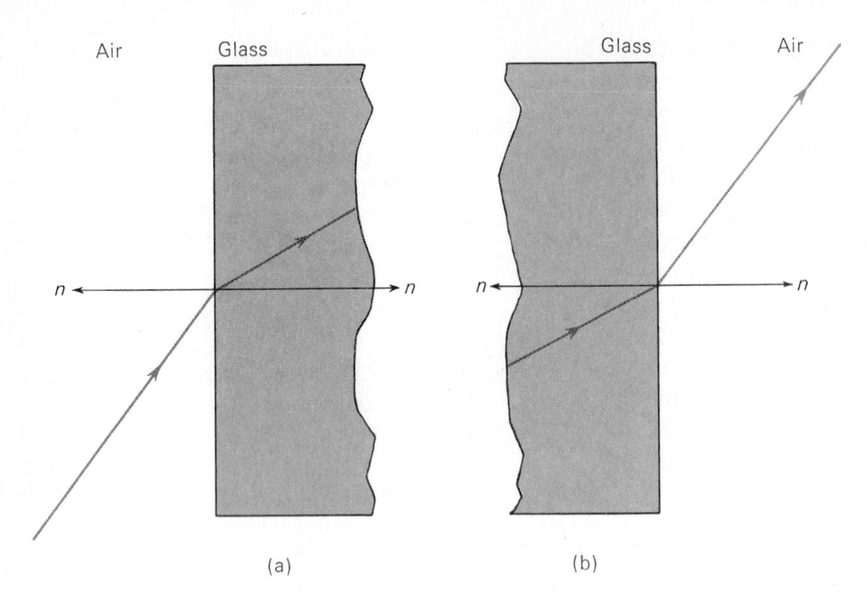

Figure 8.8 Refraction of light between glass and air. (a) When light passes from air to glass. (b) When light passes from glass to air.

horizon, not larger. The apparent enlargement of the Moon or sun when seen near the horizon is a purely psychological effect that has been the subject of much discussion and investigation by psychologists.

(c) Dispersion

Dispersion of light is the manner in which white light, a mixture of all wavelengths of visible light, can be decomposed into its constituent wavelengths or colors when it passes from one medium into another. The phenomenon of dispersion occurs because the index of refraction of a transparent medium is greater for light of shorter wavelengths. Thus, whenever light is refracted in passing from one medium into another, the violet and blue light of shorter wavelengths is bent more than the orange and red light of longer wavelengths.

Figure 8.10 shows the way in which light of different wavelengths is bent different amounts in passing from air into glass. Figure 8.11 shows how light can be separated into different colors with a prism, a triangular piece of glass. Upon entering one face of the prism, light is refracted once, the violet light more than the red, and upon leaving the opposite face, the light is bent again, and so is further dispersed. Even greater dispersion can be ob-

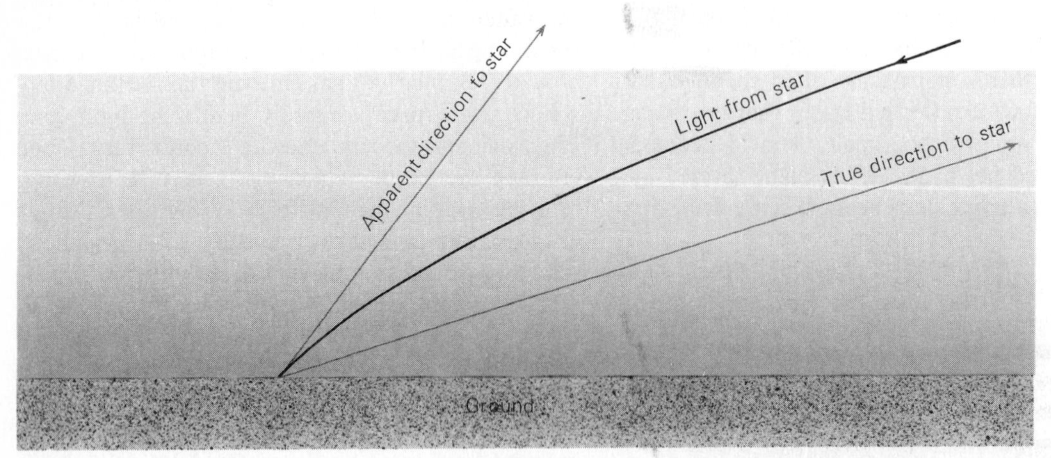

Figure 8.9 Atmospheric refraction. Light is bent upon entering the Earth's atmosphere in such a way as to make stars appear at a higher altitude than they actually are. (The effect is grossly exaggerated in this figure.)

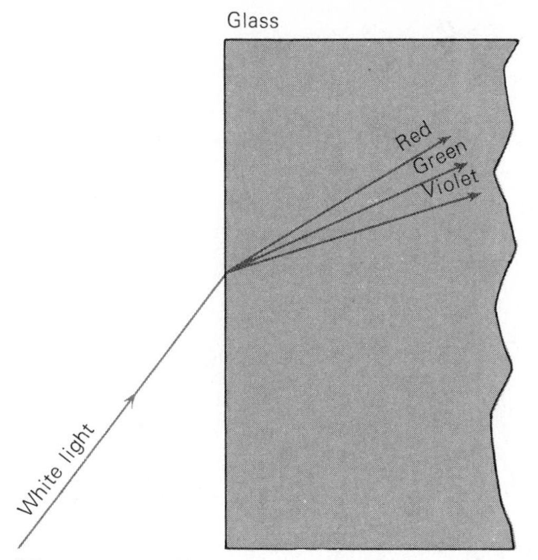

Figure 8.10 Dispersion at the interface of two media.

tained by passing the light through a series of prisms. If the light leaving a prism is focused upon a screen, the different wavelengths or colors that compose white light are lined up side by side. The array of colors is a spectrum.

It has long been known that colors are produced when white light passes through glass, but before Newton's time it was believed that the glass itself produced the colors. Newton produced a spectrum with a glass prism and thereby isolated individual rays of nearly pure color. He found that these monochromatic rays were not further dispersed with a second prism, but merely refracted, which indicated that the glass prism had not produced the colors but had merely separated them out from white light.

Atmospheric refraction occurs in varying amounts for different colors, also. This atmospheric dispersion combined with turbulence in the air often produces colorful effects when a bright star is seen low above the horizon, and its different colors are bent by the air in differing amounts. When astronomers measure the exact positions of stars in the sky, they must not only take account of the distortion of the position of the star by atmospheric refraction but also take care to note the color of the light with which the star is observed, because the correction for refraction is different for different colors.

(d) Weather Optics

Nature provides an excellent example of the dispersion of light in the production of a rainbow. Raindrops, tiny spherical droplets of water in the air, act like prisms. Light from the sun entering a raindrop is bent, the blue and violet light being bent the most. This bent light strikes the inside rear surface of the drop, and some of it is reflected back toward the front surface. This reflected light leaves the raindrop by passing through the same side that it enters. But when it leaves the drop, it is again refracted, and again dispersed, just as when light leaves a glass prism. Thus sunlight is spread into the rainbow of colors—the rainbow is nothing more than the spectrum of sunlight. Figure 8.12 shows how a raindrop produces a spectrum.

The light that emerges from a raindrop is most intense at an angle of about 42° from the direction at which it enters. Thus (Figure 8.13), to see a rainbow, the observer must have the sun behind him, at an altitude of less than 42° above the horizon. The rainbow then appears as an arc, with an angular radius of 42°,

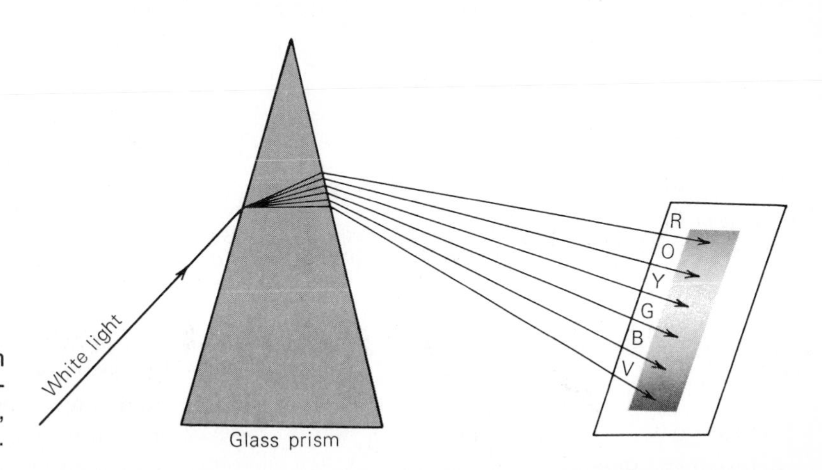

Figure 8.11 Dispersion by a prism and formation of a spectrum. The letters R, O, Y, G, B, and V stand for red, orange, yellow, green, blue, and violet.

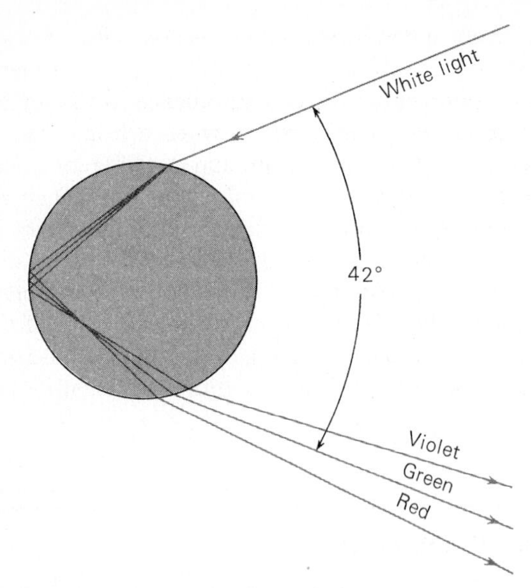

Figure 8.12 Dispersion in a raindrop.

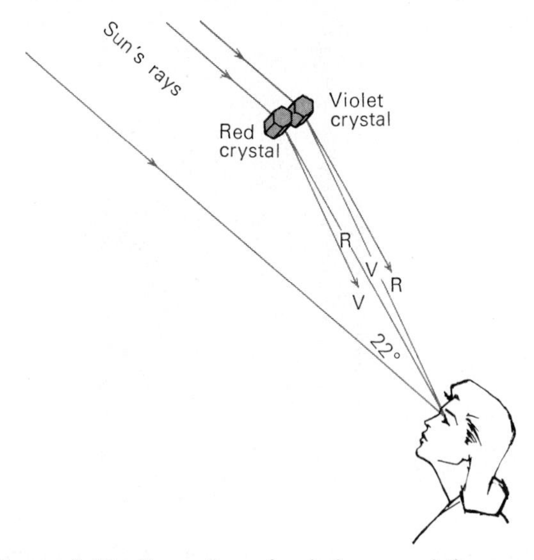

Figure 8.14 Formation of a halo around the sun or Moon by ice crystals in cirrus clouds.

centered about a point exactly opposite the sun. Since the sun must be above the horizon to illuminate the raindrops, the center of the rainbow must be below the horizon; an observer on the ground can never see a rainbow as more than half a complete circle (observers in airplanes or on mountains may occasionally see more).

Although the arc of a rainbow has a radius of approximately 42°, the different colors of sunlight are refracted and reflected by the droplets back to the observer in slightly different directions, so the band of color has a finite width. It may be seen in Figure 8.13 that from the upper drops it is the red light, bent the least, that enters the observer's eye, and from the lower drops it is the violet light, bent the most, that the observer sees. Thus, the top, or outside, of the arc of the rainbow appears red, and the bottom, or inside, of the arc appears violet. The other colors come from the drops in between.

A little of the light is reflected a second time before leaving the drops. This light emerges at an angle of about 51° from that of the incoming light and produces a fainter secondary rainbow in an arc of 51° radius, thus lying above the primary rainbow. In the secondary rainbow (less frequently seen), the red is on the inside of the arc and the violet on the outside, or upper part.

Another natural phenomenon that involves the refraction and dispersion of light is a *halo* about the sun or Moon. A halo is a faint ring of light, of angular radius 22°, caused by the bending of light as it passes through the tiny ice crystals that form cirrus clouds at altitudes of more than 7000 m. As in the rainbow, the violet light is refracted most in passing through the crystals. Consequently, the outer edge of a solar or lunar halo appears violet, the inner edge red (Figure 8.14).

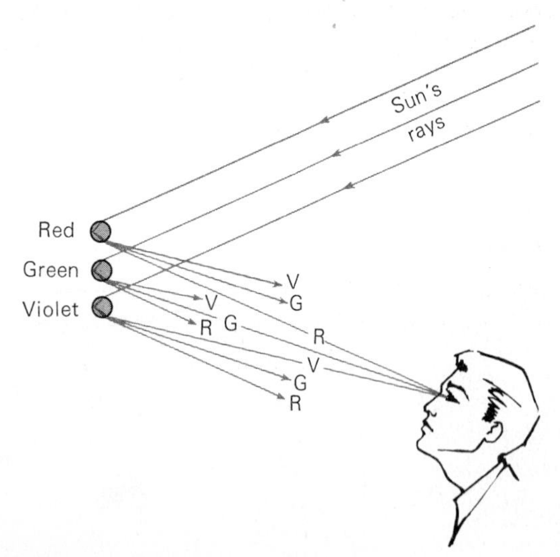

Figure 8.13 Formation of a rainbow.

8.3 SPECTROSCOPY

The light from a star or other luminous body can be decomposed into its constituent wavelengths, producing a spectrum. A device used to record the spectrum of a light source is called a *spectrometer*. Also used are the terms *spectroscope*, for an instru-

Figure 8.15 Three kinds of spectra: (a) continuous, (b) bright line, (c) dark line. The spectra are shown as they appear on photographic negatives, from which astronomers generally work.

ment used for visual observation of a spectrum, and *spectrograph*, for one used to photograph a spectrum. It will be explained in Chapter 9 how a spectrometer is constructed and attached to a telescope so that we may study the spectra of astronomical objects.

(a) Kirchhoff's Laws of Spectroscopy

If the spectrum of the white light from the sun and stars were simply a continuous rainbow of colors, astronomers would not have such intensive interest in the study of stellar spectra. To Newton, the solar spectrum did appear as just a continuous band of colors. However, in 1802, William Wollaston (1766–1828) observed several dark lines running across the solar spectrum. He attributed these lines to natural boundaries between the colors. Later, in 1814 and 1815, the German physicist Joseph Fraunhofer (1787–1826), upon a more careful examination of the solar spectrum, found about 600 such dark lines. He noted the specific positions in the spectrum, or the wavelengths, of 324 of these lines. To the more conspicuous lines he assigned letters of the alphabet, with the letters increasing from the red to the violet end of the spectrum. Today we still refer to several of these lines in the solar spectrum by the letters assigned to them by Fraunhofer.

Subsequently, it was found that such dark spectral lines could be produced in the spectra of artificial light sources by passing their light through various transparent substances or gases. On the other hand, the spectra of the light emitted by certain glowing gases were observed to consist of several separate bright lines. A preliminary explanation of these phenomena (although grossly oversimplified in light of modern knowledge) was provided in 1859 by Gustav Kirchhoff (1824–1887) of Heidelberg. Kirchhoff's explanation is often given in the form of *three laws of spectral analysis:*

FIRST LAW: A luminous solid or liquid emits light of all wavelengths, thus producing a continuous spectrum.

It is often stated that a highly compressed gas also emits a continuous spectrum, but the statement is an oversimplification. It is usually true if the gas is opaque.

SECOND LAW: A rarefied luminous gas emits light whose spectrum shows bright lines, and sometimes a faint superimposed continuous spectrum.

THIRD LAW: If the white light from a luminous source is passed through a gas, the gas may abstract certain wavelengths from the continuous spectrum so that those wavelengths will be missing or diminished in its spectrum, thus producing dark lines.

We distinguish, then, among three types of spectra (Figure 8.15). A *continuous* spectrum is an array of all wavelengths or colors of the rainbow. A *bright line* or *emission* spectrum appears as a pattern or series of bright lines; it is formed from light in which only certain discrete wavelengths are present. A *dark line* or *absorption* spectrum consists of a series or pattern of dark lines—missing wavelengths—superimposed upon the continuous spectrum of a source of white light.

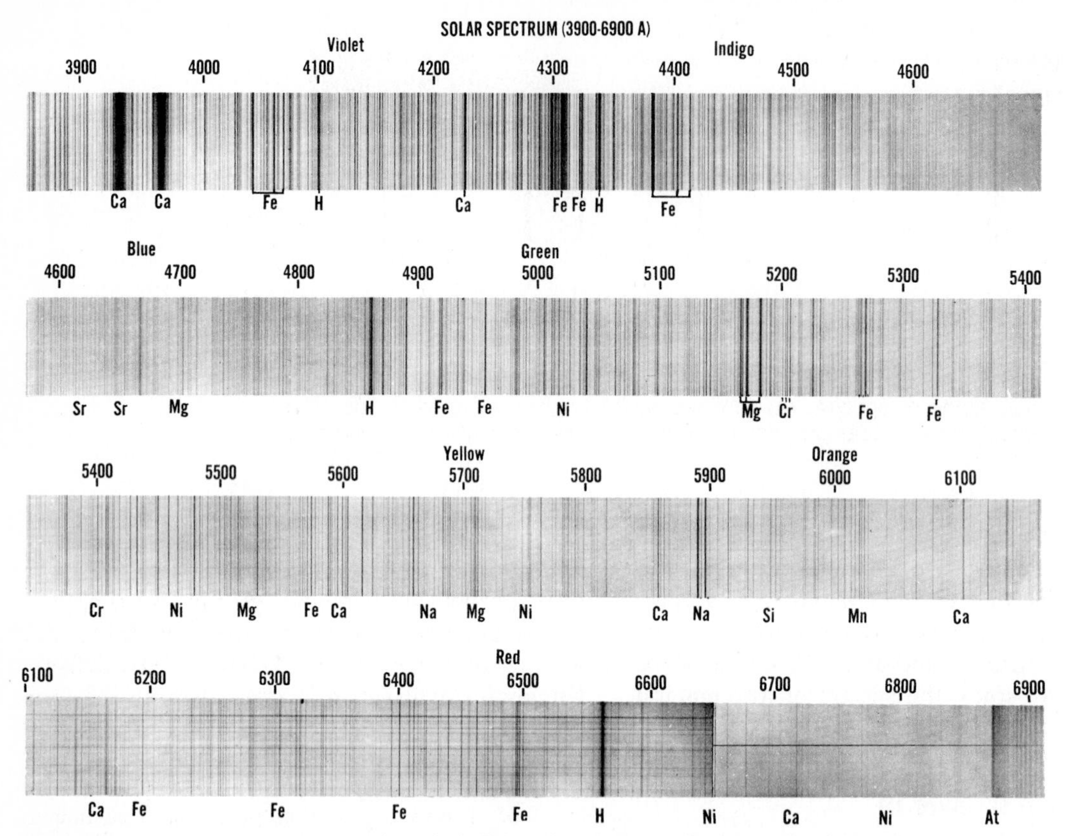

Figure 8.16 The solar spectrum. Labels indicate the elements in the sun's photosphere that cause some of the dark lines. The wavelengths in angstroms and the colors of the different parts of the spectrum are also labeled. (Mount Wilson and Las Campañas Observatories)

The great significance of Kirchhoff's laws is that each particular chemical element or compound, when in the *gaseous form,* produces its own characteristic pattern of dark or bright lines. In other words, each particular gas can absorb or emit only certain wavelengths of light, peculiar to that gas. The presence of a particular pattern of dark (or bright) lines characteristic of a certain element is evidence of the presence of that element somewhere along the path of the light whose spectrum has been analyzed.

Thus the dark lines (Fraunhofer lines) in the solar spectrum give evidence of certain chemical elements between us and the sun, absorbing those wavelengths of light. It is easy to show that most of the lines must originate from gases in the outer part of the sun itself (Figure 8.16).

The wavelengths of the lines produced by various elements are determined by laboratory experiment. Most of the thousands of Fraunhofer lines in the sun's spectrum have now been identified with most of the known chemical elements.

Dark lines are also found in the spectra of stars and in stellar systems. Much can be learned from their spectra, in addition to evidence of the chemical elements present in a star. A detailed study of its spectral lines indicates the temperature, pressure, turbulence, and physical state of the gases in that star; whether or not magnetic and electric fields are present, and the strengths of those fields; how fast the star is approaching or receding from us; and many other data. Other information about a star can be obtained by studying its continuous spectrum.

The study of the spectra of celestial objects is the most powerful means at the astronomer's disposal for obtaining data about the universe.

(b) The Doppler Effect

In 1842 Christian Doppler (1803–1853) pointed out that if a light source is approaching or receding from the observer, the light waves will be, respectively, crowded closer together or spread out. The

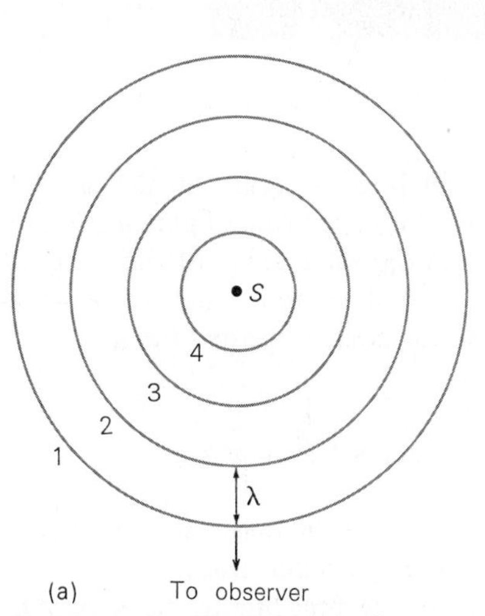

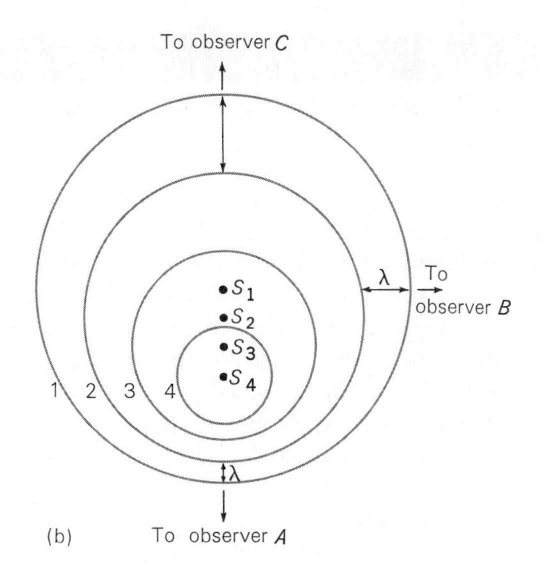

Figure 8.17 The Doppler effect.

principle, known as the *Doppler principle* or *Doppler effect,* is illustrated in Figure 8.17. In (a) the light source is stationary with respect to the observer. As successive wave crests 1, 2, 3, and 4 are emitted, they spread out evenly in all directions, like the ripples from a splash in a pond. They approach the observer at a distance λ behind each other, where λ is the wavelength of the ripple. On the other hand, if the source is moving with respect to the observer, as in (b), the successive wave crests are emitted with the source at different positions, S_1, S_2, S_3, and S_4, respectively. Thus, to observer A, the waves seem to follow each other more closely, at a decreased wavelength and increased frequency, whereas to observer C they are spread out and arrive at an increased wavelength and decreased frequency. To observer B, in a direction at right angles to the motion of the source, no effect is observed. The effect is produced only by a motion *toward* or *away* from the observer, a motion called *radial velocity.* Observers between A and B and between B and C would observe some shortening or lengthening of the light waves (increase or decrease in frequency), respectively, for a component of the motion of the source is in their line of sight.

The Doppler effect is also observed in sound. We have all heard the higher than normal pitch of the whistle of an approaching train and the lower than normal pitch of a receding one. The wavelengths are shortened if the distance between the source and observer is decreasing, and they are lengthened if the distance is increasing.

If the relative motion is entirely in the line of sight, the exact formula for the Doppler shift is

$$\frac{\Delta\lambda}{\lambda} = \frac{\sqrt{1 + v/c}}{\sqrt{1 - v/c}} - 1,$$

where λ is the wavelength emitted by the source, $\Delta\lambda$ is the difference between λ and the wavelength measured by the observer, c is the speed of light, and v is the relative line of sight velocity of the observer and source, which is counted as positive if the velocity is one of recession and negative if it is one of approach. If the relative velocity of the source and observer is small compared to the speed of light, however, the formula reduces to the simple form, which is usually used:

$$\frac{\Delta\lambda}{\lambda} = \frac{v}{c}.$$

Solving this last equation for the velocity, we find

$$v = c\frac{\Delta\lambda}{\lambda}.$$

If a star approaches or recedes from us, the wavelengths of light in its continuous spectrum appear shortened or lengthened, as well as those of the dark lines. However, unless its speed is tens of thousands of kilometers per second, the star does not appear noticeably bluer or redder than normal. The Doppler shift is thus not easily detected in a

Figure 8.18 Series of Balmer lines in the spectrum of hydrogen.

continuous spectrum (except for very remote galaxies—see Chapter 37) and cannot be measured accurately in such a spectrum. On the other hand, the wavelengths of the absorption lines can be measured accurately, and their Doppler shift is relatively simple to detect. Generally, when the spectrum of a star or other object is photographed at the telescope, sometime during the exposure the light from an iron arc or some other emission-line source is allowed to pass into the same spectrograph, and the spectrum of the arc is then photographed just beside that of the star. The known wavelengths of the bright lines in the spectrum of the arc (or other laboratory source) serve as standards against which the wavelengths of the dark lines in the star's spectrum can be accurately measured. Further illustrations of the Doppler effect are given in later chapters.

(c) The Spectrum of Hydrogen

As explained in Chapter 1, atoms are the smallest particles into which a chemical element can be subdivided and still retain its chemical identity. An atom consists of two parts: a nucleus of protons and neutrons, each proton of which carries a small unit of electric charge,* and a system of electrons, each with a negative charge equal to the positive one on a proton. In its ordinary neutral state, an atom has as many electrons as its has protons in its nucleus. The neutral hydrogen atom, for example, has one proton and one electron. A helium atom has a nucleus of two protons (and two neutrons) and two electrons. The various other elements have larger numbers of protons in their nuclei (and electrons outside): 8 for oxygen, 16 for iron, and 92 for uranium.

A clue to the structure of atoms came from the study of the spectrum of hydrogen, whose atoms are the simplest in nature. The dark lines of hydrogen that can be observed in the spectra of many stars occur in an orderly spaced series of wavelengths. The bright lines of hydrogen that are observed in the laboratory spectrum of glowing hydrogen are observed in the same series of wavelengths. The Swiss physicist Balmer found that these wavelengths could be represented by the formula

$$\frac{1}{\lambda} = \frac{1}{911.8}\left(\frac{1}{2^2} - \frac{1}{n^2}\right),$$

where λ is the wavelength in angstroms and n is an integer that can take any value from 3 on. If $n = 3$, the wavelength of the first line in the *Balmer series* of hydrogen is obtained (at 6563 Å in the red). For $n = 4, 5$, and so on, the wavelengths of the second, third, and higher Balmer lines are obtained. As n approaches larger and larger values, the wavelengths of the successive Balmer lines become more and more nearly equal. The lines of hydrogen in stellar spectra are observed to do just this; they approach a limit at about 3650 Å (Figures 8.18 and 8.19), corresponding to a value of $n = \infty$.

After Balmer's work, other series of hydrogen lines were found. The *Lyman series,* in the ultraviolet, approaches a limit at about 912 Å. The *Paschen series,* in the infrared, approaches a limit at about 8200 Å. Still farther in the infrared are found the *Brackett series,* the *Pfund series,* and so on. All these series (including the Balmer series) can be predicted by the more general formula, known as the *Rydberg formula:*

$$\frac{1}{\lambda} = R\left(\frac{1}{m^2} - \frac{1}{n^2}\right),$$

where m is an integer and n is any integer greater than m. The *Rydberg constant,* R, has the value 109,678 if λ is measured in centimeters. For the Lyman series, $m = 1$; $m = 2, 3, 4, \ldots$ for the Balmer, Paschen, Brackett, and other series.

(d) The Bohr Atom

The Danish physicist Niels Bohr (1885–1962) suggested that the hydrogen spectrum can be explained if the assumption is made that the electrons in the hydrogen atom revolve about the nucleus in orbits,

* The amount of charge is 4.8×10^{-10} electrostatic units.

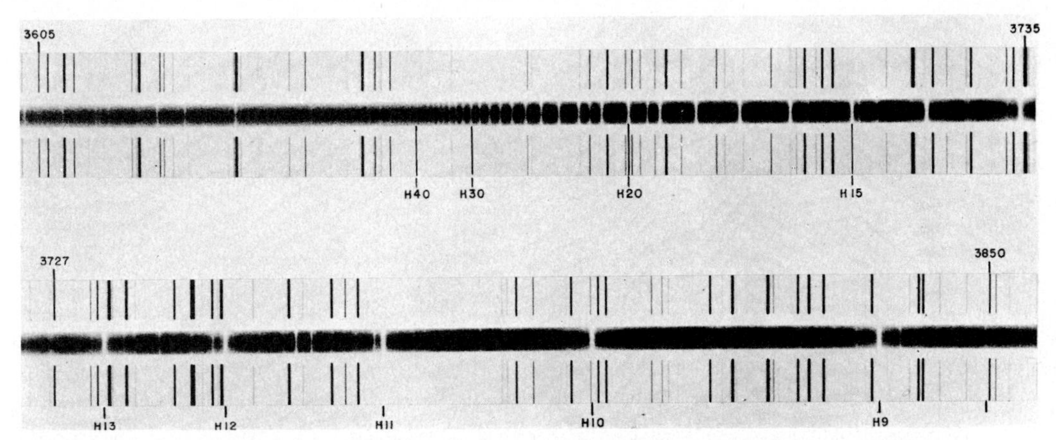

Figure 8.19 Photograph of part of the spectrum of the star HD 193182, showing many lines of the Balmer series. Several of the lines are identified by number. The spectrum has been split and is shown in two segments for convenience in reproduction on the page. (Mount Wilson and Las Campañas Observatories)

but that only orbits of certain sizes are possible. By specifying those permissible sizes for the electron orbits, Bohr was able to compute the values of energy, corresponding to the orbital motion of the electron, that are possible for an individual atom. He assumed that an atom can change from one allowed state of energy to another state of higher energy if its electron moves from a smaller to a larger allowed orbit. Conversely, according to the hypothesis, if the electron moves from a larger to a smaller orbit, the atom changes from a higher to a lower state of energy. One way in which an atom can gain or lose energy is by absorbing or emitting light. Since light is composed of *photons* whose energies depend on their wavelengths (the energy of a photon is hc/λ), the only wavelengths of light that could be absorbed or emitted are those corresponding to photons possessing energies equal to differences between various allowed energy states of the hydrogen atom. Those energy states are given by a simple formula that Bohr derived from the orderly progression of wavelengths in each series of lines in the hydrogen spectrum. In other words, by assuming that the allowed sizes of the electron orbits and thus the possible energies of the hydrogen atom are *quantized,* Bohr was able to account for the spectrum of hydrogen.

For example, suppose a beam of white light (which consists of photons of all wavelengths) is passed through a gas of atomic hydrogen. A photon of wavelength 6563 Å has the right energy to raise an electron in a hydrogen atom from the second to the third orbit, and can be absorbed by those hydrogen atoms that are in their second to lowest energy states. Since the energy of a photon is inversely proportional to its wavelength, the shorter the wavelength of a photon, the higher its energy. Photons with higher energies corresponding to the other successively shorter wavelengths in the Balmer series, therefore have the right energies to raise an electron from the second orbit to the fourth, fifth, sixth and larger orbits, and can also be absorbed. Photons with intermediate wavelengths (or energies) cannot be absorbed. Thus, the hydrogen atoms absorb light only at certain wavelengths, and produce the spectral *lines.* Conversely, hydrogen atoms in which electrons move from larger to smaller orbits emit light—but again only light of those energies or wavelengths that correspond to the energy differences between permissible orbits. The transfer of electrons giving rise to spectral lines is shown in Figure 8.20.

A similar picture can be drawn for kinds of atoms other than hydrogen. However, since they ordinarily have more than one electron each, the energies of the orbits of their electrons are much more complicated, and the problem of their spectra is much more difficult to handle theoretically.

(e) Energy Levels of Atoms and Excitation

Bohr's model of the hydrogen atom was one of the beginnings of quantum theory and was a great step

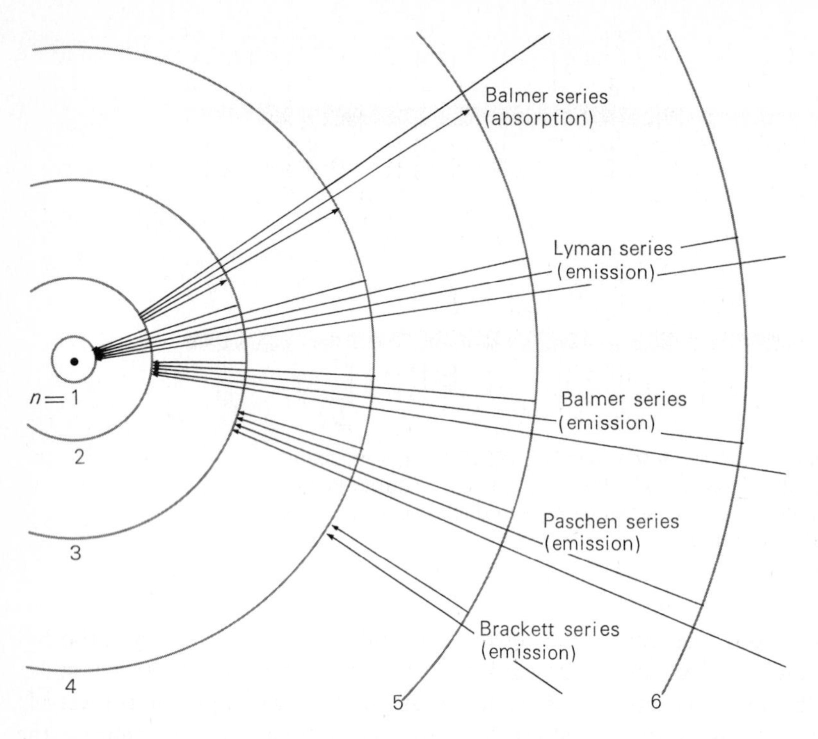

Figure 8.20 Emission and absorption of light by the hydrogen atom according to the Bohr model. Several different series of spectral lines are shown, corresponding to transitions of electrons from or to certain allowed energy levels. These series are named for the various physicists who studied them.

forward in the development of modern physics and in our understanding of the atom. However, we know today that atoms cannot be represented by quite so simple a picture as the Bohr model. Even the concept of discrete orbits of electrons must be abandoned, because according to the modern quantum theory it is impossible at any instant to state the exact position and the exact velocity of an electron in an atom simultaneously. Nevertheless, we still retain the concept that only certain discrete energies are allowable for an atom. These energies, called *energy levels,* can be thought of as representing certain mean or average distances of an electron from the atomic nucleus.

Ordinarily, the atom is in the state of lowest possible energy, its *ground state,* which, in the Bohr model, would correspond to the electron being in the innermost orbit. However, an atom can absorb energy that raises it to a higher energy level (corresponding, in the Bohr picture, to the movement of an electron to a larger orbit). The atom is then said to be in an *excited state.* Generally, an atom remains excited only for a very brief time; after a short interval, typically a hundred-millionth of a second or so, it drops back down to its ground state, with the simultaneous emission of light, unless it chances to absorb another photon first and go to a still higher state. (In the Bohr model, this corresponds to a jump by the electron back to the innermost orbit.) The atom may return to its lowest state in one jump, or it may make the transition in steps of two or more jumps, stopping at intermediate levels on the way down. With each jump, it emits a photon of the wavelength that corresponds to the energy difference between the levels at the beginning and end of that jump. An energy-level diagram for a hydrogen atom and several possible *atomic transitions* are shown in Figure 8.21; compare this figure with the Bohr model, shown in Figure 8.20.

Energy differences in atoms are usually expressed in *electron volts.* An electron volt (abbreviated eV) is the small amount of energy acquired by an electron after being accelerated through a potential difference of one volt. The energy needed to raise the hydrogen atom from its ground state to its first excited state is about 10.2 eV. One eV $= 1.602 \times 10^{-12}$ ergs. One erg is very roughly the energy required by a fly to do a pushup. Ten million ergs of energy expended in 1 s is a *watt* of power. Typical electric light bulbs use about 100 watts.

Because atoms that have absorbed light and have thus become excited generally de-excite themselves and emit that light again, we might wonder why dark lines are ever produced in stellar spectra. In

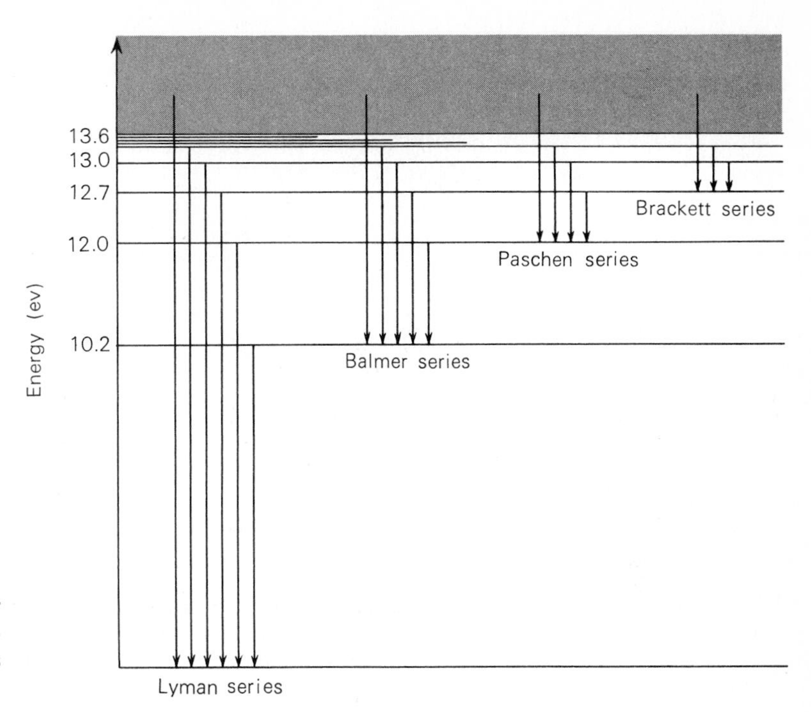

Figure 8.21 Energy-level diagram for hydrogen. The shaded region represents energies at which the atom is ionized (see Section 8.3g).

other words, why doesn't this re-emitted light "fill in" the absorption lines? Some of the re-emitted light actually *is* received by us and this light does partially fill in the absorption lines, but only to a slight extent. The reason is that the atoms re-emit the light they absorb in random directions. Now the absorption of light we would otherwise observe is of that light "coming our way" from deeper, hotter levels in the star's atmosphere, while the light they re-emit goes off in all directions, and not toward us. To be sure, other atoms can absorb light *not* coming our way and then re-emit it in our direction, but many of these atoms are in a higher, cooler layer of the star's atmosphere, so the light that they thus scatter toward our telescope is not intense enough to replace that from the hot gas that is absorbed. Therefore, the dark lines persist. We can observe the re-emitted light as emission lines only if we can view the absorbing atoms from a direction from which no light with a continuous spectrum is coming—as we do, for example, when we look at gaseous nebulae (Chapter 27). Figure 8.22 illustrates the situation.

We can calculate from theory the allowable energies of the simplest kinds of atoms and thus the wavelengths of the absorption lines that can be produced by those atoms. For the more complicated atoms, however, the wavelengths of the spectral lines are determined empirically by laboratory experiment.

Atoms in a gas are moving at high speeds and continually colliding with each other and with electrons. They can be excited and de-excited, therefore, by these collisions as well as by absorbing and emitting light. The mean energy of atoms in a gas depends upon its temperature (actually, *defines* the temperature), and if we know the temperature of the gas, we are able to calculate what fraction of its atoms, at any given time, will be excited to any given energy level. In the photosphere of the sun, for example, where the temperature is in the neighborhood of 6000 K, only about one atom of hydrogen in 100 million is excited to its second energy level by the process of collision. The Balmer lines of the hydrogen spectrum arise when atoms in this second energy level absorb light and rise to higher levels. At any given time, therefore, most atoms of hydrogen in the sun cannot take part in the production of the Balmer lines.

(f) The Uncertainty Principle

A particular energy state is one of the more likely energies an atom can have, but there is almost as good a

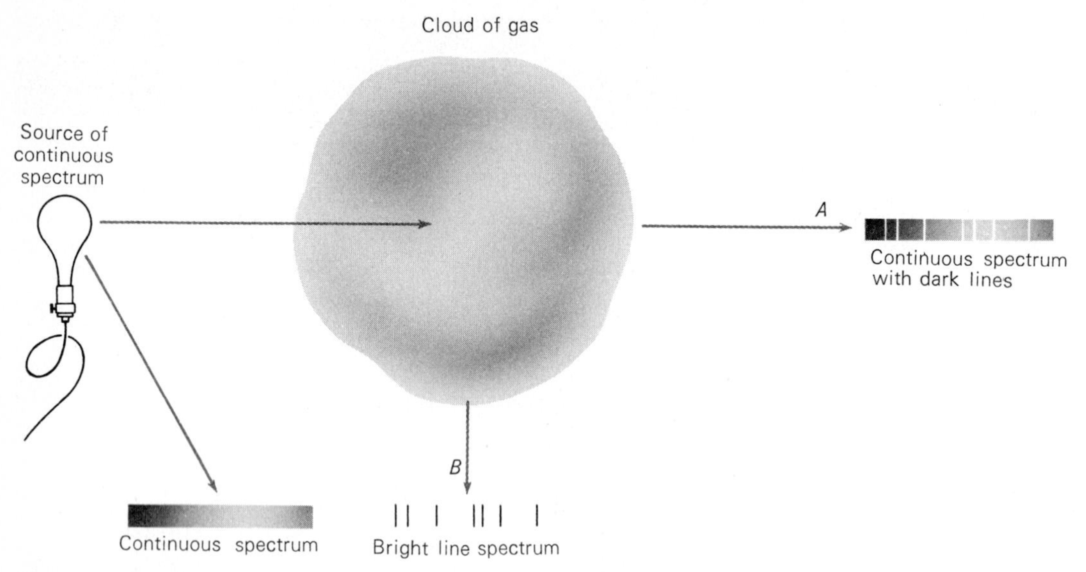

Figure 8.22 The atoms in the gas cloud produce absorption lines in the continuous spectrum of the white light source when viewed from direction *A,* but they produce emission lines (of the light they re-emit) when viewed from direction *B.* The spectra are shown as they appear on photographic negatives.

chance that it will have a very slightly different energy. So the energy levels themselves are fuzzy, and therefore so are the spectral lines. Those lines are not *absolutely* sharp, but just pretty sharp.

The vagueness of where an electron is in an atom arises from the inherent uncertainty and statistical nature of fundamental units of matter touched on in the foregoing scene. The German physicist Werner Heisenberg considered an idealized experiment in which one tries to measure with perfect precision the simultaneous position and momentum of a particle. He showed that even with theoretically optimum circumstances, there are certain inherent limitations (such as the finite speed of light) to the experimental procedures that prevent those two quantities from ever being known better than to a certain limiting accuracy. There is always an uncertainty in the position, and one in the momentum of the particle; the more accurately one of the two is known, the less accurately is the other. The product of those two uncertainties is ultimately never less than a certain small number (Planck's constant, *h,* divided by 2π). The result is called the *Heisenberg uncertainty principle.*

The uncertainty principle applies to everything, large and small. However, a large object like a bowling ball has enough mass that even a tiny uncertainty in its velocity results in the momentum being uncertain enough that its position (so far as this quantum-mechanical limitation is concerned) can be specified to a far higher accuracy than we could ever hope to measure. Suppose, for example, the speed of the bowling ball is known to within 0.1 mm/s—in other words, darned accurately. Then, the uncertainty principle says that theoretically the ball's position can never be known to better than 0.0000000000000000000000000000015 cm (more easily written 1.5×10^{-29} cm)—not a very interesting limitation to the average bowler.

But an electron has a tiny mass, and if its velocity is uncertain by that same amount, its position is uncertain by more than one meter!

It may seem that we are just talking about our inability to make a precise measurement, and that the real momentum and real position *exist* to perfect precision. But how can we define something like perfect precision if there is no possible experiment, even theoretically, by which we can determine it? According to the rules of science something that cannot be defined by means of a measurement or experiment (that is, operationally) does not exist—at least not in the realm of science. Since the limitation is theoretical, we must regard the uncertainty principle as *inherent* to nature. The *electron* cannot simultaneously know its precise position and momentum—those precise quantities do not exist at all! There is, in fact, very ample evidence to prove that such is the case.

(g) Ionization

We have described how certain discrete amounts of energy can be absorbed by an atom, raising the atom to an excited state, and moving one of its electrons farther from its nucleus. If enough energy is absorbed, the electron can be removed completely from the atom. The atom is then said to be *ionized* (Figure 8.21). The minimum amount of energy required to ionize an atom from its ground state is called its *ionization energy* or *ionization potential*. The ionization potential for hydrogen is about 13.6 eV. Still greater amounts of energy must be absorbed by the ionized atom (called an *ion*) to remove a second electron. The minimum energy required to remove this second electron is called the *second ionization energy* or *potential*. The third, fourth, and fifth ionization potentials are the successively greater energies required to remove the third, fourth, and fifth electrons from the atom, and so on. If enough energy is available (in the form of very short wavelength photons or in the form of a collision with a very fast-moving electron or another atom), an atom can become *completely ionized*, losing all of its electrons. A hydrogen atom, having only one electron to lose, can be ionized only once; a helium atom can be ionized twice, and an oxygen atom, eight times. Any energy over and above the energy required to ionize an atom can be absorbed also, and appears as energy of motion (kinetic energy) of the freed electron. The electrons released from atoms that have been ionized move about in the gas as free particles, just as the atoms and ions do.

An atom that has become ionized has lost a negative charge—that carried away by the electron—and thus is left with a net positive charge. It has, therefore, a strong affinity for a free electron, and eventually will capture one and become neutral (or ionized to one less degree) again. During the capture process, the atom emits one or more photons, depending on whether the electron is captured at once to the state corresponding to the lowest energy level of the atom, or whether it stops at one or more intermediate levels on the "way in." Any energy that the electron possessed as kinetic energy before capture can be emitted. Absorption or emission of light over a continuum of wavelengths therefore accompanies the process of ionization or recapture, at least over those wavelengths corresponding to energies higher than the ionization energy of the atom. Atoms of various kinds, being ionized and deionized, and absorbing and emitting light at various wavelengths, account for much of the continuous opacity and continuous spectrum of the sun and stars.

Just as the excitation of an atom can result from a collision with another atom, ion, or electron (collisions with electrons are usually most important), so also can ionization. The rate at which such collisional ionizations occur depends on the atomic velocities and hence on the temperature of the gas. The rate of recombination of ions and electrons also depends on their relative velocities, that is, on the temperature, but in addition it depends on the density of the gas; the higher the density, the greater the chance for recapture, because the different kinds of particles are crowded closer together. From a knowledge of the temperature and density of a gas, it is possible to calculate the fraction of atoms that have been ionized once, ionized twice, and so on. In the photosphere of the sun, for example, we find that most of the hydrogen and helium atoms are neutral, whereas most of the atoms of calcium, as well as many other metals, are once ionized.

The energy levels of an ionized atom are entirely different from those of the same atom when it is neutral. In each degree of ionization, the energy levels of the ion, and thus the wavelengths of the spectral lines it can produce, have their own characteristic values. In the sun, therefore, we find lines of *neutral* hydrogen and helium, but of *ionized* calcium. Ionized hydrogen, of course, having no electron, can produce no absorption lines.

There is an additional mechanism by which ionized atoms can absorb and emit light in the continuous spectrum. When an electron passes near an ion, it is attracted by that ion's positive charge (because opposite electrical charges attract each other). If the electron is not captured, it will pass the ion in a hyperbolic orbit, much like two stars passing in interstellar space. While the electron is passing, however, the ion can absorb or emit a photon with a corresponding increase or decrease in the kinetic energy of the hyperbolic motion of the electron. The process is called *free-free* absorption or emission (because the electron is "free" of the ion both before and after the encounter), or *bremsstrahlung*. Photons of any wavelength can be absorbed or emitted in bremsstrahlung.

(h) Spectra of Molecules

Molecules, combinations of two or more atoms, can also absorb or emit light. In addition to undergoing electronic transitions, molecules can also rotate and

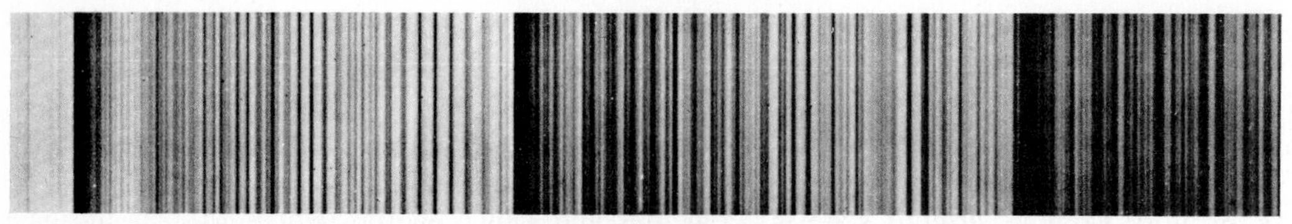

Figure 8.23 Series of closely spaced lines comprising bands in the spectrum of the compound made up of molecules of titanium oxide. Note how the bands coalesce to form band heads. (Mount Wilson and Las Campañas Observatories)

vibrate, all of which involve energy. The quantum theory predicts that the energy of vibration and rotation of molecules is quantized, like the energy of atoms. The vibrational and rotational energies are generally low, but they add to or subtract from the energies corresponding to electronic transitions. Consequently, in place of each atomic energy level there is a series of closely spaced levels, each one corresponding to a different mode of vibration or rotation of the molecule. Many more different transitions between energy levels are possible, therefore, differing from each other only slightly in energy or wavelength. Molecules, in other words, produce series of closely spaced lines known as *molecular bands* (Figure 8.23). In spectra of those stars in which molecules exist, these many molecular lines within a band are often not resolved as separate, and only a single broad absorption feature is observed.

(i) Stimulated Emission, Masers, and Lasers

In our discussion of the emission and absorption of radiation, we deferred until now mention of a phenomenon called stimulated emission. There are two ways in which excited atoms radiate photons in transitions to the ground state. Usually electrons jump down and emit photons spontaneously. However, if an atom in an excited level is passed by a photon of exactly the excitation energy, that photon can stimulate (or induce) the atom to emit an identical photon and transfer to the ground state. The two photons travel in the same direction and have the same phase as the one that was passing by (that is, the crests and troughs of their waves coincide). Emission by this process is called *stimulated* (or *induced*) *radiation*. The two photons that have identical direction of motion and phase are said to be *coherent*.

Photons of stimulated emission are usually quickly absorbed by other atoms in the ground state, exciting them to that same level. Thus we would normally not expect to see emission lines from a gas due to stimulated emission. However, in some circumstances, a larger fraction of the atoms of a gas are in a particular excited state than in the ground state. Suppose, for example, that the atoms that are in the ground state are continually excited to high energy levels by, say, intense radiation or by an electric field. Now suppose that these excited atoms undergo downward transitions to a lower excited level from which there is a relatively low probability of a spontaneous emission of a photon with a transition the rest of the way to the ground state. Thus the atoms that linger in that lower excited state can outnumber those in the ground state. Then those photons emitted spontaneously from the few excited atoms that do jump to the ground state, passing other excited atoms, can induce them to emit. Now, because there are not enough ground-state atoms to absorb all of these photons of stimulated emission, those not absorbed can induce still other atoms to emit. The radiation at that wavelength is thus amplified to a far greater intensity than would ordinarily occur.

Stimulated emission is the basis of the maser, at radio (microwave) wavelengths, and of the laser, at optical wavelengths. The word *maser* is an acronym for *microwave amplification by stimulated emission of radiation*, and *laser* for *light amplification by stimulated emission of radiation*. In these devices atoms in an enclosure, subjected to intense radiation or an electric field, are excited or ionized to energies from which they undergo downward transitions to levels from which stimulated emission can produce amplification. The energy supplied externally to produce the excitation or ionization is called *pumping* (Figure 8.24).

In the optical laser the light beam of stimulated radiation is reflected back and forth through the gas (sometimes a solid, as in the ruby laser) many times by mirrors at either end of the enclosure. With each pass the beam induces more and more radiation to

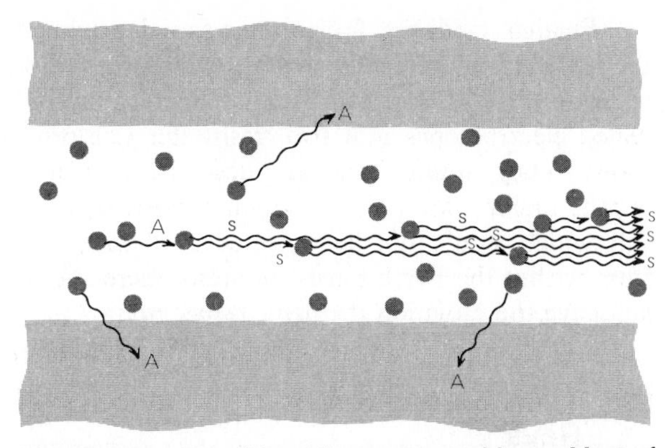

Figure 8.24 Principle of the maser and laser. Most of the atoms in an enclosure are excited from the ground state by a pumping process. The highly excited atoms transfer to lower energy levels, ending up in an excited state from which a few spontaneously jump to the ground state, emitting photons marked *A*. Some of these photons stimulate other excited atoms to emit photons, marked *S*. Because there are too few atoms in the ground state to reabsorb all of these photons from stimulated emission, they are able to induce still further emissions. Those that happen to travel along the length of the enclosure may be reflected back and forth many times (by mirrors in a laser), thus amplifying the radiation into an intense, coherent beam.

reinforce or amplify itself. The mirror at one end is half silvered to reflect part of the beam back for further amplification, and to allow the rest to pass out. Because the emerging beam of photons is coherent, it is narrow and intense, and can be transmitted over long distances.

Masers evidently occur naturally in some interstellar gas clouds. The radio emission lines from some molecules observed in interstellar space (Chapter 27) are far more intense than would be expected unless their energy were reinforced by the maser process. The pumping of the interstellar molecules to higher excited levels is due to infrared radiation from nearby stars.

(j) Summary of Emission and Absorption Processes

Atoms are characterized by energy levels that correspond to various distances of their electrons from their nuclei. By absorbing or emitting radiant energy, an atom can move from one to another of these levels, thus raising or lowering its energy. Since only certain discrete energy levels exist for each kind of atom, the absorbed or emitted radiation occurs only at certain energies or wavelengths, producing dark or bright spectral lines.

An atom is said to be *excited* if it is in any but its lowest allowable energy level. It is said to be *ionized* if, by the absorption of energy, it has lost one or more of its electrons. It can be excited or deexcited by collisions as well as by the absorption and emission of radiation; it can be ionized by collision or by absorbing radiation. Any wavelength of light corresponding to an energy greater than the ionization energy of an atom can be absorbed or emitted in the process of ionization or of deionization when the ion captures an electron.

Ionized atoms *(ions)* can also absorb or emit light at continuous wavelengths, when free electrons pass near them. The relative energy of the ion and passing electron changes by the same amount as the energy absorbed or emitted.

8.4 COSMIC RAYS

Our bodies are constantly being subjected to a rain of invisible high-energy particles passing through them. These bullets of radiation, undetected by our senses, result from the entrance of some 10^{18} atomic nuclei into the Earth's atmosphere each second at speeds near that of light. The total rate of influx of energy to the Earth from these particles is comparable to the rate at which the Earth receives energy that comes from starlight.

The physicist has learned, through the study of this phenomenon, of kinds of subatomic particles (for example, *muons* and *positrons*) hitherto unobserved. The incoming particles of extraterrestrial origin interest the astronomer because most are believed to come from beyond the solar system, and the discovery of the origin of these cosmic rays may provide us with new clues about the nature of the universe.

(a) Discovery of Cosmic Rays

For more than a century it has been known that the air has a slight electrical conductivity, because a charged body exposed to the air slowly loses its charge. Some of the atoms in the air, therefore, must be ionized. The electrons released from the atoms that have been ionized are attracted to a positively charged body, and the ions themselves are attracted to a negatively charged body; in either case, the charge on the body is reduced. The physicists J. Elster and H. Geitel investigated the conductivity of the air with an electroscope in 1899 and

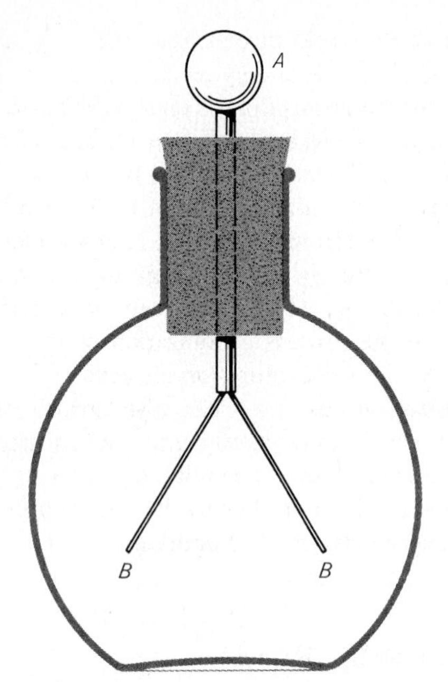

Figure 8.25 A simple electroscope.

1900 and found that the ionized particles in the air were continually being replenished.

An *electroscope* is a very simple device for detecting the presence of charged particles. An elementary homemade type of electroscope is illustrated in Figure 8.25. It operates on the principle that like charges repel. If the ball, A, at the top of the instrument is charged with electricity, part of that charge is conducted into the metal-foil leaves, B. Since the leaves have the same charge, they repel each other and separate; a measure of their separation is a measure of the total charge on the instrument. Now, if particles of the opposite charge are attracted to the ball, A, the charge on the electroscope is gradually neutralized, and the leaves come together. Thus, the rate of closing of the leaves measures the rate at which the electric charge leaks from the instrument as a result of ionization (that is, the presence of charged particles) in the air.

In August 1912 the Austrian physicist Victor Hess carried an electroscope aloft in a balloon. He found that except near the ground the conductivity of the air *increased* with altitude. In 1914 this surprising result was confirmed by D. Kolhörster, who showed that the increase in conductivity with altitude continued to more that 8000 m. Apparently the radiation that ionized the air came either from high in the atmosphere or from beyond the Earth.

Further evidence for extraterrestrial origin of the mysterious radiation was provided in 1928, when R. A. Millikan and G. H. Cameron lowered sealed electroscopes into two freshwater California lakes. They found that at successively greater depths under water, the radiation decreased, a result which would not be expected if it originated from within the Earth's crust or atmosphere. Millikan gave the radiation the name *cosmic rays*.

(b) The Charged Nature of Cosmic Rays

At first, cosmic rays were believed to be very high energy photons, that is, electromagnetic energy of wavelengths even less than those of gamma rays. In 1927, however, the Dutch physicist J. Clay found that the intensity of the ionizing radiation (cosmic rays) varies with latitude, being least near the geomagnetic equator (the circle halfway between the geomagnetic poles), and increasing as the geomagnetic poles are approached.* Clay's observations have been confirmed with many subsequent experiments. It is difficult to understand why photons, which have no electrical charge, should be in any way affected by the magnetic field of the Earth as they approach it.

It is well known, on the other hand, that charged particles move on curved paths through a magnetic field. Charged particles entering the Earth's magnetic field, therefore, would be deflected by the field; if their deflections, that is, the curvatures of their paths, are great enough, they will never strike the Earth's atmosphere. Near the geomagnetic equator, only those particles of very high energy can reach the atmosphere. At higher geomagnetic latitudes, however, the lines of force of the Earth's field curve toward its surface, and particles of lower energy can penetrate the field. At the magnetic poles, where the lines of force are perpendicular to the surface of the Earth, particles of all energies reach the upper atmosphere. Thus, if the cosmic rays were charged particles, rather than pho-

---

* The *geomagnetic poles* are in line with the ends of a hypothetical ideal bar magnet whose magnetic field most nearly matches that of the Earth. The actual field of the Earth, however, is somewhat irregular, and the Earth's *magnetic poles*, where the actual lines of force are perpendicular to the surface, deviate by some hundreds of kilometers from the idealized geomagnetic poles.

tons, it would be easily understood why cosmic-ray intensity is greater nearer the magnetic poles. Today, it is well established that this explanation is, indeed, the correct one.

The realization only a few decades ago that the ionization noticed in the Earth's atmosphere is due to charged particles striking the Earth from space was one of the important discoveries of the 20th century. Investigation of these particles became, in the 1940s, a major effort of modern physics.

(c) Detection of Cosmic Rays

The first evidence for cosmic rays came from observations with electroscopes. More sophisticated instruments, however, are used today. A *radiation counter* (for example, a *Geiger counter*) consists of a gas-filled chamber across which an electric field is provided by oppositely charged electrodes on either side. When a high-energy charged particle enters the chamber, it ionizes some of the gas, so that the gas becomes momentarily conducting. This pulse of current flowing through the tube is amplified and recorded on a meter, or detected by means of a loudspeaker. A counter, therefore, can detect and record those individual charged particles that pass through it with enough energy to ionize the gas within the chamber. (It can detect gamma rays as well, by the same process.)

A *cloud chamber* is a chamber filled with a gas that is saturated with the vapor of water or of some other liquid. The chamber is so designed that its volume can be enlarged suddenly, usually by moving a rubber diaphragm or a tightly fitting piston. As it is enlarged, the gas contained within it lowers in density and cools. If the chamber were enlarged sufficiently, the gas would cool enough for the liquid to precipitate. In actual use the cloud chamber is not enlarged quite enough for the precipitation to occur by itself. If, however, a charged particle should happen to pass through the chamber and ionize some of the atoms of the gas just as it is cooling, those ions will serve as condensation nuclei, and a line of droplets will form, marking the path of the ionizing particle. The track can be observed visually or it can be photographed.

A more useful device for many experiments is the *bubble chamber,* similar in function to the cloud chamber. The bubble chamber contains a liquid that is superheated when the chamber is suddenly expanded. A charged particle moving through the liquid when it is in this condition causes it to boil along the track of the particle, leaving a string of tiny bubbles that can be observed or photographed.

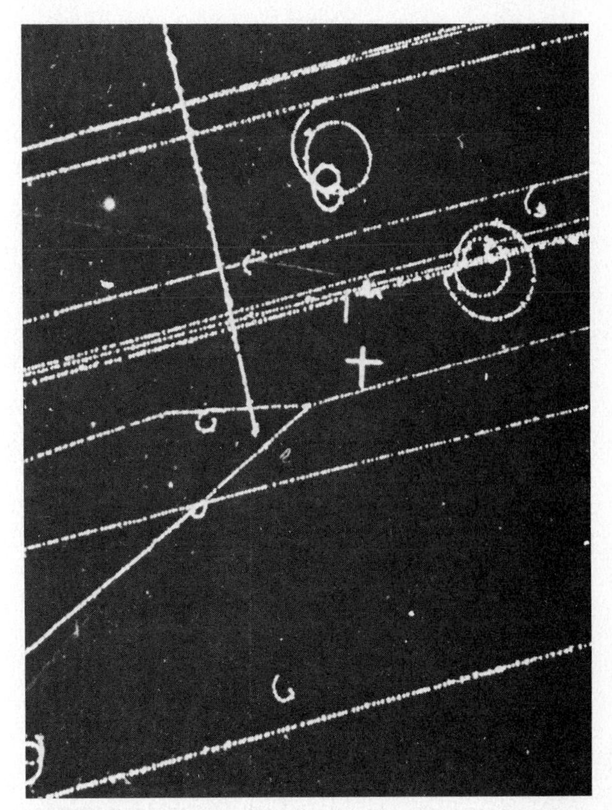

Figure 8.26 Bubble chamber photograph. (Courtesy H. Ticho, U.C.L.A.)

In either the cloud or bubble chamber, additional information is obtained by applying a magnetic field through the chamber, because the path of a charged particle in a magnetic field is curved. The amount of curvature depends on the field strength, on the charge on the particle, and on its momentum. Cloud and bubble chambers with magnetic fields, therefore, not only show the tracks of the impinging particles but indicate something about their charges and momenta.

Figure 8.26 shows tracks of charged particles through a bubble chamber containing liquid hydrogen. Most of the tracks are of the short-lived subatomic particles π-mesons, moving so fast that the magnetic field curves their trajectories only slightly. The branching near the middle of the picture is where a π-meson, coming from the right, struck a proton and produced two other charged nucleons. The upper one (a sigma-particle) quickly decayed into a neutral particle that left no track and another π-meson, whose track is visible in a still different direction. If all the particles, charged and neutral, are taken into account, the momentum is always conserved in such collisions and decays. The small spiral tracks are of low-energy electrons ejected from hydrogen atoms that are ionized by the high-speed mesons. The paths of these electrons are curved more and more by the magnetic field as they lose energy moving through the liquid.

Tracks of charged particles can also be recorded in *photographic emulsions*. The grains in an emulsion are made capable of being developed not only by photons of light but also by ionization by charged particles. Ordinarily, a photographic emulsion is a very thin coating on a piece of celluloid or glass. For detecting particle radiation, however, many layers of emulsion are often piled on top of one another, forming a thick emulsion *stack*. A charged particle passing through the stack leaves a track of developable grains behind it. After development the emulsion layers are separated and the track of the particle through each of them is measured; its course is thus determined.

Certain solid substances (for example, zinc sulphide) when bombarded by subatomic particles emit flashes of light or *scintillations*. A *scintillation counter* utilizes such a scintillating phosphor to detect particles; the light flashes produced are amplified with a photomultiplier and recorded. Certain plastics fluoresce when hit by energetic particles, and so can also be used as scintillating materials. Large disks of such plastic spread out over a substantial area of the ground (for example, in the M.I.T. experiments at Volcano Ranch, New Mexico) have been used to detect great numbers of cosmic-ray particles.

One technique by which particles from a particular direction can be isolated is to arrange two or more counters in a line, and to record only pulses that occur almost simultaneously in all of them. Such coincidences are almost always the result of single particles that pass successively through each counter, and hence that approach along (or nearly along) the direction in which the counters are lined up. If sheets of lead (or other materials) of various thicknesses are placed between the counters, the penetrating power of a particle, and thus its energy, can be determined.

Electroscopes, cloud and bubble chambers, and Geiger counters are no longer (or at best, rarely) used in cosmic-ray research, and emulsion stacks are used less and less in recent years, but these devices were important in the discovery and early investigation of cosmic rays. Space does not permit a discussion here of the many modern instruments that, in addition to scintillation counters, are in wide use today. At least, however, we have seen some ways in which fast-moving subatomic particles can be detected.

(d) Primary and Secondary Particles

Analysis of the primary cosmic-ray particles shows that most of them are high-speed protons (nuclei of hydrogen atoms), that most of the rest are alpha particles (nuclei of helium atoms), and that a few are nuclei of the still heavier atoms. A primary particle traverses, on the average, only about one-tenth of the Earth's atmospheric gases, however, before colliding with the nucleus of an air molecule.

When such a collision occurs, the nucleus in the air molecule breaks into several smaller subatomic particles. If the primary particle has high energy, each of these secondary particles is also given considerable energy. It, in turn, collides with still another nucleus in an air molecule, producing more secondary particles. In this way an original primary particle moving with high speed dissipates its energy in a great many secondary particles, producing many of the particles that are recorded at intermediate and low altitudes in the atmosphere. A large proliferation of particles by successive collisions following the impingement of a primary particle of very high energy is called a *shower*.

Perhaps it is well that we do not see these primary and secondary particles or feel them as they strike us. Otherwise, it might be discomforting to see repeated skyrocketlike bursts high above our heads, and then observe many of the burst particles passing into one side of our bodies and out through the other. Those few that are intercepted by the atoms of our bodies contribute to the natural radiation our bodies are constantly receiving, and probably are responsible for some mutations and even cancer.

From tracks in cloud chambers and through magnetic fields, physicists have learned the energies, charges, and masses of these secondary cosmic-ray objects. Most of them are charged particles with masses intermediate between those of protons and electrons. Such particles of intermediate mass are called *mesons*. At sea level the most common mesons in secondary cosmic rays are called *mu mesons* (or *muons*); each carries either a positive or negative charge that is numerically equal to the charge on the electron and has a mass equal to 207 electron masses.

Muons are very unstable particles and disintegrate in average periods of only about 1.5 microseconds. When a muon disintegrates (or *decays*) it forms an electron (if it had a negative charge) or a *positron* (if it had a positive charge). A positron is a particle equivalent to an electron but carrying a positive charge that is numerically equal to the negative charge on an ordinary electron. The excess mass of a muon, over that of an electron (or positron), is converted into energy and into *neutrinos*—particles that have energy but no mass. We shall return later (Chapter 29) to some of these unusual subatomic particles; it is worth noting, however, that a few of them were first observed in cosmic rays.

(e) Energies of Primary Cosmic-Ray Particles

The energies of most cosmic-ray primaries are near 10^9 eV (a thousand million electron volts, abbreviated GeV), but a small percentage have energies in excess of 10^{18} eV. The relative number of particles of various energies is called the cosmic-ray *energy spectrum*. The energy spectrum is now known fairly well for cosmic-ray particles with energies in the range 10^8 to 10^{19} eV. In that range the numbers of primaries of successively higher energy drops rapidly; the total number of particles, $N(E)$, with total energies greater than E is represented, approximately, by the empirical formula

$$N(E) = \frac{\text{constant}}{E^{1.6}}.$$

The constant in the above equation is different for the different kinds of atomic nuclei that make up the primary cosmic-ray particles. The conclusion is that particles of higher energy are much less common than those of lower energy.

For energies above 10^{19} eV, the energy spectrum is much less well established. It is known, however, that a few primaries of extremely high energy exist. Spectacular showers have been detected for several decades with many techniques. For example, with the large fluorescent plastic disks at their Volcano Ranch site, the M.I.T. physicists have observed very large showers of secondary particles since about 1960. Since primary particles of the highest energy produce the largest showers, the energy of a primary particle can be inferred from the number of shower particles it produces. The M.I.T. group has found that for every million showers of 1 million particles each, there are 1000 showers of 10 million particles each, and only one shower of 100 million particles.

Nevertheless, occasional showers of thousands of millions of particles are observed. A few are produced by primary particles with energy as great as 10^{20} eV; this is an amount of energy that could keep a one-watt electric light burning for one second.

The energies observed for some primary cosmic-ray particles are many times those that can be obtained in laboratory accelerators (cyclotrons, synchrotrons, betatrons, and so on). The total energy density of cosmic rays in space is estimated at about one eV/cm^3. Some of the possible origins of this energy will be considered in later chapters.

(f) Composition of the Primary Particles

As has been stated, the primary cosmic-ray particles are mostly atomic nuclei. The majority are protons (nuclei of hydrogen atoms). Most of the rest—about 15 percent—are alpha particles (nuclei of helium). The remaining 1 percent or so are nuclei of heavier elements; those of elements as heavy as iron are moderately abundant, and heavier ones have been observed. Very roughly, the relative abundances of the various atomic species in cosmic rays resemble those of elements elsewhere in the universe. There are exceptions, however. The cosmic-ray abundance of nuclei heavier than helium seems to be several times higher than the general cosmic abundance of those elements. The nuclei of the elements lithium, beryllium, and boron, in fact, are thousands of times as abundant in cosmic rays as in stars. Lithium, beryllium, and boron, however, are unstable at temperatures of a few million Kelvins, and in stars they would undergo nuclear transformations to other elements; their comparatively low abundances in stars, therefore, is not surprising. The high abundance of these nuclei in cosmic rays may result from the breakup of heavier nuclei. Finally, there is a small percentage (about 1 percent) of primary particles that are electrons.

EXERCISES

1. How many times brighter or fainter would a star appear if it were moved to (a) twice its present distance? (b) ten times its present distance? (c) half its present distance?

2. "Tidal waves," or *tsunami*, are waves of seismic origin that travel rapidly through the ocean. If tsunami traveled at the speed of 600 km/hr, and approached a shore at the rate of one wave crest every 15 min-

utes, what would be the distance between those wave crests at sea?

3. Stars are fairly good approximations to black bodies. Explain why they do not look black.

4. Suppose the sun radiates like a black body. Explain how you would calculate the total amount of energy radiated into space by the sun each second. What solar data would you need to make this calculation?

5. If the emitted infrared radiation from Pluto has a wavelength of maximum intensity at 0.005 cm, what is the temperature of Pluto?

6. What color would you expect light to appear that was equally intense at all visible wavelengths?

7. What is the temperature of a star with a wavelength of maximum light of 2.897×10^{-5} cm?

8. What is the energy of a photon of wavelength 3 cm?
Answer: 6.62×10^{-17} erg

9. Suppose that a spectral line of some element, normally at 5000 Å, is observed in the spectrum of a star to be at 5001 Å. How fast is the star moving toward or away from the Earth?
Answer: About 60 km/s away from the Earth

10. How could you measure the rotation rate of the sun by photographing the spectrum of light coming from various parts of the sun's disk?

11. How could you measure the Earth's orbital speed by photographing the spectrum of a star at various times throughout the year?

*12. Refer to the Rydberg formula and indicate to what series each of the following transitions in a hydrogen atom corresponds. State whether the spectral line produced is one of absorption or emission. List the lines in order of *increasing wavelength*.
a) $m = 3$ to $n = 8$
b) $m = 1$ to $n = 2$
c) $m = 1$ to $n = 4$
d) $m = 2$ to $n = 13$
e) $n = 17$ to $m = 2$
f) $n = 14$ to $m = 5$

13. Most hydrogen atoms in the sun are in their lowest state of energy, so what series of absorption lines would hydrogen produce most strongly in the sun?

14. Describe how the observed cosmic-ray intensity at the surface of the Earth would vary with the energy of the primary particle and with latitude if (a) the Earth had no magnetic field, (b) the Earth had a magnetic field millions of times stronger than its actual field.

15. Where on Earth would you like to live to have the best chance of avoiding radiation resulting from primary cosmic-ray particles striking the Earth at energies of 100 GeV? Why?

*16. A cosmic ray primary has an energy of 2.08×10^{11} eV. What is its energy in ergs? How many such particles would be required each second to yield the power equivalent of a 100-watt light bulb?
Answer to second part: About 3×10^9 such particles

George Ellery Hale (1868–1939) was one of the founders of the new science of astrophysics at the beginning of the 20th century. He also had the vision and leadership to initiate the construction of the world's largest telescope no less than four times! The 5-m telescope on Palomar Mountain is named in his honor. (California Institute of Technology)

9 OPTICAL ASTRONOMY

Electromagnetic radiation from space comes at all wavelengths, from gamma rays to radio waves. Unfortunately for astronomical observations (but fortunately for biological organisms) much of the electromagnetic spectrum is filtered out by the terrestrial atmosphere. There are two spectral windows in the atmospheric filter through which we can observe cosmic radiation (Figure 9.1). One of these is the *optical window,* which includes the near ultraviolet (wavelengths longer than about 3000 Å or 0.3 μm, where the notation μm means 10^{-6} meters) and portions of the infrared (with wavelengths up to about 40 μm or 0.04 mm). The other is the *radio window,* which includes radio waves that range in length from about a millimeter to about 20 m (the long wavelength cutoff depends somewhat on the variable conditions of the ionosphere).

In this chapter, we consider the optical window, together with the adjacent infrared (to about 3 μm). In this part of the infrared, optical systems

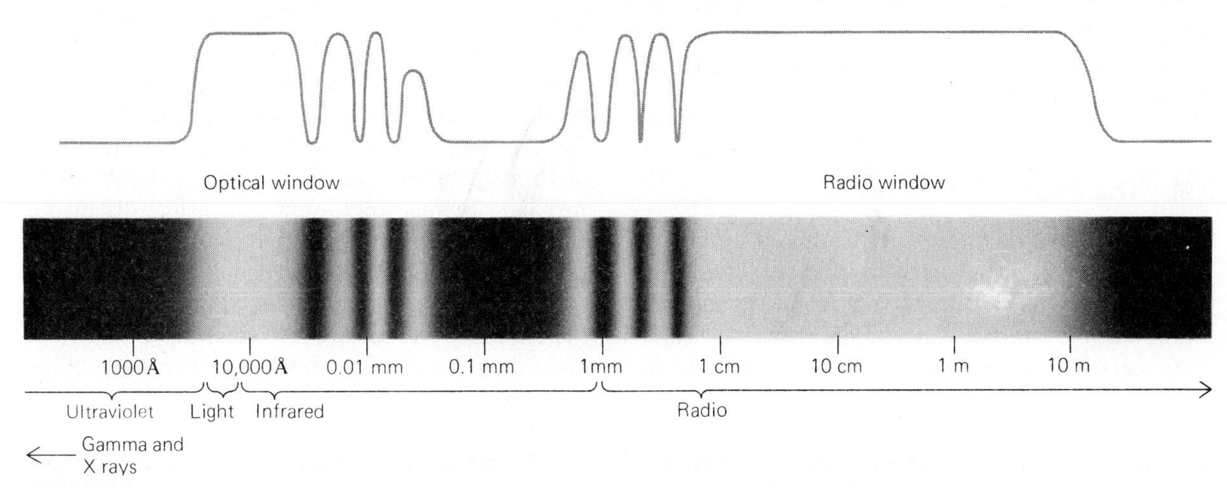

Figure 9.1 A portion of the electromagnetic spectrum, showing those regions (windows) where the Earth's atmosphere is transparent. The dark regions are those where the atmosphere is opaque. The upper graph is a plot of the transparency of the atmosphere.

and detectors are generally similar to those developed for visible light. In the next chapter we will cover longer wavelength infrared, where the situation is complicated by emission from the Earth's atmosphere and the telescope optics themselves.

9.1 OPTICAL INSTRUMENTS

Archaeoastronomy, which is the study of the practice of astronomy in the ancient world, has shown that the construction of major astronomical observatories has been a continuing concern of civilizations of diverse cultural values and levels of sophistication. Of course, ancient observatories were not used to study the heavens with any thought of trying to understand the processes that control the celestial phenomena that we observe. The Chinese, for example, were concerned exclusively with astrology, not astronomy, and in China the right to study the heavens was hereditary. Since the stars were thought to influence terrestrial events, knowledge of astronomy—or rather astrology—in the wrong hands might carry with it the power to overthrow the ruling dynasty.

Telescopes were a surprisingly late addition to the complement of instruments housed within observatories: surprising because the technology for making glass is an ancient one. The Mesopotamians were the first to fuse sand and ash to form glass, and by 1200 B.C. the Babylonians knew how to make objects of blown glass. Chinese mirrors, including curved mirrors and burning mirrors, were manufactured long before the birth of Christ. The Chinese did not, however, make either a telescope or a microscope until after the information on how to do so reached them from Europe.

(a) The First Telescopes

The telescope was almost surely invented by a manufacturer of eyeglasses, who by chance aligned and looked through two lenses, one concave and the other convex. This was an accident that took nearly 300 years to happen, since spectacles were used in Europe by the year 1300, primarily to correct the farsightedness that is a characteristic of advanced age. Once the first telescope was made, word traveled quickly throughout Europe, so quickly that no one was able to establish unambiguous claim to having been the inventor.

Galileo himself claimed after the fact that he had invented the telescope on the basis of profound considerations of perspective and of the doctrine of refraction. In fact, Galileo had done no research in optics, and his telescope was the product of trial and error combined with knowledge that such a device was feasible and had already been constructed by others. Galileo was, of course, not the first—and surely not the last—scientist to claim that results achieved by accident or serendipity or luck were, in fact, the product of a deep understanding of the implications of a new theory. On the other hand, knowledge of theory is many times not enough to produce practical results. In 1604, Kepler published a book that presented a precise explanation of the properties of lenses, but he did not build a telescope.

On August 25, 1609, Galileo demonstrated one of his first telescopes, which had a magnification of a factor of 9, to officials of the Venetian government. By a magnification of 9×, we mean that the linear dimensions of the object being viewed appeared 9 times larger, or equivalently, that the objects appeared 9 times closer than they really were. There were obvious military advantages associated with a device for seeing distant objects. For his invention Galileo's salary was nearly doubled and he was granted lifetime tenure as a professor. His colleagues were outraged, particularly since the invention was not even an original one.

In yet another example of how it often takes a very long time to do what in hindsight seems so obvious, Galileo did not use his telescope for astronomy for several months. First, he had to devise a stable mount, and he also improved the optics to provide a magnification of 30×. Galileo also had to acquire confidence in the telescope. At that time, the eyes were believed to be the final arbiter of the truth about sizes, shapes, and colors. Lenses, mirrors, and prisms were known to distort distant images by enlarging them, reducing them, or even inverting them. Galileo undertook repeated experiments to convince himself that what he saw through the telescope was identical to what he saw up close. Only then could he begin to believe that the miraculous phenomena that were revealed in the heavens were real. While Galileo was convinced of the validity of what he saw, others were not. One unbelieving colleague said that he ". . . tested this instrument of Galileo's in a thousand ways, both on things here below and on those above. Below, it

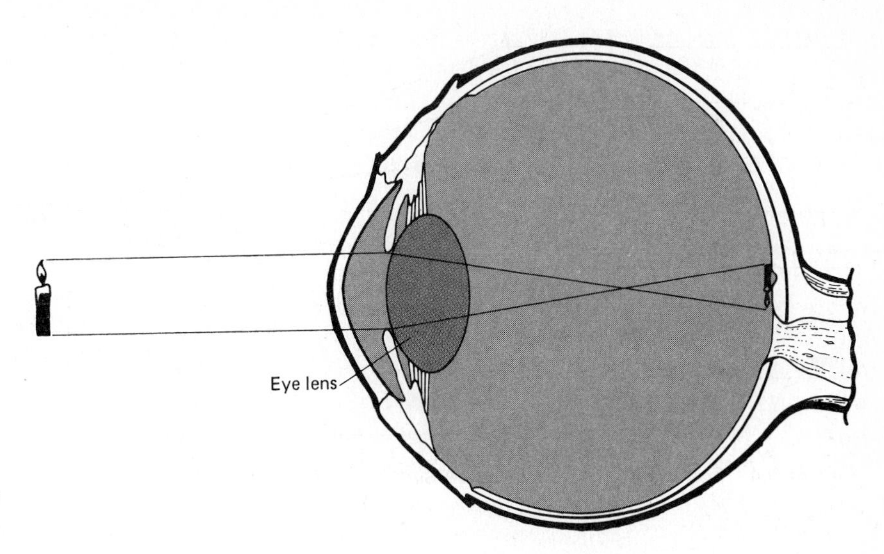

Figure 9.2 Production of an image on the retina of the eye.

works wonderfully; in the sky it deceives one, as some fixed stars are seen double." Another scholar refused even to look through the telescope because doing so gave him a headache.

We are all familiar with what Galileo discovered (Chapter 3), but stop and think how profoundly those discoveries, all made by the end of January, 1610, only five months after he constructed his first telescope, changed our view of the universe forever. Observation of the four moons of Jupiter and the subsequent detailing of their motions demonstrated unambiguously that the Earth was not the center of all orbital movement and established empirically that one body (a moon) could orbit a second (Jupiter) while that body in turn orbited a third (either the sun or the Earth, depending on whether one subscribed to the Copernican or Ptolemaic point of view). The discovery that the Moon had surface relief and geological features not altogether dissimilar to those seen on Earth raised the possibility that celestial phenomena could be studied and understood by applying the same ideas and ways of thinking that were successful in studying physical processes on Earth. The realization that the Milky Way was composed of apparently countless individual stars raised the possibility that the universe was vast beyond comprehension—perhaps even infinite in its extent.

(b) Formation of an Image by a Lens or a Mirror

In order to study astronomical objects, it is useful to form an *image* of a source of radiation of known direction in the sky. The image can then be detected, measured, reproduced, and analyzed in a host of ways. One image-forming device is the eye (Figure 9.2). It consists of a *lens,* which focuses light into an optical image on the *retina* at the back of the eye. There the image is detected by the light-sensitive retinal nerve endings, which transmit the information (through the optic nerve) to the brain for analysis. The retinal nerves can sense only that kind of electromagnetic energy we call light, but images can be made of other electromagnetic radiation as well. Ground-based optical astronomy is concerned with the forming of images in the near ultraviolet, visible light, and infrared, but the image-making principles are the same for even shorter and longer wavelengths.

Having examined the laws of geometrical optics (Section 8.2), we are in a position to understand the formation of images by optical systems. Images were first produced by simple convex lenses. To illustrate the principle, let us imagine two triangular prisms, base to base, as in Figure 9.3. Now suppose

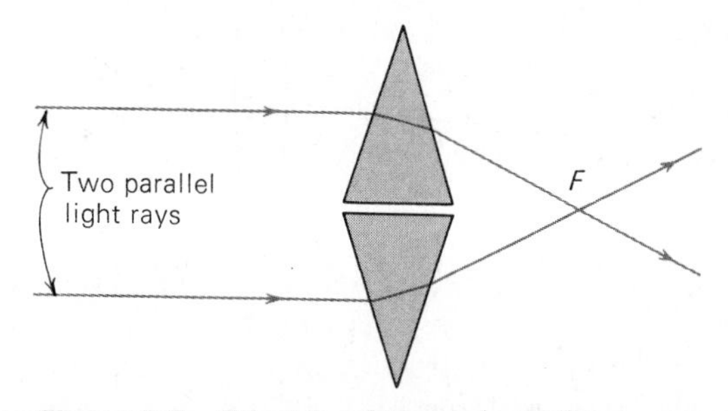

Figure 9.3 Principle of image formation.

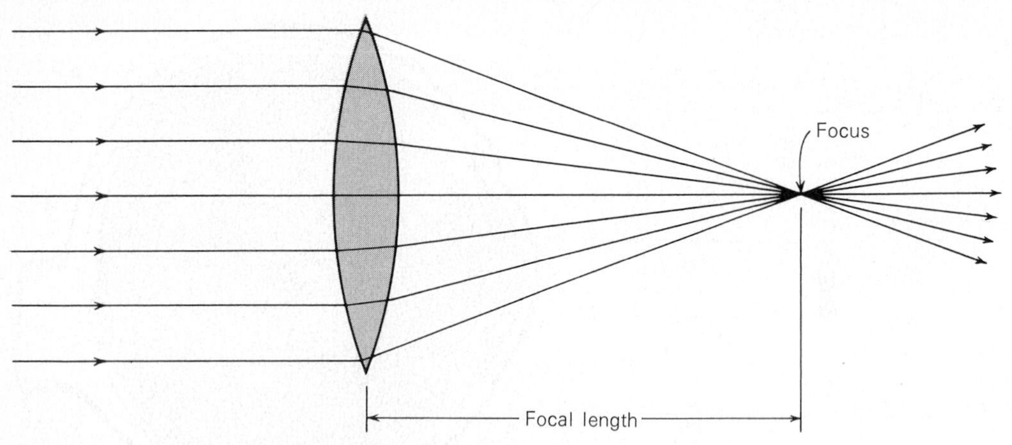

Figure 9.4 Formation of an image by a simple convex lens.

we select two of the parallel rays of light from a distant object and allow one ray to enter each prism. The light rays are refracted by the prisms and meet at a point F.

This is the principle underlying the formation of an image by a lens. In Figure 9.4 a simple convex lens is shown. Parallel light from a distant star or other light source is incident upon the lens from the left. A *convex* lens is thicker in the middle than at the edges. In cross section it is no more than a series of segments of prisms piled one on another, with slightly different slopes to their sides. If the curvatures of the surfaces of the lens are correct, light passing through the lens will be refracted in such a way that it converges toward a point. Convex lenses whose surfaces are portions of spherical surfaces are easiest to manufacture. Such lenses will refract a parallel beam of light to a point as shown in Figure 9.4, if the curvature of the surfaces is slight.

The point where light rays come together is called the *focus* of the lens. At the focus, an *image* of the light source appears. The distance of the focus, or image, behind the lens is called the *focal length* of the lens. A lens or other device that forms an image is called an *objective*.

We have seen how an image can be formed of a point source, say, a star. However, the image itself in that case is just a point of light. In Figure 9.5 we see how an image is formed of an extended source, for example, the Moon. From each point on the Moon, light rays approach the lens along parallel lines. However, from different parts of the Moon, the parallel rays of light approach the lens from different directions. The light from each point on the Moon strikes all parts of the lens (or *fills* the lens);

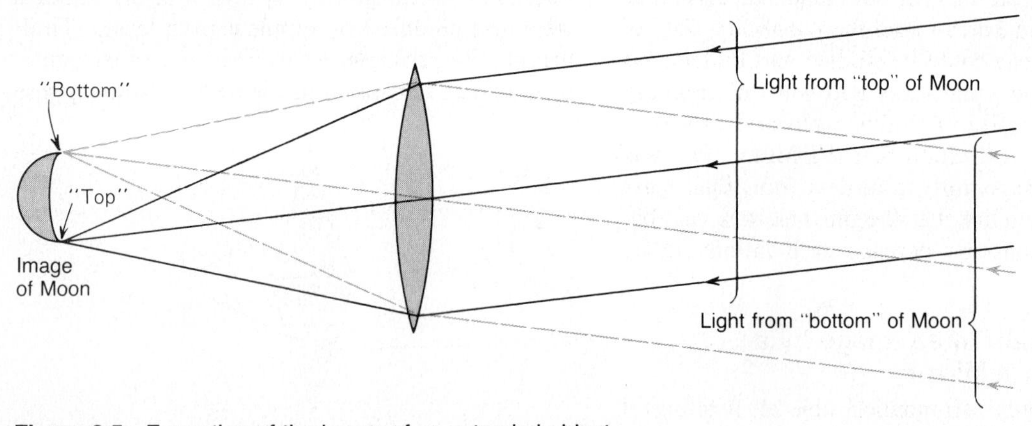

Figure 9.5 Formation of the image of an extended object.

these rays of light are focused at a point at a distance behind the lens equal to the focal length of the lens. If a screen, such as a white card, is placed at this distance behind the lens, a bright spot of light appears, representing that point on the Moon. Light from other points on the Moon similarly focuses at other points, producing bright spots on the card in different places. Thus, an entire image of the Moon is built up at the focus of the lens. The plane in which the image is formed is called the *focal plane*.

(We discuss here only the case where the object whose image is formed is so distant that light from any point on it can be regarded as approaching the lens along parallel rays. This is always true when any astronomical body is observed. Nearby terrestrial objects may be so close that the assumption is not valid. Then the image is formed at a point farther from the lens than the focal length.)

Note that if part of the lens is covered up, or if the middle is cut out, or if ink or mud is splattered over it, as long as part of the lens is still transparent to light, an entire image will be formed. All parts of the lens contribute to each part of the image. Covering up part of the lens cuts down the total amount of light that can strike each portion of the image and thus makes the image fainter, but nevertheless the whole image is formed. An ordinary camera lens produces an image at the focal plane (where the film is placed) just as is shown in Figure 9.5. Every photographer knows that it is possible to cover up part of a camera lens by "stopping it down" with an iris diaphragm. Doing so will cut down the *brightness* of the image (and hence the effective exposure on the film), but the other parts of the image will not be removed. The part of the lens that remains uncovered still produces the entire image.

Finally, we note that the image formed is always inverted and reversed (upside down and left to right) with respect to the object. The eye lens forms inverted images on the retina, but the brain interprets the image so that it appears upright. The eye is equipped with an adjustable iris diaphragm that automatically enlarges or contracts to allow more or less light to enter the eye, so that an image of the optimum brightness can be produced on the retina. An ordinary camera is a nearly complete analogy to the eye.

Rays of light can also be focused to form an image with a *concave* mirror—one hollowed out in

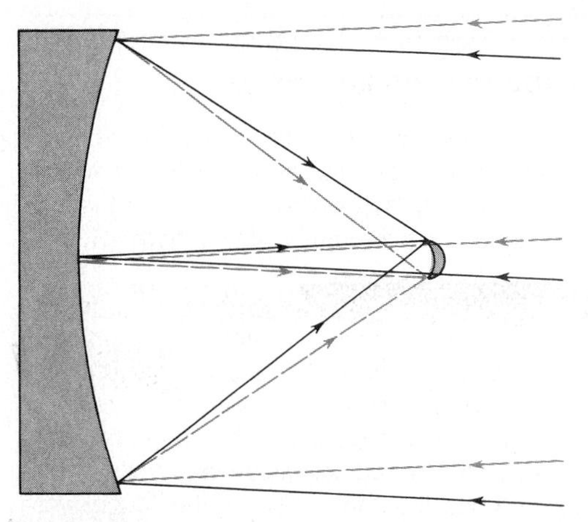

Figure 9.6 Formation of an image by a concave mirror. The dashed lines are from the "top" of the Moon.

the middle. Parallel rays of light, as from a star, fall upon the curved surface of the mirror (Figure 9.6), which is coated with silver or aluminum to make it highly reflecting. Each ray of light is reflected according to the law of reflection. If the mirror has the correct concave shape, all the rays are reflected back through the same point, the focus of the mirror. At the focus appears the image of the star. As in a lens, the distance from the mirror to the focus is called its focal length.

Rays of light from an extended object are focused by a mirror, exactly as they are by a lens, into an inverted image of the object. The principal difference between image formation by a lens and by a mirror is that the mirror reflects the light back into the general direction from which it came, so that the image forms in front of the mirror. The image can be inspected, as with the lens, by allowing the light to illuminate a screen, such as a white card, held at the focus of the mirror. The card, of course, will block off part of the incoming light, but since, as in the lens, all parts of the mirror contribute to the formation of all parts of the image, the presence of the card will not produce a "hole" in the image but will merely reduce its brightness.

If the shape of the mirror is part of a concave spherical surface, it will produce a fair-quality image, provided that the entire mirror constitutes only a very small part of a sphere, so that its size is small compared to its focal length. We shall see that the curves of telescope mirrors are usually parabolic rather than spherical in cross section.

(c) Properties of an Image

The most important properties of an image are its *scale* or size, its *brightness,* and its *resolution.* We shall consider these in turn. In this section, we ignore the limitations imposed by the Earth's atmosphere (Section 9.4a).

The *scale* of an image is a measure of its size. In all astronomical applications we are dealing with objects whose sizes can be expressed in angular units, and it is generally convenient to express the scale of an image as the linear distance in the image that corresponds to a certain angular distance in the sky. For example, suppose that an image of the Moon were produced that is exactly 1 cm across. The Moon has an apparent or angular size of ½°, that is, it subtends ½° in the sky. The scale of the image is thus ½° per centimeter, or 2 cm per degree.

The scale of an image depends only on the focal length of the lens or mirror that produces it. Numerically, the distance s in an image corresponding to 1° in the sky is given by the equation

$$s = 0.01745f,$$

where f is the focal length of the lens or mirror. For example, the 5-m (200-inch) mirror of the Hale telescope on Palomar Mountain has a focal length of 16.764 m. It produces images with a scale of 0.292 m per degree, which corresponds to 12.3 arcseconds per millimeter. Most astronomical telescopes have lenses or mirrors that give image scales ranging from 10 to 200 arcseconds per millimeter.

The *brightness* of an image is a measure of the amount of light energy that is concentrated into a unit area of the image, such as a square millimeter. The brightness of an image determines whether it is above the threshold of visibility or, alternatively, how long a time would be required to record the image photographically.

The brightness of an image of an extended object, such as the Moon, a planet, a nebula, a galaxy, or the faint illumination of the night sky, is greater the greater the amount of light flux that passes through the objective (lens or mirror) to form the image and is less the larger the image area over which that amount of flux must be spread. The amount of flux reaching the image is proportional to the area of the objective and hence to the square of its diameter or aperture (the area of a circular region of radius R or diameter D is πR^2 or $\frac{1}{4}\pi D^2$). The area the flux is spread over is proportional to the square of the focal length of the objective, for as we have seen the image diameter varies directly with the focal length. Hence the brightness B of an extended image is given by

$$B = \text{constant} \times \left(\frac{a}{f}\right)^2,$$

where a is the diameter or aperture and f is the focal length of the lens or mirror. The constant of proportionality is a number whose value depends on the units chosen to measure the various quantities, and also on the amount of light actually leaving each unit area of the object.

The quantity f/a is called the *focal ratio* or simply the *f ratio* of the lens. In common notation, if the focal length is, say, eight times the aperture of the lens, the focal ratio is written *f*/8 which should be interpreted as $a/f = 1/8$. (Note that *f*/8 means $a = f/8$). Every photographer is familiar with the concept of focal ratio when applied to a camera lens. In all but the simplest cameras the clear aperture of the lens can be increased or decreased by adjusting an iris diaphragm, thereby changing the focal ratio of the lens. A typical 35-mm camera, for example, might have focal ratio adjustments varying from *f*/1.7 to *f*/16.

Stars are in effect point sources, for even through the largest telescopes the stars still appear too small to show any apparent disks. Therefore a lens or mirror of good quality concentrates the starlight into a "point" image regardless of the focal length. For a point-source object such as a star, the amount of light in the image thus depends only on the amount of light gathered by the lens or mirror, and hence is proportional to the square of the aperture.

Resolution refers to the fineness of detail inherently present in the image. Even if the lens or mirror is of perfect optical quality, it cannot produce perfectly sharp and detailed images. Because of the phenomenon of diffraction and interference, a point source does not form an image as a true point but as a minute spot of light surrounded by faint, concentric, evenly spaced rings. The angular size of that central spot of light, called the central *diffraction disk* of the image, is inversely proportional to the aperture of the lens or mirror and directly proportional to the wavelength of the light observed. No detail can be resolved in the image if that detail is smaller than the diffraction disk. For example, the diameter predicted by geometrical optics for the image of a star produced by the lens of a telescope is much smaller than the size of the diffraction disk. Therefore, we do not see the geometrical image of a star itself but only a *diffraction pattern* that the telescope lens produces with the star's light. If a star is viewed telescopically under good conditions, the diffraction pattern, consisting of the bright central disk and faint surrounding rings, is clearly visible.

In the image of an extended source, the diffraction patterns of the various parts of the image, all overlapping each other, wash out the finest details. A feature on the surface of the Moon or Mars that is smaller than the diffraction disk produced by the telescope lens may be visible, but its true size and shape will not be distinguishable. If two stars are so near each other that their images are closer together than the size of the diffraction disk of either, they will blend together to produce a single spot, or perhaps a slightly elongated diffraction pattern. The ability of an optical system (lens or mirror) to distinguish fine detail in an image it produces, or to produce separate images of two close stars, is called its *resolving power.* In astronomical practice, the resolving power of a lens or mirror is described in terms of the smallest angle between two stars for which separate recognizable images are produced. Two stars that lie less than 1 arcminute from each other cannot be separated with the human eye. With a 15-cm lens of good quality, separate images are produced of two stars only 1 arcsecond apart. The 15-cm lens thus has higher resolving power than the eye.

We have said that the angular size of the central diffraction disk of a point source depends on the wavelength of light used and the diameter of the lens or mirror. If the wavelength and aperture are measured in the same units (for example, both in centimeters), the smallest angle (α) in seconds of arc that can be resolved by a lens or mirror of aperture d is given by the equation,

$$\alpha = 2.1 \times 10^5 \times \frac{\lambda}{d} \text{ seconds of arc,}$$

where λ is the wavelength. If λ is chosen as 5.5×10^{-5} cm (5500 Å)—near the middle of the visible spectrum— and if d is measured in centimeters, the formula becomes

$$\alpha = \frac{11.6}{d} \text{ seconds of arc.}$$

The above formula, called *Dawes' criterion,* represents a compromise or average of empirical studies of the ability of telescopes of various apertures to resolve double stars. It does not hold for the eye because the coarse structure of the retina limits the resolution ideally obtainable with the eye lens. Also, as we shall see, atmospheric turbulence usually further degrades the actual resolving power of large telescopes.

(d) Aberrations of Lenses and Mirrors

An image produced by any optical system always has imperfections. These are called *aberrations*. We describe briefly a few of the more important kinds of aberrations. Aberrations are always most serious in telescopes of low focal ratios (high speeds).

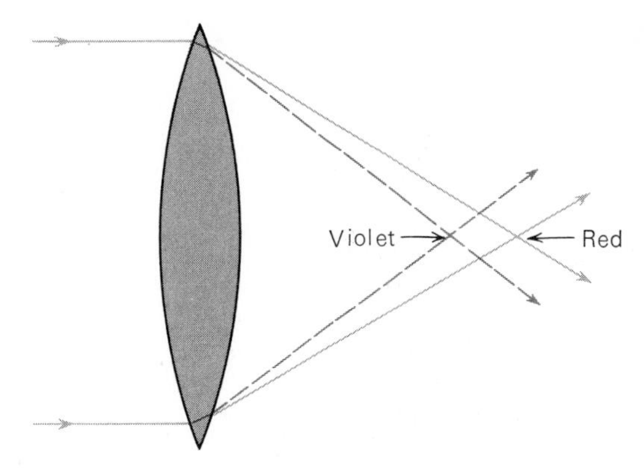

Figure 9.7 Chromatic aberration.

Chromatic or *color aberration* is a consequence of dispersion. Consider the simple lens shown in Figure 9.7. Suppose parallel rays of white light from a remote star are incident upon the left side of the lens. Because the different wavelengths that comprise white light are refracted in different amounts upon entering and leaving the lens, they do not all focus at the same place. The shorter wavelengths (violet and blue light) are bent the most and focus nearest the lens, while the longer wavelengths of orange and red light focus farther from it. The effect of this chromatic aberration is to produce color fringes in the image.

Chromatic aberration is less serious in lenses of large focal ratio. Some telescopes constructed in the late 17th century, therefore, utilized lenses of extremely long focal length to obtain as large a focal ratio as possible for a given aperture. Telescopes more than 30 m long were in common use, and ones with lengths of up to 180 m are said to have been built. Such extremely long telescopes, however, were so difficult to use that they were, for the most part, impractical.

Chromatic aberration is now corrected by constructing a lens of two pieces of glass of different types (usually crown and flint glass) and thus of different indices of refraction. One of the pieces of glass, or *elements* of the lens, is a *concave* lens, that is, it is thinner in the middle than at the edges, so that it diverges the light rather than converging it. (The concave element of the lens diverges light only enough to correct for the chromatic aberration; the combined lenses still converge incident light to a focus.) The light of shorter wavelengths is diverged the most, just as it is converged the most in the usual convex lens. Dispersion, like refraction, is greater in the glass of higher index. It is possible to make the chromatic aberration completely cancel out at any two given wavelengths, if exactly the right curvatures are chosen for the surfaces of the two lenses (Figure 9.8). Lenses can be designed so that this aberration is nearly canceled out, or is

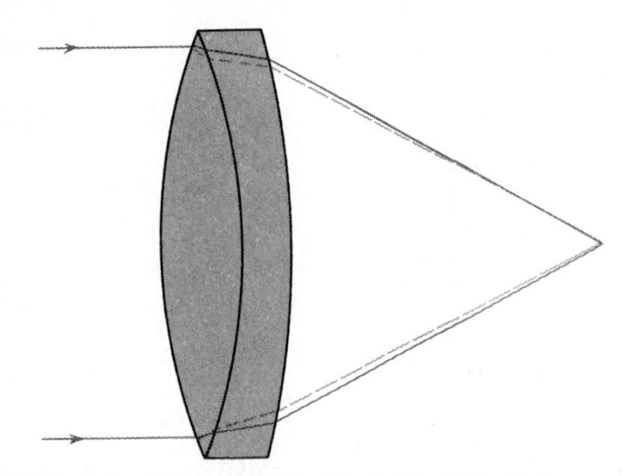

Figure 9.8 Correction of chromatic aberration.

greatly reduced, over a considerable range of wavelengths.

When an image is produced by a mirror, the light never has to pass through any glass, so it is not dispersed. Thus an image produced by a mirror does not suffer chromatic aberration.

Spherical surfaces are the most convenient to grind and polish, whether they be the concave or convex surfaces of a lens or mirror. Unfortunately, however, a simple lens or mirror with spherical surfaces suffers from the imperfection called *spherical aberration* (Figure 9.9). Light striking nearer the periphery focuses closer to the lens or mirror; light striking near the center focuses farther away.

In a lens, spherical aberration can be corrected or greatly reduced by constructing the lens of two pieces of glass or elements of different indices of refraction, just as in the correction of chromatic aberration. The elements are so designed that the spherical aberration introduced by one is canceled out by the other. Most lenses designed for astronomical purposes consist of two elements. The curvatures of the spherical surfaces of the elements are designed to reduce both the spherical and chromatic aberrations to a tolerable amount.

For a mirror, spherical aberration can be eliminated by grinding and polishing the surface not to a spherical shape, but to a *paraboloid of revolution,* that is, to a surface whose cross section is a parabola (Figure 9.10). A paraboloid has the property that parallel light rays striking all parts of the surface are reflected to the same focus. Similarly, light leaving the focus of a paraboloid is reflected at the surface in a parallel beam. An automobile headlight or searchlight, for example, has a parabolic reflector with the light source at the focus. Mirrors designed for astronomical telescopes usually have parabolic surfaces.

Coma is an aberration that distorts images formed by *off-axis* light rays that do not strike the lens or mirror "square-on" (that is, those light rays from objects away from the center of the field of view). This off-axis distortion is particularly serious in the images formed by the parabolic mirrors used in the largest existing telescopes. Photographs taken with such telescopes are reproduced in the latter part of this book. If you carefully inspect one of these, you will find that the star images near the center of the picture appear as sharp round dots, whereas those near the corners, which are formed by light entering the telescope off axis, are distorted into tiny "tear drops" or "commas" pointing toward the center of the photograph. The comet-like shape of these images accounts for the name "coma" given to this particular aberration.

In modern telescopes, special *field-correcting lenses* are often placed in the light beam before it comes to a focus. Coma can be substantially reduced by these lenses, thereby greatly increasing the useful field of view of the telescope.

Astigmatism is an aberration produced in an optical system when rays of light approaching the lens in different planes do not focus at the same spot. In the example shown in Figure 9.11, the rays in a vertical plane focus farther from the lens than those in a horizontal plane. A geometrically perfect lens produces astigmatism only for off-axis rays.

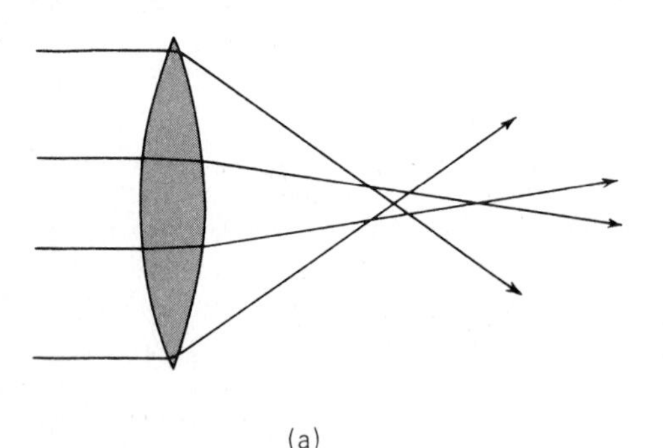

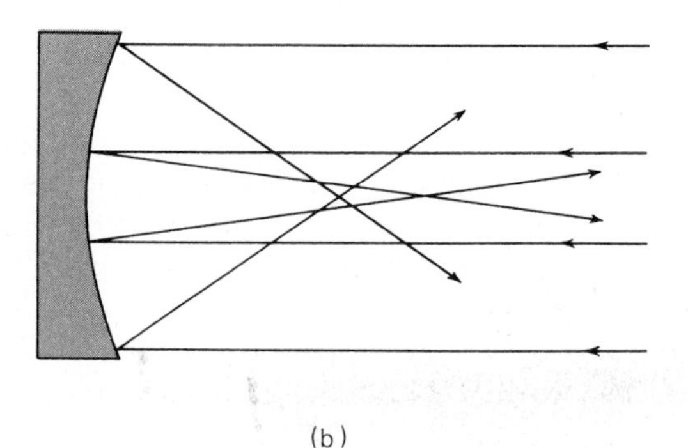

Figure 9.9 Spherical aberration. (a) In a lens. (b) In a mirror.

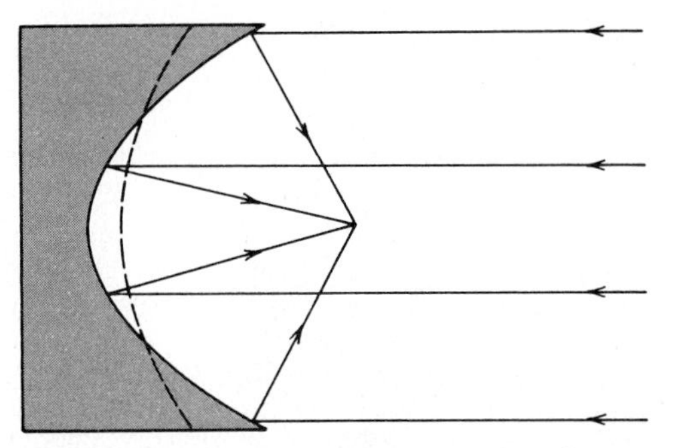

Figure 9.10 Correction of spherical aberration with a concave parabolic mirror. The dashed line is where the surface of a spherical mirror of the same focal length would lie.

(e) Magnification

When an extended celestial object is viewed through a telescope equipped with an eyepiece, it appears enlarged, that is, closer than when viewed naturally. The factor by which an object appears larger (or nearer) is called the *magnifying power* of the telescope. For example, the Moon appears to subtend an angle of ½° when viewed with the naked eye. If, when viewed through a particular telescope, the Moon appears to subtend 10°, the magnifying power of the telescope is 20.

Everyone knows that to read fine print one must use a magnifying glass of higher power than when reading the print in a newspaper. A higher power magnifying glass is one of shorter focal length. Similarly, eyepieces of different focal lengths, used in conjunction with the same telescope objective, produce different image magnifications. It is the purpose of the objective of a telescope to produce an image; it is the purpose of the eyepiece to magnify the image to the point where details in it can be viewed. In principle, any desired magnification can be obtained if an eyepiece of sufficiently short focal length is used. Therefore, it does not make sense to ask an astronomer what the "power" of a telescope is. The power can be changed at will by using different eyepieces. The term magnifying power loses its significance completely when, as in most modern telescopes, the image is not viewed through an eyepiece at all, but is displayed on a television screen.

(f) The Schmidt Optical System

An ingenious optical system that utilizes both a mirror and a lens was invented by the Estonian optician Bernhard Schmidt of the Hamburg Observatory. We noted in Section 9.1d that images formed by a mirror do not suffer from chromatic aberration. To avoid spherical aberration, parabolic rather than spherical mirrors are used. The principal disadvantage of the parabolic mirror is that it produces good images over only a relatively small field of view, that is, for light that approaches the mirror very nearly on axis.

On the other hand, for a spherical surface, any line reaching the surface through its center of curvature (that is, through the center of the sphere of which the surface is a part) is perpendicular, that is, "square on" to the surface, and hence is on axis. The Schmidt optical system, utilizing this principle, employs a spherical mirror that is allowed to receive light only through an opening located at its center of curvature (Figure 9.12). Thus there can be no off-axis aberration. The only trouble is that a spherical mirror, suffering as it does from spherical aberration, produces generally poor images for light coming from any direction. Schmidt solved the problem by introducing a thin correcting lens at the aperture at the center of curvature of the mirror. The lens is of the proper shape to correct the spherical aberration introduced by the spherical mirror but not thick enough to introduce appreciable aberrations of its own. Thus, the Schmidt optical system produces excellent images over a large angular field. A disadvantage of the Schmidt system is that the focal surface is not a plane but a sphere concentric with the spherical mirror. The resulting curved image can be photographed directly by forcing the photographic plate or film into a curved shape, but if electronic detectors are used, an additional lens must be inserted to create a flat image.

Today, the Schmidt optical system and modifications of it invented by Maksutov, Wright, and others are widely used in science and industry.

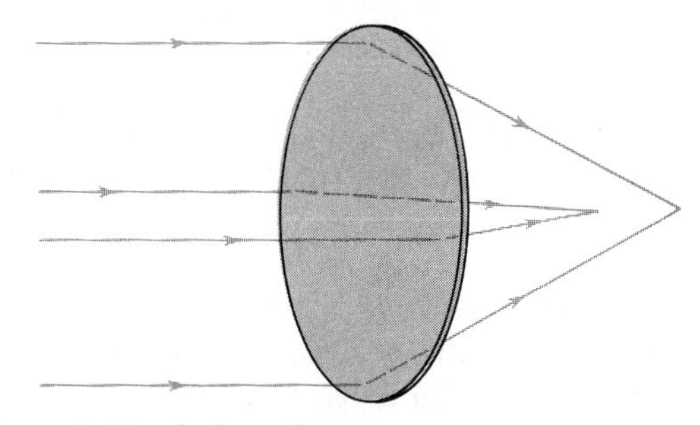

Figure 9.11 Astigmatism.

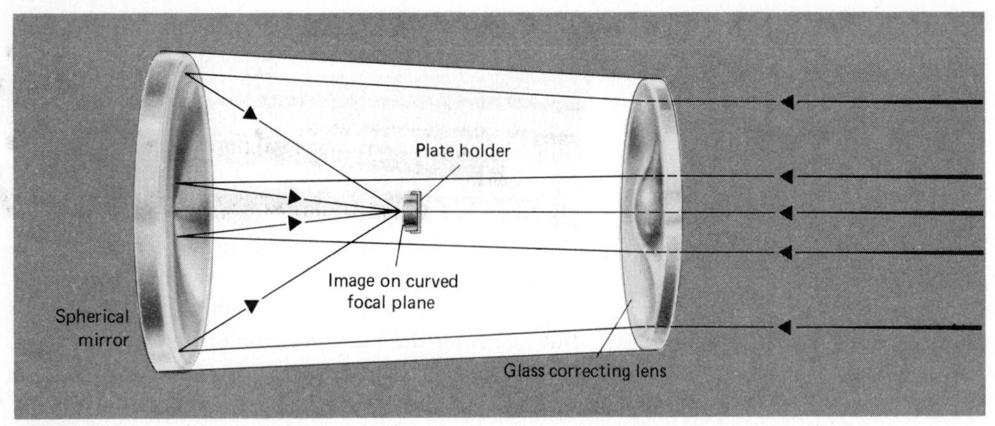

Figure 9.12 Schmidt optical system. The mirror is generally larger than the lens, so images of unreduced intensity can be produced over a wide field of view.

(g) The Complete Telescope

Now that we have seen how an image is produced, we are able to understand the operation of a telescope. Today there are two general kinds of optical astronomical telescopes in use: (1) *refracting* telescopes, which utilize lenses to produce images, and (2) *reflecting* telescopes, which utilize mirrors to produce images. As we have seen, Schmidt telescopes, and modifications thereof, are primarily reflecting telescopes but have refracting correcting lenses.

The refracting telescope is probably the most familiar, although it is rarely used in modern research observatories. This is the kind of telescope that we can literally "look through." Ordinary binoculars are two refracting telescopes mounted side by side. A lens, generally consisting of two or more elements, is usually mounted at the front end of an enclosed tube. The tube is not really essential—its purpose is merely to block out scattered light; an open framework would suffice. In a refracting telescope, the objective, the optical part that produces the principal image, is the lens at the front of the tube. The image is formed at the rear of the tube, where various devices inspect, photograph, or otherwise utilize it.

All of the large research telescopes in the world are reflecting telescopes, and this design is also more popular among amateur astronomers. The reflecting telescope was first conceived by James Gregory in 1663, and the first successful model was built by Newton in 1668. Here a concave mirror (usually a paraboloid) is used as an objective. The mirror is placed at the *bottom* of a tube or open framework. The mirror reflects the light back up the tube to form an image near the front end. Because of the difficulty of precisely producing the reflecting surface, the reflecting telescope did not become an important astronomical tool until the time of William Herschel, a century after its invention.

With a reflecting telescope the problem of image accessibility arises, because a concave mirror produces the image in front of the mirror, in the path of the incoming light. For some types of image analysis, bulky equipment must be placed at the telescope focus. There are, however, various arrangements for getting at the focus; which one is adopted depends on the type of telescope and on the purpose for which the image is to be used.

The place where the image is formed by the mirror, shown in Figure 9.13a, is the *prime focus*. If the image is to be recorded, a photographic film holder or electronic detector can be suspended at the prime focus in the middle of the mouth of the telescope tube. The detector system generally blocks out only a small fraction of the incoming light, so that the brightness of the image is only slightly dimmed. The larger the telescope, the larger the package of prime focus instruments can be before it blocks a significant fraction of the light. Therefore, the prime focus is most often used with telescopes having an aperture larger than 3 m.

Figure 9.13 illustrates a variety of other focus arrangements for reflecting telescopes. In the Newtonian design, which is commonly used on smaller amateur telescopes, the problem is solved by a flat mirror mounted diagonally in the middle of the tube so that it intercepts the light just before it

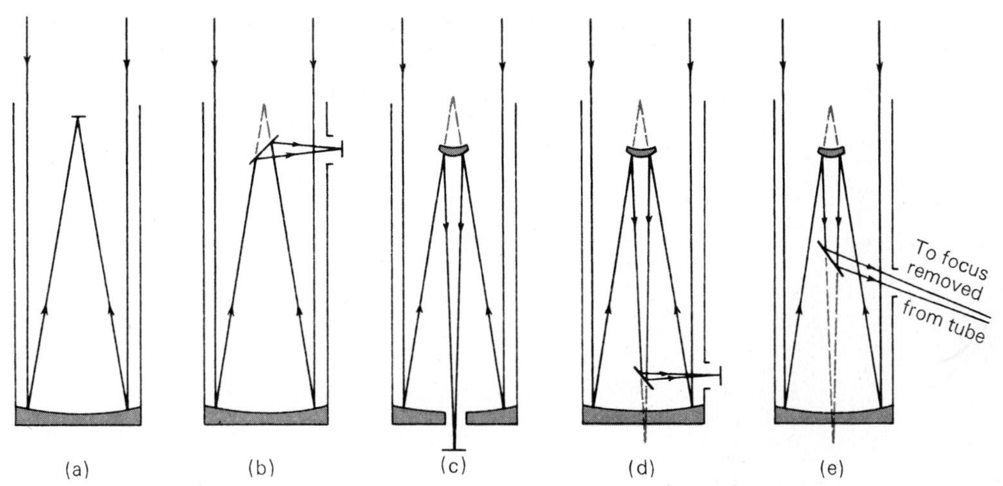

Figure 9.13 Various focus arrangements for reflecting telescopes: (a) prime focus, (b) Newtonian focus, (c) and (d) two types of Cassegrain focus, (e) Coudé focus.

reaches the focus, diverting it to an eyepiece outside the tube (Figure 9.13b).

The most popular design for modern research telescopes is the Cassegrain system, in which a small convex mirror rather than a flat mirror is suspended in the telescope tube. The convex mirror intercepts the light before it reaches the prime focus and reflects it back down the tube of the telescope. Usually a hole is provided in the center of the objective mirror so that the light reflected from the convex mirror can form an image behind the objective (Figure 9.13 c and d).

In the *coudé* arrangement, Figure 9.13e, a convex mirror intercepts the light just before the prime focus is reached. The light is reflected back down the tube until it reaches one of the pivot points about which the telescope tube can be rotated to point to various parts of the sky. There it is intercepted by a flat mirror that reflects the light outside the tube to a fixed observing station. Because the station is not attached to the moving part of the telescope, heavy equipment can be used there.

Most large reflectors in use in astronomical observatories are equipped with auxiliary flat and convex mirrors so that several of the possible focus arrangements can be used.

Schmidt telescopes are used for wide-angle photography in astronomy. The image is formed directly in front of the spherical mirror about halfway from the mirror surface to the correcting lens. Most often the entire system is enclosed in a light-tight tube, and the photographic emulsion, curved to fit the spherical focal surface, is inserted at the focus of the mirror in the center of the tube.

(h) Telescope Mounts

For any type of telescope, the tube must be mounted so that it can be pointed toward any direction in the sky and can move to follow the apparent motion of the source under study. Until very recently, almost all astronomical optical telescopes have had *equatorial mounts*. An equatorial mount (Figure 9.14) allows the telescope to turn to the north and south about one axis and to the east and west about another. The axis for the east-west motion of the telescope is parallel to the axis of the Earth's rotation, and the other axis, about which the telescope can rotate to north or south, is perpendicular to this axis.

The two axes of motion of a telescope with an equatorial mount allow the telescope to be turned directly in right ascension and declination, the celestial coordinates in which astronomical positions are generally tabulated. An important advantage of the equatorial mount is that a simple slow motion of the telescope about its axis parallel to the Earth's axis—the *polar axis* of the telescope—is sufficient to compensate for the apparent motions of the stars across the sky that result from the Earth's rotation. Even when a moving source such as a planet or asteroid is being tracked, only small adjustments with respect to the basic motion around the polar axis are needed.

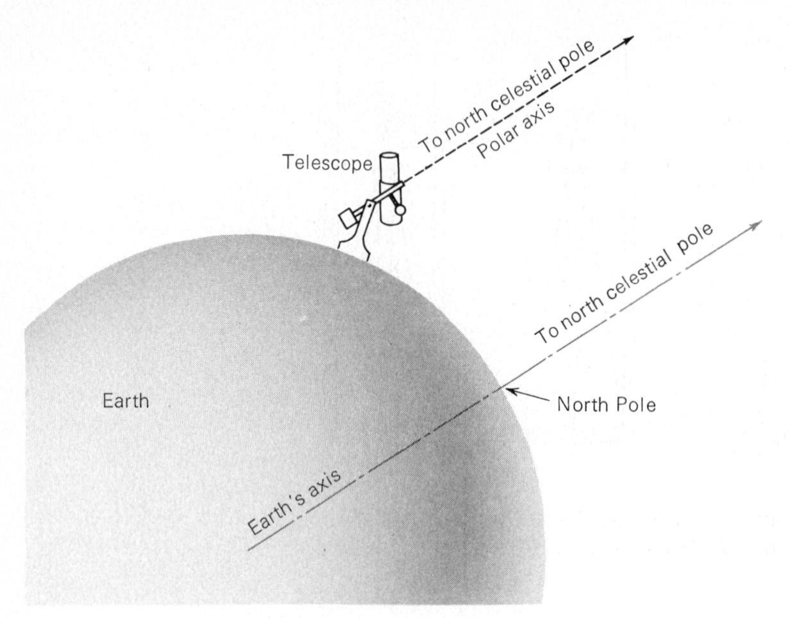

Figure 9.14 Equatorial mount.

An equatorial mount, however, is a disadvantage for a very large telescope because it is difficult to adjust for the gravitational stress on its heavy mirror. Consequently, the world's largest optical telescope, the 6-m reflector in the Soviet Union, has an *altitude-azimuth mount* in which the telescope rotates simply about one vertical and one horizontal axis. Other even larger telescopes now under design will be similarly mounted, and so are large radio telescopes (Chapter 10). It is more complicated for them to compensate for the Earth's rotation and track the stars, but today that problem can be handled with sophisticated computer-controlled driving mechanisms, while the advantages of reduced size and mechanical simplicity are considerable.

In order to point and track with the high accuracy required by modern telescopes, the motion of the telescope is controlled by a computer, which senses the position of the telescope and provides commands to electric motors connected to both axes. Many telescopes today can be pointed, on command, to an accuracy (in right ascension and declination) of a few arcseconds, thus greatly speeding the location of faint sources. Once in position, the computer-controlled drives will maintain pointing to within a fraction of an arcsecond.

The telescope computer stores the coordinates of thousands of stars in its memory, and it can calculate at any moment the apparent positions of solar system objects. In practice (and if the system is working properly), the astronomer needs only to enter the name of the object to be studied, and a minute or so later it will appear in the center of the field of view, ready for measurements to begin.

(i) Housings for Astronomical Telescopes

The conventional housing for an astronomical telescope is a hemispherical *dome*. The dome usually has an oblong window or slot on one side, extending from the base to the top of the dome. There is a *shutter*, which can close over the slot to protect the telescope during daylight and bad weather, when it is not in use. The dome is generally mounted on rails so that the slot can be turned to any direction. In a modern observatory, the dome usually turns automatically to keep the slot always oriented in the direction in which the telescope points. The domes of the largest telescopes are usually well insulated against heat, so that the interior can be maintained at nighttime temperatures, thus preventing rapid temperature changes from distorting the critical shape of the telescope mirrors or lenses.

(j) Advantages of Various Kinds of Telescopes

Refracting, reflecting, and Schmidt telescopes all have their special advantages. The choice of the kind of telescope to build or use depends on the type of project to be worked on.

Refracting telescopes were constructed primarily in the 18th and 19th centuries, before the technology of building large reflectors had been developed. They were usually constructed with long focal lengths and relatively slow speeds (*f*/12 to *f*/16). In such instruments chromatic aberration and distortion can be kept small, resulting in a larger field of view for these telescopes than for typical reflectors. On the other hand, refractors suffer from chromatic aberration, and can be constructed to perform well only for a limited spectral region. Because a refracting lens must contain at least two elements to produce good images, there are at least four surfaces of glass to be ground or polished. Furthermore, a lens can be supported only along its periphery; a very large lens sags of its own weight and distorts its shape. The largest refractor has an aperture of only 40 inches (about 1 m).

A reflecting telescope, on the other hand, utilizes a mirror that has only one optical surface to be perfected. Because the light does not go through the mirror, this need not be made of optically perfect glass; furthermore, a mirror is easier to support, for it can be braced at all points along its back. Thus it is feasible to make much larger mirrors than lenses. Optical-quality mirrors of aperture up to 6 m have been fabricated, and current work at the National Optical Astronomy Observatories and the University of Arizona has the goal of producing a new generation of telescope mirrors 7.5 m in aperture. Because it is easier to construct reflecting telescopes, most homemade telescopes built by amateur astronomers are of this type.

Because the ordinary reflecting telescope is free of chromatic aberration, it is better suited for spectroscopy. Reflecting telescopes of good optical quality can be made with relatively short focal lengths—that is, with high optical speeds—so they can photograph faint sources in shorter times than refracting telescopes can. Primarily, however, the advantage of reflectors lies in their large apertures, which collect much more light and permit the astronomer to study fainter and more distant objects. Since the early part of the 20th century, large reflecting telescopes have been the instruments of choice for most optical and infrared astronomy.

The Schmidt telescope is a compromise between reflectors and refractors. Schmidts can be made to produce excellent images over a large field of view. The nearly identical 124-cm (48-inch) Schmidts on Palomar Mountain and at Siding Spring, Australia, can record an area of the sky the size of the bowl of the Big Dipper on a single photograph, whereas with the 5-m reflector, also at Palomar, about 400 photographs are required to cover the same area. Schmidt telescopes are consequently especially useful for surveying. However, the correcting lens of a standard Schmidt system has nonspherical surfaces and is difficult to make.

9.2 OPTICAL DETECTORS

The popular view of the astronomer is of a person in a cold observatory peering through a telescope all night. Most astronomers do not live at observatories but near the universities or laboratories where they work. A typical astronomer might spend a total of only a few weeks each year observing at the telescope, and the rest of the time measuring or analyzing his or her data. Many astronomers work only with radio telescopes or with space experiments. Still others work at purely theoretical problems and never observe at a telescope of any kind. Even optical astronomers seldom inspect telescopic images visually except to center the telescope on a desired region of the sky or to make adjustments. On the contrary, photographic plates or electronic detectors are used to record the image permanently for detailed analysis after the observations are completed. Typically, one successful night at the telescope yields enough data to keep an astronomer busy for weeks of analysis and interpretation.

(a) Telescopic Photography

Through most of the 20th century, photographic emulsions have served as the prime astronomical detectors, whether for direct imaging or for photographing spectra. To photograph an astronomical object, the image of that object is allowed to fall on a light-sensitive coating that, when developed, provides a permanent record of the image—one that can be measured, studied, enlarged, published, and inspected by many individuals. When used for photography, a telescope becomes nothing more than a large camera; the lens or mirror of the telescope serves as the camera lens.

One important advantage of photography is that photographic emulsions can accumulate luminous energy and build up an image during a long exposure. Most astronomical objects of interest are remote, and hence the light we receive from them is feeble. However, long time exposures can be made. Until electronic detectors became available, astronomical exposures often would run hours in length, and occasionally over several successive nights. The longer the exposure, the more faint light gradually accumulates to help build up the photographic image. Objects can be detected that are more than one hundred times too faint to see by just looking through a telescope. The layperson is often disappointed by a first look through an astronomical telescope, for the sight is nothing compared to the spectacular photographs, such as those reproduced in this book, that are the result of long time exposures.

A photographic emulsion consists of a thin layer of gelatin, usually mounted on a base of celluloid or glass. Within the emulsion are suspended silver compounds that undergo a chemical reaction when activated by light, resulting in the formation of metallic silver. When the emulsion is immersed in an appropriate chemical solution (developed), more silver is formed where light started the process, and those molecules of silver compounds that were not activated by light are chemically removed, leaving grains of deposited silver in the parts of the emulsion that were exposed.

The photographic plate is a superb device for collecting a large amount of information, and plates have been used by astronomers for more than a century. Photographic plates do, however, have limitations. The photographic emulsion, for example, is nonlinear—that is, equal differences in exposure do not produce equal differences in the blackening of the emulsion. Moreover, the photographic plate saturates (turns black) with long enough exposure, and no additional data can be recorded. Consequently, it is far from ideal for photometry—the measuring of the brightnesses of astronomical objects. In addition, photographic plates are much less sensitive than electronic detectors.

There are ways of increasing the sensitivity of photographic plates. For example, they can be baked at 60°C or soaked in ammonia, nitrogen, or hydrogen. But at best the quantum efficiency of a photographic plate is only about 1%; this means that it requires, typically, a hundred incident photons falling on a small portion of the plate to produce a measurable blackening when the plate is developed. It has long been desirable, therefore, to find a more ideal detector—one that is linear and does not saturate, one that can yield higher resolution, and one that has high quantum efficiency.

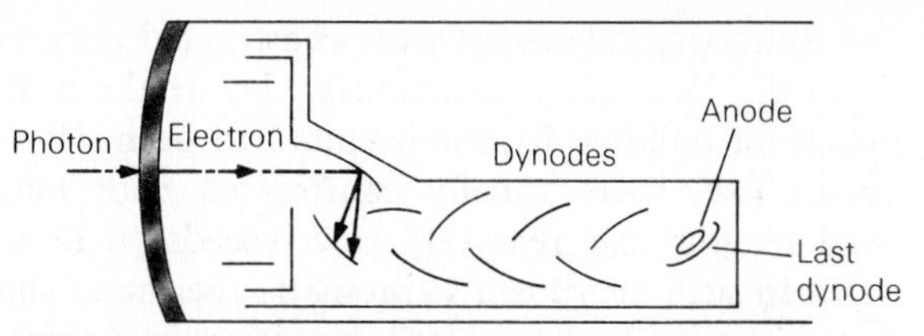

Figure 9.15 Photomultiplier. When a photon strikes the sensitive surface (photocathode) an electron is released. This electron then strikes a dynode, releasing several electrons (two of which are shown). Each of these electrons releases still more electrons when it strikes the next dynode. As all of these electrons move down the chain of dynodes, the original electron is amplified to a pulse of the order of a million electrons, which can be detected electronically.

(b) Electronic Detectors

Many of the limitations of photographic plates have been overcome through the use of electronic detectors. There are two basic types of electronic detectors that are widely used by optical astronomers. The first type is the so-called *photoemissive* detector, in which a photon strikes a light-sensitive surface with enough energy to free an electron completely from it. One example of such a device is a *photomultiplier* tube (Figure 9.15). The primary application of photomultipliers in astronomy is the precise measurement of the brightness of stars and galaxies. Light from a star is gathered by the telescope and is focused on a light-sensitive surface just inside the glass envelope of the photomultiplier tube. A photon striking this *photocathode* dislodges an electron, which is attracted to a positively charged element of the tube called a *dynode*. When the speeding electron strikes the dynode, four or five secondary electrons are dislodged. These in turn are attracted to a second dynode, where each secondary electron again dislodges four or five more electrons. After passing through ten or eleven dynodes, the original photoelectron has been amplified to a *pulse* of the order of a million electrons. This pulse is detected electronically and recorded as a *count*. The photomultiplier can thus be used to count the number of incident photons; the number recorded in a specific time is a measure of the star's apparent brightness.

Until very recently, photoemissive devices like photomultipliers have been very useful in measuring the energy emitted by astronomical objects, but they have not been designed in such a way that it was possible to form an image of that object. Instead astronomers have borrowed technology from the television studio, and have made extensive use of the kinds of light-sensitive devices used in television cameras.

The most promising technology for astronomical imaging at the present time is the *charge-coupled device* (CCD), and it is an example of a *photoconductive* detector. Such a device contains a photosensitive surface made of silicon. When a photon is absorbed by this surface, it frees an electric charge, which is stored in the absorbing region. The amount of stored charge is thus proportional to the number of photons that have struck the surface. The amount of stored charge is measured only after the exposure is complete, and in this ability to accumulate the signal from faint sources over long periods of time, CCDs resemble photographic plates. In order to construct an image of the object observed, it is necessary to record not only the *amount* of charge stored but also *where* on the CCD it is stored. The process of recording this information is called *reading out* the CCD. During readout, the charge in each small area or *pixel* is passed on to the next one in the same row, and then to the next, and so on, like a bucket brigade. As each pixel's charge reaches the end of the row it is transferred to a low-noise amplifier, and the amplifier signals are converted to digital codes and stored in the memory unit of a computer.

The CCD offers several advantages over photographic plates. For example, CCDs are more sensitive and record as many as 60 to 70 percent of all the photons that strike them. Photographic plates are much less sensitive to faint astronomical sources. A 36-inch telescope equipped with a CCD can actually detect fainter objects than the 200-inch telescope equipped with a photographic plate. A second advantage is that CCDs are most sensitive to red and near infrared radiation, while photographic plates work best in the blue region of the spectrum. With CCDs we can now measure spectral lines in galaxies and other faint sources that were previously undetectable. A third major advantage is that the amount of charge stored by a CCD is directly proportional to the amount of light falling on it. Therefore, CCDs provide more accurate measurements of the brightness of astronomical objects than do photographic plates.

On the other hand, CCDs have certainly not displaced photography entirely. The largest CCDs currently being manufactured are a factor of 100 smaller in area than the largest photographic plates. If one wants to record an image of a large area of the sky, then photographic plates are more efficient. Photographic plates are also low in cost and easy to use. No other detector now on the horizon can approach the ability of plates to record large quantities of data, inexpensively, during a single exposure.

(c) Spectroscopy

Spectroscopy (or *spectrometry*) occupies at least half of the available observing time of most large telescopes. The technique of spectroscopy is made possible by the phenomenon of dispersion of light into its constituent wavelengths. We saw in Chapter 8 that white light is a mixture of all wavelengths and that these can be separated by passing the white light through a prism, producing a spectrum.

A spectrum can also be formed by passing light through, or reflecting it from, a *diffraction grating.* These are of two kinds: A *transmission grating* is a transparent sheet or plate (such as glass) that is ruled with many fine, parallel, closely spaced scratches (typically, thousands per inch), so that light can pass through the material only between the scratches. A *reflection grating* consists of a shiny reflecting surface, say a mirror, upon which are ruled similar fine scratches or grooves, so that light can be reflected only from the places on the surface between the scratches. In either kind of grating, light emerges as if from a series of long, parallel, finely spaced slits. Because of diffraction, the light fans out in all directions as it passes through these "slits." The waves of light from different slits interfere with each other as described in Section 8.1. At any given point beyond the grating, the waves cancel each other out except at one specific wavelength. With an appropriate lens system, successive wavelengths of light can be focused at successive points along a line parallel to the surface of the grating, and in a direction perpendicular to the scratches in it. A spectrum can thus be produced by a grating as well as by a prism.

A *spectrograph* or *spectrometer* is a device with which the spectrum of a light source can be photographed or otherwise recorded. Attached to a telescope, the spectrograph can be used to record the spectrum of the light from a particular star. The

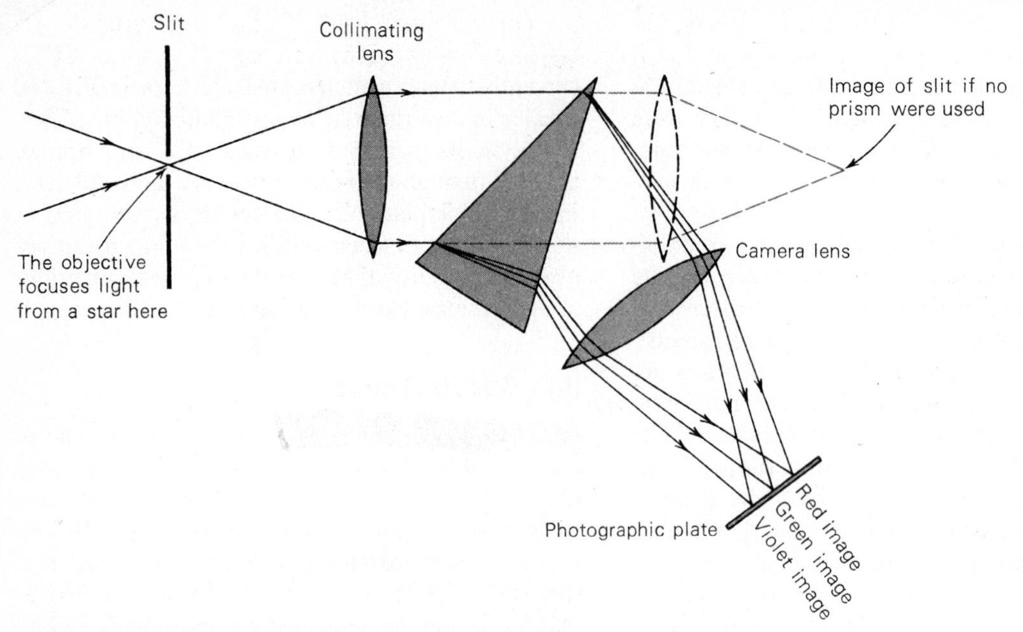

Figure 9.16 Construction of a simple prism spectrograph.

construction of a simple spectrograph is illustrated in Figure 9.16. Light from the source enters the spectrograph through a narrow slit and is then collimated (made into a beam of parallel rays) by a lens. In Figure 9.16 the collimated light is shown entering a prism, but it can just as well pass through or be reflected from a grating. Different wavelengths of light leave the prism (or grating) in different directions, because of dispersion (or diffraction). A second lens placed behind the prism forms an image of the spectrum on the photographic plate. If the light from the collimating lens had been passed directly into the second lens, rather than through the prism first, the second lens would simply produce an image of the slit through which the light entered the spectrograph (dashed lines in Figure 9.16). However, because of the dispersion introduced by the prism, light of different wavelengths enters the second lens from slightly different directions, and consequently the lens produces a different image of the slit for each different wavelength. The multiplicity of different slit images of different colors lined up at the photographic emulsion is the desired spectrum. In modern practice, the spectrum is often measured and recorded electronically rather than photographically, but the optical principle is the same.

An image of the spectrum shows the star's light spread into a streak with short wavelengths (violet) at one end and long wavelengths (red) at the other. Absorption lines in the spectrum manifest themselves as narrow ranges of wavelength where the light intensity is low. What we need to know is exactly where in terms of wavelength the dark lines and other spectral features occur. Thus it is also necessary to record some wavelength reference. Astronomers usually do this several times while recording the star's spectrum. They temporarily interrupt the exposure on the star and allow the light from an arc lamp or other laboratory source (whose spectrum consists only of bright emission lines) to pass through the slit. The spectrum of the laboratory reference source thus appears as a series of bright lines. The wavelengths of these lines are precisely known, and the wavelengths of stellar spectral lines can be determined precisely relative to the lines of the laboratory source.

9.3 ASTRONOMICAL OBSERVATORIES

(a) Optical Telescopes in the 20th Century

The giant among telescope builders in this century was surely George Ellery Hale. Not once but four times he initiated projects that led to construction

COLOR PLATE 5a The 1-m (40-inch) refracting telescope of Yerkes Observatory in Wisconsin. (Yerkes Observatory)

COLOR PLATE 5b The 1.2-m (48-inch) Schmidt telescope at Palomar. (California Institute of Technology/Palomar Observatory)

COLOR PLATE 5c The George Ellery Hale 5-m (200-inch) telescope. (California Institute of Technology/Palomar Observatory)

COLOR PLATE 5d The observer's cage at the prime focus of the Hale 5-m telescope. (California Institute of Technology/Palomar Observatory)

COLOR PLATE 6a The Multi-Mirror Telescope (MMT) of the Fred L. Whipple Observatory in Arizona, with an equivalent aperture of 4.4 m. (University of Arizona)

COLOR PLATE 6b The 4-m telescope of the Inter-American Observatory at Cerro Tololo, Chile. (National Optical Astronomy Observatories)

COLOR PLATE 6c Kitt Peak National Observatory in Arizona, with the dome of the 4-m Mayall telescope in the foreground. (National Optical Astronomy Observatories)

COLOR PLATE 7a Model of the proposed 15-m National New Technology Telescope (NNTT), with a design based on four mirrors, each of 7.5-m aperture. (National Optical Astronomy Observatories)

COLOR PLATE 7b The 100-m radio telescope near Bonn, Germany. (Max-Planck Institut für Radioastronomie)

COLOR PLATE 8a Several elements of the Very Large Array (VLA) radio telescope near Socorro, New Mexico. (National Radio Astronomy Observatory)

COLOR PLATE 8b The millimeter-wave interferometer at Owens Valley, California. (California Institute of Technology)

of what was at the time the world's largest telescope.

Hale's training and early research were in solar physics. In 1892, at age 24, he was named Associate Professor of Astral Physics and Director of the astronomical observatory at the newly founded University of Chicago. At that time, the largest telescope in the world was the 36-inch refractor at the Lick Observatory near San Jose, California. Still larger 40-inch blanks had been manufactured by Alvan G. Clark for the University of Southern California. The plan was to build an observatory that would surpass Lick. The blanks were to be paid for by a gift of land from a private donor, but just as the blanks were finished, real estate prices in California crashed, and the University of Southern California could no longer afford the lenses. Hale saw an enormous opportunity. If only he could raise about $300,000, he could build the world's largest telescope.

One prospective donor was Charles T. Yerkes, who, among other things, ran the trolley system in Chicago. Hale wrote a letter to Yerkes trying to persuade him to support construction of the giant telescope by saying, ". . . and the donor could have no more enduring monument. It is certain that Mr. Lick's name would not have been nearly so widely known today were it not for the famous observatory established as a result of his munificence."*

At one point in the discussions about the project, Yerkes said, "Here's a million dollars, if you want more, say so. You shall have all you need if you'll only lick the Lick."† In the end, the dealings with Yerkes proved difficult, sometimes acrimonious, and his gift for the new telescope did not approach a million dollars. Nevertheless, the telescope was completed in May 1897.

Yerkes Observatory is located in southern Wisconsin on the shores of Lake Geneva, and, not altogether surprisingly, clouds obscured the sky completely the first night the telescope was ready for use. The next night was clear—fortunately so, because William Rainey Harper, President of the University of Chicago, along with many trustees and faculty, had arrived to look through the new telescope. The 40-inch performed superbly well, much better, according to Hale, than the 36-inch refractor at Lick.

Close on the heels of this success came near disaster. At the end of May, just after the commissioning of the telescope, a contractor, ready to begin his day's work, was walking toward the dome that housed the 40-inch telescope. From about 200 yards away, he heard a great crash. The huge elevator floor that surrounded the telescope had plunged to the ground and was lying in ruins. It ultimately turned out that two cables holding counterbalance weights had given out. Without the floor, there was no way to reach the telescope. Hale was very concerned that perhaps the 40-inch lens had been damaged. He rushed to the topmost balcony in the dome and peered at the lens. He was horrified to see what appeared to be a network of fine cracks in the glass. Fortunately, when Hale was finally able to reach the telescope and examine the lens closely, the cracks proved to be a spider's web.

Even before the completion of the Yerkes refractor, Hale was not only dreaming of building a still larger telescope but also was taking concrete steps to achieve that goal. In the 1890s there was a major controversy about the relative quality of refracting and reflecting telescopes for astronomy. Hale realized that the 40-inch lens was close to the maximum feasible aperture for refracting telescopes. If telescopes larger in aperture by a factor of two or more were to be built, then they would have to be reflecting telescopes with mirrors as the primary optical element. This judgment proved correct—the 40-inch at Yerkes remains the largest refracting telescope in the world.

In the 1890s one certainly could not look to either the federal or state governments for support for construction of a telescope. Private philanthropy was the key, and Hale found a supporter in his own family. His father purchased a 60-inch blank for him. The glass arrived at Yerkes the year before the completion of the 40-inch refractor, and the optical polishing of the mirror was carried out at Yerkes.

Then as now, the cost of the mirror is but a small fraction of the cost of a telescope. Fourteen years were to elapse between the time Hale received the 60-inch mirror and the time that the mirror was mounted and usable for astronomical observations.

Hale's first step in the long process of building the 60-inch reflector was to identify a site for it. His thoughts turned to Mt. Wilson in southern Califor-

---

*Wright, H. *Explorer of the Universe*. New York: E. P. Dutton, Inc., 1966, p. 96.

†Ibid., p. 98.

nia, the location originally selected by the University of Southern California for the 40-inch refractor. He borrowed money from his brother and uncle and used personal funds as well to begin development of Mt. Wilson for solar research. In 1904 he received funds from the Carnegie Foundation to establish the Mount Wilson Solar Observatory.

At age 36, Hale had thus become founder and director of a second great observatory, one that was destined soon to become the world's largest. The 60-inch mirror was placed in its mount in December 1908.

Yet two years earlier, in 1906, Hale had already approached John D. Hooker, who had made his fortune in hardware and steel pipes, with a proposal to build a 100-inch telescope. The technological risks were substantial. The 60-inch was not yet complete, and the utility of large reflectors for astronomy had not yet been demonstrated. His brother called George Ellery Hale "the greatest gambler in the world." Once again, Hale was successful in obtaining funds, and the 100-inch blank arrived in Pasadena the day the 60-inch was set in place on its mount. With additional funding from Carnegie, the 100-inch telescope was completed in November 1917. Hale's first view through the telescope was of Jupiter, and he was appalled to see six or seven overlapping images. Was the new telescope a failure? Or, since workmen had left the dome open during the day, could the sun's heat have distorted the mirror? The only alternative was to wait while the mirror cooled. Hale returned to the telescope several hours later, and found that the image of the star Vega was sharp and very beautiful indeed.

Hale was not through dreaming. In 1926 he wrote an article in *Harper's* magazine about the scientific value of a still larger telescope. This article came to the attention of the Rockefeller Foundation, which granted six million dollars for the construction of a 200-inch telescope. Hale died in 1938, and the 200-inch Hale telescope on Palomar Mountain was dedicated ten years later.

(b) New Technology for Ground-Based Telescopes

Since 1948, only one telescope—the Russian 6-m—has exceeded the 200-inch Hale telescope in size (Figure 9.17). What does the future hold? In the past quarter of a century astronomical observations from space have become a reality. We have now viewed the universe with X-ray, gamma ray, and infrared eyes, discovering in the process bizarre objects, violent events, and physical processes that were totally unexpected. Does the future of astronomy belong to the world of spacecraft, of manned space stations with huge telescopes orbiting companionably nearby, of astronauts maneuvering silently in the black vacuum of space to change and repair equipment?

In part, of course, this must be the future of astronomy. But far from supplanting Earth-bound telescopes, space instruments have only increased the demand for complementary observations from the ground, where telescopes can be built and operated much more cheaply than in space. As a consequence, construction of ground-based telescopes is proceeding at an unprecedented pace. Worldwide there are now—give or take one or two—28 telescopes in operation that have apertures exceeding 2.0 m. Approximately two-thirds of those were put into operation since 1970, and still more are being planned. Most of these telescopes were constructed by following engineering designs similar to those developed for the 200-inch Palomar telescope. New technologies are now available, however, and astronomers believe it is possible to construct instruments that are two to three times greater in diameter than the 200-inch telescope and with, correspondingly, four to nearly ten times the light-gathering power.

The primary factor that drives the cost of building a telescope is its weight, and the weight of a telescope is determined by the weight of the key element, namely, the mirror. This is true not because the weight of the mirror is a significant fraction of the entire weight of the moving parts of a telescope but rather because the heavier the mirror, the more massive all of the elements that support it and preserve its shape must be, so that the images it produces remain absolutely sharp whether the telescope is pointed at the zenith or near the horizon.

The key technical breakthrough that has occurred is that we are learning how to make ever-thinner mirrors, and correspondingly we can reduce the weight of all the moving parts of the telescope. In addition, the ready availability of computers means that we can now use altitude-azimuth mounts, which are much more compact and cheaper to build than equatorial mounts. Innovations in dome design have also reduced building

Figure 9.17 The Special Astrophysical Observatory of the Academy of Sciences of the U.S.S.R., on Mount Pastukhov in the Caucasus, home of the 6-m telescope, the world's largest telescope. (Fotokhronica TASS-Sovfoto)

costs. The consequence of these technical developments is that we can now build 3- to 4-meter class telescopes for less than 10 million dollars, or less than one-third the cost of a telescope of the same aperture built with the conventional techniques developed for the 200-inch and used for most subsequent large telescopes. It should be possible to build a 10-meter—that is, a 400-inch—telescope for 100 million dollars. That sounds like an enormous sum, but it is no more than three times the cost of a conventional 3 to 4 meter telescope and is even less expensive than was the 200-inch if inflation is taken into account. For comparison, the cost of the Hubble Space Telescope, which has a 96-inch (2.4 m) mirror, is more than 10 times more than the estimated cost of a ground-based 10-meter telescope.

What will a 10-m telescope look like? Not at all like the classical telescope that we are accustomed to. For example, it will surely not be made of a single piece of glass. Engineering studies are analyzing two possibilities. The first is the so-called segmented mirror approach. Instead of a single mirror, the reflecting surface is composed of hexagonal segments fitted together to form a continuous mirror. The trick, obviously, is to find a way to control the orientation of these individual mirrors so precisely that they form images as sharp as those produced by a monolithic mirror. This is the approach adopted by the University of California and the California Institute of Technology. These two universities are planning to build a 10-m telescope on Mauna Kea on the island of Hawaii, and the project

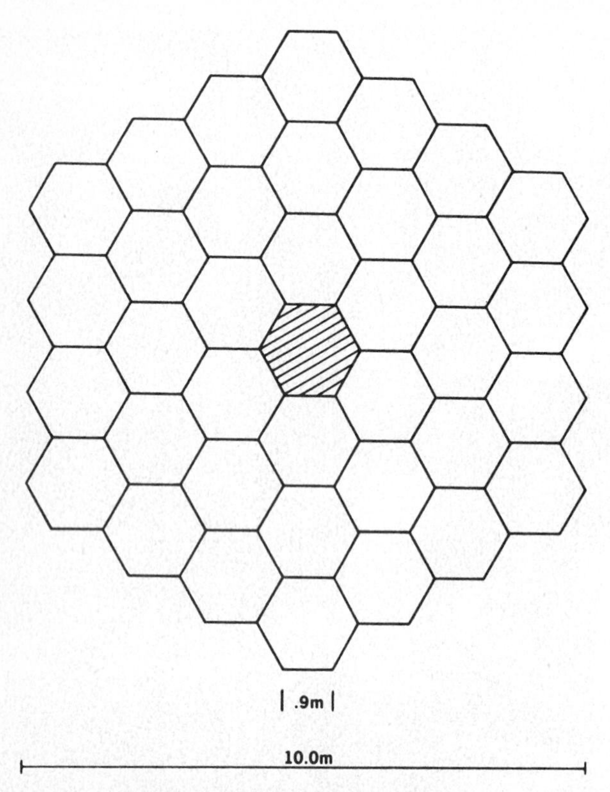

Figure 9.18 One way to build very large mirrors for telescopes is to make use of many smaller mirrors. This figure shows the design that will be used to make a 10-m mirror for the Keck Telescope. (CARA–California Association for Research in Astronomy)

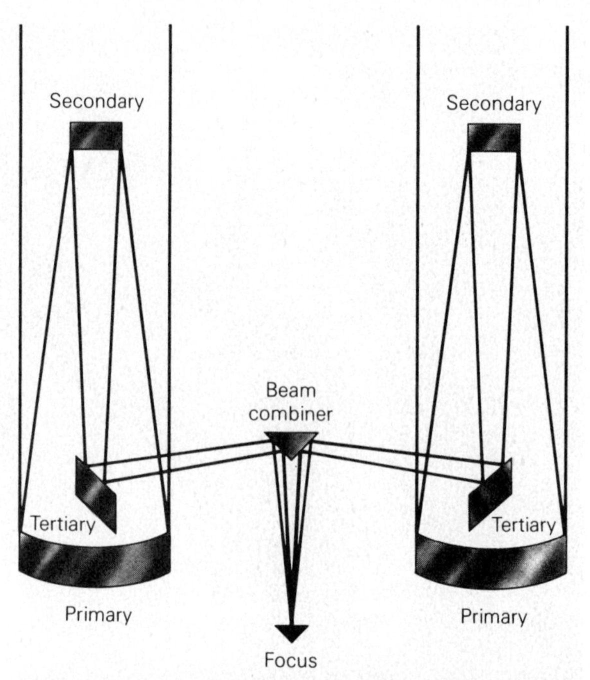

Figure 9.19 Another way to build a very large telescope is to combine the beams from smaller independent Cassegrain telescopes all carried on the same mount. This diagram shows how the beams from two of the six telescopes of the Multiple Mirror Telescope are combined. (MMT Observatory, Smithsonian Institution/University of Arizona)

has received 70 million dollars from the Keck Foundation (Figure 9.18).

The second approach is based on the design of a new technology telescope that has been in operation since 1979 on Mt. Hopkins in Arizona. This telescope consists of not one but six individual mirrors mounted together and adjusted in such a way that the light from all six mirrors comes to a focus at the same point. Each of the six mirrors is 1.8 m (72 inches) in diameter, but used together, they provide the same light gathering power as a single 4.5-m mirror. Because of its design, this telescope is called the Multiple Mirror Telescope (MMT) (Figure 9.19). Astronomers are now exploring the feasibility of building a telescope that consists of four 7.5-m mirrors on a single mount. This combination is equivalent in aperture and power to a 15-m monolithic mirror (Color Plate 7a).

How many of these gargantuan new technology telescopes will be built? The answer to that question is more political and financial than scientific and technical. There seems little doubt that either design could be successfully implemented. The problem is to find the resources to turn the dreams of astronomers and the technical feasibility studies of engineers into reality. The California project has identified the bulk of the funds required, the design is well along, and thus the Keck Telescope should become fully operational in the early 1990's (Figure 9.20). Several other universities, including Texas, Arizona, and Ohio State, as well as the Carnegie Foundation are actively pursuing plans—and funding—to build 7- to 8-meter class telescopes. The National Optical Astronomy Observatories are developing plans for a 15-m class telescope based on the MMT design with four 7.5-m mirrors. Their hope is to build the world's most powerful telescope and make it available to the whole community of astronomers, but realization of that hope depends on funding from the federal government.

What would be the primary use of such telescopes? The obvious advantage of a larger aperture is that it offers greater light-gathering power and

Figure 9.20 Scale model of the 10-m Keck Telescope, now under construction by the University of California and Caltech. The mirror is a mosaic of 36 hexagonal segments. Although larger than any other existing optical telescope in aperture, its stubby construction will allow it to be housed in a dome smaller than that of the 5-m telescope on Palomar Mountain. (CARA–California Association for Research in Astronomy)

allows us to observe ever-fainter and more distant objects. If all we wanted to do was obtain images of the most distant galaxies and quasars, however, we would probably not choose to build telescopes much larger than the ones we have now. With new electronic detectors, we can already detect 26th-magnitude objects from the ground, and the Hubble Space Telescope should allow us to see objects more than ten times fainter yet. The problem is that we must also observe faint objects spectroscopically if we want to learn about their distances, compositions, and motions, and about the violent activity that seems so often to be associated with the earliest stages of the evolution of galaxies. So many of the questions that we wish to answer about the formation and evolution of galaxies, about the large-scale structure of the universe, about the synthesis of the chemical elements, and about the mechanisms that control star formation seem just beyond the capability of existing instruments.

9.4 LOCATION OF OBSERVATORIES

The performance of a large optical telescope is determined not only by the size of its mirror but also by where it is located. One site considered originally for the Lick Observatory, which is now more than 100 years old, was Market Street in downtown San Francisco. The world's largest refractor is located in southern Wisconsin, not a region noted for its clear skies. When a site was selected for Kitt Peak National Observatory in the 1950s, no mountain peaks over 8000 feet in altitude were seriously considered. Now major observatory complexes are located in Chile, in the Canary Islands, and on Mauna Kea in Hawaii, a mountain that is 13,700 feet high. What has led astronomers to seek ever more remote locations for observatories? Why have they chosen to abandon the convenience of having telescopes near their universities for sites located often halfway around the world and, in the case of Mauna Kea, at an altitude well beyond that at which human beings work most efficiently?

(a) Atmospheric Limitations

The Earth's atmosphere, so vital to life, is the biggest headache to the observational astronomer. In at least five ways the air imposes limitations upon the usefulness of telescopes.

1. The most obvious limitation is weather—clouds, wind, rain, and the like. Even on a clear, windless night, however, the atmospheric conditions may render a telescope virtually useless for many types of work.
2. Even for visible wavelengths the atmosphere filters out a certain amount of starlight and dims the stars slightly. For wavelengths from about 20 μm in the infrared out to the radio wavelength of about 1 cm, the air is almost opaque, and it is opaque also to all wavelengths shorter than 2900 Å. In the critical infrared region between 1.0 μm and about 40 μm, the opacity is due primarily to water vapor, and a great many observations are possible at a dry site that cannot be made if the humidity is higher.

TABLE 9.1 Some Major Optical Observatories

LOCATION	INSTITUTION	LARGEST TELESCOPES
Mt. Pastukhov, Caucasus, U.S.S.R.	Soviet Special Astrophysical Observatory	6-m (236-inch) reflector
Palomar Mountain, California	Palomar Observatory (California Institute of Technology)	5-m (200-inch) Hale reflector 1.5-m (60-inch) reflector 1.2-m (48-inch) Schmidt
Canary Islands	La Palma Observatory (United Kingdom)	4.2-m (165-inch) Herschel reflector 2.5-m (98-inch) Newton reflector
Kitt Peak, Arizona	U.S. National Observatory	4-m (158-inch) Mayall reflector 2.1-m (84-inch) reflector 1.5-m (60-inch) McMath solar telescope
	University of Arizona	2.3-m (90-inch) reflector
Cerro Tololo, Chile	Inter-American Observatory	4-m (157-inch) reflector 1.5-m (60-inch) reflector
Siding Spring, Australia	Anglo-Australian Observatory	3.9-m (155-inch) reflector 1.2-m (48-inch) Schmidt
Mount Hopkins, Arizona	U. of Arizona and Smithsonian Astrophysical Observatory	Six 1.8-m (70-inch) mirrors in the multiple-mirror telescope (MMT)
Mauna Kea, Hawaii	U.K. Science Research Council (operated by Royal Observatory, Edinburgh)	3.8-m (150-inch) infrared telescope
	Canada-France-Hawaii Telescope Corporation	3.6-m (142-inch) reflector
	NASA (operated by the U. of Hawaii)	3-m (120-inch) infrared telescope
	U. of Hawaii	2.2-m (88-inch) reflector
La Silla, Chile	European Southern Observatory	3.6-m (142-inch) reflector 0.9-m (36-inch) Schmidt
Mount Hamilton, California	Lick Observatory	3-m (120-inch) Shane reflector 1-m (40-inch) reflector 0.9-m (36-inch) Crossley reflector 0.9-m (36-inch) refractor
Mount Locke, Texas	McDonald Observatory (U. of Texas)	2.7-m (107-inch) reflector 2.1-m (84-inch) reflector
Crimea	Crimean Astrophysical Observatory (U.S.S.R.)	2.6-m (102-inch) reflector
Armenia	Byurakan Observatory	2.6-m (102-inch) reflector
Las Campañas, Chile Mount Wilson, California	Mount Wilson and Las Campañas Observatories	2.6-m (101-inch) du Pont reflector (L. C.) 2.5-m (100-inch) Hooker reflector (M. W.) 1.5-m (60-inch) reflector (M. W.)
Wyoming	Wyoming Infrared Observatory	2.3-m (92-inch) infrared telescope
Williams Bay, Wisconsin	Yerkes Observatory (U. of Chicago)	1-m (40-inch) refractor

3. In the daytime, the air molecules scatter so much sunlight that the resulting blue sky hides all celestial objects except the sun, Moon, and Venus. At night the sky is darker, but never completely dark. Near cities the air scatters about the glare from city lights, producing an illumination that hides the faintest stars and limits the depths that can be probed by telescopes as well as by the naked eye. Even starlight scattered by the air contributes to the brightness of the night sky.
4. To make matters worse, the air emits light of its own. Charged particles from the sun and beyond, funneled into the atmosphere in the polar regions by the Earth's magnetic field, set the air aglow and cause auroras. In addition, at night there is an airglow all over the Earth, for atoms in the upper atmosphere are ionized by ultraviolet photons from the sun and fluoresce. The brightness of this airglow varies from time to time. During times of maximum solar activity, the night sky brightness may be two or three times normal (see Chapter 28). The faint illumination of the night sky limits the time that a telescopic photograph can be exposed without fogging and thus also limits the faintness of the stars that can be recorded.
5. Finally, when the air is unsteady, star images are blurred. In astronomical jargon, the measure of the atmospheric stability is the *seeing*.

When the seeing is good, the stars appear as sharp points and fine detail can be seen on the Moon and planets. When the seeing is bad, images of celestial objects are blurred and distorted by the constant twisting and turning of light rays by turbulent air. It is the variations in density of the upper atmosphere that causes stars to "twinkle."

(b) "Seeing"

Because stars present too small an angular size for existing telescopes to resolve their disks, their images should be geometrical points or rather tiny diffraction patterns. But in bad seeing, a star may appear as a "mothball" or as a dancing firefly. The star image usually looks "bigger" in bad seeing, but not because the telescope has magnified it, only because the air has distorted it. Use of higher magnification only magnifies these atmospheric disturbances. Under typical observing conditions, a star image appears to be from 1 to 3 arcseconds in diameter. This is called the "seeing disk" of the star, and of course has nothing to do with the star's actual size. Under the very best conditions obtainable, star images seldom have diameters less than about ¼ arcsecond.

The cells of turbulence in the atmosphere that cause seeing may be as small as a few centimeters—occasionally even less. However, they are rarely, if ever, larger than about 75 cm. If a star is viewed through a telescope whose aperture is less than the size of the turbulent seeing cells, its image may appear sharp and display the diffraction pattern characteristic of a telescope of that size, but simply dance around in the field of view as the moving air currents refract the starlight first one way and then the other. This phenomenon is called "hard" seeing. If the seeing cells are smaller than the telescope aperture, on the other hand, the telescope must see the star in several directions at once, so that its image is smeared out into that "mothball" referred to above; starlight passing through different seeing cells produces different parts of the mothball image. This is called "soft" seeing. Because seeing cells are generally less than 75 cm across, large telescopes always have soft seeing, and the size of the disk of soft seeing is seldom less than 1 arcsecond in diameter. The 5-m telescope at Palomar, which can theoretically resolve 0.05 arcsecond, thus usually falls short by 20 times of achieving its potential resolving power.

In view of the limitations set by the Earth's atmosphere, the advantages of the Hubble Space Telescope (Chapter 10) are obvious. Although many telescopes have been operated in space, as free-flying satellite observatories or from the Space Shuttle, most of these have been designed to exploit access to infrared and ultraviolet spectral regions not observable from the ground. The Space Telescope will be the first instrument optimized to take advantage of the freedom from atmospheric seeing that is realized in space. Meanwhile, balloons have carried people and instruments, or instruments alone, to altitudes of 30,000 m for the purpose of obtaining astronomical observations. At those altitudes, most of the atmosphere is below the observer, and the seeing is much improved.

(c) Speckle Interferometry

One rather modern (since about 1970) method can circumvent the seeing limitation of a large telescope. The technique is *speckle interferometry*. Suppose a very short time exposure (0.01 sec or less) is made of a highly magnified image of a star with a telescope of large aperture, and through a filter that admits only a very narrow range of wavelengths, so that the image is nearly monochromatic. In this case the photograph will reveal not one image but a multitude of images, arranged roughly in a circular pattern, with a greater concentration of images toward the center. The ensemble of images is called a *speckle pattern*. Each minute image in the pattern is that produced by the full aperture of the telescope of the starlight through a single seeing cell. That image may be too faint to reveal a useful picture, but in principle it contains information at the limit of resolution of the telescope. The technique of speckle interferometry involves the combining of many such short exposure photographs and a reconstruction of the true resolution-limited image by computer analysis.

With speckle interferometry, a considerable amount of information has been learned about the orientations of the pairs of stars in certain close binary star systems (Chapter 25). The stars in such a system have an angular separation great enough that they could theoretically be resolved by a large telescope, but that is still far smaller than the sizes of typical seeing disks. The technique has also provided information about a very few stars that are near enough and large enough that their disks would, in principle, be resolved by existing large

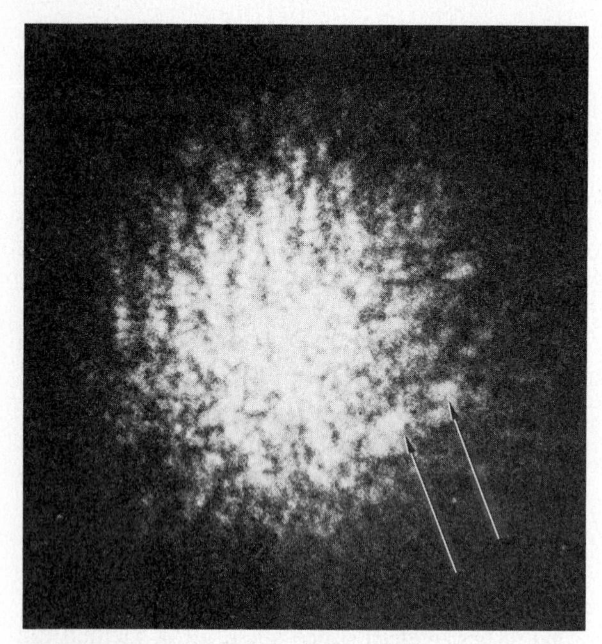

Figure 9.21 Photograph of a speckle pattern of the double star Kappa Ursae Majoris. There are really two nearly identical speckle patterns, one of each star, that are slightly displaced from each other and overlap. On a normal time exposure, the pattern would blur into the familiar seeing disk. The eye can actually pick out the speckle pairs, one of which is marked with arrows. It is more accurate and convenient, however, to measure the system by using a composite transform made with an optical processor that passes a laser beam through the negatives of many such speckle photographs taken in rapid succession. The procedure yields what is called an autocorrelation image like that shown in Figure 9.22.

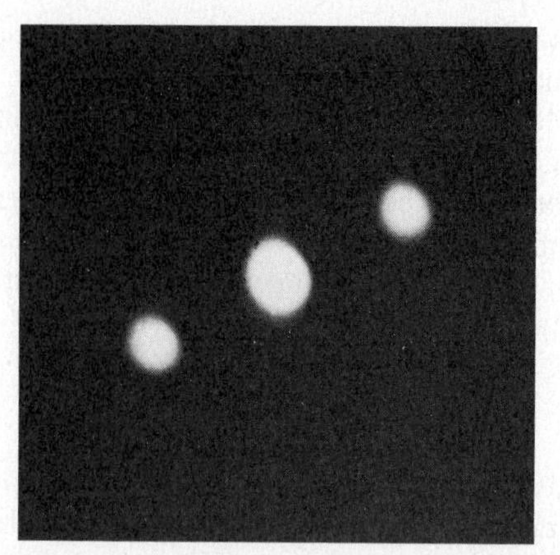

Figure 9.22 The autocorrelation image of the double star Kappa Ursae Majoris, reconstructed from 50 speckle photographs. In effect, each individual speckle image is placed at the position of the bright central spot, which serves as origin. Since each image is that of one of the two members of the binary system, the position of the image of its companion star must lie on one side or the other of the central spot (depending on which star happened to be selected as being at the origin). The companion star's position is thus represented as one of the two fainter spots flanking the central one. The separation of the two flanking spots (or the distance of either from the central spot) reveals the angular separation of the two stars in the sky, and their orientation gives the direction of one star from the other in the sky. (Figures 9.21 and 9.22 courtesy of Harold McAlister, Georgia State University; the research was carried out at Kitt Peak National Observatory)

telescopes but whose angular diameters are very much smaller than seeing disks. Already we have learned something about the distribution of light in a few giant stars, and as the speckle technique is perfected, we may be able to detect large irregularities in the surface brightnesses of some stars—such as gigantic starspots. In particular, speckle interferometry is a promising technique for determining the diameters of large nearby stars (Chapter 25).

(d) Observatory Site Selection

Despite the new observations from balloons, rockets, satellites, and space probes, ground-based observatories will continue to be the workhorses of astronomy for the foreseeable future. But as we continue to improve our observational techniques, we require ever better sites so that we can exploit fully the advances in instrumentation.

What then are the requirements for a first-class observing site? The first is that the site must have clear skies most of the time. Many people think that telescopes can see through clouds, but optical ones cannot. The best observing sites are completely clear 50 to 60 percent of the time. An additional 25 to 30 percent of the time, there may be only thin cirrus clouds, and so certain kinds of observations are possible.

The second requirement is that the seeing must be good. That is, the atmosphere must be as free of turbulence as possible so that the size of a stellar image is small and so that fine details can be seen on planets and other extended images. Tests made at various locations around the world suggest that one can expect to find the best seeing in coastal mountain ranges and on isolated peaks surrounded by ocean. Specific examples of good locations are Mauna Kea and Haleakala in the middle of the Pacific, the mountain peaks of the Canary Islands in the Atlantic, and the coastal range of mountains in Chile.

Figure 9.23 A satellite photograph of the nighttime United States. Note the strong illumination from populated areas. (Kitt Peak National Observatory and National Optical Astronomy Observatories)

A third requirement is that the sky must be dark. We are all familiar with how few stars we can see from the center of a large city. City lights can make the sky measurably brighter even at observatories that are more than 100 miles away. The problem of light pollution has become increasingly severe with the development of acutely sensitive electronic detectors. It is because of the problems of light pollution, combined with the requirement of clear skies, that the only good sites remaining in the continental United States are located in the sparsely populated southwest, most notably Arizona and New Mexico (Figure 9.23).

Fortunately, many communities in the southwest have recognized their responsibility to protect observatories from light pollution. Tucson, for example, has led the way in adopting ordinances controlling city lighting, and many other cities, includ-

Figure 9.24 Composite photograph of Tucson, Arizona, from Kitt Peak National Observatory. Upper portion photographed in 1959 and lower portion photographed in 1980 (Kitt Peak National Observatory and National Optical Astronomy Observatories)

ing Hilo in Hawaii and San Diego and San Jose in California have followed suit. A major advance has been the development of low-pressure sodium lamps. These lights emit essentially all of their light in the pure yellow color that corresponds to the two sodium lines near 5890 Å. While observations made in this particular spectral region will be contaminated by city lights, all of the rest of the spectrum will be unaffected by artificial illumination. Fortunately, low-pressure sodium lights are also extremely energy efficient, and so communities that agree to install them not only protect astronomy but save money as well.

A special requirement for an observing site is imposed by infrared astronomy. Water vapor in the Earth's atmosphere absorbs much infrared radiation. By placing telescopes at high altitude, astronomers can get above much of this water vapor, and so can detect infrared radiation that could not be measured at sites near sea level. Mauna Kea, which offers good seeing by virtue of its mid-ocean site and is an excellent infrared site because of its altitude, is the best all-around observing site developed to date. Astronomers are continuing to search, particularly in Chile and in the American southwest, for new sites that would be suitable for astronomical observations.

EXERCISES

1. Only the Balmer series of hydrogen lines is ordinarily observed in the spectra of stars. Can you suggest an explanation for why the Lyman series, for example, is hard to observe? What about the other series of hydrogen lines?

*2. If the Moon can be photographed in one second with a telescope of 50-cm aperture and 100-cm focal length, how long a time would be required to photograph the Moon with a telescope of 100-cm aperture and 400-cm focal length?

*3. Suppose that all stars emitted exactly the same total amount of light and that stars were distributed uniformly throughout space. Show that the depth in space to which stars could be observed would (ideally) be proportional to the aperture of the telescope used.

4. What is the smallest angle that could theoretically be resolved by the 5-m telescope at a wavelength of 5000 Å?

5. What kind of telescope would you use to take a color photograph entirely free of chromatic aberration? Why?

6. Ordinary 7 × 50 binoculars magnify seven times (magnifying power = 7) and have objective lenses of 50-mm aperture (5 cm). For light of 5000 Å, what is the smallest angle that can be resolved by the lenses of binoculars? Could two stars separated by this angle actually be seen as separate stars when viewed through 7 × 50 binoculars? Why?

7. If the 6-m telescope were used as a radio telescope to observe radio waves of 20 cm wavelength, how accurately could it "pinpoint" the direction of a radio source?

8. The 3-m reflector at the Lick Observatory can be operated as an *f*/5 prime-focus telescope or as an *f*/17.2 Cassegrain telescope. (a) In which of these modes can an extended source be exposed fastest, and by what factor? (b) In which of these modes can a point source be exposed fastest and by what factor?

Grote Reber (b. 1911), an amateur astronomer and electronics expert who built in his own backyard (out of wooden two-by-fours and galvanized iron) the first radio telescope specifically designed to observe radio waves from space. From 1937 until after World War II, he was the world's only active radio astronomer. (Ohio State University)

10 NEW WINDOWS ON THE UNIVERSE

The first half of the 20th century was a very exciting one for science—especially so for physics. Einstein's special and general theories of relativity had provided new insight into the meaning of space and time; physicists had begun to make real progress in understanding the workings of atoms; and with the realization of the equivalence of mass and energy, we began to learn something of the forces that bind together the atomic nucleus itself. Great discoveries were made in astronomy as well—especially the structure of our own Galaxy, the knowledge that it is but one of many in the universe, and that the universe is expanding. We were, in other words, pushing the frontier of science to the realms of the very tiny and the very large.

In the middle of the century, a fantastic change occurred that placed astronomy squarely at the forefront of science. Many physicists and scientists of other specialties changed their research interests to astronomy, and many are still doing so. What was that dramatic development? The opening of new windows in the electromagnetic spectrum. Before 1931 we recorded information from space only in optical wavelengths of electromagnetic radiation. Radio astronomy had its humble beginning in that year, but did not really develop until after World War II.

Consult again Figure 9.1. The two windows of wavelengths for which the atmosphere is transparent are the optical-infrared spectrum and part of the radio spectrum. The optical-infrared window is limited to wavelengths from about 4000 Å (0.4 μm) to 30 μm, although it is broken up by several opaque sections on its long-wave side. Radio waves can be observed from the ground over a range of wavelengths from a few millimeters to several tens of meters—more than a dozen octaves. But there's more—far more—radiation from space, ranging to the shortest gamma rays. That broad short-wave window in the electromagnetic spectrum did not open until rockets and space vehicles could carry our receptors above the atmosphere.

In less than three decades our vision, formerly responding only to light, had broadened to encompass the entire electromagnetic spectrum. It was predictable that there would be new discoveries, but no one could have foreseen just what those discoveries would be, or how far-reaching. Surely we do not yet know all of the laws of physics. Many scientists expect that some of those laws awaiting discovery could leave clues in the cosmos, and that our new eyes on the universe may be bringing us to the verge of new physics. The new astronomy of the latter half of the 20th century occupies substantial

Figure 10.1 Karl B. Jansky, the Bell Laboratories engineer who first detected radio radiation from space and correctly identified it as coming from the Milky Way. (Bell Laboratories)

parts of the chapters to follow; here we summarize briefly how we observe through those new windows to hitherto invisible radiation.

10.1 RADIO ASTRONOMY

(a) Early History

In 1931 Karl G. Jansky (1905–1950), an American radio engineer at the Bell Telephone Laboratories, built a rotating radio antenna array designed to operate at a wavelength of 14.6 m (Figure 10.2). With this array he attempted to investigate the sources of shortwave interference. In addition to temporary intermittent interference due to such phenomena as thunderstorms, he encountered a steady hiss-like static coming from an unknown source. He discovered that this radiation came in strongest about four minutes earlier on each successive day and correctly concluded that since the Earth's sidereal rotation period is four minutes shorter than a solar day, the radiation must be originating from some region of the celestial sphere. Subsequent investigation showed that the source of the radiation was the Milky Way.

It was well over a decade before the astronomical community turned serious attention to Jansky's important discovery. Jansky has, however, received belated honors; today the standard unit of radio flux received from space is the *Jansky (Jy)*. (One Jansky is 10^{-26} watts striking a square meter of the Earth's surface in one unit of frequency.)

Actually, the new radio astronomy did not go *totally* unnoticed, thanks to the American amateur astronomer, electronics engineer, and radio ham,

Figure 10.2 The rotating radio antenna used by Jansky in his serendipitous discovery of radio radiation from the Milky Way. (Bell Laboratories)

Figure 10.3 Grote Reber's original radio telescope, now reconstructed on the grounds of the National Radio Astronomy Observatory in Greenbank, West Virginia. (National Radio Astronomy Observatory)

Grote Reber (b. 1911). In 1936 Reber built the first *radio telescope*—an antenna specifically designed to receive cosmic radio waves (Figure 10.3). He constructed it in his back yard in Wheaton, Illinois, of wooden two-by-fours and galvanized iron. Subsequently he built other improved antennas and remained active in the field for more than 30 years. During the first decade, Reber worked practically alone. By 1940 he confirmed Jansky's conclusion that the Milky Way is a source of radio radiation, and in 1944 he published in the *Astrophysical Journal* the first contour maps of the radio brightness of the Milky Way as it appears at a wavelength of 1.87 m. He also discovered discrete sources of radio emission in the galactic center, Cygnus, and Cassiopeia, as well as radio waves from the sun. From 1937 until after World War II, Reber was the world's *only* active radio astronomer.

Meanwhile, in 1942 radio radiation from the sun was picked up by radar operators in England. After the war, the technique of making astronomical observations at radio wavelengths developed rapidly, especially in Australia, the Netherlands, England, and later in the United States. Radio waves have now been received from many astronomical objects—the sun, Moon, some planets, gas clouds in our Galaxy, other galaxies, and many other objects. The technique of radio astronomy has become an integral and vastly important tool in observational astronomy.

(b) Detection of Radio Energy from Space

First, it is important to understand that radio waves are not "heard"; they have nothing whatever to do with sound. Although in commercial radio broadcasting, radio waves are modulated or "coded" to carry sound information, the sound itself is not transmitted. The radio waves merely carry the information that a radio receiver must "decode" and convert into sound by means of a loudspeaker or earphones. Sound is a physical vibration of matter; radio waves, like light, are a form of electromagnetic radiation. We can also code visible light to carry sound information, as is done, for example, by the sound track on a movie film.

Many astronomical objects emit all forms of electromagnetic radiation—radio waves as well as light, infrared and ultraviolet radiation, and so on. The radio waves we can receive from space are those that can penetrate through the ionized layers of the Earth's atmosphere—those with wavelengths in the range from a few millimeters to about 20 m. The human eye and photographic emulsions are not sensitive to radio waves; we must detect this form of radiation by different means. Radio waves induce a current in conductors of electricity. An antenna is such a conductor; it intercepts radio waves which induce a feeble current in it. The current is then amplified in a radio receiver until it is strong enough to measure or record.

If we lay a photographic plate out on the ground in daylight, it will be exposed by sunlight, indicating the presence of a light source in the sky. We can place various color filters in front of such plates and detect the presence of various colors in the light that exposes them. Such an experiment, however, does not indicate the direction in the sky of the light source.

Similarly, a radio antenna can be strung up outside, and currents induced in it indicate the presence of a source of radio radiation. Electronic filters in the radio receiver can be "tuned" to amplify only a certain frequency at a time, and thus can determine what frequencies or wavelengths are present in the radio radiation. The earliest astronomical observations of radio energy were detected in this way. As in the case of a photographic plate laid out in sunlight, however, a single antenna does not indicate the direction of the source.

(c) Radio Reflecting Telescopes

Radio waves are reflected by conducting surfaces just as light is reflected from an optically shiny surface according to the same law of reflection. A radio reflecting telescope consists of a parabolic reflector, analogous to a telescope mirror. The reflecting surface can be solid metal, or a fine mesh such as chicken wire. In the professional jargon, the reflecting paraboloid is called a "dish" (Figure 10.4). Radio dishes are usually mounted so that they can be steered to point to any direction in the sky and gather up radio waves just as an optical reflecting telescope can be directed in any direction to gather up light. The radio waves collected by the dish are reflected to the focus of the paraboloid, where they form a radio image. In an optical telescope, a photographic plate, image tube, photomultiplier, spectrograph, or some other device is placed at the focus to utilize the image. In a radio telescope, an antenna or wave guide is placed at the focus of the dish. Radio waves focused on the antenna induce in it a current. This current is conducted to a receiver, not unlike ordinary home receivers in principle, where the current is amplified. Receivers can be tuned to select a single frequency, but today it is more common to use sophisticated data processing techniques to allow thousands or more of separate frequency bands to be detected. Thus the modern radio receiver operates much like a spectrometer on an optical telescope. The signals, after computer processing, are recorded on magnetic disk or tape.

One advantage of astronomical observations at radio wavelengths is that some of the important atmospheric effects discussed in Section 9.4a are not bothersome. In particular, radio observations are not affected by atmospheric seeing (although there is a similar *scintillation* effect due to clouds of ions in interplanetary space). They are less affected by weather and sky brightness, and at some wavelengths observations can even be made throughout the entire 24-hour day. Artificial radio interference, however, is a serious problem.

The ability of a radio telescope to gather radiation depends on its size. The radio energy received from most astronomical bodies is very small compared with the energy in the optical part of the electromagnetic spectrum. Hence radio dishes are usu-

Figure 10.4 The 100-m radio telescope at Effelsburg, near Bonn, West Germany. (Max-Planck Institut für Radioastronomie)

ally built in large sizes; few are under 6 m across. At first thought, the problem of constructing a large parabolic reflecting surface to sufficient accuracy might seem prohibitive. The 5-m mirror of the Hale telescope has a surface accurate to about five-millionths of a centimeter, about one-eighth of the wavelength of visible light. However, a radio dish designed to receive radio waves of a length of 25 cm, for example, need be accurate to only about 3 cm to achieve the same resolution. That is why an open wire mesh can be used as a reflecting surface.

(d) Resolving Power of a Radio Telescope

With a radio reflecting telescope, radio radiation can be detected from a particular direction, and the direction of the source in the sky can be determined. We have seen that optical telescopes can resolve images of very small angular size. A 15-cm telescope can determine the location of a star to within about 1 arcsecond. The formula given in Section 9.1c for the smallest resolvable angle,

$$\alpha = 2.1 \times 10^5 \frac{\lambda}{d},$$

where d is the aperture of the telescope, holds for radio telescopes as well as optical telescopes. Thus we can easily compute the resolving power of a radio telescope.

The main difficulty with radio telescopes is that the wavelength of radio radiation, λ in the above formula, is far greater than for visible light, so the resolving power for a telescope of a given size is correspondingly less. Radio waves of 20 cm, for example, are some 400,000 times longer than waves of visible light, so to resolve the same angle, a radio telescope would have to be 400,000 times larger than an optical telescope. To resolve 1 arcsecond at 20 cm wavelength, a radio telescope would have to be nearly 40 km across. The largest steerable radio telescopes in use today are only about 100 m in diameter. At a wavelength of 20 cm they are capable of resolving two points about 800 km apart on the Moon. If it were not for atmospheric "seeing," the 5-m telescope could resolve, in visible light, two points 40 m apart on the Moon. The human eye can resolve points on the Moon separated by about 100 km. Thus the largest radio telescopes have far poorer resolving power than even the human eye. Consequently, it is a special problem to determine accurately the positions of radio sources in the sky.

A radio astronomer commonly speaks of the *beam width* of the telescope. The *beam* is the cone-shaped bundle of radiation from space of such an angular size that within it features are smeared out

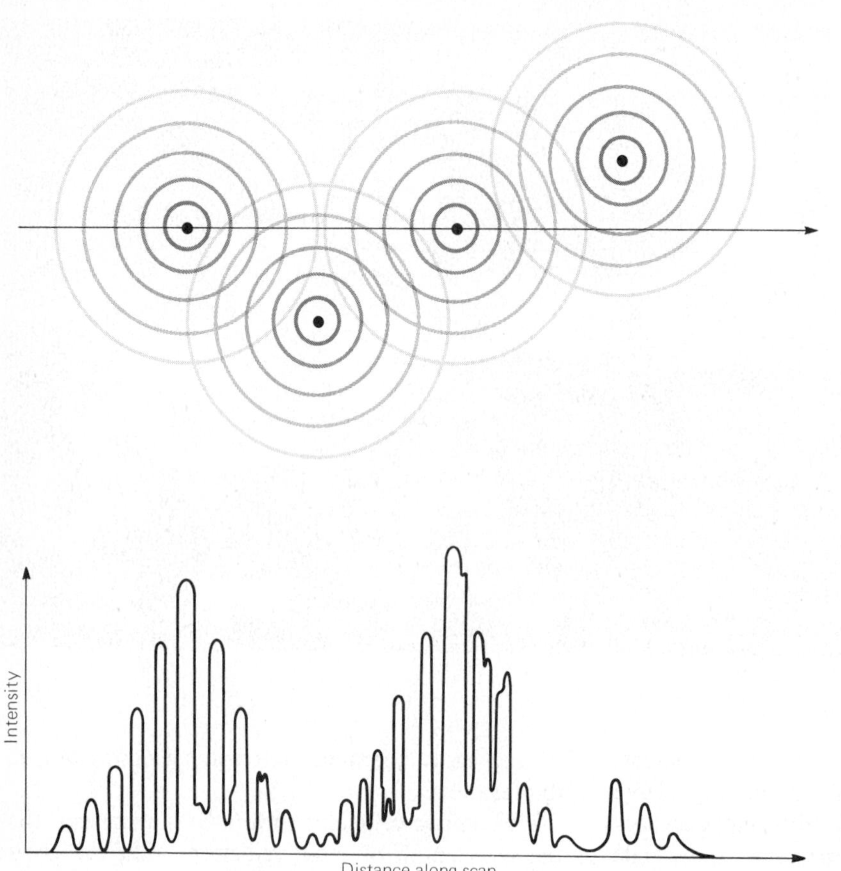

Figure 10.5 (*Top*) The overlapping diffraction patterns of five hypothetical point radio sources, as produced by a single antenna of circular aperture. *(Bottom)* The sort of signal that would be obtained by scanning through the overlapping images along the line shown.

and cannot be resolved. The beam width would correspond in optical astronomy to the angular size of the central diffraction disk of a star image. For a large optical telescope, that size might be a few hundredths of a second of arc, but for a typical radio telescope, the beam width is several minutes of arc. The beam width depends, of course, on the aperture of the telescope and on the wavelength of the radiation being detected. Now, the central disk in the optical diffraction pattern of a star is surrounded by concentric diffraction rings. Similarly, the central part of the image of a point radio source is surrounded by concentric rings of radio radiation. Suppose a radio telescope is slowly moved so that the field of view of the antenna at the focus of its dish (that is, the beam) slices across the image of such a point source of radio waves (or, equivalently, the telescope can remain stationary, and the turning Earth can cause the source to move by). In that case, the beam scans across the diffraction pattern, detecting first a series of successively brighter pieces of the diffraction rings it crosses, then the central image of the source, and finally another series of successively fainter segments of the rings on the other side. Each image, in other words, is recorded as a bright central spot of size equal to the telescope beam width flanked by fainter images on either side; these are called *side lobes*. With a large optical telescope the entire diffraction pattern of a star is usually small compared with the seeing disk caused by the Earth's atmosphere, so the star appears as a single image and many separate stars can be seen or photographed as separate images. With a radio telescope, however, the diffraction pattern is generally thousands of times larger, and the patterns of adjacent radio sources often overlap each other, creating general confusion (Figure 10.5). To make sense of the observations, the radio astronomer must sort out the mess. Fortunately, with modern computer technology, there are techniques for accomplishing this.

(e) Radio Interferometry

Angular resolution at radio wavelengths can be greatly enhanced with the technique of *interferome-*

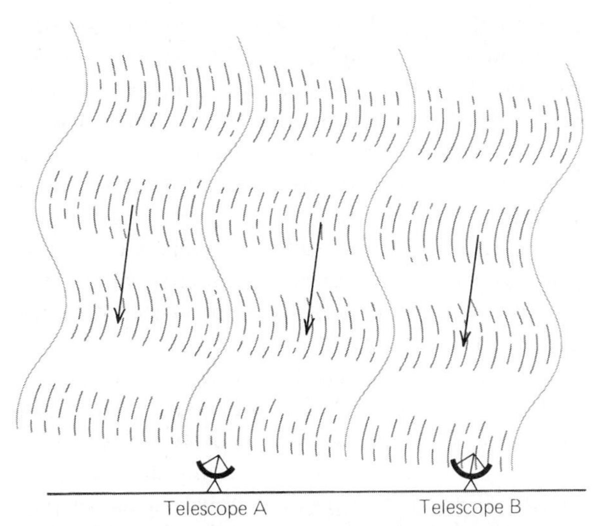

Figure 10.6 The principle of the radio interferometer. Two radio telescopes simultaneously observe the same source, which is not in a direction perpendicular to the line connecting the two telescopes. Thus the waves reach telescope A slightly out of phase with (behind) those reaching telescope B. The phase lag depends on the direction of the source, and hence determines it.

try, which uses two or more separate radio telescopes. Suppose, for example, two radio dishes are placed some distance apart. Unless the source of radio radiation happens to lie along a perpendicular bisector of the line between the antennas, the radio waves will strike one antenna a brief instant before the other, so that the two antennas will receive the same waves at slightly different times and thus become "out of phase" with each other, that is, the antennas receive different parts of a given wave (Figure 10.6). The difference in phase between the waves detected at the two antennas can be measured electronically. Because this phase difference depends on the angle the direction to the source makes with the line between the antennas, that angle can be determined. If the two antennas are due east and west of each other, an observation of this sort gives a much more accurate measure of the east-west position of the source in the sky than could be obtained with a single radio telescope. If the antennas are placed due north and south of each other, the other coordinate of the source can be found.

The farther apart the components of an interferometer are placed (the longer the *baseline*), the more sensitive is the phase discrimination of slightly off-axis sources, and the more accurately can we pinpoint the direction of the source. Most major radio astronomy observatories, therefore, have several telescopes so that pairs of them can be operated together as interferometers. The overall sensitivity to radio radiation is then the sum of the collecting areas of the telescopes so combined, and the resolution (or beam width) in a direction along the line between the telescopes is that of a single telescope of total aperture equal to that separation.

Interferometry need not be limited to two telescopes. Entire arrays can be used together, and the radiation detected at each can be compared, by computer, with that from each other one in the array. In this way an image of the source can be reconstructed as it would appear under the resolution of a single telescope of aperture equal to the maximum separation in the array. The technique is called *aperture synthesis* and was largely pioneered by astronomer Martin Ryle at Cambridge, England.

Figure 10.7 The twin 30-m radio telescopes of the Owens Valley Radio Observatory of the California Institute of Technology. (California Institute of Technology)

Figure 10.8 A portion of the Cambridge 5-km Array. (G. O. Abell)

There are three arrays at the radio observatory at Cambridge, the largest consisting of eight dishes that can be separated to a maximum of 5 km by moving four of them along a section of the old Oxford/Cambridge railway track, on which they are mounted (Figure 10.8). When operating at a radio wavelength of 2 cm, the 5-km array can obtain a resolution of 1 arcsecond, comparable to the usual resolution permitted by atmospheric seeing at optical wavelenths. Another large array at Westerbork, Netherlands, has 12 telescopes, each 25 m in diameter, spaced over a total baseline of 1.6 km.

By far the world's most impressive radio telescope array is the *Very Large Array (VLA)* near Socorro, New Mexico. The VLA, operated by the National Radio Astronomy Observatory (financed by the National Science Foundation), was essentially completed and dedicated in 1980. It has 27 telescopes, each of aperture 25 m, that can all be moved along rails laid out in a large "Y" configuration with a total span of about 36 km (Figures 10.9 and 10.10). The VLA normally operates at three wavelengths: 3 cm, 6 cm, and 21 cm. By electronically combining the signals from all of its individual telescopes, this array permits the radio astronomer for the first time to make "pictures" of the sky at radio wavelengths that are comparable to those obtained with an optical telescope. The resolution of this instrument was intentionally matched to that of ground-based optical telescopes in order to facilitate comparison of images of the same object at optical and radio wavelengths.

By tracking a single area of the sky for the 8 to 12 hours it is above the horizon, the VLA can produce a radio map or image of 1 arcsecond resolution over a field of view of several arcminutes. Approximately three such images can be obtained per day. Alternatively, and again like an optical telescope, the VLA can be used for shorter exposures of less sensitivity. Much of our modern understanding of astrophysical phenomena, particularly high-energy processes such as those associated with active galaxies, is now being derived in part from observations made with this spectacular facility (Color Plate 8a).

(f) Very Long Baseline Interferometry

Now that time standards can be coordinated to high precision, we can extend the interferometer principle to very long baseline interferometry (VLBI). Two different radio telescopes, thousands of kilometers apart, can simultaneously observe the radio waves from the same source and record them on tape, along with marks from a very accurate time standard (like the "ticks" of a very accurate clock). Later, these two tapes can be analyzed with a computer to find the phase difference between the radio radiation at the two stations, and hence the direction of the source. Baselines as long as from California to Parkes (in Australia) and from Greenbank, West Virginia, to Crimea have been used. The resulting angular resolution of the sources observed is as great as a few ten-thousandths of a second of arc—far surpassing the angular resolution of optical telescopes.

Each pair of telescopes in a VLBI acts as a two-element interferometer, yielding high-resolution data on a single source but not generating enough information to produce an image. Naturally, astronomers would like to be able to utilize an *array* of these baselines, just as they have constructed arrays

Figure 10.9 Part of the Y-shaped Very Large Array (VLA) near Socorro, New Mexico. (National Radio Astronomy Observatory)

of ordinary radio telescopes in order to generate radio images at arcsecond resolution. The objective would be a Very Long Baseline Array (VLBA) made up of 10 to 20 individual telescopes built all over the world and linked together into one gigantic instrument of nearly planetary dimensions. Such a VLBA could form astronomical images of the highest possible resolution, as high as a few hundred micro-arcseconds, where a micro-arcsecond is 10^{-6} arcseconds. At these resolutions, features as small as 10 astronomical units (AU) could be distinguished at the center of our Galaxy, and as small as a few hundred AU in many other galaxies.

Experimental combinations of several VLBI instruments have been tried in early efforts to approximate the capability of a worldwide VLBA. Meanwhile, the construction of the additional telescopes required to create a full-scale VLBA is one of the highest priority projects of astronomers, who are asking the National Science Foundation to finance the U.S. contribution to such a system. The eventual addition of one or more telescopes in space has also been suggested as a means to increase further the resolution of the ground-based VLBA, beyond the limits presently set by the dimensions of the Earth. The U.S.S.R. has announced plans to orbit such a receiver in the early 1990s and has invited the participation of astronomers from other nations in this project.

Figure 10.10 One of the 27 25-m antennas of the VLA. (G. O. Abell)

Figure 10.11 The 300-m (1000-foot) dish of the Arecibo Observatory, Puerto Rico, operated by the National Astronomy and Ionosphere Center (Cornell University) and sponsored by the National Science Foundation. (Cornell University)

(g) Radar Astronomy

Radar is the technique of transmitting radio waves to an object and then detecting the radiation that the object reflects back to the transmitter. The time required for the radio waves to make the round trip can be measured electronically, and because they travel with the known speed of light, the distance of the object is determined. The value of radar in navigation, whereby surrounding objects can be detected and their presence displayed on a screen, is well known.

In recent decades the radar technique has been applied to the investigation of the solar system. Radar observations of the Moon and of the planets have yielded our best knowledge of the distances of these worlds and played an important role in navigating spacecraft throughout the solar system. In addition, as will be discussed in later chapters, radar observations have determined the rotation periods of Venus and Mercury, probed the tiny Earth-approaching asteroids and the nuclei of comets, analyzed the rings of Saturn, and investigated the surfaces of Mercury, Venus, Mars, and the large satellites of Jupiter. Radar is particularly critical for the study of Venus, since it can penetrate the otherwise opaque atmosphere and clouds. Extensive surface mapping of Venus is being carried out by ground-based radar and by present and proposed U.S. and Soviet spacecraft in orbit around the planet.

One of the special advantages of radar as an astronomical tool is that, unlike other techniques, the scientist can control the properties of the transmitted beam. In most of astronomy, our role is entirely passive; all we can do is to detect and try to understand the radiation that nature sends to us. In contrast, the transmitted radar signal is entirely within our control, to vary its strength, frequency, polarization, and other properties. In effect, radar astronomy is an *experimental* rather than an observational science. The radar beam serves as an extension of ourselves, with which we can probe remote members of the planetary system.

(h) Famous Radio Telescopes

The largest single radio telescope, in fact, the largest area-focusing telescope of any kind in the world, is the 1000-foot (305-m) bowl near Arecibo, Puerto Rico, a facility of the National Astronomy and Ionosphere Center, funded by the National Science Foundation and operated by Cornell University (Figure 10.11). The telescope was first built in 1963, and in 1974 its reflecting surface was rebuilt with such high precision that it can be operated at radio wavelengths as short as 7 cm. Similar to several other much smaller radio "bowls" throughout

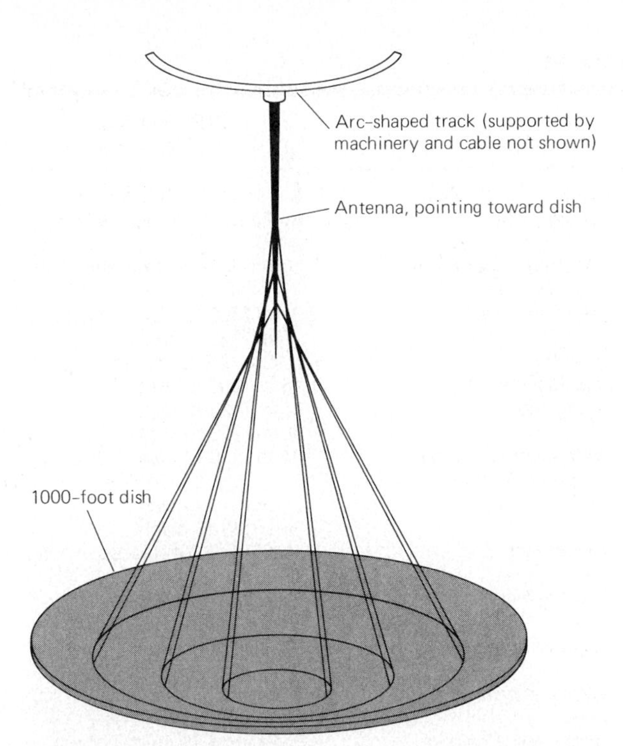

Figure 10.12 Because of spherical aberration produced by the spherical reflecting surface of the Arecibo telescope, different places along the needle-shaped antenna receive radiation reflected from different concentric zones of the 1000-foot dish.

the world, the Arecibo telescope is not movable but consists of a fine wire mesh stretched with astonishing accuracy into a spherical surface over a natural valley in the Puerto Rican mountains. Because a spherical surface has no preferential axis, it produces comparable images over a wide field. A movable antenna is supported by a large arc-shaped track at the focus of the bowl, and the track itself can rotate about a vertical axis, allowing the antenna to move in any direction; it can thus record objects as far as about 20° away from the zenith. The rotation of the Earth assures that all objects within a 40-degree belt around the sky come into view of the telescope at one time or another. The reflector has spherical aberration, of course, so a rather long antenna, pointing toward the bowl, is used. Different points along the antenna thus pick up radiation reflected from different ring-shaped zones of the bowl, and all these radiations are combined electronically (Figure 10.12). The 1000-foot dish is used not only to record extremely feeble radio signals from the remote universe but also to study the Earth's atmosphere. Moreover, it is used

Figure 10.13 Closeup of the reflecting surface of the Arecibo radio telescope. The debris that collects on the wire mesh does not cause appreciable interference and is not removed frequently in order to avoid unnecessary risk of bending the dish out of shape. (G. O. Abell)

as a very powerful radar telescope to explore the surfaces of planets. Frank Drake, former director of the facility, has pointed out that the volume of the bowl could contain the world's annual consumption of beer. Our own reaction is that the surface, as is, would be a mecca for skateboard enthusiasts (Figure 10.13).

Large movable dishes include the 100-m (328-ft) telescope at Bonn, the 91-m (300-ft) telescope of the National Radio Astronomy Observatory at Greenbank, West Virginia, the 76-m (250-ft) at Jodrell Bank, England, the 64-m (210-ft) dish at Goldstone, California (operated by the Jet Propulsion Laboratory in Pasadena), and the 64-m (210-ft) Parkes dish in Australia. The VLA and some other large arrays have already been mentioned.

Many arrays do not use separate parabolic dishes. For example, the Molonglo Radio Telescope, near Canberra, Australia, operated by the University of Sydney, consists of wire mesh stretched along two large troughs, each a mile long and crossing at right angles at their centers. Designed by B. Y. Mills, it is of the type known as the "Mills cross." Each trough is parabolic in cross section, so rather than concentrating the radiation at a single point focus, it reflects the radio waves to a long antenna running down the length of the trough at the focus of its cross-section parabola. The different signals striking different parts of the cross are used to reconstruct a radio image of the source observed by an aperture synthesis technique.

TABLE 10.1 Some Famous Radio Astronomy Observatories

OBSERVATORY	DATE FOUNDED	LOCATION	IMPORTANT TELESCOPES
Arecibo	1963	Arecibo, Puerto Rico	305-m (1000-ft) bowl
Very Large Array	1980	Socorro, NM	27 25-m (82-ft) dishes in 36 km array
Max-Planck Institut für Radioastronomie	1971	Effelsburg, near Bonn, W. Germany	100-m (328-ft) steerable dish
National Radio Astronomy Observatory	1958	Greenbank, WV	91-m (300-ft) dish 43-m (140-ft) dish 3 26-m (85-ft) dishes
Jodrell Bank	1949	near Manchester, England	76-m (250-ft) dish 38-m (125-ft) dish 66-m (218-ft) dish
Parkes National Radio Observatory	1961	North Goobang, near Parkes, Australia	64-m (210-ft) dish
Goldstone Tracking Station (Jet Propulsion Laboratory)	1958	near Barstow, CA	64-m (210-ft) dish 26-m (85-ft) dish
Westerbork	1955	Netherlands	12 25-m (82-ft) dishes in 1.6-km array
Owens Valley (Caltech)	1958	near Big Pine, CA	2 27-m (90-ft) dishes 40-m (130-ft) dish
CSIRO Radioheliograph (Solar radio telescope)	1967	Culgoora, Australia	96 13.7-m (45-ft) dishes in 3-km diam. circular array
Molonglo Radio Observatory (U. of Sydney)	1967	near Canberra, Australia	2 crossed arms 1.6 km (1 mi) long and 12 m wide
Mullard Radio Astronomy Observatory	1945	Cambridge, England	5-km array (8 15-m dishes) 1.6-km array (3 18-m dishes) 0.8-km array (4 9-m dishes)
Haystack telescope	1956	Westford, MA	36-m (120-ft) dish 25-m (84-ft) dish

Of the many other important radio observatories, a few of the most famous are described in Table 10.1.

10.2 INFRARED AND SUBMILLIMETER ASTRONOMY

(a) The Infrared Window

Infrared is a term applied to a very broad wavelength band, from about 1μm (= 10,000 Å) out to the submillimeter and millimeter wavelengths at the short end of the radio window. Over much of this span of nearly 10 octaves the Earth's atmosphere is opaque, more-or-less, depending primarily on the amount of water vapor it contains. Within this range of wavelengths, a variety of astronomical sources emit radiation, and many different kinds of techniques have been developed to detect and analyze this radiation.

From the visible band out to about 2.5 μm, the techniques of infrared astronomy are very similar to those described in Chapter 9 for optical astronomy. The same telescopes can be used, and for the most part the Earth's atmosphere is transparent. Although telescope instrumentation and detectors are not identical, many are actually modifications of those developed previously for use in the visible, such as photographic plates and CCD detector arrays. The sun and most stars are strong sources of infrared radiation in this *near-infrared* range, and the planets and other members of the solar system can also be observed by reflected sunlight.

To most astronomers, the term *infrared astronomy* refers primarily to observations made at wavelengths from about 2.5 μm out to 40 μm, where the Earth's atmosphere becomes too opaque to observe from even the driest ground-based sites. This is sometimes called the *intermediate infrared.* This is the spectral region in which the dominant sources of radiation have temperatures between 100 and

1000 K, including interstellar matter, regions of star formation, dust clouds surrounding the nuclei of active galaxies, and the planets and other members of the solar system. Many special observational techniques have been developed to work in this spectral region, as described in Sections 10.2b and 10.2c.

The *far infrared* extends from about 40 μm out to several hundred μm. Since the Earth's atmosphere is opaque over this span of wavelengths, observations must be carried out from high-flying instrumented airplanes or balloons, or from space. Only recently have we begun to probe this part of the spectrum. The first all-sky survey in the far infrared was not made until 1983, when the NASA Infrared Astronomy Satellite (IRAS) was placed into orbit for that purpose.

The *submillimeter* is that part of the spectrum at wavelengths from a few hundred μm to a few millimeters. Because the atmosphere is largely opaque, and no spacecraft have been launched with a capability for submillimeter astronomy, this spectral region has not been explored in detail. Several new submillimeter telescopes have been built recently, however, so this situation should improve in the near future.

(b) Infrared Telescopes

Superficially, infrared telescopes are very similar to optical telescopes. They all use reflecting optics, since infrared radiation cannot be focused with glass lenses. Their mountings and optical designs are also similar to the telescopes described in Chapter 9. Indeed, most large optical telescopes are now used part of the time for observations in the infrared, especially in the near infrared, at wavelengths of 2.5 μm or less.

At wavelengths longer than about 2.5 to 3.0 μm, however, the differences between infrared and optical systems become important. Typical temperatures on the Earth's surface are near 300 K, and the atmosphere through which observations are made is only a little cooler. This means that the telescope, the observatory, and even the sky are radiating infrared energy with a peak wavelength at about 10 μm. To infrared eyes, everything is brightly aglow. The problem is to detect faint cosmic sources against this sea of light. The infrared astronomer must always contend with a situation such as would face a visible-light observer working in broad daylight with a telescope and optics lined with bright fluorescent lights.

The first problem is to protect the infrared detector from this radiation, just as you would shield photographic film from bright daylight. Since anything that is warm radiates infrared energy, the detector must be isolated in very cold surroundings. Even in the near infrared, detectors are typically cooled by liquid nitrogen to about −200° C. At longer wavelengths, they are more commonly cooled with liquid helium, the gas with the lowest known boiling temperature (about 4° above absolute zero). Even greater cooling can be achieved by using a vacuum pump to reduce the pressure over the liquid helium, bringing its temperature down below 2K. Therefore, the detector is held in a metal container surrounded by liquid helium with only a tiny window through which it can view the hot world outside.

The second problem is radiation from the telescope structure and optics. To reduce the emission from the optics, they are kept very clean, since every bit of dust is an infrared source. The telescope is itself designed so that the detector does not see its structure, but only looks out through the optics at the sky. In ordinary visible-light astronomy it is common to use black baffles to reduce stray light, but no black (and therefore strongly emitting) surfaces are allowed to intrude into the optical path of a telescope that has been optimized for infrared observations. This is a fundamental distinction that forces the astronomer to chose whether to build an optical or infrared telescope, since what is best for one spectral region is worst for the other.

Finally, it is necessary to contend with the infrared emission of the atmosphere. Ideally, infrared observations should be made from outside the atmosphere, or at least from as high an elevation as possible (Section 10.2c). They always require absolutely cloud-free skies, since even the thinnest wisp of cirrus cloud emits a great deal of infrared radiation.

In addition, it is generally the practice in infrared astronomy to observe a source and compare it with the adjacent sky, detecting only the *difference* in radiation between the two directions. The more rapidly this comparison can be made, the better. In infrared telescopes this is done by tilting a secondary mirror back and forth rapidly, but very

Figure 10.14 Infrared and sub-millimeter observations require dry observing sites, such as the 4.2-km high summit of Mauna Kea in Hawaii. The telescopes seen here include the 3.8-m UK Infrared Telescope (the largest in the world), the 3.0-m NASA IR telescope, and the 15-m James Clerk Maxwell submillimeter telescope operated by the UK and Netherlands. In the background is another shield volcano, Mauna Loa. (C.R. Chapman/University of Hawaii)

slightly, so that the source's image produced by the primary mirror is alternately on and off the detector, usually at frequencies of 10 to 50 cycles per second. The alternating current produced by this periodic change of radiation intensity reaching the detector is a measure of the infrared brightness of the source.

Infrared detectors were originally single, tiny chips of semiconductor material connected electrically to sensitive amplifiers, and this is still a common way to observe. However, it has the disadvantage of allowing measurement of the signal from only one point source at a time, or from only a single wavelength of a spectrum. Efforts have therefore been devoted to developing *one-dimensional arrays*—rows of individual detectors that can observe a series of points on the sky, or a number of wavelengths in a spectrum, simultaneously. Even more recently, technologies have been found to manufacture *two-dimensional arrays* of infrared detectors. With such arrays, it is possible for the first time to obtain images of the infrared sky. As of this writing, such infrared arrays are still small, providing no more than 100 × 100 pixels (picture elements), but they are improving all the time, with the future promise of greatly enhancing our capability to observe in this part of the spectrum.

The best sites for infrared observatories are located where the air is as dry as possible. Throughout this spectral region, water vapor is the primary

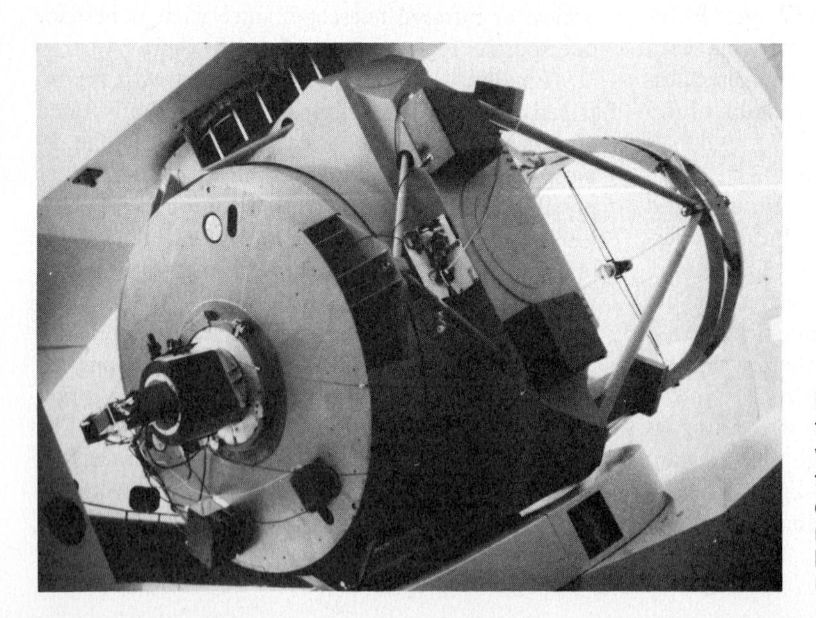

Figure 10.15 NASA's 3-m infrared telescope (IRTF) on Mauna Kea, Hawaii. The center section, supported by the large yoke mounting, contains the optics. The 3-m–diameter primary mirror is inside the circular section at the lower end of the telescope. (NASA/University of Hawaii)

Figure 10.16 The NASA IRTF (foreground) and the 3.6-m Canada–France–Hawaii telescope on Mauna Kea. (D. P. Cruikshank/University of Hawaii)

source of opacity in the Earth's atmosphere. Even when the sky is otherwise beautifully clear, a site with high humidity can be nearly useless for wavelengths longer than 12 μm, or even for observations near the water vapor bands in the near infrared. Low humidity can be found at very cold locations such as the Antarctic in winter, but these have other disadvantages as observatory sites. Generally, the best way to escape atmospheric water vapor is to go to high altitudes, and today nearly all infrared telescopes are located at elevations above 3000 m, and preferably above 4000 m. Mauna Kea in Hawaii, at 4200 m, is the home of the two largest telescopes specifically designed for infrared work, the United Kingdom 3.8-m telescope and the NASA/University of Hawaii 3.0-m Infrared Telescope Facility. Another major instrument used for infrared work is the 4.5-m multi-mirror telescope (MMT) of the Smithsonian Institution and University of Arizona, located on Mt. Hopkins near Tucson at an elevation of 2600 m.

(c) Airborne and Space Infrared Telescopes

Water vapor, the main source of atmospheric interference throughout the infrared, is concentrated in the lower part of the Earth's atmosphere. That is why a gain of even a few hundred meters in elevation can make an important difference to the quality of a site. Given the limitations of high mountains, most of which attract clouds and violent storms, it is natural to investigate the possibility of observing from airplanes and ultimately from space.

The first airborne infrared observations were made in the 1960s by Frank J. Low of the University of Arizona. NASA equiped a small Lear Jet airplane with a 12-inch (30-cm) telescope that looked out through an open hole cut in the side of the plane. Based on the success of these flights, NASA decided to construct and operate a larger airborne observatory, flying regularly out of the Ames Research Center south of San Francisco.

The Gerard P. Kuiper Airborne Observatory—named for one of the pioneers of infrared astronomy, who practically invented the field in the years after World War II—consists of a 0.9-m Cassegrain reflector mounted in a Lockheed C-141 (Figure 10.17). The telescope views the sky through a large

Figure 10.17 The NASA Kuiper Airborne Observatory (KAO) consists of a 1-m telescope mounted in a modified C-141 aircraft. The KAO observes from the bottom of the stratosphere, at an altitude of between 40,000 and 44,000 feet. (NASA)

Figure 10.18 IRAS carried out the first all-sky survey in the infrared in 1983. The entire optical section of the telescope is cooled to just a few degrees above absolute zero, permitting very much greater sensitivity than is possible with any ground-based infrared telescope.

hole in the side of the airplane. To minimize vibrations, it is mounted on an air bearing and controlled by an elaborate computerized system on the plane. A crew of about a dozen people, in addition to the astronomers, is required to fly the plane and operate the telescope.

The KAO (as it is called) flies near the top of the troposphere at elevations of 12 to 13 km, where it is above 99 percent of the atmospheric water vapor. Although the atmosphere above this elevation is not completely transparent at all infrared wavelengths, observations can be made throughout the critical far-infrared range from 40 to 100 μm. This telescope is used on 60 to 100 nights per year for astronomical work. NASA is now considering building a second infrared telescope, at least twice as large, to fly in a modified Boeing 747 aircraft.

Taking the next step, to observations from space itself, has important advantages for infrared astronomy. First, of course, is the elimination of all interference from the atmosphere, including any thermal radiation that it emits into the telescope. But equally important, it provides the opportunity to cool the entire optical system of the telescope so as to reduce, and nearly eliminate, infrared radiation from this source as well. If we tried to cool a telescope within the atmosphere, it would quickly become coated with condensing water vapor and other gases, making it useless. Only in the vacuum of space can optical elements be cooled to hundreds of degrees below freezing and still remain operational. This is a tremendous advantage for space observations in this part of the spectrum.

Although a number of small experimental payloads had been flown previously, the first real infrared observatory was IRAS, the Infrared Astronomy Satellite built as a joint project among the U.S., the Netherlands, and Britain (Figure 10.18). IRAS was equipped with a 0.6-m telescope cooled to a temperature of less than 10 K, with a series of detectors sensitive to wavelengths of 10, 25, 60, and 100 μm. The shorter wavelength bands overlapped with ground-based telescopes, but the longer two extended our capability into a largely unexplored spectral region.

Because it operated at such a low temperature, the IRAS telescope was much more sensitive at all wavelengths than much larger telescopes on the ground. For the first time the infrared sky could be seen at night, as it were, rather than through a bright foreground of atmospheric and telescope emission. Therefore, IRAS was well suited to carry out a rapid but thorough survey of the entire sky, sweeping the fields of view of its detectors over huge areas rather than pointing at one spot at a time. Launched early in 1983, IRAS just managed to complete this survey before its liquid helium coolant was exhausted ten months later (Color Plates 22 and 23).

The next step in infrared observations from space will require larger cooled telescopes that will operate, like ground-based facilities, in a pointed rather than a survey mode. The European Space Agency plans to launch such a facility, called ISO (Infrared Space Observatory), in the early 1990s. Later in the same decade, NASA hopes to complete a 1-m class telescope called SIRTF (Space Infrared Telescope Facility) to be operated as part of the Space Station or an associated space science orbital platform. Since SIRTF will be serviceable by astronauts, its liquid helium can be resupplied, so there is no reason that it cannot continue operation indefinitely, unlike IRAS and ISO.

(d) Submillimeter Observations

The submillimeter represents one of the few remaining unexplored regions of the electromagnetic spectrum. Radio astronomers have worked to wavelengths as short as about 1 millimeter, where there are a few spectral intervals where the Earth's atmosphere is reasonably transparent. In addition, ground-based observers at Mauna Kea have found that there is sufficient transparency on dry nights to work in two other wavelength bands, at about 300 and 700 μm (0.3 and 0.7 mm). Submillimeter observations have also been pioneered by Caltech astronomers at their Owens Valley observatory in California (Color Plate 8b).

At this writing (1986) there are two new telescopes under construction on Mauna Kea that are optimized for submillimeter observations. One, with a 10-m aperture, is being built by Caltech; the other, with 15-m aperture, is a joint project of the Netherlands and Britain. A survey from space of the submillimeter sky is one of the projects that the European Space Agency would like to carry out, and NASA also has long-term plans for a space-based facility called the LDR (Large Deployable Reflector). However, these observatories are not likely to be operational before well into the next century.

10.3 HIGH-ENERGY ASTRONOMY

(a) The Short-Wave Window

The atmosphere is completely opaque to electromagnetic radiation of wavelength less than about 3000 Å. Consequently, ultraviolet, X-ray, and gamma ray observations must be made from space. Such observations first became possible in 1946, when the U.S. acquired several V2 rockets captured from the Germans. Herbert Friedman (b. 1916) and his colleagues from the U.S. Naval Research Laboratory instrumented these rockets for a series of pioneering flights, used initially to detect far-ultraviolet radiation from the sun. Subsequently many rockets have been launched to make X-ray and ultraviolet observations of the sun, and later of other celestial objects as well.

Beginning in the 1960s Earth satellites have been launched to carry out astronomical observations. These have included the orbiting solar observatories (OSOs), the orbiting astronomical observatories (OAOs), and the high-energy astronomy observatories (HEAOs). Scientist astronauts also made astronomical observations at short wavelengths from Skylab, and more recently from Spacelab, a European Space Agency (ESA) facility for Earth-orbital observations.

(b) The Orbiting Ultraviolet Observatories

Ultraviolet telescopes are similar to optical telescopes, except that their optical surfaces need special coatings that have high ultraviolet reflectivity, and, of course, the telescopes must be taken outside the atmosphere. In addition to the small ultraviolet instruments flown in rockets and those operated on the Moon and from Skylab, the United States has launched a highly successful series of satellites especially designed as ultraviolet observatories.

A series of eight satellites, launched during the years 1962 through 1975, were primarily for solar observations. These orbiting solar observatories (OSOs) have obtained thousands of ultraviolet spectra of the sun. The last and most successful observatory, OSO 8, was launched in June 1975 and carried eight instruments. Two of these (one American and one French) are for obtaining ultraviolet spectra of light from tiny regions of the sun (down to one arc second across) and six are X-ray and gamma ray detectors for exploring sources of this high-energy radiation from other directions in space.

The only major solar space observatory since the OSO series is the Solar Maximum satellite, launched in 1980 but disabled after a few months by electronic failures. In 1984, Space Shuttle astronauts docked with the errant satellite, brought it on board, and made the necessary repairs before releasing it back into an independent orbit. At this writing (1986) the Solar Max satellite is continuing its interrupted mission successfully.

The first successful orbiting astronomical observatory, OAO 2, was launched December 7, 1968, and carried instruments developed by the University of Wisconsin and by the Smithsonian Astrophysical Observatory at Harvard for obtaining ultraviolet spectra of astronomical objects in the spectral range 1200 to 4000 Å. During the useful life of the observatory (to February 1973), more than a thousand observations were made. The next important ultraviolet observatory was OAO Copernicus, launched in August 1972, which carried a 0.8-m ultraviolet telescope and three small X-ray telescopes. Hundreds of astronomers served as guest investigators to obtain observations with it.

The most successful orbiting ultraviolet observatory to date (still in operation at this writing—early 1986) is the International Ultraviolet Explorer (IUE) satellite, launched in January 1978 and carrying instruments developed by NASA and by the United Kingdom's Science Research Council. The IUE can obtain high-quality spectra in the range 1150 to 3200 Å and uses a vidicon image converter that can allow exposure times of up to 15 hours. The IUE is in a synchronous orbit (that is, its period of revolution about the Earth is equal to the period of the Earth's rotation), so that it is always in view of its control headquarters at NASA's Goddard Space Flight Center, in Greenbelt, Maryland. Astronomers from all over the world go to Greenbelt as guest investigators. They must, of course, have their applications for observing time on the IUE approved, and the competition is great. Careful preparation must be made to be able to orient the satellite so that its telescope will point to the desired object. The guest observer at Greenbelt can actually see a TV-displayed image of the field of view through IUE's 0.45-m Cassegrain telescope, as it is televised live to Earth, and can compare what she sees with her star charts and make final adjustments as needed. When the exposure on the object is over, the observer sees the spectral data displayed on the TV screen; a printout on paper or magnetic tape is provided for the astronomer to take back to her home institution, to be analyzed at leisure.

(c) X-Ray Observations

X-rays are electromagnetic radiation of wavelength less than about 100 or 120 Å. The shorter the wavelength, the higher the energy of the photons (Section 8.1), and X-ray and gamma ray astronomers often speak of the energies of the photons they observe rather than of their wavelengths. A photon of wavelength 12.4 Å has an energy of 1000 eV, abbreviated 1 keV. Those of lower energy are called *soft* X-rays, and those of higher energy (shorter wavelength), *hard* X-rays. Radio waves are characteristic of the energy emitted from cold bodies, and visible light and the ultraviolet are characteristic of that from hot bodies of temperature 10^3 to 10^5 K (like the surfaces of stars). X-rays are emitted from gas at very high temperatures—10^7 to 10^9 K. Thus in different spectral regions we preferentially observe parts of the universe that are at different temperatures.

X-rays from space were first observed with instruments flown on balloons and rockets. By 1967 about 30 discrete sources of X-rays had been discovered. X-ray astronomy made a sudden advance in December 1970, when the first orbiting X-ray observatory was launched from Kenya through an international program. That observatory was named *Uhuru,* the Swahili word for "freedom." Uhuru systematically scanned the sky for X-ray sources and charted more than 200 of them during its lifetime (to 1973). The convention for naming these sources has been numerically according to the constellation in which they are found; for example, the famous source Cygnus X-1 (Chapter 33) is the first X-ray source to be found in Cygnus.

In 1977 the United States launched the first of its series of high-energy astronomy observatories, HEAO 1. At this writing HEAO 1 is still active and has recorded many more X-ray sources and measured their intensities. Now, up to 1978, X-ray telescopes did not produce images that could be inspected but could merely record the presence of X-rays of certain intensity from a particular direction. The detectors most commonly used were gas ionization detectors. By putting in front of the detector a shield or baffle of metal (for example, lead) that is opaque to the X-rays, the telescope could be made to record radiation only from such a direction as to pass through an aperture in the shield. One kind of shield consisted of many banks of parallel wires, one bank behind the other; the whole apparatus would have to be aligned in a critical way for the spacing between the wires to be lined up from bank to bank so that the radiation could pass through to the detector. With such arrangements it was possible to determine the directions of the X-rays' sources, but not their shapes or much about their sizes.

X-ray astronomy received a truly spectacular boost with the launching of HEAO 2, the Einstein Observatory, in November 1978 (Figure 10.19). The Einstein X-ray telescope was the culmination of a nearly 20-year dream of Harvard-Smithsonian scientist Ricardo Giacconi, a pioneer in X-ray astronomy and now Director of the Space Telescope Science Institute.

Although X-rays are easily absorbed in ordinary optical systems, they can be reflected from polished surfaces that they strike at a grazing angle—like stones skipping across water. The Einstein satellite has an X-ray telescope consisting of a complex set of concentric parabolic and hyperbolic cylindrical surfaces that use the grazing reflection principle to focus X-rays into an actual X-ray image that can be detected electronically and transmitted to Earth. The size of the telescope aperture is 58 cm (23 inches), but because of the grazing angles of the reflecting surfaces to the incoming X-ray photons, the actual mirror surface is equivalent to that of the 2.5-m (100-inch) telescope on Mount Wilson. The Einstein telescope was designed to record X-rays of wavelength from 3 to 50 Å and had a field of view of about 1°. In the first few months of its operation, it was clear that it was an unqualified and spectacular success, with a sensitivity for detecting weak sources 1000 times as great as anything that preceded it. This is equivalent to changing from a small amateur telescope to the 200-inch Hale reflector on Palomar. In later chapters we shall have more to say about the thousands of X-ray sources discovered with the Einstein telescope and what they mean.

The third satellite in the HEAO series, launched in September 1979, is used to detect high-energy particles from space (cosmic rays).

(d) Gamma-Ray Astronomy

Gamma rays were first discovered among the radiation emitted during the decay of radioactive ele-

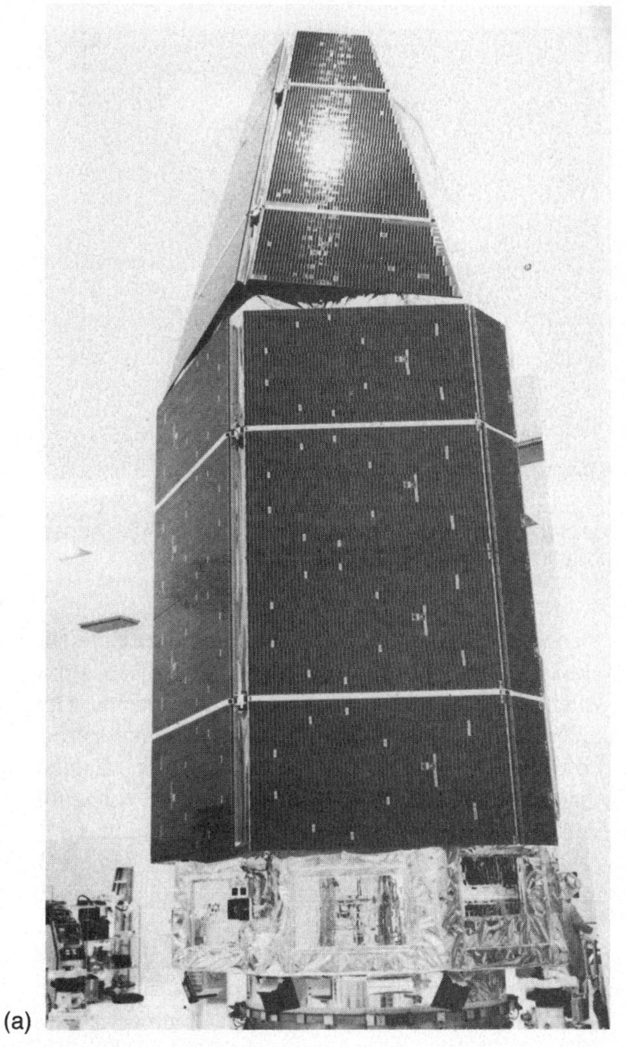

Figure 10.19 (a) The Einstein X-ray Observatory (HEAO 2) before being launched with an Atlas Centaur rocket from Cape Canaveral in November 1978 (b) Cutaway drawing showing the instrument array in the Einstein orbiting X-ray observatory. (NASA/Center for Astrophysics)

Figure 10.20 Artist's rendering of AXAF, the next generation of large X-ray telescopes that NASA hopes to launch during the 1990s. (NASA)

ments. We know today that atomic nuclei have excited energy states, analogous to those of an atom when it changes its configuration of electrons. The differences in energy between the nuclear states, however, are very much higher, and the photons that can be absorbed or emitted by nuclei changing states are gamma rays. Many physicists, in fact, *define* gamma rays as electromagnetic radiation involved in nuclear transitions and call all other high-energy photons X-rays. Although this gamma radiation is of far higher energy than that of common X-rays, there is no gap in the electromagnetic spectrum, and indeed photons can be produced by other mechanisms that have higher energy than the gamma rays emitted from many nuclear transitions. Physicists, in other words, sometimes call the same kind of photons by different names (gamma rays or X-rays) according to their origin. There are cosmic sources of electromagnetic radiation of all energies, however, so astronomers tend to ignore the nuclear definition and call *all* photons of high energy gamma rays. The boundary between X-rays and gamma rays is, of course, arbitrary, but is usually taken as about 0.1 Å (or the order of 100 keV).

Gamma rays of cosmic origin were first discovered by the *Vela* satellites. The Vela satellite system conducts worldwide surveillance for possible explosions of nuclear bombs, which would emit gamma rays. In 1967 they detected bursts of gamma radiation that investigation showed could not originate from within the solar system. Such bursts have also been detected by Uhuru and other orbiting observatories equipped with detectors of high-energy radiation, including some of the OSOs. A few dozen bursts have been observed altogether, sometimes single and sometimes complex and multiple. Each lasts anywhere from about a tenth of a second to several seconds. The sources of these bursts, not yet identified, seem to be scattered more or less at random about the sky.

The bursts are of relatively low energy, as gamma rays go. Sources of far higher energy gamma rays—greater than 10 million eV (10 MeV)—have been found subsequently. The first certain detection of high-energy gamma rays of cosmic origin was with an experiment on OSO 3, in which the plane of our Galaxy was detected in the light of gamma rays of energy more than 50 MeV. Other satellites, especially the second Small Astronomy Satellite, SAS 2, have provided a far better picture of the gamma-ray sky and supplement that obtained from lower energy gamma rays detected from other satellites and from experiments flown in high balloons. In high-energy gamma radiation, the Milky Way appears somewhat as it does to the unaided eye, only narrower—a belt around the sky about 2° wide. There are also discrete gamma-ray sources, mostly along the Milky Way. About 30 localized sources had been found by 1980, and two were definitely identified with old supernovae. There are other probable identifications, at least one being a quasar.

Among the devices that can be used to detect gamma rays are *scintillation counters* and *spark chambers*. Scintillation counters use a material that fluoresces—emits a flash of visible light—when it absorbs a gamma ray. The flash is generally recorded photoelectrically. A spark chamber consists of a stack of slightly separated conducting plates with high electrical voltages between them. When an incoming gamma ray is absorbed in one of the plates, its energy is converted to an electron-positron pair, and these charged particles are attracted by the voltage differences between the plates, creating a current that manifests itself as a series of sparks between the plates. The light from the sparks is recorded photoelectrically, and the direction of the sparks gives some indication of the incoming photon's direction.

(e) Observatories in Space

Most of the orbiting observatories we have described were launched in the 1970s and operated for a few years at most. During the 1980s there has been a substantial gap in this sequence of space observations, with only the reliable IUE continuing its operations. (A similar gap exists in planetary exploration, with at most two spacecraft launches during the entire decade of the 1980s, compared with an average of almost one per year going to the Moon and planets during the previous two decades.) This interruption can be attributed in large part to the development of the Space Shuttle, which absorbed a disproportionate share of the NASA funding during the 1970s. The problem became worse after the explosion of the Shuttle *Challenger* in 1986, just when the Shuttle was becoming operational and scientific flights were about to resume. It is also true, however, that as the cost of individual science missions goes up, their frequency of flight must decline unless there are major infusions of new funding into the NASA science program.

For the period of the late 1980s and into the 1990s, scientists and NASA officials hope to reverse this decline and establish a series of *space observatories*. Unlike previous shorter-lived spacecraft, these observatories will be serviceable by astronauts, either from the Shuttle directly or from the Space Station (expected to be operational by the mid-1990s). They should therefore have very long operational lifetimes, just like ground-based telescopes. Further, it will be possible to modify and improve their instrumentation from time to time, thus utilizing advances in technology to increase their power.

The new generation of space-based telescopes will not be attached to the Space Station, since they operate best when isolated from the vibrations, electrical interference, and pollution generated by a large manned facility. Some will be "free-flyers" in their own independent orbits. Others may be grouped together in a space science orbiting platform. But all will still be operated remotely from the ground, by scientists sitting at control consoles in NASA centers or even in their offices distributed around the world.

The first of the new observatories is the Hubble Space Telescope (HST). This optical-ultraviolet telescope, with its aperture of 2.4 m, will be launched (as we now understand) in late 1988 or early 1989, and operated from the NASA-funded Space Telescope Science Institute in Baltimore (Figure 10.21). It is initially instrumented for direct imaging and spectroscopy in the visible and ultraviolet, but in the mid 1990s one or more of these instruments will be replaced, and it is expected that the HST will then also work in the near infrared part of the spectrum. The primary objective of the HST is to take advantage of the absence of atmospheric "seeing" to increase its resolution and thereby to probe deeper into space than is possible with any ground-based telescope. The HST also takes advantage of the absence of atmospheric absorption to operate in the ultraviolet, at wavelengths too short to penetrate to the surface of the Earth. U.S. and European astronomers, who have collaborated to develop the HST, are looking forward to this telescope as the premier astronomical instrument of the 1990s.

The second of the space observatories is a more specialized facility called the Gamma Ray Observatory (GRO). Now under development by NASA, GRO is scheduled for launch in 1988 or 1989. Its objectives are to conduct an all-sky survey for gamma-ray sources as well as to study individual objects in detail at a variety of energies. The potential for new discoveries from this instrument is great, since it will provide us with the first in-depth look at the gamma-ray universe (Figure 10.22).

The third of the observatories, the Advanced X-Ray Astrophysics Facility (AXAF), has not yet formally begun development. If given approval at the end of the 1980s, it could be launched by about 1995. AXAF is the logical successor to the highly successful Einstein X-ray satellite, with much larger imaging optics and advanced detector systems. Like the HST, AXAF is expected to be a long-lived, adaptable, user-oriented observatory facility.

The fourth and final spectral region that needs to be covered from space is the infrared. The proposed U.S. infrared observatory is SIRTF (the Space Infrared Telescope Facility), mentioned in Section 10.2c. This 1-m class cooled telescope could be in operation by the mid- to late 1990s.

While the U.S. is pursuing these large, long-lived observing facilities in space, the European Space Agency also has an ambitious program in space astronomy built around somewhat smaller satellites to be launched with the Ariane rocket. For the 1980s these include a German-built X-ray satellite called Rosat, a satellite named Hipparcos designed for precision measurement of stellar positions (astrometry), and the previously mentioned ISO infrared satellite. For the 1990s, ESA objectives include development of a large X-ray spectroscopy observatory to complement the AXAF X-ray imaging telescope, and a new spacecraft to operate in the far-infrared and submillimeter regions of the spectrum. In addition, of course, there are the plan-

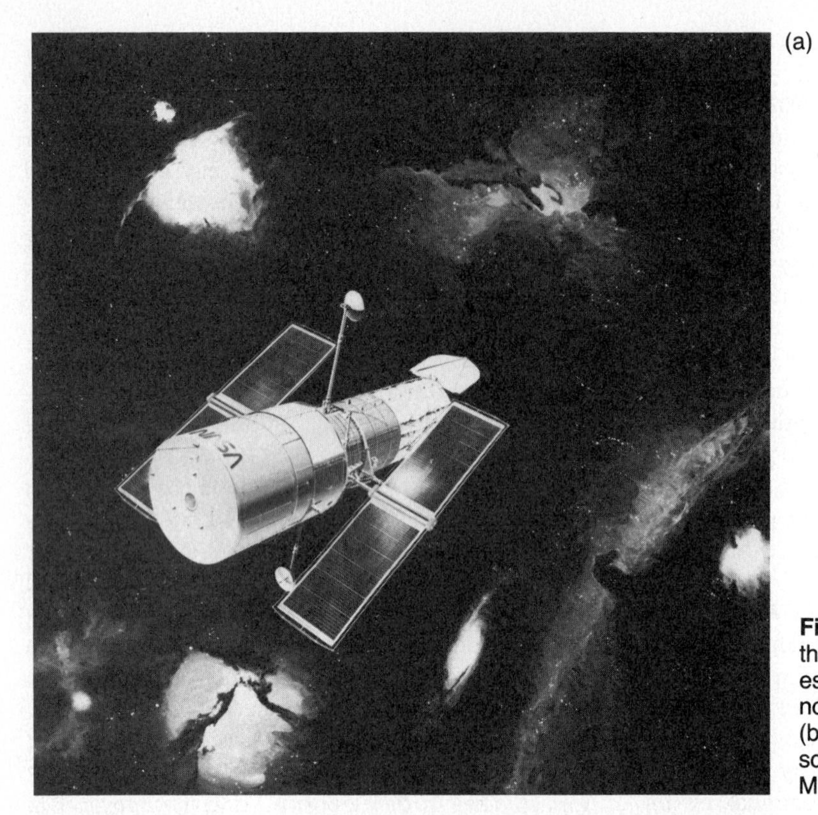
(a)

Figure 10.21 (a) Artist's rendering of the Hubble Space Telescope, the largest and most sophisticated orbital astronomical observatory ever built. (NASA) (b) The completed Hubble Space Telescope undergoing testing. (Lockheed Missiles and Space Company)

(b)

Figure 10.22 The Gamma Ray Observatory, a large NASA satellite to be launched in about 1988. (NASA)

etary exploration missions being planned by the U.S., the U.S.S.R., and ESA. If the funds can be found to support construction and operation of these many space-observing facilities, the outlook for astronomy from above the atmosphere looks very bright indeed for the 1990s and beyond.

EXERCISES

1. When the 1000-foot radio telescope at Arecibo is used to observe an object that is not precisely overhead, less than the full aperture of the telescope is effective. Why?
2. Suppose there were a stationary array of radio telescopes lined up in an east-west direction and located at the equator of the Earth. Would the system determine more accurately the right ascension or the declination of a cosmic radio source? Why?
*3. Suppose at some northern latitude a straight-line array of radio telescopes is laid out in an arbitrary direction (not necessarily north-south or east-west, but not excluding these possibilities). Explain how the use of the array as an interferometer at different times during the day (or night) can yield two-dimensional information about the structure of the source and its direction in the sky.
4. Suppose the VLA to have a maximum baseline of 36 km. What is the angular resolution possible with the system operating at a wavelength of 6 cm?
5. If, with a baseline of 6000 km, a VLB interferometer just resolved an angle of 0.0042 arcseconds, at what wavelength would the observations have been made? ***Answer:*** 12 cm
*6. If a radio telescope array has four separate antennas, how many different pairs can be selected among them for interferometry? Actually, the number of different pairs among n objects is $n(n - 1)/2$. How many pairs exist among the 27 antennas in the VLA? ***Answer:*** 351
7. Radio and radar observations are often made with the same antenna, but otherwise they are very different techniques. Compare and contrast radio and radar astronomy, in terms of the equipment needed, the methods used, and the kind of results that are obtained.
8. What is the wavelength of 50 MeV gamma-ray photons? ***Answer:*** about 2.5×10^{-4} Å
9. Compare infrared telescopes with those built primarily for observations in the visible part of the spectrum. What are the main differences? Should it be possible to construct a telescope so that it can operate in an optimum fashion in both wavelength regions?

10. Infrared detectors are usually operated at very low temperatures, typically 4 K. Explain the advantages of using cold detectors. In space, the entire telescope can also be cooled. What additional advantages does this bring? Why can't we also cool the telescope optics on a ground-based telescope?
11. What are the advantages of putting a telescope in an airplane, such as the NASA Kuiper Airborne Observatory? How does the 1-m KAO telescope compare in capability with larger ground-based telescopes, for both infrared and visible observations?
12. Suppose you are looking for a site for an optical observatory, an infrared observatory, a submillimeter observatory, and a radio observatory. List the main criteria of excellence for each of these cases. What sites on Earth are actually thought to be the best today?
13. Astronomers who hope to see the Hubble Space Telescope, GRO, AXAF, and SIRTF all operating by the end of this century refer to an era of the "great observatories in space." If all four of these instruments were in operation at the same time, together with ground-based instruments, what would be the coverage of the electromagnetic spectrum? Are any important wavelength regions left out?

James Clerk Maxwell (1831–1879), the great Scots physicist, unified electricity and magnetism into a coherent theory, much as Newton had unified celestial and terrestrial mechanics. Maxwell's theory was the cornerstone on which Einstein built his theory of special relativity. (American Institute of Physics, Niels Bohr Library)

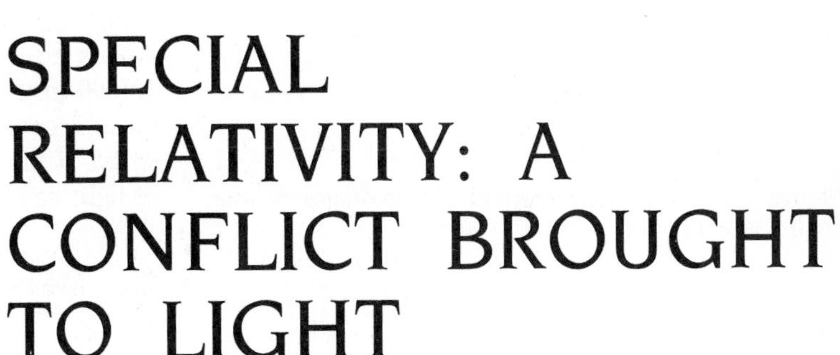

11 SPECIAL RELATIVITY: A CONFLICT BROUGHT TO LIGHT

Newton unified science in 1687, and his theories of mechanics and gravitation dominated scientific thought for nearly two centuries. But by the middle of the 19th century, there was a change in the wind—a change set in motion by the Scottish physicist James Clerk Maxwell (1831–1879), who unified electricity and magnetism and showed that they are really different manifestations of the same thing. In the vicinity of an electric charge, another charge feels a force of attraction or repulsion, depending on whether the two charges have the opposite or the same sign, respectively. But in the vicinity of a *moving* charge another charge is acted on by a *magnetic* force. Maxwell spoke of an *electric field* around a static charge, and a *magnetic field* around a moving charge (as in an electric current).

Maxwell's theory of electromagnetism was codified in 1873 in his four famous equations, which describe electric and magnetic fields, and also how a change in one always induces a change in the other. If a charge moves back and forth, for example, as in an alternating current, the magnetic field set up is continually breaking down, and reforming with the opposite polarity. These changes in the magnetic field induce a constantly changing electric field, which in turn induces a changing magnetic field, and so on. These rapidly alternating fields are in the form of a disturbance that propagates as waves away from the moving charge. Maxwell's equations also predict that this disturbance should move with a very definite speed, equal to the ratio of the electromagnetic and electrostatic units of electricity. Maxwell recognized that this speed is remarkably similar to the speed that had been measured for light, and he suggested that light must be one form of this electromagnetic radiation. He even suggested that there were other forms of electromagnetic radiation, and of course he was right, but it was more than 20 years before radio waves were generated for the first time by the German physicist Heinrich Hertz.

11.1 THE SPEED OF LIGHT

Long before Maxwell's time it was known that light travels with a finite speed, and by the time of his prediction of electromagnetic radiation that speed had even been rather accurately measured. The speed of light is found by measuring the time required for it to travel an accurately known distance.

Galileo suggested a way to measure the speed of light with two experimenters, separated by a mile or more, and each equipped with a lantern that can be covered. The first opens his lantern, and the sec-

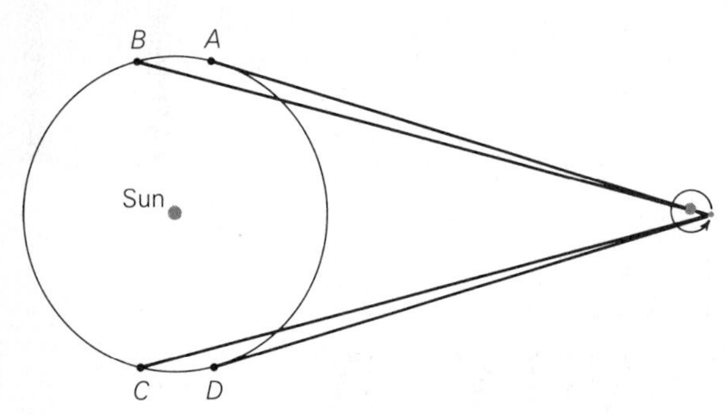

Figure 11.1 Roemer's method of demonstrating the finite speed of light.

ond, on seeing the light from the first, uncovers his. The time that elapses between the time that the first experimenter opens his lantern and the time that he sees the light of his associate's lantern, after correction for the human reaction time, is how long light spends making the round trip. It is not clear whether Galileo actually conducted this experiment, but he correctly concluded that the speed of light is too great to be measured by so crude a technique.

(a) Roemer's Demonstration of the Finite Speed of Light

The first demonstration that light travels at a finite speed was provided by observations of the Danish astronomer Olaus Roemer (1644–1710) in 1675. Jupiter's inner satellites are regularly eclipsed when they pass into the shadow of the planet. Thus a convenient way of timing the period of revolution of a satellite is to note the interval between successive eclipses. Roemer was doing just this when he noted that the period of a satellite seems longer when the Earth, in its orbit, is moving away from Jupiter than it is six months later when the Earth is approaching Jupiter. Roemer correctly attributed the effect to the time it takes light to travel through space.

Suppose the Earth is at A (Figure 11.1) when we note an eclipse, indicating that a cycle of revolution of the satellite has begun. Before the revolution is completed, the Earth has moved somewhat further from Jupiter, and light from Jupiter has farther to travel to catch up with the Earth, at B, to bring us news of the next eclipse. Six months later the situation is reversed; the Earth then approaches Jupiter, and thus advances to meet the oncoming light. If the first eclipse is then observed with the Earth at C, the second is observed at D a little ahead of the time it would have been seen had the Earth remained at C. Observations like Roemer's indicate that light takes about 16½ minutes to cross the orbit of the Earth. Because the Earth revolves about the circumference of its orbit in one year, it is easy to see that light must have a speed about 10,000 times that of the Earth. Later, when the distance from the sun to the Earth, and hence the speed of the Earth, was well determined, the speed of light could be deduced in km/s.

(b) Measuring the Speed of Light

In 1849, the French physicist Hippolyte Fizeau (1819–1896) invented a laboratory method of measuring the speed of light. His procedure was to send a beam of light to a distant mirror. On the way, the light had to pass through the gap between two teeth in a toothed wheel. If the wheel was stationary, or rotating only slowly, the mirror would reflect the light beam back through the same gap. If, however, the wheel was set in rapid rotation, by the time the reflected light beam reached the wheel a tooth would have moved into the location occupied by the gap when the light first passed the wheel. Fizeau's procedure was to find at what speed he had to rotate the wheel in order that the reflected light beam would be so eclipsed. Knowing the speed of rotation of the wheel, he could calculate the time required for a tooth to move into the position occupied by its adjacent gap. This was the same time that the light spent traversing the distance from the wheel to the mirror and back. In Fizeau's experiment, the mirror was a little over 8 km from the toothed wheel, so that the light had to travel about 17.2 km. His result for the speed of light was accurate to within about 4 percent. Cornu later applied the same method with an improved apparatus and measured the speed of light to an accuracy of 1 percent.

A superior procedure was developed independently in 1850 by the French physicist Jean Foucault (1819–1868), also famous for his pendulum experiment. The principle of Foucault's method is shown in Figure 11.2. Light from a source, S, is reflected from a rapidly rotating mirror, M. Each time that the mirror turns through the correct orientation, the reflected beam strikes a stationary mirror B some distance away. A second reflection sends the beam from B back to M, but while light is

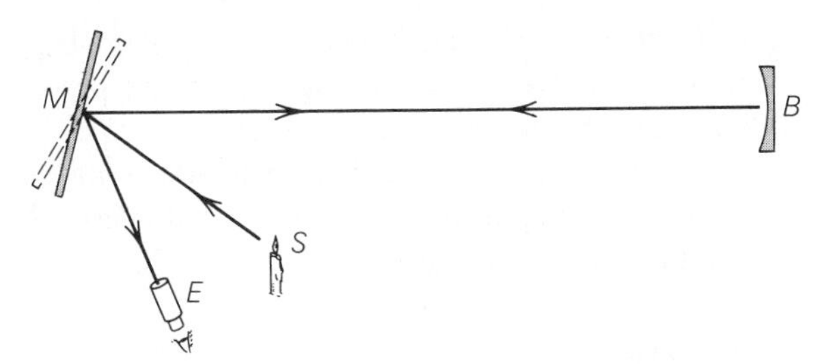

Figure 11.2 Principle of Foucault's method of measuring the speed of light.

making that round trip, M, because of its rapid rotation, has turned a bit (dashed line in Figure 11.2), so that the light does not reflect directly back to S, but off to the side, to E, where it can be observed with an eyepiece. From the geometry of the setup, it is easy to calculate the angle through which the mirror M must rotate to send the light into the eyepiece at E. Thus by adjusting the rotation speed of the mirror until he could see the final reflected beam at E, Foucault could determine how long it took the mirror to turn through the requisite angle while light was making the round trip from M to B and back.

Foucault's experiment was performed entirely within the laboratory, and his total light path was only about 20 m; even so, he found the speed of light to an accuracy better than 1 percent. An adaptation of Foucault's method was later applied by Albert A. Michelson (1852–1931), the first American physicist to win the Nobel prize. Michelson measured the speed of light a number of times between 1878 and 1926. His most accurate determinations (1924 to 1926) were made by passing the light beam from a rotating eight-sided mirror on Mount Wilson, in Southern California, to a stationary mirror on Mount San Antonio, 35 km away. Michelson arranged to have the distance between the two peaks surveyed with the help of the United States Navy and the National Bureau of Standards to a precision of better than 1 cm; it was, at that time, the most precise land survey ever attempted.

Michelson's best result at Mount Wilson gave, for the speed of light in air, the value 299,729 km/s. But light is slowed in passing through transparent media; in air it is retarded by nearly 70 km/s. Michelson began the construction of a mile-long vacuum tube on the Irvine Ranch in Southern California, through which he had hoped to measure the speed of light in empty space. Unfortunately he died before the experiment could be completed, and without his expert touch the results were not as good as had been hoped.

Today electronic timing techniques can measure the speed of light with high accuracy completely in the laboratory, eliminating the necessity of sending light over long distances. The modern accepted value for the speed of light in a vacuum, c, is 299,792.458 km/s.

(c) The Special Nature of the Speed of Light

The speed of light is very unlike other speeds that we normally encounter. First, of course, it is very great. Indeed, the speed of light is an absolute barrier. Nothing can go faster than light; in fact, no material body can ever travel quite at the speed of light. But in addition to these properties, the speed of light is always the same for all observers, irrespective of how they may be moving with respect to each other or to the source of the light. This absoluteness of the speed of light (and other electromagnetic radiation) is predicted by Maxwell's equations. But it is not something one would be likely to expect from "common sense." It is, indeed, a most remarkable thing.

Consider, for example, a material body, such as a bullet fired from a pistol. There is a certain muzzle velocity of the bullet, with respect to the gun, but the speed of the bullet with respect to an observer also depends on the speed with which the gun is moving when it is fired. If the bullet is discharged from a moving car, for example, its speed would certainly have the speed of the car added to its own speed. Alternately, one can, in principle, catch up with a speeding bullet, or even outrun it. (Astronauts in orbit, for example, travel much faster than bullets.) None of this is so for light! No matter how fast you approach or recede from the source, the speed of light, with respect to you, is always the same as if you and the source had no relative motion at all. Even if you could race from the Earth in a spaceship at 99 percent the speed of light, and a colleague on Earth were to send you a light signal, when that light caught up with your ship, and entered the rear port, its speed, with respect to the ship, would still be c. The speed of light depends in no way on the speed of the source.

But neither does the speed of sound, which is a compressional wave that travels through the atmosphere with a (rather slow) speed that depends on the temperature and other characteristics of the air. Similarly, ocean waves travel across the surface of the water as a transverse wave—the water level rising and falling vertically while the wave moves forward—and their speed depends on the wind and water conditions. But you can outrun sound waves, as is done in a supersonic airplane, and you can swim into ocean waves, increasing their speed with respect to you. Not so for light! No matter how fast you move, or in what direction, light waves approach you with that same speed—c. You can race forward to meet the waves of light, like the swimmer in the ocean, and, to be sure, that light will reach you sooner than if you were stationary, just as Roemer found for the times of eclipses of Jupiter's satellites, but the speed of that light when it reaches you is nevertheless the same, with respect to you, as if you were not moving; the *speed* of the light from Jupiter is the same all year round.

Light (and other electromagnetic radiation) behaving as predicted by Maxwell's theory, therefore, does not act the way one would expect from considering only the science of Newton. Does light point out a conflict between Newtonian mechanics and Maxwellian electrodynamics?

The existence of just such a contradiction was realized in 1895 by a 16-year-old schoolboy in the Luitpold Gymnasium in Munich. The boy was regarded as backward and indifferent by his teachers and was advised to leave the gymnasium without a diploma, because he "would never amount to anything and his indifference was demoralizing." But his thought experiment at that time directed his thinking along lines that 10 years later would lead him to his special theory of relativity.

Albert Einstein (1879–1955) reasoned that it should be possible to catch up with any uniformly moving object, after which the relative velocity of the two would be zero. But if you could catch up with a light beam—an electromagnetic wave—you would find it still oscillating back and forth in time, and varying in intensity in space, but not moving! But not only is no such electromagnetic field known, it is impossible according to Maxwell's equations, which say it must be moving with a speed c. Here is surely a contradiction; either Maxwell's equations must be wrong, or our fundamental Newtonian concepts must be wrong. Yet all the predictions of Maxwell's theory that could be tested turned out to be correct. The electronic technology available to us today certainly attests to the power and success of electromagnetic theory.

But to young Einstein there was also a strong philosophical reason for suspecting that Maxwell was right: the *principle of relativity*.

11.2 THE PRINCIPLE OF RELATIVITY

The principle of relativity states that there is no physical experiment by which one can detect his state of uniform relative motion. What this means is that if two observers, moving uniformly with respect to each other, perform the identical experiment in their own moving environment they will obtain identical results, so neither can say, from anything the experiment told him, that he was or was not moving, or how fast he was moving.

For example, two people standing in the aisle of an airliner going 1000 km/hr can play catch exactly as they would on the ground. On that same airplane you can drop a heavy and light object together, and they will hit the cabin floor at the same time, and they fall at the same rate as they would if you had dropped them on the ground (provided that the airplane is moving uniformly—in a straight line at a constant speed). You can play table tennis quite normally on a moving ship on a calm sea. You can swing pendulums in an automobile (so long as it is not turning or accelerating in some other way) and they will swing in the same way, with the same periods, obeying the same pendulum laws as do pendulums in the laboratory. When you are moving uniformly, you experience no physical sensation of speed, or any other sensation that will tell you that you are in motion. You can, of course, look out the window and see the ground moving by, but if you were stationary and the *ground* were moving you would feel the same and see the same thing. It is common to sit in a train in a station, and momentarily wonder whether it is your train or the one on the next track that starts to move. For that matter, none of us can feel the motion of the Earth carrying us about the sun with its orbital speed of 30 km/s; so emphatically do we not feel it, that scarcely three centuries have elapsed since it has been generally accepted that, indeed, the Earth *does* move.

But does this principle of relativity apply only to mechanics, or does it apply to electromagnetic

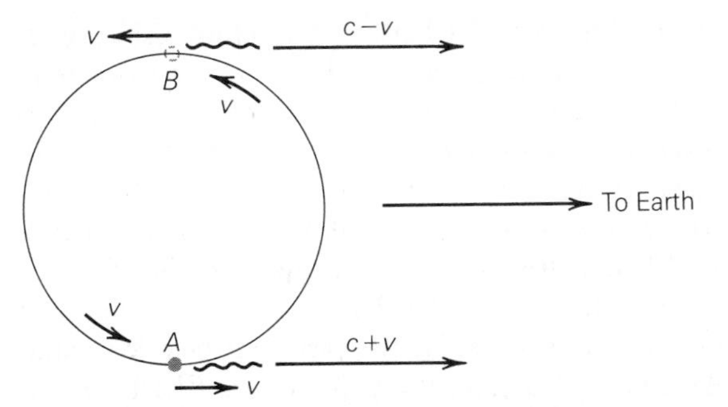

Figure 11.3 If the speed of light depended on the speed of the source, the light emitted by a binary star in a binary system would have a speed toward the Earth that depended on the location of the star in its orbit.

phenomena as well? All experiments indicate that it does apply to electromagnetic experiments, and that the principle of relativity is quite general. Do not radios and tape recorders work the same on an airplane as in the house? You can pick up iron filings with a magnet just as easily in a moving automobile as in a classroom.

On the other hand, if the speed measured for light depended in any way on the velocity of the observer, then relatively simple experiments could be performed that would reveal the observer's motion. All such experiments invariably fail. Like Einstein, we are forced to the conclusion that you cannot catch up with a light beam; light will always have the same speed (in a vacuum)—c—and the prediction of Maxwell's electromagnetic theory is correct. The principle of relativity holds, and there is no experiment—whether mechanical, or involving electricity and magnetism, or light, or anything else—by which we can detect our state of uniform motion.

The principle is profound, for if it is impossible to detect uniform motion, the idea of *absolute motion*—motion with respect to absolute space, as envisioned by Newton—can have no meaning. There can be no absolute reference frame or coordinate system that is guaranteed to be at rest in the universe, and with respect to which other motion can be referred. All we can define is *relative* motion with respect to something else. Thus the Earth moves 30 km/s with respect to the sun; the sun moves 20 km/s with respect to the average of its stellar neighbors; these nearby stars all move at about 250 km/s with respect to the center of our Galaxy; and so on. But there is no way to know how fast we are "really" moving, with respect to absolute space.

But it is a bizarre thing that two observers, one stationary with respect to a light source, and one moving rapidly away from (or toward) it, will still measure the light from it to be approaching them with the same speed. All the seemingly strange results of special relativity come about because of that bizarre fact, and once we can swallow it, everything else in relativity makes perfect sense. Before proceeding, therefore, let's make sure we all understand that it is the real world we are talking about, and not fantasy. We describe briefly therefore, a few (of very many) observations and experiments that demonstrate the absoluteness of the speed of light.

(a) c Cannot Depend on the Motion of the Source

A simple experiment to show that the speed of light cannot depend on its source is provided by nature herself, and was pointed out by the Dutch astronomer Willem de Sitter (1872–1934) early in the century. Many stars are found in double-star systems—in which the two stars revolve about their common center of mass. Take one star, say, the brighter, in such a system whose orbit lies roughly edge on to our line of sight (Figure 11.3). Suppose the orbital speed of that star about the center of mass of the system is v, and consider what would happen if the speed of the light it emits included the speed of the star itself, just as the speed of a bullet fired from a moving automobile has the speed of that car added to it. Then when the star approaches us, at point A in Figure 11.3, light from it should be traveling toward the Earth at a speed of $c + v$, and when the star is moving away from us in its orbit, at B in the figure, its light should approach Earth with a speed of $c - v$. To be sure, the speeds of the stars in binary systems (v) are very small compared to the speed of light (c), but the stars are very far away and over the many years it takes their light to reach us, the faster beam, traveling at $c + v$, can gain considerably over the slower beam, traveling only at $c - v$. If the distance to the binary system were just right, we could be receiving light from the star at position A at the same time as the light sent to us at a slower speed at an earlier time, when the star was at position B. A little thought will show that under some circumstances we could be seeing the same star in a double-star system at many different places in its orbit at once, and analysis of the orbit would end in hopeless confusion. But we have actually analyzed the orbital motions of stars in thousands of double-star system with distances ranging from a few light years to many hundreds of light years, and the orbital motions are all well-behaved, with the stars moving in accordance with Newton's laws. The speed of light from them therefore cannot include the speeds of the stars themselves.

Further proof that the speed of light is independent of the speed of its source comes from the nuclear physics laboratory. In nuclear accelerators, subatomic particles moving at nearly the speed of light are often observed to change form *(decay)* and emit photons, but these photons are always observed to move with the normal speed of light, c, with respect to the laboratory.

(b) c Cannot Depend on the Motion of the Observer

The most famous experiment showing that the speed of light does not depend on the motion of the observer was performed in 1887 in Ohio by A. A. Michelson and E. W. Morley, but its results have been confirmed by hundreds if not thousands of other experiments since then. Michelson and Morley, unaware at that early date of the principle of relativity, were attempting to measure the absolute speed of the Earth through space by measuring the speed of light in two different directions.

To illustrate the idea of the Michelson-Morley experiment, consider three hypothetical astronauts, Able, Baker, and Charley, as shown in Figure 11.4(a), all stationary in space. Suppose further (contrary to experiment, as we shall see) that the speed of light is constant through absolute space. Baker and Charley are each 4 light years (LY) away from Able, but in directions at right angles to each other. Able sends radio signals (which travel with the speed of light) to Baker and Charley at the same time, and those signals reach their destinations 4 years later; immediately Baker and Charley respond, and Able receives their answers simultaneously, 8 years after his original transmission.

Now suppose the three astronauts maintain their relative positions, but all three are moving at 60 percent the speed of light, and in the direction from Able toward Charley. As before, Able sends out the two messages, and, by our supposition, those signals move at the same speed, c, but with respect to stationary absolute space. This time, however, Charley is moving away from the point where the signal was emitted at $0.6c$ [Figure 11.4(b)], so waves from the signal approach Charley at only $0.4c$, and take 2.5 times as long to reach him as before—that is, 10 years. On the other hand, Able is approaching the point where Charley sends back his response, at $0.6c$, so Able moves forward to meet Charley's transmission at a relative speed of $1.6c$; thus those waves take only 2.5 years to span the 4 LY, and reach Able 12.5 years after his original transmission.

But how about the message going from Able to Baker and back? The radio waves that reach Baker from Able must be directed *ahead* of Baker's position *(B)* as seen from Able *(A)* at transmission time, just as a hunter must "lead" his running prey [see Figure 11.4(c)]. Thus the message that reaches Baker from Able travels an oblique path, and since Baker's speed is 60 percent that of the radio waves, Baker has traveled 0.6 times as far as the message has when it catches him at P. The three points—A, B, and P—make a right triangle. Since BP and AP are in the ratio 3 to 5, the theorem of Pythagoras tells us that AB must be $4/5$ of AP; but AB is 4 LY, so the radio message, traveling at the speed of light, took 5 years to reach Baker. The same geometry holds for the return message, so Able receives Baker's reply 10 years after his original transmission, and 2.5 years ahead of hearing from Charley!

We selected a speed and separation for the astronauts of 60 percent c and 4 LY to make the arithmetic easy. What is the effect, in general, for a speed of v and separation L? The signal from Able to Charley has speed $c - v$, and the return one from Charley to Able has speed $c + v$. Thus the total round-trip light travel time in this direction parallel to the motion of the astronauts is

$$T_{\parallel} = \frac{L}{c - v} + \frac{L}{c + v} = \frac{2L}{c} \frac{1}{(1 - v^2/c^2)}.$$

Now the effective speed of the light from Able to Baker (and return) in a direction perpendicular to the motion is $\sqrt{(1 - v^2/c^2)}$, from the theorem of Pythagoras, so the round-trip time for that perpendicular signal is

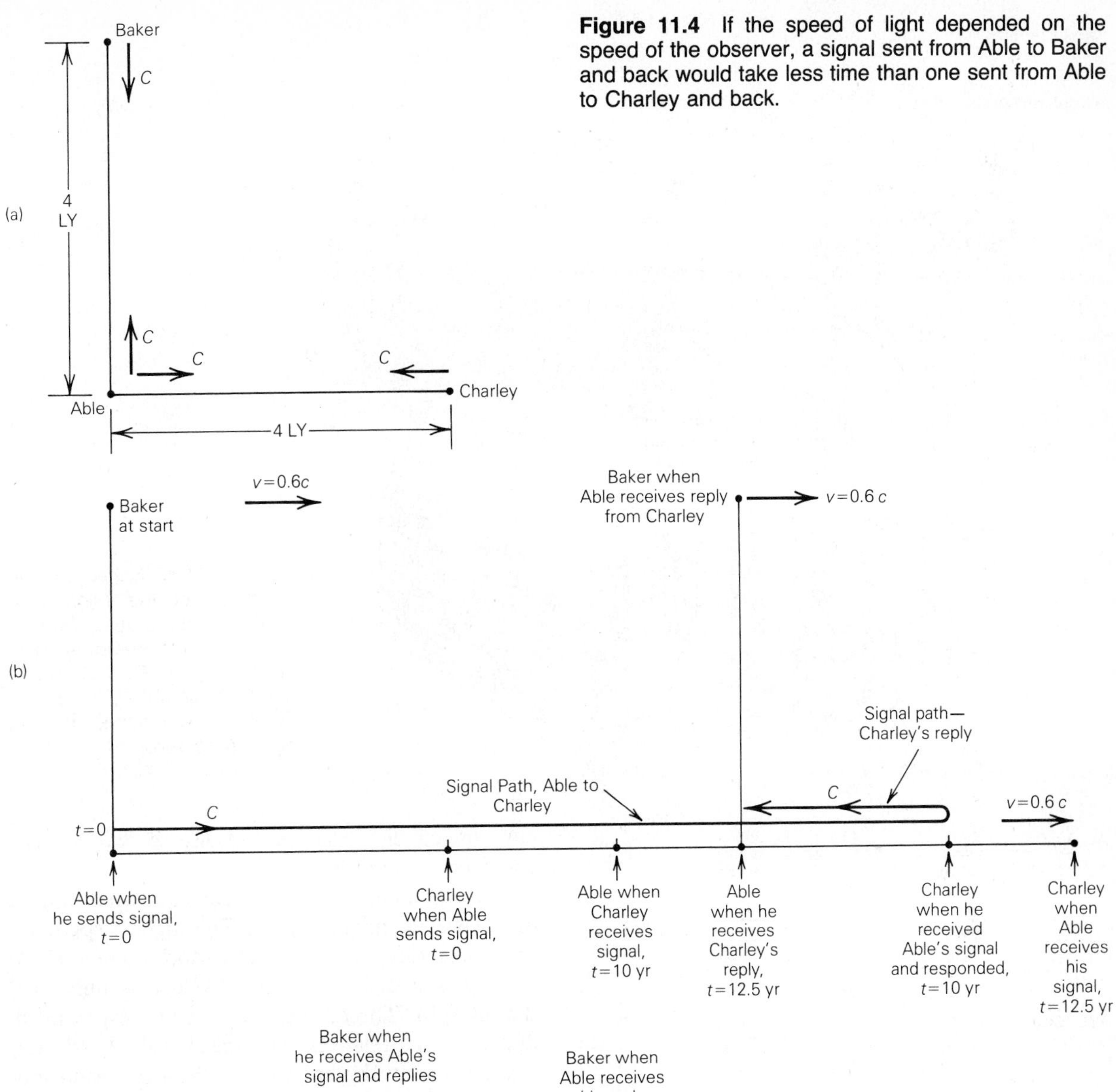

Figure 11.4 If the speed of light depended on the speed of the observer, a signal sent from Able to Baker and back would take less time than one sent from Able to Charley and back.

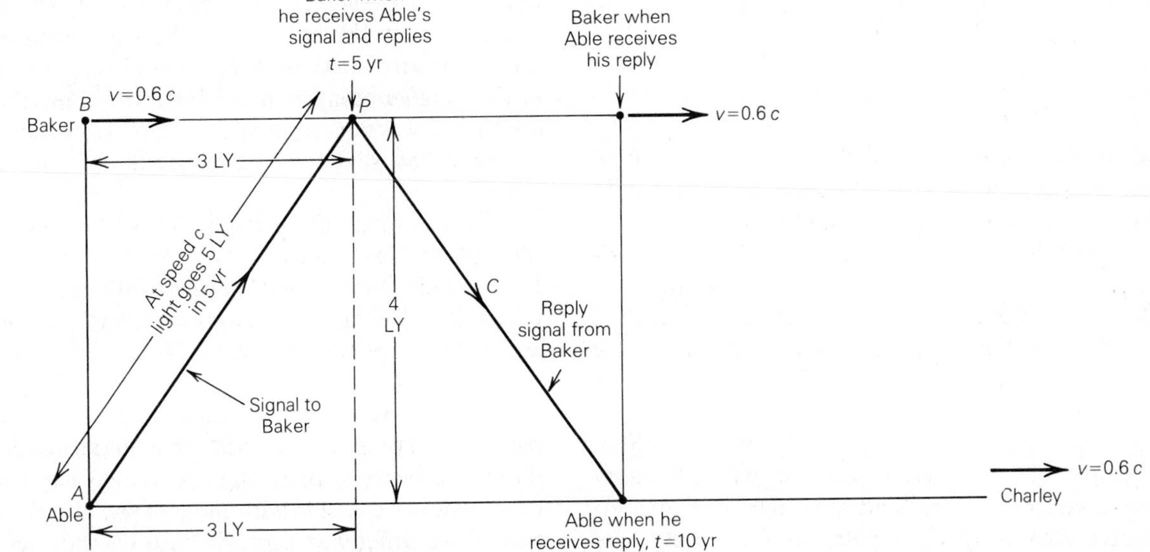

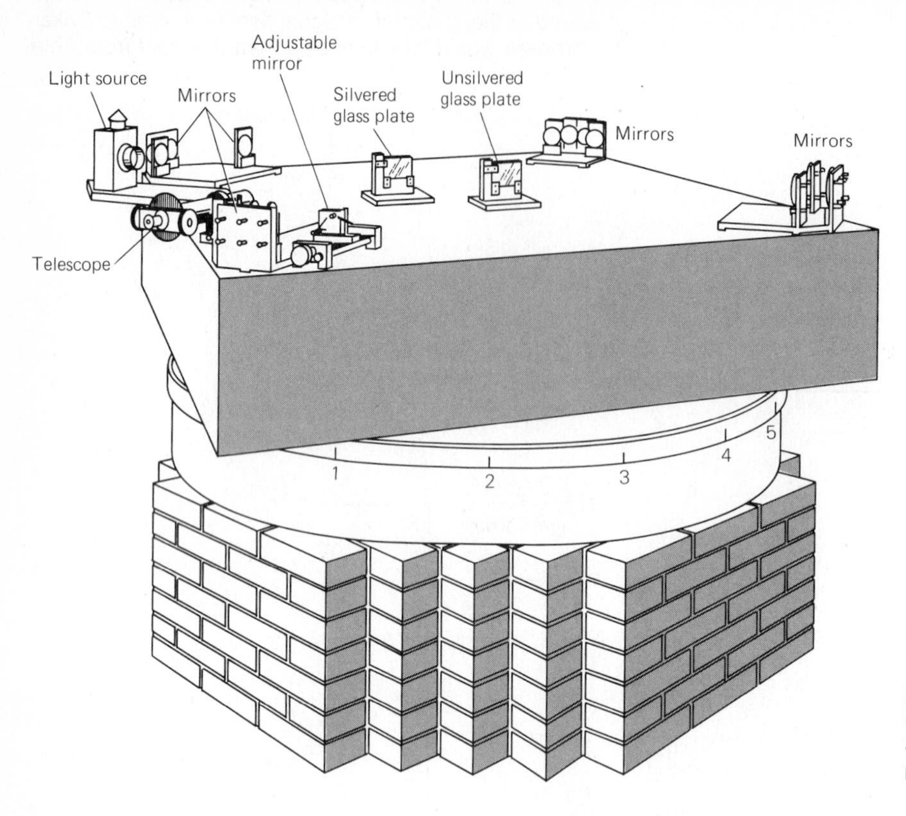

Figure 11.5 Schematic illustration of the Michelson-Morley apparatus. (From "The Michelson-Morley Experiment" by R. S. Shankland. Copyright © 1964 by Scientific American, Inc. All rights reserved.)

$$T_{\perp} = \frac{L}{c\sqrt{(1 - v^2/c^2)}} + \frac{L}{c\sqrt{(1 - v^2/c^2)}} = \frac{2L}{c}\frac{1}{\sqrt{(1 - v^2/c^2)}}.$$

Hence,

$$T_{\parallel} = \frac{T_{\perp}}{\sqrt{(1 - v^2/c^2)}} = \gamma T_{\perp},$$

where we define γ by $1/\sqrt{(1 - v^2/c^2)}$. Thus if Able had not known his speed through space, he could have deduced it from the ratio of $T_{\parallel}$ to $T_{\perp}$.

Michelson and Morley performed their experiment in a basement laboratory, but the principle is the same. They hoped to determine the absolute speed of the Earth through space by measuring the difference in times required for light to travel across distances in the laboratory that were at right angles to each other. The two light paths were set up by multiple reflections between mirrors on the horizontal surface of a heavy sandstone slab. The total effective round-trip light path for each beam was about 22 m, and the slab floated on a layer of mercury to reduce vibration, and also so that the slab could be rotated easily.

At this point, we should consider what kind of difference we might expect. The highest speed for the Earth that Michelson and Morley knew about for sure was its 30 km/s orbital velocity—only 10^{-4} that of light. The ratio of the travel times, γ, introduced by that relatively tiny speed of the Earth, differs from unity by only 5 parts in a thousand million. The perpendicular beam should return ahead of the parallel beam by only about 10^{-5} cm, about a fifth of a wavelength of visible light! How can one determine so slight a lead of one beam over the other?

Here is where the ingenuity of Michelson came into play. He arranged that the two returning beams come together and interfere with each other, producing that familiar pattern of alternately dark and light fringes (Section 8.1). Now, obviously no one could assume that the light paths of the two beams were precisely of the same length in the first place, so they will return with some unknown phase difference between them (that is, the crests of one beam will not coincide with those of the other), and that phase difference depends both on the slightly unequal lengths of the light paths and on the effect

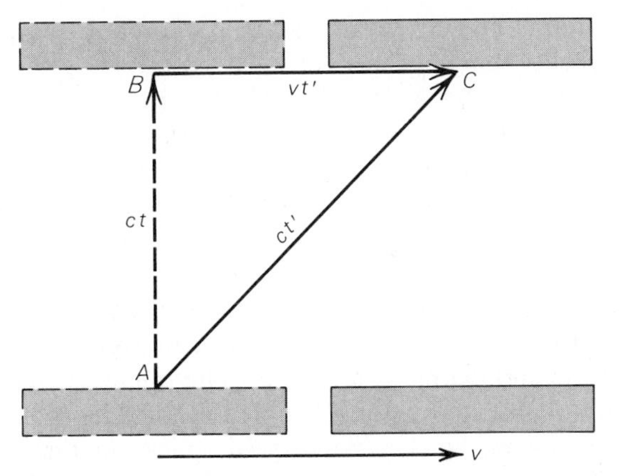

Figure 11.6 The light path in a moving observer's "ideal clock."

due to the Earth's motion that was being sought. Consequently, the observation of the fringe pattern does not, in itself, reveal directly the difference in light travel time. That was the reason for being able to rotate the slab supporting the apparatus. Turning it through 90° interchanges the roles of the beams; if one happened, at the outset, to be parallel to the direction of motion of the apparatus due to the velocity of the Earth and the other was perpendicular to that direction, after rotation the first beam would be perpendicular and the second parallel to the motion. This would change the difference in arrival times of the beams by exactly twice the amount Michelson and Morley were trying to measure. The pattern of fringes would thus shift, and the amount of shift would reveal the speed of the Earth.

Michelson and Morley made their observations at noon and at 6:00 P.M. on several consecutive days beginning with July 8, 1887. During that six-hour interval the direction of the earth's motion must change by 90° relative to the laboratory. During each observation the slab was rotated slowly, and the fringe position was measured at 16 different positions around the circle. The Earth's orbital velocity alone should cause a shift in the interference pattern of two-fifths of a fringe width. And what did they observe?

Nothing! Rotating the slab made absolutely no difference in the light travel time along the two beams at right angles to each other. Their experiment was accurate enough to detect a small part of the Earth's orbital motion, but there was no difference in light travel times. It was as if the Earth were absolutely stationary, but Copernicus could not be wrong, for the success of gravitational theory shows that the Earth has to be moving. Michelson and Morley thought their experimental setup was at fault, so they repeated the experiment with even greater accuracy. Still no result, and there has never been any in all the many, many times this and comparable experiments have been repeated. The only conclusion is that the speed of light does *not* depend on the motion of the observer; it is always c with respect to him.

Strange as this result seems, however, it is completely consistent with Maxwell's theory and with the principle of relativity.

11.3 THE SPECIAL THEORY OF RELATIVITY

How can we understand the bizarre properties of the propagation of light? Einstein gave the solution in his special theory of relativity. He showed that different observers in uniform relative motion (moving at constant velocity with respect to each other) perceive space and time differently. There are two assumptions on which the special theory is based: the principle of relativity and, embodied within it, the absolute constancy of the speed of light.

(a) Time Dilation

Let us imagine the construction of an ideal clock. Of many possible designs, we shall choose a clock consisting of two parallel mirrors, and a pulse of light reflecting back and forth perpendicularly between them. Each time the pulse passes from one mirror to the other, we shall count it as a "tick" of the clock. Because light travels at an absolutely constant rate, by carefully standardizing the spacing of the mirrors, we can agree that all such clocks should keep identical time.

On the other hand, what if an observer is moving very rapidly to the right with respect to us, carrying his two-mirror clock with him? Further, suppose his direction of motion with respect to us is parallel to the surfaces of the mirrors (see Figure 11.6). As far as he is concerned, his clock, in his own system, is at rest, for there is no experiment by which he can detect his own motion; consequently, as far as he is concerned his clock is operating normally, with the light pulse reflecting perpendicularly back and forth between the mirrors. But as *we* see the situation, because of his clock's rapid motion to the right, the light pulse is not bouncing simply back and forth along a single line, but is following a slanting path. In other words, we see the moving observer's light pulse traveling farther between ticks than he sees it traveling. But he and we agree on the *speed* of the light pulse, so *we* must conclude that the interval between *his* pulses is longer than it is between ours; that is to say, his seconds are too long, and his clock is running slowly. On the other hand, he, aware of no motion on his part, argues that it is *we* who are moving to the left, and that it is in *our* clock that light travels obliquely, and that it is *our* clock that runs slowly. Each of us insists that the other's clock is slow.

By isolating a triangle in Figure 11.6 we can see by how much we disagree on the rate of passage of time. As far as our moving friend is concerned, he is stationary, and the light pulse has traveled vertically from A to B at a speed c, and the time it has taken to do so is t; since distance equals rate times time, the distance from A to B must be ct. But we see the light taking the slanting path AC, and requiring, at the same speed c, a longer time, t', to do so; thus we say that the pulse traveled a distance ct'. Meanwhile, our friend with his moving clock has gone from B to C, and if his speed relative to us is v, that distance must be vt'. The theorem of Pythagoras tells us that $c^2t^2 = c^2t'^2 - v^2t'^2$, from which we find, upon solving for t',

$$t' = \frac{t}{\sqrt{1 - v^2/c^2}} = \gamma t.$$

Thus what the moving observer thinks is an interval t, we see to be a longer interval t', and it is longer by the factor $1/\sqrt{1 - v^2/c^2}$. He, of course, regards his time intervals as normal and *ours* as too long by the same factor, γ.

Which of us is right? We both are. Time really *does* move at different rates in two different systems in uniform relative motion. We simply perceive time differently: Time is not absolute; each of us has his own private time.

(b) Reality of the Time Dilation

We must not think of this time dilation (the stretching out of time) between observers in uniform relative motion as some artifact of the clock we choose to construct. It is a very real thing. All processes slow down in moving systems; moving observers actually age more slowly than we do.

Nature provides a spectacular example of time dilation. The upper atmosphere of the Earth is continually being bombarded by cosmic rays—atomic nuclei moving at very nearly the speed of light. When a cosmic ray particle strikes a molecule in the upper air, it breaks the molecule into a number of subatomic particles. Common among these are particles called *muons*. A muon is rather like an electron, but has about 200 times the mass. Muons spontaneously *decay*, turning into electrons, and emitting certain other radiation, in an average time of 1.5 millionths of a second. The muons formed by collisions of cosmic rays high in the atmosphere are moving at very high speeds, close to that of light. But even if they moved *at* the speed of light, they could travel, on the average, only about a half kilometer before decaying. Now the muons are formed at altitudes of 10 to 20 km; yet they rain down to the surface of the Earth in enormous numbers. In fact, the muon is the principal kind of cosmic ray particle observed at ground level. If they are formed 15 km above the ground and decay before having time to travel as far as 1 km, how can we observe them at the bottom of the atmosphere?

As far as the muons are concerned, they *do* decay on the average in 1.5 millionths of a second, but because of their high speeds, as we observe them their time has slowed down, so that they have time to go very much farther than 0.5 km before decaying, and, in fact, most of them survive all the way to the Earth's surface. Muons are also observed to live very much longer when they are accelerated to high speeds in the nuclear physics laboratory. In 1976 at CERN, the international laboratory of the European Council of Nuclear Research in Geneva, muons were accelerated to a speed of $0.9994c$; the formula for time dilation predicts that their time should slow down by a factor of 30 at that speed. Indeed, the average lifetime of those high-speed muons was 44 millionths of a second, 30 times the 1.5 millionths of a second they survive at rest. Note that we are not speaking here of light pulses bouncing between mirrors, but of muons waiting to disintegrate; time dilation is not just a strange property of our light clock, but a fundamental property of time itself!

Would people live longer if they were rapidly moving? Not to their own way of thinking, of course, for they would have no sensation of moving. But relative to *us* they most certainly would age more slowly. In principle, long space trips could be made by astronauts if they were moving near enough to the speed of light. If we were to send a manned spaceship at a speed of $0.98c$ on a round trip to a star 100 LY away, the round trip would take just over 200 years of our time, but time for the astronauts would slow down by a factor of about 5, and on their return they would be only 40 years older than when they left. (Although such relativistic space travel is theoretically possible, the virtually prohibitive energy requirements make it unfeasible in practice—see Chapter 38.)

(c) Contraction of Length and Distance

Let us return to those astronauts traveling to a star 100 LY away at 98 percent the speed of light. If they make the trip in 20 years of their time (40 years round trip) does that mean that they have traveled at 5 times the speed of light? No, for lengths (and distances) as perceived by different observers in uniform relative motion are also different. As perceived by the astronauts moving $0.98c$ with respect to the Earth, the Earth and star are moving at that same speed in the opposite direction. The astronauts see the separation of the Earth and star to be very much less than as perceived by earthlings; in fact, they find the distance from Earth to star to be just under 20 LY. If a system is in uniform motion with respect to us, we see all dimensions in that system that lie along the direction of relative motion to be *shorter* than as perceived by an observer in that moving system. He, on the other hand, sees lengths in *our* system (that lie parallel to the direction of relative motion) to be shorter than we see them. All objects in a moving system, in other words, appear foreshortened in the direction of motion.

We can see how this foreshortening must come about by reconsidering the astronauts Able, Baker, and Charley, who are lined up at right angles to each other. Suppose, as before, they are moving with respect to us along the direction from Able to Charley. We have already seen that there is no way that they can detect their own motion; thus if Able sends signals to Baker and Charley and receives simultaneous replies, he must conclude that they are equidistant from him. But not so for us, for we see Charley moving away from Able's signal until it catches up with him, and then Able rushing forward to meet the return signal from Charley. And we see the signal from Able to Baker, and Baker's return signal, traveling on slanting paths, as shown in Figure 11.4. As we found before, if Baker and Charley are equidistant from Able, we should see Able receive Baker's reply first. Thus if we see the two signals return to Able at the same time, we must conclude that Charley is *closer* to him than Baker is. As judged by Able, Baker and Charley are the same distance from him, but to us the moving system of astronauts is foreshortened in the direction of motion. Only a bit of algebra is needed to show that the factor of foreshortening is just the same factor by which time intervals in the moving system are too long. That is, a distance in the moving system, along the direction of motion, that the moving observer would say is D, we would say is only $D\sqrt{1 - v^2/c^2}$. Of course the moving observer sees *our* distances as foreshortened, not his own. *Length* is just as private a matter as time is!

(d) The "Twin Paradox"

There is a famous story, the so-called "twin paradox" (which is not really a paradox, as we shall see), describing a debate between two identical twins, Peter and Paul. Peter, the astronaut, travels to a remote star and returns, aging less than his brother, Paul, who stays on Earth. But Peter claims that he was actually stationary all the time and that it was the Earth that rushed away from him, and then came back again, carrying Paul. Thus, argues Peter, it is really Paul who has aged less. Who is right? It turns out to be Peter who ages less, for he had to change direction to return to Earth and hence is *not* an equivalent observer. To see how this comes about, we have adapted the discussion by N. David Mermin in Chapter 16 of his book, *Space and Time in Special Relativity*.

Imagine that Peter makes a trip to the star Vega, 25 LY distant, at a speed that is 99 percent that of light ($0.99c$). To keep track of what is happening, both Peter in the spacecraft and Paul on Earth are equipped with a flasher that sends out a brilliant signal exactly every second.

What does Paul see when Peter is on his way to Vega? Since Peter is moving at $0.99c$, his time, relative to Paul, is slow by a factor $\gamma = 1/\sqrt{(1 - v^2/c^2)} = 1/\sqrt{(1 - 0.99^2)} = 7.09$. This fact alone would slow the frequency of Peter's flashes, as Paul sees them, to one every 7.09 s. But in addition, since Peter is moving away at $0.99c$, each successive flash has farther to go to reach Paul. In the 7.09 s between Peter's flashes (as Paul sees them), Peter moves 7.09 s $\times$ $0.99c$ = 7.02 light seconds farther away, and it takes light 7.02 s to go that extra distance, so Paul sees Peter's signals coming every 7.09 + 7.02 = 14.11 s.

Peter, moving away from Paul, sees exactly the same effect, for Paul is moving relatively away from Peter, and Paul's flashes reach the spacecraft every 14.11 s.

Now Peter, moving at $0.99c$, judges the distance to Vega to be $25/\gamma = 25/7.09 = 3.53$ LY, and he arrives there in $3.53/0.99 = 3.56$ years. During that time, Peter has received flashes from Earth every 14.11 s, but he knows that those flashes were emitted one each second of Earth time, so he reasons that when he reaches Vega only $3.56/14.11 = 0.25$ years had passed on Earth while those pulses were emitted.

Immediately upon reaching Vega, Peter turns around and starts his return to Earth at $0.99c$. The distance home is the same as the distance out, of course, so Peter arrives at Earth after another 3.56 years, or after a total journey (of his time) of 7.12 years. During his return trip, how much time does he calculate to have passed on Earth?

While Peter approaches Earth he perceives, as before, that Earth time is running slower than his by the factor $\gamma = 7.09$. Thus he would expect Paul to be emitting pulses every 7.09 s (of Peter's time), just as during the outward journey. But this time Peter *approaches* Paul at $0.99c$, so during the interval of 7.09 s between the emission of those pulses, Peter has closed the gap by $7.09 \text{ s} \times 0.99c = 7.02$ light seconds. Thus Peter's approach toward Paul saves 7.02 s of the 7.09 s that would otherwise separate the arrival of Paul's pulses at the spacecraft, so Peter receives the pulses every $7.09 \text{ s} - 7.02 \text{ s} = 0.07$ s, or at the rate of 14.11 per second (what may appear to be a small arithmetic error here and elsewhere arises because we have rounded all of the figures to two decimals; we have, however, carried out the calculation to higher precision, so the answers given here are actually the correct ones). In other words, Peter's movement toward Paul at $0.99c$ more than makes up for the slowing of Paul's time as perceived by Peter, and he receives Paul's flashes about 14 times as frequently as he would if he and Paul were not moving with respect to each other.

During the 3.56 years of Peter's time that he is returning home, he receives signal flashes from Paul at the rate of 14.11 per second. But he knows that Paul emitted them, in Paul's own time, at the rate of one each second. Consequently, Peter realizes that the interval of time that has elapsed on Earth must be 14.11 times as long as the 3.56 years of his time in the spacecraft. Peter realizes, therefore, that while he is returning home Paul has aged $14.11 \times 3.56 = 50.25$ years. The total amount that Paul has aged during Peter's round trip to Vega is therefore $50.25 + 0.25 = 50.50$ years. Meanwhile, Peter has aged just 7.12 years.

Now let's explore things from the point of view of Paul, on Earth. He received flashes from Peter every 14.11 s while Peter was en route to Vega, as we have seen above. However, Paul does not receive that last flash emitted by Peter at Vega, just before he began his return trip home, for 25 more years, because Vega is 25 LY away. On the other hand, Paul knows that Peter, traveling at $0.99c$, takes $25/0.99 = 25.25$ years to reach the star. Thus Paul easily calculates that by the time he has received that last flash emitted by Peter at Vega, Peter has been gone from Earth for $25.25 + 25 = 50.25$ years. In fact, by that time, Peter is already only $2 \times 25.25 - 50.25 = 0.25$ years from home and is closing fast! Now during those 50.25 years of Paul's time, Peter had been emitting a pulse of light every second of his own time, but those flashes were received by Paul at the rate of one every 14.11 s. Thus Paul finds that Peter, while en route to Vega, must have aged $50.25/14.11 = 3.56$ years—exactly what Peter himself had calculated.

Finally, during those 0.25 years between the time Paul received Peter's last flash from Vega and the time of Peter's splashdown in the Gulf of Iran, Paul receives all of those flashes that Peter emitted on his way home. They arrive at Earth at the rate of 14.11 per second—just the rate that Peter received flashes from Paul during his return to Earth. Paul, of course, knows that those flashes were emitted at one per second by Peter's time, so Peter must have aged, on his way home, by Paul's reckoning, 0.25 times 14.11, or 3.56 years, making a total aging for Peter on his round trip of 7.12 years—again, exactly in agreement with what Peter himself found. Meanwhile, Paul ages $2 \times 25/0.99 = 50.50$ years, again in agreement with Peter's calculations.

Note that Peter and Paul each received all of the pulses emitted by the other—all flashes are accounted for. Moreover, Paul and Peter, fully understanding how relative motion affects time, are in complete agreement on how old each of them is. There is no paradox; Paul and Peter are not equivalent observers. Peter, on his outward trip, received flashes every 14.11 s, and then *changed direction,* henceforth receiving flashes at the rate of 14.11 per second. Paul, who never moved, received flashes from Peter at the rate of one every 14.11 s during all but the last quarter-year of Paul's 50½-year absence; during that last 3 months, Peter received 14.11 flashes per second.

Now, although the traveling astronaut came back having aged far less than his brother at home, it is not correct to say that he bought any additional longevity by taking the journey. To be sure, he would likely outlive his brother after returning to Earth, but in his own time he didn't get any extra years of life; he only *lived* those 7 years he was gone. He brother at home had a full life of 50 fun-filled, adventurous (perhaps even amorous) years!

(e) Relativistic Mass Increase

If different observers in uniform relative motion disagree on length and time, they must also disagree on velocity, which is distance covered in a given time. Thus they must, in turn, disagree on such things as momentum and energy, which depend on velocity. But they do agree on the laws of physics and the results of physical experiments—such as the conservation of momentum.

Suppose Jane and Mary are astronauts in space, moving together so that their relative velocity is zero [Figure 11.7(a)]. At a given instant each fires an elastic missile, such as a billiard ball, toward the other. The two balls are identical and are fired at identical speeds. They meet halfway between the spaceships at C, rebound, and return to the ships from which they were launched. The balls had equal but opposite momenta before the impact (since they were moving in opposite directions), and since each was turned about, they had equal and opposite momenta after the collision, so the total momentum is conserved, as it must be.

Now suppose that Jane and Mary are moving with equal speeds (with respect to us) but in opposite directions. As before, Jane and Mary discharge missiles toward each other, but because of their rel-

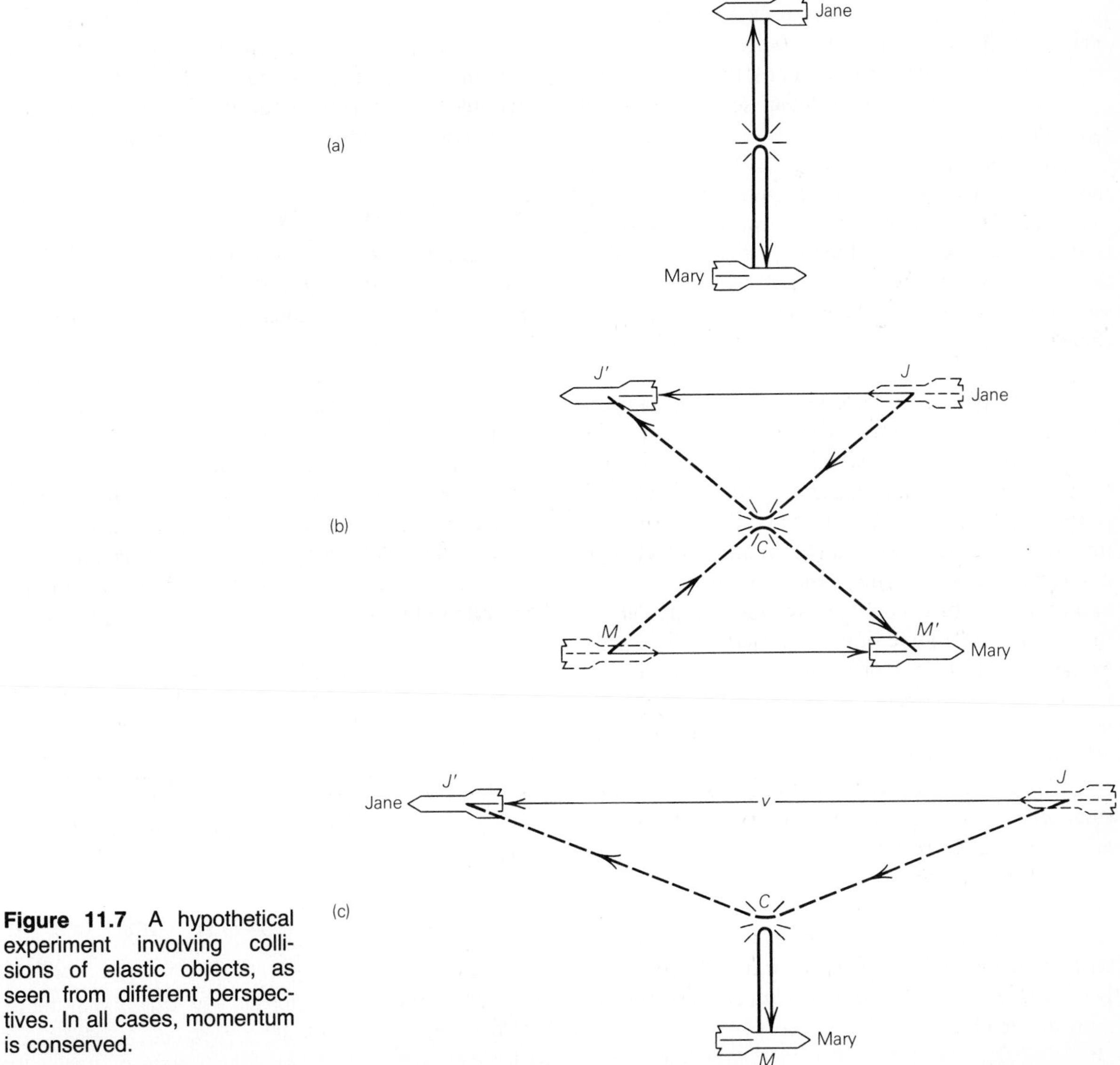

Figure 11.7 A hypothetical experiment involving collisions of elastic objects, as seen from different perspectives. In all cases, momentum is conserved.

ative motion they fire the balls at $\mathcal{J}$ and M, respectively, and in directions perpendicular to their relative velocity. Because the balls move forward with the spaceships, they follow the dashed paths shown in Figure 11.7(b), meet at C, rebound, and return to their own ships at $\mathcal{J}'$ and M', respectively. Again, each is reversed in a symmetrical way and momentum is conserved.

But let's look at the last experiment from the point of view of Mary [Figure 11.7(c)]. Now Mary is stationary (in her own system), so her missile moves straight out perpendicular to the path of Jane's ship. But Jane's missile is released when she was way back at $\mathcal{J}$. As before, the two missiles meet at C, rebound, and Mary's missile returns to her ship, while Jane's returns to hers at $\mathcal{J}'$, as must happen, since it is the identical experiment we described in the last paragraph. Both Mary and Jane must agree that momentum is conserved (if not, one of them would be able to detect something about her own motion).

But now there is a problem, because Jane is moving rapidly with respect to the stationary Mary, hence Jane's time passes more slowly. Similarly, all physical processes in Jane's system must slow down, including the component of velocity with which Jane's missile is fired toward Mary, perpendicular to the direction of their relative motion. But if Jane's missile is moving more slowly than Mary's, we would expect it to have less momentum perpendicular to the direction of motion as well, since the balls are of the same mass. But then how can each turn the other around, conserving momentum? We would expect Mary's missile, with the greater momentum, to suffer less change in the impact, and not return to Mary's ship. But from our own vantage point [Figure 11.7(b)], we saw that it *did* return, and that momentum *is* conserved. The only explanation is that Jane's missile, as observed by Mary, must have greater mass to compensate for its lower velocity in the direction perpendicular to that of their relative motion.

If two observers are in uniform relative motion, each will say that the masses of objects in the other's system are greater than they would be if they were at rest. The factor by which mass is increased is exactly the same as the factor by which time is slowed. If an object has a mass m_o when it is at rest, when it is moving with a speed v its effective mass is $m_o/\sqrt{1 - v^2/c^2}$. The quantity m_o is called the *rest mass* of the object.

The increase in mass of rapidly moving objects is not illusory; it is real. We observe it commonly in nuclear accelerators. As subatomic particles are sped up to nearly the speed of light, their masses increase manyfold, and enormously more power is required to provide them additional acceleration. Note that if the speed of a body were equal to the speed of light, v/c would be 1, and $\sqrt{1 - v^2/c^2}$ would be zero. Anything (except zero) divided by zero is infinite, so the mass of a particle moving with the speed of light would be infinite, which, of course, is impossible. Thus no material body (a body with nonzero rest mass) can ever travel at quite the speed of light. Here is the physical explanation of the fact that the speed of light is an absolute barrier that no body can cross. To accelerate a body of appreciable mass to a speed even very close to that of light would require absolutely tremendous amounts of energy; so far we have succeeded in making only objects of the mass of subatomic particles reach speeds close to that of light.

(f) Mass and Energy

All material bodies in motion possess energy of motion called *kinetic energy*. With a little algebra, it can be shown that the increase in mass of an object caused by its motion is its kinetic energy divided by the square of the speed of light, or equivalently, its kinetic energy equals its mass increase times c^2. Thus there is an equivalence between the mass and energy of a moving body. Einstein postulated that even when a body is at rest there is an energy equivalence to its rest mass, so that its total energy is equal to its total mass times c^2, a concept made famous by that equation that is the hallmark of special relativity,

$$E = mc^2.$$

With a little algebra, we can derive the above relation. We have

$$m = \frac{m_o}{\sqrt{(1 - v^2/c^2)}} = m_o(1 - v^2/c^2)^{-1/2}.$$

We can expand the factor $(1 - v^2/c^2)^{-1/2}$ with the binomial theorem, and obtain

$$m = m_o\left(1 + 1/2\frac{v^2}{c^2} + \text{terms of order } \frac{v^4}{c^4}\right).$$

For v very much smaller than c, v^4/c^4 is so small that it can be ignored, and

$$m = m_o + 1/2\, m_o v^2 \frac{1}{c^2},$$

or,

$$(m - m_o)c^2 = 1/2\, m_o v^2.$$

We call $m - m_o$ the increase in mass, Δm, and note that $1/2 m_o v^2$ is the kinetic energy due to the motion of the body. Thus Δmc^2 is the extra energy gained by motion, which is ΔE. But if kinetic energy has an equivalence to mass gained by virtue of its motion, what if the velocity is zero, and the kinetic energy is zero? Einstein postulated that there must be energy associated with mass even at rest, and proposed:

$$E = mc^2,$$

where E is the total energy, including that associated with the rest mass, and m is the total mass of a moving body. In particular, if $v = 0$, $m = m_o$, and the rest energy is

$$E_o = m_o c^2.$$

The equivalence of mass and energy stated in the above equation suggests that matter can be converted into energy and vice versa. Indeed, conversions in both directions are commonly observed in experiments with subatomic particles. For example, the electron has a twin called a *positron,* which has the opposite charge of the electron but identical mass. When a positron and electron come into contact, they annihilate each other, turning into photons of energy equal to the combined mass of the positron and electron times the square of the speed of light. Energetic photons can also combine to produce a positron and electron pair.

Because c^2 is a very large quantity, the conversion of even a small amount of mass results in a very great amount of energy. For example, the mutual annihilation of one gram of electrons and one gram of positrons (about 1/14 ounce in all) would produce as much energy as 30,000 barrels of oil. Here is the source of nuclear energy. Commercial nuclear power plants do not, however, involve the complete conversion of the nuclear fuel, but only a small fraction of it. In the hoped-for hydrogen reactor of the future, hydrogen is converted to helium with the destruction of a little under half of one percent of the original hydrogen. Still, the conversion of only 15 kg (about 33 lb) of hydrogen into helium per hour annihilates enough matter to produce energy at the rate of the current United States oil consumption. We are still a long way from the technology to accomplish this, but the sun and stars derive their energy by a similar process, as we shall describe in Chapter 29.

The fact that mass can be converted into energy and vice versa means that the old concepts of conservation of mass and conservation of energy are not strictly correct. However, the total of mass and energy equivalence *is* conserved; that is, if we calculate the energy equivalence of all mass by multiplying that mass by c^2, the resulting figure, added to the total energy, is conserved. Of course, we could also divide the total energy by c^2 and add that to the total mass to obtain a quantity that is conserved.

(g) How Velocities Add

Let an object move with a speed U' in the system of coordinates of an observer moving at a speed v with respect to another observer, taken to be at rest. Let U'_x be the component of U' along the direction of relative motion of the two observers, and let U'_y be one of the components of U' perpendicular to the direction of relative motion. It can be shown that the corresponding components of U, U_x and U_y, in the system of the stationary observer, are

$$U_x = \frac{U'_x + v}{1 + \frac{v}{c^2}U'_x},$$

and

$$U_y = \frac{U'_y}{\gamma\left(1 + \frac{v}{c^2}U'_x\right)}.$$

With Newtonian reasoning, one would expect the velocity of the object in the moving system and the velocity of that system itself to simply add—that is, $U_x = U'_x + v$—but we see that the velocity is "corrected" by the terms in the denominator.

As an example, suppose $U'_x = 0.9c$, and $v = 0.9c$. Newton would have said $U_x = 1.8c$, but relativity says:

$$U_x = \frac{1.8c}{1 + 0.81} = \frac{1.8c}{1.81} = 0.9944c.$$

If $U'_x = c$, we have

$$U_x = \frac{c + v}{1 + \frac{vc}{c^2}} = \frac{c + v}{(c + v)/c} = c;$$

that is, the speed of light is the same in both systems (as we have seen it must be!)

(h) Faster than Light?

It has been speculated that there could exist particles that always must move faster than light, and can never slow down to the speed c. Such hypothetical particles have been called tachyons, and experiments have been performed to search for them, to date with negative results. But as Nobel Laureate physicist Julian Schwinger (b. 1918) has pointed out, there is excellent reason for believing that tachyons cannot exist. If there *were* particles that could travel faster than light, then we could, at least in principle, use them to transmit signals at a faster rate than light can. But then we could build an ideal clock that ticks at more nearly the same rate for different observers in uniform relative motion, and all the special relativity effects we have described would not be correct; in fact if we could communicate with infinite speed (instantaneously), there would be no special relativity at all. Michelson and Morley's experiment would have given a positive result, muons would not arrive at the ground, electrons and protons would not gain mass in accelerators, and all of the many thousands of extremely accurate tests of relativity would not have turned out the way they did. In particular, E would *not* equal mc^2, and we would not have nuclear bombs and reactors. One could, of course, hypothesize that tachyons exist but are totally unobservable, but then we can never know of them nor detect their existence, directly or indirectly, and their existence would have no practical significance on the real world. Most physicists now discard the tachyon hypothesis.

But this does not stop many people from feeling that somehow science and technology will somehow find a way to "break the light barrier." Perhaps they read the wrong science fiction authors. Anyway, irrespective of Captain Kirk's taking the *Enterprise* to "warp II," it is impossible for a material body to ever reach the speed of light. It is not a technological problem, but a fundamental principle of nature. As we have seen, the mass of such a body would become infinite; it would become the entire universe itself.

Nor is there any need to travel faster than light, for, at least in principle, a person can travel at a speed as close to that of light as he wishes (given enough energy), and the closer his speed is to light's, the smaller all distances ahead of him become. Our hypothetical astronauts going only 98 percent the speed of light could reach a star 100 LY away in 20 years, but by going even closer to the speed c they could make the trip in a far shorter time. As one approaches c, his time slows and distances shrink so that he can go anywhere in as short a time as he likes.

If one *could* travel at the speed of light, riding on a photon as it were, his time would stand still and he would be everywhere in the universe at once. Of course only a photon or other body of no rest mass can really do that—a body of pure energy. Massless bodies—photons and neutrinos*—can only travel at the speed of light; material bodies may approach that speed but never attain it.

11.4 CONCLUSION

Special relativity can be understood with very little effort in mathematics. Yet the concepts are very difficult to grasp. The mathematics is easy enough, but the ideas are totally alien to our experience, and present no easy conceptual hurdle. Why is this so? Because we have all grown up in a world where speeds around us are very small compared to the

*Neutrinos are particles released in certain nuclear reactions. They carry energy and have other properties, and can (with difficulty) be detected, but until recently were thought to have no rest mass. Some recent (1980) experiments suggest that a neutrino may have a very tiny rest mass, in which case it could not move with the speed of light. At this writing, none of the experiments is considered very definitive, and the question of whether or not neutrinos have rest mass is still open.

TABLE 11.1 Gamma ($1/\sqrt{1 - v^2/c^2}$) for Various Speeds

MOVING OBJECT	v	v/c	GAMMA (γ)
Automobile	100 km/hr	0.00000009	1.000000000
Concorde SST	2000 km/hr	0.000002	1.000000000
Rifle bullet	1 km/s	0.000003	1.000000000
Earth escape speed	11 km/s	0.000037	1.000000001
Orbital speed of Earth	30 km/s	0.0001	1.000000005
10% light's speed	30,000 km/s	0.1	1.005
		0.5	1.155
		0.9	2.294
		0.98	5.025
		0.99	7.089
		0.999	22.37
Muons in CERN experiment		0.9994	28.87
		0.9999	70.71
		0.999999	707.1
		0.999999999	22360.7

speed of light. All of the relativistic effects we have discussed depend on that factor $1/\sqrt{1 - v^2/c^2}$, usually denoted by the Greek letter gamma (γ). Values of gamma corresponding to several values of v/c are given in Table 11.1. Values in the table show by how much masses increase, lengths shrink, and clocks slow in moving systems. Until v/c is a pretty good-sized fraction, gamma is essentially equal to 1. In such low-velocity systems, Newton's laws of motion apply with admirable precision. Even the Earth's speed about the sun—30 km/s—is only $0.001c$, and gamma is equal to unity within one part in a hundred million. We have become used to the low-velocity world, and it has prejudiced our ideas of "common sense."

On the other hand, how about a hypothetical civilization living on another world in an environment where speeds close to that of light are commonplace? Relativity would not seem strange to them. They, like us, given enough time, would discover the laws of physics, but not in the same order. As Schwinger has put it, "They would have their Maxwell and their Einstein, but alas, no Newton."

EXERCISES

1. Do you think that two observers on systems that are rotating with respect to each other would find that all physical laws are the same in their two systems? Could they tell which observer was rotating more rapidly? If so, how?
2. Suppose a ball is thrown forward at 60 km/hr from an automobile moving at 100 km/hr. How fast is the ball moving with respect to an observer on the roadside? What if the ball is thrown toward the rear of the car with the same speed?
3. Compare the ways in which different observers compare the speed of the ball in the last exercise with how they compare the speed of light.
4. From the fact that light takes 16½ minutes to cross the orbit of the Earth, show that the speed of light is about 10,000 times that of the Earth.
5. Prepare a pendulum consisting of a small weight at the end of a string exactly 40 cm long. Start the pendulum swinging and time it for 10 complete oscillations (one oscillation is to and fro). What is your result? Now take the pendulum into an automobile, try to arrange that the car drive as smoothly and at as constant a speed as possible, and repeat the experiment. Now what is your result?

*6. Suppose a star in a binary system revolves about its companion with an orbital speed of 90 km/s (three times the Earth's orbital speed). Now suppose (which, as we have seen, is incorrect) that the speed of light emitted by the star toward us *did* depend on how fast the star was approaching us or receding from us.
   a. How much faster, in km/s, would light be approaching us that is emitted when the star ap-

proaches us in its orbit, than when it is moving away (assume that the orbit is edge-on to our line of sight)?

b. One year is approximately 3×10^7 s. If the double star system were 200 LY away, by how many kilometers would the faster approaching light, by the time it reached the Earth, have gained on the slower light, emitted when the star was on the opposite side of its orbit?

c. How long (approximately) would it take the slower light to make up this distance?

d. How far does the star move in its orbit in this time? If the star's orbit were the size of the Earth's, what fraction of its orbit would the star have traversed?

★7. Show that in the Michelson-Morley experiment the predicted lead of the faster light beam over the slower one (due to the Earth's orbital revolution) is 10^{-5} cm.

★8. Refer to Table 11.1. What would we measure for the mass of a 100-kg body moving past us with a speed of 90 percent that of light?

★9. According to Newton's laws the ordinary kinetic energy of a body of mass m moving at a speed v is $\frac{1}{2}mv^2$. Calculate the kinetic energy of a body of mass 1 g moving with a speed of 10^6 cm/s (about one third the orbital speed of the Earth). Now calculate the energy associated with the rest mass of the same body. How do the two energies compare?

★10. By what factor does time slow for an astronaut moving 99.99 percent the speed of light? How long would it take an astronaut going that fast to make a round trip to a star 100 LY away: (a) according to people who stayed behind on Earth? (b) according to his own time?

★11. By what factor is the mass of an object increased if it moves: (a) 3/5 the speed of light? (b) 99.99 percent the speed of light?

Gerald P. Kuiper (1905–1973), Dutch-born American astronomer, made many contributions to the theory of the origin of the solar system, discovered the atmosphere of Titan, carried out pioneering work in infrared astronomy, founded several observatories, initiated the NASA airborne astronomy program, and was an influential architect of the early NASA program of lunar and planetary exploration. (University of Arizona)

STRUCTURE AND ORIGIN OF THE SOLAR SYSTEM

The ancient observer, who considered the Earth to be central and dominant in the universe, regarded the sun, Moon, and planets as luminous orbs that moved about on the celestial sphere through the zodiac.

Our solar system is indeed dominated by one body, but it is the sun, not the Earth. Our sun, so important to us, is merely an ordinary, "garden-variety" star. Only careful scrutiny at close range would reveal the tiny planets to an imaginary interstellar visitor. First Jupiter, the largest, would be seen; then Venus and Saturn; and perhaps only with the greatest difficulty, the Earth and other planets. Almost 99.9 percent of the matter in the system *is* the sun itself; the planets comprise most of what is left—the Earth scarcely counts among them. The countless millions of other objects in the solar system, mostly unknown to the ancients, would probably remain unnoticed by a casual traveler passing through the solar neighborhood.

We turn now to those worlds of the solar system. They will be considered individually in detail in later chapters. Here we take only a brief look at some of the general characteristics of the solar system and remark on a few of the properties that its constituent worlds have in common.

12.1 Inventory of the Solar System

The solar system consists of the sun and a large number of smaller objects gravitationally associated with it. The other objects are the planets, their satellites and rings, and a great deal of smaller debris, including asteroids, comets, meteoroids, and dust. It is believed that most of these objects formed at about the same time as the sun, as the result of processes generally associated with star formation. The relative prominence of the various members of the solar system can be seen from Table 12.1, which lists the approximate distribution of mass in the solar system.

(a) The Sun

The sun, practically, *is* the solar system (Figure 12.1). It is a typical star—a great sphere of luminous gas. It is composed of the same chemical elements that compose the Earth and other objects of the universe, but in the sun (and other stars) these elements are heated to the gaseous state. Tremendous pressure is produced by the great weight of the sun's layers. The high temperature of its inte-

TABLE 12.1 Mass of Members of the Solar System

OBJECT	PERCENTAGE OF MASS
Sun	99.85
Jupiter	0.10
All other planets	0.04
Comets	0.01(?)
Satellites and rings	0.00005
Asteroids	0.0000002
Meteoroids and dust	0.0000001(?)

rior and the consequent thermonuclear reactions keep the entire sun gaseous. There is no distinct "surface" to the sun; the apparent surface we observe is optical only—the level in the sun at which its gases become opaque, preventing us from seeing deeper into its interior. The temperature of that region is about 6000 K. Relatively sparse outer gases of the sun extend for millions of kilometers into space in all directions. The visible part of the sun is 1,390,000 km across, which is 109 times the diameter of the Earth. Its volume is 1⅓ million times that of the Earth. Its mass of 2×10^{33} g exceeds that of the earth by 330,000 times. The sun's energy output of 4×10^{33} ergs/s, or about 5×10^{23} hp, provides all the light and heat for the rest of the solar system. The sun derives this energy from thermonuclear reactions deep in its interior, where temperatures exceed 14 million K.

In addition to its heat and light, the sun is the origin of most of the thin gas that fills the solar system. A *solar wind* of ionized atoms and electrons constantly streams away from the hot upper atmosphere of the sun at speeds of several hundred kilometers per second. The solar wind strikes the planets, interacting with their atmospheres and magnetic fields. The solar wind also causes comet tails to stream away from the sun, no matter which way the comet itself is moving.

We shall describe the sun, a typical star, in Chapter 28. In this and the following chapters our attention is focused instead on the smaller bodies: *the planetary system,* which includes (in addition to the planets themselves) satellites, rings, comets, asteroids, and the other solid matter associated with the sun.

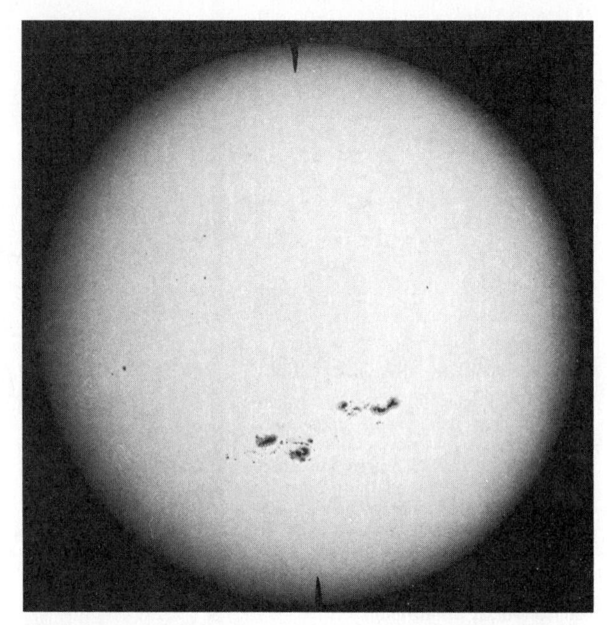

Figure 12.1 The sun, which contains more than 99 percent of the mass of the solar system. (Mount Wilson and Las Campañas Observatories)

(b) The Planets

Most of the material in the planetary system is concentrated in the planets, and particularly in the largest planet, Jupiter (see Table 12.1). The nine planets include the Earth, the five other planets known to the ancients (Mercury, Venus, Mars, Jupiter, and Saturn), and the three discovered since the invention of the telescope (Uranus, Neptune, and Pluto). These planets circle the sun in approximately the same plane in orbits of small eccentricity (Figure 12.2). Their sizes range from Jupiter, with a mass of about 1/1000 that of the sun, down to tiny Pluto, which is smaller than our Moon.

The four innermost planets (Mercury through Mars) are called the *inner planets* or *terrestrial planets;* often the Moon is also discussed as a part of this group, bringing the total of terrestrial bodies to five. The outer five planets (Jupiter through Pluto) are called the *outer planets.* Of these, the four largest planets (Jupiter, Saturn, Uranus, and Neptune) are referred to as *giant* or *Jovian planets.* Pluto is neither a terrestrial nor a Jovian planet; it is similar to the icy satellites of the outer planets.

The terrestrial planets are rocky worlds, composed largely of silicate rocks and metals. Venus, Earth, and Mars each have atmospheres that es-

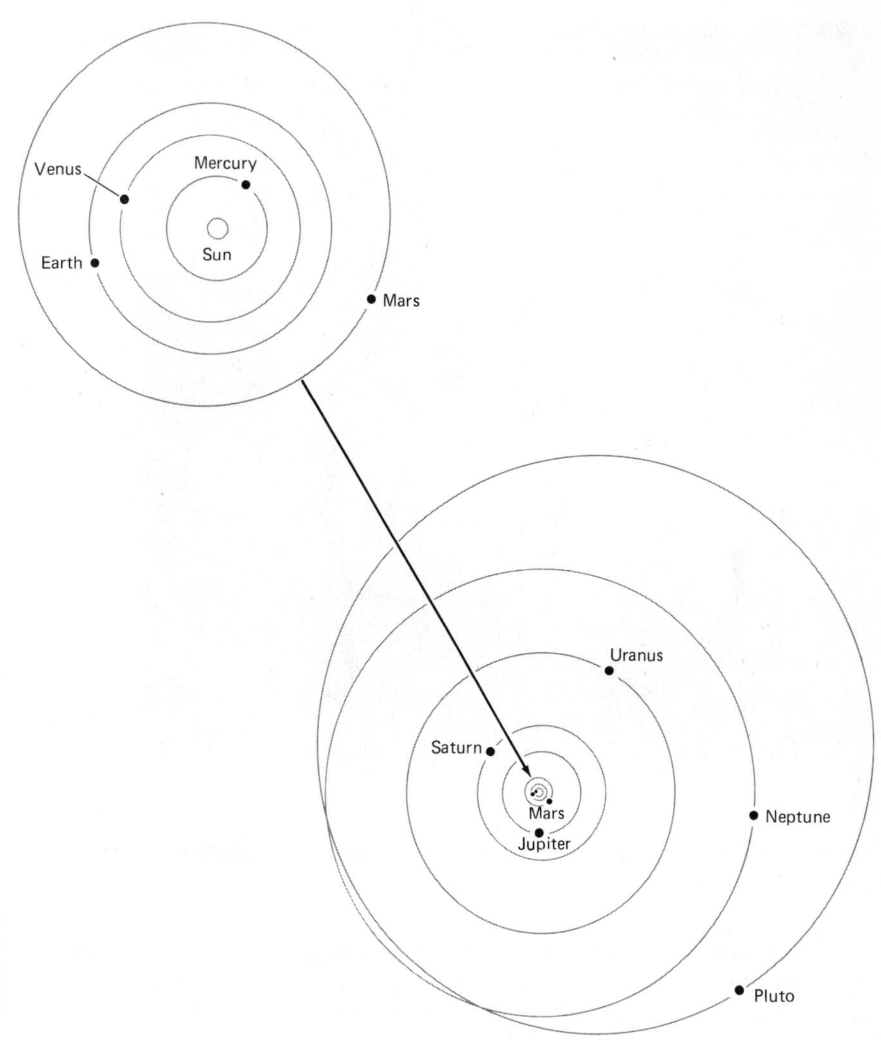

Figure 12.2 The relative sizes of the orbits of the planets. Two different scales are used to show the outer and inner planets.

caped from their interiors; the Moon and Mercury, however, are too small to retain atmospheres. In dramatic contrast, the Jovian planets lack solid surfaces; their composition is primarily light elements (hydrogen, helium, argon, carbon, oxygen, and nitrogen) in gaseous or liquid form (Figure 12.3). Thus, the giant planets have compositions somewhat similar to that of the sun and stars, while the terrestrial planets consist almost entirely of the rarer elements, primarily silicon, iron, and oxygen.

All of the planets *revolve* about the sun in the same counterclockwise direction as viewed from the north, at distances (semimajor axis of the orbit) that range from 0.39 AU (58 million km) for Mercury to 39.5 AU (5900 million km) for Pluto. Their periods of orbital revolution range from 88 days for Mercury to 248 years for Pluto; the corresponding average orbital velocities range from 48 km/s to 5 km/s.

Each of the planets also *rotates* about an axis running through it, and in most cases the direction of rotation is the same as that of revolution about the sun. The most obvious exception is Venus, which rotates backward (that is, in a *retrograde* direction). Uranus and Pluto each rotate about an axis tipped nearly on its side.

(c) Satellites and Rings

Most of the planets are accompanied by one or more satellites; only Mercury and Venus are alone.

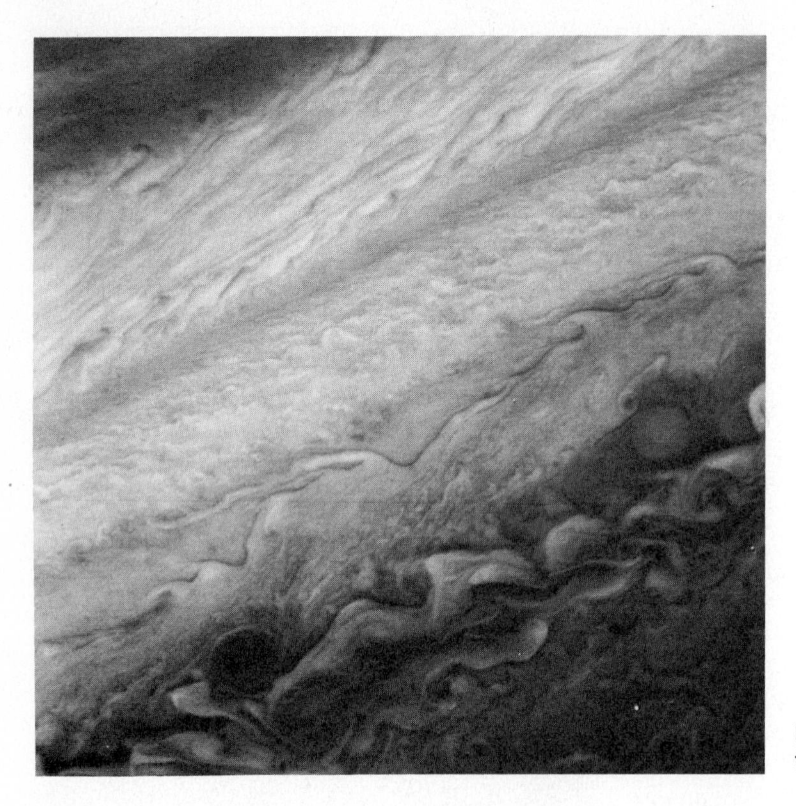

Figure 12.3 Mid-latitudes of Jupiter, photographed by Voyager 1. (NASA/JPL)

Most of the satellites, as well as the ring systems, are in the outer solar system, associated with the giant planets.

There are 56 known satellites, and undoubtedly many other very small satellites remain undiscovered. Saturn has 19, Jupiter 16, Uranus 15, Neptune and Mars 2, and Earth and Pluto 1 each. Six of these satellites are comparable in size to our Moon or to the planet Pluto. The largest is Ganymede in the Jovian system, which has a diameter almost as large as that of Mars. A close second is Saturn's Titan, the only satellite with a large, cloudy atmosphere.

Satellites in the inner solar system are of a rocky composition like the terrestrial planets. Most of the satellites reside in the outer solar system, however, where lower temperatures allow the preservation of water ice. Thus the typical satellite consists of a substantial fraction—sometimes more than half—of ice. Pluto shares this ice-dominated composition.

Most satellites revolve in nearly circular orbits in the plane of the planet's equator, much as the planets orbit the sun. Presumably these *regular satellites* were formed with the planet in a manner analogous to the formation of the solar system. Other satellites are in eccentric, highly inclined, or even retrograde orbits; these *irregular satellites* are thought to have been captured subsequent to the formation of the regular satellite systems. Jupiter has 8 regular and 8 irregular satellites; Saturn has 17 regular and 2 irregular satellites; and all 15 known satellites of Uranus are regular.

Each of the Jovian planets has a ring system, but each is unique, bearing rather little similarity to the others. By far the largest, and certainly the best known, are the rings of Saturn. The rings of Uranus were discovered in 1977, those of Jupiter in 1979, and those of Neptune not until 1985.

(d) Comets and Asteroids

The comets and asteroids are small bodies orbiting the sun. They are undoubtedly remnants of the original population of solid bodies that ultimately formed the planets, as we will describe in Section 12.4. The great majority of such bodies either impacted the planets or were ejected from the solar

Figure 12.4 Comet Kohoutek, photographed December 7, 1973. (NASA)

system, with less than one-millionth of the original mass surviving as today's comets and asteroids.

Comets differ from asteroids in both composition and orbits. The comets are chunks of frozen gases, ice, and dust that revolve about the sun in orbits of very high eccentricity. Although they are only a few kilometers in diameter, they become visible to us when they are heated by the sun to produce sometimes spectacular atmospheres and tails (Figure 12.4). The asteroids, in contrast, are rocky objects that orbit the sun primarily between the orbits of Mars and Jupiter. They are never spectacular, being visible only through a telescope. Comets and asteroids share common traits in that they are small, are older than the planets, and have managed to survive since the formation of the solar system.

When a comet is close to the sun, it can form a tenuous, extended atmosphere up to 100,000 km across, called the *head* of the comet. Under pressure from sunlight and the solar wind, a part of the gas and dust in the head is usually forced away from the sun to form the comet's *tail,* which can be tens of millions of kilometers long. In contrast to these huge structures, the tiny solid *nucleus* is usually invisible.

More than a thousand comets have been observed, and some five to ten new ones are discovered each year. About once a decade a comet is bright enough to attract public attention. The observed comets, however, are only the tip of the iceberg; as discussed in Chapter 20, a vastly larger number are believed to exist in a cloud surrounding the sun and extending about one-quarter of the distance to the nearest stars.

The asteroids, discussed in Chapter 19, range in size from Ceres, with a diameter of just under 1000 km, down to the limits of detection at about 1 km. There are tens of thousands of asteroids between Mars and Jupiter, of which about 4000 of the larger objects have well-determined orbits. The asteroids are also called *minor planets.*

(e) Meteoroids and Dust

The number of small objects revolving about the sun that are too small to observe with telescopes is great indeed and increases for objects of smaller and smaller size. These tiny astronomical bodies, with diameters less than a kilometer, are called *meteoroids.* The smallest meteoroids are called micrometeoroids or, more commonly, simply *dust.*

When a meteoroid collides with the Earth, it is heated by friction with the atmosphere and at least partially vaporized. Looking up at the night sky, we see a brief flash of light and perhaps a trail of luminous vapor that persists for a few seconds before fading back into blackness. This "falling star" or "shooting star" is properly called a *meteor.* On a typical clear dark night, about half a dozen meteors can be seen per hour from any given place on Earth. The typical meteor represents the death of a grain of material no larger than a pea. Bigger meteoroids produce the rarer *fireballs,* or very bright meteors. If a part of the meteoroid should survive its fiery plunge through the atmosphere to reach the surface, we call it a *meteorite.*

Meteorites are extraordinarily valuable to scientists because they represent actual samples of cosmic material that can be analyzed in detail in the laboratory. Many of these fragments preserve information about the earliest history of the solar system, and much of our present understanding of the formation of our system is derived from the study of meteorites. Other meteorites have been identified, at least tentatively, as fragments from the Moon, from Mars, and from the asteroid Vesta.

Cosmic dust—the smallest meteoroids—is thought to be derived primarily from comets and secondarily from the asteroids. Its presence can be detected by faint glows in visible and infrared light that pervade the inner solar system. This dust is also constantly striking the other bodies in the solar system. It contributes hundreds of tons of material daily to the upper atmosphere of the Earth. Cosmic dust also erodes the surfaces of spacecraft exposed to interplanetary space for long periods.

12.2 SOME BASIC IDEAS IN PLANETARY SCIENCE

The study of the planetary system has gone through three distinct phases. From the most ancient times through the 18th century, the mystery of the *motions* of the planets (and their possible effects on human life) dominated astronomy. As we have seen, Tycho's observations of planetary positions led to Kepler's determination of the laws of planetary motion and thus to Newton's brilliant synthesis and the foundation of modern science. This emphasis on celestial mechanics persisted until the evolution of the telescope led to the second, or *astronomical* phase. For about the last century planetary astronomers have concentrated their efforts on the study of the physical and chemical nature of the planets as determined by telescopic observation. The third phase is that of *direct exploration* of the planetary system by spacecraft, in which the traditional interests are supplemented by new fields of planetary geology, planetary meteorology, and space physics. We are now experiencing the exciting early part of this third phase.

In the following chapters we will consider the results of both telescopic and space-probe exploration of the individual planets and their satellite and ring systems. But first we introduce, in this and the following section, some basic ideas that are needed to understand the processes at work in the planetary system and the interrelationships among the various planets.

(a) Basic Characteristics of Planets

The most fundamental characteristics of a planet are its mass, its chemical composition, and its distance from the sun. From these data alone, we can predict to some extent many of its other characteristics, such as its surface temperature or whether or not it is likely to have an atmosphere. We will discuss chemical composition in Section 12.3, but other basic ideas are reviewed here.

(b) Rotation of the Planets

All the planets are observed to rotate. Of the number of techniques employed to determine rotation rates, the following are the most useful.

1. The most direct method is to watch permanent surface features move across the disk. Even naked-eye observations are sufficient to show that the Moon always keeps the same face toward the Earth, that is, that the rotation period of the Moon is equal to its period of revolution about the Earth. Mars' rotation period of a little over 24 hours was observed telescopically more than a century ago, and today most of the known rotation rates for planets and satellites have been derived by this technique, using either telescopic or spacecraft data. If the object is too small to show individual surface features, one can look for a periodic variation in brightness as it rotates; this approach has given us our value for the rotation period of Pluto, and it yielded rotation periods for many of the satellites of Jupiter and Saturn decades before the arrival of the first spacecraft.
2. Radar was first used to derive the rotation periods of Mercury and Venus, although the results were later verified using method 1. Radar (radio) waves at a single frequency are beamed to a planet and the reflected signal is measured. These reflected waves are Doppler-shifted according to the line-of-sight velocity of each part of the target with respect to the transmitter on Earth. Waves reflected from the approaching edge are thus shifted to shorter wavelengths than those that are reflected from the center of the disk, and the signal from the receding part of the planet's surface is shifted to longer wavelengths. Measurement of the range of wavelengths in the reflected signal (its bandwidth) thus gives a measure of the rotation rate of the target.
3. The gaseous giant planets do not reflect radar waves, and observations of the motions of markings yield only the wind velocities in the atmosphere. The true rotation rate of the underlying planet can be found, however, if there

is a magnetic field generated in the core, since this field rotates with the core. The rotation rates of Jupiter, Saturn, and Uranus are defined by the measured rotation rates of their magnetic fields.

(c) Temperatures of Planets

Since the planets obtain most of their energy from sunlight, their temperatures depend on their distance from the sun; the closer to the source of energy, the warmer the planet should be. It is relatively straightforward to calculate a characteristic or *effective temperature* for any particular distance from the sun, although as we will see below and in the following chapters, the actual temperatures on a planet may differ greatly from the calculated values.

The effective temperature corresponds to a balance between the sunlight striking a planet and the emission of infrared radiation back into space. At the Earth (a distance of 1 AU from the sun), the amount of energy from the sun—the *solar constant*—is 1.37×10^6 erg/s striking each square centimeter of surface. The corresponding temperature for a dark, perfectly absorbing material to radiate an exactly equal quantity of energy is just under 400 K. However, the Earth is not a black surface oriented toward the sun, but a real spherical planet with oceans and atmosphere. It reflects part of the sunlight back into space, and it distributes heat from the subsolar regions to the cooler poles and to the night side. The average surface temperature on the Earth is actually a little above the freezing point of water, or about 280 K.

The incident sunlight varies with distance according to the inverse-square law for the propagation of electromagnetic energy. As we saw in Section 8.1, the emission of energy from a hot surface is proportional to the temperature to the fourth power (when the temperature is expressed in Kelvins). A little algebra should convince you, therefore, that the effective temperature varies as the *inverse square root* of the distance from the sun.

We can apply this simple relationship between temperature and distance from the sun to calculate what sort of temperatures should be characteristic of other planets, starting from the average Earth value of 280 K. All we need to do is divide 280 by the square root of the distance from the sun, expressed in AU. The results are illustrated in Table 12.2.

TABLE 12.2 Calculated Characteristic Planetary Temperatures

PLANET	DISTANCE (AU)	TEMPERATURE (K)
Mercury	0.39	450
Venus	0.72	330
Earth	1.00	280
Mars	1.52	230
Jupiter	5.2	125
Saturn	9.6	90
Uranus	19.	65
Neptune	30.	50
Pluto	40.	45

In practice, there are a great many effects that produce departures from the estimates in Table 12.2. First, of course, is the dependence of temperature on location on a planet; it is colder at the poles than on the equator, and colder on the night side than facing the sun. Depending on the period of rotation of a planet and the nature of its atmosphere, these temperature values may vary widely. Second, the atmosphere itself can elevate the surface temperature by the greenhouse effect (Chapter 16), and within the atmosphere temperatures can vary greatly depending on the absorption of sunlight, especially ultraviolet radiation. In particular, the upper atmospheres of planets are usually much hotter than would be expected from Table 12.2. Third, the giant planets Jupiter, Saturn, and Neptune all have significant internal heat sources that elevate temperatures considerably above the values that would be expected on the basis of sunlight alone. Fourth, temperatures can be effected by the evaporation of ices, resulting in much lower surface temperatures in comets, for instance, than would be expected for non-icy objects. And finally, small dust grains do not follow the pattern in Table 12.2, but tend to reach much higher equilibrium temperatures because they are too small to radiate their energy efficiently into space.

(d) Finding the Ages of Rocks

In order to trace the history of a planet or of the solar system as a whole, we must have a way to determine the ages of individual rocks. Before the discovery of radioactivity at the end of the 19th century, there was no way to do this. It is not surprising, therefore, that 18th and 19th century es-

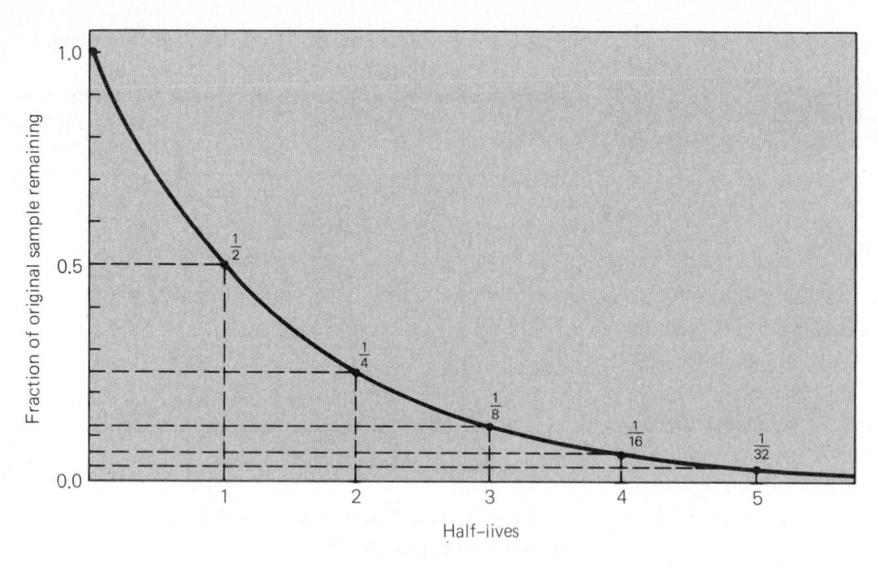

Figure 12.5 Radioactive decay and half-life.

timates of the age of the Earth, for example, ranged all the way from a few thousand years to the opposite extreme, an infinitely old planet. Now we can date the formation of our planet with substantial precision, and similar techniques are used to determine the geological history of the Moon and of the parent bodies from which the meteorites are derived.

Radioactive nuclei are unstable and spontaneously convert to other nuclei with the emission of particles such as electrons or of radiation in the form of gamma rays. For any given nucleus the decay process is random, and it might happen at any time. But for a very large collection of identical radioactive atoms there is a specific time period, called its *half-life,* during which the chances are fifty-fifty that decay will occur. A particular nucleus may last a shorter or longer time then its half-life, but in a large sample almost exactly half will have decayed after a time equal to one half-life, and half of those remaining (three-quarters in all) will have decayed in two half-lives. After three half-lives, only one-eighth of the original sample remains, and so on (Figure 12.5). Thus radioactive elements provide accurate nuclear clocks; by comparing the relative abundance of a radioactive element (the *parent*) with that of the element it decays into (the *daughter*), we can learn how long the process has been going on and hence the age of the sample.

Most natural rocks contain several radioactive elements with half-lives that provide suitable clocks. Among these are potassium-40, which decays to argon-40 with a half-life of 1.3 billion years; rubidium-87, which decays to strontium-87 with a half-life of 50 billion years; and uranium-238, which decays to lead-106 with a half-life of 4.5 billion years. In practice as many as five separate parent-daughter pairs are used to date a rock, and the process is carried out for several different mineral grains in the same rock so as to further eliminate uncertainties. The final age, checked by the agreement of the different methods, is usually accurate to within a few percent, that is, to a few tens of millions of years in a rock several billion years old.

In order to interpret such measurements, it is important to recognize what the age represents. A radioactive age is the time during which the parent has been decaying into the daughter undisturbed, so that both parent and daughter are still present in each mineral grain. In most cases it can be thought of as the *solidification age* of the rock: the time since it cooled from the molten state. Thus, for example, it is straightforward to use the radioactive ages measured for a rock from a lunar lava flow to determine when volcanic activity took place on the Moon. In addition, however, it is possible to use age measurements from a wide variety of rocks to find out when an entire planet formed, and such results date the formation of both the Earth and the Moon at between 4.5 and 4.6 billion years ago.

(e) Planetary Atmospheres

All of the planets except Mercury (and possibly Pluto) are surrounded by appreciable gaseous atmospheres. Some of the constituents of those at-

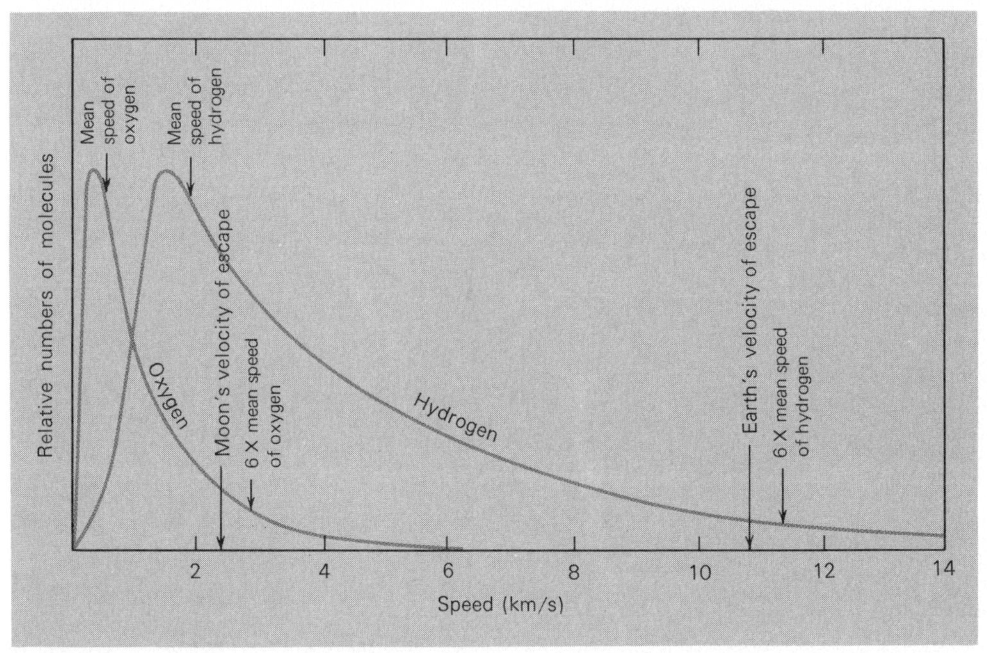

Figure 12.6 Distribution of speeds of oxygen and hydrogen molecules at a temperature appropriate for the Earth and Moon.

mospheres have been known for many years from Earth-based spectral analysis of sunlight reflected from the planets. This sunlight has traversed a part of the planet's atmosphere, and the molecules of the gas in that atmosphere absorb certain wavelengths, leaving dark lines in the spectrum that are not present in the spectrum of sunlight alone. Proper identification of those lines reveals the identity of the atmospheric gases that produce them, and it may also indicate the pressure and temperature in the planetary atmosphere.

New constituents continue to be identified in the spectra of planetary atmospheres, particularly as new regions of the ultraviolet and infrared spectrum are studied. This information has been greatly extended, however, by spacecraft measurements. In the cases of Mars and Venus, direct analyses have been carried out by entry probes and by instruments landed on the surface. An entry probe is part of the Galileo mission payload, to be deployed into the atmosphere of Jupiter about 1994, and other similar probes are under study for both Saturn and its cloudy satellite Titan.

The gases present in a planet's atmosphere depend on the chemical constituents out of which the planet formed, a topic we return to in the next section. Even more important, however, is the mass of the planet. The terrestrial planets formed without extensive atmospheres, and their present atmospheres represent a combination of gas that has escaped or *outgassed* from their interiors over geological time, together with a component derived from the impacts of comets. The atmospheres of these planets are in equilibrium between outgassing and loss, either by escape of gas to space or by chemical reaction with the surface rocks. Further modifications take place as the result of a variety of chemical reactions with ultraviolet sunlight in the upper atmosphere, with the crust of the planet, or (in the case of the Earth) with life.

The massive giant planets have entirely different kinds of atmospheres, dominated by the lightest and most common gases, especially hydrogen and helium. Evidently, these planets were able to retain the cosmically abundant gas present in the solar system at the time of their formation. Indeed, we can calculate today that the gravitational attractions of the giant planets are sufficient to hold these gases. In contrast, hydrogen and helium escape from the Earth and smaller planets. The individual molecules of a gas are always in a rapid motion (Figure 12.6); if their speed exceeds the velocity of escape of a planet the atmosphere gradually "evaporates" into space. Even heavy gases cannot be retained by an

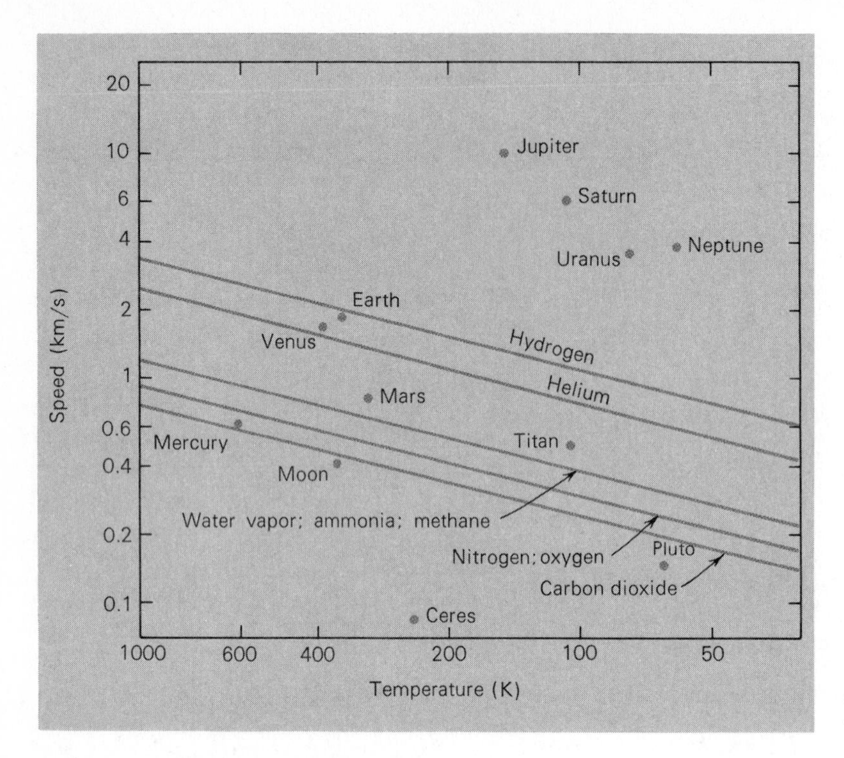

Figure 12.7 Molecular speeds of common gases at various temperatures.

object as small as the Moon, which is why there is no lunar atmosphere.

(f) Kinetic Theory of the Escape of Atmospheres

Gases, like all matter, are composed of units of matter called *molecules.* Molecules are composed, of course, of the still smaller *atoms.* Molecules of pure elements consist of one or more atoms of that element. Molecules of a chemical compound consist of two or more atoms of two or more different kinds bound together. For example, a molecule of the compound *water* consists of one oxygen atom and two hydrogen atoms. The atomic structure of the molecules determines the identity of the substance they compose.

Molecules of a gas are in rapid motion, darting this way and that, frequently colliding with each other. At sea level on the Earth there are some 10^{19} such molecules bouncing about in each cubic centimeter of air. The *kinetic energy* (or energy of motion) of a moving object is defined as $\frac{1}{2}mv^2$, where m is the mass of the object and v is its speed. Ordinary temperature, *kinetic temperature,* is a measure of the mean (or average) kinetic energy of molecules. Specifically, the absolute temperature of a gas, T, is related to the mean energy of its individual molecules by the formula

$$\overline{\tfrac{1}{2}mv^2} = \tfrac{3}{2}kT,$$

where the bar over the term on the left indicates that it is the average kinetic energy of the gas molecules, and k is Boltzmann's constant (in metric units, $k = 1.38 \times 10^{-16}$ erg/deg). If the above formula is solved for the mean speed of molecules in a given kind of gas, there results

$$\bar{v} = \sqrt{\frac{3kT}{m}}.$$

Thus we see that the mean speed of molecules in any particular gas is proportional to the square root of the temperature and inversely proportional to the square root of the mass of a single molecule of that gas. The higher the temperature, the faster the molecules move; at a given temperature, molecules of greater mass move slower, on the average, than those of smaller mass.

The temperature is measured from absolute zero; at $T = 0$, the mean energy of the molecules is zero. Here is the meaning of absolute zero: As gases are cooled, their molecules move more and more slowly, and at absolute zero all molecular motion ceases. This occurs at -273°C (-459°F). Absolute, or Kelvin, temperature is thus the Celsius temperature + 273K. (See Appendix 5.)

At any given time, some molecules move at less than the average speed, and others move faster. A few are moving at several times the average speed. An individual molecule, suffering frequent collisions, constantly exchanges energy with other molecules. Sometimes it moves relatively slowly; at other times it may get a good jolt in a collision and move far faster than the average. From kinetic theory we can calculate the relative numbers of molecules moving at various speeds if we know the kind of gas (and hence the mass of each molecule) and the temperature. It is these fast-moving atoms in the highest regions of a planet's atmosphere that gradually escape, leading to depletion of light elements over the age of the solar system.

12.3 CHEMISTRY OF THE PLANETARY SYSTEM

Unlike most objects studied by astronomers, the planetary system is composed primarily of matter in solid and liquid form, generally at temperatures far lower than those encountered in the sun and stars. Matter at high temperatures is relatively easy to understand, since it consists of individual atoms or fragments of atoms in the gaseous state. But under planetary conditions, atoms interact with each other to produce molecules and minerals, and we must deal with the complexities of *chemistry*.

When we look at the individual bodies in the planetary system we quickly see that their compositions are also much more diverse than the compositions of stars, which are almost all made predominantly of just two gases, hydrogen and helium. Most solar system objects, in contrast, have lost much of their hydrogen and helium and other light elements. Their chemical makeup is both a reflection of the processes that formed them and a major factor in determining their evolution throughout the 4.6-billion year history of the solar system.

(a) Five Kinds of Matter

In a quick overview of planetary compositions, we can simplify the situation by considering just five characteristic kinds of matter:

1. *Gas.* The gaseous form of matter is well known to us. It forms the atmospheres of planets, and we often speak of the giant planets as composed primarily of the gases hydrogen and helium. At the pressures found within the giant planets, however, these gases undergo a transformation to the liquid state. Most of the matter in the planetary system is either gas or liquid.
2. *Plasma.* As we saw in Chapter 8, a plasma is a dilute hot gas composed of ionized atoms: basically, positively charged ions and negatively charged electrons. Unlike electrically neutral gas, a plasma's motion is responsive to magnetic and electric fields. The solar wind that streams through interplanetary space is a plasma, as are the charged atomic particles trapped in the magnetic fields of planets.
3. *Ice.* The cosmically abundant elements hydrogen, oxygen, carbon, and nitrogen all form simple compounds that freeze into solids at the temperatures of the outer solar system. The most important of these is water ice (H_2O), but significant quantities are also expected of the ices of NH_3, CH_4, CO_2, and CO. Ices form the main building blocks of comets and outer planet satellites, and they may also contribute most of the mass to the cores of the giant planets.
4. *Rock.* The next most abundant class of materials after the ices are the rocks, which consist of more complex compounds of silicon, oxygen, magnesium, calcium, sulfur, carbon, iron, and other elements. Rocks are the main building blocks of the inner planets and the asteroids, which formed in parts of the solar system where temperatures were too high for the accumulation of ices.
5. *Metal.* Most metallic elements readily form compounds with oxygen and thus contribute to the rocky material. However, there are places in the planetary system, particularly in the cores of planets, where metal is separated from rocks and exists in the pure state. The two most abundant metals in the core of the Earth are iron and nickel.

As we look at individual planetary bodies we will see that they are composed of various proportions of metal, rock, ice, and gas, all surrounded by a sea of interplanetary plasma. One of the main challenges to any theory of solar system origin is to explain the observed distribution of these materials.

(b) Oxidized and Reduced Environments

Because oxygen and hydrogen are both abundant and very chemically reactive elements, they tend to dominate the chemistry of the solar system. Much of the chemical evolution of a body therefore depends on the relative proportions of these two elements.

On the Earth, hydrogen is relatively rare, since this light gas escapes easily from the upper atmosphere (Section 12.2f). Oxygen therefore dominates, and we live in an *oxidized* environment. Most of the rocks that make up the crust of the Earth are composed of various compounds of oxygen, and there is oxygen to spare in our atmosphere. If hydrogen or hydrogen-rich compounds are introduced on Earth, they are quickly broken down by chemical interactions with oxygen. Pure metal is also unstable and oxidizes, as we see by the rusting of iron, the tarnishing of silver, or the rapid transformation of shiny copper to a blue-green oxide. (One of the reasons gold has always been so highly valued is that it is one of the few metals that does not form an oxide.)

All of the terrestrial planets are chemically oxidized to various degrees, but only the Earth has free oxygen in its atmosphere. As we will see in Chapter 13, the presence of oxygen is the direct result of photosynthetic life on our planet.

When there are two or more hydrogen atoms present for each atom of oxygen, the hydrogen dominates and the chemical environment is said to be *reducing*. Any available oxygen combines with hydrogen to produce water (H_2O), and the leftover hydrogen combines with other elements to produce an entirely different set of compounds. Among these are ammonia (NH_3) and the hydrocarbons (compounds of hydrogen and carbon).

The giant planets all have chemically reduced atmospheres, with plentiful free hydrogen. Their visible clouds are composed of ammonia (NH_3) crystals, and their spectra show the presence of many hydrocarbons (CH_4, C_2H_2, etc.) Thus most of the material in the planetary system is characterized by a reducing chemistry.

(c) Rocks and Minerals

Rocks are mixtures of compounds composed in part of the elements silicon and oxygen. They are made up of assemblages called *minerals*. Unlike a mineral, which consists of a single compound, a rock is typically much more heterogeneous. Examples of minerals include a silicate, quartz (SiO_2); a metallic oxide, hematite (Fe_2O_3); a sulfide, iron pyrite (FeS_2); and a carbonate, calcite ($CaCO_3$).

Figure 12.8 Eruption of the Kilauea volcano in Hawaii. (J.D. Griggs, USGS)

Rocks can be classified by their minerals, but a much simpler system is to categorize them according to their history. On Earth, three such classes are commonly used:

1. *Igneous rock,* which formed by the cooling of molten material (Figure 12.8). The two most abundant kinds of igneous rock on the surface of the Earth are *basalt,* the lava that makes up the ocean floors, and *granite,* the most common continental rock. Much of the lunar crust is made of a different class of igneous rock, the anorthosites.
2. *Sedimentary rock,* which formed by the deposition of fragments of igneous rock or of living organisms. On Earth, these include the common sandstones, shales, and limestones. We have a great deal of sedimentary rock on our planet, as a result of the constant weathering and erosion of the primary igneous rock.

3. *Metamorphic rock,* produced by the chemical and physical alteration of igneous or sedimentary rock at high temperature and pressure. Metamorphic rock is produced on Earth because geological activity carries surface rock to considerable depths and then brings it back up to the surface. On less active planets, metamorphic rock should be rare.

There is a fourth very important category of rock not represented on the Earth or Moon that can tell us much about the early history of the planetary system:

4. *Primitive rock,* which has largely escaped modification by heating. Primitive rock represents the original material out of which the planetary system was made. There is no primitive material left on the terrestrial planets because these planets were heated above the melting point of rock early in their history. To find primitive rock, we must look to smaller objects: comets, asteroids, and small planetary satellites. Fragments of primitive material also reach the Earth in the form of some meteorites.

A piece of marble on Earth is composed of materials that have gone through all four of these stages: beginning as primitive material, it was heated in the early Earth to form igneous rock, subsequently eroded and redeposited (perhaps many times) to form sedimentary rock, and finally transformed several kilometers below the Earth's surface into the hard white metamorphic stone we see today.

(d) Planetary Interiors

The study of the interiors of planets is a difficult subject. Our knowledge of the interior of the sun is far more advanced than our knowledge of the interior of our own planet. It is easier to explore the surfaces of the other planets than to penetrate even a few kilometers toward the interior of the Earth.

There are indirect ways, however, to probe the interiors of our own and other planets. One of the simplest clues to the composition of the interior, for example, is given by a calculation of the density of the planet (mass/volume). Consider the five kinds of matter discussed in Section 12.3a. Ices have densities typically near 1.0 g/cm^3. Rocks are much more dense, usually in the range 2.5 to 3.5 g/cm^3. And metals are denser yet, often more than 8 g/cm^3. A knowledge of the density of a planet may be enough to estimate the relative proportions of these three components.

As an example, we can look at the densities of the Earth, the Moon, and the planet Pluto. The Moon's density is 3.3 g/cm^3, within the range for rock. Since we know the Moon is too warm to contain ice, we conclude that it is predominantly composed of rock, with at most a small metal core. The Earth's higher density (5.5 g/cm^3) indicates a mixture of rock and metal, consistent with the presence of a large iron-nickel core. (Incidentally, this difference in composition between the Earth and Moon is one of the great unsolved mysteries of the planetary system.) Finally, Pluto has a density of only about 1.5 g/cm^3, so it must be made primarily of ice, with some rock or metal.

The structure of the material in the interior of a planet can be probed in considerable detail if we can measure the transmission of *seismic waves* through it. Such waves are produced by natural earthquakes or by artificial impacts or explosions. The seismic waves travel through a planet rather like sound waves through a bell, and the response of the planet to various frequencies is characteristic of interior structure just as the sound of a bell reveals its size and construction. So far, only the Earth and Moon have been investigated by this technique, which has clearly shown the presence of a metal core on the Earth and the absence of any substantial core on the Moon.

Why should the Earth have a metal core? And more generally, why should any planet be separated into layers like the crust, mantle, and core structure of the Earth (Figure 12.9)? The answer goes back to the early history of the solar system, when the larger bodies in the planetary system *differentiated. Differentiation* is the name given to the process by which a planet organizes its interior into layers of different composition. It results from heating of the planet, either during its formation or subsequently by the release of energy through natural radioactivity. Once the planet becomes molten, the heavier metal tends to sink to form a core, while the lightest minerals float to the surface to form a crust. Later, when the planet cools, this layered structure is preserved. In order for a rocky planet to differentiate, it must be heated to the melting point of rocks, typically above 1100 K. But an object composed in large part of water ice will differentiate as soon as

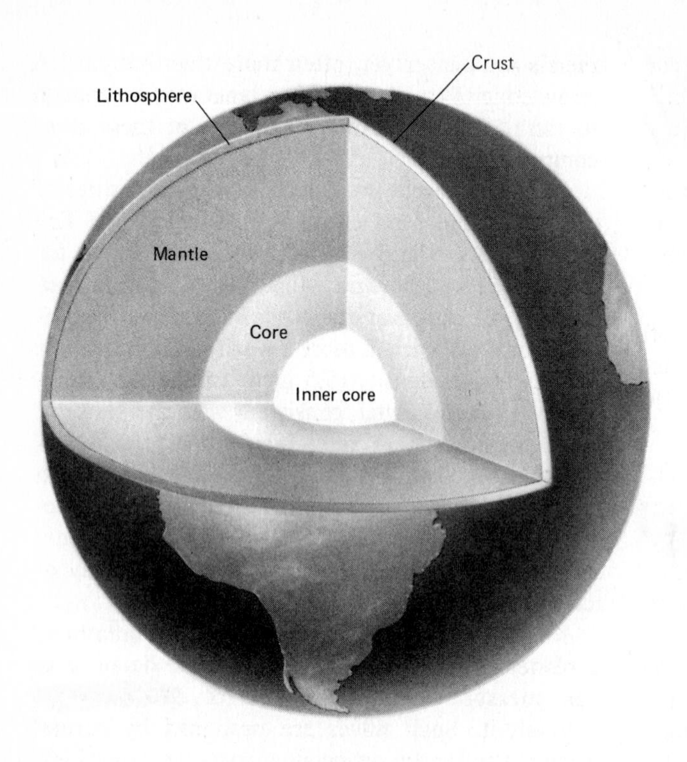

Figure 12.9 Structure of the Earth's interior.

its temperature rises above the melting point of water, at 273 K. Only very small objects are likely to have escaped differentiation.

12.4 Origin of the Solar System

The origin of the solar system is a complicated subject, involving the formation of both the sun and the planetary system that orbits it. Here, our emphasis is on the solid material that formed with the sun and evolved into the planets and other members of the system. In Chapter 30 we will discuss the more general question of star formation, and in Chapter 38, at the end of this book, we will return to questions of planetary formation in the context of a search for other planetary systems beyond our own.

(a) Observational Constraints

There are certain basic properties of the planetary system that any theory of formation should explain. These may be summarized under three categories: dynamical constraints, chemical constraints, and age constraints.

DYNAMICAL CONSTRAINTS

The planets all move around the sun in the same direction and approximately in the plane of the sun's own rotation. In addition, most of the planets share this same sense of rotation, and most of the satellites also move in counterclockwise orbits. With the exception of the comets, the members of the system define a disk shape. On the other hand, exceptions are possible in the form of retrograde rotation, like that of Venus, or of the roughly spherical shape of the cloud of comets that accompanies the sun.

CHEMICAL CONSTRAINTS

The planets Jupiter and Saturn have approximately the same *cosmic composition* as the sun and stars, dominated by hydrogen and helium. Each of the other members is to some degree lacking in the light elements. A careful examination of the composition of solar system objects shows a striking progression from the metal-rich inner planets through predominantly rocky materials out to ice-dominated composition in the outer solar system. The comets are also icy objects, while the asteroids represent a transitional rocky composition that is rich in dark, carbon-rich materials. This general chemical pattern can be interpreted as a temperature sequence, with the inner parts of the system strongly depleted in materials that do not form stable solids at high temperatures. Again, however, there are important exceptions to the general pattern. In particular, it is difficult to explain the presence of water on Earth and Mars if these planets had formed in a region where the temperature was too hot for ice to condense, unless the ice or water was brought in later from cooler regions.

AGE CONSTRAINTS

Radioactive age dating demonstrates that there are rocks on the surface of the Earth that have been present for at least 3.8 billion years and lunar samples that are 4.2 billion years old. In addition, the primitive meteorites—those that do not appear ever to have been melted—all have radioactive ages of 4.5 to 4.6 billion years. The age of these unaltered building blocks is considered the age of the planetary system. This age of about 4.6 billion years is in excellent agreement with theoretical calculations of the probable age of the sun, suggesting that the sun and planets formed together. These dates also tell us that the planets formed and their crusts cooled within a few hundred million years, at most, of the beginning of the solar system. Further, detailed examination of primitive meteorites indicates that they all represent material that condensed out of a hot gas; no identifiable fragments or grains survived from before this hot vapor stage 4.6 billion years ago.

(b) The Solar Nebula

All of the above constraints lead to the conclusion that the solar system formed 4.6 billion years ago out of a rotating cloud of hot vapor of approximately cosmic composition. This cloud is called the *solar nebula*. The general idea of such an origin appears to have been first suggested by the German philosopher Immanuel Kant (1724–1804), and it was developed into a specific model by the French astronomer Marquis Pierre Simon de Laplace (1749–1827) in his *Système du Monde* in 1796. The Kant-Laplace idea is known as the *nebular hypothesis*.

The modern concept of the solar nebula, with the detailed constraints outlined previously, dates from work carried out in the 1940s and 1950s by the German theoretical physicist Carl von Weizsacker (b. 1912), the Dutch-American astronomer Gerard P. Kuiper (1905–1973), and the American Nobel-prize–winning chemist Harold Urey (1893–1981). In Chapter 30 we will return to the question of how such a nebula might form in the first place. Here we begin our discussion with such a rotating nebula in place, surrounding a central condensation that would evolve into the sun.

As the solar nebula collapsed, it was heated by its own gravitational energy and its rotation speed increased as a consequence of the conservation of angular momentum (Figure 12.10).

As temperatures rose, any solid material that was originally present was vaporized. Its rapid spin caused the nebula to evolve into a disk shape, with most of the material confined to a thin spinning sheet. At the center, continuing collapse ultimately led to self-sustaining nuclear reactions, and the new star we call the sun was born. The existence of this disk-shaped rotating nebula explains the primary dynamical properties of the solar system as described in the previous section.

There is a great deal of difference of opinion today on the total mass of the solar nebula. Some calculations suggest that it might have been only a few percent more massive than the sun and planets; others believe that the total amount of material present was equal to as much as twice the current mass of the sun. In either case, the mass in the spinning disk was much larger than the total mass of the planetary system today, so that the material we see now is no more than a small remnant of the hot vapor originally present.

(c) Condensation and Accretion

Picture the solar nebula at the end of the collapse phase, when it is at its hottest. With no more gravitational energy to heat it, the nebula begins to cool. In the center, however, the newly formed sun keeps the temperatures up. The nebula therefore develops a *temperature gradient*, with the temperature varying with distance roughly as indicated in Table 12.2. As the nebula cools, the gases interact chemically to produce compounds, and eventually these compounds *condense* into liquid droplets or solid grains.

The *chemical condensation sequence* (Figure 12.11) in the cooling nebula was calculated in the 1970s by geochemists such as John Lewis of M.I.T.

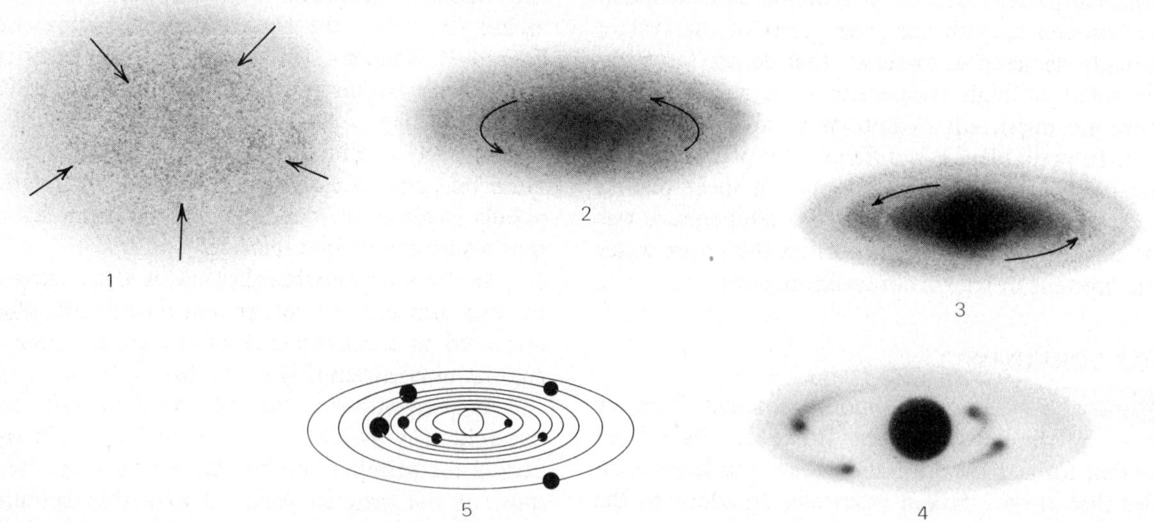

Figure 12.10 Schematic representation of the formation of the solar system. (1) The solar nebular condenses from the interstellar medium and contracts. (2) As the nebular shrinks, its rotation causes it to flatten, until (3) the nebular is a disk of matter with a concentration near the center, which (4) becomes the primordial sun. Meanwhile, solid particles condense as the nebula cools. These (5) accrete to form the cores of the planets. The pressure of radiation and the solar wind blow the solar system clean of most of the matter in the disk that did not form into planets. The five drawings are not to the same scale; the original solar nebula had to contract greatly before its rotation produced appreciable flattening.

and Laurence Grossman of the University of Chicago. The first materials to form grains are the metals and various rock-forming silicates and other minerals. As the temperature dropped, these were joined by sulfur compounds and by carbon- and water-rich silicates such as those now found abundantly among the asteroids. However, in the inner parts of the nebula the temperature never dropped low enough for these materials to condense, so they are lacking on the innermost planets. Still further out, where temperatures fell below about 200 K, the oxygen combined with hydrogen and condensed in the form of water ice. Beyond the orbit of Saturn, carbon and nitrogen combined with hydrogen to condense as additional ices such as methane (CH_4) and ammonia (NH_3). This chemical condensation sequence explains the basic age constraint that all primitive material seems to have formed at

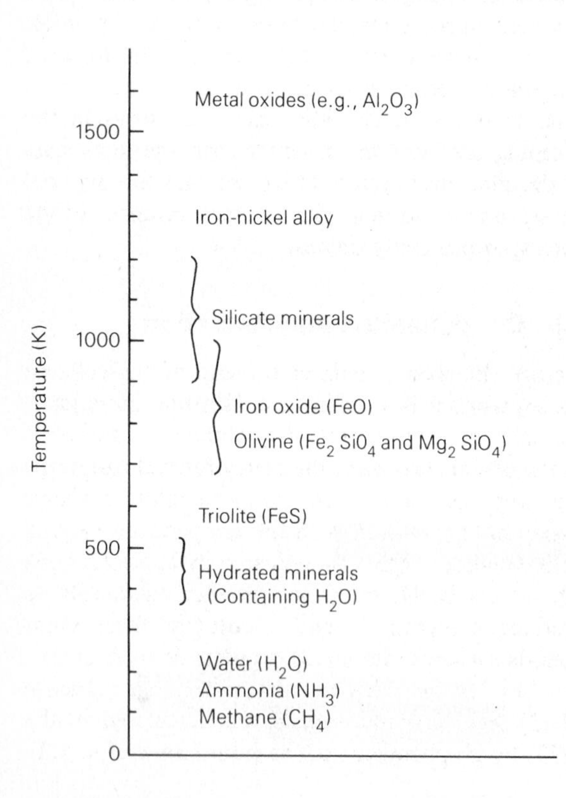

Figure 12.11 The chemical condensation sequence in the solar nebula, showing the primary chemical species that would be expected to form in a cooling gas cloud of cosmic (solar) composition under equilibrium conditions. (Adapted from diagrams published by John Lewis, University of Arizona)

the same time, and it also shows how the observed pattern of planetary compositions originated.

By the time the temperatures in the solar nebula approached the values given in Table 12.2, the nebula contained solid grains, sorted chemically by distance from the sun, mixed with the still-abundant hydrogen and helium gas. It is believed that these grains rather quickly formed into larger and larger aggregates, until most of the solid material was in the form of *planetesimals* a few kilometers to a few tens of kilometers in diameter. The Soviet theorist V. A. Safronov (b. 1917) studied the formation and properties of planetesimals in the late 1960s, and his theory of formation of the planets is sometimes called the *planetesimal hypothesis*.

Some planetesimals were large enough to attract their neighbors gravitationally and thus to grow by the process called *accretion*. While the intermediate steps are not well understood, ultimately there seem to have developed four large centers of accretion in the inner solar system, with perhaps several dozen more objects of at least the size of the Moon still whizzing about among them. In the outer solar system, where the building blocks included ices as well as silicates, much larger bodies grew, with masses of 10 to 20 times the mass of the Earth.

(d) Formation of the Giant Planets

In the inner solar system, the rocky objects continued to grow by accretion, but they had little interaction with the residual gas of the solar nebula. In contrast, the proto-planets of the outer solar system became so large that they were able to attract the surrounding gas. As the hydrogen and helium rapidly collapsed onto their cores, the giant planets were heated by the energy of contraction, just as the contraction of the solar nebula ultimately ignited the nuclear fires of the sun. But these giant planets were far too small to achieve the central temperatures and pressures necessary to initiate self-sustaining nuclear reactions. After glowing dull red for a few thousand years, they gradually cooled to their present state.

The collapse of nebular gas onto the cores of the giant planets explains how these objects came to have about the same hydrogen-rich composition as the sun itself. The process was most efficient for Jupiter, so that its composition is most nearly "cosmic." Much less gas was captured by Uranus and Neptune, which is why these two planets have compositions dominated by the icy building blocks that made up their large cores, rather than by hydrogen and helium.

Some time after the formation of the giant planets, the newly formed sun, like other very young stars, went through a stage in which it developed a very strong solar wind. Blasts of hot plasma flowed away from its atmosphere, sweeping through the remains of the solar nebula. Although this intense solar wind had little effect on the planets and other solid material, it interacted strongly with the gas still present, driving it out of the system. The solar nebula was dissipated, leaving a new star surrounded by a few planets and a disk of smaller bodies. As we will see in Chapter 38, the dissipation of nebulae by strong stellar winds and the presence of disks of solid material have all been observed in association with other young stars, so the processes we have described are not unique to our own system.

(e) The Dynamical Evolution of the System

All of the processes described above, from the collapse of the solar nebula to its dissipation by the solar wind, took place within at most a few million years, and possibly much faster yet. However, the story of the formation of the solar system is not complete at this stage—there remains the fate of the planetesimals and other debris that did not initially accumulate to form the planets.

Throughout the system, the remaining solid matter continued to interact gravitationally with the planets. Each time a fragment of debris came close to a planet, its orbit was altered, and eventually it found itself on a near-collision course with a larger object. These close encounters could result either in a direct impact on the planet or in ejection as the small body was accelerated by the gravitation of the large one. Calculations show that these two possibilities are about equally probable. Thus about half the material should have been ejected from the system on hyperbolic orbits and about half smashed into the surfaces of the planets. As much as several percent of the masses of the terrestrial planets may have come from this late stage of accretion.

The impacting material added to the planets during this phase could have come from almost anywhere within the solar system. Unlike the previous

Figure 12.12 Two rocks with very different ages. On the left, a three-week old Hawaiian basalt, produced as a result of our planet's intense geological activity; on the right, a 4.6-billion year old fragment of the Allende primitive meteorite, one of the earliest remnants of the solar nebula out of which the solar system formed. (David Morrison)

stage of accretion, therefore, this new material did not represent just a narrow range of compositions as specified by the initial temperatures in the solar nebula. Much of the debris striking the inner planets, for example, was ice-rich material that had condensed in the outer part of the solar nebula. As this comet-like bombardment progressed, the Earth accumulated the water and various organic compounds that would later be critical to the formation of life. Mars and Venus should also have acquired water and organic materials from the same source. Thus this late bombardment, bringing in material condensed throughout the planetary nebula, can explain the presence of water and other chemical peculiarities of the inner planets, properties that would otherwise contradict the theory of the chemical condensation sequence.

Gradually, as the planets swept up the remaining debris, most of the leftover planetesimals disappeared. In the region between Mars and Jupiter, however, there existed stable orbits where small bodies could avoid impacting the planets or being ejected from the system. The remnant of objects that survives in this special location is what we now call the *asteroids.* A much larger number of icy planetesimals in the outer solar system were gravitationally ejected, not completely, but into a large spherical volume of space surrounding the solar system. This distant reservoir now contains the *comets,* perhaps as many as ten million million objects, each typically a few kilometers in diameter. Because the comets were ejected into this cloud as a result of encounters with the giant planets rather than forming there originally, their orbits have a random distribution, and they alone of the members of the solar system do not retain any memory of the original rotation of the solar nebula. This is why new comets can arrive from any direction, as discussed in Chapter 20.

No one knows what the largest objects were that may have been available to impact the newly formed planets, but there is increasing evidence that they may have been comparable in size to the planet Mars. The impact of a Mars-sized object with the Earth or Venus would, of course, have been an extremely dramatic event. Indeed, such an impact would have released almost enough energy to break apart the target planet.

The evidence for such impacts is circumstantial. Remember that we mentioned the retrograde rotation of Venus as a dynamical constraint that would not be expected for a planet forming by the aggregation of planetesimals in a spinning solar nebula. Similar unexplained peculiarities include the rotational orientations of Uranus and Pluto, both of which are tipped on their sides. All of these departures from the general rotational symmetry of the solar system could be understood if the final stages of planetary accretion were marked by *giant impacts.* In effect, these final, essentially random events struck three of the nine planets so hard that their rotation axes were knocked on their sides or reversed. As we will see in Chapter 14, a similar argument has been made that a giant impact on Earth gave rise to our Moon.

(f) Summary

All of these violent events terminated by about 4.4 billion years ago, which is when the oldest rocks in the lunar crust solidified. Since then there has been a continuing sweep-up by the planets of remaining debris, but at a much slower rate than during the accretionary period. Planetary orbits have remained stable, and no errant lunar or Mars-sized objects remain to threaten planetary collisions. There was a particularly intense period of bombardment about 4 billion years ago (described in Chapter 14), but even this is minor in comparison with the events of the early history of the planetary system.

The foregoing account of the solar system's beginnings is probably close to what actually took place, but future research can be expected to fill in many details and show others to be incorrect. In addition, there are several problem areas not addressed at all in this outline of events, including the formation of the satellite and ring systems of the outer planets. An especially "hot topic" in current research deals with the origin of our Moon, and as this problem is addressed, the solutions may have much wider implications concerning the origins of other members of the solar system.

In Chapters 13 through 20 we will be looking in more detail at the individual members of the planetary system, and from time to time we will return to the questions of the origin and early evolution of our system. Later, when we examine the more general problem of star formation, we will see that all stars do not form the way we have described for the sun. Often the collapsing nebula develops two central condensations, giving rise to a double star; other times the cloud breaks up into whole clusters of stars. We don't know how often planetary systems form. At most it is only about half of the time, because half the stars around us are members of double-star systems, and planetary orbits are unstable in the presence of two stars. We return to the topic of other planetary systems in the final chapter of this book.

EXERCISES

1. From the data given in this chapter, show that Kepler's third law holds for Pluto.
2. What would be the period of a comet whose orbit has a semimajor axis of 10,000 AU?
3. What are the main distinctions between (a) the terrestrial planets, (b) the giant planets, and (c) the outer planet satellites (and Pluto)?
4. Why is density an important characteristic of a planet? What would you conclude if a planet were discovered around another star and found to have a density of 1.0 g/cm^3? Or 10 g/cm^3? Or 100 g/cm^3?
5. Suppose a planet were found orbiting a star that has 100 times the luminosity of the sun. If the distance of the planet from the star is 100 AU, what would you expect its characteristic temperature to be? At about what wavelength would you want to observe to detect thermal radiation from the planet?
6. Suppose a lunar rock is carefully age dated by measuring the parent and daughter elements in a variety of its mineral grains. The uranium-lead age is 3.2 billion years; the rubidium-strontium age is 3.3 billion years; and the potassium-argon age is 54 million years. When do you think the rock solidified from the molten state? Can you think of any reason why the different ages do not all agree?
7. The velocity of escape of Mars is only a little greater than that of Mercury. Why then does Mars have an appreciable atmosphere, while Mercury does not?
8. Suppose there were a planet 100 AU from the sun with mass and diameter similar to the Earth's. Would it be expected to have helium in its atmosphere? What about ammonia? Explain in each case.
9. Explain the difference between a rock and a mineral. Is ice a rock or a mineral? What about a piece of lava? Or a lunar breccia?
10. Planets of rock and planets with large amounts of ice can both differentiate. Describe the process of differentiation in each case. What are the main differences?
11. Describe the chemical building blocks that are thought to have been available in the grains that condensed from the solar nebula. If each planet formed in place from these grains, what would be the chemical composition of objects at 0.3 AU, 1.0 AU, 5.0 AU, and 25 AU from the sun?
12. What is the main difference between the giant planets and the terrestrial planets? How can this difference be understood in terms of the theory of the formation of the solar system described in this chapter?
13. How do we know when the solar system formed? Usually we say that the solar system is about 4.5 billion years old. What does this age correspond to? Are there parts of the solar system that might be substantially older than this?

Alfred Wegener (1880–1930), German meteorologist, suggested the idea of "continental drift" in 1912 from a study of geological similarities between the two sides of the Atlantic Ocean. His arguments, although not accepted then, are now regarded as precursors of the theory of plate tectonics—the most important development in Earth sciences of the 20th century. (Historical Pictures Service, Chicago)

13 EARTH AS A PLANET

The Earth, our own planet, is a special place. Only during the past four centuries have people even realized that the Earth is a planet, and the scientific comparison of the Earth with the other worlds of the solar system has taken place only during recent decades. Our planet has now become a critical benchmark for understanding other members of the system. In a complementary way, we also realize that the study of other planets—the subject often called *comparative planetology*—provides important insights concerning the workings of the Earth.

Of the dozens of planets and major satellites we have studied, only Earth has conditions suitable for life as we know it. But ours is a fragile environment, and not only do we have the power to alter it, we have already begun to do so in significant ways. The special conditions on Earth that make life possible, and also the ways in which life has itself modified our planet, are of particular interest to us. We cannot say what the future of our world as a habitable planet may be, but if we care about it, it behooves us to look carefully and attempt to understand its workings.

Many of the major properties of the Earth were described in the Chapter 12, and some of these are summarized in Table 13.1.

13.1 INTERIOR OF THE EARTH

(a) Composition

The Earth is a terrestrial planet, with a bulk composition of what we have called rock and metal. How do we know this? Our only direct experience is with the outermost skin of the Earth's crust, a layer no more than a few kilometers in depth. All of the information we have about the bulk properties of the Earth has been deduced indirectly. It is important to remember that we know less about our

TABLE 13.1 Properties of the Earth

Semimajor axis	1.00 AU
Period	1.00 yr
Mass	5.98×10^{27} g
Diameter	12,756 km
Density	5.5 g/cm^3
Uncompressed density	4.5 g/cm^3
Surface gravity	980 cm/s^2
Escape velocity	11.2 km/s
Rotation period	23h 56m 4.1s
Surface area	5.1×10^8 km
Atmospheric pressure	1.00 bar

Figure 13.1 The first view of the Earth taken by a spacecraft from the vicinity of the Moon; Lunar Orbiter 1 photograph. (NASA)

own planet a few kilometers beneath our feet than we do about the surfaces of Venus or Mars.

The surface rocks of the Earth have densities mostly in the range of 2.5 to 3.0 g/cm^3. However, the density of the planet as a whole is 5.5 g/cm^3. Because of this high overall density, we conclude that the interior of the Earth is very dense indeed, and therefore that its composition is probably quite different from that of the crust.

The above argument is not sufficient, however, to demonstrate that the interior composition of the Earth is different from that of the observable crust. The weight of the various layers of the Earth causes the pressure to increase inward, compressing the materials in the interior and increasing their density. It is necessary, therefore, to use experimental measurements of the physical properties of rocks to see if they can be compressed sufficiently to explain the apparent increase in density of the Earth with depth.

Such studies show that rock is *not* sufficiently compressible, and they lead us to an interesting property called the *uncompressed density* of the Earth. The uncompressed density, which is equal to 4.5 g/cm^3, is the density that an average piece of our planet would have if it were not under high pressure. It is the uncompressed density that should be compared with the densities of various materials to estimate composition. To achieve the *average* uncompressed density of 4.5 g/cm^3, the interior of the Earth must include high-density material as well as rock. Since the cosmically most abundant such material is metallic iron, we conclude that the interior of the Earth is enriched in iron and perhaps other metals relative to crustal rocks.

(b) Structure

The structure of the planet can be studied using the transmission of seismic waves through its interior. As we will see in Chapter 28, similar methods are being used to investigate the interior structure of the sun. On Earth, seismic waves are produced by earthquakes, which generate vibrations from the sudden slippage of parts of the crust. Some of these vibrations travel along the surface; others pass directly through the interior.

The study of these waves that originate in earthquakes is *seismology*. The seismic vibrations are recorded by delicate instruments and analyzed to determine the paths and velocities of the waves through the Earth's interior. Such studies have shown that most of the interior of the Earth is solid, and that it consists of several distinct layers of different composition.

The major part of the Earth is called the *mantle,* which stretches from the base of the crust down to a depth of 2900 km. The density in the mantle increases downward from about 3.5 g/cm^3 to more than 5 g/cm^3; however, its uncompressed density is everywhere about 3.5 g/cm^3. Its composition is believed to be igneous silicate rocks. Samples of upper mantle material are occasionally ejected from volcanoes, permitting a detailed analysis of its chemistry.

Above the mantle is the *crust,* the part of the Earth we know best. The oceanic crust, which covers 55 percent of the surface, is typically about 8 km thick and is composed of basalts. The continental crust, which covers 45 percent of the surface, is from 20 to 70 km thick and is predominantly granitic. On both ocean and continents, these igneous rocks are often buried by sedimentary and metamorphic rocks produced by weathering and erosion of the surface material. The crust makes up only about 0.3 percent of the mass of the Earth.

Within the mantle is the *core* of the Earth, a high-density region with a diameter of 7000 km, substantially larger than the planet Mercury. The outer part of the core acts like a liquid, for it does not transmit certain kinds of seismic waves. The innermost part of the core (about 2400 km in diameter) is extremely dense and probably solid. The primary constituent of both parts of the core is believed to be iron, probably also containing substantial quantities of nickel, sulfur, and other cosmically abundant elements.

At the base of the mantle, the temperature is 4500 K and the pressure is 1.3 million bars (where one *bar* is defined as 10^6 dyne, or approximately the atmospheric pressure at the surface of the Earth). At the center of the planet, the pressure is nearly 4 million bars, but the temperature is about the same, approximately 5000 K. The primary heat source in the interior is the decay of radioactive elements in the mantle and crust. In the mantle, this heat is carried upward by *convection,* the slow movement of currents of hotter material. In the crust, heat escapes upward to the surface by *conduction* through the solid rock or by the release of molten lava in volcanic eruptions. We will examine some consequences of this release of energy from the interior in Section 13.2.

(c) The Earth's Magnetism

Additional clues concerning the interior are provided by the Earth's magnetic field. This field is similar to that produced by a bar magnet. Nearly everyone is familiar with the way iron filings align themselves along the lines of force that extend between the north and south poles of such a magnet. Between the magnetic poles of the Earth stretch similar lines of force along which compass needles align. The magnetic poles, however, are not coincident with the rotational poles, but are tilted by a few degrees. The overall strength of the Earth's magnetic field is fairly weak, averaging about one-half gauss at the surface.

The Earth's magnetism results from electric currents moving in the core of the planet. Being composed of metal, the core is electrically conducting. The rotation of the Earth generates slow motions in the metallic core that act like a giant dynamo, generating the observed field. Because these motions are turbulent, the strength and alignment of the field varies, and the magnetic poles wander about with respect to geographic position.

In addition, the polarity of the field reverses itself completely from time to time. We detect these changes in polarity from the direction of magnetism preserved in igneous rocks. When the rock is molten, its iron-bearing compounds are weakly magnetized by the Earth's field; when it solidifies, this alignment is "frozen in." We measure the strength and polarity of magnetism in rocks of different ages to trace the magnetic history of the Earth, and find that its field has reversed polarity about 100 times in the past 50 million years. The cause of these field reversals is not well understood.

The Earth's magnetic field extends into surrounding space. Above the atmosphere, this field is able to trap small quantities of plasma, creating the Earth's *magnetosphere* (Figure 13.2). The magnetosphere is defined as that region surrounding the planet within which our magnetic field dominates over the weak interplanetary field that is carried along with the solar wind. Its size depends on the strength of the Earth's field relative to the strength of the solar wind, so it can expand or contract depending on the current value for the Earth's field strength and the level of solar activity. Typically, the Earth's magnetosphere extends about 60,000 km, or ten Earth radii, in the direction of the sun.

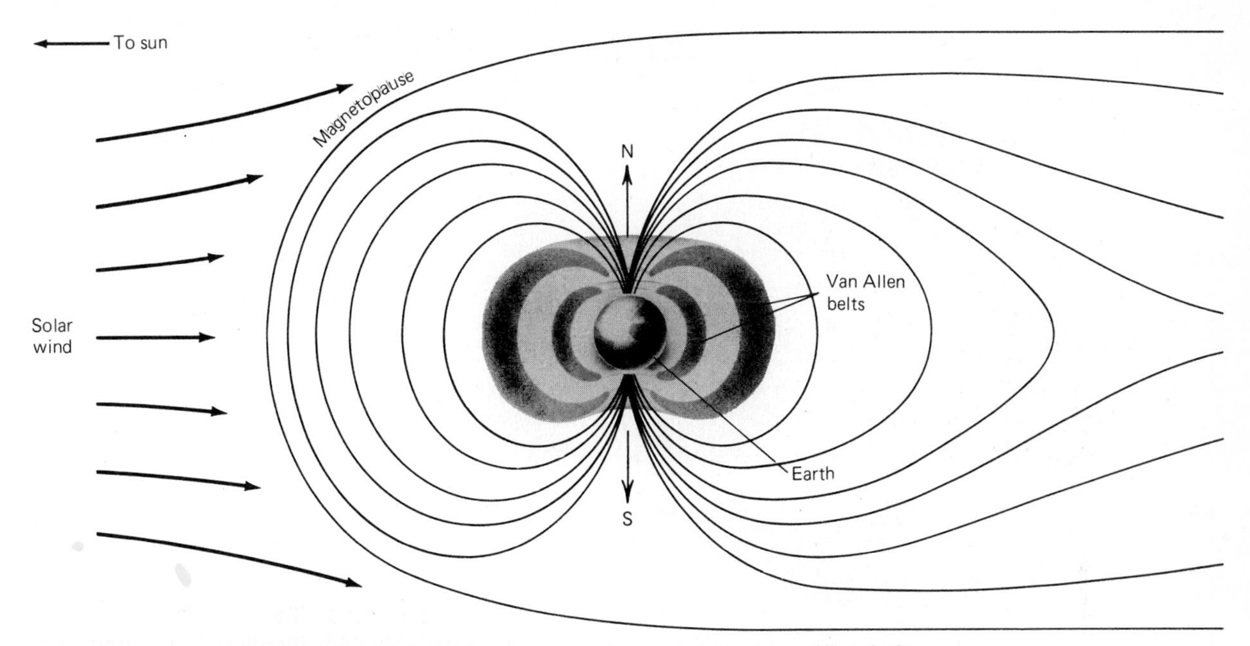

Figure 13.2 Cross section of the Earth's magnetosphere and the Van Allen belts, as revealed by numerous spacecraft.

It is shaped like a wind sock pointing away from the solar wind. In the downstream direction the magnetosphere can reach as far as the orbit of the Moon.

The existence of the magnetosphere was discovered in 1958 by instruments on the first U.S. Earth satellite, Explorer 1, which recorded the plasma trapped in its inner part. The regions of high-energy trapped plasma are often called the *Van Allen belts* in recognition of the State University of Iowa professor who built the scientific instrumentation for Explorer 1 and correctly interpreted the satellite measurements. Since 1958, hundreds of satellites have explored various regions of the magnetosphere, and a scientific discipline called *space plasma physics,* which is devoted to understanding magnetospheric phenomena, has developed. We will discuss planetary magnetospheres in more detail in Chapter 17, when we look at the much larger magnetospheres of Jupiter and Saturn.

13.2 THE CRUST OF THE EARTH

(a) Geological Processes

The study of the Earth's crust and the processes that modify it is called *geology*. Most geological structures can trace their origin to the effects of heat escaping from the interior of the planet. Geologists call these *endogenic* processes, meaning that their source is from the inside. The dominance of endogenic geological processes is a special property that sets the Earth apart from most other planetary bodies, which are much more influenced by *exogenic,* or external processes. The most important exogenic features on other planets are *impact craters,* produced by collisions with comets, asteroids, or other space debris.

Impact craters are also produced on Earth. If the Earth's level of internal geological activity were as low as that of the Moon, for instance, it would have as many impact scars as does our cratered satellite. The primary reason there are few identifiable impact craters on Earth is its high level of endogenic geological activity. This activity has destroyed all evidence of craters produced early in the history of the planet, while even younger craters are pretty well erased within a few million years of their formation by a combination of weathering and endogenic forces.

The most prominent impact crater on the Earth is Meteor Crater, a mile-wide scar in the arid plains of northern Arizona. Meteor Crater was produced 50,000 years ago by the impact of an iron meteroid

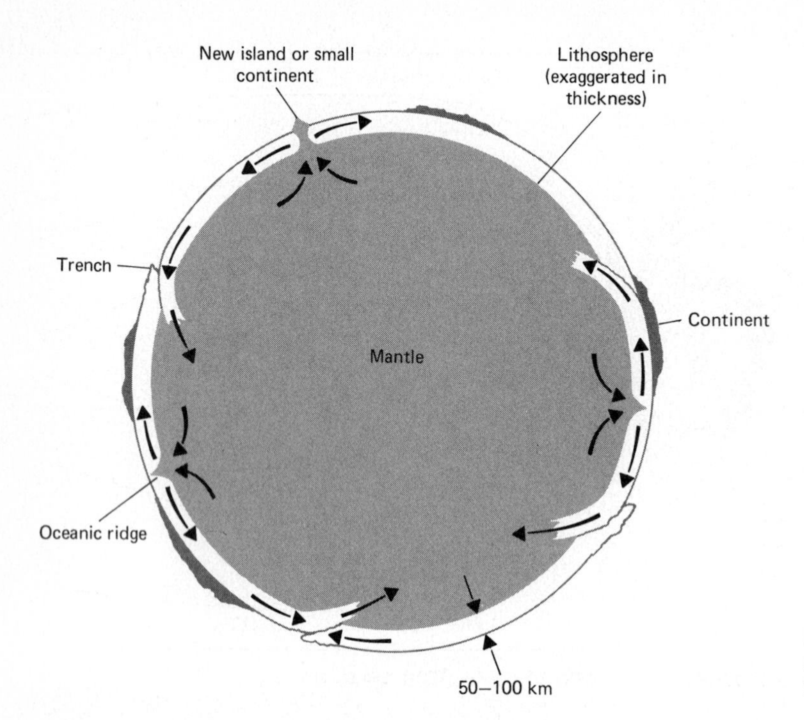

Figure 13.3 The lithosphere (of much exaggerated thickness), showing the emergence and immersion of plates.

with a mass of about a million tons, creating an explosion equivalent to a 20-megaton nuclear bomb. Larger craters are older and more eroded. Altogether, several hundred craters have been found on Earth, but most are so badly eroded that they can be identified only by an expert geologist.

In contrast, we see around us a multitude of landforms produced by endogenic processes: mountains, valleys, volcanoes, even the continents and ocean basins themselves. We will defer the detailed discussion of impact cratering to the next chapter, when we investigate the Moon, and concentrate here on the internally driven geology that is so characteristic of our planet.

(b) Plate Tectonics: A Geological Revolution

Lots of schoolchildren, in studying maps or globes of the Earth, notice that North and South America, with a little juggling, look as if they could almost be nestled up against Europe and Africa; it is as if these great land masses were once together but somehow tore apart. The same idea also occurred to the German meteorologist Alfred L. Wegener (1880–1930) early in the 20th century. He looked at the matter in considerable detail, making a good empirical case that the continents had drifted apart, basing his arguments on detailed geological similarities between the east and west shores of the Atlantic Ocean. At that time, however, the mantle and crust of the Earth were known to be solid, and Wegener could propose no plausible mechanism by which the continents could move. His ideas of *continental drift* were therefore dismissed and even ridiculed by nearly all of the scientific community, even though additional evidence accumulated over the following decades that pointed to connections tens of millions of years ago between land masses now widely separated.

Wegener had proposed that the continents somehow moved through the fixed, underlying basaltic crust. Continental drift could only be accepted when a new theory of the crust of the Earth was developed in the 1960s called *plate tectonics*. In this theory, we recognize that the oceanic crust also moves, driven by the slow convection currents that transport heat within the mantle. The crust and upper mantle (to a depth of about 60 km) are divided into about a dozen major *plates* that fit together like the pieces of a jigsaw puzzle, but that are also capable of moving slowly with respect to each other. This mobile part of the Earth is called the *lithosphere* (Figure 13.3)

The direct evidence for moving lithospheric plates was first supplied in the Atlantic Ocean,

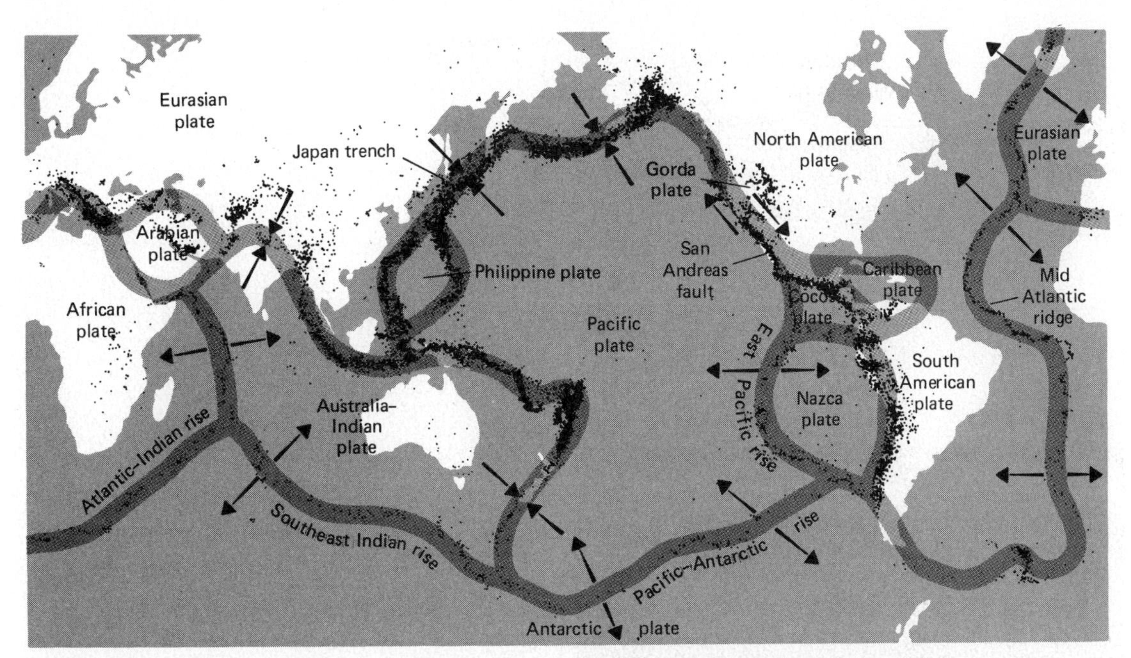

Figure 13.4 Tectonic plates on the Earth. The dots indicate regions of seismic activity, where the boundaries of the plates generally lie. The major plates that have been identified are labeled, and the arrows indicate the direction of motion of the plates.

where geologists demonstrated that fresh lava was being injected along the Mid-Atlantic Ridge, a line of volcanic mountains running approximately north and south along the center of the ocean basin. At the same time, they found that on either side of the ridge, the ocean floor was gradually separating at a speed of a few centimeters per year. At this rate, the entire Atlantic Ocean, which is about 4000 km (400 million cm) across, could have formed within 100 million years, in good agreement with the evidence for geological continuity across the Atlantic accumulated by Wegener and others. Very quickly, other plate boundaries were located where similar sea-floor spreading was taking place, and a great many previously unconnected geological phenomena were seen to make better sense in the context of the new plate tectonics (Figure 13.4). Within a few years, a fundamental change in our geological perspective took place—a true "scientific revolution."

But how can a system of interlocking plates covering the entire surface of the Earth move about? If plates in the Atlantic and elsewhere are spreading apart, they must be jamming together somewhere else. The analysis of the interactions of moving plates provides the basis for understanding most large-scale geological activity on Earth.

Four basic kinds of interactions between plates are possible: (1) they can pull apart (2) they can slide alongside each other, (3) they can jam together, or (4) one plate can burrow under another. Each of these activities is important in determining the geology of the Earth.

(c) Rift Zones

Plates pull apart from each other along *rift zones*, such as the Mid-Atlantic Ridge, driven by upwelling convection currents in the mantle (Figure 13.5). A few rift zones are found on land, the best known being the central African rift, an area in which the African continent is slowly breaking apart. Most rift zones, however, are in the oceans. The new material that rises to fill the space between the receding plates is basaltic lava, the kind of igneous rock that forms most of the ocean basins.

The clinching evidence for the existence of rift zones was derived during the 1960s from studies of the ocean floor on either side of the Mid-Atlantic rift. Samples of sea-bottom basalt displayed a clear sequence of ages, youngest at the ridge and increasing in age with distance east or west. This pattern is made even clearer when the magnetism of the Atlantic basalts is measured. Consider what happens

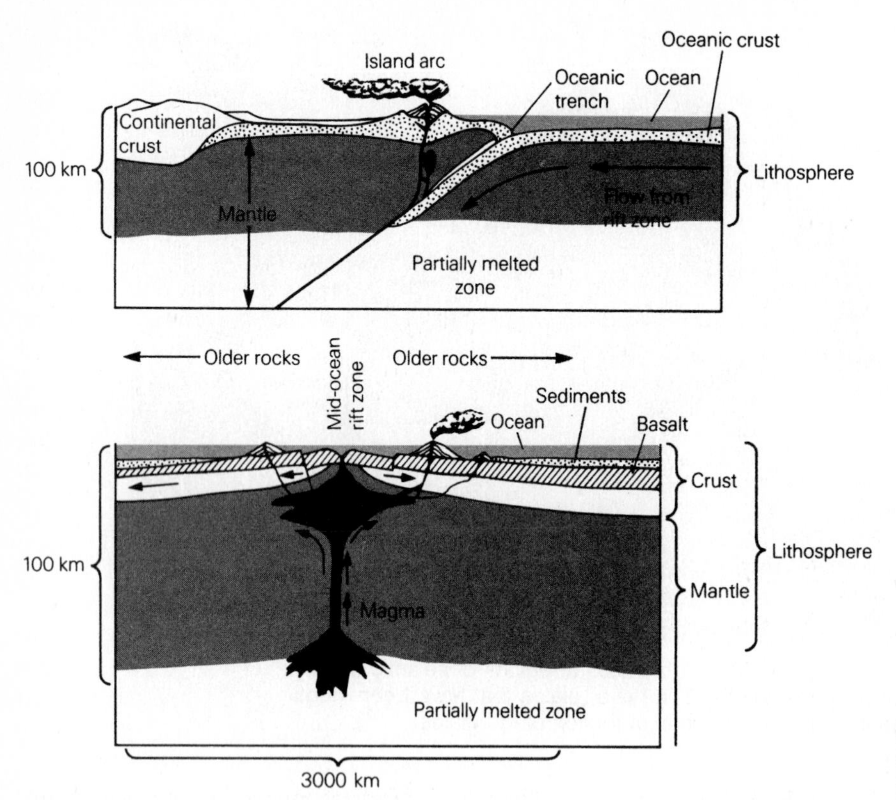

Figure 13.5 Rift zones and subduction zones in the Earth's crust.

when molten rock is injected along the mid-ocean rift. It spreads out, hardens, and becomes part of the spreading plates. As it hardens, it preserves a record of the magnetism of the Earth; however, as we have seen, the Earth's magnetic field reverses periodically. By measuring the magnetism of the rocks on the ocean floor with delicate magnetometers towed behind ships, we can see how the polarity switches back and forth many times as the ship moves away from the Mid-Atlantic Ridge, crossing over basalts of steadily greater age.

Considerable heat is released at the mid-ocean rifts, not all of it in the form of volcanic eruptions. One of the most dramatic discoveries about our planet in recent years has been the existence of colonies of remarkable lifeforms that cluster around the hot, mineral-laden springs that result when seawater circulates through the superheated rift rocks. These hydrothermal vents in the dark ocean depths represent one of the few places on our planet where life has learned to obtain its nutrients and energy directly from geological activity, without dependence on sunlight or the organic products of other living creatures.

From a knowledge of sea-floor spreading, we can calculate the average age of the oceanic crust. About 60,000 km of active rifts have been identified, with average separation rates of about 4 cm per year. The new area added to the Earth each year is therefore 60,000 km × 0.00004 km/yr or just over 2 square kilometers. The total area of ocean crust is 260 million km^2. Dividing the total area by the new area added each year, we obtain about 100 million years as the average age of the crust. This is a very short interval in geological time, less than 3 percent of the age of the Earth. The present oceans are among the youngest features on the planet!

(d) Fault Zones

Along much of their lengths, the crustal plates slide along parallel to each other. These plate boundaries are marked by cracks or *faults*. Along active fault zones, the motion of one plate with respect to the other is several centimeters per year, about the same as the spreading rates along rifts.

One of the most famous faults is the San Andreas Fault, lying on the boundary between the Pa-

Figure 13.6 The San Andreas Fault, a very active region where one crustal plate is sliding sideways with respect to the other. The fault is marked by the valley running up the center of this photo; the dark line to the left is tumbleweeds piled along a fence line. (USGS)

cific Plate and the North American Plate. This fault runs from the gulf of California in the south in the Pacific Ocean just west of San Francisco in the north (Figure 13.6). The Pacific Plate, to the west, is moving northward, carrying Los Angeles, San Diego, and parts of the southern California coast with it. In several million years, Los Angeles will be an island off the coast of San Francisco.

Unfortunately for us, the motion along most fault zones does not take place smoothly. The creeping motion of the plates against each other builds up stresses in the crust that are released in sudden, violent slippages, generating *earthquakes*. Since the average motion of the plates is constant, the longer the interval between earthquakes, the greater the stress and the larger the energy released when the surface finally moves. For example, the part of the San Andreas Fault near the central California town of Parkfield has been sliding about every 22 years, moving an average of about 1 meter (5 cm/yr × 22 years). In contrast, the average interval between major earthquakes in the Los Angeles region is about 140 years, and the average motion is about 7 meters. The last time the San Andreas slipped in this area was in 1857; tension has been building ever since, and sometime soon it is bound to be released.

Charles Richter (of the Richter scale of earthquake size) once said: "Some people have an irrational fear of cats, and others an irrational fear of earthquakes; the former should not have cats for pets, and the latter should not live in California." Yet millions of people do live in California, and the San Andreas Fault passes through both Los Angeles and San Francisco. On the other hand, some Californians relish the opportunity to live where they do, arguing that it is exciting to witness the grand changes that shape the face of the Earth. If we could see a time-lapse motion picture of the Earth's surface, with millions of years collapsed to a few seconds, we would find it a tremendously alive place, with its entire surface constantly rearranging itself; the motions of plates at this scale would seem continuous. It is only on our tiny time scale, confined to barely more than a snapshot of the Earth's cosmic existence, that things seem so abrupt and chaotic. In witnessing an earthquake, we are witnessing a tiny sliver of the evolution of the Earth.

(e) Mountain Building

When two continental masses are brought together by the motion of the crustal plates, they are forced against each other under great pressure. The Earth buckles and folds, forcing some rock deep below the surface and raising other folds to heights of many kilometers. This is the way most of the mountain ranges on Earth were formed; as we will see, however, quite different processes produced the mountains on other planets.

The highest mountain range on Earth, the Himalayas, is still being formed as the Indian subcontinent is forced against the Asian mainland. India was formerly a large island near Australia in the Indian Ocean, but during the past few million years it has been forced up against the Eurasian Plate. The

Figure 13.7 The Alps, a young region of the Earth's crust where sharp mountain peaks are being sculpted by glaciers. (David Morrison)

Alps are the result of the African Plate bumping into Europe (Figure 13.7). In North America, the Rocky Mountains probably also are the product of pressure between two plates, but this boundary is not very active at present, and there is some debate among geologists concerning exactly what is happening here.

At the same time a mountain range is being formed by upthrusting of the crust, its rocks are subject to the erosional force of water and ice. The sharp peaks and serrated edges that are characteristic of our most beautiful mountains have little to do with the forces that make them, but are instead the result of the processes that tear them down. Ice is an especially effective sculptor of mountains. In a planet without moving ice or running water, mountains will tend to remain smooth and dull. This is exactly the case on our Moon, as we will see in the next chapter.

(f) Subduction Zones

The final way that moving plates can interact is by one sliding under another. The region where one plate dives down beneath another is called a *subduction zone* (Figure 13.5). Generally, continental masses cannot be subducted, but the thin oceanic plates can be rather readily forced down into the upper mantle. Often a subduction zone is marked by an oceanic trench, a fine example being the deep Japan Trench along the coast of Japan. Approximately the same total area—2 square kilometers—is subducted each year as is created by sea-floor spreading.

The subducted plate is forced down into regions of high pressure and temperature, eventually melting several hundred kilometers below the surface. Its material is recycled back into a downward-flowing convection current, ultimately balancing the material that rises along rift zones.

A part of the subducted material reaches the surface more directly through volcanic eruptions. All along the subduction zone, earthquakes and volcanoes mark the death throes of the plate. Some of the most destructive earthquakes in history have taken place along subduction zones. These include the 1923 Yokahama earthquake and fire, which killed 100,000; the 1966 quake that leveled the Chinese city of Tangshan and killed more than half a million people; and more recently the 1985 Mexico City quakes with their toll of more than 10,000 lives.

(g) Volcanoes

Terrestrial volcanoes mark the locations where molten rock, called *magma,* rises to the surface. One example is the mid-ocean ridges, which are volcanic features formed by mantle convection currents at plate boundaries. A second, major kind of volcanic activity is associated with subduction zones, and volcanoes sometimes also appear in regions where continental plates are colliding.

Another location for volcanic activity on our planet is found above so-called *mantle hot spots,* areas far from plate boundaries where heat is nevertheless rising from the interior of the Earth. Perhaps the best known such hot spot is under the island of Hawaii, where it supplies the energy to maintain three currently active volcanoes, two on land and one under the ocean. The Hawaii hot spot has been active for at least a hundred million years, and it has generated an entire 3500-km–long chain of volcanic islands (Figure 13.8). The tallest Hawaiian volcanoes are among the largest single volcanic features on Earth, up to 100 km in diameter and rising 9 km above the ocean floor; however, as we shall see, other planets have volcanoes that considerably surpass these size records.

The fact that the Hawaiian volcanoes form a long chain of islands rather than one gigantic vol-

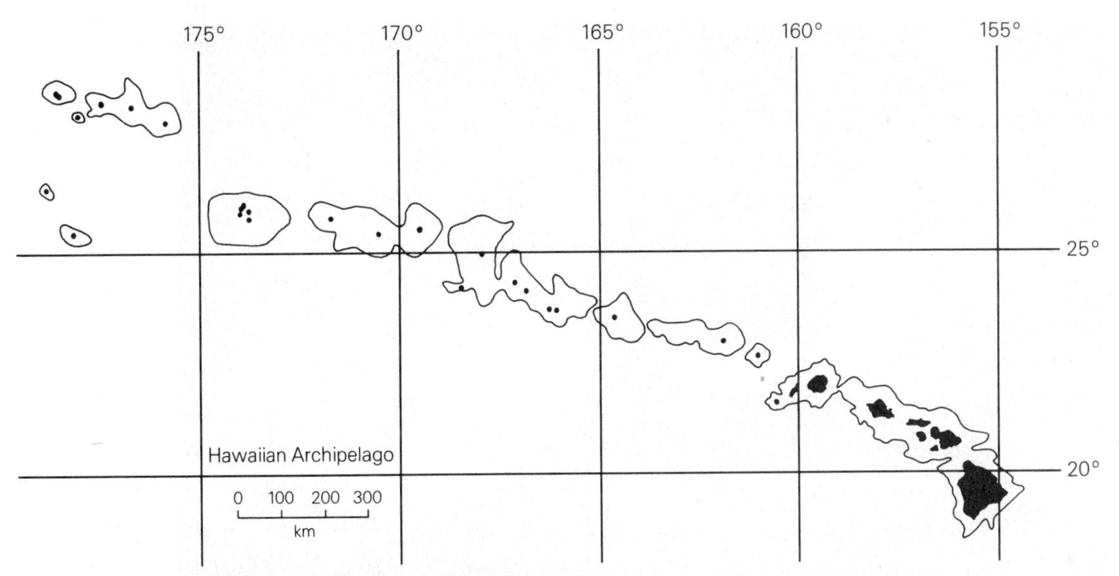

Figure 13.8 The Hawaiian Islands, a chain of volcanic mountains formed during the past 100 million years by movement of the Pacific plate over a mantle hot spot. The figure shows islands and the contour at 10,000-foot depth. Three volcanoes are currently active at the southeast end of the chain.

cano is the direct result of the motion of the Pacific plate, which slides over the hot spot at a rate of several centimeters per year. In one or two million years, the plate moves a distance equal to the average separation of the Hawaiian islands; thus, the island chain represents a time sequence, with each major island a few million years older than the next. A couple of million years from now, all the currently active volcanoes will have fallen silent, and a new volcanic island will be forming to the southeast of the present island of Hawaii.

Volcanic eruptions on Earth can take several forms. The Hawaiian volcanoes (Figure 10.14) are called *shield volcanoes* because of their shape, characterized by long, gradual slopes built up by successive flows of fluid lavas (Figure 13.9). In contrast, there are other more explosive volcanoes (Fuji in Japan, Vesuvius in Italy, and Mt. St. Helens in

Figure 13.9 Highly fluid basaltic lava that characterizes the thin overlapping flows of shield volcanoes. (David Morrison)

Figure 13.10 Paricutin, a cone-shaped volcano in Mexico formed by fire-fountains of lava erupted under great pressure. (Tad Nichols)

Washington State are well-known examples) that are characterized by steep-sided *cinder cones* created by the fall-back from violent fire-fountains of molten magma (Figure 13.10). Both shield volcanoes and cinder cones have been identified on other planets as well.

A third major type of volcanism does not produce a volcanic mountain at all, but rather lays down vast plains of basalt called flood basalts. These eruptions are the largest in terms of volume of lava produced, and they also have their counterparts on other planets, particularly in the formation of the lunar lava plains. A terrestrial example of such an eruption is the 60-million-year-old Columbia River flood basalts that cover eastern Washington State and have a total volume estimated at 400,000 cubic kilometers.

(h) Continental Drift

One consequence of plate tectonics is the constant rearrangement of the continental masses. Figure 13.4 is a map of the world showing the outlines of the major plates. The many tiny dots along those boundaries (and a few elsewhere) are the sites of moderate and major earthquakes in recent years, while the arrows show the location of major rift and subduction zones.

If we extrapolate the plates' motions backward we arrive, some 200 million years ago, at the supercontinent of *Pangaea*. At that time the continents of the Earth were all together, as shown in Figure 13.11. The map may not be absolutely accurate but shows how we believe the continents lined up. Separation along many of the presently active rift zones began between 150 and 200 million years ago, although new rifts (such as the African Rift) continue to develop.

The configurations of the continents before Pangaea is less well understood. It is clear that plate tectonics were at work, from the existence of old mountains (such as the Urals in the U.S.S.R. and the Appalachians in the U.S.) that were formed more than 200 million years ago. Recent studies indicate that plate tectonic activity in what is now North America can be traced back 1.5 billion years.

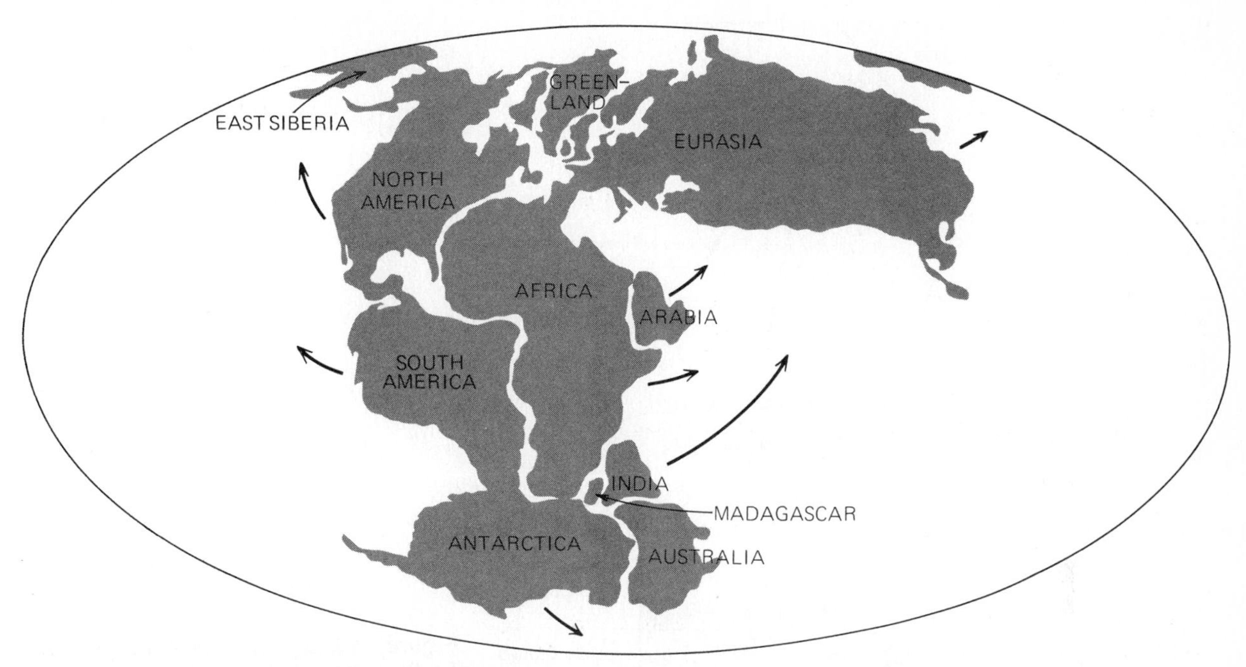

Figure 13.11 A map of the Earth approximately as it appeared 150 million years ago. At that time the continents were clustered together in a supercontinent called Pangaea.

We are not certain, however, if plate tectonics have been a part of the Earth's geology for all of its existence. Beyond a billion years ago, a mere 20 percent of our planet's history, most of the geological record becomes sparse and confused. A few continental rocks survive from earlier epochs, but in general the crust of the Earth is very young on the cosmic time scale, and we must look to other planets for records of the earlier history of the solar system.

(i) Erosion and Sedimentation

While most terrestrial landforms can ultimately trace their origin to the deep internal processes discussed above, their details are often the result of local forces of weathering and erosion. Water, wind, and ice all modify the surface of our planet, as does the chemical interaction with the atmosphere and oceans. In particular, almost all sharply sculpted features, from mountain peaks to deep canyons, are the product of water and ice erosion.

Water and ice do their work in arid climates as well as wet ones. Indeed, the lack of vegetation in arid lands often makes the ground more susceptible to erosion when occasional big rainstorms or floods occur. Do not be surprised, therefore, that the spectacular rock formations of the American Southwest or the Arabian Desert are primarily the result of water erosion. Wind rarely has a major effect on rock, although it can create beautiful landscapes of its own in the sand dunes of the great sand seas of the world.

The fine material eroded away from the mountains of the Earth or produced by chemical weathering of the crust must go somewhere. A part ends up in the desert sand seas, but most is carried by water into low-lying areas or into the sea itself. These *sediments* cover much of the igneous rock on both the continents and the oceans, and eventually they fuse into new sedimentary rock. Typically, the sedimentary deposits on the ocean floors are more than a kilometer thick. Of course, this sediment is recycled along with the oceanic basalt in the subduction process, some to replenish the mantle and some to reappear on the surface by volcanic eruption.

13.3 THE EARTH'S ATMOSPHERE

(a) Structure of the Atmosphere

We live at the bottom of the ocean of air that envelops our planet. The atmosphere, weighing down upon the surface of the Earth under the force of

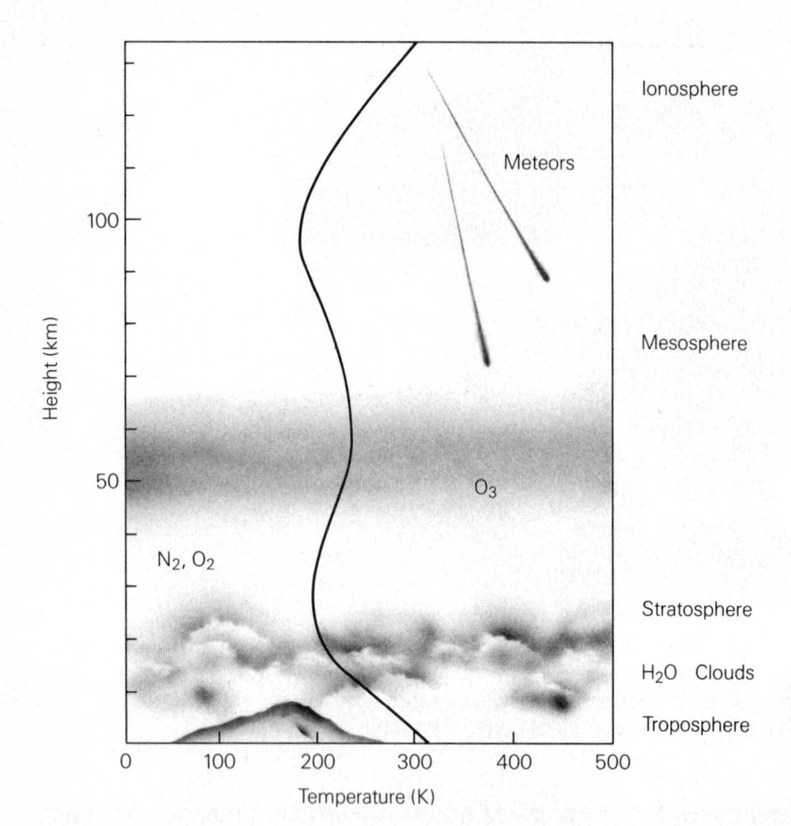

Figure 13.12 The structure of the Earth's atmosphere. The curving line shows the temperature measured at various altitudes. Temperature is higher where sunlight is absorbed, in the ozone layer and in the ionosphere.

gravitation, exerts a pressure at sea level of one *bar*, which equals the weight of 1.03 kg over each square centimeter. If the mass of the air over 1 square centimeter is 1.03 kg, the total mass of the atmosphere may be found by multiplying this figure by the surface area of the Earth in square centimeters. We find, thus, that the total mass of the atmosphere is about 5×10^{15} tons, or about one-millionth of the total mass of the Earth.

At higher and higher altitudes the air thins out more and more until it disappears into the extremely sparse gases of the magnetosphere, at an altitude of several hundred kilometers (Figure 13.12). In the thin upper reaches of the atmosphere we can observe *auroras* or polar lights, which are electrical discharges caused when ions and electrons from the magnetosphere bombard the upper air. Most auroras are found to occur at heights of from 80 to 160 km, but a few are as high as 1000 km (Figure 13.13).

Most of the atmosphere is concentrated near the surface of the Earth, within about the bottom 10 kilometers. That is where clouds form and airplanes fly. This region, called the *troposphere,* is characterized by convection currents produced as warm air, heated by the surface, rises and is replaced by descending currents of cooler air. The convection generates clouds and other manifestations of weather. Within the troposphere, temperature drops rapidly with increasing elevation to values near 50°C below freezing at its upper boundary. On other planets, the term troposphere is used as on Earth, to designate the lower, convective part of the atmosphere.

Above the troposphere, and extending to a height of about 80 km, is the *stratosphere.* Most of the stratosphere is cold and free of clouds, but in its upper part the temperature rises again. The hot layer is due to the presence of *ozone* in that level of the atmosphere. Ozone (abreviated O_3) is a heavy form of oxygen, having three atoms per molecule instead of the usual two. It has the property of being a good absorber of ultraviolet light. In absorbing the sun's short-wavelength ultraviolet light, the ozone is heated up and warms the parts of the atmosphere where it is present. Incidentally, the protective ozone layer helps to prevent some of the sun's dangerous ultraviolet radiation from penetrat-

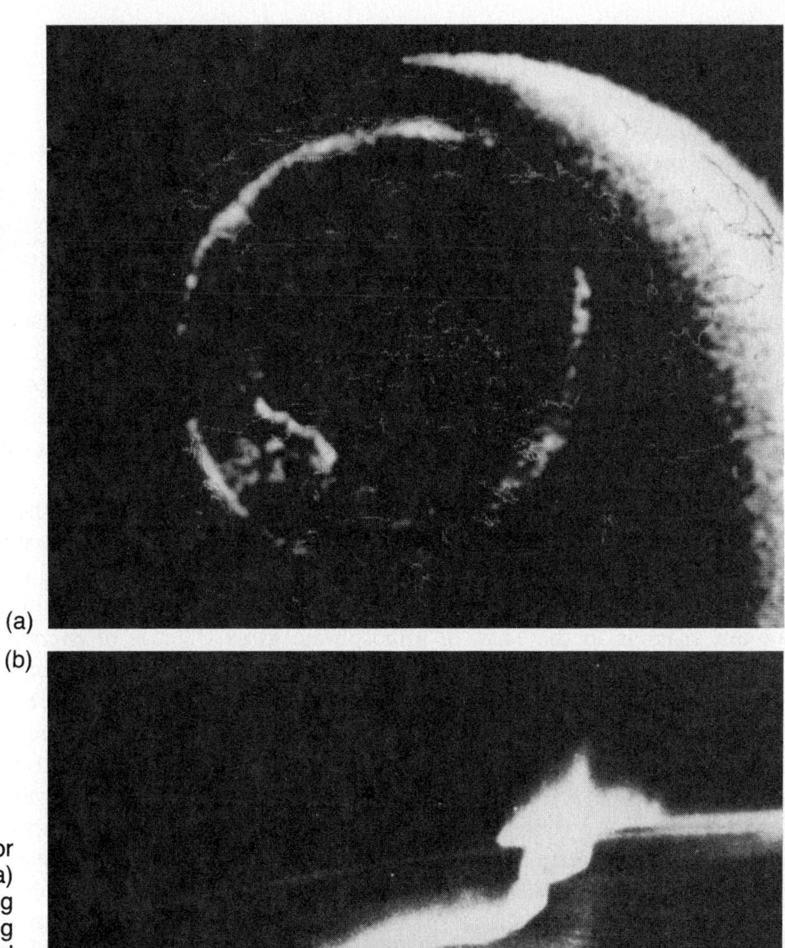

Figure 13.13 The Earth's aurora, or northern lights, as seen from space. (a) View from the Dynamics Explorer during Northern-Hemisphere winter, showing that most auroral activity is concentrated in a narrow band encircling the North Magnetic Pole (NASA/L. Frank, Iowa State University). (b) Detailed view from Earth orbit showing a single curtain of auroral activity. (NASA)

ing to the Earth's surface. The region of the stratosphere extending upward from the ozone layer is sometimes called the *mesosphere*.

From 65 to 80 km, the temperature drops to below −50°C again. Above 80 km the temperature rises rapidly through a region called the *thermosphere*, and at 400 to 500 km reaches values above 1000°C. The highest layer of the atmosphere, above about 400 km, is called the *exosphere*. In the upper atmosphere, the thermosphere and exosphere, molecules of oxygen and nitrogen break up into individual atoms of those elements. Ultraviolet radiation from the sun ionizes many of these atoms. Therefore, part of the upper part of the atmosphere (above 80 km) is also called the *ionosphere*. Because it is electrically conducting, the ionosphere reflects long-wave radio, making it possible to receive AM radio broadcasts over very large areas.

(b) Composition

The chemical composition of the Earth's atmosphere is determined by quantitative chemical analysis. At the Earth's surface, the constituent gases are 78 percent nitrogen, 21 percent oxygen, and 1 percent argon, with traces of water (in the gaseous form), carbon dioxide, and other gases. At lower elevations, variable amounts of dust particles and water droplets are also found suspended in the air.

The gases nitrogen and argon are relatively inert chemically. It is oxygen that sustains most animal life on Earth, by allowing animals to oxidize their food to produce energy. Oxygen is also required for all forms of combustion (rapid oxidation) and thus is necessary for most of our heat and power production. In the process of photosynthesis, green plants absorb carbon dioxide and release oxygen, which helps to replenish the oxygen consumed by humans and other animals.

The Earth's atmosphere is the result of outgassing of compounds originally accreted to the planet in the form of rocky planetesimals, together with the gas that has been contributed by later impacts with comets or comet-like icy planetesimals. Given the close proximity of the Earth to the sun, it should be no surprise that only about one-millionth of our mass is in the form of atmospheric gases, even if the interior of the planet is by now completely degassed.

A complete census of the Earth's atmosphere, however, should look at more than the gas now present. Suppose, for example, that our planet were heated the way Venus is now; the water in the oceans would vaporize to water vapor and become a part of the atmosphere. To estimate how much water vapor would be released by boiling away the oceans, we first note that there is enough water to cover the entire Earth to a depth of about 3000 meters. The pressure exerted by 10 meters of water is about equal to 1 bar, as any scuba diver knows. Thus the average pressure at the ocean floor is about 300 bars (3000 meters divided by 10 meter/bar). Since water weighs the same whether it is in liquid or vapor form, the atmospheric pressure of water if the oceans boiled away would also be 300 bars. Water would therefore greatly dominate the Earth's atmosphere, with nitrogen and oxygen reduced to the status of trace constituents.

On a warmer Earth another source of additional atmosphere would be found in the sedimentary carbonate rocks of the crust. These minerals contain abundant carbon dioxide, which is readily released when the rocks are heated. At the temperature of Venus, the Earth's atmosphere would be expected to contain about 70 bars of carbon dioxide, far more than the current CO_2 pressure of only 0.0004 bars. Thus the atmosphere of a warm Earth would be dominated by water vapor and carbon dioxide, with a surface pressure of close to 400 bars.

(c) Weather and Climate

The convection of the troposphere, and particularly the evaporation and condensation of water, give rise to the phenomena we call *weather*. The energy that powers the weather is derived primarily from the sunlight that heats the surface. As the planet rotates, and also as slower seasonal changes take place in the deposition of sunlight, the atmosphere and oceans try to redistribute the heat from warmer to cooler areas. Weather on any planet represents the response of its atmosphere to changing inputs of energy from the sun.

The weather on Earth is closely tied to the presence of the oceans and of water vapor in the atmosphere. As water evaporates, it stores large quantities of energy, which can be released later through condensation. The violence of a summer thunderstorm or the awesome power of a hurricane are the result of energy suddenly released by the condensation of atmospheric water. On a planet without abundant water, the weather should be calmer and more predictable.

The rotation of the Earth also has a major influence on the circulation of the atmosphere. On a slowly rotating planet, the simplest circulation pattern would involve warm air rising near the equator and cool air descending near the poles; this is approximately the kind of circulation observed on Venus. The Earth's rotation, however, breaks up this simple flow pattern to create the large cyclonic weather systems that dominate the temperate regions of the planet (Figure 13.14). On an even more rapidly spinning planet such as Jupiter, we will see that north-south motion of the atmosphere is almost impossible, and the circulation is dominated by extremely strong eastward-blowing winds.

Climate is a term used to refer to the effects of the atmosphere on a longer time scale, measured in decades or centuries. Changes in climate are difficult to detect, but as they accumulate their effect can be devastating. Modern farming practice, in particular, is highly sensitive to the temperature and rainfall. Calculations show, for instance, that a drop of only 2°C throughout the growing season would cut the wheat production of Canada and the U.S. in half. In Section 13.4 we will return to the

Figure 13.14 A large tropical storm, marked by clouds swirling in a cyclonic direction around a low-pressure region. (NASA)

problem of climate change, while in Chapter 28 we will discuss evidence for association of climate changes with the activity level of the sun.

(d) The Ice Ages

The best-documented changes in the climate of the Earth are the great ice ages, which have periodically lowered the temperature of the Northern Hemisphere of the Earth over the past million years or so. During an ice age, snow accumulates over the continental masses of North America, Europe, and Asia, eventually building up great ice deposits or glaciers, at the same time lowering the level of the oceans as more and more water builds up on land. Such conditions can persist for tens of thousands of years. Today we are in a relatively warm period, interpreted by many scientists as a fairly short-lived interglacial interval between major ice ages.

It is generally believed today that the ice ages are the result of changes in the tilt of the Earth's rotation axis (its *obliquity*) produced by the perturbative gravitational influences of other planets, as originally calculated in 1920 by the Serbian scientist Milutin Milankovitch (1879–1958). With modern computing techniques, these changes can be predicted for the Earth and for other planets. As we will see in Chapter 15, one of the exciting discoveries about Mars is that it also seems to experience periodic ice ages. Soon we may have data on both the Earth and Mars to use in testing and refining our understanding of the astronomical causes of ice ages.

13.4 LIFE AND THE CHEMICAL EVOLUTION OF THE EARTH

(a) The Uniqueness of Life

Earth is the only inhabited planet in the solar system. We can now make this dramatic assertion with some confidence. About a century ago, it was widely believed that Mars harbored not only life, but also intelligent creatures with an advanced civilization. As recently as the 1960s, most astronomers thought that life was probable on Mars. But by now the data from numerous space probes have pretty well ruled out the presence of life of any sort we are familiar with on other planets in our solar system. Of course, there are likely to be many planets elsewhere in the galaxy, including those with intelligent inhabitants, but that is another story that we save for the final chapter of this book.

Terrestrial life forms an important part of the story of our planet. Life arose early in Earth's history, and living organisms have been interacting with their environment for billions of years. We all recognize that lifeforms have evolved to adapt themselves to the environment on Earth, and now we are beginning to realize that the Earth itself has been changed in important ways by the presence of living matter.

There is an interesting thing about the organic molecules in living organisms on the Earth. The basic molecules are helical—corkscrew shaped—and the more complex ones are built up of these simpler ones. They are all with the right-hand thread; that is to say, in all living organisms, the helical molecules twist the same way. It's easy enough to understand why this should be so. In the early development of life, some molecule with the ability to reproduce itself got formed; it had to be of either right-hand or left-hand thread—it couldn't be both at once. It just happened to have a right-hand thread; it then fed and reproduced, and fed and reproduced, and had mutations, and was modified by natural selection, and eventually ended up with something holding an astronomy book in its hand. If life had sprung up independently all over the Earth, we might expect by chance to find some forms with right-hand helices and others with left-hand helices in their chromosomes. But all of us, from the lowliest bacterium to the biggest elephant, are made of organic molecules twisting the same way, and all are based on the same genetic code. If other, independent lifeforms originated during the long history of the Earth, their descendants do not survive today.

(b) Origin of Life

The record of the birth of life on Earth has been lost in the restless motions of the crust. By the time the oldest surviving rocks were formed, life already existed. And by 3.5 billion years ago, only about a billion years after the formation of the planet, life had evolved to the stage of photosynthetic colonies of micro-organisms whose fossils can still be seen in ancient rocks. Any theory of the origin of life must therefore be partly speculative, since there is little direct evidence to go on.

An important clue to the origin of life is found from attempts to simulate, in the laboratory, the conditions under which the appropriate chemical building blocks of life might have formed. Such studies indicate clearly that an oxygen-rich atmosphere like that of the Earth today was not suitable. However, if the experiments start with a reducing atmosphere (no free oxygen) and abundant water, it turns out to be easy to produce a wide variety of organic molecules of the sort that life is made of. These include such complex compounds as the amino acids and simple proteins. In such an environment, organic matter must have been abundant.

Is it reasonable to expect Earth to have had a reducing atmosphere at the time life was forming? The answer is clearly yes, for free oxygen turns out to be a very unusual gas in the solar system. We have oxygen in our atmosphere today only because it is produced by photosynthetic plants. In the absence of life, any free oxygen quickly combines with surface rocks to produce oxides. Before life began, our atmosphere presumably was dominated by carbon dioxide and perhaps carbon monoxide, like the atmospheres of Venus and Mars today.

To form abundant organic compounds, however, we require more than the absence of oxygen. Also apparently needed are such hydrogen-rich compounds as methane (CH_4) and ammonia (NH_3). The fact that life arose here suggests that an atmosphere containing methane and ammonia existed on the early Earth. Alternatively, it may be possible that the necessary organic materials came from elsewhere, carried to the Earth as part of the early cometary bombardment. No one knows for sure.

For hundreds of millions of years after its formation life (perhaps little more than large molecules like the viruses of today) probably existed in warm, nutrient-rich seas, living off the accumulated organic matter. Eventually, however, as this easily accessible food became depleted, life began the long evolutionary road that led to the proliferation of different organisms on Earth today. As it did so, life influenced the chemical evolution of the atmosphere.

(c) Evolution of the Atmosphere

If methane and ammonia were originally abundant in the Earth's atmosphere, they probably did not last long. Ultraviolet light easily breaks such compounds apart, and in the low gravity of the Earth hydrogen would have escaped, so that such compounds could not be put back together again. Thus there was an inevitable transition from NH_3 to N_2, and from CH_4 to CO. The period during which organic material could easily be produced on Earth was probably short-lived.

Studies of the chemistry of ancient rocks show that the atmosphere of the Earth remained reduced until at least two billion years ago, in spite of the presence of plants releasing oxygen by photosynthesis. Apparently the oxygen gas was removed by chemical reactions with the crust as quickly as it formed. Slowly, however, the increasing evolutionary sophistication of life led to a growth in plant population, finally reaching the point where oxygen was produced faster than it could be removed, and the atmosphere became more and more oxidizing.

The appearance of free oxygen between one and two billion years ago eventually led to the formation of the Earth's ozone layer, which protects the surface from lethal solar ultraviolet light. Before that, it was unthinkable for life to venture outside the protective oceans, and the land masses of Earth were barren. The presence of oxygen thus allowed the colonization of the land. It also made possible a tremendous proliferation of dependent creatures called animals, who lived off of the organic materials produced by plants and the oxygen dissolved in

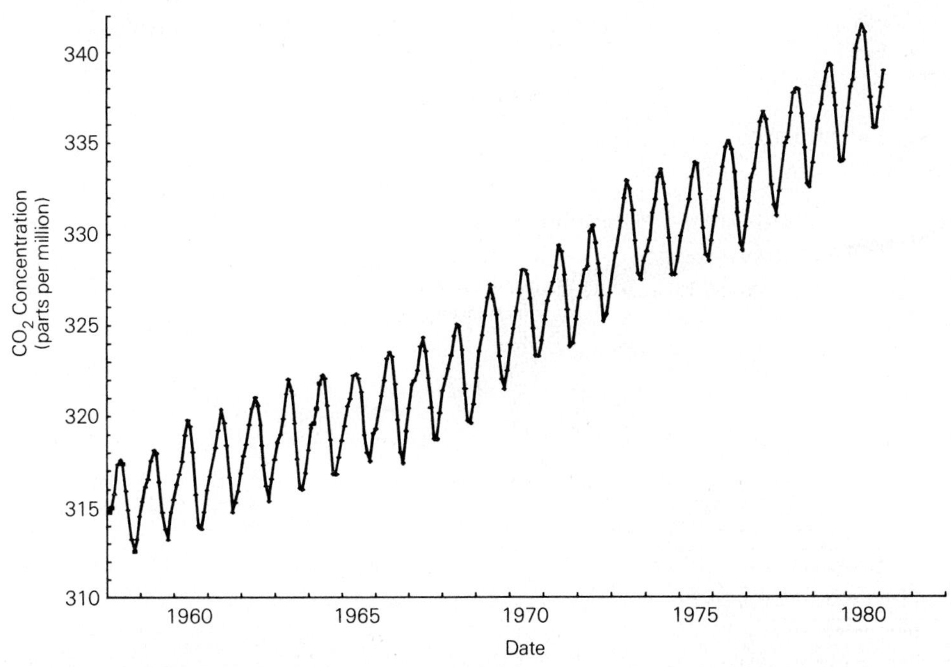

Figure 13.15 Variation with time of atmospheric carbon dioxide content, which has increased by about 20 percent during the past century. (Adapted from data obtained at the Mauna Loa Observatory of NOAA)

the seas. Then the animals too moved to the land, where they adapted techniques for breathing oxygen directly from the atmosphere.

(d) Carbon Dioxide in the Atmosphere

One of the most important consequences of life has been a reduction in atmospheric carbon dioxide. In the absence of life we would probably have a much more massive atmosphere dominated by carbon dioxide. But life has effectively stripped us of most of this gas.

Most of the carbon dioxide is locked up in sediments composed of carbonate minerals. Carbonates are an essential material for the shells of marine creatures, which have evolved techniques for extracting CO_2 from the water and manufacturing intricate carbonate structures for their protection. When they die, their shells sink to the ocean floor. We have all seen the great beds of limestone that were formed in this way on the continents, and the ocean carbonate sediments are much greater. The CO_2 remains trapped on the ocean floor until it is subducted, when much of it returns to the atmosphere in volcanic eruptions. But evidently the marine organisms quickly cycle it back into sediments, so that at any time only a minute fraction of the Earth's CO_2 is present in the atmosphere.

Another way that life removes CO_2 is by producing deposits of fossil fuels, predominately coal and oil. These substances are primarily carbon, for the most part extracted from atmospheric CO_2 hundreds of millions of years ago, when the first great forests populated the land.

We have a special interest in the CO_2 content of the atmosphere because of the special role this gas plays in retaining heat from the sun. In Chapter 16 we will describe this situation, called the *greenhouse effect,* in more detail. Basically, a larger quantity of CO_2 results in a more effective blanket and a higher average surface temperature. Even the tiny amount of this gas in our atmosphere today is sufficient to raise the temperature several degrees higher than it would otherwise be.

Modern industrial society depends on energy extracted from burning fossil fuels. As these ancient coal and oil deposits are oxidized, additional carbon dioxide is released into the atmosphere. So far in this century, the amount of CO_2 in the atmosphere has increased by about 20 percent, and it is continuing to rise at 0.4 percent per year (Figure 13.15).

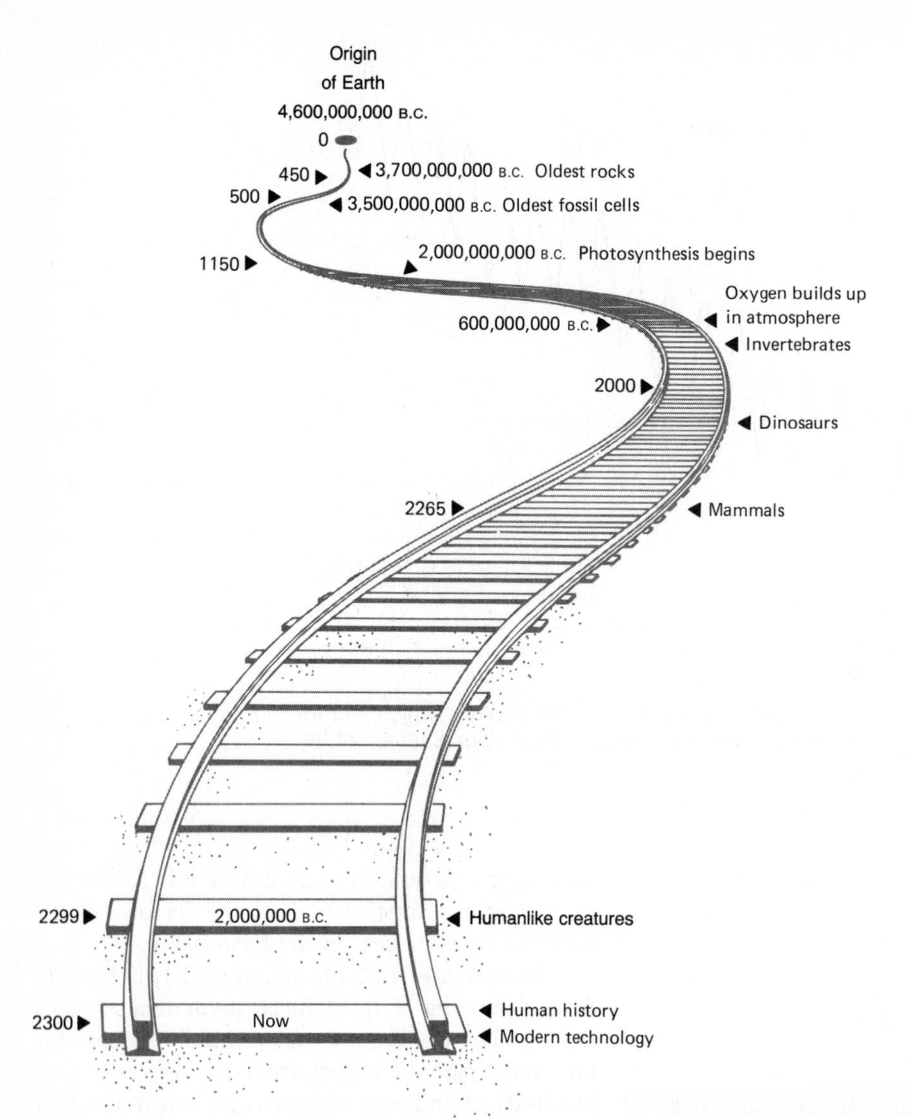

Figure 13.16 A symbolic railway through time.

By early in the 21st century, the CO_2 level is predicted to reach twice its preindustrial value. Because of the lag in responsiveness of the Earth, the consequences on the climate will not become apparent until 20 years or so later.

Unless other effects intervene, we can expect a significant warming of the Earth, with possibly disastrous effects. The problem is that no one really knows how to estimate the size of these effects, and even if we did, there is no way known to bring the carbon dioxide content back down to normal levels.

(e) Conclusion

Figure 13.16 is a figurative railway track, built as the Earth has aged. Each tie in the track corresponds to 2 million years. There are 2300 ties so far, and events in the history of the Earth are marked at various ties along the way. Suppose typical railway ties are about 1 m apart and that the last tie is being laid as we write these words. The second to last one, 1 m back, was laid when humanlike creatures first stood upright on Earth. Only a few centimeters from the front end of the last tie is where the history of humans begins, as gleaned from their drawings and crude writing. Modern history begins a few millimeters back. Now bear in mind that the entire length of track is more than 2 km. Yet our technological society goes back only 200 micrometers—0.2 mm—just the thickness of that grain of dust on the front side of the tie that was just laid.

In the time represented by that grain of dust, humankind has drastically changed the environment, plundering resources and polluting waters. We could probably solve those problems, but in the past few years some of us have begun to realize that we may have permanently altered, to the disastrous worse, that fragile, thin, atmospheric envelope that protects us from the outside universe and makes our own existence possible.

But to the Earth itself, chugging along with its plates burrowing under each other, switching its magnetic field around chaotically, having ice ages come and go, and whose existence is pretty much assured for another 5 billion years, what humans do to themselves is of no consequence whatsoever.

EXERCISES

1. Why do we include a chapter on the Earth in an astronomy text? Explain whether you think more, or less, space should be devoted to our own planet.
2. What fractions of the volume of the Earth are occupied by the inner core, the entire core, the mantle, and the crust?
3. What are the compositions of the core and mantle of the Earth? Explain how we know, and indicate what the uncertainties in your conclusions might be.
4. Explain what is meant by uncompressed density. Why is it a useful concept? Why is the uncompressed density of the Moon almost the same as its actual density, while for the Earth the uncompressed density is substantially less than the measured density?
5. When the first rockets and satellites were sent into near-Earth space, they found that the density of charged atomic particles was greatest at an elevation of a few thousand kilometers (in the Van Allen belts), that this density declined with altitude, and that above a certain well-defined point there were no ions or electrons at all circling the Earth. Explain why there should be this upper cutoff. What defines this boundary?
6. Why does the Earth have so few impact craters, relative to the Moon and many other planets and satellites?
7. If you wanted to live where the chances of a destructive earthquake are small, would you pick a location near a fault zone, near a mid-ocean ridge, near a subduction zone, or on a volcanic island like Hawaii? What are the relative risks of earthquakes at each of these locations?
8. If the Hawaiian island chain is 3500 km long and dating of the volcanoes shows that on the average a million years is added for each 100 km along the chain, what is the speed of motion of the Pacific plate, and how long has this hot spot been active?
9. What is the volume of new oceanic basalt that is added to the Earth's crust each year? Assume that the thickness of the new crust is 5 km, that there are 60,000 km of active rifts, and that the average speed of plate motion is 4 cm per year. What fraction of the entire volume of the Earth does this annual addition of new material represent?
10. From your answer to the previous question, also estimate the volume of material subducted each year.
11. Suppose that the next slippage along the San Andreas Fault in southern California takes place in the year 2000, and that it completely relieves the accumulated strain in this region. How much slippage will take place?
12. Why are volcanoes often associated with mountains that form by uplift, where one plate is running into another? (Examples occur in the Andes of South America, in the Japanese islands, and elsewhere.)
13. Assume that atmospheric pressure decreases by a factor of one-half with every increase of 5500 m in altitude above the Earth's surface. At what altitude would the pressure be $\frac{1}{16}$ that at sea level? What is the pressure near the summit of Mt. Everest at about 8 km altitude?
14. What is the probable origin of the main gases in the atmosphere of the Earth (nitrogen, oxygen, argon, water vapor, carbon dioxide)?
15. Is there evidence of changes in climate in your area over the past century? How would you distinguish a true climate change from the random variations in weather that take place from one year to the next?
16. If all life were destroyed on Earth, would new life eventually form to take its place? Explain how conditions would have to change for life to start again on our planet.
17. At the present rate of burning fossil fuels, the amount of carbon dioxide in our atmosphere will double before the end of the first half of the 21st century—within most of your lifetimes. What do you expect the consequences to be for our planet? Is there any way to stop this increase from taking place?

Grove K. Gilbert (1843–1918), a founding member and Chief Geologist of the U.S. Geological Survey, was among the first scientists to recognize the importance of impact cratering on the Earth and Moon. In the 1890s his careful arguments for the impact origin of the lunar craters, although decades ahead of their time, laid the foundations for the modern science of lunar geology and the development of a chronological sequence for the evolution of the terrestrial planets. (USGS)

14

CRATERED WORLDS: MOON AND MERCURY

The Moon is the nearest large celestial object to the Earth and the only one to have been visited by humans. Unlike our planet, the Moon is geologically dead, a world that has long since exhausted its internal energy sources. As such, it is a window on the earlier eras of solar system history. The planet Mercury is in many ways similar to the Moon, which is why the two are discussed together. Among the terrestrial bodies, these two are both relatively small, lacking in atmospheres, deficient in geological activity, and dominated by the effects of impact cratering from outside.

14.1 GENERAL PROPERTIES OF THE MOON

(a) Some Basic Facts

The Moon has only 1/80 of the mass of the Earth, and a surface gravity too low to retain an atmosphere. If, early in its history, the Moon outgassed an atmosphere from its interior or collected a temporary envelope of gases from impacting comets, such an atmosphere was lost before it could leave any recognizable evidence of its short existence. All signs of water are similarly absent. Indeed, the Moon is dramatically deficient in a wide range of *volatiles,* those elements and compounds that evaporate at moderate temperatures. Something in its past removed whatever volatiles might have been accumulated initially from the solar nebula or from cometary impacts.

Some of the properties of the Moon are summarized in Table 14.1, along with comparative values for Mercury.

TABLE 14.1 Properties of the Moon and Mercury

	MOON	MERCURY
Mass (Earth = 1)	0.0123	0.055
Diameter (km)	3476	4878
Density (g/cm^3)	3.34	5.43
Uncompressed density (g/cm^3)	3.2	5.3
Surface gravity (Earth = 1)	0.17	0.38
Escape velocity (km/s)	2.4	4.3
Rotation period (days)	27.3	58.6
Surface area (Earth = 1)	0.07	0.15

Figure 14.1 The Moon in a gibbous phase as seen mostly from the far side. Mare Crisium is visible at the edge of the Moon in the upper left. The photograph was taken by the Apollo 16 crew. (NASA)

(b) Exploration of the Moon

Most of what we know about the Moon today derives from the Apollo program, which sent nine manned spacecraft to our satellite between 1968 and 1972 and landed 12 human astronauts on its surface. Before the era of spacecraft studies, astronomers had mapped the side of the Moon that faces the Earth to a best telescopic resolution of about 1 kilometer, had measured the surface temperatures with infrared detectors, and had used spectroscopy to verify the absence of an atmosphere or surface water and to make some preliminary deductions about the composition of the surface rocks. But lunar geology was only beginning, and the fields of lunar geochemistry and geophysics were nonexistent. All that changed beginning in the early 1960s.

Initially, the U.S.S.R. dominated in sending spacecraft to the Moon. In 1959 Luna 2 became the first vehicle to reach the lunar surface and Luna 3 returned the first photos of the lunar far side. By 1966, the Soviets were able to land a spacecraft (Luna 9) on the surface and transmit pictures and other data to Earth. At about the same time, the U.S. achieved success with a series of five Lunar Orbiters that mapped the entire surface and with five Surveyor spacecraft that landed and conducted experiments on the surface. All of these craft helped pave the way for the first Apollo landing, on July 20, 1969 (Figure 14.2).

Table 14.2 summarizes the nine Apollo flights—six landings and three other missions that circled the Moon but did not land. The initial landings were on flat plains selected out of considerations of safety (Figure 14.3), but with increasing experience and confidence NASA targeted the last three missions to more geologically interesting locales. The level of scientific exploration also increased with each mission, as the astronauts spent longer on the Moon and carried more elaborate equipment. Finally, on the last Apollo landing, NASA included one scientist-astronaut, geologist Harrison Schmitt (Color Plate 9).

In addition to landing and photographing the lunar surface at close hand, the Apollo missions accomplished three objectives of major importance for lunar science. First, and most important, the astro-

Figure 14.2 Launch of one of the Apollo/Saturn V rockets carrying astronauts to the Moon. (NASA)

Figure 14.3 Apollo 12 astronaut on Moon, visiting the Surveyor 3 spacecraft that had landed several years earlier. Note the footprints in the soft lunar soil and the presence of rocks ejected from nearby impact craters. (NASA)

nauts collected nearly 400 kg of samples to return to Earth for detailed laboratory analysis. These samples, which are still being extensively studied, have probably revealed more about the Moon and its history than all other lunar studies combined. Second, each Apollo landing after the first deployed an ALSEP, or Apollo Lunar Surface Experiment Package, which continued to operate for years after

TABLE 14.2 Apollo Flights to the Moon

FLIGHT	DATE	LANDING SITE	ACCOMPLISHMENTS
8	Dec 68	Orbiter	First human circumlunar flight.
10	May 69	Orbiter	First lunar orbit rendezvous.
11	Jul 69	Mare Tranquillitatis	First human landing. 22 kg samples returned.
12	Nov 69	Oceanus Procellarum	First ALSEP. Visit to Surveyor 3.
13	Apr 70	Mission aborted due to explosion in Command Module	
14	Jan 71	Mare Nubium	First "rickshaw."
15	Jul 71	Imbrium/Hadley	First "rover." Visit to Hadley Rille. 24 km traverse.
16	Apr 72	Descartes Highlands	Only highland site. 95 kg samples returned.
17	Dec 72	Taurus Mts.	Only geologist present. 111 kg samples returned.

the astronauts departed. (The ALSEPS were turned off by NASA in 1978 as a cost-cutting measure.) And third, the orbiting Apollo Command Modules carried a wide range of instrumentation to photograph and analyze the lunar surface from above.

The last human left the Moon in December 1972, just a little more than three years after Neil Armstrong took his "giant step for mankind." The program of lunar exploration was cut off in midstride as the result of political and economic pressures. It had cost just about 100 dollars per American, spread over ten years—the equivalent of one pizza and a six-pack of beer per year. The giant Apollo rockets built to travel to the Moon have been left to rust on the lawns of NASA centers at Cape Canaveral and Houston. Today no nation on Earth has the technology and the rockets to return to the Moon. Having reached our nearest neighbor in space, humans retreated back to our own planet. How long before we will venture out again into the solar system?

The scientific legacy of Apollo remains, however, as we will see in the following sections of this chapter.

(c) The Composition of the Moon

The uncompressed density of the Moon, as shown in Table 14.1, is only 3.2 g/cm^3, compared with 4.5 g/cm^3 for the Earth. This one fact clearly tells us that the bulk compositions of the two bodies are different, a conclusion that has been amply verified by returned lunar samples. As we will see in Section 14.4, these differences in composition provide important constraints on any theory of the origin of the Moon.

A density of 3.2 g/cm^3 is characteristic of many kinds of silicate rocks such as those that make up the crust and mantle of the Earth, as well as the lunar samples. The fact that the average density of the Moon is not very different from that of its common surface rocks shows that it does not display the large increase in density with depth that the Earth does. Its interior composition must also be dominated by rocky material, and there is at most a very small metallic core.

The Moon therefore differs from the Earth primarily in being *depleted* in iron and other metals. We also know from the study of lunar samples that it is depleted in water and other volatiles. On the other hand, its rocky materials are very similar to those of the Earth. It is as if the Moon were composed of the same basic silicates as the Earth's mantle and crust, with the core metals and the volatiles selectively removed.

(d) Interior of the Moon

Probes of the interior carried out with seismometers carried to the Moon as part of the Apollo program confirm the absence of a large lunar core. The crust of the Moon is much thicker than that of the Earth, ranging from about 60 km on the side facing the Earth up to at least 150 km on the far side. If there is a metallic core, it can be no more than a few hundred kilometers in diameter.

The level of seismic activity on the Moon is much less than on the Earth, as expected for a smaller and less active body. No moonquake measured by any of the Apollo seismometers could have been *felt* by a person standing on the Moon, and the total energy released by moonquakes is a hundred billion times less than that on our planet. In the absence of major movements of the crust, the tremors that were measured were caused by impacts of meteoroids (and of the Apollo rocket boosters aimed for that purpose) and by small internal adjustments triggered by tidal stresses.

Additional information on the lunar interior is provided by Apollo measurements of heat flow from beneath the crust. On Earth, temperature increases rapidly with depth, and a substantial flow of heat outward from the mantle can be measured. The lunar heat flow is much less and is consistent with local sources of radioactive heating in the crust rather than with heat escaping from a hot interior.

The Moon also lacks a global magnetic field like that of the Earth, a result consistent with the absence of a metal core. There are some very weak local fields, however, indicating slight magnetism in some of the surface rocks. We do not know how these rocks became magnetized, but one explanation might be the presence of a general lunar field billions of years ago, when the Moon was younger and more active.

14.2 THE LUNAR SURFACE

(a) General Appearance

Through the telescope (or photographed from a spacecraft), the lunar crust is seen to be covered by impact craters of all sizes. We will discuss the for-

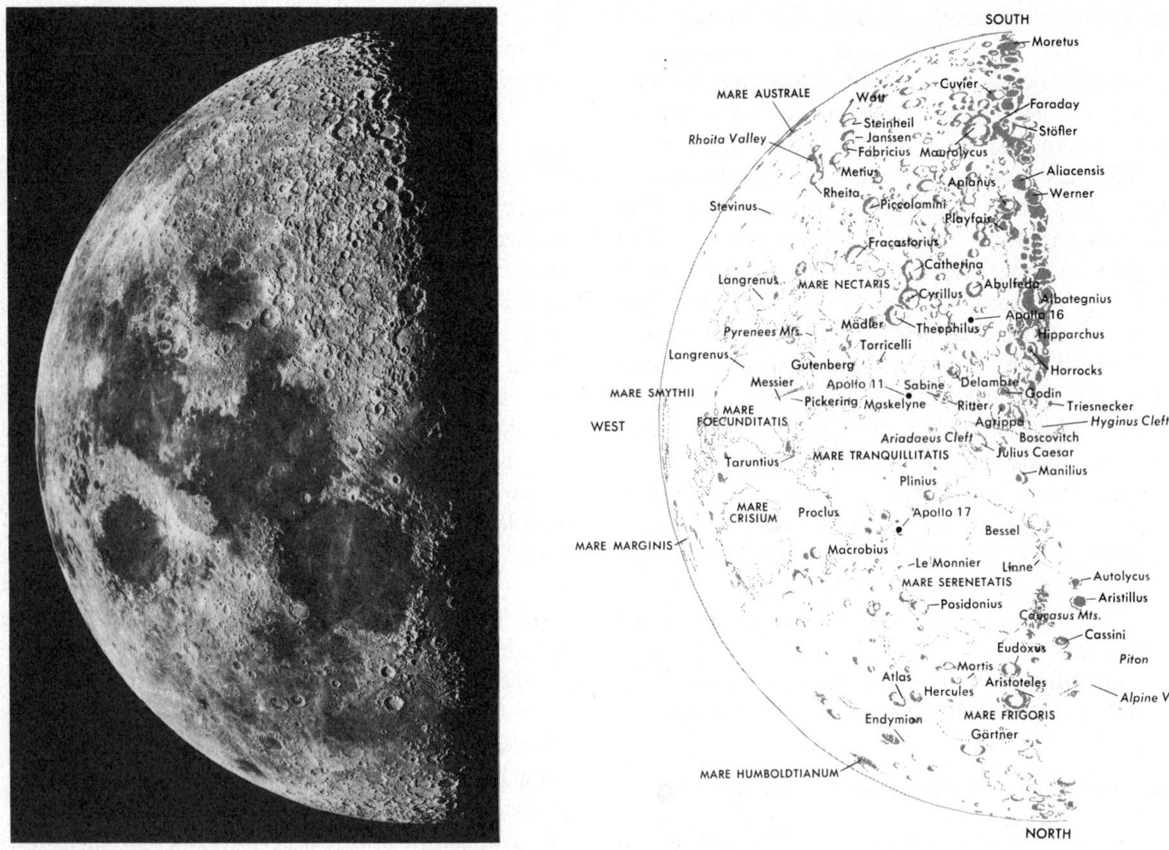

Figure 14.4 The western hemisphere of the Moon. The numbered symbols on the identification chart show the locations of Apollo landings. (Lick Observatory)

mation of impact craters in some detail in Section 14.3. None of these craters or other topographic features are large enough, however, to be seen without optical aid. The few features that can be distinguished without a telescope reflect the division of the lunar surface into two kinds of rocks, one much darker than the other.

The most conspicuous of the Moon's surface features—those composing what we familiarly call "the man in the Moon"—are due to splotches of darker material of volcanic origin. When Galileo first looked at the Moon through a telescope, he saw mountains, craters, and valleys, in addition to these dark regions which he thought were seas. The idea developed that the Moon might be a world, perhaps not unlike our own (Figures 14.4 and 14.5). It is interesting to note that, before the era of spacecraft, no observations of resolution comparable to Galileo's first glimpses of the Moon were available for any other planet.

Many other lunar observers followed Galileo's lead. In 1647 John Hevel of Danzig (1611–1687) published a comprehensive treatise on the Moon, *Selenographia*. In a series of carefully prepared plates, Hevel identified many of the Moon's features and gave them names, in many cases in honor of similar features on the Earth. The names he gave the mountain ranges on the Moon survive today. Later, cartographers adopted the convention still in use of naming lunar craters for great scientists and philosophers.

The early lunar observers regarded the Moon as having continents and oceans and as being a possible abode of life. We know today, however, that the resemblance of lunar features to terrestrial ones is superficial. Even when they look somewhat similar, the origins of lunar features such as craters and mountains may be very different from their terrestrial counterparts. The Moon's relative lack of internal activity, together with the absence of air and

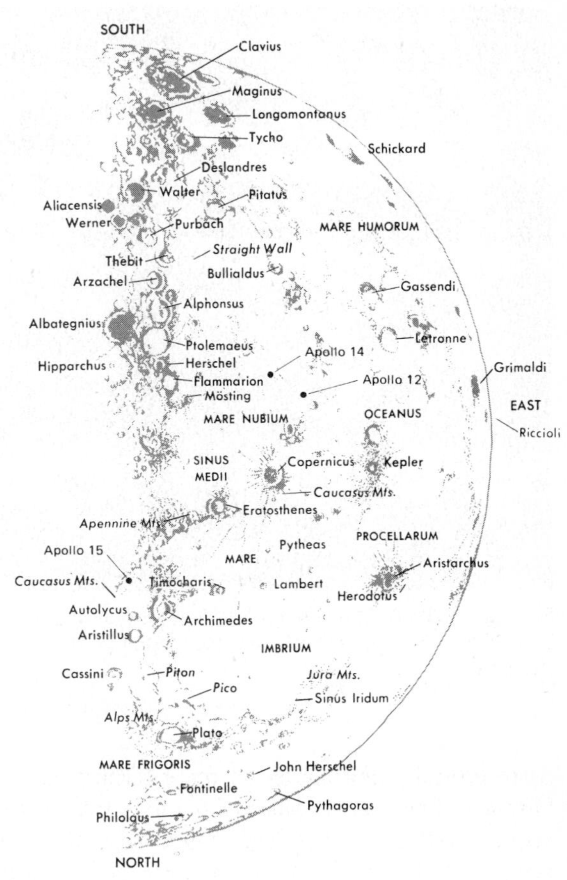

Figure 14.5 The eastern hemisphere of the Moon. The numbered symbols on the identification chart show the locations of Apollo landings. (Lick Observatory)

water, makes most of its geological history unlike anything we know on Earth.

(b) The Lunar Highlands

Most of the crust of the Moon (83 percent) is heavily cratered and consists of relatively light-colored silicate rocks called anorthosites. In the past, this dominant terrain on the Moon was called the "terra," or "land," in contrast with the "seas" discussed below. Today, in recognition of its higher elevation, it is known more simply as the *lunar highlands*.

The highlands (Figures 14.6 and 14.7) are the oldest surviving part of the lunar crust. They represent material that solidified on the crust of the cooling Moon within a hundred million years or so of its formation. Like slag floating on the top of a smelter, the anorthositic rocks are the lowest-density part of the Moon.

Because they are so old, the highlands are also extremely heavily cratered. Early in its history, the solar system was filled with planetesimals and other debris from its formation, as discussed in Section 12.4. Impact rates were therefore much higher than at present, a topic we will discuss in more detail in Section 14.3. The cumulative effect of these impacts over more than 4 billion years has been sufficient to distribute craters one on top of another in the highlands, until the surface has reached a state of *saturation*. Once the craters are shoulder to shoulder, any new impacts destroy older craters as rapidly as they create new ones.

Although none of the Apollo landings took place directly in the highlands, several were close to highland regions, and all brought back samples of highland rock distributed over the Moon as ejecta from cratering impacts. A few of the oldest highland samples solidified from molten rock 4.4 billion years ago, although ages of 4.2 billion years are

Figure 14.6 A densely cratered region in the lunar highlands, photographed by the Apollo 16 crew. (NASA)

Figure 14.7 Apollo 16 photograph of its landing site in the region of the lunar crater Descartes. (NASA)

more typical. Information on these earliest periods is sparse, however, because of the intensity of subsequent cratering. Although much older than the surface of the Earth, even the lunar highlands do not take us back to the earliest periods of solar system history, as some scientists had hoped before Apollo.

(c) The Lunar Seas

The largest of the lunar features are the so-called seas, still called *maria* (Latin for "seas"). These dark plains (Figures 14.8 and 14.9), which are much less cratered than the highlands, cover just 17 percent of the lunar surface, mostly on the side facing the Earth. The maria are of course dry land. They are volcanic plains, laid down in eruptions billions of years ago.

The largest of the 14 lunar maria on the Moon's visible hemisphere is Mare Imbrium (the "Sea of Showers"), about 1100 km across. The other maria have equally fanciful names, such as Mare Nubium ("Sea of Clouds"), Mare Nectaris ("Sea of Nectar"), Mare Tanquillitatis ("Tanquil Sea"), Mare Serenitatis ("Serene Sea"), and so on. Most of the maria are roughly circular in shape, although many of them are interconnected or overlap slightly. Typically, they are several kilometers lower than the surrounding highlands, and their borders are marked by curving arcs of mountains.

The lunar maria are all composed of basalt, very similar in composition to the oceanic crust of

Figure 14.8 A closeup of a part of a lunar mare photographed by the Apollo 16 crew. (NASA)

Figure 14.9 A typical lunar mare landscape seen from the window of the Apollo 11 landing module. (NASA)

the Earth or to the lavas erupted by many terrestrial volcanoes. The dark color and uniformity of this basalt produces the striking pattern of the maria, visible even to the unaided eye. Analysis of samples of mare basalts reveals that they were erupted between 3.3 and 3.8 billion years ago, and that their sources were deep in the lunar mantle.

It is evident from a study even of Earth-based photographs of the Moon that the formation of the maria was a two-step process. Their low elevation and roughly circular shapes reveal them to be *impact basins*—great scars in the lunar crust formed by the impact of planetesimals up to 100 km in diameter. Examination in the laboratory of material ejected from these giant impacts shows that the last of these impacts took place about 3.9 billion years ago. But the impacts themselves did not produce the lava that later filled these basins.

Within the maria are a number of large craters that are flooded by lava. These used to be called "ghost craters." They represent impact craters that formed subsequent to the excavation of the great basins and before the volcanic eruptions that later filled the basins. Both a study of these flooded craters and the direct dating of the eruptions show that an interval of more than a hundred million years years elapsed between the basin-forming impacts and the major era of volcanic eruptions.

The lunar eruptions that formed the maria probably resembled the terrestrial activity that formed *flood basalts,* such as the Snake River Plain in Washington State (Section 13.2g). A series of such eruptions laid down smooth flows typically a few meters thick but extending over distances of hundreds of kilometers. Eventually, these flows filled in the lowest parts of the basins to form the mare surfaces we see today.

Volcanic activity may have begun very early in the Moon's history, although most evidence of the first half-billion years is lost. What we do know is that the major mare volcanism, which involved the release of highly fluid basalts from hundreds of ki-

Figure 14.10 Mt. Hadley on the edge of Mare Imbrium, photographed by the Apollo 15 astronauts. (NASA)

lometers below the surface, was largely confined to the interval between 3.8 and 3.3 billion years ago. After that the Moon's interior cooled, and by 3.0 billion years ago all volcanic activity ceased. Since then, our satellite has been a geologically dead world, changing only slowly as the result of random impacts.

(d) Mountains on the Moon

There are several mountain ranges on the Moon, generally along the edges of the maria. Most of them bear the names of terrestrial ranges—the Alps, Apennines, Carpathians, and so on—but their mode of origin is entirely different from their namesakes.

The major lunar mountains are all the result of our satellite's history of impacts. The long, arc-shaped ranges that border the maria are simply ejecta from the impacts that formed these giant basins. The impact explosion shattered and lifted the surrounding rock, forming outcrops such as Mt. Hadley, photographed by the Apollo 15 astronauts (Figure 14.10). Other, isolated peaks are little more than huge boulders thrown about the Moon from these impacts. A very few inconspicuous low mountains may be volcanoes from the last stages of lunar internal activity.

The lunar mountains have low, rounded profiles that resemble old, eroded mountains on Earth. But appearances can be deceiving. The mountains of the Moon have not been eroded, except for the effects of meteoritic impacts. They are rounded because that is the way they formed, and there has been no water or ice to carve them into cliffs and sharp peaks.

Even from Earth the heights of isolated lunar mountains above the surrounding plains can be determined by measuring the lengths of the shadows they cast. Figure 14.11a illustrates one of the possible procedures for determining the elevation of a lunar feature. A mountain peak, *M*, near the terminator and near the center of the Moon's disk, casts its shadow over the distance *AB* from its base toward the terminator. The length of the shadow *AM* and the projected distance of the mountain from the terminator, *T'B*, can be measured in seconds of arc, and these measures can be converted to kilometers.

The geometry of the problem is shown in Figure 14.11b, in which the Moon is seen from a direction at right angles to the line from the Moon to Earth. The line *AM* is perpendicular to *OT*, and *AB* is perpendicular to *OM*. Therefore, the two right triangles *OT'B* and *ABM* are similar, and we have the proportion

$$\frac{BM}{AM} = \frac{T'B}{OB'}$$

Since *OB* is the known radius of the Moon, and *T'B* and *AM* are measured quantities, the height of the mountain, *BM*, can be calculated.

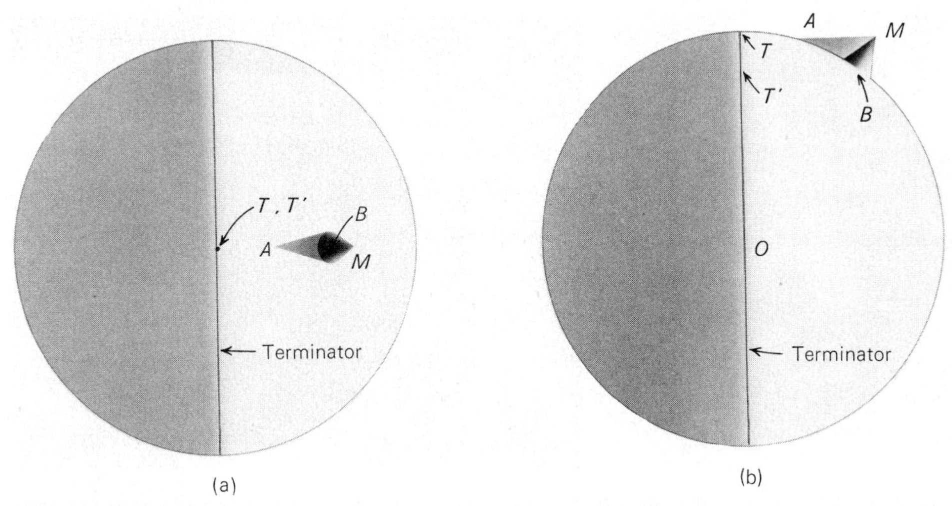

Figure 14.11 Measuring the height of a lunar mountain. (a) The apparent disk of the Moon as it appears in the sky, (b) the geometry of the situation seen at right angles to the line of sight from the Earth.

(e) The Lunar Soil

The surface of the Moon is everywhere covered with a fine-grained soil of tiny, shattered rock fragments. The dark basaltic dust of the lunar maria was kicked up by every astronaut footstep, and this dust eventually worked its way into all of their equipment. The upper layers of the surface are porous, consisting of loosely packed dust into which their boots sank several centimeters (Fig. 14.12). In the absence of weathering and terrestrial sources of erosion, where did this ubiquitous dust come from?

The lunar dust, like so much else on our satellite, is the product of impacts. Each cratering event, large or small, breaks up the rock of the lunar surface and scatters the fragments about. The impacts also melt tiny droplets of rock, producing spherules of *impact glass* that are found throughout the soil. Especially important are the multitudes of very small impacts by *micrometeorites,* grains of interplanetary dust that never strike the Earth's surface because they are filtered out by our atmosphere.

The larger impacts have broken and shattered the crust to a considerable depth to produce a layer called the *regolith.* The material of the regolith consists of rock fragments and small glass spheres produced in the heat of impacts (Figure 14.13). The depth of the regolith is least on younger surfaces and greatest on the oldest parts of the Moon, which have experienced the most impacts. On the maria, the regolith is typically about 10 meters deep. On the border of the highlands, Apollo 17 measured thicknesses of about 40 meters. And on the older parts of the highlands, it is estimated that the regolith is hundreds of meters deep. We have no similar layer on Earth because water and various chemical

Figure 14.12 Apollo photo of bootprint on Moon. (NASA)

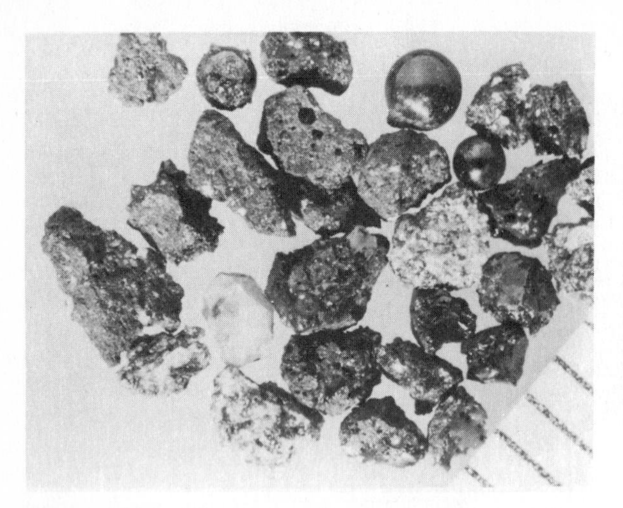

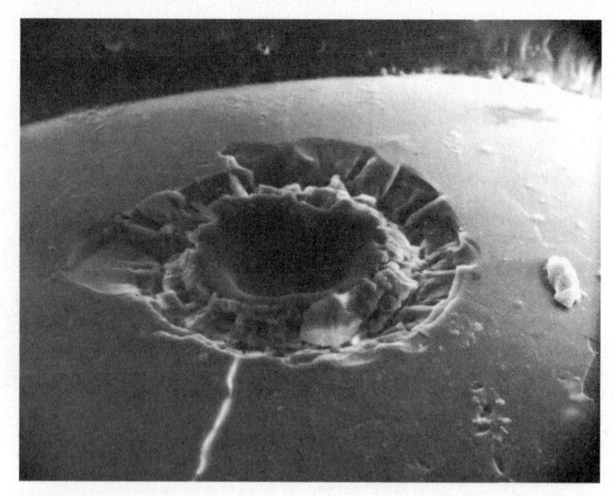

Figure 14.13 The lunar soil seen through a microscope. (a) Rock fragments and small glass spheres typically about 1 mm in size. (b) Highly magnified view of a glass sphere with a tiny microcrater, produced by the impact of an interplanetary dust grain. (NASA)

processes cement loose material together to form new sedimentary rocks.

A special kind of sedimentary rock can be produced on the Moon, however, as a direct result of impacts. The force of large impacts fuses some loose rock together to form recemented rocks called *breccias*. Almost all highland rock samples are breccias, providing further evidence of their long history of violent impacts.

In the absence of any air, the lunar surface experiences much greater temperature extremes than we do on Earth, even though we are at the same distance from the sun. Near local noon, the temperature of the dark lunar soil rises to just above the boiling point of water, while during the long lunar night it drops to about 100 K (−173°C). The extreme cooling is a result not only of the absence of air, but of the porous nature of the dusty soil, which cools more rapidly than would solid rock.

14.3 IMPACTS AND AGES

The Moon provides an important benchmark for understanding the history of the planetary system. Most solid bodies show the effects of impacts often extending back to the era when a great deal of debris from the formation process was still present. On the Earth this long history has been erased by our active geology. On the Moon, in contrast, most of the impact history is preserved. If we can understand what has happened on the Moon, we may be able to extrapolate this knowledge to the other cratered planets and satellites.

(a) Volcanic Versus Impact Origin of Craters

Until relatively recently, the question of interpreting the impact history of the Moon or other planets would have been meaningless, because few scientists believed that impacts had played any significant role in lunar geology. This attitude, which persisted into the mid-20th century, is an example of the projection of our ideas about the Earth to other planets, where they may not apply.

The difficulty was, quite simply, that impact craters do not play a major role in the geology of the Earth. Indeed, until about 1940 even the one prominent terrestrial impact crater, Meteor Crater in Arizona, was widely thought to be volcanic in origin. On the other hand, there are plenty of volcanic craters on our planet, so, it was reasoned, why shouldn't the lunar craters have a similar origin?

One of the first geologists to address this question was G. K. Gilbert, a scientist with the U.S. Geological Survey in the 1890s. Gilbert pointed out that the large lunar craters, which are broad-ringed features with floors generally below the level of the surrounding plains, are quite different in form from terrestrial volcanic craters, which tend to be smaller, deeper, and almost always occur at the tops of volcanic mountains. Further, Gilbert argued that in the absence of erosion, very ancient impact craters would survive on the Moon, while their terrestrial counterparts might well disappear. Now that we understand the role of plate tectonics in rework-

Figure 14.14 Meteor Crater in Arizona, the best-preserved terrestrial impact crater. (American Meteorite Museum)

ing the surface of the Earth, we can make this argument even more strongly.

Gilbert believed that the lunar craters were of impact origin, but he still had difficulty explaining why all of them are circular. Impacts of the sort we are most familiar with, such as throwing a stone into a sandbox, only make circular features when falling straight down; otherwise the outline of the crater is more-or-less elliptical. The solution to this problem is readily seen, however, when we note the speed with which projectiles approach the Earth or Moon. Attracted by the gravity of the larger body, the projectile strikes with at least escape velocity, which is 11 km/s for the Earth and 2.4 km/s for the Moon. The corresponding energy of impact leads to an *explosion*, and we know from experience with bomb and shell craters on Earth that explosion craters are always essentially circular. Thus recognition of the similarity between *impact craters* and *explosion craters* removed the last important objection to the impact theory for the origin of lunar craters (Figure 14.14).

(b) The Cratering Process

When an impacting projectile strikes at a speed of several kilometers per second, it penetrates two or three times its own diameter before stopping. During these few seconds, its energy of motion is transferred into a shock wave, which spreads through the target body, and into heat, which vaporizes most of the projectile and some of the surrounding target. The shock wave fractures the rock of the target, while the superheated silicate vapor generates an explosion not too different from that of a nuclear bomb detonated at ground level (Figure 14.15).

The size of the crater that is excavated depends primarily on the speed of impact. On Earth, with its larger escape velocity, the crater is typically 20 times the diameter of the projectile. On the Moon, the ratio of crater to projectile size is about 10. In neither case does it make much difference what the impacting body was made of; an icy comet is just as effective at producing craters as a stony asteroid.

An impact explosion of the sort described above leads to a characteristic kind of crater. We will describe here a crater 20 or 30 kilometers in diameter, the sort that would be produced by a small asteroid or comet impact, but most of the same features are also present for both larger and smaller craters.

The central cavity is initially bowl shaped ("crater" comes from the Greek word fro "cup"), but the gravitational rebound of the crust partially fills it in, producing a flat floor and sometimes creating a *central peak*. Around the rim, landslides create a series of *terraces*. Terraces, flat floors, and central peaks (Figure 14.17) are all characteristics of impact craters that distinguish them from most volcanic craters.

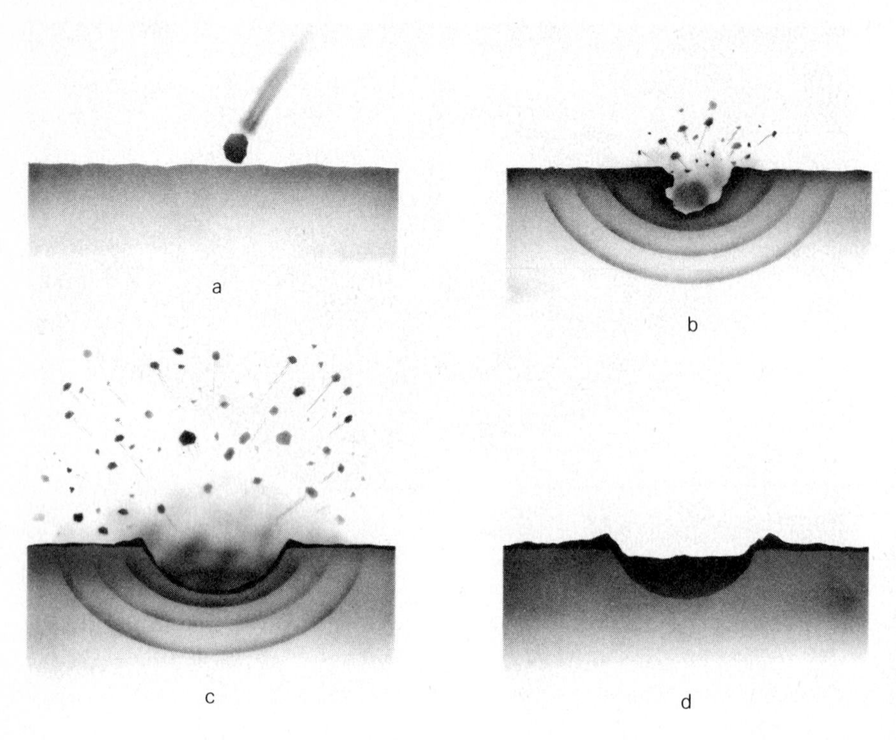

Figure 14.15 Stages in the formation of an impact crater: (a) the impact; (b) the projectile vaporizes and a shock wave spreads through the lunar rock; (c) ejecta are thrown out of the crater; and (d) most of the ejected material falls back to form secondary craters, rays, and the ejecta blanket.

The rim of the crater is turned up by the force of the explosion, so it rises above both the floor and the adjacent terrain. Surrounding the rim is an *ejecta blanket*, consisting of material thrown out by the explosion that falls back to create a rough, hummocky apron, typically about one crater diameter wide. Additional, higher-speed ejecta fall at greater distances from the crater, often digging small *secondary craters* where they strike the surface. Some of these streams of ejecta can extend for hundreds or even thousands of kilometers from the crater, creating on the Moon the bright *crater rays* that are so prominent in photos taken near full moon. The brightest lunar crater rays are associated with large young craters such as Kepler and Tycho. Rays formed long ago from other craters have been obscured by the erosional effects of subsequent impacts. On Earth, a part of the ejecta from a large impact are shaped by atmospheric friction into a spherical or teardrop form and are called *tektites*.

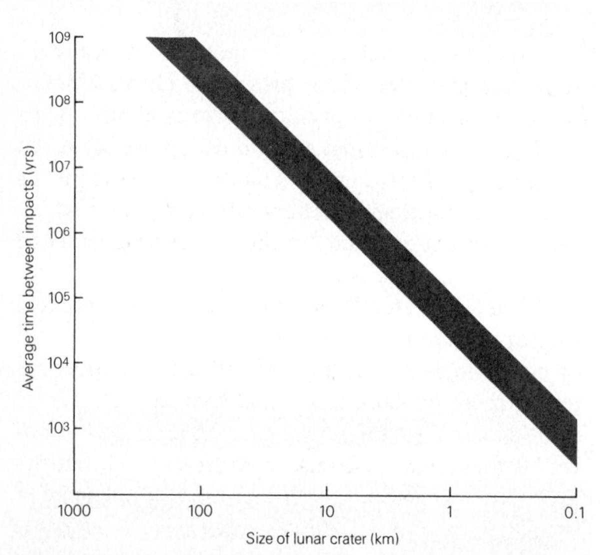

Figure 14.16 Rate of formation of impact craters on the Moon and Earth, based on our current best interpretation of numbers of Earth-crossing comets and asteroids.

(c) Using Crater Counts to Date Planetary Surfaces

If a planet has little erosion or internal activity, like the Moon during the past three billion years, it is possible to use the numbers of impact craters counted on its surface to estimate the age of that surface. By "age" we mean the time since there was a major disturbance such as the volcanic eruptions that produced the lunar maria.

Figure 14.17 King Crater on the far side, a fairly recent lunar crater 75-km in diameter, clearly showing most of the features associated with large lunar impact craters. (NASA)

This technique works because the rate at which impacts have occurred has been roughly constant for several billion years (see Section 14.3d). Thus, in the absence of forces to eliminate craters, the number of craters is simply proportional to the length of time the surface has been exposed.

Estimating ages from crater counts is a little like the experience you might have walking along the sidewalk in a snowstorm, when the snow has been falling steadily for a day or more. You may notice that in front of some houses the snow is deep, while next door the sidewalk may be almost clear. Do you conclude that less snow has fallen in front of Mr. Jones's house than Ms. Smith's? No, of course not. Instead, you conclude that Jones has recently swept the walk clean while Smith has not. Similarly, the numbers of craters indicate how long since a planetary surface was last "swept clean."

Except for the Moon and Earth, we have no planetary samples to permit exact ages to be calculated from radioactive decay (Section 12.2d). Without samples, the planetary geologist cannot tell if a surface is a million years old or a billion. But the number of craters provides an important clue. On a given planet, the more heavily cratered terrain will always be the older (as we have defined age above). And if we can calibrate the relationship between crater numbers and ages using what we know about the Moon, we may be able to estimate the ages of surfaces on other cratered planets, such as Mars or Mercury.

(d) Cratering Rates

The rate at which craters are being formed in the vicinity of the Earth and Moon cannot be measured directly, since the interval between large crater-forming impacts is longer than the span of human history. Remember that Meteor Crater is 50,000 years old. However, the cratering rate can be estimated from the number of craters on the lunar maria, or it can be calculated from the numbers of potential projectiles (asteroids and comets) present in the solar system today (Figure 14.16). Fortunately, both lines of reasoning lead to about the same answers.

For the entire land area of the Earth, these calculations indicate that a crater of 1-km diameter should be produced every few tens of thousands of years, about one or two 10-km craters every million years, and about one 100-km crater every 50 million years. Comets and asteroids appear to contribute about equally to these statistics. For the Moon the numbers are about one-tenth as great, as a consequence of the Moon's smaller total area and the lower velocity of impacts there.

If these cratering rates are extrapolated back in time, they lead to the conclusion that the craters on the lunar maria should accumulate to presently observed values in several billion years. Since the maria are 3.3 to 3.8 billion years old, this agreement suggests that comets and asteroids in approximately their current numbers have been impacting planetary surfaces for at least 3.8 billion years.

Earlier than 3.8 billion years ago, however, the impact rates must have been a great deal higher. This conclusion is immediately evident when we compare the numbers of craters on the lunar maria with those on the highlands. Typically, there are ten times as many craters on the highlands as on a similar area of maria, yet as we have seen the highlands are only a little older than the maria, typically 4.2 billion years rather than 3.8 billion years. If the rate of impacts had been constant, the highlands would have had to be at least ten times older. They would therefore have formed 38 billion years ago—long before our universe began.

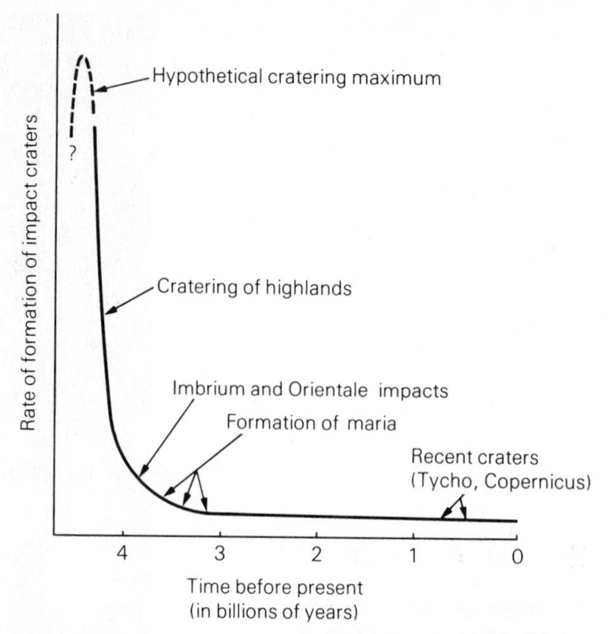

Figure 14.18 Schematic diagram of cratering flux history for the Moon, covering the past 4.3 billion years.

(e) A Theory of Planetary Cratering

A detailed examination of the lunar and terrestrial cratering record has led in recent years to a theory of planetary cratering history that can be used to estimate ages on other planets and to develop a scenario for the early history of the solar system. Much of this theory is the work of two geologists, Laurence Soderblom and Eugene Shoemaker, both at the U.S. Geological Survey's center for astrogeology at Flagstaff, Arizona (not far from Meteor Crater), although many other planetary geologists have contributed.

The first part of this theory is fairly straightforward. It notes that cratering on Earth and Moon during the past 3 billion years or so can be accounted for by comets and asteroids in about equal numbers. The cratering rates for the other terrestrial planets can then be calculated, and they turn out to be similar to those for the Moon. In the outer solar system the comets are still present, but the asteroids are not, leading to rates roughly half as great as those on planets inside the orbit of Jupiter. In both cases, the age of any lightly or moderately cratered surface (like the lunar maria) can be determined from a count of its impact craters.

Earlier than about 3.8 billion years ago the situation becomes more complicated (Figure 14.18). For the Moon, at least, the impact rates were hundreds, even thousands, of times higher. Was this a local event, or did similar high impact rates apply throughout the solar system? A partial answer is provided by Voyager pictures of the satellites of Jupiter and Saturn, which reveal some surfaces to be as heavily cratered as the lunar highlands. Since it is impossible to accumulate this many craters *at the present rate* within the lifetime of the solar system, there must have been a period of high bombardment in the outer solar system as well.

The hypothesis is made that the heavily cratered surfaces all correspond to the same intense bombardment period approximately 4.0 billion years ago. (Earlier impact rates, perhaps still larger, do not enter into this argument because their effects were obliterated by the 4-billion-year-old impacts.) It is then possible to use the measured lunar rates to calibrate the impacts throughout the solar system. This is an interesting hypothesis, and one that lets geologists estimate surface ages on more than two dozen bodies in the planetary system. However, it may not be true. Just because each object experienced a high impact rate early in its history does not ensure that these periods of high bombardment were simultaneous.

As it turns out, it does not matter for most purposes whether the heavy bombardments were really simultaneous. The main point is that they all occurred a long time ago, so that any heavily cratered surface can be assigned an age going back to the first 700 million years of solar system history. Even in this form, without trying to determine the exact cratering rates early in solar system history, the theory of a common origin for impact cratering on all the planets and satellites is a powerful unifying idea that has contributed greatly to establishing a chronology for planetary evolution. We will make use of these ideas in the following chapters as we discuss the interpretation of spacecraft photos of other planets.

14.4 THE ORIGIN OF THE MOON

(a) Three Theories

Understanding the origin of the Moon has proved to be extremely difficult for planetary scientists. Part of the problem is simply that we know too much about our satellite. There is a great wealth of data, particularly on the details of the elemental and isotopic composition of the Moon, that present a challenge to any simplified theory of lunar origins.

Most of the various theories for the Moon's origin follow one of three general ideas: (1) that the Moon was once part of the Earth but separated from it early in their history (the fission theory); (2) that the Moon formed together with (but independent of) the Earth, as we believe many satellites of the outer planets formed (the sister theory); and (3) that the Moon formed elsewhere in the solar system and was captured by the Earth (the capture theory).

Unfortunately, there are fundamental problems with each of these ideas. Perhaps the easiest theory to reject is the *capture theory*. The difficulty is primarily that no one knows of any way that the early Earth could have captured a large satellite from elsewhere. Recall (Section 5.1h) that one body approaching another cannot go into orbit around it without a substantial loss of energy—this is the reason that spacecraft destined to orbit other planets are equipped with retrorockets. Further, if such a capture did take place, it would be into a very eccentric orbit, rather than the nearly circular orbit the Moon occupies today. And finally, there are too many compositional similarities between the Earth and the Moon, particularly an identical fraction of the major isotopes of oxygen, to justify seeking a completely independent origin.

The first possibility, that the Moon separated from the Earth, was suggested in the late 19th century by George Darwin, the astronomer son of naturalist Charles Darwin. It is often called the *fission theory*, since it suggested that the early, liquid Earth was spinning so fast that the Moon separated from it by fission. More modern dynamical calculations have shown that fission, as imagined by Darwin, is impossible. Further, it is difficult to understand how a Moon made out of terrestrial material in this way could have developed the many distinctive chemical differences that are now known to characterize our satellite.

Scientists are therefore left with the *sister theory*, that the Moon formed alongside the Earth, or with some modification of the fission theory that finds a more acceptable way for the lunar material to be separated from the Earth.

(b) Finding an Acceptable Hypothesis

THE SISTER THEORY

In the sister theory the Earth and Moon formed together out of the solar nebula. Growing in the same region, they presumably began with nearly identical composition, namely, the composition of the planetesimals that orbited at about 1 AU from the protosun. This idea has the great virtue that it is not unique to the Earth-Moon system. It is widely believed that the regular satellites of the outer planets also formed from solid material orbiting them in miniature versions of the solar nebula itself, as we will discuss further in Section 18.1.

The primary problems with the sister theory arise when we look at the differences in composition between the Moon and the Earth, especially the depletion of iron and other metals in our satellite. How could the Moon, forming out of the same raw material as the Earth, avoid accumulating its share of metals? No one knows. Nearly as great a problem is raised by the absence of water and other volatiles on the Moon. Some gases would escape because of the lower lunar gravity, but it is difficult to see how so many volatiles were lost, or how other differences arose in the abundances of individual elements.

One way out of these difficulties is to make the Moon out of the same material as the Earth's *mantle*, after the planet had differentiated and the met-

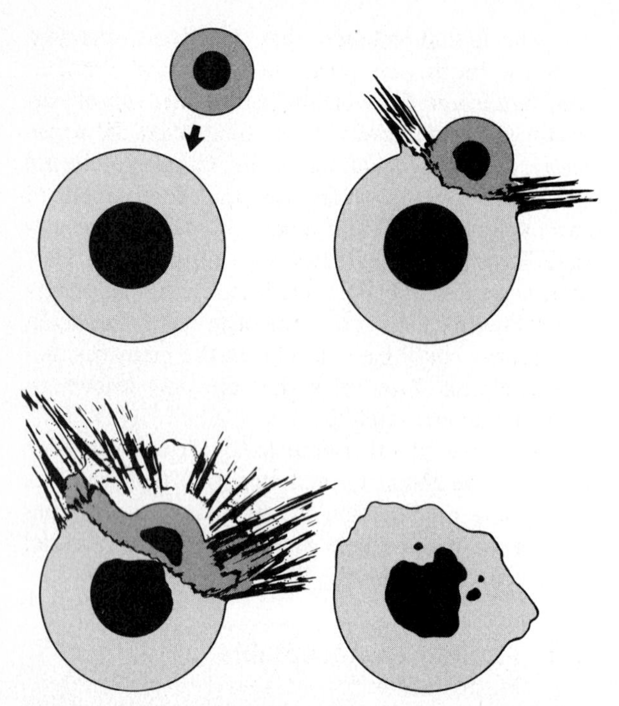

Figure 14.19 Schematic stages in the impact of a Mars-sized projectile with the Earth, adapted from calculations carried out by J. Melosh of the University of Arizona. Such a catastrophic impact may have resulted in the formation of the Moon.

als settled into its core. But then we find ourselves edging toward the fission theory again. Out of these thoughts has come a possible synthesis, called the *giant impact theory* of the origin of the Moon.

THE GIANT IMPACT THEORY

The giant impact theory is in part a response to the problems of lunar chemistry posed by the sister theory, and in part a reflection of an increasing interest among planetary scientists in the role of impacts in the late stages of planet growth. As described in Section 12.4, there is increasing evidence for a few very large impacts as part of the final clearing out of debris from between the planets. What if one of these very large projectiles struck the Earth? Some of the answers are being provided by calculations carried out by Jay Melosh of the University of Arizona, David Stevenson of Caltech, and George Wetherill of the Carnegie Institution of Washington, D.C.

The giant impact theory envisions the Earth being struck obliquely by an object of about one-tenth the Earth's mass, or about the size of Mars. This is very nearly the largest impact the Earth could experience without being shattered. Calculations with supercomputers (Figure 14.19) show that such an impact would disrupt much of the Earth, penetrating into the core, ejecting a vast amount of material into space, and releasing almost enough energy to break the planet apart.

Material totalling several percent of the mass of the Earth could be ejected in such an impact. Most of the material would be from the mantles of the Earth and the impacting body, and it would initially be ejected at high speed in the form of hot vapor. Perhaps—and the calculations are not really clear on this point—much of this hot vapor could condense into a ring of material in orbit around the Earth. Alternatively, it might form a sort of superheated silicate atmosphere in rapid rotation around the planet. Either way, it is suggested that this ejected material ultimately condensed to form the Moon, as described in the sister theory.

The giant impact hypothesis offers potential solutions to most of the major problems raised by the chemistry of the Moon. First, since the raw material for the Moon is derived from the mantles of the Earth and the projectile, the absence of metals is easily understood. Second, most of the volatile elements could have been lost during the high-temperature phase following the impact. Yet by making the Moon primarily of terrestrial mantle material, it is also possible to understand similarities such as identical oxygen isotopic abundances.

No one knows whether the giant impact hypothesis or some variant to be derived from it really describes the origin of the Moon. We present it here primarily as an illustration of the complexity of many current problems in planetary science. As a result of space missions we now have a great deal of data on the planets, but in order to understand fully the implications of those data will require much additional work and more than a little inspiration.

(c) Subsequent Evolution of the Moon

If the Moon formed from a ring of particles in orbit around the Earth, it initially was much closer and revolved around the Earth in much less than a month. Being closer to our planet, the Moon was subject to strong tidal forces (Section 12.2). The effect of these forces would have been to slow the ro-

Figure 14.20 A large boulder on the Moon with many cracks. It is one of many lunar features described in considerable detail by Apollo 17 Astronaut Harrison H. Schmitt. (NASA)

tation of both the Earth and the Moon, eventually bringing the Moon into its current configuration, always keeping the same side toward the Earth.

From the principle of the conservation of angular momentum, the orbital momentum of the Earth-Moon system must have increased as the rotational momentum of these two bodies decreased. An increase of orbital momentum involves an expansion of the Moon's orbit, increasing its distance from Earth and lengthening its period of revolution. Thus we can understand how the Moon's orbit evolved to its present state. Sensitive laser ranging devices show that the Moon's orbit is still expanding, compensating for the gradual slowing of the rotation of the Earth by lunar tides.

The Moon should have formed from solid material, but it was probably heated from the process of accretion and from the continuing bombardment by other debris still populating the terrestrial neighborhood. Sometime within the first hundred million years of its formation the Moon's temperature rose high enough for it to differentiate. The anorthositic materials rose to the surface, where they cooled to form the original lunar crust. This crust was fully formed by 4.4 billion years ago, and perhaps even sooner.

Probably continuing heavy bombardment of both Earth and Moon resulted in heavy cratering and considerable recycling of surface material for several hundred million years. Little evidence for this early period remains on either planet, however. All we know for sure is that heavy bombardment of the Moon continued up to about 4 billion years ago, or at least half a billion years after the Moon's formation. Then, between 4.0 and 3.8 billion years ago, the cratering rate dropped to essentially its current value.

As a part of the final burst of bombardment, a dozen or so large impact basins were formed on the Moon (Figure 14.20). Later, internal heating by radioactivity produced partial melting of the mantle. Basaltic lavas worked their way to the surface on the side of the Moon facing the Earth, where elevations are lower than on the far side, flooding most of these basins to produce the maria. This volcanic activity persisted until a little more than 3 billion years ago, when, as we have seen, the Moon settled down into a premature old age.

14.5 MERCURY

Mercury is one of the brightest objects in the sky. At its brightest it is inferior in brilliance only to the sun, the Moon, the planets Venus, Mars, and Jupiter, and the star Sirius. Yet, most people—including even Copernicus, it is said—have never seen Mercury. The planet's elusiveness is due to its proximity to the sun. Its orbit is only about 40 percent the size of Earth's; it can never appear further from the sun than about 28°. It is visible to the unaided eye for a period of only a week or so, at times when it is near eastern elongation, appearing above the western horizon just after sunset, and also when it is near western elongation, rising in the east shortly before sunrise.

Mercury was well known to the ancients of many lands. The earliest observers, however, did not recognize it as the same object when it appeared as an evening star and as a morning star. The early Greeks, for example, called it Hermes (Mercury) in the evening and Apollo when it was seen in the morning twilight.

(a) Mercury's Orbit

Mercury is the nearest to the sun of the nine major planets and, in accordance with Kepler's third law, has the shortest period of revolution about the sun (88 of our days) and the highest average orbital speed (48 km/s). It is appropriately named for the fleet-footed messenger god of the Greeks and Romans.

The semimajor axis of the orbit of Mercury, that is, the planet's median distance from the sun, is 58 million km or 0.39 AU. However, because Mercury's orbit has the high eccentricity of 0.206, its actual distance from the sun varies from 46 million km at perihelion to 70 million km at aphelion. Pluto is the only major planet with a more eccentric orbit. Furthermore, the 7° inclination of the orbit of Mercury to the plane of the ecliptic is also greater than that of any other planet except Pluto.

Figure 14.21 The planet Mercury in a mosaic of photographs made by Mariner 10 during its approach to the planet. (NASA/JPL)

(b) General Properties and Structure

Table 14.1 compared some basic properties of Mercury with those of the Moon. Mercury's mass is 1/18 that of the Earth; Pluto is the only planet with a smaller mass. Mercury is also the second smallest of the planets, having a diameter of only about 4880 km, less than half that of the Earth. Its density is 5.5 g/cm^3, about the same as that of the Earth. Being much smaller than the Earth, however, Mercury has an uncompressed density of 5.3 g/cm^3, as compared with Earth's 4.5 g/cm^3. Thus the composition of Mercury is fundamentally different from either the Earth or the Moon.

The higher uncompressed density of Mercury implies a larger proportion of metals than in the Earth, and of course much more than in the Moon. The most likely models for the interior suggest an iron-nickel core with a mass amounting to 60 percent of the total, with the rest of the planet made up primarily of silicates. The core has a diameter of 3500 km and extends to within 700 km of the surface. We could think of Mercury as a metal ball the size of the Moon surrounded by a rocky crust 700 km thick. Lighter materials, including many silicates, must have been depleted where Mercury formed, compared with the Earth's position in the solar nebula.

The escape velocity is too low and the surface temperature too high for Mercury to retain any substantial atmosphere. In 1985, however, an extensive but extremely thin atmosphere of sodium was detected spectroscopically. Apparently atoms of this metal are ejected from the surface through bombardment by the solar wind.

One of the surprises of the Mariner 10 mission in 1974 was the discovery of a magnetic field on Mercury. Although less than 1 percent as strong as the Earth's field, Mercury's magnetic field is likewise intrinsic to the planet. Such planetary magnetism was not expected because the interior of Mercury was thought to be solid, so that a field would not be generated. It is difficult to see how such a small planet could retain enough interior heat to have a liquid metal core. No really satisfactory resolution of this seeming contradiction has been achieved in the years since the Mariner 10 flybys.

(c) Rotation

Visual studies of Mercury's indistinct surface markings were once thought to indicate that the planet kept one face to the sun, and for many years it was widely believed that Mercury's rotation period equalled its period of revolution about the sun of 88 days.

Radar observations of Mercury in the mid-1960s, however, showed conclusively that Mercury does rotate with respect to the sun. Its sidereal period of rotation (that is, with respect to the distant stars) is about 59 days. The Italian dynamicist Giuseppe Colombo (1920–1984) first pointed out that this is very nearly two-thirds of the planet's period of revolution, and subsequently it was found that there are theoretical reasons for expecting that Mercury can rotate stably with a period of exactly two-thirds that of its revolution—58.65 days. If Mercury were not perfectly spherical, but were deformed so that one dimension through the equator were longer than the others, the sun's force on that bulge should force the long axis of the planet to point to the sun when it is at perihelion, where the sun's differential force on Mercury is strongest. This condition would be met if the planet rotated with its revolution period, but also with certain other rotation periods, the most likely being two-thirds the revolution period, so that at successive perihelions alternate ends of the long axis of the planet are pointed toward the sun.

(d) Temperature

Although the temperature of Mercury had been measured from the Earth, better data came from the Mariner 10 flyby in 1974. The daylight temperature on the surface ranges up to about 700 K at noontime. Just after sunset, however, the temperature drops quickly to about 150 K, and then slowly descends to about 100 K just before dawn. The range in temperature on Mercury is thus 600 K, more than on any other planet.

(e) The Surface of Mercury

The first closeup look at Mercury came on March 29, 1974, when the Mariner 10 spacecraft passed 9500 km from the surface of the planet. On that date, and for several days before and after the encounter, Mariner 10 televised more than 2000 photographs to Earth, revealing details with a resolution down to 150 m. Mariner 10, in a planned elliptical orbit about the sun, passed Mercury again on September 21, 1974, and finally on March 16, 1975, obtaining over 2000 more pictures.

Mercury strongly resembles the Moon in appearance (Figures 14.21 and 14.22). It is covered with thousands of craters up to hundreds of kilometers across, and larger basins up to 1300 km in diameter. There are also scarps (cliffs) over a kilometer high and hundreds of kilometers long, as well as ridges and plains. Some of the brighter craters are rayed, like Tycho and Copernicus on the Moon, and many have central peaks.

Most of the Mercurian features have been named in commemoration of artists, writers, composers, and other contributors to the arts and humanities, in contrast with the scientists commemorated on the Moon. One of us (Morrison) is the head of the International Astronomical Union commission that assigns names to features on Mercury. Among the most prominent craters are Bach, Shakespeare, Tolstoy, Mozart, and Goethe. A large basin has been named Caloris (Figure 14.23), for the sun shines directly down on the basin when Mercury is at perihelion, and Caloris is probably thus the place on the planet with the highest noon temperature.

The larger basins resemble the lunar maria, in both size and appearance. They show evidence of much flooding, as do the maria, from lava flows, possibly (unlike the Moon) released during the impacts that produced the features. However, these flows, if they are real, do not have the distinctive dark color that characterizes the lunar maria. Since Mercury is so difficult to study telescopically and has been visited by only one spacecraft, we actually know very little about the chemistry of its surface or the details of its geological history.

Unlike Earth there is no evidence of faulting on Mercury; plate tectonics evidently does not occur there. Mercury's distinctive long scarps, however, which sometimes cut across craters, seem to be due to crustal compression. Apparently this planet shrank, wrinkling the crust, and it did so after most of the craters on its surface were formed. If the standard cratering chronology applies to Mercury, this shrinkage must have taken place during the past 4.0 billion years. If we understood better the interior structure and composition of this planet, we could probably calculate how internal changes in temperature might have led to this global compression, which has no counterpart on the Moon or the other terrestrial planets.

Figure 14.22 Cratered terrain of Mercury photographed by Mariner 10. The large flat-floored crater at the right is about 100 km in diameter—about the size of the lunar crater Copernicus. Many of the smaller craters are probably caused by particles thrown out by the impact that formed the large crater. (NASA/JPL)

Figure 14.23 In the left half of this mosaic is the ring basin Caloris, 1300 km in diameter. It is the largest structural feature on Mercury seen by Mariner 10 in its first pass of the planet. (NASA/JPL)

EXERCISES

1. One of the primary scientific objectives of the Apollo program was the return to Earth of samples of lunar material. Why was this so important? What have we learned from the samples, and are they still of value now?
2. Apollo astronaut David Scott dropped a hammer and a feather together on the Moon, and both reached the ground at the same time. There are two reasons why this experiment on the Moon had distinct advantages over the same experiment performed on the Earth. What are these advantages?
3. What would a 100-kg person weigh on the Moon?
4. Compare the bulk composition of the Moon with that of the Earth. From the differences, what might you conclude about the origin of the Moon? Is there any way to get from terrestrial-type material to the composition of the Moon?
5. Galileo thought the lunar maria were seas of water. If you had no better telescope than Galileo's, could you prove that the maria are not composed of water?
6. The lunar highlands have about ten times more craters on a given area than do the maria. Does this mean that the highlands are ten times older? Explain your reasoning.
7. Explain why the presence of flooded craters (the "ghost craters") in the lunar maria is evidence for a long time interval between the formation of the mare basins and the filling of those basins by lunar magma.
8. In any one mare there are a variety of rock ages, spanning typically about a hundred million years. The individual lava flows as seen in Hadley Rille by the Apollo 15 astronauts were about 4 m thick. Estimate the average interval between lava flows if the total depth of the lava in the mare is 2 km.
9. Why are the lunar mountains smoothly rounded rather than having sharp, pointed peaks (the way they had almost always been depicted before the first lunar landings)?

*10. Suppose an isolated mountain is observed on the Moon 100 km from the terminator. Its shadow is 40 km long. How high is the mountain?

*11. Suppose a mountain peak on the night side of the Moon rises just high enough to catch some of the rays of the rising sun and shine like a bright spot of light. If the mountain is just 100 km from the terminator, what is its height?

12. If the typical depth of the regolith on the lunar maria is about 10 m, how much new regolith accumulates, on the average, each century?
13. Why did it take so long for geologists to recognize that the lunar craters had an impact origin rather than being volcanic?
14. Explain the evidence for a period of heavy bombardment on the Moon about 4 billion years ago. What might have been the source of this high flux of impacting debris?
15. How would a crater made by the impact of a comet with the Moon differ from a crater made by the impact of an an asteroid?
16. The suggestion that a giant impact that nearly destroyed the Earth was necessary for the formation of the Moon seems like a drastic hypothesis. Indicate why planetary scientists are invoking such a seemingly implausible idea for the origin of the Moon.
17. What are the main geological differences between the Moon and Mercury? (Remember that geology deals with surface topography and composition, not the deep interior.)
18. Give several reasons why Mercury would be a particularly unpleasant place to live.
19. Only one spacecraft, Mariner 10, has been sent to Mercury. Can you think of any reasons why there have not been additional missions to follow up on the discoveries of Mariner 10?

MARS: THE MOST HOSPITABLE PLANET

Thomas A. Mutch (1931–1980) of Brown University was one of the first geologists who sought to carry out comparative studies of the planets. In 1976, as the Team Leader for the Viking lander imaging experiment, he served as an articulate and enthusiastic advocate for the exploration of Mars. Following the completion of the Viking program, Mutch became the NASA Associate Administrator for Space Science, but his leadership of the U.S. space science program ended little more than a year later when he was killed in a mountain climbing accident in the Himalayas. (Brown University)

15.1 GENERAL PROPERTIES OF MARS

Mars is the most favorably situated of the planets for observation from Earth and has excited more interest and comment than any other. It has long been suspected, and is now abundantly clear, that Mars is also the only other place in our planetary system that resembles the Earth closely enough to be considered an abode for life. While Mars supports no life today, it is not impossible that it may have done so in the past. Even more exciting, Mars could become inhabited in the future, as a result of human colonization.

(a) Basic Facts

The median distance of Mars from the sun is 227 million km, but its orbit is somewhat eccentric (0.093), and its heliocentric distance varies by 42 million km. The sidereal period of revolution is 687 days, and the synodic period is 780 days. Thus approximately once every 26 months Mars is at opposition, when it is above the horizon all night and is most favorably placed for observation. The same 26-month interval separates the best launch opportunities for sending spacecraft to Mars.

Seen through a telescope, Mars is both tantalizing and disappointing. The planet is distinctly red in color, due (as we now know) to the presence of iron oxides in its soil. At its nearest, Mars subtends about 25 arcseconds, and the best resolution obtainable is about 100 km, or about the same as the Moon seen with the unaided eye. At the resolution of 100 km, however, no hint of topographic structure can be detected: no mountains, no valleys, not even impact craters (Figure 15.1). On the other hand, the bright polar caps of the red planet can easily be seen, together with dusky surface markings that gradually change in outline and intensity from season to season. Mars alone of all the planets has a surface that can clearly be made out from Earth, and this surface exhibits changes that bespeak a dynamic atmosphere and that were once considered evidence for life.

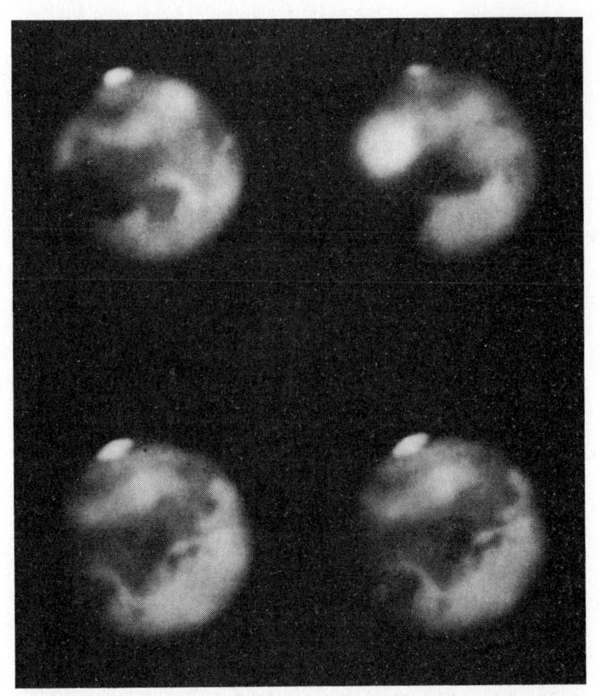

Figure 15.1 These are among the best Earth-based photos of Mars, taken in 1971 by the NASA-Lowell Observatory Planetary Patrol station at Mauna Kea Observatory. (Lowell Observatory Planetary Patrol)

The presence of permanent surface markings enables us to determine the rotation period of Mars with great accuracy; its sidereal day is $24^h\ 37^m\ 23^s$, just a little greater than the rotation period of the Earth. This high precision is not obtained by watching Mars for a single rotation, but by noting how many turns it makes in a long period of time. Good observations of Mars date back for more than 200 years, a period during which tens of thousands of Martian days have passed. The rotation period is known now to within a few hundredths of a second.

The rotational pole of Mars has a tilt or obliquity of about 25°, very similar to that of the Earth's pole. Thus Mars experiences seasons very much like those on Earth. Because of the longer Martian year, however, seasons there each last about six of our months. Bright polar caps of water or carbon dioxide ice form at each pole during the winter and largely evaporate during spring and summer.

As a planet, Mars is rather small, with only 11 percent the mass of the Earth. It is larger than either the Moon or Mercury, however, and unlike them it retains a thin atmosphere. This atmosphere

TABLE 15.1 Properties of Mars

Semimajor axis	1.52 AU
Period	1.88 yr
Mass (Earth = 1)	0.11
Diameter	6794 km
Density	3.9 g/cm^3
Uncompressed density	3.8 g/cm^3
Surface gravity (Earth = 1)	0.38
Escape velocity	5.0 km/s
Rotation period	24.6 hrs
Surface area (Earth = 1)	0.28
Atmospheric pressure	0.006 bars

is composed almost entirely of carbon dioxide, but its pressure is less than 1 percent that of Earth. Mars is also large enough to have supported considerable geological activity, some apparently persisting to the present day. On the whole, it is a much more interesting and attractive place than either of the objects studied in the last chapter.

Table 15.1 summarizes some of the basic data for Mars.

(b) The Canal Controversy

Approximately 100 years ago Mars became an object of great public interest and the subject of a controversy that persisted for decades. At the favorable opposition of 1877, the Italian astronomer Giovanni Schiaparelli (1835–1910) announced the discovery of long, faint, straight lines on Mars that he called *canale*, or channels. In English-speaking countries, the term canale was translated as *canals*, a word that implies an artificial origin. After the existence of these markings was confirmed by other observers at the opposition of 1879, there was widespread speculation in both scientific and public circles that artificial waterways existed on Mars, presumably the work of an advanced race of Martians. The public interest in (and concern over) Martians was considerably enhanced in 1898 with the publication of H. G. Wells's popular novel *War of the Worlds*, which depicts an invasion of Earth by aliens from Mars.

The most effective proponent of intelligent life on Mars was Percival Lowell (1855–1916), self-made American astronomer and member of the great Lowell family of Boston (Figure 15.2). Lowell was fascinated by the discovery of the canals, and he decided to devote his talents and fortune to pur-

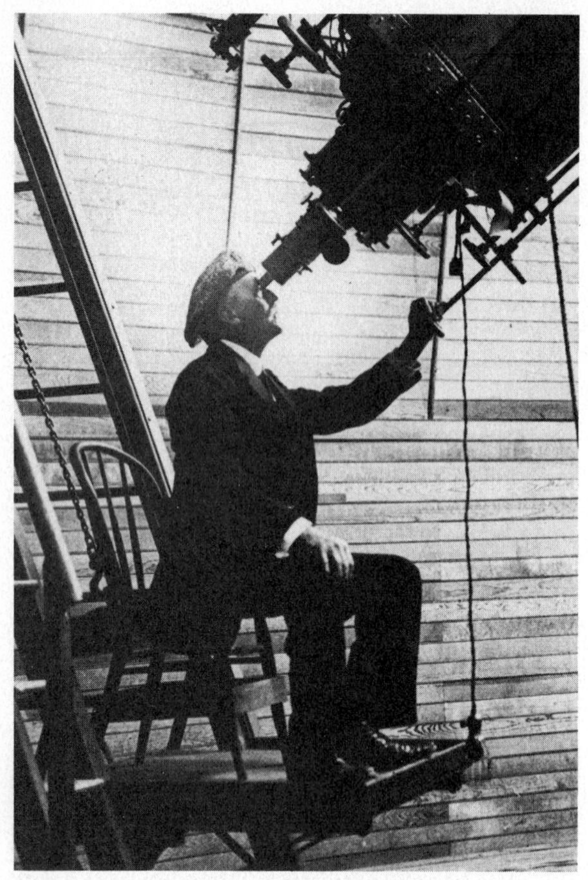

Figure 15.2 Percival Lowell in about 1910 observing with his 24-inch telescope at Flagstaff. (Lowell Observatory)

suing studies of the red planet. After an extensive search for the best available site, in 1894 he constructed the Lowell Observatory in Flagstaff, in the Territory of Arizona. An ardent observer, Lowell saw hundreds of Martian canals through his telescopes, and even spotted similar linear markings on Venus and the large satellites of Jupiter. In addition, however, Lowell was an effective author and public speaker, and during the first decade of the 20th century he made a convincing case for intelligent Martians, who he believed had constructed huge canals to carry water from the polar caps in an effort to preserve their existence in the face of a deteriorating climate.

The Martian canals were always difficult to study, glimpsed only occasionally as "seeing" caused the tiny image of Mars to shimmer in the telescope. Lowell saw them everywhere, but many other observers were skeptical. When the new, larger telescopes constructed at Mt. Wilson failed to confirm the existence of canals, the skeptics seemed to be vindicated. Astronomers had given up the idea by the 1930s, although it persisted in the public consciousness until the first spacecraft photographs clearly showed that there were no Martian canals. Now it is generally accepted that the canals were an optical illusion, the result of the human mind's tendency to see order in random features glimpsed dimly at the limits of the eye's resolution.

Lowell and the canal controversy did have an effect on astronomy, however, that extended long beyond his death in 1916. He stimulated a public awareness of Mars that was influential 50 years later in the decision to send the Viking spacecraft to search for life on the red planet. But at the same time the sensationalism of the subject alienated many professional astronomers and contributed to a marked decline in planetary studies. Not until the dawn of the space age did planetary astronomy again become an active and productive discipline.

(c) The Exploration of Mars

Although astronomical studies of Mars revived somewhat in the 1950s, it was not until the first spacecraft visited Mars that both popular and scientific interest was again focused on this planet. On July 22, 1965, the U.S. Mariner 4 spacecraft flew past Mars and radioed 22 closeup pictures to Earth. These first photos showed an apparently bleak planet with abundant impact craters. In those days craters were unexpected, and perhaps people who were romantically inclined still expected to see canals; in any case, the Mariner 4 results represented something of a shock. Mariners 6 and 7 followed in 1969, returning much more data and apparently confirming that Mars was cratered, with little evidence of internal geological activity.

A major step forward was achieved in 1971 when Mariner 9 became the first spacecraft to orbit another planet. In the same year the U.S.S.R. successfully landed two spacecraft on the surface, but both failed before they could return useful data. The Mariner 9 orbiter was spectacularly successful, mapping the entire planet at a resolution of about 1 km and discovering a great variety of geological features missed by the previous flybys, including volcanoes, huge canyons, intricate layers on the polar caps, and channels that appeared to have been cut by running water. Mariner 9 set the stage for Viking (Figure 15.3), the most ambitious and successful of all U.S. planetary missions.

Figure 15.3 The Viking lander spacecraft, identical to the two craft that reached the surface of Mars in 1976. (NASA/JPL)

The Viking missions, which flew to Mars in 1976, consisted of two orbiters and two landers. The orbiters surveyed the planet at higher resolution than Mariner 9 and served to relay communications for the two landers. After an exciting and often frustrating search for a safe landing spot, Viking 1 touched down on the surface of Chryse Planitia (the Plains of Gold) on July 20, 1976, exactly seven years after Neil Armstrong's historic first step on the Moon. Two months later Viking 2 landed with equal success in another plain farther north, called Utopia.

Most of the information discussed in this chapter is derived from the four Viking spacecraft. The two landers each conducted a search for evidence of Martian micro-organisms, as discussed in Section 15.4. Instruments analyzed the atmosphere and surface material, and "weather stations" were established to radio back meteorological data to Earth. Detailed photography from the orbiters eventually achieved a resolution of just a few meters, and close flybys of the two Martian moons were successfully completed. The orbiters continued to operate until their control gas was exhausted, after which the surface stations were commanded directly from Earth. Not until November 5, 1982, did the mission end, when Viking 1 sent its last message after almost 6½ Earth years on the surface. Even then, the mission was terminated by a human programming error on Earth, not by failure of the robot spacecraft on Mars.

In 1981, NASA transferred ownership of the Viking 1 lander to the National Air and Space Museum. It is the only museum exhibit located on another planet! Perhaps someday it will be brought back for public display in Washington (or Moscow), but even more appropriate would be to leave it on Mars, as a symbol for the human colonists who will someday make the red planet their home.

(d) Global Properties

Mars has a diameter of 6787 km, just over half that of the Earth, resulting in a surface area very nearly equal to the continental area of our planet. Its mass is 11 percent of the Earth's, and its uncompressed density is 3.8 g/cm^3, about midway between that of the Earth (4.5) and the Moon (3.2). Thus Mars apparently has a composition between that of the Earth and the Moon, consisting primarily of silicates but with the possibility of a substantial metal core. Tracking of spacecraft in orbit around Mars confirms this suspicion of a core, which is estimated to consist primarily of iron sulfide (FeS) and to have a diameter of about 2400 km. The planet has no detectable magnetic field, suggesting that this core is not liquid.

From its size, it is clear that Mars (like Earth, Moon, and Mercury) is a differentiated planet. Any scenario for its origin would have permitted it to heat sufficiently to separate into a core, mantle, and crust. However, since no seismic studies have been carried out, any conclusions about the exact interior structure of Mars are somewhat speculative.

In an overview of the properties of Mars, it is appropriate to discuss the large-scale features of the

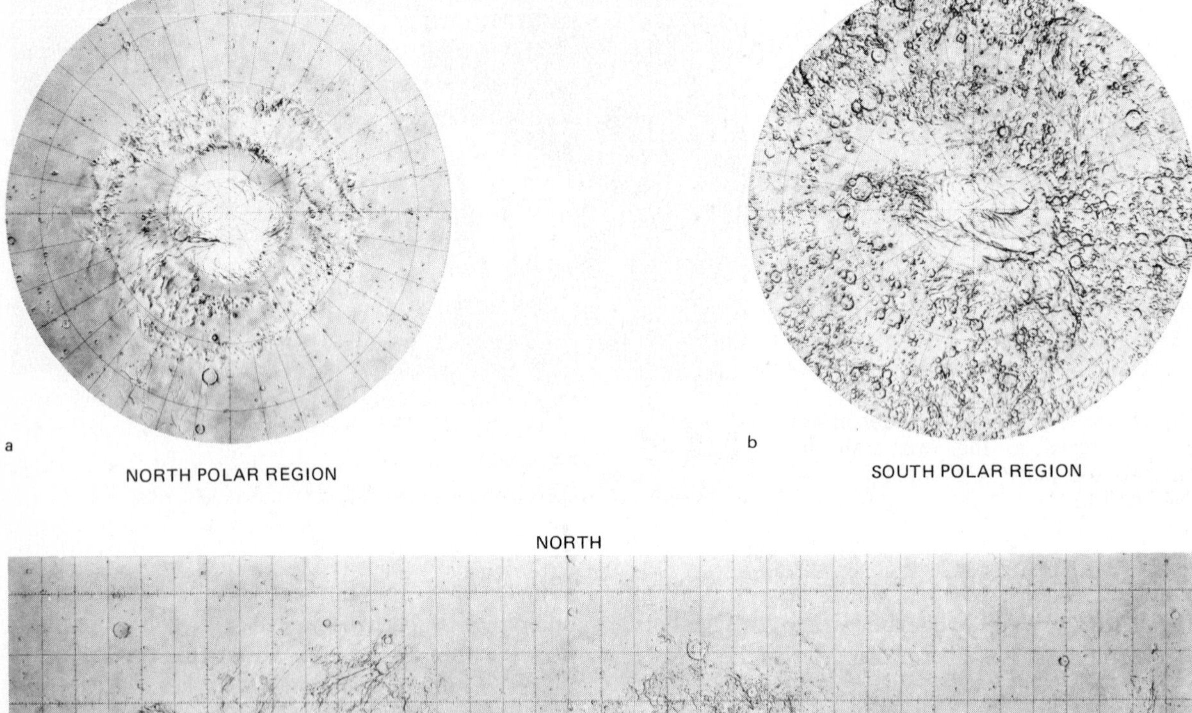

Figure 15.4 A topographical map of the planet Mars: (a) the north polar region, (b) the south polar region, and (c) the equatorial and temperate regions. (NASA/JPL)

surface topography. Does Mars have continents and ocean basins like the Earth, indicating a possible history of plate tectonics? Or does it have large impact basins like the lunar maria? Or is it different from either Earth or Moon? The answers, as revealed from spacecraft mapping, are a partial "yes" to all three questions.

The surface of Mars divides into two major terrains, superficially analogous to the way the Earth might have looked 200 million years ago, when all the continents were together (Figure 15.4). Approximately half the planet, lying primarily in the southern hemisphere, consists of ancient cratered terrain. The other half, which is primarily in the north,

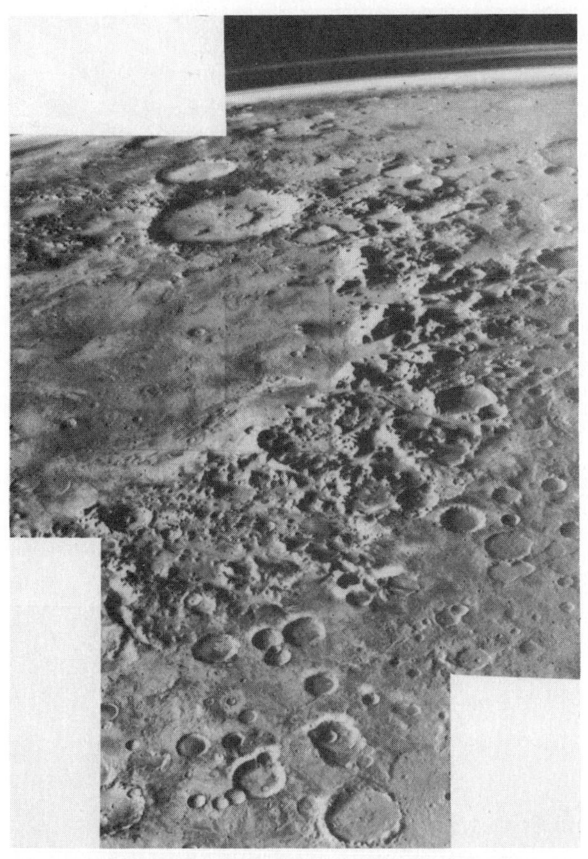

Figure 15.5 The Argyre impact basin, about 700 km in diameter, photographed by the Viking orbiter. (NASA/JPL)

contains younger volcanic plains. These volcanic plains, like the terrestrial ocean basins or the lunar maria, are at an average elevation several kilometers lower than the older southern uplands. In many places the boundary between the two terrains is relatively sharp, with the surface sloping down 4 or 5 km in a span of a few hundred km. There are no mountain rings to mark the borders of the lowland plains, however, and they did not originate like the lunar mare basins in catastrophic impacts. Some other process, still unknown, destroyed the ancient crust over half the planet and lowered the surface levels by about 5 km on the average.

Mars does have impact basins, but they lie in the old southern uplands. The largest basin, called Hellas, is about 1800 km in diameter and 6 km deep, larger than the Imbrium basin on the Moon. Smaller but better preserved is the Argyre basin (Figure 15.5), with a diameter of 700 km. Both basins are surrounded by mountains formed from the impact explosion.

In addition to the north-south division of the planet into old uplands and younger volcanic plains, Mars displays an impressive east-west asymmetry. On one side of the planet, straddling the boundary between uplands and plains, is an immense bulge the size of North America that rises nearly 10 km above its surroundings. This is the Tharsis bulge, a volcanically active region crowned by four great volcanoes that rise another 15 km into the pink Martian sky. We will discuss Tharsis and its volcanoes in some detail in Section 15.2.

The total range in elevations on Mars is very large, amounting to 30 km between the tops of the highest volcanoes and the bottom of the Hellas basin. In comparison, the maximum elevation range on Earth (from Mt. Everest to the deep ocean trenches) is only about 17 km. As we will see in Chapter 16, Venus has about the same elevation range as the Earth. Why are the mountains of Mars so much larger? We will answer this question in Section 16.4.

(e) The Moons of Mars

Centuries ago, Kepler, hearing of Galileo's discovery of four satellites of Jupiter, speculated that Mars should have two satellites. The speculation was based on numerological, not scientific considerations. In 1726, Jonathan Swift in his satire *Gulliver's Travels* described Gulliver's fictional visit to the land of Laputa, where he found scientists engaged in many interesting investigations. In one of them, Gulliver reported, Laputian astronomers had discovered

> ". . . [two] satellites, which revolve about Mars, whereof the innermost is distant from the centre of the primary planet exactly three of the diameters, and the outermost five; the former revolves in the space of ten hours, and the latter in twenty one and a half; so that the squares of their periodical times are very near in the same proportion with the cubes of their distance from the centre of Mars, which evidently shows them to be governed by the same law of gravitation that influences the other heavenly bodies."

It is an interesting coincidence that in 1877, 150 years later, Asaph Hall (1829–1907) of the U.S. Naval Observatory actually discovered two small sat-

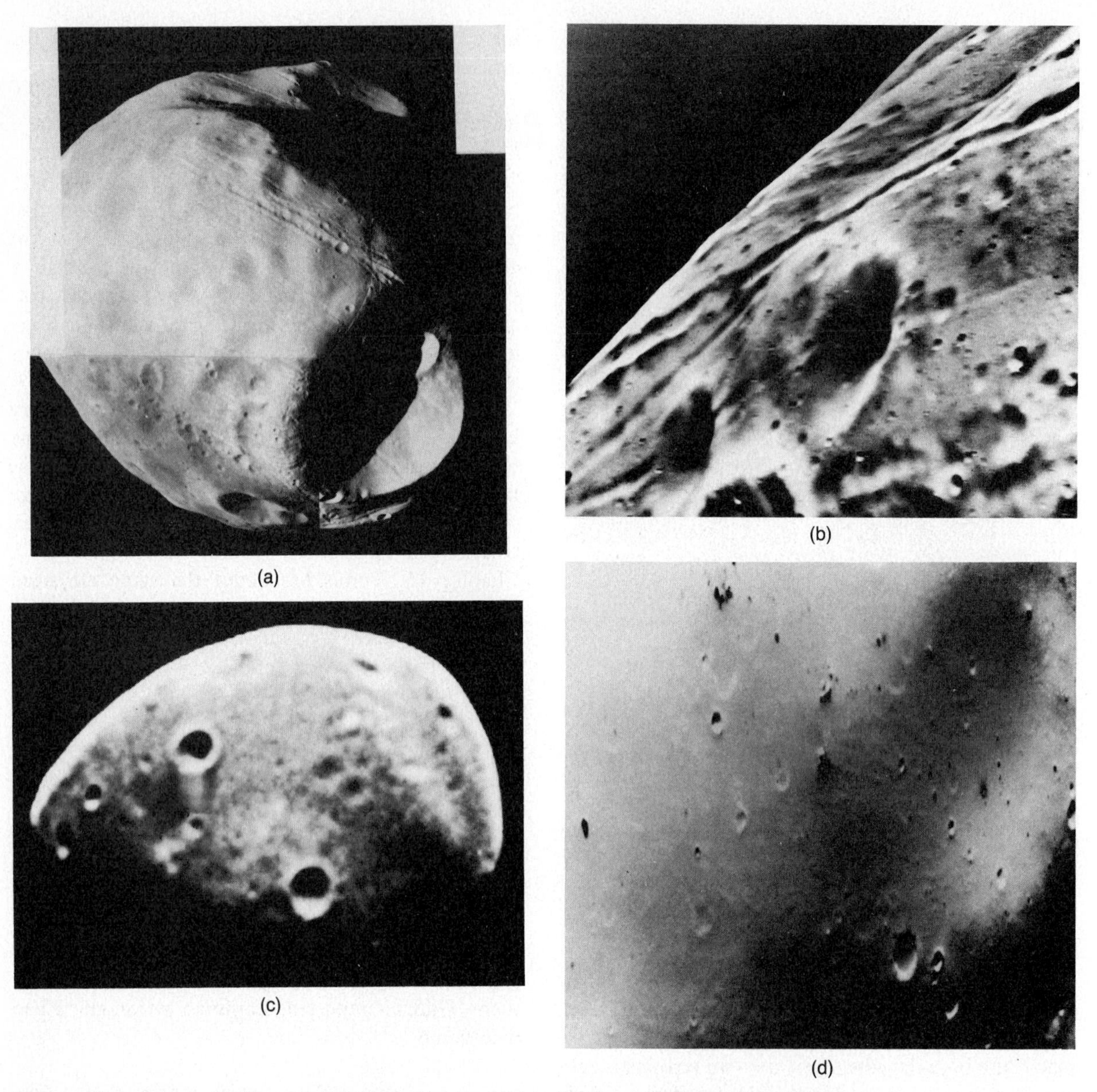

Figure 15.6 Mars' satellites, photographed by the Viking Orbiters. (a) Phobos; the large (10-km diameter) crater at the upper right (partly in shadow) is Stickney. (b) Phobos at a range of 120 km. The photograph covers an area of 3.6 × 3 km, and the largest crater visible is 1.2 km across. (c) Deimos, from a range of 3300 km. The satellite is seen in the gibbous phase; the illuminated portion facing the camera has an area of about 12 × 8 km. (d) Deimos, photographed by Viking Orbiter 2 at a range of only 30 km. The picture covers an area of 1.2 × 1.5 km and shows features as small as 3 m across, including strewn boulders of sizes 10 to 20 m. (NASA/JPL)

ellites of Mars that closely resemble those described by Swift. (More significant than Swift's "prediction" of the Martian moons, perhaps, is that he was aware of and understood Kepler's third law. Swift was not a scientist. How many intellectuals of today are knowledgeable in natural philosophy?)

The satellites are named Phobos and Deimos, meaning "fear" and "panic"—appropriate companions of the god of war, Mars. Phobos is 9380 km from the center of Mars, and revolves about it in 7^h 39^m; Deimos has a distance of 23,500 km and a period of 30^h 18^m. The "month" of Phobos is less than the rotation period of Mars; consequently, Phobos would appear to an observer on Mars to rise in the west.

Phobos and Deimos were studied at close range by the Viking orbiters (Figure 15.6). They are both rather irregular, somewhat elongated, and heavily cratered. The largest diameters of Phobos and Deimos are about 25 km and 13 km, respectively. Each is a dark brownish grey in color, and spectral analysis suggests that each is composed of dark materials similar to those out of which most asteroids are made. Apparently these two satellites are *chemically primitive;* that is, they represent material that is little changed since it condensed in the solar nebula 4.6 billion years ago. For this reason, some scientists consider them to be almost as interesting as the planet Mars itself.

Since they are primitive bodies closely related to the asteroids, it seems likely that these satellites are captured objects. Presumably they were captured very early in solar system history, when the frictional drag of an extended Martian atmosphere was sufficient to decrease their energy and trap them into stable orbits. Their origin is thus most likely not at all like that of Earth's Moon.

Phobos and Deimos have often been suggested as the target for space missions. The Viking orbiters carried out special maneuvers to come within 100 km of Phobos and down to just 23 km from Deimos. In 1988, the U.S.S.R. will undertake an ambitious pair of flights, the first to Phobos and the second (if the first is successful) to Deimos. In each case the spacecraft will match orbits with the satellite, and it is expected to come within 50 to 100 m of the surface. Many people consider this mission to be a precursor of human flights to Mars, which might use these two satellites as ready-made space stations.

15.2 VOLCANOES AND OTHER GEOLOGICAL FEATURES

(a) The Martian Shield Volcanoes

About a dozen very large volcanoes have been found on Mars, most of them associated with the

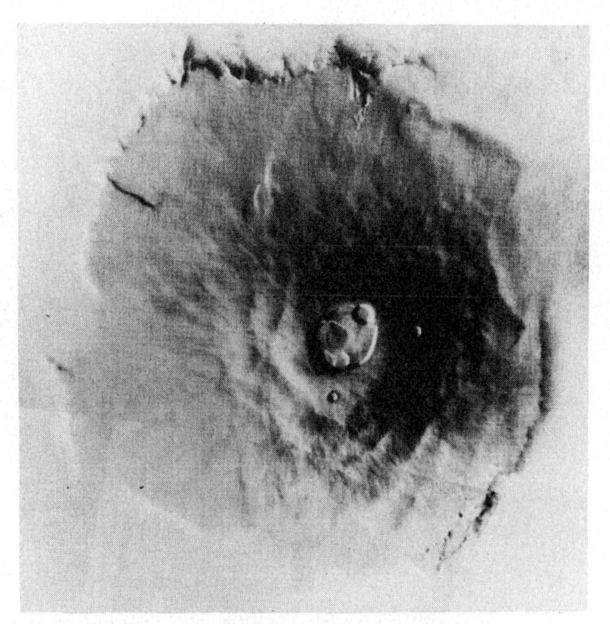

Figure 15.7 The large Martian shield volcano Olympus Mons, rising about 26 km high and more than 500 km across at its base. (NASA/JPL)

Tharsis bulge and its immediate surroundings. Many more small volcanoes also dot the surface, primarily in the younger northern half of the planet. Three of the most dramatic of these volcanoes lie along the crest of the Tharsis bulge. Each is about 400 km in diameter, and all rise to the same height. The fourth and largest volcano, Olympus Mons (Mount Olympus), is on the northwest slope of Tharsis (Figure 15.7 and Color Plate 10a). It is more than 500 km in diameter, with a summit 25 km above the surrounding plains. The volume of this immense volcano is nearly 100 times greater than the largest terrestrial volcano, Mauna Loa in Hawaii. The only comparable feature in the solar system is the Maxwell Mountains on Venus (Section 16.2), which are about as large across but only half as high as Olympus Mons.

These great Martian mountains are *shield volcanoes,* similar in shape to their terrestrial counterparts in such locations as Hawaii or the Galapagos Islands. Presumably they consist of many overlapping flows of fluid basaltic lavas. Indeed, the detailed patterns of these flows can be traced in high-resolution Viking photos of their slopes. The summits of shield volcanoes are marked by *calderas,* which are broad, flat-floored collapse craters. The caldera on Olympus Mons is 80 km across, while

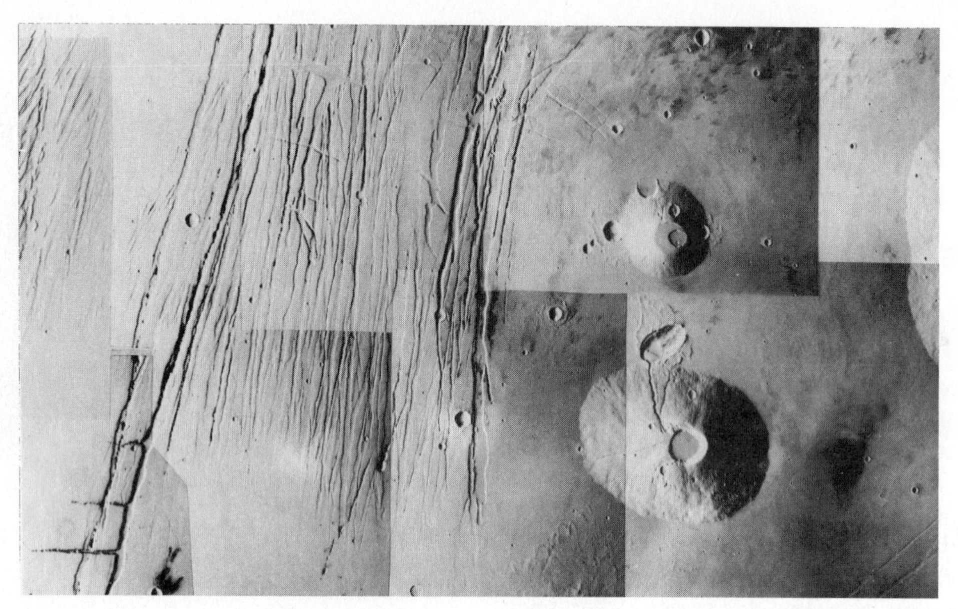

Figure 15.8 Tectonic features in the Tharsis region of Mars, produced by tension in the crust, and two large volcanoes. (NASA/JPL)

that of Arsia Mons, another Tharsis volcano, is 120 km in diameter.

The Viking imagery permits a detailed examination of the calderas and slopes of these volcanoes to search for impact craters. As described in Section 14.3c, the number of such impact craters is a direct measure of the age of a surface. Many of the volcanoes show fair numbers of such craters, suggesting that they ceased activity a billion years or more ago. However, Olympus Mons has very, very few impact craters. Its present surface cannot be more than about a hundred million years old, and it could be much younger yet. Some of the fresh-looking lava flows we see might have been formed a hundred years ago, or a thousand, or a million, but geologically speaking they are young. It is quite probable that these great volcanoes remain intermittently active today.

(b) Canyons and Tectonic Features

The Tharsis bulge consists of more than a collection of huge volcanoes. In this part of the planet the surface itself has bulged upward, forced by great pressures from below. Such crustal forces are called *tectonic forces*, and Tharsis is a *tectonic feature* (Figure 15.8). Other tectonic features on Mars include extensive cracks in the crust that criss-cross and surround the Tharsis area, produced by the same forces that raised the bulge.

On the Earth we see abundant evidence of *plate tectonics*, in which the crust responds to internal forces by dividing into plates that move with respect to each other. There is no evidence of plate tectonics on Mars, however. For example, there is no Martian equivalent of a San Andreas Fault, where one piece of the crust slides alongside another. Instead, there is the single feature of Tharsis, forced upward but not induced to shift sidewise. It is as if crustal forces began to act but then subsided before full-scale plate tectonics could begin.

The most spectacular tectonic features on Mars make up a great canyon system called the Valles Marineris, which extends for about 5000 km (nearly a quarter of the way around Mars) along the slopes of the Tharsis bulge (Figure 15.9). The main canyon is about 7 km deep and up to 100 km wide. It is so large that the Grand Canyon of the Colorado River would fit comfortably into one of its side canyons (Color Plate 11).

The term canyon is somewhat misleading, because the Valles Marineris canyons were not cut by running water. They have no outlets. They are basically cracks, produced by the same crustal ten-

Figure 15.9 The great Valles Marineris canyon system as photographed by Viking. These valleys are primarily tectonic in origin; that is, they represent splitting of the crust rather than erosion. (NASA/JPL)

sions that caused the Tharsis uplift. However, water is believed to have played a later role in shaping the canyons, primarily through the undercutting of the cliffs by seepage from deep springs. This undercutting led to landslides, gradually widening the original cracks into the great valleys we see today. The material from recent landslides is clearly visible on the valley floors, while earlier landslides have probably been eroded away by wind storms sweeping down the canyons.

The Valles Marineris has been compared to the African Rift Valley on Earth, which is the result of mantle convection pulling apart the crust. It may be that similar forces were at work on Mars 2 to 3 billion years ago, which is when we estimate both Tharsis and the canyons were formed. But again we must conclude that full-scale plate tectonics never developed on Mars, even though the planet may have been headed in that direction early in its geological history.

(c) The Cratered Uplands

Most of the Martian southern hemisphere consists of cratered upland plains. The Martian craters are clearly of impact origin, although they have been modified more than lunar craters by subsequent erosion. There are no rays, and most craters have also lost their ejecta blankets. Some of the oldest appear to have been partially filled in, perhaps by windblown dust.

One of the most curious features of the equatorial plains are multitudes of small sinuous (twisting) channels that appear to have been cut by running water. Typically these channels are simple valleys, a few meters deep, some tens of meters wide, and perhaps 10 or 20 km long. They are called *runoff channels,* because they appear to have carried the surface runoff of ancient rainstorms (Figure 15.10). As we will discuss in the next section, there are no rainstorms today on Mars. These runoff channels seem to be telling us that the planet had a very different climate long ago. How can we tell when the rain might have last fallen on Mars?

The best that we can do is to count impact craters and use the theory described in Section 14.3. Such crater counts show that the uplands of Mars are more cratered than the lunar maria but less so than the lunar highlands. Thus they are older than the maria, presumably at least 3.9 billion years.

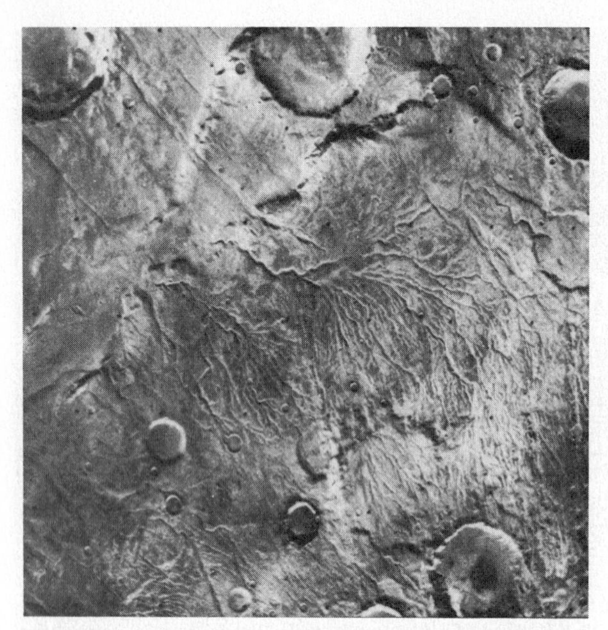

Figure 15.10 Runoff channels in the Martian highlands, interpreted as the valleys of ancient rain-fed rivers. The width of this frame is about 250 km. (NASA)

Most scientists would guess that they were formed during the final heavy bombardment about 4.0 billion years ago. Interestingly, there are no runoff channels in the younger terrain to the north.

(d) Volcanic Plains

The northern hemisphere of Mars is characterized by rolling, volcanic plains that lie several kilometers lower in elevation than the uplands. These plains look very much like the lunar maria, and they have about the same numbers of impact craters. Like the lunar maria, they probably formed between 3.0 and 4.0 billion years ago. Apparently Mars experienced extensive volcanic activity at about the same time the Moon did, producing similar plains of flood basalt. That much is evident from the Viking photos. What is not understood are the forces that removed the old crust and dropped the elevation level in this hemisphere of the planet.

Although basically volcanic in origin, the Martian plains have been modified by erosion and sedimentation. Some of them form basins into which drain huge river channels from the uplands (to be discussed in Section 15.3). They may once have contained shallow seas. Others, nearer the north pole, are covered today by extensive sand dunes.

Both Vikings landed in such volcanic plains. The atmospheric entry technique required a low elevation site, excluding both the uplands and the Tharsis bulge from consideration. In addition, it was thought that lower regions, especially near former river mouths, might be more likely to harbor Martian micro-organisms.

(e) The View from the Surface

Viking 1 landed in Chryse at a latitude of 22 N, on a 3-billion-year-old windswept plain near the lowest point of a broad basin. Its desolate but strangely beautiful surroundings included numerous angular rocks, some more than a meter across, interspersed with dune-like deposits of fine-grained soil. On the horizon, the low profiles of several distant impact craters could be seen (Figure 15.11). At the Viking 2 site in Utopia, at latitude 48 N, the surface was somewhat similar, but with substantially greater numbers of rocks. Probably Viking 2 landed on ejecta from a 90-km-diameter crater about 200 km distant (Figure 15.12 and Color Plate 12b). In both cases, it is believed that winds had stripped the surface of loose, fine material to leave the numerous rocks exposed.

Each lander peered at its surroundings through color stereo cameras, sniffed the atmosphere with a variety of analytical instruments, and poked at nearby rocks and soil with its mechanical arm. As part of its primary mission of searching for Martian life, each lander collected soil samples and brought them on board for analysis, as will be described in more detail in Section 15.4. The soil was found to consist of clays and iron oxides, as had long been expected from the red color of the planet.

Each lander also carried a seismometer in the hopes of detecting marsquakes to use to probe the planet's interior. Unfortunately, the Viking 1 seismometer failed to operate, and the Viking 2 instrument picked up nothing more than some vibrations of the lander in strong gusts of wind. Thus we know nothing as yet about the level of seismic activity on Mars.

Each lander also carried a weather station to measure and record the temperature, pressure, and wind speed and direction at the site, at a height of 1.3 m above the ground. The surface pressure at each location was about 0.007 bar, which although less than 1 percent of the sea-level pressure on Earth is actually fairly high for Mars. This pressure

Figure 15.11 Viking 1 view of its landing site in Chryse. (NASA/JPL)

was much more variable than on our planet, however, because the main atmospheric gas, carbon dioxide, is also the primary component of the polar caps. Therefore, the pressure changes with the seasons, as the polar caps grow and recede. At the Viking 1 site, this pressure ranged from 0.0067 to 0.0088 bar (Figure 15.13). The temperatures also vary much more on Mars than on Earth, due to the absence of moderating oceans and clouds. Typically, the summer maximum was 240 K (−33°C), dropping to 190 K (−83°C) at the same location just before dawn. The lowest air temperatures, measured farther north by Viking 2, were about 173 K (−100°C). During the winter, Viking 2 also photographed frost deposits on the ground (Color Plate 12b).

Most of the winds measured at the Viking sites were low to moderate, typically from 2 m/s at night up to 7 m/s in the daytime. However, Mars is capable of great windstorms, which can shroud the entire planet in dust, as we will discuss further in Section 15.3. At such times the sun was greatly dimmed at the Viking sites, and the sky turned a dark red in color.

(f) Martian Samples

One of the most striking differences in our knowledge of Mars, compared with that of the Moon, results from the absence of samples of Martian material to be analyzed in the laboratory. Much of what

Figure 15.12 Viking 2 view of its landing site in Utopia, on ejecta from the crater Mie. (NASA/JPL)

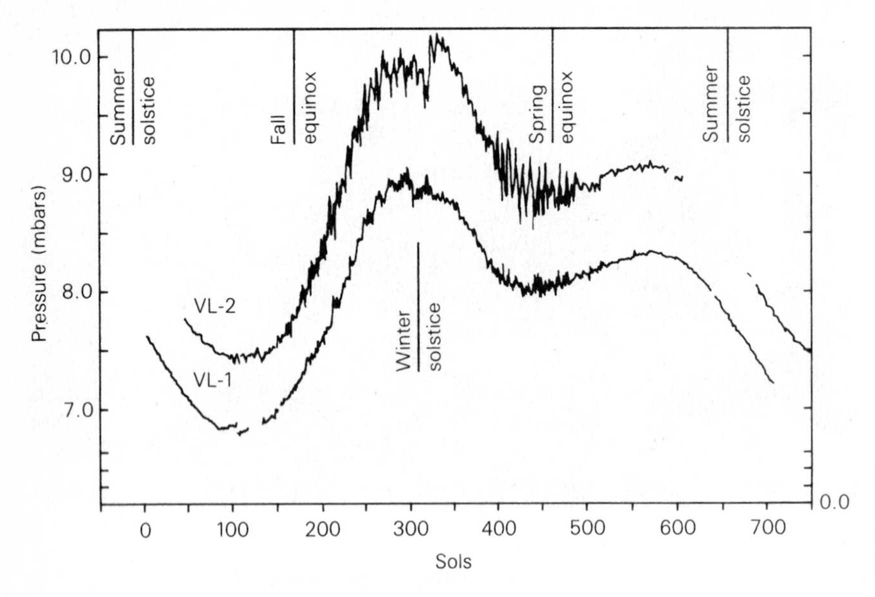

Figure 15.13 The changing surface pressure of the Martian atmosphere as the CO_2 polar caps grow and shrink. (NASA/JPL)

we know of the Moon, including the circumstances of its origin, comes from sample studies. Radioactive age dating is also the only dependable way we know to establish a chronology of the geological history of a planet.

It is with great interest, therefore, that scientists have recently concluded there may be samples of Martian material available for our study on Earth. These candidate Martian rocks are a rare class of meteorites called *SNC meteorites*. Four SNC meteorites have been in our collections for a number of years, and they have recently been supplemented by four new fragments found in Antarctica in an excellent state of preservation (Figure 15.14).

The most obvious special characteristic of these meteorites is that they are *basalts,* and they are relatively *young,* with ages of 1.3 billion years. We know they are not from the Moon, and in any case there was no lunar volcanic activity as recently as 1.3 billion years ago. By process of elimination, the only reasonable origin seems to be Mars, where the Tharsis volcanoes were certainly active at that time. This guess seems to have been confirmed by recent analysis of tiny bubbles of gas trapped in one of these meteorites, which match almost perfectly the atmospheric properties of Mars as measured directly by Viking.

If these are pieces of Mars, they were presumably ejected in some large impact on one of the Tharsis volcanoes. The scar from that impact must be visible, but of course we do not know which crater it was. As to helping us understand Mars, the work on these meteorites has just begun, and we do not know how much they will tell us. This is an intriguing chapter in Martian science that is only beginning to be written.

15.3 POLAR CAPS, ATMOSPHERE, AND CLIMATE

(a) The Martian Atmosphere

The atmosphere of Mars is composed primarily of carbon dioxide (95 percent), with about 3 percent nitrogen and 2 percent argon. The predominance of carbon dioxide over nitrogen is not too surprising when you remember (Section 13.3b) that the Earth's atmosphere would also be mostly carbon dioxide if the CO_2 were not locked up in marine sediments. Water vapor is not expected to be a major component of the Martian atmosphere, since temperatures are almost always well below the freezing point. Table 15.2 lists the exact abundances as measured by the Viking landers.

Like the Earth, Mars has a troposphere, in which convection takes place, and above it a more stable cold stratosphere. The Martian troposphere typically has a height of about 10 km during the daytime, when it is heated from below by sunlight absorbed at the surface. At night or over the polar regions, however, the surface temperature is much lower, and the troposphere disappears.

Figure 15.14 One of the SNC meteorites, which are believed to be samples from the surface of Mars. (NASA)

Several types of clouds can form in the atmosphere. First there are dust clouds, raised by winds, which can sometimes grow to cover a large fraction of the surface. Second are water ice clouds similar to those on Earth. These often form around mountains, just as happens on our planet (Color Plate 10a); low-lying fog also is possible near dawn, when the surface temperature drops to 100° below freezing. Finally, the carbon dioxide of the atmosphere can itself condense at high altitudes to form hazes of dry ice crystals. The carbon dioxide clouds have no counterpart on Earth, since temperatures never drop low enough here for this gas to condense.

Most of the time, the atmosphere of Mars is much clearer than that of the Earth, with relatively few clouds. The exceptions to this rule are associated with the major dust storms that occur during southern hemisphere summer. Once begun, these storms seem to feed upon themselves, spreading outward across the surface and carrying more and more dust high into the atmosphere. On occasion, such storms can envelop the entire planet. When this happens, sunlight is absorbed primarily in the dust clouds rather than at the surface, and the atmosphere is heated from above. The troposphere disappears, temperatures all over the planet tend to become uniform, and the circulation of the atmosphere changes. Eventually, after a month or two, the dust settles out and the atmosphere returns to its normal state.

TABLE 15.2 Composition of the Martian Atmosphere

Carbon dioxide	95.3%
Nitrogen	2.7
Argon	1.6
Oxygen	0.15
Water	0.03 (variable)
Neon	0.0003
Krypton	0.00003
Xenon	0.000003

With the exception of the great dust storms, Martian weather is largely predictable. The thinness of the atmosphere and the absence of water lead to a situation in which the atmosphere responds very quickly to changing surface temperature, and the wind patterns tend to repeat almost exactly each day. The situation can vary substantially from one surface location to another, however, influenced by local topography. Martian conditions provide an interesting and useful test of models developed to predict weather on the Earth, since the Martian case is simpler. If a meteorological computer program fails for Mars, it is unlikely to work properly for Earth.

Although there is water vapor in the atmosphere and occasional clouds of water ice can form, *liquid* water is not stable under present conditions on Mars. Part of the problem is the low temperatures on the planet. But even if the temperature on a sunny summer day rises above the freezing point, liquid water cannot exist. At an atmospheric pressure of 0.006 bars, only the solid and vapor forms are possible. In effect, the boiling point is as low or lower than the freezing point, and water changes directly from solid to vapor without an intermediate liquid state. Thus Mars is truly a dry planet, even though there may be abundant water ice available in the polar caps or frozen in the soil.

(b) The Polar Caps

Through a telescope, the most prominent surface features on Mars are the bright polar caps, which change in extent with the seasons. These *seasonal* caps are similar to the seasonal snow cover on Earth. We do not usually think of the winter snow as a part of our polar caps. But seen from space, the thin snow would blend with the thick permanent ice caps to create a situation much like that seen on Mars.

The seasonal caps on Mars are composed not of ordinary snow, but of frozen carbon dioxide (dry

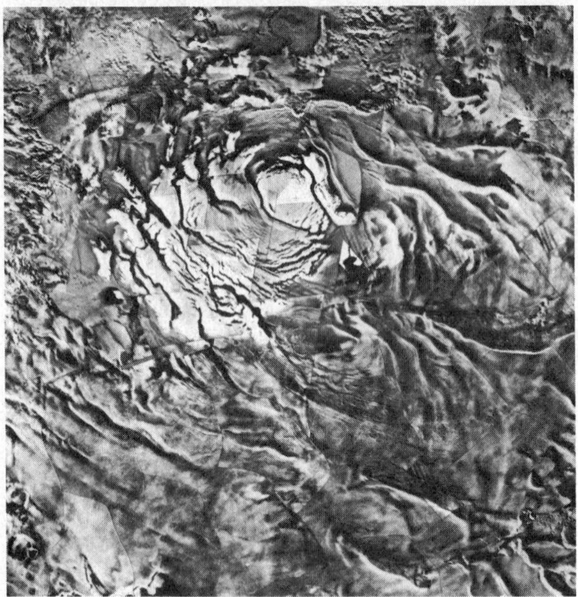

Figure 15.15 Photo maps of the north and south residual (permanent) polar caps of Mars. (NASA/USGS)

ice). These deposits condense directly from the atmosphere when the surface temperature drops below about 150 K. The caps develop during the cold Martian winters, extending down to about 55° in the south and to 45° in the north. The difference in extent of the two caps is a reflection of the greater severity of the southern winter, which takes place when the planet is near aphelion. The northern winter is shorter and warmer, since Mars is then closer to the sun.

Quite distinct from these thin seasonal caps of carbon dioxide are the *permanent,* or *residual,* caps that are always present near the poles (Figure 15.15). As the seasonal cap retreats during spring and early summer, it reveals a brighter, thicker cap beneath. The southern permanent cap has a diameter of 350 km and is composed of thick deposits of frozen carbon dioxide. Throughout the southern summer, it remains at the freezing point of carbon dioxide, 150 K, and this cold reservoir is thick enough to survive the summer heat intact.

The northern permanent cap is different. It is much larger, never shrinking below a diameter of 1000 km, and it is composed of ordinary water ice. We do not know the thickness of the cap, but it may be as much as several kilometers. In any case, this permanent cap represents a huge reservoir of water, in comparison with the very small amounts of water vapor in the atmosphere.

The explanation for the different composition of the two caps is complex, since the southern summers are hotter than the northern summers, yet it is in the south that carbon dioxide survives. Probably the explanation is associated with the major dust storms, which always take place during northern winter; the northern cap therefore becomes dusty and dark, so that it absorbs more sunlight and heats more rapidly when spring arrives.

The terrain surrounding the permanent polar caps and seen in ice-free areas within the caps is remarkable (Figure 15.16). At latitudes above 80° in both hemispheres, the surface consists of recent, layered sedimentary deposits that entirely cover the older cratered ground below. These layers show up especially well on eroded slopes, which appear to be terraced. Individual layers are typically a few tens of meters in thickness, marked by alternating light and dark bands of sediment. Probably the material in the polar deposits is dust carried by wind from equatorial regions. This dust is apparently mixed with water or CO_2 ice.

Calculations suggest that as much as 1 mm/yr of dust might be deposited in this way. In that case the time necessary to build up a 10-m layer would be 10,000 years. Thus each major layer corresponds to an interval of tens of thousands of years. In Section 15.3d we will return to a discussion of ways the Martian climate might change on this kind of time scale.

COLOR PLATE 9 Apollo 17 Astronaut Harrison Schmitt working beside a large boulder on the Moon. The lunar roving vehicle is at the left. (NASA)

COLOR PLATE 10a The large Martian shield volcano, Olympus Mons, rising high above surrounding clouds. (NASA/JPL)

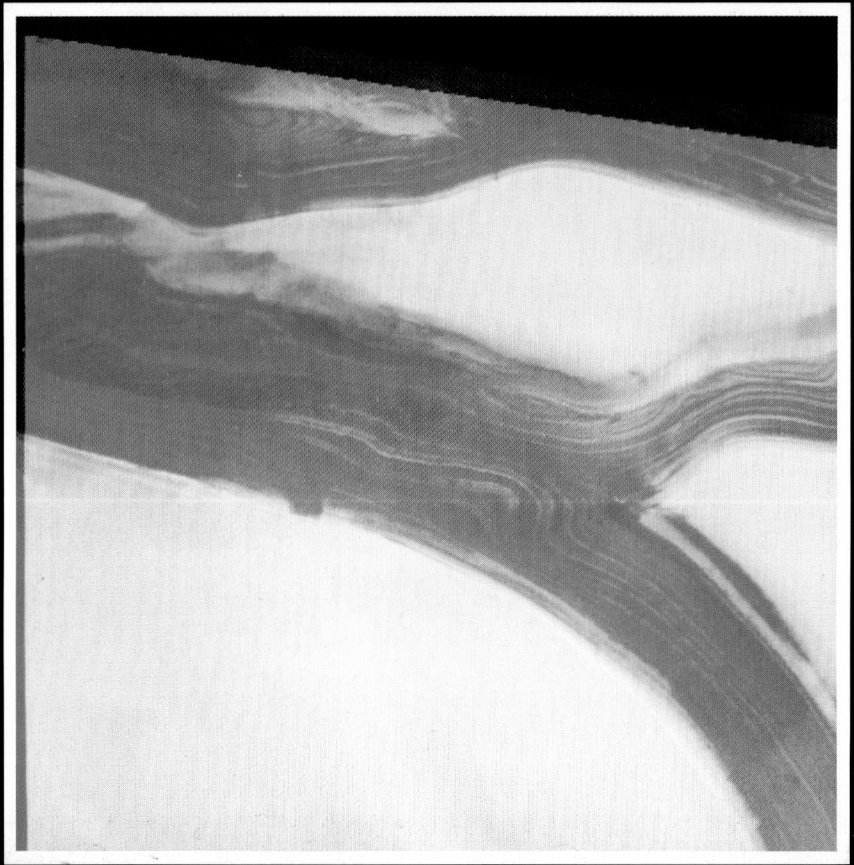

COLOR PLATE 10b Part of the residual north polar cap of Mars with reddish terraced slopes showing through the carbon dioxide ice. (NASA/JPL)

COLOR PLATE 11 High-resolution view of the Candor Chasma region of the Martian canyonlands. (NASA/U.S. Geological Survey; Courtesy Alfred McEwen, Arizona State University)

COLOR PLATE 12a The Martian surface photographed from the Viking 1 lander. (NASA/JPL)

COLOR PLATE 12b Frost near th Viking 2 lander, photographed dur ing Martian spring. (NASA/JPL)

Figure 15.16 Martian layered terrain in the north polar region, preserving a record of past climate variations. (NASA/JPL)

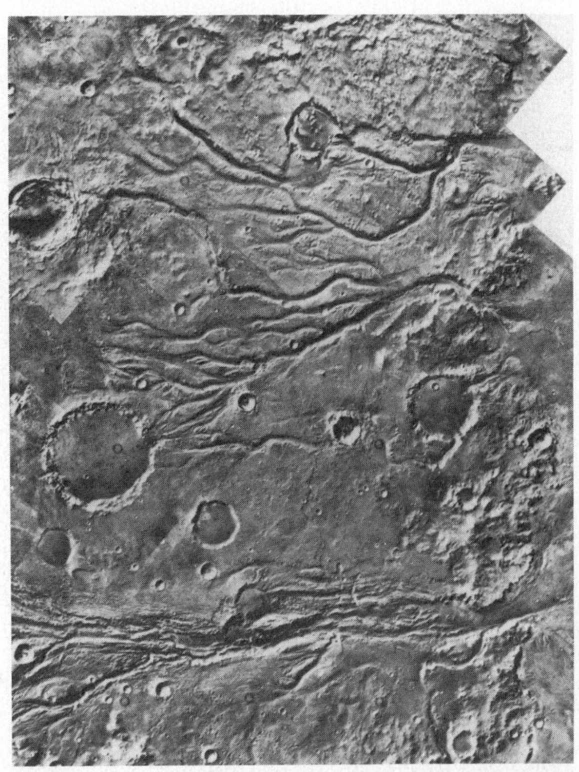

Figure 15.17 Large outflow channels photographed by Viking. These features appear to have been formed in the distant past from massive floods of water. (NASA/JPL)

(c) Catastrophic Floods

There is a variety of evidence suggesting that the climate of Mars is variable. We have already noted the layered polar deposits, corresponding to time scales of tens of thousands of years. At the other extreme, the runoff channels in the older cratered terrain represent conditions existing probably about 4 billion years ago. Now we turn to another indication of liquid water on Mars, provided by the huge dry river channels that drain from the uplands into the northern volcanic plains.

These *outflow channels* (Figure 15.17) are much larger than the older runoff channels. The largest of these, which drain into the Chryse basin where Viking 1 landed, are ten or more kilometers in width and hundreds of kilometers long. Over much of their length, the channels are multiple, with branches that separate from the main channel, run parallel to it for a while, and then rejoin. They have few if any tributaries, apparently flowing at full force from their origin in the cratered uplands to their mouths.

Many features of these outflow channels have convinced geologists that they were carved by huge volumes of running water (Figure 15.18). Such floods could not have been sustained by rainfall, but must have some other source. The closest terrestrial analog is provided by the system of flood channels in eastern Washington State called the Channeled Scablands. These were carved 18,000 years ago when a large glacial lake in Montana suddenly burst its natural dam and emptied within a matter of hours, generating flows 120 m deep and cutting new channels up to 10 km wide through the bedrock.

Presumably the Martian outflow channels were also formed by sudden floods. But where did the water come from? As far as we can tell, the source regions contained abundant water frozen in the soil as *permafrost*. Some local source of heating must have released this water, leading to catastrophic flooding. Perhaps this heating was associated with the formation of the volcanic plains, which apparently happened at about the same time the channels formed.

The two kinds of Martian channels thus provide evidence of two periods in the past when water was present in liquid form. The first, about 4 billion years ago, was a time when rain fell and the

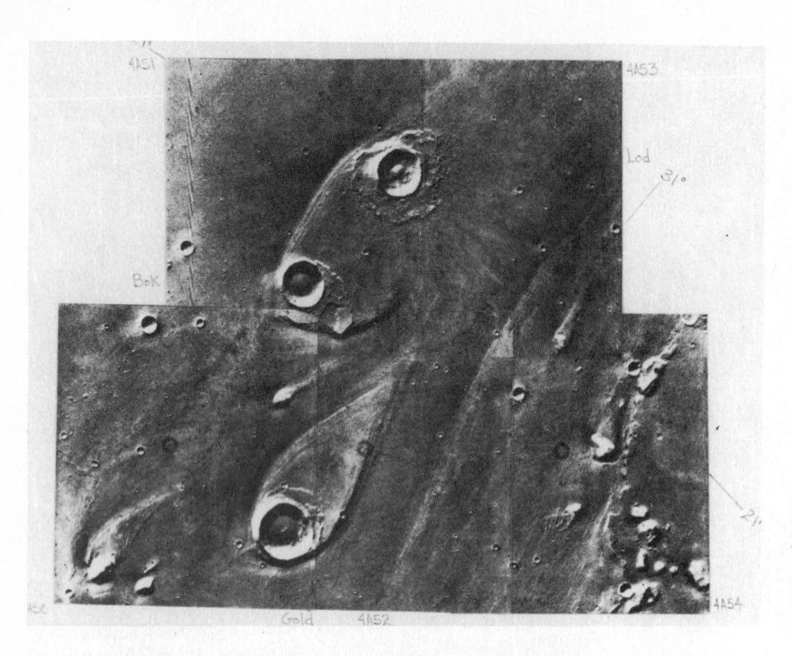

Figure 15.18 Flow features in Chryse, near the place where several major outflow channels emptied into the basin. (NASA/JPL)

atmosphere was probably much larger and warmer than it is at present. The second, perhaps a billion years later, represented the release of frozen groundwater by volcanic heating. Since then, the planet has probably been as cold and dry as we see it today, although occasional shorter periods of warmer and wetter climate are not impossible in more recent times.

(d) Climate Change

The evidence cited above suggests that the climate of Mars has varied on at least two different time scales. Billions of years ago temperatures were warmer, rain fell, and the atmosphere must have been much more substantial than it is today. And much more recently there appears to have been a cyclic variation as recorded in the layered deposits of the Martian polar regions.

The long-term cooling of Mars and loss of its atmosphere are a result of both its small size (and low escape velocity) relative to the Earth and its greater distance from the sun. Presumably Mars formed with a much thicker atmosphere, and the atmosphere maintained a higher surface temperature due to the *greenhouse effect,* discussed in more detail in Section 16.3. Escape of the atmosphere to space, however, gradually lowered the temperature. Eventually it became so cold that the water froze out of the atmosphere, further reducing its ability to retain heat. The result is the cold, dry planet we see today. Probably this loss of atmosphere took place within a few hundred million years; from the absence of runoff channels in the northern plains, it seems that rain has not fallen for at least 3 billion years on Mars.

The climate variations recorded in the polar regions are much more recent. As noted in Section 15.3b, the time scales represented by the polar layers are tens of thousands of years. Apparently the Martian climate experiences periodic changes with this frequency, which is similar to the intervals between ice ages on the Earth. Calculations indicate that the causes are probably also similar: variations in the orbit and obliquity of the planet induced by the gravitational perturbations of other planets.

15.4 THE SEARCH FOR LIFE ON MARS

(a) Early Ideas

Even after scientists rejected the Martian canals as evidence of intelligent life, there was still a widespread feeling in the astronomical community that some simpler kind of life was possible, or even probable, on Mars. Of all the planets, Mars had conditions most similar to those on Earth, with surface temperatures not much lower than those in the polar regions of our planet. Its atmospheric pressure was thought by scientists to be about $\frac{1}{10}$ of the terrestrial value, rather than 1/100 as we now know. In addition, there was direct evidence that argued for plant life on that planet. Observed changes in the dusky markings on Mars seemed to follow a seasonal pattern. In many cases these markings darkened with the coming of spring, as if in response to higher temperatures and the availability of water vapor from the shrinking polar cap. Some observers argued that the markings were green in color, suggesting the presence of photosynthetic plant life. There were even hints of spectral features that matched those of sunlight reflected from green leaves on Earth.

With the advantages of hindsight, we now recognize that much of this attitude represented wishful thinking. The spectral features turned out to have been an observational error, and later measurements showed that the dark markings are orange, not green, in color. The apparent seasonal changes, while real, are the result of the deposition and stripping away of light-colored dust from darker terrain, not the growth of plant life. Further, the Mariner 4 spacecraft revealed in 1965 a desolate cratered surface and showed that the atmosphere was too inconsequential to permit liquid water to exist on the Martian surface.

Although the new data from the first spacecraft and from more sophisticated telescopic studies were not encouraging for the idea of widespread plant life on Mars, they by no means excluded the possibility of life of some kind. Conditions on Mars were still the most favorable to be found outside of the Earth. Most scientists felt that the only way to settle the issue was to go to Mars and look.

(b) The Viking Life-Detection Experiments

The primary objectives of the Viking mission were to land on the Martian surface and search for life, using an automated, miniaturized biological laboratory. The entire spacecraft was carefully sterilized to ensure that no terrestrial micro-organisms were accidentally transported to Mars, thus protecting the integrity of the Martian environment as well as guarding against a false-positive signal from the spacecraft experiments.

The Viking strategy was to search for micro-organisms, on the grounds that these were likely to be the most abundant lifeforms as well as the easiest to detect. In any terrestrial environment there are abundant microscopic creatures; a handful of even the most inhospitable desert soil contains millions of them. The approach therefore was for the spacecraft to scoop up Martian soil and distribute this sample to three different life-detection experiments, each designed to test for a different chemical indication of life processes (Figure 15.19). As far as possible, these experiments were designed to minimize the natural bias toward terrestrial biological systems, although it was understood from the beginning that the only kind of life we knew how to search for was basically similar, on a chemical level, to the carbon-based life we know on Earth.

The three experiments flown to Mars each began with a sample of soil, which was introduced into a small test chamber. Any micro-organisms within the sample were then given an opportunity to interact with their environment in these closed, controlled conditions, and the chemical consequences of this interaction were measured.

In the *gas-exchange* experiment, the environment included water and a variety of nutrients that would appeal to terrestrial organisms. If the organisms liked the food and were able to metabolize it, they should alter the composition of the gas in the chamber in a measurable way. The *labeled-release* experiment followed a similar strategy, but it used nutrients that were tagged with radioactive atoms and measured the release of these atoms into the atmosphere of the test chamber. Both of these experiments had the drawback that they would give a positive result only if the Martian micro-organisms metabolized the offered food and released products into the atmosphere.

The third approach—the *pyrolitic-release* experiment—followed a more general strategy, maintaining the sample in essentially a Martian environment. After all, Martian organisms might be killed by the water and nutrients supplied by the first two experiments. In this case, the organisms were offered only a carbon dioxide atmosphere, suitably tagged with radioactivity. They were then given an opportunity to take some of this carbon dioxide into their own bodies before the chamber was cleared of gas and the sample soil analyzed. If radioactive carbon showed up in the soil, it was taken as evidence that microbial activity had taken place.

(c) The Viking Results

Much to the surprise of the scientists who designed these experiments, all three initially gave strong positive results! Apparently the Martian samples re-

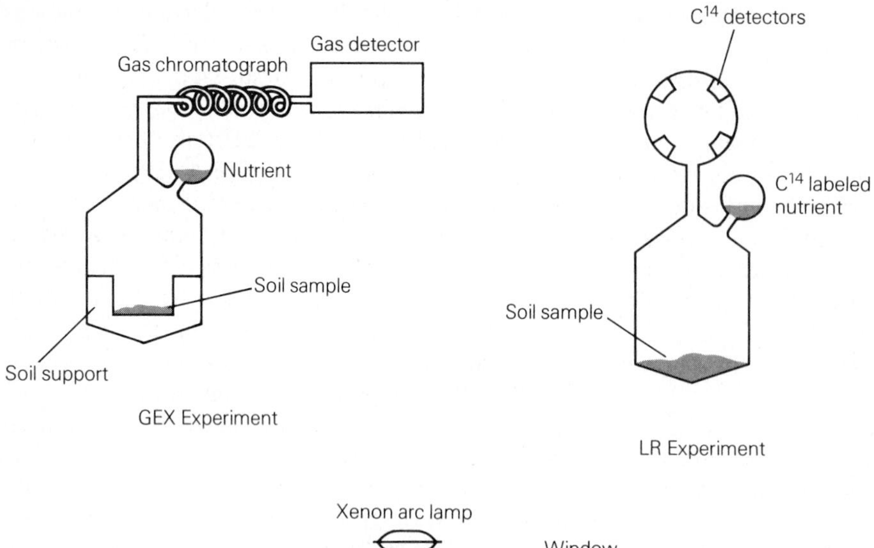

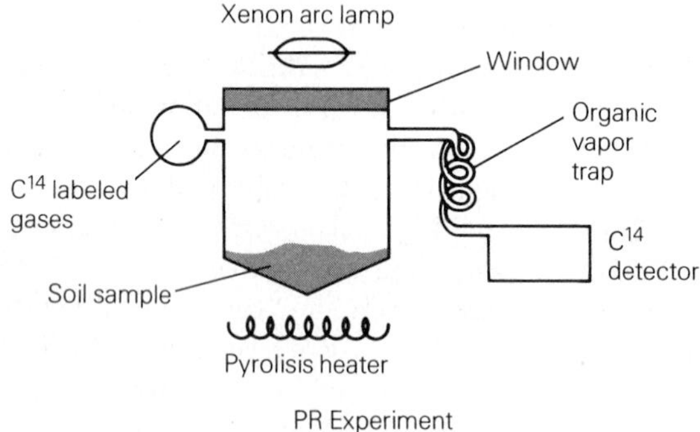

Figure 15.19 Schematic representation of the three Viking life-detection experiments.

acted strongly with the nutrients and gas in the test chambers. However, the general character of this reaction was different from that expected for living organisms, since the immediate activity was high but declined with time, just the opposite of that usually encountered with growing things. Were the Martian organisms gorging themselves as soon as they were placed in the test chambers, and then quickly dying? Or might there be a nonbiological explanation of the results?

There was another reason to be skeptical about the positive results of the life-detection experiments. The Viking landers each also carried a very sophisticated chemical analysis device called a GCMS, which was used to determine the chemical composition of the atmosphere and soil. To everyone's surprise, the GCMS found no organic chemicals in the soil, down to the parts-per-billion level. How could there be life without organic chemicals?

The resolution of this complex set of results required many exciting months of effort. Each of the Viking spacecraft repeated its tests under varying experimental conditions, using soil collected from different places within reach of its sampler arm. Finally it became clear that life was not being detected. Instead, the soil itself was much more chemically active than terrestrial soil. The explanation for this activity is the same as the explanation for the absence of organic matter in the soil: solar ultraviolet light.

The thin Martian atmosphere does not protect the surface from ultraviolet light the way the Earth's atmosphere, and particularly our ozone layer, does. This ultraviolet radiation breaks apart carbon molecules and sterilizes the soil, as seen by the GCMS analysis. It also produces compounds called super-oxides in the soil, which are highly reactive when brought into contact with water or even with the walls of the Viking test chambers. These nonbiological reactions produced the false-positive signals from the Viking life-detection experiments.

Is there life on Mars? Given the ultraviolet radiation and the absence of organic material in the soil, probably not. Certainly there does not seem to have been life as we know it at the two Viking landing sites. The possibility always exists of protected "oases" somewhere on the planet where life has adapted to the harsh environment. On the basis of only two landings, we can hardly claim to have explored the whole planet. But most scientists now feel that the long-held dream of life on Mars has finally reached its end.

EXERCISES

1. Why does Mars have the longest synodic period of any planet but a sidereal period of only 687 days?
2. Compare Mars with Mercury and the Moon in terms of its bulk properties. What are the main similarities and differences?
3. Why is Mars red? Why aren't the Moon or Earth the same color?
4. Why is Mariner 9 often quoted by NASA planners as the kind of mission they would like to fly to each planet? What was particularly significant about Mariner 9?
5. Explain the major division of the surface of Mars into two distinct kinds of topography. Compare with the distinction on the Moon between the highlands and the maria.
6. How do the mountains on Mars compare with those on Earth and Moon?
7. Show that the satellites of Mars obey Kepler's third law.
8. Discuss why Olympus Mons and other major volcanoes on Mars are so much larger than terrestrial shield volcanoes. Think about why there are no mountain chains (such as Hawaii) on Mars.
9. Explain how the theory that all of the terrestrial planets had similar impact histories can be used to date the formation of the Martian uplands, the Martian basins, and the Tharsis volcanoes. How certain are the ages derived for these features?
10. Describe the weather patterns experienced by the Viking landers. Why was the Martian weather more repeatable from day to day than that usually encountered on the Earth?
11. What is the evidence that leads scientists to suspect that the SNC meteorites are from Mars? (*Hint:* consider where else they might have originated.)
12. Why are the two polar caps of Mars so different from each other? The Earth also has polar areas that are different from each other; are any of the reasons similar?
13. Where is the water on Mars? Try to estimate how much water might be present in various forms.

*14. If the period for a shift of the great seasonal dust storms from the southern hemisphere to the northern and back again is 25,000 Earth-years and the layers in the polar caps that represent this shift are 10 m thick, how thick is the seasonal deposit of dust produced on the polar cap each Martian year?

15. List the main differences among the Martian canals, the runoff channels, and the outflow channels.
16. Suppose the sun increased its luminosity by 50 percent. How would you expect conditions on Mars to change?
17. Why is there so little organic material in the soil of Mars?
18. Describe a possible "oasis" on Mars where life might exist today. How likely do you think it is that there are such places on Mars?

Carl Sagan (b. 1934) of Cornell University is probably the best-known scientist in the U.S. today, as a consequence of his popular writings and television appearances. Sagan's earlier research work concentrated on Venus; he developed the first quantitative models for the atmosphere of that planet and established that its high surface temperature was the result of a greenhouse effect. Later he extended his studies to Mars, where he played a leading role in the development of the Viking program. As founder and president of the Planetary Society, he is a major advocate of the scientific exploration of the planets. (Brent Peterson, *Parade*)

16

VENUS: THE HOTTEST PLANET

16.1 GENERAL PROPERTIES OF VENUS

Venus, named for the goddess of love and beauty, is sometimes called Earth's "sister," for of all the planets it is most like the Earth in mass and size. As we have learned during the past 25 years, however, conditions on the two worlds could hardly be more different. Nor would visitors find Venus appropriately named, for its surface environment of furnace-hot temperatures and 90-bar pressures is certainly not conducive to romance.

Understanding the divergent evolution of Venus and Earth is one of the challenging problems of planetary science. How did two planets, so alike in most ways, develop such different surface environments? Is there a chance that Earth could end up the way Venus is today? Or could Venus be changed to transform it from a hot, sterile planet into one more like the Earth? We are only beginning to be able to answer such questions.

(a) Basic Facts

Venus approaches the Earth more closely than any other planet—at its nearest it is only 40 million km away. Like Mercury, Venus is an inferior planet (nearer the sun than the Earth); consequently, it appears to swing back and forth in the sky, during its synodic period, from one side of the sun to the other. It appears sometimes as an "evening star" and sometimes as a "morning star." Because Venus is farther than Mercury from the sun, it reaches much greater eastern and western elongations, and can appear as far as 47° from the sun. The early Greeks, thinking it to be two objects, called it Phosphorus when it was in the morning sky and Hesperus when it was in the evening sky. Pythagoras, in the sixth century B.C., is credited with being the first to recognize that Phosphorus and Hesperus were one and the same planet.

The median distance of Venus from the sun is 108 million km. Its orbit is the most nearly circular

TABLE 16.1 Properties of Venus

Semimajor axis	0.72 AU
Period	0.61 yr
Mass (Earth = 1)	0.82
Diameter	12,104 km
Density	5.24 g/cm^3
Uncompressed density	4.4 g/cm^3
Surface gravity (Earth = 1)	0.91
Escape velocity	10.4 km/s
Rotation period	−243 days
Surface area (Earth = 1)	0.94
Atmospheric pressure	90 bar

of any of the planetary orbits, having an eccentricity of only 0.007. Venus has a mass 0.82 times the mass of the Earth, and a radius only about 300 km less than Earth's. The uncompressed density calculated for Venus is 4.4 g/cm^3, almost the same as the Earth's, suggesting nearly identical composition and probably very similar interior structure.

Table 16.1 summarizes some of the basic properties of Venus.

(b) Appearance of Venus

Venus is one of the most beautiful objects in the night sky. Its brilliance is exceeded only by that of the sun and the Moon; at night it can cast a shadow, and it can even be seen in broad daylight if one knows exactly where to look for it. Indeed, when Venus happens to be conspicuous in the evening sky in December, people, surprised at its brilliance, often call observatories to inquire whether it is the star of Bethlehem.

Venus goes through phases like the Moon, as discovered by Galileo in 1610. At full phase, however, Venus is not observable from the Earth, because then it is at superior conjunction—on the other side of the sun from the Earth—and is too nearly in line with the sun. Even if it could be seen, however, Venus would not show its greatest brilliance at full phase because at that time it is also at its greatest distance from the Earth and subtends an angle of only 10 arcseconds. Venus has its largest angular size of 64 arcseconds when it is nearest the Earth, but then it is at the new phase (at inferior conjunction) and, again, is unobservable. Venus is at its greatest brilliance when it is a crescent, with an elongation of about 39°; this occurs about 36 days before and after inferior conjunction.

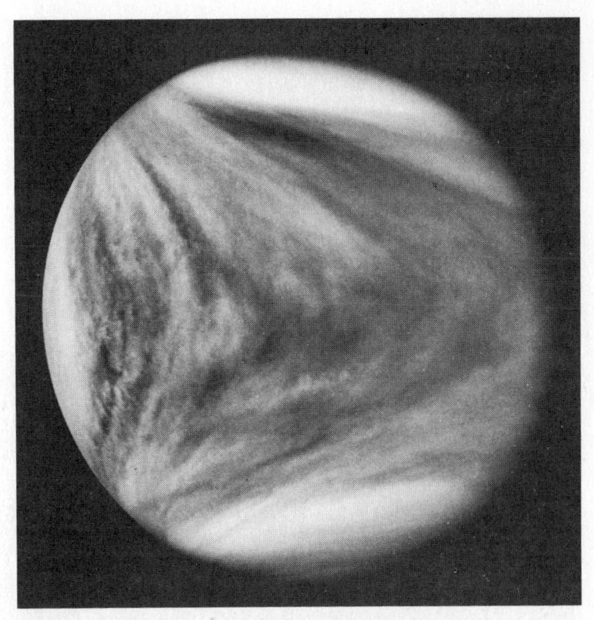

Figure 16.1 Venus, photographed by Pioneer Venus orbiter at a distance of 58,000 km. (NASA)

The surface of Venus is not visible because it is shrouded by dense clouds (Figure 16.1). These clouds reflect sunlight very efficiently; measures of the albedo of the planet show that more than 70 percent of the incident visible sunlight is reflected, a circumstance that contributes greatly to the planet's brightness. Some rather indistinct dark markings can be observed on Venus from Earth; they show up best when the planet is photographed in violet light. Ultraviolet photography from spacecraft has been used to determine the motion of the upper atmosphere, but radar is required to penetrate the clouds and examine the surface.

(c) Rotation

The rotation period of Venus, like that of Mercury, was determined by radar. The first radar observations of the planet's rotation were made in the early 1960s. Surprisingly, they showed Venus to rotate from east to west—in the reverse direction from the rotation of most other planets—in a period of about 250 days. Subsequently, topographical surface features were identified on the planet that show up in the reflected radar signals. The rotation period of Venus, determined from the motion of such features across its disk, is 243.08 days retrograde (east to west). The rotation period of Venus is about 19

Figure 16.2 Pioneer Venus 2 multiprobe *(foreground)* and Pioneer Venus 1 orbiter. The entry probes can be seen mounted on the bus, which itself carried instruments into the Venerian atmosphere. (NASA)

days longer than its period of revolution about the sun. The length of a day on Venus—the time between successive noons—is 116.67 Earth days.

The synodic period of Venus is 243.16 days, a value very close to the rotation period of 243.08 days. This coincidence creates a problem for observers, because it means that very nearly the same face of Venus is turned toward Earth at each inferior conjunction, when radar is most useful for mapping the surface. As a result, we have good radar maps only of this one hemisphere.

Because Venus rotates so slowly, it was not expected to have a strong magnetic field, for there is no driving force to set up a current of moving charges in its interior. The prediction has been confirmed by the various space probes sent to the planet; no general magnetic field has been detected.

(d) The Exploration of Venus

Venus and Mars are the two planets that have been most extensively explored by spacecraft. In the case of Venus, the U.S.S.R. has taken the lead, sending many more probes and developing a number of innovative techniques for studying both the atmosphere and the cloud-shrouded surface beneath.

The first U.S. interplanetary spacecraft, Mariner 2, flew by Venus in 1962, followed by Mariner 5 in 1967. Meanwhile, the U.S.S.R. began its series of missions called *Venera,* the Russian word for Venus. The early Venera entry probes were crushed by the high pressure of the atmosphere before they could reach the surface, but in 1970 Venera 7 successfully landed and broadcast data from the surface for 23 minutes before succumbing to the 730 K surface temperature. Additional successful probes and

Figure 16.3 Soviet scientists have taken the lead in the exploration of Venus through their extensive series of Venera spacecraft, which have included orbiters, atmospheric probes, and landers. (a) V. Barsukov, Director of the Vernadsky Institute. (b) Yu. A. Surkov, the designer of many of the instruments flown to Venus. (David Morrison)

landers culminated in 1978 with Veneras 11 and 12, which photographed the surface of Venus.

In the 1980s, the Venera landers have sent additional data from the surface, including both color pictures and chemical analyses of the soil. At the same time, the Soviet scientists have expanded into new areas. Veneras 15 and 16 were orbiters that arrived at Venus in 1983 and have carried out detailed surface mapping of the northern hemisphere of the planet using powerful radars to penetrate the clouds and thick atmosphere. And in 1985 the VEGA spacecraft (VEGA = Venera + Halley) successfully deployed two French-built balloons into the atmosphere, where they transmitted data to Earth for several days. Some of the instruments on these balloons were built by U.S. experimentors.

In contrast to the U.S. planetary exploration program, the Soviet effort has been characterized by conducting many flights using modifications of a well-proven spacecraft and lander system. Having developed this capability, the Soviet scientists have used it repeatedly to carry out a long-term scientific investigation. The U.S., on the other hand, tends to build expensive spacecraft to be used for just one or two flights. Once these are successful, the U.S. goes on to new challenges rather than returning to the same planet. It is not clear which is the better strategy; however, the Russians have certainly been very successful with their approach to the exploration of Venus.

The only U.S. mission to Venus in the past decade was Pioneer Venus, which consisted of an orbiter and multiprobe sent to the planet in 1978 (Figure 16.2). The multiprobe sent five separate instrument packages into the Venerian atmosphere: a large probe, three small probes, and the bus—the vehicle on which the other four instruments were mounted before deployment. The large probe descended by parachute, while the smaller probes descended without parachute and took 55 minutes to fall the 200 km from entry down to the surface. Unlike the Veneras, none of the Pioneer Venus probes was designed to survive impact; nevertheless, one of them did transmit data for 67 minutes after landing.

The surface environment of Venus is an extremely difficult one for spacecraft designers. So far, no lander has operated for more than a few hours, and in the U.S., at least, there is no prospect for better performance with present technologies. Much remains to be done from orbiting spacecraft, however, and in the analysis of the atmosphere. The next U.S. mission is an orbital radar mapper called Magellan scheduled for launch in 1988, designed to map the entire planet with a resolution of a fraction of a kilometer. Magellan is similar in concept to the Venera 15 and 16 radar orbiters, but with its improved resolution it should provide a major advance in our understanding of the geology of Venus.

16.2 THE SURFACE REVEALED

(a) Probing Through the Clouds

The clouds of Venus are opaque, totally blocking our view of the surface beneath. Thus even though Venus comes closer than any other planet to Earth, we have only recently obtained any information on the topography of the planet. Geological studies of Venus are in their infancy compared to those of the Earth, the Moon, or even Mars.

The information that has recently been acquired on the surface of Venus comes from four sources: the Venera landers; Pioneer Venus; the Arecibo radar system; and orbital imaging radar on Veneras 15 and 16. First and most directly, there are the pictures and soil analyses obtained by the various Venera landers, but of course these provide us only a limited, worm's-eye view of a tiny fraction of the surface. For a global perspective, we must rely on penetration of the clouds by radar.

The only comprehensive radar survey of the entire Venerian surface was carried out by the Pioneer Venus orbiter between 1978 and 1981. This artificial satellite carried a radar transmitter that beamed pulses straight down and measured the time required for them to be reflected back up from the surface. In this manner, the orbiter gradually built up a map of the elevation of the surface. The vertical resolution in this technique is very good, better than a hundred meters, but the horizontal resolution is much more limited. Even with the most careful processing of the data, the best that can be achieved is about 25 km, about the same as the resolution of the Moon obtained with Galileo's first low-powered telescopes.

Considerably better resolution has been obtained over part of the surface with Earth-based radar, particularly employing the great 300-m radar telescope at Arecibo, Puerto Rico. When Venus is close to the Earth, this radar system can map the surface with a resolution of a few kilometers, comparable to normal telescopic resolution of the Moon. A major problem with Earth-based radar studies, however, is their limitation to under a quarter of the surface area of Venus, which is a consequence of the coincidence that Venus always presents the same face toward Earth near inferior conjunction, when it is closest to us (Section 16.1c).

The best, but also the most expensive, approach is to map the planet from orbit using high-resolution imaging radar. Imaging radar systems use highly sophisticated data processing with a powerful radar beam to reconstruct a picture of the surface, very similar in appearance to a satellite photograph of the Earth or Mars. Such systems are flown on the Space Shuttle for scientific studies of the Earth, and on U.S. and U.S.S.R. spy satellites (since they are unaffected by weather conditions). The Venera 15 and 16 spacecraft obtained an imaging resolution of about 2 km over the northern hemisphere above latitude 30°.

(b) Large-Scale Topography

The general appearance of Venus as revealed by the global Pioneer Venus map (Figure 16.4 and Color Plate 13) is very different from that of either Earth or Mars. On Earth about 45 percent of the crust is continental, while approximately half the surface area of Mars is uplands. Venus, in contrast, consists mostly of low, rolling, relatively flat terrain. Only about 10 percent of the surface is highlands, which might (or might not) be similar to the terrestrial continents.

The largest highland area on Venus, called Aphrodite, is about the size of Africa. Aphrodite stretches along the equator nearly halfway around the planet. Next in size is the northern region called Ishtar, which is about the size of Australia. Ishtar contains the highest region on the planet, the Maxwell Mountains, which rise about 11 km above the surrounding lowlands. Fortunately for us, Ishtar and the Maxwell Mountains are on the side of Venus that faces Earth near inferior conjunction as well as being within the area mapped by Veneras 15 and 16. Therefore we have relatively high-resolution radar data on this interesting region.

There are no other continent-size highland areas on Venus. There are several rises, however, that appear to be either huge individual mountains or mountain ranges. The most prominent of these, called Alpha and Beta, are south of Ishtar. Beta appears to be divided by a broad valley, while another valley cuts across part of Aphrodite. Otherwise,

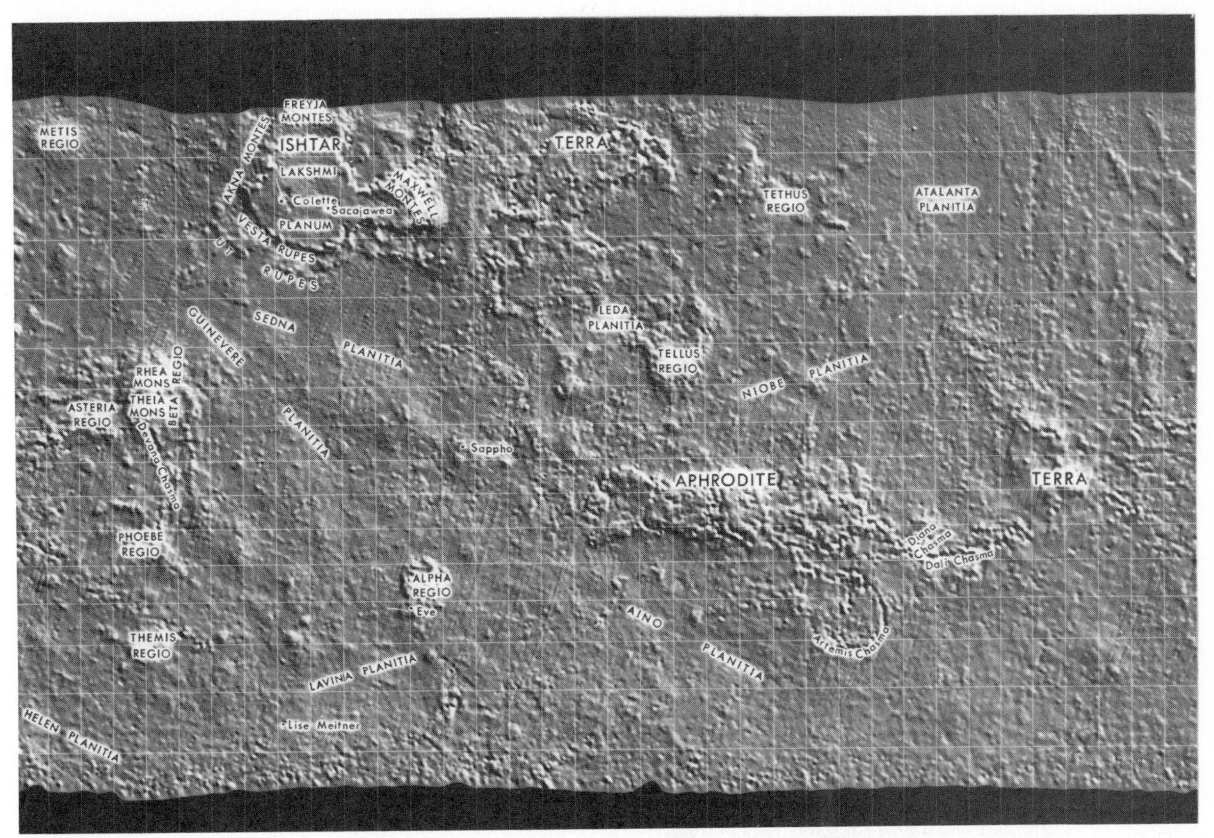

Figure 16.4 Pioneer Venus map of Venus showing landing sites of the Venera probes. (NASA/USGS)

low-resolution maps such as that obtained by Pioneer Venus reveal a relatively flat and dull surface.

As far as can be determined from existing radar data, Venus is not experiencing plate tectonic activity like the Earth. However, there are a number of suggestions of active volcanism, as will be discussed below, as well as the possibility that compression of the crust was involved in raising the Aphrodite and Ishtar highland regions. Future, higher-resolution images are likely to provide evidence of a great deal of interesting geological activity.

(c) Geology of Venus

In the best radar images, whether obtained from Arecibo or the Veneras, a wealth of additional detail becomes apparent. Unfortunately, however, the resolution is still too poor for a definitive geological interpretation of most of these features.

One of the first questions to ask from the new data is what is the age of the surface of Venus. Is this planet old, like the Moon or most of the surface of Mars, or is it young, like the Martian volcanoes or the Earth's ocean basins? As we have seen in the previous two chapters, such questions can be answered by counting impact craters. You might think that the thick atmosphere of Venus would protect the surface from impacts, but this is true only for relatively small projectiles. For impacts of the size that produce craters larger than a few kilometers in diameter, the atmosphere of Venus will not have provided any protection to the surface, so crater counts should be as significant here as on other bodies with much less atmosphere.

The Venera images indicate that the numbers of craters on the plains of Venus are only 10 to 20 percent of the lunar mare values, indicating a surface age considerably less than that of either the lunar maria or the volcanic plains of Mars. These results suggest that the surface of Venus is less than a billion years old, and they strongly argue for a planet with persistent geological activity. Thus, al-

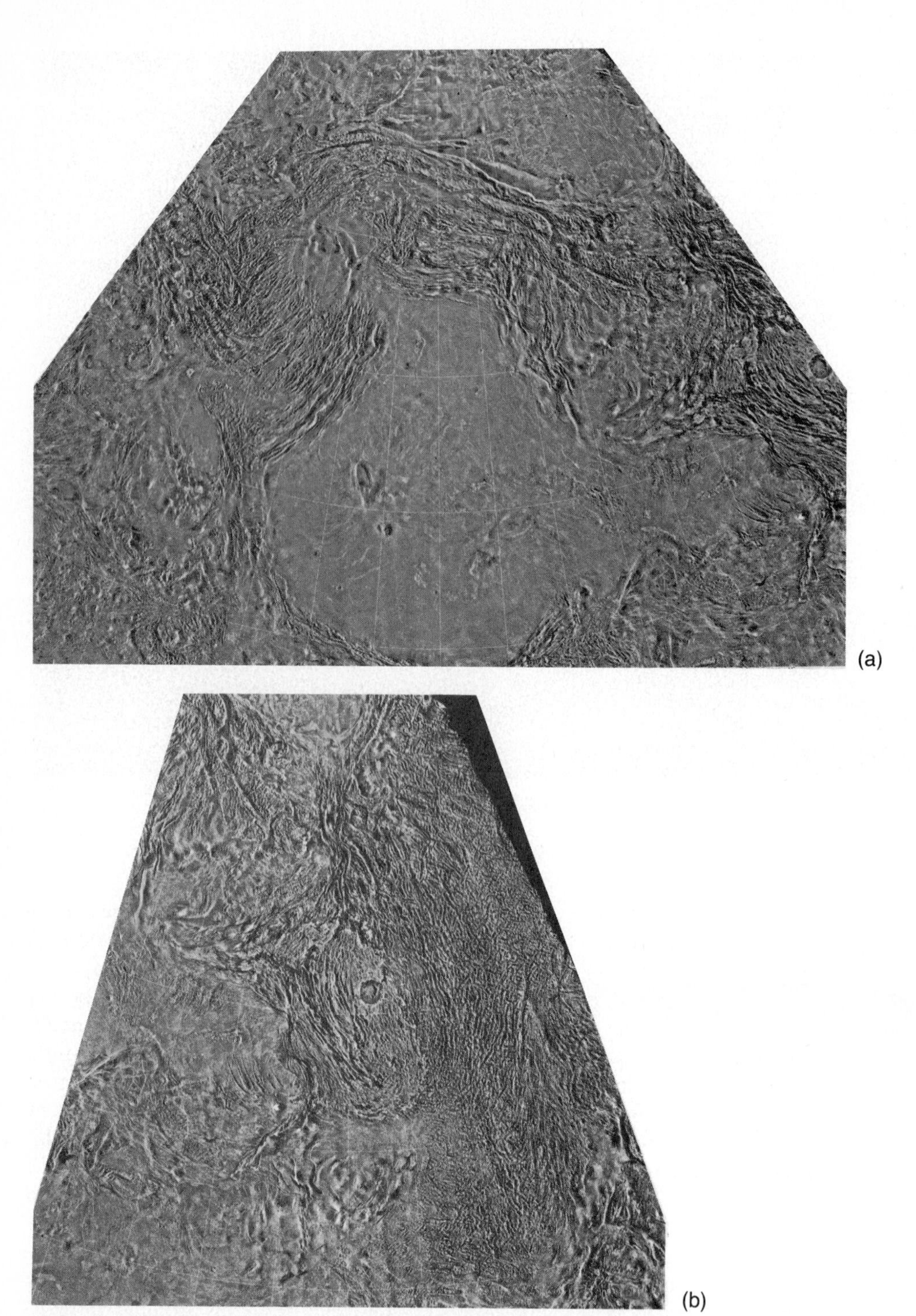

Figure 16.5 Two Soviet photomosaics of the surface of Venus at a resolution of about 2 km obtained with the Venera 15 and 16 imaging radar system. (a) Lakshmi Planum in the Ishtar highlands with its surrounding mountain ridges, produced from images collected between December 1, 1983, and January 25, 1984. (b) The Maxwell Mountains, the highest region of Venus, including the 80-km wide volcanic caldera Cleopatra, produced from images collected between December 30, 1983, and February 1, 1984. (Institute of Radioengineering and Electronics, U.S.S.R. Academy of Science)

though apparently less active than Earth, Venus seems to have retained more of its internal heat than Mars, a result that is not surprising in view of Venus' large size.

The Maxwell Mountains are the highest feature on Venus, and the best radar images (Figure 16.5) show that this range generally consists of long parallel ridges spaced about 20 km apart, similar to many folded mountains on Earth. Some scientists argue that such mountains were produced by uplift associated with motion of crustal plates on Venus. However, the 85-km-diameter circular depression called Cleopatra near the summit of these mountains appears to be a volcanic caldera, based on the Venera radar images. If Cleopatra is a caldera, the Maxwell Mountains themselves are more likely to represent a large eroded volcano, about the same size as Olympus Mons on Mars.

Among the most interesting features on the planet are the two mountainous regions, Alpha and Beta. As described below, the Venera landers have analyzed some surface material in the vicinity of these features and found it to be basalt, suggesting that these are large volcanoes. In addition, the Pioneer Venus orbiter repeatedly detected atmospheric electrical discharges from this region of the surface. We know from terrestrial experience that lightning often accompanies volcanic eruptions, and the same thing may be happening on Venus. There is therefore a growing suspicion that there are currently active volcanoes in this region of the planet. It is with great anticipation that we await high-resolution images of Alpha and Beta from the Magellan radar mapper mission.

(d) On the Surface

The successful Venera landers have found themselves on an extraordinarily inhospitable planet, with a surface pressure of 90 bars and a temperature of 730 K. (In more familiar American units, this is about 860°F. For comparison, the highest temperature in a kitchen oven is about 500°F.) The surface of Venus is hot enough to melt lead and zinc.

Despite these unpleasant conditions, the Soviet spacecraft have photographed their surroundings and collected surface samples for chemical analysis. Measured levels of the radioactive elements are suggestive of terrestrial igneous rocks, probably basalts, although in one case a granite seems more likely. More sophisticated chemical studies carried out by Veneras 13 and 14 near the bases of the Alpha and Beta mountains confirm the presence of basalt and support the hypothesis that these are volcanic features. Veneras 17 and 18 (the landers associated with the VEGA mission) landed on the Aphrodite continent in 1985 and found evidence of nonvolcanic igneous rocks, suggesting that this continental area may represent a part of the early crust of Venus, like the highlands of the Moon.

Examples of the Venera photographs are shown in Figure 16.6. Each of these is a wide-angle image built up out of many individual scans, just as were the Viking pictures of the Martian surface. In the foreground, the camera looks down to see the base of the lander and the adjacent soil, while toward either side the image extends up to the horizon. Each picture shows a desolate, flat landscape with a variety of rocks, somewhat like the two Viking landing sites on Mars. Some of these rocks may be ejecta from impacts. Other areas show flat, layered rock, perhaps exposed lava flows. Unfortunately, there are no chemical studies of sites that have been photographed, and no photographs of sites studied chemically, so direct comparisons are not possible.

The sun cannot shine directly through the heavy, opaque clouds, but the surface is fairly well lit by diffused light. The illumination is actually about the same as that on Earth under a very heavy overcast, but with a strong red tint, since the massive atmosphere blocks blue light. The weather is unchanging, with a temperature of 730 K and winds of less than 2 m/s. Because of the heavy blanket of clouds and atmosphere, one spot on the surface of Venus is pretty much like any other as far as weather is concerned. The reason for the uniform conditions and blistering temperatures is to be found in the atmosphere.

16.3 THE MASSIVE ATMOSPHERE

(a) Composition and Structure

The first gas detected in the atmosphere of Venus, from telescopic observations in 1932, was carbon dioxide. Now that the chemical composition of the atmosphere has been measured accurately by the Venera and Pioneer probes, we can confirm that CO_2 accounts for 96 percent of the atmosphere. The second most abundant gas is nitrogen (Table 16.2).

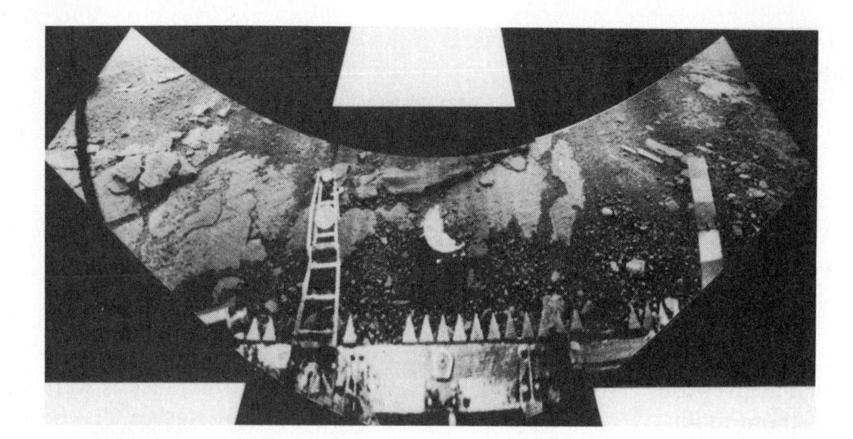

Figure 16.6 Views of the surface of Venus from the Venera landers. (U.S.S.R. Academy of Science/Brown University)

These proportions are actually very similar to those for Mars, but in other respects the atmosphere of Venus is dramatically different. With its surface pressure of 90 bars, the Venerian atmosphere is more than 10,000 times more massive than its Martian counterpart.

In addition to these gases, there are measurements of sulfur dioxide (SO_2) in the middle atmosphere, a subject we will return to when we look at the clouds of Venus. There have also been conflicting measurements of a very small amount of water vapor. However, the extreme dryness of the atmosphere is striking. Unlike the case of Mars, there is no possibility that Venus' water is frozen in polar caps or beneath the surface. This planet really is dry, and the absence of water is one of the important characteristics that distinguishes Venus from Earth.

The Venerian atmosphere (Figure 16.7) has a huge troposphere that extends up to at least 50 km above the surface. Within the troposphere, the gas is heated from below and circulates slowly, rising near the equator and descending over the poles. With no rapid rotation to break up this flow, the atmospheric circulation is highly stable. In addition, its very size maintains atmospheric stability. Being at the base of the atmosphere of Venus is something like being a kilometer below the ocean surface on the Earth. There, also, the mass of water evens out temperature variations and results in a uniform, stable environment.

TABLE 16.2 Composition of the Atmosphere of Venus

Gas	
Carbon dioxide	96.5%
Nitrogen	3.5
Argon	0.006
Oxygen	0.003
Carbon monoxide	0.002
Neon	0.001

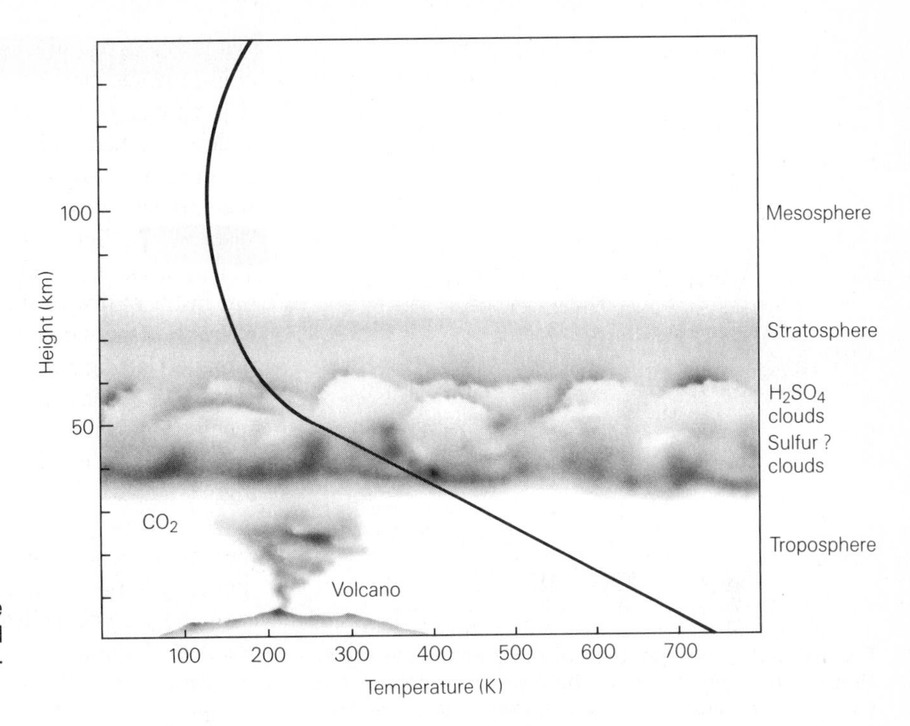

Figure 16.7 Structure of the atmosphere of Venus, based on data from the Pioneer entry probes.

The nature of the thick clouds of Venus was a problem that eluded astronomers for many years, in spite of the fact that light reflected from these clouds was repeatedly analyzed spectroscopically. There simply were no identifiable spectral features in the visible part of the spectrum. It was not until the 1970s that study of the spectrum in the infrared, together with a number of less direct arguments, demonstrated that the visible clouds were composed of sulfuric acid droplets. Sulfuric acid (H_2SO_4) is formed from the chemical combination of sulfur dioxide (SO_2) and water (H_2O).

Spacecraft measurements have confirmed the presence of both sulfuric acid and sulfur dioxide in the clouds, and have indicated the presence of other solid particles in addition. These are probably grains of elemental sulfur, but we are not sure. In the atmosphere of the Earth, sulfur dioxide is one of the primary gases emitted by volcanoes, but it is quickly diluted and washed out of the atmosphere by rainfall. In the absence of water, this unpleasant substance is apparently stable in the atmosphere of Venus. If we are right about the presence of active volcanoes on the surface, the source of sulfur dioxide is readily understood.

The clouds lie in the upper troposphere, between 30 and 60 km above the surface. Below 30 km, the air is clear. In the middle of the cloud layer, at the 53-km altitude where the French-Soviet balloons floated for 46 hours each in 1985, the conditions are quite Earth-like. The pressure here was 0.5 bar and the temperature was a comfortable 305 K, just a little warmer than in the room where you are reading this book. Were it not for the absence of oxygen and the nasty sulfuric acid clouds, this would not be too bad a place to visit. Certainly, it beats conditions on the surface of our "sister" planet.

At the top of the clouds the wind speeds are very high, blowing from east to west at about 100 m/s (360 km/hr). These winds can be tracked by motions of the clouds photographed from above. At the speeds of the Venerian "jet streams," cloud patterns are carried clear around the planet in just over four days, giving the appearance of a four-day rotation period. Of course, this four-day period has nothing to do with the actual rotation of the surface far below.

(b) The Greenhouse Effect

The high surface temperature of Venus was discovered by radio astronomers in the late 1950s and confirmed by Mariner 2 observations and by the early Venera probes. It was not easy to understand, however, how this planet could be so much hotter than

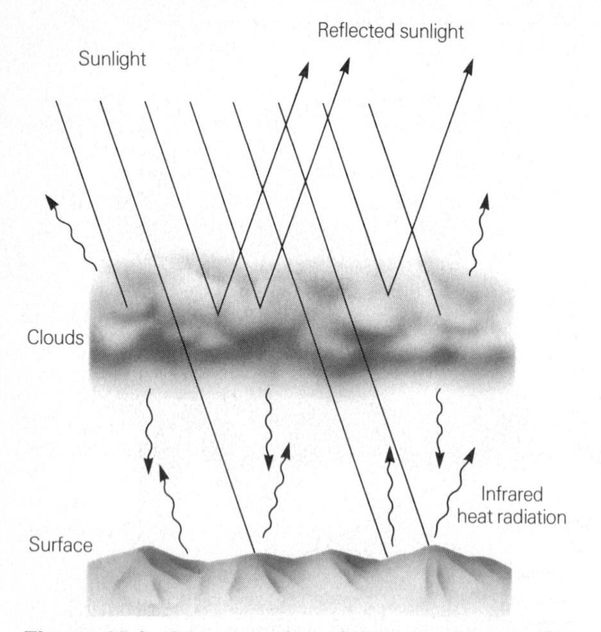

Figure 16.8 The operation of the greenhouse effect. Sunlight that penetrates to the lower atmosphere and surface is reradiated in the infrared, where the high opacity of the atmosphere restrains this heat from escaping. The result is a much higher surface temperature than would be present without the blanketing effect of the atmosphere.

would be calculated from solar heating. A major question therefore became: what is heating the surface of Venus to a temperature above 700 K?

Before trying to answer this question, it is important that we understand exactly what it is that we are trying to explain. Venus is hot, but it is *not* emitting more energy than it receives from the sun. The temperature at the cloud tops is about 250 K, just what would be calculated for a planet with Venus' albedo and distance from the sun. Therefore, we should not look for an explanation in terms of a strong internal heat source. Rather, we should ask how the surface can be so thermally isolated from its surroundings that it can remain hot without losing its heat to space. The answer to this question is found in the blanketing effect of the massive, cloudy atmosphere.

Imagine the fate of the sunlight that diffuses through the atmosphere of Venus and strikes the surface. It is absorbed, heats the surface, and is reradiated in the infrared part of the spectrum. However, carbon dioxide, which is a colorless, transparent gas in the visible, is opaque in the infrared. As a result, it acts as a blanket, making it very difficult for the infrared radiation to leak back to space. In consequence, the surface heats up, until eventually it is emitting so much infrared that an energy balance is reached with incoming sunlight.

The process whereby an atmosphere blankets the surface by its infrared opacity is called the *greenhouse effect* (Figure 16.8). It is similar to the heating of a gardener's greenhouse or the inside of a car left out in the sun with the windows rolled up. In these examples, the window glass plays the role of carbon dioxide, letting sunlight in but impeding the outward flow of infrared radiation. You all know the result: a greenhouse or car interior much hotter than would otherwise be expected from solar heating.

Calculations of the greenhouse effect for Venus were first carried out in about 1960 by Carl Sagan, then a graduate student at the University of Chicago. Sagan and colleague James Pollack later developed elaborate computer models for the Venerian atmosphere, finally demonstrating that even the extreme level of surface heating seen on Venus was the result of this basically simple process. Similar but much smaller greenhouse heating occurs on the Earth and even on Mars. As noted in Section 13.4d, the size of the greenhouse effect on our planet is very sensitive to the amount of carbon dioxide in the atmosphere, and will probably increase as more fossil fuels are burned, adding more CO_2 to our atmosphere.

(c) The Runaway Greenhouse Effect

In order to evaluate the possible consequences of increasing the carbon dioxide in Earth's atmosphere, it might be useful to try to trace the past history of Venus. Venus may once have had a much more Earth-like environment. If so, how could it have evolved to its present state?

Imagine that the Earth or an Earth-like Venus is heated, for example, by a small rise in the energy output of the sun or by an increase in atmospheric carbon dioxide. One consequence is a further increase in atmospheric carbon dioxide and water vapor, as a result of increased evaporation from the oceans and release of gas from surface rocks. These two gases would in turn produce a stronger greenhouse effect, further raising the temperature and leading to still more CO_2 and H_2O in the atmosphere. An unstable situation quickly arises, called the *runaway greenhouse effect.*

The runaway greenhouse is not just a larger greenhouse effect; it is a process whereby an atmosphere can evolve rapidly from a state where the greenhouse effect is small, such as on the Earth, to one with a much larger effect, such as we see today on Venus. Once the larger greenhouse conditions develop, the planet establishes equilibrium, and reversing the situation is difficult if not impossible.

If large bodies of water are available, the runaway greenhouse leads to their evaporation, creating an atmosphere of hot water vapor, which itself is a major contributor to the greenhouse effect. Water vapor is not stable in the presence of solar ultraviolet light, however, which tends to *dissociate,* or break apart, the molecules of H_2O into their constituent parts, oxygen and hydrogen. As we have seen, hydrogen can escape from the atmospheres of the terrestrial planets, leaving the oxygen to combine chemically with surface rock. The loss of water is therefore an *irreversible process;* once the water is gone, it cannot be restored.

There is evidence that this is exactly what happened to the water once present on Venus. This evidence comes from isotopic analysis of a sulfuric acid cloud particle carried out by the Pioneer Venus large probe as it descended through the atmosphere. Thomas Donahue of the University of Michigan and his colleagues argued that if large quantities of hydrogen escaped from the atmosphere of Venus, the hydrogen that remains should be enriched in its heavy isotope, deuterium, which has twice the mass of an ordinary hydrogen atom. Being heavier, deuterium escapes more slowly. Donahue found that the ratio of deuterium to hydrogen was about a hundred times greater on Venus than on Earth, indicating that enough water had been lost to account for modest-size oceans in the distant past.

This result is particularly interesting, since it allows the possibility that Venus and Earth both once had similar amounts of water. If, as many scientists believe, much of the water on the terrestrial planets arrived after planet formation in the form of icy planetesimals from the outer solar system, then Venus, Earth, and Mars each should each have accumulated this substance. We know there is water on Earth and ice on Mars. If Venus had never had water, however, the theory would have been suspect. Thanks to the Pioneer Venus result, this theory for the origin of water in the inner solar system is strengthened.

16.4 PLANETARY EVOLUTION: A COMPARISON OF THE TERRESTRIAL PLANETS

This chapter completes the discussion of the terrestrial planets. In Chapter 18 we will encounter a number of large satellites of the outer planets that are similar in size to the terrestrial planets, but most of these objects are different in composition, having formed in cold regions where water ice was a major building block. These icy bodies will be compared with the terrestrial planets later, but more appropriate now is a summary of the similarities and differences of Earth, Venus, Mars, Mercury, and the Moon.

(a) Composition and Origin

All of the inner planets formed in a region of the solar nebula where the primary condensates were rocky and metallic. It is therefore no surprise that the compositions of all five of the terrestrial bodies are dominated by these same materials. Venus and Earth are nearly identical in composition, each with a large metal core and a mantle of silicates. Mercury, forming closer to the sun, was depleted in some of the silicates, and therefore formed with a proportionately larger metal core. Mars, being farther from the center of the solar nebula, formed from slightly lower-temperature condensates, in which more of the metal was incorporated into minerals and less was available to form a core; its uncompressed density is therefore significantly lower than that of Venus or Earth.

The progression in composition from Mercury to Mars is readily understood in terms of the temperatures in the solar nebula at the time the planetesimals formed. The Moon, however, does not fit this simple picture. It appears to be made of terrestrial-type material from which both metals and volatiles were selectively removed. It has thus been necessary to seek an alternative explanation for the Moon's origin, involving a giant impact with the Earth or some other mechanism for depleting the metals and volatiles in the material out of which the Moon formed.

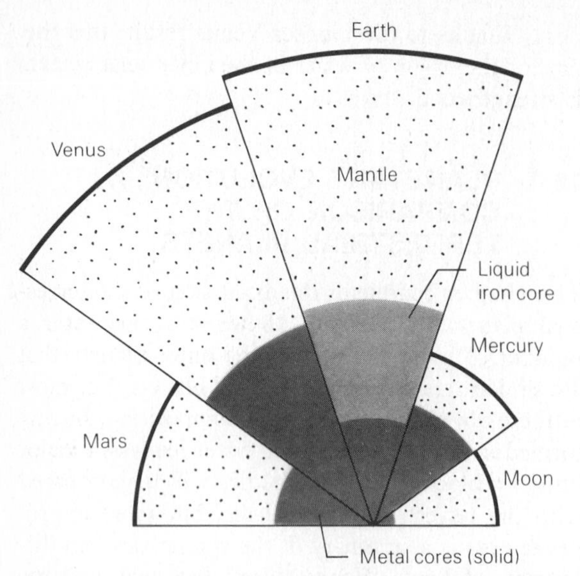

Figure 16.9 The interior structure of the terrestrial bodies.

In the inner part of the solar nebula, relatively little water in any form is expected; yet there is abundant water on the Earth and evidence of past water on both Venus and Mars, perhaps in comparable amounts. Presumably this water, and probably other low-temperature volatiles and organic matter as well, were added late in the planetary accretion process by impacting icy bodies from the outer solar system. This influx of water took place before the Moon formed, since the Moon is depleted in volatiles.

Early in their history, each of the five terrestrial bodies differentiated. The heat to initiate melting may have been supplied by the energy of impacts as the planets accreted, or it may have been generated internally by radioactivity. The sources of heating may have been different for different bodies. But we know that the Moon had a solid crust 4.4 billion years ago, and there is indirect evidence that the Earth is equally old. The heavily cratered surfaces of both Mars and Mercury also suggest that their present crusts were in existence by 4.0 billion years ago, if not much earlier.

As a result of differentiation, the terrestrial bodies all now have similar internal structures, consisting of core, mantle, and crust (Figure 16.9). Where there is a combination of a liquid metal core and rapid rotation, an internal magnetic field is generated, as we see on the Earth. Little or no magnetic field is present on Moon or Mars, presumably because they lack a liquid metal core. Venus probably has such a core, but it rotates so slowly that the absence of a magnetic field can also be understood. However, the weak magnetic field of Mercury remains something of a mystery, since we would not expect such a small planet to have retained enough internal heat to permit a liquid core. The theory seemingly breaks down here, suggesting that the process of generation of a planetary magnetic field is not very well understood after all.

(b) Geological Activity

Internal sources of geological activity require energy, either in the form of primordial heat left over from the formation of a planet or from decay of radioactive elements in the interior. The larger the planet, the more likely it is to retain its internal heat, and therefore the more we expect to see evidence on the surface of continuing geological activity.

The amount of radioactive heat generated in a planet is simply proportional to its mass. Doubling the amount of material doubles the energy released. To escape, however, this heat must work its way to the surface by a combination of conduction and convection, and it must then be radiated to space. Doubling the mass of a planet increases the thickness of blanketing material, and instead of doubling the surface area it increases it only by a factor of 1.6; thus larger objects cannot rid themselves easily of their internal heat, while smaller objects of the same composition tend to cool down more rapidly.

For the most part, the level of geological activity seen on the terrestrial planets conforms to the expectation of this simple theory (Figure 16.10). The Moon, the smallest of these objects, was internally active until about 3.3 billion years ago, when the major mare volcanism ceased. Since that time the mantle has cooled and become solid, and today even internal seismic activity has declined almost to zero. The Moon is a geologically dead world. Although we know much less about Mercury, it seems likely that this planet too ceased most activity about the same time that the Moon did.

Mars represents an intermediate case, and it has been much more active than the Moon. The southern hemisphere crust had formed by 4 billion years ago, and the northern hemisphere volcanic plains seem to be contemporary with the lunar maria. However, the Tharsis bulge formed somewhat later, and volcanic activity in the large Tharsis volcanoes

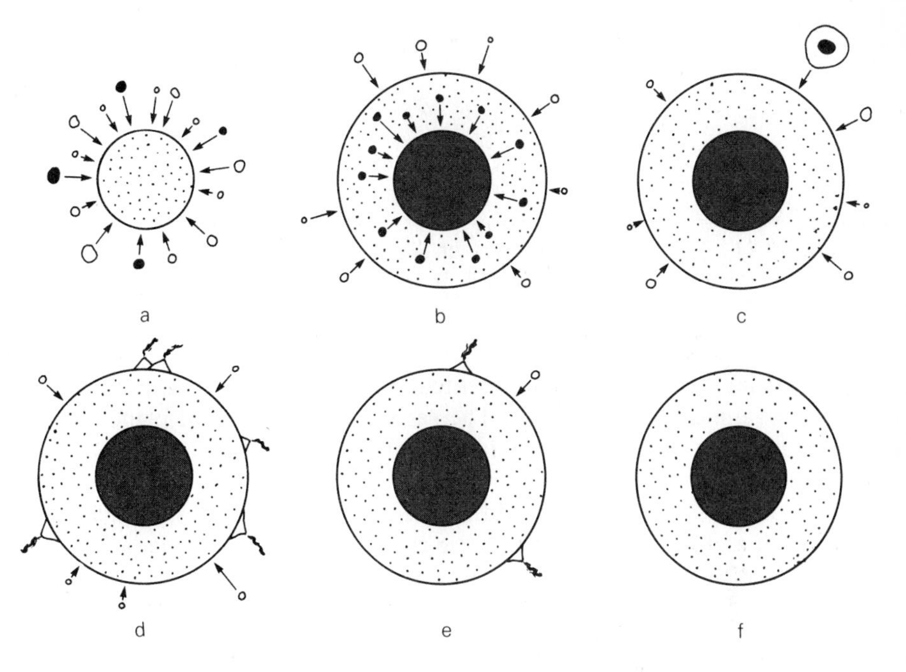

Figure 16.10 Schematic history of the growth and early evolution of a terrestrial planet. (a) Accretion of planetesimals of various compositions. (b) While accretion continues, the interior begins to melt and differentiate. (c) Final stages of accretion, which include impacts by a few very large objects. (d) Generation of heat by radioactivity results in extensive volcanism, while the final heavy bombardment is still taking place. (e) Volcanism declines as the interior begins to cool. (f) Finally, the planet settles down into a mature state with low flux of impacts and reduced volcanic activity.

has apparently continued intermittently to the present. Although there has been no large-scale plate tectonic activity, hot spots in the interior must still be carrying significant quantities of heat to the surface.

The Earth is the largest and the most active terrestrial planet, with its global plate tectonics driven by mantle convection. As a result, our surface is continually reworked, and over most of the planet the age of the surface material is less than 200 million years. Venus, with its similar size, should have similar levels of activity, but the limited data now available suggest that its surface is somewhat older and that it may not be experiencing plate tectonics. At the same time there is evidence of currently active volcanoes on Venus. A better understanding of the geological differences between Venus and Earth will have to await better radar data.

(c) Elevation Differences

The mountains on the terrestrial planets have very different origins. On the Moon and Mercury the major mountains are just ejecta thrown up by the large basin-forming impacts that took place billions of years ago. On Mars and perhaps Venus, the largest mountains are huge shield volcanoes, produced by repeated eruptions of lava from the same vents. Finally, the highest mountains on Earth are the result of collisions of one continental plate with another, although shield volcanoes nearly 10 km high are also present.

It is interesting to compare the maximum heights of the volcanoes on Earth, Venus, and Mars (Figure 16.11). On both of the larger planets, the maximum elevation differences between these mountains and their surroundings are about 10 km. Olympus Mons, in contrast, towers 26 km above its surroundings, and nearly 30 km above the lowest elevation areas on Mars.

One reason that Olympus Mons is so much larger than its terrestrial counterparts is that the lithospheric plates on Earth never stop long enough to let a really large volcano grow. Instead, the moving plate creates a long row of volcanoes, like the Hawaiian islands. On Mars (and perhaps Venus), the crust remains stationary with respect to the underlying hot spot, and the volcano can continue to grow for hundreds of millions of years.

A second difference relates to the force of gravity on the three planets. The surface gravity on Venus is nearly the same as on Earth, but on Mars it is only about one-third as great. In order for a mountain to survive, its internal strength must be great enough to support its weight against the force of gravity. Volcanic rocks have known strengths, and it is apparent that on Earth 10 km is about the limit; for instance, when new lava is added to the

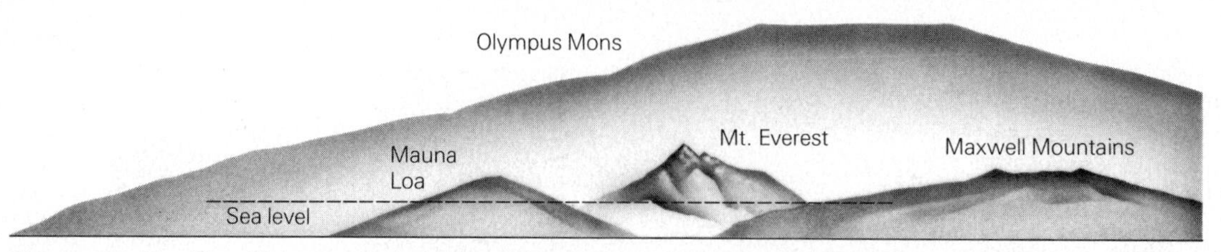

Figure 16.11 The highest mountains on Mars, Venus, and Earth. Mountains can rise taller on Mars because the surface gravity is less. The vertical scale is exaggerated 3:1 in this sketch.

top of Mauna Loa, the mountain eventually slumps downward under its own weight. The same height limit applies on Venus, where the force of gravity is the same. On Mars, however, with its lesser surface gravity, much greater elevation differences can be supported, and it should be no surprise that Olympus Mons is more than twice as high as the mountains of Venus or Earth.

The same kind of calculation that illustrates the limiting height of a mountain can be used to determine the largest body that can have an irregular shape. All of the planets and larger satellites are nearly spherical, as a result of the force of their own gravity. But the smaller the object, the greater the departures from spherical shape that can be supported by the strength of its rocks. For silicate bodies, the limiting diameter is about 400 km; larger objects will always be approximately spherical, while smaller ones can have almost any shape. We will see later that this prediction is supported by the observations of satellites (Chapter 18) and asteroids (Chapter 19).

(d) Atmospheres

The atmospheres of the terrestrial planets were formed by a combination of outgassing from their interiors and the impacts of volatile-rich planetesimals from the outer solar system. Each of the four must have originally had similar atmospheres, but Mercury was too small and too hot to retain its gas. The Moon probably never had an atmosphere, since the material out of which it is made was depleted in volatiles.

The predominant atmospheric gas on the terrestrial planets is now carbon dioxide, but initially there were probably also hydrogen-containing gases. In this more chemically reduced environment there should have been large amounts of carbon monoxide, and perhaps also ammonia and methane. Solar ultraviolet light dissociated the reducing gases, however, and eventually the hydrogen escaped to leave behind the oxidized atmospheres seen today on Earth, Venus, and Mars.

The fate of water was different for each of these three planets, depending on its size and distance from the sun. Early in its history Mars apparently had a thick atmosphere with abundant liquid water, but it could not retain these conditions. The carbon dioxide necessary for a substantial greenhouse effect was lost, the temperature dropped, and eventually the water froze. On Venus the reverse process took place, with a runaway greenhouse effect leading to the permanent loss of water. Only on Earth was the delicate balance maintained that permits liquid water to persist.

With the water gone, Venus and Mars each ended up with an atmosphere that is about 96 percent carbon dioxide and a few percent nitrogen. On Earth, the presence first of water and then of life led to a very different kind of atmosphere. The carbon dioxide was removed to be deposited in marine sediment, while proliferation of photosynthetic life eventually led to the release of enough oxygen to overcome natural chemical reactions that are eliminating this gas from the atmosphere. As a result, we on Earth find ourselves with a great deficiency of carbon dioxide and with the only planetary atmosphere containing free oxygen.

EXERCISES

1. Venus requires 440 days to move from greatest western elongation to greatest eastern elongation, but only 144 to move from greatest eastern elongation to greatest western. Explain why. (A diagram will help.)
2. At its nearest, Venus comes within about 40 million km of the Earth. How distant is it at its farthest?
3. Would astronomers be likely to learn more about the Earth from observatories on Venus or Mars? Explain.
4. Explain some of the problems that would be encountered in trying to build a spacecraft that could operate on the surface of Venus for a full Venus year.
5. Discuss the differences between the Soviet program to explore Venus and the U.S. program to explore Mars. What are the advantages of each approach?
6. What are the advantages of using radar imaging rather than ordinary cameras to study the topography of Venus? What are the relative advantages of these two approaches to mapping the Earth or Mars?
7. Discuss the main ways in which the geology of Venus is similar to that of the Earth, and the main differences between the two planets.
8. Explain how the greenhouse effect results in such a high surface temperature on Venus. Is there any possible alternative explanation for the high temperature?
9. Why is there so much more carbon dioxide in the atmosphere of Venus than in that of the Earth? Why so much more than in the Martian atmosphere?
10. How could Venus have avoided a runaway greenhouse effect? (In formulating your answer, consider the effects of changing its distance from the sun.)
11. Venus is now a very dry planet. Why is this troubling? Discuss why it seems likely that Venus should once have had more water, and what evidence there is from Pioneer Venus for the existance of more water in the past.

*12. We have seen how Mars can support greater elevation differences than Earth or Venus. According to the same arguments, the Moon should have higher mountains than any of the other terrestrial planets, yet we know it does not. What is wrong with this line of reasoning?

James Van Allen (b. 1914), American physicist from the Iowa State University, has given his name to the largest feature of planet Earth, the belts of charged atomic particles that make up the inner magnetosphere. One of the originators of the discipline of space plasma physics, Van Allen has extended his studies from the Earth to the larger and more complex magnetospheres of the outer planets as an experimenter on the Pioneer spacecraft, and his strong advocacy was critical in approval of the Galileo mission to Jupiter planned for the late 1980s.

17

THE GIANT PLANETS

17.1 FOUR GIANTS: AN OVERVIEW

The members of the outer solar system are very different from the inner planets, which have been discussed in the previous four chapters. Most obviously, the scale of the system is greatly expanded. The planets themselves are much larger, distances between them are vastly increased, and they are accompanied by extensive systems of satellites and rings. From the perspective of an objective observer, the outer solar system is where the action is, and the giant planets are the important members of the sun's family. In contrast, the little cinders of rock and metal that orbit closer to the sun seem like an insignificant afterthought.

In spite of the size and complexity of the outer solar system, we devote only two chapters to its discussion. The reason for this seeming discrimination is simply lack of detailed information on these distant worlds. They are far from the Earth, difficult to study telescopically, and much more demanding technologically to explore with spacecraft. From our perspective, living on one of the small inner planets, the giant planets and their systems seem distant and exotic, and only very recently have we begun to focus much attention on them.

(a) A Census of Objects

There are five planets in the outer solar system; Jupiter, Saturn, Uranus, Neptune, and Pluto. The first four of these are often called the giant or Jovian planets, while Pluto is a small world, by far the smallest in the planetary system. Pluto is physically similar to the satellites of the giant planets, and it is therefore discussed with them in Chapter 18.

Jupiter and Saturn (Figure 17.1) are so large that in spite of their great distances, they appear as bright as the brightest stars. They were both well known to ancient peoples, who revered them as important gods. In contrast, both Uranus and Neptune were *discovered*, one by accident and the other as the result of detailed theoretical calculations. Pluto is also a discovered planet, the product of a careful, systematic telescopic search.

Figure 17.1 The two largest planets, Jupiter and Saturn, shown to scale. (NASA/JPL)

Each of the giant planets has a system of orbiting satellites and rings. Many of these have been discovered recently as a result of the first spacecraft studies of these systems, and undoubtedly our present census is incomplete. As of this writing, however, Saturn appears to have the most extensive system, consisting of its beautiful and complex rings plus 19 known satellites. Jupiter is a close second, with 16 satellites and a very faint ring. Uranus has an intricate system of narrow, dark rings and five telescopically discovered satellites; ten additional satellites were discovered by the Voyager 2 spacecraft in 1986, bringing the total to 15. Finally, Neptune has only fragmentary rings and two satellites, according to telescopic studies; the first spacecraft will not reach this distant region until 1989.

Most of the mass in the planetary system is found in the giant planets, as described in Chapter 12. Jupiter alone accounts for more than half of this mass. Saturn is about a quarter as massive as Jupiter, and both Uranus and Neptune are much smaller yet. Uranus and Neptune are also different from the larger giants in both composition and structure, as described below.

(b) Chemistry in the Outer Solar System

In moving outward beyond Mars and the asteroid belt, we enter a region of very different planetary composition. Much of this difference, as noted in Section 12.3, results simply from the temperatures in the solar nebula. Beyond about 4 AU from the protosun, water ice was able to form grains and to accrete, together with rocky and metallic compounds, into planetesimals. Since the atoms that constitute water—hydrogen and oxygen—are among the most abundant in the universe, a great deal of water ice formed, resulting in planetesimals with about half their mass in the form of H_2O. Beyond 10 AU, additional ices could also condense, but none of these is nearly as plentiful as water ice.

The second major chemical distinction in the outer solar system is a result of the larger spacing between planets and the accumulation of more massive cores of rock and ice. The developing cores of Jupiter and Saturn apparently grew large enough before the dissipation of the gaseous solar nebula to attract and hold the hydrogen and helium from large volumes of space. Uranus and Neptune captured some hydrogen and helium, but much less; this is why these two planets are both smaller and different in composition than Jupiter and Saturn.

With so much hydrogen available, the chemistry of the outer solar system is *reducing*. Most of the oxygen present is chemically combined with hydrogen to form water, and it is therefore unavailable to form many oxidized compounds with other elements. The compounds detected in the atmospheres of the giant planets are thus hydrogen-based gases,

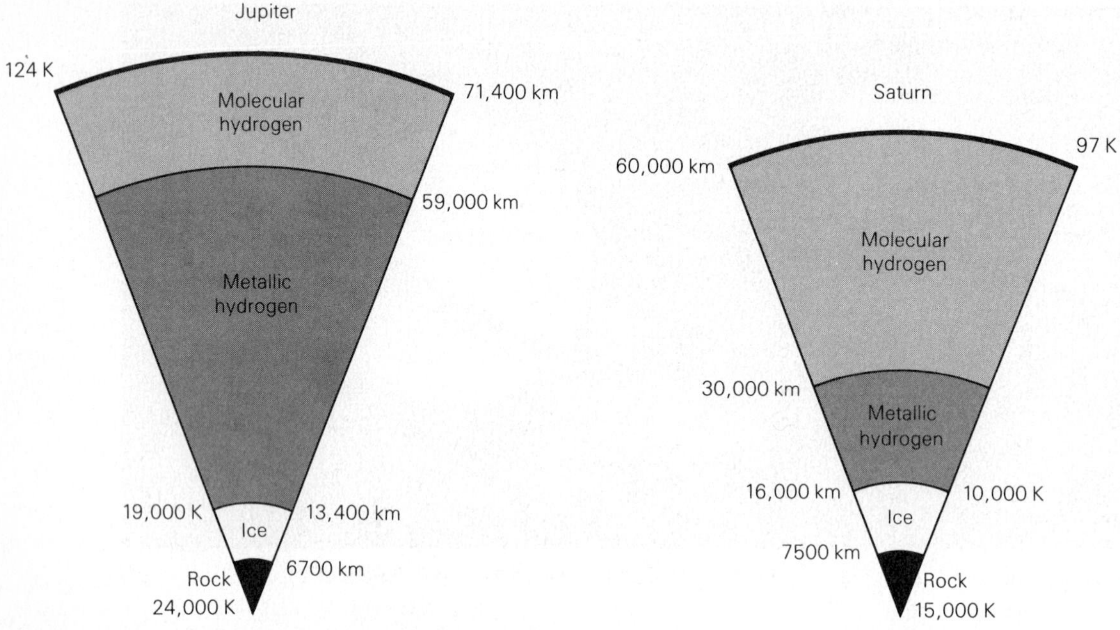

Figure 17.2 Sizes and interior structures of the giant planets.

such as methane (CH_4) and ammonia (NH_3), or more complex hydrocarbons such as ethane (C_2H_6) and acetylene (C_2H_2).

The internal structures of the giant planets are very different from those of the rocky inner planets (Figure 17.2). Each has an extensive atmosphere, as we would expect from the large quantities of gas they acquired from the solar nebula. However, at depths of only a few thousand kilometers below their visible clouds, pressures become so large that hydrogen changes from a gaseous to a liquid state. Still deeper, this liquid hydrogen acts like a metal. The equivalent of a mantle for these planets is thus a huge region composed primarily of liquid hydrogen. The so-called "gas giants" are therefore really primarily liquid.

Each of these planets has a core composed of heavier materials, as demonstrated by detailed analyses of their gravitational fields. Presumably these cores are the original rock-and-ice body that formed before the capture of gas from the surrounding nebula. The cores exist at pressures of tens of millions of bars, compared with a pressure of 4 million bars at the center of the Earth. While scientists speak of the giant planet cores being composed of rock and ice, we can be sure that neither rock nor ice assumes any familiar forms at such pressures and the accompanying high temperature. In actuality, we really know very little about conditions at the centers of the giant planets.

(c) Internal Heat Sources

Because of their large sizes, each of the giant planets was strongly heated during its own formation by the collapse of surrounding nebular gas onto its core. Jupiter, being the largest, was by far the hottest. In addition, it is possible for giant, largely gaseous planets to generate heat after formation by slow contraction. A shrinkage of as little as 1 millimeter a year can liberate substantial gravitational energy. Nevertheless, the discovery in 1969 by Frank J. Low of the University of Arizona of an internal source of energy for Jupiter came as a considerable surprise to the astronomical community. Only later, after a similar source had been found for Saturn, did theorists carry out the calculations required to show that such energy sources were to be expected for the giant planets.

Jupiter has the largest internal energy source, amounting to 4×10^{17} watts. It is glowing with the equivalent of 4 million billion hundred-watt light bulbs. This is about the same as the total solar energy absorbed by Jupiter. The atmosphere of Jupiter is therefore somewhat of a cross between a normal planetary atmosphere, which obtains most of its energy from the sun, and the atmosphere of a star, which is entirely heated from below. Most of the internal energy of Jupiter is primordial heat, left over from the formation of the planet 4.6 billion years ago.

Saturn has an internal energy source about half as large as that of Jupiter, which means (since its mass is only about one-quarter as great) that it is producing twice as much energy per ton of material as does Jupiter. Since Saturn is expected to have much less primordial heat, there must be another source at work generating most of this 2×10^{17} watts of power. This source is believed to be the separation of helium from hydrogen in the interior. In the liquid hydrogen mantle, the heavier helium forms drops that sink toward the core, releasing gravitational energy. In effect, Saturn is still differentiating. This precipitation of helium is possible in Saturn because it is cooler than Jupiter; at the temperatures in Jupiter's interior, hydrogen and helium remain well mixed.

Uranus and Neptune are different. Neptune has a small internal energy source, while Uranus does not emit enough internal heat for it to have been measured. No one knows why these two planets differ in this respect.

(d) Exploration of the Giant Planets

Four spacecraft, all from the U.S., have penetrated beyond the asteroid belt to initiate the exploration of the outer solar system. The challenges of probing so far from Earth are considerable. Flight times to the outer planets are measured in years to decades, rather than the few months required to reach Venus or Mars. Spacecraft must be highly reliable, and they must also be capable of a fair degree of independence and autonomy, since the light-travel time between Earth and the spacecraft is several hours. If a problem develops near Saturn, for example, the spacecraft computer must deal with it directly; to wait hours for the alarm to reach Earth and instructions to be routed back to the spacecraft could spell disaster. These spacecraft also must carry their own electrical energy sources, since sunlight is too weak to supply energy through solar cells. Heaters are required to keep instruments at proper operating temperatures, and spacecraft must have more powerful radio transmitters and large antennas if their precious data are to be transmitted to receivers on Earth a billion kilometers or more distant.

The first spacecraft to the outer solar system were Pioneers 10 and 11, launched in 1972/73 as pathfinders to Jupiter. Their main objectives were to determine whether a spacecraft could navigate through the asteroid belt without collision with small particles, and to measure the radiation hazards in the magnetosphere of Jupiter. Both spacecraft passed through the asteroid belt without incident, but the energetic plasma associated with Jupiter nearly wiped out their electronics, providing information necessary for the design of subsequent missions. Pioneer 10 flew past Jupiter in 1974, after which it sped outward toward the limits of the solar system. Pioneer 11 undertook a more ambitious program, using the gravity of Jupiter during its 1975 encounter to divert it toward Saturn, which it reached in 1979. Both Pioneers are still operating, each leaving the solar system at sufficient velocity to escape entirely from the sun.

Following Pioneer, the primary scientific missions to the outer solar system have been Voyagers 1 and 2, launched in 1977. These are larger and more capable spacecraft, launched with the same powerful Titan-Centaur rocket that was used for the Vikings sent to Mars. The Voyagers carry 11 scientific instruments, including cameras and spectrometers as well as devices to measure the magnetic field and plasma characteristics of planetary magnetospheres. Voyager 1 reached Jupiter in 1979 and used a gravity assist from that planet to take it on to Saturn in 1980. The second Voyager, arriving four months later at Jupiter, followed a different path to reach Saturn in 1981 and Uranus in 1986. Voyager 2 is now en route to Neptune, which it will encounter in 1989. Most of the information in this chapter and in Chapter 18 is derived from the Voyager missions.

Voyager 2 is following a trajectory made possible by the alignment of the four giant planets on the same side of the sun that took place around 1980. About once per century these planets are in a position where a single spacecraft can visit them all, using gravity-assisted flybys to adjust its course for the next encounter. We are very fortunate that this opportunity was seized. Because of this "grand tour" alignment, every planet in the outer solar system except Pluto will have been visited by spacecraft by 1990; otherwise, it would probably have been well into the next century before this basic reconnaissance of the planetary system was accomplished.

The next step in the exploration of the outer solar system is an extended study of Jupiter and its satellites. This mission, called Galileo, is awaiting launch at this writing (in 1986). When it reaches Jupiter in the early 1990s, it will deploy an entry probe into the planet for direct studies of the atmosphere, before beginning a three-year orbital tour, during which there will be repeated close flybys of the four large Galilean satellites. A similar mission to Saturn, called Cassini, is under discussion as a possible cooperative venture between NASA and the European Space Agency.

Table 17.1 summarizes the encounter dates for the Pioneer and Voyager missions to the outer solar system.

TABLE 17.1 Missions to the Outer Planets

PLANET	SPACECRAFT	ENCOUNTER DATE
Jupiter	Pioneer 10	Dec 73
	Pioneer 11	Dec 74
	Voyager 1	Mar 79
	Voyager 2	Jul 79
Saturn	Pioneer 11	Sep 79
	Voyager 1	Nov 80
	Voyager 2	Aug 81
Uranus	Voyager 2	Jan 86
Neptune	Voyager 2	Aug 89

17.2 JUPITER AND SATURN

(a) Basic Facts

Next to the sun, Jupiter (Figure 17.3 and Color Plate 14a) is the largest and most massive object in the solar system, so it is well named for the master of the gods. Saturn (Color Plate 15a) is the second-largest planet in the solar system and was the most remote one known in antiquity. It is named for the Greco-Roman god of agriculture and the father of Jupiter. Its beautiful ring system makes it one of the most impressive telescopic objects, and it has long been a favorite object for public viewing.

Jupiter and Saturn have similar compositions and structures, in spite of the fact that Jupiter is nearly four times more massive. Both planets have also been studied to approximately the same level of detail by the Voyager 1 and 2 spacecraft. Therefore it is appropriate to consider them together. Table 17.2 summarizes some basic information about both bodies.

(b) Orbits and Rotations

The median distance of Jupiter from the sun is 778 million km, 5.2 times that of the Earth; its period of revolution is just under 12 years. Saturn is about twice as far away as Jupiter, at an average distance from the sun of 1427 million km, or 9.6 AU. Saturn completes one sidereal revolution in very nearly the standard human generation of 30 years. Its slow cycle around the sky provided the longest natural time interval available to ancient peoples.

Distinct details in the cloud patterns on Jupiter allow us to determine the rotation rate of the atmo-

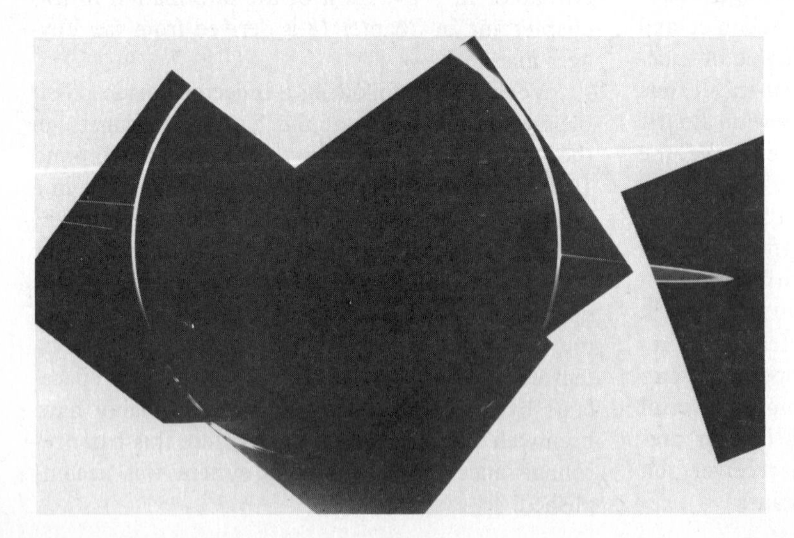

Figure 17.3 A mosaic of photographs obtained by Voyager 2, showing the night side of Jupiter, with the ring silhouetted against the dark sky. (NASA/JPL)

TABLE 17.2 Properties of Jupiter and Saturn

	JUPITER	SATURN
Semimajor axis (AU)	5.2	9.6
Period (yrs)	11.9	29.5
Mass (Earth = 1)	318	95
Diameter (km)	142800	120660
Density (g/cm^3)	1.3	0.7
Escape velocity (km/s)	60	36
Rotation (hrs)	10	11

sphere at the cloud level, although as we noted in the case of Venus (Section 16.4), such an apparent rotation of the atmosphere may have little to do with the spin of the underlying planet. Much more fundamental is the rotation of the mantle and core, as indicated by periodic variations in the magnetic field. This period of $9^h\ 56^m$ gives Jupiter the shortest "day" of any planet. Atmospheric rotation periods depend on latitude.

The rotation period of Saturn is determined in the same way as that of Jupiter. The fundamental period of $10^h\ 40^m$ is derived from variations in its radiation at radio wavelengths (which are in turn linked to its magnetic field), while motions in the upper atmosphere are determined from cloud observations. Like Jupiter, Saturn appears to rotate more slowly at higher latitudes. Because of its rapid rotation and low density, Saturn is the most oblate of all the planets; its equatorial diameter is about 10 percent greater than that through its poles.

The axis of rotation of Jupiter is tilted by only 3°, so there are no seasons to speak of; however, Saturn does have seasons, since its axis of rotation is inclined at 27° to the perpendicular to its orbit. Because the planet's rings lie in its equatorial plane, they are tipped by this same angle to its orbit. During part of Saturn's orbital revolution we look obliquely at one face of the rings, and when the planet is on the opposite side of the sun we see the other face. At intermediate points we view the rings edge on, at which times they disappear because the ring system is so thin—only a few tens of meters (see Section 18.5).

(c) Large-Scale Physical Properties

Jupiter is 318 times as massive as the Earth, a value that is very close to 1/1000 the mass of the sun. Its diameter is about 11 times the Earth's diameter and about 1/10 the diameter of the sun. Jupiter's density is 1.33 g/cm^3, much lower than that of any of the terrestrial planets. The mass of Saturn is 95 times that of the Earth, and its density is only 0.7 g/cm^3—the lowest of any planet. In fact, Saturn would be light enough to float, if an ocean large enough to contain it existed.

Jupiter has the highest escape velocity of any planet, 60 km/s; it can therefore retain all kinds of gases in its atmosphere. Saturn has an escape velocity of 36 km/s, also great enough to hold the lightest gases.

The interiors of both planets are composed primarily of hydrogen and helium. Of course, these gases have been measured only in their atmospheres, but calculations first carried out nearly 50 years ago by Yale University astronomer Rupert Wildt (1905–1976) have shown that they are the only possible materials out of which a planet with the observed masses and densities of Jupiter and Saturn could be constructed. There remain some uncertainties in calculating models, however, primarily because of our incomplete knowledge of the compressibility of liquid hydrogen and helium at the various temperatures and pressures that exist inside Jupiter and Saturn, and to the unknown abundances of other elements there. The best Jupiter models predict a central pressure of over 100 million bars, and a central density of about 31 g/cm^3.

It is interesting to note that Jupiter has very nearly the maximum possible size for a body of "cold" hydrogen, that is, one that is not generating energy. Less massive bodies than Jupiter would occupy a smaller volume. More massive bodies, by virtue of their greater gravitation, would also be compressed to a smaller volume than Jupiter's. In the latter case, the extreme internal pressures would lead to ionization. The separated atoms and electrons then would constitute a gas of a sort, but it would not obey the laws of physics appropriate to ordinary gases. Such an object, with a mass considerably larger than Jupiter's but not large enough to maintain nuclear reactions, is called a *brown dwarf* (Chapter 38).

(d) Composition and Structure of the Atmosphere and Clouds

Methane was first discovered in the atmospheres of Jupiter and Saturn more than 50 years ago, fol-

TABLE 17.3 Atmospheric Compositions of Jupiter and Saturn*

GAS	JUPITER	SATURN
H_2	1	1
He	0.12	0.06
CH_4	2×10^{-3}	2×10^{-3}
NH_3	2×10^{-4}	$>2 \times 10^{-5}$
C_2H_2	8×10^{-7}	1×10^{-7}
C_2H_6	4×10^{-5}	8×10^{-6}
PH_3	4×10^{-7}	3×10^{-6}

* Numbers are relative to hydrogen.

lowed by ammonia. Neither gas is nearly as abundant as hydrogen or helium, but both are more easily detected spectroscopically in visible light. It was not until the Voyager spacecraft measured the far-infrared spectra of Jupiter and Saturn that a reliable abundance could be found for the elusive helium on either planet. Table 17.3 summarizes the compositions of these two atmospheres.

At the temperatures and pressures of the upper atmospheres of Jupiter and Saturn, methane remains a gas, but ammonia can condense, just as water vapor condenses in the Earth's atmosphere, to produce clouds. When we look down on these planets, what we see primarily are the tops of the ammonia clouds.

The clouds of Jupiter are one of the most spectacular sights in the solar system, much beloved by makers of science fiction films. They range in color from white to orange to red to brown, swirling and twisting in a constantly changing kaleidoscope of patterns (Color Plate 14b). Saturn shows similar but very much subdued cloud activity; instead of vivid colors, its clouds are a nearly uniform butterscotch hue (Color Plate 15a). The beauty of Saturn lies not in the planet, but in its surroundings rings.

The primary clouds that we see when we look at either planet, whether from a spacecraft or through a telescope, are composed of crystals of frozen ammonia. The ammonia cloud deck marks the upper edge of the convective troposphere; above it is the cold stratosphere. The temperature near the cloud tops is about 140 K (only a little cooler than the polar caps of Mars). On Jupiter this cloud level is at a pressure of about 0.1 bar, but on Saturn it occurs at about 1 bar pressure.

The Galileo probe will enter the atmosphere of Jupiter and descend to a pressure level of between 10 and 20 bars before its battery power is exhausted or the probe is crushed by increasing atmospheric pressure. Below the thin ammonia clouds it should pass through a clear region, but at a pressure of about 3 bars we expect the probe to enter another thick deck of condensation clouds, this time composed of ammonium hydrosulfide (NH_4SH). These clouds have never been detected directly, but there is strong cricumstantial evidence for their existence. The ammonium hydrosulfide clouds probably also contain some sulfur particles, which color them a darker yellow or brown.

As it descends to a pressure of 10 bars and ever-higher temperatures, the Galileo probe should pass next into a region of frozen water clouds, and below that into clouds of liquid water droplets perhaps similar to the common clouds of the terrestrial troposphere. This region corresponds almost to a "shirt sleeve" environment, in which astronauts could exist quite comfortably if they carried scuba gear for breathing. But with no solid surface to stop it, the probe will continue to descend, penetrating to dark regions of higher and higher pressure and temperature. No matter how strongly it was built, eventually the probe will be crushed and swallowed in the black depths, where the great pressures finally transform the atmospheric hydrogen into a hot dense liquid.

Above the visible ammonia clouds the atmosphere of Jupiter is clear and cold, reaching a minimum temperature of near 120 K. At still higher altitudes temperatures rise again, just as they do in the thermosphere of the Earth, due to the absorption of solar ultraviolet light. In this region *photochemical reactions* create a variety of fairly complex hydrocarbon compounds by the action of solar radiation on the gases in the atmosphere. A thin layer of organic aerosols—a *photochemical smog*—lies far above the visible clouds. Jupiter also has an ionosphere, and at very high altitudes its upper atmosphere blends into the inner edge of its magnetosphere (Section 17.4).

The atmospheric structure on Saturn is very similar. Temperatures are somewhat colder, and the atmosphere is more extended as the result of Saturn's lower surface gravity, but qualitatively the same atmospheric regions, and the same condensation clouds and photochemical reactions, should be present (Figure 17.4). The composition of the two atmospheres is also similar, except that on Saturn there is only about half as much helium—the result of the precipitation of helium that also contributes to Saturn's internal energy source (Section 17.1c).

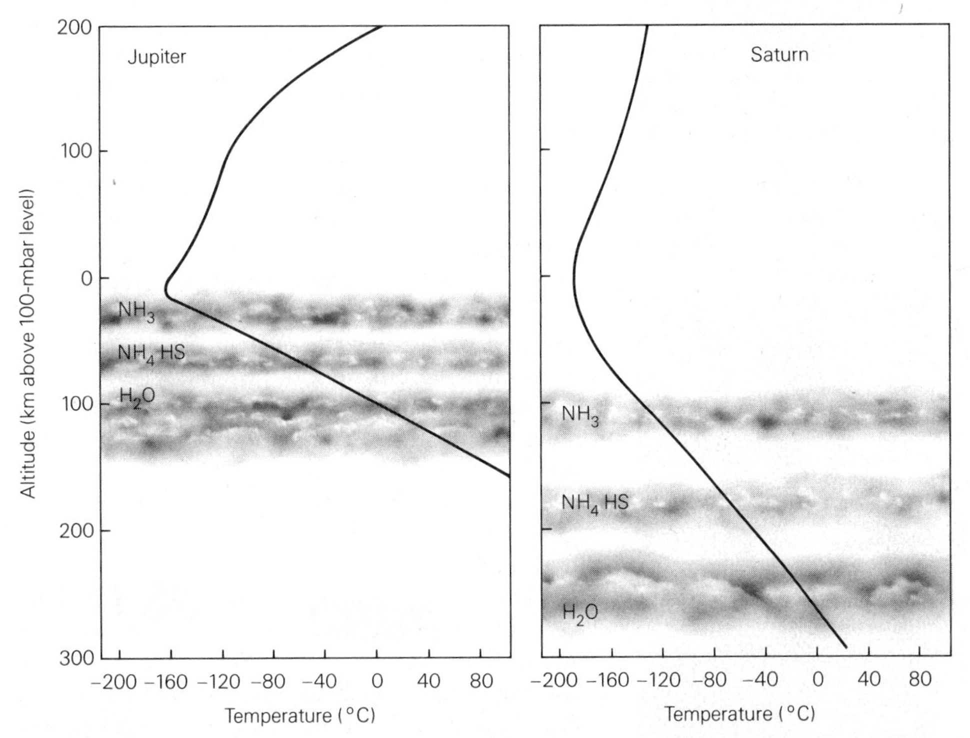

Figure 17.4 The atmospheric structure of Jupiter and Saturn as revealed by the Voyager spacecraft.

(e) Circulation of the Atmosphere

The atmospheric dynamics on Jupiter and Saturn differ fundamentally from those of the terrestrial planets. There are three primary reasons for these differences: (1) these planets have much deeper atmospheres, with no solid lower boundary; (2) they spin faster than the terrestrial planets, suppressing north-south circulation patterns and accentuating east-west airflow; and (3) major internal heat sources contribute about as much energy as sunlight, forcing the atmospheres into deep convection to carry the internal heat outward.

There are no weather stations in the atmospheres of Jupiter and Saturn, and no probe has investigated these atmospheres directly. However, the circulation of the atmosphere near the top of the troposphere can be mapped by observing the motions of the clouds. From a knowledge of these superficial wind patterns and of the energy balance that must be maintained deeper down, it is possible to calculate models of atmospheric circulation. These models have major uncertainties, however, especially for Saturn (Figure 17.5), where less is known about even the high-level wind patterns. Giant planet meteorology is a young subject, with fewer than a dozen people knowledgeable in the field.

The main features of the visible clouds of Jupiter are alternating dark and light bands that stretch around the planet parallel to the equator. Traditionally the lighter bands are called *zones* and the darker ones *belts*. The belts and zones are semipermanent features, although they shift in intensity and position from year to year. Consistent with the small obliquity of Jupiter, there are no seasonal effects detectable.

More fundamental than the zones and belts are the underlying east-west wind patterns in the atmosphere, which do not appear to change at all, even over many decades (Figure 17.6). The main such feature on Jupiter is an eastward-flowing equatorial jet stream with a speed of 300 km/hr, similar to the speed of jet streams in the Earth's upper atmosphere. At higher latitudes there are alternating east- and west-moving streams, with each hemisphere an almost perfect mirror image of the other. Saturn shows a similar pattern, but with a much stronger equatorial flow at a speed of 1300 km/hr.

Generally, the light zones are regions of upwelling air, capped by white ammonia cirrus clouds.

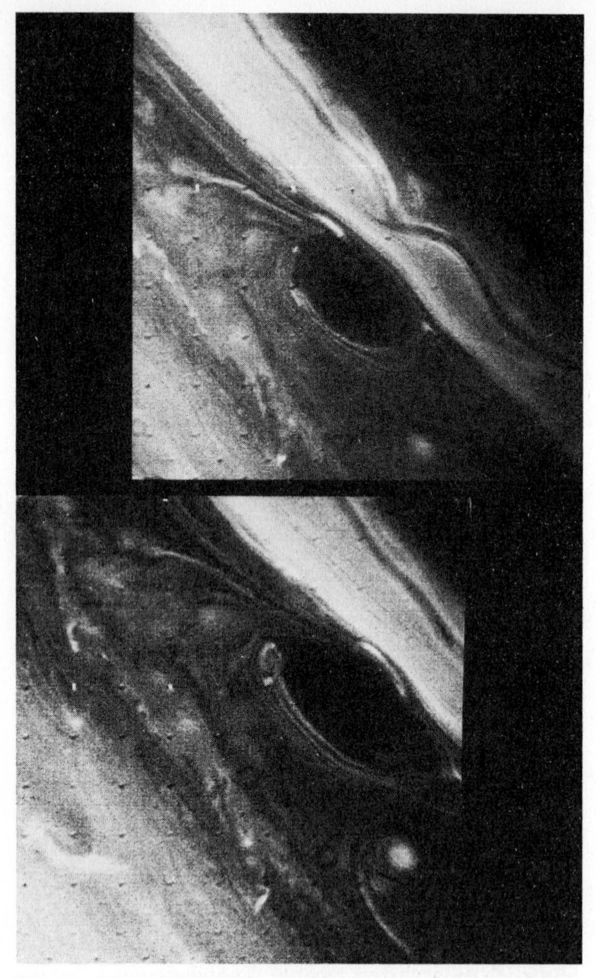

Figure 17.5 Two images of Saturn's atmosphere near the north polar region, taken by Voyager 2 from distances of 2.7 and 2.3 million kilometers, respectively. Note the circulation around the large brown spot. (NASA/JPL)

They apparently represent the tops of upward-moving convection currents. The darker belts are regions where the cooler atmosphere moves downward, completing the convection cycle; they are darker because there are less ammonia clouds and it is possible to see deeper in the atmosphere, perhaps down to the ammonium hydrosulfide clouds. This convection, driven by the heat escaping from the interior, gives rise on these rapidly spinning planets to eddies that transfer energy and momentum to the east- and west-flowing air currents, maintaining the major circulation pattern of the atmosphere.

Superimposed on the regular pattern described above are many local disturbances—weather systems or storms, to borrow terrestrial terminology. The most prominent of these are large oval high-pressure regions, some of which can even be seen from the Earth.

The largest and most famous "storm" on Jupiter is the *Great Red Spot,* or *GRS,* a reddish oval in the southern hemisphere that is almost 30,000 km long—big enough to hold two Earths side by side (Figure 17.7 and Color Plate 14b). First seen 300 years ago, the GRS is clearly much longer-lived than storms in our own atmosphere. The GRS also differs from terrestrial storms in being a high-pressure region characterized by anti-cyclonic motion. The counterclockwise rotation has a period of six days. Three similar but smaller disturbances on Jupiter formed about 1940, called the "white ovals"; these are only about 10,000 km across.

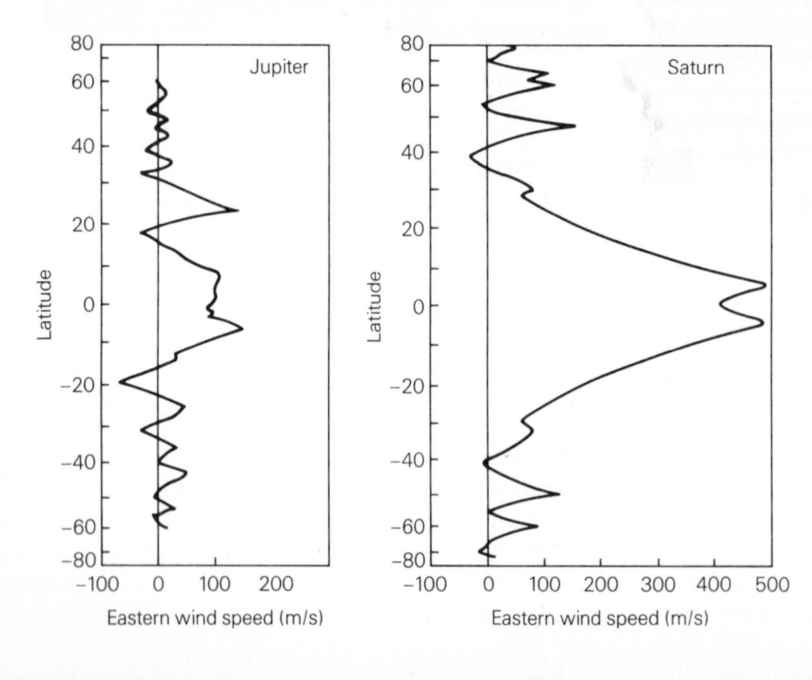

Figure 17.6 Zonal (east-west) winds on Jupiter and Saturn as measured by Voyager. The wind speeds are referred to the rotation of the core as determined from magnetic field and radio measurements. (NASA/JPL)

We don't know what causes the GRS or the white ovals, but it is possible to understand how they can last so long once they do form. On Earth, a large oceanic hurricane or typhoon typically has a lifetime of a few weeks, or even less when it moves over the continents and encounters friction with the land. On Jupiter, there is no solid surface to slow down an atmospheric disturbance, and furthermore the sheer size of these features lends them stability. It is possible to calculate that on a planet with no solid surface, the lifetime of anything as large as the GRS should be measured in centuries, and as large as the white ovals in decades. These time scales are consistent with the observed lifetimes of Jovian storms.

There is one other mystery of the Jovian clouds that should be mentioned, and that concerns their colors. All of the condensation clouds identified on the planet should be white, like water clouds on Earth, yet we see beautiful and complex patterns of red, orange, and brown (Color Plate 14b). Some additional chemical or chemicals must be present to lend the clouds such colors, but we do not know what they are. Various photochemically produced organic compounds have been suggested, as well as sulfur and red phosphorus. But there are no firm identifications, nor any immediate prospects of solving this mystery.

17.3 URANUS AND NEPTUNE

(a) Basic Facts

Uranus and Neptune make an appropriate pair. They have about the same size and composition, and they differ considerably from Jupiter and Saturn. It is an oversimplification to speak of the four "giant planets" or the "Jovian planets" as if they were all alike. At the least, Uranus and Neptune should be separated from their bigger cousins, but no one has invented an appropriate term for these two "small giants."

Uranus has an axis of rotation that is tilted by 98° with respect to the north direction. This unusual tilt creates very strange seasons, with each pole alternately tipped toward the sun for about 40 years at a time.

Uranus and Neptune are also interesting for the contrasting stories of their discoveries. Because these were considered events of major importance in their times, we describe these discoveries in some detail.

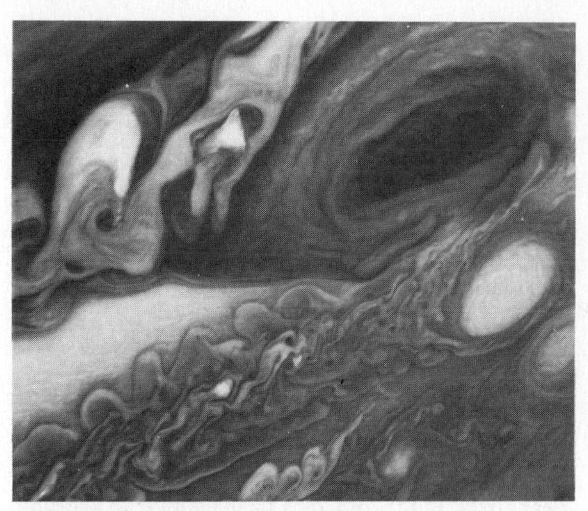

Figure 17.7 The Great Red Spot of Jupiter, photographed by Voyager. (NASA/JPL)

Table 17.4 summarizes some of the basic data on these two distant planets.

TABLE 17.4 Properties of Uranus and Neptune

	URANUS	NEPTUNE
Semimajor axis (AU)	19.2	30.1
Period (yrs)	84	165
Mass (Earth = 1)	15	17
Diameter (km)	50,800	50,500
Density (g/cm^3)	1.3	1.5
Escape velocity (km/s)	21	24
Rotation period (hrs)	−17	16

(b) Discovery of Uranus

Uranus is a planet that must surely have been seen by the ancients and yet was unknown to them. It was discovered on March 13, 1781, by the German-English musician and amateur astronomer William Herschel, who was making a routine telescopic survey of the sky in the constellation of Gemini. Herschel noted that through his telescope the planet did not appear as a stellar point but seemed to present a small disk. He believed it to be a comet and followed its motion for some weeks. Later, a solution for its orbit was computed and the orbit was found to be nearly circular, lying beyond that of Saturn; Herschel's "comet" was unquestionably a new planet.

TABLE 17.5 The Titius-Bode Law

PLANET	CALCULATED DISTANCE	ACTUAL DISTANCE
Mercury	(0+4)/10 = 0.4	0.39
Venus	(3+4)/10 = 0.7	0.73
Earth	(6+4)/10 = 1.0	1.00
Mars	(12+4)/10 = 1.6	1.5
	(24+4)/10 = 2.8	
Jupiter	(48+4)/10 = 5.2	5.2
Saturn	(96+4)/10 = 10	9.6
Uranus	(192+4)/10 = 19.6	19.2
Neptune		30.1
Pluto	(384+4)/10 = 38.8	39.5

Uranus can be seen by the unaided eye on a clear, dark night, but it is near enough to the limit of visibility that it is indistinguishable from a very faint star. It is so inconspicuous that its motion escaped notice until after its telescopic discovery. However, it turned out that Uranus had been plotted as a star on charts of the sky on at least 20 previous occasions between 1690 and 1781. These earlier observations were useful later in determining how perturbations were altering the planet's orbit.

Herschel proposed to name the newly discovered planet Georgium Sidus, in honor of George III, England's reigning king. Others suggested the name Herschel, after the discoverer. The name finally adopted, in keeping with the tradition of naming planets for gods of Greek and Roman mythology, was Uranus, father of the Titans and grandfather of Jupiter.

The discovery of Uranus brought Herschel great fame and a full-time career in astronomy. It also brought delight to the German astronomer Johann Bode (1749–1826), because it fit beautifully into a sequence of numbers announced in 1766 by Daniel Titius that describe the approximate distances of the planets from the sun. Bode had been so impressed with Titius' progression that he published it in his own introductory astronomy text. This sequence is known as the *Titius-Bode Law.* It is obtained by writing down the numbers: 0,3,6,12, . . . , each succeeding number in the sequence (after the first) being double the preceding one. If 4 is added to each number and the sum divided by 10, the resulting numbers are the approximate radii of the orbits of the planets in astronomical units, as can be seen in Table 17.5. The rule breaks down completely for Neptune and Pluto, but these planets were, of course, unknown in Bode's time.

The fact that Uranus fit so well into this scheme suggested to Bode that the progression was a law of nature, which led him to expect an unknown planet at 2.8 AU. Most of the asteroids have orbits near 2.8 AU, so we shall have more to say about this law in Chapter 19; it also played a role in the discovery of Neptune.

(c) Discovery of Neptune

Whereas the discovery of Uranus was quite unexpected, Neptune was found as the result of mathematical prediction. By 1790, an accurate orbit had been calculated for Uranus based on observations of its motion in the decade following its discovery. By early in the 19th century, however, it was realized that all was not well with Uranus. Even after allowing for the perturbing effects of the known planets, it was not possible to account for the motion of Uranus over the entire period for which it had been observed—including the observations back to 1690. The discrepancies between the observed and predicted positions grew worse and worse, until by 1844 they amounted to 2 arcminutes—an angle almost large enough to be resolved without optical aid. Since this difference was totally unacceptable in gravitational theory, it seemed clear that there must be an unknown planet providing gravitational perturbations on Uranus in addition to those of the known planets.

Most astronomers concluded that it would require many additional decades of observation to obtain enough data to predict the size and position of this unknown planet. Still, the problem was intriguing, and it appealed to two relatively unknown young mathematicians, each without any knowledge of the other. They were Urbain Jean Joseph Leverrier (1811–1877) in France and John Couch Adams (1819–1892) in England (Figure 17.8).

Both Adams and Leverrier saw that the peculiar way in which Uranus had been first ahead of and then behind its predicted position was an important clue to finding where a hypothetical perturbing planet would have to be. They both solved the problem and obtained very nearly identical results. They both had been misled into believing that the unknown planet would be as far from the sun as is suggested by the next planet in the series in the Titius-Bode law—39 AU. The perturbing planet was actually closer to the sun by about 21 percent, but fortunately this error almost cancels

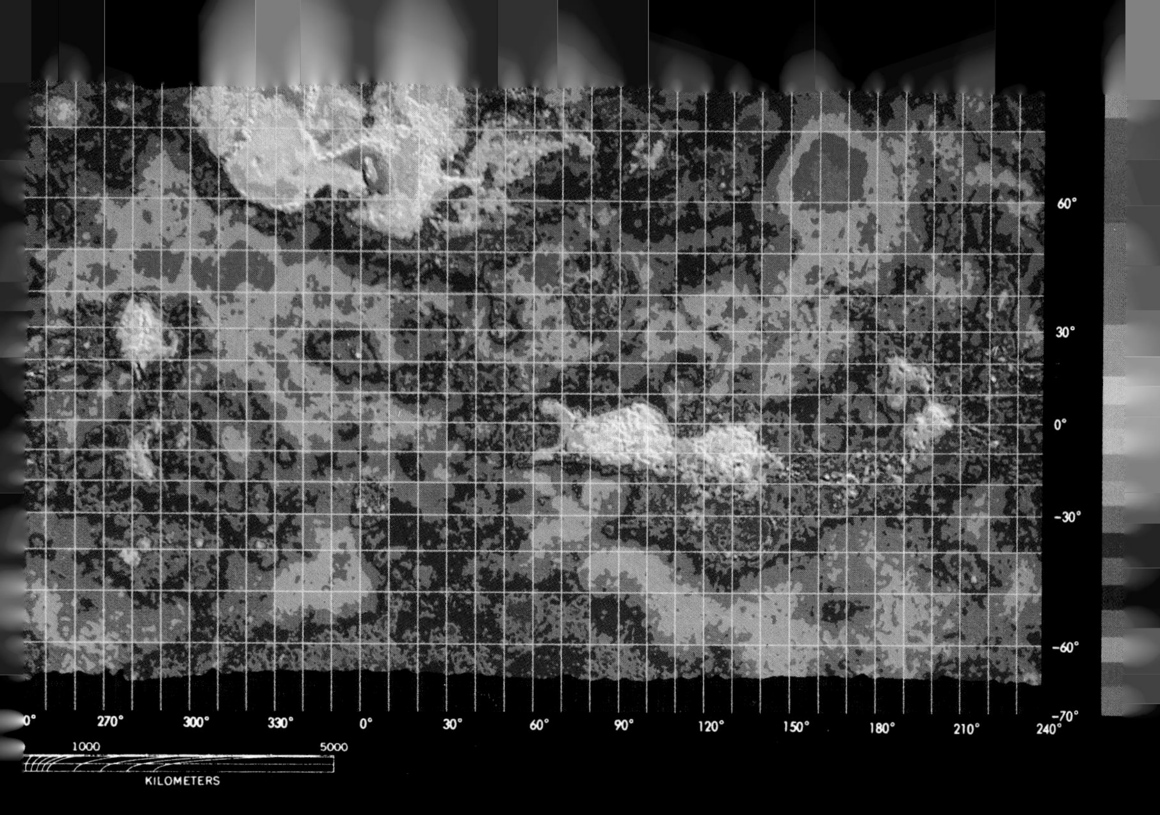

60°
30°
0°
-30°
-60°
-70°
0°
270°
300°
330°
0°
30°
60°
90°
120°
150°
180°
210°
240°
1000
5000
KILOMETERS

VENUS

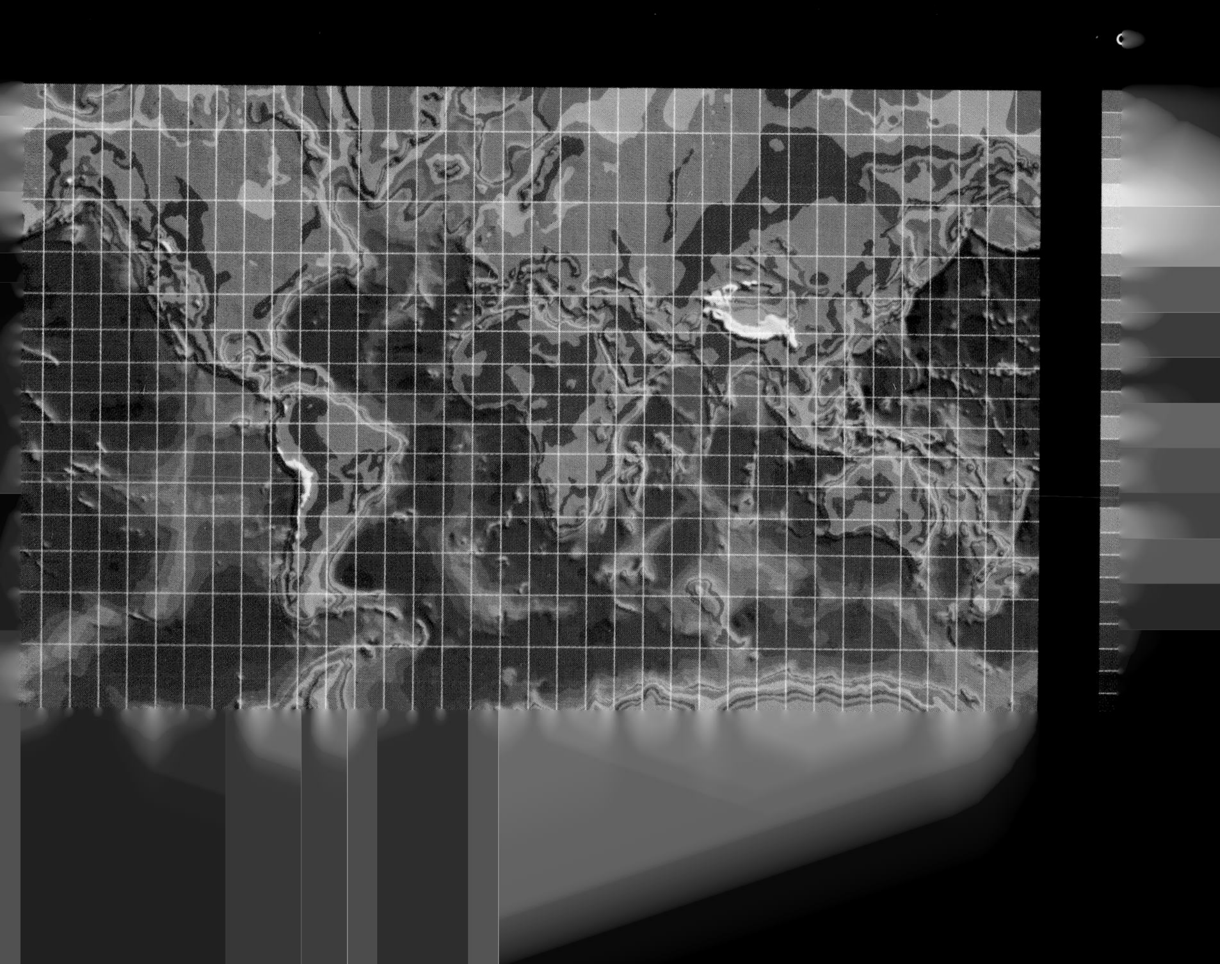

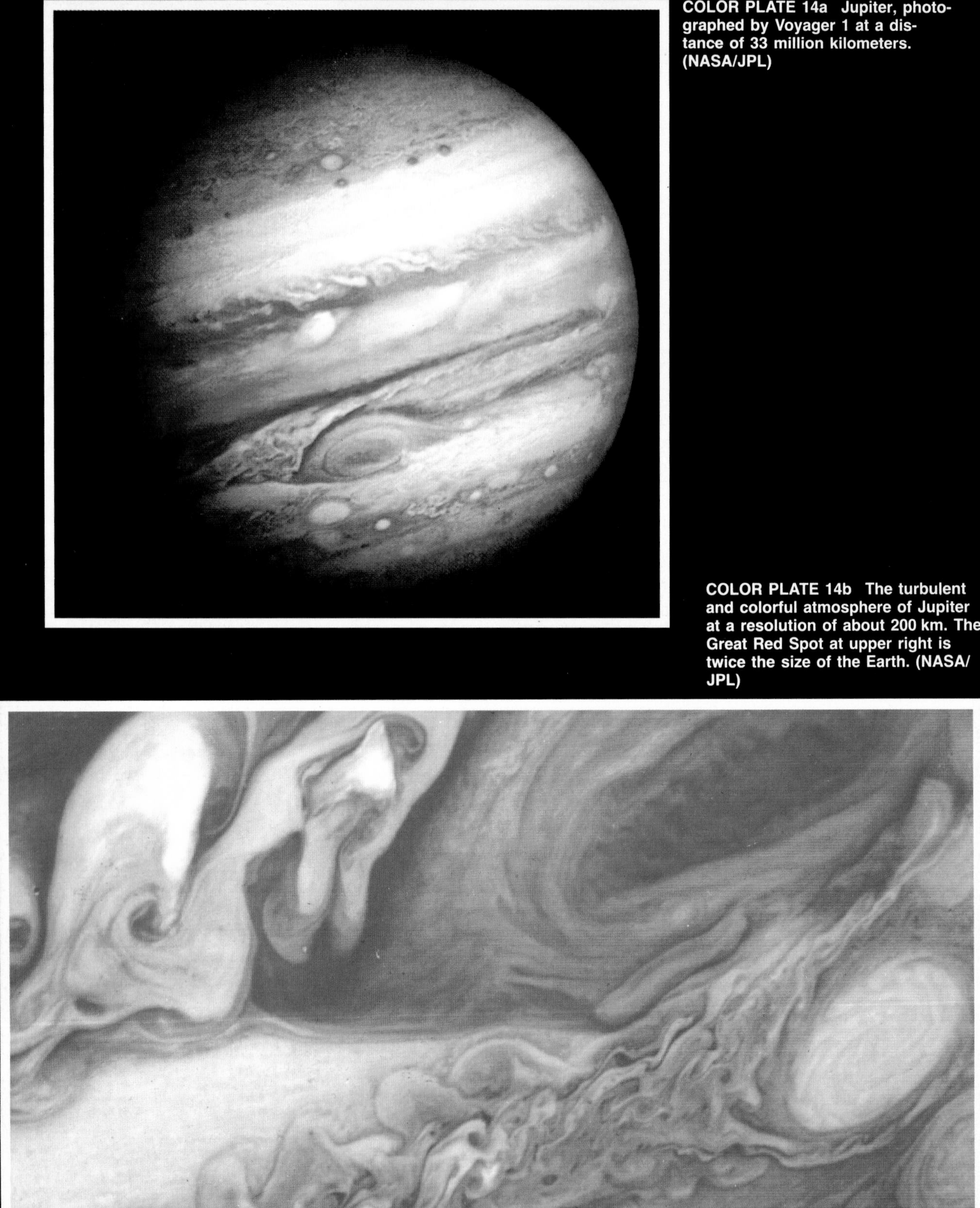

COLOR PLATE 14a Jupiter, photographed by Voyager 1 at a distance of 33 million kilometers. (NASA/JPL)

COLOR PLATE 14b The turbulent and colorful atmosphere of Jupiter at a resolution of about 200 km. The Great Red Spot at upper right is twice the size of the Earth. (NASA/JPL)

COLOR PLATE 15a Saturn, photographed by Voyager 1 four days after passing the planet. (NASA/JPL)

COLOR PLATE 15b Uranus, photographed by Voyager 2 a few hours after closest approach. (NASA/JPL)

COLOR PLATE 16a Jupiter and two of its Galilean satellites: Io on the left and Europa on the right. (NASA/JPL)

COLOR PLATE 16b Jupiter's largest satellite, Ganymede, photographed by Voyager 1 at a resolution of about 100 km. (NASA/JPL)

Figure 17.8 The mathematicians who share credit for the discovery of Neptune: John Couch Adams and Urbain J. J. Leverrier. (Yerkes Observatory)

out in the solution for the direction of the planet in the sky, so both Adams and Leverrier predicted correctly where to look for the new object.

Unfortunately, the actual discovery of Neptune was not so straightforward. Adams finished his calculations first and had attempted to deliver them in person to the Astronomer Royal, Sir George Biddell Airy, in late September 1845. Airy, at the time, was on the Continent and had not received word of Adams' call. Adams tried again on October 21, but on that morning Airy was out of his office. Adams left a brief summary of his results, and a message that he would call later that same day. Regrettably, the message about the later call was never delivered to Airy. When Adams came again, Airy was at dinner, and his overprotective butler would not allow him to be disturbed. Adams, feeling rebuffed, left.

When Airy got around to reading Adams' summary, he was skeptical because he was not familiar with the detailed theory himself, and Adams was still young and unproven. Thus when Airy replied to Adams, on November 5, he attempted to put the young man to the test by asking him about what seemed to Adams to be an almost irrelevant and trivial problem concerning the earlier observations of Uranus. Probably annoyed, Adams did not reply until more than a year later, by which time the discovery of the new planet was history.

Meanwhile, on November 10, Leverrier had completed the first part of his own calculations and had published them in the appropriate French journal. Airy saw Leverrier's paper and was most impressed with its clarity and style. During the winter and spring of 1846, Leverrier completed his analysis, and his published solution reached Airy in late June 1846. Airy noted that Leverrier's prediction for the position of the new planet agreed to within 1° with Adams', and he immediately sent Leverrier the same question about the orbit of Uranus that he had sent Adams. Airy received a prompt and authoritative reply from Leverrier on July 1 (the mails must have been more efficient in those days!) that convinced him that the new planet must exist.

Airy sent two letters to James Challis, Director of the Cambridge Observatory, pointing out that the observatory's 12-inch Northumberland telescope was ideal for the search for the suspected planet. (That instrument is one of the oldest working telescopes in the world and is still used on clear nights by Cambridge undergraduates.) Airy's second letter, dated July 13, expresses the urgency with which Airy finally regarded the search for the new planet. "In my opinion," he wrote, "the importance of this inquiry exceeds that of any current work, which is of such a nature as not to be totally lost by delay." Accordingly, Challis began a systematic search of the area of the sky indicated by Airy, based on the predictions of Adams and Leverrier. The Cambridge astronomers did not possess up-to-date star charts against which to compare suspected planets, so their procedure was to survey systematically the designated part of the sky on dates separated by several days and then to compare the records of the positions of the observed stars to see if one, by virtue of its motion, had revealed itself as a planet. Challis began trial observations on July 29.

Leverrier completed his final calculations and presented his paper to the French Academy on August 31. In that paper he gave detailed information on where to look for the new planet. Leverrier did not have Adams' problem of establishing his credibility; his work was warmly received. On the other hand, Leverrier had no success whatsoever convincing the French astronomers that they should bother to look for the new planet. In desperation, he wrote to a young assistant at the Berlin Observatory, Johann Gottfried Galle, from whom he had had some correspondence a year earlier. Galle was much excited about the prospect of discovering a new planet, and after much effort finally received permission from the Observatory Director, Johann Franz Encke, to try a search the very night he received Leverrier's letter—September 23. Galle agreed to allow a young student astronomer, Heinrich Louis d'Arrest, to join him.

The 9-inch Fraunhofer refractor, pride of the observatory, was turned to the predicted position; right ascension $22^h\ 46^m$, declination $-13°\ 24'$. At first they found nothing, but Galle and d'Arrest located new star charts of that region of the sky, and while Galle looked through the telescope and described the stars he saw, d'Arrest compared Galle's comments with the stars on the chart. After checking out a few stars, Galle described an eighth-magnitude star at right ascension $22^h\ 53^m\ 25^s$, and d'Arrest excitedly pointed out that that star was not on the map; it was, at last, the eighth planet!

When word reached Cambridge of the Berlin discovery of the new planet, Challis checked through his old records and found that the planet had actually been observed, but not recognized, six weeks earlier at Cambridge. It appeared in the observing records of August 12 as a marginal comment, which Challis added later. It points out, "This is the planet. It does not appear in this place on page 3." On page 3 of the observing notebook we find the observations of the same part of the sky made 12 days earlier, on July 30th.

Today the honor for the discovery of Neptune is justly shared by those two brilliant and tenacious mathematicians, Adams and Leverrier, who successfully predicted its existence. That prediction ranks high in the history of science, for it was an extraordinary triumph of Newtonian gravitational theory. But by the time it was announced, the discovery was not so much of a surprise as a fulfillment of expectations. On September 10, 1846, 13 days before Galle's observation of the planet to be named Neptune, John Herschel, son of the discoverer of Uranus, remarked in his valedictory address as president of the British Association, "We see it as Columbus saw America from the shores of Spain. Its movements have been felt, trembling along the far-reaching line of our analysis with a certainty hardly inferior to ocular demonstration."

A search of old records revealed two prediscovery observations of Neptune—on May 8 and 10, 1795, by Joseph Lalande. These observations were used in 1847 to calculate an improved orbit for Neptune. It was then for the first time that it was realized that Neptune is only about 30 AU from the sun and that the Titius-Bode law does not work for all the planets. Table 17.6 compares some of the orbital elements calculated by Adams and Leverrier with the modern values.

TABLE 17.6 Comparison of Elements for Neptune's Orbit

	ADAMS	LEVERRIER	MODERN
Semimajor axis (AU)	37.25	36.15	30.134
Eccentricity	0.12062	0.10761	0.005
Period (years)	227.3	217.387	164.74
Mass (sun = 1.0)	1/6666	1/9300	1/19,350

It is now believed that Neptune was also observed a full 234 years before its "official" discovery by no less a person than Galileo. Astronomer Charles Kowal of Caltech and historian Stillman Drake of Toronto investigated close conjunctions of planets that occurred in historical times and found that Neptune approached very close to Jupiter in January 1613, about the time Galileo was observing Jupiter and its satellites. A check of Galileo's notebook showed that he had indeed seen an eighth-magnitude star near Jupiter on several occasions and had even noticed that it seemed to move. He evidently was not sure enough of the motion to realize the significance of what he had observed, however.

(d) Current Knowledge of Uranus and Neptune

Although Uranus and Neptune are near twins in terms of their size and density, the two are certainly not identical, just as important differences exist between the "twins" Earth and Venus. Since we know much more about Uranus as a result of the Voyager flyby of January 1986, we will discuss it first.

Unlike Jupiter and Saturn, the atmosphere of Uranus is almost entirely featureless as seen at wavelengths that range from the ultraviolet to the infrared. Not since Venus have we encountered such a dull-looking planet (Color Plate 15b). Yet calculations indicate that the basic atmospheric structure of this planet should resemble that of Jupiter and Saturn. The primary difference in appear-

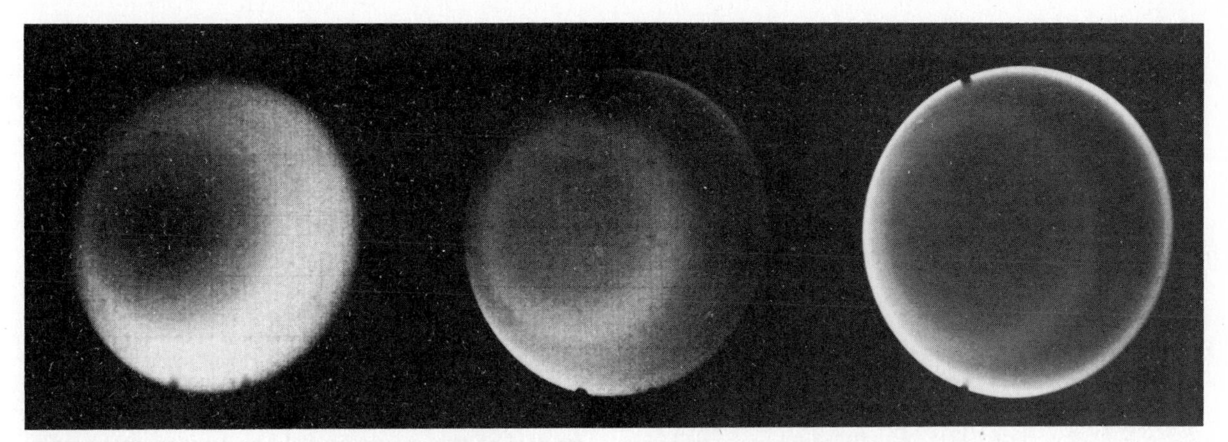

Figure 17.9 Three photos of Uranus obtained in 1986 by the Voyager 2 spacecraft. They show the appearance of the planet seen through filters that transmit violet light, orange light, and red light in a deep absorption band of methane. Note the polar haze visible in the violet. (NASA/JPL)

ance results from the colder temperatures in the atmosphere of Uranus, which cause its clouds to form at lower elevations. As a result, the troposphere and its associated activity are hidden from our view by a deep, cold, hazy stratosphere.

Underneath, the clouds of Uranus may be as colorful and dynamic as those of Jupiter, but looking in from the outside we cannot see them. Consequently, it is very difficult to determine the nature of the atmospheric circulation on Uranus. Figure 17.9 illustrates some structure visible at different wavelengths, including a polar haze. Other Voyager photos reveal a few small clouds that could be tracked to derive a rotation period of approximately 16 hours.

In spite of the strange seasons induced by the 98° tilt of its axis, Uranus' basic circulation is east-to-west, just as it is on Jupiter and Saturn. The mass of the atmosphere and its capacity to store heat are so great that the alternating 40-year periods of sunlight and darkness have little effect; in fact, Voyager measurements show that the atmospheric temperatures are a few degrees higher on the dark, winter side than on the hemisphere facing the sun.

Although the bulk composition of the planet is different from that of Jupiter and Saturn, the composition of the atmosphere of Uranus is very similar to that of its larger cousins. Hydrogen and helium predominate, with a ratio similar to that of Jupiter and the sun. There is no indication of depletion of helium, as we found in the atmosphere of Saturn.

Although Neptune is farther away than Uranus and consequently even more difficult to study telescopically, it actually shows more structure in its atmosphere (Figure 17.10). Large-scale atmospheric features can be photographed from the Earth, and

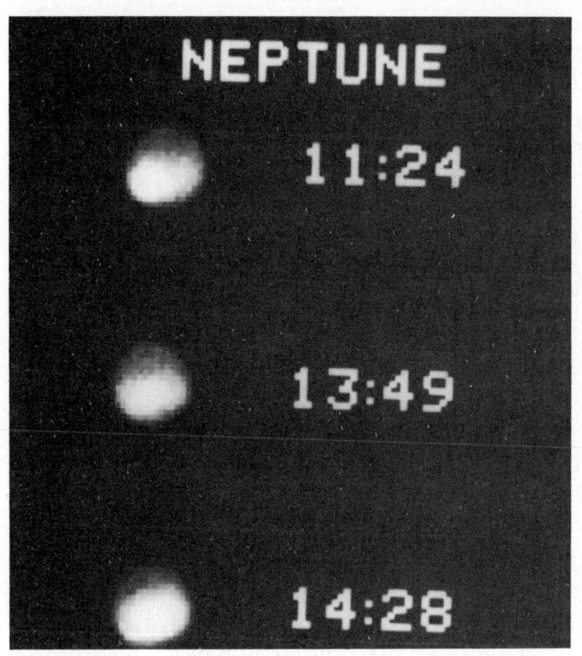

Figure 17.10 Three images of Neptune taken at Mauna Kea Observatory on May 20, 1986 through a filter centered at 8900 Å wavelength. The apparent motion of the cloud structure is due to rotation of the planet. (University of Hawaii; courtesy H. Hammel)

infrared observations have shown that high-altitude clouds or hazes can form or dissipate over nearly the entire planet in times as short as a few weeks. Whatever clouds may exist in the deep atmosphere of Neptune, it is clear that there is also a cloud-forming material very high in the stratosphere, where we can see its comings and goings. Tracking the motions of these high-altitude clouds has also yielded an apparent rotation period of Neptune of about 16^h. In terms of its atmospheric structure and circulation, Neptune should be a much more interesting target for Voyager than was Uranus.

Voyager discovered that Uranus has a magnetic field and magnetosphere, both somewhat smaller than those of Saturn. The rotation of the magnetic field yields the true internal rotation period of the planet, just as it does for Jupiter and Saturn: the Voyager result is 17^h 14^m. The magnetosphere is compared with those of Jupiter and Saturn in Section 17.4.

17.4 MAGNETOSPHERES

Among the most dramatic features of the outer planets are their *magnetospheres*. Like the magnetosphere of the Earth, these regions are defined as the large cavities within which the planet's magnetic field dominates over the interplanetary magnetic field. Inside the magnetosphere higher-density plasma can be contained, and ions and electrons can be accelerated to very high energies. These physical processes are similar to those dealt with by astrophysicists in many distant objects, from pulsars to quasars. The magnetospheres of Jupiter, Saturn, Uranus, and the Earth provide the only nearby analogs of these cosmic processes that can be studied directly. (Neptune may also have a magnetosphere, but we won't be able to find out until Voyager arrives in 1989.)

(a) Planetary Magnetic Fields

In the late 1950s, radio energy was observed from Jupiter that is more intense at longer than at shorter wavelengths—just the reverse of what is expected from thermal radiation. It is typical, however, of the radiation emitted by electrons accelerated by a magnetic field, called *synchrotron radiation* (Chapter 34). Later observations showed that the radio energy originated from a region surrounding the planet whose diameter is several times that of Jupiter itself. Furthermore, the radio energy was found to be polarized, another characteristic of synchrotron radiation from accelerated electrons. The evidence suggested, therefore, that there are a vast number of charged atomic particles circulating around Jupiter, spiraling through the lines of force of a magnetic field associated with the planet. This phenomenon is like the Van Allen belt that had recently been discovered around the Earth (Section 13.1).

The Pioneer and Voyager spacecraft supplemented these indirect measurements with direct studies of the magnetic field and magnetosphere of Jupiter (Figure 17.11). They found Jupiter's surface magnetic field to be from 20 to 30 times as strong as the Earth's field. Because of Jupiter's great size, moreover, its total magnetic energy is enormous compared with the Earth's. The Jovian magnetic axis, like that of the Earth, is not aligned exactly with the axis of rotation of the planet, but is inclined at some 15°. It also does not pass through the planet's center, but is displaced by about 18,000 km. In addition, the Jovian field has the opposite polarity from the Earth's current value, but of course the Earth's field is known to reverse polarity from time to time, and the same may be true of Jupiter.

Saturn does not emit strong synchrotron radiation, because its magnetosphere is depleted in electrons by collisions between electrons and its rings and inner satellites. It does have a substantial magnetic field, however, as discovered by Pioneer 11 and the Voyagers. Unlike the Earth and Jupiter, Saturn's field is almost perfectly aligned with its rotation axis. Its polarity is the same as Jupiter's and the opposite of Earth's.

The magnetic field of Uranus was not discovered until the Voyager flyby in 1986. The size of the field is comparable to that of Saturn, about what would be expected from the size of the planet, but in its orientation the field of Uranus is remarkable (Figure 17.12). Like Jupiter's field, it is offset from the center of the planet by about one-third of the planet's radius. In addition, however, the field of Uranus is tilted by 60° with respect to the axis of rotation—the extreme opposite case from that of Saturn.

Presumably the magnetic fields of these three outer planets are generated in much the same way as the field of the Earth. All of these planets spin

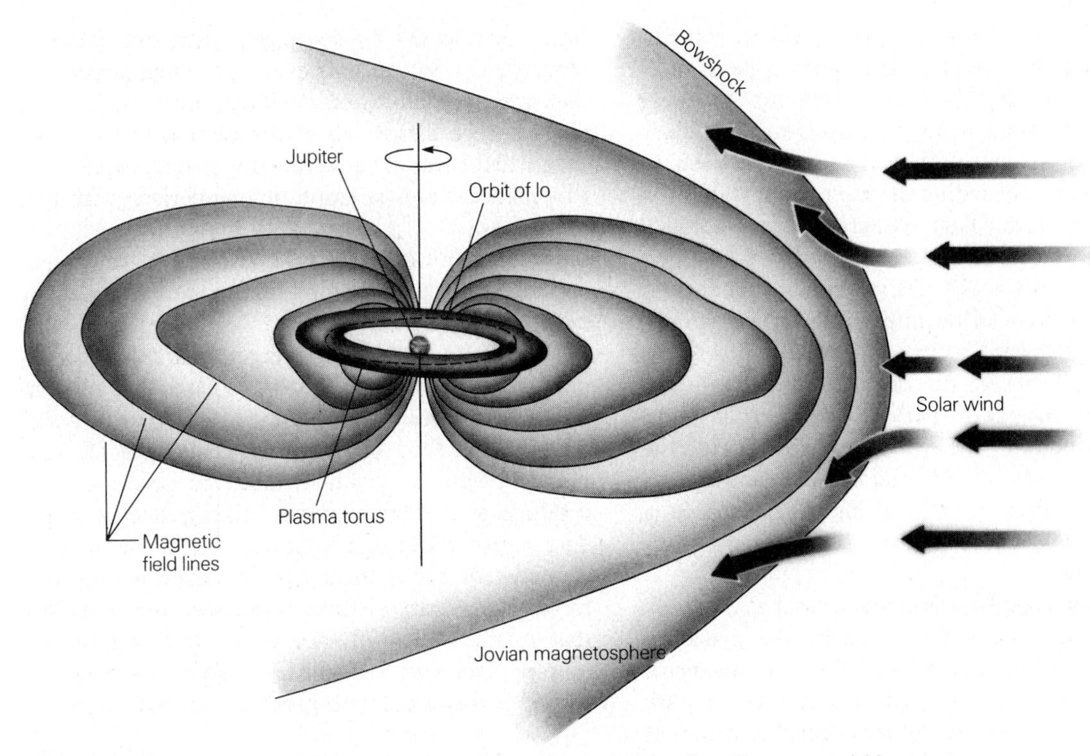

Figure 17.11 The magnetosphere of Jupiter as revealed by the Pioneer and Voyager probes.

rapidly, so there is a ready source of energy to power their internal magnetic generators. Jupiter and Saturn have large interior regions of metallic liquid hydrogen that act like the liquid iron core of the Earth. In the case of Uranus, however, the metallic region may be in the hydrogen-water mantle, possibly accounting for the large offset of the field from the center of the planet. Although the detailed mechanisms may not be well understood, these planets seem to meet the conditions required for the generation of a planetary magnetic field by self-maintaining dynamo action.

(b) Magnetospheres of Jupiter, Saturn, and Uranus

The Jovian magnetosphere is one of the largest features in the solar system. It is actually much larger than the sun and completely envelops the innermost satellites of Jupiter. If we could see the magnetosphere, it would appear the size of our Moon. The physical dimensions of Saturn's magnetosphere are about one-third as great, and that of Uranus is still smaller, approximately in proportion to the sizes of the planets themselves.

On its upstream side (facing toward the solar wind), each magnetosphere is bounded by a pressure balance between the plasma inside and the solar wind streaming toward it at about 400 km/s. At the outer planets, the magnetic fields are stronger

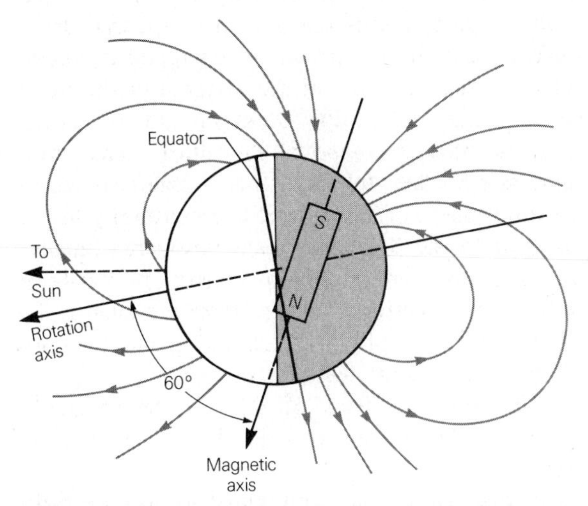

Figure 17.12 The magnetic field of Uranus as revealed by Voyager in 1986; note the large offset from the center of the planet and the 60° tilt with respect to the planet's axis of rotation.

than Earth's, and the solar wind is weaker, contributing to the large size of their magnetospheres. The actual borders vary, however, with the changing pressure of the solar wind. The upstream boundary of the Jovian magnetosphere is usually between 50 and 100 times the radius of Jupiter, or roughly 5 million kilometers. That of Saturn is at 20 to 40 Saturn radii, or about 2 million kilometers, and that of Uranus is at near 20 Uranus radii, or about 0.5 million kilometers. One interesting aspect of Saturn's magnetosphere is that its large satellite Titan, while normally inside the magnetosphere, can find itself on the outside in periods of high solar wind pressure, which was the case for example during the Pioneer 11 encounter. On the downstream side all three magnetospheres extend much farther in a *magnetotail,* and the boundary is even more variable.

The magnetosphere is characterized as much by the plasma trapped within it as by the planetary magnetic field. In this respect the four magnetospheres of Earth, Jupiter, Saturn, and Uranus are quite different from each other. Each has different sources and sinks of atomic particles.

The two primary sources of plasma in the Earth's magnetosphere are solar wind protons and electrons that leak in from the outside, and atmospheric particles that escape upward from the planet. On Jupiter both of these sources also apply, but they are supplemented by a much stronger source on the large Galilean satellites. Unlike our Moon, these satellites are enveloped by the magnetosphere. Some atoms are ejected or *sputtered* from their surfaces by the impact of energetic magnetospheric ions, but most of the material in the inner magnetosphere of Jupiter is oxygen and sulfur ions from the active *volcanoes* of Io. Saturn lacks a volcanic satellite like Io, but it has a major source in the atmosphere of Titan, which is constantly losing nitrogen to the Saturnian magnetosphere. Finally, the magnetosphere of Uranus consists primarily of protons and electrons, derived (like the ions in the Earth's magnetosphere) from the planet's atmosphere and the solar wind.

The ultimate fates of ions in the magnetospheres of these four planets also differ. For all four, some magnetospheric particles escape (mostly down the magnetotail) and some are lost by collision with the planet's atmosphere (where they generate auroral discharges). Jupiter also loses some by collision with its satellites, but in general more new ions are released by sputtering than old ones destroyed. On Saturn, however, the much larger surface areas of its inner satellites and especially its rings absorb almost all of the plasma in the inner magnetosphere. Therefore, the magnetosphere of this planet is almost empty in comparison with that of Jupiter.

The ions and electrons within the magnetosphere are accelerated by the spinning magnetic field of the planet, eventually reaching extremely high energies. It is these energetic particles that radiate at radio wavelengths by processes that are similar, whether they take place at Jupiter or in a distant galaxy. One of the main challenges of *space plasma physics* is to understand the processes that produce such high energies. Experience with the data acquired during recent decades on the magnetospheres of the planets has demonstrated the extreme complexity of these processes, and it is clear that much additional theoretical work will be required before we can interpret observations of astrophysical sources with great confidence.

(c) The Io Plasma Torus

One of the major features of the magnetosphere of Jupiter is associated directly with its innermost Galilean satellite, Io. As we will see in the next chapter, Io is volcanically active, erupting large quantities of sulfur and sulfur dioxide into the space surrounding it. While most of this material falls back to the surface, it is estimated that about ten tons per second is lost to the magnetosphere. These ions of oxygen and sulfur form a donut-shaped *plasma torus* surrounding Jupiter approximately at Io's orbit, at a distance of 5 Jupiter radii from the planet. In addition to being investigated directly by spacecraft, the Io plasma torus has been imaged telescopically from Earth in the glow emitted by oxygen and sulfur atoms as they recapture electrons from the plasma (Figure 17.13).

The energetic ions of the plasma torus and its surroundings would be very dangerous to both spacecraft and humans, if any should ever venture close to Jupiter. When these ions strike a solid surface, they damage it directly and also generate lethal X-rays. On or near Io, a human could survive for only a few minutes. Special shielding would be required for spacecraft electronics, and no spacecraft has been built that could last more than a few hours

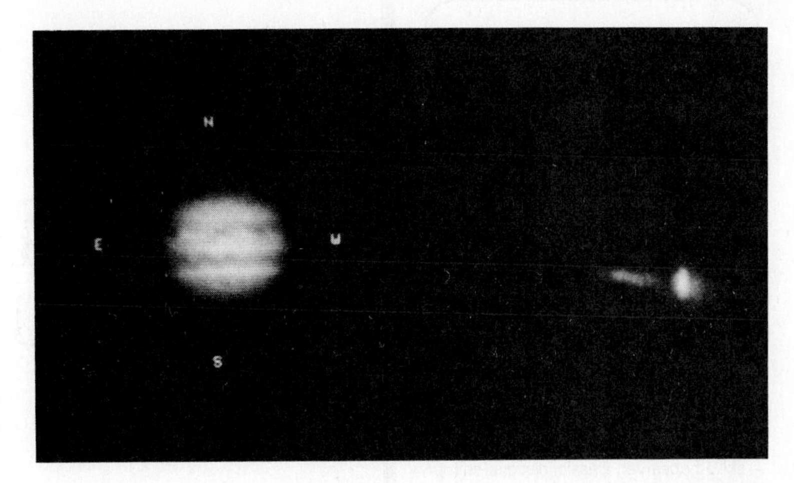

Figure 17.13 The Io plasma torus as photographed in the light of glowing sulfur ions. (University of Hawaii)

in this environment. Thus an Io lander or orbiter is far beyond our present capabilities, and it is probably safe to predict that human exploration of the inner Jovian system will never be possible.

As Io orbits about Jupiter in this sea of energetic plasma, it generates an electric current that flows along magnetic field lines between the satellite and the planet. This *magnetic flux tube* acts like a wire carrying a current estimated at 5 million amperes. Where it reaches the atmosphere of Jupiter, this current stimulates strong radio noise emissions. These emissions were picked up by radio astronomers many years before spacecraft reached Jupiter, and their association with the orbital position of Io was established. Only recently, however, have we come to understand the remarkable phenomenon that generates these bursts.

Ions from the Io plasma torus also strike the upper atmosphere of Jupiter at high latitudes, producing auroral glows that were measured by the Voyager spacecraft and should be observable by the Hubble Space Telescope.

EXERCISES

1. We often speak of the giant planets having approximately solar or cosmic composition. Is this strictly true? To what degree does each of these planets depart from cosmic composition?
2. Make a table showing the oxidized (oxygen-rich) and reduced (hydrogen-rich) compounds of some common elements.
3. If Saturn's mass is 95 times that of the Earth, how does it happen that its surface gravity is so nearly the same as that of the Earth?
4. Describe the different processes that lead to substantial internal heat sources for Jupiter and Saturn. Since these two objects generate much of their energy internally, should they be called stars instead of planets? Justify your answer.

*5. Estimate how frequently all four giant planets are approximately in alignment (for example, with longitudes differing by no more than 60°), permitting a "grand tour" trajectory like that of Voyager 2.

6. Compare the atmospheric circulation (weather) for Jupiter, Saturn, and Uranus.
7. Make a table comparing the magnetospheres of Earth, Jupiter, Saturn, and Uranus. List their sizes and indicate the main sources and sinks of charged atomic particles.
8. The inclination of the equatorial plane of Uranus to its orbit (its obliquity) is given as 98° rather than 82°. Can you suggest why?
9. Describe the seasons of Uranus.

Jean Dominique Cassini (1625–1712) was an Italian-born astronomer who founded and later directed the Paris Observatory. A skilled observer, Cassini discovered four satellites of Saturn—Rhea, Iapetus, Dione, and Tethys—and he first suggested that Iapetus had one dark side and one light side. He also discovered the gap in the rings of Saturn that bears his name, and he was one of the first to argue that these rings were made up of individual small moons rather than being a solid disk. (Yerkes Observatory)

18

RINGS, SATELLITES, AND PLUTO

18.1 RING AND SATELLITE SYSTEMS

The giant planets are centers of systems of satellites and rings that orbit about them like planets in a miniature solar system. Although small in comparison to the giant planets they accompany, a number of these satellites are larger than the planet Mercury, and many of them show evidence of a surprising degree of geological and atmospheric evolution. Little Pluto, although technically classed as a planet, is similar to the larger of these satellites. Equally interesting in the outer solar system are the ring systems, which provide fascinating examples of physical processes with applications ranging from the formation of planetary systems to the spiral structure of galaxies.

(a) General Properties

The rings and satellites of the outer solar system are chemically distinct from objects in the inner solar system, as is to be expected from the fact that they formed in regions of lower temperature. The primary difference, as we have frequently noted, was in the availability of large quantities of water ice as a building block for bodies beyond the asteroid belt. Notable in addition is the presence of dark, organic compounds that also seem to be the product of condensation in the solar nebula at fairly low temperatures. Mixed with the ice that is present in these objects, this dark primitive material often results in low reflectivities, which was something of a surprise to those who had expected ice-rich objects to be bright. There is even evidence that the icy comets are dark objects.

Most of the satellites in the outer solar system are in *direct* or *regular* orbits; that is, they revolve about their parent planet in an east to west direction and very nearly in the plane of the planet's equator. Ring systems are also in direct revolution in the planet's equatorial plane. Such objects probably formed at about the same time as the planet by processes similar to those that formed the planets in orbit around the sun. Because of this analogy, it is really quite appropriate to speak of the regular satellite systems of Jupiter, Saturn, and Uranus (Figure 18.1) as miniature solar systems.

In addition to the regular satellites, there are *irregular* satellites that orbit in a *retrograde* (west to east) direction or else have orbits of high eccentric-

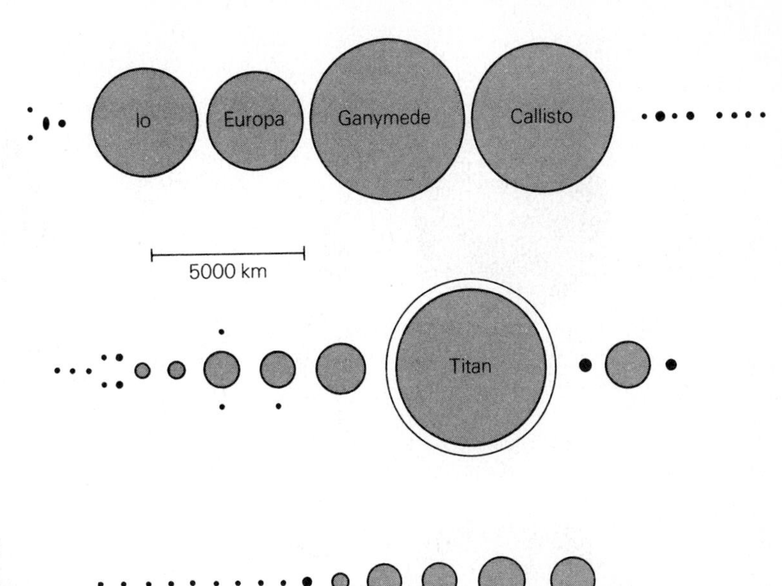

Figure 18.1 Comparative sizes and distributions of the satellites of Jupiter, Saturn, and Uranus.

ity or inclination. These are usually smaller satellites, located relatively far from their planet. These are generally thought to be captured planetesimals, like the two satellites of Mars (Section 15.1e).

We have said that ice is a major component of most of these objects, but how do we know this to be the case? There are two ways, and they tend to complement each other. First, we can analyze the spectrum of reflected sunlight to determine if any of the prominent infrared ice absorption bands are present. Using this approach, many satellites and the rings of Saturn have been found to have icy surfaces. However, such spectra refer only to the surface, and may not be representative of the bulk composition of the satellites.

Estimating the composition of planetary interiors is never easy, since they cannot be sampled directly. But as we saw for the inner planets, a useful indication of bulk composition is provided by a measurement of density. Most of the measured densities for outer planet satellites are less than 2.0 g/cm^3, in contrast to larger densities of the terrestrial bodies. Since water ice is the main low-density material expected from condensation in the solar nebula, it is natural to try to reproduce these densities with combinations of rock and ice. Most of these measured densities are matched very well by compositions that are approximately half ice by mass.

Planet-sized objects composed of equal parts ice and rock are unlikely to experience the same sort of geological evolution as the rocky worlds studied in Chapters 13 to 16. One difference results from the low melting temperature of ice. Only a little internal heating is required to melt these objects, resulting in rapid differentiation. Ice also expands and contracts differently from rock as its temperature changes, producing different stresses in the planetary crust. Even before the Voyager flights, there was reason to expect some geological surprises in the outer solar system.

(b) The Jupiter System

Jupiter has 16 satellites and a faint ring. The 16 satellites include the four large Galilean satellites (Figure 18.2) discovered in 1610 by Galileo: Callisto, Ganymede, Europa, and Io. The smallest of these, Europa and Io, are about the size of the Moon. The largest, Ganymede and Callisto, are larger than Mercury and only a little smaller than Mars. Io is the most volcanically active body in the solar system.

The other 12 Jovian satellites are much smaller. They divide themselves conveniently into three groups of four each. The inner four all circle the planet inside the orbit of Io; one of these, Amalthea, has been known for about a century, but the other three were discovered by Voyager. The outer satellites consist of four in direct but highly inclined orbits and four farther out in retrograde orbits. These eight are believed to be captured objects.

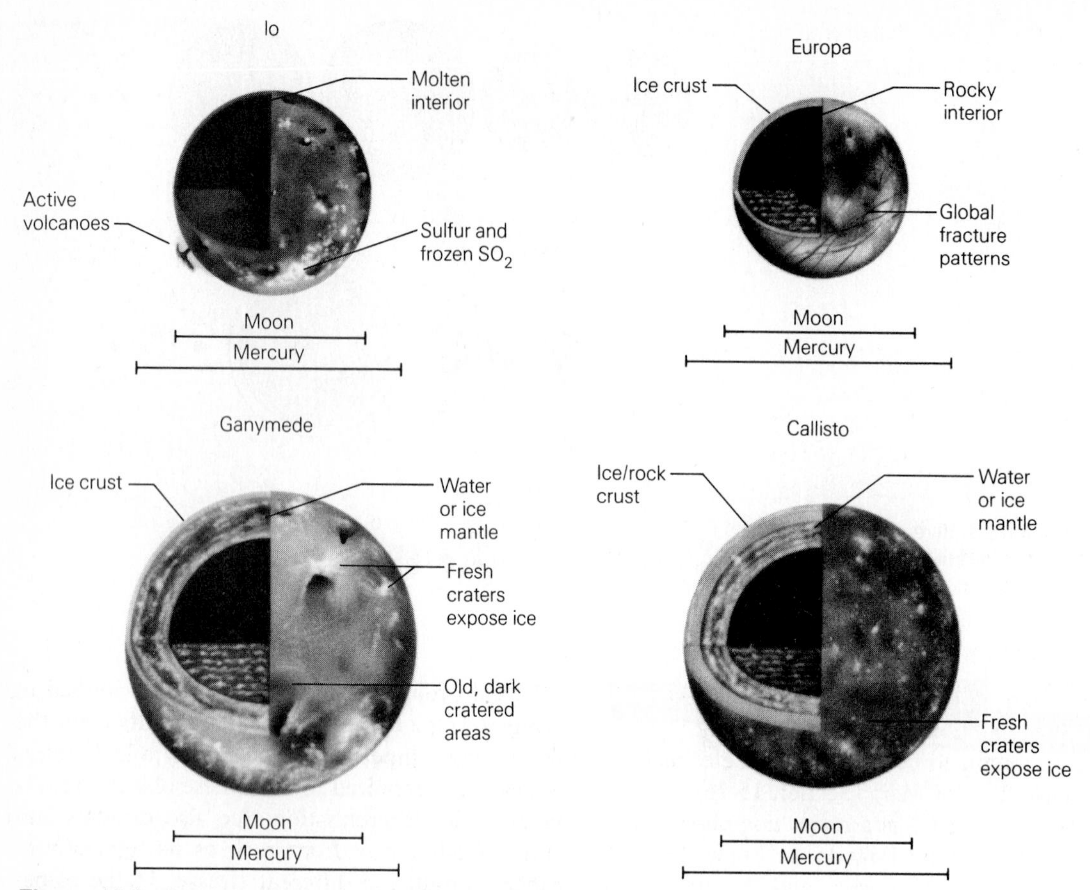

Figure 18.2 Interior structures of the Galilean satellites of Jupiter. Callisto and Ganymede are about half ice and half rock, while both Io and Europa are largely rocky bodies like the Moon. (NASA/JPL)

The two groupings may indicate two parent bodies that were broken up in a collision early in the history of the Jovian system. These eight outer satellites are dark, apparently primitive objects probably similar to the Trojan asteroids (Chapter 19).

(c) The Saturn System

Saturn has 19 known satellites in addition to its magnificent rings. The largest of the satellites, Titan, is almost as big as Ganymede in the Jovian system. Titan is the only satellite with a substantial atmosphere; the atmospheric pressure there is greater than at the surface of the Earth.

There are no other Saturn satellites comparable to the Galilean satellites of Jupiter, but there are six regular satellites with diameters between 400 and 1600 km, each composed about half of water ice.

Saturn has two distant irregular satellites, one of which (Phoebe) is in a retrograde orbit. It also has a fascinating collection of small, irregular-shaped, icy satellites circling close to the planet and the rings. Most of these smaller satellites were discovered in 1980, in part by Voyager and in part with Earth-based telescopes. Gravitational interactions between these satellites and the rings are responsible for much of the detailed ring structure observed by Voyager.

(d) The Uranus and Neptune Systems

The regular ring and satellite system of Uranus shares the 98° tilt of the planet. It consists of 11 rings (two discovered by Voyager in 1986) and 15 satellites, all revolving in the same direction as the rotation of Uranus itself. The five largest satellites are similar in size to the regular satellites of Saturn, with diameters from 500 to 1600 km. In contrast with these larger icy satellites, the 10 smaller satel-

TABLE 18.1 The Largest Satellites

NAME	PERIOD (days)	DIAMETER (km)	DENSITY (g/cm^3)
Ganymede (J3)	7.2	5280	1.9
Titan (S6)	15.9	5150	1.9
Callisto (J4)	16.7	4840	1.8
Io (J1)	1.8	3640	3.5
Moon (E1)	27.3	3476	3.3
Europa (J2)	3.6	3130	3.0
Triton (N1)	5.9	3000 (?)	?

lites and the ring particles are very dark, reflecting only a few percent of the sunlight that strikes them. There are no known irregular satellites in the Uranian system.

Neptune has only two known satellites, both of them irregular. Triton, the largest, has an atmosphere and is similar in size to Pluto. Its orbit is retrograde. Much farther from the planet is the smaller Nereid, in an extremely eccentric direct orbit. There are also rings or more likely fragments or arcs of ring material, but the ring observations are contradictory and difficult to interpret.

18.2 LARGE SATELLITES

In this section we discuss the six large satellites of the outer solar system. Table 18.1 summarizes their properties, with the Moon listed for comparative purposes.

(a) The Three Largest Satellites

The three largest satellites are Ganymede and Callisto in the Jovian system and Titan in the Saturn system. All three of these have about the same diameter (from 5280 km for Ganymede down to 4840 km for Callisto) and nearly identical densities of 1.8 to 1.9 g/cm^3. (Their uncompressed densities are about 1.3 g/cm^3.) Each therefore appears to have the same composition, and we would expect them to have experienced parallel, and perhaps nearly identical, evolution. It is thus with considerable interest that we note that the three are different in several fundamental ways.

Since they have the same size and composition, these three largest satellites probably have the same general interior structure. M.I.T. chemist John Lewis first argued in about 1970 that such objects are almost surely differentiated into a central, Moon-sized core of rock and mud, surrounded by a thick mantle of ice or, possibly, liquid water. The crust would be hard, brittle ice at the temperature prevalent on these satellites.

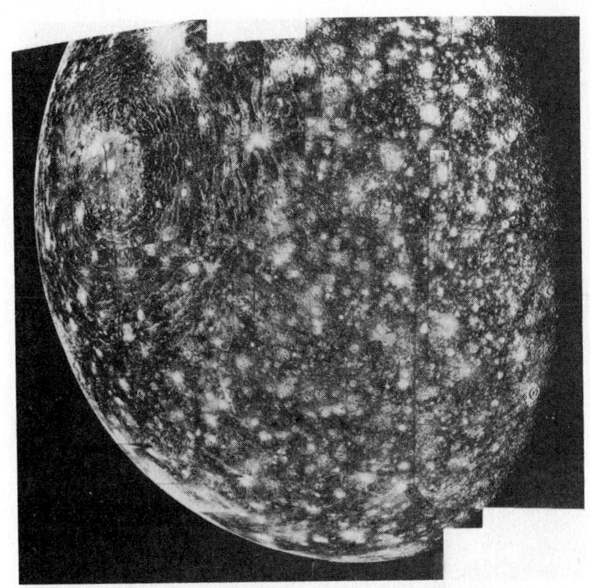

Figure 18.3 The heavily cratered surface of Jupiter's outermost Galilean satellite, Callisto. (NASA/JPL)

Both Callisto and Ganymede apparently conform to this expectation. As we will describe below, each has regions that are heavily cratered, suggesting that their crusts stabilized long ago. Titan may be similar, but unfortunately we have no knowledge of its surface, since Titan, like Venus, is hidden under opaque clouds and a thick atmosphere. One of the most obvious questions to ask when comparing these three objects is why Titan developed an extensive atmosphere while Callisto and Ganymede have none. First, however, let us look at the geological record preserved on the surfaces of Ganymede and Callisto.

(b) Geology of Ganymede and Callisto

Callisto and Ganymede provide an excellent introduction to the geology of icy worlds. We begin with Callisto, the simpler of the two. The entire surface of Callisto is covered with impact craters, like the lunar highlands (Figure 18.3). The existence of this heavily cratered surface tells us three important things not known before Voyager: (1) an icy planet

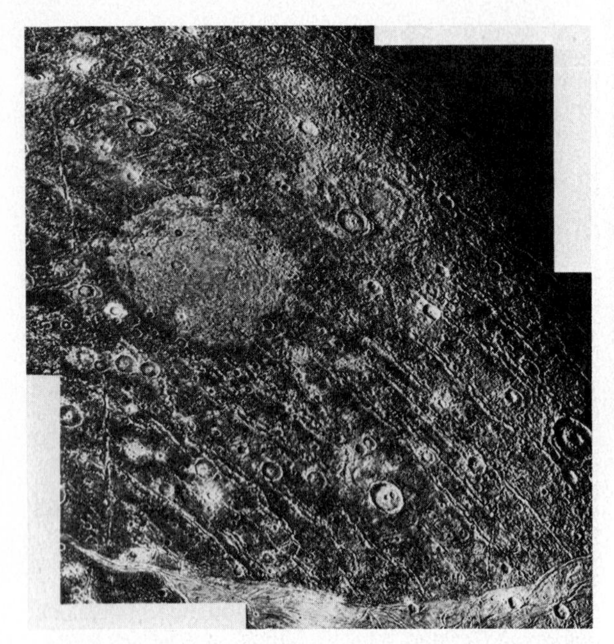

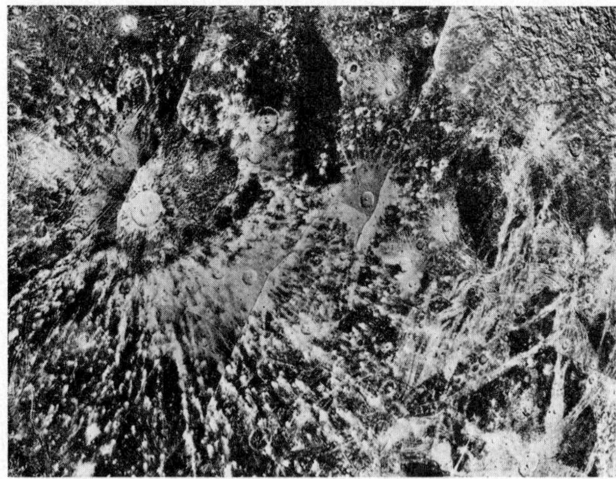

Figure 18.4 Cratered terrain on Jupiter's largest satellite, Ganymede. (NASA/JPL)

can form and retain impact craters in its surface, (2) there was a heavy bombardment by debris in the outer solar system as well as nearer the sun, and (3) Callisto has experienced little if any geological activity other than impacts for a long time—probably billions of years.

Calculations suggest that impacts occurring now in the outer solar system are primarily from comets, and that the impact rates should be roughly half as great as in the inner solar system. If correct, these calculated current rates could not accumulate as many craters over the entire history of the solar system as are seen on Callisto. Therefore, Callisto tells us that there was a *heavy bombardment* in the outer solar system too, probably at about the same time that the heavily cratered surfaces of Moon, Mars, and Mercury formed.

The craters of Callisto do not look exactly like their counterparts in the inner solar system. They tend to be much flatter, as if the surface did not have the strength to support much vertical relief. Such subdued topography is to be expected for an ice crust at the temperatures of 130 to 140 K measured near local noon on Callisto, since ice loses some of its strength as it is warmed. Farther from the sun, in the Saturn system, temperatures are so low that ice is as strong as rock.

Ganymede, the largest satellite in the solar system, is also cratered, but less so than Callisto. About one-third of its surface seems to be contemporary with Callisto; the rest formed later, after the end of the heavy bombardment period. This younger terrain on Ganymede is probably about as old as the lunar maria or the Martian volcanic plains, judging from crater counts.

The younger terrain on Ganymede (Figure 18.4) was produced when internal forces (probably due to expansion of ice in the mantle) cracked the crust, flooding many of the craters with water from the interior and forming extensive parallel mountain ridges. This mountainous terrain, with its ridges evenly spaced about 15 km apart, covers more than one-quarter of the surface. There is even evidence that blocks of the older, heavily cratered terrain may have rotated or slipped at the time the younger crust was forming, providing a surprising analog of Earth's plate tectonics on this icy satellite.

Thus we see that Ganymede experienced expansion and consequent resurfacing during the first billion years after its formation, while Callisto did not. Apparently the small difference in size between the two led to this difference in their evolution. Meanwhile, the third of these satellite triplets, Titan, was outgassing and retaining an extensive atmosphere.

(c) The Atmosphere of Titan

Titan was found in 1655 by Christian Huygens (1629–1695); it was the first satellite discovered since Galileo had seen the four moons of Jupiter that bear his name. The presence of an atmosphere on this satellite was discovered spectroscopically in 1944 by another Dutch astronomer, Gerard P. Kuiper (1905–1973), one of the few professional astronomers who worked in the area of planetary studies during the middle part of this century. His spectra, obtained at the McDonald Observatory in Texas, showed the presence of methane. The next step, based on polarization measurements made 25 years later by Joseph Veverka of Cornell University, was the demonstration that Titan's atmosphere also contained dense clouds, obscuring the surface from our view. This opened the possibility that the atmosphere beneath the clouds might be much more extensive than had seemed possible previously. In the late 1970's, Donald M. Hunten of the University of Arizona calculated models for the atmosphere in which he suggested that nitrogen, although then undetected, was the primary constituent of Titan's atmosphere.

The Voyager 1 flyby of Saturn was designed to yield as much information as possible about Titan, even though a close encounter with this satellite precluded a trajectory that would take the spacecraft on to Uranus and Neptune. NASA planners thus considered data on Titan to be more valuable than that on the two outer planets. Voyager 1 passed within 4000 km, and it also flew behind the satellite as seen from the Earth, producing an *occultation*. In this way, its radio signal traversed successive paths through Titan's atmosphere, generating data from which scientists could reconstruct the atmospheric profile all the way down to the invisible surface of the satellite.

The Voyager occultation determined that the surface pressure on Titan is 1.5 bar, higher than that on any of the terrestrial planets except Venus. Since the surface gravity on Titan is less than that on the Earth, it actually requires much more gas to yield the same pressure, so the amount of atmosphere above each point is nearly ten times greater on Titan than on the Earth. The composition is primarily nitrogen, another respect in which Titan's atmosphere resembles Earth's. Methane amounts to only a few percent, and there is probably also a few percent argon. Exact abundances are unlikely to be determined until the atmosphere is probed directly, a mission (called Cassini) that NASA would like to carry out during the late 1990s in collaboration with the European Space Agency.

A variety of additional compounds have been detected spectroscopically in Titan's upper atmo-

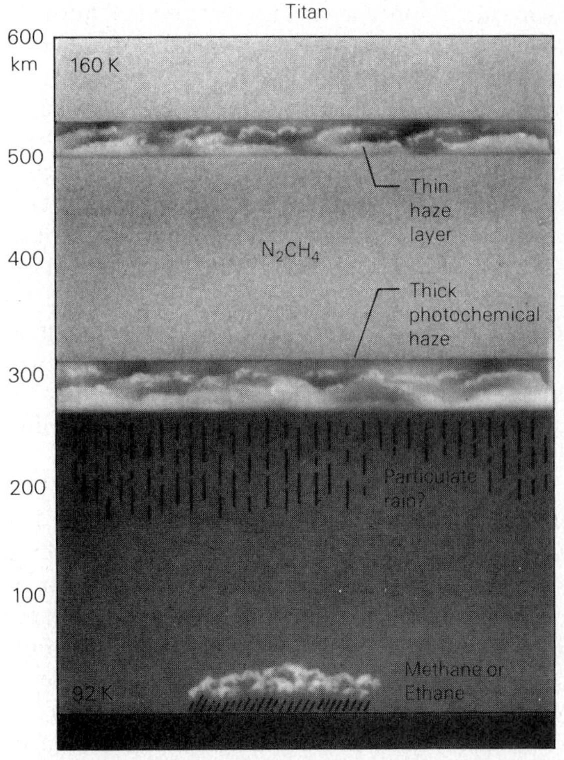

Figure 18.5 Atmospheric structure of Titan as derived from Voyager observations.

sphere, including carbon monoxide (CO), hydrocarbons such as ethane (C_2H_6) and propane (C_3H_8), and nitrogen compounds such as hydrogen cyanide (HCN), cyanogen (C_2N_2), and cyanoacetylene (HC_3N). These indicate an active *photochemistry,* in which sunlight interacts with atmospheric nitrogen and methane to create a rich organic mix, in much the way we believe organic compounds were formed on the Earth when it still had a reducing atmosphere. The discovery of HCN was particularly interesting, since this molecule is the starting point for formation of some of the components of DNA, the fundamental genetic molecule essential to life on Earth.

There are multiple cloud layers on Titan (Figure 18.5). The lowest clouds are in the troposphere, within the bottom 10 km of the atmosphere; these are condensation clouds composed of methane. Methane plays the same role in Titan's atmosphere as water does on Earth; the gas is only a minor constituent of the atmosphere, but it condenses to form the major clouds in the troposphere. Much higher, photochemical reactions have produced a dark reddish haze or smog consisting of complex organic chemicals. Formed at an altitude of several hundred kilometers, this aerosol slowly settles downward, where it presumably has built up a deep layer of tar-like organic chemicals on the surface of Titan.

Titan's surface temperature is about 90 K, held uniform by the blanketing atmosphere. At such a low temperature, there may be seas of liquid methane and ethane. Organic compounds are chemically stable, unlike the situation on the warmer, oxidizing Earth; therefore Titan's surface probably records a chemical history that goes back billions of years. Many people believe that this satellite will provide more insights into the early history of Earth's atmosphere, and even into the origin of life, than any other object in the solar system.

Why does Titan have an atmosphere while Ganymede and Callisto do not? Part of the answer lies in Titan's greater distance from the sun, producing cooler temperatures that slow down the molecules in the atmosphere and decrease their rate of escape. But the primary reason must be that Titan outgassed from its interior more gas than was ever present on the two Jovian satellites. This difference once again relates to the changing composition of the solar nebula condensates with distance from the sun. Where Titan formed, small but significant amounts of methane and ammonia were present, while Ganymede and Callisto apparently had none. Subsequently, photochemical reactions dissociated most of the ammonia (NH_3) to release nitrogen, while the hydrogen escaped from Titan's atmosphere to produce the conditions seen today.

(d) Europa and Io

Europa and Io, the inner two Galilean satellites, are not icy worlds like most of the satellites of the outer planets. With densities and sizes similar to our Moon, they appear to be predominantly rocky objects. How did they fail to acquire the ice that must have been plentiful at the time of their formation? The most probable cause is Jupiter itself, which became hot and radiated a great deal of infrared energy during the first few million years after its formation. Temperatures therefore rose in the disk of material near the planet, and the ice evaporated, leaving Europa and Io with compositions more appropriate to bodies in the inner solar system.

In spite of its mainly rocky composition, Europa (Color Plate 18c) has an ice-covered surface. In this way it is like the Earth, which also has global

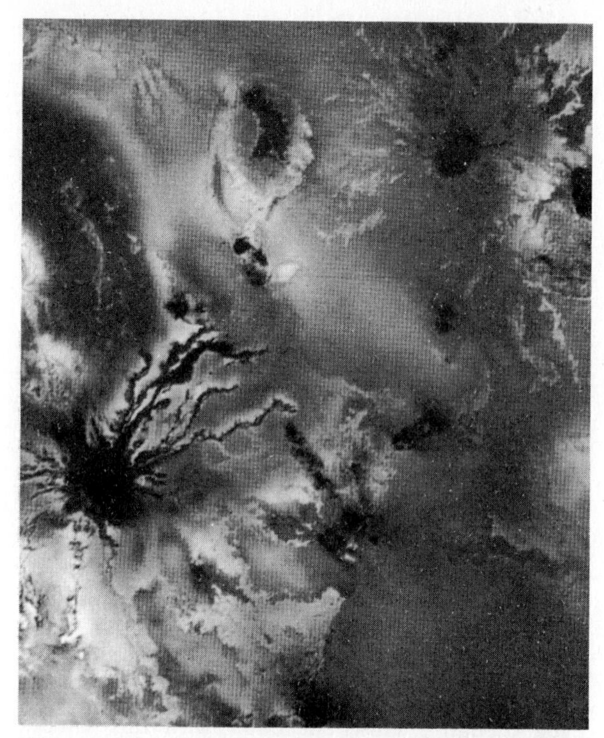

Figure 18.6 Jupiter's satellite Io, showing the complex volcanic flows that dominate its highly active surface and make it unlike any other planet or satellite. (NASA/JPL)

oceans of water, except that Europa's ocean is frozen. There are very few impact craters, indicating that the surface of Europa has been capable of some degree of self-renewal. Additional indications of continuing internal activity are provided by an extensive network of cracks in its icy crust.

Io (Color Plate 17a), the innermost of Jupiter's large satellites, might have been expected to be a twin of Europa. Instead, it displays a high level of volcanic activity, setting it off from all the other objects in the planetary system.

(e) Volcanoes of Io

The discovery of active volcanism on Io was the most dramatic event of the Voyager flybys of Jupiter. Eight volcanoes were seen erupting when Voyager 1 passed in March 1979, and six of these were still active four months later when Voyager 2 passed. These eruptions consisted of graceful *plumes* that extended hundreds of kilometers into space (Figures 18.6 and 18.7). The material erupted is not lava or steam or carbon dioxide, all of which are

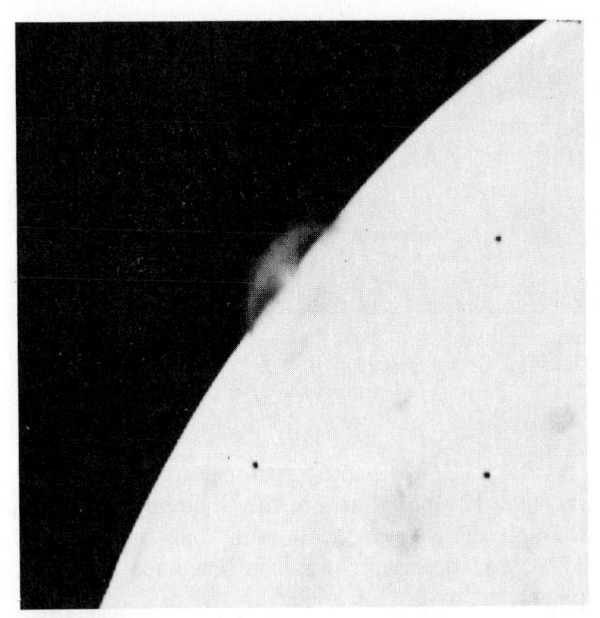

Figure 18.7 A portion of the limb of Io, showing a particularly striking plume from an erupting volcano. (NASA/JPL)

vented by terrestrial volcanoes, but sulfur and sulfur dioxide. As shown in detailed calculations by Susan Kieffer of the U.S. Geological Survey, both of these can build up to high pressure in the crust of Io and then be ejected to tremendous heights. As the rising plume cools, the sulfur and sulfur dioxide recondense as solid particles, which fall back to the surface in gentle "snowfalls" that extend as much as a thousand kilometers from the vent. The sulfur dioxide snow is white, while sulfur forms red and orange deposits. The surface of Io is slowly buried in these deposits, which accumulate at an average rate of a millimeter or so per year. Over millions of years, this is sufficient to cover any impact craters, so it is no surprise that no such craters have been seen on Io's surface.

Io displays other types of volcanic activity in addition to the spectacular plume eruptions. Images of its surface show numerous shield volcanoes, calderas, and twisting lava flows hundreds of kilometers long. From their bright colors, these lava flows are thought to be sulfur. Further volcanic activity is indicated by *hot spots,* surface areas that are hundreds of degrees warmer than their frigid surroundings. (Note that on Io, where the average daytime temperature is only 130 K, even a 300 K area, the surface temperature of the Earth, would classify as a "hot spot.") The largest of these hot spots is a type of "lava lake" 200 km in diameter near the Loki eruption. The "lava" in this case is probably liquid sulfur. Telescopic observations made at Mauna Kea Observatory in Hawaii show that this Loki hot spot has been active for at least ten years, and it accounts for about half of the total volcanic energy released by Io.

The sulfur dioxide and other gases belched out by Io's volcanoes form a tenuous atmosphere. Io, however, orbits deep within the Jovian magnetosphere, and its surface is subject to a tremendous bombardment by energetic ions of sulfur and oxygen, as discussed in Section 17.4c. The molecules in Io's thin atmosphere are dissociated and ionized by these charged particles. Once ionized, they are swept up in Jupiter's magnetic field to form the Io plasma torus. Thus the volcanic eruptions on this satellite have a major influence on the huge magnetosphere of Jupiter.

How can Io maintain this remarkable level of volcanism, which exceeds that of much larger planets such as Earth and Venus? The answer lies in *tidal heating* of the satellite by Jupiter. Io is about the same distance from Jupiter as is our Moon from the Earth, yet Jupiter is more than 300 times more massive than Earth, causing tremendous tides on Io. These tides pull the satellite into an elongated shape, with a several-kilometer–high bulge extending toward Jupiter. Now if Io always kept exactly the same face turned toward Jupiter, this tidal bulge would not generate heat. However, Io's orbit is not exactly circular, because of gravitational perturbations from Europa and Ganymede. In its slightly eccentric orbit, Io twists back and forth with respect to Jupiter, at the same time moving nearer and farther from the planet on each revolution. The twisting and flexing of the tidal bulge heats Io, much as repeated flexing of a wire coathanger heats the wire. In this way, the complex interaction of orbit and tides pumps energy into Io, melting its interior and providing power to drive its volcanic eruptions.

After billions of years, this tidal heating has taken its toll on Io, driving away water and carbon dioxide and other gases, until now sulfur and sulfur compounds are the most volatile materials remaining. The inside is entirely melted, and the crust itself is constantly recycled by volcanic activity. Although Io was well mapped by Voyager, we expect that when reimaged by the Galileo spacecraft, its surface will wear a partly unfamiliar face.

Figure 18.8 Two photographs of Pluto, showing its motion among the stars in a 24-hour period, photographed with the 200-inch telescope. (Caltech/Palomar Observatory)

18.3 PLUTO, CHARON, AND TRITON

In this section we examine Pluto and its satellite Charon. For comparative purposes we also look at Triton, the large satellite of Neptune, which appears to be physically very similar to Pluto.

(a) Discovery of Pluto

Pluto was discovered through a careful, systematic search, not (like Neptune) as the result of a position calculated from gravitational theory. Nevertheless, the history of the search for Pluto began with indications of departures of Uranus and Neptune from their predicted orbits. Early in the 20th century several astronomers became interested in this problem, most notably Percival Lowell, then at the peak of his fame as an advocate of intelligent life on Mars (Section 15.1b).

At the time Lowell made his calculations, Neptune had moved such a short distance since its discovery that it could not be used effectively to search for perturbations by an unknown ninth planet. Therefore Lowell and his contemporaries based their calculations primarily on the minute remaining irregularities of the motion of Uranus. His computations indicated two places where a perturbing planet could be, the more likely of the two being in the constellation of Gemini. Lowell predicted a mass for the planet intermediate between that of the Earth and that of Neptune (his calculations gave about 6.6 Earth masses). Other astronomers, however, obtained other solutions, including one that indicated *two* exterior planets. Lowell searched for the unknown planet at his Arizona observatory from 1906 until his death in 1916, without success. Subsequently, Lowell's brother donated to the observatory a 33-cm photographic telescope that could record a 12° by 14° area of the sky on a single photograph. The new camera went into operation in 1929, and the search was continued for the ninth planet.

Unfortunately, Gemini lies near the Milky Way, and some 300,000 star images were recorded on each exposure. It was an immense task to compare all the star images on each of two or more photographs of the same field in the hope of finding one image that changed position with respect to the rest, revealing itself as the new planet. The job was facilitated by the invention of the *blink microscope,* a device for comparing two different photographs of the same region of the sky. The operator's vision is automatically shifted back and forth between corresponding parts of the two photographs. If the star patterns are the same on the two plates, the observer sees a constant, although flickering, picture. However, if one object has moved slightly in the interval between the times the two plates were taken, the image of that object appears to jump back and forth as the view alternates between the two plates. In this way, moving objects can quickly be picked out from among the many thousands of star images.

In February 1930, Clyde Tombaugh (b. 1906), comparing photographs made on January 23 and 29 of that year, found an object whose motion appeared to be about right for a planet far beyond the orbit of Neptune (Figure 18.8). It was within 6° of the position Lowell predicted for the unknown planet. Announcement of the discovery was made on Lowell's birthday, March 13, 1930. The new planet was named for Pluto, the god of the underworld. (Appropriately, the first two letters of Pluto are the initials of Percival Lowell; this is about as

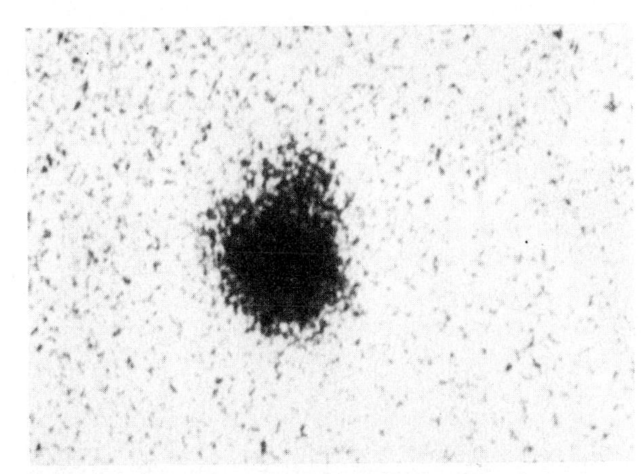

Figure 18.9 Highly enlarged image of Pluto on photograph made at the U.S. Naval Observatory, Flagstaff Station. The "bump" on the upper right is Pluto's satellite, Charon. (U.S. Naval Observatory)

close as one can come to naming a planet for a person.)

Although at the time the discovery of Pluto appeared to be a vindication of gravitational theory similar to the triumph of Adams and Leverrier, we now know that Lowell's calculations were wrong. When the mass of Pluto was finally measured, it was found to be much less than that of the Moon. It could not possibly have exerted any measurable pull on either Uranus or Neptune. To the degree that the discrepancies in the orbits of these two planets are real, they must be due to some other cause. A survey of the entire sky in the infrared carried out in 1983 by the Infrared Astronomical Satellite (IRAS), however, has revealed no hidden "Planet X," and today it is generally accepted that the supposed perturbations of Uranus and Neptune are not, and never were, real. But even if his calculations were wrong, it was certainly Lowell's faith and enthusiasm that ultimately led to the discovery of Pluto.

Pluto's orbit has the highest inclination to the ecliptic (17°) of any planet and also the largest eccentricity (0.248). Its median distance from the sun is 40 AU, or 5.9 billion km, but its perihelion distance is under 4.5 billion km, within the orbit of Neptune. Pluto is now closer to us than Neptune, and it will remain so until 1999. Even though the orbits of these two planets cross, there is no danger of collision; because of its high inclination, Pluto's orbit clears Neptune's by 385 million km. Pluto completes its orbital revolution in a period of 248.6 years; since its discovery in 1930, the planet has traversed less than one-quarter of its long path around the sun.

(b) Discovery of Charon

Pluto's satellite was discovered in 1978 by James W. Christy of the U.S. Naval Observatory. He noticed a peculiarity on the images of Pluto obtained with the observatory's 1.5-m astrometric telescope at Flagstaff. The images of Pluto were slightly elongated, while those of stars on the same plate were not (Figure 18.9). A check of the observatory plate files revealed that some of the other Pluto images taken under excellent seeing conditions also showed this image distortion, although most did not. Such an effect could take place if there were a second, unresolved source in a periodic orbit around Pluto.

Christy followed up on this hunch and found that all of the observations could be matched if the satellite had a period of 6.387 days and a maximum separation from Pluto of just under 1 arcsecond. This is the same as Pluto's rotation period, showing that the planet keeps the same side always turned toward its satellite. Since the satellite also almost surely has the same rotational period, we have here the only fully tidally evolved planet-satellite system, in which both members are tidally locked together.

The exact nature of the satellite's orbit was not confirmed until 1985, when the system had turned to the point at which the satellite and the planet began to occult each other on each satellite orbit. These observations confirmed that the satellite was in a retrograde orbit and indicated that it had a diameter of about 1200 km, nearly half the size of Pluto itself. It was given official status and a name—Charon—by the International Astronomical Union in 1985.

Confirmation of the orbit of Charon also clarified what had been suspected for some time—that Pluto's own rotation is retrograde. The obliquity of Pluto's rotational axis is 112°, similar to that of Uranus. At present, the equator of Pluto faces approximately toward the sun, but in about 60 years the pole will nearly point to the sun, the way Uranus' pole does now.

(c) Physical Nature of Pluto and Triton

Since Neptune's large, retrograde-orbiting satellite Triton appears to be very similar to Pluto, we discuss them together. Of course, neither has ever been visited by spacecraft, and they are both so faint that they are difficult objects for ground-based studies, requiring use of the largest telescopes in the world. Thus our knowledge is limited, and some of what we report below may turn out to be incorrect.

Pluto and Triton are each somewhat smaller than the Moon, with diameters between 2600 and 3000 km. The mass of Pluto, measured by tracking its satellite Charon, is 1/400 the mass of the Earth. The corresponding density is about 1.0 g/cm^3, lower than that of any solid planetary object encountered previously. Such a low density cannot be matched with combinations of water ice and rock, unless the amount of rock is extremely small, which is thought to be unlikely. It is more probable that a third component, with density less than 1.0 g/cm^3, is mixed with the ice and rock.

The most likely candidate is *methane ice,* which is expected to have condensed in the solar nebula at temperatures below about 50 K. But new observations by D. Tholen and M. Buie, made as this text went to press, indicate that Pluto is smaller than was previously thought, and that its density is actually about 1.5 g/cm^3, similar to a number of the icy satellites. If this is correct, the argument for the presence of methane ice is weakened. The mass of Triton has not been measured, so we cannot use its density as a clue to its composition.

Both Pluto and Triton have highly reflective surfaces, indicative of the presence of fairly fresh ice of some kind. One of us (Morrison) was part of a team effort in 1975 that first detected methane ice on the surface of Pluto, using infrared spectroscopic observations with the 4-m telescope of Kitt Peak National Observatory. Subsequently, methane was also found on Triton. Equally interesting, Dale P. Cruikshank of the University of Hawaii has recently found spectral evidence in observations made at Mauna Kea Observatory of *liquid* nitrogen on the surface of Triton. He suggests that a part of the surface of the satellite is covered with pools or seas of this substance. The surface temperature on Pluto is expected to range from about 60 K near perihelion to below 50 K near aphelion; that of Triton is probably always near 60 K. Pluto and Triton each also have thin atmospheres of methane gas, in equilibrium with the frozen methane on their surfaces.

Probably the similarities between Pluto and Triton are simply the result of their formation at about the same distances from the sun. However, it has been suggested that they were both once satellites of Neptune, and that Pluto was later ejected in a catastrophic collision of some kind. Such an event seems dynamically unlikely, especially now that we know that Pluto has a satellite of its own. However, the overlap of the orbits of Pluto and Neptune is intriguing, as are the physical similarities between Triton and Pluto.

Figure 18.10 Saturn's satellite Rhea, with its heavily cratered surface, photographed by Voyager 1. (NASA/JPL)

18.4 SMALLER SATELLITES

(a) Regular Satellites of Saturn

Saturn has six regular satellites with diameters between about 400 and 1600 km, a size range not represented in the Jovian system. All were discovered

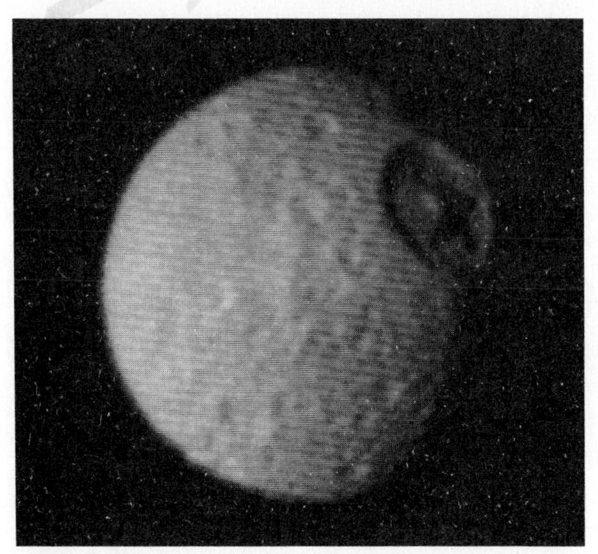

Figure 18.11 Saturn's satellite Mimas, photographed by Voyager 1. (NASA/JPL)

Figure 18.12 Saturn's satellite Dione, photographed by Voyager 1. (NASA/JPL)

long ago, four of them by the Italian-born French astronomer Jean Dominique Cassini (1625–1712). Five of these are close to the planet, inside the orbit of Titan. These are, in order from the inside out: Mimas, Enceladus, Tethys, Dione, and Rhea. Iapetus, the sixth of these satellites, occupies a somewhat inclined orbit far outside that of Titan.

Each of these six satellites has a surface that displays the spectral signature of water ice. Further, each has a density of about 1.3 g/cm^3, close to the expected uncompressed density of an object composed half of water ice. (The uncompressed densities of Titan, Ganymede, and Callisto are also about 1.3 g/cm^3.) Unlike in the Jovian system, there is no indication of a systematic variation of density and composition with distance from the planet. Evidently Saturn was never hot enough to eliminate water ice from its inner satellites, as Jupiter seems to have done with Io and Europa.

The largest in this group of satellites is Rhea (Figure 18.10), with a diameter of 1530 km, just half as large as Europa, the smallest Galilean satellite. Rhea, and also its smaller cousins Mimas, Dione, and Tethys, are all heavily cratered worlds with bright surfaces of relatively clean water ice (Figures 18.11 and 18.12). Although there are indications of some tectonic cracking and resurfacing early in their histories, all four of these objects seem to have stabilized geologically billions of years ago. In general, they behave as might be expected for icy objects of this size at this distance from the sun, where any internal activity should have ceased early as the body cooled, and where the icy crusts can retain impact craters as well as rock does on the Moon.

The other two of this group of six Saturn satellites are more unusual. Iapetus (Figure 18.13) is nearly as large as Rhea, and on one side it looks very much like Rhea, with a bright, heavily cratered surface of water ice. The other hemisphere is entirely different, however. The side of Iapetus that

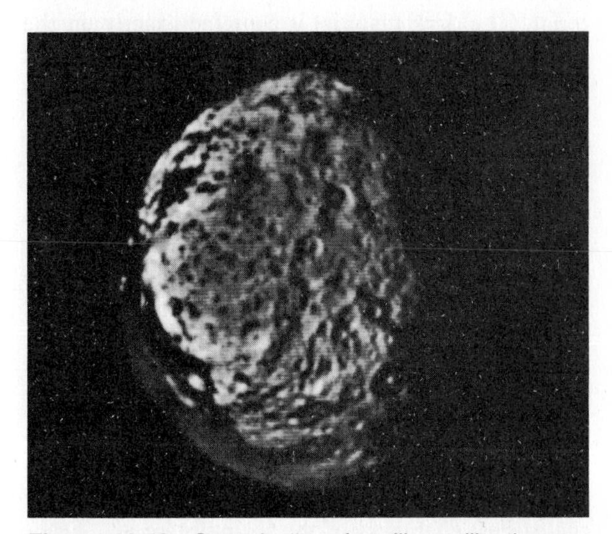

Figure 18.13 Saturn's "two-faced" satellite Iapetus, seen by Voyager 2, showing the strong contrast between its bright and its dark regions. (NASA/JPL)

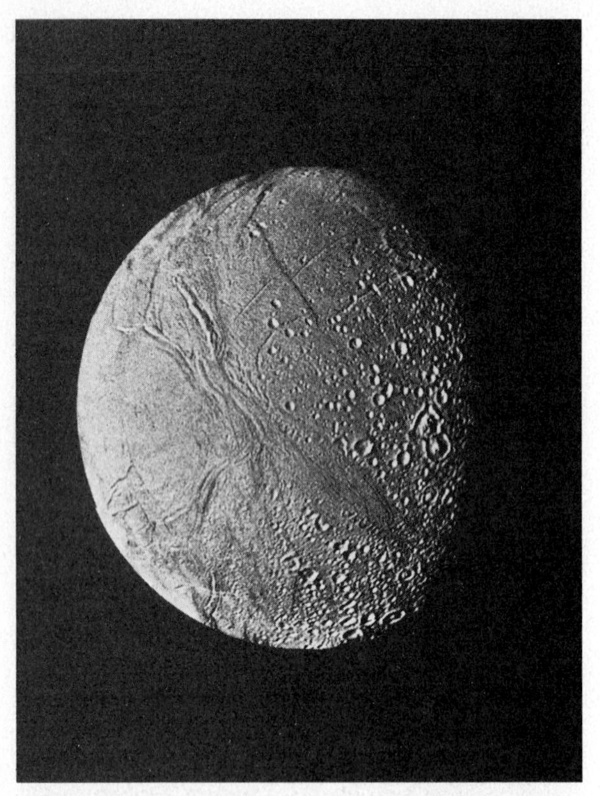

Figure 18.14 Saturn's volcanically modified satellite Enceladus, seen by Voyager 2. (NASA/JPL)

faces forward in its orbit—like all of these satellites Iapetus always keeps the same side toward the planet—is covered with a very dark carbon-rich material. The dark material is centered exactly on the forward hemisphere, in the form of a huge oval spot, and it reflects less than 1/10 as much light as the surface on the trailing (backward-facing) hemisphere. The contrast between the two kinds of surface material on this satellite is as great as that between a black asphalt pavement and freshly fallen snow.

From its orientation, this dark spot must have an external cause. Possibly impacts have vaporized enough ice on this side to concentrate near the surface the dark primitive dust that was originally mixed with the ice when Iapetus formed. If so, then this satellite may never have differentiated. Its surface, especially on the dark side, might be an extremely interesting place to search for evidence of the earliest chemistry of the solar system.

The other peculiar satellite is Enceladus (Figure 18.14). With a diameter of only about 500 km, Enceladus is a near-twin of heavily cratered Mimas, yet about half of its surface is nearly crater-free. There is evidence of powerful internal activity in the geologically recent past—within the past few hundred million years. In addition, the surface of Enceladus is the most highly reflective of any planet or satellite, suggesting that it is covered with fine particles of fresh crystalline ice, like the glass beads on a projection screen. Finally, this satellite also seems to have associated with it a very tenuous ring, called the E Ring, circling Saturn, also made up predominantly of small (1 μm) ice particles. Something very strange is going on here! Has a recent volcanic eruption or impact sprayed out water droplets to freeze and form the E Ring, and subsequently to coat the surface of Enceladus with bright material? Is such an event related to the crater-free areas on the surface? And most puzzling, what could possibly maintain internal activity on a body as small as Enceladus, which should have cooled down very quickly after its formation? No one knows the answers to such questions.

(b) The Strange Small Satellites of Saturn

We have discussed Titan and the six intermediate-sized Saturn satellites. The outermost satellite, Phoebe, appears to be a captured object. Finally, now, we turn to the 11 known small satellites. In many ways, these small bodies are what one would expect to find: heavily cratered, icy in composition, with irregular shapes, they appear to be fragments of once-larger parent bodies. What makes them interesting, however, is the variety of orbits they occupy, and the close interaction many of them have with the rings of Saturn (Section 18.5d).

One of these satellites, named Hyperion, has a very strange kind of rotation. Unlike all of the other satellites we have discussed, Hyperion does not appear to keep the same side toward its planet. In fact, it does not even have a well-defined period of rotation. Theorists such as Stanton Peale of the University of California at Santa Barbara believe that Hyperion is in a state of chaotic rotation, in which gravitational interactions with Titan cause it to exchange angular momentum between its orbit about Saturn and its rotation.

Two of these satellites, Janus and Epimethius, are *co-orbital*—that is, they occupy nearly, but not quite the same orbit around Saturn. If they had exactly the same period, they could avoid each other, but such a state is dynamically impossible. Instead, their orbits differ in radius by about 50 km, corresponding to a difference in orbital periods of 1 part in 2080. The inner co-orbital, following Kepler's laws, therefore catches up with the outer at a relative speed of about 9 m/s. Since these satellites are more than 100 km across, they cannot pass. Fortunately, they interact gravitationally before they get too close, exchanging orbits. Their relative motion therefore reverses, and they pull apart. This intricate orbital maneuver repeats about once every four years.

Other strange satellite orbits in the Saturn system include several cases where small satellites occupy the *Lagrangian points* associated with a larger satellite (Section 5.2). In the next chapter we discuss a similar situation in which some asteroids are in Lagrangian orbits with respect to Jupiter.

Figure 18.15 Uranus' satellite Ariel, with its deep valleys and other tectonic features, as photographed by Voyager 2 in 1986. (NASA/JPL)

(c) Satellites of Uranus

The 15 known satellites of Uranus are conveniently divided into two groups: the five larger bodies that had been discovered telescopically; and 10 smaller moons, all relatively close to the planet, that were discovered by Voyager. The 10 new satellites are all small and dark, with the largest less than 200 km in diameter. They may be related both chemically and gravitationally to the dark rings of Uranus, discussed in the next Section. Here we will focus our attention on the five larger satellites, all named after characters from Shakespeare or from Alexander Pope's poem "The Rape of the Lock."

The larger satellites of Uranus have diameters from about 1600 km down to 500 km, just the same range as the regular satellites of Saturn. Their densities (1.3 to 1.6 g/cm^3) are also similar, although a little greater, indicating that there is a slightly smaller proportion of ice, relative to rock, in the Uranian satellites. Like the satellites of Saturn, their surfaces show the spectral signature of water ice, although their reflectivities are also generally lower (ranging from 20 to 40 percent), suggesting that their surfaces are "dirtier." They are not nearly so dark as the newly discovered small Uranus satellites, however, which reflect only about 5 percent of incident sunlight.

As might be expected from our previous examination of the satellites of Jupiter and Saturn, these objects are all more or less heavily cratered. Presumably most of this cratering took place during the first billion years of solar system history, at a time when there was more cometary debris present than there is today. If we apply the standard theories for cratering to these distant objects, we conclude that they are relatively inactive geologically, with surfaces that have been stable for billions of years. There is no young, active object like Io or Enceladus in the Uranus system.

Although all five of these satellites of Uranus are similar in being ice–rock mixtures and having heavily cratered surfaces, there are still striking differences in their individual geological histories. Of the five, only Umbriel shows no sign of internal activity. The two largest, Titania and Oberon, both have tectonic cracks or valleys rather like those of the Saturn satellites, Dione and Tethys. However, the most geologically interesting are the two inner satellites, Ariel and Miranda.

Ariel (Figure 18.15), which has a diameter of 1160 km, is characterized by deep valleys that seem to be the result of stretching of the crust. There are also relatively young flows that may represent a period of water "volcanism" in the history of this satellite. Since pure water seems unlikely to have been a fluid at the temperatures (well below 100 K) found in the Uranian system, it is speculated that

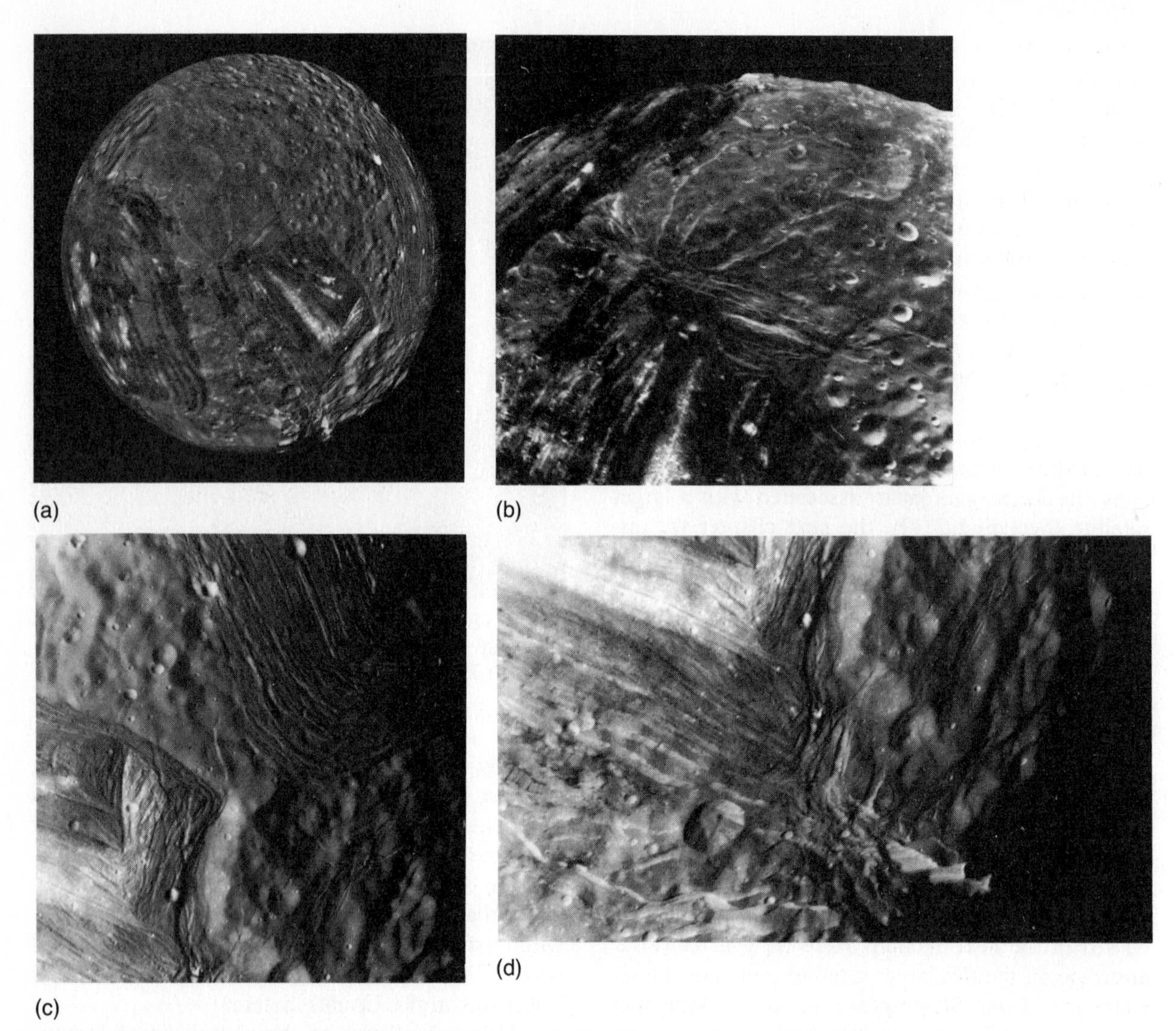

Figure 18.16 Four images of Uranus' remarkable satellite Miranda, as photographed by Voyager 2 at a resolution of about 1 km. (a) Overall view of the southern hemisphere. (b, c, and d) Detailed images of terrain heavily modified by tectonic forces. (NASA/JPL)

ammonia–water mixtures, or fluids involving carbon monoxide or methane, may have constituted the "lava" in this case.

Miranda, the smallest (484 km) and innermost of the five pre-Voyager satellites, is the most geologically diverse and mysterious of the Uranian moons. Its surface (Figure 18.16), like that of Ganymede in the Jovian system, consists of both older heavily-cratered terrain and widespread younger structures that nearly defy description. These include great valley systems with gorges as much as 10 km deep and complex oval or trapezoidal ranges of mountains. At this time (1986) planetary geologists are trying to understand the origin of these features, many of which have no counterparts on other planetary bodies. Also mysterious are the reasons why this small satellite should have displayed such a high degree of geological activity, in comparison to the larger members of the Uranian system.

18.5 PLANETARY RINGS

Three of the outer planets have well-developed ring systems, and even Neptune seems to have some ring-like material in orbit about it. Each ring sys-

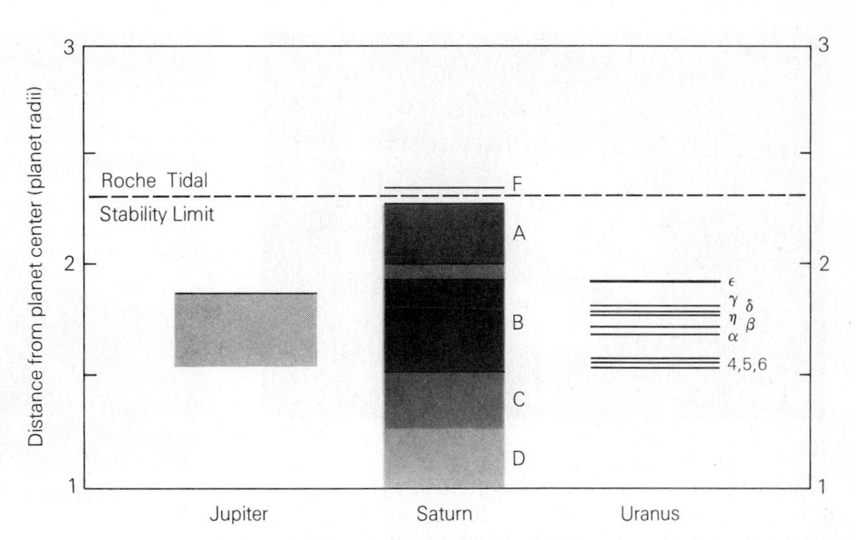

Figure 18.17 Comparative sizes of the known planetary ring systems. Also shown are the Roche tidal stability limits for each system.

tem consists of billions of small particles or moonlets orbiting close to their planet, and each displays complex structure apparently related to interactions between the ring particles and the larger satellites. However, the outer planet ring systems are also very different from each other. Saturn's system, which is by far the largest, is made up primarily of small icy particles spread out into a vast flat ring with a great deal of fine structure. The Uranus rings are nearly the reverse, consisting of larger, very dark particles confined to a few narrow rings, with broad empty gaps in between. The Jupiter ring is merely a transient dust band, constantly renewed by erosion of dust grains from its inner two satellites, while the Neptune "ring" is too poorly documented to permit much analysis. In this section, we will concentrate on the larger and better studied rings of Saturn and Uranus.

(a) Ring Origin and Dynamics

A ring is a collection of vast numbers of particles, each obeying Kepler's laws as it follows its own orbit around the planet. Thus the inner particles orbit faster than those farther out, and the ring as a whole does not rotate as a solid body. In fact, it is better not to think of a ring *rotating* at all, but rather to consider the *revolution* of its individual moonlets. If the particles were widely spaced, they would move independently, like separate small satellites. However, in the rings of Saturn and Uranus the particles are close enough to each other to exert mutual gravitational influence, and occasionally even to rub together or bounce off of each other in low-speed collisions. Because of these interactions, phenomena such as waves can be produced that move across the rings, like water waves moving over the surface of the ocean.

There are two basic theories of ring origin. First is the breakup theory, which suggests that the rings are the remains of a shattered satellite. The second theory, which takes the reverse perspective, suggests that the rings are made of particles that were unable to come together to form a satellite in the first place.

In either theory, an important role is played by tidal forces. As we saw in Section 5.4f, tides are very sensitive to distance, with the tidal force varying as the inverse cube of the separation between two bodies. If objects approach too closely, their tidal bulges become so large that they are torn apart. The same effect takes place for some double stars, which can become so tidally distorted that one star leaks material over onto its companion.

Around each planet there exists a *tidal stability limit,* often called the Roche limit after the French mathematician who first calculated it. This is the distance within which a satellite with no internal strength (like a pile of gravel) would be disrupted. Alternatively, we may think of it as the distance within which the individual particles in a disk cannot attract each other to form a satellite. As discussed in Section 5.4f, for most objects in the outer solar system the limit is at about 2.5 planetary radii from the center of a planet. All three known ring systems lie within the tidal stability limits for their respective planets (Figure 18.17).

(a)

(b)

Figure 18.18 (a) Saturn, photographed by Voyager 1 from a distance of 18 million km. (b) Image of the crescent Saturn transmitted from 1.5 million km by Voyager 1 as the spacecraft left the Saturn system. (NASA/JPL)

This stability limit applies only to a satellite with no intrinsic strength of its own; a solid object held together by its own strength will not necessarily break up inside the limit. If the satellite is large enough, however, its intrinsic strength becomes less important in comparison to the differential tidal forces, and breakup is more likely. Also, a satellite within the limit will break up if it is fractured by a large impact, while if it is outside the limit it is likely to fall back together under its own gravitation after such an impact.

In the breakup theory of ring formation, we can imagine a satellite or even a passing comet coming too close and being torn apart by tidal forces. A more likely variant of this idea suggests that a small satellite near the stability limit might be broken apart in a collision, with the fragments then dispersing into a disk. The third possibility is that the rings represent primitive material left over from the time of formation of the planet and its satellite system. We do not know which of these three explanations holds for the rings of Saturn and Uranus.

The rings of Jupiter are composed of very small dust particles apparently being eroded from the surfaces of the two known inner satellites, and perhaps from smaller unknown bodies within the rings. They represent an equilibrium between production and loss of the dust. The smallest particles in other rings as well have short lifetimes, and are presumably being renewed from sources within the rings.

(b) The Rings of Saturn

The rings of Saturn circle the planet in its equatorial plane, which is tilted by 27° to the planet's orbit plane. As Saturn revolves about the sun, we see one side of the rings for about 15 years, followed by the other side for the same period. The three brightest rings of Saturn, visible from Earth, are labeled (from outer to inner) the A, B, and C Rings. The outer radius of the A Ring is 136,780 km, while the inner edge of the C ring is just 12,900 km above the cloud tops of Saturn. The rings are illustrated in Figure 18.18, and Table 18.2 summarizes the dimensions of Saturn's main rings.

The B Ring is the brightest and has the most closely packed particles, while the A and C Rings are translucent. The total mass of the B Ring, which is probably close to the mass of the entire ring system, is about equal to that of a 250-km diameter icy satellite. The B and A Rings are sepa-

TABLE 18.2 Rings of Saturn

NAME	OUTER EDGE (R_S)	OUTER EDGE (km)
D Ring	1.235	74,510
C Ring	1.525	92,000
B Ring	1.949	117,580
Cassini Div.	2.025	122,170
A Ring	2.267	136,780
F Ring	2.324	140,180

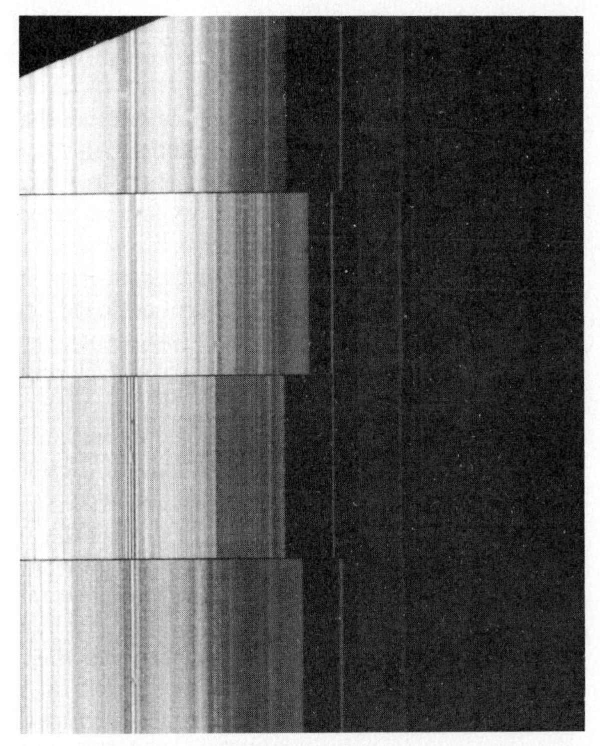

Figure 18.19 Fine structure within the Cassini Division in the rings of Saturn in four photos taken at different times. (NASA/JPL)

rated by a gap easily visible from the Earth, discovered in 1675 by J. D. Cassini and called the Cassini Division. Although it looks like a gap from the Earth, the Cassini Division actually contains many ring particles with considerable structure, including several truly empty gaps, within it.

The Pioneer and Voyager spacecraft revealed additional rings not visible from Earth. A faint D Ring lies inside the C Ring, and a very narrow F Ring, of radius 140,180 km, lies outside the A Ring. As we will describe below, the F Ring is one of the most interesting features of the Saturn system, and it is the one Saturn ring that is similar in many ways to the rings of Uranus. There is a suggestion of a G Ring with a radius of 150,000 km, and we have already noted the broad, tenuous E Ring that is associated with the satellite Enceladus.

The rings of Saturn are very broad in width but very thin. The width of the main rings is 70,000 km, yet their thickness is only about 20 m. If we made a scale model of the rings out of paper the thickness of the sheets in this book, we would have to make the rings a kilometer across—about eight city blocks. On this scale, Saturn itself would loom as high as an 80-story building.

The ring particles are composed primarily of water ice, and they span a range of sizes from grains of sand up to house-sized boulders. An insider's view of the rings would probably resemble a bright cloud of floating snowflakes and hailstones, with a few snowballs and larger objects, many of these loose aggregates of smaller particles.

As revealed by Voyager, the rings of Saturn have a great deal of complex structure, including about a dozen empty gaps, each a few hundred km wide. Two of these gaps are in the C Ring and two in the A Ring, but most of them are associated with the 4800-km wide Cassini Division between the B and A Rings (Figure 18.19). Some of these gaps contain peculiar eccentric ringlets, that is, ribbons of particles that do not share the circular orbits of the other ring particles. For the ring as a whole to be eccentric, it is not sufficient that the individual particles have eccentric orbits; in addition, the major axes of these orbits must be aligned in space. Some of the gaps have wavy edges, and one of the gap ringlets is kinky.

The isolated F Ring is also eccentric. This ring (Figure 18.20) is only about 100 km wide, and it shows a great deal of remarkable structure. Within its 100-km width there are many ringlets, including a double bright ring with two components just a few hundred meters wide. In some places the F Ring breaks up into two or three parallel strands, which sometimes show bends or kinks. Two of the Voyager pictures also show what appears to be a braiding of these multiple strands.

The major B Ring has no gaps. The Voyager data, however, indicate intricate structure, partly in the form of waves. Each wave corresponds to alternating ringlets where the ring particles are bunched together or spread more thinly. Photographed from the spacecraft, these waves, which are typically separated by 100 km or so, look like the grooves in a phonograph record. The A Ring has even more of this wave-like structure.

(c) The Rings of Uranus

The rings of Uranus are narrow and black, making them almost invisible from the Earth. They were discovered accidentally on March 19, 1977 during

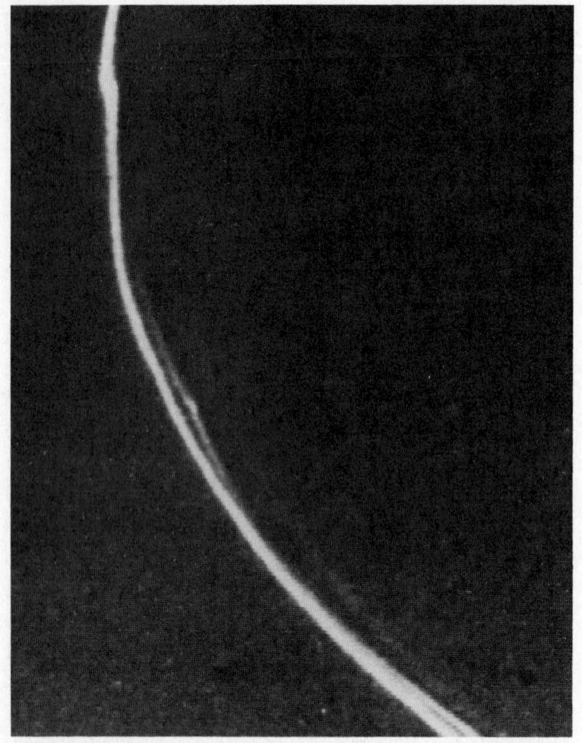

Figure 18.20 Voyager photograph of the complex F Ring of Saturn. (NASA/JPL)

observations of the *occultation* of a bright star by Uranus. Since this occultation could be observed only from the Indian Ocean and its surroundings, several teams of astronomers planned to measure it from observatories in Australia, China, India, and South Africa. In addition, Cornell University astronomer James Elliot led a group that observed from above the middle of the Indian Ocean using the NASA 1-m airborne observatory (Section 10.2d).

About 20 minutes before the predicted occultation of the star by the planet, Elliot and his colleagues saw a series of brief dimmings of the star, as it disappeared behind successive narrow rings. This pattern of ring occultations was repeated later, as the opposite side of each arc passed in front of the star (Figure 18.21). Several ground-based observers also recorded some of these ring events, permitting the identification of nine individual rings (Table 18.3).

Since the 1977 discovery, many more occultations have been observed to map out the Uranus rings in detail, and in January 1986 the Voyager 2 spacecraft was able to study them at close range. In spite of their low reflectivity—typically about 5 percent—they could be photographed by the spacecraft cameras (Figure 18.22). In addition, Voyager used observations of occultations of stars by the rings, and an occultation of the spacecraft itself as it passed behind the planet, to probe the structure of the rings in greater detail than is possible from telescopes on the Earth. Voyager discovered two additional faint rings, one of them broad and diffuse, to bring the total to 11.

The broadest and outermost of the rings of Uranus is called the Epsilon Ring. It is about the width of the Saturn F Ring and has an eccentricity of 0.008. Its thickness is probably no more than 100 m, and from probes with the spacecraft radio system it appears that most of the particles are relatively large—several meters or more in diameter.

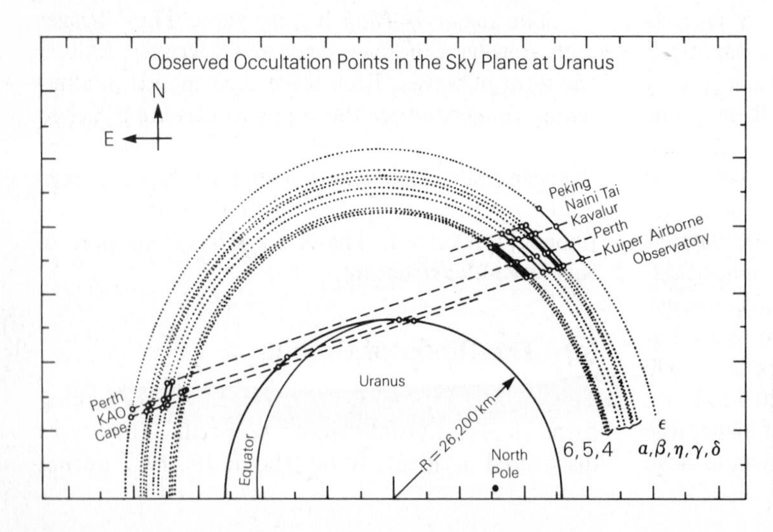

Figure 18.21 The rings of Uranus as revealed by measurements of occultations carried out from the Earth. (Adapted from a summary prepared by J. Elliot of Cornell University)

TABLE 18.3 Rings of Uranus

NAME	DISTANCE (km)	WIDTH (km)	ECCENTRICITY
Epsilon	51,160	22–93	.0079
1986U1R	50,040	1–2	?
Delta	48,310	3–9	0
Gamma	47,630	1–4	0
Eta	47,180	2	0
Beta	45,670	7–11	.0004
Alpha	44,730	8–11	.0008
4 Ring	42,580	2	.0011
5 Ring	42,240	2–3	.0019
6 Ring	41,850	1–3	.0010
1986U2R	38,000	2500	?

The Epsilon Ring circles Uranus at a distance of 51,000 km, or 2.2 Uranus radii—near the position of the tidal stability limit. This ring (shown in Figure 18.23 at a resolution of about 20 m) probably contains as much mass as all of the other 10 rings combined. With one exception, all of the other rings are narrow ribbons less than 10 km in width—just the reverse of the broad rings of Saturn.

Figure 18.22 The dark, narrow rings of Uranus as seen by the Voyager spacecraft in 1986. (NASA/JPL)

(d) Satellite–Ring Interactions

Most of the structure in the rings of Uranus and Saturn owes its existence to the gravitational effects of satellites. If there were no satellites, the rings

Figure 18.23 Detailed cross sections of the epsilon ring of Uranus, showing structure as small as 10–20 meters across, reconstructed from occultation measurements obtained from the Voyager spacecraft. (NASA/JPL).

Figure 18.24 Voyager photo of one of the "shepherd satellites" with Saturn's narrow F ring. (NASA/JPL)

Figure 18.25 Spiral density waves and bending waves in the A ring of Saturn as photographed by Voyager. (NASA/JPL)

would be flat and featureless. Indeed, if there were no satellites, there would probably be no rings at all, since left to themselves, thin disks of matter gradually spread out and dissipate. The sharp edges as well as the fine structure of rings are due to the satellites.

Most of the gaps, and the location of the outer edge of the Saturn rings, apparently result from gravitational *resonances* with Mimas and the two co-orbital inner satellites. A resonance takes place when two objects have orbital periods that are exact ratios of each other, such as ½, ⅓, etc. For example, any particle in the gap at the inner side of the Cassini Division would have a period exactly equal to one-half that of Mimas. Such a particle would be nearest Mimas in the same part of its orbit every second revolution, and the repeated gravitational tugs of Mimas, acting always in the same direction, would perturb it, forcing it into a new orbit that does not represent a resonance with a satellite.

Resonances can apparently form gaps by ejecting material with periods that are an exact multiple of satellite periods, but they can also lead to circumstances where the boundary of a ring is stabilized by these gravitational effects. Such is the case for the sharp outer edge of the A Ring of Saturn, which is in a 7:6 resonance with the two co-orbital satellites, Janus and Epimetheus.

Small satellites that orbit very close to rings can also control their shape and position. This certainly seems to be the case for Saturn's F Ring, which is bounded by the orbits of Pandora and Prometheus (Figure 18.24). These two small objects are referred to as "shepherd satellites," since their gravitation serves to "shepherd" the ring particles and keep them confined to a narrow ribbon. A similar situation applies to the Epsilon Ring of Uranus, which is shepherded by the satellites provisionally called 1986U6 and 1986U7. These two shepherd satellites, each about 50 km in diameter, orbit about 2000 km inside and outside the ring.

Theoretical calculations suggest that the other narrow rings in the Uranian system should also be controlled by shepherd satellites, but none have been located. The calculated diameter for such shepherds—about 10 km—was just at the limit of detectability for the Voyager cameras, so it is impossible to say if they are present or not.

A great deal of the Saturn ring structure photographed by Voyager, particularly in the A Ring, represents waves of higher and lower density that are propagating across the rings. Many of these features are tightly wound spirals, like the grooves in a phonograph record (Figure 18.25). These *spiral density waves* are produced at distances from Saturn corresponding to satellite resonances. The effect is very similar to that which is thought to generate the spiral arms of galaxies, with the role of individual stars being played here by the ring particles.

Recently, waves seen in the Voyager images have been used to locate two new satellites, imbedded in the rings of Saturn. Even though the satellites themselves have not been seen, the evidence from the waves they set up seems convincing enough to add them to the census of Saturn satellites.

Studies of planetary rings and of the gravitational interactions between rings and small satellites have come a long way in the past few years, but many problems in understanding these complex phenomena remain unsolved. Without additional spacecraft missions to the outer planets, however, it may be difficult to obtain the data needed to test the new dynamical theories now being developed.

EXERCISES

1. Why do the outer planets have such extensive systems of rings and satellites, while the inner planets do not?
2. Which would have the greater period, a satellite 1 million km from the center of Jupiter or a satellite 1 million km from the center of Earth? Why?
3. Ganymede and Callisto were the first icy objects to be studied from a geological point of view. Summarize the main differences between their geology and that of the rocky terrestrial planets.
4. Compare the properties of the atmosphere of Titan with those of the Earth's atmosphere.
5. Compare the properties of the volcanoes of Io with those of terrestrial volcanoes.
6. Where did the nitrogen in Titan's atmosphere come from? Compare with the origin of the nitrogen in our atmosphere.
7. Two of the differences between the Saturn and Jupiter systems are (a) the presence of icy rings at Saturn, and (b) the absence of a dependence among the Saturn satellites of density on distance. Can you think of a common explanation for both of these differences?
8. Compare the satellites of Uranus with those of Saturn.
9. Explain why a large satellite is more likely to break up inside the tidal stability limit than a small satellite.
10. Make a table comparing the rings of Saturn with those of Uranus.
11. Saturn's A, B, and C Rings extend from a distance of about 75,000 to 137,000 km from the center of the planet. What is the approximate variation factor for the periods of time for various parts of the rings to revolve about the planet?
12. Three possibilities were suggested in the text for the origin of the rings of Saturn. List them and briefly summarize the arguments in favor of each.

19

THE ASTEROIDS

Eugene Shoemaker (b. 1928) of Caltech and the U.S. Geological Survey has contributed in many ways to our understanding of planetary geology. He began his career with a definitive study of the impact process at Meteor Crater in Arizona, and he has played a leading role in the Apollo exploration of the Moon and the Voyager flybys of the satellites of the outer planets. Shoemaker is best known, however, for his analysis of the dynamics of Earth-approaching asteroids and the role these bodies play in the cratering of the Earth and other planets. (Ramona Boudreau)

The asteroids or *minor planets* are small objects in orbits similar to those of the major planets. They differ from the planets primarily in size. Distinguishing them from their cousins the comets is a little more difficult. A comet is defined as a small object with a visible transient atmosphere—the coma or tail. If there is no atmosphere, a small body is called an asteroid. This practical definition reflects primarily a difference in composition between comets and asteroids: comets contain water ice and other volatiles that vaporize when heated by the sun, while asteroids are rocky objects with little volatile material.

All comets and most asteroids are *primitive bodies*, which have been relatively little altered chemically or physically since the formation of the solar system. They thus provide a window on our earliest history, the period before the planets formed. That is why they are of such great interest to scientists, in spite of their small sizes and insignificant contribution to the total mass of the solar system. Indeed, it is precisely because they are small that the asteroids have been able to preserve a record of the early chemical history of the solar system.

19.1 DISCOVERY AND ORBITS

(a) Discovery of the Asteroids

Most of the asteroids have orbits between those of Mars and Jupiter. From the time of Kepler it was recognized that this region of the solar system represented a gap in the spacing of planetary orbits. We have already seen that the Titius-Bode law predicts a planet at 2.8 AU from the sun (Table 17.5). After the discovery of Uranus seemed to confirm this "law," many astronomers felt there should be a concerted effort to locate this missing planet.

The view that there was another planet was not held by everyone. The noted philosopher Georg Wilhelm Hegel, for instance, argued that there could be only seven planets because, among other things, there are only seven openings in the human head. Nevertheless, in 1800 a methodical search for the missing planet was organized by Baron Francis Xavier von Zach. The plan was to divide the zodiac—that band around the sky centered on the ecliptic and through which the planets' orbits lie—into 24 sections and to assign each section to a different astronomer to search. One of these was the Sicilian astronomer Giuseppe Piazzi (1749–1826).

Piazzi, however, had not yet been informed of his role in the search and had not received the charts of the region of the zodiac he was to survey. He was, therefore, working independently on an entirely different project when, on January 1, 1801—the first night of the 19th century (1800 is the last year of the 18th century)—he discovered a new object not on his star charts. The next night he observed it again and thought that it had moved; on the third night he was sure of it. Piazzi, like Herschel in his discovery of Uranus, thought he had found a comet. He observed the new object regularly until February 11, when his work was interrupted by a severe illness.

On January 24, however, Piazzi had written to report his new comet to others, including Bode in Berlin. The news created a great deal of enthusiasm, and von Zach even published a report that the missing planet between Mars and Jupiter had been discovered! But by the time the astronomers in northern Europe were aware of the discovery, the new object was too close to the sun to be observed, and it would be September before it should again be visible.

By summer of 1801, Bode realized that the new planet was lost. Piazzi had observed it for only six weeks, too short a time to calculate its orbit by techniques then available. Fortunately, the discovery was saved by a 23-year-old genius from Brunswick, Germany: Karl Friedrich Gauss (1777–1855). Gauss was intrigued by an account of the discovery he had read in a newspaper. Putting his other work aside, he devised a new method of orbit calculation to deal with observations made over only a short arc of the total orbit. He finished his calculations in November 1801, and sent the results to von Zach.

Von Zach caught a glimpse of a possible object on December 7, but bad weather prevented him from confirming the observation until December 31, the last night of the year of the object's discovery; there it was, in the constellation of Virgo, almost precisely where Gauss had predicted it would be. At Piazzi's request, the object was named Ceres, for the Roman goddess of agriculture and protecting goddess of Sicily. The fame earned by Gauss for his efforts in the problem eventually led, in 1807, to his appointment as director of the Goettingen Observatory, where he remained the rest of his life.

Ceres was widely assumed to be the missing planet predicted by the Titius-Bode law. It came as a complete surprise, therefore, when in March 1802 Heinrich Olbers discovered a second moving star-like object—the minor planet to be named Pallas. It was a natural speculation that if there was room for two minor planets, there could be room for others as well, and a search for such objects began in earnest. The discovery of Juno followed in 1804 and of Vesta in 1807. But it was 1845 before Karl Hencke discovered the fifth minor planet, after 15 years of search. Subsequently, new ones were found with increasing frequency, until by 1890 more than 300 were known.

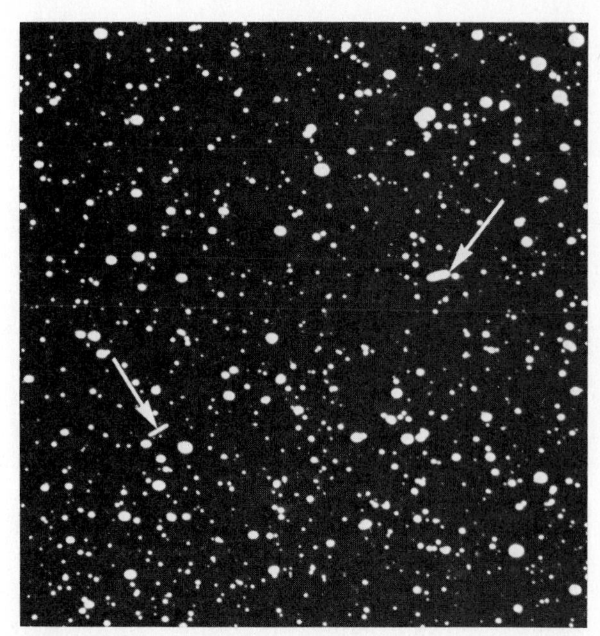

Figure 19.1 Time exposure showing trails left by two minor planets (marked by arrows). (Yerkes Observatory)

In 1891 Max Wolf (1863–1932) of Heidelberg introduced the technique of astronomical photography as a means of searching for asteroids. The angular motion of a minor planet is large enough (especially if it is near opposition) so that during a long time exposure its image will form a trail. The object appears on the photograph, therefore, as a short dash rather than a star-like point image (Figure 19.1). Brucia, the 323rd asteroid to be discovered, was the first to be found photographically.

Today discoveries of asteroids are usually accidental; they most often occur when the tiny objects leave their trails on photographs that are taken for other purposes. Literally thousands of minor planet trails appear on the photographs taken for systematic surveys of the sky, such as those with the

1.2-m Schmidt telescopes at Palomar and at Siding Spring, Australia. The majority of these trails are of objects that have never been catalogued. Most of them have been ignored, however, and the searches carried out today usually focus on special groups of asteroids, such as those in orbits that bring them close to the Earth. More than 3000 asteroids have well-determined orbits at this writing.

(b) Naming the Asteroids

By modern custom, after a newly found asteroid has had its orbit calculated, and has been observed again after another circuit of the sun since its first discovery, it is given a name and a number. The number is a running index that indicates the order of discovery among the minor planets. The discoverer is customarily given the honor of supplying the name. The full designation of the asteroid contains both the number and the name, thus: 1 Ceres, 2 Pallas, 433 Eros, 1566 Icarus, and so on.

Originally, the names were chosen from goddesses in Greek and Roman mythology. These, however, were soon used up. After exhausting the names of heroines from Wagnerian operas, discoverers chose names of wives, friends, flowers, cities, colleges, pets, and even favorite desserts. One is named for a computer (NORC). The 1000th minor planet to be discovered was named Piazzia, and number 1001 is Gaussia. Other asteroids bear the names Washingtonia, Hooveria, and Rockefellia. Asteroid 2410 is named Morrison, for one of the authors of this text.

(c) Statistical Studies

It would be a formidable task to discover, determine orbits for, and catalogue all the minor planets bright enough to be observed with modern telescopes. Nevertheless, the total number of such objects can be estimated by systematically sampling regions of the sky.

Several investigators have estimated the number of asteroids by photographing selected regions of the zodiac. The total number of asteroids bright enough to leave trails on photographs taken with an efficient instrument such as the Palomar Schmidt telescope is estimated to be about 100,000. Such a survey includes essentially all objects down to a diameter of about 1 kilometer.

In 1983 the Infrared Astronomical Satellite (IRAS) carried out an all-sky survey from orbit that was particularly sensitive to asteroids, which are strong emitters of thermal radiation. More than 10,000 individual observations were made, corresponding to perhaps 5,000 separate objects. The results, which are still being analyzed, should contribute significantly to our statistical knowledge of these objects.

The largest asteroid is 1 Ceres, with a diameter of just under 1000 km. Two have diameters near 500 km, and about 30 are larger than 200 km. The number of asteroids increases rapidly with decreasing size; there are nearly 1000 more objects 10 km across than the number of asteroids 100 km across.

It is estimated that our census of asteroids is 98 to 99 percent complete for objects down to 100 km diameter, and at least 50 percent complete down to 10 km. It is possible to estimate the total mass of the asteroids, which is about 1/2000 the mass of the Moon. Ceres alone accounts for nearly half of this total mass. If the asteroids represent surviving planetesimals from the early days of the solar system, they are only a very tiny remnant of the original population.

(d) Orbits of Asteroids

The asteroids all revolve about the sun in the same direction as the principal planets (from west to east), and most of them have orbits that lie near the plane of the Earth's orbit; the mean inclination of their orbits to the plane of the ecliptic is 10°. A few, however, have orbits inclined more than 25°; the orbit of 2102 Tantalus is the most inclined (64°) to the ecliptic. The *main asteroid belt* contains minor planets with semimajor axes in the range 2.2 to 3.3 AU, with corresponding periods of orbital revolution about the sun from 3.3 to 6 years. The mean value of the eccentricities of the main belt asteroid orbits is 0.15, not much greater than the average for the orbits of the planets.

Some asteroids have orbits rather far outside the main asteroid belt. A few with semimajor axes around 1.9 to 2.0 AU are called the Hungarias (for the prototype 434 Hungaria), and a few with larger orbits, near 4 AU, are called the Hildas (for 153 Hilda). There are asteroids with even more extreme orbits, some of which cross the orbit of the Earth, and a few that cross the orbit of Jupiter. We shall return to these objects later.

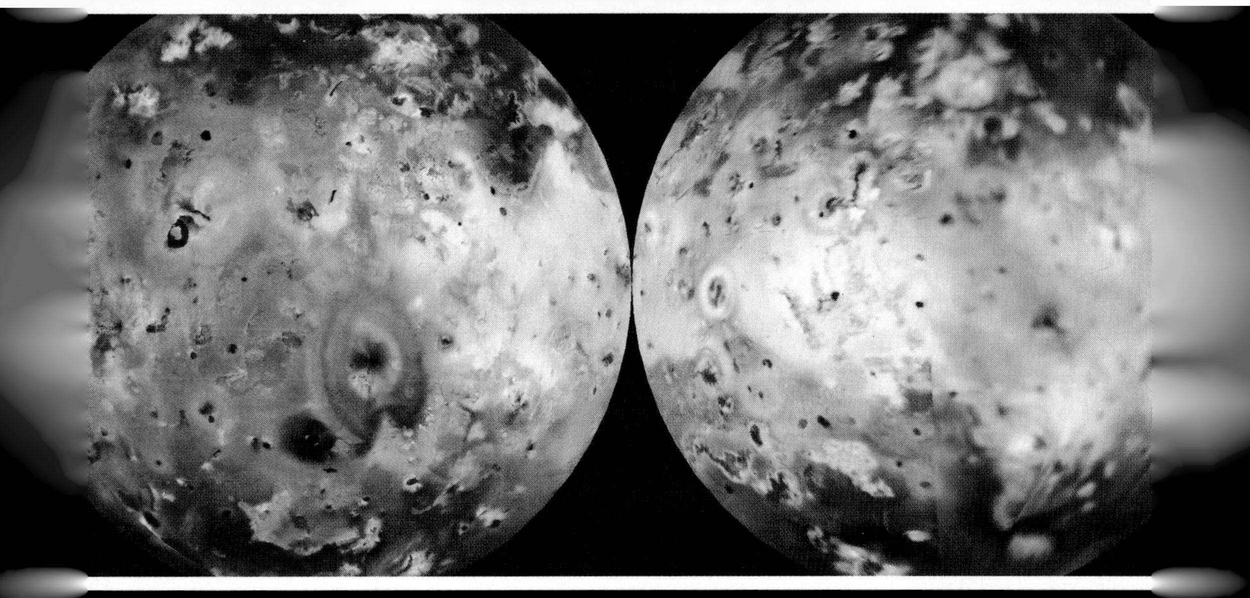

PLATE 17a Photomap of Io reconstructed from Voyager 1 images. (U.S. Geological
Courtesy Alfred McEwen, Arizona State University)

COLOR PLATE 17b Numerous volcanic features in the northern hemisphere of Io. (U.S.
Geological Survey; Courtesy Alfred McEwen, Arizona State University)

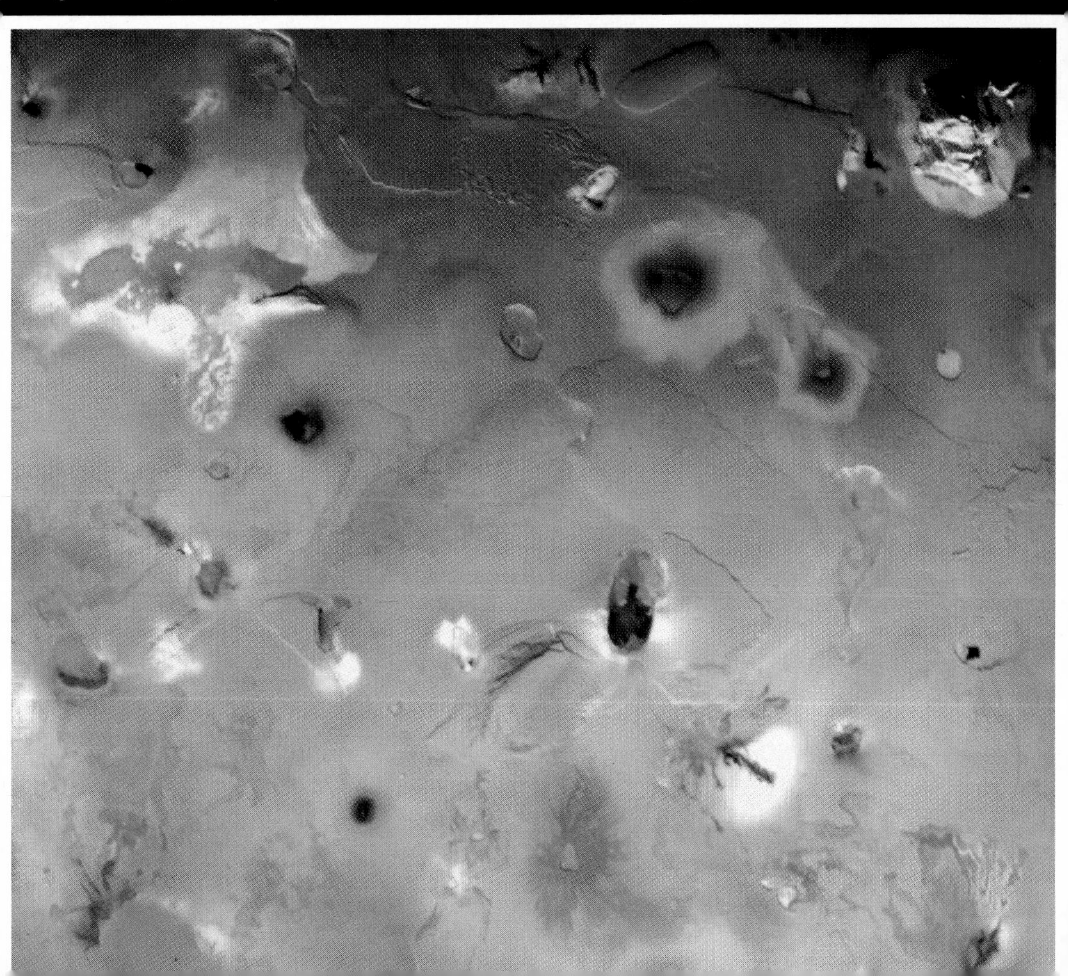

COLOR PLATE 18a The Pele volcano of Io, with a part of its erupting plume projecting above the limb of the satellite. (U.S. Geological Survey; Courtesy Alfred McEwen, Arizona State University)

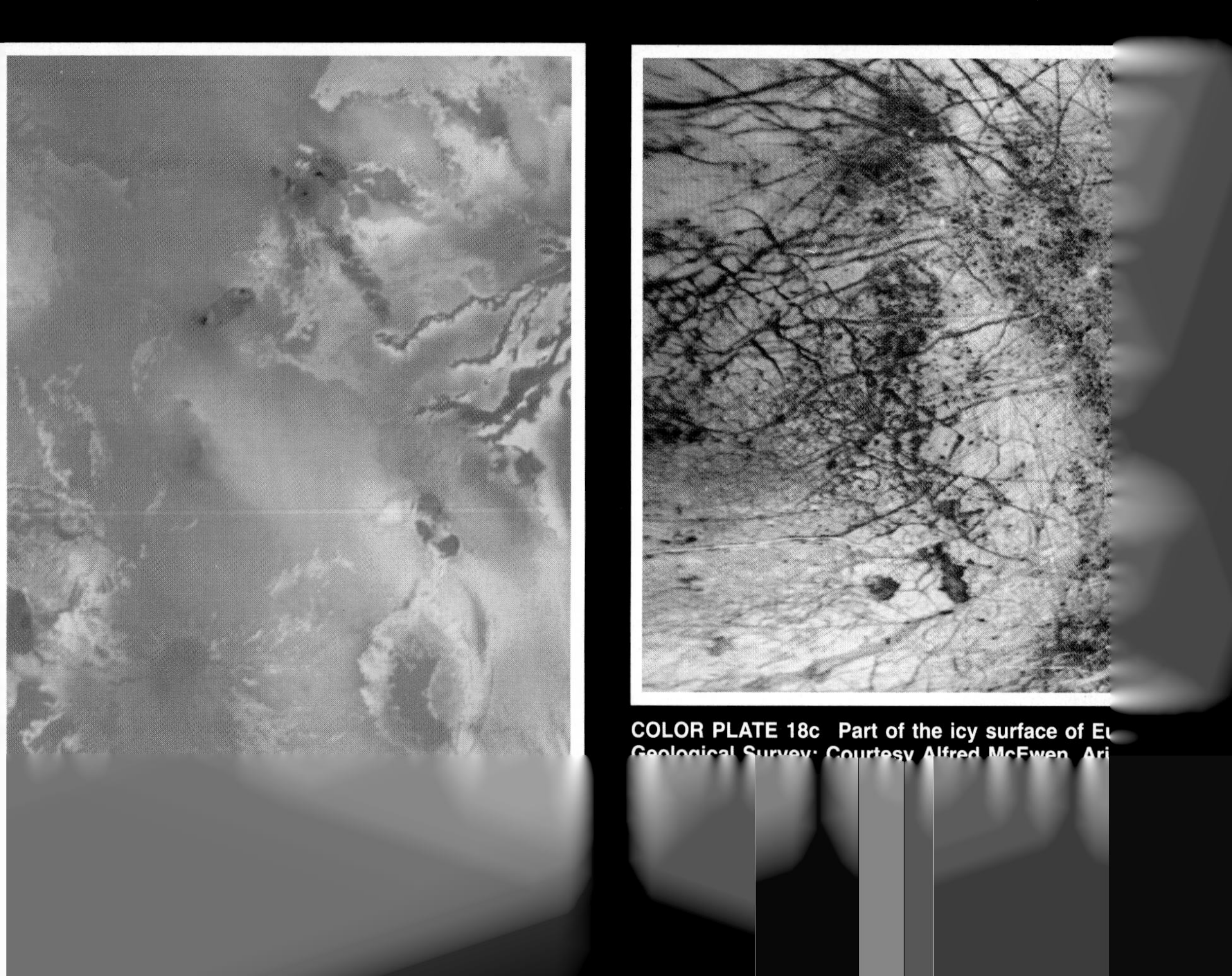

COLOR PLATE 18c Part of the icy surface of Eu
Geological Survey; Courtesy Alfred McEwen, Ari

COLOR PLATE 19a Comet Ikeya-Seki, photographed by J.B. Irwin in Chile, October 1965.

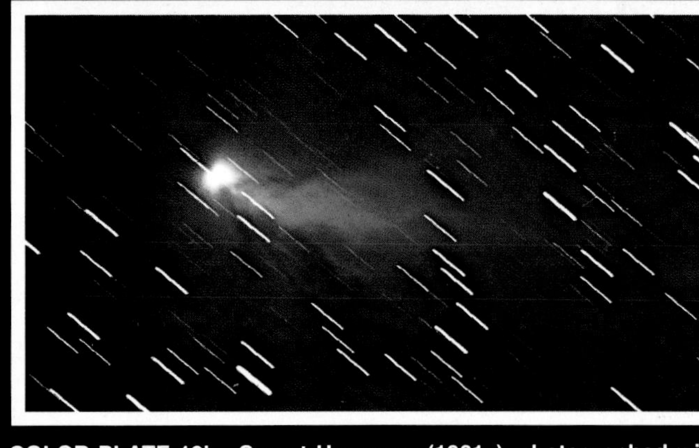

COLOR PLATE 19b Comet Humason (1961a), photographed with the Palomar Schmidt telescope. (California Institute of Technology/Palomar Observatory)

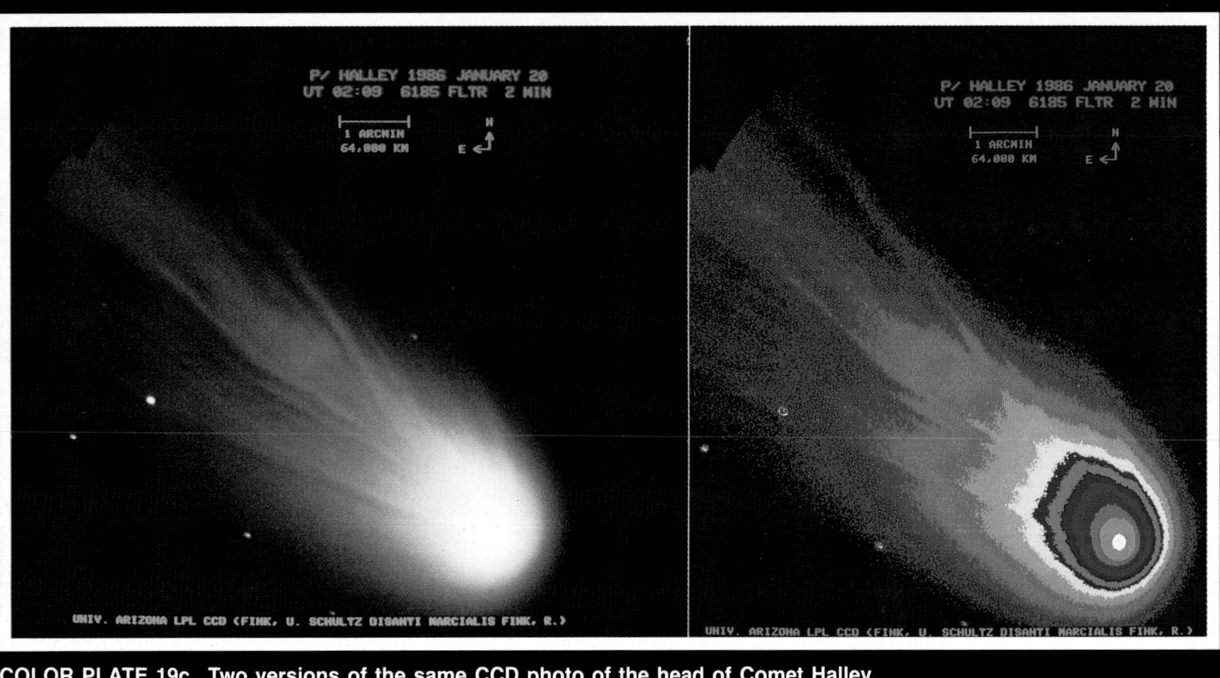

COLOR PLATE 19c Two versions of the same CCD photo of the head of Comet Halley, taken January 20, 1986, in Tucson. False color is frequently used in modern image processing to display the wide range in brightness that is recorded in images obtained with CCD cameras. (University of Arizona; Courtesy Uwe Fink)

COLOR PLATE 20a Collecting an iron meteorite from the Antarctic ice. (NASA)

COLOR PLATE 20b Antarctic meteorite ALHA81005, which is actually a fragment of lunar material ejected from the Moon in an ancient impact. (NASA)

COLOR PLATE 20c Antarctic meteorite EETA79001, one of the SNC basaltic meteorites, is widely suspected to be a similar ejection fragment from the surface of Mars. (NASA)

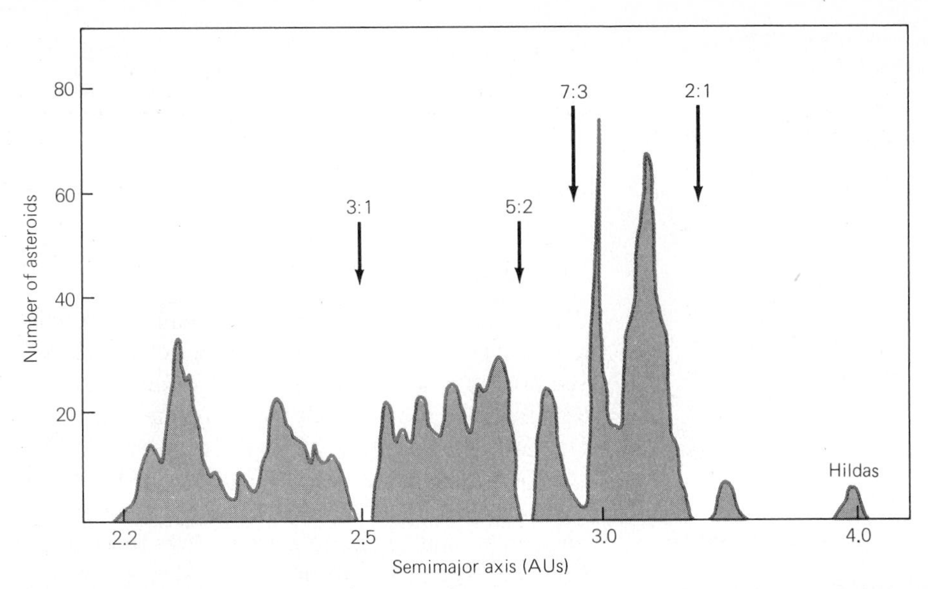

Figure 19.2 The distribution of the number of asteroids at various distances from the sun in astronomical units. Some of the resonances are indicated at gaps (Kirkwood's gaps) where the period of an asteroid would be a simple fraction of the period of Jupiter. For example, an asteroid at about 2.5 AU from the sun would have a period of one-third that of Jupiter's.

An interesting characteristic in the distribution of asteroid orbits is the existence of several clear areas or gaps (Figure 19.2), corresponding to orbital periods that the asteroids seem to avoid. These unoccupied periods were interpreted in 1866 by the American astronomer Daniel Kirkwood (1814–1895) as a resonance phenomenon caused by gravitational perturbations from Jupiter. For example, a minor planet at about ⅝ of Jupiter's distance from the sun (3.3 AU) would have a period exactly half that of Jupiter, and every two times around the sun it would find itself relatively near the planet. The repeated attractions toward Jupiter, always in the same direction, would eventually perturb the orbit of such an asteroid, just as resonances with satellites produce gaps in the rings of Saturn (Section 18.5d).

In 1917 the Japanese astronomer K. Hirayama found that a number of the asteroids fall into *families* or groups with similar orbital characteristics. He hypothesized that each family may have resulted from an explosion of a larger body or from the collision of two bodies. Slight differences in the initial velocities given the fragments account for the small spread in orbital characteristics now observed for the different asteroids in a given family. There are several dozen such families, and for the larger ones observations have shown that their individual members are physically similar, as if they were fragments of a common parent. The existence of these families testifies to the frequency of asteroid collisions in the past.

(e) Spacing and Collisions

The asteroids in the main belt have the kind of distribution of sizes that characterizes a population of fragments. With the exception of the largest objects (diameters greater than about 100 km), therefore, the asteroids are probably mostly broken or shattered remnants of larger *parent bodies.* The Hirayama families are the most obvious result of such collisions, but the majority of the asteroids not in families may also have been involved in collisions further in the past. It has also been suggested by some that a very large collision or group of collisions in the asteroid belt about 4 billion years ago might have been responsible for the late heavy bombardment experienced by the Moon and the other terrestrial bodies.

It is easy to estimate the spacing of the asteroids today and to calculate how often collisions take place. We have seen that there are about 100,000 objects in the belt with diameters of 1 km or more. These asteroids occupy a torus about 1 AU across and 0.1 AU thick, corresponding to a volume of:

$$v = \text{circumference} \times \text{width} \times \text{height}$$
$$v = 1\ \text{AU}^3 = 3 \times 10^{24}\ \text{km}^3$$

Dividing by the number of asteroids gives an average of one asteroid per $3 \times 10^{19}\ \text{km}^3$. The spacing between objects is found by taking the cube root, yielding about 3 million km as the typical asteroid separation in the main belt.

We can now calculate the frequency of collisions. Consider an asteroid with a diameter of 5 km. Observations have shown that the typical *relative* speed of one asteroid with respect to others it approaches is 4 km/s. The cross section of our typical asteroid is πr^2, or about 75 km^2. The volume of space it moves through in a given time is its cross section times its speed:

$$\text{volume swept out} = 300\ \text{km}^3/\text{s} = 1 \times 10^{10}\ \text{km}^3/\text{yr}.$$

To find the frequency with which this object collides with another, we compare this volume of space swept out in a year with the volume of space associated with each asteroid, which we calculated above:

$$\frac{\text{volume per object}}{\text{volume swept out}} = \frac{3 \times 10^{19}}{1 \times 10^{10}} = 3 \times 10^9\ \text{years}.$$

This result tells us that an individual 5-km asteroid can expect about one collision in 3 billion years. Thus in the 4.6-billion-year lifetime of the solar system, nearly every object will have had at least one such collision, but few will have had a large number of them. This calculation supports our suggestion that most of the smaller asteroids are collisional fragments.

In a similar way we can estimate the probability that a spacecraft will impact an asteroid while crossing the belt. The spacecraft cross section is about 3 m^2 and its speed is typically 20 km/s with respect to the asteroids. If the craft takes six months to cross the belt, the volume it sweeps out is about 3000 km^3. Comparing this volume to the typical volume associated with each asteroid of 3×10^{19}, we get the a probability of collision:

$$\text{probability} = \frac{3 \times 10^3}{3 \times 10^{19}} = 10^{-16}.$$

Thus there is only one chance in 10 million billion that a randomly aimed craft will collide with an asteroid of 1 km or larger diameter.

Finally, we can use this same kind of arithmetic to calculate how often a collision takes place somewhere in the belt. This is just the average time between collisions for any one asteroid divided by the number of asteroids, or:

$$\text{time between collisions} = \frac{3 \times 10^9}{1 \times 10^5} = 30{,}000\ \text{years}.$$

19.2 THE ASTEROID BELT

(a) Geography of the Belt

The main asteroid belt stretches from its inner edge at 2.2 AU out to a rather sharp cutoff at 3.3 AU, the 2:1 resonance with Jupiter. Probably more than 90 percent of the asteroids are in the main belt. The belt is divided by the Kirkwood gaps into several subregions, separated by resonances with Jupiter. It should be realized, however, that these "gaps" do not correspond to empty lanes such as those in the rings of Saturn, but to missing *periods*. Because of their substantial orbital eccentricity, the physical distribution of asteroids across the belt is fairly uniform.

Although there are probably about 100,000 belt asteroids larger than 1 km diameter, the asteroids are not closely spaced. The volume of the belt is actually very large, and typical spacing between objects is about 3 million km (Section 19.1e).

Not all asteroids are alike; in fact, their histories are surprisingly varied. Most of them are primitive objects, composed of silicates mixed with a variety of dark carbon-bearing compounds and with metallic grains. These appear to be unaltered condensates from the solar nebula. Other asteroids, however, have undergone an extensive thermal evolution, including examples of objects that fully differentiated. For the most part, the more primitive objects are in the outer part of the belt, while the more rare evolved objects are closer to the sun, but there is considerable mixing of these populations. Before describing them in further detail, however, it is appropriate to look at how we measure the sizes and chemical properties of such small objects.

(b) Asteroid Sizes

There are several methods of determining the sizes of minor planets. A few are near and large enough to present measurable disks in the telescope; in those cases we can, at least in principle, measure their true (linear) diameters from the measurements of their angular diameters. The largest such angular

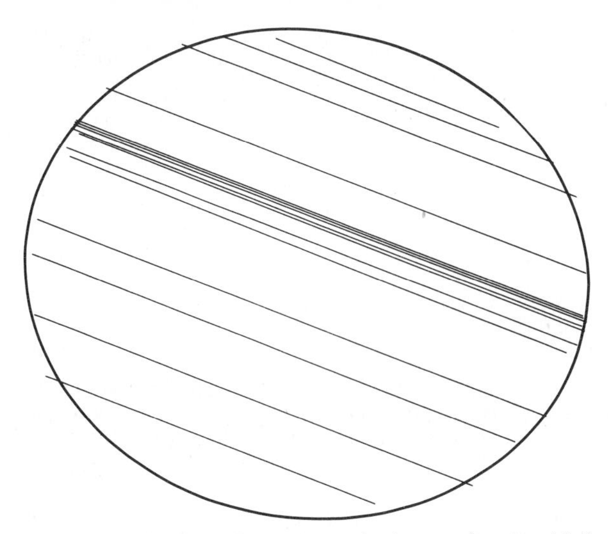

Figure 19.3 The diameter and shape of asteroid 3 Juno as determined from occultation measurements. The straight line segments represent the duration of the occultation timed from different observing sites. (R. Millis, Lowell Observatory)

TABLE 19.1 The Largest Asteroids

Name	Diameter (km)	Semimajor Axis of Orbit (AU)	Class
1 Ceres	940	2.77	C
2 Pallas	540	2.77	*
4 Vesta	510	3.36	*
10 Hygeia	410	3.14	C
704 Interamnia	310	3.06	C
511 Davida	310	3.18	C
65 Cybele	280	3.43	C
52 Europa	280	3.10	C
87 Sylvia	275	3.48	C
3 Juno	265	2.67	S
16 Psyche	265	2.92	M**
451 Patientia	260	3.06	C
31 Euphrosyne	250	3.15	C
15 Eunomia	240	2.64	S
324 Bamberga	235	2.68	C
107 Camilla	230	3.49	C
532 Herculina	230	2.77	S
48 Doris	225	3.11	C
29 Amphitrite	225	2.55	S

* Two of the largest asteroids do not fit into the common C or S classes. Pallas has a dark surface but shows some evidence of modification by heating. The surface of Vesta is covered with basalt from a time when this asteroid was volcanically active.

** The M class is apparently metallic; Psyche is the largest such asteroid.

diameters are less than 1 arcsecond, however, so the results are not reliable. Some improvement is possible with the technique of speckle interferometry (Section 9.4), and it has been tried on a few asteroids.

The most accurate technique for measuring the size of a minor planet is timing how long it takes to pass in front of a star. Several of the larger minor planets occult stars from time to time, and considerable effort has been made to coordinate observations over a large geographical area to observe such occultations. From different places on Earth, different parts, or chords, of an asteroid will appear to pass in front of the same remote star because the various observers see the asteroid in slightly different directions compared with the direction to the star. In effect, the asteroid casts its own moving shadow on the Earth. The combination of many observations of an occultation permits an accurate determination of the size and shape of the asteroid (Figure 19.3). About a dozen asteroid diameters have been measured by this technique, and for three, Ceres, Pallas, and Juno, values so derived are accurate to within a few percent.

The method that works best to estimate the sizes of large numbers of asteroids is to compare their brightnesses in visible light (which is reflected sunlight) with the light they emit in the infrared, which energy, of course, they have previously absorbed from the sun. For a particular asteroid, we know its distance from the sun and therefore how much sunlight falls on each area of its surface. The total sunlight it intercepts is equal to its cross-sectional area times the incident flux. Of that intercepted light, the asteroid reflects part, a fraction A (the *albedo*), and hence absorbs the rest, the fraction $1-A$. The asteroid re-emits the energy it absorbs at infrared wavelengths appropriate to its temperature. If we measure the infrared radiation coming from the minor planet, we are recording how much energy it must have absorbed from the sunlight falling upon it. If we compare this measure with that of the light reflected from the object, we find what its albedo—reflecting power—is. Then we can easily calculate what its size must be to account for the amount of light it reflects to us, that is, its observed brightness.

Table 19.1 gives the diameters and semimajor axes of the largest asteroids. Also listed is the compositional type, discussed below.

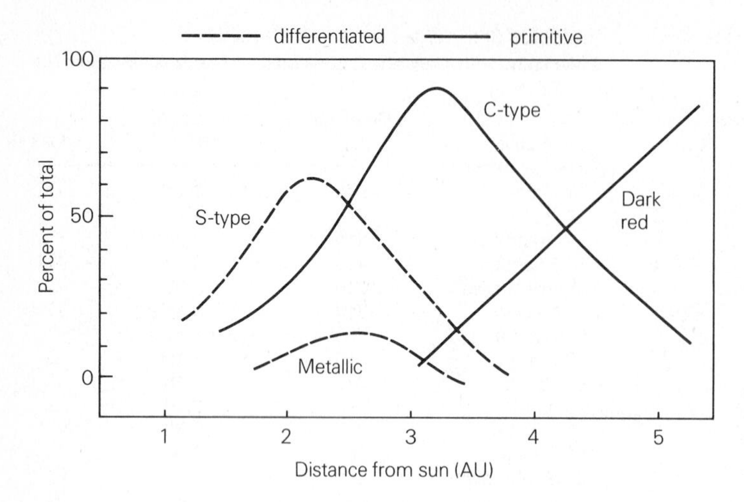

Figure 19.4 The distribution of asteroid compositional types with distance from the sun. (Adapted from work carried out by D. Tholen, J. Gradie, and E. Tedesco)

(c) Composition and Classification

When the reflectivities or albedos of many asteroids are compared, a great diversity is seen. The majority of minor planets are very dark, with reflectivities of only 3 or 4 percent, about as dark as a lump of coal. However, there is another large group with typical reflectivities of about 15 percent, a little brighter than the Moon, and still others with reflectivities as high as 60 percent. From these reflectivity measurements alone we can see that there are different kinds of asteroids, but further classification requires spectral information as well.

Spectroscopy is of limited use in the identification of the chemistry of solids and liquids. Almost all gases have their own unique pattern of sharp spectral lines that can be used for the analyses of starlight, for example. But solids tend to have only a few broad *spectral bands,* and their interpretation is often ambiguous. The common ices (H_2O, CO_2, NH_3, CH_4) are easy to distinguish by their infrared spectra, but the more complex silicate minerals that make up the surfaces of asteroids present the astronomer with a difficult challenge. In spite of these problems, however, spectra have been used together with reflectivity measurements for a compositional classification of the asteroids, beginning with the work of Thomas B. McCord and Clark R. Chapman in the early 1970s.

The majority of the asteroids are revealed from these studies to be *primitive bodies,* composed of silicates mixed with dark organic carbon compounds, whose presence reduces the asteroids' reflectivities to the 3 to 4 percent level observed. Many of these objects also include some water chemically bonded to the silicates. Two of the largest asteroids, 1 Ceres and 2 Pallas, are primitive, as are almost all of the objects in the outer third of the belt. Most of the primitive belt asteroids are classed *C asteroids,* where the C stands for "carbonaceous," but recently several other classes of primitive objects with different silicate minerals have also been identified.

The second most populous asteroid group is the *S asteroids,* where S stands for "silicaceous." In these asteroids the dark carbon compounds are missing, resulting in higher reflectivities and clearer spectral signatures of silicate minerals. The minerals present are common ones, similar to those that make up many meteorites, but the exact compositions of the S asteroids remain in dispute. In particular, we are unable to answer the basic question of whether these asteroids are primitive bodies or whether they are differentiated. Most workers suspect that they are primitive, representing material that condensed in the inner part of the asteroid belt where dark carbonaceous materials were not present, but we really do not know if this is the case.

Spectral observations have identified a few asteroids, not more than 5 percent of the total, that are clearly differentiated objects. These include asteroids suspected to be made largely of metal (presumably the cores of differentiated parent bodies that were shattered in collisions), and asteroids that have basaltic surfaces like the volcanic plains of the Moon and Mars. The large asteroid 4 Vesta is in the latter category. Apparently some of the asteroids were heated early in the history of the solar system, but why these, and why only a small percentage of the total number, we do not know.

TABLE 19.2 Densities of Asteroids

NAME	DIAMETER (km)	DENSITY (g/cm^3)
1 Ceres	940	2.7 ± 0.2
2 Pallas	540	2.8 ± 0.3
4 Vesta	510	3.6 ± 0.4
Phobos	24	2.0 ± 0.3
Deimos	13	2.0 ± 0.3

Detailed surveys of the population of the belt have revealed that the different compositional classes are at different distances from the sun. Thus we may think of the belt as made up of overlapping rings of similar kinds of objects, with each ring having a width of about half an AU (Figure 19.4). The presence of this structure tells us that the asteroids must still be in approximately the positions where they formed; the belt has not been totally mixed and homogenized. Therefore, if we could sample the compositions of the different kinds of primitive asteroids, we could map out in some detail the nature of the solar nebula out of which they formed.

Measurements of reflected sunlight provide compositional information on asteroid surfaces only. If densities were also known, we could draw some conclusions concerning their internal properties. Since asteroids do not have satellites, however, and no spacecraft have yet visited them, their masses are very difficult to determine. Only for the largest—Ceres, Pallas, and Vesta—have the masses been estimated from the perturbations they produce on other asteroids. Table 19.2 lists the best values now available, with the satellites of Mars, Phobos and Deimos, listed for comparison. The densities of Ceres and Pallas are about what one would expect for primitive bodies with little or no water ice. Vesta apparently has a higher density, presumably the result of loss of volatiles when it differentiated.

(d) Vesta: A Volcanic Asteroid

Minor planet 4 Vesta is one of the most interesting of the asteroids. Vesta orbits the sun with a semimajor axis of 2.4 AU, and its relatively high reflectivity of almost 30 percent makes it the brightest of the main belt objects, clearly visible to the unaided eye if you know just where to look. But its real claim to fame is the fact that its surface is covered with basalt, indicating that Vesta was once volcanically active in spite of its small size (diameter about 500 km). At this writing, Vesta is the only asteroid with a clearly identified basaltic surface, although a couple of very small objects (less than 10 km diameter) may be similar.

Figure 19.5 Photograph of one of the eucrite meteorites, believed to be fragments from the crust of asteroid 4 Vesta. (NASA)

Greatly adding to the importance of Vesta is the fact that we apparently have samples of its surface to study directly in the laboratory. Meteorites (Chapter 21) have long been suspected to come from the asteroids, but there is generally no way to identify the particular source of a given meteorite that strikes the Earth. In Vesta's case, however, this identification seems fairly firm.

The meteorites that are believed to come from Vesta are called the *eucrites,* a group of about 30 basaltic meteorites of very similar composition (Figure 19.5). Chemical analysis of the eucrites has shown that they cannot have come from the Earth, Moon, or Mars. On the other hand, their spectra (measured in the laboratory) match perfectly the spectra of Vesta obtained telescopically. The age of the lava flows from which the eucrites derived has been measured at 4.5 billion years, just about 100 million years after the formation of the solar system. This age is consistent with what we might expect for Vesta; whatever process heated such a small object was probably intense and short-lived.

19.3 OTHER ASTEROIDS

(a) The Trojans

The *Trojan asteroids* are objects located far beyond the main belt, orbiting the sun at about the same

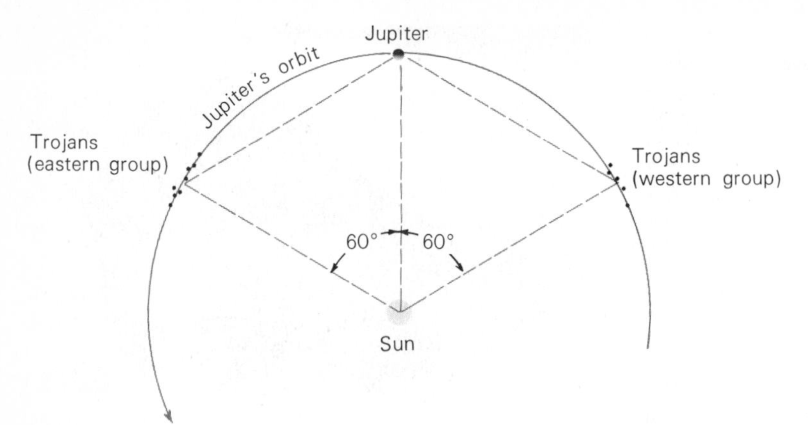

Figure 19.6 Locations of the Trojan asteroids near the Lagrangian points of the orbit of Jupiter.

distance as Jupiter, 5.2 AU. Lagrange's solution to a particular case of three bodies (Section 5.2) shows that there should be two points in the orbit of Jupiter near which an asteroid can remain almost indefinitely. These are the two points which, with Jupiter and the sun, make equilateral triangles (Figure 19.6). Between 1906 and 1908, four such asteroids were found; the number has now increased to about two dozen. These asteroids are named for the Homeric heroes from the *Iliad* and are collectively called the Trojans. Those that are preceding Jupiter (that is, ahead of it in its orbit) are named for the Greek heroes (with the exception of the Trojan spy, 624 Hektor), and those following Jupiter are named for the Trojan warriors (with the Greek spy, 617 Patroclus, among them).

To a first approximation, the Trojan asteroids circle the sun with Jupiter's period of 12 years, one-sixth of a cycle ahead of or behind the planet. Their detailed motion, however, is very complicated; they slowly oscillate around the points of stability found by Lagrange, some of their oscillations taking as long as 140 years.

Measurements of the reflectivities and spectra of the Trojans show that they are all very dark, presumably primitive objects like those in the outer part of the asteroid belt. They appear faint because they are so dark and far away, but actually the larger Trojans are quite sizable. Four of them — Hektor, Diomedes, Agamemnon, and Patroclus — have diameters between 150 and 200 km. Hektor is about twice as long as it is wide, leading to the suggestion that it is a *double asteroid,* with two similar objects orbiting in contact with each other.

A recent search for Trojan asteroids has revealed that there are actually many of them—far more than have been catalogued. Current estimates are that there must be close to 700 Trojan asteroids in the region preceding Jupiter (the Greek camp) that are at least 15 km in diameter and about 200 in the region following Jupiter (the Trojan camp).

(b) Hidalgo and Chiron

There are two asteroids known with orbits that carry them far beyond Jupiter. In 1920 Walter Baade of Mt. Wilson Observatory discovered 944 Hidalgo. With its semimajor axis of 5.85 AU and a very large eccentricity of 0.66, Hidalgo has an aphelion outside the orbit of Saturn.

The most distant asteroid was found in 1977 by Charles Kowal of Caltech. This object, 2060 Chiron, has a semimajor axis of 13.7 AU; its orbit (eccentricity 0.38) carries it from just inside the orbit of Saturn at perihelion out almost to the orbit of Uranus. Its diameter is greater than 100 km, and may be as large as 300 km if Chiron is a dark object, as some scientists suspect (Figure 19.7).

When we look at Hidalgo and Chiron we face the question of the relationship between such distant asteroids and the comets. Neither of these objects comes close enough to the sun to develop a cometary atmosphere and tail, even if it is made of the same volatile-rich materials as the comets. Probably, if they were diverted into the inner solar system, we would find them transformed into spectacular comets.

(c) Earth-Approaching Asteroids

There is an important class of asteroids with orbits that either come close to or cross the orbit of the

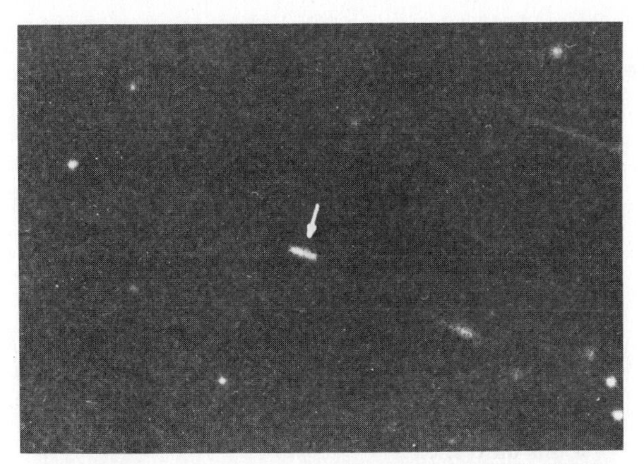

Figure 19.7 The most remote known asteroid, Chiron, photographed on 19 October 1977 with the Palomar Schmidt telescope. The bright streak in the center (marked with arrow) is the trail left by Chiron during the 75-minute time exposure. The faint, irregular streak below and to the right of Chiron is an edge-on galaxy. At the upper right is a faint trail of a "normal" asteroid. (Caltech/Palomar Observatory)

Earth. Some of these are the nearest approaching celestial objects, save the Moon and meteorites. A few even collide with the Earth at intervals of millions of years, producing major impact craters. These are known collectively as *Earth-approaching asteroids*.

The Earth-approaching asteroids are divided into three groups on the basis of their orbits. The innermost are the *Atens,* named for 2062 Aten, which have orbits with semimajor axes less than 1.0 AU. The best-known group are the *Apollos* (for 1862 Apollo, discovered in 1948), which cross the orbit of the Earth (or nearly do so) but have semimajor axes greater than 1.0 AU. The Apollos include 1566 Icarus, which has its perihelion inside the orbit of Mercury, and 1620 Geographos, named for the National Geographic Society, which financed the Palomar Sky Survey on which this little world was discovered in 1951. The outer group are the *Amors* (for 1221 Amor), which are Mars-crossing asteroids with perihelion distances between 1.017 and 1.400 AU.

At present about 50 Earth-approaching asteroids have been located. The largest, 433 Eros, is about 20 km across, similar in size to the satellites of Mars; most of the known Earth-approaching asteroids, however, are only 1 or 2 km in diameter. Physically, they seem representative of the main belt asteroids, with several different compositional classes identified. Extensive searches for additional Earth-approaching asteroids under way at both the University of Arizona and Caltech are presently leading to the discovery of two or three new objects each year. Radar is also proving to be a powerful tool for investigating these small objects, particularly for determining their shapes. As befits small objects that are probably fragments of larger precursors, the Earth-approachers are very irregular. One, 1627 Ivar, is 2½ times longer than it is wide.

Eugene Shoemaker of the U.S. Geological Survey in Flagstaff leads one of the searches for new Earth-approaching asteroids and has carried out extensive calculations of the dynamics of the entire population. He estimates that there are between 1000 and 2000 Earth-approaching asteroids down to a diameter of ½ kilometer. About half of these are Apollos and half Amors, with only a few Atens.

A few of the Earth-approaching asteroids hold the distinction of being among the easiest objects in the solar system for round-trip travel from the Earth. In the best cases, the energy required for such a space flight is actually less than the amount needed for a round trip to the Moon. Therefore, these little objects may become targets for human exploration, if and when we are prepared to venture once again beyond low-Earth orbit.

As Shoemaker and others have shown, the orbits of Earth-approaching asteroids are unstable. These objects will meet one of two fates: either they will impact one of the terrestrial planets, or they will be ejected from the inner solar system as the result of a near encounter with a planet. The probabilities for these two outcomes are about the same. The time scale for impact or ejection is only about 100 million years, very short in comparison with the age of the solar system.

If the current population of Earth-approaching asteroids will be removed by impact or ejection in 100 million years, there must be a continuing source of new objects. Some of these apparently come from the main asteroid belt, where collisions can eject fragments into Earth-crossing orbits. Others may be dead comets that have exhausted their volatiles. Possibly as many as half of the Earth-approachers are the solid remnants of former comets. Recall that similar calculations (discussed in Section 14.3) suggest that about half the impacts on the terrestrial planets come today from comets and half from asteroids. We will discuss the fate of old comets in Chapter 20.

EXERCISES

1. Asteroids can be discovered by the trails they leave on astronomical photographs. Fainter asteroids could be recorded, however, if their images were points rather than trails. Explain how you might make a photographic search for very faint asteroids.

2. If you weigh 100 kg on the Earth, by what factor would your weight change on the surface of an asteroid that has a mass 1/10,000 that of the Earth and a diameter 1/20 that of the Earth?

3. Many small bodies, including the asteroids, follow a size distribution in which there are almost 1000 times more objects for each decrease of a factor of 10 in diameter. For this size distribution, calculate the numbers of objects and their total mass for diameters of 100, 10, 1, 0.1, and 0.01 km, starting with the fact that there are about 100 asteroids with diameters near 100 km. Then carry out the same calculation for a size distribution in which the number increases by a factor of only 100 for each decrease of a factor of 10 in diameter. Note the dramatic difference in the two cases in the number of small objects and in the total mass of the asteroids.

*4. Use the method given in Section 19.1e for finding the probability of a spacecraft colliding with an asteroid, together with the results from the preceding exercise, to estimate the chances of collision with objects of different size. Which size asteroid poses the main danger to a spacecraft crossing the belt?

5. Describe the main differences between C asteroids and S asteroids.

6. At times the Earth-approaching asteroid 433 Eros fluctuates in brightness as it rotates by a factor of about five. At other times its light fluctuates by a smaller amount. Can you offer an explanation for this difference?

7. If 4 Vesta is not the parent body of the eucrite meteorites, what might be an alternative source for these objects?

8. Compare the properties of the Martian satellites Phobos and Deimos (described in Chapter 15) with the asteroids as discussed in this chapter. Do you expect that most asteroids will look like Phobos and Deimos? Explain.

9. There is a great deal of interest today in the discovery of additional Earth-approaching asteroids. Can you think of several reasons for this high interest?

Fred L. Whipple (b. 1906) of Harvard University, formerly director of the Smithsonian Astrophysical Observatory, is one of the leading students of comets in the world. In 1950 he published his "dirty snowball" model of the cometary nucleus and provided the first explanation of the nongravitational forces that modify the orbits of comets—two contributions that still form the basis for our understanding of the nature of these celestial visitors.

20

COMETS: NATURE'S SPACE PROBES

Comets have been observed from the earliest times; accounts of spectacular comets are found in the histories of virtually all ancient civilizations. Yet, until comparatively recently, comets were not generally regarded as celestial objects. Moreover, they have had a special connotation in Western culture as harbingers of evil. John Milton, for example, characterized Satan as a comet that "from its horrid hair/Shakes pestilence and war."

A typical comet that is bright enough to be conspicuous to the unaided eye has the appearance of a rather faint, diffuse spot of light, somewhat smaller than the Moon and many times less brilliant. There may be a very faint nebulous tail, extending for a length of several degrees away from the main body of the comet. Like the Moon and planets, comets slowly shift their positions in the sky from night to night, remaining visible for periods that range from a few days to a few months. Unlike the planets, however, most comets appear at unpredictable times, perhaps explaining why more fear and superstition have been attached to them than to any other astronomical objects (Figure 20.1).

Today we recognize comets as the best preserved, most primitive material available in the solar system. Stored in the deep freeze of space, these icy objects are messengers from the distant past, providing us unique access to the initial condensates from the solar nebula.

20.1 DISCOVERY AND ORBITS

(a) Early Investigations

The first "modern" investigation of a comet, based on careful observation, was Tycho Brahe's diligent study of the brilliant comet that appeared in 1577 (Section 3.2). Had the comet been inside the Earth's atmosphere, as was then generally supposed, changes in its apparent direction would easily have been detectable to an observer who changed position by several kilometers. Brahe, not being able to detect any such changes, realized that the comet was a celestial object. Moreover, his failure to detect any *diurnal parallax*—the apparent change in the comet's position that should accompany the observer's motion as the Earth rotates—led him to the conclusion that the comet was at least three times as distant as the Moon and that it probably revolved around the sun.

Figure 20.1 Comet Halley in 1066, as depicted on the Bayeux tapestry. (Yerkes Observatory)

Kepler described in detail the comet of 1607 (later known as Comet Halley). He held the view that this and other comets are celestial bodies that travel in straight lines through the solar system.

When Newton applied his law of gravitation to the motions of the planets, he wondered whether comets might similarly be gravitationally accelerated by the sun. If so, their orbits should be conic sections. If comets, like planets, had nearly circular orbits, they would be visible at regular and frequent intervals. On the other hand, if a comet moved in an elongated elliptical orbit of large size, it would spend most of its time far from the sun and would be visible only rarely, near perihelion. Furthermore, the periods of comets moving in such orbits would be very long. A comet that had been seen for a short time, and then was seen again many tens or hundreds of years later when it next appeared near perihelion, would naturally be mistaken for a new object.

One end of an ellipse, of very great length and of eccentricity near 1.0, is nearly indistinguishable from a parabola; thus the motion of a comet moveing on a parabolic orbit would closely resemble that of a comet whose orbit is a very long ellipse. Newton concluded that comets are gravitationally attracted to the sun and move about it in orbits that are either very long ellipses or parabolas.

Edmund Halley (1656–1742), the British astronomer who oversaw the publication of Newton's *Principia*, greatly extended Newton's studies of the motions of comets (Figure 20.2). In 1705 he published calculations relating to 24 cometary orbits. In particular, he noted that the elements of the orbits of the bright comets of 1531, 1607, and 1682 were so similar that the three could well be the same comet, returning to perihelion at average intervals of 76 years. If so, he predicted that the object should return about 1758.

Alexis Clairaut calculated the perturbations that this comet should experience in passing near Jupiter

Figure 20.2 Edmund Halley. (Yerkes Observatory)

and Saturn and predicted that it should first make its appearance late in 1758 and should pass perihelion within about 30 days of April 13, 1759. The comet was first sighted on Christmas night, 1758, and it passed perihelion on March 12, 1759. The comet has been named Comet Halley, in honor of the man who first recognized it to be a permanent member of the solar system. Subsequent investigation has shown that Comet Halley has been observed and recorded on every passage near the sun at intervals from 74 to 79 years since 239 B.C. The period varies somewhat because of perturbations upon its orbit produced by the Jovian planets. In 1910 the Earth was brushed by the comet's tail (Figure 20.3), and Comet Halley last appeared in 1985/86, with perihelion passage on February 9, 1986.

Figure 20.3 Comet Halley in 1910. (Yerkes Observatory)

(b) Discovery and Nomenclature

Observational records exist for about a thousand comets. Today, new comets are discovered at an average rate of between five and ten per year. Some of these are found accidentally on astronomical photographs taken for other purposes. Many are discovered by amateur astronomers.

Most of the new comets found each year never become conspicuous and are visible only on photographs made with large telescopes. Every few years, however, a comet may appear that is bright enough to be seen easily with the unaided eye. In 1957 there were two conspicuous naked-eye comets, Comet Arend-Roland and Comet Mrkos. Other recent bright comets were Comet Bennett in 1970, Comet Kohoutek in 1973, and Comet West in 1976. About two or three times each century, there appear spectacular comets that reach naked-eye visibility even in daylight. The first "daylight comet" to appear since the brilliant comet of 1910 (which preceded the more famous Comet Halley by a few months), was Comet Ikeya-Seki in 1965.

A newly found comet is named for its discoverer or discoverers (there are often more than one). It is also given a temporary designation consisting of the year of its discovery followed by a lowercase letter indicating its order of discovery in that year. For example, Comet 1975a is the first comet discovered in 1975, 1975b is the second, and so on. Later, when the orbits of all recently observed comets have been calculated, each of the comets is given a permanent designation consisting of the year in which it passed nearest the sun, followed by a Roman numeral designating its order among the comets that passed perihelion during that particular year. For example, Comet 1975 I would be the first comet to pass perihelion in 1975, 1975 II, the second, and so on.

(c) Membership in the Solar System

Because the orbits of most comets appear to be nearly parabolic, and a few are even slightly hyperbolic, the question arises whether all comets are members of the solar system or whether some might be accidental intruders from interstellar space. The evidence is conclusive, however, that comets have always been members of the solar system.

If comets were intruders from interstellar space, their orbits should nearly all be markedly hyperbolic. An elliptical orbit is possible only for a body permanently revolving about the sun, not for

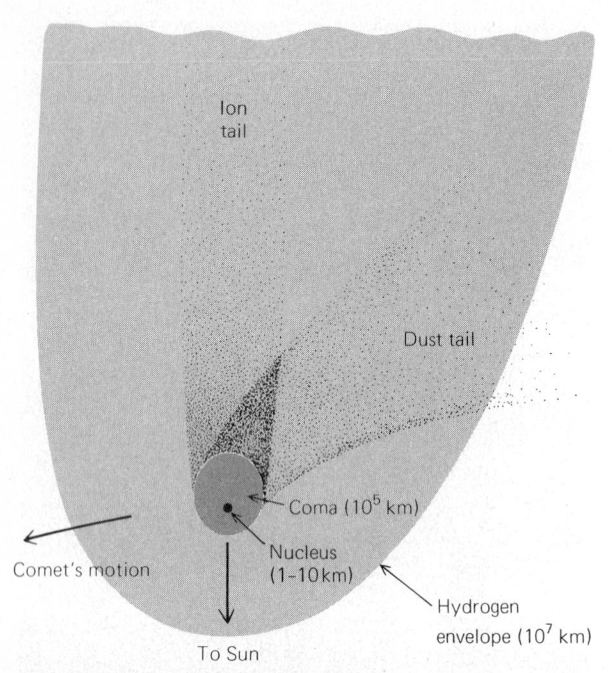

Figure 20.4 Parts of a typical comet.

one coming from outside the solar system. Those few comets that do appear to have had slightly hyperbolic orbits are believed to have approached the sun on long ellipses that were perturbed by Jupiter or another planet. Although a comet, through planetary perturbations, may thus *escape* from the solar system, there is no case where a comet is known to have *approached* the sun on a significantly hyperbolic orbit.

Moreover, the orbits of comets, unlike those of planets, are oriented at random in space. As many comets appear to approach the sun from one direction as from another. If comets were interstellar objects, there should be a preponderance of them approaching from the direction of the star Vega, toward which the sun is moving with a speed of about 20 km/s (see Chapter 22). It is generally accepted, therefore, that comets are members of the solar system, and that they approach the sun on elliptical orbits, most of which are extremely long. The aphelia of new comets are usually at about 50,000 AU from the sun, and their periods are a million years or more.

(d) Periodic Comets

A new comet, entering the inner solar system on a nearly parabolic orbit, is frequently perturbed by Jupiter into an orbit of smaller eccentricity and shorter period. Only those comets whose orbits are well determined and that differ rather substantially from parabolas are technically classed as *periodic comets*. The comets of longest definitely established periods are Pons-Brooks (71 years), Halley (76 years), and Rigollet (151 years). The comet of the shortest known period is Comet Encke (3.3 years).

There are about 50 known comets whose orbits are inclined at less than 45° to the ecliptic, that travel from west to east (like planets), and whose aphelion distances are near Jupiter's mean distance from the sun. They comprise the *Jupiter family* of comets, with periods that range from five to ten years. The Jupiter-family comets are objects that were strongly perturbed by Jupiter during one of their passes through the inner solar system. Because their orbits make them more accessible, almost all of the suggested targets for space missions to rendezvous (match orbits) have been Jupiter-family comets, including Kopff, Temple 2, and Wild 2.

20.2 THE PHYSICAL NATURE OF COMETS

No two comets are alike, but they all have (by definition) an extended atmosphere that forms the comet's *head*, appearing as a round, diffuse, nebulous glow. Many comets also develop long *tails* that stream away from the head. But the most important part of the comet is rarely seen: its small solid *nucleus* (Figure 20.4).

(a) The Nucleus

When we look at a comet we see its transient atmosphere of gas and dust, illuminated by sunlight. Since the escape velocity from such small bodies is very low, the atmosphere we see is rapidly escaping; therefore it must be coming from somewhere. The source is in the heart of the comet's head, usually hidden in the glow of its atmosphere; there is found the small solid *nucleus*. The nucleus is the *real* comet, the fragment of primitive material preserved since the beginning of the solar system.

Until about the past decade, no comet nucleus had been seen or measured; the very existence of the nucleus as a single solid object was speculative. Now, most of the studies being made of comets are directed toward a better understanding of the nucleus.

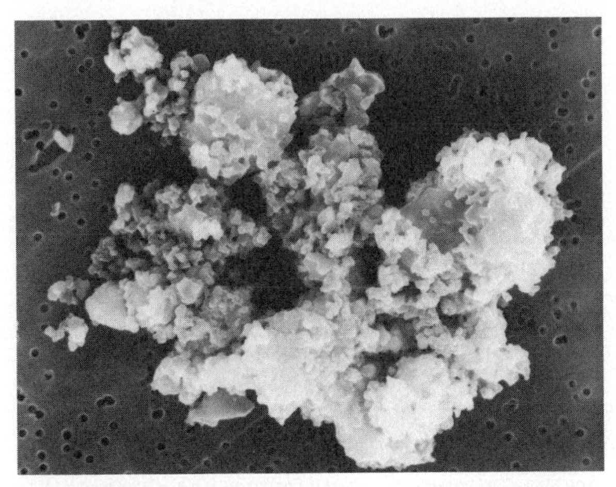

Figure 20.5 A *Brownlee particle,* believed to be a tiny fragment of cometary dust. (D. Brownlee, University of Washington)

The modern theory of the physical and chemical nature of comets was first proposed by Harvard astronomer Fred L. Whipple (b.1906) in a 1950 paper on Comet Encke. Before Whipple's work many persons had thought that the nucleus of a comet might be a loose agglomeration of solids of meteoritic nature—a sort of orbiting "gravel bank." Whipple proposed instead that the nucleus was a solid object a few kilometers across composed in substantial part of water ice, mixed with silicate grains and dust. This proposal became known as the "dirty snowball" model for the nucleus of a comet.

Later observations detected large quantities of water vapor and its dissociation products in cometary atmospheres, confirming the importance of this material. Other ices may also be present, including CO_2, CO, NH_3, and CH_4. These are, of course, exactly the same molecules that are expected to have condensed in low-temperature regions of the solar nebula, as well as the materials we have noted to be in the satellites of the outer planets.

We are somewhat less certain of the non-icy component of the nucleus, the material that when released contributes to the comet's dust tail. No fragments of solid matter from a comet have ever survived passage through the Earth's atmosphere to be studied as meteorites. Some very fine, microscopic grains have been collected in the Earth's upper atmosphere by Donald Brownlee of the University of Washington, however, and dust collectors have also flown in Earth orbit on the NASA Long-Duration Exposure Facility (Figure 20.5). The spacecraft that encountered Comet Halley in March 1986 also carried dust detectors. From these various investigations it seems that much of the "dirt" in the dirty snowball is in the form of tiny bits of dark, primitive hydrocarbons and silicates rather like the material thought to be present on the distant asteroids and dark satellite surfaces such as the trailing hemisphere of Saturn's Iapetus.

The presence of this dark carbon-rich material gives the comet nucleus a low reflectivity, particularly when the evaporation of ice concentrates the non-volatile material at the surface. The darkness of the nucleus makes it faint and contributes to the problem of studying it telescopically. A recent innovation in the study of the nucleus is to use radar to probe through the surrounding gas and dust, just as radar penetrates the clouds of Venus to yield images of its surface. Several comets have been studied by radar, which has also permitted direct measurement of the size of the nucleus.

The typical diameter derived by radar for the nucleus of several faint comets seems to be 2 to 5 km. The more active Comet Halley, as measured in 1986, has dimensions of about 12 by 8 km. The corresponding comet masses, although not measured directly, must be about 10^{-9} to 10^{-11} Earth masses.

(b) Activity on the Nucleus

The spectacular activity of comets that gives rise to their atmospheres and tails results from the evaporation of the cometary ices when they are heated by sunlight. Beyond the asteroid belt, where even short-period comets spend most of their time, the ices are tightly frozen. But as a comet approaches the sun, it begins to warm up. If water is the dominant ice, significant quantities will outgas as temperatures rise toward 200 K, somewhat beyond the orbit of Mars. The evaporating water in turn releases the dust that was mixed with the ice. Since the comet nucleus is so small, its gravity offers no resistance to either the gas or the dust, both of which flow away into space at speeds of about 1 km/s.

The comet continues to absorb energy from the sun as it approaches perihelion. More and more of this energy goes into the evaporation of its ice, however, rather than heating of the surface. If the surface of the nucleus were primarily made of water ice, its temperature would stabilize at about 230 K.

Figure 20.6 The nucleus of Comet Halley on March 14, 1986, photographed by the Giotto spacecraft. (Max Planck Institut für Aeronomie)

However, observations of many comets indicate that the evaporation is not uniform, and that most of the gas is released in sudden spurts, perhaps confined to a few areas of the surface. Such jets were observed directly on the surface of Comet Halley by the VEGA and Giotto spacecraft in 1986, where they resembled volcanic plumes or geysers (Figure 20.6). Most of the surface was apparently inactive, with the ice buried under a layer of black silicates and carbon compounds.

On Comet Halley, the same two major plumes were photographed by each of the three spacecraft, indicating that these centers of activity persisted with little change over a period of a week. However, we do not know if they represented deep-seated eruptive centers, or if perhaps the plume regions migrate from one part of the surface to another on time scales of several weeks or more. To answer such questions, it will be necessary to send a spacecraft to orbit with a comet and observe it for longer periods of time.

(c) Nongravitational Forces

The escaping jets of gas from an active comet can have an effect on its orbit. It was the problem of accounting for observed changes in cometary orbits that originally led Whipple to develop the dirty snowball model for the nucleus. His 1950 paper was primarily concerned with interpreting orbital variations seen in Comet Encke.

As the gas streams away from the nucleus of an active comet, it exerts a reaction force on the nucleus, just like a rocket engine. If the nucleus were very large, this force would not be important. Also, if the gas were escaping from all directions equally, all of the forces would cancel out. However, Whipple hypothesized that Comet Encke had a small nucleus rotating in a period of a few hours. The side of the nucleus facing the sun would be the warmest, and in particular the highest temperatures would be on the "afternoon" quadrant of the nucleus. The strongest forces would therefore be concentrated in one direction, and acting continuously over weeks they could measurably influence the comet's orbit. Such reaction effects of cometary activity are called *nongravitational forces*.

Whipple used the observed changes in the orbit of Comet Encke to estimate the size and rotation period of its nucleus. Subsequently he, Zdenik Sekanina of the Caltech Jet Propulsion Laboratory, and others have used similar reasoning to derive rotation periods and even orientations of the pole for a number of cometary nuclei. Comets, like asteroids, typically have rotation periods that range from a few hours to a few days. One of the longer periods is that associated with Comet Halley, which rotates in two days and five hours.

(d) The Cometary Atmosphere

The atmosphere of a comet is composed of the gas released from the nucleus together with the dust and other solid material being carried along with it. Expanding at a speed of about 1 km/s, the atmosphere can reach an enormous size. The diameter of the comet's head is usually as large as Jupiter, and it often approaches 1 million kilometers. The head of a comet is also called its *coma*.

When it is far from the sun, the comet's spectrum is simply that of sunlight reflected from the nucleus. The spectrum changes, however, when the comet comes within about 3 AU and its atmosphere begins to develop. Bright emission lines (actually, molecular bands—see Section 8.3h) of the molecules and radicals carbon (C_2), cyanogen (CN), and hydroxyl (OH) are generally present, and often NH, NH_2, and other molecules as well. If a comet comes very close to the sun, as did Ikeya-Seki in 1965, the temperature in its inner atmosphere can become high enough for bright emission lines produced by atoms of various metals to appear, including sodium, iron, silicon, and magnesium.

Ultraviolet light from the sun dissociates the molecules in the head of a comet to produce the *radicals*—molecular fragments—whose spectra we commonly observe. The *parent molecules*—the original molecules that were dissociated to form the radicals we actually measure—are presumed to be water (H_2O), methane (CH_4), and ammonia (NH_3). Direct spacecraft measurements of the inner atmosphere of Comet Halley revealed many additional gases, including hydrocarbons. Hydrogen is released in large quantities from comets, not only from the dissociation of the parent molecules (especially H_2O) into radicals, but eventually from the further dissociation of OH as well. Since its spectral features can only be seen from above the Earth's atmosphere, this hydrogen was not observed in comets until 1970, when it was detected around Comet Bennett with the Orbiting Astronomical Observatory. Generally the hydrogen cloud is at least ten times the size of the coma on normal photographs; Mariner 10 instruments detected radiation from hydrogen from a region around Comet Kohoutek at least 40 million km across. Observations of Comet Kohoutek also showed that water is dissociating into hydrogen and the OH radical within the inner 15,000 km around the nucleus, and that OH is further dissociating into hydrogen and oxygen beyond 45,000 km from the nucleus.

(e) Comet Tails

Many comets develop tails as they approach the sun (Figure 20.7). The tail of a comet is an extension of its atmosphere, consisting of the same gas and dust that make up the head. As early as the 16th century observers realized that comet tails point away from the sun, not back along the comet's orbit. Newton attempted to account for comet tails by a repulsive force of sunlight driving particles away from the comet's head, an idea that is close to our modern view.

Actually the primary repulsive force that causes tails of comets to point away from the sun is the solar wind. This outward-streaming plasma interacts with the plasma in the comet tail and carries it along at a speed of about 400 km/s. Decades before the solar wind was detected directly by spacecraft its existence was inferred by Ludwig Biermann and other astronomers from the behavior of comet tails. The solar wind interacts strongly only with ionized gas, however; other forces, primarily radiation pressure, are important in repelling dust grains, as Newton had suggested.

Figure 20.7 Comet Seki-Lines photographed from Frazier Mountain, California, August 9, 1962. (A. McClure, Los Angeles)

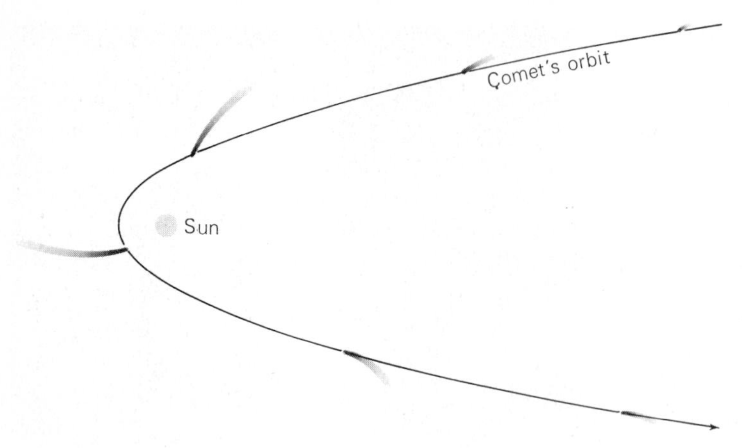

Figure 20.8 Shape of a typical comet tail as the comet passes perihelion.

Tails generally grow in size as comets near the sun; some comet tails have reached lengths of more than 150 million km. Once the tail material has left the vicinity of the comet's head, the only forces acting upon it are radial forces toward and away from the sun (gravity and the force of repulsion). As the material recedes farther from the sun, its angular velocity about the sun decreases, in accord with Kepler's laws, and the material lags behind the comet. In general, therefore, comet tails lie in the plane of the comet's orbit, pointing more or less away from the sun, but curving somewhat backward, away from the direction of the comet's motion (Figure 20.8).

The acceleration produced by the force of repulsion from the sun varies according to the size of the particles it acts upon and according to whether the particles are gas molecules or solid grains. Material of different types, therefore, is accelerated away from the sun by different amounts. If the repulsive force exceeds the gravitational force by 20 or 30 times for a certain particle, particles of that type are driven outward from the sun so rapidly that they form a tail that is nearly straight. On the other hand, if the repulsive force is only 2 or 3 times the gravitational force, the particles move outward more slowly; the difference between the orbital motions of the comet and tail material becomes apparent before the latter has receded very far from the comet's head, and the tail appears noticeably curved.

Most comets have two tails that are very different from each other. The most prominent is the *plasma tail;* it is usually nearly straight, and its spectrum shows emission by ions such as CO^+, N_2^+, and H_2O^+. The other tail is a *dust tail,* broad and generally curved, displaying the spectrum of reflected sunlight.

The nearly straight plasma tails sometimes have recognizable features or knots that can be seen, from one day to the next, moving away from the comet (see the photographs of Comet Mrkos in Figure 20.9). Measures of such structures show that the streaming ions have speeds that increase from several km/s near the head of the comet to hundreds of km/s toward the end of the tail. From time to time, sudden solar outbursts (flares) result in a greater than average flow of plasma from the sun; these variations in the solar wind intensity, and variations in the ejection of material from the comet's nucleus, probably account for the irregular features in its ion tail.

The dust tails are produced by solar radiation pressure. Each photon of electromagnetic radiation has associated with it a small amount of momentum, which is numerically equal to the energy of the photon divided by the velocity of light. If a photon strikes a particle, or is absorbed by it, the momentum of the photon is transferred to the particle, thereby accelerating the latter. On particles of moderate size, the acceleration is entirely negligible in comparison to that produced by the gravitational forces acting on them. On the very small dust particles in comets, however, the force of solar radiation can be greater than the sun's gravitational attraction.

Radiation pressure is most effective on particles whose sizes are near that of the average wavelength of sunlight. If they have a density near 1 g/cm^3 (like ice crystals), the force of radiation on such particles can exceed that of gravitation by about five times. Particles several times larger than the wavelength of

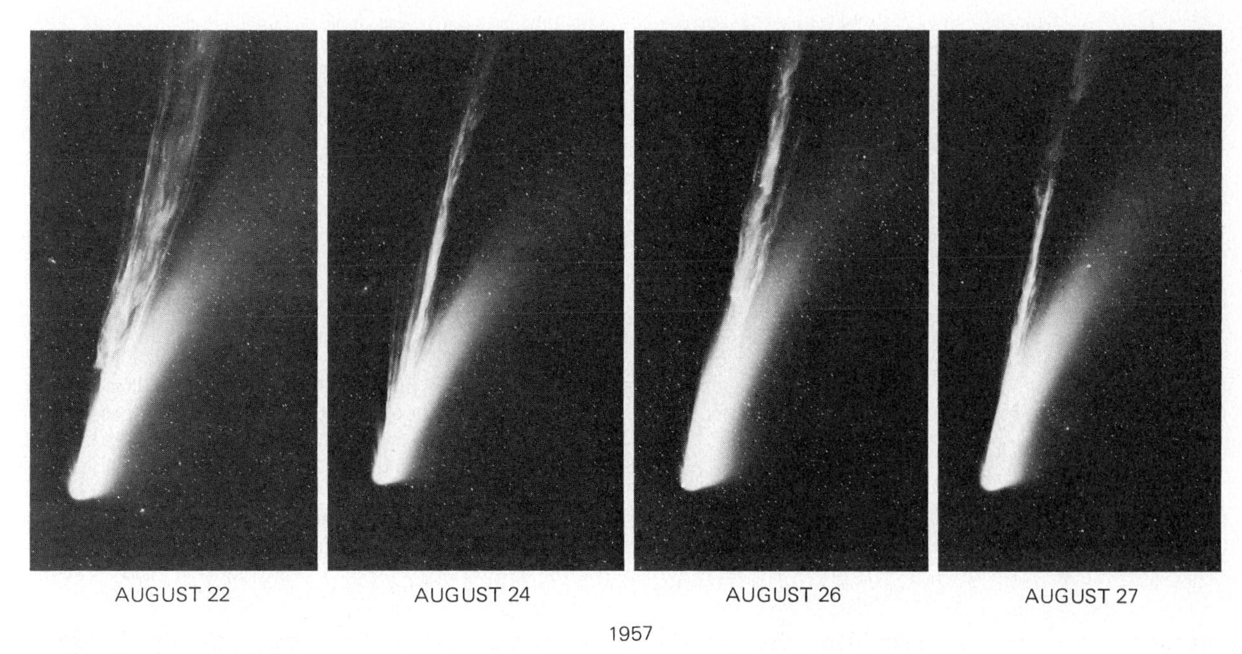

Figure 20.9 Comet Mrkos, photographed with the Palomar Schmidt telescope. (Caltech/Palomar Observatory)

the incident light have masses great enough so that the force of gravitational attraction between them and the sun exceeds the outward force of radiation pressure. The effect of radiation pressure on smaller particles depends on their chemical composition; some types of smaller particles are repulsed. Many kinds of molecules also absorb radiation and are driven away from the sun.

The dust tails of comets are often broad or multiple because the particles of different sizes are accelerated by different amounts; radiation pressure generally imparts smaller speeds to particles of greater mass, so that they trail farther behind the comet. Sometimes the dust tails are distinctly fan-shaped, the trailing edge of the fan consisting of the largest particles and curving back the most.

(f) Spacecraft Encounters with Comets

In 1985 the first spacecraft encounter with the tail of a comet took place. An Earth satellite called the International Sun-Earth Explorer was diverted in a complicated set of maneuvers devised by Robert Farquahar of the NASA Goddard Space Flight Center so that it left our planet and intercepted Comet Giacobini-Zinner, passing 7800 km behind the comet nucleus on September 11, 1985. Among the results were the discovery of an unexpectedly large and energetic region of interaction between the plasma tail and the solar wind, and a lower than anticipated density of dust (only about 50 hits measured). This spacecraft was not instrumented, however, to measure the nucleus; that was left to the flotilla of spacecraft that encountered Comet Halley the following March.

Six spacecraft made measurements of Comet Halley in March of 1986, as the comet crossed the plane of the Earth's orbit a month after perihelion. Two of these—the U.S. ICE spacecraft mentioned above and a small Japanese craft named Sakagaki—served primarily to monitor the comet and the interplanetary medium from a distance of several million kilometers, while a second Japanese spacecraft—Suisei—passed about 1 million km from the comet. The primary exploration tasks, however, were undertaken by three craft targeted for the nucleus itself.

Because of its retrograde orbit, Comet Halley can only be approached head-on, at speeds of about 70 km/s. The encounter period is therefore very short—of the order of an hour or two—and the nucleus itself is passed in just a fraction of a second. Hence all three craft were designed for autonomous

Figure 20.10 Leaders of the teams that sent spacecraft through Comet Halley confer in Moscow at the time of the VEGA 1 encounter. *Left:* Roger Bonnet of the European Space Agency; *right:* Roald Sagdeev of the USSR Space Research Institute. (David Morrison)

operation at encounter. The high speed of the spacecraft relative to the comet also created extra risks, since even a tiny grain of dust can pierce the spacecraft like a bullet at 70 km/s.

The Soviet VEGA 1 and VEGA 2 were the first craft to arrive, on March 6 and 9, 1986. Each plunged deeply into the inner atmosphere and dust cloud of the comet, passing within about 8000 km of the nucleus. In addition to making many direct measurements of the gas and dust, they photographed the dust-shrouded nucleus, but they were able to see little beyond the bright plumes of material jetting out from the two most active regions on its surface. Both VEGA craft were severely damaged by dust impacts, losing most of their solar cells and suffering the loss of several instruments at the time of closest approach.

The trajectory data for the VEGA craft were provided to the European Space Agency to allow them to target their Giotto spacecraft (Figure 20.11) for an even closer encounter on March 14, 1986—just 605 km from the comet's nucleus. Giotto also carried out many measurements of the near environment of the nucleus, confirming and extending the Soviet results. Among the most exciting discoveries was the fact that much of the dust is in the form of very small particles consisting largely of carbon and hydrocarbon compounds, rather than silicates. Thanks to the pathfinder data from VEGA, Giotto was able to image the nucleus directly before its camera was destroyed by impacts a few seconds before closest approach (Figure 20.6). As we have noted above, the nucleus of Halley is large and very dark (reflectivity 2–3 percent), with the activity concentrated in two bright jets of gas and dust.

Halley is the only bright comet with a predictable orbit that can be targeted for spacecraft investigation. To take the next step, mission planners are turning to the smaller Jupiter-family comets. The U.S. has proposed Comet Tempel 2 as a target for a mission they would like to launch in 1993. Their plans call for the spacecraft to match orbit with the comet near aphelion, when it is inactive, and to examine the nucleus at close range, including the dropping of instruments directly on its surface. The spacecraft would then stay with the comet as it approaches the sun, monitoring its activity through perihelion. By thus extending the observation period from a few minutes to several months, this mission should provide much more information on cometary activity than the Halley flybys.

20.3 THE ORIGIN AND EVOLUTION OF COMETS

(a) The Oort Comet Cloud

In Section 20.1c we explained that comets are a part of the solar system, not interlopers from interstellar space, but that they come initially from very great distances. Observationally, the aphelia of new comets typically have values near 50,000 AU. This clus-

Figure 20.11 The Giotto spacecraft, which encountered Comet Halley on March 14, 1986. (ESA)

tering of aphelion distances was first noted by Dutch astronomer Jan Oort (b. 1900), who proposed in 1950 a scheme for the origin of the comets that is still accepted today.

It is possible to calculate that the gravitational *sphere of influence* of a star—the distance within which it can exert sufficient gravitation to hold onto orbiting objects—is about ⅓ of the distance to the nearest other stars. In the vicinity of the sun, stars are spaced about three light years (200,000 AU) apart. Thus the sun's sphere of influence extends only a little beyond 50,000 AU, and at such distances objects in orbit about the sun will be perturbed by the gravitation of passing stars. Oort suggested, therefore, that the new comets were objects orbiting the sun with aphelia near the edge of its sphere of influence, and that the perturbing effects of other nearby stars modified their orbits to bring them close to the sun where we can see them. This region of space from which the new comets are derived is called the *Oort comet cloud*.

Several new comets are discovered each year, most of them probably approaching the sun for the first time. If comets have been entering the inner solar system at this same rate for its entire history of 4.6×10^9 years, then the source region must have originally held at least 10^{10} comets. Using similar but more detailed arguments Oort calculated that there were at least 10^{11} comets in the cloud, while recent estimates place the number closer to 10^{12}.

If a typical comet has a mass of about 10^{-11} Earth masses (Section 20.2a), we can see that the entire mass of this comet cloud is equivalent to only a few Earths, and much less than that of the giant planets. It is still much greater, however, than the total mass of the asteroids.

Just because most new comets have aphelia near 50,000 AU, we should be careful not to conclude that the Oort cloud consists of billions of comets in roughly circular orbits at this distance, like a shell around the sun. There may be some comets in nearly circular orbits, but if so, we have no evidence of them. What we see are comets that are perturbed into the inner solar system from orbits that must have been already of very large eccentricity. A passing star can cause only slight changes in orbital elements of the comet, so if it has an orbit with a perihelion of, say, 1 AU when we discover it, it must have had a perihelion of no more than a few AU on its previous unobserved passage near the sun. For every comet that comes close enough to be seen, there must be many more skimming invisibly through the orbits of the outer planets.

(b) Formation of the Comet Cloud

There are two theories for the formation of the Oort cloud: either it is made of material that condensed in place, tens of thousands of AU from the sun, or else the comets were formed closer to the sun and subsequently ejected to these large distances.

It would be easy to imagine that the comets are condensates from the outer fringes of the solar nebula, were it not for two problems. The mass of the comets in the cloud is as large as that of the terrestrial planets, yet models suggest that the solar nebula thinned out rapidly with increasing distance from the sun. It is difficult to understand how any

Figure 20.12 Comet Halley in March 1986, photographed from Australia. (Royal Observatory, Edinburgh)

substantial amount of material could have condensed at such huge distances, or how small grains might have accumulated into bodies several kilometers across. The second problem is associated with the randomly inclined orbits of new comets, which tell us that the distribution of comets in the Oort cloud is spherical, not disk-like, as it should be if the comets formed directly from the spinning solar nebula.

A more likely hypothesis is that the comets were ejected into the Oort cloud from initial orbits in the disk, probably near the present orbits of Uranus and Neptune. Calculations indicate that if many icy planetesimals were still present after the massive cores of the giant planets formed, a substantial fraction of them would have been ejected by gravitational encounters with these bodies. If this is correct, then the comets are leftovers from the building blocks of the outer planets, preserved for 4.6 billion years in the Oort cloud.

The same process that ejected comets to the present Oort cloud must have supplied an even larger number of cometary bodies to regions closer to the sun. In fact, if the ejection hypothesis is correct, the vast majority of these objects now have orbits between that of Pluto and the observed Oort cloud. Unfortunately, there is no way to detect these objects in what is sometimes called the *inner Oort cloud*. All we are aware of is the outer fringe, where stellar perturbations place a tiny fraction of the objects in orbits that bring them close to the sun. Orbits in the much larger inner Oort cloud would be stable against such perturbations.

(c) The Fate of Comets

Once a comet enters the inner solar system, its previously uneventful life history begins to accelerate. It may, of course, survive its initial passage near the sun and return to the cold reaches of space where it spent the previous 4.6 billion years. At the other extreme, it may impact the sun or pass so close that it is utterly destroyed on its first perihelion passage. Observations from space indicate that at least one comet collides with the sun every year, something that had never been expected from Earth-based observations. Frequently, however, the new comet does not come this close to the sun, but instead it interacts with one or more of the planets.

A comet coming within the gravitational influence of a planet has three possible fates. (1) it can impact that planet, ending the story at once; (2) it can be ejected on a hyperbolic trajectory, leaving the solar system forever; or (3) it can be perturbed into a shorter period. In the last case, its fate is sealed. Each time it approaches the sun, it will lose part of its material, and it still also has a significant chance of collision with a planet. Once in a short-period orbit, the comet's lifetime is measured in thousands, not billions, of years.

Measurements of the amount of gas and dust in the atmosphere of a comet permit an estimate of the total losses during one orbit (Figure 20.12). Typical loss rates are up to a million tons per day from an active comet near the sun, adding up to some tens of millions of tons per orbit. This is equivalent to stripping off the top several meters from the nucleus. Comparing these loss rates with the total mass of the comet, we see that they amount to about 0.1 percent per orbit. At that rate, the comet will be gone after a thousand orbits.

Whether the comet evaporates away completely is not known. If the gas and dust are well mixed, we would expect the nucleus to shrink each time around the sun until it has entirely disappeared. However, there remains the suggestion that many of the Earth-approaching asteroids are extinct comets. If there is a silicate core in the comet, or if the dirty snowball includes large blocks of nonvolatile material that are held gravitationally to the surface, then there could be a substantial solid residue after the ices are gone. We simply do not know which of these alternatives is correct.

Some comets end catastrophically. Even if they avoid impacting a planet or being perturbed into an orbit that collides with the sun, they can break apart for reasons that are not well understood. In 1846, for example, the nucleus of Comet Biela split into two parts, and on its next return in 1852 it appeared as two comets, separated by 2 million km. In 1866, the next perihelion year, nothing appeared; Comet Biela had simply and totally disappeared. A more recent example of breakup is provided by Comet West; shortly after its 1976 perihelion passage its nucleus split into four components, which drifted apart at a rate of several hundred kilometers per day. The smallest fragment survived only a few days, but the other three retained their identities until the comet became invisible with increasing distance from the sun.

20.4 IMPACTS OF COMETS AND ASTEROIDS WITH THE EARTH

We have frequently discussed crater-forming impacts with the Earth and other planets. Chapter 12 introduced the concept of the accretion of planets from planetesimals. In Chapter 14 we suggested that a giant impact had been responsible for the formation of the Moon, and we found that the lunar cratering history can be used to establish a chronology for the geology of the terrestrial planets, which we used in Chapters 15 and 16 to interpret the evolution of Mars and Venus. Chapter 18 described how the heavily cratered surfaces of the outer planet satellites allowed us to generalize terrestrial planet impact history to the entire solar system, and we suggested that ring systems might be the result of the impact fracturing of satellites near the tidal stability limit. In Chapter 19 we showed that the smaller asteroids represented a population of impact fragments of larger parent bodies, and we introduced in the Earth-approaching asteroids some of the objects that are present today to collide with the terrestrial planets. This chapter has been devoted to the comets, another group of potential projectiles for collision with the planets.

Let us now look again at the history of collisions with our own planet, the Earth. We will see increasing evidence that the impacts of comets and Earth-approaching asteroids have had important effects, particularly in the unexpected field of biological evolution.

(a) Historic Impacts

Twice in the 20th century large objects have collided with the Earth. The first such event took place on June 30, 1908, near the Tunguska River in Siberia. In this desolate region and witnessed by only a handful of people, a remarkable explosion took place in the atmosphere about 8 km above the surface. The shock wave flattened more than a thousand square kilometers of forest; herds of reindeer and other animals were killed; and a man at a trading post 80 km from the blast was thrown from his chair and knocked unconscious. The blast wave spread around the world, recorded by instruments designed to measure changes in atmospheric pressure. Yet in spite of this violence, no craters were formed by the explosion.

While we do not know exactly what caused the Tunguska event, it certainly represented the disintegration of an impacting body weighing approximately 100,000 tons. The force of the blast was equivalent to about a 10-megaton nuclear bomb. The material may have been cometary, and Comet Encke has even been suggested as a source. In any case, the projectile did not have the strength to survive its plunge to the surface, but rather gave up its energy of motion in the atmosphere, creating the equivalent of an "air burst" in nuclear weapons jargon.

The second impact event also took place in Siberia, near Vladivostok. On February 12, 1947, observers saw a fireball "as bright as the sun." The impact produced 106 craters and pits ranging in size up to 28 m across. Trees were felled radially around each of the large craters. More than 23 tons of iron meteorite fragments have subsequently been recovered from the area, demonstrating that the impacting bodies here were a swarm of iron meteorites.

Either of these impacts was large enough to do substantial harm if it had occurred in a populated area. Perhaps more important, explosions of this magnitude would be immediately detected by the various surveillance and defense systems around the world, and they could easily be mistaken for the beginning of a nuclear attack. Can you imagine the reaction if the 10-megaton Tunguska event took place in the U.S.S.R. today, especially if it were near a major city or military base?

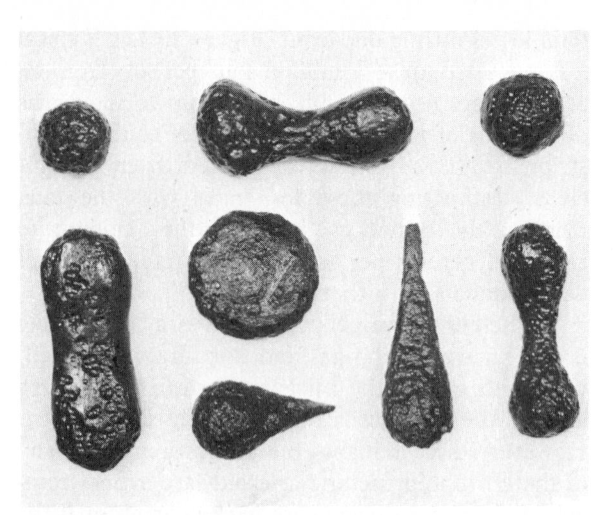

Figure 20.13 Tektites: melted rock from terrestrial impact craters. The largest is about 8 cm long. (R. Oriti, Griffith Observatory)

(b) Terrestrial Impact Craters

Neither the Tunguska nor the Vladivostok impact created large craters of the sort seen on the Moon. Major crater-forming events on Earth are much rarer, probably taking place only once in several thousand years on the average. Meteor Crater in Arizona (see Figure 14.14) is the only substantial young crater that has been identified on the land areas of the Earth; it is 50,000 years old. Even the Meteor Crater projectile, however, was no more than a hundred meters across. Impacts of objects the size of the smallest known Earth-approaching asteroids or comet nuclei are much rarer yet.

More than one hundred major impact scars have been identified on the Earth, most of them in the very old rocks of the continental shields. Rarely do they resemble the craters on other planets, however. Erosion and sedimentation have generally removed such characteristic features as the rim and the ejecta blanket. Often the only remaining indication of the impact is a circular region of shocked and shattered rock below the original location of the crater.

An interesting and relatively recent such impact feature is the Ries structure near the town of Nordlingen in southern Germany. This crater, originally 27 km in diameter and 5 km deep, was formed about 15 million years ago. The impacting object was of rocky composition and had a mass of at least a billion tons; presumably it was an Earth-approaching asteroid with a diameter of 2 to 3 km. As the asteroid plunged through the Earth's atmosphere, it left a partial vacuum behind it, allowing part of the vaporized ejecta to arc out into space. Upon re-entering the atmosphere, this ejected material was molded into teardrop-shaped lumps of black glass, called *tektites* (Figure 20.13). Thousands of tons of tektites from the Ries impact are strewn over half of Europe.

(c) Extinction of the Dinosaurs

An impact the size of the Ries event must have been felt over all of central Europe, but it did not have serious consequences for the entire Earth. Larger impacts, however, can disturb the whole planet and have a major influence on the course of evolution.

The best-documented such impact took place 65 million years ago, at the boundary between the Cretaceous and Tertiary periods of geological history. It is called the *K/T event* (K rather than C because Cretaceous is spelled with a K in German). This break in the Earth's history is marked by a *mass extinction,* when as many as half of the species on our planet became extinct. While there are a dozen or more mass extinctions in the geological record, the K/T boundary has always intrigued paleontologists because it marks the end of the age of the dinosaurs. For tens of millions of years these great reptiles had ruled the world. Then, suddenly, they disappeared, and thereafter the mammals be-

Figure 20.14 TIME magazine cover depicting the extinction of the dinosaurs. (Copyright 1985 Time Inc. All rights reserved. Reprinted by permission from TIME)

gan their development and diversification (Figure 20.14).

The body that impacted the Earth at the end of the Cretaceous period was an asteroid with a mass of more than a trillion tons and a diameter of at least 10 km. We know this because of a worldwide layer of sediment deposited from a dust cloud that enveloped the planet after the impact. First identified in 1980, this sediment layer is enriched in the rare metal iridium and other elements that are relatively abundant in an undifferentiated asteroid but very rare in the crust of the Earth. Even diluted by the terrestrial material excavated from the crater, this asteroidal component is easily identified. The impact site, however, has not been located. This is not really too surprising, since about half of the Earth's ocean floor and some of the continental areas consist of crust formed during the 65 million years since the impact.

The K/T impact must have released energy equivalent to 5 billion Hiroshima-size nuclear bombs, excavating a crater 200 km across and deep enough to penetrate through the Earth's crust. It can be compared with the lunar crater Tycho, the youngest major crater on the Moon. The explosion lofted about 100 trillion tons of dust into the atmosphere, as can be determined by measuring the thickness of the sediment layer formed when this dust settled to the surface. Such a quantity of material would have blocked sunlight completely from reaching the surface, plunging the Earth into a period of cold and darkness that lasted at least several weeks, and more likely several months. Presumably it is the darkness and cold that were responsible for the mass extinction, including the death of the dinosaurs.

Several other mass extinctions in the geological record have been tentatively identified with large impacts, although none is so dramatic as the K/T event. But even without such specific documentation, it is clear that impacts of this size do occur and that their effects can be catastrophic for life. What is a catastrophe for one group of living things, however, may create opportunities for another group. Following each mass extinction, there is a sudden evolutionary burst as new species develop to fill the ecological niches opened by the event.

(d) Nuclear Winter

Research into the K/T extinction and the behavior of major dust storms in the Martian atmosphere has recently led atmospheric scientists into a consideration of the effects of a nuclear war on the Earth's atmosphere and climate. The analog of the period of darkness that followed the K/T impact is called in this case the *nuclear winter,* a term coined in 1983 by Paul J. Crutzen. Detailed calculations of the effects of a nuclear war on the atmosphere were first undertaken by Richard P. Turco, Brian Toon, and their colleagues, and these scientists have since been joined by a worldwide effort involving, among others, the Academies of Science of both the U.S. and the U.S.S.R.

A nuclear war of the sort planned for by the military of both the U.S. and the U.S.S.R. involves the explosion of from 1000 to 10,000 warheads, at least some of which are targeted for cities or for military bases located near population centers. The total energy yield of such a war might be a thousand

Figure 20.15 Aftermath of the Tunguska impact. This photo, taken 21 years after the blast, shows a part of the forest that was destroyed by an explosion that must have registered at least 10 megatons. (Novosty)

megatons, which is much less than the energy of impact of an asteroid or comet. However, the effects on the atmosphere are comparable because of the added contribution of smoke from burning cities. Smoke and soot are much stronger absorbers of sunlight than is the naturally occurring dust ejected into the atmosphere by an asteroidal impact. The situation is further complicated by the toxic gases released when the products of modern civilization are consumed by fire. Calculations show that the burning of even a hundred cities, added to the smoke from forest and range fires triggered by blasts, could create a situation on Earth nearly as severe as that following the K/T impact.

The discovery of the nuclear winter phenomenon, which is a result of computer programs and other research tools first developed for the investigation of other planets, may be one of the most important research results of our time. Carl Sagan and other scientists have argued that it changes the basic assumptions of nuclear strategy. If the climate of the entire planet is disrupted by any nuclear war, independent of who starts it or against which nation state it is waged, then any such war becomes untenable. To wage war is to commit mass suicide, and the notion of a winner or even a survivor does not apply. Politicians and military strategists are facing this problem now, while research is continuing to define better the exact consequences for the planet of different scenarios for nuclear war.

EXERCISES

1. Suggest reasons why comets have been associated in Western culture with disaster and evil.
2. Find the period of a comet that at perihelion just grazes the sun, and whose aphelion distance from the sun is: (a) 200 AU; (b) 2000 AU; (c) 20,000 AU; (d) 200,000 AU.
3. On the assumption that a periodic comet can survive 1000 perihelion passages, find the lifetimes of the first two comets in the previous exercise.
4. Suppose a comet is discovered next year approaching the sun on a distinctly hyperbolic orbit. List the possibilities for its previous history, and evaluate how likely it is that this comet is truly a visitor from interstellar space rather than a member of the solar system.
5. It has been suggested that comets may have formed from material near the orbits of Uranus and Neptune, or that they may be from the outer edges of the solar nebula far beyond the orbits of the planets. Can you suggest any way we might decide between these alternatives if we could make really precise measurements of the composition of the nucleus of a comet?
6. Comets are considered to be the "most primitive" solid bodies in the solar system. What does this statement mean?

*7. The force due to radiation pressure on a particle is proportional to the amount of radiation that the particle intercepts. Show that the ratio of the force of radiation pressure to the solar gravitational force on a spherical particle is independent of the distance of the particle from the sun.

8. The 1986 reappearance of Comet Halley was a disappointment to many people, most of whom were unable to see it. Suggest at least three reasons why it did not live up to the layperson's expectations.
9. How did the 1986 spacecraft encounters with Comet Halley change our ideas about the nature of comets?
10. If the Oort comet cloud contains 10^{12} comets and 10 new comets are discovered each year, what percentage of the comets have been used up since the beginning of the solar system?
11. Suppose a comet is discovered approaching the sun and it is found to be on an orbit that will cause it to collide with the Earth 20 months later, after perihe-

lion passage. (This is approximately the situation described in the popular science fiction novel *Lucifer's Hammer*.) What could we do? Is there any way to protect ourselves from a catastrophe?

*12. Read the novel *Lucifer's Hammer*. Does it present an accurate picture of the nature of comets and of the consequences if one impacted the Earth?

13. Make a table comparing the consequences of a nuclear war (the nuclear winter) with the consequences of impact of a large comet or asteroid on the Earth.

Harold C. Urey (1893–1981) won the Nobel Prize for chemistry in 1934 for the discovery of deuterium. Later in his career he became interested in the origin of the Earth and planets, and during the 1950s and 1960s his recognition of the role of primitive meteorites as remnants from the birth of the solar system laid the foundation for much of the modern interest in both the meteorites and the broader study of cosmochemistry.

21

METEOROIDS, METEORITES, AND METEORS

With the study of the meteorites and their little cousins the meteors, we come full circle in our discussion of the solar system. Chapter 12 described the origin of the system as an introduction to the planets. However, much of what we understand about these beginnings is derived from the study of the meteorites, a scientific field known as *meteoritics*. During the past two decades, this branch of research has steadily increased in the sophistication of its laboratory techniques and the impact of its conclusions. Indeed, the great power of laboratory analysis has now become so widely recognized that the primary objectives of future space missions to the planets are likely to involve the return of samples for study using the techniques developed for the investigation of meteorites.

21.1 METEORS

Although the layperson often confuses comets and meteors, these two phenomena could hardly be more different. Comets can be seen when they are many millions of miles away from the Earth, and may be visible in the sky for weeks, or even months, slowly shifting their positions from day to day. They rise and set with the stars, and during a single night appear motionless to the casual glance. *Meteors,* on the other hand, are small solid particles that enter the Earth's atmosphere from interplanetary space. Since they move at speeds of many kilometers per second, the high friction they encounter in the air vaporizes them. The light caused by the luminous vapors formed in such an encounter appears like a star moving rapidly across the sky, fading out within a few seconds. Meteors are commonly called "shooting stars."

On rare occasions, an exceptionally large particle may survive its flight through the Earth's atmosphere and land on the ground. Such an occurrence is called a *meteorite fall,* and if the particle is later recovered, it is known as a *meteorite.* Before a particle encounters the atmosphere of the Earth, it is called a *meteoroid.* Thus, in summary: the particle, when it is in space, is called a meteoroid; the luminous phenomenon caused when the particle vaporizes in the Earth's atmosphere is a meteor; and if it survives and lands on the ground, the particle is a meteorite.

(a) Phenomenon of a Meteor

On a typical dark, moonless night an alert observer can see half a dozen or more meteors per hour. To

Figure 21.1 Photograph of the trail of a bright meteor that happened to cross the field of view of the telescope photographing the Andromeda galaxy. (F. Klepesta)

be visible, a meteor must be within 150 to 200 km of the observer; over the entire Earth, the total number of meteors bright enough to be visible must total about 25 million per day. Faint meteors are far more numerous than bright ones; the number of meteors that are potentially visible with binoculars or through telescopes, and that range down to a hundred times fainter than unaided-eye visibility, must number more than 10^{11} per day.

More meteors can generally be seen in the hours after midnight than in the hours before, since we are then on the leading edge of the Earth as it moves through space. Moreover, there are certain times when meteors can be seen with much greater than average frequency—up to 60 or more per hour. These unusual displays are called *meteor showers*. Meteor showers occur when the Earth encounters swarms of meteoroids moving together in space.

Occasionally an exceptionally bright meteor is reported by many observers. These bright meteors are called *fireballs* (Figure 21.1). It is estimated that tens of thousands of fireballs appear every day over the entire Earth. Some fireballs have been visible in broad daylight; some are as bright as the full moon. Sometimes fireballs break up in mid-air with explosions that are audible from the ground. Such exploding fireballs are called *bolides*.

Fireballs sometimes leave luminous trails or trains behind them, which may persist for periods ranging from one second to half an hour. The velocities of upper atmospheric winds are sometimes revealed by the twisting and distortion of meteor trains.

(b) Observations of Meteors

The paths of bright fireballs through the atmosphere can occasionally be traced from an analysis of lay sightings. The untrained observer, to be sure, does not often provide reliable accounts: fireballs usually seem many times closer than they actually are; also, their angular altitudes and speeds are frequently grossly exaggerated. Nevertheless, the paths of spectacular fireballs have been derived by comparing reports from many observers scattered over areas of hundreds of kilometers.

Far better data on the paths of meteors come from special photographic patrols and radar observations (Figure 21.2). In the former, arrays of specially designed, high-speed, wide-angle meteor cameras are placed many kilometers apart, and are directed to the same region of the sky. When a meteor chances to pass through the common field of view of two of these cameras, it is recorded by both. At each camera, the exposure is interrupted by a rotating propeller-type shutter, so that the trail of the meteor on the film consists of a series of dashes rather than a continuous streak.

When trails of the same meteor are identified on photographs obtained simultaneously by the two cameras, the meteor's elevation and direction of motion through the atmosphere can be computed by triangulation. Since the rotation rate of the propeller shutters is known, the spacing of the dashes that constitute the meteor's trailed image indicates the velocity of the meteor. During the 1960s, the Smithsonian Astrophysical Observatory operated the Prairie Network, a system of 16 such meteor cameras in the midwestern United States, which covered an area of more than a million square kilometers. Similar networks have operated in Canada and in Czechoslovakia. Since the early 1970s, however, interest in such studies has declined and the sky patrols have closed down.

Meteors are detected by radio and radar as well as visually and photographically. In the early 1940s

Figure 21.2 Photograph *(left)* obtained with a Baker Super-Schmidt camera *(right)*. Trails of three meteors are visible; they are interrupted by a timing device in the telescope. (Harvard Observatory)

it was found that meteors caused brief interruptions in high-frequency broadcasting reception; these interruptions took the form of "whistles," which usually fell quickly in pitch. Since then, the rate at which meteors occur has been determined by counting the occurrences of such whistles.

Meteoroids themselves are too small to reflect radar waves back to the ground. These waves are reflected, instead, by the ionized gases that are formed when meteoroids vaporize in the air. The radio or radar waves travel, of course, with the speed of light; consequently, the time required for them to travel from the ground to the meteor and back again indicates the distance of the meteor from the radar station. The motion of the meteoroid toward or away from the radar station can be detected from the rate at which the distance changes, which in turn gives the speed of the meteoroid in the line of sight. Simultaneous observations from two radar stations give the height, speed, and direction of the meteoroid.

It is found that meteoroids produce meteors at an average height of 95 km. The highest meteors form at heights of 130 km. Nearly all meteoroids completely disintegrate, and their luminous paths end, by the time they reach altitudes of 80 km. A few meteoroids doubtless skim out of the atmosphere, returning to space before they are completely burned up.

The typical bright meteor is produced by a meteoroid with a mass of less than 1 gram—no larger in size than a pea. Of course, the light you see comes from the much larger region of ionized gas surrounding this little grain of interplanetary material, not from the meteoroid itself. A meteoroid the size of a marble produces a fireball when it strikes the atmosphere, while one as big as your fist has a fair chance of surviving its fiery entry to become a meteorite. The total mass of meteoritic material entering the Earth's atmosphere is estimated to be about 100 tons per day.

(c) Meteor Showers

Most meteors are *sporadic,* that is, they come from no particular direction and represent random bits of interplanetary material. More interesting are the *meteor showers* that occur when the Earth encounters swarms of particles that produce spectacular meteor displays.

Meteors belonging to a shower all seem to radiate or diverge away from a single point on the celestial sphere; that point is called the *radiant* of the shower. Recurrent showers are named for the

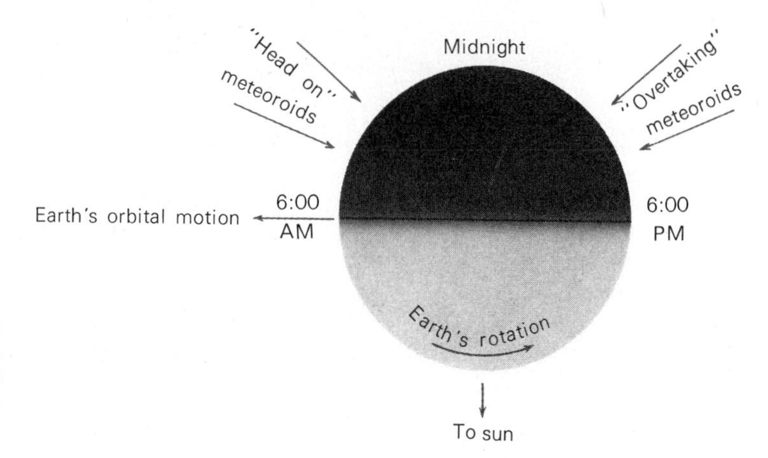

Figure 21.3 Meteors seen before midnight have overtaken the Earth. Meteors seen after midnight are produced mostly by particles approaching the Earth from the east.

constellation within which the radiant lies or for a bright star near the radiant.

The seeming divergence of shower meteors from a common point is easily explained. The meteoroids producing a meteor shower are members of a swarm; they are all traveling together in closely spaced parallel orbits about the sun. When the Earth passes through such a swarm, it is struck by many meteoroids, all approaching it from the same direction. As we, on the ground, look toward the direction from which the particles are coming, they all seem to diverge from it. Similarly, if we look along railway tracks, those tracks, although parallel to each other, seem to diverge away from a point in the distance (Figure 21.4).

Actually, a meteor radiant is not a perfectly sharp point. The meteors seem to radiate away from a small region of the celestial sphere. In some show-

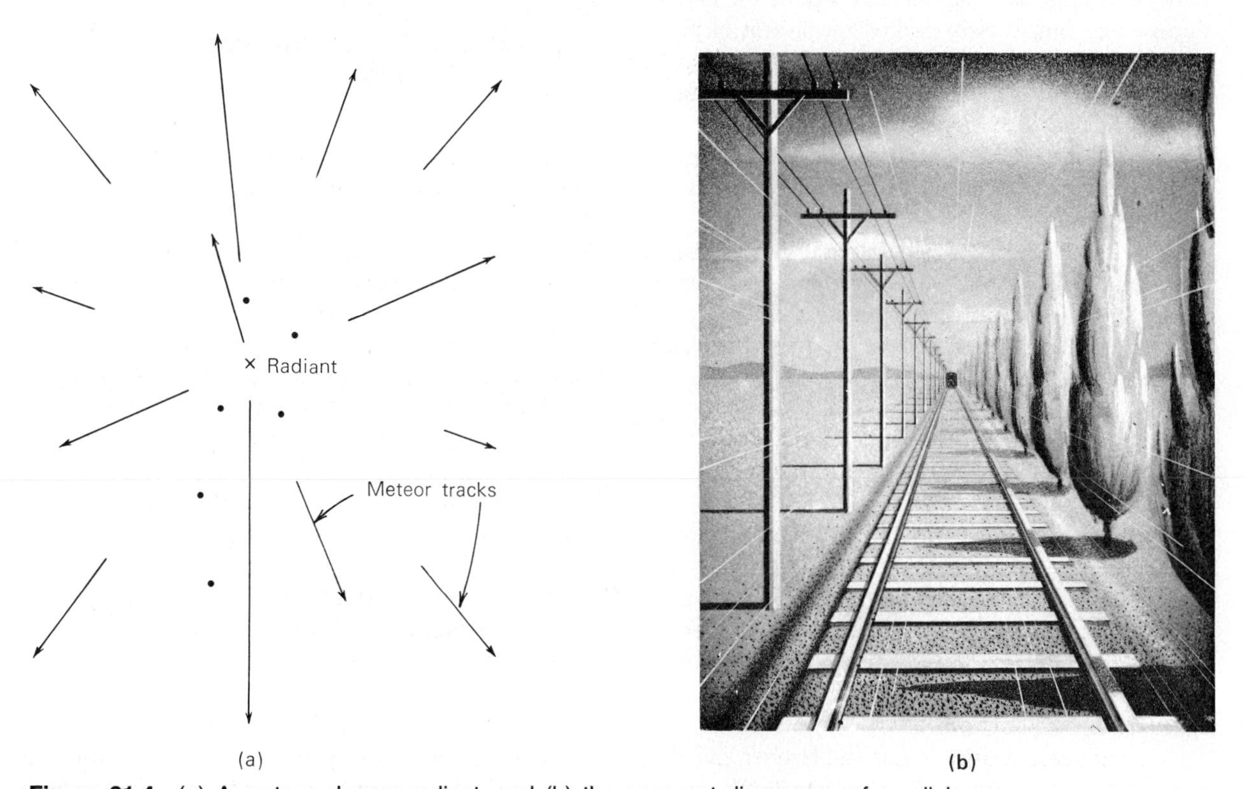

Figure 21.4 (a) A meteor shower radiant, and (b) the apparent divergence of parallel lines.

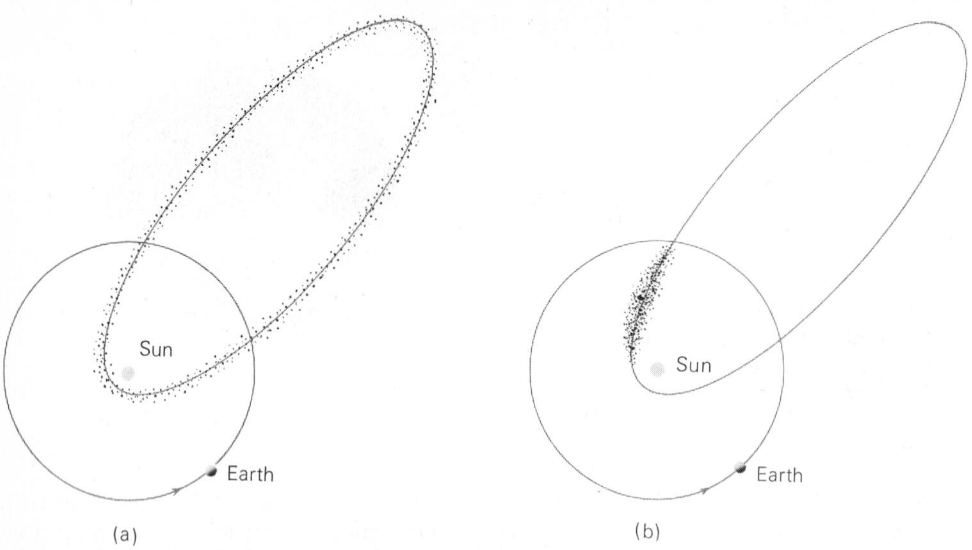

Figure 21.5 Meteoroids in a swarm may be strewn more or less uniformly along their orbit (a), or bunched up (b).

ers, that region is very small—as little as 3 arcminutes in diameter. In other showers, the radiant may be as large as 1° across. The size of the radiant depends on how closely packed the swarm of particles is. The more closely packed the swarm, the more spectacular the meteor shower it produces. The spectacle also depends on how closely the Earth approaches the densest part of the swarm.

Many meteors classed as sporadic may actually belong to showers that are so inconspicuous that they go unnoticed. Scarcely a night passes that a patient observer cannot see three or more meteors moving in the sky in directions away from a common point. Many of these may be members of minor showers.

21.2 INTERPLANETARY DUST

The meteors represent one form of interplanetary dust. Other ways to detect the fine-grained material between the planets are by observing its faint glow in reflected sunlight or its rather stronger thermal emission at infrared wavelengths, or by detecting it directly as it impacts spacecraft on their flights through the solar system.

Interplanetary dust is not stable over the lifetime of the solar system; it must constantly be replenished. Probably the major source for this material is the comets, as shown by the association between comets and meteor showers.

(a) Association of Showers and Comets

The direction to the radiant of a meteor shower indicates the direction in which the swarm of particles that produces it is moving through space with respect to the Earth. The velocity of the swarm is found from the velocity of the meteors. These are enough data to specify completely the orbit of the swarm.

On about August 11 of each year, the Earth passes through a swarm of particles that approach from the direction of the constellation Perseus. In 1866 it was observed that the particles producing this Perseid shower travel in an orbit that is almost identical to that of Comet 1862 III. It was then realized that those meteoroids encountered each August are debris from the comet that has spread out along the comet's orbit.

Subsequently, it has been found that the elements of the orbits of many other meteoroid swarms are similar to those of the orbits of known comets. Not all meteor showers have yet been identified with individual comets, but it is presumed that all showers have had a cometary origin. These swarms of debris, provided by the gradual disintegration of comets, give further evidence of the flimsy nature of comets.

Twice each year the Earth passes through a swarm of particles moving in the orbit of Comet Halley. Debris from that famous comet gives us the Eta Aquarids in May and the Orionids in October.

Sometimes, if a meteoroid swarm is old enough, its particles are strewn more or less uniformly along their entire orbit, as shown in Figure 21.5a; an example is the swarm of particles that produces the Perseid shower. Consequently, the Earth meets about the same number of particles each time it passes through the orbit of the swarm, and the Perseid shower is of about equal intensity each year. More often, a swarm is bunched as in Figure 21.5b, and we experience a spectacular display only when the Earth intercepts the meteoroids as they reach that same point in their revolution about the sun.

A good example is provided by the Leonid meteors in the last century, when the Earth met the debris of Comet 1866 I in 1833 and again in 1866 (after an interval of 33 years—the period of the comet). Those were among the most spectacular showers ever recorded. As many as 200,000 meteors could be seen from one place within a span of a few hours. The last good Leonid shower was on November 17, 1966, when in some southwestern states up to 140 meteors could be observed per second. Even in such dense swarms the individual particles of the swarm are separated by distances of 30 km or more; in most meteoroid swarms, the particles are more than 100 km apart.

The best meteor shower that can be depended on at present is the Perseid shower, which appears for about three nights near August 11 each year. In the absence of bright moonlight, meteors can be seen with a frequency of about one per minute during a typical Perseid shower. It is estimated that the total combined mass of the particles in the Perseid swarm is near 5×10^8 tons; this gives at least a lower limit for the original mass of Comet 1862 III. The orbit of this comet (and hence of the Perseids) was nearly perpendicular to the ecliptic plane; thus the meteoroids are little perturbed by the planets, accounting for the reliability of the Perseid shower.

One spectacular shower of recent decades was the Draconid shower that reached maximum display on October 9, 1946 (Figure 21.6). On that date the Earth reached the point that Comet Giacobini-Zinner had passed 15 days earlier. Debris from the comet produced meteors that could be counted from points in the southwestern United States at a rate of two per second, even though the Moon was full at the time.

Figure 21.6 Photograph showing trails of many meteors during the shower of October 1946. Note the divergence of the trails from a radiant (off the field to upper right). Other streaks are star trails.

The characteristics of some of the more famous meteor showers are summarized in Table 21.1. Other spectacular meteor showers can occur, however, at almost any time, just as some bright comets appear unexpectedly.

(b) Physical Properties of Shower Meteors

No shower meteor has ever survived its flight through the atmosphere and been recovered for laboratory analysis. However, there are other ways to investigate the nature of these particles.

Analyses of the photographic tracks of meteors show that most of them are very light or porous, with densities typically less than 1.0 g/cm^3. Estimates for some of the meteors associated with Comet Giacobini-Zinner suggest that the metoroids

TABLE 21.1 Major Annual Meteor Showers

SHOWER	DATE OF MAXIMUM DISPLAY	VELOCITY (km/s)	ASSOCIATED COMET	PERIOD OF COMET (yr)
Quadrantid	Jan 3	43	—	7.0
Lyrid	Apr 21	48	1861 I	415.00
Eta Aquarid	May 4	59	Halley	76.0
Delta Aquarid	Jul 30	43	—	3.6
Perseid	Aug 11	61	1862 III	105.0
Draconid	Oct 9	24	Giacobini-Zinner	6.6
Orionid	Oct 20	66	Halley	76.0
Taurid	Oct 31	30	Encke	3.3
Andromedid	Nov 14	16	Biela	6.6
Leonid	Nov 16	72	1866 I	33.0
Geminid	Dec 13	37	Phaethon	1.6

are so fragile that a 1-kilogram lump, if you placed it on a table, would fall apart under its own weight. Such particles break up very easily in the atmosphere, probably accounting for the failure of even relatively large shower meteoroids to produce meteorites. Comet dust, apparently, is fluffy, rather inconsequential stuff.

In spite of their fragility, these meteors yield products that can be studied in the laboratory. Centimeter-sized meteoroids can produce millimeter-sized droplets of metal and other nonvolatile materials, and these tiny spherules have been collected in large numbers from deep-sea cores and from the Arctic icefields of the Earth. Donald Brownlee of the University of Washington and his colleagues have analyzed these melt-products and find that their chemistry matches closely that of the most primitive meteorites, the carbonaceous meteorites (Section 21.4b). Thus, apparently, the dust of comets is similar to the material that condensed in the solar nebula near the outer part of the asteroid belt.

Brownlee has also collected even smaller but less modified meteor-derived particles from the upper atmosphere. These *Brownlee particles* (see Figure 20.5) are micrometeorites—fragments of comet dust less than 0.1 mm across. They too indicate a primitive composition, with from 5 to 10 percent carbon. So far, the Brownlee particles are the nearest we have been able to come to samples of cometary material for laboratory study.

(c) Zodiacal Dust

The *zodiacal light* is a faint glow of light along the zodiac (or ecliptic). It is the brightest along those parts of the ecliptic nearest the sun and is best seen in the west in the few hours after sunset or in the east before sunrise. Under the most favorable circumstances, the zodiacal light rivals the Milky Way in brilliance. It is sometimes called the "false dawn" because of its visibility in the morning hours before twilight actually begins.

The zodiacal light has the same spectrum as the sun, which shows it to be reflected sunlight. Gas molecules and atoms cannot be numerous enough in space to scatter enough sunlight to contribute appreciably to the zodiacal light. Not only are molecules inefficient scatterers of light, but what light they do scatter (that is, reflect helter skelter) is mostly blue and violet light of short wavelength—the blue daylight sky comes from the scattering of sunlight by air molecules. The zodiacal light represents rather sunlight reflected by particles of dust, which are concentrated in the plane of the ecliptic. The dust reflects enough sunlight to produce the faint glow along the zodiac.

The total mass of the material responsible for the zodiacal light is estimated at about 10^{13} tons. Because of the rate at which these particles are spiralling into the sun due to the Poynting-Robertson effect (see Section 21.2d below), there must be a continual replacement from the comets and from collisional fragmentation of objects in the asteroid belt, in the amount of about 10 tons per second.

The zodiacal dust has also been detected in the infrared. In 1983 the IRAS satellite carried out an all-sky survey at wavelengths from 10 to 100 μm, in which the presence of this material is obvious (Figure 21.7). IRAS also detected previously unrecognized structure in the emission, indicating the

Figure 21.7 The central asteroidal dust belt and two fainter dust lanes from Comets Tempel 2 and Encke as photographed from thermal emission in 1983 by IRAS at a wavelength of 60 μm. (NASA/University of Arizona)

presence of three distinct dust bands. These bands may represent the debris from asteroid collisions, and they have tentatively been associated with two of the largest asteroid families, the Eos and Themis families.

IRAS also located dust bands along the orbits of comets, thus confirming directly the association between comets and meteor showers. Two of the identified bands are from Comets Tempel 2 and Encke (Figure 21.7). Data from this orbiting observatory also led to the discovery of an asteroid (3200 Phaethon) that seems to have the same orbit as the Geminid meteor stream—the first association of an asteroid, rather than a comet, with a meteor shower.

Spacecraft that have explored beyond the asteroid belt confirm that most of the interplanetary dust is located within a few AU of the sun. This is to be expected, whether the dust originates mostly in comets or whether the asteroids also make an important contribution.

(d) The Poynting-Robertson Effect

For particles the size of micrometeorites, the sun's attractive gravitational force exceeds the repulsive force of the sun's radiation pressure. Like the planets, therefore, the micrometeorites revolve about the sun in Keplerian orbits. If the force of radiation pressure acting on such a particle were *exactly radial* (away from the sun), its only effect would be to reduce slightly the sun's attraction, resulting in a somewhat increased period of orbital revolution. However, the direction of the force of radiation pressure is not exactly radial but rather has a small component in a direction opposite to that of the particle's motion.

Suppose one runs through the rain. Even if the rain is falling vertically, it appears to strike the runner's face obliquely, because *relative to him* the raindrops have a horizontal component of motion—namely, a velocity equal and opposite to his velocity with respect to the ground. Similarly, to objects revolving about the sun, photons do not appear to come from an exactly radial direction, away from the sun, but have a slight component of velocity in a direction opposite to that of the objects' own motions. Thus, radiation pressure produces a slight "backward" force upon them. The effect of radiation pressure is negligible on particles of large mass (like the planets), but small particles, such as micrometeorites, are appreciably perturbed. The "backward" force acts like a "drag" on the orbital motion of such a particle, first making its orbit more and more circular and then gradually causing the orbit to diminish in size until the particle ultimately spirals into the sun. A particle 1 mm in diameter that orginates in the region of the minor planet belt (between the orbits of Mars and Jupiter) spirals into the sun in only 10 million years. A particle even as large as 10 cm in diameter at the Earth's distance from the sun will spiral into the sun in about 10^8 years. The fact that we find small particles around the Earth is evidence that they are either newly formed or have newly arrived in our part of the solar system.

Attention was first directed to the effect by J. H. Poynting in 1903; it was confirmed by a rigorous application of relativistic theory by H. P. Robertson in 1937. It is thus called the *Poynting-Robertson effect.*

21.3 METEORITES: STONES FROM HEAVEN

Occasionally, a meteoroid survives its flight through the atmosphere and lands on the ground; this happens with extreme rarity in any one locality, but over the entire Earth hundreds of meteorites fall each year. These rocks from the sky carry a remarkable record of the formation and early history of the solar system.

(a) Extraterrestrial Origin of Meteorites

While occasional meteorites have been recovered throughout history, their extraterrestrial origin was not accepted by scientists until the beginning of the 19th century. Before that, these strange stones were either ignored or else treated with supernatural respect.

The earliest recovered meteorites are lost in the fog of mythology. A number of religious texts speak of stones from heaven, which sometimes arrive at opportune moments to smite the enemies of the authors of the texts. At least one sacred meteorite has survived in the form of the Ka'aba, the holy black stone in Mecca that is revered by Islam as a relic from the time of the Patriarchs.

Ancient people apparently found practical uses for iron meteorites at a time when this metal was difficult or impossible for them to refine from available terrestrial ores. The legendary sword Excalibur of the Arthurian legend in ancient Britain may have been made of extraterrestrial iron. So probably is the iron dagger found in the 3500-year-old burial tomb of the Egyptian Pharaoh Tutankhamun in the Valley of the Kings. The Greek myth of the gift of iron from the gods to Prometheus and similar stories in Japan and in other cultures may all have their roots in falls of meteoritic iron.

The modern scientific history of the meteorites begins in the late 18th century, when a few scientists suggested that some of the strange stones that had been found around the world were of such peculiar composition and structure that they were probably not of terrestrial origin. Their ideas revived interest in the stories of falling stones. The general acceptance that indeed "stones fall from the sky" occurred after the French physicist Jean Baptiste Biot (1774–1862) described the circumstances of a fall in the Orne village of l'Aigle on April 26, 1803, in which many witnesses observed the explosion of a bolide, after which many meteoritic stones were found, reportedly still warm, on the ground.

A fall of meteorites may represent a group of meteoroids that were moving together in space before they collided with the Earth, but more likely the different stones are fragments of an original meteoroid that broke up during its violent passage through the atmosphere. It is important to remember that such a *shower of meteorites* has nothing to do with a *meteor shower*. As we have noted, no meteorites have ever been recovered in association with meteor showers. Whatever the ultimate source of these objects, they do not appear to come from the comets or their associated meteoroid streams.

(b) Orbits

One way to investigate the source of meteorites is to determine the orbit of the meteoroid while it is still in space, before it encounters the Earth's atmosphere. Once the meteorite has fallen, of course, its former orbit cannot be reconstructed. However, there have been three cases during about 20 years of searches where photographic patrols have yielded the path of a fireball from which a meteorite was later recovered.

The first successful orbit was calculated for the Pribram meteorite that fell in Czechoslovakia in 1959, followed by Lost City (U.S., 1970) and Innisfree (Canada, 1977). All three of these meteoroids proved to have been on eccentric orbits that carried them from the main asteroid belt to the Earth, similar to the orbits of many Apollo asteroids (Figure 21.8). While not conclusive, these results suggest the asteroids as the source of at least some of the meteorites.

(c) Meteorite Falls and Finds

Today, meteorites are found in two ways. First, sometimes bright fireballs are observed to penetrate the atmosphere to very low altitudes. A search of the area beneath the point where the fireball was observed to burn out may reveal one or more remnants of the meteoroid. Observed *falls*, in other words, may lead to the recovery of fallen meteorites.

Second, unusual-looking "rocks" are occasionally discovered that turn out to be meteoritic. These are termed *finds*. Now that the public has become meteorite conscious, many suspected meteorites are sent to experts each year. The late F. C. Leonard, a specialist in the field, referred to these objects as "meteorites" and "meteorwrongs." Outside of Antarctica (see below), genuine meteorites are turned up at an average rate of 25 per year. Of these, from five to ten are recovered from observed falls.

About a thousand meteorites have been collected, both falls and finds, where the many fragments from a single fall are counted as just one meteorite. Most of them are to be found today in

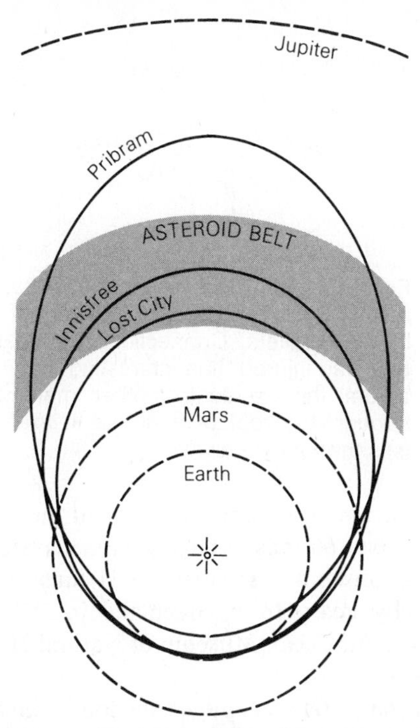

Figure 21.8 Calculated orbits of 3 recovered meteorites: Pribram (1959), Lost City (1970), and Innisfree (1977).

natural history museums or specialized meteoritical laboratories throughout the world. During the past decade, however, a new source of meteorites from the Antarctic has dramatically increased this total and is greatly enriching our knowledge of these objects.

The large number of meteorites recovered from the Antarctic are the result of the low precipitation and peculiar motion of the ice in some parts of that continent. Meteorites that fall in regions where ice accumulates are buried and then carried slowly, with the motion of the ice, to other areas where the ice is gradually worn away. After thousands of years the rock again finds itself on the surface, along with other meteorites carried to these same locations. The ice thus concentrates the meteorites from both a large area and a long period of time. Since there are no other exposed rocks on the Antarctic ice, any rock spotted in these areas is almost sure to be a meteorite (Figure 21.9 and Color Plate 20a).

Several thousand Antarctic meteorites have been collected during the past decade, most of them by Japanese scientists at sites near the Yamato Mountains. The primary U.S. site is the Allan Hills, a 20-km long mountain range in Victorialand.

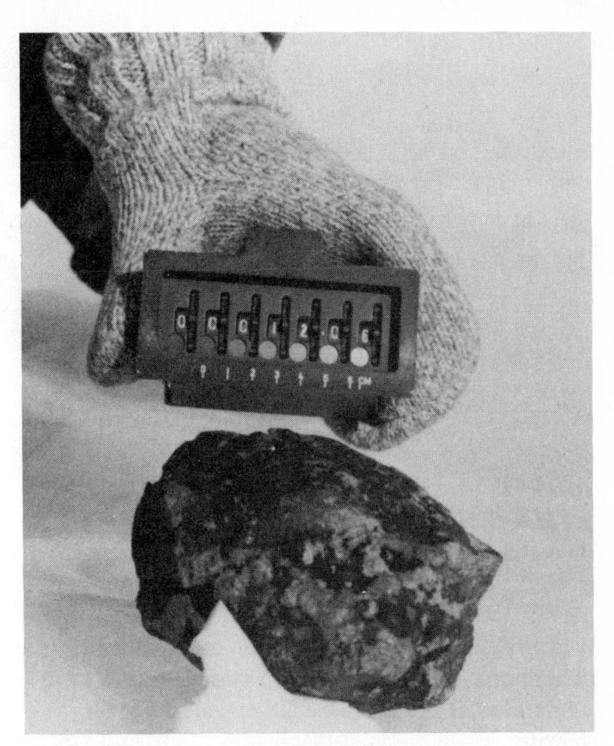

Figure 21.9 An Antarctic meteorite lying on the ice just before it was added to our collections. (NASA)

Most of the meteorites collected are small stones, weighing less than a kilogram. Often later analysis indicates that many of these small pieces are fragments from a single fall, so that the actual number of new falls represented by the Antarctic meteorites may be only a thousand or so. Even in this case, however, the Antarctic meteorites have more than doubled the total size of our meteoritic collections, and in addition these samples are being treated with much more care to avoid terrestrial contamination than was common in the past.

(d) Classification and Nomenclature

The meteorites in our collections include a wide range of compositions and histories, but traditionally they have been placed into three broad classes. First, there are the *irons,* which are composed of nearly pure metallic nickel-iron. Second are the *stones,* which is the term used for any silicate or rocky meteorite. Third are the much rarer *stony-irons,* which are (as the name implies) made of mixtures of stony and metallic iron materials.

Figure 21.10 On November 9, 1982, a small meteorite crashed through the roof of a home in Wethersfield, Connecticut. Fortunately, no one was injured. It is interesting that 11 years before this incident another meteorite tore through the roof of a house in Wethersfield less than a mile away.

Of these three types, the irons are the most obviously extraterrestrial in origin because of their metallic content. Native, or unoxidized, iron almost never occurs naturally on Earth. This metal is always found here as an oxide or other mineral ore. Therefore, if you ever come across a chunk of metallic iron, it is sure to be either manmade or a meteorite.

The stones are much more common than the irons, but harder to recognize. Often a laboratory analysis is required to demonstrate that a particular sample is really of extraterrestrial origin, especially if it has lain on the ground for some time and has been subject to weathering. The most scientifically valuable stones are those that are collected immediately after they fall or the Antarctic samples that have been preserved in a nearly pristine state by the ice.

Meteorities have traditionally been named for the town nearest to the place where they are found. For example, the large fall that took place in 1969 near the village of Pueblito de Allende in northern Mexico has given us the Allende meteorite. Since there are no towns in the Antarctic, the meteorites collected there are designated by a combination of letters and numbers. An example is the first meteorite of lunar origin, found near the Allan Hills and known as ALHA 81005. The numbers indicate that this meteorite was found in 1981, and that it was the fifth sample collected in that year at the Allan Hills site.

(e) Some Meteorite Trivia

The largest meteorite ever found on the Earth is Hoba West, near Grootfontein, Namibia. It has a volume of about 7 cubic meters and an estimated mass of about 60 tons. The largest meteorite on display in a museum has a mass of 31 tons; it was recovered by Peary from Greenland in 1897 and is now at the American Museum of Natural History in New York.

The largest fallen meteorite found in a single piece in the United States was discovered in a forest near Willamette, Oregon, in 1902; it has a mass of 13 tons. (The discoverer spent about three months hauling the meteorite to his own property, where he put it on display for an admission price. Among those interested in the new exhibit were the attorneys of the Oregon Iron and Steel Company, owner of the land on which the meteorite was found. After litigation, it was decided that the company was the rightful owner of the object.)

There is no authenticated case of the killing of a human being by a meteorite fall; however, there have been some close calls. The Nahkla meteorite that fell in Egypt in 1911 is reported to have killed a dog. On September 29, 1938, a woman in an Illinois town heard a crash in the back yard. Later, it was found that a meteorite had pierced the roof of a nearby garage. A car was in the garage at the time, and the meteorite was found buried in the cushion of the car seat. In 1954 an Alabama woman was struck and injured by a ricocheting meteorite—the only modern case of injury.

21.4 REMNANTS OF CREATION

It was not until the ages of meteorites were measured and techniques developed for the detailed analysis of their chemistry and isotopic composi-

Figure 21.11 An iron meteorite cut and polished to show the crystal patterns known as Widmanstaetten figures. (R. Oriti, Griffith Observatory)

tions that their true significance as the oldest and most primitive materials available for direct study in the laboratory was appreciated. One of the earliest and most influential advocates of the importance of meteorite studies was chemist Harold Urey (1893–1981) of the University of California. His interest during the 1950s and 60s in using the primitive meteorites as probes of conditions at the birth of the solar system was largely responsible for the foundation of the modern science of *cosmochemistry*.

(a) Ages of Meteorites

The ages of stony meteorites can be determined from the careful measurement of the parent/daughter ratios of various radioactive decay reactions (Section 12.4) For irons this technique does not work, unless small silicate grains containing the required radioactive decay products can be identified in the metal. The ages of stony-iron meteorites are derived from their silicate fraction.

Almost all meteorites have radiometric ages between 4.5 and 4.6 billion years. The few exceptions are mostly igneous rocks—basalts—that will be discussed separately in Section 21.4c. The average age for all of the old meteorites, calculated using the best data and the most accurate values now available for the radioactive decay constants, is 4.55 billion years, with an uncertainty of less than 0.1 billion years. This value is taken to represent the *age of the solar system*—the time since the first solids condensed from the solar nebula and began to accrete into larger bodies.

(b) Compositions of Meteorites

The traditional classification of meteorites into irons, stones, and stony-irons is easy to use because it is obvious from inspection which category a meteorite falls into (although it may be much harder to distinguish a meteoritical stone from a terrestrial rock). Much more significant, however, is a system that classes meteorites in terms of their parent objects. The simplest such classification, paralleling the classes of asteroids (Section 19.2c), involves the distinction between *primitive* and *differentiated* meteorites.

A primitive meteorite is one that is made of materials that have never been subject to great heat or pressure since their formation from the solar nebula. The fiery passage of the meteorite through the air takes place so rapidly that the interior (below a burned crust a few millimeters thick) never even becomes hot, so a primitive meteoroid is still primitive (in the sense in which we use the word) after it lands on the Earth. Most, but not all, stone meteorites are primitive.

Differentiated meteorites are fragments of differentiated parent bodies. Like the igneous terrestrial rocks, they have been heated above their melting points and subjected to a degree of chemical reshuffling. The irons (Figure 21.11), which are derived from the metallic cores of their parent bodies, and the stony-irons (Color Plate 21), which probably represent the interface between iron core and silicate mantle, are all differentiated meteorites. Some stones are also differentiated, including sev-

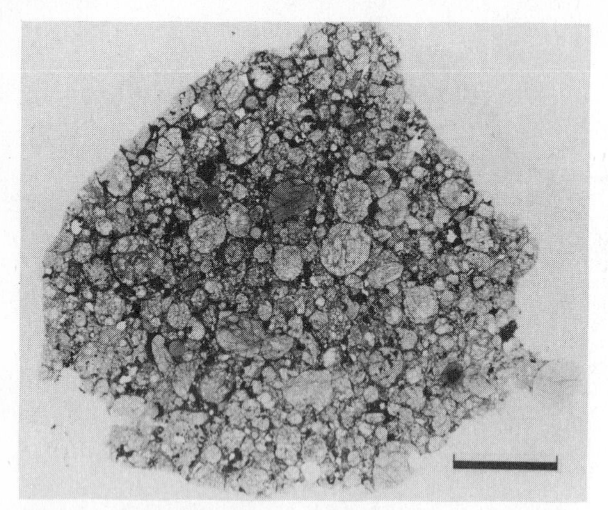

Figure 21.12 Enlarged cross-section of a primitive meteorite showing many chondrules, which are material from the solar nebula. Chondrules are typical of primitive meteorites, which are also called chondrites. (NASA)

eral groups composed of basalt that must have originated on volcanically active parent bodies.

The great majority of the meteorites that reach the Earth are primitive stones. The primitive meteorites are also called *chondrites,* because most of them contain small rounded granules or *chondrules* (Figure 21.12). The ordinary chondrites are mostly composed of light-colored grey silicates with some metallic grains mixed in, but there is also an important group of darker stones called *carbonaceous meteorites.* As their name suggests, these meteorites contain carbon, various complex organic compounds, and often chemically bound water; they are also depleted in metallic iron. The carbonaceous meteorites are presumably related to the dark, carbonaceous asteroids, which we saw (Figure 19.4) were concentrated in the outer part of the asteroid belt. Carbonaceous meteorites probably come from a source region more distant from the sun than that of the lighter-colored ordinary primitive stones.

Differentiated meteorites are much rarer, amounting to only about 10 percent of observed *falls.* In this respect the meteorites parallel the asteroids, which are predominantly primitive objects. However, irons make up nearly half of all *finds,* for the simple reason that they are much easier to recognize than the stones. Irons are also prominent in museum displays of meteorites because of their obviously extraterrestrial appearance.

TABLE 21.2 Frequency of Occurrence of Different Meteorite Classes

	FALLS	FINDS	ANTARCTIC
Primitive stones	87%	52%	85%
Differentiated stones	9	1	12
Irons	3	42	2
Stony-irons	1	5	1

Table 21.2 summarizes the frequencies of the different classes of meteorites among falls, finds, and the Antarctic meteorites.

(c) Basaltic Meteorites

The basaltic meteorites are samples from the surfaces of differentiated parent bodies that have experienced active volcanism. Because most asteroids are too small to have retained the internal heat necessary for volcanic eruptions, they are not likely to be the source of such meteorites. Any asteroid that has experienced volcanism is automatically interesting. But equally intriguing is the possibility that some of these meteorites are derived from the major, not the minor, planets.

The first basaltic meteorite to yield a definitive identification of its parent body was ALHA 81005, which was found at the Allan Hills Antarctic site in 1981. This meteorite is clearly lunar, similar in many ways to the samples returned in the Apollo program. More recently, two additional lunar samples have been identified by the Japanese among their collection of Antarctic meteorites. The presence of these lunar fragments demonstrates that cratering impacts can eject material with high enough velocity to escape from the Moon and impact the Earth (Color Plate 20b).

Another closely related group of basaltic meteorites is the SNC meteorites, mentioned in Chapter 15, which have solidification ages of 1.3 billion years (see Figure 15.14). These eight stones, including four from the Antarctic, are now generally believed to represent samples of the Martian surface, presumably from the Tharsis area. However, the mechanisms by which fragments could be ejected from Mars at escape velocity without being melted or even severely shocked are not well understood, and considerable effort is going into this problem at present (Color Plate 20c).

The third and largest group of basaltic meteorites with an identified parent body is the eucrites, mentioned in Chapter 19 (Figure 19.5). Largely by a process of elimination of alternatives, these approximately 30 stones are thought to represent samples of the surface of the asteroid 4 Vesta. With solidification ages of nearly 4.5 billion years, the eucrites represent a period of volcanism at the very beginning of solar system history.

(d) The Allende and Murchison Primitive Meteorites

The carbonaceous meteorites are the most primitive materials available for laboratory study, excepting the tiny Brownlee particle micrometeorites from comets. Two large carbonaceous meteorites that fell within a few months of each other have proved particularly valuable in probing the birth of the solar system.

The Allende meteorite fell in Mexico and the Murchison meteorite in Canada. Arriving in 1969 at the same time that many laboratories were preparing for analyses of the first Apollo lunar samples, these two meteorites were widely studied from the beginning. Indeed, Allende served as a "dry run" for the Apollo 11 samples (Figure 21.13).

Figure 21.13 The Allende carbonaceous meteorite. Some of the tiny grains in this meteorite were among the first solids to condense from the cooling solar nebula nearly 4.6 billion years ago. (NASA)

Murchison is best known for the variety of organic, or carbon-bearing, chemicals that it has yielded. Most of the carbon compounds in these meteorites are complex, tar-like substances that defy exact analysis. However, Murchison also contains 16 amino acids, 11 of which are rare on Earth. Unlike terrestrial amino acids, which are formed by living things, the Murchison chemicals include equal numbers with right-handed and left-handed molecular symmetry. The presence of these amino acids and other complex organic compounds in Murchison demonstrates that a great deal of interesting chemistry must have taken place in the solar nebula. Perhaps some of the molecular building blocks of life on Earth were actually derived from the primitive meteorites and comets.

The Allende meteorite has proved to be a rich source of information on the solar nebula because it contains many individual grains with varied chemical histories. Among the discoveries from Allende have been isotopic evidence of separate supernova explosions associated with the formation of the sun, identification of shortlived radioactive elements that may have contributed to the early heating of solid bodies immediately after the solar nebula condensed, and hints of the possible survival of a few pre–solar-system interstellar grains that were not vaporized in the formation of the solar nebula. Nearly 20 years after its fall, this plain grey stone continues to provide new insights into the formation and evolution of the solar system.

EXERCISES

1. Meteors apparently come primarily from comets, while the meteorites are thought to be fragments of asteroids. This may seem contradictory. Explain why we do not believe meteorites come from comets, or meteors from asteroids.
2. Two meteor cameras are located 190 km apart on an east-west line. A meteor is recorded by both cameras. The western camera shows its trail to begin 45° above the east point of the horizon, while the eastern camera shows its trail to begin 45° above the west

point on the horizon. At what altitude did the meteor trail begin?

3. Comets that have been associated with meteor showers are all periodic comets. Why do you suppose that showers have not been identified with comets having near-parabolic orbits?
4. Two meteoroids of the same mass enter the Earth's atmosphere at the same instant. Both produce observable meteors. One is moving 70 km/s, the other 35 km/s. Which meteor gives out more light? Estimate how many times as much light it produces as the other meteor.
5. For a long time astronomers speculated whether the interplanetary dust that causes the zodiacal light was concentrated near the Earth or whether it stretched to the asteroid belt or even beyond. Can you think of ways to distinguish between these hypotheses?
6. Explain why iron meteorites represent a much higher percentage of finds than of falls.
7. Why is it more useful to classify meteorites according to whether they are primitive or differentiated, rather than into stones, irons, and stony-irons?
8. Consider the differentiated meteorites. We think the irons are from the cores, the stony-irons from the interface between mantle and core, and the stones from the mantles of their differentiated parent bodies. If these parent bodies were like the Earth, what fraction of these meteorites would you expect to consist of irons, stony-irons, and stones? Is this consistent with the observed numbers of each?
9. Which meteorites are the most useful for defining the age of the solar system? Why?
10. We have suggested that the SNC meteorites are fragments from Mars. Suppose we are wrong. What other body might be the parent to these meteorites?
11. Suppose a new primitive meteorite is discovered and analysis shows that it contains a trace of amino acids, all of which show the same rotational symmetry (unlike the Murchison meteorite). What might you conclude from this finding?

Sir William Herschel (1738–1822), a German musician, emigrated to England to avoid service in the Seven Years' War. While composing and giving music lessons, he built the first large reflecting telescopes, which he used to survey the skies. He attempted the first quantitative measurements of star brightnesses and was the first to derive the motion of the sun through space. (Yerkes Observatory)

22 SURVEYING THE HEAVENS

Nearly all measurements of astronomical distances, as well as of distances on the Earth, depend directly or indirectly on the principle of triangulation. Some of the concepts of triangulation were known to the Egyptians; the art was developed by the Greeks.

22.1 DISTANCES IN THE SOLAR SYSTEM

(a) Triangulation

Six quantities describe the dimensions of a triangle: the lengths of the three sides and the values of the three angles. It is a well-known theorem in elementary geometry that any three of these quantities in succession around the perimeter of the triangle (for example, two sides and an included angle or two angles and an included side) determine the triangle uniquely.

As an example, suppose that in the triangle ABC (in Figure 22.1) the side AB and the angles A and B are all known. The triangle can then be constructed without ambiguity, for the side AB can be laid out and the lines AE and BD can be drawn at angles A and B, respectively, to the line AB. The two lines intersect at C, which completes the construction.

Suppose, in the triangle in Figure 22.1, that the point C represents an inaccessible object—say, a remote mountain peak, or an island in a large river, or the Moon. The distance to C can be determined by setting up two observation stations at A and B, separated by the known distance AB. This known distance (AB) is called a *base line*. At station A the angle A is observed between the directions to B and C. At station B, the angle B is observed between the directions to A and C. Enough information is

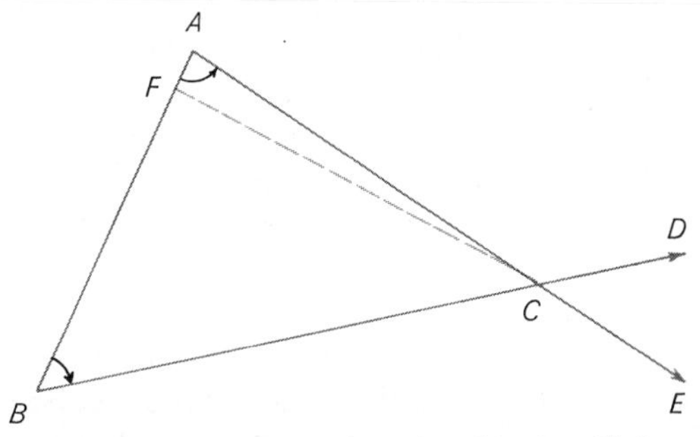

Figure 22.1 Construction of a triangle with two angles and an included side given.

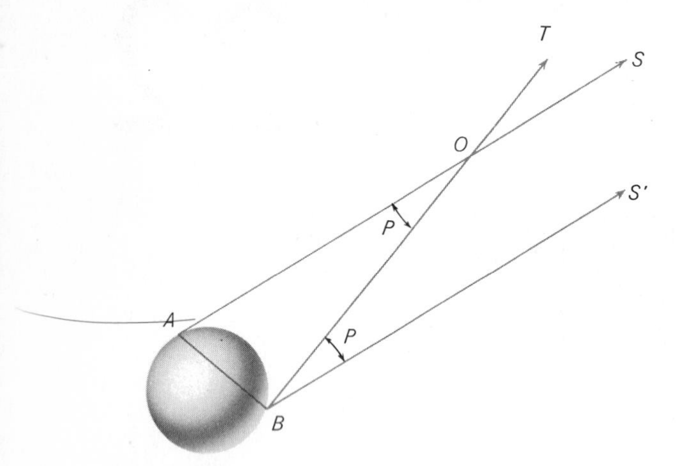

Figure 22.2 "Skinny" triangle.

now available to construct a scale drawing of the triangle ABC. The base line (AB) is first laid out on the drawing at some convenient scale. Lines AE and BD are then constructed at angles A and B to the base line, and their intersection is at C. The distance of C from any point on the base line (for example, A, or B, or F) can now be measured in the drawing; since the scale of the drawing is known, the real distances to C are determined.

The preceding two paragraphs describe a *geometrical* method for solving a triangle. In practice the triangle can be solved more simply and more accurately by numerical calculation. The solution of triangles by calculation rather than by geometrical construction is the subject of *trigonometry*.

(b) The "Skinny" Triangle

In astronomy we frequently have to measure distances that are very large in proportion to the length of the available base line, and the triangle to be solved is thus long and "skinny." Suppose (Figure 22.2) that it is desired to measure the distance to the Moon or a minor planet, located at O. Two observation stations, A and B, are set up on the Earth. The base line AB is known, since the size of the Earth and the geographical locations of A and B are known. From A, O appears in the direction AS, in line with a distant point, such as some very remote star. Because of the great distance of that star, observers at A and B would look along parallel lines to see it; thus from B the same star is in direction BS'. Observer B sees O, on the other hand, in direction BT, at an angle p away from direction BS'. This angle, p, is the same as the angle at O between the lines AO and BO (because the line BO intercepts the parallel lines AS and BS'). Note that p is the difference in the directions of O as seen from A and B; this is the apparent displacement (or *parallax*) of an object seen from two different points.

Now the problem is to find the distance to O, which lies somewhere along the line AS. This can be done in two ways. One way is to determine the angles OAB and OBA and then determine the distance to O by solving the triangle OAB as described in Section 22.1a. However, because the sides AO and BO are so nearly parallel, the two angles would have to be measured with very great precision. It would be very difficult to obtain adequate precision, because at each station (A and B) it would be necessary to compare the direction of O with that of the other terrestrial observation point. An alternative procedure for finding the distance AO is to measure the parallax p and then calculate how distant O must be in order for the base line AB to subtend that angle. Such a calculation is illustrated in Section 22.1c.

Because observers at A and B both see O projected against remote stars on the celestial sphere, each can measure O's direction among those stars with great accuracy; the difference between those directions, the parallax, can be found, therefore, with considerable precision. In astronomical practice the greatest emphasis is thus placed on measuring the parallax, rather than the angles at the end of the base line. Even the parallax, however, can be determined only as accurately as the direction to O can be measured with respect to the background of stars on the celestial sphere. The more distant the object, the smaller its parallax and the greater the uncertainty of its value in comparison to its size, that is, the greater its *percentage error,* and the greater the uncertainty in the distance to the object that is derived from its parallax. It is advantageous, therefore, that the base line be made as long as possible, so that the parallax will be as large as possible in comparison to the error of its measurement.

Even for the nearest stars, measured parallaxes are usually only a small fraction of a second of arc. In this text, we will write one second of arc as either 1 arcsec or $1''$. One minute of arc will be written as 1 arcmin or $1'$. One arcminute is, of course, equal to 60 arcseconds, and there are 60 arcminutes in one degree. A coin the size of a quarter would ap-

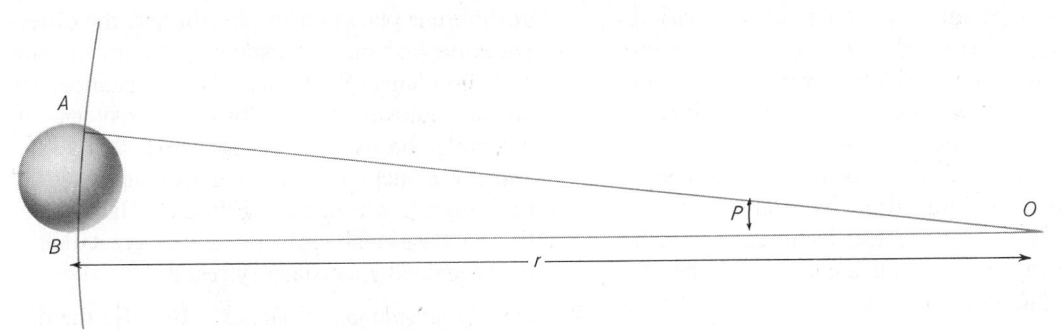

Figure 22.3 Solution of the "skinny" triangle.

pear to have a diameter of one arcsecond if it were at a distance of a little over three miles.

Even though the parallaxes of stars are very small indeed, we can under the most favorable circumstances measure parallaxes to within a few thousandths of an arcsecond with ground-based telescopes. It is hoped that the Hubble Space Telescope, with its capability of measuring much smaller images unblurred by the atmosphere, will achieve parallaxes of one-thousandth of an arcsecond or better. In addition, a special-purpose satellite called Hipparchos, to be built and operated by West Germany, may measure parallaxes for tens of thousands of stars brighter than about 12th magnitude over much of the sky to accuracies of at least two-thousandths of an arcsecond.

(c) An Example of the Solution of the "Skinny" Triangle

Suppose it is found that the displacement in direction of an object (at *O*, in Figure 22.3) viewed from opposite sides of the Earth is the angle *p*; *p*, then, is the angle at *O* subtended by the diameter of the Earth. Imagine a circle, centered on *O*, that passes through points *A* and *B* on opposite ends of a diameter of the Earth. If the distance of *O* is very large compared to the size of the Earth, then the length of the chord *AB* is very nearly the same as the distance along the arc of the circle from *A* to *B*. This arc is in the same ratio to the circumference of the entire circle as the angle *p* is to 360°. Since the circumference of a circle of radius *r* is $2\pi r$, we have

$$\frac{AB}{2\pi r} = \frac{p}{360^\circ}.$$

By solving for *r*, the distance to *O*, we find

$$r = \frac{360^\circ}{2\pi}\frac{AB}{p}.$$

If *p* is measured in seconds of arc, rather than in degrees, it must be divided by 3600 (the number of seconds in 1°) before its value is inserted in the above equation. After such arithmetic, the formula for *r* becomes

$$r = 206{,}265\,\frac{AB}{p(\text{in arcseconds})}.$$

As an example, suppose *p* is 18 seconds of arc (about what would be observed for the sun). Since *AB*, the Earth's diameter, is 12,756 km,

$$r = 206{,}265\,\frac{12{,}756}{18} = 1.46 \times 10^8\ \text{km}.$$

In Exercise 2 the Moon's distance is to be calculated by the same procedure.

(d) The Astronomical Unit

The planets and other bodies that revolve about the sun in the solar system are so far away that at any time each one is seen in almost the same direction by all terrestrial observers. The parallaxes of these bodies are therefore very small and difficult to measure accurately; a larger base line than the diameter of the Earth is desirable. As the Earth moves about the sun, however, it carries us across a base line that can be as large as the diameter of the Earth's orbit—about 300 million km. Thus, we can determine accurate distances to the other members of the solar system by observing them at times when the Earth is in two different places in its orbit. Kepler arranged that the different observations of the same body were made at intervals of its sidereal period when it was at the same place in its orbit (Section 3.3). This is not necessary, however, for we can take into account the motion of the object during the interval between the sightings of it. It is convenient, of course, not to have to restrict our observations of a body to intervals of its sidereal period. There are various mathematical techniques (which will not be gone into here) for unscrambling the effects of the combined motions of the body and the Earth.

Measures described above—measures within the solar system—are not obtained directly in kilometers, however; they are found in terms of the *astronomical unit,* the semimajor axis of the Earth's orbit. The foregoing procedure for surveying the distances to the planets provides us with an accurate map of the solar system, but to find the scale of the map—that is, to evaluate the astronomical unit—we must find the distance to some object that revolves about the sun, both in astronomical units and in kilometers (or some other terrestrial units).

The earliest known attempt to find the length of the astronomical unit was made by Aristarchus in the third century B.C. His value of 200 Earth diameters for the distance to the sun (the Earth's size was measured by Aristarchus' contemporary, Eratosthenes—Section 2.3) was later corrected to 600 Earth diameters by Hipparchus. The latter value survived until the 17th century, when Kepler estimated that it was too small by a factor of at least three because of the absence of a diurnal parallax of Mars.

Today the astronomical unit is known from radar measures of the distances to planets with high precision. Before describing the modern procedure for evaluating it, however, we shall review some of the historical approaches, which in themselves are very instructive.

OLDER METHODS OF DETERMINING THE ASTRONOMICAL UNIT

Before radar some of the fairly accurate methods of measuring AU were the following:

1. *Direct triangulation of the sun.* It is difficult to observe the sun's direction in the sky with respect to the stars. Occasionally, however, Mercury or Venus can be seen in transit across the sun's disk. The exact times of the start and finish of such a transit are different for observers at different places on the Earth, and the differences depend on the distances of both the sun and the planet. Since the relative distances are already known, transit observations can, in principle, be used to triangulate both the sun and the planet. This method was suggested by the English astronomer Edmund Halley, but the observational problems are so great that it never yielded a satisfactory result.
2. *Direct triangulation of planets.* Because the distances to the planets are known directly in AU, there is no need to attempt to measure the sun's parallax directly; better, we can triangulate the distance to Mars, which at its closest is only ⅓ AU away and is favorably situated for observations during an entire night. (We cannot use the Moon, however, for although its distance is found directly with considerable accuracy in terrestrial units, say, kilometers, it cannot be surveyed in AU from different sides of the sun, for as the Earth moves in its orbit, the Moon follows along with us.)
3. *Triangulation of asteroids.* Earth-approaching asteroids (Section 19.3c) come even closer than Mars. In 1932 the British astronomer Harold Spencer Jones coordinated a world wide effort to triangulate the distance to the asteroid Eros, which passed only 22 million km from Earth and had a parallax of about 2′.
4. *Gravitational perturbations of asteroids.* As an asteroid (for example, Eros) comes near the Earth, its orbit is changed by the Earth's gravitational effect on it. We can calculate the shape and size of its orbit, in AU, by surveying from various parts of our own orbit, both before and after the encounter, thereby obtaining an accurate measure of how the Earth's gravitational perturbation on the asteroid changed its orbit. The Earth's gravitational influence that produced this change, however, depends on how far in kilometers from the center of the Earth the asteroid must have passed. Since we know how far the asteroid passed in astronomical units, the number of kilometers per AU is determined. Because the gravitational pull of the Earth drops off with the *square* of the distance from Earth, the actual error in the derived distance is proportional to only the *square root* of the errors in the observations of the orbital changes, so this is a more sensitive determina-

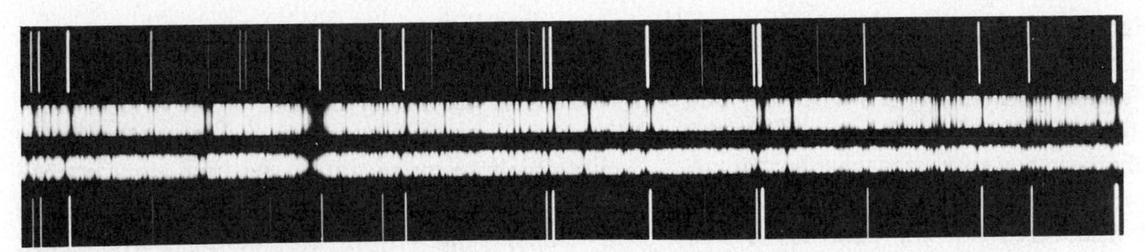

Figure 22.4 Two spectrograms of Arcturus, taken six months apart. On July 1, 1939 (top), the measured radial velocity was +18 km/s; on January 19, 1940, it was −32 km/s. The difference of 50 km/s is due to the orbital motion of the Earth. (Caltech/Palomar Observatory)

tion than is that of direct triangulation. Astronomer Eugene Rabe used this technique to find the distance to Eros when it passed in 1932.

5. *Doppler shifts.* Stars near the ecliptic show periodic changes in radial velocity during the year because of the Earth alternately approaching and receding from them. These changes, measured from the changing Doppler shifts of the stars' spectral lines (Section 8.3 and Figure 22.4), tell us how fast the Earth is moving in its orbit. This speed, in km/s, multiplied by the number of seconds in a year, gives us the circumference of the Earth's orbit and hence the length of the astronomical unit.
6. *Aberration of starlight.* The 20″.5 displacement of stars in the direction of the Earth's motion (Section 22.2a) indicates the speed of the Earth in terms of that of light. In 1862 the French physicist Jean Foucault used his measurement of the speed of light to find the Earth's speed and hence the distance to the sun.

MODERN DETERMINATION OF THE ASTRONOMICAL UNIT

We have already seen how measurement of the round-trip travel time of the radio waves beamed to the Moon and reflected back gives the Moon's distance accurately. Radar observations of Venus, the nearest approaching major planet, began in 1958. Today radar observations are routinely carried out on all but the most remote planets. Since we know the distance to each planet in astronomical units, measurement of the time required by radar signals to travel from the Earth to a planet and return establishes directly the size of an astronomical unit in kilometers. In order to convert this time to a distance, we must also know the speed of light, c. The Jet Propulsion Laboratory finds 1 AU = (499.004784s) × c. For c = 299,792.458 km/s, the value of the astronomical unit is 149,597,870.7 km, with an uncertainty of about 1 km. The use of radar now replaces direct triangulation.

The *solar parallax* is defined as the angle subtended by the radius of the Earth at a distance of 1 AU. The modern radar determination of the AU gives a value for the solar parallax of 8″.7940; before 1961 the best value was 8″.798, obtained by Rabe from the perturbations of Eros when it passed the Earth in 1932.

22.2 SURVEYING DISTANCE TO STARS

(a) Early Efforts

In principle one should be able to detect the parallax of a star—a change in its apparent direction—as the Earth moves from one side of its orbit to the other—over a baseline of 2 AU. The stars are so remote, however, that none has a parallax perceptible to the unaided eye. This fact alone, by the way, shows the stars to be self-luminous. At their great distances, to reflect enough sunlight back to us to account for their apparent brightnesses they would have to be extremely large and show observable disks, which they do not.

In the century and a half following the publication of Newton's *Principia,* by which time the motion of the Earth was generally conceded, a number of investigators had tried to detect stellar parallaxes telescopically. One of them, the German-English astronomer William Herschel, catalogued many instances of two stars appearing close together in the sky. Herschel thought that careful observation of the brighter, presumably nearer star with respect to the fainter, presumably more remote one might reveal the parallax of the former. What the observations showed, however, was that the stars in a pair are in mutual revolution about each other; Herschel had discovered not the parallax he was searching for but rather that many stars are members of binary star systems. As we shall see in Chapter 25, Herschel's discovery was even more important than parallax, for by analyzing binary stars we learn a great deal about stellar masses and sizes; in addition, Newton's laws were shown to apply to these stars as well as to the planets.

Another serendipitous discovery was made during an attempt to detect stellar parallax by James Bradley in 1729. Bradley mounted a vertical telescope in his chimney and measured the positions of stars as the rotation of the Earth carried them past an illuminated reticle or scale in the field of view of his telescope. He had hoped that some relatively nearby stars, because of parallax, would pass the reticle in different places at different times of the year. What he found instead is that *every* star shifts its direction periodically during the year by 20″.5 in either direction. This phenomenon is *aberration of starlight*.

Parallaxes were eventually detected, but even the nearest star shows a total annual displacement of only about 1″.5. The first observation of the parallax of a star is usually credited to the German astronomer Friedrich Bessel, who measured the parallax of 61 Cygni in 1838 and obtained a value only about 6 percent too large. Later in the same year Thomas Henderson, at the Cape of Good Hope, and Friedrich Struve, in Russia, reported measures of the parallaxes of the stars Alpha Centauri and Vega, respectively. At last quantitative measurements had extended beyond the solar system, opening the new era of stellar astronomy.

(b) The Parallactic Ellipse

As the Earth moves about its orbit, the place from which we observe the stars continually changes. Consequently, the positions of the comparatively near stars, projected against the more remote ones, are also continually changing. If a star is in the direction of the ecliptic, it seems merely to shift back and forth in a straight line as the Earth passes from one side of the sun to the other. A star that is at the pole of the ecliptic (90° from the ecliptic) seems to move about in a small circle against the background of more distant stars, as we view it from different positions in our nearly circular orbit. A star whose direction is intermediate between the ecliptic and ecliptic pole seems to shift its position along a small elliptical path during the year. The eccentricity of the ellipse ranges from that of the Earth's orbit (nearly a circle) for a star at the ecliptic pole to unity (a straight line) for a star on the ecliptic (see Figure 22.5). This small ellipse is called the *parallactic ellipse*.

(c) Stellar Parallax and Stellar Distances

The angular semimajor axis of the parallactic ellipse is called the *stellar parallax* of the star. Since the major axis of the ellipse is the maximum apparent angular deflection of the star as viewed from opposite ends of a diameter of the Earth's orbit, the stellar parallax, the semimajor axis of this ellipse, is the angle, at the star's distance, subtended by 1 AU perpendicular to the line of sight.

Until now we have implied that the star whose parallax is observed is motionless with respect to the sun. Actually, the stars are all moving at many kilometers per second (Section 22.3). The effects of the relative motion of the star and the sun can be separated from the effects of the Earth's motion (the star's parallax) by observing the star's direction at the same time of the year several years apart. Any observed change in its direction then must be due to its own motion relative to the sun. The change indicates the corrections that must be applied to observed total changes in the direction of the star in order to obtain its true stellar parallax.

(d) Units of Stellar Distance

If a line of length D subtends an angle of p seconds of arc as seen from a distant object, the distance r of that object is given by the formula derived in Section 22.1c:

$$r = 206{,}265\,\frac{D}{p}.$$

Since the parallax of a star is the angle, in arcseconds, subtended by 1 AU at the star's distance, the distance of a star, in astronomical units, is $206265/p$. The length 206,265 AU is defined as a *parsec* (abbreviated pc). One parsec is, therefore,

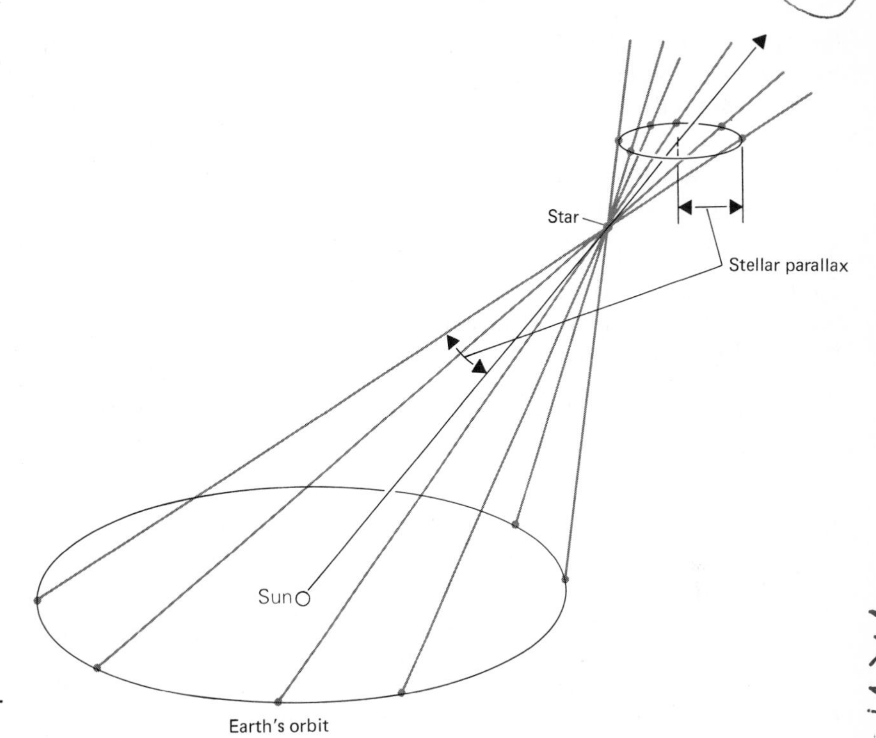

Figure 22.5 The parallactic ellipse.

the distance of a star that would have a *par*allax of one *sec*ond. The distance of any star, in parsecs, is thus given by

$$r = \frac{1}{p},$$

where p is the parallax of the star. A star with a parallax of ½″, for example, has a distance of 2 pc; one with a parallax of ⅒″ has a distance of 10 pc. Parallaxes of stars are usually measured in arcseconds, so the parsec is a convenient unit of distance—a star's distance in parsecs is simply the reciprocal of its parallax. One parsec is 3.086×10^{13} km. Since 1 pc is 206,265 AU, the sun's distance is 1/206,265 pc.

Another unit of stellar distance is the *light year,* which is the distance traversed by light in one year at the rate of 299,792.5 km/s. One light year (abbreviated LY) is 9.46×10^{12} km; 1 pc contains 3.26 LY.

(e) The Nearest Stars

The nearest stellar neighbors to the sun are three stars that make up a multiple system. To the naked eye the system appears as a single bright star, Alpha Centauri, which is only 30° from the south celestial pole and hence is not visible from the mainland United States. Alpha Centauri itself is a double star—two stars revolving about each other that are too close together to be separated by the naked eye. Nearby is the third member of the system, a faint star known as *Proxima Centauri.* Proxima is slightly closer to us than the other two stars of the system. All three have a parallax of about 0.″76, and a distance of 1/0.76, or about 1.3 pc (4.3 LY). The nearest star, except for the sun, visible to the naked eye from most parts of the United States is the brightest-appearing of all the stars, *Sirius.* Sirius has a distance of 2.6 pc, or about 8 LY. It is interesting to note that light reaches us from the sun in eight minutes and from Sirius in eight years.

Parallaxes have been measured for thousands of stars. Only for a fraction of them, however, are the parallaxes large enough (about 0.″05 or more) to be measured with a precision of 10 percent or better. The 1969 edition of Gliese's catalogue of nearby stars lists 1049 within 20 pc, but the total number of such stars, including those not yet discovered may be near 4000. Of those stars within about 20 pc, most are invisible to the unaided eye and actually are intrinsically less luminous than the sun. Most of the stars visible to the unaided eye, on the other hand, have distances of hundreds or even thousands of parsecs and are visible not because they are relatively close, but because they are intrinsically very luminous. The nearer stars are described more fully in Chapter 26.

(f) Other Methods of Measuring Stellar Distances

The vast majority of all known stars are too distant for their parallaxes to be measured, and we must resort to other methods to determine their distances. Most of these methods are either statistical or indirect; they are discussed in later chapters. For completeness, however, some of the more important procedures, other than parallax measurement, for determining stellar distances are listed:

1. *Stellar motions (Section 22.5b).* All stars are in motion, but only for comparatively nearby ones are the angular motions perceptible. Statistically, therefore, the stars that have large apparent motions are the nearer ones; we can estimate the average distance to stars in a large sample from the average angular motions of those stars.
2. *Moving clusters (Section 22.5c).* In a few cases the direction of motion through space of a cluster or swarm of stars can be determined from the apparent convergence or divergence of the directions of motions of the individual stars in that cluster. In such a case, an analysis of the apparent motions and radial velocities of the member stars gives the distance to the cluster.
3. *Inverse-square law of light (Chapter 23).* The apparent brightness of a star depends on both its intrinsic luminosity and its distance (through the inverse-square law of light). Very often it is possible to infer the intrinsic brightness of a star from its spectrum (Chapter 26) or because it is a recognizable type of variable star (Chapter 23). Then its distance can be calculated from a knowledge of its observed brightness.
4. *Interstellar lines (Chapter 27).* The space between the stars throughout much of space contains a sparse distribution of gas. Sometimes this interstellar gas leaves absorption lines superposed upon the spectrum of a star whose light must shine through the gas to reach us. The amount of the star's light absorbed by these interstellar lines indicates the total mass of the gas that must lie in the light path from the star. If we can estimate the approximate density of the gas in space (as we often can), we can tell what the total path length of the star's light must be, and hence the distance of the star.
5. *Galactic rotation (Chapter 34).* The sun and its neighboring stars are part of a vast system of stars—our Galaxy. The Galaxy is rotating; the stars that compose it all revolve about its center much as the planets revolve about the sun. The speeds with which distant stars in the Galaxy approach us or recede from us as a result of this galactic rotation depends on the directions and distances of these stars. Observations of their radial velocities, therefore, can lead to an estimate of their distances.

22.3 STELLAR MOTIONS

The ancients distinguished between the "wandering stars" (planets) and the "fixed stars," which seemed to maintain permanent patterns with one another in the sky. The stars are, indeed, so nearly fixed on the celestial sphere that the apparent groupings they seem to form—the constellations—look today much as they did when they were first named, more than 2000 years ago. Yet the stars are moving with respect to the sun, most of them with speeds of many kilometers per second. Their motions are not apparent to the unaided eye in the course of a single human lifetime, but if an ancient observer who knew the sky well—Hipparchus, for example— could return to life today, he would find that several of the stars had noticeably changed their positions relative to the others. After some 50,000 years or so, terrestrial observers will find the handle of the Big Dipper unmistakably "bent" more than it is now (Figure 22.6). Changes in the positions of the nearer stars can be measured with telescopes after an interval of only a few years.

(a) Proper Motion

The *proper motion* of a star is the rate at which its direction in the sky changes, usually expressed in seconds of arc per year. Proper motion is a consequence of the intrinsic motion of the star relative to the sun. (Parallax, in contrast, is a measure of the change in apparent direction to a star caused by the orbital motion of the Earth.) It is almost always an

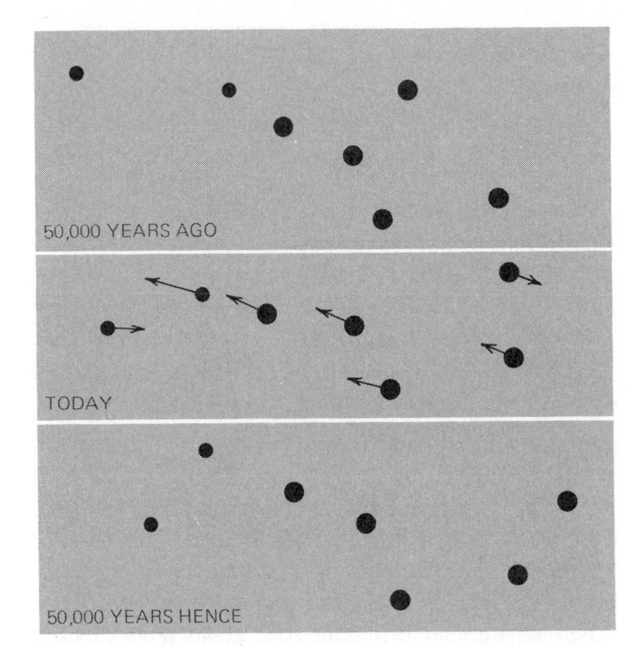

Figure 22.6 Appearance of the Big Dipper over 100,000 years.

angle that is too small to measure with much precision in a single year; in an interval of 20 to 50 years, on the other hand, many stars change their directions by easily detectable amounts. The modern procedure for determining proper motions is to compare the positions of the star images on two different images of the same region of the sky taken at least a decade apart. Most of the star images on such photographs do not appear to have changed their positions measurably; these are, statistically, the more distant stars that are relatively "fixed," even over the time interval separating the two photographs. With respect to these "background" stars, the motions of a few comparatively nearby stars can be observed. Modern practice is to measure proper motions of stars with respect to remote galaxies, which can show no measurable proper motions themselves (Chapter 35).

The star of largest known proper motion is *Barnard's star*, which changes its position by 10.″25 each year (Figure 22.7). This large proper motion is partially due to the star's relatively high velocity with respect to the sun but is mostly the result of its relative proximity. Barnard's star is the nearest known star beyond the triple system containing Alpha Centauri; its distance is only 1.8 pc, but it emits less than 1/2500 as much light as the sun and is about 25 times too faint to see with the unaided eye. There are several hundred stars with proper motions greater than 1.″0 per year. The mean proper motion for all naked-eye stars is less than 0.″1; nevertheless, the proper motions of most stars are larger than their stellar parallaxes.

The complete specification of the proper motion of a star includes not only its angular rate of motion, but also its *direction* of motion in the sky.

(b) Radial Velocity

The *radial* velocity (or line-of-sight velocity) of a star is the speed with which it approaches or recedes from the sun. This can be determined from the Doppler shift of the lines in its spectrum (Section 8.3). Unlike the proper motion, which is observable only for the comparatively nearby stars, the radial velocity can be measured for any star that is bright enough for its spectrum to be photographed. The radial velocity of a star, of course, is only the component of its actual velocity that is projected along the line of sight, that is, that carries the star toward or away from the sun. Radial veloc-

Figure 22.7 Two photographs of Barnard's star, showing its motion over a period of 22 years. (Yerkes Observatory)

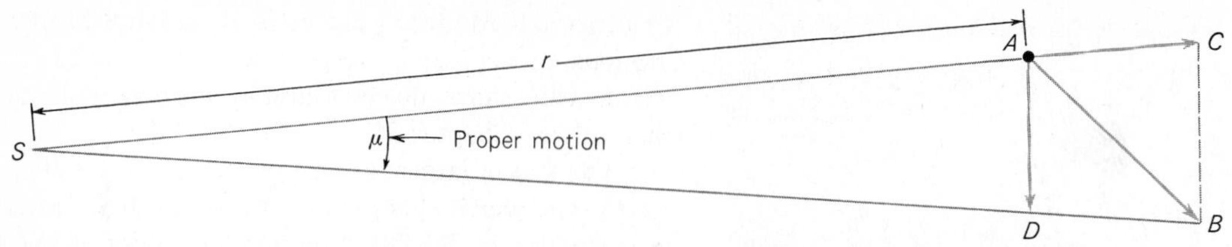

Figure 22.8 Relationship of proper motion, radial velocity *(AC),* and tangential velocity *(AD).*

ity is usually expressed in kilometers per second and is counted as *positive* if the star is moving *away* from the sun, and *negative* if the star is moving *toward* the sun. Since motion of either the star or the observer (or both) produces a Doppler shift in the spectral lines, a knowledge of the radial velocity alone does not enable us to decide whether it is the star or the sun that is "doing the moving" (indeed, as we saw in Chapter 11, it does not even make sense to ask which is moving). What we really measure, therefore, is the speed with which the distance between the star and sun is increasing or decreasing, that is, the star's radial velocity *with respect to the sun.*

Actually, the radial velocity of a star with respect to the sun is not obtained directly from the measured shift of its spectral lines, because we must observe the star from the Earth, whose rotational and orbital motions contribute to the Doppler shift. Since the direction of the star is known, however, as are the speed and direction in which the moving Earth carries the telescope at the time of observation, it is merely a problem in geometry (albeit a slightly complicated one) to correct the observed radial velocity to the value that would have been found if the star had been observed from the sun.

(c) Tangential Velocity

Radial velocity is a motion of a star along the line of sight, while proper motion is the angular motion produced by the star's motion *across*, or at right angles to, the line of sight. Whereas the radial velocity is known in kilometers per second and is independent of distance, the proper motion of a star does not, by itself, give the star's actual *speed* at right angles to the line of sight. The latter is called the *tangential* or *transverse* velocity. To find the tangential velocity of a star, we must know both its proper motion and *distance*. A star with a proper motion of 1".0, for example, might have a relatively low tangential velocity and be nearby, or a high tangential velocity and be far away.

The relation between tangential velocity and proper motion is illustrated in Figure 22.8. As seen from the sun *S*, a star *A*, at distance *r*, is in direction *SA*. During one year it moves from *A* to *B* and then appears in direction *SD*, at an angle μ (the proper motion) from *SA*. The star's radial motion is *AC*—it has moved a distance *AC* farther away during the year; its *tangential* motion is *AD*—it has moved that distance across the line of sight. The motions shown in the figure are grossly exaggerated; actual stars do not move enough to change their distances by an appreciable percentage, even in 100 years (see Exercise 14).

The tangential motion *AD* can be approximated very accurately by a small arc of a circle of radius *r* centered on the sun. The arc *AD* is the same fraction of the circumference of the circle, $2\pi r$, as the proper motion is of 360°. The proper motion is expressed in seconds of arc per year, so we have

$$\frac{\mu}{1{,}296{,}000} = \frac{AD}{2\pi r}$$

(there are 1,296,000" in 360°). If we solve for *AD* in km/yr, and note that *AD* is the product of the star's tangential velocity in km per second, *T*, and the number of seconds in a year (3.16×10^7), we obtain

$$T = \frac{\mu r}{6.52 \times 10^{12}} \text{ km/s},$$

where *r* must be in kilometers. If *r* is expressed in parsecs, we must multiply the right-hand side of the above equation by 3.086×10^{13}, the number of kilometers per parsec. The formula for the tangential velocity then becomes

$$T = 4.74\ \mu r = 4.74\ \frac{\mu}{p}\ \text{km/s},$$

where p is the stellar parallax, which is merely the reciprocal of the distance in parsecs (Section 22.2d). The above equations show how the tangential velocity of a star is related to its proper motion and distance (or parallax).

(d) Space Velocity

The *space velocity* of a star is its total velocity in kilometers per second with respect to the sun. The radial velocity is the distance the star moves toward or away from the sun in one second; the tangential velocity is the distance it moves at right angles to the line of sight in one second. The space velocity, therefore, the total distance the star moves in one second, is simply the hypotenuse of the right triangle whose sides are the radial and tangential velocities (Figure 22.9). It is found immediately by the theorem of Pythagoras,

$$V^2 = V_r^2 + T^2$$

where V and V_r are the space and radial velocities, respectively.

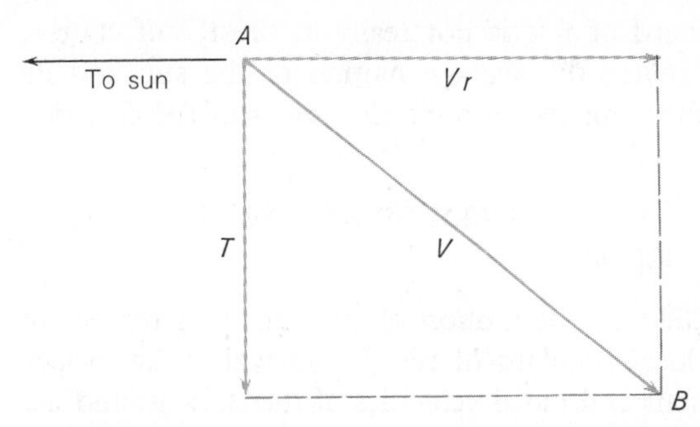

Figure 22.9 Space velocity; the star moves from *A* to *B* in one second.

22.4 MOTIONS OF THE SUN AND STARS

It might be expected that the sun, a typical star, is in motion, just as the other stars are. We turn now to the motion of the sun, and how it affects the apparent motions of the stars.

(a) The Local Standard of Rest

As we know, the sun is a member of the Milky Way Galaxy, a system of a hundred thousand million stars. The most luminous stars in the Galaxy lie in a disk, which is flat, like a pancake, and is rotating. The sun, partaking of this general rotation of the Galaxy, moves with a speed of about 250 km/s to complete its orbit about the galactic center in a period of about 200 million years. At first thought it might seem that the galactic center is the natural reference point with which to refer the stellar motions. However, our observations of the proper motions and radial velocities of the stars that surround the sun in space, all in our own so-called local "neighborhood" of the Galaxy, do not give us directly the motions of these stars about the galactic center. The reason is that the stars' orbits around the galactic center and their orbital velocities are both nearly the same as those of the sun. The motions we observe are merely small residual or *differential* motions of these stars with respect to the sun. These small residual motions arise because our neighboring stars' orbits about the galactic center are not absolutely identical to our own. We are overtaking and passing some stars, while others are passing us; the slightly different eccentricities and inclinations of our respective orbits bring us closer to some stars and carry us farther from others. We can study these residual motions without knowing anything about the actual motions of stars around the center of the Galaxy. Our situation is analogous to that of our driving an automobile on a busy highway. All the cars around us are going the same direction and at roughly the same speed, but some are changing lanes and others are passing each other. More or less like the highway traffic, the residual motions of the stars around us seem to be helter-skelter.

Astronomers have defined a reference system (that is, a coordinate system) within which the motions of the stars in the solar neighborhood—within a hundred parsecs or so—average out to zero. In other words, those stars in our neighborhood appear, on the average, to be at rest in this system; it is thus called the *local standard of rest.* The local standard of rest is not really at "rest," of course, but shares the average motion of the sun and its neighboring stars around the center of the Galaxy.

(b) The Solar Motion and Solar Apex

We deduce the motion of the sun with respect to the local standard of rest by analyzing the proper motions and radial velocities of the stars around us. The easiest way to understand how the sun's motion is found is to consider the effect it has on the apparent motions of the other stars.

First, consider the radial velocities of stars with respect to the sun. Note that we have defined the local standard of rest as being stationary with respect to the average of the motions of the stars in the solar neighborhood. Therefore, if we could correct the observed space motions of the stars to those values they would have if the sun were not moving in the local standard of rest, they would then average out to zero. Now it is clear that in a direction that is at right angles to the direction in which the sun is actually moving, the solar motion cannot affect the observed radial velocities of the stars. In those directions, indeed, we find as many stars approaching us as receding from us—their radial velocities *do* average zero. On the other hand, if we look in the direction *toward* which the sun is moving, we find that most of the stars are approaching us, because, of course, we are moving forward to meet them. The only stars in that direction that have radial velocities of recession are those that are moving in the same direction we are going, but at a faster rate, so that they are pulling away from us. The observed radial velocities of all the stars in the direction toward which the sun is moving do not average to zero, but to −20 km/s, showing that we are moving toward them at about 20 km/s. Similarly, stars in the opposite direction have an average radial velocity of about +20 km/s, because we are pulling away from them at that speed.

Now consider the proper motions of stars. Part of a star's proper motion, in general, is due to its own motion and part is due to the sun's motion. However, the sun's motion can contribute nothing to a star's proper motion if the star happens to lie in the direction toward which the sun is moving. Therefore, if we look at stars that lie in a path along the direction of the solar motion, as many of their proper motions should be in one direction as any other; the average of the motions of many stars in those directions, therefore, should be zero. On the other hand, the maximum effect on the proper motions of stars should occur in directions that are at right angles to the direction of the solar motion. If the stars were at rest, they would all show a backward drift due to our forward motion. As it is, the stars have motions of their own, but only those moving in the same direction we are and at a faster rate appear to have "forward" proper motions—the rest, by far the majority, *do* appear to drift backward.

William Herschel was the first to attempt to detect the direction of the solar motion from the proper motions of stars. In 1783 he analyzed the proper motions of 14 stars and deduced that the sun was moving in a direction toward the constellation Hercules—a nearly correct result.

Modern analysis of the proper motions and radial velocities of the stars around the sun has shown that the sun is moving approximately toward the direction now occupied by the bright star Vega in the constellation of Lyra. The value found for the sun's speed depends somewhat on what stars are observed to determine it. Analysis of most of the stars in the standard catalogues gives the *standard solar motion*, which is 19.7 km/s (4.14 AU/yr). The direction in the sky toward which the sun is moving is called the *apex* of solar motion,* and the opposite direction, away from which the sun is moving, is called the *antapex*.

(c) Peculiar Velocities of Stars

The velocity of a star with respect to the local standard of rest is called its *peculiar velocity*. The *space velocity* of a star, its motion with respect to the sun, is made up of both the star's peculiar velocity and a component due to solar motion. Since the solar motion is known, the peculiar velocity can be calculated for a star of known space velocity.

Consider, for example, the special case of a star that is not moving with respect to the local standard of rest and which therefore has zero peculiar velocity. The entire space motion of that star, then, is merely a reflection of the solar motion; its space ve-

---

* The equatorial coordinates of the *solar apex* are (1900) $\alpha = 18^h 4^m \pm 7^m$, $\delta = +30° \pm 1°$.

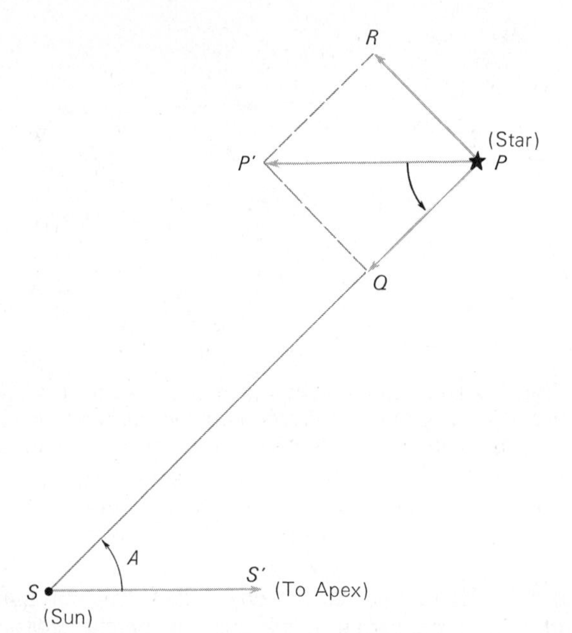

Figure 22.10 How the solar motion affects a star's space motion.

locity is a vector of magnitude exactly equal to the solar velocity but directed toward the *solar antapex*. In Figure 22.10, the vector SS' and PP' represent the solar velocity and that star's space velocity, respectively; A is the angle at the sun between the directions to the star and to the solar apex. Vector PQ, the projection of PP' onto the line of sight from the sun to the star, is the star's radial velocity. Vector PR, the component of PP' that is perpendicular to the line of sight, is the star's tangential velocity. Since this hypothetical star has no peculiar velocity, PQ and PR are due entirely to the solar motion and are, therefore, the corrections that must be applied to the observed radial and tangential velocities of *any* star in that same direction to obtain, respectively, its *peculiar radial velocity* and *peculiar tangential velocity*. The angle A can always be observed; consequently, the desired corrections to the radial and tangential velocities can always be found geometrically or calculated trigonometrically.* The correction to the star's observed proper motion that is required to obtain its *peculiar proper motion* can be found from the relation between proper motion and tangential velocity given by the formula in Section 22.3c. The peculiar proper motion can be found only for stars of known distance.

* The corrections to the observed radial and tangential velocities are $V_0 \cos A$ and $V_0 \sin A$, respectively, where V_0 is the solar velocity (about 19.5 km/s).

22.5 DISTANCES FROM STELLAR MOTIONS

The proper motions of stars can be expected to be largest, statistically, for the nearest stars. If, for example, a star is only a few parsecs away, its proper motion will almost certainly be observable after a few years (but see Exercise 20). The proper motion of a very distant star, on the other hand, may only be detectable after a long time, and then only if the star has a very great space velocity. Searches for nearby stars, therefore, are usually conducted by searching for stars of large proper motion. Conversely, remote stars can be identified by their lack of observable proper motions; they serve as standards against which we can measure the parallaxes of the nearby stars to determine their distances. The proper motion of an individual star does not in itself indicate its distance uniquely. However, proper motions and distances of stars are inversely correlated, and investigations of such motions do give some information about stellar distances.

(a) Secular Parallaxes

Trigonometric parallaxes of stars are based on triangulation with a base line of only 2 AU, the diameter of the Earth's orbit. The solar motion, however, amounts to 4.1 AU/yr. If we wait long enough between observations, the motion of the sun will carry us over as long a base line as we desire. For example, after 20 years we move 82 AU; with such a base line we should be able to detect parallaxes of stars 41 times as distant as we can with a base line of 2 AU. By this method, accurate distances to many stars could be determined if they did not have peculiar velocities of their own. Unfortunately, without knowing a star's distance in advance, we have no way of knowing how much of its proper motion is due to the solar motion and how much to its own peculiar velocity. On the other hand, the peculiar velocity of a large number of stars, spaced well about the sky, should average out to zero, for peculiar velocities are measured with respect to the local standard of rest, which is *defined* to have zero average velocity with respect to stars in the sun's neighborhood. Consequently, the *average* of the space velocities for a large sample of stars must be due entirely to the solar motion. The proper motion corresponding to the average space velocity depends upon the average of the distances to those stars. This motion is an average angular drift of the stars in the sample toward the solar antapex; it results from the fact that the sun carries us over a long base line. Since we know how far the sun carries us in any interval of time, we can calculate the average of the distances for the stars in question (more accurately, we calculate the *mean parallax*). The procedure is called the method of *secular parallaxes*.

At first thought, it might not seem particularly useful to know only the average of the distances to a large number of stars, because some of those stars will be much more distant than others, and we have no way of knowing which are which. Suppose, however, that we have chosen the sample in such a way that we have reason to believe that all the stars in it have more or less equal luminosities (as determined, say, from their spectra). In that case, the brighter-appearing stars would be the nearer and the fainter-appearing stars the more distant. The inverse-square law of light, in other words, allows us to calculate the relative distances of stars of similiar luminosities. A knowledge of their mean distance, found from the method of secular parallaxes, then enables us to find both their actual distances and luminosities. (The relation between the apparent brightness, luminosity, and distance of a star will be further explained in Chapter 23.) The method of secular parallaxes has proved very useful for determining the luminosities of relatively rare stars of certain types that are not represented by examples near enough for direct parallax measurement.

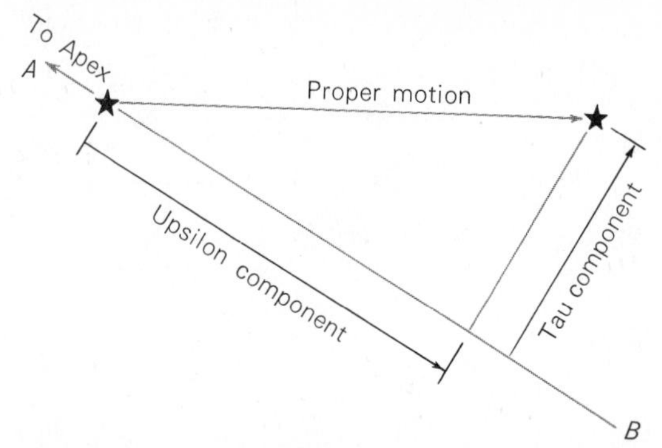

Figure 22.11 Components of proper motion; the figure is in the plane of the sky, perpendicular to the line of sight. The line *AB* is part of a great circle running through the star and the solar apex and antapex.

(b) Statistical Parallaxes

The method of secular parallaxes is based on the detection of the average backward drift, due to solar motion, in a large sample of stars. Now, because the sun moves toward the apex, that part of a star's motion due to the solar motion is a drift toward the antapex. However, the solar motion cannot affect the component of a star's proper motion that is in a direction *perpendicular* to a line between the apex and antapex. It is convenient, therefore, to regard the proper motion of a star as having two components: one—the *upsilon component*—along a great circle on the celestial sphere from the star through the antapex, that is, parallel to the direction to the solar apex, and another—the *tau component*—perpendicular to the direction of the apex (Figure 22.11). The upsilon component of the proper motion of a star can be augmented or diminished by the solar motion, but the tau component can involve only the star's peculiar motion—in fact, only that part of it that contributes to the star's tangential velocity perpendicular to the direction of the solar apex. If we had some way of knowing the value of this perpendicular component of the star's tangential velocity, $T_\perp$, we could find its parallax from the tau component, τ, of its proper motion, by applying the formula of Section 22.3c:

$$T_\perp = 4.74 \frac{\tau}{p} \text{ km/s.}$$

There is no way, in general, of knowing $T_\perp$ for an individual star. The stars in a large sample, however, should be moving at random in the local standard of rest—as many are going in any one direction as in any other. We should expect, therefore, that in such a sample the average value of $T_\perp$ should be the same as the average value of the *peculiar radial velocities,* and so $T_\perp$ can be replaced by the latter (without regard to sign) in the above equation. If the quantity τ is then replaced by the average value of the many individual values of τ for the stars in the sample, the average value of the parallaxes of those stars can be computed. This procedure is called the method of *statistical parallaxes.*

Like secular parallaxes, statistical parallaxes do not indicate distances of individual stars but only the average distances for large numbers of stars. The method is very useful, however, if applied to a sample of stars that can all be expected to have the same intrinsic brightness. It gives greater accuracy than does the method of secular parallaxes, when applied to stars whose peculiar velocities are statistically larger than the sun's velocity.

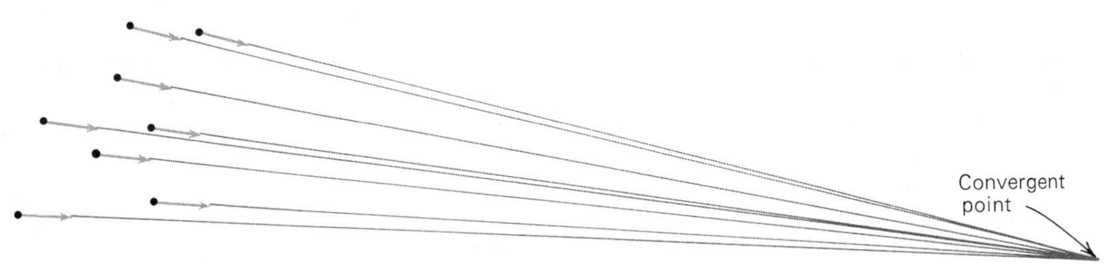

Figure 22.12 Proper motions of the stars in a moving cluster. Note that here we see the cluster as seen from the Earth. We see the projections of the stellar motions on the plane of the sky. Although the stars are moving parallel to one another in space, their proper motions appear to converge toward a point infinitely far away. The direction from the Earth to that convergent point is also parallel to the space motions of the cluster stars.

(c) Moving Clusters

The stars that belong to a group or cluster, moving more or less together through space, all have about the same space velocity. If such a cluster is close enough to the sun, however, the proper motions of its member stars may not all be exactly parallel. If, for example, the cluster is approaching us, its stars appear to radiate away from a distant point in the direction from which the cluster is coming. Conversely, if the cluster recedes from us, the stars appear to converge toward the direction in which the cluster is moving. In either case, the proper motions of the member stars indicate the direction in which the cluster is traveling. The situation for a hypothetical cluster is illustrated in Figure 22.12, which shows the apparent convergence of the proper motions of its stars toward a distant "convergent point," and in Figure 22.13, which shows the actual space velocities of the stars. In these diagrams the cluster is viewed from two different directions.

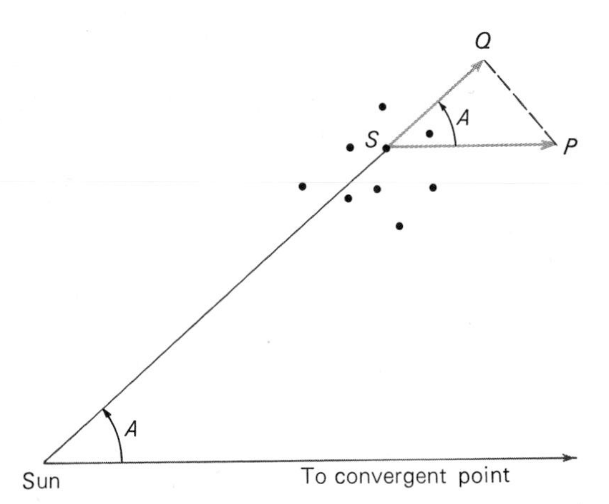

Figure 22.13 Space velocities of the stars in a moving cluster.

In studying a moving cluster for which the direction of convergence or divergence can be observed, we have the great advantage of knowing the angle A (Figure 22.13) between that direction of motion and the line of sight from the sun to the cluster. Now consider any star in the cluster, such as star S. Its space velocity, radial velocity, and tangential velocity are SP, SQ, and QP, respectively. Angle SQP is a right angle, and angle PA and the star's radial velocity are both observed. Thus the tangential velocity QP can be calculated by solving the triangle.* Then the parallax (or distance) of the star S, and thus that of the cluster, can be found from the observed proper motion of the star and from the formula relating tangential velocity and proper motion:

$$QP = 4.74\mu r = 4.74 \frac{\mu}{p} \text{ km/s.}$$

There are several groups and clusters that are near enough for us to observe the convergence or divergence of the proper motions of their member stars. The three best-known examples are the *Hyades* in Taurus, the *Ursa Major* group (which contains most of the naked-eye stars in the Big Dipper), and the *Scorpio-Centaurus* group. These three clusters contain, in all, more than 500 stars. This is a sizable fraction of the number of stars whose parallaxes can be directly observed to an accuracy of 10 percent or better. Thus, this method of using moving clusters to obtain stellar distances is very important and fruitful. It has contributed greatly to our knowledge of fundamental data about stars.

* $QP = SQ \tan A$.

EXERCISES

1. At a distance of 3500 m from the base of a vertical cliff, the top of the cliff is observed to have an angular altitude of 45°. What is its height?
2. The Moon's position among the stars was observed at moonrise. Six hours and 12 minutes later the Moon happened to be exactly at the zenith. Its position among the stars by then had shifted to the east by 148′. On the other hand, because of the Moon's orbital motion, it is known to move eastward on the celestial sphere at the rate of 33′ per hour. Find the distance to the Moon. Compare your result with the value given in Section 7.1.
3. At a particular opposition of Mars, the planet was simultaneously observed from two points on the equator—one where it was rising and one where it was setting. Its direction among the stars from the two observing stations differed by 41″. What was the distance to Mars in astronomical units?
   Answer: 0.43 AU
4. Why would observations such as those described in Exercise 3 be very difficult to make?
5. When the minor planet Geographos is only 10 million km away, how large an angular displacement in direction can be observed for it from different places on Earth?
6. Describe how you might organize a program to determine the length of the astronomical unit from observations of Geographos. Explain what kinds of observations would be needed, how many would be desired, and what use you would make of them.
7. A star lying on the ecliptic and exactly 90° to the west of the sun is observed to have a radial velocity of exactly 48 km/s toward the Earth. Six months later the star's radial velocity is 12 km/s away from us. What is the star's radial velocity with respect to the sun? What is the Earth's orbital velocity? Assume the Earth's orbit to be circular and use its orbital velocity to evaluate the astronomical unit in kilometers.
8. Suppose that when Mars is exactly 56 million km away, radar waves are beamed toward it. Exactly 4.5 minutes later a return "beep" is picked up that the radar operator thinks is a radar echo reflected from Mars. Do you agree with him? Why?
9. Show that a light year contains 9.46×10^{12} km.
10. Show that 1 pc equals 3.086×10^{13} km and that it *also* equals 3.26 LY. Show your calculations.
11. Make up a table relating the following units of astronomical distance: kilometer, Earth radius, astronomical unit, light year, parsec.
12. Another convenient unit of angular measure is the radian, defined such that 2π radians = 360°. Calculate the number of degrees in a radian and the number of seconds of arc in a radian.
13. In 50 years a star is seen to change its direction by 1′40″. What is its proper motion?
14. Suppose a star at a distance of 10 pc has a radial velocity of 150 km/s. By what percentage does its distance change in 100 years?
15. Find the tangential velocities of the following stars:
    a) Proper motion = 1.5″; distance = 20 pc.
    b) Proper motion = 0.01″; distance = 1000 pc.
    c) Proper motion = 0.01″; distance = 20 pc.
    d) Proper motion = 0.01″; parallax = 0.001″.
16. The wavelength of a particular spectral line is normally 6563 Å; in a certain star, the line appears at 6565 Å. How fast is that star moving in our line of sight? Is its motion one of approach or recession?
17. On June 22 a star is observed that is exactly in the direction of the autumnal equinox; its radial velocity appears to be +36 km/s. On December 22, the same star appears to have a radial velocity of −24 km/s. What is the radial velocity of the star as seen from the sun?
18. Show by a diagram how two stars can have the same radial velocity and proper motion but different space motions.
19. A star has a proper motion of 3.00″, a parallax of 0.474″, and a radial velocity of 40 km/s. What is its space motion?
    Answer: 50 km/s
20. Under what conditions would the proper motion of a nearby star be zero: (a) if it were in the direction of the solar apex? (b) if it were *not* at the solar apex?
21. What is the space velocity of a star with a distance of 10 pc, a radial velocity of −8 km/s, and a proper motion of 0.5″?
22. A certain star is 90° from the solar apex. The observed radial velocity is +12 km/s, and the observed tangential velocity is 24.5 km/s in the direction of the solar antapex. What is the peculiar velocity of this star?
23. Review the conditions that a moving cluster must satisfy in order that we can find its distance by the method described in Section 22.5c.

Henrietta Swan Leavitt (1868–1921) joined the staff of the Harvard College Observatory in 1892. While studying variable stars in the Magellanic Clouds, she discovered the period-luminosity relation for cepheid variable stars, which later made it possible for Edwin Hubble to demonstrate that our Galaxy is but one among billions in the universe. (Harvard College Observatory)

23

THE LIGHT FROM STARS

The most casual glance at the sky shows that stars differ from one another in apparent brightness. They differ not only because of actual differences in their output of luminous energy but also because they are at widely varying distances. Measures of the amounts of light energy, or *luminous flux*, received from stars are among the most important and fundamental observational data of astronomy; they are used in estimating both the distances and the actual output of energy of stars.

23.1 STELLAR MAGNITUDES

In the second century B.C., Hipparchus compiled a catalogue of about a thousand stars. He classified these stars into six categories of brightness, which are now called *magnitudes*. The brightest-appearing stars were placed by him in the *first magnitude;* the faintest naked-eye stars were of the *sixth magnitude*. The other stars were assigned intermediate magnitudes. This system of stellar magnitudes, which began in ancient Greece, has survived to the present time, with the improvement that today magnitudes are based on precise measurements of apparent or total luminosity rather than arbitrary and uncertain eye estimates of star brightness.

(a) The Magnitude Scale

In 1856, after early methods of making quantitative measurements of stellar brightness had been developed, Norman R. Pogson proposed the quantitative scale of stellar magnitudes that is now generally adopted. He noted, as did William Herschel before him, that we receive about 100 times as much light from a star of the first magnitude as from one of the sixth, and that, therefore, a difference of five magnitudes corresponds to a ratio in luminous flux of 100:1. Pogson suggested, therefore, that the ratio of light flux corresponding to a step of one magnitude be the fifth root of 100, which is about 2.512. Thus, a fifth-magnitude star gives us 2.512 times as much light as one of sixth magnitude, and a fourth-magnitude star, 2.512 times as much light as a fifth, or 2.512×2.512 times as much as a sixth-magnitude star. From stars of third, second, and first magnitude, we receive 2.512^3, 2.512^4, and 2.512^5 ($= 100$) times as much light as from a sixth-magnitude star. By assigning a magnitude of 1.0 to the bright stars Aldebaran and Altair, Pogson's new scale gave magnitudes that agreed roughly with those in current use at the time.

Table 23.1 gives the approximate ratios of light flux corresponding to several selected magnitude

TABLE 23.1 Magnitude Differences and Light Ratios

DIFFERENCE IN MAGNITUDE	RATIO OF LIGHT
0.0	1:1
0.5	1.6:1
0.75	2:1
1.0	2.5:1
1.5	4:1
2.0	6.3:1
2.5	10:1
3.0	16:1
3.5	25:1
4.0	40:1
4.5	63:1
5.0	100:1
6.0	251:1
10.0	10,000:1
15.0	1,000,000:1
20.0	100,000,000:1
25.0	10,000,000,000:1

differences. Note that a given ratio of light flux, whether between bright or faint stars, always corresponds to the same magnitude interval. Further, note that the numerically *smaller* magnitudes are associated with the *brighter* stars; a numerically *large* magnitude, therefore, refers to a faint star.*

The so-called first-magnitude stars are not all of the same apparent brightness. The brightest-appearing star, Sirius, sends us about ten times as much light in the visual region of the spectrum as the average star of first magnitude, and so has a magnitude of 1.0 − 2.5 (see Table 23.1), or of about −1.5. Several of the planets appear even brighter; Venus, at its brightest, is of magnitude −4. The sun has a magnitude of −26.5. Some magnitude data are given in Table 23.2. It is of interest to note that the brightness of the sun and Sirius differ by 25 magnitudes—a factor of 10 billion (10^{10}) in light energy, or flux—and that we also receive 10^{10} times as much light from Sirius as from the faintest stars that can be photographed with the 5-m telescope. The entire range of light flux represented in Table 23.2 covers a ratio of about 10^{20} to 1. The magnitudes in this table are visual magnitudes, which measure the apparent brightnesses of

TABLE 23.2 Some Visual Magnitude Data

OBJECT	MAGNITUDE
Sun	−26.5
Full moon	−12.5
Venus (at brightest)	−4
Jupiter, Mars (at brightest)	−2
Sirius	−1.5
Aldebaran, Altair	1.0
Naked-eye limit	6.5
Binocular limit	10
15-cm telescope limit	13
4-m CCD limit	26
Hubble Space Telescope	28–29

stars in the wavelength region to which the eye is most sensitive.

(b) Measuring Starlight—Photometry

One modern method of measuring the brightness of stars, *photoelectric photometry*, has come into widespread use since World War II. The light from a star, coming to a focus at the focal plane of the telescope, is allowed to pass through a small hole in a metal plate and thence onto the photosensitive surface of a photomultiplier (Section 9.2b). The electric current generated in the photomultiplier is amplified and recorded and provides an accurate measure of the light passing through the hole. Because even the darkest night sky is not completely dark, not all the light striking the photomultiplier is due to the star, but some is provided by the light of the night sky, which is, of course, also gathered by the telescope and passed through the hole. Consequently, the hole is next moved to one side of the star image, so that only light from the sky passes through and is recorded; the difference between the first and second readings is a measure of the star's light entering the telescope.

A photomultiplier is useful for measuring the brightness of a single star. During the last decade, electronic devices that record the precise brightness of all of the stars in an entire image have become available to astronomers. One of these types of detectors, the charge-coupled device (CCD), was described in Section 9.2b. Imaging detectors allow astronomers to measure with a single exposure the individual luminosities of every object that lies within the field of view. All of the stars in a cluster

* If m_1 and m_2 are the magnitudes corresponding to stars from which we receive light flux in the amounts l_1 and l_2, the difference between m_1 and m_2 is defined by

$$m_1 - m_2 = 2.5 \log\frac{l_2}{l_1}.$$

Figure 23.1 Both telescope aperture and exposure time determine the limiting magnitude of the faintest stars that can just barely be detected. These photographs illustrate the effects of telescope aperture and exposure time. *(Top left)* 36 cm, 1 min. Stars to 12th magnitude are visible. *(Top right)* 152 cm, 1 min. Stars to 15th magnitude. *(Lower left)* 152 cm, 27 min. Stars to 18th magnitude. *(Lower right)* 152 cm, 4 hr. Stars to 20th magnitude. (Mt. Wilson and Las Campañas Observatories)

can be measured at once, as can the distribution of brightness across the surface of an extended object like a galaxy, a cloud of luminous gas, or a planet. The brightness of the sky between the stars or surrounding an extended source of light is also recorded during this single exposure. Photometry with imaging detectors is therefore much more efficient than with photomultipliers, and more and more astronomers are choosing to use CCDs rather than photomultipliers in order to measure the energy output of astronomical objects.

(c) Magnitude Standards

Magnitudes, as explained above, are defined in terms of ratios of the light received from stars (or other celestial objects); thus magnitude differences between objects indicate the *relative* amounts of luminous flux received from them. To set the magnitude scale unambiguously, however, it is necessary to pick out some standard or standards with respect to which all other stars are to be compared. Accurate photoelectric measurements have been used to establish precise apparent magnitudes for a small number of stars in all parts of the sky, and these stars are used as standards against which all other measurements can be compared. The magnitude values assigned to these standards have been chosen in such a way that the brightest-appearing stars in the sky, in accordance with historical tradition, are still reasonably near first magnitude. The apparent brightnesses of several astronomical objects were listed in Table 23.2

23.2 THE "REAL" BRIGHTNESSES OF STARS

Even if all stars were identical, and even if there were no interstellar dust to dim their light, stars would not all appear to have the same brightness, because they are at different distances from us, and the light that we receive from a star is inversely proportional to the square of its distance (Section 8.1).

(a) Absolute Magnitudes

The sun gives us thousands of millions of times as much light as any of the other stars, but on the other hand, it is hundreds of thousands of times as close to us as any other star is. One way to compare the intrinsic luminous outputs of the sun and of other stars is to determine what magnitude the sun would have if it were at a specified distance, typical of the distances of the other stars. Suppose we choose 10 pc as a more or less representative distance of the nearer stars. Since 1 parsec is about 200,000 AU, the sun would be 2,000,000 times as distant as it is now if it were removed to a distance of 10 pc; consequently, it would deliver to us $(1/2{,}000{,}000)^2$ or $1/(4 \times 10^{12})$ of the light it now does. A factor of 4×10^{12} corresponds to about 31½ magnitudes (which can be verified by raising 2.512 to the 31.5 power). The sun, therefore, if removed to a distance of 10 pc, would appear fainter by some 31½ magnitudes than its present magnitude of −26.5; the sun would then appear as a fifth-magnitude star.

Similarly, we can use the inverse-square law of light to calculate how luminous all other stars of known distance would appear if they were 10 pc away. Suppose, for example, that a tenth-magnitude star has a distance of 100 pc. If it were only 10 pc away, it would be 10 times closer to the sun and hence 100 times as bright—a difference of five magnitudes. At 10 pc, therefore, it too would appear as a fifth-magnitude star.

We now define the *absolute magnitude* of a star as the magnitude that star would have if it were at the standard distance of 10 pc (about 32.6 LY). The absolute magnitude of the sun is about +5. The absolute magnitudes of stars are measures of how bright they really are; they provide a basis for comparing the amount of light emitted by stars.

For historical reasons, optical astronomers continue to use the magnitude scale, particularly when describing the brightness of stars. Measurements in the ultraviolet, infrared, and radio regions of the electromagnetic spectrum may be stated in terms of magnitudes but are usually given in terms of luminosity. *Apparent luminosity* is a measure of the amount of electromagnetic radiation that reaches the Earth. Intrinsic luminosity, or simply *luminosity* with no qualifying adjective, is the rate at which electromagnetic energy is radiated into space by an astronomical object. Luminosity is usually expressed in units of ergs per second. Two stars that differ by a factor of 100 in intrinsic luminosity differ by 5 magnitudes in absolute magnitude. The extreme range of absolute magnitudes observed for normal stars is −10 to nearly +14, a range of almost a factor of 10^{10} in intrinsic light output.

(b) Distance Modulus

The absolute magnitude of a star is the magnitude that the star *would* have if it were at a distance of 10 pc, and therefore is a measure of the actual rate of emission of visible light energy by the star, which, of course, is independent of the star's actual distance. On the other hand, the magnitude of a star (sometimes called the *apparent magnitude* to avoid confusion with *absolute magnitude*) is a measure of how bright the star *appears* to be, and this depends on both the star's actual rate of light output and its distance. The *difference* between the star's apparent magnitude, symbolized m, and absolute magnitude, symbolized M, can be calculated from the inverse-square law of light and from a knowledge of how much greater or less than 10 pc the star's distance actually is. The difference $m - M$ therefore depends only on the distance of the star and is called its *distance modulus*. The distance modulus of a star, then, is a measure of its distance, from which the actual distance, in parsecs, can be calculated.* For example, suppose the distance

---

* Let $l(r)$ be the observed light of a star at its actual distance, r, and $l(10)$ the amount of light we would receive from it if it were at a distance of 10 pc. From the definition of magnitudes, we have

$$m - M = 2.5 \log[l(10)/l(r)]$$

and from the inverse-square law of light,

$$l(10)/l(r) = (r/10)^2$$

Combining the above equations, we obtain

$$m - M = 5 \log(r/10)$$

The quantity, $5 \log (r/10)$, then, is the distance modulus.

modulus of a star is 10 magnitudes. Ten magnitudes (see Table 23.1) corresponds to a ratio of 10,000:1 in light. Thus, we actually receive from the star 1/10,000 of the light that we would receive if it were 10 pc away; it must, therefore, be 100 times as distant as 10 pc, or at a distance of 1000 pc. In succeeding chapters we shall see that the absolute magnitude of a star can often be inferred from its spectrum or from some other observable characteristic. Since the apparent magnitude of a star can be observed, a knowledge of its absolute magnitude is equivalent to a knowledge of its distance.

Most of us make use of this same principle, subconsciously, in everyday life. Every experienced motorist has an intuitive notion of the actual brightness of a stoplight. If, while driving down the highway at night, he sees a stoplight, he judges its distance from its apparent faintness. In other words, the difference between the light's *apparent* and *real* brightness indicates its distance. The computation of a star's distance from its distance modulus is analogous.

The calculation of a stellar distance by the inverse-square law, as described above, is actually valid only if the space between the star and us is completely transparent. We shall see (Chapter 27) that the light from many stars is attenuated by absorbing dust in interstellar space. Fortunately, however, we can usually tell when cosmic dust is dimming the light from a star and can correct for it.

(c) Colors of Stars

Until now, this discussion has ignored the fact that stars have different colors and that all colors do not produce an equal response in the human eye, photographic film, or CCDs. The apparent brightness of a star can depend to some extent, therefore, upon its color.

Every device for detecting light has a particular color or spectral sensitivity. The human eye, for example, is most sensitive to green and yellow light; it has a lower sensitivity to the shorter wavelengths of blue and violet light and to the longer wavelengths of orange and red light. It does not respond at all to ultraviolet or to infrared radiation. The eye, in fact, responds roughly to the same kind of light that the sun emits most intensely; this coincidence is probably not accidental—the eye may have evolved to respond to the kind of light most available on Earth.

Another detecting device is the photographic plate (or film). The basic photographic emulsion is sensitive to violet and blue; dyes must be added to make it sensitive to longer wavelengths. Electronic detectors like CCDs are most sensitive to red and near-infrared radiation.

Suppose, now, that the total amount of light energy entering a telescope from each of two stars is exactly the same if light of all wavelengths is considered, but that one star emits most of its light in the blue spectral region and the other in the yellow spectral region. If these stars are observed visually (that is, by looking at them through the telescope), the yellow one will appear brighter, that is, will have a numerically smaller magnitude, because the eye is less sensitive to most of the light emitted by the blue star. If the stars are photographed on a blue-sensitive photographic plate, however, the blue star will produce the more conspicuous image; measures of the photographic images will show the blue star appearing brighter and having the smaller magnitude. Consequently, when a magnitude system is defined, it is necessary also to specify how the magnitudes are to be measured, that is, what detecting device is to be used.

(d) Color Indices

Since about 1960, when the use of photoelectric photometry became widespread, the most commonly measured magnitudes have been U (ultraviolet), B (blue), and V (visual). The U and B magnitudes are obtained from measures of the flux from stars through certain standardized ultraviolet and blue filters with a common type of photomultiplier. The visual magnitude (V) is measured through a filter that approximates the response of the human eye. The *difference* between any two of these magnitudes, say between blue and visual magnitudes ($B - V$), is called a *color index*. Thus the standard U, B, V system provides two independent color indices, $U - B$, and $B - V$.

Since the inverse-square law of light applies equally to all wavelengths, the color index of a star would not change if the star's distance were changed. The color index of a star, therefore, can be defined in terms of either its apparent or absolute magnitudes; for example,

$$B - V = M_B - M_V.$$

A very blue star measures brighter through a blue filter than through a yellow one; its blue magnitude, then, is *less* positive than its visual magnitude, and its $B - V$ color index is *negative*. A yellow or red star, on the other hand, has a brighter (smaller) visual magnitude, and a *positive* color index. Color indices, therefore, provide measures of the *colors* of stars. Colors, in turn, indicate the temperatures of stars. According to Wien's law (Section 8.1), the hotter an object, the shorter the wavelength at which it emits the most radiation and, therefore, the bluer its color. Ultraviolet, blue, and visual magnitudes are adjusted to be equal to each other, so that they give a color index of zero to a star with a temperature of about 10,000 K (a spectral type A star—see Chapter 24). The $B - V$ color indices of stars range from -0.4 for the bluest and hottest to more than $+2.0$ for the reddest and coolest.

(e) Bolometric Magnitudes and Luminosities

All the magnitude systems so far discussed refer only to certain wavelength regions. A magnitude system that measures *all* the electromagnetic energy reaching the Earth from the stars would seem to be more fundamental. Magnitudes so based are called *bolometric magnitudes*, m_{bol}, and the bolometric magnitudes that stars would have at a distance of 10 pc are *absolute bolometric magnitudes*, M_{bol}.

The bolometric luminosity of an object is equal to the total energy that it radiates in all the wavelength regions of the electromagnetic spectrum. Unfortunately, bolometric luminosities are difficult to observe directly because some wavelengths of electromagnetic energy do not penetrate the Earth's atmosphere. It is true that most of the radiation from stars with temperatures comparable to that of the sun does reach the Earth's surface, and for such stars bolometric luminosities can be determined approximately. A large part of the energy from stars that are substantially hotter or cooler than the sun, however, lies in the far ultraviolet or in the infrared, both of which are blocked by the Earth's atmosphere and cannot be observed from the ground; for those stars, bolometric luminosities can only be directly measured from rockets or satellites or calculated from theoretical models.

The difference between the visual and bolometric magnitudes of a star is called its *bolometric correction, BC:**

$$V - m_{bol} = M_V - M_{bol} = BC$$

A bolometric correction, then, is a measure of the difference between the intrinsic visual luminosity of a star and the total energy that it emits in all wavelengths. Bolometric magnitudes are scaled so that bolometric corrections are nearly zero for stars like the sun. They are positive quantities for stars that are much hotter or cooler than the sun because these stars would appear brighter (would have numerically smaller bolometric than photovisual magnitudes) if all of their radiant energy could be observed. For the hottest stars, calculated bolometric corrections reach as much as eight magnitudes.

(f) The Luminosity of the Sun

The first step in determining the luminosity of the sun is to measure the rate at which its radiation falls upon the Earth. There are various devices for measuring this quantity. In one method, the sun's radiation is allowed to fall upon the blackened surface of some metal (for example, platinum or silver), which absorbs this radiant energy and rises in temperature. The rate of the temperature rise can be measured by the increased resistance that the metal offers to an electric current. A simpler device measures the rate at which the temperature of water rises as the sun's radiation is absorbed by it. The measures obtained of the sun's radiation must, in any case, be corrected for the attenuation of the sunlight passing through the Earth's atmosphere. (Measures of stellar magnitudes must similarly be corrected.)

It is found that a surface of 1 cm^2 just outside the atmosphere and oriented perpendicular to the direction of the sun receives from the sun 1.37×10^6 ergs/s. This value is the solar constant.

The total energy that leaves the sun during an interval of one second diverges outward, away from the sun, in all directions. Since one astronomical unit is 1.49×10^{13} cm, the area of the spherical surface over which the solar radiation has spread by the time it reaches the Earth's distance from the sun (about eight minutes later) is 2.8×10^{27} cm^2. The solar constant of 1.37×10^6 ergs/s/cm^2 is the energy that crosses just one of those square centimeters. The total energy that

* Sometimes bolometric corrections are defined as $m_{bol} - V$, in which case they are always negative, rather than positive as in the convention used in this book.

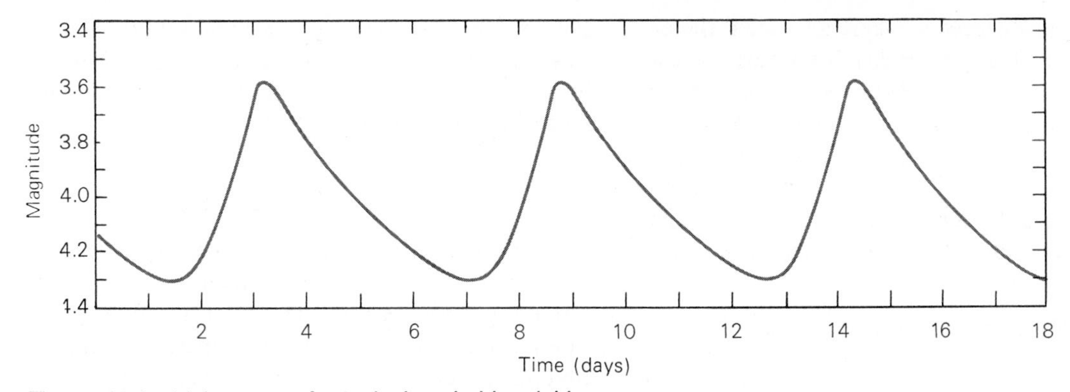

Figure 23.2 Light curve of a typical cepheid variable.

leaves the sun in one second—its luminosity—is thus $1.37 \times 10^6 \times 2.8 \times 10^{27} = 3.8 \times 10^{33}$ ergs/s. The corresponding absolute bolometric magnitude of the sun is +4.6.

A dramatic illustration of the magnitude of that amount of energy is obtained by imagining a bridge of ice 3 km wide and 1.5 km thick and extending over the 150 million-km span from the Earth to the sun. If all the sun's radiation could be directed along the bridge, it would be enough to melt the entire column of ice in one second.

23.3 STARS THAT VARY IN LIGHT

Many stars vary in brightness. There are three basic types of variable stars: (1) pulsating variables, (2) eruptive variables, and (3) eclipsing variables. *Pulsating variables* are stars that periodically expand and contract, pulsating in size, temperature, and luminosity. *Eruptive variables* (including novae and supernovae—see Chapter 32) are stars that show sudden, usually unpredictable, outbursts of light, or, in some cases, diminutions of light. *Eclipsing variables* (or eclipsing binaries) are binary stars whose orbits of mutual revolution lie nearly edge-on to our line of sight and that periodically eclipse each other. Eclipsing variables are not, of course, intrinsically variable; they will be discussed in Chapter 25.

(a) Designation

Variable stars are designated in order of time of discovery in the constellation in which they occur. If a star that is discovered to vary in light already has a proper name or a Greek-letter designation, it retains that name; examples are Polaris, Betelgeuse (α Orionis), Algol, and δ Cephei. Otherwise, the first star to be recognized as variable in a constellation is designated by the capital letter R, followed by the possessive of the Latin name of the constellation (for example, R Coronae Borealis). Subsequently discovered variables in the same constellation are designated with the letters S, T, . . . , Z, RR, RS, . . . , RZ, SS, ST, . . . , SZ, and so on, until ZZ is reached. Then the letters AA, AB, . . . , AZ, BB, BC, . . . , BZ, and so on, are used up to QZ (except that the letter J is omitted). This designation takes care of the first 334 variable stars in any one constellation. Thereafter, the letter V followed by a number is used, beginning with V 335. Examples are V 335 Herculis and V 969 Ophiuchi.

(b) The Light Curve

A variable star is studied by analyzing its spectrum and by measuring the variation of its light with lapse of time. Some stars show light variations that are apparent to the unaided eye. Generally, however, the apparent brightness of a variable star is determined by telescopic observation.

A graph that shows how the magnitude of a variable star changes with time is called a *light curve* of that star. An example is given in Figure 23.2. The *maximum* is the point on the light curve where the maximum amount of light is received from the star; the *minimum* is the point where the least amount of light is received. If the light variations of a variable star repeat themselves periodically, the interval between successive maxima is called the *period* of the star. The *median light* of a variable star is the amount of light it emits when it is halfway between its maximum and minimum brightness. The *amplitude* is the difference in light (usually expressed in magnitudes) between the maximum and the minimum. The amplitudes of variable stars range from less than 0.1 to several magnitudes.

(c) Long-Period Variables

The largest group of pulsating stars consists of *Mira*-type stars; these are named for their prototype, Mira, in the constellation of Cetus. Other large groups of pulsating stars are the *RR Lyrae* variables, the *semiregular* variables, and the *irregular* variables.

The Mira or *red variables* are giant stars that pulsate in very long or somewhat irregular periods of months or years. Because they are not highly predictable, an important service is provided by amateur astronomers who keep track of the magnitudes of these stars. It would require far too much of the time of professional astronomers to maintain constant vigil on all of them. The *American Association of Variable Star Observers* has a well-planned program of careful surveillance of these long-period and irregular variables and has been gathering valuable data on them for years.

(d) Cepheid Variables

Although relatively rare, the cepheid variables are very important in astronomy because, as we shall see, it is possible to determine how far away they are simply by measuring their periods. Cepheids are large yellow stars named for the prototype and first known star of the group, δ Cephei. The variability of δ Cephei was discovered in 1784 by the young English astronomer John Goodricke just two years before his death at the age of 21. The visual magnitude of δ Cephei varies between 3.6 and 4.3 in a period of 5.4 days. The star rises rather rapidly to maximum light and then falls more slowly to minimum light (see Figure 23.2).

More than 700 cepheid variables are known in our Galaxy. Most cepheids have periods in the range 3 to 50 days and absolute magnitudes (at median light) from -1.5 to -5. Cepheids are thus highly luminous stars that radiate as much as 10,000 times more energy than the sun. The amplitudes of cepheids range from 0.1 to 2 magnitudes.

Polaris, the *North Star*, is a small-amplitude cepheid variable that varies by one-tenth of a magnitude, or by about 10 percent in visual luminosity, in a period of just under four days.

(e) The Period-Luminosity Relation

The importance of cepheid variables lies in the fact that a relation exists between their periods of pulsation (or light variation) and their absolute magnitudes at median light. Simply by measuring the period of a cepheid, we can estimate its intrinsic luminosity and, since its apparent magnitude can be determined, we can calculate how far away it is (Figure 23.3).

The relation between period and luminosity was discovered in 1912 by Henrietta Leavitt, an astronomer of the Harvard College Observatory. Some hundreds of cepheid variables had been discovered in the Large and Small Magellanic Clouds, two great stellar systems that are actually neighboring galaxies (although they were not known to be galaxies in 1912—see Chapter 35).

Many photographs of the Magellanic Clouds had been taken at the Southern Station of the Harvard College Observatory in South America. (These stellar systems are too far south in the sky to be observed from the United States.) On some of the photographs, Leavitt identified 25 cepheids in the smaller cloud and plotted light curves for them. She found that the periods of the stars were related to their relative brightness; the brighter-appearing ones always had the longer periods of light variation. Since the 25 stars are all in the same stellar system, they are all at about the same distance—the distance to the Small Magellanic Cloud—and their relative apparent brightnesses, therefore, indicate their actual relative luminosities. Leavitt, in other words, had found that the median luminosities, or absolute magnitudes, of the cepheid variables were correlated with their periods. Subsequent investigation showed that this relation exists between all the cepheids in the Large and Small Magellanic Clouds.

Now, the cepheid variables known in our Galaxy are intrinsically very luminous stars. If those cepheids in the Magellanic Clouds are the same kind of object, they must also be highly luminous. They appear very faint, however, of 15th or 16th apparent magnitude, which indicates that these stellar systems must be very remote. In 1912 the actual distances to the Magellanic Clouds were not known.

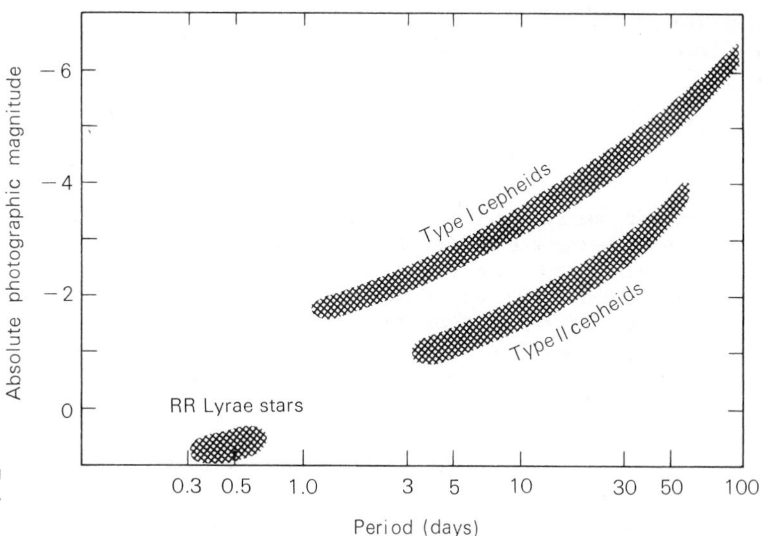

Figure 23.3 Period-luminosity relation for type I cepheids. (Other types of cepheids will be discussed in Chapter 31.)

While it was known then that the periods of cepheids are correlated with their absolute magnitudes, it was not known what those absolute magnitudes are; only *differences* between absolute magnitudes of cepheids of different periods could be determined. For example, if one cepheid had a period of three days and another of 30 days, it was known that the one of longer period was about two magnitudes, or six times, brighter than the star of shorter period. Unfortunately, however, the actual absolute magnitude of neither star was known, beyond the fact that they were both very luminous; hence, the true distances to the stars remained inaccessible.

If the distances to only one or a few cepheid variables could be determined by independent means, however, the absolute magnitudes of those few stars could be found, and if it is assumed that the period-luminosity relation applies to cepheids in the Galaxy, as well as in the Magellanic Clouds, the scale of the relation would be set. Thereafter, the distance could be determined to any system or cluster of stars in which a cepheid variable had been identified. Harlow Shapley was one of the astronomers who recognized the importance of cepheids as distance indicators, and he pioneered the work of determining distances to some of them in our Galaxy. His work was extended by others, and by 1939 it appeared that the distances to these stars were well determined. Unfortunately, however, the determination of distances to cepheid variables is a very difficult observational problem. In the first place, there is not a single cepheid near enough so that its parallax can be measured (Section 22.2c); therefore, statistical methods involving the proper motions and radial velocities of these stars must be employed (Section 22.5). In the second place, most of the cepheid variables in our Galaxy lie close to the plane of the galactic system, where clouds of interstellar dust heavily obscure their light (see Chapter 27). Corrections must be applied therefore to the measured apparent magnitudes of these cepheids. Nevertheless, today we use cepheid variables as fundamental indicators of distance to other galaxies (Chapter 35). We shall return to the absolute magnitudes of cepheid variables in Chapter 31.

EXERCISES

1. Find the absolute magnitude of each of the following stars of apparent magnitude m and distance r:
   a) $m = 7$; $r = 10$ pc
   b) $m = 20$; $r = 100$ pc
   c) $m = 0$; $r = 100$ pc
   d) $m = 3.5$; $r = 5$ pc
   e) $m = 17.5$; $r = 20$ pc
   f) $m = -5$; $r = 1/10$ pc (there is no such star as this)

2. What are the distances of stars of the following apparent magnitudes m and absolute magnitudes M?
   a) $m = 10$; $M = 5$
   b) $m = 5$; $M = 10$
   c) $m = 13.5$; $M = 15$
   d) $m = 20$; $M = 10$
   e) $m = -26.5$; $M = 5$

3. Suppose obscuring matter between a star and an observer dims the star so much that only 10 percent of the star's light he would otherwise receive reaches him. If the observer were unaware of the obscuration, what would be the error in the distance modulus he would assign to the star?
   Answer: His assigned modulus would be 2.5 magnitudes too great.

4. If a star has a color index of $B - V = 2.5$, how many times brighter in visual light does it appear than in blue light?

5. What would be the designations of the 4th, 40th, and 400th discovered variable stars in the constellation Ophiuchi? (Ignore proper names and Greek letter designations.)
   Answer: U Ophiuchi, VV Ophiuchi, V 400 Ophiuchi

6. Suppose a type I cepheid variable is observed in a remote stellar system. The cepheid has a period of 50 days and an apparent magnitude of +20. What is the distance to the stellar system? (See Figure 23.3.)

7. Draw a diagram showing how two stars of equal bolometric magnitude, one of which is blue and the other red, would appear on photographs sensitive to blue and to yellow light.

8. What is the distance of a star for which $B = 12.4$, $M_V = 6.8$, and $B - V = 0.6$?

9. If a star has a bolometric correction of six magnitudes, what fraction of its radiant energy is observed in its visual magnitude?

10. The apparent visual magnitude of a star is 10.4. Its bolometric correction is 0.8. Its parallax is 0"001. What is its luminosity?
    Answer: 3.8×10^{35} ergs/s

11. By what steps do you suppose the absolute bolometric magnitude of the sun is found?

Annie Jump Cannon (1863–1941), who during her long career at Harvard College Observatory personally classified more than 500,000 stellar spectra and arranged them in the spectral sequence: O, B, A, F, G, K, M. She received many honors for her work, which laid the groundwork for modern stellar spectroscopy, so fundamental to our understanding of the stars. (Harvard College Observatory)

24

ANALYZING STARLIGHT

About 1665 Newton showed that white sunlight is really a composite of all colors of the rainbow (Section 8.3), and that the various colors, or wavelengths, of light could be separated by passing the light through a glass prism. William Wollaston (Section 8.3) first observed dark lines in the solar spectrum, and Joseph Fraunhofer catalogued about 600 such dark lines. As early as 1823, Fraunhofer observed that stars, like the sun, also have spectra that are characterized by dark lines crossing a continuous band of color. Sir William Huggins, in 1864, first identified some of the lines in stellar spectra with those of known terrestrial elements.

24.1 CLASSIFICATION OF STELLAR SPECTRA

When the spectra of different stars were observed, it was found that they differed greatly among themselves. In 1863 the Jesuit astronomer Angelo Secchi classified stars into four groups according to the general arrangement of the dark lines in their spectra. Secchi's scheme subsequently was modified and augmented, until today we recognize seven such principal *spectral classes*.

(a) The Spectral Sequence

As we have seen (Section 8.3), each dark line in a stellar spectrum is due to the presence of a particular chemical element in the atmosphere of the star observed. It might seem, therefore, that stellar spectra differ from each other because of differences in the chemical makeup of the stars. Although some stars do have abundance anomalies, the differences in stellar spectra are due mostly to the widely differing temperatures in the outer layers of the various stars. Hydrogen, for example, is by far the most abundant element in the outer layers of all stars, except probably in those at an advanced stage of evolution (Chapter 32). In the atmospheres of the very hottest stars, however, hydrogen is completely ionized. Since the electron and proton are separated, ionized hydrogen can produce no absorption lines. In the atmospheres of the coolest stars hydrogen is neutral and can produce absorption lines, but in these stars practically all of the hydrogen atoms are in the lowest energy state (unexcited), and can absorb only those photons that can lift them from that first energy level to higher ones; the photons so absorbed produce the *Lyman series* of absorption lines (Section 8.3), which lies in the ultraviolet part

TABLE 24.1 Spectral Sequence

SPECTRAL CLASS	COLOR	APPROXIMATE TEMPERATURE (K)	PRINCIPAL FEATURES	STELLAR EXAMPLES
O	Violet	>25,000	Relatively few absorption lines in observable spectrum. Lines of ionized helium, doubly ionized nitrogen, triply ionized silicon, and other lines of highly ionized atoms. Hydrogen lines appear only weakly.	10 Lacertae
B	Blue	11,000–25,000	Lines of neutral helium, singly and doubly ionized silicon, singly ionized oxygen and magnesium. Hydrogen lines more pronounced than in O-type stars.	Rigel Spica
A	Blue	7,500–11,000	Strong lines of hydrogen. Also lines of singly ionized magnesium, silicon, iron, titanium, calcium, and others. Lines of some neutral metals show weakly.	Sirius Vega
F	Blue to white	6,000–7,500	Hydrogen lines are weaker than in A-type stars but are still conspicuous. Lines of singly ionized calcium, iron, and chromium, and also lines of neutral iron and chromium are present, as are lines of other neutral metals.	Canopus Procyon
G	White to yellow	5,000–6,000	Lines of ionized calcium are the most conspicuous spectral features. Many lines of ionized and neutral metals are present. Hydrogen lines are weaker even than in F-type stars. Bands of CH, the hydrocarbon radical, are strong.	Sun Capella
K	Orange to red	3,500–5,000	Lines of neutral metals predominate. The CH bands are still present.	Arcturus Aldebaran
M	Red	<3,500	Strong lines of neutral metals and molecular bands of titanium oxide dominate.	Betelgeuse Antares

of the spectrum that is unobservable from the ground. In a stellar atmosphere with a temperature of about 10,000 K, many hydrogen atoms are not ionized; nevertheless, an appreciable number of them are excited to the second energy level, from which they can absorb additional photons and rise to still higher levels of excitation. These photons correspond to the wavelengths of the *Balmer series,* which is in the optical part of the spectrum and is readily observable. Absorption lines due to hydrogen, therefore, are strongest in the spectra of stars whose atmospheres have temperatures near 10,000 K, and they are less conspicuous in the spectra of both hotter and cooler stars, even though hydrogen is, roughly, equally abundant in all the stars. Similarly, every other chemical element, in each of its possible stages of ionization, has a characteristic temperature at which it is most effective in producing absorption lines in the observable part of the spectrum.

Once we have ascertained how the temperature of a star can determine the physical states of the gases in its outer layers, and thus their ability to produce absorption lines, we need only to observe what patterns of absorption lines are present in the spectrum of a star to learn its temperature. We can therefore arrange the seven classes of stellar spectra in a continuous sequence in order of decreasing temperature. In the hottest stars (temperatures over 25,000 K) only lines of ionized helium and highly ionized atoms of other elements are conspicuous. Hydrogen lines are strongest in stars with atmospheric temperatures of about 10,000 K. Ionized metals provide the most conspicuous lines in stars with temperatures from 6000 to 8000 K. Lines of neutral metals are the strongest in somewhat cooler stars. In the coolest stars (below 4000 K), absorption bands of some molecules are very strong. The most important among the molecular bands are those due to titanium oxide, a tenacious chemical compound which can exist at the temperatures of the cooler stars. The sequence of spectral types is summarized in Table 24.1 and Figure 24.1. (Hot stars—types O, B, A—are sometimes referred to as

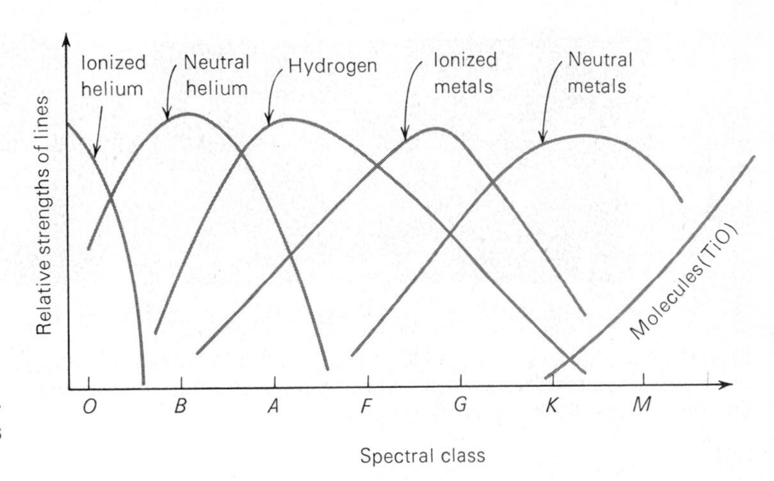

Figure 24.1 Relative intensities of different absorption lines in stars at various places in the spectral sequence.

having *early* spectral types, and cool stars—G, K, M—as having *late* spectral types. This jargon derived from old ideas of stellar evolution that are now discarded, but the terminology is still in wide use.)

The spectral classes of stars listed in Table 24.1 can be subdivided into tenths; thus a star of spectral class A5 is midway in the range of A-type stars, that is, halfway between stars of type A0 and F0. The sun is of spectral class G2—two-tenths of the way from class G0 to K0.

The spectral sequence, ranging smoothly from O to M with decreasing temperature (see Figure 24.2 for sample spectra), was established through the classification of hundreds of thousands of stellar spectra in the years 1918 to 1924 by Harvard astronomer Annie Cannon. The famous American astronomer Henry Norris Russell proposed a scheme by which every student can remember the order of classes in the spectral sequence: The class letters are the first letters in the words, "Oh, Be A Fine Girl, Kiss Me!"

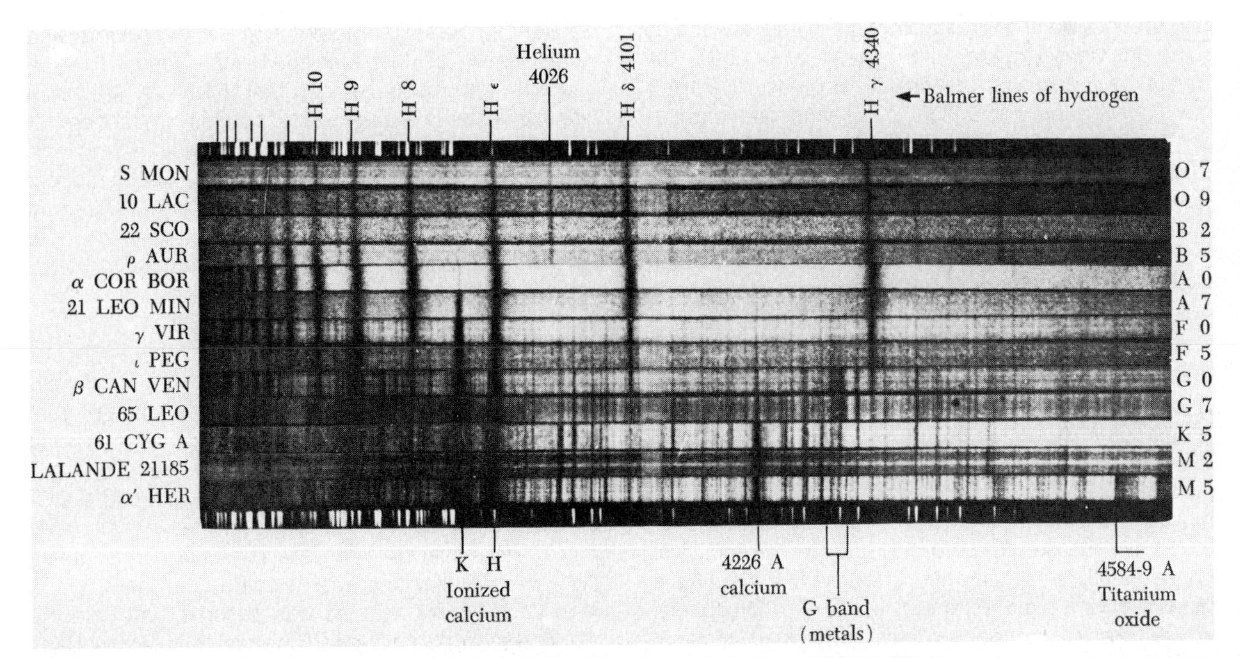

Figure 24.2 Spectra of several stars of representative spectral classes. (UCLA Observatory)

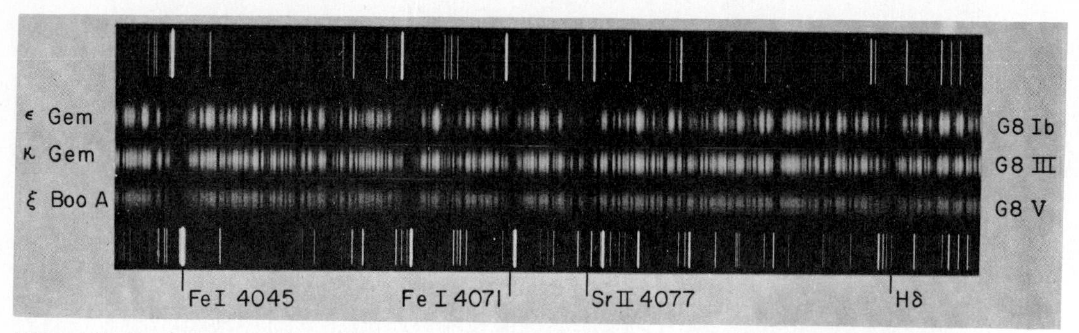

Figure 24.3 Spectra of three G8 stars; from top to bottom, a supergiant, a giant, and a main-sequence star. Note the differences in spectra due to the lower surface gravities of the larger stars. (Lick Observatory)

(b) The Role of Pressure

Although temperature is the most important factor that determines the characteristics of a stellar spectrum, other factors are present. Two stars whose atmospheres have the same temperatures but different pressures, for example, will have somewhat different spectra. The degree to which atoms of a particular kind are ionized in a gas depends on the rate at which those atoms can recapture electrons, and this, in turn, depends on how closely the atoms, ions, and electrons in the gas are packed together. If the density of the gas is high, the particles are close together; it is easier for ions to capture electrons, and the fraction of atoms that are ionized at any instant of time is then lower than if the gas density is low. At any given temperature, the density of a gas is proportional to its pressure *(Boyle's law);* consequently, in a star whose atmospheric gases are at a high pressure, a smaller fraction of each kind of atom will be ionized than in a star of the same temperature whose atmospheric gases are at a low pressure.

Atmospheric pressure, either on a planet or a star, results from the *weight* of the atmosphere, that is, from the gravitational attraction exerted on the atmosphere. The weight of a unit mass upon the surface of a spherical body is called the *surface* gravity of that body; it is proportional to the mass of the body and is inversely proportional to the square of its radius (Section 4.2). We shall see in Chapter 25 that stars differ enormously from one another in radius, but only moderately in mass. Two stars that differ very greatly in radii may thus have masses that are only slightly different, and consequently these stars must have very different surface gravities (see Exercise 4). The photosphere, where most of the spectral lines are formed, may extend in depth through a different total mass of material in a large star than in a small one, but the difference is generally small compared to the difference in surface gravity. Consequently, the weight of the atmosphere of the larger star, and hence its pressure, is less than in the smaller star. Even if the two stars are at the same temperature, therefore, the gases in the larger one must be at a higher degree of ionization than those in the smaller one. The spectrum of a larger star thus resembles that of a smaller star of greater temperature.

The resemblance is not, however, perfect. The lower atmospheric pressure of the larger star partially compensates for its lower temperature, but the compensation is not exactly the same for all elements; there are subtle differences in the fractions of the atoms of the various elements that are ionized in the two stars (Figure 24.3). Spectroscopic evidence for these differences was first noted in 1913 by Adams and Kohlschütter at the Mount Wilson Observatory. A trained spectroscopist can tell from its spectrum whether a star is a giant with a tenuous atmosphere, or a smaller, more compact star of somewhat higher temperature. The importance of the subtle spectroscopic differences between large and small stars will become clear when we consider the method of spectroscopic parallaxes in Chapter 26.

(c) Special Spectral Classes

The majority of all known stars fit into the spectral sequence outlined in Section 24.1a. Some relatively rare stars, however, require special classification. Among the most important of these groups are the following:

The Wolf-Rayet stars. The Wolf-Rayet stars are O-type stars that have broad emission lines in their spectra. These emission lines are presumed to originate from material that has been ejected from the star at a high velocity. The ejected gas absorbs light from the star and re-emits it as emission lines; the different parts of the envelope, which are moving at different velocities relative to an observer on Earth, absorb light at

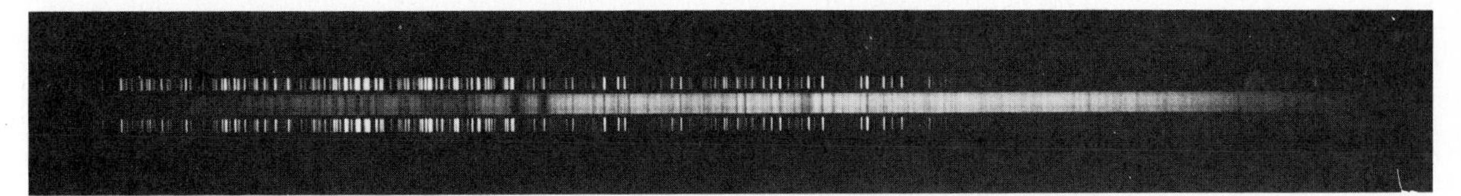

Figure 24.4 A typical stellar spectrogram. The bright streak along the middle, crossed by dark lines, is the spectrum of the star. The bright lines on either side form the comparison (emission) spectrum of iron.

varying wavelengths because of the Doppler effect. Each emission line is, therefore, a broad band that covers a relatively wide range of wavelengths.

Early emission-line stars. The spectra of Of, Be, and Ae stars display bright emission lines of hydrogen. These lines, like those of the Wolf-Rayet stars, are presumed to come from extended gaseous envelopes surrounding the stars.

Peculiar A stars. The peculiar A stars are stars of spectral type A that show abnormally strong lines of certain ionized metals. The intensities of these absorption lines sometimes vary periodically as the star rotates. The variation is caused by an unequal distribution of the chemical elements over the star's surface.

R stars. R stars are stars with spectral characteristics of K stars, except that molecular bands of C_2 and CN (carbon and cyanogen, respectively) are present.

N stars. The N stars are like M stars, except that bands of C_2, CN, and CH are strong, rather than those of titanium oxide. R and N stars are also called *carbon stars* and are often grouped together as a carbon sequence.

S stars. The S stars are like M stars except that molecular bands of zirconium oxide and lanthanum oxide are present in addition to, or instead of, bands of titanium oxide.

24.2 SPECTRUM ANALYSIS AND THE STUDY OF THE STELLAR ATMOSPHERES

Perhaps half of the observing time of many large telescopes is assigned to the study of stellar spectra. Analyses of stellar spectra yield an enormous amount of information about the stars. Unfortunately, spectrum analysis is a very complicated subject; in the following sections the approach is only briefly outlined. Sections 24.2a, b, and c can be skipped over by those not interested in the underlying physics. Some of the more important data that can be gleaned from the detailed study of stellar spectra are summarized at the end of this section.

(a) How Stellar Spectra Are Observed

The design of a spectrometer or spectrograph and how spectra are observed are described in Section 9.2c.

If the spectrum is observed with a modern detecting device, the intensity at various wavelengths is recorded electronically and can be displayed by a computer or plotted out. If a stellar spectrum has been photographed, the blackness of the image of the spectrum at various wavelengths indicates the intensity of the starlight at those wavelengths (Figure 24.4). The intensity of the light drops down, for example, where there is an absorption line. These variations in intensity along a stellar spectrum, and through its various absorption lines, can be exhibited conveniently if the spectrogram (the photograph of the spectrum), which is recorded on glass or transparent film, is scanned with a *microphotometer* (or *microdensitometer*). In this device a narrow beam of light is passed through the spectrogram. The intensity of the transmitted beam is a measure of the blackness of the image; it can be recorded as various parts of the spectrogram are passed along the light beam so that the beam scans along different wavelengths. Figure 24.5 illustrates the appearance of a recording of a portion of a spectrogram made with a microphotometer. The same sort of plot results from a computer display of a spectrum recorded electronically.

We note that the spectral lines are not perfectly sharp but have a non-zero width (some of the reasons will be explained below). Because the shapes or profiles of different lines are not identical, it is convenient to define some measurable quantity that can be used to calculate the total amount of light energy that is subtracted from the spectrum by a line. The most widely used measure is the *equivalent width,* the width of a hypothetical line with rectangular profile of zero inten-

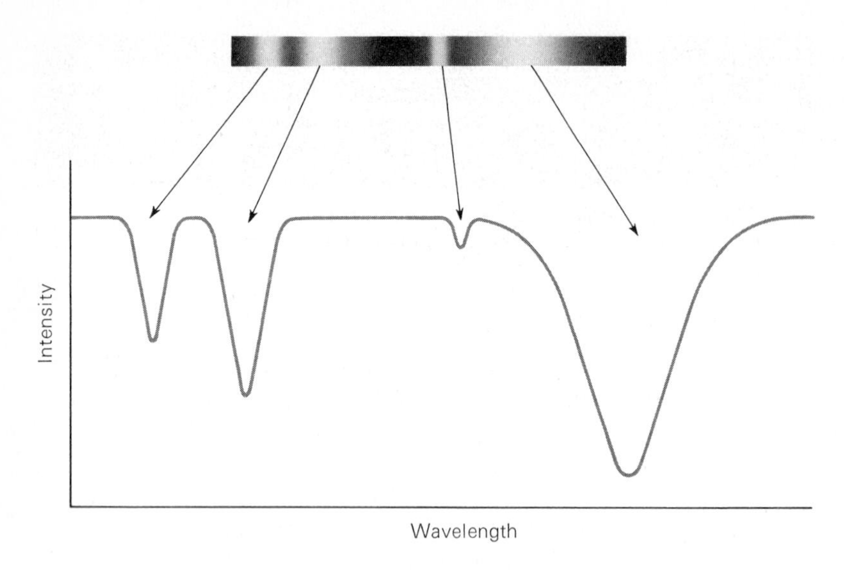

Figure 24.5 Appearance of the recording of a microphotometer scan of a portion of the negative of a spectrogram.

sity along its entire width. The equivalent width represents the same subtraction of light from the stellar spectrum as is removed by the actual line (Figure 24.6).

(b) Analysis of the Continuous Spectrum

The continuous spectrum of a star consists of radiation of a wide range of wavelengths that escapes from a relatively shallow layer of gases in the outer part of the star, called the photosphere. One of the problems of astrophysics is to understand the atomic processes that give rise to this emission of light.

In any given region in a stellar photosphere the gas atoms must reach equilibrium and emit radiation at the same rate that they absorb it. The problem of understanding the emission of light in stellar photospheres is thus equivalent to that of understanding how the photospheric gases absorb light, that is, what their source of opacity is.

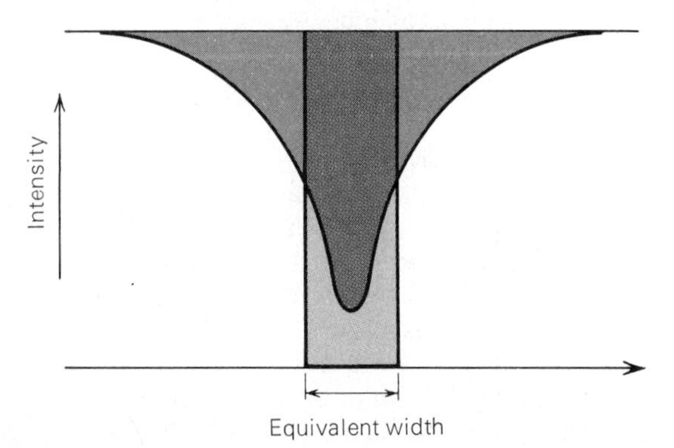

Figure 24.6 Equivalent width of a line. The regions of two different shadings have the same area.

Calculations show that only hydrogen is sufficiently abundant in the atmospheres of stars of most spectral classes to contribute appreciably to the opacity of their outer layers. We have seen (Section 8.3) that hydrogen atoms can absorb light over a range of wavelengths in the process of becoming ionized. The ionization of hydrogen atoms by the absorption of photons, then, is a possible source of opacity in a stellar photosphere. On the other hand, in stars as cool as the sun, or cooler, most atoms of hydrogen are in their lowest energy states. They can be ionized only if they absorb the invisible radiation of wavelengths less than 912 Å (which possesses energy greater than the ionization energy of hydrogen). Opacity at visible wavelengths can result only from the ionization of hydrogen atoms that are already excited to their third or higher energy levels, so that less energetic photons, of longer wavelengths, can be absorbed. Those excited hydrogen atoms are so rare in the solar photosphere that if they were the only source of opacity, the photosphere should be far more transparent than it actually is; we should be able to see into far hotter and deeper layers of the sun than we do.

The problem was solved by the realization that in stars of temperatures near that of the sun and lower, an occasional hydrogen atom can capture and temporarily hold a passing electron. Such a hydrogen atom, possessing *two* electrons, is called a *negative hydrogen ion*. The negative hydrogen ion can absorb visible light and dissociate into an ordinary hydrogen atom and a free electron again. Only about one hydrogen atom in 100 million in the solar photosphere has an extra electron at any given time. This number is

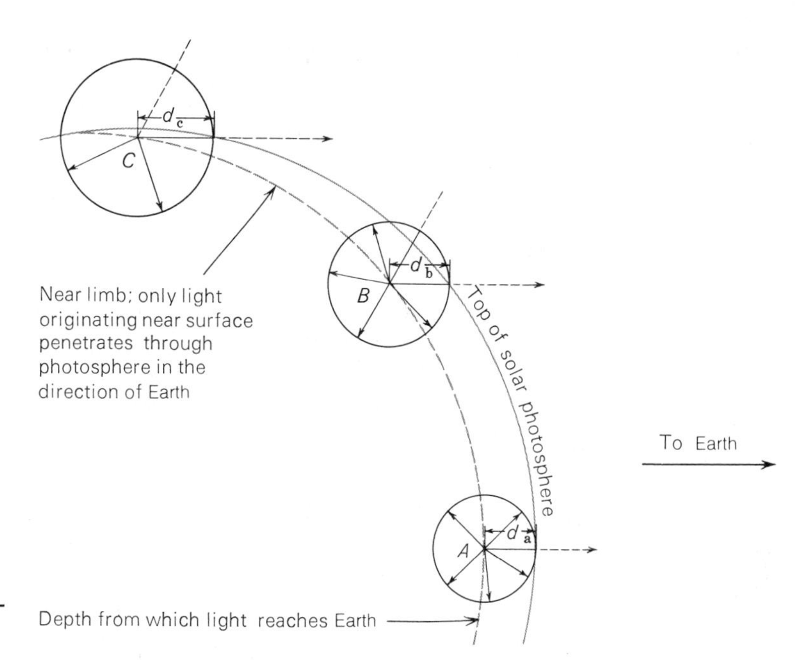

Figure 24.7 Illustration of limb darkening.

enough, however, to account for the opacity of the sun's atmospheric gases in the visible spectrum. Negative hydrogen ions re-form at the same rate that they dissociate, and in doing so emit the visible continuous spectrum that we observe.

We can probe the outer layers of the sun by observing to different depths of its photosphere. This is possible because of the phenomenon of *limb darkening*. The *limb* of the sun (or of any celestial body) is its apparent edge as it is seen in the sky. In visible light, the limb of the sun appears darker than the center of the disk, because when we look at the center of the disk we see into deeper, hotter layers of the sun's photosphere than when we look at the limb. The explanation for this phenomenon is seen in Figure 24.7. Point A is relatively deep in the photosphere and is beneath the exact center of the sun's disk. Suppose photons that are emitted at point A travel, on the average, a distance d_a before being absorbed. The only photons that have an even chance of escaping from the photosphere are those that are moving radially outward, a direction that happens to be toward the Earth. We see, then, sunlight emerging from points as deep as point A only at the center of the sun's disk. Suppose that at point B, photons travel, on the average, a distance d_b before being absorbed. To get into the line of sight to the Earth they must travel outward obliquely, rather than radially, and hence can reach the Earth only if point B is closer to the surface of the sun than point A. To reach the Earth from point C, photons would have to traverse an extremely long path through the solar photosphere unless point C were practically at the outermost photospheric level. The depths to which we see into the sun at various distances from the center of its disk are indicated (although not to scale) by the dashed line.

Once we know the source of opacity in a stellar photosphere, we can calculate how the temperature, pressure, and density of its gases increase with depth. One clue is provided by the condition that the pressure of the gases in a photosphere must be just great enough to balance the weight of the overlying layers of gases (the condition of *hydrostatic equilibrium*—see Chapter 28). The laws of gases (also see Chapter 28) tell us how density, pressure, and temperature are related. Finally, the known opacity of the gases tells us how they impede the flow of radiation through them. It is the opacity of the gases, blocking in the radiation, that maintains the increase of temperature inward through the stellar photosphere. Thus, a knowledge of the opacity enables us to calculate the temperature at each depth in the photosphere. When the clues are put together, the astrophysicist can calculate what is called a *model photosphere* or *model atmosphere* for a star. In practice, the calculations are difficult and the results obtained for the match of pressure, temperature, and density through a photosphere are somewhat uncertain. Model photospheres do indicate, however, at least roughly, the physical nature of the outer layers of a star.

(c) Analysis of the Line Spectrum

The analysis of the absorption-line spectrum (sometimes called the *Fraunhofer spectrum*) of a star is even more difficult than that of the continuous spectrum, but it yields a great deal of additional information about the star. The object of the analysis is to account in detail for the appearance of each spectral line and to extract such information as is possible concerning the physical and chemical characteristics of the star.

Before we can account for the shape, or profile, of a spectral line, we must consider those processes that *broaden* it, that is, that keep it from being an infinitely sharp line of one precise wavelength only. The principal sources of *line broadening* are the following.

Natural broadening. We saw in Section 8.3 that atoms have discrete energy levels. The levels are not *perfectly* sharp, however. The nominal energy levels of an atom, and the wavelengths of the photons it can absorb, are actually only the most probable energies (or wavelengths). There is a small range of energy about each nominal energy that the atom can have.* There is, therefore, a small range of wavelengths that can be absorbed or emitted. The resulting (usually) slight range of wavelengths over which an atom can absorb radiation in the vicinity of a line is called the *natural width* of the line.

Doppler broadening. Photospheric atoms are in rapid motion because of the high temperatures of the outer layers of stars. Atoms, moving at various speeds with respect to the Earth, "see" the wavelengths of the photons they encounter as different from the lengths we would measure on Earth, because of the Doppler shift. At any instant of time, some atoms approach us and others recede from us. Some, therefore, absorb photons of wavelengths that we measure to be a little shorter than the laboratory or rest value for an absorption line, and others absorb photons that we measure to have slightly longer wavelengths. The result is that the line is smeared out into a finite width. This *Doppler broadening* is the main source of broadening of most "weak" lines, that is, lines in which only a little energy is subtracted from the spectrum.

Collisional broadening. The energy levels of an atom can be *perturbed* by the presence of other atoms and ions that pass near it or that collide with it. These perturbations of the energy levels shift them to slightly different energies, so that the perturbed atoms then absorb at wavelengths slightly different from their usual ones. As in Doppler broadening, the effect is to smear out each absorption line. *Collisional broadening* is the most important source of line broadening for the majority of "strong" lines.

Were it not for the broadening, very little energy could be removed from the continuous spectrum by line absorption. Since, however, atoms can absorb radiation over a small range of wavelengths near each line, they can subtract enough energy from the continuous spectrum so that the lines can be observed easily and analyzed in considerable detail.

The strength of an absorption line depends not only on the total abundance of the relevant atomic species, but also on the fraction of those atoms that are in the right state of ionization and excitation to produce the line. From a model photosphere for a star, we have the temperatures and pressures as functions of depth in its atmosphere. For each depth in the photosphere, therefore, we can calculate what fraction of the atoms are in the relevant ionization and excitation state, and also how that particular absorption line is broadened at each depth. Since an absorption line arises from atoms at all depths throughout the photosphere, we must consider how the contributions from atoms at all layers combine to predict the observed profile of the line. We can then predict how the profile depends on the total number of atoms of the element in question. It is found that for a relatively low abundance, the equivalent width of a line is approximately proportional to the total number of atoms of the element present. For a great enough number of atoms, the line "saturates" and, with the addition of a still larger number of atoms, its strength increases only moderately. If the abundance of the element is very great, however, broad "wings" appear on the line as the result of collisional broadening. Figure 24.8 shows how the profiles of successively stronger lines appear.

(d) What Is Learned from Stellar Spectra

Having described some of the procedures by which stellar spectra are analyzed, we now summarize some of the more important kinds of data that can be obtained from spectrum analysis.

TEMPERATURE

The kind of temperature measured by an ordinary thermometer is called *kinetic temperature;* it is a measure of the energies associated with the motions of the atoms or molecules. Since we cannot place a thermometer in the photosphere of a star, we cannot measure its kinetic temperature directly but

* According to the Heisenberg uncertainty principle, the mean range in energy of an energy level, ΔE, and the average time an atom spends in that level, Δt, are related by the formula

$$\Delta E \Delta t = h,$$

where h is Planck's constant (6.6×10^{-27} erg·s). A typical value of Δt is 10^{-8} second; the corresponding value of ΔE is 10^{-18} erg.

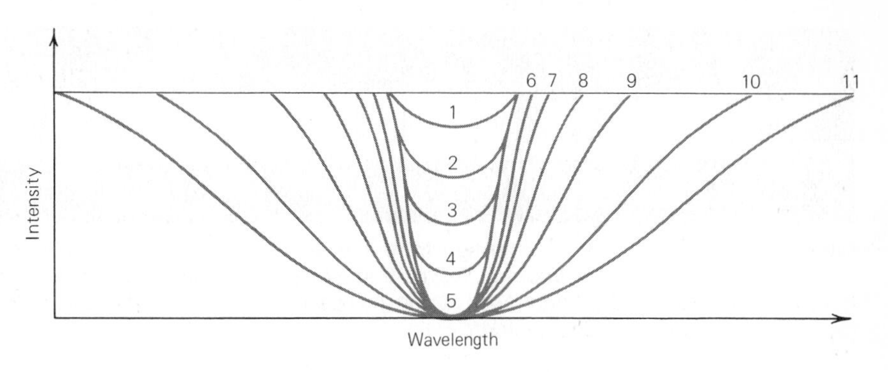

Figure 24.8 Profiles of spectral lines of different equivalent widths. The profiles designated by larger numbers correspond to larger numbers of atoms being present in the stellar photosphere in the right atomic energy level to produce the absorption line.

must infer its temperature from the quality of the radiation that we receive from the star. Because that radiation emerges from a variety of depths in the stellar photosphere, it cannot be expected to indicate a unique temperature that corresponds to any one layer in the star. The temperature that we derive is a representative temperature, corresponding to some representative depth in the photosphere. What that depth is depends on the manner in which the temperature is measured. If the star's continuous spectrum, or a portion of it, is fitted to the spectral energy distribution of a black body (perfect radiator), for example, the temperature of that black body is called the star's *color temperature*. A temperature that is estimated from the extent to which lines of ionized atoms appear in the spectrum is called an *ionization temperature*. A temperature that is estimated from the extent to which lines of certain atoms in certain states of excitation are present is called an *excitation temperature*. The temperature of a perfect radiator that would emit the same total amount of radiant energy per unit area as the star does is the *effective temperature* of the star; the effective temperature cannot be determined directly from the spectrum alone, but is found, rather, from the luminosity and radius of a star if they are known (see Chapter 25). If a star were a perfect radiator and were in complete thermodynamic equilibrium, all these temperatures would agree. As it is, the different kinds of temperatures usually do agree with each other fairly well—a circumstance that results from the fact that the observable radiation from most stars emerges from a relatively small range of depth and temperature in their photospheres.

PRESSURE

We have already discussed how the pressure in a stellar photosphere can affect its spectrum and how stars of high photospheric pressure and high temperature can be distinguished from those of low photospheric pressure and a somewhat lower temperature. It is generally possible, therefore, to make a rough estimate of the pressures of the gases in a stellar photosphere.

CHEMICAL COMPOSITION

Dark lines of a majority of the known chemical elements have now been identified in the spectra of the sun and stars. Of course, the lines of *all* elements are not observable in the spectrum of a star, nor are lines of any one element visible in the spectra of *all* stars. As we have seen, because of variations among the stars in temperature and pressure of the photosphere, only certain of the prevailing kinds of atoms are able to produce absorption lines in any one star. The absence of the lines of a particular element, therefore, does not necessarily imply that that element is not present. On the other hand, spectral lines of an element, in the neutral state or in one of its ionized states, certainly imply the presence of that element in the star (Figure 24.9).

Once due allowance has been made for the prevailing conditions of temperature and pressure in a star's photosphere, analyses of the equivalent widths of absorption lines in its spectrum can yield information regarding the relative abundances of the various chemical elements whose lines appear. It is found that the relative abundances of the different chemical elements in the atmospheres of the sun and of most stars (as well as in most other regions of space that have been investigated) are approximately the same. Hydrogen seems to comprise from 60 to 80 percent of the mass of most stars. Hydrogen and helium together comprise from 96 to 99 percent of the mass; in some stars they comprise more than 99.9 percent. Among the 4 percent or less of the mass that is made up of "heavy ele-

Figure 24.9 A portion of the solar spectrum and a comparison spectrum of iron photographed with the same spectrograph. Notice that many of the dark lines in the solar spectrum are matched by the bright lines in the comparison spectrum, showing that iron is present in the sun. (Lick Observatory)

ments," neon, oxygen, nitrogen, carbon, magnesium, argon, silicon, sulfur, iron, and chlorine are among the most abundant. Generally, but not invariably, the elements of lower atomic weight are more abundant than those of higher atomic weight.

RADIAL VELOCITY

The radial or "line-of-sight" velocity of a star can be determined from the Doppler shift of the lines in its spectrum (Section 8.3). William Huggins, an English astronomer, made the first radial-velocity determination of a star in 1868. He observed the Doppler shift in one of the hydrogen lines in the spectrum of Sirius and found that the star was receding from the solar system. The space motion of a star towards or away from us shifts the stellar lines in wavelength but does not broaden the lines.

ROTATION

As a star rotates, unless its axis of rotation happens to be directed exactly toward the sun, one of its limbs approaches us and the other recedes from us, relative to the star as a whole. For the sun or a planet, we can observe the light from one limb or the other and measure directly the Doppler shifts that arise from the rotation. A star, however, appears as an unresolved point of light, and we are obliged to analyze the light from its entire disk at once. Nevertheless, part of the light from a rotating star, including the spectral lines, is shifted to shorter wavelengths and part is shifted to longer wavelengths. Each spectral line of the star is a composite of spectral lines originating from different parts of the star's disk, all of which are moving at different speeds with respect to us. The effect produced by a rapidly rotating star is that all its spectral lines are broadened so that their profiles have a characteristic "dish" shape (see Exercise 8). Fortunately, this dish shape is highly characteristic and usually can be distinguished from broadening produced by other sources. The amount of rotational broadening of the spectral lines, if observable, can be measured, and a lower limit to the rate of rotation of the star can be calculated (see Exercise 9).

TURBULENCE

If large masses of a star's photospheric layers have vertical turbulent motion, with some blobs of gas rising and others falling, this motion, like rotation, causes a broadening of the spectral lines.

MAGNETIC FIELDS

Magnetic fields cause the energy levels in atoms to split into two or more levels, which are shifted from the nominal levels by amounts that depend on the strength of the field. The spectral lines are corre-

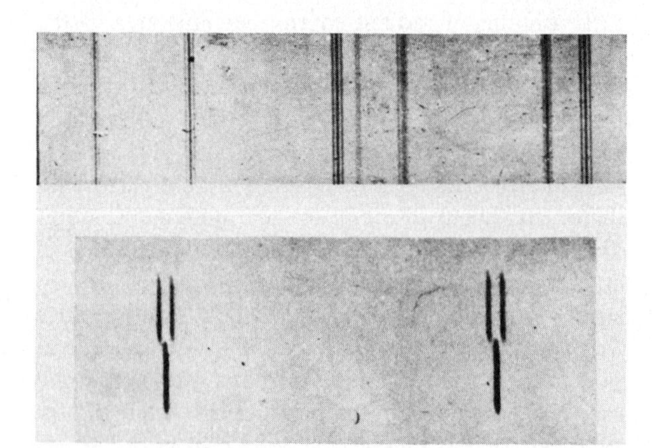

Figure 24.10 The Zeeman effect. *(Above)* The splitting of several iron lines in a magnetic field; *(below)* two mercury lines in the presence *(upper half)* and in the absence *(lower half)* of a magnetic field. (Yerkes Observatory)

spondingly split into two or more components, and this splitting is referred to as the *Zeeman effect* (Figure 24.10). The actual splitting of the spectral lines, for example, can be observed in light coming from the magnetic regions associated with spots on the sun (Chapter 28). However, in the spectra of those stars that possess appreciable magnetic fields, the lines are usually only broadened, because the separate components of a line are not resolved. Zeeman broadening can still be differentiated, however, because the light that is not absorbed in the components of the split lines is polarized. The amount by which a spectral line is broadened by a magnetic field can be detected, therefore, by photographing the star's spectrum through a polarizing filter. About 10 percent of the stars with temperatures between 8000 K and 25,000 K have very strong magnetic fields. In most cases, these field strengths are variable. The explanation is that the field on such a star is localized, and rotation of the star carries the face with the magnetic field alternately into and out of our view. Magnetic fields, probably associated with starspots, the stellar equivalent of sunspots, have been detected in many cool stars.

SHELLS AND EJECTED GASES

In the spectra of some stars, absorption lines are observed that do not appear to originate in the photospheres. Sometimes such lines can be associated with shells or rings of material ejected by the star. If enough material is ejected, and if the star has radiation of high enough energy to excite or ionize the gas, the latter produces emission lines instead of absorption lines, superposed on the stellar spectrum. Examples are Wolf-Rayet stars and hot emission-line stars. Ejection of matter may play a crucial role in determining the fate of dying stars (see Chapter 32).

EXERCISES

1. What are the probable approximate spectral classes for stars whose wavelengths of maximum light have the following values (see Section 8.1)?
   a) 2.9×10^{-5} cm
   b) 0.5×10^{-5} cm
   c) 6.0×10^{-5} cm
   d) 12.0×10^{-5} cm
   e) 15×10^{-5} cm
2. What are the probable approximate spectral classes of stars described as follows?
   a) Balmer lines of hydrogen are very strong; some lines of ionized metals are present.
   b) Strongest lines are those of ionized helium.
   c) Lines of ionized calcium are the strongest in the spectrum; hydrogen lines show with only moderate strength; lines of neutral and ionized metals are present.
   d) Strongest features are lines of C_2, CN, and CH; titanium oxide is absent.
3. The spectrum of a star shows lines of ionized helium and also molecular bands of titanium oxide. What is strange about this spectrum? Can you suggest an explanation?
4. Most stars have masses in the range 0.1 to 10 times the sun's mass. Stellar radii range, however, from 0.01 to 1000 times that of the sun. What is the ratio of the surface gravities of two stars, each of one solar mass, but one with a radius of 0.01 times the sun's radius, and one with a radius of 1000 times the sun's radius?
5. * Sketch how the microphotometer tracing of a spectrogram might appear if the spectrum showed *both* absorption and emission lines.
6. * At what part of the solar disk would the temperature of the visible gases correspond to the "boundary temperature" of the sun? How could you define a "boundary"?
7. * Explain why extremely little light could be subtracted from a continuous spectrum if there were no line broadening.
8. Explain (with a diagram) how stellar rotation broadens spectral lines.
9. Why is it that only a lower limit to the rate of stellar rotation can be determined from rotational broadening, rather than the actual rotation rate?
10. * Consider a small star and a large star of the same mass and spectral type. Would you expect the large or the small star to have sharper (narrower) lines in its spectrum? Why?
11. * Outline the steps by which you would analyze the spectrum of a star to determine the relative abundances of the chemical elements in its outer layers.
12. Draw a diagram showing a shell of gas that has been ejected from a star. Indicate on the diagram the direction of the Earth. Also show which parts of the shell give rise to emission lines superimposed on the star's spectrum, and which part or parts can produce absorption lines in the star's spectrum. (*Hint:* See Chapter 32.)

Edward Charles Pickering (1846–1919), American astronomer, was a pioneer in the study of stellar spectra, and especially in the study of the spectra of binary stars. He was the first to demonstrate the existence of spectroscopic binaries, and he also carried out fundamental work in the investigation of eclipsing binary stars—so important to our knowledge of the fundamental properties of stars. (Yerkes Observatory)

25

DOUBLE STARS: WEIGHING AND MEASURING THEM

At least half the stars around the sun are found in pairs (binary stars) or in systems of three or more, ranging up to clusters of thousands—each star moving under the combined gravitational influence of the other stars in the same system. The circumstance is fortunate, because analyses of these systems provide us with our best means of learning stellar masses and sizes. Although by direct methods we have accurate data for relatively few stars in such systems, they make it possible for us to calibrate indirect methods that we can use to estimate the masses and radii of many others. Because binary stars, the simplest systems, have lent themselves best to analysis, this chapter is especially concerned with them.

25.1 DETERMINATION OF THE SUN'S MASS

The masses of stars must be inferred from their gravitational influences on other objects. We can measure the force exerted by the sun, for example, on the Earth and other planets, and the mass of the sun can be determined more reliably than that of any other star.

The most direct way to calculate the mass of the sun is from the acceleration of the Earth. For the sake of illustration, let us assume that the orbit of the Earth is circular. Then the acceleration required to keep the Earth in its orbit is the centripetal acceleration (Section 4.1):

$$a = \frac{v^2}{R},$$

where v is the Earth's orbital speed and R is the radius of its orbit. This acceleration must be provided by the gravitational attraction of the sun on a unit mass at the Earth's distance; that is,

$$a = \frac{v^2}{R} = \frac{GM_s}{R^2},$$

where G is the universal gravitational constant and M_s the desired mass of the sun. Solving the above equation for the sun's mass, we obtain

$$M_s = \frac{v^2R}{G}.$$

Both v and R are known from observation and G is determined from laboratory measurements (Section 4.2). In metric units, $v = 3 \times 10^6$ cm/s, $R = 1.49 \times 10^{13}$ cm, and $G = 6.67 \times 10^{-8}$. Substituting these values into the above formula, we find for the mass of the sun,

$$M_s = 2 \times 10^{33}\ \text{g} = 2 \times 10^{27}\ \text{tons}.$$

Because the Earth's orbit is nearly circular, the value thus found for the mass of the sun is very nearly the correct one. Of course, the mass of the sun can be found more accurately by using the instantaneously correct acceleration of the Earth and its exact distance from the sun at a given moment in its elliptical orbit.

A simple way to calculate the mass of the sun relative to that of the Earth is to apply Kepler's third law, as it was refined by Newton (Section 4.4), to two systems, the sun-Earth system in which the sun is the accelerating mass, and the Earth-Moon system in which the Earth is the accelerating mass. Since the Earth's mass is negligible compared with the sun's, Kepler's third law, when applied to the Earth-sun system, becomes

$$M_s P_{es}^2 = K a_{es}^3,$$

where P_{es} is the period of revolution of the Earth about the sun (sidereal year), a_{es} is the semimajor axis of the relative orbit of the Earth and sun (the length of the astronomical unit), and K is a known constant of proportionality whose value depends on the units used to measure mass, time, and length. The same formula may be applied to another pair of mutually revolving bodies, the Earth-Moon system:

$$(M_e + M_m) P_{em}^2 = \frac{82.3}{81.3} M_e P_{em}^2 = K a_{em}^3,$$

where M_e is the Earth's mass, which is 81.3 times the mass of the Moon (M_m), P_{em} is the period of revolution of the Moon about the Earth (the sidereal month), and a_{em} is the semimajor axis of the relative orbit of the Earth-Moon system. If the first of the above equations is divided by the second, the result is

$$\frac{M_s}{M_e} = \frac{82.3}{81.3}\left(\frac{P_{em}}{P_{es}}\right)^2 \left(\frac{a_{es}}{a_{em}}\right)^3.$$

The sidereal month is about 27.32 days, or about 1/13.4 sidereal years, and the astronomical unit is about 389 times the Moon's distance from the Earth. If these figures are inserted in the above equation,

$$\frac{M_s}{M_e} = 33{,}000.$$

That is, the sun is about one-third of a million times as massive as the Earth. Since the Earth's mass is 6×10^{27} gm (Section 4.2), the sun's mass is 2×10^{33} gm, as found above.

25.2 BINARY STARS AND THE DETERMINATION OF STELLAR MASSES

We have not yet been able to observe directly planets revolving around other stars, but we do observe many double stars and cases of three or more stars that belong to the same dynamical system.

(a) Discovery of Binary Stars

In 1650, less than half a century after Galileo turned a telescope to the sky, the Italian Jesuit astronomer Giovanni Baptista Riccioli observed that the star Mizar, in the middle of the handle of the Big Dipper, appeared through his telescope as *two* stars. Mizar was the first *double star* to be discovered. In the century and a half that followed, many other closely separated pairs of stars were discovered telescopically.

One famous double star is Castor, in Gemini. The telescope reveals Castor to be two stars. They were separated by an angle of nearly 5″ in 1804, when William Herschel had noted that the fainter component of Castor had changed, slightly, its direction from the brighter component. Here, finally, was observational evidence that one star was moving about another; it was the first evidence that gravitational influences exist outside the solar system.

If the gravitational forces between stars are like those in the solar system, the orbit of one star about the other must be an *ellipse*. The first to show that such is the case was Felix Savary, who in 1827 showed that the relative orbit of the two stars in the double system ξ *Ursae Majoris* is an ellipse, the stars completing one mutual revolution in a period of 60 years. An example of a double star system is shown in Figure 25.1.

Another class of double stars was discovered by E. C. Pickering, at Harvard, in 1889. He found that the lines in the spectrum of the brighter component of Mizar (the first double star to be discovered) are usually *double*, but that the spacing in wavelength of the components of the lines varies periodically, and at times the lines even become single. He correctly deduced that the brighter component of Mizar (Mizar A) itself is really *two* stars that revolve about each other in a period of 104 days. When one star is approaching us, relative to the center of mass of the two, the other star is receding from us; the radial velocities of the two stars, and therefore the Doppler shifts of their spectral lines, are different,

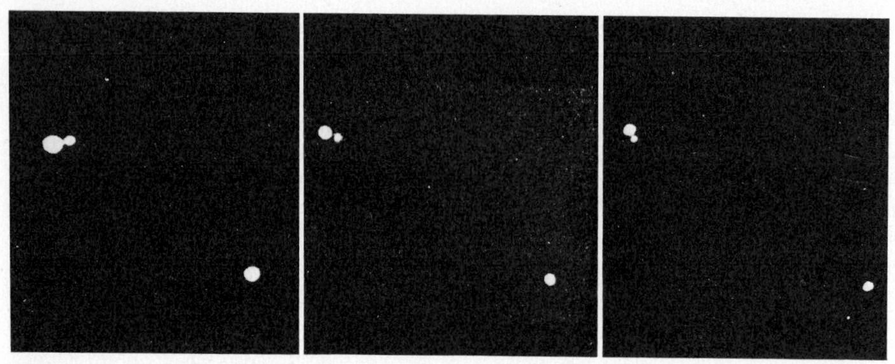

Figure 25.1 Three photographs, covering a period of about 12 years, which show the mutual revolution of the components of the visual binary star Kruger 60. (Yerkes Observatory)

so that when the composite spectrum of the two stars is observed, each line appears double. When the two stars are both moving across our line of sight, however, they both have the same *radial* velocity (that of the center of mass of the pair), and hence the spectral lines of the two stars coalesce. An example is shown in Figure 25.2.

Stars like Mizar A, which appear as single stars when photographed or observed visually through the telescope, but which the spectrograph shows really to be double stars, are called *spectroscopic binaries;* systems that can be observed visually as double stars are called *visual binaries*. In 1908 Frost found that the fainter component of Mizar, Mizar B, is also a spectroscopic binary. Mizar and nearby Alcor, the faint companion visible to the naked eye, form an optical double. These stars are members of a group of stars, all having nearly the same space velocities, but which do not orbit each other.

Almost immediately following Pickering's discovery of the duplicity of Mizar A, Vogel discovered that the star Algol, in Perseus, is a spectroscopic binary. The spectral lines of Algol were not observed to be double, because the fainter star of the pair gives off too little light compared with the brighter for its lines to be conspicuous in the composite spectrum. Nevertheless, the periodic shifting back and forth of the wavelengths of the lines of the brighter star gave evidence that it was revolving about an unseen companion; the lines of both components need not be visible in order for a star to be recognized as a spectroscopic binary (Figure 25.3).

The proof that Algol is a double star is significant for another reason. In 1669 Montonari had noted that the star varied in brightness; in 1783 John Goodricke established the nature of the variation. Normally, Algol is a second-magnitude star, but at intervals of $2^{d}20^{h}49^{m}$ it fades to one-third of its regular brightness; after a few hours, it brightens to normal again. Goodricke suggested that the variations might be due to large dark spots on the star, turned to our view periodically by its rotation, or that the star might be eclipsed regularly by an invisible companion. Vogel's discovery that Algol is

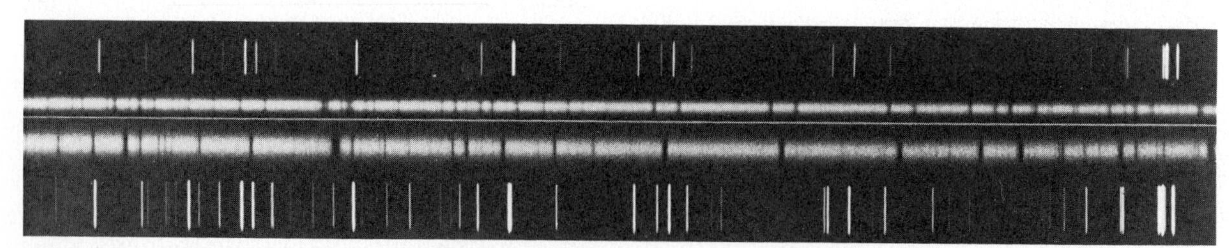

Figure 25.2 Two spectra of the spectroscopic binary κ Arietis. When the components are moving at right angles to the line of sight *(bottom)*, the lines are single. When one star is approaching us and the other receding *(top)*, the spectral lines of the two stars are separated by the Doppler effect. (Lick Observatory)

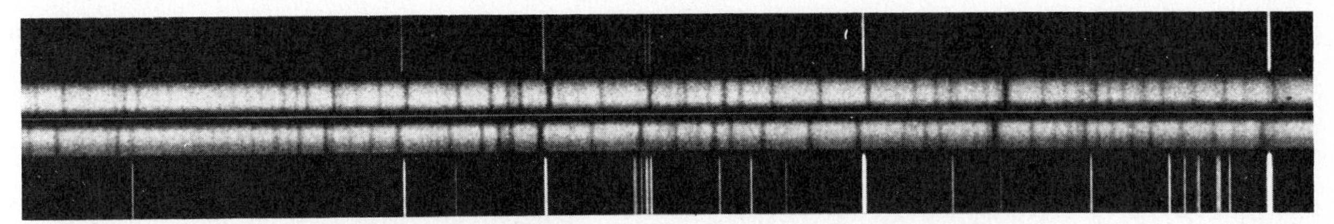

Figure 25.3 Two spectrograms of the single-line spectroscopic binary α^1 Geminorum. The upper part shows the spectrum of the brighter star when it is moving away from us at nearly its maximum orbital speed, and the lower one when it is moving toward us. (Lick Observatory)

a spectroscopic binary verified the latter hypothesis. The plane in which the stars revolve is oriented nearly edgewise to our line of sight, and each star is eclipsed or occulted once by the other during every revolution. The eclipse of the fainter star is not very noticeable because the part of it that is covered contributes little to the total light of the system; the eclipse of the brighter star can, however, be observed. A binary such as Algol, in which the orbit is nearly edge-on to the Earth so that the stars eclipse each other, is called an *eclipsing binary*.

Binary stars are now known to be very common; they may be the rule, not the exception. In the stellar neighborhood of the sun, somewhere between one-half and two-thirds of all stars are members of binary or multiple star systems.

(b) Binary Star Orbits and Stellar Masses

Observations of binaries provide the most direct measurements of the masses of stars. In the case of visual binaries, measurement of the separation of the two stars and of the direction of an imaginary line drawn between them as both change with time can, if the distance to the double star system is known, yield the masses of the two individual components. For spectroscopic binaries, if the spectra of both stars can be observed, we can derive the ratio of the masses of the stars, and we can also determine minimum values of the mass of each star. The actual values of the masses of the individual stars cannot be determined from spectroscopic observations of the orbital velocities alone. If the line spectrum of only one of the stars is visible, then we can derive only an estimate for the lowest mass that the *unseen* star could have. While this number may not seem to be very useful, it can be a very important guide in searching for candidates for black holes (Section 33.6).

Most useful for determining stellar masses are spectroscopic binary systems in which the two stars also eclipse each other. For such double star systems, even if we do not know how far away they are, we can determine the actual masses of the two individual stars, their diameters, and their effective temperatures. Among the thousands of known binaries, however, only a few dozen are so favorably disposed for observation that all of the necessary data can be obtained. There are only about 100 individual stars in eclipsing systems and fewer than three dozen more in visual binary systems for which we have derived masses that are reliable to 15 to 20 percent.

In the following sections, we show how observations of binaries can be used to derive stellar masses.

(c) Mass Determinations of Visual Binary Stars

The two stars of a binary pair revolve mutually about their common center of mass (or barycenter), which in turn moves among the neighboring stars. Each star, therefore, describes a wavy path around the course followed by the barycenter. With careful observations it is possible to determine these individual motions of the member stars in a visual binary system. It is far more convenient, however, simply to observe the motion of one star (by convention the fainter), about the other (the brighter). The observed motion shows the *apparent relative orbit*. The periods of mutual revolution for visual binaries range from a few years to thousands of years, but only for those systems with periods less than a few hundred years can the apparent relative orbits

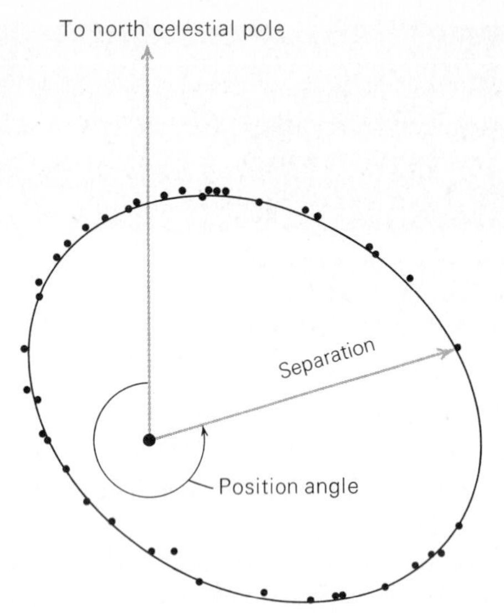

Figure 25.4 Separation and position angle in a visual binary star.

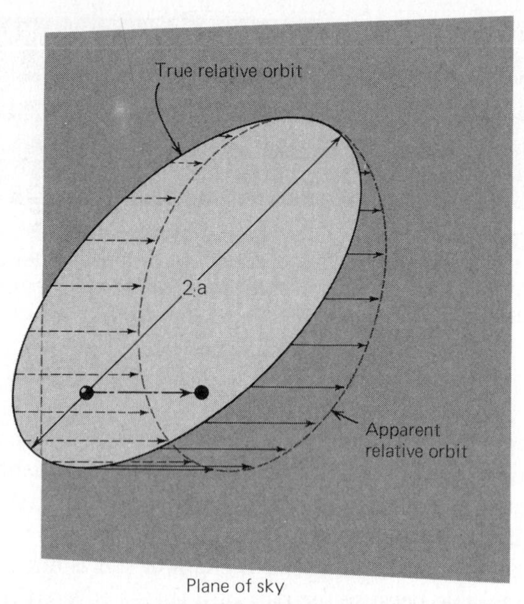

Figure 25.5 Relation between a true and an apparent relative orbit, showing the apparent displacement of the focus.

be determined with much precision; even then, a long series of observations covering a number of decades is usually necessary. The data observed are the angular separation of the stars and the *position angle,* which is the direction, reckoned from the north, around toward the east, of the fainter star from the brighter one. These data can be measured on a photograph if the separation of the stars is not too small; in any case, they can be measured directly at the telescope. A typical apparent relative orbit, assembled from many observations of separation and position angle, is shown in Figure 25.4.

The true orbit of the binary system does not, in general, happen to lie exactly in the plane of the sky (that is, perpendicular to the line of sight). Consequently, the apparent relative orbit is merely the *projection* of the *true relative orbit* into the plane of the sky. Now it is easy to show that when an ellipse in one plane is projected onto another plane (that is, is viewed obliquely), the projected curve is also an ellipse, although one of different eccentricity. Moreover, the foci of the original ellipse do *not* project into the foci of the projected ellipse. Therefore, the brighter star, although it is located at one focus of the *true* relative orbit, is not at a focus of the *apparent* relative orbit. This circumstance makes it possible to determine the inclination of the true orbit to the plane of the sky. The problem is simply one of finding the angle at which the ellipse of the true relative orbit must be projected in order to account for the amount of displacement of the brighter star from the focus of the apparent relative orbit (Figure 25.5). There are several techniques for solving this geometry problem, so that the shape and angular size of the true orbit can be found. The period of mutual revolution, of course, is observed directly if the system has completed one revolution during the interval it has been under observation; otherwise, the period must be calculated from the rate at which the fainter star moves in the relative orbit.

If the semimajor axis of the true relative orbit (that is, the semimajor axis the orbit would appear to have if it were seen face on) has an angular length, in seconds of arc, of a'', and if the system is at a distance of r pc, the semimajor axis, in astronomical units, is $r \times a''$ (see Exercise 7). The sum of the masses of the two stars, in solar units, is given by Kepler's third law:

$$m_1 + m_2 = \frac{(r \times a'')^3}{P^2}.$$

To find what share of the total mass belongs to each star, it is necessary to investigate the individual motions of the stars with respect to the center of mass of the system. The distance of each star from the barycenter is inversely proportional to its own mass.

As an example, consider the visual binary Sirius A and Sirius B. The semimajor axis of the true relative orbit is about 7½″ and the distance from the sun to Sirius is 2.67 pc. The period of the binary system is 50 years. The sum of the masses of the stars, then, is

$$m_1 + m_2 = \frac{(2.67 \times 7.5)^3}{(50)^2} = 3.2 \text{ solar masses.}$$

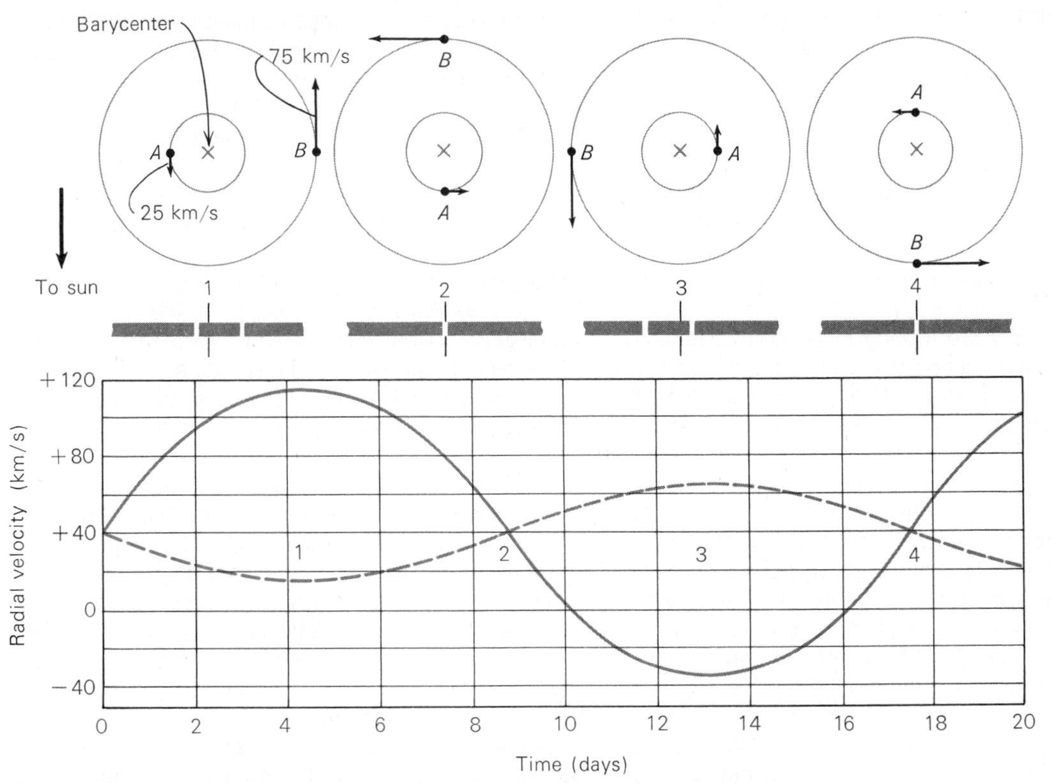

Figure 25.6 Hypothetical spectroscopic binary system with circular orbits.

Sirius B, the fainter component, is about twice as far from the barycenter as Sirius A, and so has only about half the mass of Sirius A. The masses of Sirius A and Sirius B, therefore, are about 2 and 1 solar masses, respectively.

(d) Spectroscopic Binaries

If the two stars of a binary system have a small linear separation, that is, if their relative orbit is small, there is little chance that they will be resolved as a visual binary pair. On the other hand, they have a shorter period, and their orbital velocities are relatively high, compared with the stars of a visual binary system; unless the plane of orbital revolution is almost face-on to our line of sight, there is a good chance that we will be able to observe radial velocity variations of the stars due to their orbital motions. In other words, they constitute a spectroscopic binary system.

Most spectroscopic binaries have periods in the range from a few days to a few months; the mean separations of their member stars are usually less than 1 AU. If the two stars of a spectroscopic binary are not too different in luminosity, the spectrum of the system displays the lines of both stars, each set of lines oscillating in wavelength in the period of mutual revolution. More often, lines of only one star are observed. A graph showing the radial velocity of a member of a binary star system plotted against time is called a *radial velocity curve,* or simply, a *velocity curve.*

Here, we shall only consider, as an example of the analysis of a spectroscopic binary, the simple case in which the orbits of the stars are circular, and in which the spectral lines of both stars are observed. Figure 25.6 shows the two stars in their orbits, the Doppler shifts of a hypothetical spectral line, and the two radial velocity curves. Because the orbits of the stars in this example are circular, the velocity curves are symmetrical; they have shapes that are known, technically, as *sine curves.* A complete cycle of variation of radial velocity—the period of the system—is 17.5 days. At position 1, star *A* has its maximum possible component of velocity toward the sun, and star *B* has its maximum possible component of velocity away from the sun. The conditions are reversed at position 3. At positions 2 and 4, both stars are moving across our line of sight and neither has a component of velocity in our line of sight due to orbital motion; both have the same radial velocity as that of the center of mass of the system, 40 km/s. The radial velocity of star *B* ranges from +115 to −35 km/s, a range of 150 km/s. The maximum difference between the radial velocities of star *B* and that of the barycenter, then, is 75 km/s. The corresponding value for star *A* is only 25 km/s. Because both stars have the same period and since star *A* moves only one-third as fast as star *B*, with respect to the barycenter, star *A* must have only one-third as far to go to get around its orbit; its orbit must be one-third the size of that of star *B*, and its mass, therefore, must be three times as great.

Stars *A* and *B* are moving in opposite directions with respect to their center of mass; thus the maximum radial velocity (as observed from the solar system) of star *B* with respect to star *A* must be the sum of the radial velocities of the two stars with respect to the barycenter, or 100 km/s. If the orbital plane of the system were in our line of sight, this would be the relative orbital velocity of one star with respect to the other. As it is, however, the orbital plane is tilted at some unknown angle to our line of sight, and the 100 km/s is only the maximum *radial component* of the relative velocity; the actual relative velocity will be greater by an unknown factor. Now, the distance around the relative orbit—its circumference—is the relative orbital velocity multiplied by the period—the time one star takes to get around the other. The distance between the stars, *a*, is the circumference of the orbit divided by 2π; that is,

$$a = \frac{V \times P}{2\pi},$$

where V is the relative velocity. If we substitute the observed lower limit to the relative velocity—in our example, 100 km/s—for V in the above equation, we will obtain a lower limit to the separation of the stars. If this lower limit to a is applied to Kepler's third law, we obtain a lower limit to the sum of the masses of the stars;

$$m_1 + m_2 = \frac{a^3}{P^2}.$$

The calculation for the numerical example given in Figure 25.6 may be illustrated as follows: A velocity of 100 km/s is equivalent to about 20 AU/yr, and a period of 17.5 days is about 0.048 yr; thus a lower limit to a is given by

$$a = \frac{20 \times 0.048}{2\pi} = 0.153 \text{ AU}.$$

A lower limit to the sum of the masses, then, is

$$m_1 + m_2 = \frac{0.153^3}{0.048^2} = 1.6 \text{ solar masses}.$$

Since star *A* is three times as massive as star *B*, it has 75 percent of the total mass; lower limits to the individual masses of the stars are therefore

$$m_A \geq 1.2 \text{ solar masses}$$

and

$$m_B \geq 0.4 \text{ solar mass},$$

where the symbol "$\geq$" means "greater than or equal to."

The analysis is more difficult in the general case of elliptical, rather than circular, orbits, but it can nevertheless be carried out. The results are similar except that a is then the semimajor axis of the relative orbit instead of the constant separation between the stars that exists if the orbits are circular. In either case, individual masses of the stars are not found, only lower limits to their masses.* If the spectral lines of only one star are visible, we do not even find lower limits to the masses of the individual stars but only a relation between their masses, known as the *mass function*.†

The mass function does, however, indicate a lower limit to the mass of the star whose spectral lines are *not* observed in terms of the one whose spectral lines *are* seen. If the spectrum of the visible star gives an indication of its mass (as is often the case), we have at once an estimate of the lower limit to the mass of the fainter companion. We shall see (Chapter 33) that this analysis can be very important for finding candidates for black holes.

The analysis of spectroscopic binaries, we have seen, yields only lower limits to stellar masses. To find the masses of individual binary systems we must have some way of determining the angle at which the orbital plane of the system is inclined to our line of sight (or to the plane of the sky); we can find this angle only when the orbit is almost edge-on, so that the spectroscopic binary is also an eclipsing binary (or in a few cases, when it is also a visual binary).

(e) Eclipsing Binaries

To simplify our discussion of eclipsing binaries, we shall assume that the two stars in such a system re-

---

* What is actually found for each star is $m \sin^3 i$, where i is the inclination angle of the plane of orbital revolution to the plane of the sky.

† The *mass function* is

$$\frac{m_2^3 \sin^3 i}{(m_1 + m_2)^2},$$

where m_2 is the mass of the star whose spectrum is not observed.

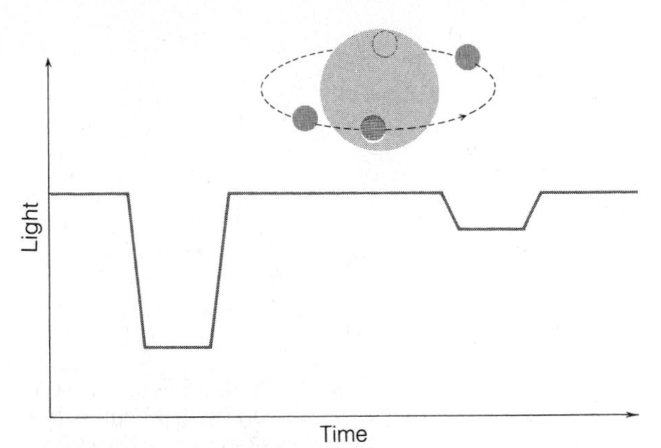

Figure 25.7 Schematic light curve of a hypothetical eclipsing binary star with alternating total and annular eclipses.

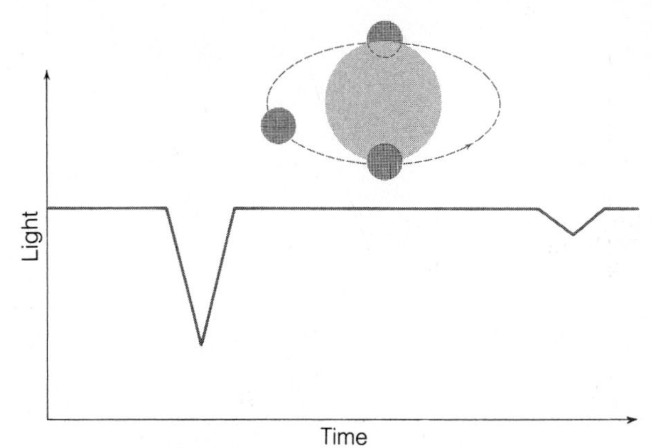

Figure 25.8 Schematic light curve of a hypothetical eclipsing binary star with partial eclipses.

volve about each other in circular orbits. Actually the assumption is not so bad. Most stars in binary systems that are seen to eclipse each other are relatively close to each other. (Here we use the term *eclipse* for the passage of either star in front of the other, without distinguishing between *transits* and *occultations*—see Section 7.4.) Their mutual proximity enhances the effects of their perturbations on each other's motions, and these perturbations tend to produce circular, or nearly circular orbits.

Let us consider an eclipsing binary in which the two stars are of unequal diameters. During the period of revolution, there are two times when the light from the system diminishes—once when the smaller star passes behind the larger one and is occulted, and once when the smaller star transits in front of the larger one and eclipses part of it. If the smaller star goes completely behind the larger one, that eclipse is *total* and the other eclipse half a period later is *annular* (see Figure 25.7). If the smaller star is never completely hidden behind the larger star, both eclipses are partial (Figure 25.8). Each interval during an eclipse when the light from the system is farthest below maximum brightness is called a *minimum*. Both minima are not, in general, equally low in light. The same area of each star is covered during the time it is eclipsed; for example, if the eclipses are total or annular, an area equal to the total cross-sectional area of the smaller star is eclipsed at each minimum. The relative amount of light drop at each minimum, however, depends on the relative *surface brightnesses* of the two stars, and hence on their temperatures. (Remember that, according to Planck's law, hotter stars emit more total energy per square centimeter, and therefore have a greater surface brightness—see Section 8.1.) *Primary minimum* occurs when the hotter star is eclipsed (whether it is a total, an annular, or a partial eclipse), and *secondary minimum* occurs when the cooler star is eclipsed. A graph of the light from an eclipsing binary system, plotted against time through a complete period, is called a *light curve.*

The most important data that are obtained from the analysis of the light curve of an eclipsing binary system are the sizes of the stars, relative to their separation, and the inclination of the orbit to our line of sight. To illustrate how the sizes of the stars are related to the light curve, we may consider a hypothetical eclipsing binary in which the stars are very different in size, in which the orbit is exactly edge-on, so that the eclipses are *central* (Figure 25.9), and in which the or-

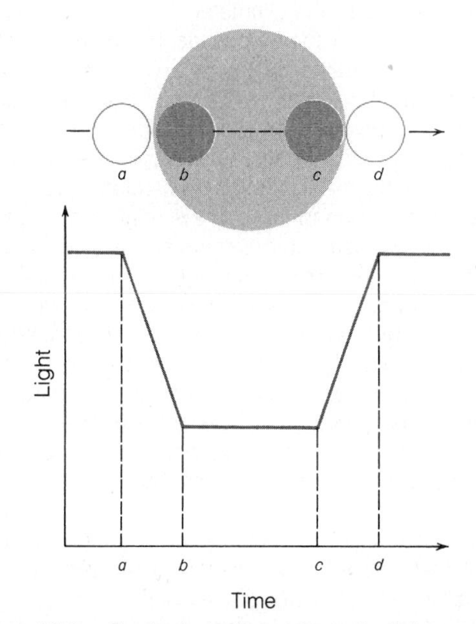

Figure 25.9 Contacts in the schematic light curve of a hypothetical eclipsing binary with central eclipses.

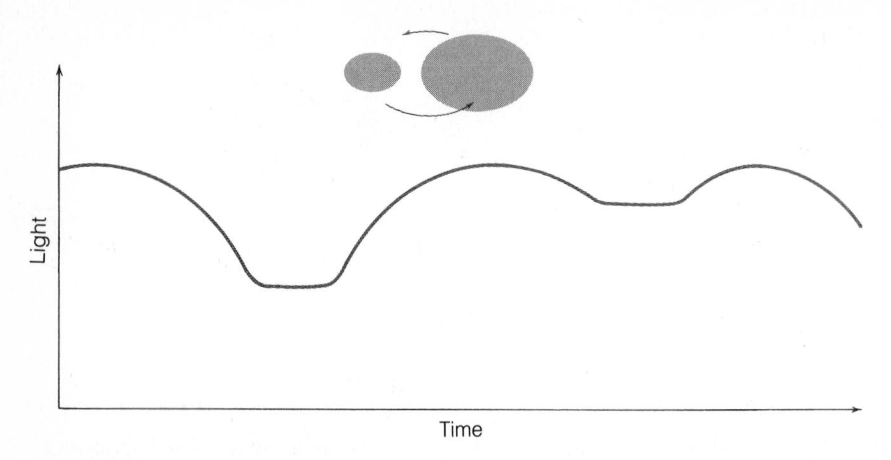

Figure 25.10 Effect of tidal distortion on the light curve in a binary system.

bit is large enough that we can assume the stars to move in approximately straight lines during the eclipses. When the small star is at point *a (first contact),* and is just beginning to pass behind the large star, the light curve begins to drop. At point *b (second contact),* the small star has gone entirely behind the large one and the total phase of the eclipse begins. At *c (third contact)* it begins to emerge, and when the small star has reached *d (fourth* or *last contact)* the eclipse is over. During the time interval between first and second contact (or between third and last contacts), the small star has moved a distance equal to its own diameter. That time interval is in the same ratio to the period of the system as the diameter of the small star is to the circumference of the relative orbit. During the time interval from first to third contacts (or from second to last contacts), the small star has moved a distance equal to the diameter of the large star; that time interval is to the period as the diameter of the large star is to the circumference of the relative orbit. We see, therefore, that the light curve alone gives the sizes of the stars in terms of the size of their orbit. If the lines of both stars are visible in the composite spectrum of the binary, both radial velocity curves can be observed. Then the size of the relative orbit can be found, and we can determine the actual (linear) radii of the stars. In other words, the velocity of the small star with respect to the large one is known, and, when multiplied by the time intervals from first to second contacts and from first to third contacts, gives, respectively, the diameters of the small and large stars.

In actuality the orbits are not, generally, exactly edge-on, and the eclipses are not central. However, it is a relatively simple geometry problem, at least in principle, to use measurements of the depths of the minima and the exact instants of the various contacts to calculate both the inclination of the orbit and the sizes of the stars relative to their separation.

The foregoing discussion applies only to eclipses that are total and annular. If they are partial, the analysis is far more difficult, although it can still be accomplished.

There are various other complications which have been ignored here. For example, stars, like the sun, exhibit limb darkening (Section 24.2b), which affects the rate of the drop of light during eclipse. Also, the two stars in a short period eclipsing binary may be so close together that they suffer severe tidal distortion and have shapes more like footballs than spheres. The light from such systems is not constant, even outside of eclipse, but is greatest when the stars' longest dimensions are turned "broadside" to us, and is less just before and just after a minimum (Figure 25.10). There are many other complications as well, which the specialist must be concerned with.

To summarize: From the analysis of the light curve of an eclipsing binary we can find the inclination of the orbit and the sizes of the stars relative to their separation. If, in addition, we can measure the Doppler shifts of the spectral lines of both stars during their period of revolution, we can obtain their velocity curves. The analysis of the velocity curves, as described in the preceding section, leads to a determination of lower limits to the masses of the stars. The knowledge of the inclination of the orbit now allows us to convert these minimum values for the masses to actual masses for the individual stars. We can also convert the lower limit to the separation of the stars (or the semimajor axis of their relative orbit) to the actual value when the inclination is known; since the sizes of the stars relative to this separation are found from the light curve, we find their actual diameters or radii. Finally, from the relative depths of the primary and secondary minima, we can calculate the relative surface brightnesses of the stars and hence their effective temperatures (the surface brightness of a star is proportional to the fourth power of its effective temperature—Sections 8.1h and 24.2d). Note that we do not need to know the distance to an eclipsing-spectroscopic binary to determine its mass, as we do in the case of a visual binary. Among the

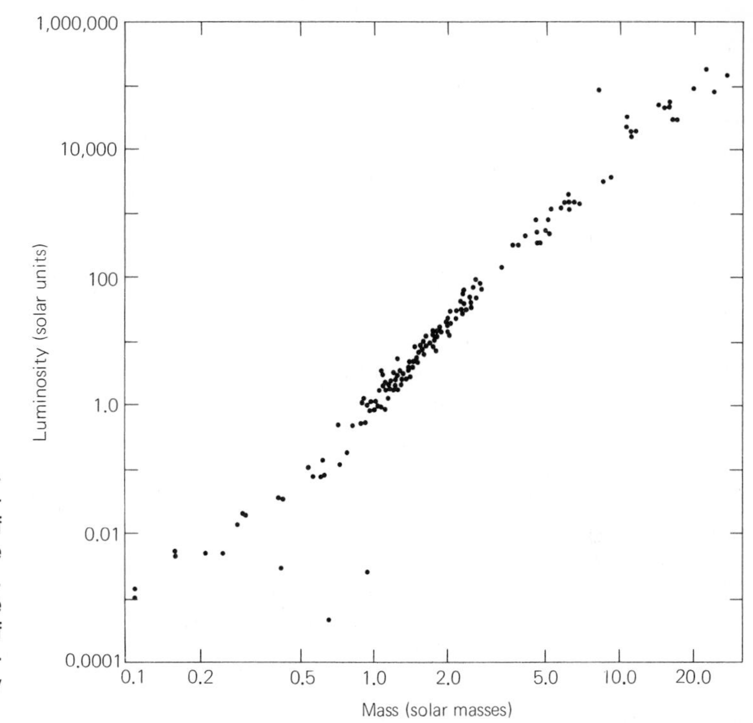

Figure 25.11 Mass-luminosity relation. The plotted points show the masses (abscissas) and luminosities (ordinates) of stars for which both of these quantities are known to an accuracy of 15 to 20 percent. The open circle represents the sun. The three points lying below the sequence of points are all white dwarf stars (see Chapter 26). (Adapted from data compiled by D. M. Popper)

thousands of known eclipsing binaries, however, only a few dozen are so favorably disposed for observation that all the necessary data can be obtained.

(f) Mass Exchange in Close Binaries

We saw in Section 5.2 that in Lagrange's solution to the three-body problem, in which two stars revolve about each other in circular orbits, there are equilibrium points where gravitational forces are balanced and a low-mass body can remain stationary with respect to the two stars. Should the small body be shoved closer to one of the revolving stars, however, it is drawn to it by gravitational attraction. We shall also see that as stars age, the generation of nuclear energy in their interiors causes them to distend their outer layers greatly, so that those stars become giants (Chapter 30). If such a star is a member of a close binary system, the atoms in its expanding outer layers may reach and pass through one of the points (the *inner Lagrangian point*) where they do not belong to either star. Subsequently the matter from the expanding star can flow to the other star.

Mass exchange is believed to occur between many stars in close binary systems. This exchange of mass can have profound effects on the evolution of the stars in a system. Not only may matter stream from one star to another in a close binary system, possibly accounting for such phenomena as novae and supernovae, but it can also form a large circumstellar disk or ring of material around the binary system, and even be ejected from the system altogether. It also may be involved in the creation of neutron stars and black holes (Chapters 32 and 33).

(g) The Mass-Luminosity Relation

Studies of binary stars have provided knowledge of the masses of over a hundred individual stars. When the masses and luminosities of those stars for which both of these quantities are determined are compared, it is found that, in general, the more massive stars are also the more luminous. This relation, known as the *mass-luminosity relation*, is shown graphically in Figure 25.11. The mass-luminosity relation provides a useful means of estimating the masses of stars of known luminosity that do not happen to be members of visual or eclipsing binary systems. We shall return to the physical explanation of the relation in Chapter 30.

It should be noted how very much greater the range of stellar luminosities is than the range of stellar masses. Luminosities of main-sequence stars are roughly proportional to their masses raised to the 3.5 to 4.0 power. Most stars have masses between one-tenth and 50 times that of the sun; according to the mass-luminosity relation, however, the corresponding luminosities of stars at either end of the range are respectively less than 0.001 and about 10^6 solar luminosities. If two stars differ in mass by a factor of 2, their luminosities would then be expected to differ by a factor of 10 to 15.

25.3 DIAMETERS OF STARS

The eclipsing binary systems, as we have seen, provide one means of determining the diameters of stars, but of course only of those stars that happen to be members of eclipsing systems for which the necessary analysis can be carried out. It would be convenient if the angular sizes could be measured directly for many stars of known distances; then their linear diameters could be calculated just as they are for the Moon or planets. The sun is unfortunately the only star whose angular size can be resolved visually and whose diameter can be calculated simply. There are many other stars, however, whose angular sizes are only slightly beyond the limit of resolution of the largest telescopes and can be measured with special techniques. These techniques include speckle interferometry (Section 9.4c) and high-speed photoelectric observations of the dimming of light from stars being occulted by the Moon or planets.

In all, about a hundred angular stellar diameters have been measured by one or more of these methods. There are some difficulties with each technique (for example, in connection with limb darkening—see Section 24.2b). Still, these data give us some confidence in the correctness of the less direct determinations of stellar diameters from radiation laws.

(a) Finding Stellar Radii from Radiation Laws

For most stars we must use an indirect method by which we can calculate their radii from theory. The theory involved is the Stefan-Boltzmann law (Section 8.1); we calculate the radius of a spherical perfect radiator that has the same luminosity and effective temperature that a star does.

The luminosity of a star can be obtained by the procedure discussed in Chapter 23, and the effective temperature of a star can be obtained in various ways, as from its color or its spectrum (Section 24.2d). Now, the energy emitted per unit area of a star (given by the Stefan–Boltzmann law), multiplied by its entire surface area, gives the star's total output of radiant energy, that is, its luminosity. Since the surface area of a sphere of radius R is $4\pi R^2$, the luminosity of a star is given by

$$L = 4\pi R^2 \times \sigma T^4$$

The above equation can be solved for the radius of the star.

Note that the temperature appearing in the above formula is raised to the fourth power; if it is in error, therefore, the computed value of the star's radius can be substantially incorrect. In particular, because stars are *not* perfect radiators (that is, not true black bodies), values of stellar temperatures as determined by different methods do not all agree precisely. Different kinds of stellar temperatures were discussed more fully in Section 24.2d. Despite this uncertainty, observations indicate that we can use the Stefan–Boltzmann law to find the sizes of most stars with an accuracy of 10 to 20 percent.

We shall illustrate the use of Stefan's law for the computation of stellar radii with two examples. Consider, first, a star whose red color indicates that it has a temperature of about 3000 K, roughly half the temperature of the sun. Each square centimeter of the star, therefore, emits only 1/16 as much light as the sun (for the light emitted is proportional to the fourth power of the temperature). Suppose, however, that the star is, nevertheless, 400 times as luminous as the sun. It must be many times larger than the sun to emit more light despite its much lower surface brightness. We can find its radius, in terms of the sun's, by noting that L is proportional to R^2T^4, and thus (since the constants of proportionality cancel in each of the ratios),

$$\frac{R_\star}{R_\odot} = \sqrt{\frac{L_\star}{L_\odot}\left(\frac{T_\odot}{T_\star}\right)^2} = \sqrt{400} \times 4 = 80.$$

(The subscripts $\star$ and $\odot$ refer to the star and sun, respectively.) This star has 80 times the sun's radius; if the sun were placed at its center, the star's surface would reach past the orbit of Mercury.

Next, consider a star whose blue color indicates a temperature of about 12,000 K—twice the sun's temperature. Suppose, however, that this star has a luminosity of only 1/100 that of the sun. Now we find, for the star's radius,

$$\frac{R_\star}{R_\odot} = \sqrt{\frac{L_\star}{L_\odot}}\left(\frac{T_\odot}{T_\star}\right)^2 = \sqrt{\frac{1}{100}}\left(\frac{1}{2}\right)^2 = \frac{1}{40}.$$

The star has only 1/40 the sun's radius—less than three times the radius of the Earth. These two examples are by no means extreme cases, but are more or less typical of *very red giants* and *white dwarfs*, respectively (Chapter 26).

(b) Summary of Stellar Diameters

The few score good geometrical determinations of stellar radii come from (1) direct measure of the sun's angular diameter, (2) measures of the angular diameters of about 100 stars by special techniques including speckle interferometry and lunar occultations, and (3) analyses of the light curves and radial velocity curves of eclipsing binary systems. All other determinations of stellar radii make use of the radiation laws; the validity of this indirect method is verified by noting that it gives the correct radii for those stars whose sizes can also be measured by geometrical means.

EXERCISES

1. Many eclipsing binaries can be observed which do not appear in catalogues of spectroscopic binaries. Can you suggest an explanation?
2. A few stars are both visual binaries *and* spectroscopic binaries (their radial velocity variations can be detected). Why do you suppose such stars are rare? Can you suggest a way of determining the distance to such a system? (*Hint:* Consider the method of determining the parallax to a moving cluster, Section 22.5c.)
3. Describe the apparent relative orbit of a visual binary whose true orbital plane is edge-on to the line of sight. Describe the apparent motions of the individual stars of the system among the background stars in the sky.
4. What, approximately, would be the periods of revolution of binary star systems in which each star had the same mass as the sun, and in which the semimajor axes of the relative orbits had these values? (a) 1 AU; (b) 2 AU; (c) 6 AU; (d) 20 AU; (e) 60 AU; (f) 100 AU.
5. ★ In each of the binary systems in Exercise 4, at what distance would the two stars appear to have an angular separation of 1″? (Assume circular orbits.)
6. Why do most visual binaries have relatively long periods and most spectroscopic binaries relatively short periods? Under what circumstances could a binary with a relatively long period (over a year) be observed as a spectroscopic binary?
7. Show that the semimajor axis of the true relative orbit of a visual binary system, in astronomical units, is equal to its angular value, in seconds of arc, times the distance of the system, in parsecs.
8. The true relative orbit of ξ Ursae Majoris has a semimajor axis of 2″.5, and the parallax of the system is 0″.127. The period is 60 years. What is the sum of the masses of the two stars in units of the solar mass?
   Answer: 2.1 solar masses
9. In a particular visual-spectroscopic binary the maximum value of the radial velocity of one star with respect to the other is 60 km/s, the inclination of the orbital plane to the plane of the sky 30°, and the period is 22 days. If the stars in the system have circular orbits, what is the sum of their masses?
   Answer: 3.8 solar masses
10. ★ The observed component of a hypothetical astrometric binary system is found to move in an elliptical orbit of semimajor axis 1 AU about the barycenter of the system in a period of 30 years. If the visible star is assumed to have a mass equal to that of the sun, what is the mass of its unseen companion?
    Answer: About 0.11 solar masses
11. A hypothetical spectroscopic-eclipsing binary star is observed. The period of the system is three years. The maximum radial velocities, with respect to the center of mass of the system, are as follows: Star A, $\frac{4}{3}\pi$ AU/year; Star B, $\frac{2}{3}\pi$ AU/year.
    a) What is the ratio of the masses of the stars?
    b) Find the mass of each star (in solar units). Assume that the eclipses are central.
12. In an eclipsing binary in which the eclipses are exactly central, and in which a small star revolves about a considerably larger one, the interval from first to second contacts is one hour and from first to third contacts is four hours. The entire period is three days. The centers of the stars are separated by

11,460,000 km. What are the diameters of the stars?
Answer: 1,000,000 and 4,000,000 km

13. Although the periods of known eclipsing binaries range from 4^h39^m to 27 years, the average of their periods is less than the average period of all known spectroscopic binaries. Can you suggest an explanation?

14. How many times as massive as the sun would you expect a star to be that is 1000 times as luminous? What if it were 10,000 times as luminous? (Assume that the mass-luminosity relation holds for these stars.)

*15. Measured angular diameters of several stars measured by Michelson and Pease are given in the following table:

STAR	ANGULAR DIAMETER	DISTANCE *(pc)*	LINEAR DIAMETER *(In Terms of Sun's)*
Betelgeuse	0″.034*	150	500
(α Orionis)	0.042		750
Aldebaran (α Tauri)	0.020	21	45
Arcturus (α Bootis)	0.020	11	23
Antares (α Scorpii)	0.040	150	640
Scheat (β Pegasi)	0.021	50	110
Ras Algethi (α Herculis)	0.030	150	500
Mira (o Ceti)	0.056*	70	420

* Variable in size.

Which of the stars are larger than the orbit of the Earth? Are any larger than the orbit of Mars? of Jupiter?

16. Show how the measured angular diameters and observed energy fluxes of stars can be used to measure their effective temperatures. (*Hint:* Use the inverse-square law of light, and recall the Stefan–Boltzmann law.)

17. What is the radius of a star (in terms of the sun's radius) with the following characteristics:
a) Twice the sun's temperature and four times its luminosity?
b) Eighty-one times the sun's luminosity and three times its temperature?

18. Assume the wavelength of maximum light of the sun to be exactly 5000 Å, its temperature exactly 6000 K, and its absolute bolometric magnitude exactly 5.0. Another star has its wavelength of maximum light at 10,000 Å. Its apparent visual magnitude is 15.5, its bolometric correction is 0.5, and its parallax is 0″.01. What is its radius in terms of the sun's?
Answer: $R/R_{\odot} = 0.4$

26

THE STELLAR POPULATION: A CELESTIAL CENSUS

Ejnar Hertzsprung (1873–1967), Danish astronomer, spent most of his productive career in Göttingen (Germany) and Leiden (the Netherlands). He specialized in the study of binary stars and star clusters and the properties of stars, and he was the first to recognize the distinction between giant and dwarf stars. (Nordisk Pressefoto A/S and the American Institute of Physics)

Henry Norris Russell (1877–1957), American astronomer, was professor at Princeton University and director of the observatory. His many interests included the study of stellar evolution. Hertzsprung and Russell independently discovered the main sequence of stars, best illustrated on the famous diagram now known as the Hertzsprung-Russell diagram. (Princeton University Archives)

Having described the methods by which we are able to obtain basic information about individual stars, we now summarize the data that have been gathered about the stellar population in our own Galaxy.

26.1 THE NEAREST AND THE BRIGHTEST STARS

Let us consider first our most conspicuous stellar neighbors, the brightest-appearing stars in the sky. Appendix 14 lists some of the properties of the brightest 20 stars. Many of these are double or triple star systems; data are given, in such cases, for each component.

(a) The Brightest Stars

The most striking thing about the brightest-appearing stars is that they are bright not because they are nearby, but because they are actually of high intrinsic luminosity. Of the 20 brightest stars listed in Appendix 14, only six are within 10 pc of the sun. The absolute magnitude of a star (Section 23.2a) is the apparent magnitude that it would have if it were at a distance of 10 pc. Since the brightest 20 stars are of apparent magnitude 1.5 or brighter, the 14 of them that are more distant than 10 pc must have absolute magnitudes *less* (that is, brighter) than 1.5. Even among the approximately 3000 stars with apparent magnitudes less than 6.0, only about 60 are

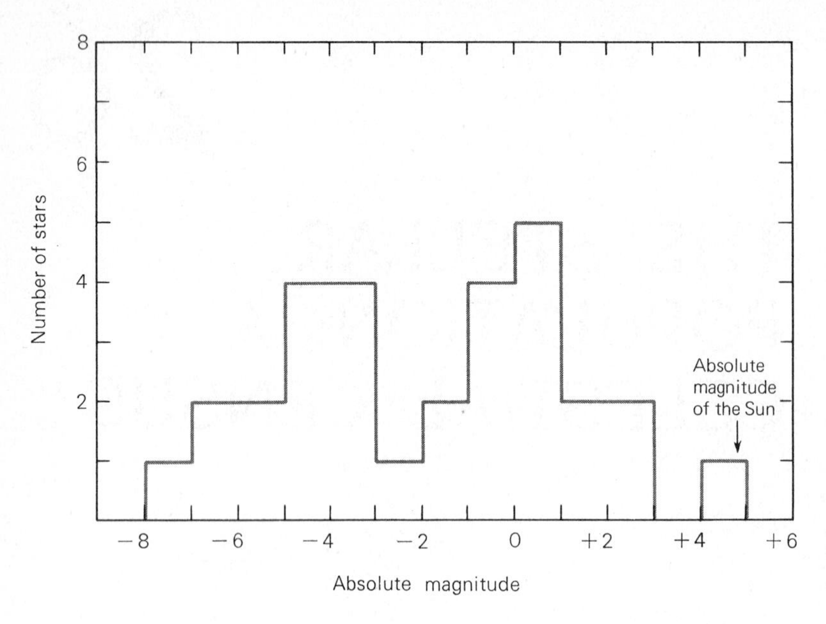

Figure 26.1 Distribution among absolute magnitudes of the 30 brightest-appearing stars. The units are the numbers of stars per unit of absolute magnitude.

within 10 pc. Most naked-eye stars are tens or even hundreds of parsecs away and are many times as luminous as the sun. Indeed, among the 6000 stars visible to the naked eye, at most 50 are intrinsically fainter than the sun. Figure 26.1 is a histogram showing the distribution among various absolute visual magnitudes of the 30 brightest-appearing stars (the absolute visual magnitude of the sun is +4.8).

From Appendix 14 or Figure 26.1, we might gain the impression that the sun is far below average among stars in luminosity. This is not so. Most stars, as we shall see in the next subsection, are much less luminous than the sun is. They are too faint, in fact, to be conspicuous unless they are nearby. Stars of high luminosity are rare—so rare that the chance of finding one within a small volume of space, say, within 10 pc of the sun, is very slight. Why, then, are the most common, intrinsically faint stars not among the most common naked-eye stars, while rare, highly luminous stars are?

The question is best answered with the help of some numerical examples. The sun, whose absolute visual magnitude is +4.8, would appear as a very faint star to the naked eye if it were 10 pc away. Stars much less luminous than the sun would not be visible at all at that distance. Stars with absolute magnitudes in the range +10 to +15 are very common, but a star of absolute magnitude +10 would have to be within 1.6 pc to be visible to the naked eye. Only Alpha Centauri and its companions are closer than this. The intrinsically faintest star observed has an absolute visual magnitude of about +20. For this star to be visible to the naked eye, it would have to be within 0.025 pc, or 5200 AU. The star could not be photographed even with the 5-m telescope if it were more distant than 100 pc. It is clear, then, that the vast majority of nearby stars, those less luminous than the sun, do not send enough light across interstellar distances to be seen without optical aid. For example, the star closest to the sun is Proxima Centauri, a companion of Alpha Centauri; Proxima Centauri has an absolute visual magnitude of +16, and it is invisible to the naked eye.

In contrast, consider the highly luminous stars. Stars with absolute magnitudes of 0 have luminosities of about 100 times that of the sun. They are far less common than stars less luminous than the sun, but they are visible to the naked eye even out to a distance of 160 pc. A star with an absolute magnitude of −5 (10,000 times the sun's luminosity) can be seen without a telescope to a distance of 1600 pc (if there is no dimming of light by interstellar dust—see Chapter 27). Such stars are very rare, and we would not expect to find one within a distance of only 10 pc; the volume of space included within a distance of 1600 pc, however, is about 4 million times that included within a distance of only 10 pc. Hence many stars of high luminosity are visible to the unaided eye.

(b) The Nearest Stars

Evidently, the brightest naked-eye stars do not provide a representative sample of the stellar population in the neighborhood around the sun. Let us turn then to the nearest known stars. Appendix 13 lists the 44 known stars within 5 pc (some are double or multiple systems) from data provided by the U. S. Naval Observatory. (Additional nearby stars are discovered from time to time; the total number of stars within 5 pc may be double the number listed. Moreover, the measurements of distances, luminosities, and so on for nearby stars are being continually refined, but the table does indicate the general characteristics of the sun's nearest stellar neighbors.)

First the table shows that only three of the 43 stars (other than the sun) are among the 20 brightest stars: Sirius, Alpha Centauri, and Procyon. This fact is further confirmation that the nearest stars are not the brightest-appearing stars. The nearby stars also tend to have large proper motions, as would be expected (Section 22.5). In fact, the large proper motions of many of these stars led to the discovery that they are located nearby. Another interesting observation is that 13 of the 44 stars are really binary or multiple star systems; the table thus contains, actually, 59 rather than 44 stars. Twenty-eight of these 59 stars, or nearly half, are members of systems containing more than one star. Moreover, two or three other stars in the list are suspected of having astrometric companions.

The most important datum concerning the nearest stars is that most of them are intrinsically faint. Only ten of the nearest stars are individually visible to the unaided eye. Only three are as intrinsically luminous as the sun; 43 have absolute magnitudes fainter than +10. If the stars in our immediate stellar neighborhood are representative of the stellar population in general, we must conclude that the most numerous stars are those of low luminosity; in a random sample, only about one star in 60 is as intrinsically luminous as the sun.

An estimated lower limit can be established for the mean density of stars, i.e., the number of stars per cubic parsec, in the solar neighborhood. There are at least 59 stars within 5 pc (counting the members of binary and multiple star systems and the sun). We would expect eight times as many stars within 10 pc (for that distance includes a volume of space eight times as large), and about 64 times as many stars within 20 pc. W. Luyten estimates there to be more than 500 stars within 10 pc and some of the fainter stars within that distance must have escaped discovery. Within 20 pc there may be 4000 or more stars. A sphere of radius 5 pc has a volume of $\frac{4\pi(5)^3}{3}$, or about 524 pc^3. Since this volume of space contains at least 59 stars, the density of stars in space is at least one star for every 9 pc^3; the actual stellar density, of course, can be greater than this figure if there are undiscovered stars within 5 pc. The mean separation between stars is the cube root of 9, or about 2.1 pc. If the matter contained in stars could be spread out evenly over space, and if a typical star has a mass 0.4 times that of the sun, the mean density of matter in the solar neighborhood would be about 3×10^{-24} g/cm^3.

The nearest stars constitute a much more nearly representative sample of the stellar population in the vicinity of the sun than do the brightest stars. We are still not sure, however, that we have identified all of the faintest stars in the solar neighborhood. Moreover, there do not happen to be any stars of high luminosity in this "tiny" volume of space. Yet we can identify all the luminous stars, with a reasonable degree of completeness, out to a much greater distance. If we make allowance for the different volumes of space that we must survey to catalogue large samples of stars of different intrinsic luminosities, we can gain some indication of their relative abundances. For example, within 10 pc there are about 12 known stars brighter than absolute magnitude +4, while within 5 pc there are about 57 known stars fainter than absolute magnitude +4. We would expect, however, some 8 × 57, or 456 stars fainter than absolute magnitude +4 within 10 pc; therefore, the ratio of stars with absolute magnitude greater (fainter) than +4 to the number of more luminous stars is about 456 to 12, or 38:1. This calculation is only an example and may not indicate the precise ratio that exists in the stellar population.

(c) The Luminosity Function

Once the numbers of stars of various absolute magnitudes or intrinsic luminosities have been found, the relative numbers of stars in successive intervals of absolute magnitude within any given volume of space can be established. This relationship is called

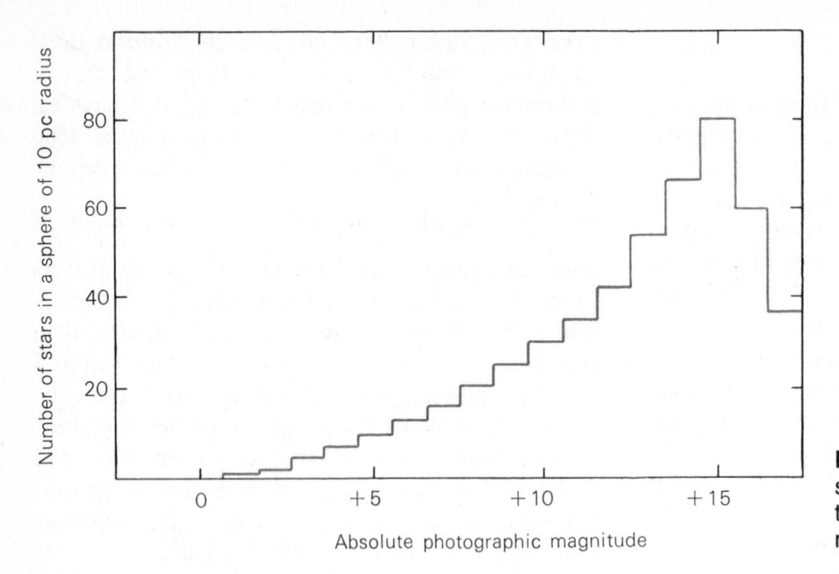

Figure 26.2 Luminosity function of stars in the solar neighborhood. Note that faint stars are much more common than bright ones.

the *luminosity function*. Figure 26.2 shows the luminosity function for stars in the solar neighborhood, as it has been determined by W. J. Luyten of the University of Minnesota. Compare Figure 26.2 with Figure 26.1.

The sun, we see, is more luminous than the vast majority of stars. Most of the stellar mass is contributed by stars that are fainter than the sun. On the other hand, the relatively few stars of higher luminosity than the sun compensate for their small numbers by their high rate of energy output. It takes only ten stars of absolute magnitude 0, which is 100 times the luminosity of the sun, to outshine 1000 stars fainter than the sun, and only one star of absolute magnitude -5 to outshine 10,000 stars fainter than the sun. Most of the starlight from our part of space, it turns out, comes from the relatively few stars that are more luminous than the sun.

26.2 THE HERTZSPRUNG–RUSSELL DIAGRAM

In 1911 the Danish astronomer E. Hertzsprung compared the colors and luminosities of stars within several clusters by plotting their magnitudes against their colors. In 1913 the American astronomer Henry Norris Russell undertook a similar investigation of stars in the solar neighborhood by plotting the absolute magnitudes of stars of known distance against their spectral classes. These investigations by Hertzsprung and by Russell led to an extremely important discovery concerning the relation between the luminosities and surface temperatures of stars. The discovery is exhibited graphically on a diagram named in honor of the two astronomers—the *Hertzsprung–Russell* or *H–R diagram.*

(a) Features of the H–R Diagram

Two easily derived characteristics of stars of known distances are their absolute magnitudes (or luminosities) and their surface temperatures. The absolute magnitudes can be found from the known distances and the observed apparent magnitudes. The surface temperature of a star is indicated either by its color or its spectral class. Before the development of yellow- and red-sensitive photographic emulsions and, of course, photoelectric techniques, spectral classes of stars were usually used to indicate their temperatures. Now that stellar colors can be measured with precision, the color index is more often employed, even though the use of spectral classes is still of great value.

If the absolute magnitudes of stars are plotted against their spectral classes, an H–R diagram like that of Figure 26.3 is obtained. The most significant feature of the H–R diagram is that the stars are not distributed over it at random, exhibiting all combinations of absolute magnitude and temperature, but rather cluster into certain parts of the diagram. The majority of stars are aligned along a narrow sequence running from the upper left (hot, highly luminous) part of the diagram to the lower right (cool, less luminous) part. This band of points is

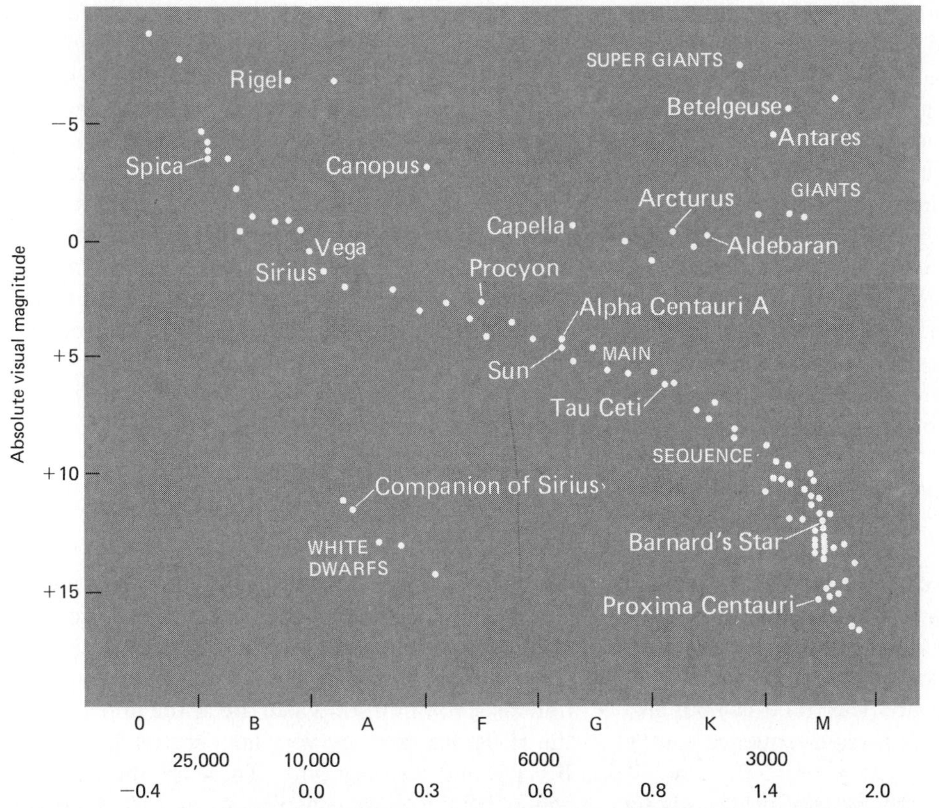

Figure 26.3 Hertzsprung-Russell diagram for a selected sample of stars. Note that the stars do not have all possible values of temperature and luminosity, but rather that most are found along the *main sequence,* a band that stretches from upper left to lower right in the diagram, or from high temperature and high luminosity to low temperature and low luminosity. A few stars are red giants (high luminosity, low temperature) or white dwarfs (high temperature, low luminosity).

called the *main sequence*. A substantial number of stars, however, lie above the main sequence on the H–R diagram, in the upper right (cool, high luminosity) region. These are called *giants*. At the top part of the diagram are stars of even higher luminosity, called *supergiants*. Finally, there are stars in the lower left (hot, low luminosity) corner known as *white dwarfs*. To say that a star lies "on" or "off" the main sequence does not refer to its position in space, but only to the point that represents its luminosity and temperature on the H–R diagram.

An H–R diagram, such as Figure 26.3, that is plotted for stars of known distance does not show the relative proportions of various kinds of stars, because only the nearest of the intrinsically faint stars can be observed. To be truly representative of the stellar population, an H–R diagram should be plotted for all stars within a certain distance (see Exercise 11). Unfortunately, our knowledge is reasonably complete only for stars within a few parsecs of the sun, among which there are no giants or supergiants. It is estimated that about 90 percent of the stars in our part of space are main-sequence stars and about 10 percent are white dwarfs. Less than 1 percent are giants or supergiants. We shall return to the theoretical interpretation of the distribution of stars on the H–R diagram in Chapter 30.

(b) Method of Spectroscopic Parallaxes

One of the most important applications of the H–R diagram is in the determination of stellar distances. Suppose, for example, that a star is known to be a spectral class G2 star on the main sequence. Its absolute magnitude could then be read off the H–R diagram at once; it would be about +5. From this absolute magnitude and the star's apparent magnitude, its distance can be calculated (Section 23.2b).

In general, however, the spectral class alone is not enough to fix, unambiguously, the absolute magnitude of a star. The G2 star described in the last paragraph could have been, for example, a main-sequence star of absolute magnitude +5, a giant of absolute magnitude 0, or a supergiant of still higher luminosity. We recall, however (Section 24.1b), that pressure differences in the atmospheres of stars of different sizes result in slightly different degrees of ionization for a given temperature. It will be seen in the next subsection that giant stars are larger than main-sequence stars of the same spectral class and that supergiants are larger still. In 1913 Adams and Kohlschütter, at the Mount Wilson Observatory, first observed the slight differences in the degrees to which different elements are ionized in stars of the same spectral class but different luminosities (and therefore different sizes). It is thus possible to classify a star by its spectrum, not only according to its temperature (spectral class) but also according to whether it is a main-sequence star, a giant, or a supergiant.

The most widely used system of classifying stars according to their luminosities is that of W. W. Morgan and his associates at the Yerkes Observatory. In favorable cases it has been found possible to divide stars of a given spectral class into as many as six categories, called *luminosity classes*. These luminosity classes are:

- Ia Brightest supergiants
- Ib Less luminous supergiants
- II Bright giants
- III Giants
- IV Subgiants (intermediate between giants and main-sequence stars)
- V Main-sequence stars

A small number of stars that may lie below the normal main sequence are called *subdwarfs* (Sd). The white dwarfs are much fainter. Main-sequence stars (luminosity class V) are often termed "dwarfs" to distinguish them from giants. The term "dwarf," when applied to a main-sequence star, should not be confused with its use as applied to a white dwarf. The full specification of a star, including its luminosity class, would be, for example, for a spectral-class F3 main-sequence star, F3 V. For a spectral class M2 giant, the specification would be M2 III. Figure 26.4 illustrates the approximate mean positions of stars of various luminosity classes on the H–R diagram. The dashed portions of the lines represent those spectral classes (for a given luminosity class) for which there are very few or no stars.

With both its spectral class and luminosity class known, a star's position on the H–R diagram is uniquely determined. Its absolute magnitude, therefore, is also known, and its distance can be calculated. Distances determined this way, from the spectral and luminosity classes, are said to be obtained from the *method of spectroscopic parallaxes*.

(c) Extremes of Stellar Luminosities, Radii, and Densities

The most massive stars have absolute magnitudes of −6 to −8. A few stars are known that have absolute bolometric magnitudes of −10; they are a million times as luminous as the sun. These super-luminous stars, most of which are at the upper left on the H–R diagram, are very hot spectral-type O and B stars, and are very blue. These are the stars that would be the most conspicuous at very great distances in space.

Consider now the stars at the upper right corner of the H–R diagram. A typical red, cool supergiant has a surface temperature of 3000 K and an absolute bolometric magnitude of −5. This star has 10,000 times the sun's luminosity but only half its surface temperature. Since each unit area of the star emits only 1/16 as much light as a unit area of the sun (Section 8.1), its total surface area must be greater than the sun's by 160,000 times. Its radius, therefore, is 400 times the sun's radius. If the sun could be placed in the center of such a star, the star's surface would lie beyond the orbit of Mars.

Red giant stars have extremely low mean densities. The volume of the star described in the last paragraph would be 64 million times the volume of the sun. The masses of such giant stars, however, are probably at most only 50 solar masses, and very likely much less. If we assume that a supergiant star with 64 million times the sun's volume has only 10 times its mass, we find that it has just over 1 ten-millionth the sun's mean density, or only about 2 ten-millionths the density of water; the outer parts of such a star would constitute an excellent laboratory vacuum.

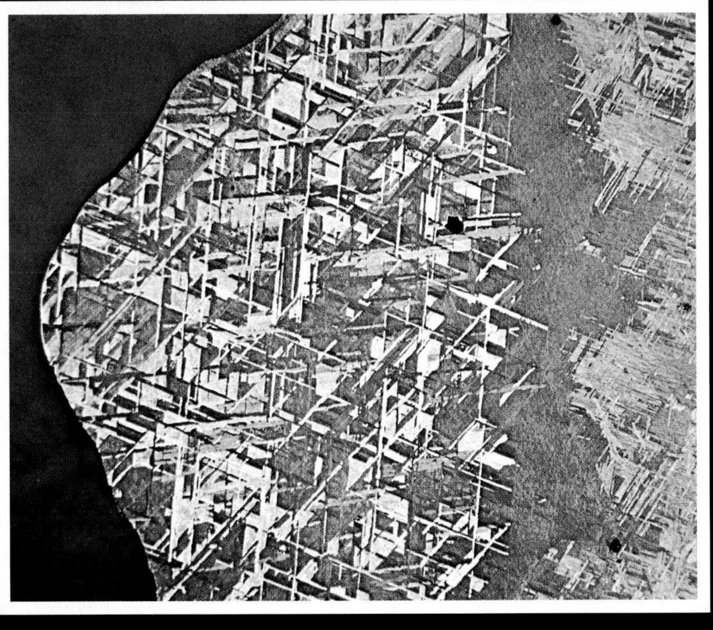

COLOR PLATE 21a Slice of the Kamkas iron meteorite, cut and polished and then etched with acid to show the crystal pattern in the metal. (Photo by Ivan Dryer)

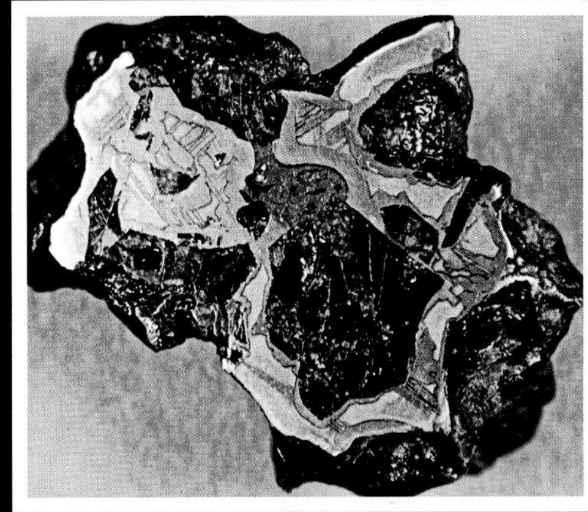

COLOR PLATE 21b Stony-iron meteorite, Glorieta Mountains, New Mexico. (Photo by Ivan Dryer)

COLOR PLATE 21c Polished slice of the Albin, Wyoming, stony-iron meteorite. This type of meteorite, called a Pallasite, consists of nickel-iron mixed with crystals of the green mineral olivine. (Photo by Ivan Dryer. All meteorite specimens on this page are from the collection of Ronald Oriti and are reproduced with his kind permission.)

COLOR PLATE 22a The IRAS (Infrared Astronomy Satellite), launched in 1983, provided the first all-sky survey in the infrared. This cooled, orbiting telescope was a joint U.S./U.K./ Netherlands project. (NASA)

COLOR PLATE 22b The central region of the Milky Way Galaxy as imaged by IRAS in thermal infrared radiation. Most of the radiation in this image is emitted by interstellar dust at temperatures of 20–200 K. (NASA)

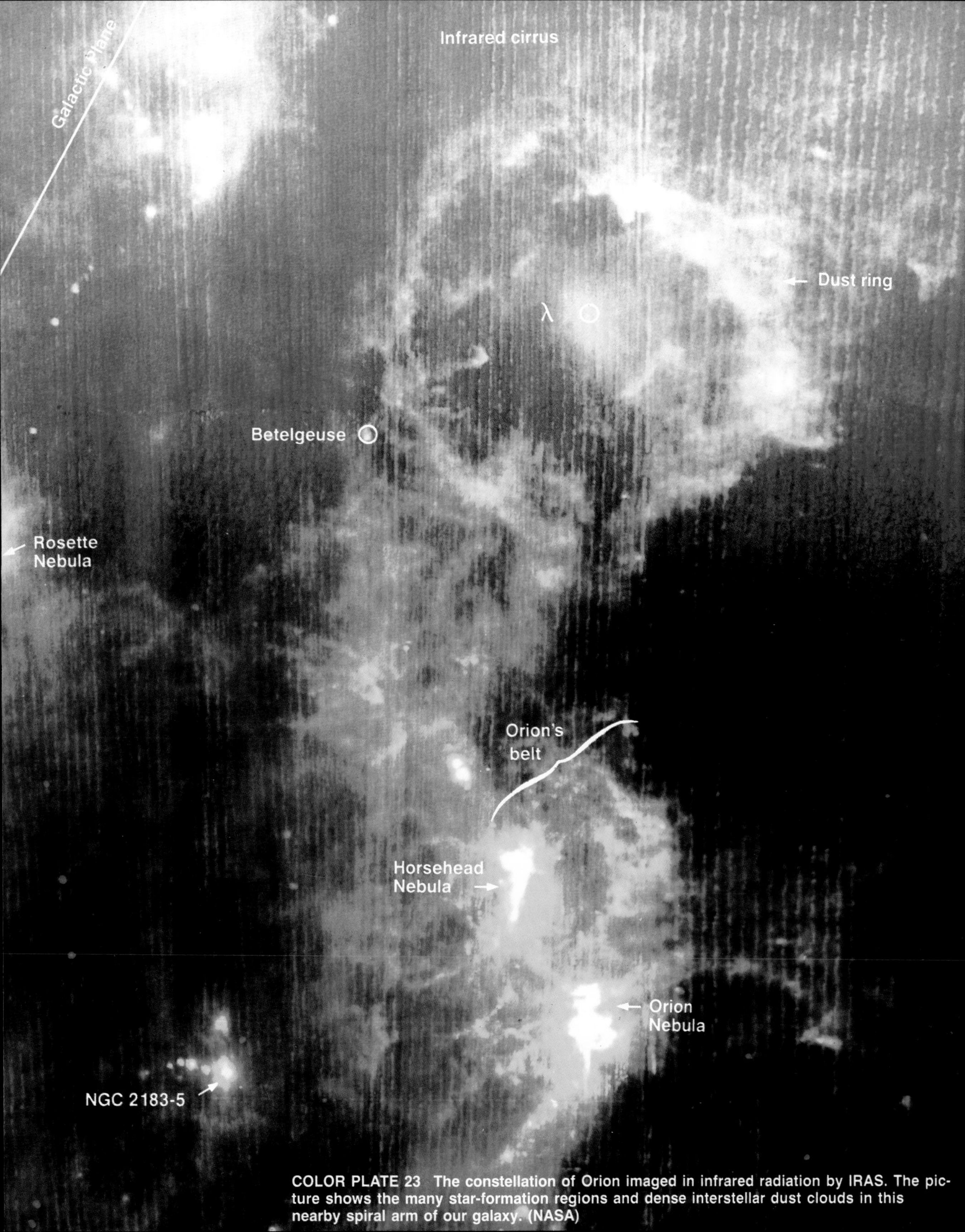

COLOR PLATE 23 The constellation of Orion imaged in infrared radiation by IRAS. The picture shows the many star-formation regions and dense interstellar dust clouds in this nearby spiral arm of our galaxy. (NASA)

COLOR PLATE 24a The Lagoon Nebula, in the constellation Sagittarius. The Lagoon Nebula glows with the red light of hydrogen excited by the radiation of very hot stars buried inside it. Deep within the cloud, there are dark filaments of obscuring matter. The nebula is at a distance of 6500 LY and has a 60-LY diameter. (Kitt Peak National Observatory 4-m Mayall telescope photograph/National Optical Astronomy Observatories)

COLOR PLATE 24b Eagle Nebula, an irregular gaseous nebula in the constellation Serpens. The many bright-edged features visible are shock fronts that mark collisions between gas clouds in different states of ionization. The dark areas consist of opaque dust and gas. (Kitt Peak National Observatory 4-m Mayall telescope photograph/National Optical Astronomy Observatories)

COLOR PLATE 24c Trifid Nebula, in the constellation Sagittarius. This colorful nebula consists of clouds of hydrogen and helium which glow from the radiation of stars within the nebula. Radiation pressure and stellar winds from stars in the overexposed central area create a shock wave, pushing the gases outward. Dark lanes are opaque regions of dust and gas. The 3000 LY distant nebula is approximately 30 LY in size. (Kitt Peak National Observatory 4-m Mayall telescope photograph/National Optical Astronomy Observatories)

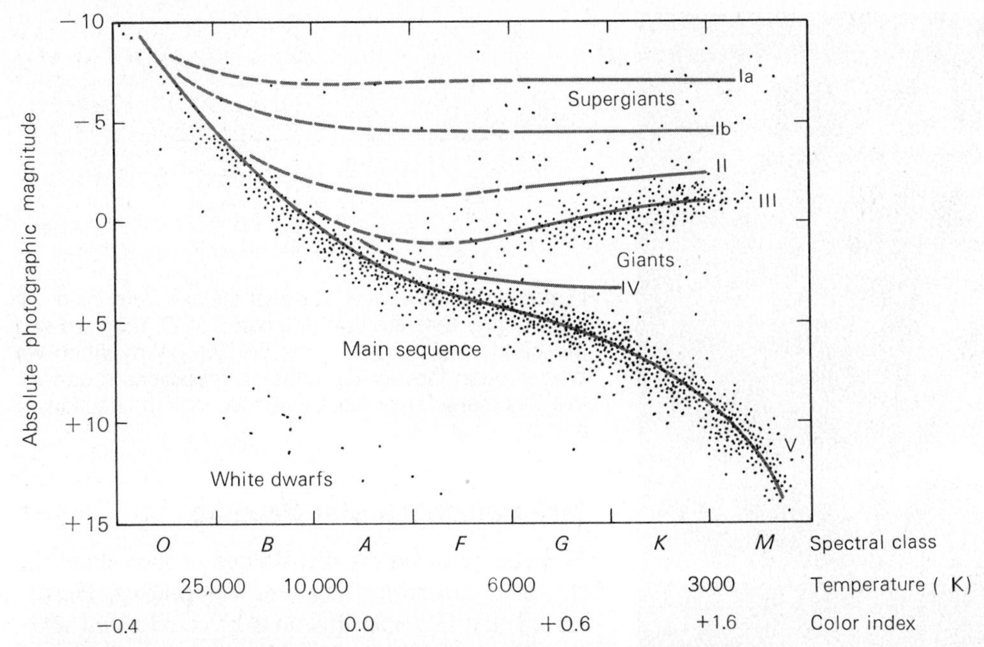

Figure 26.4 Luminosity classes on the Hertzsprung-Russell diagram.

In contrast, the very common red, cool stars of low luminosity at the lower end of the main sequence are much smaller and more compact than the sun. As an example, consider such a red dwarf, the star Ross 614B, which has a surface temperature of 2700 K and an absolute bolometric magnitude of about +13 (1/2000 of the sun's luminosity). Each unit area of this star emits only 1/20 as much light as a unit area of the sun, but to have only 1/2000 the sun's luminosity, the star must have only about 1/100 the sun's surface area, or 1/10 its radius. A star with such a low luminosity also has a low mass (Ross 614B has a mass about 1/12 that of the sun), but still would have a mean density about 80 times that of the sun. Its density must be higher, in fact, than that of any known solid found on the surface of the Earth.

The faint red main-sequence stars are not the stars of the most extreme densities, however. The white dwarfs, at the lower left corner of the H–R diagram, have the highest densities of the normal stars known to be common.

(d) The White Dwarfs

The first white dwarf to be discovered is the companion to Sirius, the brightest-appearing star in the sky. From its wavy proper motion, Sirius was known to have a companion since 1844 (Section 25.2c); it was first seen telescopically in 1862. Sirius is the brightest star in the constellation Canis Major, Orion's big dog. Procyon, the brightest star in Orion's other dog, Canis Minor, also has a white dwarf companion.

The companion of Sirius has a mass about 5 percent greater than that of the sun. From its temperature and luminosity, however, we find its diameter to be only about 1 percent of the sun's, or about twice that of the Earth. The white dwarf has a mean density more than 100 thousand times that of the sun, and a sixth of a million times that of water. Some white dwarfs have much higher mean densities, and many have central densities in excess of 10^7 times that of water. A teaspoonful of such material would weigh nearly 50 tons.

The British astrophysicist Sir Arthur Eddington described the first known white dwarf this way: "The message of the companion of Sirius, when decoded, ran: I am composed of material three thousand times denser than anything you've come across. A ton of my material would be a little nugget you could put in a matchbox. What reply could one make to something like that? Well, the reply most of us made in 1914 was, 'Shut up; don't talk nonsense.'"

We shall describe the theory of white dwarfs in Chapter 32, where we shall also learn that there are stars millions of times denser yet!

Figure 26.5 The Milky Way in Sagittarius. (Yerkes Observatory)

26.3 THE DISTRIBUTION OF THE STARS IN SPACE

In the immediate neighborhood of the sun, the stars seem to be distributed more or less at random (except for their tendency to form small clusters). The larger the volume of space we survey, the more stars we find, and if allowance is made for the fact that the faintest stars become invisible at larger distances, it is found that the number of stars we can count is roughly proportional to the cube of the distance to which we look. Eventually, however, the stars do thin out more rapidly in some directions than in others. The way they thin out is a clue to the nature of the stellar system to which the sun belongs. The idea that the sun is a part of a large system of stars was suggested as early as 1750 by Thomas Wright in his *Theory of the Universe*. Immanuel Kant, the great German philosopher, suggested the same hypothesis five years later. It was the German-English astronomer William Herschel, the discoverer of Uranus, who first demonstrated the nature of the stellar system.

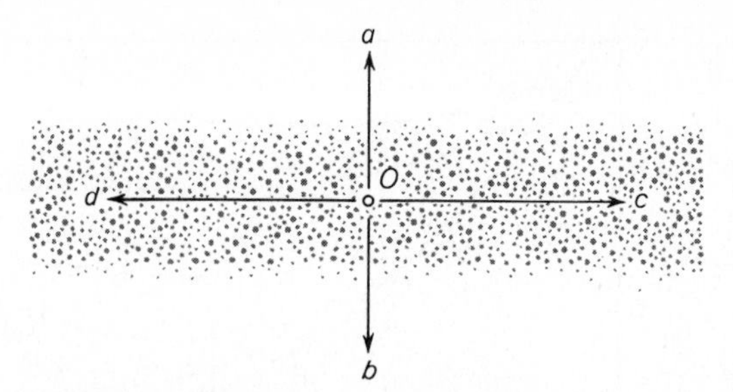

Figure 26.6 We view our own Galaxy from inside. If we imagine that the sun is located at *O*, then we see the band of stars that forms the Milky Way when we look through the Galaxy edge-on (directions *c* and *d*). We see many fewer stars when we look in directions *a* and *b*.

(a) Herschel's Star Gauging

Herschel sampled the distribution of stars about the sky by a procedure he called *star gauging*. He observed that in some directions he could count more stars through his telescope than in other directions. In 1784 and 1785 he presented two papers giving the results of gauges or counts of stars that he was able to observe in 683 selected regions scattered over the sky. While in some of these fields he could see only a single star, in others he was able to count nearly 600. Herschel reasoned that in those directions in which he saw the greatest numbers of faint stars, the stars extended the farthest, and in other directions they thinned out at relatively shorter distances. As a result of his star gauging, Herschel arrived at the conclusion (only partially correct, as we shall see) that the sun is inside a great sidereal system, and that the system is disk-shaped, with the sun near the center.

(b) The Phenomenon of the Milky Way

All of us who have looked at the sky on a moonless night away from the glare of city lights are aware of the Milky Way (Figure 26.5), a faint, luminous band of light that completely encircles the sky. Galileo turned his telescope on the Milky Way and saw that it really consists of a myriad of faint stars. Herschel's grindstone explains why the Milky Way appears as a band all the way around the sky.

It must be recalled that we view our sidereal system from the inside. Figure 26.6 shows an idealized portion of the "grindstone," viewed edge-on. The sun's position is at *O*. If we look from *O* to-

Figure 26.7 A copy of a diagram by Herschel, showing a cross section of the Milky Way system of stars. The large circle shows the location of the sun.

ward either face of the wheel, that is, in directions *a* or *b*, we see only those stars that lie between us and the nearest boundary of the stellar system. In these directions in the sky, therefore, we see only scattered stars. On the other hand, if we look edge-on through the wheel, say in directions *c* or *d*, we encounter so many stars along our line of sight that we get the illusion of a continuous band of light. Since the greatest dimensions of the grindstone extend in all directions along its flat plane, the band of light extends completely around the sky. This band of light is the Milky Way; it is simply the light from the many distant stars that appear lined up in projection when we look edge-on through our own flattened stellar system. Figure 26.7 is a copy of one of Herschel's diagrams, showing a cross section of the stellar system, as he derived its shape.

(c) The Galaxy

We call our stellar system the *Galaxy*, or sometimes the *Milky Way Galaxy*, or even just the *Milky Way*. The Galaxy is far more complicated than Herschel's image of the grindstone. It is a vast, wheel-shaped system of some 10^{11} stars, with a diameter that probably exceeds 30,000 pc (100,000 LY). The flattened shape of the Galaxy is a consequence of its rotation. The sun, about 8500 parsecs from the center of the Galaxy, moves at a speed of about 250 km/s to complete its orbital revolution about the galactic center in some 200 million years. (Unlike a wheel, the Galaxy does not rotate as a solid body—see Chapter 34.

At the hub of the Galaxy is a huge nuclear bulge of stars, where the stars are somewhat closer together than they are in the solar neighborhood (although still light-months or light-years apart). Extending outward from the nucleus, and winding through the disk of the Galaxy, like the spirals of light in a gigantic pinwheel, are the *spiral arms*. The spiral arms consist of vast clouds of gas and cosmic dust—the interstellar medium. Associated with these gas and dust clouds are many young stars, a few of which are very hot and luminous. In the interstellar gas and dust clouds of the spiral arms star formation is taking place.

In addition to individual stars and clouds of interstellar matter, the Galaxy contains many *star clusters*—groups of stars, presumably having a common origin and age. The most common star clusters, numbering in the thousands, are the *open* or *galactic clusters*. Typically, an open cluster consists of a few hundred stars, loosely held together by their mutual gravitation, and moving together through space. The open clusters are located in the main disk of the Galaxy and are usually in or near spiral arms. Besides the open clusters, there are over a hundred *globular clusters*—beautiful, spherically symmetrical clusters, each containing hundreds of thousands of member stars (Figure 26.8). Most of the globular clusters are scattered in a roughly spherical distribution about the main wheel of the Galaxy, grouped around it rather like bees around a flower. They

Figure 26.8 Omega Centaurus globular star cluster. (Kitt Peak National Observatory)

form a more or less spherical *halo* surrounding the main body of the Galaxy.

Our Galaxy is not alone in space. Today we know that far beyond its borders there are countless millions of other galaxies, extending as far as we can see in all directions. Our Galaxy, interstellar matter, star clusters, and other galaxies are the subjects of some of the later chapters.

EXERCISES

1. (a) At what distance would a star of absolute magnitude +15 appear as a fifth-magnitude star? (b) At what distances would a star of absolute magnitude −10 and one of absolute magnitude +15 appear brighter than the fifth apparent magnitude?
2. If the brightest-appearing star, Sirius, were three times its present distance, would it still make the list of "Twenty Brightest Stars" (Appendix 14)? What about the second brightest star, Canopus?
3. Would any of the stars within five parsecs (Appendix 13) be naked-eye stars at a distance of 100 parsecs? If so, which ones?
4. Describe an everyday situation that is analogous to the fact that most naked-eye stars are of far more than average stellar luminosity.
5. If stars of all kinds were uniformly distributed through space, what would the approximate luminosity function have to be in order for intrinsically faint stars to be the most common among naked-eye stars?
6. Given the luminosity function, how would stars have to be distributed in space in order for the intrinsically faint ones to be most common among naked-eye stars?
7. ★ Verify that the mean density of stellar matter in the solar neighborhood is about 3×10^{-24} g/cm^3.
8. Suppose that within 10 pc there were 11 stars of absolute magnitude +5, and that within 30 pc there were 11 stars of absolute magnitude 0. Estimate the true ratio of the number of stars of absolute magnitude +5 to the number of absolute magnitude 0 in a given volume of space.
9. From the data in Appendix 13 (the nearest stars), plot the luminosity function for stars nearer than 5 pc. Compare your plot with Figure 26.2.
10. Why are most faint-appearing stars blue?
11. Plot a Hertzsprung-Russell diagram for the stars within 5 pc of the sun. Use the data of Appendix 13. How does this H–R diagram differ from the one in Figure 26.3? Explain the reasons for these differences.
12. Find the distances to the following stars (see Figure 26.4):
    a) $m = +10$, spectral designation A0 Ib;
    b) $m = +5$, spectral designation K5 III;
    c) $m = 0$, spectral designation G2 V.
13. Consider the following data on five stars:

STAR	m	SPECTRUM
1	15	G2 V
2	20	M3 Ia
3	10	M3 V
4	15	B9 V
5	15	M5 V

(a) Which is hottest? (b) coolest? (c) most luminous? (d) least luminous? (e) nearest? (f) most distant? In each case, give your reasoning.

14. Suppose you had data on the apparent magnitudes and colors of several hundred stars in a cluster. Explain how you could use these data to determine the distance to the cluster.
15. For normal stars what is the approximate range (in order of magnitude) of (a) effective temperature; (b) mass; (c) radius; (d) luminosity?
16. Suppose you weigh 70 kg on the Earth. How much would you weigh on the surface of a white dwarf star, the same size as the Earth, but having a mass of 300,000 times the Earth's (nearly the mass of the sun)?
17. Why do you suppose that most visual binaries are stars of low luminosity?
18. Sometimes our Galaxy is called, simply, the "Milky Way." Why is this poor terminology? Where, exactly, *is* the Milky Way? Does the question make sense? Why?
19. Suppose the Milky Way were a band of light extending only halfway around the sky (that is, in a semicircle). What then would you conclude about the sun's location in the Galaxy?

Bengt Georg Daniel Strömgren (b. 1908), Danish astronomer, spent much of his productive career in the United States, especially as Professor and Director of the Yerkes Observatory (University of Chicago). He has received many honors for his fundamental work in the study of the structure and evolution of stars, and especially for his pioneering investigation of the physics of the interstellar medium. (John B. Irwin)

27

BETWEEN THE STARS: GAS AND DUST IN SPACE

By earthly standards the space between the stars is empty, for in no laboratory on Earth can so complete a vacuum be produced. Yet throughout large regions of space this "emptiness" consists of vast clouds of gas and tiny solid particles. Sometimes these tenuous clouds are visible, or partially so, in the form of *nebulae* (Latin for "clouds"). More often they are invisible, and their presence must be deduced.

The conditions in this tenuous matter that lies between the stars, which astronomers refer to as the interstellar medium, vary widely. There are dense clouds with temperatures as low as 10 degrees Kelvin and low density regions with temperatures of a million degrees. The interstellar medium is dynamic. Clouds form, collide, coalesce, and fragment to form stars. An understanding of the origin of interstellar matter and of the physical processes that control its evolution is critical to understanding where and how stars form.

27.1 THE INTERSTELLAR MEDIUM

The primary components of the interstellar medium are gas and dust, and the gas is composed mainly of hydrogen and helium. About 1 percent by mass of interstellar material is in the form of solid material, frozen particles of dust that are sometimes called interstellar grains.

Interstellar material is concentrated between the stars in the spiral arms of our own and other galaxies. The density of the interstellar matter in the arms of our Galaxy in the neighborhood of the sun, for example, is estimated to be from 3 to 20 times that of the interarm regions. The gas and dust are not distributed uniformly, however, but have a patchy, irregular distribution, being denser in some areas than in others, hence forming "clouds." In the spiral arms, on the average, there is about one atom of gas per cubic centimeter in interstellar space, and from a few hundred to a few thousand tiny particles or grains, each less than a micrometer (one-thousandth of a millimeter) in diameter, per cubic kilometer. In some of the denser clouds, the density of gas and dust may exceed the average by as much as a thousand times or even more, but even this density is more nearly a vacuum than any attainable on Earth. In air, for contrast, the number of molecules per cubic centimeter at sea level is on the order of 10^{19}.

While the density of interstellar matter may be very low, its total mass is substantial. Stars occupy only a small fraction of the volume of the Milky Way Galaxy. For example, it takes only about 2 seconds for light to travel a distance equal to the radius of the sun but more than four years to travel from the sun to the nearest star. Even if the density of gas and dust surrounding the sun and lying between it and the nearest stars is small, the volume of space filled by this low-density material is so large that the total mass of gas and dust contained within a sphere centered on the sun is equal to a few percent of the mass of the sun. The mass of the interstellar matter in the Milky Way Galaxy is probably equal to about 10 percent of the mass now contained in stars. The total mass of interstellar gas and dust in our Galaxy therefore amounts to several billion times the mass of the sun.

Figure 27.1 A portion of the Milky Way in Cygnus, showing the "dark rift." (Caltech/Palomar Observatories)

27.2 COSMIC "DUST"

The tiny solid grains in interstellar space, commonly called interstellar dust, are manifested in the following ways: (1) dark nebulae, (2) general obscuration, (3) reddening of starlight, (4) reflection of starlight, (5) polarization of starlight, (6) infrared radiation from circumstellar and interstellar clouds, (7) interstellar molecules, which we theorize must have formed on dust grains, and (8) absorption features due to silicates, amorphous solid carbon, water ice, and solid CO in the spectra of highly reddened stars.

(a) Dark Nebulae

Relatively dense clouds of the solid grains produce the *dark nebulae,* the opaque-appearing clouds that are conspicuous on any photograph of the Milky Way. Even in the densest clouds, the particles are very sparse, but the clouds extend over such vast regions (measured in parsecs) that they absorb or scatter a considerable portion of the starlight passing through them. Such concentrations of dust often have the appearance of dark curtains, greatly dimming or completely obscuring the light of stars behind them.

The "dark rift," running lengthwise down a long part of the Milky Way and appearing to split it in two, is an excellent example of a collection of such obscuring clouds (Figure 27.1). The obstruction of light from the stars located behind it is so great that a century ago astronomers thought that it was a sort of "tunnel" through which they could see beyond the Milky Way, into extragalactic space. Today, we know that the dark rift is not such a tunnel; the Galaxy extends far beyond such observable dark nebulae (Chapter 34).

The conspicuous obscuring dust clouds are relatively close to us, within 1000 pc. The more distant opaque clouds are difficult to discover because the large number of stars lying between them and us reduces the contrast produced by their own obscuration of starlight.

In addition to the large dark clouds, many very small dark patches can be seen on Milky Way photographs, silhouetted against bright backgrounds of star fields or glowing gas clouds. Many of these patches, called *globules,* are round or oval and have angular diameters of down to a few seconds of arc. They probably range from a few thousand to a hundred thousand astronomical units across. The high opacity of the globules (they dim background objects by five magnitudes or more) implies that they must be very dense compared with the usual interstellar material. It has been suggested that the globules may be condensations of matter that may ultimately form into stars.

(b) The General Obscuration

Although the distribution of the interstellar dust is spotty, and dense clouds produce conspicuous dark nebulae, some of the dust is thinly scattered more or less evenly throughout the spiral arms of the Galaxy. As a result, some absorption of starlight occurs even in regions where dark clouds are not apparent. Unfortunately, the presence of such sparse absorbing matter is not obvious, and it has been the cause of considerable difficulty in the determination of stellar distances.

It has been shown (Section 23.2b) that the distance to a star can be calculated from comparison of its apparent and absolute magnitudes. If light from a star has had to pass through interstellar dust to reach us, however, it is dimmed, much as a traffic light is dimmed by fog. We therefore underestimate the true apparent brightness of the star—that is, we assign to it too large (that is, faint) an apparent magnitude, and the distance we calculate for the star, corresponding to its known (or assumed) absolute magnitude, is too large. Analogously, a motorist may overestimate the distance to a stoplight that he views through fog.

Astronomers once tried to calculate the extent of the Galaxy by counting the numbers of visible stars in various directions and calculating the distances required to account for their observed apparent magnitudes. Because of the general obscuration, however, they overestimated the distances of the stars, and the error increased with increasing actual distances (and hence increasing obscuration). In other words, interstellar absorption of light produces an *apparent* thinning out of stars with distance. The early investigators arrived at the erroneous conclusion that the Galaxy was centered on the sun, and thinned out to its "edge" at a distance of only about 10,000 LY. In actual fact, we do not even see (in visible light) as far as the Galaxy's brilliant central nucleus. Were it not for the obscuring dust in space, we would be able to read at night by the light of the Milky Way.

The presence of the general interstellar obscuration can be demonstrated in several simple ways. We now have independent knowledge that the Galaxy does not thin out a few thousand parsecs from the sun (Chapter 34). The numbers of stars counted in any given region of the sky would be expected to increase, therefore, as the limiting brightness to which stars are counted is decreased (see Exercise 2). As we count to fainter and fainter magnitudes in any direction along the Milky Way, we count more and more stars, but their numbers do not increase as rapidly as we would expect for a more or less uniform stellar distribution because of the dimming of the stars by the dust. It is possible, in fact, to estimate the average absorbing power of the interstellar material by making such star counts in directions at different angles to the plane of the Galaxy.

Interstellar obscuration also affects the apparent distribution of external galaxies. We believe that external galaxies are distributed uniformly in space. Near the galactic plane, however, we *see* practically no galaxies. Hubble called this region the "zone of avoidance." More and more galaxies can be observed as one turns away from the Milky Way, the greatest number at roughly 90° from the plane of the Galaxy (see Chapter 36). The distribution of interstellar dust is the cause of this variation in the number of observable galaxies. Because of the disk-like shape of the interstellar matter in our Galaxy, we encounter more absorption in the direction of the Milky Way (in the plane of the system) than at right angles to it (out either face of the disk). Near the galactic plane, the absorption is so heavy that practically no external galaxies show through. Light from a direction perpendicular to the plane of the Galaxy is probably dimmed by less than 30 percent, and so we can see many galaxies. However, the varying numbers of galaxies in various directions also indicate that there is a spotty, irregular distribution of the absorbing material.

(c) Interstellar Reddening

It is a fortunate circumstance for observational astronomy that the interstellar obscuration is *selective;* that is, light of short wavelengths is obscured more readily than that of long wavelengths. Seventy years ago, astronomers were puzzled by the existence of stars whose spectral lines indicate that they are intrinsically hot and blue, of spectral-type B, although they actually appear as red as cool stars of spectral-type G. We know today that the light from these stars has been reddened by the interstellar absorbing material; most of their violet, blue, and green light has been obscured, leaving a greater percentage of their orange and red light, of longer wavelengths, which penetrates through the obscuring dust. This *reddening* of starlight by interstellar dust not only shows that the stars are dimmed, but also provides a means of estimating the amount of obscuration they have suffered.

The manner in which the dimming or *extinction* of starlight depends on wavelength can be evaluated by comparing, at various wavelengths, the relative brightnesses of two appropriate stars. Stars are chosen whose spectral lines show them to be approximately identical. One, however, is dimmed and reddened by interstellar dust, while the other is not, being in a direction in the sky that is relatively free of interstellar obscuration. As an illustration, suppose the nearer of the two stars, in the absence of obscuration, would be brighter than the other by half a magnitude at all wavelengths. If this nearer star were dimmed by dust, then, at a wavelength of 10,000 Å it might be, say, only 0.2 magnitude brighter than the unobscured star; at 5000 Å it would appear about half a magnitude fainter, and at 3300 Å it would be about one magnitude fainter than the more distant, unobscured star. From a study of several pairs of such obscured and unobscured stars, the extinction of the interstellar material at various wavelengths has been determined. Over the visible spectral region, it turns out that the extinction, expressed in magnitudes, is roughly inversely proportional to wavelength. The extinction properties of interstellar dust are more complicated, however, outside the visible spectrum. Infrared observations show absorption features (bands) characteristic of silicates at 9.7 μm (1 μm = 1 micrometer = 10^4 Å); a few clouds show bands at 3.1 μm due to water ice. Ultraviolet observations from satellites show an absorption feature at 2200 Å characteristic of small particles, which are probably made of carbon. We shall return shortly to the significance of these observations.

We can estimate the total amount by which a star is dimmed from the amount that it is reddened. The extinction of the light from a star increases its apparent color index (the redder the star, the greater the color index—see Section 23.2d). The difference between the *observed* color index and the color index that the star *would have* in the absence of obscuration and reddening is called the *color excess*. The $B - V$ color excess, for example, is the amount by which the difference between the blue and visual magnitudes of a star is increased by reddening. In most directions in the Galaxy, the total absorption, in visual magnitudes, is found empirically to be about three times the $B - V$ color excess.

The calculation of the distance to a star that is dimmed by obscuration may be illustrated with a numerical example.* Suppose a spectral-type G2 star is observed to have an apparent blue magnitude of 14.4 and an apparent visual magnitude of 12.8. Its *observed* color index, therefore, is +1.6. Now it is known that this type of star has an absolute visual magnitude of +4.8, an absolute blue magnitude of +5.4, and therefore, an intrinsic color index of +0.6. Its color excess is 1.6 − 0.6 = 1.0, and its obscuration, in visual magnitudes, is about three times its color excess, or about 3.0 magnitudes. In the absence of obscuration, therefore, we estimate that its apparent visual magnitude would be 12.8 − 3.0 = 9.8, just five magnitudes fainter than its absolute magnitude (its distance modulus is 5 magnitudes, which corresponds to a distance of 100 parsecs—see Section 23.2b). In the absence of interstellar dimming, in other words, the star would appear five magnitudes, or 100 times, fainter than it would appear at a distance of 10 pc; we estimate the true distance of the star, therefore, at 100 pc.

(d) Reflection Nebulae

Until now, the term "absorption" has been used loosely. The tiny interstellar grains actually absorb some of the starlight they intercept, but at least half of it they merely scatter—that is, they redirect it helterskelter in all directions. Since the starlight that is scattered, as well as that which is truly absorbed, does not come directly to us, the effect is the same as if the loss were all due to actual absorption. It is this whole process that is termed *interstellar extinction*. The scattered or reflected light comes to us from the direction of the scattering dust, thus betraying its presence. Consequently, even the darkest dark nebulae are not completely dark but are illuminated by a faint glow of scattered starlight that can actually be measured. It is estimated that about one-third of the light of the Milky Way is diffused starlight, scattered by interstellar dust.

---

* The apparent distance modulus of a star dimmed by interstellar obscuration is given by

$$V - M_v = 5 \log \frac{r}{10} + 3CE,$$

where V and M_v are its apparent and absolute visual magnitudes, r its distance in parsecs, and CE its blue minus visual color excess.

Figure 27.2 Reflection nebula about the star Merope in the Pleiades. (Caltech/Palomar Observatories)

The light scattered by a particularly dense cloud of dust around a luminous star may be bright enough to be seen or photographed telescopically. Such a cloud of dust, illuminated by starlight, is called a *reflection nebula*. One of the best known examples is the nebulosity around each of the brightest stars in the Pleiades cluster (Figure 27.2 and Color Plate 25c).

Blue light is scattered more than red by the dust. A reflection nebula, therefore, usually appears bluer than its illuminating star. A reflection nebula could be red only if the star that is the source of its light were very red. However, it takes a bright star to illuminate a dust cloud sufficiently for it to be visible to us, and in the regions of the Galaxy where dust clouds are found, the brightest stars are usually blue main-sequence stars. Most of the reflection nebulae that are conspicuous, therefore, are very blue, since they are illuminated by blue stars. Sometimes, of course, an intrinsically blue star illuminating a reflection nebula may appear much redder than it actually is because of interstellar extinction and reddening.

From the measured brightness and size of a reflection nebula we can calculate how much light it actually reflects. We find that this is usually a large fraction of the starlight intercepted by the dust. Thus, much of the obscured starlight is scattered (or reflected) by the dust and is not truly absorbed. The interstellar dust, in other words, has a rather high reflecting power or *albedo*. The fraction of the light that is truly absorbed by the dust heats the dust and eventually is reradiated in longer wavelengths.

(e) Interstellar Polarization

Molecules of gas scatter light and polarize it at the same time. The light of the blue daylight sky, for example, is highly polarized. The dust particles of interstellar space also polarize light, but not so much. The light from reflection nebulae is observed to be polarized typically by 10 to 20 percent, and occasionally as much as 50 percent, which means that the brightness of a reflection nebula as observed through a polaroid filter varies by that percentage as the filter is rotated through various angles. The light of stars dimmed by interstellar dust is also slightly polarized. Evidently, the dust grains preferentially scatter light in a particular plane of vibration.

The interstellar polarization can be understood if the particles are elongated in shape and at least partially aligned with each other. The mechanism by which the elongated particles become aligned, while not understood in detail, is undoubtedly associated with the presence of interstellar magnetic fields (see also Chapter 34).

(f) Circumstellar Dust

Infrared surveys of the sky reveal a large number of stars that appear very much brighter at infrared wavelengths than in visible light. It is now thought that many of these objects are red giant stars surrounded by circumstellar dust clouds. Visible and ultraviolet light from such a star is greatly dimmed by the dust, but the absorbed radiation heats up the dust itself, which then reradiates the energy at infrared wavelengths. Typically, the temperature of the dust is in the range 10 K to 1000 K, corresponding to maximum infrared emission between 2 μm and 400 μm. Additional support for this model comes from the fact that infrared spectra of some red giant stars show emission or absorption at 9.7 μm, and this spectral feature is produced in rock-like material (silicates). To emit as strongly as it does at the shorter infrared wavelengths, the dust must absorb considerable energy from the star, which means it must be quite near it, generally within a few astronomical units.

(g) Nature of the Interstellar Grains

The preceding paragraphs have described various phenomena associated with the interstellar dust. These phenomena reveal something of the nature of the solid grains themselves.

First, it is found that the absorption of light is accomplished by *solid particles*, and not by interstellar gas. Except for specific spectral lines, atomic or molecular gas is almost transparent. Consider the Earth's atmosphere; despite its incredibly high density compared with interstellar gas, it is so transparent as to be practically invisible. The absorbing power of the interstellar medium exceeds that of an equal mass of gas by more than 100,000 times. The quantity of gas that would be required to produce the observed absorption in space would have to be many thousands of times the amount that can possibly exist. The gravitational attraction of so great a mass of gas would produce effects upon the motions of stars that would be easily detected; such effects, however, are not observed.

Moreover, molecules or atoms (of a gas) scatter light quite differently from the interstellar material. Scattering by gas molecules, as happens in the Earth's atmosphere, is called *Rayleigh scattering*. Rayleigh scattering discriminates very strongly among colors, blue light being scattered very efficiently, as we see from the brilliant blue of the daytime sky; the scattering efficiency is inversely proportional to the fourth power of the wavelength. Interstellar particles, too, scatter selectively, but not so strongly as the Earth's atmosphere. Interstellar extinction, as we have seen, is approximately inversely proportional to the first power of the wavelength.

Whereas gas can contribute only negligibly to absorption of light, we know from our everyday experience that tiny solid or liquid particles can be very efficient absorbers. Water vapor in the air is quite invisible. When some of that vapor condenses into tiny water droplets, however, the resulting cloud is opaque. Dust storms, smoke, and smog furnish other familiar examples of the opacity of solid particles.

Based on these arguments, we must conclude that widely scattered solid particles in interstellar space are responsible for the observed dimming of starlight. What then are these particles made of? And how did they form? The answers to those questions are far from certain, and no single grain composition, shape, or size can explain all of the observations. We can, however, make a number of general statements about what the grains must be like.

From measurements of the densities of interstellar dust, that is, of the total amount of dust within a given volume of space, we know that about one of every two atoms heavier than helium must be locked up in the grains. The grains cannot, therefore, be made of rare elements but rather must be composed primarily of the most abundant elements in the universe. After hydrogen and helium, the most abundant elements are oxygen, carbon, and nitrogen, and these three elements, along with magnesium, silicon, iron, and perhaps hydrogen, are thought to be the most important components of interstellar dust.

Observations support this line of reasoning; many heavy elements, including iron, magnesium, and silicon, are less abundant in interstellar *gas* than they are in the sun and young stars. These heavy elements are assumed to be missing from the interstellar gas because they are condensed into solid particles of interstellar *dust*. A characteristic dust grain contains 10^6 to 10^9 atoms and has a diameter of 10^{-5} to 10^{-6} cm (0.1 to 0.01 μm).

It seems most likely that interstellar grains consist of a core of rock-like material (silicates), including such common minerals as olivine and enstatite, both frequently found in terrestrial igneous rocks and in meteorites. Other grains appear to be nearly pure carbon (graphite). The nuclei of the grains are probably formed in shells of cooling gas ejected by red giants and other stars, including some novae and even possibly supernovae, that are nearing the end of their evolution (Chapter 32). These grain nu-

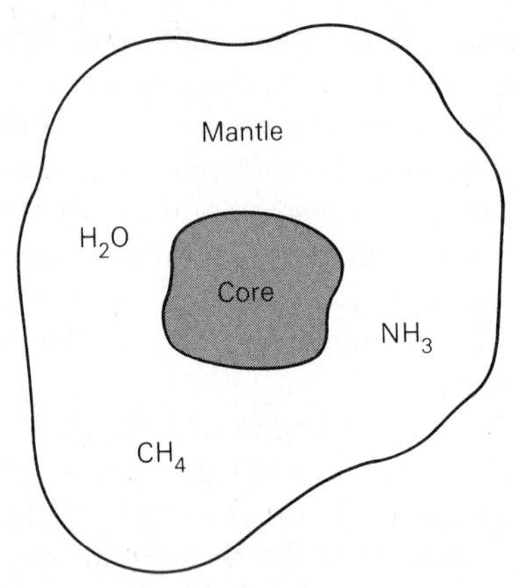

Figure 27.3 A typical interstellar grain is thought to consist of a core of rocky material (silicates), graphite, or possibly iron surrounded by a mantle of ices.

clei may then subsequently be incorporated into an interstellar cloud, where they can grow by accreting other atoms. The most widely accepted model pictures the grains as consisting of rocky cores with icy mantles (Figure 27.3). The most common ices are water (H_2O), methane (CH_4), and ammonia (NH_3).

The growth of the grains is not uninterrupted. Several processes tend to destroy the grains. If a high-energy ion or photon collides with a grain, it can knock off atoms, and this process is called *sputtering*. If a grain is heated to too high a temperature, atoms can evaporate from its surface. The temperatures required are not very high. In a typical interstellar cloud, the vaporization temperature is 20 K for CH_4, 60 K for NH_3, and 100 K for H_2O. Finally, if two grains collide at a speed of several km/s, both will vaporize.

Ultimately, those grains that survive may be incorporated into a new generation of stars, where they will be broken apart by the star's heat into individual atoms to begin the cycle over again.

27.3 INTERSTELLAR GAS

Although interstellar gas is estimated to be a hundred times as abundant by mass, on the average, as the dust, because of its high transparency it is not visible by reflected starlight, nor does it contribute to the general interstellar absorption. It does manifest itself, however, in several other ways. Because of the process of *fluorescence*, clouds of gas near hot stars often shine brightly. The gas also produces narrow absorption lines superposed upon the spectra of stars whose light passes through the gas. Moreover, the gas is responsible for the emission of radio waves over a broad range of wavelengths and for many emission and absorption lines at radio wavelengths, including the important line at 21 cm, and those of dozens of molecules in space. Finally, a small fraction of the gas is probably very hot (10^6 K) and emits low-energy X-rays.

(a) General Gas in Space

Interstellar gas is distributed generally throughout the regions of the spiral arms of the Galaxy, and to a lesser extent in other regions. The very hot gas (that emits X-rays) is found throughout the disk of the Galaxy and even on either side of the disk, sometimes as far as 10^3 pc from the galactic plane. Hydrogen makes up about three-quarters of the gas, and hydrogen and helium together compose from 96 to 99 percent of it by mass. Most of the gas is cold and nonluminous. Near very hot stars, however, it is ionized by the ultraviolet radiation from those stars. Since hydrogen is the main constituent of the gas, we often characterize a region of interstellar space according to whether its hydrogen is neutral—an "H I region"—or ionized—an "H II region."

The gas in the H II regions glows by the process of fluorescence. The light emitted from these regions of ionized gas consists largely of emission lines, so they are also called *emission nebulae*. Those emission nebulae in which the gas happens to be much denser than average (it occasionally reaches densities of 10^3 or 10^4 atoms per cubic centimeter—still an extremely high vacuum on Earth) are especially conspicuous. The best-known example is the Orion nebula, which is barely visible to the unaided eye, but easily seen with binoculars, in the middle of the sword of the hunter. Other famous emission nebulae are the North America nebula in Cygnus and the Lagoon nebula in Sagittarius (Figure 27.4).

(b) Fluorescence in H II Regions

All ultraviolet radiation of wavelength 912 Å or less can be absorbed by neutral hydrogen, and in the process the hydrogen is ionized (Section 8.3). An appreciable fraction of the energy emitted by the hottest stars lies at wavelengths shorter than 912 Å. If such a star is embedded in a cloud of interstellar

Figure 27.4 The Lagoon nebula in Sagittarius, photographed in red light with the 5-m telescope. (Caltech/Palomar Observatories)

gas, the ultraviolet radiation from that star ionizes the hydrogen in the gas, converting it into a plasma of positive hydrogen ions (protons) and free electrons. These detached protons and electrons are then a part of the gas, each of them acting like a free particle. Protons in the gas are continually colliding with electrons and capturing them, becoming neutral hydrogen again. As the electrons cascade down through the various energy levels of the hydrogen atoms on their way to the ground states, they emit light in the form of emission lines. Lines belonging to all the series of hydrogen (Section 8.3) are emitted—the Lyman series, Balmer series, Paschen series, and so on—but the lines of the Balmer series are most easily observed from the surface of the Earth because of the opacity of our atmosphere to most wavelengths. Part of the invisible ultraviolet light from the star is thus transformed into visible light in the Balmer emission lines of hydrogen. After an atom has captured an electron and emitted light, it loses that electron again almost immediately by the subsequent absorption of another ultraviolet photon from the star. Thus, although neutral hydrogen absorbs and emits light in H II regions, almost all the hydrogen, at any given time, is in the ionized state.

The interstellar gas, of course, contains other elements besides hydrogen. Many of them are also ionized in the vicinity of hot stars and are capturing electrons and emitting light, just as the hydrogen does. Of these, only helium is abundant enough to contribute an appreciable amount of light to an emission nebula by the process of electron capture that we have described.

(c) Intermixture of Interstellar Gas and Dust

Gas and dust are generally intermixed in space, although the proportions are not everywhere exactly the same. The presence of dust is apparent on many photographs of emission nebulae (Figure 27.5). Clouds of dark material can be seen silhouetted on the Orion nebula, actually hiding a large part of the H II region from our view. Foreground dust clouds produce the lagoon in the Lagoon nebula, and the Atlantic Ocean and Atlantic coastline in the North America nebula. Although the dust is most conspicuous when it is in front of an emission nebula and is silhouetted against it, the dust is also intermixed with the gas. Spectra of H II regions often reveal the faint continuous spectrum (with absorption lines) of the central star, whose light is reflected to us by the dust associated with the gas. In other words, emission nebulae are generally superimposed upon *reflection nebulae*.

Figure 27.5 Messier 16, nebula in Serpens. Note that gas and dust appear together in the same region in space. Photographed in red light with the 5-m telescope. (Caltech/Palomar Observatories)

Both the emission component (due to the gas) and the reflection component (due to the dust) are brighter the hotter and more luminous the central star. The brightness of an emission nebula, however, is far more sensitive to the kind of central, exciting star than is that of a reflection nebula. Stars cooler than about 25,000 K have so little ultraviolet radiation of wavelengths shorter than 912 Å (that is, which can ionize hydrogen) that the reflection nebulae around such stars outshine the emission nebulae. The dust around a star with a surface temperature of only 10,000 K scatters more than 5000 times as much visible light as is emitted by the gas around the same star, although even the reflection component would probably not be conspicuous unless the star were a supergiant. Stars hotter than 25,000 K emit enough ultraviolet energy so that the emission nebulae produced around them generally outshine the reflection nebulae. The dust around a star of 50,000 K reflects less than one-tenth the amount of light that is emitted by the gas that is expected to be present.

(d) Forbidden Radiation

It was explained in Section 8.3 that an atom or ion can be excited in either of two ways: by absorbing radiation or by collision with another particle. Atoms of singly ionized nitrogen and singly and doubly ionized oxygen all contain energy levels that correspond to low energies above their ground states. The ions are easily excited to these "low-energy" levels by collisions with free electrons in the H II regions (most of these electrons have been freed from hydrogen atoms by ionization). Ordinarily, observed emission lines originate from atoms or ions that have remained excited for only a very brief period—on the order of a hundred millionth to a ten millionth of a second—before becoming de-excited by the emission of radiation. The levels to which oxygen and nitrogen ions are excited by collision, however, are said to be *metastable* levels, because the ions will normally remain in them for periods of hours before radiating energy and dropping to their ground states. Consequently, emission lines corresponding to such atomic transitions are exceedingly weak compared with other lines and are not usually observed in the laboratory; they are known as *forbidden lines*.

In the interstellar gas, however, the cards are stacked in favor of forbidden radiation. At the temperatures of H II regions many of the free electrons have just the right kinetic energy to excite oxygen and nitrogen ions to their metastable levels. Although transitions from these levels are slow to occur, so many ions are excited to them at any time that many such "forbidden" transitions or de-excitations occur in an H II region. Now the gas is transparent to visible light, so the photons emitted through the entire depth of an H II region contribute to visible emission lines; indeed, the forbidden radiation often comprises half or more of the observable light from H II regions. The most important forbidden lines in the spectra of emission nebulae are two green lines (5007 and 4959 Å) due to doubly ionized oxygen. Other important forbidden lines are two ultraviolet lines (near 3727 Å) due to singly ionized oxygen, two red lines (6584 and 6548 Å) due to singly ionized nitrogen, and two ultraviolet lines (3867 and 3968 Å) due to doubly ionized neon.

When the green forbidden oxygen lines were first observed in the spectra of emission nebulae, their origin was a mystery. For a time, they were ascribed to an unknown element, *nebulium,* named for its apparent prevalence in gaseous nebulae. The correct explanation of the "nebulium lines" was provided by the American physicist I. S. Bowen in 1927.

(e) Size and Brightness of an H II Region

If a cloud of gas surrounding a hot star is not very extensive, some of the ultraviolet radiation emitted by the star that is capable of ionizing hydrogen may leak out through the gas, and the apparent boundary of the emission nebula will be the actual edge of the gas cloud. Such a nebula is said to be *optically thin,* and the H II region is said to be *gas* or *density bounded.* Usually, however, an H II region is *optically thick,* which means that all of the star's ultraviolet radiation (of wavelengths less than 912 Å) is absorbed within the gas, and the H II region is said to be *radiation bounded.* In this case, the boundary of the H II region is merely the limiting distance through the gas to which the star's ultraviolet radiation penetrates.

If the interstellar gas were distributed with absolute uniformity, and if it and the stars were all at rest, every emission nebula would be a spherical H II region exactly centered on a hot star. Because the distribution of the gas is patchy and irregular, and because the ionizing stars are often moving through it, actual H II regions are only approximately spherical, with irregular boundaries corresponding to the irregularities in the gas density; where the gas is dense, the ionizing radiation is consumed, and the H II region ends closer to the star. Further irregularities may result when two or more stars are responsible for the radiation, and their H II regions overlap. The theory of H I and H II regions was worked out in detail by the Danish astronomer B. Strömgren in 1939. The more or less spherical emission regions are therefore sometimes called *Strömgren spheres.*

The linear size of an H II region depends on two things: (1) the "ultraviolet" luminosity of the central star, that is, how much energy it emits per second in wavelengths less than 912 Å; and (2) the density of the gas in the nebula. The higher the density, the more circumstellar hydrogen per unit volume there is to ionize, and the shorter is the distance through the gas that the ultraviolet energy can penetrate before it is completely absorbed. If the gas density is very low, and if the star is very hot and luminous, the H II region can be very large; if the density is one atom per cubic centimeter, a main-sequence spectral-type O6 star can ionize a region more than 100 pc in diameter. Main-sequence stars of types B0 and A0 would produce in the same gas H II regions having diameters of 40 and 1 pc, respectively.

(f) Temperatures of H I and H II Regions

When we speak of the temperature of a gas, we usually mean its *kinetic temperature*, which is a measure of the energy with which typical particles in the gas are moving. To understand the temperature of the interstellar medium, we must first consider the processes that heat the gas and then those that cool it. Finally, we calculate the equilibrium temperature that exists when these two processes—the heating and the cooling of the gas—exactly balance each other.

In H II regions the heating is principally by ionization of hydrogen, which is by far the most abundant element in the gas. The neutral hydrogen atom, in becoming ionized, can absorb *any* photon of wavelength shorter than 912 Å. The energy absorbed by the atom, in excess of that required for its ionization, is converted to kinetic energy of the freed electron. By collisions, the free electrons gradually share their energy with the other particles of an H II region. Excess energy absorbed in the process of ionization of hydrogen, therefore, is slowly converted into heat in the gas.

Electrons can lose energy by collisions with ions of oxygen and nitrogen, because these ions can be excited by such collisions and can eventually radiate the energy away in the form of "forbidden" emission lines. This mechanism, in other words, takes energy *from* the gas and therefore *cools* it. Calculations show that the heating (by ionization) and cooling (by the emission of forbidden lines) should balance each other at an equilibrium temperature in the range of 7000 to 20,000 K. If the temperature should drop much below 10,000 K, the rate of collisional excitations of oxygen and nitrogen would decrease (because the electrons would be moving more slowly) and the gas would heat up. If the temperature rose much above 20,000 K, the collisional excitations would become so numerous that the gas would cool rapidly. The heating and cooling mechanisms, therefore, act like a thermostat keeping the gas in the H II region at a relatively even temperature.

We may measure the temperature of H II regions by comparing the intensities of two or more lines originating from different collisionally excited levels of the same kind of atom or ion. Measures of intensities of the forbidden lines of doubly ionized oxygen (the "nebulium lines") are especially useful for this purpose. Temperatures of H II regions can also be estimated from the intensities of their radio emission at different wavelengths. Most such measurements suggest that H II regions have temperatures between 8000 and 10,000 K.

The temperature of H I regions is also dictated by a balance of heating and cooling mechanisms, but the processes are more complicated, and they vary in effectiveness from place to place, depending, among other things, on the shielding of the matter from starlight by dust obscuration. Heating can occur by ionization of atoms of such heavier elements as carbon and silicon and by cosmic rays. Cooling mechanisms include collisional excitation of carbon and other atoms, and collisions between atoms and solid grains. Measures mostly derived from radio observations indicate that some cold clouds have temperatures below 20 K, while other regions are at temperatures of more than 125 K.

In any case we can conclude that H II regions are hot, with temperatures on the order of 10,000 K, and that H I regions are cold, with temperatures on the order of 100 K, or less.

(g) Expansion of H II Regions

It will be seen in Chapter 30 that the very hot, luminous stars—the type that are generally responsible for producing H II regions—are relatively short-lived. H II regions, therefore, are temporary phenomena; most existing emission nebulae have been formed within the past few million years. Consider, now, what happens when a very hot, luminous star is first formed, and rather suddenly ionizes the gas surrounding it in space, thereby producing an emission nebula. The ionized gas, although initially of the same density as the unionized gas in the adjacent H I region, is about 100 times hotter than the cool gas. Since the pressure in a gas is proportional to the product of its density and temperature (see Section 27.4a), the H II region has a pressure that is also about 100 times greater than that of the H I region. Consequently, the H II region expands, decreasing its own density and pushing outward, away from the central star, against the cold gas in the H I region. At the same time, the radiation pressure exerted by the light from the star upon the interstellar dust particles pushes them outward, away from the star, much as sunlight pushes out the dust tail of a comet.

The expanding front of an emission nebula—the interface between the H I and H II regions—may move at first very rapidly. As a result, there is often a buildup in the density of the gas and dust at the boundary of the nebulae. The bright edges visible in many emission nebulae constitute observational evidence for the higher densities of the gas at the expanding front.

Eventually, the increasing density of the material in the H I region surrounding the hot, expanding emission nebula offers sufficient resistance to slow down the expansion. Sometimes points of higher than average density in the H I region break through the expanding front of hot gas and appear as "intrusions" into the emission nebula. Such intrusions generally look like dark cones with bright edges pointing toward the star at the center of the nebula; the emission nebula Messier 16 in Serpens is an excellent example. The dark intrusions are sometimes referred to as "comet-tail" or "elephant-trunk" structures. They are not, of course, actually moving inward; the hot gas expands *outward*, around them. Dark globules are often found near, or as dissociated extensions of, such elephant-trunk structures (Figure 27.5).

The fact that stars exist that we have reason to believe are very young (Chapter 30) implies that stars must be continually forming. The only material for them to form from is the interstellar gas and dust. Regions within the interstellar medium, therefore, must be the "birthplaces" of new stars. Places where the densities of the interstellar gas are unusually high may be future sites of star formation.

(h) Interstellar Absorption Lines

The cold interstellar gas—that in the H I regions—is not visible by reflected or emitted light, nor does it appreciably dim the light of stars shining through it. Yet it often reveals its presence by leaving dark absorption lines superposed upon the spectra of stars that lie beyond it. There are several ways of knowing that the interstellar lines seen in the spectrum of a star do not originate in the star itself. Since the interstellar gas is cold, most of its atoms are neutral and are practically all in the state of lowest energy; the lines they produce, therefore, are generally not the same as the ones that are produced by the atoms of the hot gases in stellar photospheres. Moreover, the lines of the cold interstellar gas are very narrow, while those formed in stellar photospheres show the characteristic broadening associated with the spectra of hot gases at relatively high pressure (Section 24.2c). Finally, an interstellar gas cloud does not, in general, move with the same radial velocity as the star whose spectrum is observed, and the Doppler shifts of the interstellar

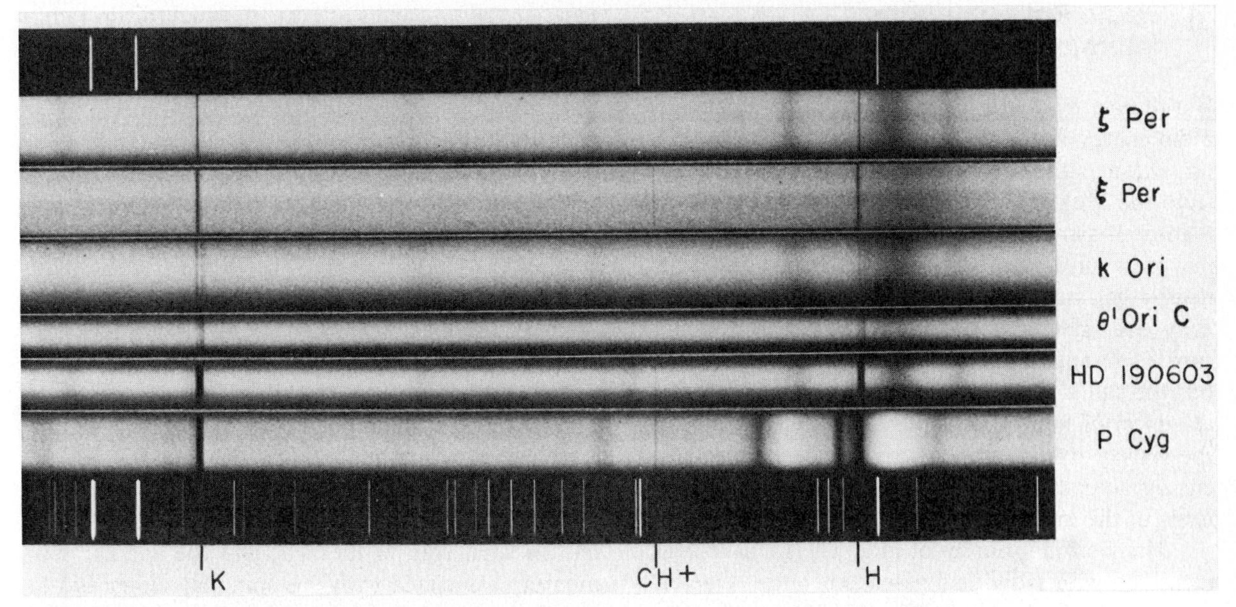

Figure 27.6 Interstellar H and K lines of ionized calcium and a line of the CH radical showing in the spectra of several stars. (Lick Observatory)

lines are thus different from those of the stellar lines.

Interstellar lines have been found of most of those elements for which observable lines would be expected. The most conspicuous optical interstellar lines are produced by sodium and calcium (Figure 27.6). Lines are also observed of some other common elements as well as bands of CN, CH, and CH^+. Satellite ultraviolet observations have detected lines of carbon, hydrogen, oxygen, nitrogen, and other elements; of molecular hydrogen; and of CO (carbon monoxide). The strengths of interstellar lines lead to estimates of the relative abundances of the elements that produce them. For some elements such estimates do not differ markedly from their relative abundances in the sun and other stellar photospheres. For others the relative abundance is noticeably lower, especially for elements that most easily condense into solids at relatively high temperatures (notably aluminum, calcium, and titanium, as well as iron, silicon, and magnesium).

Sometimes the strength of an interstellar line seen in the spectrum of a star provides an indication of the distance to that star. A very strong interstellar line, for example, indicates that the starlight has traversed a considerable amount of interstellar gas. Because the density of gas in space is very low, the starlight would have to travel a long distance to encounter that much material; that is, the star would have to be very far away. Sometimes the interstellar lines are double or even multiple, which indicates that the starlight has traversed two or more gas clouds, moving with respect to each other, whose different radial velocities produce different Doppler shifts.

(i) Molecular Hydrogen in Space

It was once thought that in H I regions, all of the hydrogen was in the atomic form. Recently, however, observations from rockets and satellites have made it possible to observe the ultraviolet bands of the hydrogen molecule H_2, which occur in the ultraviolet part of the spectrum. These observations show that in dense clouds, 10 percent or more of the hydrogen is in molecules. In diffuse, low-density clouds, however, only one out of every hundred thousand H nuclei is to be found as part of an H_2 molecule.

Ultraviolet radiation from stars easily dissociates the hydrogen molecule into two hydrogen atoms. Thus the H_2 is found mostly in interstellar clouds that are shielded from ultraviolet photons by interstellar dust or even by the H_2 molecules themselves. We shall see that the dust particles are probably necessary for the formation of molecular hydrogen as well as for its protection.

(j) Radio Emission from Interstellar Gas

We receive from the Milky Way a large amount of radio energy in the frequency range 10 to 300 MHz (or wavelength range 1 to 30 m). This emission from the Milky Way was, in fact, the first radio radiation of astronomical origin to be observed. Similar radiation at radio frequencies is received from nearby galaxies. The radio energy from the Milky Way does not originate in stars. The sun, to be sure, is an apparently strong source of radio waves, but the sun is very close to us. If its radio emission is typical of that of the other stars, all the stars in the Galaxy would emit less than 10^{-9} of the radio energy actually observed. The radio energy originates in the interstellar gas.

Many strong sources of radio energy have been identified with individual gaseous nebulae. One example is the Crab nebula (Chapter 32); another is a group of faint gaseous filaments in Cassiopeia (known as the Cassiopeia A source). The optical spectra of the different filaments in the Cassiopeia A source show Doppler shifts that indicate that some blobs are moving with respect to one another with speeds in excess of 4000 km/s. Both the Crab nebula and Cassiopeia A are the remnants of supernova outbursts (Chapter 32)—gas shells blown out by the explosions of dying stars. Other discrete radio sources are associated with the more conspicuous nebulae (like the Orion nebula). Superimposed on them all is the general background of radio radiation from those regions where interstellar gas prevails.

(Thousands of individual radio sources have been catalogued that are in directions in the sky away from the Milky Way. Most of these are extragalactic sources, and will be discussed in more detail in Chapter 35).

One cause of galactic radio emission is free-free transitions (bremsstrahlung—Section 8.3) in H II regions. A free electron, passing near an ionized hydrogen atom (proton), can change from its original hyperbolic orbit to another hyperbola, emitting or absorbing in the process a photon of a definite wavelength. The many photons of different wavelengths arising from such free-free transitions in the interstellar gas constitute continuous radiation, a good part of which lies in the radio part of the spectrum. This is called *thermal radiation*—the normal emission from hot gas. On the other hand, some galactic radio sources are due to synchrotron radiation (Chapter 34), which results when electrons spiral around magnetic lines of force with speeds near that of light. In particular, the radio sources associated with old supernovae (such as the Crab nebula) are produced by the synchrotron mechanism.

(k) Radio Lines: The 21-cm Line of Hydrogen

Some of the most important radio observations carried out in the interstellar medium are of the spectral line of hydrogen at the radio wavelength of 21.11 cm. A hydrogen atom possesses a tiny amount of angular momentum by virtue of the axial spin of its electron and the electron's orbital motion about the nucleus (proton). In addition, the proton has an axial spin of its own, and the angular momentum associated with this spin may either add to or subtract from that of the electron, depending on the direction of the spin of the nucleus, with respect to that of the electron. If the spins of the two particles oppose each other, the atom as a whole has a very slightly lower energy than if the two spins are aligned. If the requisite minute amount of the energy is imparted to an atom in the lower energy state, however, the spins of the proton and electron can be aligned, leaving the atom in a slightly *excited state*. If it loses that same amount of energy again, the atom returns to its ground state. The amount of energy involved is that associated with a photon of 21-cm wavelength. An atomic transition of this type, which involves the spin of the nucleus of an atom, is called a *hyperfine transition*.

Neutral hydrogen atoms in H I regions can be excited to this 21-cm level by collisions with electrons and other atoms. Such collisions are extremely rare in the sparse gases of interstellar space; an individual atom may wait many years before such an encounter reverses its nuclear spin, that is, aligns it with the electron spin. Nevertheless, over many millions of years a good fraction of the hydrogen atoms are so excited. An excited atom can then lose its excess energy either by a subsequent collision or by radiating a photon of 21-cm wavelength. It happens, however, that hydrogen atoms are extremely loath to do the latter; an excited atom will wait, on the average, about 10 million years before emitting a photon and returning to its state of lowest energy (it is a highly forbidden transition). On the other hand, there is a definite chance that the atom will radiate before a second collision carries away its energy of excitation. In 1944 the Dutch astronomer H. C. van de Hulst predicted that enough atoms of interstellar hydrogen would be radiating photons of 21-cm wavelength to make this radio emission line observable. Equipment sensitive enough to detect the line was not available until 1951. Since that time, "21-cm astronomy" has been a very active field of astronomical research.

Observations at 21 cm show that the neutral hydrogen in the Galaxy is confined to an extremely flat layer, most of it in a sheet less than 100 pc thick, extending throughout the plane of the Milky Way. The measurement of the Doppler shifts of the 21-cm radiation from various directions also helps us detect the spiral structure of our galactic system (Chapter 34).

Since the discovery of the 21-cm line, other radio spectral lines have been observed. Among them is a multiplet of four closely spaced lines due to the OH radical—a bond of one oxygen and one hydrogen atom. Also, very high-level hydrogen emission lines are observed. The various series of hydrogen lines are described in Section 8.3. Transitions from energy level 110 to 109 of the hydrogen atom, for example, give rise to a radio line of 6-cm wavelength; those from level 157 to 156 produce a line at 17 cm.

(I) Molecules in Interstellar Space

The first molecules in interstellar space were discovered spectroscopically many years ago. The molecules CN and CH and the radical CH^+ all have spectral lines in the optical part of the spectrum. Ultraviolet observations from space revealed H_2 and CO as well. Radio observations proved that OH is present, and radio lines of many polyatomic molecules have also been detected. All of this work has led to the exciting new field of interstellar chemistry.

The radio radiation from OH near 18 cm seemed at first to be anomalous, in that it is enormously stronger than would be expected from the temperature of interstellar gas. The explanation is that the intensity of this radiation is greatly amplified by maser action. Large numbers of the molecules are excited by stellar radiation to a metastable energy level from which they undergo spontaneous transitions to their lowest (ground) energy states only very slowly; thus, a large concentration of the molecules in that metastable level is built up. However, radiation from the occasional molecules that do jump to their ground states induces or *stimulates* other excited molecules to make the same transitions by emitting photons of the same wavelength and direction as the ones stimulating them. This process of stimulated emission repeats until the radiation is amplified into a powerful beam. Meanwhile, molecules in the ground state are continually re-excited or "pumped" back to the metastable level by the radiation from surrounding stars. Water molecules produce another such *interstellar maser*. (Masers are described more fully in Chapter 8).

More than 60 kinds of molecules and radicals have been identified in space. Common atoms of hydrogen, oxygen, carbon, nitrogen, and sulfur make up molecules of water, carbon monoxide, ammonia, hydrogen sulfide, and such common organic molecules as formaldehyde, hydrogen cyanide, methyl cyanide, and simple alcohols.

Like the hydrogen molecule (H_2), most of these more complex molecules are dissociated by short-wave stellar radiation. Hence they are found in relatively dense, dark interstellar clouds where dust shields the region from ultraviolet starlight. Such clouds are estimated to have a thousand to a million times the mass of the sun and to be several light years across. The molecules in the clouds are excited to various states of vibration and rotation by collisions among themselves; they subsequently transform to lower energy states with the emission of the infrared and radio radiation by which we observe them. In the process, energy is removed from the dark clouds, which causes them to cool, contract, and become denser. Particularly cold clouds, rich in interstellar molecules, are found behind the Orion nebula, near the galactic center, and elsewhere, and are believed to be prime sites of star formation (Chapter 30).

Relatively heavy molecules, such as HC_9N, are found in some cold clouds, and ethyl alcohol, C_2H_5OH, is quite plentiful—up to one molecule for every cubic meter in space. The largest of the cold clouds have enough ethyl alcohol to make 10^{28} fifths of 100-proof liquor. Wives or husbands of future interstellar astronauts, however, need not fear that their spouses will become interstellar alcoholics; even if a spaceship were equipped with a giant funnel 1 km across and could scoop through such a cloud at the speed of light, it would take about a thousand years to gather up enough alcohol for one standard martini!

Dust grains play a crucial role in the formation of H_2 molecules and probably are important in helping to form more complex molecules. According to a model by G. Field, atoms such as those of hydrogen, oxygen, carbon, and nitrogen strike the grains. The last three tend to stick and unite with the impinging hydrogen atoms, building up a mantle of ice and other molecules. Some of the hydrogen forms H_2 molecules, which are easily dislodged from the grains and escape into space again as molecular hydrogen. Ultraviolet photons from stars striking the grains break up some of the molecules there. These molecular fragments may unite with other molecules or fragments to form some of the more complex molecules observed in space. Further chemical changes can occur through interactions of cosmic rays with the grains or with the molecules in space.

The cold interstellar clouds also contain cyanoacetylene and acetaldehyde, generally regarded as starting points for the formation of amino acids necessary for living organisms. The presence of these organic molecules does not, of course, imply the existence of life in space. On the other hand, as we learn more about the processes by which they are produced, we gain an increased understanding of similar processes that must have preceded the beginnings of life on the primitive Earth some thousands of millions of years ago.

27.4 A MODEL OF THE INTERSTELLAR MEDIUM

(a) Structure and Distribution of Interstellar Clouds

We have now described the basic constituents of interstellar matter. There are cold dense clouds in which molecules abound. There are clouds so hot that molecules cannot survive and in which atoms are mainly ionized. Interstellar matter may appear in the form of gas or dust. Can we now assemble from the observations a model of the interstellar medium that tells us where we might expect to find these various types of clouds, what the structure of an individual cloud is like, and how the clouds change as time passes?

Several important clues have already been mentioned, either implicitly or explicitly, in this chapter. First, the interstellar material is distributed in a patchy way. There are regions where the interstellar material is concentrated into clouds and regions of very low density between the clouds. We can see the patchy structure of dark dense clouds simply by looking at the Milky Way, but even low-density material is also distributed in clouds. For the gas, we can detect the individual clouds by looking at the interstellar absorption lines that appear in the spectra of distant stars. In many cases, we can see several absorption lines, each formed in a separate cloud moving with its own distinct radial velocity, superposed on the spectrum of a single distant star.

The dust is also distributed nonuniformly throughout space. Stars in certain directions may be highly reddened, while other stars in sometimes only slightly different directions but at the same distance from the sun may show little or no reddening.

A second clue to the nature of the interstellar medium is that dust and gas are found together. For a long time astronomers argued about whether or not there were two separate types of clouds, one composed of dust and the other of gas. This argument has now been settled by measuring the total amount of gas and the total amount of dust along the line of sight from us to many, many individual stars. For most stars, the ratio of the amount of gas to the amount of dust is the same within a factor of two or so. This result can only be explained if the gas and dust occur together within the same clouds.

The third vital clue to the structure of the interstellar medium was the discovery, made when satellite observations of far-ultraviolet spectra became possible, that much of interstellar space is filled with a very high-temperature, low-density gas. Most hot stars exhibit absorption lines at 1038 and 1032 Å, and these lines are produced by oxygen atoms in the interstellar medium that have been ionized five times. To strip five electrons from their orbits around an oxygen nucleus requires a lot of energy. In fact, these observations imply that the temperature of the interstellar medium where these atoms occur must be approximately one million degrees.

What causes such high temperatures in the interstellar gas far from any star? These high temper-

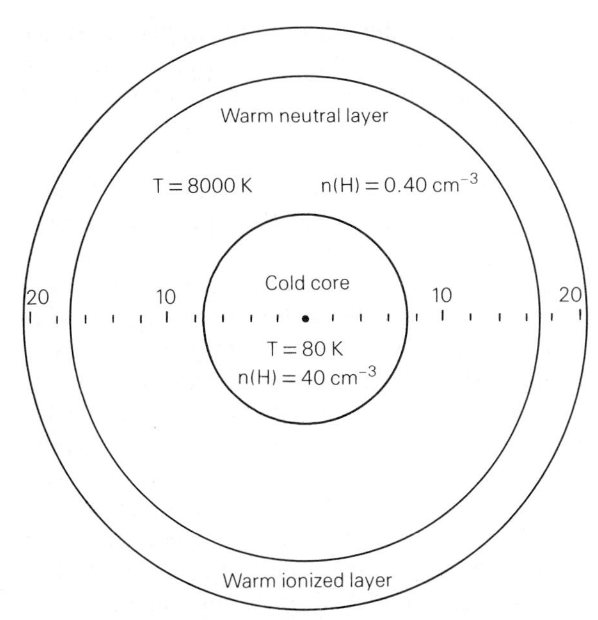

Figure 27.7 A typical interstellar cloud consists of a dense cold core surrounded by a warm envelope. The horizontal scale shows the radius in light years.

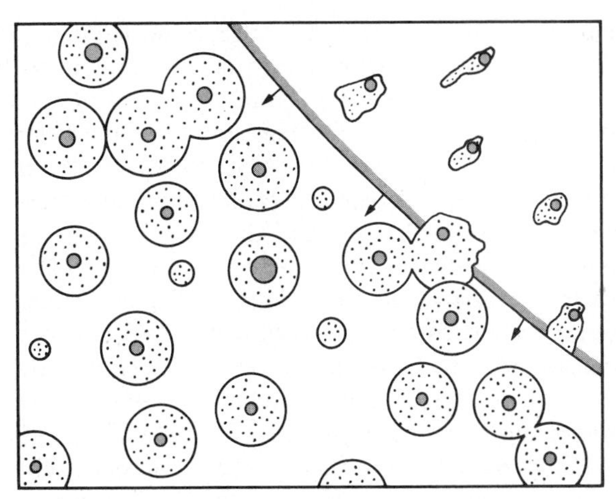

Figure 27.8 Interstellar clouds are embedded in hot low-density gas, which is heated to a temperature of several hundred thousand degrees by supernova explosions. In the upper right, a supernova remnant is shown sweeping through interstellar space. (Diagram is taken from work published in the *Astrophysical Journal* by C. McKee and J. Ostriker)

atures are almost certainly produced by the explosive force of supernovae. Stars nearing the ends of their lives (Chapter 32) explode and send high-temperature gas, moving at velocities of thousands of kilometers per second, out into interstellar space. This high-speed gas will sweep away any low-density gas, compressing it in the process and so perhaps creating some of the raw material required for the formation of cold clouds. Astronomers estimate that there is about one supernova explosion somewhere in the Galaxy every 25 years, and so on the average, the hot gas from a supernova will sweep through any given point in the Galaxy about once every two million years. At this rate, the sweeping action is continuous enough to keep most of the space between clouds filled with gas at a temperature of one million degrees.

The final constraint in developing a model of the structure of the interstellar medium is that the clouds and the gas between the clouds must be at approximately the same pressure. Suppose they were not. If the cloud pressure were higher, the cloud would expand until its pressure matched that of its environment. If, on the other hand, the pressure of the hot gas were greater than that of a cloud embedded in it, the hot gas would compress the cloud and force it to shrink in size until its pressure became high enough to resist further compression (Section 27.3g).

The pressure in a gas is proportional to temperature (T) and to the number of particles per cubic centimeter in the gas (n) (Section 28.4a). Specifically, the pressure (P) of a gas can be calculated according to the formula

$$P = nkT,$$

where k is a constant (and is equal to 1.38×10^{-16} erg deg^{-1}).

If the pressure of a gas cloud is to be equal to the pressure of the intercloud gas that surrounds it, then

$$\frac{n(\text{cloud})}{n(\text{intercloud})} = \frac{T(\text{intercloud})}{T(\text{cloud})}.$$

In words, this equation says that if the gas pressures in these two regions are equal, then the region at higher temperature must have fewer particles per cubic centimeter, i.e., it must have a lower density.

Figures 27.7 and 27.8 show in a schematic way what we think the interstellar clouds look like. Individual clouds are scattered at random throughout the galaxy. The typical cloud may be a few tens of light years in diameter. Since this cloud is embedded in gas with a temperature of one million degrees or so, the outer portions of the cloud are heated by conduction. The temperature of the outer portion of the cloud is typically about 8000 K. If the cloud is large enough, it can shield its innermost core from being heated, and the core may have a temperature that is 100 times lower and a density correspondingly 100 times higher. Typical values for the cloud core are a temperature of 80 K and a density of 40 hydrogen atoms per cubic centimeter.

The sun itself seems to be located in a region where the interstellar gas density is only about 0.1 particles per cubic centimeter, and the temperature of the nearby gas is about 10,000 K. The sun is definitely not located in the cold core of a typical interstellar cloud.

(b) Evolution of Interstellar Clouds

We have now described a picture of the interstellar medium as it appears on the average at any given time. The individual clouds do, however, change with time. Clouds collide with one another, and such collisions may cause clouds to coalesce or, if the collision is a violent one, to fragment into many smaller clouds.

Schematically, we think that clouds are formed initially from the expanding gas around supernovae or hot stars. These clouds are relatively small and do not exceed 100 times the mass of the sun. The clouds then grow through collisions with other clouds. Ultimately, this process may lead to the formation of giant molecular clouds with diameters of 200 light years and masses that exceed 100,000 times the mass of the sun. It is in these molecular clouds that the most vigorous star formation occurs. Only a small fraction of the mass of a giant molecular cloud is converted to stars. These stars then evolve and become supernovae and in the process eject gaseous material, thus starting the cycle over again.

EXERCISES

1. Identify several dark nebulae on photographs of the Milky Way in this book. Give the figure numbers of the photographs, and specify where the dark nebulae are to be found on them.

2. Suppose all stars had the same absolute magnitude, and that they were distributed uniformly through space. Show that the number of stars from which we receive an amount of light b is inversely proportional to b raised to the $3/2$ power. Now suppose that stars of many absolute magnitudes exist, but that the numbers of stars of various absolute magnitudes exist in the same relative proportions everywhere in space. Is the proportionality between the number of stars appearing brighter than b and $b^{-3/2}$ changed? Explain.

★ 3. Suppose stars are counted to increasingly fainter limiting brightnesses. Make a sketch showing how the numbers of stars counted will increase as b is decreased, both in the absence and in the presence of general interstellar obscuration.

4. A spectral-type A star normally has a color index of 0.0. The blue and visual magnitudes of an A star are $B = 11.6$ and $V = 10.8$, respectively.
   a) What is the color excess of the star?
   b) What is the total absorption of its light in visual magnitudes?
   c) What is the total absorption of its light in blue magnitudes?

5. The sun is observed from a distant star to have an apparent blue magnitude of 14.4 and an apparent visual magnitude of 12.8. How far away, approximately, is the star? Assume that the sun's color index is $B - V = 0.6$, and that its absolute visual magnitude is $+4.8$.

6. Suppose a bright reflection nebula appears yellow. What kind of star probably is producing it?

7. The red color of the sun, when seen close to the horizon, and the blue color of the daytime sky provide analogies to the reddening of starlight and the blue color of reflection nebulae. Discuss this analogy more fully, and also explain how it breaks down.

★ 8. The amount of light emitted in the hydrogen emission lines of an emission nebula depends on the amount of ionized hydrogen. This in turn depends on the amount of ultraviolet radiation, capable of ionizing hydrogen, that is emitted by the central star. The star's brightness in the visible part of the spectrum can be observed and compared with the amount of light in the hydrogen lines of the nebula. Does this suggest a way by which the temperature of the star might be estimated? Explain. (See Chapters 8 and 24.)

★ 9. Explain why "nebulium lines" are not observed in the solar spectrum. Might they be possible in the spectrum of the very rarefied upper atmosphere of the Earth?

*10. Describe in detail the appearance of the spectrum of an emission nebula such as the Orion nebula. Would you expect any continuous spectrum to be present? Explain.

*11. Suppose you examined the spectrum of some nebulosity surrounding a main-sequence spectral type O star and found that it contained no emission lines, only the continuous spectrum of the star. What conclusions could you draw about the nature of the interstellar material around that star?

*12. If an H II region at a temperature 10,000 K is in pressure equilibrium with an adjacent H I region at a temperature of 100 K, what can you say about the relative densities of the two regions?

Sir Arthur Stanley Eddington (1882–1944), British mathematician and astrophysicist, organized two expeditions to observe the total solar eclipse of 1919 in order to test a prediction of Einstein's general theory of relativity. Eddington is best known among astronomers for his development of theoretical methods of investigating the internal structure of the sun and stars.

28

THE SUN: AN ORDINARY STAR

The stars are faint and very far away. None can be resolved so that we can study their surfaces in any detail. Fortunately, we have near at hand (astronomically speaking!) an ordinary star, which we call the sun. By observing the sun closely, we can learn a great deal about the physical processes that must determine the structure and evolution of other stars. Of even more immediate importance is the fact that the sun's warmth, and the constancy of the heating it supplies, is vital to life on Earth.

Some basic data for the sun are summarized in Table 28.1.

28.1 OUTER LAYERS OF THE SUN

The only parts of the sun that can be observed directly are its outer layers, collectively known as the sun's *atmosphere*. There are three general regions, each having substantially different properties: the *photosphere*, the *chromosphere*, and the *corona*.

(a) The Solar Photosphere

What we see when we look at the sun with our unaided eyes is the solar photosphere (Figure 28.1). As stated in Section 24.2b, the photosphere is not a discrete surface but covers the range of the depths from which the solar radiation escapes. Most of the absorption of visible light in the solar photosphere, as we have seen, is done by negative hydrogen ions. As we look toward the edge or limb of the sun, our line of sight enters the photosphere at a grazing angle, and the depth below the outer surface of the photosphere to which we can see is even less than at the center of the sun's disk. The light from the limb of the sun, therefore, comes from higher and cooler regions of the photosphere. Analysis of this *limb darkening* can be used to determine the variation of temperature with depth in the photosphere, as described in Section 24.2. From the data obtained from observations of limb darkening, using our knowledge of the physics of gases and the way in which atoms absorb and emit light, we can calculate a *model solar photosphere*. Such a model solar photosphere is given in Table 28.2.

It is evident from Table 28.2 that within a depth of about 260 km the pressure and density increase by a factor of 10, while the temperature climbs from 4500 to 6800 K. At a typical point in the photosphere, the pressure is only a few hundredths of sea-level pressure on the Earth, and the density is about one ten-thousandth of the Earth's atmospheric density at sea level.

TABLE 28.1 Solar Data

DATUM	HOW FOUND	TEXT REFERENCE (Section)	VALUE
Mean distance	Radar reflection from planets	22.1d	1 AU 149,597,892 km
Maximum distance from Earth			1.521×10^8 km
Minimum distance from Earth			1.471×10^8 km
Mass	Acceleration of Earth	25.1	333,400 Earth masses 1.99×10^{33} g
Mean angular diameter	Direct measure	7.1h	31′59″.3
Diameter of photosphere	Angular size and distance	7.1h	109.3 times Earth diameter 1.39×10^{11} cm
Mean density	Mass/volume	4.1g	1.41 g/cm^3
Gravitational acceleration at photosphere (surface gravity)	$\frac{GM}{R^2}$	4.2c	27.9 times Earth surface gravity 27,300 cm/s^2
Solar constant	Measure with instrument such as bolometer	23.2f	1.96 cal/min/cm^2 1.368×10^6 ergs/s/cm^2
Luminosity	Solar constant times area of spherical surface 1 AU in radius	23.2f	3.8×10^{33} ergs/s
Spectral class	Spectrum	24.1a	G2V
Effective temperature	Derived from luminosity and radius of sun	25.3a	5800 K
Visual magnitude			
Apparent	Received visible flux	23.1a	−26.7
Absolute	Apparent magnitude and distance	23.2a	+4.8
Rotation period at equator	Sunspots, and Doppler shift in limb spectra	28.1e	24^d16^h
Inclination of equator to ecliptic	Motions of sunspots	28.1e	7°10′.5

(b) Chemical Composition

More than 60 of the elements known on the Earth have now been identified in the solar spectrum. Those that have not been identified in the sun either do not produce lines in the observable spectrum or are so rare on the Earth that they cannot be expected to produce lines of observable strength on the sun unless, proportionately, they are far more abundant there. Most of the elements found in the sun are in the atomic form, but more than 18 types of molecules have been identified. Most of the molecular spectra are observed only in the light from the cooler regions of the sun, such as the sunspots.

The relative abundances of the chemical elements in the sun are similar to the relative abundances found for other stars. About three-quarters of the sun (by weight) is hydrogen, and about 98 percent is hydrogen and helium. The remaining few percent is made up of the other chemical elements, in approximately the amounts that are described in Section 24.2 (See also Appendix 19).

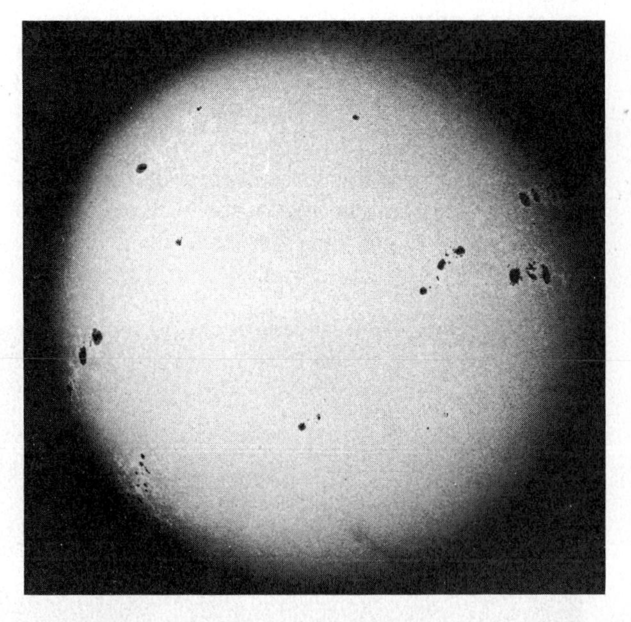

Figure 28.1 The sun, photographed under excellent conditions, showing a large number of sunspots, on September 15, 1957. (Mt. Wilson and Las Campañas Observatories)

Table 28.2 Model Solar Photosphere

DEPTH BELOW SURFACE *(km)*	PERCENT OF LIGHT THAT EMERGES FROM THAT DEPTH	TEMPERATURE *(K)*	PRESSURE *(Earth Atmospheres)*	DENSITY *(g/cm³)*
0	100	4500	1.0×10^{-2}	2.8×10^{-8}
50	95	4800	1.7×10^{-2}	4.2×10^{-8}
100	91	5000	2.6×10^{-2}	6.2×10^{-8}
140	82	5300	3.8×10^{-2}	8.7×10^{-8}
170	67	5600	5.4×10^{-2}	11.5×10^{-8}
225	37	6200	8.3×10^{-2}	16.0×10^{-8}
260	13	6800	11.2×10^{-2}	20.0×10^{-8}

(c) The Chromosphere

There is a change in the physical state of the gases just above the photosphere. The photosphere ends at about the place where the density of negative hydrogen ions has dropped to a value too low to result in appreciable opacity. Gases extend far beyond the photosphere, but they are transparent to most radiation. The region of the sun's atmosphere that lies immediately above the photosphere is the chromosphere.

Until this century the chromosphere was best observed when the photosphere was occulted by the Moon during a total solar eclipse. In the 17th century several observers described what appeared to them as a narrow red "streak" or "fringe" around one limb of the Moon during a brief instant after the sun's photosphere had been covered. In 1868 the spectrum of the chromosphere was first observed; it was found to be made up of bright lines, which showed that the chromosphere consists of hot gases that are emitting light in the form of emission lines. These bright lines are difficult to observe against the bright light of the photosphere but appear in the spectrum of the light from the extreme limb of the sun just after the Moon has eclipsed the photosphere. They disappear within a few seconds, when the Moon has covered the chromosphere as well. Because of the brief instant during which the chromospheric spectrum can be photographed during an eclipse, its spectrum, when so observed, is called the *flash spectrum* (Figure 28.2). The element *helium* (from *helios,* the Greek word for "sun") was discovered in the chromospheric spectrum before its discovery on Earth in 1895.

Today it is possible to photograph both the chromosphere and its spectrum outside of eclipse. One instrument used for this purpose is the *coronagraph,* a telescope in which a black disk at the focal plane occults the photosphere, producing an artificial eclipse. The chromosphere can also be photographed in the light of its strongest emission lines through monochromatic filters, which selectively transmit a narrow range of wavelengths. In addition, the chromosphere is regularly observed in ultraviolet, X-ray, infrared, and radio wavelengths.

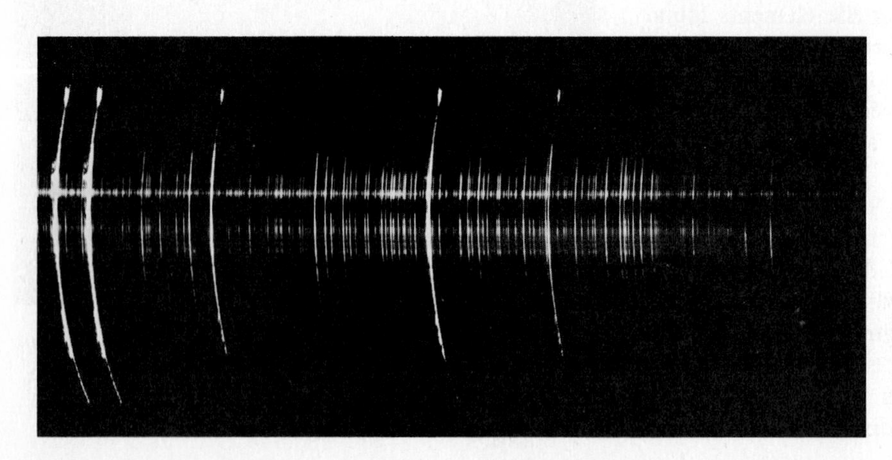

Figure 28.2 Flash spectrum of the eclipse of 2 February 1968. The bright arcs are monochromatic images of the solar limb. The irregularities are prominences. (Courtesy Sacramento Peak Observatory, Air Force Cambridge Research Laboratories)

The chromosphere is about 2000 to 3000 km thick, but in its upper region it breaks up into a forest of jets (called *spicules*), so the position of the upper boundary of the chromosphere is somewhat arbitrary. Its reddish color (whence comes its name) arises from one of the strongest emission lines in the visible part of its spectrum, the bright red line due to hydrogen (the Hα line— first line in the Balmer series—Section 8.3c). The density of the chromospheric gases decreases upward above the photosphere, but spectrographic studies show that the temperature *increases* upward through the chromosphere, from 4500 K at the photosphere to 100,000 K or so at the upper chromospheric levels. The processes that heat the corona (see below) evidently also heat the chromosphere. Far-ultraviolet observations of the chromosphere, such as those made from the Skylab, have been especially helpful in revealing information about its structure.

(d) The Corona

The chromosphere merges into the outermost part of the sun's atmosphere, the corona. Like the chromosphere, the corona was first observed only during total eclipses, but unlike the chromosphere, the corona has been known for many centuries; it is referred to by Plutarch and was discussed in some detail by Kepler. The corona extends millions of miles above the photosphere and gradually thins to a sparse wind of ions and electrons flowing outward through the entire solar system (the *solar wind;* Section 28.3). The corona emits half as much light as the full Moon; its invisibility, under ordinary circumstances, is due to the overpowering brilliance of the photosphere. Like the chromosphere, the corona can now be photographed, with the coronagraph and other instruments, under other than eclipse conditions.

The corona is also observed at radio wavelengths. In Great Britain in 1942, unexpected noise was picked up on radar receivers. It was subsequently learned that the source of this noise was the sun. Since World War II, radio observations of the sun have been made regularly at many radio astronomical observatories. Shortwave radio energy (near 1 cm wavelength) can escape the sun from the lower chromosphere. The corona is more and more opaque, however, to longer and longer radio wavelengths. Those of 15 m escape the sun only if they originate high in the corona. Thus, by observing the sun at different radio wavelengths, we observe to different depths in the corona and chromosphere and can determine the heights in the solar atmosphere at which various disturbances giving rise to radio emission occur.

Not only do radio waves originate in the corona, but the corona produces scintillations in distant radio sources when they are observed through its outer part. The phenomenon is somewhat analogous to the scintillation of the light from stars caused by the Earth's atmosphere. Scintillations of remote radio sources observed 90° away from the sun in the sky show that the corona actually reaches out beyond the Earth. Coronal atoms are also regularly detected in the vicinity of the Earth by space vehicles.

The visible light from the corona is about 10^{-6} that of the photosphere of the sun. It has three components. The spectrum of the first is simply that of reflected sunlight that shows the same dark Fraunhofer lines as the solar photospheric spectrum; this part is called the F (Fraunhofer) corona. It extends far from the sun and gradually fades into interplanetary space. The F corona is believed to be caused by interplanetary dust particles. Superimposed on the F corona is the K corona, whose light dominates in the corona's inner, brighter region, less than two solar radii from the center of the sun. The K corona is photospheric light reflected by free electrons. However, all of the absorption lines are washed out, leaving the K corona with a pure continuous spectrum (the "K" designation comes from the German word *Kontinuum*). The washing out of the spectral lines is due to the very high random speeds of the electrons, redistributing the reflected light, thereby filling in the absorption lines by many different Doppler shifts. The inferred high speeds of the electrons provide evidence that the corona is extremely hot—in the millions of Kelvins.

The third spectral component of the corona consists of bright emission lines superimposed on the light of the K and F coronas. In 1942 the Swedish physicist B. Edlen identified these lines as forbidden lines of calcium, iron, and nickel (see Section 27.3d for a discussion of forbidden radiation). There are also many ordinary permitted emission lines, but most of these lie in the far ultraviolet, where they have been mapped by instruments aboard spacecraft.

Analysis of the coronal spectral lines confirms the K-corona inference that the corona is very

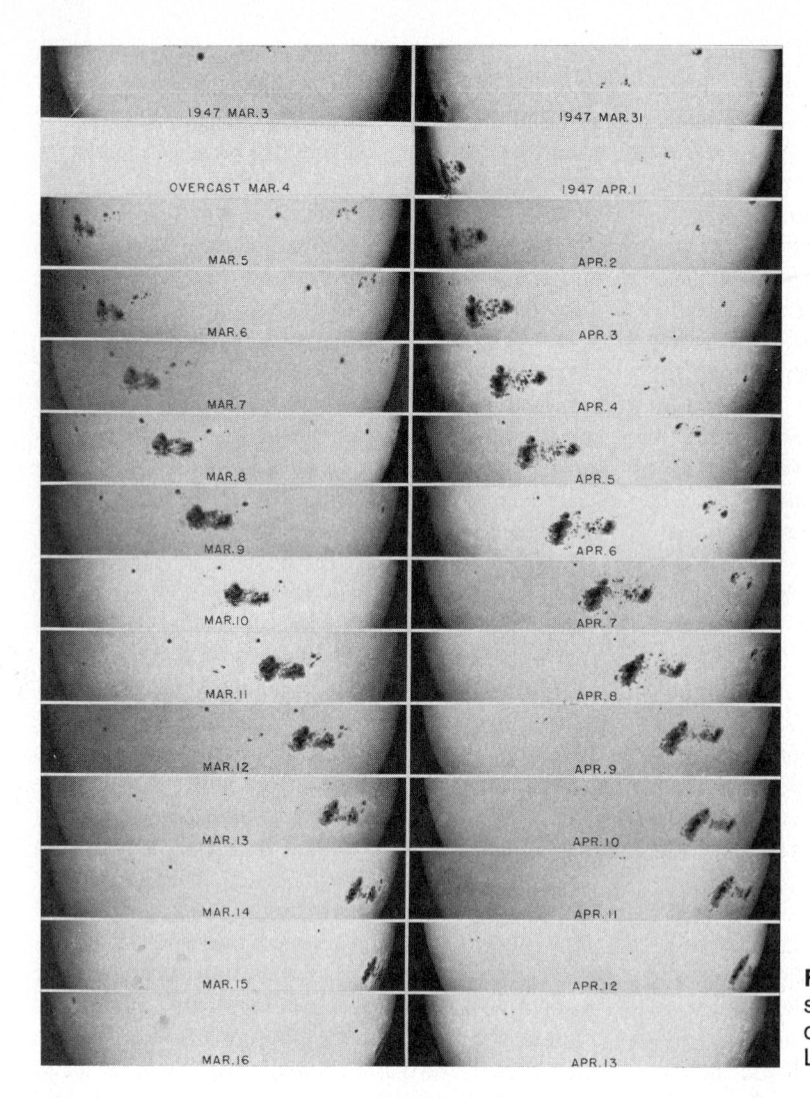

Figure 28.3 Series of photographs showing the motions of sunspots, indicating the solar rotation. (Mt. Wilson and Las Campañas Observatories)

hot—millions of Kelvins. The atoms are all very highly ionized—for example, there are ultraviolet lines of iron ionized 16 times. The density, however, is very low. At the base of the corona there are about 10^9 atoms/cm^3, compared with 10^{16}/cm^3 in the upper photosphere and 10^{19}/cm^3 at sea level in the Earth's atmosphere. Thus, despite the high temperature of the corona (a measure of how fast the particles are moving), its density is so low that the actual heat (energy content per cubic centimeter) is very low; in that near-vacuum it would take a long time for the hot coronal gases to warm up a cup of coffee (of course, the radiant energy from the nearby photosphere would do the job in a hurry).

Observations from Skylab in 1973 revealed that there are sometimes large regions of the corona that are relatively cool and quiet. These *coronal holes* are places of extremely low density and are usually (but not always) found in the polar regions of the sun. They cause the empty spaces that can be seen on some of the eclipse photographs of the solar corona (see Section 7.3).

(e) Solar Rotation

Galileo first demonstrated that the sun rotates on its axis by recording the apparent motions of the sunspots as the turning sun carried them across its disk (Figure 28.3). He found that the rotation period of the sun is a little less than one month. In 1859 Richard Carrington found that in its equatorial re-

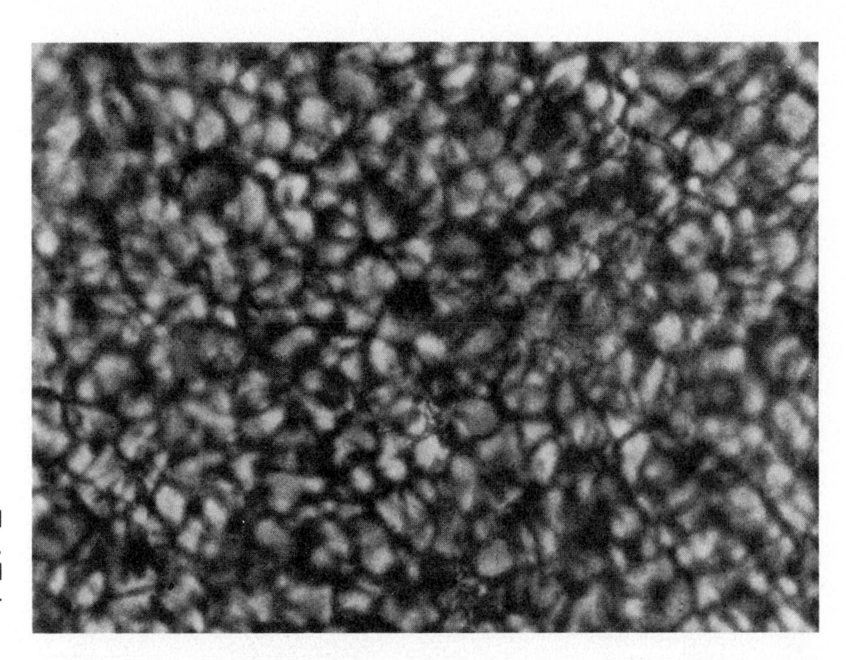

Figure 28.4 A highly magnified photograph of solar granulation. (National Solar Observatory and National Optical Astronomy Observatories)

gions the sun rotates in about 25 days, but that at a latitude of 30° the period is about 2½ days longer.

The sun's rotation rate can be determined also from the difference in the Doppler shifts of the light coming from the receding and approaching limbs (Section 24.2d). Very sensitive Doppler shift measurements show the period to average 25.8 days at the equator, 28.0 days at latitude 40°, and 36.4 days at latitude 80°, in the direction west to east (like the orbital motions of the planets). There are, however, variations in the rotation rate over time at any given latitude, corresponding to speeds of up to a few meters per second. Those variations appear to be correlated with the solar cycle (next section). The rotation may be quite different beneath the surface, as well. Of course the sun, being a fluid gas, need not rotate as a solid body.

The apparent motions of sunspots are not usually straight lines across the sun's disk but rather slight arcs, because the axis of rotation of the sun is not exactly perpendicular to the plane of the ecliptic. The angle of inclination of the solar equator to the ecliptic is about 7°.

28.2 PHENOMENA OF THE SOLAR ATMOSPHERE

In its gross characteristics the sun is quite stable, but the detailed features of its atmosphere are constantly changing.

(a) Photospheric Granulation and Supergranulation

Direct telescopic observation and photography show that the photosphere is not perfectly smooth but has a mottled appearance resembling rice grains—this structure of the photosphere is now generally called *granulation* (Figure 28.4). Typically, granules are 700 to 1000 km in diameter; the smallest observed are about 300 km across. They appear as bright areas surrounded by narrow darker regions.

The motions of the granules can be studied by the Doppler shifts in the spectra of gases just above them. It is found that the granules themselves are columns of hotter gases arising from below the photosphere. As the rising gas reaches the photosphere it spreads out and sinks down again. The darker intergranular regions are the cooler gases sinking back. The centers of the granules are hotter than the intergranular regions by 50 to 100 K. The vertical motions of gases in the granules have speeds of about 2 or 3 km/s. Individual granules persist for about 8 minutes. The granules, then, are the tops of convection currents of gases rising through the photosphere.

The granules themselves form part of a structure of still larger scale called *supergranulation*. Supergranules are cells that average about 30,000 km in diameter, within which there is a flow of gases

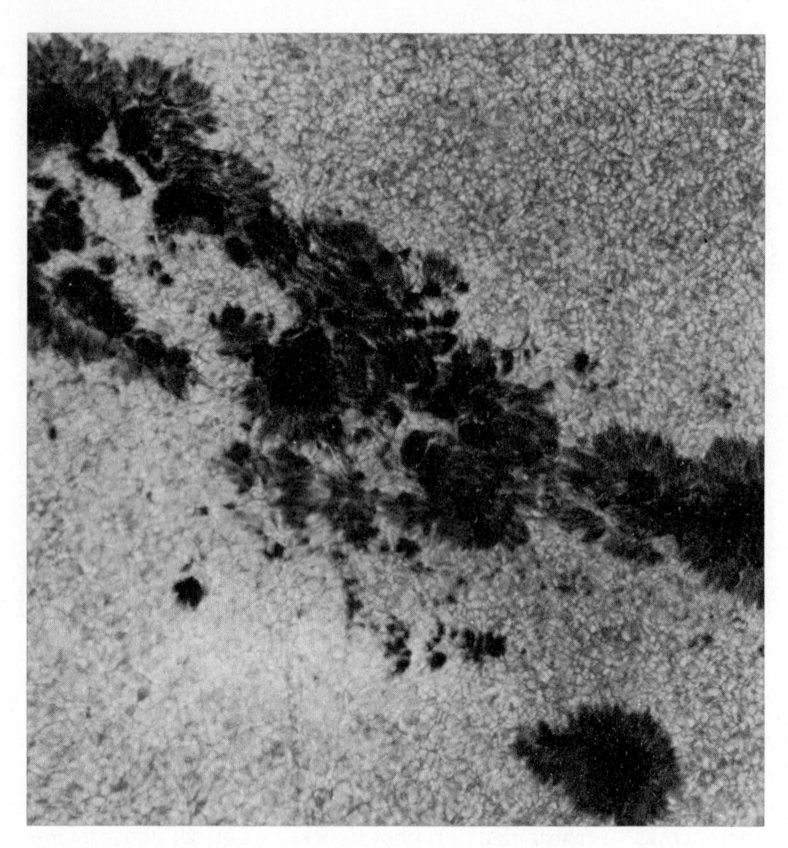

Figure 28.5 An excellent photograph of sunspots and solar granulation in the surrounding photosphere. (National Solar Observatory and National Optical Astronomy Observatories)

from center to edge. The structure that outlines the supergranules persists throughout the upper photosphere and chromosphere. Moreover, magnetic fields are concentrated at supergranule boundaries.

In addition to the vertical currents of gases in the granules and the center-toward-edge flow in the supergranules, in each region of the sun (typically up to tens of thousands of kilometers across) the gases rhythmically pulse up and down with speeds of about ⅓ km/s, taking about 5 minutes for a complete cycle—a phenomenon known as the *five-minute oscillation.*

(b) Sunspots

The most conspicuous of the photospheric features are the *sunspots* (Figure 28.5). Occasionally, spots on the sun are large enough to be visible to the naked eye. Galileo first showed that sunspots are actually on the surface of the sun. In 1774 the Scot Alexander Wilson suggested that spots were "holes" through which we could see past the photosphere into a cooler interior of the sun. William Herschel held a similar view; he imagined that the sun had a cool, probably inhabited, interior that was surrounded by two cloud layers.

Actually the spots are regions of the photosphere where the gases are up to 1500 K cooler than those of the surrounding photosphere. Sunspots are nevertheless hotter than the surfaces of many stars. If they could be removed from the sun, they would be seen to shine brightly; they appear dark only by contrast with the hotter, brighter surrounding photosphere.

Individual sunspots have lifetimes that range from a few hours to a few months. They are first seen as small dark "pores" somewhat over 1500 km in diameter. Most of them disappear within a day, but a few persist for a week or occasionally much longer. If a spot lasts and develops, it is usually seen to consist of two parts: an inner darker core, the *umbra,* and a surrounding less dark region, the *penumbra.* Many spots become much larger than the Earth, and a few have reached diameters of 50,000 km. Frequently spots occur in groups of from two to 20 or more. If a group contains many spots, it is likely to include two large ones, one approximately east of the other, and many smaller spots clustered

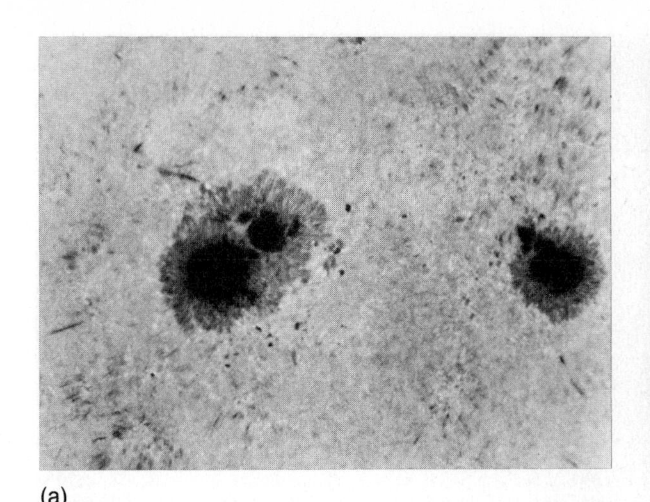

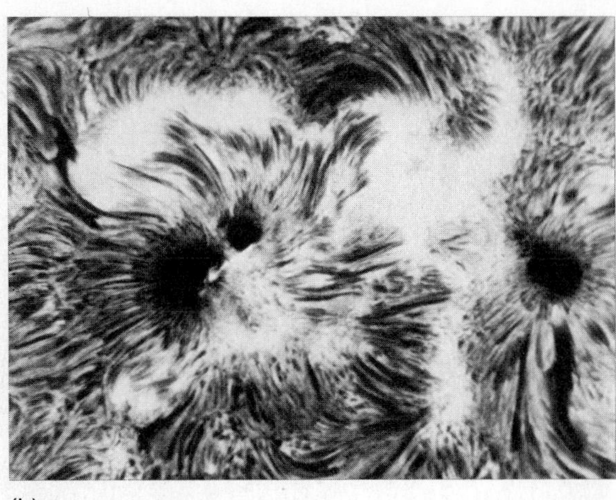

(a) (b)

Figure 28.6 Bipolar sunspot group photographed May 21, 1972, at Big Bear Solar Observatory, (a) in white light; (b) in the red light of the hydrogen Hα line. (California Institute of Technology)

around the two principal ones (Figure 28.6). The principal spot to the east is most often the largest one of the group. The largest groups are very complex and may have over a hundred spots. Like storms on the Earth, sunspots may move slowly on the surface of the sun, but their individual motions are slow when compared with the solar rotation, which carries them across the disk of the sun.

(c) The Sunspot Cycle

In 1851 Heinrich Schwabe, a German apothecary and amateur astronomer, published an important conclusion he had reached as a result of his observations of the sun over the previous decade. He found that the number of sunspots visible, on the average, varied with a period of about ten years. Since Schwabe's work, the *sunspot cycle* has been clearly established. Although individual spots are short-lived, the total number of spots visible on the sun at any one time is likely to be very much greater during certain periods, the periods of *sunspot maximum*, than at other times, the periods of *sunspot minimum* (Figure 28.7). Sunspot maxima have occurred at an average interval of 11.1 years, but the intervals between successive maxima have ranged from as little as eight years (from 1830 to 1838) to as long as 16 years (from 1888 to 1904). During sunspot maxima, more than 100 spots can often be seen on the sun at once. During sunspot minima, the sun sometimes has no visible spots. The last maximum (at this writing) was in 1979; the next is expected in 1991.

At the beginning of a cycle, just after a minimum, a few spots or groups of spots appear at latitudes of about 30° on the sun. As the cycle progresses, the successive spots occur at lower and lower latitudes, until, at the maximum of the cycle, their average latitude is about 15°. Near minimum, the last few spots of a cycle appear at about 8° latitude. About the same time, the next cycle begins with a few spots occurring simultaneously at higher latitudes. Sunspots almost never appear at latitudes greater than 40° or less than 5°. The locations of sunspots on the sun in both the northern and southern hemispheres are related to the sunspot cycle in the same way; however, sunspot activity in one hemisphere may dominate for long periods. We shall return later to the interpretation of the sunspot cycle.

(d) Magnetic Fields on the Sun

As stated in Section 24.2d, a spectral line is usually split up into several components in the presence of a magnetic field, the phenomenon known as the Zeeman effect. In 1908 the American astronomer George E. Hale observed the Zeeman effect in the spectrum of sunspots and found evidence for strong magnetic fields. The magnetic fields observed in sunspots range from 100 to nearly 4000 gauss. This is as great as the field of a good alnico magnet and,

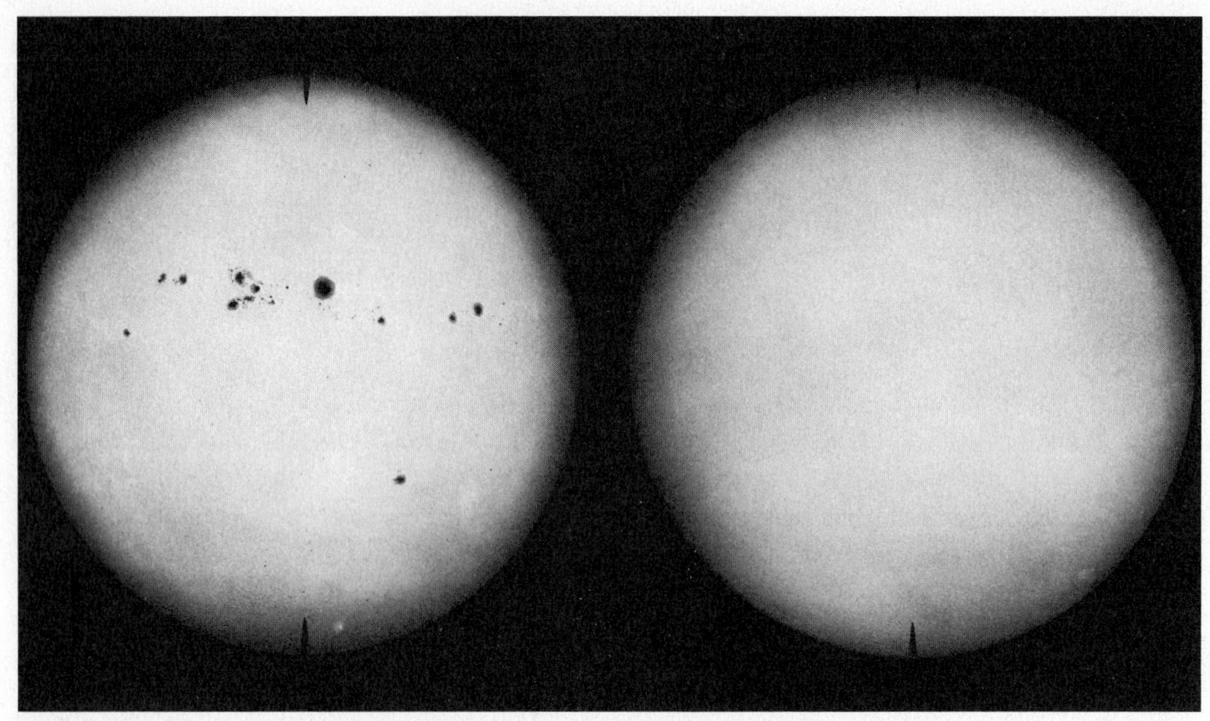

Figure 28.7 Direct photographs of the sun near the time of sunspot maximum *(left)* and near sunspot minimum *(right)*. (Mt. Wilson and Las Campañas Observatories)

moreover, is spread over a region tens of thousands of kilometers across. The magnetic field is present in the region surrounding a spot as well as in the spot itself and persists even after the spot has disappeared.

Whenever sunspots are observed in pairs or in groups containing two principal spots, one of the spots usually has the magnetic polarity of a north-seeking magnetic pole, and the other has the opposite polarity. Moreover, during a given cycle, the leading spots of pairs (or leading principal spots of groups) in the northern hemisphere all tend to have the same polarity, while those in the southern hemisphere all tend to have the opposite polarity. During the next sunspot cycle, however, the polarity of the leading spots is reversed in each hemisphere. For example, if during one cycle the leading spots in the northern hemisphere all had the polarity of a north-seeking pole, the leading spots in the southern hemisphere would have the polarity of a south-seeking pole; during the next cycle, the leading spots in the northern hemisphere would have south-seeking polarity and those of the southern hemisphere would have north-seeking polarity. We see, therefore, that the sunspot cycle does not repeat itself as regards magnetic polarity until *two* 11-year maxima have passed. The magnetic cycle of the sun is therefore sometimes said to last 22 years, rather than 11.

The strong magnetic fields on the sun are associated with sunspots. When the spots die, their fields spread out towards the poles, contributing to a more general, but far weaker, solar magnetic field. H. W. and H. D. Babcock, of the Mount Wilson Observatory and the Hale Solar Observatory, investigated magnetic fields on the sun for many years, and in the 1950s first established the existence of the general solar magnetic field with a strength of a few gauss. Because the sunspot polarity reverses with each cycle, the general magnetic field of the sun also reverses, but in a rather irregular way.

The solar magnetic field is not uniform. The motions of gas in the supergranules, for example, cause the field strength to build up to strengths of 1500 to 2000 gauss at the supergranule boundaries. Magnetism is still not a dominant force in the photospheric layers, but in the chromosphere and corona, where the gas density and pressure are enor-

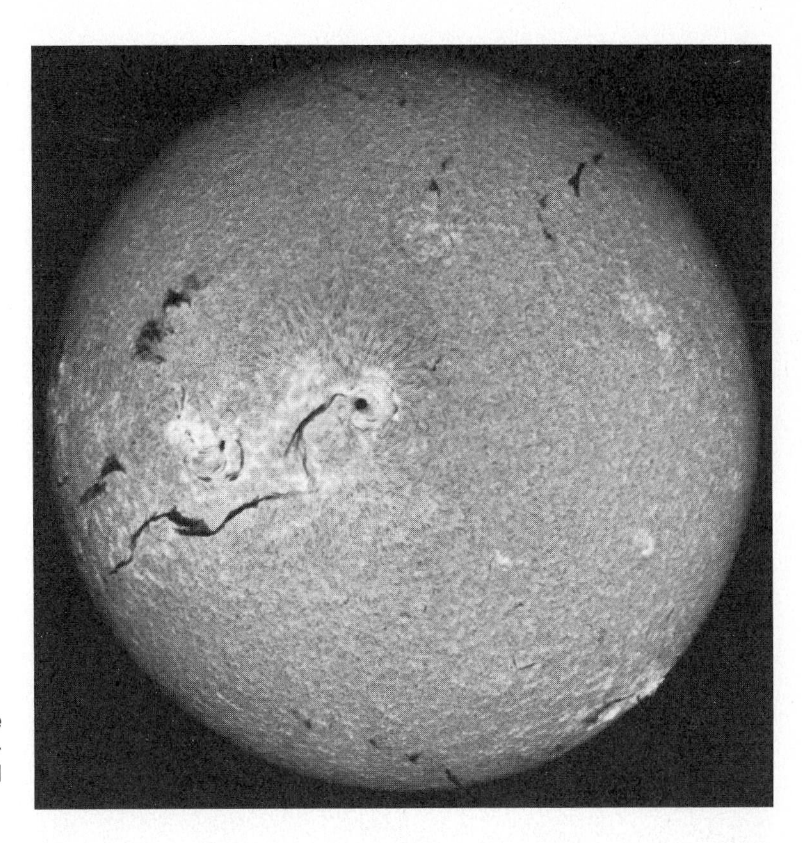

Figure 28.8 The sun in the light of the first Balmer line of hydrogen (Hα). (National Solar Observatory and National Optical Astronomy Observatories)

mously less, the magnetic fields are relatively strong and play an important role in influencing the motions of ionized gases. Far out in the corona, magnetic lines of force manifest themselves by organizing ionized coronal gases into streamers, which are easily seen and photographed during total solar eclipses. Low in the corona, magnetic fields guide the motions of ions in solar prominences (see below). The solar magnetic field even extends into interplanetary space and is measured in the vicinity of the Earth and other planets with magnetometers carried on space probes. It can even be deduced from changes in the Earth's field.

(e) Filtergrams

In order to see regions of the sun that lie directly above the photosphere, we may observe in spectral regions to which the photospheric gases are especially opaque—at the centers of strong absorption lines such as those of hydrogen and calcium.

It is most common to isolate one of the absorption lines of ionized calcium in the ultraviolet (the K line of ionized calcium) or the Hα line of hydrogen in the red. These spectral lines appear dark when viewed against the rest of the solar spectrum, but they are not completely dark; what light remains in the centers of the lines is emitted from atoms of calcium or hydrogen, respectively, in the chromosphere. Thus, spectroheliograms reveal the appearance of the chromosphere in the light of calcium, or hydrogen. There are special filters that pass only light in narrow spectral regions, and now astronomers routinely observe the sun through such *monochromatic* filters. These images are called *filtergrams* (Figure 28.8).

Spectacular motion pictures have been taken through monochromatic filters, or with spectroheliographs. Time-lapse photographs, in which frames are exposed every few seconds or every few minutes and then run through a projector at normal speed, show in a dramatic way changes that occur in the solar chromosphere.

(f) Plages and Faculae

Filtergrams in the light of calcium and hydrogen show bright "clouds" in the chromosphere in the

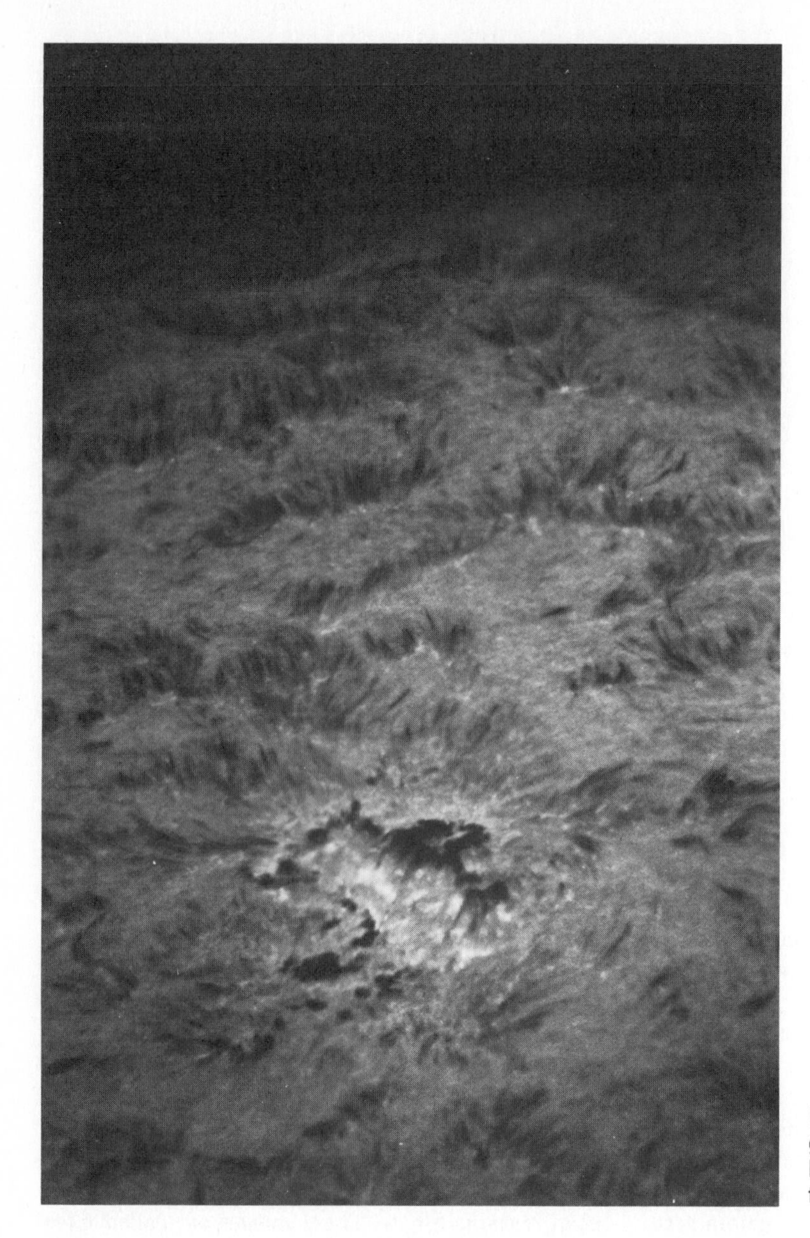

Figure 28.9 Solar spicules, photographed in the light of Hα. (National Solar Observatory and National Optical Astronomy Observatories)

magnetic-field regions around sunspots. These bright regions (formerly called *flocculi*—"tufts of wool") are known as *plages*. Calcium and hydrogen plages are also sometimes seen in regions where there are no visible sunspots, but these regions are generally those of higher than average magnetic fields.

The plages are not, of course, concentrations of calcium or hydrogen, but are regions where calcium and hydrogen happen to be emitting more light at the observed wavelengths. These elements are partially ionized throughout most of the visible chromosphere, and some of the atoms emit light as they capture electrons and become neutral (or less ionized) or as those atoms (or ions) cascade down through the various excited energy levels. The plages, then, are regions where some of the atoms of the element observed are changing their states of ionization or excitation and are emitting more light than in the surrounding areas. Plages of hydrogen and calcium usually occur in approximately the same projected regions at the same time.

Plages sometimes emit light at many wavelengths and can be seen in the direct image of the

Figure 28.10 A solar prominence 160,000 km high, photographed in the light of Hα at Big Bear Solar Observatory on June 12, 1972. (Mt. Wilson and Las Campañas Observatories)

sun. These "white-light" plages are called *faculae* ("little torches") and were first described by Galileo's contemporary Christopher Scheiner. Faculae are seen best near the limb of the sun where the photosphere is not so bright and the contrast is more favorable for their visibility.

(g) Spicules

The chromosphere also contains many jet-like spikes of gas rising vertically through it. These features, called *spicules,* occur at the edges of supergranule cells; when viewed near the limb of the sun so many are seen in projection that they give the effect of a forest (Figure 28.9). They show up best when the chromosphere is viewed in the light of hydrogen. They consist of gas jets moving upward at about 30 km/s and rising to heights of from 5000 to 20,000 km above the photosphere. Individual spicules last only 10 minutes or so. Through the spicules matter continually flows into the corona. They are now believed to be of fundamental importance in the energy and momentum balance in the solar atmosphere.

(h) Prominences

Among the more spectacular of coronal phenomena are the *prominences.* Prominences have been viewed telescopically during solar eclipses for centuries. They appear as red flame-like protuberances rising high above the limb of the sun. Prominences can now be viewed at any time on spectroheliograms and filtergrams. The features of some, the *quiescent* prominences, may remain nearly stable for many hours, or even days, and may extend to heights of tens of thousands of km above the solar surface. Others, the more active prominences, move upward or have arches that surge slowly back and forth (Figures 28.10 and 28.11). The relatively rare *eruptive* prominences appear to send matter upward into the corona at speeds up to 700 km/s, and the most active *surge* prominences may move upward at speeds up to 1300 km/s. Some eruptive prominences have reached heights of over one million km above the photosphere. When seen silhouetted against the disk of the sun, prominences have the appearance of irregular dark filaments.

Superficially, prominences appear to be material ejected upward from the sun, but motion pictures show that whereas a prominence may grow in size and rise higher and higher above the photosphere, the actual material in the prominence most often appears to move downward in graceful arcs, evidently along lines of magnetic force. Apparently, most prominences form from coronal material that cools and moves downward, even though the disturbance that characterizes the prominence may move upward.

Prominences are cool and dense regions in the corona where atoms and ions are capturing electrons and emitting light. Their origin is unknown, but it is significant that they usually originate near regions of sunspot activity and lie on the boundary between regions of opposite magnetic polarity. Quiescent prominences are supported by coronal magnetic fields, and eruptive prominences evidently result from sudden changes in the magnetic fields. Prominences seem to be further symptoms of the

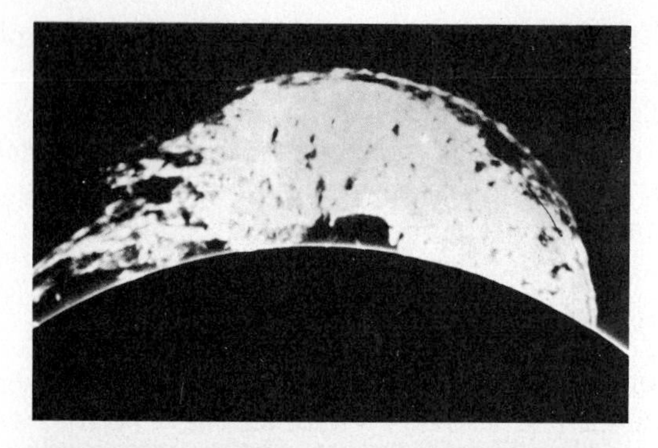

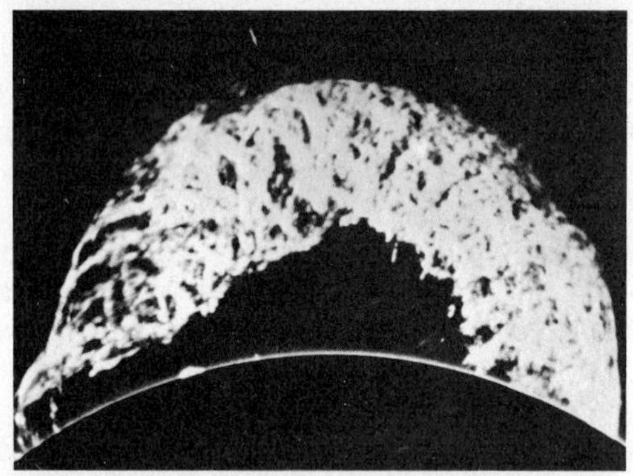

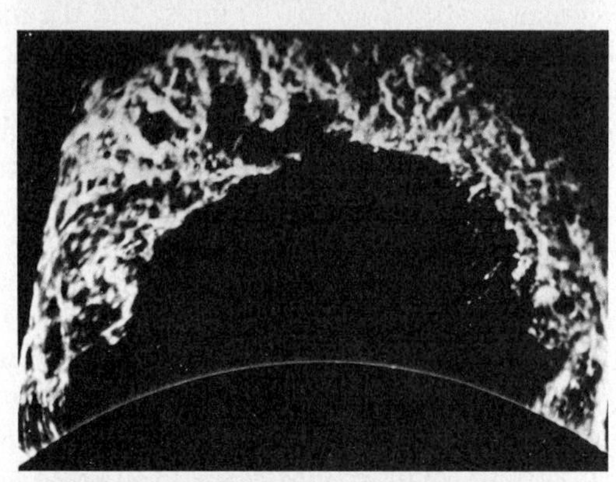

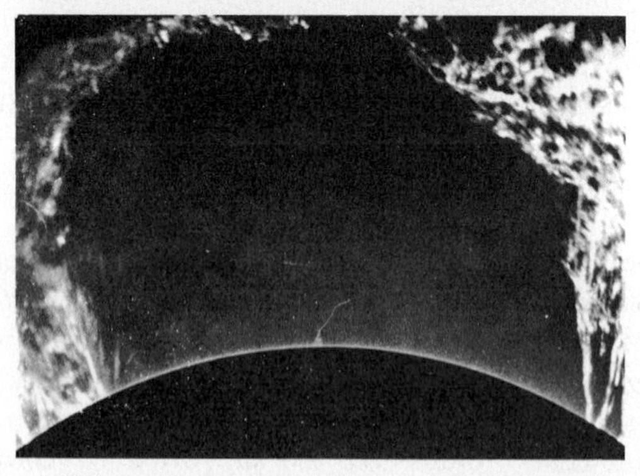

Figure 28.11 Four successive spectroheliograms of the great explosive prominence of June 4, 1946, taken with a coronagraph. The total elapsed time between the first *(top)* and the last *(bottom)* picture was 1 hour. (Harvard Observatory)

same general disturbances that produce spots and plages, that is, local magnetic fields.

(i) Flares

Occasionally, the chromospheric emission lines (the Hα and ionized calcium lines in particular) in a small region of the sun abruptly brighten up to unusually high intensity. Such an occurrence is called a *flare* (Figure 28.12). A flare appears as an intensely bright spot a few thousand or tens of thousands of kilometers in diameter. Very rarely, the continuous spectrum of that part of the solar surface affected also brightens during a flare. These white-light flares are among the most intense observed. A flare usually reaches maximum intensity a few minutes after its onset. It fades out slowly, and after an interval ranging from a few minutes to a few hours, it disappears. Near sunspot maximum, small flares occur several times per day and major ones may occur every few weeks.

During major flares, an enormous amount of energy is released. The visible light emitted is, of course, very small compared to that from the entire sun, but a flare covering only a thousandth of the solar surface can actually outshine the sun in the ultraviolet. X-rays and gamma rays are emitted as well and are observed from space probes and satellites. In addition, matter is thrown out at speeds of 500 to 1000 km/s.

Radio radiation from the sun shows sudden bursts at times of flares. Some of this radiation is nonthermal, showing that high-speed electrons are ejected into coronal magnetic fields, but thermal energy also is emitted from coronal regions near flares. From the intensity of the thermal X-ray radiation we calculate that those regions of the corona have been heated to temperatures near 20 million K. Initially the energy of radio bursts is received only at short wavelengths, but at successively greater intervals of time after the start of the burst, energy is received over longer and longer wavelengths, which shows that the source of the radio energy is rising higher and higher in the corona; the outward velocity of such a source may reach 1500 km/s. About 50 hours later the matter ejected reaches the Earth.

Flares are most frequent in the regions of complex sunspot groups with complicated magnetic field structure. Material thrown into the corona from a flare expands and cools and, as ions and electrons recombine, it may emit light and be seen as a loop prominence over the active region of the sun.

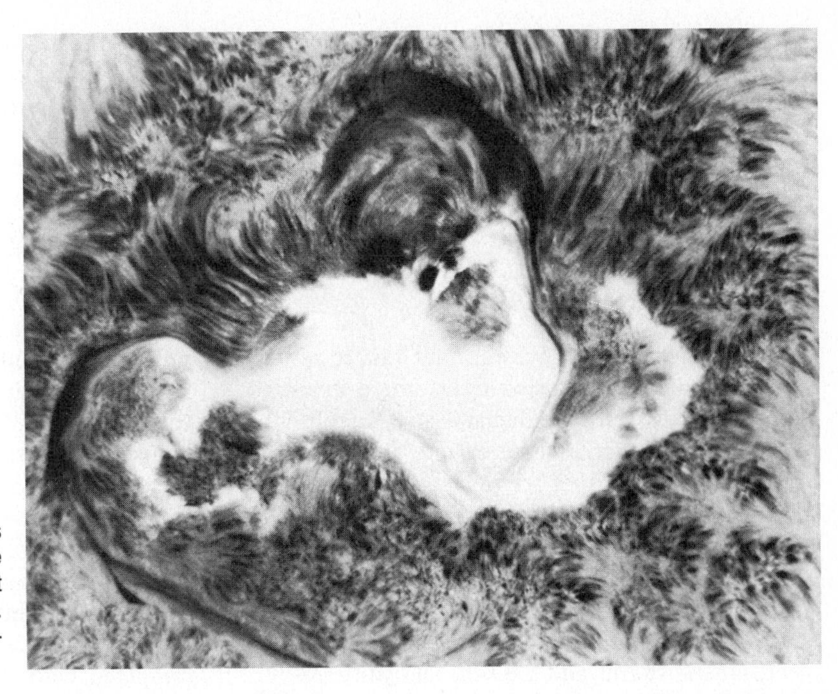

Figure 28.12 A solar flare at its peak intensity, photographed in the light of Hα on August 7, 1972, at Big Bear Solar Observatory. (Mt. Wilson and Las Campañas Observatories)

Loop prominences are the hottest of all prominences and are indicative of rather violent activity in the sun's atmosphere. As is the case with almost all solar phenomena, we do not know the energy source of flares or the mechanism that triggers them, but they are somehow connected with the magnetic fields of sunspot regions.

(j) Theory of the Solar Cycle

There are many solar phenomena that seem to be associated with a particular kind of disturbance in the solar atmosphere. The disturbances are accompanied by strong magnetic fields, sunspots, plages, prominences, and solar flares. All these phenomena are related to the semiregular activity cycle of 22 years. Even the shape of the corona varies with the sunspot cycle. During sunspot maximum it is nearly spherical and distended, while at sunspot minimum it is contracted but extends to relatively greater distances near the plane of the solar equator.

The most successful theory yet advanced to account for the solar cycle is that of H. D. Babcock, E. Parker, and others. They suggest that early in the cycle the inner loops of the lines of force of the general solar field are buried a few hundred kilometers below the photosphere, with the outer loops arching through the corona. As the sun rotates, it carries the buried magnetic field with it. However, as we have seen, the sun does not rotate as a solid body, but faster at the equator. Thus at low latitudes the field lines are carried forward faster than at high latitudes, and after a large number of solar rotations, they are stretched out around the sun many times (Figure 28.13). Adjacent field lines, wound ever closer together, build up strong local fields; the lines may even be twisted together by turbulence in the outer solar layers. The matter in the region of these strong fields expands, lowers in density, and rises, carrying loops of the "cables" of field lines through the surface of the photosphere to form active regions. As the differential rotation of the sun continues to wind up the lines at lower and lower latitudes, the spots occur nearer the equator.

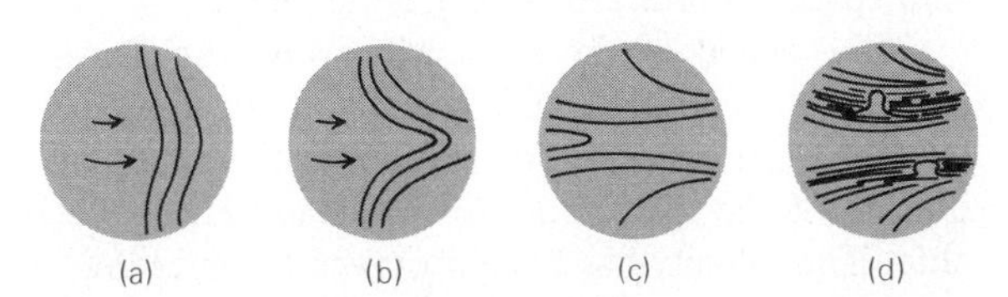

Figure 28.13 Babcock's hypothesis for the solar cycle. (a) The lower loops of the lines of force of the solar magnetic field are buried shallowly beneath the photosphere. (b) and (c) The solar differential rotation stretches the lines out, winding them around the sun. (d) Eventually, the wound-up force lines are buoyed to the surface and break through as active regions. In all figures, north is at the top.

Eventually the field lines coalesce and disperse, dissipating the strong fields, and the half cycle is over. Detailed analysis shows that as the local field neutralizes, the general field of the sun reverses, so that spots in the next half cycle are of opposite polarity from the previous one.

28.3 THE SOLAR WIND

In 1852, one year after Schwabe announced his discovery of the solar sunspot cycle, three investigators, Edward Sabine in England and Rudolf Wolf and Alfred Gautier in Switzerland, independently discovered that sunspot activity was correlated with magnetic storms on the Earth. During geomagnetic storms, the Earth's magnetic field is disturbed, and the compass needle shows fluctuations. Today we know that long-range shortwave radio interference and displays of the aurora are also correlated with geomagnetic storms and the sunspot cycle.

These effects are due to the ultraviolet and X-ray radiation from the sun and also to the *solar wind*. In addition to electromagnetic radiation, the sun emits plasma, mostly protons and electrons. This solar wind is actually an extension of the solar corona in the form of a more or less continuous outflowing of ions and electrons. The effect of the solar wind on comet tails has been known for years (Section 20.2).

In optical photographs, the solar corona appears to be much more uniform and smoother than it appears in X-ray pictures, which indicate that the corona is very patchy. Hot gas is present mainly where magnetic fields have trapped and concentrated it. The regions between these concentrations of gas, where X-ray emission is weak, are called *coronal holes*. The solar wind comes predominantly from these coronal holes, streaming through them into space unhindered by magnetic fields.

The speed of the solar wind near the Earth's orbit averages about 400 km/s, and its density is usually from 2 to 10 ions/cm^3. Both the speed and density of the solar wind, however, are highly variable. The density has been observed to be as low as 0.4 and as high as 80 ions/cm^3, and speeds in excess of 700 km/s have been measured.

During solar flares, high-energy particles are sometimes ejected from the sun into space. Some of these, with energies of 10^9 to 10^{10} eV, are in the low-energy cosmic ray range. Thus the sun (at times of flare activity) is an occasional source of weak cosmic rays. These, however, constitute only a tiny fraction of the total cosmic ray influx to the Earth.

28.4 THE SOLAR INTERIOR: STELLAR STRUCTURE

One circumstance that greatly facilitates the computation of conditions in the interiors of the sun and stars is that stars, in most cases at least, are completely gaseous throughout. Not only are the temperatures too high to permit molecules to exist in stellar interiors, but even the atoms are almost completely ionized. Consequently, the overwhelming majority of the particles of which stars are made are free electrons and atomic nuclei, and most of the latter are simple protons. When we recall (Section 1.1) how extremely tiny these particles are compared with the sizes of neutral atoms, we can see that even in stars where the gases are compressed to enormous densities, there is mostly empty space between the electrons and atomic nuclei. For this reason the idealized gas laws, described below, hold throughout the interiors of most stars to a high degree of accuracy.

(a) The Perfect Gas Law

The particles that constitute a gas are in rapid motion, frequently colliding with each other and with the walls of the container of the gas. This constant bombardment is the *pressure* of the gas. The pressure is greater, the greater the number of particles within a given volume of the gas, for of course the combined impact of the moving particles increases with their number. The pressure is also greater the faster the molecules or atoms are moving; since their rate of motion is determined by the temperature of the gas, the pressure is greater the higher the temperature.

Most students have run across these concepts in high school, in the form of Boyle's law, which states that the pressure of a gas at constant temperature is inversely proportional to the volume to which it is constrained (that is, is proportional to its density), and Charles' law, which states that the pressure (at constant volume) is proportional to the temperature of the gas. These two ideas combine to give us the *perfect gas law* (also called the *equation of state* for a perfect gas). The perfect gas law provides a mathematical relation between the pressure, density, and temperature of a perfect, or ideal, gas (one in which intermolecular or interatomic forces can be ignored) and states that the pressure is proportional to the product of the density and temperature of the gas. The gases in most stars closely approximate an ideal gas; thus, they must obey this law. The exceptions are very massive stars, where radiation pressure (Section 20.2e) can play an important role, and collapsed stars or the collapsed cores of stars, where the matter is *degenerate*. We shall describe degeneracy in Chapter 32.

(b) Hydrostatic Equilibrium

Apart from some very low-amplitude pulsations (Section 28.5), the sun, like the majority of other stars, is *stable,* that is, neither expanding nor contracting. Such a star is said to be in a condition of *equilibrium;* all the forces within it are balanced, so that at each point within the star the temperature, pressure, density, and so on, are maintained at constant values. We shall see (Chapters 30 and 32) that even these stable stars, including the sun, are changing as they evolve, but such evolutionary changes are so gradual that to all intents and purposes the stars are still in a state of equilibrium.

The mutual gravitational attraction between the masses of various regions within a star produces tremendous forces that tend to collapse the star toward its center. Yet, since stars like the sun have remained more or less unchanged for billions of years, the gravitational force that tends to collapse a star must be exactly balanced by a pressure from within. Most of it is the pressure of the gases themselves, although in some very luminous stars the pressure of radiation also contributes appreciably.

If the internal pressure in a star were not great enough to balance the weight of its outer parts, the star would collapse somewhat, contracting and building up the pressure inside. If the pressure were greater than the weight of the overlying layers, the star would expand, thus decreasing the internal pressure. Expansion would stop, and equilibrium would be reached, when the pressure at every internal point again equaled the weight of the stellar layers above that point. An analogy is an inflated balloon, which will expand or contract until an equilibrium is reached between the excess pressure of the air inside over that of the air outside and the tension of the rubber. This condition is called *hydrostatic equilibrium.* Stable stars are all in hydrostatic equilibrium; so are the oceans of the Earth, as well as the Earth's atmosphere. The pressure of the air keeps the air from falling to the ground.

(c) Minimum Pressures and Temperatures in the Solar Interior

For mathematical purposes, we can regard the sun (or a star) as being composed of a large number of concentric spherical shells (like the layers in an onion). The sun is not actually stratified, of course; we speak of these shells in the same sense that we speak of levels in the ocean. Now suppose we knew how matter is distributed within the sun—that is, what fraction of its mass is included within each shell. Since the weight of a shell is the gravitational attraction between it and all the underlying layers, we could then calculate the weight of each shell. From the condition of hydrostatic equilibrium, we could next calculate how the pressure must increase downward through each shell to support its weight. At the surface of the sun, where there are no overlying layers of stellar matter, the pressure is zero. By simply adding up the increases of pressure through the successive layers inward, we would be able to find the pressure at each point within the sun, in particular at its center. Using the pressures and densities thus determined at all points along the radius of the sun, we could then find the corresponding temperatures from the perfect gas law. In other words, if we only knew how the material within the sun is distributed, we would be able to calculate the density, pressure, and temperature at all its internal points.

It is not known in advance, of course, how the matter in a particular star is distributed. On the other hand, some ways that it is *not* distributed can be specified. Internal gravity, for example, must force the gases composing the sun into higher and higher compression at deeper and deeper levels in its interior. The material is expected to show high central concentration; the densities of outer layers would certainly not exceed those of inner layers. To assume that the matter in a star is distributed with uniform density, therefore, would certainly be to underestimate its central compression, and the values calculated for its internal pressures and temperatures would certainly be lower than the true values. Here, then, is a method by which *lower limits* can be found for the pressures and temperatures of the solar interior.

Thus, with only the assumption of hydrostatic equilibrium and a knowledge of the perfect gas law, it is possible to learn something of the conditions inside the sun. We find that the mean pressure is at least 500 million times the sea-level pressure of the Earth's atmosphere, that the central pressure is at least 1.3×10^9 times that of the Earth's atmosphere, and that the mean temperature is at least 2.3 million Kelvins. Since these pressures and temperatures would exist if the sun were uniform in density, the actual values must be much higher. Under such conditions all elements are in the gaseous form, and the atoms cannot be combined into molecules. Moreover, most of the atoms are almost completely ionized—that is, stripped of their electrons (Section 8.3g). These electrons, freed from their parent atoms, become part of the gas itself, moving about as individual particles.

(d) Thermal Equilibrium

From observation we know that electromagnetic energy flows from the surfaces of the sun and stars. According to the second law of thermodynamics, heat always tries to flow from hotter to cooler regions. Therefore, as energy filters outward toward the surface of a star, it must be flowing from inner hotter regions. The temperature cannot ordinarily decrease inward in a star, or energy would flow in and heat up those regions until they were at least as hot as the outer ones. We conclude that the highest temperature occurs at the center of a star and that temperatures drop to successively lower values toward the stellar surface.* The outward flow of energy through a star, however, robs it of its internal heat and would result in a cooling of the interior gases were that energy not replaced. There must therefore be a source of energy within each star.

If a star is in a steady state (that is, in hydrostatic equilibrium and shining with a steady luminosity), the temperature and pressure at each point within it must remain approximately constant. If the temperature were to change suddenly at some point, the pressure would similarly change, causing the star to contract suddenly, or to expand, or otherwise to deviate from hydrostatic equilibrium. Energy must be supplied, therefore, to each layer in the star at just the right rate to balance the loss of heat in that layer as it passes energy outward toward the surface. Moreover, the rate at which energy is supplied to the star as a whole must, at least on the average, exactly balance the rate at which the whole star loses energy by radiating it into space; that is, the rate of energy production in a star is equal to the luminosity. We call this balance of heat gain and heat loss for the star as a whole and at each point within it the condition of *thermal equilibrium*.

(e) Heat Transfer in a Star

There are three ways in which heat can be transported: by *conduction*, by *convection*, and by *radiation*. Conduction and convection are both important in planetary interiors, while radiation is unimportant because of the low temperatures and great opacity of planetary materials. In stars, which are so much more transparent, radiation and convection are important, while conduction can be ignored unless the gas is degenerate (Chapter 32) or is very hot, as in solar flares, the solar corona, and the interstellar medium.

Stellar convection occurs as currents of gas flow in and out through the star. While these convection currents travel at moderate speeds and do not upset the condition of hydrostatic equilibrium or result in a net transfer of *mass* either inward or outward, they nevertheless carry *heat* outward through a star very efficiently. However, convection currents cannot be maintained unless the temperatures of successively deeper layers in a star increase rapidly in relation to the rate at which the pressures increase inward. In a similar way, convection in planetary atmospheres is only important in the troposphere, which also has a temperature that decreases rapidly with height. Convection does occur, nevertheless, in certain parts of many stars, and convection currents may travel completely through some of the least luminous stars.

Unless convection occurs, the only significant mode of energy transport through a star is by electromagnetic radiation, which gradually filters out-

* The high temperature of the sun's chromosphere and corona may therefore appear to be a paradox. The actual heat energy in the sun's corona is relatively small because the corona is a highly rarefied gas. Its high temperature is believed to be maintained by shock waves or by some other process that would not exist for a gas in radiative equilibrium.

ward as it is passed from atom to atom. However, radiative transfer is not an efficient means of energy transport, because under the conditions that prevail in stellar interiors gases are very opaque—that is, a photon does not go far before it is absorbed by an atom (typically, in the sun, about 1 cm). The energy absorbed by atoms is always re-emitted, to be sure, but most of it is re-emitted in random directions. A photon that is traveling outward in a star when it is absorbed has almost as good a chance of being reradiated back toward the center of the star as toward its surface. A particular quantity of energy being passed from atom to atom, therefore, zigzags around in an almost random manner and takes a long time to work its way from the center of the star to the surface; in the sun, the time required is of the order of a few million years.

The measure of the ability of matter to absorb radiation is called its *opacity*. It should be no surprise that the gases in the sun are opaque. If they were completely transparent, we would be able to see all the way through the sun. We have discussed earlier (Section 8.3) the processes by which atoms and ions can interrupt the flow of energy—such as by becoming ionized and by bremmstrahlung (free-free transitions). In addition, individual electrons can scatter radiation helter-skelter. For a given temperature, density, and composition of a gas, all of these processes can be taken into account, and the opacity can be calculated. The computations are very complicated and thus require large computers.

Once the opacity is known, we can find how each layer or shell of the sun or a star impedes the outward flow of radiation. Of course there is such a net outward flow of the energy generated by thermonuclear reactions in the interior, or the star would have no luminosity. Thus from the opacity we calculate how the temperature must increase inward through the shell to force the observed radiation out, and thereby learn the temperature distribution throughout the interior.

If the temperature difference across some regions of a star should be high enough to support convection, convection currents, rather than radiation, carry most of the energy. Within those regions the variation of temperature with depth is determined by the expansion of outward-moving masses of gas and the contraction of inward-moving ones. Here again, knowledge of the energy transport mechanism within a star makes possible calculation of the temperature distribution.

In this section we have described some of the ways we probe the interiors of the sun and stars. The story is not yet complete, however, for the various physical variables (such as temperature, density, pressure, and composition) are interdependent, and so far we have not specified enough conditions to separate and solve for all unknowns. To do so, we need to know the source of energy of the sun and stars. In the next chapter we will explore what makes the sun and the stars shine.

28.5 THE SOLAR INTERIOR: OBSERVATIONS

For a long time, the only information available to us about the internal structure of the sun was derived from numerical calculations of the consequences of the physical processes described in the preceding section. Unfortunately, many important characteristics of the sun that are required to carry out these calculations are not known. Is the composition of the interior of the sun the same as the composition we see at its surface? Does the interior of the sun rotate at the same rate as its surface? How efficient is convection as a way to transport energy?

Figure 28.14 shows schematically what we think, based on computations, the interior of the sun looks like. We can, of course, observe none of this structure directly. All of the radiation that we receive from the sun originates in the outer layers of its atmosphere, that is, from the photosphere, chromosphere, and corona.

Recently, astronomers have come to realize that there are two kinds of observations that we can make that will tell us directly something about the interior of the sun. The first technique relies on the measurement of neutrinos emitted by the sun. This experiment will be described in more detail in Chapter 29. Neutrinos are massless, or nearly massless, particles. They rarely interact with other matter, and the neutrinos that are emitted from the center of the sun travel unimpeded to the sun's surface and on toward the Earth at the speed of light. In contrast, radiation—heat and light—generated at the center of the sun travels an erratic path, constantly being absorbed and re-emitted by the atoms in its way. So tortuous is this path, that it takes a few million years for energy generated in the center of the sun to reach its surface. The photons that we see are only those that are emitted so close to the

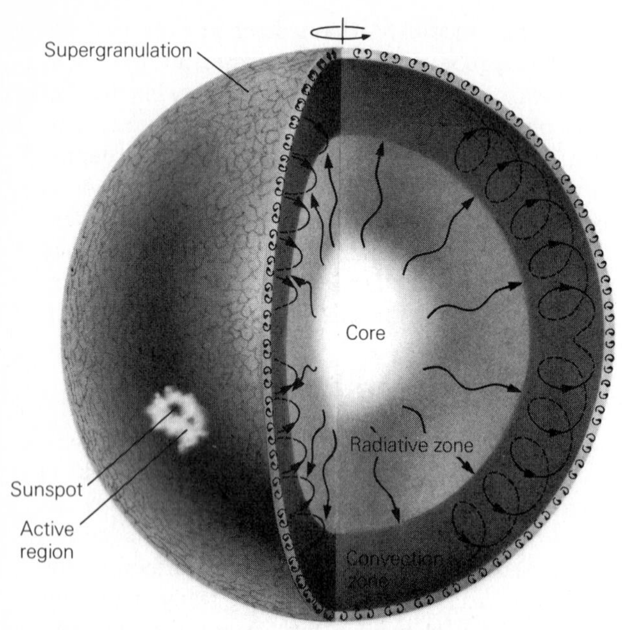

Figure 28.14 The interior structure of the sun. Energy is generated in the core by the fusion of hydrogen to form helium. This energy is transmitted outwards by radiative processes, i.e., by the absorption and re-emission of photons. In the outermost layers, energy is transported mainly by convection.

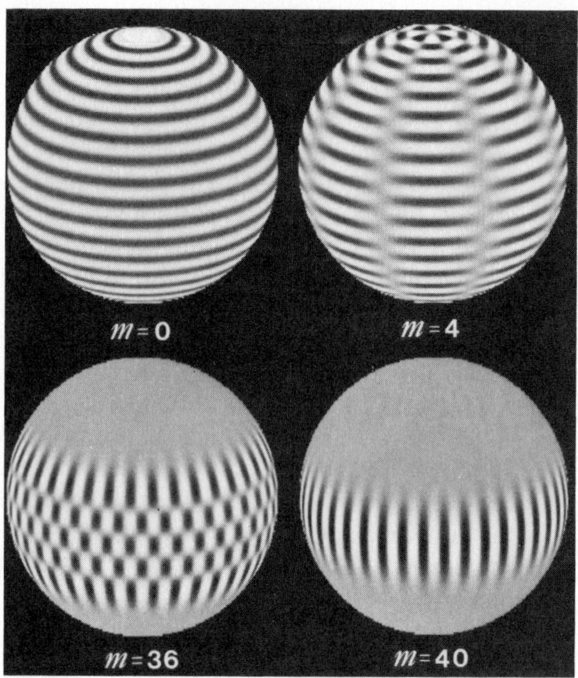

Figure 28.15 Sound waves resonating in the interior of the sun cause the sun's surface to oscillate in complex patterns. These images show some of the possible patterns. (The letters below each image are mathematical shorthand for describing the specific pattern of motion.) Dark regions are approaching the observer; light regions are receding. Gray regions have zero radial velocity. All of the patterns of oscillation shown, plus several million others, are present at the same time. The velocities measured at the solar surface are produced by the superposition or summation of all of these sound waves. (National Optical Astronomy Observatories)

surface of the sun that they can escape without being absorbed again.

Based on models of what the interior of the sun is like, we can calculate how many neutrinos should reach the Earth. Computations predict that we should detect approximately three times more neutrinos than we actually do. There is a major disagreement between theory and observation. What then is wrong with our model of the solar interior? And is there any other way that we can measure what the sun is like beneath its photosphere?

Astronomers have discovered very recently that the sun is a pulsating variable star. Measurements of radial velocity show oscillations in the sun's surface. The changes in velocity of the sun measured as a whole are very small—in some cases less than 10 cm/s. The time required to complete a cycle from maximum to minimum velocity and back again is only a few minutes, and so the total change in the apparent radius of the sun is only about 5 meters. Since the total radius of the sun is about 6.5×10^8 m, the percentage change in the size of the sun is very small indeed.

In addition to this pulsation, in which a large fraction of the visible solar surface moves as a whole, alternately in and out, measurements also show that any small portion of the surface also moves first toward and then away from the center of the sun (Figure 28.15). Regions on the solar surface with diameters of 4000 to 15,000 kilometers oscillate with a period of about 5 minutes.

The discovery of the 5-minute oscillation was soon followed by the realization that it could be used to determine empirically what the interior of the sun is like. The motion of the sun's surface is caused by waves generated deep within it, and thus study of the amplitude and period of the velocity changes produced by this motion can yield information about the temperature, density, and composition beneath the sun's surface. The situation is somewhat analogous to the use of earthquakes to infer the properties of the interior of the Earth, and correspondingly, studies of solar oscillations are referred to as *solar seismology*. It takes about an hour for these waves to traverse the sun, and so they, like neutrinos, provide information about what the solar interior is like at the present time.

The observations of velocity amplitudes and periods do not agree with the predictions of standard models of the interior structure of the sun. First of all, measurements indicate that the convection zone at the surface of the sun extends about 30 percent of the way toward the center. Calculations had suggested that the convection zone was shallower and extended only 15 to 20 percent of the distance from the surface of sun to its center. Measurements of solar oscillations also suggest, but do not yet prove, that the regions just below the surface of the sun rotate more slowly than the surface itself, while the core of the sun may rotate much more rapidly than the surface.

Even with the above changes, the computed structure of the sun still does not explain the pattern of changes in the velocity of the sun's surface. In order to resolve this discrepancy between theory and observations, theorists have explored a variety of non-standard solar models. Some models assume that the ratio of the abundance of helium to hydrogen in the solar interior is different from that of the solar surface. Other models assume that the interior of the sun rotates at a very different rate than do the surface layers. Neither of these modifications of standard solar models can account for the observed solar oscillations.

Calculations show that theory and observation can be reconciled, however, if the interior of the sun is cooler than standard models predict. The same change could also explain why the number of neutrinos detected at the Earth is as small as it is measured to be (but for an alternative explanation for the low number of detected neutrinos, see Section 29.3d). Two otherwise puzzling observations may therefore be explained by a single hypothesis—namely, that the interior of the sun is, perhaps only momentarily, cooler than we thought.

Is this conclusion correct? Much additional work, both observational and theoretical, is required to test this hypothesis, and it remains very controversial. Astronomers are now planning to place at several sites around the Earth instruments designed to measure the radial velocity of the sun and monitor its oscillations continuously. If the sun is indeed temporarily cooler than models predict, the implications may be profound. As we have seen, neutrinos travel from the center of the sun to its surface and then on to the Earth at the speed of light. Neutrinos, therefore, provide information on conditions in the sun at the present time. The time for radiation to reach the surface of the sun and to be seen by us as light and felt as heat is on the order of a few million years. The present-day luminosity of the sun, therefore, depends on conditions in the sun a few million years ago. If the sun is really cooler deep within than we would predict based on its current luminosity, is it generating less energy now than it was a few million years ago? Will it appear dimmer and provide less heat to the Earth a few million years from now? These questions are obviously of great importance, but will remain unanswered until more extensive studies of solar oscillations are completed.

28.6 IS THE SUN A VARIABLE STAR?

If direct observations of the sun cannot yet tell us whether or not the energy output of the sun changes with time, then perhaps we can examine historical records to determine whether or not the sun is variable. We already know that solar variations, if any do exist, will be subtle. The existence of life on Earth requires that there have been no major recent changes in the climate of the Earth, and so the energy output of the sun cannot vary by large amounts. There is, however, growing evidence that some characteristics of the sun do change.

(a) Variations in Solar Activity

There is considerable evidence that solar activity, as estimated from the number of sunspots, was much lower from 1645 to 1715 than it is now. This interval of quiescence in solar activity, brought to the attention of contemporary astronomers by John Eddy, of the High-Altitude Observatory in Boulder, Colorado, was first noted by Gustav Spörer in 1887 and by E. W. Maunder in 1890, and is now called the *Maunder minimum*. The incidence of sun-

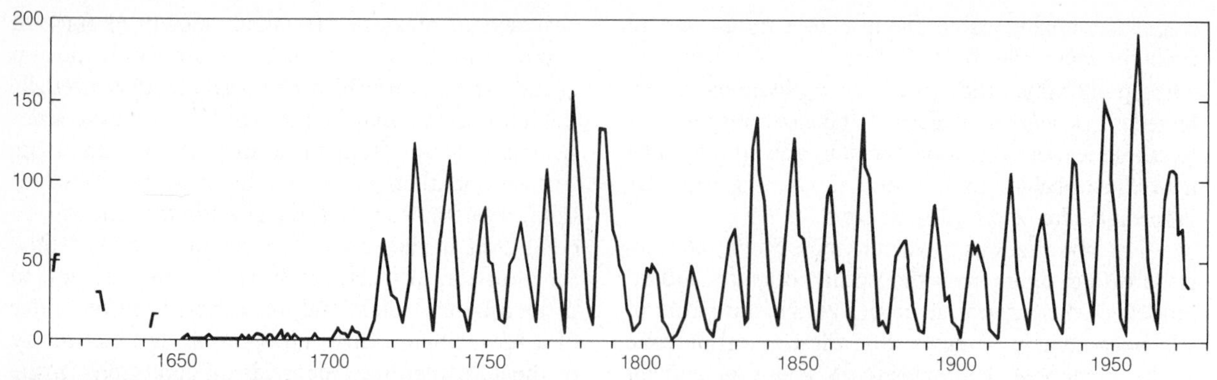

Figure 28.16 The relative numbers of sunspots, as a function of time. Note the absence of sunspots from 1645 to 1715. (Courtesy John Eddy, National High-Altitude Observatory)

spots over the past four centuries is shown in Figure 28.16.

Sunspots did occur during the time of the Maunder minimum. Chinese records, for example, indicate that there were 21 sunspot groupings that were large enough to be seen with the naked eye, and still more sunspots were recorded in Europe, where telescopes were in regular use. It seems well established, however, that the number of sunspots was markedly lower than before 1645 or in modern times. Sunspot numbers were also somewhat lower than they are now during the first part of the 19th century, and this period of time is called the little Maunder minimum.

One might question the validity of sunspot numbers as estimated by a variety of observers using a variety of techniques and sometimes relying on visual observations unaided by telescope. There may even have been times in history when reports of sunspots were suppressed for political reasons. In China, for example, where belief in astrology was strong, sunspots were often thought to signal dire consequences for the ruling dynasty. Fortunately, there is independent corroboration that solar activity was unusually low during the Maunder minimum.

The appearance of the solar corona varies markedly, depending on the level of solar activity. When the sun's activity is low, the corona is extended near the solar equator, with streamers reaching far out into space. Near the poles, we see only weak plumes that do not extend very far from the solar surface. At times of high activity, the corona is much more nearly symmetric and is filled with long, extended streamers. Eclipses of the sun have been extensively described in literature and in historical records. During the time of the Maunder minimum there are no reports of coronal streamers or of other structure in the corona, and this absence of remarks is consistent with the hypothesis that the level of solar activity was low.

There is a direct connection between solar activity and the occurrence of aurorae. Northern (and southern) lights are caused by the impact of charged particles from the sun on the Earth's magnetosphere. There is a strong correlation between sunspot number and the frequency of auroral displays. Historical accounts indicate that auroral activity was low during the Maunder minimum.

The best quantitative evidence of variations in the level of solar activity comes from studies of the radioactive isotope carbon-14 (^{14}C). Cosmic rays reach the Earth from a variety of sources beyond the solar system. When these energetic particles impact the upper atmosphere, they produce several different radioactive isotopes. Carbon-14 is produced when nitrogen is struck by high-energy cosmic rays. The rate at which cosmic rays from galactic sources reach the upper atmosphere depends on the level of solar activity. When activity levels are high, the incoming cosmic rays are prevented from reaching the Earth by irregularities in the sun's magnetic field. At times of low activity, when the sun's magnetic field is weak, cosmic rays reach the Earth in larger numbers.

The production of carbon-14 is therefore higher when the activity of the sun is lower. Some of this radioactive carbon is contained in carbon dioxide molecules (CO_2), and is ultimately incorporated into trees through photosynthesis. By measuring the amount of radioactive carbon in tree rings, we can estimate the historical levels of solar activity. Correlations with visual estimates of sunspot numbers over the past 300 years for which reliable data are available indicate that the carbon-14 estimates of solar activity are indeed valid. Because it takes about 10 years on the average for a carbon dioxide molecule to be absorbed from the atmosphere or ocean into plants, this technique cannot provide data on the 11-year solar cycle, but it can be used to look for long-term (over several decades) changes in the level of solar activity.

Estimates of the amount of carbon-14 in tree rings now extend continuously back about 8000 years into the past. Variations in solar activity levels have occurred throughout this period of time, and the sun has been at times both more and less active than it is now. The measurements confirm that the amount of carbon-14 was unusually high, and solar activity correspondingly low, during both the Maunder minimum and during the little Maunder minimum between the years 1800 and 1830. During the past thousand years, activity was also low in the years A.D. 1410–1530 and A.D. 1280–1340. Between about A.D. 1100 and A.D. 1250, the level of solar activity may have been even higher than it is now.

(b) Solar Variability and the Earth's Climate

Variations in the overall level of the sun's activity seem to be well established. Do these variations have any direct impact on the Earth or on its climate? It has long been known that the period of the Maunder minimum was a time of unusually low temperatures in Europe—so low, in fact, that this period is described as a Little Ice Age. One example does not, however, establish a causal connection, and recent measurements cast doubt on a direct link between climate and solar activity.

The most obvious way in which the sun and the Earth's climate might be linked is through variations in the output of energy by the sun. If the sun puts out less energy, then logically one might expect the Earth to become colder. At times of high levels of solar activity, a larger fraction of the sun is covered by sunspots, which are cooler than the surrounding solar surface. At the same time, plages and faculae, bright hot regions, also become more prominent. Which effect is more important? Does the luminosity of the sun decrease when it is active because sunspots block some of the radiation? Or increase, because of the radiation from plages and faculae? Or remain constant because the radiation blocked by the sunspots is balanced by the excess radiation from the plages and faculae?

The answer to this question was obtained only after measurements were made from Earth orbit. The changes in solar luminosity with activity level are too small to be measured reliably from the ground because of uncertainties in estimating how much of the sun's energy is transmitted by the Earth's atmosphere.

Precise measurements show that the luminosity of the sun varies on time scales of weeks to months by 0.1 to, in extreme cases, 0.5 percent; that the variation in luminosity can be accurately estimated simply by knowing what fraction of the surface is covered by sunspots; and that the sun is fainter when more sunspots are present. In other words, the sun does change in brightness, albeit by a very small amount, and these changes are almost entirely caused by the blocking of the sun's radiation by sunspots.

An examination of temperatures measured at various places around the Earth suggests that average temperatures in inland regions, isolated from the moderating effect of ocean waters, may vary on an 11-year cycle by a few tenths of a degree Celsius, becoming cooler in response to the lower luminosity of the sun that results from blocking of radiation by large numbers of sunspots.

These results are precisely opposite to what is required to explain the Little Ice Age at the time of the Maunder minimum. The unusually cold temperatures at this time imply a drop in the solar luminosity of about 1 percent. Yet, as we have seen, if there are few or no sunspots, the energy radiated by the sun should be higher than usual. It may be that there are other effects besides blocking by sunspots that cause variations in the solar luminosity over long time scales. After all, precise measurements of the total energy output of the sun extend only over a period of four years. Alternatively, the Little Ice Age may have been the consequence of forces unrelated to the sun. We are only beginning to understand how the sun varies. Until we know a great deal more about these variations than we do now, we will be unable to estimate how, if at all, they may affect the climate of the Earth.

EXERCISES

1. Describe the principal spectral features of the sun.
2. (a) What is the distance to the sun in parsecs? (b) What is the distance modulus of the sun? (c) How many times farther away would the sun be if it were removed to a distance of 10 pc? (d) How many times fainter would the sun appear at 10 pc?
3. How might you convince an ignorant friend that the sun is not hollow?
4. Give at least three good arguments that refute the view proposed by Herschel that the sun has a cool interior that is inhabited.
5. Suppose an eruptive prominence rises at 150 km/s. If it did not change speed, how far from the photosphere would it extend in three hours?
6. Would the material in the prominence in Exercise 5 escape the sun? Why? (See Section 4.5).
7. Suppose you were to take two photographs of the sun, one in light at a wavelength centered on a strong absorption line, and the other at a wavelength region in the continuum away from strong lines. In which photograph would you be observing deeper, hotter layers? Why?
8. From the Doppler shifts of the spectral lines in the light coming from the east and west limbs of the sun, it is found that the radial velocities of the two limbs differ by about 4 km/s. Find the approximate period of rotation of the sun.
9. If the rotation period of the sun is determined by observing the apparent motions of sunspots, must any correction be made for the orbital motion of the Earth? If so, explain what the correction is and how it arises. If not, explain why the Earth's orbital revolution does not affect the observations.
10. If the corona, which is outside the photosphere, has a temperature of 1,000,000 K, why do we measure a temperature of 6000 K for the surface of the sun?
11. Show with a diagram why the duration of totality is extremely sensitive to the distance of the observer from the edge of the eclipse path.
12. From the data in Section 28.3, find how long it takes solar wind particles, on the average, to reach the Earth from the sun.
13. Give some everyday examples of hydrostatic equilibrium. It is known that the pressure in a container of water increases with depth in the container. Is this a consequence of hydrostatic equilibrium? Explain. Compare the pressure-depth relation in water with that in the Earth's atmosphere. Why is the case much simpler for water?
14. If the atmospheric pressure were the same on two different days, but if one day were much hotter than the other, what could you say about the relative density of the air on the two days?
15. If, in a vacuum chamber, the pressure is only one millionth of sea-level pressure, how does the density of the gas in the chamber compare with the average density of air at sea level?
16. Give everyday examples of convection and radiation of heat through a gas.

Sir James Chadwick (1891–1974), British physicist. In addition to his knighthood in 1945, he received the Nobel Prize in physics for his basic contributions to our understanding of the atomic nucleus. He proved the existence of the neutron in 1932 by bombarding beryllium nuclei with alpha particles (helium nuclei) and also worked on the generation of chain reactions and nuclear fission. (American Institute of Physics)

STELLAR ENERGY: THE ATOMIC NUCLEUS

The rate at which the sun emits electromagnetic radiation into space, and thus the rate at which energy must be generated within it, is about 4×10^{33} ergs/s (Section 23.2f). Moreover, the power output of the sun has been about the same throughout recorded history and, according to geological evidence, not very different since the formation of the Earth thousands of millions of years ago. Our problem is now to find what sources can provide the gigantic amounts of energy required to keep the stars like the sun shining for so long.

29.1 THERMAL AND GRAVITATIONAL ENERGY

Two large stores of energy in a star are its internal heat, or *thermal energy,* and its *gravitational energy.* The heat stored in a gas is simply the energy of motion (kinetic energy) of the particles that compose it. If the speeds of these particles decrease, the loss in kinetic energy is radiated away as heat and light. This is how a hot iron cools after it is withdrawn from a fire (except that the atoms in a solid vibrate within a crystalline structure, rather than moving freely, as in a gas).

Because a star is bound together by gravity, it has gravitational potential energy. If the various parts of a star fall closer together, that is, if the star contracts, it converts part of its potential energy into heat, some of which can be radiated away. About the middle of the 19th century, the physicists Helmholtz and Kelvin postulated that the source of the sun's luminosity was indeed the conversion of part of its gravitational potential energy into radiant energy.

The sun cannot be infinitely old, of course, for no source of energy can last forever. Sometime in the past the sun must have formed as the solar nebula gradually fell or gravitated together, giving up its potential energy. It can be shown by thermodynamics that about half the potential energy released by a contracting star goes into radiation (or luminosity) and the other half goes into heating up its interior. Thus the internal heat or thermal energy of a star is numerically equal to about half the potential energy it has given up in its contraction.

Helmholtz and Kelvin showed that because of its enormous mass, the sun need contract only extremely slowly to release enough gravitational potential energy to account for its present luminosity. In fact, over the time span of recorded history, the decrease in the sun's size resulting from its contraction would be so negligible as to escape detection. It seemed to these researchers, therefore, that the sun's gravitational and thermal energies were sufficient to keep it shining for an extremely long time, and were certainly the source of its power.

The amount of potential energy that has been released since the presolar cloud began to contract is of the order of 10^{49} ergs. This is the amount, according to the Helmholtz and Kelvin theory, that the sun could have converted to thermal energy and luminosity. Since the present luminosity of the sun is 4×10^{33} ergs/s, or about 10^{41} ergs/yr, its contraction can have kept it shining at its present rate for a period of the order of 100 million years. It is only within the present century that it has been learned that the Earth, and hence the sun, has an age of at least several billion years, and therefore that the sun's gravitational energy is grossly inadequate to account for the luminosity it has generated over its lifetime.

Einstein's special theory of relativity shows that there is an equivalence between mass and energy and that one can be converted to the other. About 1928 it was suggested that the conversion of matter to energy might account for the sun's luminosity. Such conversion is possible through nuclear reactions. These reactions involve the fundamental particles of physics, the study of which is one of the most exciting frontiers in modern physics. Before proceeding, therefore, let us digress briefly for a quick review of that field.

29.2 THE ELEMENTARY PARTICLES AND FORCES

Our present physical interpretation of nature depends on an understanding of the interactions among the basic entities of mass and energy—the fundamental particles. The interactions are manifestations of the forces of nature, of which four are known: strong, weak, electromagnetic, and gravitational. The particles are the things on which the forces act. Exchange of certain fundamental particles can also transmit the forces of nature.

(a) The Forces of Nature

The known forces of nature were alluded to in Chapter 1. Their properties are summarized in Table 29.1. Gravitation and electromagnetism are both long-range forces whose strengths drop with the square of the distance. Gravitation, however, acts on all particles, while the electromagnetic force acts only between charged particles. Because bulk matter is usually electrically neutral, the most important manifestation of the electromagnetic force is in binding together the charged parts of atoms, or in binding atoms together in molecules or in solids, and the interacting particles are separated only by atomic dimensions (10^{-8} cm). (We will see, however, that magnetism can be important over at least the dimensions of the Galaxy.)

The strong force (or strong nuclear force) binds together the heavy particles in the atomic nucleus and is effective only over nuclear dimensions (10^{-13} cm). We shall see that most (all?) heavy particles are unstable, in that they decay (change) into other particles spontaneously. Often these decays occur over extremely short times (typically, 10^{-23} s), and these also involve the strong force. Other decays, however, are much slower (typically, 10^{-10} s, but sometimes very much slower), and involve the weak (or weak nuclear) force. These weak interactions also always accompany the formation of a neutrino. The weak force, like the strong, is effective only over nuclear dimensions.

It is now believed that each of the four forces is mediated through the exchange of certain particles. The exchange particle for the electromagnetic force is the familiar photon, and that for the gravitational force is the hypothetical *graviton* (not yet observed). Both photons and gravitons are massless. The exchange particle for the strong force is the neutral π meson, with a mass of 264 times that of the electron. The exchange particle for the weak force has not yet been observed but is believed to be very heavy—in excess of 150,000 times the mass of the electron.

(b) Elementary Particles

The most familiar of the elementary particles are the proton, neutron, and electron, which, as we have seen, are the constituent particles of ordinary atoms. But each of these can exist as a separate entity, as in cosmic rays, in the solar wind, in laboratory particle accelerators, or, for that matter, in the ionized gases that make up the bulk of stars. We have learned in the 20th century, however, that these are by no means *all* the particles that exist.

First, for each kind of particle, there is a corresponding *antiparticle*. If the particle carries a charge, its anti has the opposite charge. The anti-

TABLE 29.1 The Forces of Nature

FORCE	RELATIVE STRENGTH	RANGE	INTERACTING PARTICLES	EXCHANGE PARTICLE	IMPORTANT ASTRONOMICAL APPLICATIONS
Gravitation	10^{-38}	Universe	All	Graviton	Planets, stars, galaxies and the universe
Electromagnetic	1	Usually effective over 10^{-8} cm	All charged	Photon	Atoms, molecules, and solids
Weak	10^{-12}	10^{-13} cm	Leptons	Intermediate vector bosons (?)	Big bang; stellar energy sources
Strong	10^{2}	10^{-13} cm	Hadrons	π mesons	Big bang; stellar energy sources

electron is the *positron,* of the same mass as the electron but positively charged. The antiproton has a negative charge. The antineutron, like the neutron, has no charge, but interacts with other matter opposite to the way the neutron does. Some particles—among them the photon and the graviton—are their *own* antiparticles; for them the particles and antiparticles are identical. Whole atoms could exist of positrons, antiprotons, and antineutrons. They constitute what is called *antimatter.* Such atoms do not exist around here because when a particle comes in contact with its antiparticle the two annihilate, turning into energy. Antimatter in our world of ordinary matter, therefore, is highly unstable (in large doses, it would be mighty dangerous!), but individual antiparticles are found in cosmic rays and can be formed in the laboratory.

In addition, hundreds of other kinds of particles are formed in the nuclear reactions produced in the laboratory when particles are smashed into each other at very high speeds. The host of particles now known are classed into two groups: the *leptons* (light-weights) and *hadrons* (heavy-weights). Electrons are examples of leptons. Hadrons are further subdivided into *baryons,* which include the proton and neutron, and *mesons,* which are intermediate in mass and include a particle known as the π *meson,* or *pion,* whose mass is about 270 times that of the electron. Pions can be positive, negative, or neutral; their antis have the opposite sign, except the neutral pion, which, like the photon, is its own antiparticle. Baryons make up the nuclei of atoms.

Most hadrons are extremely unstable and decay into other particles in very short times. The neutron is stable in the atomic nucleus, but when isolated it decays with a half-life of about 11 minutes into a proton, an electron, and an antineutrino. Most hadrons decay very much more rapidly than the neutron.

In contrast to the hundreds of known hadrons, only six leptons (and their antis) are known. They consist of three kinds of electrons—the familiar one, the heavier muon, and the recently discovered, much heavier tau—and associated with each kind of electron is a neutrino. The muon and tau are unstable and decay eventually to electrons and neutrinos.

The existence of neutrinos was originally postulated in 1933 by physicist Wolfgang Pauli to account for small amounts of energy that appeared to be missing in certain nuclear reactions. They were presumed to be massless and to move with the speed of light. They interact very weakly with other matter and so are very difficult to detect; most of them pass completely through a star or a planet without being absorbed. Yet, the neutrinos associated with the electron and the muon have now been detected. The third kind of neutrino, that associated with the tau electron, has not yet been observed but is presumed to exist. Each of the three neutrinos has a corresponding antineutrino.

Recent experiments and new theoretical calculations suggest that neutrinos may have a tiny mass. This conclusion is, however, far from certain. The measured and calculated masses are not even in agreement. If neutrinos do turn out to have mass, it could have interesting consequences for cosmology (Chapters 36 and 37) and for models of the interior of the sun (Section 29.3d).

The properties of a few of the principal elementary particles are summarized in Table 29.2.

(c) Quarks

The existence of hundreds of kinds of hadrons suggests that they may not all be fundamental particles. A

TABLE 29.2 Properties of Some of the Elementary Particles

FAMILY	PARTICLE	SYMBOL	RELATIVE MASS	CHARGE	ANTIPARTICLE SYMBOL		ANTIPARTICLE CHARGE
	Graviton	—	0	0	Same		0
	Photon	γ	0	0	Same		0
Leptons	Electron	e	1	-1	e^+	(positron)	$+1$
	Electron neutrino	ν_e	0(?)	0	$\bar{\nu}_e$		0
	Muon	μ	207	-1	μ^+		$+1$
	Muon neutrino	ν_μ	0(?)	0	$\bar{\nu}_\mu$		0
	Tau	τ	3500	-1	τ^+		$+1$
	Tau neutrino	ν_τ	0(?)	0	$\bar{\nu}_\tau$		0
Hadrons Baryons	Proton	p	1836	$+1$	$\bar{p}$		-1
	Neutron	n	1839	0	$\bar{n}$		0
	+ many others						
Mesons	Pion	π^+	273	$+1$	π^-		-1
		π^0	264	0	Same		0
		π^-	273	-1	π^+		$+1$
	+ others						

substantial simplification and unification was provided by an idea suggested independently in 1963 by Murray Gell-Mann and George Zweig, both at the California Institute of Technology. These physicists postulated that all hadrons are composed of various combinations of a small number of more fundamental particles called *quarks*. (The name comes from James Joyce's *Finnegan's Wake:* "Three quarks for Muster Mark . . . .")

Originally three kinds of quarks were postulated: "up," "down," and "strange," each with its corresponding antiquark. Baryons are made of three quarks each, and antibaryons of three antiquarks. A meson is made of one quark and its own antiquark in combination. Since the introduction of the theory, evidence has been found for two additional quarks: "charm" and "bottom." It is generally expected that a sixth quark, "top," will eventually be found to bring the total number equal to the number of leptons.

The different kinds of quarks are referred to as different "flavors." In addition, each quark is presumed to come in three different "colors"—three states otherwise indistinguishable from each other. (The terms "flavor" and "color" as used here, have, of course, nothing to do with the common usage of those words.) The quarks making up each baryon must represent each of the three colors. Unlike other elementary particles, quarks have fractional charge: up and charm quarks have a charge of $+2/3$, down and strange of $-1/3$, and bottom and top (if it exists) of $+1/3$. Quarks of the many various combinations of possible colors and flavors then can account for the hundreds of known hadrons.

Isolated quarks have never been detected in the laboratory (or at least their detection has not been confirmed), but strong evidence for their existence comes indirectly from the way the hadrons interact with other particles. It is postulated that quarks are bound together by forces that increase with separation, so that they can never be separated from their hadrons—or if they were separated with the application of very great energy, a pair of new quarks would be produced, creating two hadrons from one. The hypothetical exchange particle that binds quarks is called the *gluon,* and it would provide the underlying basis for the strong force.

(d) GUTs

At the forefront of theoretical physics is the speculation that the forces of nature are different manifestations of the same thing, and that there is a basic unity to all particles and forces; the concept is an extension of Einstein's dream of finding a unified field theory. Already there has been progress in developing such a theory—in particular in achieving a partial unification of the weak and electromagnetic forces. The theories that lead to the ultimate grand unity are called *grand unified theories (GUTs).*

One of the predictions of the grand unified theories is that baryons are not really fundamental and can decay to leptons. The only baryon known to be stable to present experimental accuracy is the proton (to which other baryons decay). Modern experiments show that if the proton *does* decay, it can be only after an average time of 10^{30} years. But the lifetime for the proton predicted by GUTs is about 10^{31} years. A human would have to live well over a hundred years before a single proton in his body had an even chance of decaying—not very exciting fireworks. Yet experiments are now being carried out with which it is hoped to detect proton decay if the proton's lifetime does not exceed 10^{33} years. It will be exciting to see if the prediction of the theory is realized.

29.3 NUCLEAR ENERGY

It is hadrons (protons and neutrons) that make up an atomic nucleus. In the nuclei of most atoms, the number of neutrons is roughly equal to the number of protons, but the most common kind of atom of hydrogen contains one proton but no neutron in its nucleus. Atoms of the same chemical element (whose nuclei have the same number of protons) but with different numbers of neutrons are said to be different *isotopes* of that element. Thus *deuterium* is an isotope of hydrogen whose atoms contain nuclei of one proton and one neutron each, and *tritium* is an isotope of hydrogen whose atomic nuclei have one proton and two neutrons each.

(a) Binding Energy

Just as the gases give up gravitational potential energy when they come together to form a star, so particles release energy in uniting to form an atomic nucleus. This energy given up is called the binding energy of the nucleus.

The binding energy is greatest for atoms with a mass near that of the iron nucleus, and it is less both for the lighter and heavier atoms. Thus heavy nuclei, like those of atoms of lead and uranium, are held together with less energy per atomic mass unit than iron is, and so are the lighter nuclei, like carbon and lithium. In general, therefore, if light atomic nuclei come together to form a heavier one (up to iron), energy is released; this is called *nuclear fusion*. On the other hand, if heavy atomic nuclei can be broken up into lighter ones (down to iron), an increase in the total binding energy results, with the release of that much potential energy; this is called *nuclear fission*. Nuclear fission sometimes occurs spontaneously, as in natural radioactivity.

Mass and energy, of course, are equivalent, so the binding energy must correspond to a *mass defect*. Indeed, we find that the mass of every nucleus (other than the simple proton nucleus of hydrogen) is less than the sum of the masses of the nuclear particles that are required to build it; this slight deficiency in mass is always only a small fraction of a mass unit (the unit of mass, u, is $\frac{1}{12}$ the mass of an atom of carbon). The mass defect per nucleon (nuclear particle) is greatest for the nucleus of the iron atom and is less for both more massive and less massive nuclei. A *nuclear transformation* is a buildup of a heavier nucleus from lighter ones, or a breakup of a heavier nucleus into lighter ones. In any such nuclear transformation, if the mass defect increases, the equivalent amount of energy is released. That energy, of course, is the difference in mass defect times the square of the speed of light. On the other hand, if, in the nuclear transformation, the mass defect *decreases*, a corresponding amount of energy must be put into the system.

(b) Nuclear Energy in Stars

It was suggested about 1928 that the energy source in stars might be fusion of light elements into heavier ones. Since hydrogen and helium account for about 98 or 99 percent of the mass of most stars, we logically look first to these elements as the probable reactants in any such fusion reaction. Helium atoms are about four times as massive as hydrogen atoms, so it would take four atoms of hydrogen to produce one of helium. The masses of hydrogen and helium atoms are 1.007825 u and 4.00268 u, respectively. Let us compute the difference in initial and final mass. (Here we include the mass of the entire atoms, not just the nuclei, because the electrons are involved as well, even though, in stellar interiors, the hydrogen and helium are completely ionized. When hydrogen is converted to helium, two positrons are created in the nuclear reactions, and these annihilate with two free electrons, adding to the energy produced.)

$$
\begin{aligned}
4 \times 1.007825 = \quad & 4.03130 \text{ u (mass of initial hydrogen atoms)} \\
& \underline{-4.00268} \text{ u (mass of final helium atom)} \\
& 0.02862 \text{ u (mass lost in the transaction)}
\end{aligned}
$$

The mass lost, 0.02862 u, is 0.71 percent of the mass of the initial hydrogen. Thus if 1 g of hydrogen turns into helium, 0.0071 g of material is converted into energy. The velocity of light is 3×10^{10} cm/s, so the energy released is

$$E = 0.0071 \times (3 \times 10^{10})^2$$
$$= 6.4 \times 10^{18} \text{ ergs.}$$

This 6×10^{18} ergs is enough energy to raise the 5-m telescope 150 km above the ground.

To produce the sun's luminosity of 4×10^{33} ergs/s some 600 million tons of hydrogen must be converted to helium each second, with the simultaneous conversion of about 4 million tons of matter into energy. As large as these numbers are, the store of nuclear energy in the sun is still enormous. Suppose half of the sun's mass of 2×10^{33} g is hydrogen that can ultimately be converted into helium, then the total store of nuclear energy would be 6×10^{51} ergs. Even at the sun's current rate of energy expenditure, 10^{41} ergs/year, the sun could survive for more than 10^{10} years.

There is little doubt today that the principal source of energy in stars is thermonuclear reactions. Deep in the interiors of stars, where the temperatures range up to many millions of degrees, nuclei of atoms are changing from one kind to another with an accompanying release of energy. The most important of these changes is the conversion of hydrogen to helium.

(c) Rate of Nuclear Reactions

For a nuclear reaction to occur, the nuclei of the reacting atoms must first collide with each other. As a result of this collision, a temporary "compound" nucleus is formed that either turns into a single nucleus of a different kind of atom from those originally taking part in the collision, or breaks into two or more less massive nuclei, again of different atomic types from those of the original reactants. How rapidly such events occur depends on (1) how fast the nuclei in the gas are moving (which depends on the temperature), (2) what the effective sizes of the nuclei are (which is computed from quantum theory for the simplest kinds of nuclei and measured in the nuclear physics laboratory for the more complicated ones), (3) how close together the nuclei are (which depends on the density of the gas), and (4) the probability that a particular compound nucleus that is formed will break into the relevant kinds of new nuclei (which is also determined either from theory or laboratory experiment).

With a rigorous mathematical treatment of the factors discussed in the preceding paragraph, nuclear physicists combine theory and experimental data to derive formulas that predict the rate of nuclear-energy production. Such formulas enable astronomers to predict the rate of energy release in a given region of a star in terms of the chemical composition of that region, its temperature, and the gas density. Since the total rate of energy release from a star (its luminosity) is known, the formulas obtained from nuclear physics give new information about the physical conditions in the stellar interior, which, when combined with our knowledge of the conditions of hydrostatic and thermal equilibrium, provide additional clues about the structure of the star.

Under conditions that prevail at the centers of most stars, there are two chains of nuclear reactions possible by which hydrogen can be converted to helium. One is the *carbon cycle,* also called the *carbon-nitrogen-oxygen (CNO)* cycle, in which carbon, nitrogen, and oxygen nuclei are involved in collisions with hydrogen nuclei (protons), eventually ending with carbon again and a new helium nucleus. The CNO cycle is important at temperatures above 15×10^6 K. At temperatures that prevail in the sun and in less massive stars, most of the energy is produced by what is called the *proton-proton* chain. Here, by any of several possible branches, protons collide directly to build into helium nuclei. The details of the more important series of nuclear reactions that occur in stars are given in Appendix 8.

It happens that the proton-proton chain, very important in the sun, begins with a most improbable event: the collision of two protons resulting in the formation of a nucleus (a *deuteron*) of the heavy isotope of hydrogen called *deuterium.* Usually the formation of a compound nucleus of two protons simply breaks up into two protons again, rather than ejecting a positron and turning into a deuteron, and very many compound nuclei must form to produce appreciable amounts of deuterium. But even at the high temperatures of stellar interiors (up to 15×10^6 K in the sun), it is extremely hard for two positively charged nuclei to come together to undergo any kind of reaction.

The reason is that the nuclei (protons, for example) are not moving fast enough to overcome their mutual electrostatic repulsion (due to their like positive charges) and come in contact. One might not expect nuclear reactions to occur at all in stars. Note in Figure 29.1a that two protons on a

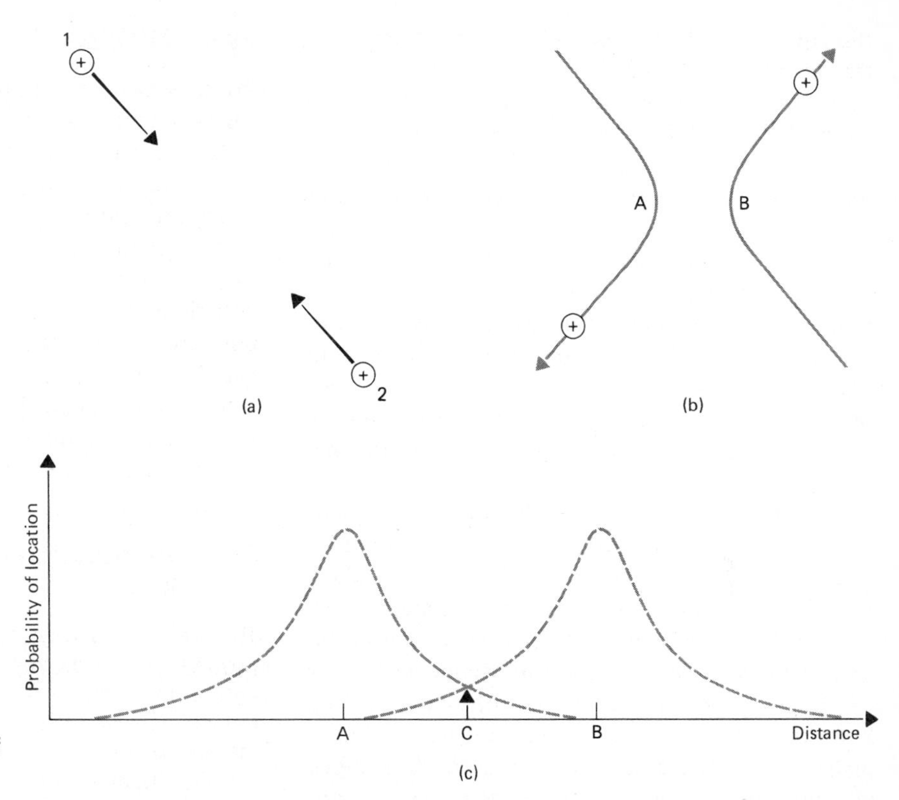

Figure 29.1 An encounter of two protons.

collision course would be expected to deflect each other without touching (b). A reaction *can* occur all the same, because of the Heisenberg uncertainty principle (Section 8.3f). In Figure 29.1c are shown (schematically) graphs of the probability of each proton having various positions when the two are at closest approach. While the most likely positions for the protons are at *A* and *B*, there is a small but finite chance that they will actually be at the same place, for example, at *C*. Only rarely does probability permit such a union of two nuclei, but there are lots of them in a stellar interior, so it nevertheless happens often enough for nuclear reactions to work and supply the stars with energy.

Most of the electromagnetic radiation released in these nuclear reactions is at very short wavelengths—in the form of X-rays and gamma rays. Nuclear reactions are important, however, only deep in the interior of a star. Before this released energy reaches the stellar surface, it is absorbed and re-emitted by atoms a very great number of times. Photons of high energy (short wavelength) that are absorbed by atoms are often re-emitted as two or more photons, each of lower energy. By the time the energy filters out to the surface of the star, therefore, it has been converted from a relatively small number of photons, each of very high energy, to a very much larger number of photons of lower energy and longer wavelength, which constitute the radiation we actually observe leaving the star.

(d) Solar Neutrinos

The proton-proton chain, which is thought to provide almost certainly most of the sun's energy, involves the emission of neutrinos. In fact, the neutrinos should constitute about 3 percent of the energy released from the sun. Their detection, however, because of their low interaction with matter, is an extremely difficult business.

Nevertheless, Raymond Davis, Jr., and his colleagues at Brookhaven National Laboratory have devised a technique by which they are detecting solar neutrinos. On rare occasions a neutrino of the energy of some of those emitted from the sun should react with the isotope chlorine-37 to transmute it to argon-37 and an electron. Davis has placed a tank containing 378,000 liters of tetrachloroethylene, C_2Cl_4 (ordinarily used as cleaning fluid), 1.5 km beneath the surface of the Earth in a gold mine at Lead, South Dakota. Even though an individual neutrino is extremely unlikely to react with the chlorine in the cleaning fluid, calculations show that about one atom of argon-37 should nevertheless be produced daily. Because argon-37 is radioactive (about half of it decaying in 35 days), it is possible to isolate and detect most of those few argon-37 atoms from the more than 10^{30} atoms of chlorine in the tank.

In analyzing dozens of experimental runs, Davis found that the flux of neutrinos from the sun is about three times less than that expected. Careful analysis of the laboratory procedure has failed to suggest obvious sources of experimental error. Attempts to explain this discrepancy between theory and observation have taken two forms. On the one hand, astrophysicists have attempted to construct some rather exotic and highly speculative models of the solar interior that vary in composition, rotation rates, and central temperature from the standard model. One such model, involving variations with time in the rate at which energy is generated in the core of the sun, was discussed in Section 28.5.

On the other hand, it may be that it is our knowledge of the properties of the neutrino that is inadequate. There are, as we have seen, three types of neutrinos. If these neutrinos have even a very small mass, then they may change from one type to another. The neutrinos created in the core of the sun are electron neutrinos, and the reaction on which Davis relies to detect neutrinos can be triggered only by electron neutrinos. If the solar neutrinos can be converted to, say muon neutrinos, sometime during their journey from the center of the sun to the surface of the Earth, then they would no longer be detectable by Davis' experiment.

Recent theoretical calculations suggest that this may be exactly what happens. The density in the interior of the sun appears to be just what is needed to convert certain high energy electron neutrinos to muon neutrinos. The "missing" neutrinos then do indeed arrive at the Earth in the appropriate numbers, but in a form that current experiments cannot detect. As this is being written (August 1986) experiments are being designed to test these new theoretical ideas about the nature of neutrinos. Even if the theory proves to be correct, there still may be problems with the standard model of the solar interior. This explanation for the missing solar neutrinos does not resolve the problem (see Section 28.5) that the standard model for the internal structure of the sun does not reproduce the solar oscillations.

29.4 MODEL STARS

We now have enough theory to determine the internal structure of a star. We must combine the principles we have described: hydrostatic equilibrium, the perfect gas law, thermal equilibrium, energy transport, the opacity of gases, and the rate of energy generation from nuclear processes. These physical ideas are formulated into mathematical equations that are solved to determine the march of temperature, pressure, density, and other physical variables throughout the stellar interior. The set of solutions so obtained, based upon a specific set of physical assumptions, is called a theoretical model for the interior of the star in question.

(a) Computation of a Stellar Model

There are many ways to formulate mathematically the physical principles that govern the structure of a star and to solve the resulting equations to obtain a stellar model. Here we shall illustrate a particular procedure that has been used widely.

Four quantities are chosen to describe the physical conditions at any distance, r, from the star's center: the pressure, $P(r)$, the temperature, $T(r)$, the mass, $M(r)$, contained within a sphere of radius r concentric with the star's center, and the contribution to the star's total luminosity, $L(r)$, that is generated within this sphere. Once the pressure and temperature are known, the density can be calculated from the perfect gas law. The opacity and rate of energy generation at each point in the star involve no new parameters, for they, like the density, can also be expressed as functions of the pressure, temperature, and chemical composition of the stellar material at that point.

The four quantities $P(r)$, $T(r)$, $M(r)$, and $L(r)$ are then combined into four equations that express the physical principles involved. Each equation describes how one of the quantities changes through a small radial distance within the star (say, across one of the imaginary layers or shells described in Section 28.4c). The change in pressure across such a layer is given by the condition of hydrostatic equilibrium. The mass of the shell is given by the density of the layer, which in turn is given by the pressure and temperature of the layer and the perfect gas law. The change in luminosity across the layer is given by the rate of energy generation within the shell. The change in temperature is governed by the mode of energy transport; if in that region of the star the energy is transported by radiation, the opacity determines the temperature variation, while if that region is in convection, the expansion and contraction of the gases determine the change in temperature. Since there are four equations that hold for each point within a star, their simultaneous solution determines the four desired quantities at that point.

The solution of the equations may begin at the surface of the star, where the physical conditions are known. The mass and luminosity included within the surface are, of course, the total mass and luminosity of the star. The pressure and temperature often can be considered zero to a sufficient approximation, since their actual values in the photospheric layers of a star are very small compared to those in the interior. If higher precision is desired, "surface" values of pressure and temperature can be taken from the solution of a model photosphere of the star (Section 24.2b).

The four equations are then used to calculate how the values of pressure, temperature, mass, and luminosity change over a short distance inward, beneath the surface of the star, thus yielding the values of these quantities at that new depth. Next, the equations are used to calculate the changes over the next short distance inward. So, step by step, the pressure, temperature, mass, and luminosity are found at successively deeper layers in the star, until the center is reached.

At the center, of course, the mass and luminosity should be zero, for no mass or luminosity can be contained within a point. This would be true if all the physical laws governing a star were precisely understood, and if the chemical composition were precisely known at each depth in the star. In practice, neither the physical details entering into the opacity and nuclear-energy generation formulas nor the chemical composition are known with absolute accuracy. Consequently, the solution of the four equations of structure may not lead to zero values of mass and luminosity at the center. The physical laws, therefore, are expressed as accurately as knowledge permits, and trial adjustments to the chemical composition are made until a set of solutions is found for which $M(r)$ and $L(r)$ do equal zero at the center. Then the runs of the pressure, temperature, mass, and luminosity throughout the star constitute a finished model for its interior.

Since a possible model is found only for a "correct" choice of chemical composition, something is learned of the distribution of the various chemical elements in a star as well as of the physical conditions in its interior. Hydrogen and helium are thus found to constitute (usually) more than 95 percent of the mass of a star, and more than two-thirds of that 95 percent is hydrogen. Unfortunately, however, the physical parameters in the opacity and nuclear physics theory are not yet known accurately enough to determine the chemical composition within a star with high precision from such studies.

The solution of the equations of structure to obtain a stellar model is a difficult and tedious business. Until the 1950s it often took as long as a year to compute a stellar model, and such a computation was a satisfactory topic for the dissertation of a student earning a Ph.D. degree. Now, however, high-speed electronic computers enable the calculation of a model in a few minutes or, in some cases, even in a few seconds.

(b) A Model for the Sun

The sun is the most studied of all stars, and models of its interior have been calculated for several decades. Each new model of the sun represents a refinement resulting from an improvement in our knowledge of physics or of computing methods or both. The general run of the physical parameters in the sun, however, was fairly well established even in the early approximate models. The temperature within the sun increases gradually toward its center and reaches a value of about 15 million Kelvins at the center. The density (like the pressure), on the other hand, increases very sharply near the center of the sun (indicating a high degree of central concentration of its material) and reaches a maximum value over 100 times the density of water.

As time goes on, the thermonuclear conversion of hydrogen to helium in the sun's central regions gradually changes its chemical composition. The exact temperature and density of the material at the sun's center that are required to account for its observed size and luminosity depend on how much the composition of that material has been changed—that is, on the age of the sun.

Series of models that trace the past history of the sun, therefore, have been calculated (more of this in Chapter 32). Table 29.3 and Figure 29.2 exhibit a model appropriate to the present-day sun that results from one such set of calculations. The model is based on the assumption that the mass of the sun was originally 73 percent hydrogen and 24.5 percent helium. According to this model, the outer layers of the sun are in convection, while the inner parts transport energy by radiation. The hydrogen abundance at the very center has been reduced (by nuclear reactions) to only 38 percent, and the present age of the sun is about 4.5×10^9 years.

(c) The Russell-Vogt Theorem and the Interpretation of the Main Sequence

If a star is in hydrostatic and thermal equilibrium, and if it derives all its energy from nuclear reac-

TABLE 29.3 Model for the Structure of the Sun*

FRACTION OF RADIUS	FRACTION OF MASS	FRACTION OF LUMINOSITY	TEMPERATURE *(Millions of K)*	DENSITY *(g/cm³)*	FRACTION HYDROGEN *(By Weight)*
0.00	0.000	0.00	15.0	148	0.38
0.05	0.011	0.10	14.2	125	0.47
0.10	0.076	0.45	12.5	86	0.59
0.15	0.19	0.78	10.7	56	0.67
0.20	0.33	0.94	9.0	36	0.71
0.30	0.61	1.00	6.5	12	0.73
0.40	0.79	1.00	4.9	4	0.73
0.60	0.95	1.00	3.1	0.5	0.73
0.80	0.99	1.00	1.3	0.1	0.73
1.00	1.00	1.00	0.0	0.0	0.73

* Adapted from R. Ulrich.

tions, then its structure is completely and uniquely determined by its total mass and by the distribution of the various chemical elements throughout its interior. This is not to imply that we have the ability to compute perfect models for such stars; the stars themselves, nevertheless, must conform to the physical laws that govern their material and will adjust themselves to unique configurations. It should not, of course, surprise us that mass and composition, the very properties that a star is "born" with, and over which it "has no control," are just those that determine its structure. This important principle is known as the *Russell-Vogt theorem.*

Suppose a cluster of stars were to form from a cloud of interstellar material whose chemical composition was similar to the sun's. All condensations that become stars would then begin with the same chemical composition and would differ from each other only in mass. Suppose now that we were to compute a model for each of these stars for the time at which it became stable, and derived its energy from nuclear reactions, but before it had had time to alter its composition appreciably as a result of these reactions. The admixture of chemical elements would be the same, then, at all points within the star; such a composition is *homogeneous.*

The models we would calculate for these stars would indicate, among other things, their luminos-

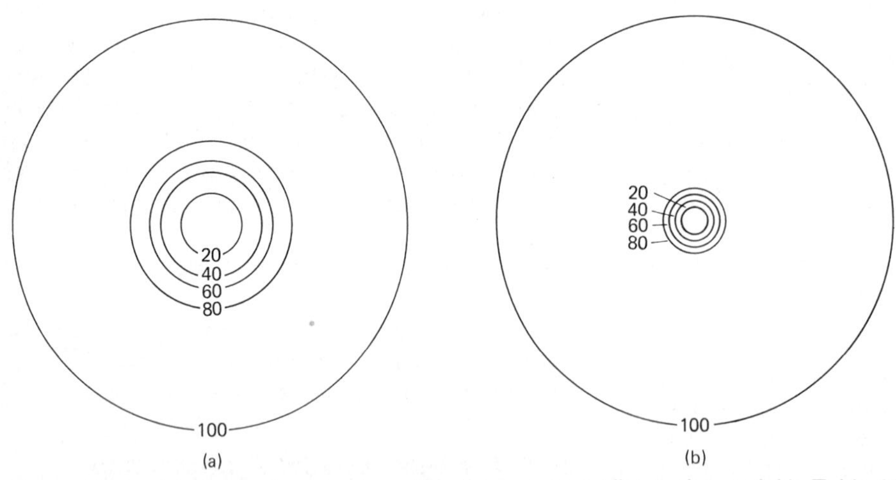

Figure 29.2 (a) Distribution of mass within the sun, according to the model in Table 29.3; numbers show what percentage of the sun's mass is included within the radial zones shown. (b) Distribution of energy generation in the sun according to the same model. Successively smaller circles show the regions within which 100, 80, 40, and 20 percent of the sun's energy is produced.

ities and radii. From Stefan's law we know that the luminosity of a star is proportional to the product of its surface area and the fourth power of its effective temperature (Section 25.3a). We can, therefore, calculate the temperature for each of the stars and plot it on the Hertzsprung-Russell diagram (Section 26.2). We would find that the most massive stars were the hottest and most luminous and would lie at the upper left corner of the diagram, while the least massive were coolest and least luminous and would lie at the lower right. The other stars would all lie along a line running diagonally across the diagram—the *main sequence*. The main sequence, then, is the locus of points on the H–R diagram representing stars of similar chemical composition but different mass. The observed fact that most stars in the Galaxy do lie along the main sequence is evidence that they have compositions similar to the sun's and are nearly chemically homogeneous. The observed scatter about the main sequence represents slight differences in the chemical compositions of stars. This explanation of the main sequence was first presented in the 1930s by the Danish astrophysicist B. Strömgren.

If we now plot the masses of the stars in our hypothetical cluster against their luminosities, we also find that the points lie along a line—the *mass-luminosity* relation. We have seen (Section 25.2g) that most real stars do, indeed, obey a mass-luminosity relation; these stars are also of similar chemical composition. The locus of points on a plot of the masses of stars against their luminosities is simply the main sequence, plotted on a different kind of diagram. (Actually, it is possible for stars of the same mass but different chemical composition to have nearly the same luminosity—although they will differ in radius and temperature. Some stars, therefore, may obey the mass-luminosity relation even though they are *not* main-sequence stars, but this circumstance is fortuitous.)

Those stars that do not lie on the main sequence in the Hertzsprung-Russell diagram (for example, red giants and white dwarfs) *must* differ somehow from the majority in their chemical compositions, or else they are not stable and are not shining by nuclear energy alone. We have seen, however, that as stars age they convert hydrogen to helium, and so change their compositions, especially near their centers. Chapters 30 and 32 will describe how most non-main-sequence stars can be interpreted either as stars that are still forming from interstellar matter and are not yet deriving all their energy from nuclear sources, or as stars that, by virtue of nuclear transformations, have altered their chemical compositions and hence their entire structures.

EXERCISES

1. In what important respect (or respects) is a neutrino very different from a neutron?
2. If the nuclear force is the strongest of the known natural forces, why do you suppose it escaped discovery until the 20th century?
3. What do you suppose are the decay products of an antineutron?
4. Consider a nucleus of 11 protons. The total electrostatic repulsion between the 11 protons is expected to be about 10 times as strong as that between two protons of comparable separation. If the strong nuclear force is about 100 times as strong as the electrical force, but acts only between adjacent particles, what is the largest number of protons you would expect a nucleus to be able to have and still remain stable? What do you know about nuclei that actually contain larger numbers of protons?
5. Which of the following transformations is fusion and which is fission: the transformation of (a) helium to carbon; (b) carbon to iron; (c) uranium to lead; (d) boron to carbon; (e) oxygen to neon?
6. Verify that some 600 million tons of hydrogen are converted to helium in the sun each second.
7. Stars exist that are as much as a million times as luminous as the sun. Consider a star of mass 2×10^{35} g, and luminosity 4×10^{39} ergs/s. Assume that the star is 100 percent hydrogen, all of which can be converted to helium, and calculate how long it can shine at its present luminosity. There are about 3×10^{7} seconds in a year.
8. Perform a similar computation for a typical star less massive than the sun, such as one whose mass is 1×10^{33} g and whose luminosity is 4×10^{32} ergs/s.
9. Why do you suppose so great a fraction of the sun's energy comes from its central regions? Within what fraction of the sun's radius does practically all of the sun's luminosity originate? (See Table 29.3). Within what radius of the sun has its original hydrogen been

partially used up? Discuss what relation the answers to these questions bear to each other.

10. Why do we not expect nuclear fusion to occur in the surface layers of stars?

*11. Let ΔP denote the pressure increase inward through a spherical shell of inner radius r and thickness Δr. Multiply ΔP by the area of the inner surface of the shell to find the total outward force on the shell. Now equate this outward force to the weight of the shell pulling it inward, and show that

$$\Delta P = \frac{GM(r)}{r^2} \rho \Delta r,$$

where $M(r)$ is the mass of the star interior to r and ρ is the density of the shell.

*12. Use the equation derived in Exercise 11 to make a very rough estimate of the pressure at the center of the sun. For this estimate suppose Δr to be the entire radius of the sun (that is, the entire sun is taken as one shell) and ΔP to be the increase in pressure from the surface to the center. For ρ use the mean density of the sun, and for $M(r)$ the entire solar mass. See Appendix 6 for needed data. Your estimate cannot be expected to be better than a few orders of magnitude, but it should give some indication of the amount of pressure.
Answer: About 3×10^{15} dynes/cm^2

*13. What is the minimum uncertainty in the momentum of a particle whose position is specified to within 1 Å (10^{-8} cm)? If the particle is a proton, what is the uncertainty in its velocity? See Appendix 6 for needed data.
Answer to Second Part: About 6×10^4 cm/s

Sir Fred Hoyle (b. 1915), the well-known British astrophysicist and cosmologist, is also well known for his science fiction and even for an opera libretto. Hoyle was one of the pioneers in the modern study of stellar evolution, and in the 1950s his brilliant deduction about the nature of the carbon nucleus enabled us to understand where the atoms of our bodies originated (Chapter 32). (Floyd Clark, California Institute of Technology)

30

STAR FORMATION AND EVOLUTION

It is only natural to think of the stars as fixed, permanent, and unchanging. Apart from an occasional nova or supernova, few stars undergo any fundamental changes in the span of a human lifetime. Yet stars are radiating energy at a prodigious rate, and no source of energy can last forever. For example, we know (Chapter 29) that the sun shines by converting hydrogen to helium. Specifically, deep in the interior of the sun, 600 million tons of hydrogen are converted to helium every second, with the simultaneous conversion of about 4 million of these tons to energy. So massive is the sun, however, that it can survive for about ten billion years before it exhausts its fuel supply.

The implications of a finite fuel supply are much more immediate for stars more massive than the sun. The most massive stars have masses that are 50 to 100 times the mass of the sun, yet their luminosities are a million times greater than that of the sun. Accordingly, these massive stars must exhaust their fuel supply, burn themselves out, and become unobservable in no more than a few million years. The brightest hot star in Orion—Rigel—cannot have been shining when the first man-like creatures walked the Earth.

Astronomers estimate that there are about 18,000 stars in our own Galaxy that have lifetimes that are measured in millions of years. We also see highly luminous stars in galaxies other than our own. It seems likely, therefore, that highly luminous stars must have been present throughout the billions of years that our Galaxy has existed. If so, then as some stars die, they must be replaced by new ones. On average, in fact, one new bright star must be formed somewhere in our Galaxy every 500 to 1000 years if the total number of highly luminous stars is to remain approximately constant.

As we shall see in this chapter, star formation is a continuous process that is going on right now. Stars of all masses, low as well as high, are being formed, and that formation process is taking place in the interiors of clouds of dust and gas that provide the necessary raw materials.

30.1 FORMATION OF STARS

Young stars are typically found in groups, not in isolation. The Pleiades consists of about 300 stars, and the total mass of the stars in this cluster is

Figure 30.1 Orion nebula. This cloud of dust and gas is one of the nearest regions where star formation is currently taking place.

Figure 30.2 A region of gas and dust where star formation has occurred recently. The stars in the cluster (NGC 2264) associated with the gas and dust are only a few million years old.

about 500 times greater than the mass of the sun. Where did the material come from to form such a cluster? The only possible source is a giant cloud of dust and gas. (The wisps of dust and gas that can be seen in the vicinity of the Pleiades are probably the result of a random encounter between the cluster and an interstellar cloud and not the remnants of the cloud from which these stars formed originally.)

(a) Molecular Clouds and Star Formation

Observations suggest that most star formation takes place in the interiors of the giant molecular clouds. A typical cloud is about 100 light years across, and the total mass of such a cloud can be as much as a million times the mass of the sun. The best studied of the stellar nurseries is the Orion region (Figure 30.1). A luminous cloud of dust and gas can be seen with binoculars in the middle of the dagger in the constellation of Orion, but associated with it is a much larger molecular cloud, which is invisible in the optical region of the spectrum. Near the center of the optically bright region of the nebula is the Trapezium cluster of stars. The most massive of these stars radiate energy 100,000 times faster than the sun, and observations show that these stars are only about 2 million years old.

The interiors of molecular clouds like the one in Orion cannot be observed with visible light. The dust in these clouds acts like a thick blanket of interstellar smog, which cannot be penetrated by visible radiation (Figure 30.2). It is only with the new techniques of infrared and millimeter radio astronomy that we have been able to measure the conditions inside these clouds and study directly the very early stages of the births of stars.

In the specific case of Orion, these new techniques show that the long dimension of the Orion molecular cloud stretches over a distance of about 100 light years and is about 50 times larger than the region that can be seen optically. The total quantity of molecular gas is about 200,000 times the mass of the sun and about 1000 times the mass of the stars in the Trapezium cluster. Typical temperatures inside the cloud are less than 100 K, and so most atoms have indeed been bound into molecules. Molecular hydrogen (H_2) is the most abundant, but other molecules including carbon monoxide (CO), CN, and ammonia (NH_3) are also present. Infrared observations have even shown that there are two star clusters embedded deep in the Orion cloud, neither of which can be seen at all in the optical region of the spectrum even though one emits about the same total amount of energy as the Trapezium cluster.

The earliest stages of star formation are still clouded in mystery. There is almost a factor of 10^{20} difference between the density of a molecular cloud and that of the youngest stars that can be observed. So far we have been unable to observe directly what happens within a cloud as material comes together and collapses gravitationally through this range of densities to form a star. We do not know what causes a single large cloud to fragment and form individual stars, nor do we know why multiple star systems and planets form. We do not even know what triggers the initial gravitational collapse of the cloud. Perhaps the complex gravitational field of the galaxy (see discussion of spiral density waves in Section 34.3e) forces the density in a portion of the molecular cloud to increase enough so that gravitational forces in the cloud can overpower gas pressure and initiate the collapse of the cloud. In other cases, the necessary compression and increase in density may result from the collision of two gas clouds. (Gas clouds, like stars, orbit the center of the Galaxy, but each has a slightly different space velocity, and collisions can occur.)

However the first stars form, there is strong observational evidence that these stars can trigger the formation of additional stars, which can lead to the formation of still more stars (Figure 30.3). The basic idea is as follows. When a massive star is formed, it emits copious amounts of ultraviolet radiation, which heats the surrounding gas in the molecular cloud. This heating increases the pressure in the gas and causes it to expand. Supernova explosions may also cause the gas to expand. The hot gases explode into the surrounding cold cloud, compressing the material in it until the cold gas is at the same pressure as the expanding hot gas. In the conditions typical of molecular clouds, the compression is enough to increase the gas density by about a factor of 100. At densities this high, stars can begin to form in the compressed gas.

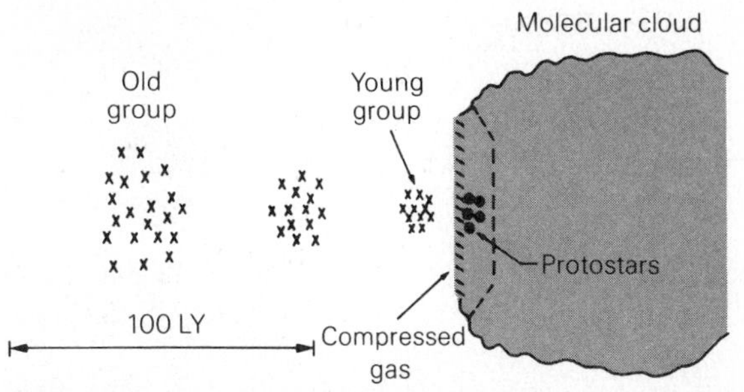

Figure 30.3 Schematic diagram showing how star formation can move progressively through a molecular cloud. The oldest group of stars lies to the left of the diagram and has expanded because of the motions of the individual stars. Eventually the stars in this group will disperse and will no longer be recognizable as a cluster. The youngest group of stars lies to the right next to the molecular cloud. This group of stars is only one to two million years old. The pressure of the hot, ionized gas surrounding these stars compresses the material in the nearby edge of the molecular cloud and initiates the gravitational collapse that will lead to the formation of more stars.

The Orion Nebula is a good example of how star formation can move progressively through a cloud. Star formation began about 12 million years ago at one edge of this molecular cloud and has slowly moved through it, leaving behind groups of stars. The oldest of these groups is the one farthest from the site of current star formation, and the ages become progressively younger the closer the groups are to the Trapezium cluster. Still younger stars are embedded in the molecular cloud and can be observed only in the infrared. The oldest groups of stars can be easily observed because they are no longer shrouded in dust and gas. In these older regions, any material not incorporated into stars has been heated either by stellar radiation or by supernova explosions (Chapter 32) and blown away into interstellar space.

(b) Gravitational Collapse and the Formation of Stars

The story of stellar evolution is the story of the competition between two forces—gravity and pressure. The force of gravity tries to make a star collapse. Pressure tries to force the star to expand.

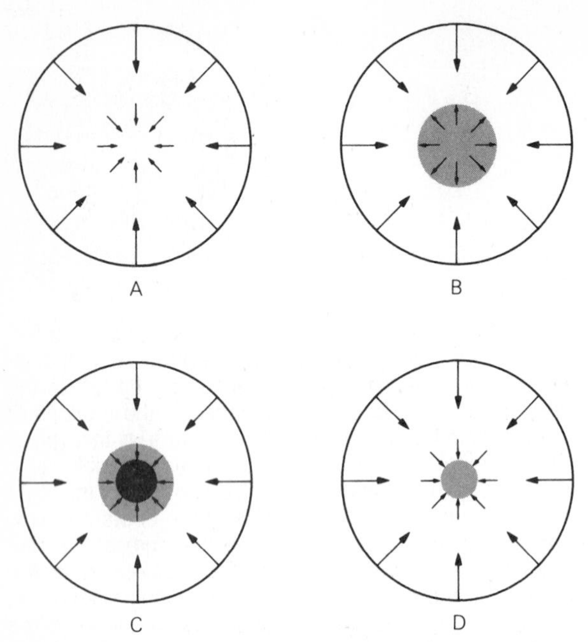

Figure 30.4 Stellar models showing various stages in the collapse of a nonrotating, spherical cloud of gas to form a star. Initially *(panel A)*, the gas and dust fall rapidly toward the center of the cloud. When the central region of the cloud becomes dense enough to be opaque to infrared radiation, heat is trapped within the cloud and both temperature and pressure build. The collapse is temporarily halted *(panel B)*. When the protostar becomes hot enough to break up molecules of hydrogen into individual hydrogen atoms, there is another phase of rapid collapse *(panel C)*. A final and smaller core forms, which continues to accrete matter and to contract slowly until its interior becomes hot enough to ignite nuclear reactions.

When these two forces are in balance, the star is stable. Major changes in the structure of a star occur when one or the other of these two forces gains the upper hand.

Because we are unable to observe directly the earliest stages of star formation, we must rely on calculations to tell us how stars are born. In the theoretical study of stellar evolution, we compute a series of models for a star, each successive model representing a later point in time. Given one model, we can calculate how the star should change (in the case of the young stars under discussion, due to gravitational contraction), and hence what the star will be like at a slightly later time. At each step we find the luminosity and radius of the star, and from these its surface temperature.

Let us now follow the evolution of a stellar condensation that is on its way to becoming a normal star with a mass similar to that of the sun (Figure 30.4). This young stellar object is not yet self-sustaining with nuclear reactions but derives its energy from its gravitational contraction—by exactly the process proposed for the sun by Helmholtz and Kelvin (Section 29.1). Initially the radius of the protostar is very large and its density is very low. It is transparent to infrared radiation, and the heat generated by gravitational contraction can be radiated away freely into space. Because no heat builds up inside the protostar, the gas pressure remains too low to counteract gravity, and the protostar collapses extremely rapidly. The temperature of the protostar remains fairly constant at only about 10 K.

During this phase of rapid collapse, the density of the protostar is increasing dramatically (by a factor of a million or more), and eventually the protostar becomes opaque to infrared radiation. Because the heat released as a consequence of gravitational contraction is now trapped, both the temperature and the pressure within the protostar increase. When the pressure becomes high enough to support the weight of the outer material that has been falling inward, hydrostatic equilibrium is reached—the collapse is halted and the protostar becomes (temporarily) stable.

The region of this collapsing gas cloud that is opaque to infrared radiation is called the core of the protostar and is perhaps 5 astronomical units in diameter. (Remember that one astronomical unit is equal to the average distance of the Earth from the sun.) The portion of the gas cloud outside the core is still transparent to radiation and continues to fall onto the core, and as a consequence the core becomes gradually hotter and denser. When it finally reaches a temperature of about 2000 K, the molecular hydrogen in the core is dissociated—separated into individual hydrogen atoms. This process soaks up energy, which is then no longer available to heat the core of the protostar and produce the pressure needed to counterbalance gravity. The protostar again undergoes a phase of very rapid collapse, which stops only when the interior temperature of the star reaches about 100,000 K.

The star is now smaller and denser but has reached a new equilibrium. It continues to contract very slowly until it reaches temperatures in excess of a million degrees and ignites nuclear reactions in its interior. When the central temperature becomes high enough to fuse hydrogen into helium, then we say that the star has reached the main sequence.

This description of the early history of a star like the sun is highly simplified. In particular, we have not considered what will happen if the gas cloud is rotating. The typical contracting cloud probably has at least some rotation, if for no other reason than that it formed from material undergoing differential rotation in the Galaxy (Section 34.3). Early on, the rotation is likely to be exceedingly slow, but to conserve angular momentum the cloud must spin faster and faster as it contracts. The angular momentum, in fact, will probably prevent the cloud from collapsing entirely to a single star. In the solar system, the nebula flattened to a disk, and planets accreted in the disk. The formation of planets may be commonplace, but often (perhaps half the time or so) the cloud must split and form two or more stars, whose orbital motions about each other contain most of the original angular momentum. It may be that formation of either a planetary system or a multiple star system are the two alternatives open to a condensing cloud.

(c) Evolution in the H–R Diagram

We have now described in words the early evolutionary history of the sun and stars like it. Theoretical calculations, however, give us quantitative information about how the luminosity, radius, and temperature of protostars change with time. We can thus find where any star (or its embryo) should be represented on the Hertzsprung-Russell diagram. By plotting the temperature and luminosity of the collapsing protostar as it changes with time, we can trace the track that the star follows on the H–R diagram.

In its early contraction phases, a star transports its internal energy not by radiation but by leisurely convection currents (Section 28.4e). The Japanese astrophysicist C. Hayashi first showed that such stars must lie in a zone on the H–R diagram extending nearly vertically from the lower main sequence to the right extreme of the regions occupied by red giants and red supergiants (shaded region in Figure 30.5). There can be no stable star such that the point representing it on the H–R diagram lies to the right of this zone. In accord with the Hayashi theory, stars in the initial stages of their evolution contract and move (on the H–R diagram) downward in the zone along *Hayashi lines.* Representative tracks for stars or stellar embryos of several masses and of chemical composition more or less like the sun's are shown in Figure 30.5.

Except for a star of low mass, after a period of some thousands or millions of years, the convection currents cease at the center of the star, and energy must be transported by radiation in those regions. The central zone in radiative equilibrium gradually grows in size, while the convection currents extend less and less deeply beneath the stellar surface. In this stage of its evolution, the star or embryo, still slowly shrinking and deriving its energy from gravitational contraction, turns sharply on the H–R diagram and moves left, almost horizontally, toward the main sequence. Eventually, as the release of gravitational energy continues to heat up the star's interior, its central temperature becomes high enough to support nuclear reactions. Soon this new source of energy supplies heat to the interior of the star as fast as energy is radiated away. The central pressures and temperatures are thus maintained and the contraction of the star ceases; it is now on the main sequence. By this time the infall of matter is complete and the star is fully formed. The small hooks in the evolutionary tracks of the stars shown in Figure 30.5, just before they reach the main sequence, are the points (according to theory) where the onset of nuclear-energy release occurs. Calculations show that stars more massive than the sun would be observable only in the infrared during most of their pre-main-sequence evolution because the light they emit is absorbed by the surrounding dust in the infalling material.

By the time stars of mass appreciably greater than the sun's have reached the main sequence, their outer convection zones have disappeared, but new cores of convection exist at their centers. Main-sequence stars of mass near that of the sun still have appreciable regions in their outer layers in convection, with their deep interiors in radiative equilibrium. Stars of rather low mass remain in complete convective equilibrium throughout and follow their Hayashi lines right down to the main sequence, where nuclear reactions finally stop their contraction. Objects of extreme low mass, on the other hand, never achieve high enough central temperature to ignite nuclear reactions. The lower end of the main sequence terminates where stars have a mass just barely great enough to sustain nuclear reactions at a sufficient rate to stop gravitational contraction; this critical mass is calculated to be near $\frac{1}{12}$ that of the sun.

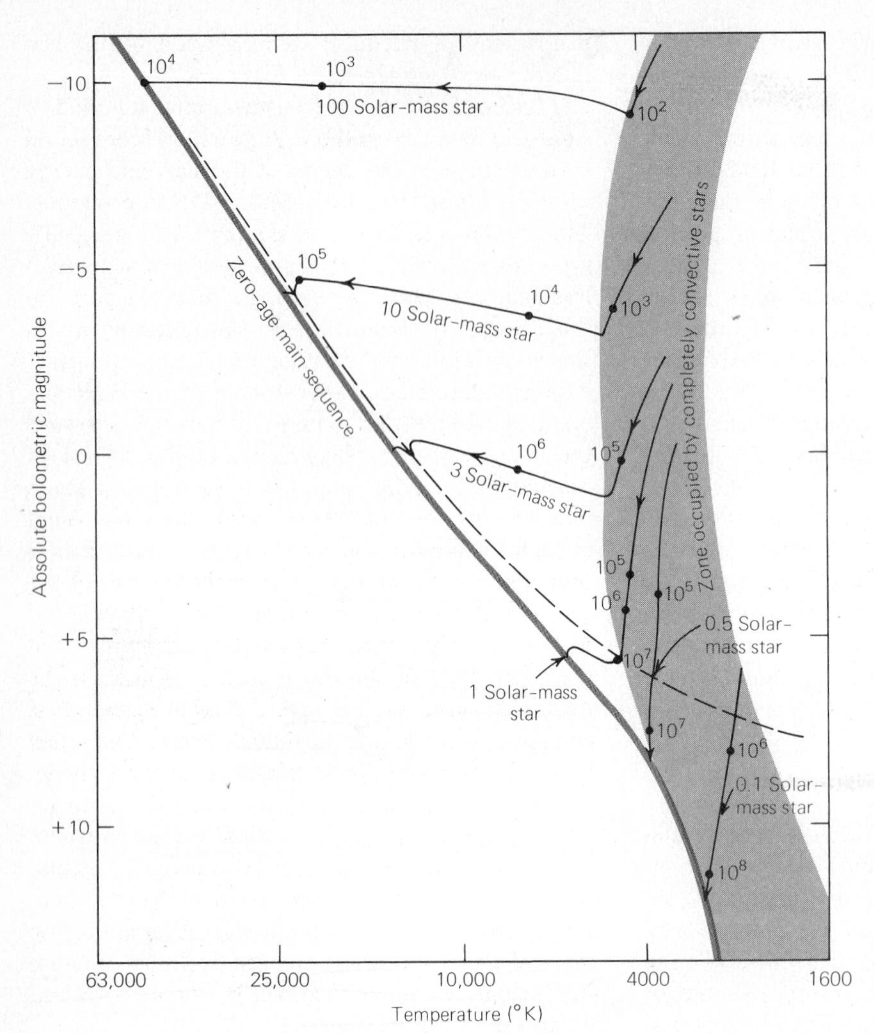

Figure 30.5 Theoretical evolutionary tracks of contracting stars or stellar embryos on the Hertzsprung–Russell diagram. According to calculations, stars or embryos lying roughly above the dashed line are still surrounded by infalling matter and would be hidden by it. (Based on calculations by R. Larson)

At the other extreme, the upper end of the main sequence terminates at the point where the mass of a star would be so high and the internal temperature so great that radiation pressure would dominate (Section 20.2e). The radiation produced from nuclear reactions would be so extreme that when absorbed by the stellar material it would impart to it a force greater than that produced by gravitation; hence, such a star could not be stable. The upper limit to stellar mass is calculated to be about 100 solar masses.

In general, the pre-main-sequence evolution of a star slows down as the star moves along its evolutionary track toward the main sequence; the numbers labeling the points on each evolution track in Figure 30.5 are the times, in years, required for the embryo stars to reach those stages of contraction. The time for the whole evolutionary process, however, is highly mass-dependent. Stars of mass much higher than the sun's reach the main sequence in a few thousand to a million years; the sun required millions of years; tens of millions of years are required for stars to evolve to the lower main sequence. For all stars, however, we should distinguish three evolutionary time scales (although the first two, as we have seen, may overlap):

1. The initial gravitational collapse from interstellar matter is relatively quick. Once the condensation is, say, 1000 AU in diameter, the time for it to reach hydrostatic equilibrium is measured in thousands of years.
2. Pre-main-sequence gravitational contraction is much more gradual; from the onset of hydrostatic equilibrium to the main sequence requires, typically, millions of years. For the stars

COLOR PLATE 25a The globular cluster M 13. (U.S. Naval Observatory)

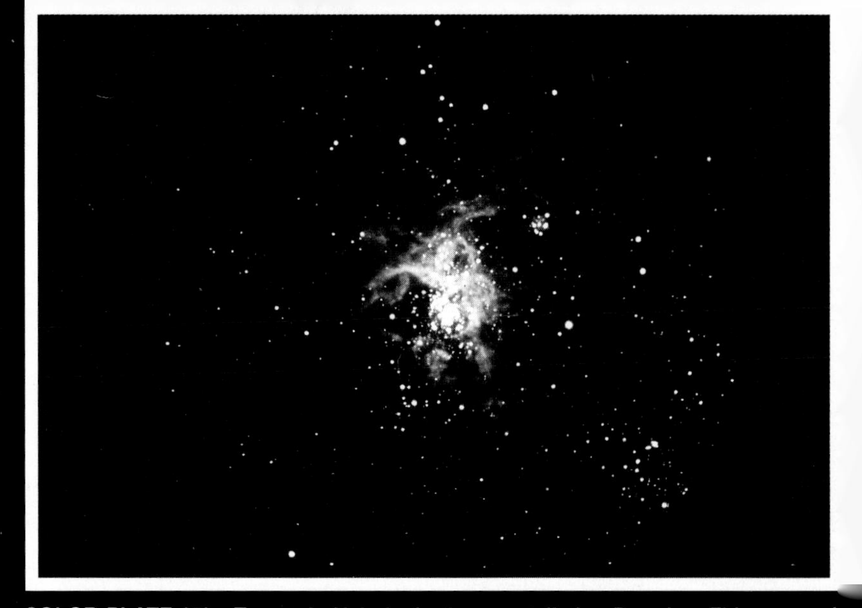

COLOR PLATE 25b Tarantula Nebula, in the constellation Doradus. This unusual object in the Large Magellanic Cloud is the brightest emission nebula known. A central cluster of blue supergiant stars excites its hydrogen gas clouds out to a radius of about 400 LY, although the radius of the entire nebula is estimated to be nearly 800 LY—with a few streamers reaching 1800 LY—making it 30 times larger than the Orion Nebula in our Galaxy. (Cerro Tololo Inter-American Observatory photo/National Optical Astronomy Observatories)

COLOR PLATE 25c The Pleiades and associated nebulosity in Taurus, photographed with the Palomar Schmidt telescope. (California Institute of Technology/Palomar Observatory)

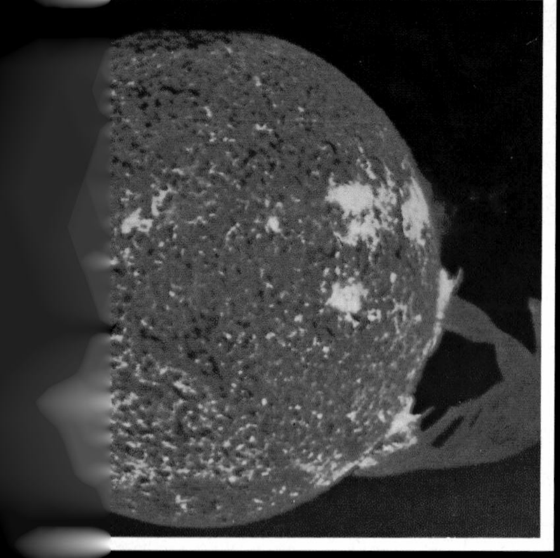

COLOR PLATE 26a Ultraviolet photograph of the entire sun, taken by Skylab, showing several flares and a large prominence. (NASA)

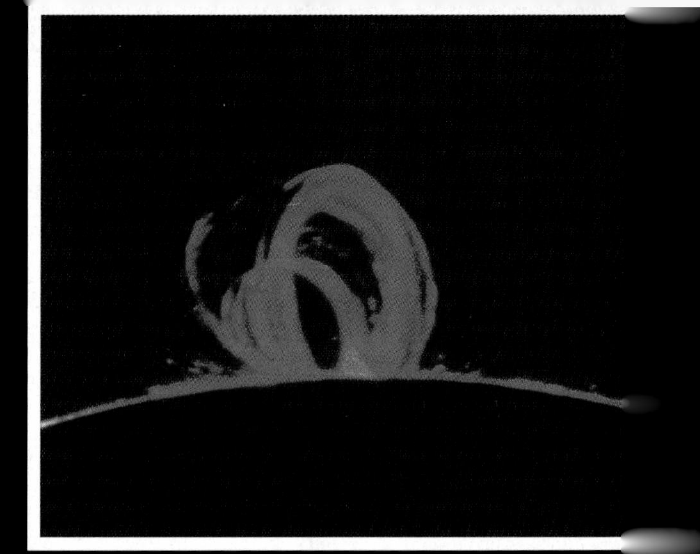

COLOR PLATE 26b A loop prominence on the sun, photographed by the National Solar Observatory at Sacramento Peak, New Mexico. The distinctive shape of the prominence results from strong magnetic fields in the region bending the hot plasma into a loop. (National Optical Astronomy Observatories)

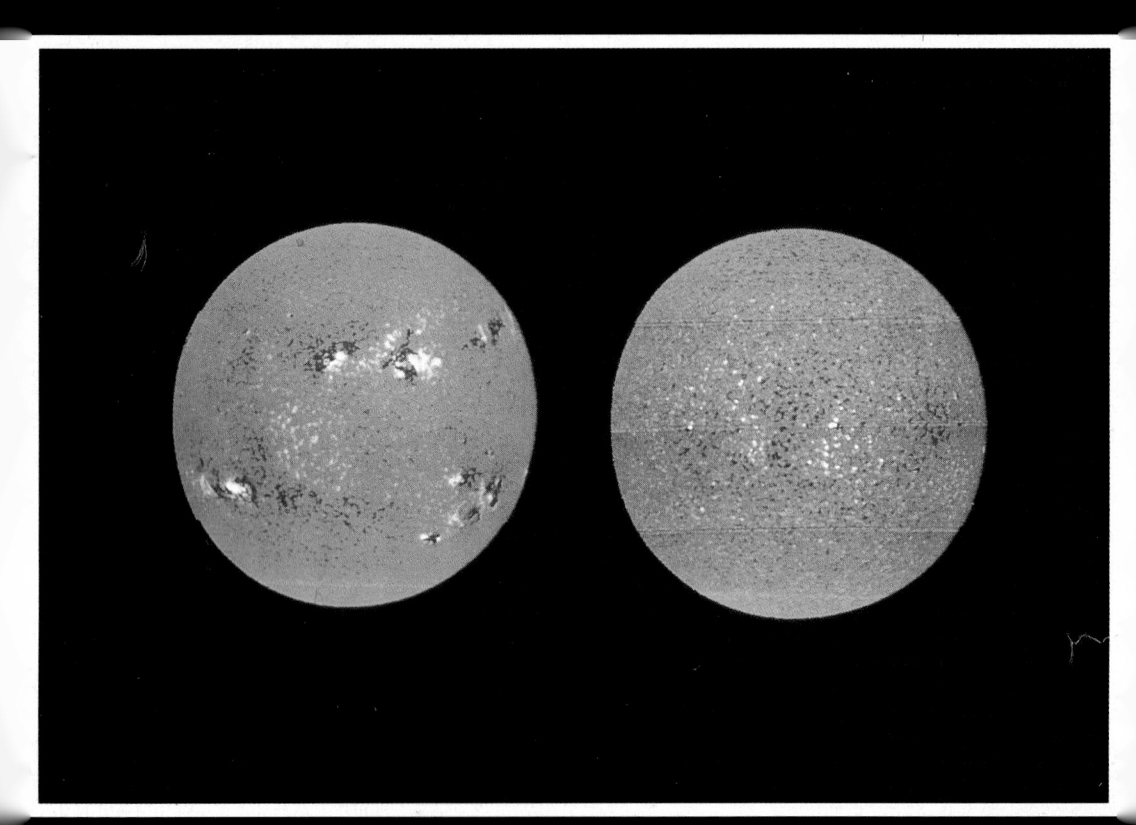

COLOR PLATE 26c These National Solar Observatory images show a comparison of magnetic activity on the quiet and active sun. The computer-generated images use yellow to indicate positive or north polarity, and blue for negative or south polarity. The image of the quiet sun was taken on 26 January 1976 with the Vacuum Solar Telescope at Kitt Peak; the image of the active sun on 3 January 1978. (National Optical Astronomy Observatories)

COLOR PLATE 27 Eta Carinae Nebula. This bright nebula, which includes a great mass of gas and intermixed dust, surrounds the peculiar variable star Eta Carinae. In 1843, Eta Carinae was the second brightest star in the sky. It has faded considerably since that time. Astronomers believe that Eta Carinae is a massive supergiant star that sheds an amount of material equal to the mass of the sun every 100 to 1000 years. The star, nearly invisible around the turn of the century, is brightening again. The nebula is 9000 LY from Earth. (Cerro Tololo Inter-American Observatory Curtis Schmidt telescope photograph/National Optical Astronomy Observatories)

COLOR PLATE 28a The Crab Nebula, in the constellation Taurus. This nebula is the remnant from a supernova explosion, which was seen on Earth in the year 1054 A.D. It is located some 6300 LY away and is approximately 6 LY in diameter, still spreading outward. This composite of images taken with the Kitt Peak National Observatory 4-m Mayall telescope shows hydrogen (red) and sulfur (blue) emission; different colors thus signify different ionization conditions and chemical compositions in various portions of the nebula. (National Optical Astronomy Observatories)

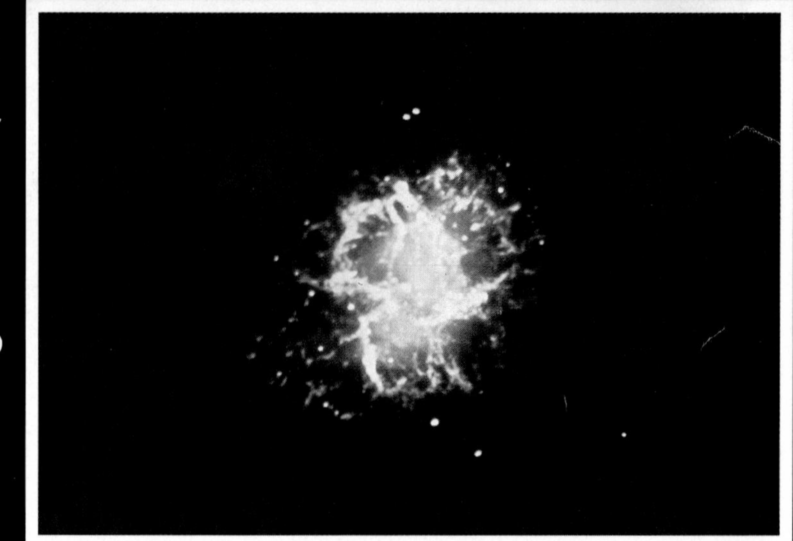

COLOR PLATES 28b and c Two planetary nebulae, the Ring Nebula (*top*) and NGC 2346. Images were taken through three filters and then superposed. The images are coded in such a way that red corresponds to forbidden emission from singly ionized nitrogen, green corresponds to forbidden emission from doubly ionized oxygen, and blue corresponds to emission from ionized helium. The helium is most difficult to ionize and its radiation is concentrated in the center of the nebulae, close to the central stars, which are the sources of the ionizing radiation. Less energy is required to ionize nitrogen, and the emission from it is most conspicuous far from the central star. Images were taken with a CCD at the 2.1-m telescope at Kitt Peak National Observatory. (Courtesy Bruce Balick, University of Washington)

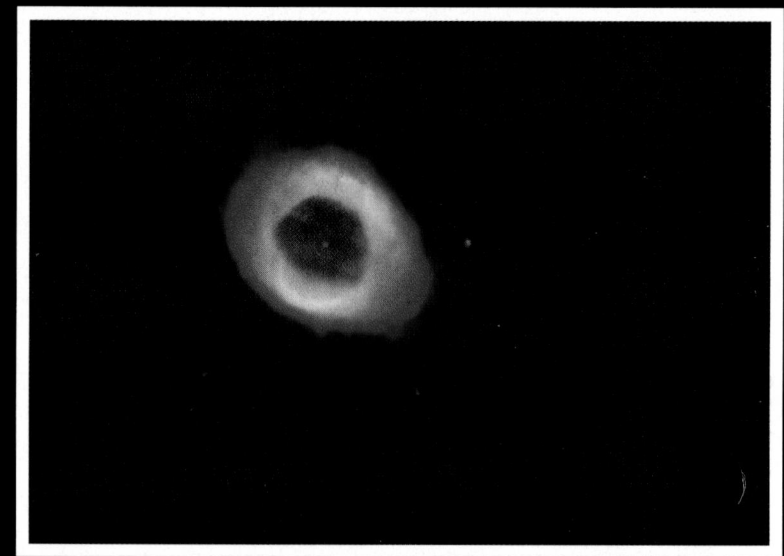

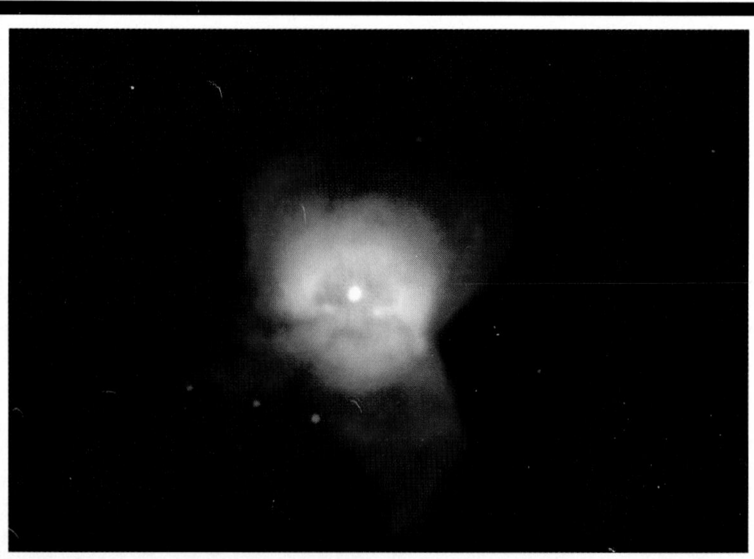

with masses just barely high enough to ignite hydrogen burning, this phase of evolution can take as long as 100 million years.

3. Subsequent evolution on the main sequence is very slow, for a star changes only as thermonuclear reactions alter its chemical composition. For a star of a solar mass, this gradual process requires billions of years. All evolutionary stages are relatively faster in stars of high mass and slower in those of low mass.

30.2 EVOLUTION FROM THE MAIN SEQUENCE TO GIANTS

As soon as a star has reached the main sequence, it derives its energy almost entirely from the thermonuclear conversion of hydrogen to helium. It remains on the main sequence for most of its "life." Since only 0.7 percent of the hydrogen used up is converted to energy, the star does not change its mass appreciably, but in its central regions, where the nuclear reactions occur, the chemical composition gradually changes as hydrogen is depleted and helium accumulates. This change of composition forces the star to change its structure, including its luminosity and size. Eventually, the point that represents it on the H–R diagram evolves away from the main sequence. The original main sequence, corresponding to stars of homogeneous chemical composition, is called the *zero-age main sequence*.

(a) Evolution on the Main Sequence

As helium accumulates at the expense of hydrogen in the center of a star, calculations show that the temperature and density in that region must increase. Consequently, the rate of nuclear-energy generation increases, despite the depletion of hydrogen "fuel," and the luminosity of the star slowly rises. A star, therefore, does not remain indefinitely *exactly* on the original zero-age main sequence. The most massive and luminous stars alter their chemical composition the most quickly. The faintest stars will scarcely change in composition or in luminosity even after billions of years. During the time that a star, even a very massive one, is burning hydrogen in its core, its luminosity increases only by a moderate amount—probably less than a factor of three—before subsequent more rapid changes alter its structure enormously.

When the hydrogen has been depleted completely in the central part of a star, a core develops containing only helium, "contaminated" by whatever small percentage of heavier elements the star had to begin with. The energy source from hydrogen burning* is now used up, and with nothing more to supply heat to the helium core, it begins again to contract gravitationally. Once more the star's energy is partially supplied by potential energy released from the contracting core; the rest of its energy comes from hydrogen burning in the region immediately surrounding the core. These changes result in a substantial and rather rapid readjustment of the star's entire structure, so that the star leaves the vicinity of the main sequence altogether. About 10 percent of a star's mass must be depleted of hydrogen before the star evolves away from the main sequence. The more luminous and massive a star, the sooner this happens, ending its term on the main sequence. Because the total rate of energy production in a star must be equal to its luminosity, the core hydrogen is used up first in the very luminous stars. The most massive stars spend only a few million years on the main sequence; a star of one solar mass remains there for about 10^{10} years, and a spectral-type M0 V star of about 0.4 solar mass has a main-sequence life of some 2×10^{11} years, a value much longer than the age of the universe. Therefore, such low-mass stars would not have had time to complete their main-sequence phase and go on to the next stage of evolution.

(b) Evolution to Red Giants

As the central core contracts, it releases gravitational potential energy, which is absorbed in the surrounding envelope, thereby forcing the outer part of the star to distend greatly. The star as a whole, therefore, expands to enormous proportions; all but its central parts acquire a very low density. The expansion of the outer layers causes them to cool, and the star becomes red. Meanwhile, some of the potential energy released from the contracting core heats up the hydrogen surrounding it to ever

---

* The term "burning" is often used to describe the depletion of an element by nuclear reactions. This "nuclear burning" is not, of course, burning in the literal chemical sense.

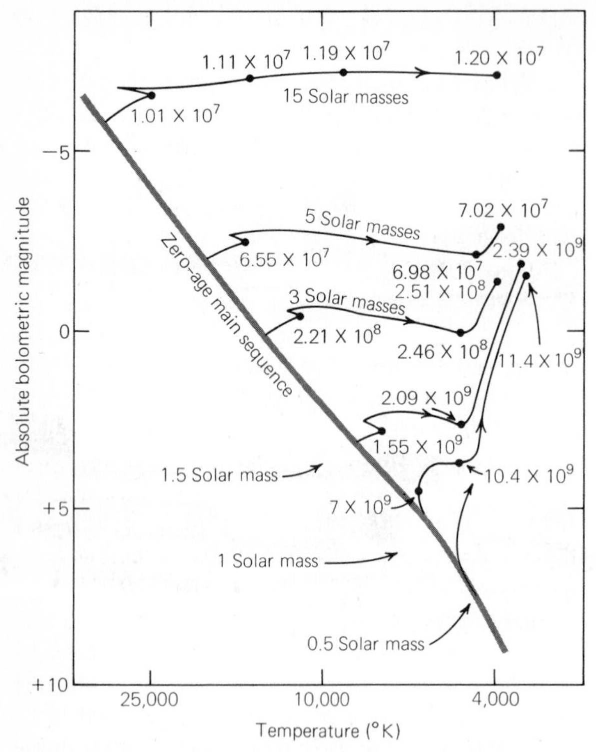

Figure 30.6 Predicted evolution of stars from the main sequence to red giants. See text for explanation. (Based on calculations by I. Iben)

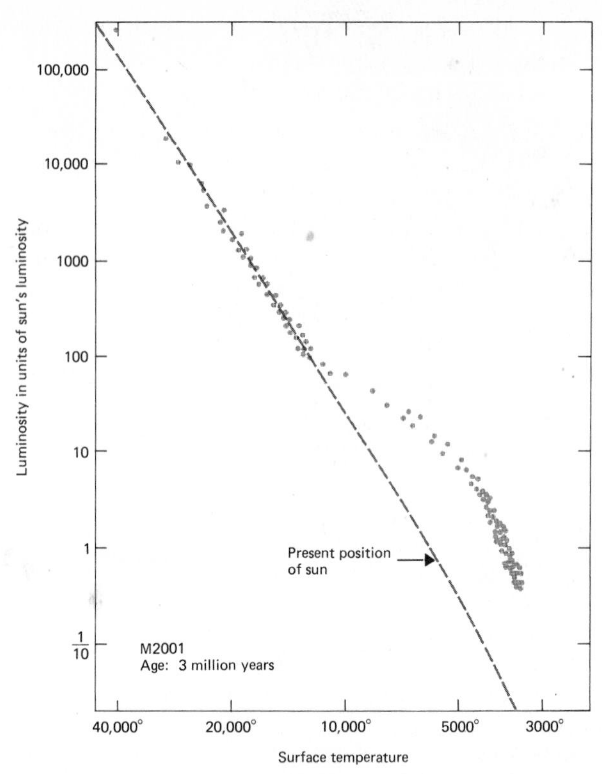

Figure 30.7 The H–R diagram of M2001 at an age of 3 million years.

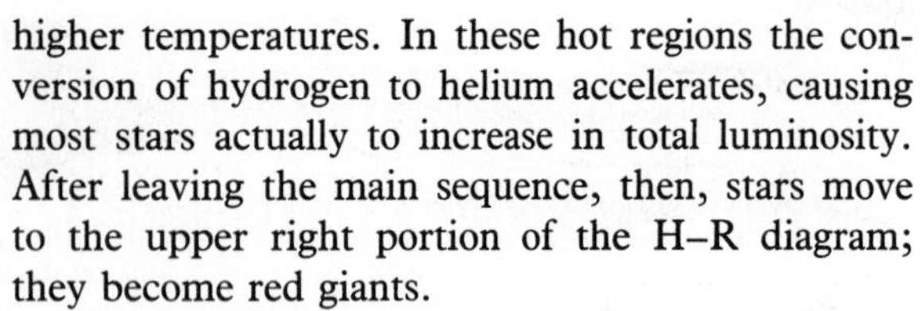

higher temperatures. In these hot regions the conversion of hydrogen to helium accelerates, causing most stars actually to increase in total luminosity. After leaving the main sequence, then, stars move to the upper right portion of the H–R diagram; they become red giants.

Figure 30.6, based on theoretical calculations by Illinois astronomer Icko Iben, shows the tracks of evolution on the H–R diagram from the main sequence to red giants for stars of several representative masses and with chemical composition similar to that of the sun. The broad band is the zero-age main sequence. The numbers along the tracks indicate the times, in years, required for the stars to reach those points in their evolution after leaving the main sequence.

30.3 CHECKING OUT THE THEORY

This description of stellar evolution is based entirely on calculations of stellar models. No star completes its main-sequence lifetime or its evolution to a red giant quickly enough for us to observe these structural changes. Fortunately, nature has provided us a way to test the calculations. Instead of observing the evolution of a single star, we can look at a group or cluster of stars that all formed at the same time but that have different masses. Since stars with higher masses evolve more quickly, we can hope to find clusters in which massive stars have already completed their main-sequence phase of evolution and have become red giants or supergiants, while lower-mass stars in the same cluster are still on the main sequence or even undergoing pre-main-sequence gravitational contraction. One main advantage of using clusters of stars rather than individual stars like those in the solar neighborhood to check the theory of stellar evolution is that the many stars in a cluster are at the same distance, so that their luminosities can be directly compared. If a group of stars is very close together in space, it is also reasonable to assume that they all formed nearly at the same time, from the same cloud, and with the same composition.

In the next chapter, we will describe the properties of star clusters and show how the observa-

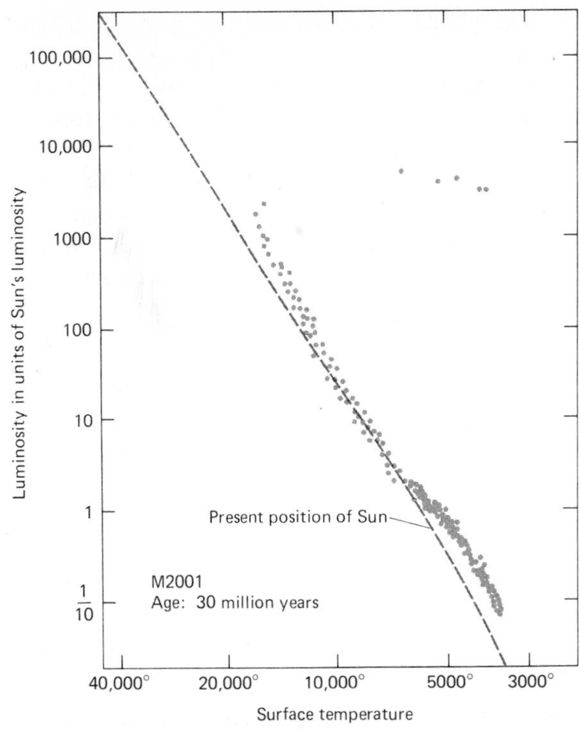

Figure 30.8 The H–R diagram of M2001 at an age of 30 million years.

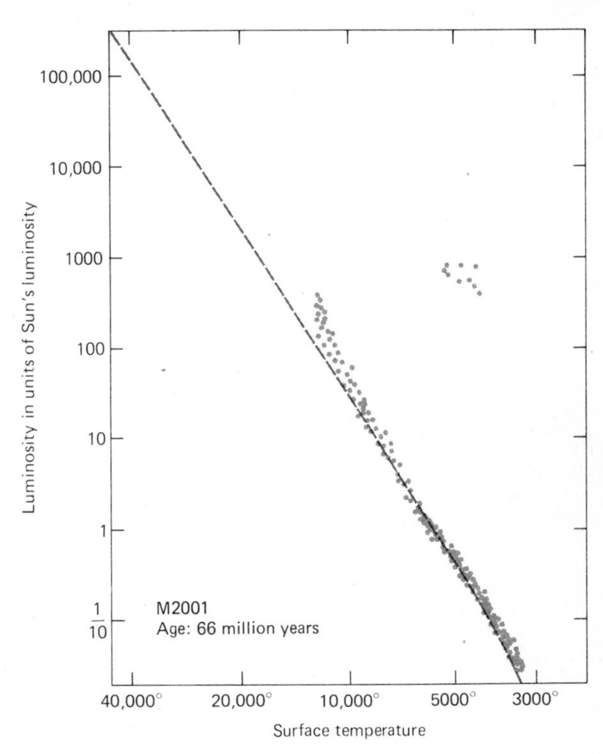

Figure 30.9 The H–R diagram of M2001 at an age of 66 million years.

tions support the theoretical models of stellar evolution. We will conclude this chapter by discussing what theory predicts that the H–R diagrams of clusters of various ages will look like. The theoretical H–R diagrams that we will use were calculated by Rudolf Kippenhahn and his associates at Munich. These diagrams allow us to follow the evolution of a hypothetical star cluster whose members have a chemical composition similar to the sun's. We will call this hypothetical cluster M2001, in honor of a well-known motion picture.

(a) Evolution of a Cluster of Stars

What should the H–R diagram be like for a cluster whose stars have recently condensed from an interstellar cloud? After a few million years, the most massive stars should have completed their contraction phase and be on the main sequence, while the less massive ones should be off to the right, still on their "way in." Figure 30.7 shows the H–R diagram of M2001 at an age of 3 million years.

After a short time—only a few million years after reaching the main sequence—the most massive stars use up the hydrogen in their cores and evolve off the main sequence to become red giants. As more time passes, stars of successively lower mass leave the main sequence to become red giants. Meanwhile, even after 30 million years, the least massive stars will not have completed their gravitational contraction to the main sequence (Figure 30.8).

Figure 30.9 shows the H–R diagram of our hypothetical cluster M2001 at an age of 66 million years. At a much older age of 4.2 billion years, M2001 has the H–R diagram shown in Figure 30.10. Stars only a few times as luminous as the sun have begun to leave the vicinity of the main sequence. The total main sequence lifetime for the sun is expected to be about 10^{10} years. Since the universe is probably no more than 2×10^{10} years old (Chapter 37), few stars less massive than the sun have had time to become red giants.

The stages of evolution of stars after they become red giants are more complicated and are less thoroughly understood. It is there, however, that we find most variable stars, eruptive stars, mass ejection, enrichment of the heavy elements in the Galaxy, and—a subject at the very frontier of mod-

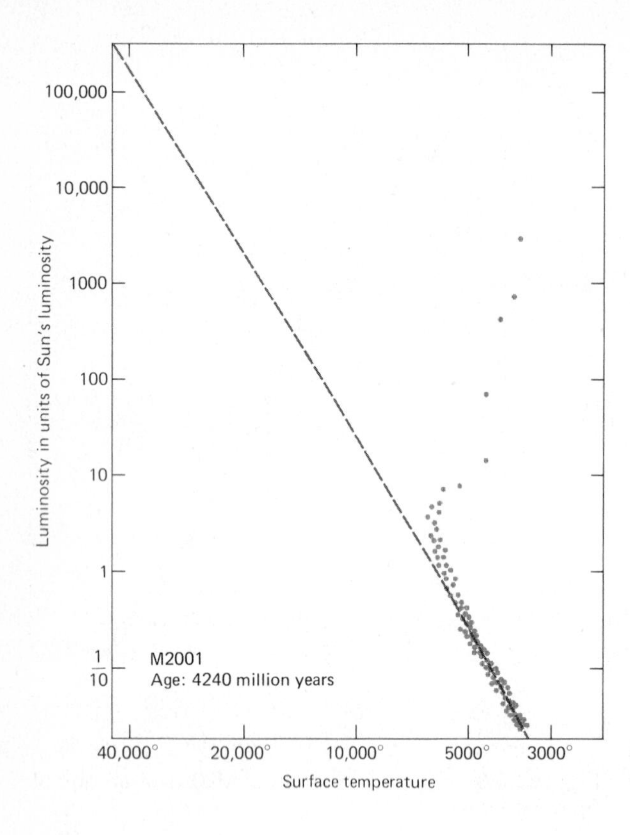

Figure 30.10 The H–R diagram of M2001 at an age of 4.24 billion years.

ern astrophysics–the final phases of a star's life, and its death.

In the next chapter we will discuss real, rather than theoretical, clusters of stars, and show that the observed H–R diagrams are very much like the theoretical diagrams presented here. In the subsequent two chapters, we will describe the often violent and altogether remarkable events that befall a star that has finally exhausted its store of nuclear energy.

EXERCISES

1. Where on the H–R diagram does a star *begin?*
2. ★ Suppose stars contracting from interstellar matter evolved exactly to the left across the H–R diagram (that is, at constant luminosity). The luminosities of main-sequence stars are approximately proportional to the cubes of their masses. Show that more massive stars would contract faster and reach the main sequence sooner than less massive stars.
3. The H–R diagram for field stars (that is, stars all around us in the sky) shows very luminous main-sequence stars and also various kinds of red giants and supergiants. Explain these features and interpret the H–R diagram for field stars.
4. In the H–R diagrams for some young clusters, stars of very low and very high luminosity are off to the right of the main sequence, while those of intermediate luminosity are on the main sequence. Can you offer an explanation? Sketch an H–R diagram for such a cluster.
5. The star Rigel has an absolute magnitude of -6.8. Its mass is uncertain but is probably no more than 50 times the mass of the sun. Estimate the lifetime of Rigel. (Assume that the lifetime of the sun is 10 billion years.)
6. Why is star formation more likely to occur in cold molecular clouds than in regions where the temperature of the interstellar medium is several hundred thousand degrees?
7. Suppose a star cluster were at such a large distance that it appeared as an unresolved spot of light on telescopic photographs. What would you expect the color of the spot to be if it were the image of the cluster immediately after it was formed? How would the color differ after 10^{10} years? Why?
8. Explain the statement in the text that the tracks of evolution on the H–R diagram of globular cluster stars from the main sequence to the red giant branch lie very nearly along the observed sequence of stars in the diagram from the main sequence to red giant.
9. Explain how you could decide whether red giants seen in a star cluster probably had evolved away from the main sequence or were still evolving along their Hayashi lines.
10. What form of energy transport is occurring in the cores and envelopes of the following main-sequence stars: (a) O9; (b) G2; (c) M8?
11. If all the stars in a cluster have the *same* age, how

can clusters be useful in studying evolutionary effects?

12. Assume the sun to be 73 percent hydrogen by mass. What mass of hydrogen will then be "burned" by the time the sun moves away from the main sequence? If the sun's luminosity were constant during this time, how long would it take for the sun to convert this amount of hydrogen into helium? (See Section 29.3b and Appendix 6.)
Answer to Second Part: 7.6×10^9 years

Harlow Shapley (1885–1972) began his career as a newspaper reporter. In his twenties he enrolled in the University of Missouri, where he searched through the catalogue for a suitable major and found Astronomy in the "A's." After earning his bachelor's degree, he went on to Princeton for his Ph.D. Subsequently, at Mount Wilson, his study of globular clusters revealed the true extent of our Galaxy. (Yerkes Observatory)

31

STAR CLUSTERS

In Chapter 30, we showed that theoretical models of stellar evolution make a number of definite predictions about how a group of stars, all formed at about the same time from prestellar material, should change as the group ages. Fortunately, nature has provided groups or clusters of such stars so that we can check the validity of the calculated models.

In this chapter, we will first describe star clusters—where they are found, how big they are, what kind of stars they contain. Then we will show that clusters of various ages do indeed differ from one another in the way that theory predicts.

31.1 DESCRIPTIONS OF STAR CLUSTERS

Clusters that contain a great many stars are said to be *rich;* those that contain comparatively few are said to be *poor*. Rich clusters are likely to be conspicuous, and their identification as genuine stellar systems is certain. Poor clusters, on the other hand, are much more difficult to pick out from among the unrelated stars that lie in front of and behind them. Sometimes a real cluster may not be identifiable as such. Other groups of stars that appear to be real systems may actually be stars at different distances seen close together in projection. Most of the clusters that are catalogued, however, contain a high enough density of stars to stand out against the nonmember stars that lie in the foreground or background, so that there is virtually no chance of their being accidental superpositions of stars at different distances. Even so, it is often difficult or impossible to say with certainty whether a given individual star is a member of the cluster or not. In general, therefore, a few of the stars studied as cluster members are actually stars in the foreground or background, that is, stars which belong to the *field*.

In our Galaxy, there are two basic types of clusters. *Globular clusters* were formed about 12 to 20 billion years ago and contain only very old stars. *Open clusters* are much younger, some being only a few million years old, and still contain massive hot stars. The properties of these two types of clusters are described in detail in the sections that follow.

(a) Globular Clusters

About a hundred globular clusters are known in our Galaxy, most of them in a spherical halo surround-

Figure 31.1 The globular cluster M13 in Hercules. (Palomar Observatory; California Institute of Technology)

ing the flat wheel-like shape formed by the spiral arms. All are very far from the sun, and some are found at distances of 60,000 LY or more from the galactic plane. A few, nevertheless, are bright enough to be seen with the naked eye; they appear as faint, fuzzy stars. One of the most famous naked-eye globular clusters is M13 (Figure 31.1 and Color Plate 25a), in the constellation of Hercules, which passes nearly overhead on a summer evening at most places in the United States. Through a good pair of binoculars the more conspicuous globular clusters resemble tiny mothballs. A small telescope reveals their brightest stars, while a large telescope shows them to be beautiful, globe-shaped systems of stars. Visual observation, however, even through the largest telescope, does not reveal the multitude of fainter stars in globular clusters that can be recorded with long exposures.

A good picture of a typical globular cluster shows it to be a nearly circularly symmetrical system of stars, with the highest concentration of stars near its own center (a few globular clusters, such as Omega Centauri, appear slightly flattened). In the central regions of the cluster, the stars are so closely packed that most of them cannot be resolved as individual points of light but appear as a fairly uniform and continuous glow. Two photographs of a globular cluster made in light of two different colors, say red and blue, show that the brightest stars are red. From measurements of the distance to globular clusters, we know that the brighter red stars must be red giants.

Distances to globular clusters are sometimes calculated from the apparent magnitudes of the RR Lyrae stars they contain (Section 31.2c). From their angular sizes (typically a few minutes of arc) their actual linear diameters are found to be from 20 to 100 pc or more. In one of the nearer globular clusters more than 30,000 stars have been counted, but if those stars too faint to be observed are considered, most clusters must contain hundreds of thousands of member stars. The combined light from all these stars gives a typical globular cluster an absolute magnitude in the range -5 to -10, or ten thousand to one million times the luminosity of the sun.

The average star density in a globular cluster is about 0.4 star per cubic parsec. In the dense center of the cluster the star density may be as high as 100 or even 1000 per cubic parsec. There is plenty of space between the stars, however, even in the center of a cluster. The "solid" photographic appearance of the central regions of a globular cluster results from the finite resolution of the telescope and seeing effects of the Earth's atmosphere. (With the Hubble Space Telescope, which will be above the blurring effects of the atmosphere, it should be possible to resolve and analyze individual stars near the centers of globular clusters.) A bullet fired on a straight line through a point near the center of a cluster would have far less than one chance in 10^{11} of striking a star. If the Earth revolved not about the sun, but about a star in the densest part of a globular cluster, the nearest neighboring stars, light-months away, would appear as points of light. Thousands of stars, however, would be scattered uniformly over the sky. The Milky Way would be hard, if not impossible, to see, and even on the darkest of nights the brightness of the sky would be comparable to faint moonlight.

The motions of globular clusters will be described in Chapter 34. They are high-velocity objects that do not partake of the general galactic rotation. They are believed to revolve about the nucleus of the Galaxy on orbits of high eccentricity and high inclination to the galactic plane (rather like the orbits of comets in the solar system). Obeying Kepler's second law, a cluster spends most of its time far from the nucleus; a typical cluster probably has a period of revolution of the order of 10^8 years.

Figure 31.2 The Pleiades. North is to the left. The wisps of gas that surround these very young stars are probably the result of an encounter between these stars and an interstellar cloud. (Lick Observatory)

Because most globular clusters lie outside the plane of the Milky Way, probably a good fraction of them have been discovered, although a few dozen, hidden by the obscuring dust clouds, may remain undiscovered in the disk and nucleus of the Galaxy.

(b) Open Clusters

In contrast to the rich, partially unresolved globular clusters, *open clusters* appear comparatively loose and "open" (hence their name). They contain far fewer stars than globular clusters and show little or no strong concentration of stars toward their own centers. Although open clusters are usually more or less round in appearance, they lack the high degree of spherical symmetry that characterizes a globular cluster; some open clusters actually appear irregular. The stars in these clusters are usually fully resolved, even in the central regions.

Open clusters, as mentioned previously, are found in the disk of the Galaxy, often associated with interstellar matter. Because of their locations, they are sometimes called *galactic clusters* rather than open clusters. They are low-velocity objects and are presumed to originate in or near spiral arms. Over 1000 open clusters have been catalogued, but many more are identifiable on good search photographs such as those of the Palomar Sky Survey. Yet only the nearest open clusters can be observed, because of interstellar obscuration in the Milky Way plane. We conclude, therefore, that we see only a small fraction of the open clusters that actually exist in the Galaxy; possibly tens or even hundreds of thousands of them escape detection.

Several open clusters are visible to the unaided eye. Most famous among them is the *Pleiades,* which appears as a tiny group of six stars (some people see more than six) arranged like a tiny dipper in the constellation of *Taurus* (Figure 31.2 and

Figure 31.3 The open star cluster M67 (NGC 2682) in Cancer. This is one of the oldest and most densely populated of the open clusters. (Caltech/Palomar Observatory)

Color Plate 25c); a good pair of binoculars shows dozens of stars in the cluster, and a telescope reveals hundreds. (The Pleiades is *not* the Little Dipper; the latter is part of the constellation of *Ursa Minor,* which also contains the North Star.) The *Hyades* is another famous open cluster in Taurus. To the naked eye, the cluster appears as a V-shaped group of faint stars, marking the face of the bull. Telescopes show that the Hyades actually contains more than 200 stars. The naked-eye appearance of the *Praesepe,* in Cancer, is that of a barely distinguishable patch of light; this group is often called the "Beehive" cluster, because its many stars, when viewed through a telescope, appear like a swarm of bees.

Typical open clusters contain several dozen to several hundred member stars, although a few, such as M67, contain more than a thousand (Figure 31.3). Compared with globular clusters, open clusters are small, usually having diameters of less than 10 pc. The RR Lyrae stars are never found in open clusters, but other kinds of variable stars, such as type I cepheids, are sometimes present.

(c) Associations

For more than 50 years it has been known that the most luminous main-sequence stars—those of spectral types O and B—are not distributed at random in the sky but tend to be grouped into what are now called *associations,* lying along the spiral arms of our Galaxy. In the decade following World War II, interest was greatly revived in these groups of hot stars, especially by the Soviet astronomer V. A. Ambartsumian, who called attention to many of them and pointed out that they must be very young groups of stars. Because the stars of an association lie in the galactic plane and are spread over tens of parsecs, each revolves about the galactic center with a slightly different orbital speed; Ambartsumian showed that the different orbital speeds of the different members of an association would completely disrupt the group after a few million years. All associations, therefore, must be very young objects on the astronomical time scale.

We distinguish between two kinds of associations: those containing luminous, massive O and B stars are called *O-associations;* the others contain only low-mass stars and are called *T-associations* because they contain numerous T Tauri stars. The T Tauri stars are low-mass stars (masses less than three times that of the sun) that have formed so recently that they are still contracting and have not yet reached the main sequence. Associations are always associated with interstellar matter and are believed to be very young stars that have just formed from it. Sometimes, as in Orion, T-associations and O-associations coexist.

An O-association appears as a group of several (say, 5 to 50) O stars and B stars, and sometimes

TABLE 31.1 Characteristics of Star Clusters

	GLOBULAR CLUSTERS	OPEN CLUSTERS	ASSOCIATIONS
Number known in Galaxy	125	1055	70
Location in Galaxy	Halo and nuclear bulge	Disk (and spiral arms)	Spiral arms
Diameter (pc)	20 to 100	<10	30 to 200
Mass (solar masses)	10^4 to 10^5	10^2 to 10^3	10^2 to 10^3?
Number of stars	10^4 to 10^5	50 to 10^3	10 to 100?
Color of brightest stars	Red	Red or blue	Blue
Integrated absolute visual magnitude of cluster	−5 to −10	0 to −10	−6 to −11
Density of stars (solar masses per cubic parsec)	0.5 to 1000	0.1 to 10	<0.01
Examples	Hercules Cluster (M13)	Hyades, Pleiades, h and χ Persei, Praesepe	Zeta Persei, Orion

Wolf-Rayet stars, scattered over a region of space some 30 to 200 pc in diameter. Because these stars are rare, it would be very unlikely for so many of them to exist by chance in so relatively small a volume of space. It is assumed, therefore, that the stars in an association are either physically associated or at least have had a common origin. Stars of other spectral types may also belong to associations, but these more common stars are not conspicuous against the general star field and do not attract attention to themselves as belonging to any particular group.

Often a small open cluster is found near the center of an association. It is presumed, in such cases, that the stars in the association are the outlying members of the cluster, and that they probably had a common origin with the cluster stars.

Some associations are expanding—their member stars moving radially outward from the association center. An example of such an association is one in Perseus, known as the *Zeta Persei association.* More than a dozen of its member stars are bright enough so that their positions have been measured accurately for more than a century. Most of the stars seem to be moving outward, from the center of the group. If we extrapolate their motions backward, we find an age for the expansion of about 1½ million years. Another famous association is in *Orion,* centered on a group of stars in the middle of the Orion Nebula. Analysis of the statistics of associations and O and B stars have led some investigators to conclude that all such stars have originated in associations.

About 70 associations are now catalogued. Like ordinary open clusters, however, they lie in regions occupied by interstellar matter, and many others must be obscured. There are probably several thousand undiscovered associations in our Galaxy.

(d) Summary of Clusters

The foregoing descriptions of star clusters are summarized in Table 31.1. The numbers of globular clusters, open clusters, and associations are taken from the second edition of the *Catalogue of Star Clusters and Associations,* published by the Hungarian Academy of Sciences. The sizes, absolute magnitudes, and numbers of stars listed for each type of cluster are approximate only and are intended as representative values.

31.2 PULSATING VARIABLES

In Chapter 23, we discussed the fact that some stars vary rhythmically in brightness. Furthermore, the period of time between successive phases of maximum brightness is longer for variable stars that are intrinsically more luminous. Variable stars are found in both open and globular clusters and are used to establish distances to them. What causes some stars to vary? And why are most stars constant in luminosity to an accuracy of 1 percent or less?

(a) The Pulsation of Cepheid Variables

Cepheid variables (Section 23.3e) provide a good example of pulsating variables. A hint of the physical nature of the variability of these stars comes from their spectra. At maximum light, their spectral classes correspond to higher surface tempera-

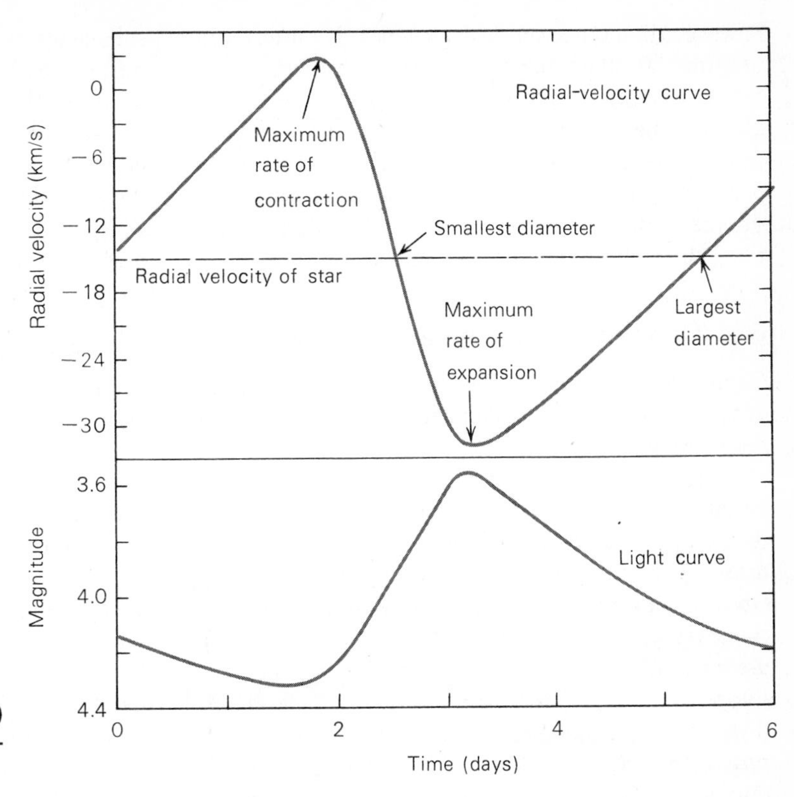

Figure 31.4 Radial-velocity curve *(top)* and light curve *(bottom)* of the star δ Cephei.

tures than at minimum light. The light variations of a cepheid, therefore, are due in part to variations in the temperature of its outer layers. Further spectroscopic evidence reveals that the temperature fluctuations are accompanied by actual pulsations in the sizes of these stars. The lines in the spectrum of a cepheid show Doppler shifts that vary in exactly the same period as that of the star's light fluctuations. Evidently, the changes in light are associated with a periodic rise and fall of the cepheid's atmosphere.

A graph that displays changes in the Doppler shifts of the spectral lines of a cepheid with lapse of time is called a *radial-velocity curve* (Figure 31.4). It is like the radial-velocity curve of a spectroscopic binary star, except that in a cepheid the Doppler shifts are due to the periodic rising and falling of its outer layers rather than to the orbital motion of the star as a whole. The mean value of the apparent radial velocity corresponds to the line-of-sight motion of the star itself; the photospheric pulsations cause variations about this mean value. When the photosphere expands, it is approaching us with respect to the rest of the star, and each spectral line is shifted to slightly shorter wavelengths than that of its mean position; when the photosphere contracts, the lines are shifted to slightly longer wavelengths. When the photosphere reaches its highest or lowest point—that is, when the star is at its largest or smallest size—the position of the spectral lines corresponds to the radial velocity of the star itself.

We can calculate the total distance through which the cepheid's photosphere rises or falls by multiplying the velocity of its rise at each point in the pulsation cycle by the time it spends at that velocity, and then adding up the products.* For δ Cephei, for example, we find by such a calculation that the photosphere pulsates up and down over a distance of somewhat under 3 million km. On the other hand, the mean diameter of δ Cephei as calculated from Stefan's law (Section 25.3a) is about 40 million km. During a pulsation cycle, therefore, the radius of the star changes by about 7 or 8 percent.

We might expect a pulsating star to be hottest when it is smallest and most compressed. δ Cephei,

---

* Or, technically, by *integrating* the velocity curve over the time of rise or fall.

however, like other cepheid variables, is hottest and brightest at about the time when its radiating surface is rushing outward at its maximum speed. Evidently, the greatest compression of the star as a whole does not correspond to the maximum temperature at its surface; the explanation is related to the mechanism by which energy is transferred outward through the outer layers of the star.

(b) Cepheids in Globular Clusters

Cepheids of the kind just described are massive young stars. Some are found in open clusters. In a very few globular star clusters, and also outside of clusters in the galactic halo (Chapter 34), there are variable stars with periods in the range of 10 to 30 days. The light curves of these stars are similar to those of the cepheid variables except that they fall from maximum to minimum light more slowly, and the stars are somewhat bluer in color. These objects are sometimes called W Virginis stars, after the prototype W Virginis, or sometimes *type II cepheids,* to distinguish them from ordinary or "classical" cepheids, which are called *type I cepheids.* Although only a few dozen type II cepheids are known, a period-luminosity relation also seems to hold for them, and it resembles that for type I cepheids, in that stars of longer period are the more luminous. Because of the similar characteristics of type I and type II cepheids, they were originally thought to be the same kind of star, and only since about 1950 has it been realized that type I and type II cepheids are actually different.

(c) RR Lyrae Stars

Among the most common variable stars are the *RR Lyrae* stars, named for RR Lyrae, best known member of the group. Nearly 4500 of these variables are known in our Galaxy. Almost all of them are found in the nucleus or the halo of our Galaxy (Chapter 34) or in globular clusters. In fact, nearly all globular clusters contain at least a few RR Lyrae variables, and some contain hundreds; these stars, therefore, are sometimes called *cluster-type* variables.

The periods of RR Lyrae stars are less than one day; most periods fall in the range of 0.3 to 0.7 day. The amplitudes of their visual light curves never exceed two magnitudes, and most RR Lyrae stars have amplitudes less than one magnitude. Several subclasses of RR Lyrae stars are recognized, but the differences between these subclasses are small and need not be considered here.

It is observed that the RR Lyrae stars occurring in any particular globular cluster all have about the same median apparent magnitude. Since they are all at approximately the same distance, it follows that they must also have nearly the same absolute magnitude. Because the RR Lyrae stars in different clusters are all similar to each other in observable characteristics, it is reasonable to assume that *all* RR Lyrae stars have about the same absolute magnitude. If we could learn what that absolute magnitude is, we could immediately calculate the distances to all globular clusters that contain these stars.

Unfortunately, as is true for cepheids, not a single RR Lyrae star is near enough to measure its parallax by direct triangulation. Like the cepheids, distances to RR Lyrae stars have also had to be determined by statistical means, that is, by analyzing their proper motions and radial velocities. The early distance investigations, carried out from 1917 to about 1940, seemed to indicate an absolute magnitude of about 0 for these important stars—they are, therefore, about 100 times as luminous as the sun, whose absolute magnitude is about +5. More recent work shows that individual RR Lyrae stars may differ from each other slightly in median absolute magnitude, and that the average value may be a little fainter than was once thought—somewhere between 0 and +1. These recent revisions do not vitiate the important results concerning the nature of our Galaxy derived from the study of RR Lyrae stars (Chapter 34).

Figure 31.5 shows the form of the period-luminosity relation for the three kinds of variable stars discussed in the last four subsections.

(d) Cause of Pulsation

In a stable star the weights of the constituent layers, bearing toward the center of the star under the influence of gravity, are just balanced by the pressure of the hot gases within (Chapter 28.4). A pulsating star, on the other hand, is something like a spring: as the star contracts, its internal pressures build up until they surpass the weights of its outer layers. Eventually, these pressures start the star pulsing outward, but because of their inertia, the outward-moving layers over-

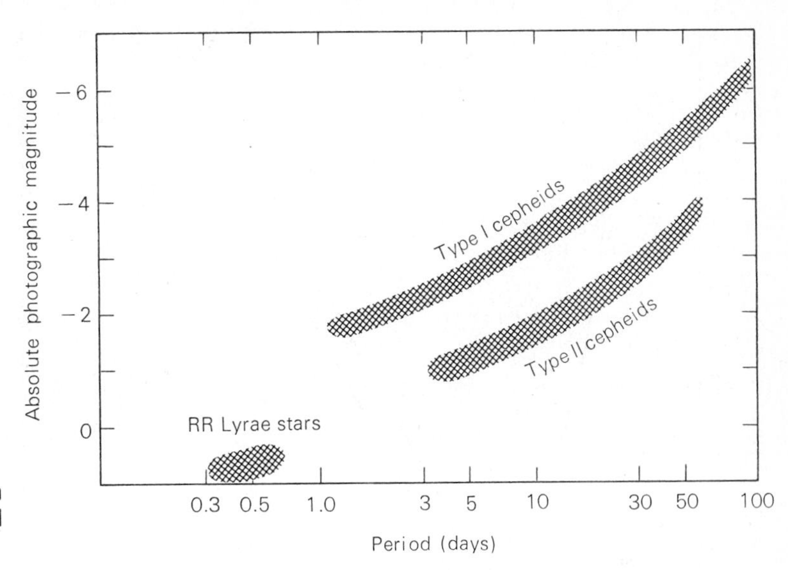

Figure 31.5 Period-luminosity relation for type I cepheids, type II cepheids, and RR Lyrae stars.

shoot the equilibrium point where their weights will just balance the internal pressures. As the star expands further, the weights of the overlying layers decrease, but the internal pressure decreases faster. Hence, the overlying layers are not supported adequately, and the star begins to contract. As it does so, it overshoots again, and this time it becomes too highly contracted. Once more the inner pressures cause the star to expand—and so the pulsation continues.

Stars would pulsate indefinitely if there were no dissipation of energy. Most stars do not pulsate in this way, however, because with each pulsation some of the energy is radiated away, converted to convection of stellar gases, or otherwise lost. Thus if a star were to start oscillating, its pulsations would quickly die out due to these damping forces unless it experienced some kind of driving force.

What is the driving force that causes some stars to pulsate? A highly simplified explanation for the cepheids and stars with similar temperature follows. It is important in reading this explanation to understand that the pulsation is confined to the outer layers of the star. The deep interior does not pulsate.

During the phase when the star is smallest, that is, when it is most compressed, the temperature becomes hot enough to strip both electrons away from the helium nuclei in the star's atmosphere. Energy absorbed by the helium atoms in excess of that required for ionization is converted to kinetic energy of the freed electrons, which in turn heat the gas in the stellar atmosphere through collisions. Energy absorbed by the ionization process is thus dammed up in the atmosphere instead of being radiated into space. The temperature increases in the atmosphere where the ionization is occurring, pressure builds, and the outer layers of the star are forced to expand. Once most of the helium is ionized, however, then radiation can once again flow freely through the atmosphere, the atmosphere cools, the pressure drops, and the weight of the outer layers of the atmosphere compress the star. As the pressure increases, electrons recombine with helium nuclei, and the process begins all over again as these helium nuclei are reionized and in the process block again the flow of energy. In some stars, the ionization of hydrogen can also drive pulsation.

All stars have helium. Why do only some of them pulsate? Again, in very simplified terms, in cool stars, the temperature becomes high enough to ionize helium so deep in the star that the pressure in this region is not high enough to lift all of the mass in the layers lying above it. In very hot stars, the ionization of helium occurs in a layer of the atmosphere so close to the outer boundary of the star that the weight of the layer above it is not great enough to compress the star.

This mechanism will, therefore, cause only stars of intermediate temperature to pulsate. Figure 31.6 shows the *instability strip,* which is that portion of the H–R diagram where the temperature is right for helium ionization to cause pulsation. Cepheids and RR Lyrae stars lie within the instability strip. Even white dwarfs that fall within the instability strip are variable.

Although most stars do not pulsate, it is still interesting to ask what the pulsation period would be for an ordinary star if it were unstable. It turns out that the period is greater for a giant star of low mean density than for a smaller compact star of higher density—just as a long piano string vibrates more slowly than a short one. It can be shown that for pulsating stars of any one type, the period of a particular star is inversely proportional to the square root of its mean density.

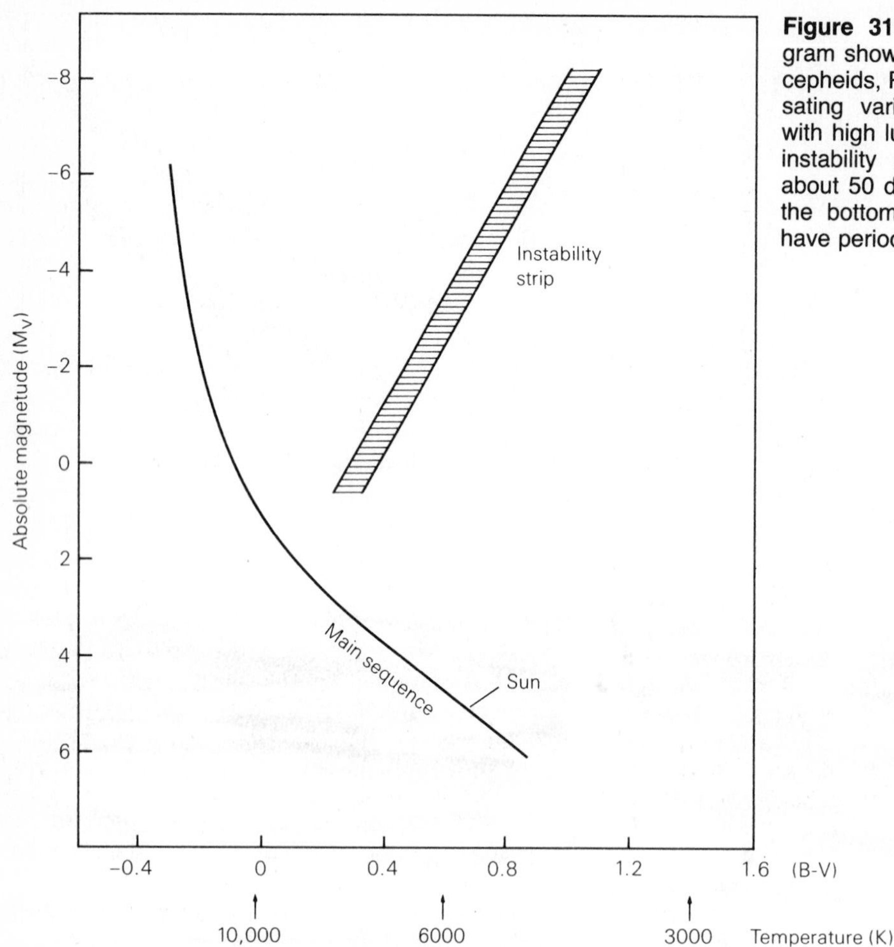

Figure 31.6 Hertzsprung-Russell diagram showing the instability strip where cepheids, RR Lyrae stars, and other pulsating variables are found. Cepheids with high luminosity lie at the top of the instability strip and have periods of about 50 days. RR Lyrae stars lie near the bottom of the instability strip and have periods of only about half a day.

We can derive the relation between the period and the mean density of a pulsating star from a consideration of Kepler's third law. When a star is at its maximum size and there is no longer enough pressure to support its outer layers, they simply fall inward under the influence of the star's gravitation. Thus the matter at the surface of such a star is temporarily in free fall toward the star's center, as if it were in an elliptical orbit about the center of the star (but a degenerate ellipse that is a straight line). The period of revolution of such material in this hypothetical orbit is of the same order of magnitude as the period with which the star would pulsate. By Kepler's third law, this period, P, is related to the mass, M, of the star and the semimajor axis of the orbit—which is proportional to the radius, R, of the star—by the equation (Section 4.4)

$$MP^2 \propto R^3.$$

Upon solving for the period, we find

$$P \propto R^{3/2}/M^{1/2}.$$

Now the mass of the star is the product of its volume, $4\pi R^3/3$, and its mean density, ρ; therefore,

$$P \propto R^{3/2}/(R^{3/2}\rho^{1/2}),$$

or

$$P\sqrt{\rho} = constant.$$

The surface of the star does not, of course, complete an actual orbit about its center. We have seen that in a cepheid, for example, the radius changes by less than 10 percent. On the other hand, if a point on the stellar surface *could* complete such a full orbit, it would still spend most of its time in the outer part of that orbit (in accord with Kepler's second law); thus the representation is not as bad as it might seem. Even if the actual pulsation period differed by five or ten times from that derived by the above procedure, it would still be the case that the ratios of pulsation periods of different stars would be roughly inversely proportional to

Table 31.2 Pulsation Periods for Various Stars of One Solar Mass

RADIUS (Solar Radii)	PERIOD	EXAMPLES
1	1 hr	Sun
1000	4 yr	Red supergiants
100	1 month	Cepheid
10	1 day	RR Lyrae star
0.1	2 min	
0.01	4 s	White dwarf
10^{-5}	10^{-4} s	Neutron star

the square roots of their mean densities. (The proportionality would be exact if the different stars were exactly similar to each other, that is, were built on the same model.)

The shortest period with which a star could rotate is, in order of magnitude, the same period with which it would pulsate. This relation is easily seen; the speed at the equator of a rotating star cannot exceed the speed of a body at that point on a circular orbit, and that circular orbit has only twice the semimajor axis of the rectilinear elliptical orbit of major axis equal to the star's radius.

In sum, the shorter the period of pulsation of a star, the higher the mean density of the star. We can estimate the periods of pulsation of other kinds of stars by comparing their mean densities. Table 31.2 shows the results of such a comparison for stars all of one solar mass but of different radii. Because most stars have masses that differ from the sun's by less than a factor of 10, their pulsation periods would differ from the tabulated values for one solar mass stars of the same radius by only at most a factor of 3.

A note of caution is in order. While Table 31.2 indicates what the period of these various stars would be if they did pulsate, additional calculations are required to determine whether or not these stars are unstable and if so, whether the amplitudes of pulsation are large enough to be detected. In the case of the sun, we know that pulsation does occur but the amplitude is very small—so small in fact, that similar pulsations could not be detected except in a few of the brightest (in terms of apparent magnitude) stars in the sky.

31.3 DYNAMICS OF STAR CLUSTERS

The problem of two bodies moving under the influence of their mutual gravitational attraction was solved by Newton. We have seen (Chapter 4) that each member of such a two-body system moves on a path (a conic section) that can be described by a simple algebraic equation. A star cluster, on the other hand, is a system of many bodies. Each star in a cluster moves under the influence of gravitational forces exerted upon it simultaneously by all the other stars. Since all the stars in a cluster are moving, the gravitational forces they exert on each other are constantly changing, and it is an enormously complicated problem to predict the future path of any one.

In principle, the motion of an individual star in a cluster could be computed in detail. In fact, the individual motions in hypothetical clusters of up to hundreds of stars have been calculated with electronic computers to study theoretically the dynamical evolution of such systems. Similar calculations must be performed in space science applications, for example, to compute the motion of a space vehicle moving in the combined gravitational fields of the Earth, Moon, and sun. The particular wanderings of an individual star in a cluster, however, are not of much interest. Far more significant is the way all the cluster members move on the average, for their average motion depends upon certain fundamental characteristics of the cluster—its mass, size, and structure. This section will describe, briefly, the statistical methods that enable us to learn some of the properties of clusters from the average motions of their member stars.

(a) The Virial Theorem and Masses of Clusters

If a stone is raised high above the ground and released, the force of the Earth's gravity upon it accelerates it downward; as it falls, it picks up more and more speed, or energy of motion. The energy associated with the motion of an object is its *kinetic energy.* The potential ability of gravity to accelerate a body and give it kinetic energy is called *gravitational potential energy.* The greater the height from which it is dropped, the greater is its potential energy before it is released.

Similarly, a star in a cluster feels itself attracted generally toward the center of the cluster, and so has gravitational potential energy. The actual amount of its potential energy depends upon the strength of the total resultant gravitational force acting on it and its distance from the cluster center. As a star moves about in a cluster, it sometimes decreases its distance from the center, and thus speeds up, converting some of its potential energy to kinetic energy. At other times the star increases its distance from the cluster center, pulling against gravity, and therefore slows down; it then converts kinetic to potential energy. The actual path of the star may be very complicated, because occasionally it may pass near another star, and its direction of motion is then deflected by the gravitational attraction between the two. Such a deflection is called a *gravitational encounter.* An "encounter" in this case is not a real collision; because of the distances separating the stars in a cluster, actual stellar collisions are extremely improbable.

There is associated with a star cluster a certain total kinetic energy, which is defined as the sum of the individual kinetic energies of its member stars. (The kinetic energy of a star of mass m and speed v is $\frac{1}{2} mv^2$.) By convention, the potential energy of a star in a cluster at a given instant is defined as the work that must be done on the star (that is, the energy that must be given it) to remove it from its location in the cluster at that instant to a point infinitely far away, pulling, in the process, against the gravitational attraction between the star and cluster. The potential energy of the entire cluster is the energy required to separate all the stars infinitely far apart. Actually, as defined, the potential energy of a star cluster is the potential energy its stars would have given up if they had all fallen together under their mutual gravitational attraction, from a configuration in which they were extremely widely separated to their present configuration in the cluster. Clearly, more potential energy would have been released if the cluster were smaller and more concentrated than it is, and less if it were more spread out.

After a cluster has existed for a long enough time (usually an interval of hundreds or thousands of millions of years) there will have been enough encounters between stars to divide the total energy of the cluster (potential plus kinetic) approximately evenly among the stars. The energy will never be distributed exactly equally; there will always be a few stars that by virtue of recent encounters have more energy than average, and others that have less energy than their share. On the average, however, each star will have its share of the total energy of the cluster; no one *kind* of star would be expected to have more energy than any other kind. The cluster is then said to be in a state of *statistical equilibrium.*

Under the above conditions of statistical equilibrium, the total potential energy of a cluster (as defined above) is twice its kinetic energy. This statement is called the *virial theorem.* An important application of the virial theorem in astronomy is in the estimation of masses of systems of stars presumed to be in statistical equilibrium. The underlying physical idea can be summarized briefly as follows: If we assume that a cluster is in equilibrium, its members are held together and their motions are determined by their mutual gravitation. The virial theorem tells us how much mass the cluster must have (and hence how much gravitation) in order for its stars to move with their observed speeds, and in such paths that the cluster has its observed shape.

(b) Stability of Star Clusters

A condition for the stability of an isolated cluster is that its total potential energy be greater than its kinetic energy. Otherwise, the average stellar speed exceeds the escape velocity from the cluster, which dissipates into space. If the potential energy of the cluster exceeds its kinetic energy, it is gravitationally bound—that is, its member stars cannot all escape. (See Exercise 11.)

Clusters, however, are not completely isolated but move in various orbits in the Galaxy. Thus, an added condition for the stability or permanence of a cluster is that it be bound together with gravitational forces that are stronger than the disrupting tidal forces of the Galaxy, or other nearby stars, upon it. The more compact a cluster, the greater is its own gravitational binding force compared to the disrupting forces, and the better chance it has to survive to old age.

Globular clusters are highly compact systems and are, consequently, very stable. Most globular clusters can probably maintain their identity almost indefinitely. Even these clusters lose some stars, however—especially those of relatively small mass. A few stars in a cluster are always moving substantially faster than average. Every now and then one of them, through an encounter, will be given enough speed to escape the cluster. Some of the stars in the galactic halo must be stars that have, in the past, escaped in this way from globular clusters.

When a star escapes, it carries off energy, leaving less energy than before for the stars remaining in the cluster. The result is that, over time, the cluster develops a tightly bound core surrounded by a rarefied halo of stars. In the dense core, stars occasionally collide and some of the debris eventually coalesces. It is predicted that this dynamical evolution will lead to the development of a massive black hole at the cluster center (Chapter 33). Meanwhile, a few stars in the outer parts of the cluster continue to escape.

The escape rate and dynamical evolution for the rich globular clusters, however, is so slow that the clusters can survive for many thousands of millions of years. The situation is analogous to the evaporation of molecules from a more or less permanent planetary atmosphere.

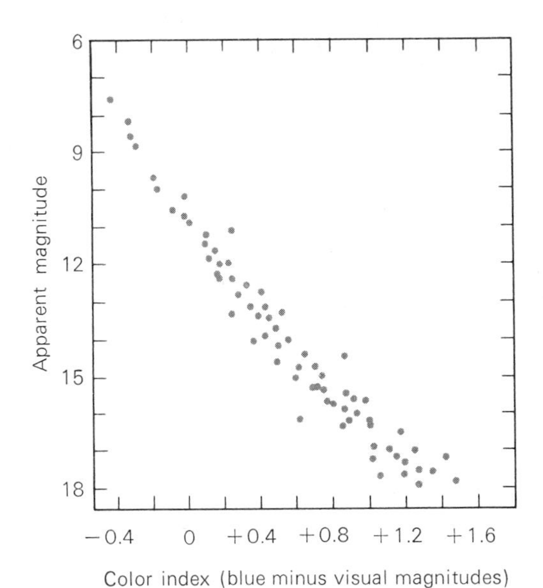

Figure 31.7 Color-magnitude diagram for a hypothetical open star cluster.

Matters are very different for most open clusters. Those that we have discovered are relatively close to the sun; at our distance from the galactic center, the tidal force of the Galaxy will shear a cluster apart in short order if the star density within it is much less than about one star per cubic parsec. The Pleiades, for example, is probably just stable in its central regions, while its outer parts are probably dissipating. The Hyades is on the verge of instability. Typical open clusters have maximum lifetimes as clusters of only a few hundreds of millions of years; a handful of the richer, denser ones, like M67, can be expected to survive for thousands of millions of years. Most open clusters, in other words, are relatively young stellar groups, and only a small fraction are very old.

The case for stellar associations, which have very low star densities, is even more extreme. These loose groups cannot possibly be permanent stellar systems. If they are actually expanding, as one or two appear to be, they would be highly unstable even in the absence of disruptive galactic tidal forces.

31.4 POPULATIONS OF STAR CLUSTERS

Hertzsprung-Russell diagrams, or color-magnitude diagrams, of star clusters are extremely useful in the study of stellar evolution. A color-magnitude diagram is a plot of the apparent magnitudes of cluster stars versus their color indices, as shown in Figure 31.7. Recall that the color index of a star depends on the star's temperature and is thus related directly to its spectral class. (Colors can be measured much more quickly than spectral types and so are frequently used in studies of large numbers of faint stars.) Therefore, a plot of apparent magnitude versus color index is, with one exception, like a Hertzsprung-Russell diagram for the cluster. The exception is that a normal H–R diagram is a plot of absolute magnitudes (rather than apparent magnitudes) of stars against their spectral classes (or color indices). All the stars in the cluster, however, are at very nearly the same distance from the sun, and the difference between the apparent and absolute magnitudes is the same for every star and is equal to the distance modulus of the cluster (Section 23.2b). In this section we shall describe some of the properties of the color-magnitude diagrams for different kinds of clusters and some other properties of the stellar populations of clusters.

(a) Color-Magnitude Diagrams of Globular Clusters

Globular clusters nearly all have very similar appearing color-magnitude diagrams. Figure 31.8 shows the appearance of the color-magnitude diagram for a typical globular cluster of known distance, for which the apparent magnitudes have been converted to absolute magnitudes. The region from a to b is the main sequence. Presumably, the main sequence would extend farther down than a if the cluster were near enough for us to observe its fainter stars. Above point b, however, the main sequence seems to terminate; in most globular clusters, this point occurs at about absolute magnitude, $M_v = +3.5$. From b to c there extends a sequence of stars that are yellow and red giants; the brightest and reddest of them (at c) at $M_v = -3$ are brighter than typical red giants in the solar neighborhood. A third sequence of stars extends from d to f; it is called the *horizontal branch* of the H–R diagram for a globular cluster. There is a gap in the horizontal branch at $M_v = 0$ (point e), where no star of constant light output is found. The stars observed in this gap are the RR Lyrae variables. The stars on the horizontal branch have already been through the red giant phase of evolution once. In Chapter 32, we will see what happens to stars after they become red giants and will discuss the nature of the horizontal branch stars.

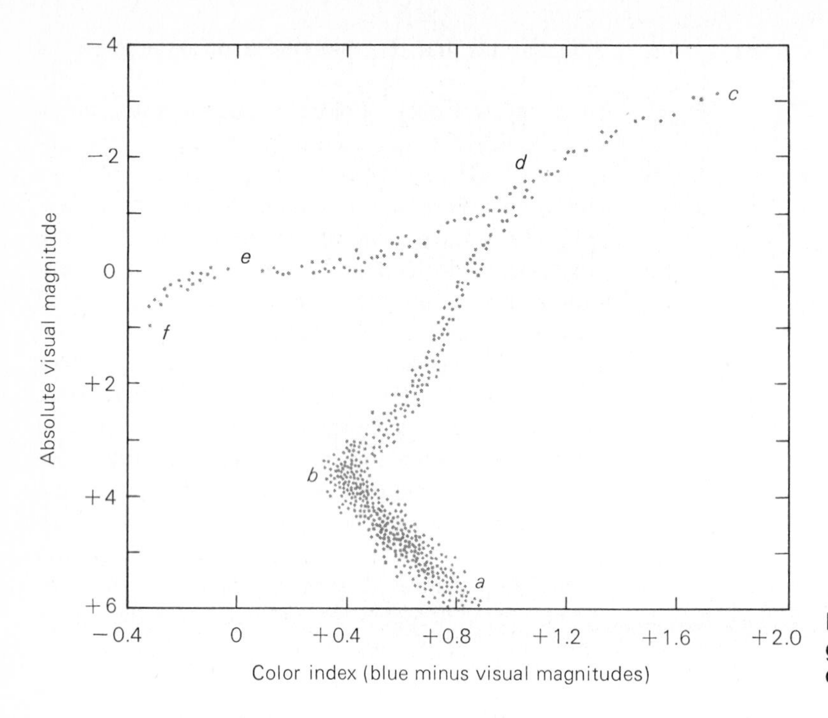

Figure 31.8 Hertzsprung-Russell diagram for a hypothetical globular star cluster.

The color-magnitude or H–R diagram of a globular cluster is more or less similar to that of any system of old stars. After a cluster forms, the most massive stars in that cluster use up the hydrogen in their cores and evolve off the main sequence to become red giants. As more time goes on, stars of successively lower mass leave the main sequence, making it seem to burn down, like a candle. The globular clusters are 10 to 16 billion years old. The only stars still remaining on the main sequence have masses comparable to that of the sun or less. All of the more massive stars have evolved away from the main sequence. Some are now red giants. Still others have completed their evolution and become white dwarfs or neutron stars (Chapter 32).

In the globular clusters, the giants are brighter than the brightest main-sequence stars and must, therefore, have increased in luminosity during their evolution from the main sequence. The red giants in globular clusters are even more luminous than are those in the oldest open clusters, such as M67. Calculations show these differences to be due to differences in chemical composition; globular-cluster stars have, on the average, lower abundances of heavier elements than do open-cluster stars (Section 31.4c).

Counts of stars of various kinds in old clusters show that the number of giants is very small compared with the number of main sequence stars, and that all the present giants in these clusters must have evolved from a very short segment of the original main sequence, just above its present termination point. (In Figure 31.8, to avoid crowding, points are plotted for only about one in ten of the main-sequence stars that lie below the giant branch turnoff.) In other words, all the giants in a single cluster are expected to have nearly the same mass and also to have had almost the same luminosity when they were on the main sequence. Those stars at the top of the giant branch on a cluster H–R diagram are, to be sure, further evolved than those, say, only halfway up to the top, but they started from only very slightly greater luminosities on the main sequence, and so had only a slight "head start." We can conclude, therefore, that the sequence of stars forming the giant branch in the H–R diagram of an old cluster lies very nearly along the evolutionary tracks of individual stars.

In any case, the red giant stage must be a relatively brief part of a star's life. In this stage of evolution, therefore, a star's nuclear fuel is consumed relatively quickly, and further evolutionary changes soon follow.

(b) Color-Magnitude Diagrams of Open Clusters

Whereas globular clusters nearly all have very similar H–R diagrams, those of open clusters differ

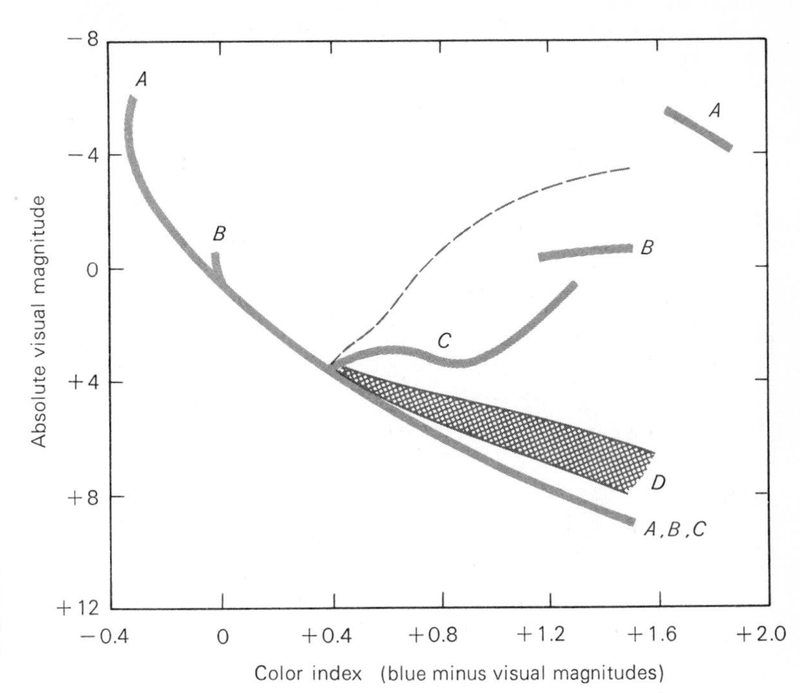

Figure 31.9 Composite Hertzsprung-Russell diagram for four hypothetical open clusters, *A, B, C,* and *D*. The dashed line shows the location of the giants in globular clusters.

widely from one another. Figure 31.9 shows, schematically, superposed H–R diagrams for four hypothetical open clusters, which will serve as examples.

In cluster *A* the main sequence extends to high-luminosity stars, and the highly luminous cool stars in the cluster are red supergiants. The double cluster *h* and χ Persei has an H–R diagram similar to that of cluster *A*. The main sequence of cluster *B* (representative of the open clusters M11 and M41—Figure 31.10) extends to less luminous stars, and the red giants of that cluster are less luminous than those in *A* as well. Cluster *C* (characteristic of the open cluster M67) has a main sequence that terminates at about the same absolute magnitude as do those of globular clusters. Cluster *C* has a sequence of yellow and red giants that resembles, but is less luminous than, that of the globular clusters (dashed line). The red giants in cluster *C* have about the same luminosity as typical red giants near the sun. At least one open cluster of the "*C* type" (M67) has a suggestion of a horizontal branch.

Note that in the three clusters discussed so far, the upper ends of the main sequences terminate at different absolute magnitudes and that the red giants (if present) usually lie off to the right of the top of the main sequence in each H–R diagram. In a few clusters (such as the Pleiades), no red giants

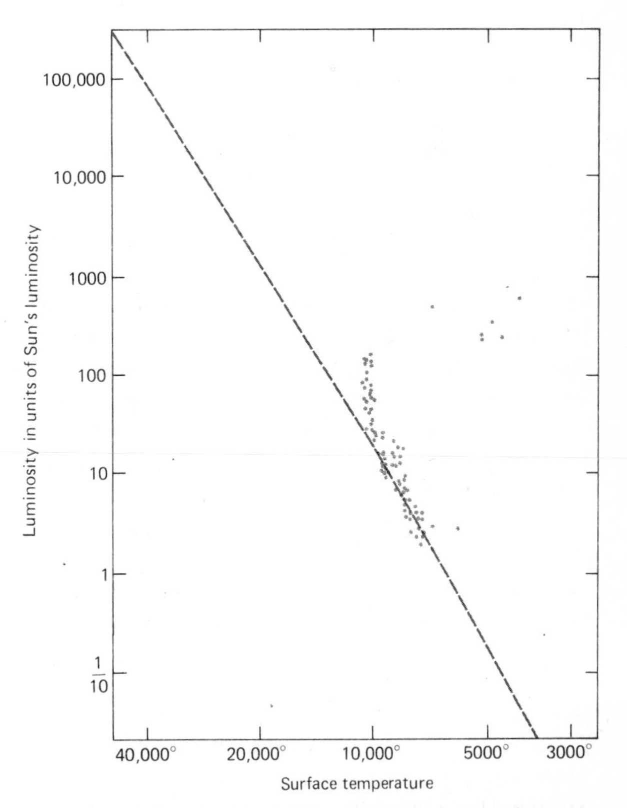

Figure 31.10 The H–R diagram of M41, from data by A. N. Cox.

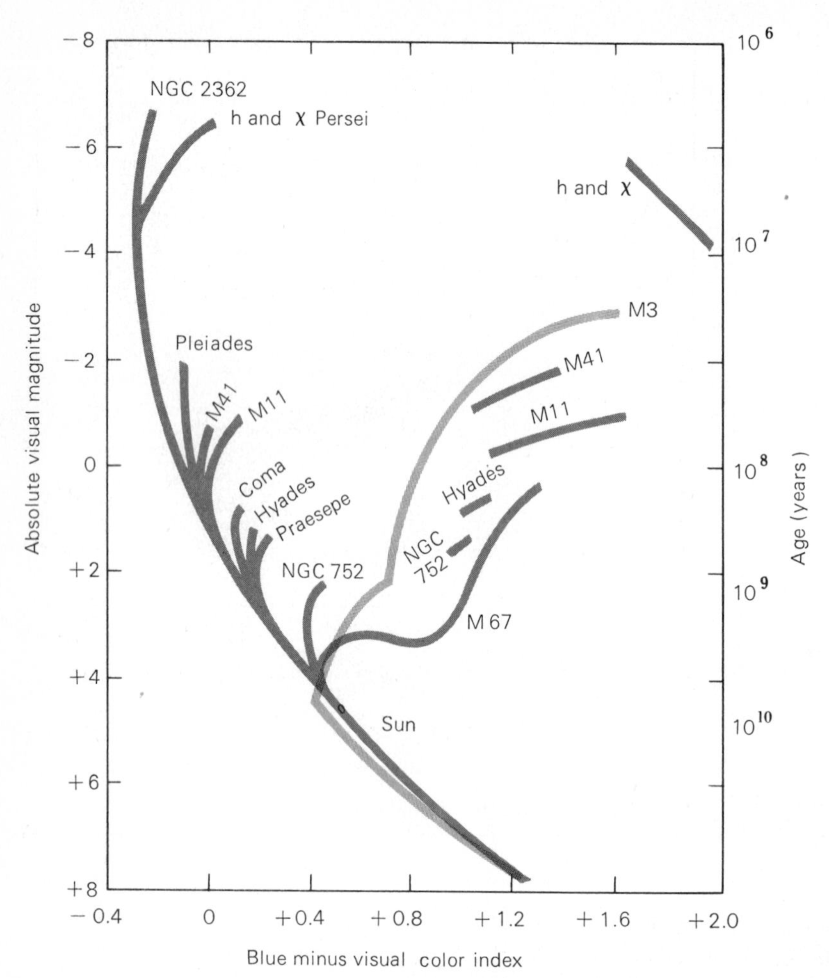

Figure 31.11 Composite Hertzsprung-Russell diagram for several star clusters of different ages. (Adapted from a diagram by A. Sandage)

are observed, but such clusters are not generally very rich. We note that there is a gap between the top of the main sequence and the red giants in the H–R diagrams of clusters A and B. This gap, called the *Hertzsprung gap,* is broadest in the color-magnitude diagrams of clusters whose main sequences extend to high luminosities, and narrows for clusters whose main sequences terminate at successively lower luminosities. The gap finally disappears in the color-magnitude diagrams of clusters whose main sequences extend only as far as that of cluster C. At this gap stars are in stages of evolution in which they are unstable and either pulsate, like the cepheids, or evolve very rapidly. We would, therefore, expect few stars to exist in this part of the H–R diagram.

The least luminous stars observed in clusters A, B, and C all lie on the main sequence. In cluster D, however, the brighter stars are on the main sequence, while the faintest observed ones lie off to the *right.* The open cluster NGC 2264 has such an H–R diagram. The faintest stars in this cluster (and others like it) are believed to be still in the process of contraction from interstellar matter.

If we now compare these observed color-magnitude diagrams with the theoretical ones in the previous chapter, we will see that the calculations agree very well with what we actually see. For example, the H–R diagram of NGC 2264 (Figure 31.12) is very similar to that of M2001 at an age of 3 million years (Figure 30.7). The cluster M41 (Figure 31.10) is like M2001 at an age of 66 million years (Figure 30.9). The H–R diagram for the old open cluster M67 (Figure 31.11) closely matches that of M2001 at an age of 4.24 billion years (Figure 30.10).

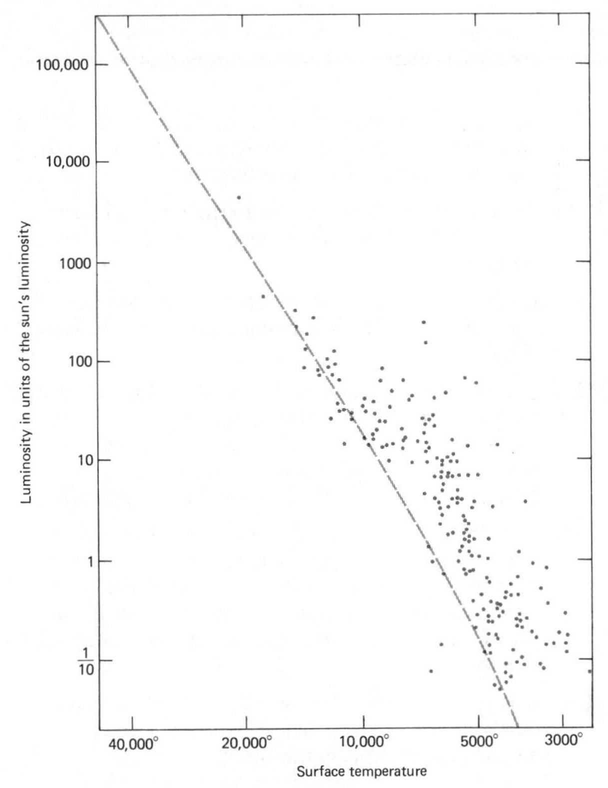

Figure 31.12 The H–R diagram of cluster NGC 2264, from data by M. Walker.

(c) Differences in Chemical Composition of Stars in Different Clusters

Hydrogen and helium, the most abundant elements in stars in the solar neighborhood, are also the most abundant constituents of the stars in all kinds of clusters. The exact abundances of the elements heavier than helium, however, vary from cluster to cluster. In the sun, and most of its neighboring stars, the combined abundance (by mass) of the heavy elements seems to be between 1 and 4 percent. The strengths of the lines of heavy elements in the spectra of stars in most open clusters show that they, too, have 1 to 4 percent of their matter in the form of heavy elements.

Globular clusters, however, are a different story. Spectra of their brightest stars often show extremely weak lines of the heavy elements. The heavy-element abundance of stars in typical globular clusters is found to range from only 0.1 to 0.01 percent, or even less. Other very old stars outside of globular clusters also have spectra that often indicate low abundances of heavy elements, although the difference between them and the sun is not usually as extreme as for some of the globular cluster stars. Differences in chemical composition are related to where and when stars were formed. The probable explanation of these phenomena is discussed in Chapter 32.

(d) Determination of Distances to Clusters

Distances to globular clusters can be determined from the period-luminosity relationship for RR Lyrae stars. In Section 22.5c we saw how distances to three moving open clusters—the Hyades, the Ursa Major Group, and the Scorpio-Centaurus Group—have been found from the apparent convergence or divergence of their member stars as they move across the sky.

If such techniques are not available, the most useful method of determining the distance to a star cluster is from a plot of the apparent magnitudes versus the color indices of its member stars, as shown in Figure 31.7. Most of the stars in a cluster lie along a *main sequence* similar to that defined by stars in the neighborhood of the sun. We find a main sequence, therefore, in the color-magnitude diagram of a cluster (see Figure 31.7), with the bluer, brighter-appearing stars (which are really the more luminous main-sequence stars in the cluster) farther up on the diagram than the redder, fainter-appearing stars. From the ordinary H–R diagram, we know what absolute magnitudes correspond to various color indices along the main sequence. The difference, at any given color index, between the apparent magnitude of the cluster stars and the absolute magnitude of known main-sequence stars of the same color is the distance modulus of the cluster.

As a numerical example, consider the hypothetical cluster whose color-magnitude diagram is shown in Figure 31.7. A main-sequence cluster star of color index +0.6 is seen to have an apparent magnitude of +15. But this is a star like the sun, whose absolute magnitude is +5. At a distance of 10 pc, therefore, this star would appear 10 magnitudes brighter than it does at the actual distance of the cluster. Since 10 magnitudes corresponds to a factor of 10,000 in light, the cluster must be 100 times as distant as 10 pc or must be 1000 pc away.

In actual practice the apparent magnitudes and color indices must first be corrected for effects of interstellar absorption and reddening (Chapter 27).

EXERCISES

* 1. The RR Lyrae stars in a particular globular cluster appear at apparent magnitude +15. How distant is the cluster? (Ignore interstellar absorption and assume that the RR Lyrae stars have absolute magnitudes of 0.)

2. Where in the Galaxy do you suppose undiscovered globular clusters may exist?

3. Suppose globular clusters have orbits about the galactic center with very high eccentricities—near unity. When a particular globular cluster is at its farthest from the center of the Galaxy, its distance from the center is 10^4 pc. What is its period of galactic revolution? (*Hint:* 1 pc = 2×10^5 AU. Assume that the mass of the Galaxy is 10^{12} solar masses.)

4. Table 31.1 indicates that stellar associations can emit even more light than a globular cluster. How is this possible if the associations have so few stars?

5. What color would a globular cluster appear? Why?

6. From the data of Table 31.1, estimate the average mass of the stars in each of the three different cluster types.

7. What is the density in solar masses per cubic parsec of the following clusters: (a) a globular cluster 50 pc in diameter containing 10^5 stars? (b) a stellar association of 100 solar masses and 20 pc in radius?
Answer: (a) 1.5 solar masses/pc^3, (b) 0.003 solar masses/pc^3

8. Why do you suppose it is sometimes said that the problem of dealing with the motions of more than two bodies interacting gravitationally has no solution? Is the statement true? Explain.

* 9. At what time of year is the Earth's potential energy greatest? At what time of year is its kinetic energy greatest?

*10. Could the virial theorem be used to compute the mass of a flock of birds flying together in formation? Why?

*11. Compare the question of the stability of a star cluster to the question of whether a body moving in the solar system has a closed elliptical, or an open hyperbolic, orbit. Is the orbital potential energy of the Earth less or greater than its kinetic energy? Why?

12. A main-sequence star of color index 0 has an absolute magnitude of about +1. In the color-magnitude diagram of a certain cluster, it is noted that stars of 0 color index have apparent magnitudes of about +6. How distant is the cluster? (Ignore interstellar absorption.)

13. It is often possible to observe fainter main-sequence stars in open clusters than in globular clusters. Why do you suppose this is the case?

14. Suppose a type I cepheid and a type II cepheid are both observed to have the same apparent magnitude and a period of 10 days. Consult Figure 31.5 and estimate how many times further away the type I cepheid is.

32.2
.8
10
12
13

Jocelyn Bell Burnell, British astronomer, discovered the first known pulsar when she was a graduate research student at Cambridge University Observatory. She analyzed 400 feet of chart recordings each week and found the sky to be heavily populated with compact radio sources, one of which was flaring. This first pulsar led to the realization that neutron stars exist. *(Sky and Telescope)*

EVOLUTION AND DEATH OF OLD STARS

Chapter 30 followed the evolution of stars up to the point where they leave the main sequence to become red giants. Already it has become apparent that the evolutionary development of a star is shaped by the balance between two forces—pressure and gravity. When these two forces are in balance, a star is stable. If pressure forces exceed the force of gravity, a star will expand. If pressure for any reason becomes inadequate to resist gravity, a star will contract.

As a star consumes its supply of hydrogen and evolves away from the main sequence, these two forces do get out of balance. The star changes the way it generates energy, and related changes in its size and internal structure are required to achieve a new equilibrium. Sometimes the changes in structure are accompanied by loss of mass. This mass loss may occur gently, in the form of a stellar wind, but often it is explosive and can destroy most or all of the star.

Each new stage of equilibrium is only temporary. At each phase of its evolution a star will ultimately exhaust the supply of whatever fuel it is using to generate energy, the balance between pressure and gravity will be destroyed, and the structure of the star will change in such a way as to restore equilibrium. In the end, however, all sources of energy will be completely consumed, and the star will cease to shine. Its ultimate fate will depend on its mass. Stars with masses similar to that of the sun will end their lives as white dwarfs. More massive stars will become neutron stars or possibly black holes or may even be completely disrupted.

In the chapter that follows, we trace the late stages of evolution first of stars like the sun and then of more massive stars. By late stages of evolution, we refer to what happens after a star first becomes a red giant. We will begin, however, with a discussion of a new kind of pressure—the pressure exerted by an electron-degenerate gas—that becomes important near the end of a star's life.

32.1 DEGENERATE MATTER

The electrons in a neutral atom occupy certain allowed *states,* each of which involves a certain energy level for the atom as a whole. We saw in Section 8.3 how an electron can change states when the atom absorbs or emits energy. When the atom is ionized, one or more of its electrons are free, and we speak of them as occupying a *continuum* of states. Even free electrons, however, do not have an infinite number of states available to them.

Electrons, protons, and neutrons belong to a class of particles (there are others) that must obey a rule of quantum mechanics, which insists that no two of them occupy the same state. What this means is that no two electrons can be in the same place at the same time doing the same thing. The rule is known as the *Pauli exclusion principle,* after the Austrian physicist Wolfgang Pauli, who enunciated it. We specify the *place* of the electron by its precise position in space, and we specify what it is doing by its momentum and the way it is spinning.

But recall from Section 8.3 that the simultaneous position and momentum of an electron (or anything else) cannot be known any more precisely than is allowed by the Heisenberg uncertainty principle. Specifically, this means that the precision of its position, or the amount by which the x-coordinate of its location is uncertain, Δx, and the precision of the x-coordinate of its momentum, Δp_x, are governed by the condition

$$\Delta x \Delta p_x \geq \frac{h}{2\pi},$$

where h is Planck's constant (6.626×10^{-27} erg · s). The state of an electron cannot be specified more precisely; the states, therefore, are fuzzy. The Pauli principle permits only one electron in each of these fuzzy states (actually two, because an electron can spin in either of two possible directions).

Imagine the free electrons in an ionized gas with a certain temperature. That temperature determines the distribution of velocities, and hence the range of momenta, for the electrons. If the temperature is high enough, the momentum range is large, and the electrons have plenty of possible momentum-position states to occupy without violating the Pauli principle. On the other hand, what if the temperature is *not* high enough? (How high is "high enough" depends on how many electrons are crowded into a given volume—that is, their density.) Then two or more electrons (of the same spin) would have to occupy the same fuzzy state, which will not happen, for the electrons will resist such crowding with overwhelming pressure. When all the available states (of position and momentum) are occupied, the electrons are said to be *degenerate,* and the gas is an *electron-degenerate gas.*

The electrons in a degenerate gas move about, as do particles in any gas, but not with freedom. A particular electron cannot change position or momentum until another electron in an adjacent state gets out of the way. It is as if all the particles were geared together. Crystalline solids are analogous; there, each electron occupies a particular state in the latticework, and all states are occupied (the lattice is also degenerate), so that one electron cannot move over until another next to it moves, producing a "hole." If electrons are removed from one side, however, others can move over, and still others behind them, producing a current through the solid. In a similar way, an electron-degenerate gas is highly conducting, and heat can flow through it with great ease.

Now what has all this to do with stars? Simply that if part or all of a star contracts enough and increases sufficiently in density it will become electron-degenerate unless the temperature also rises enough to make available a sufficient range of momentum states. The contracting star (or the contracting core of a star) does release gravitational potential energy, half of which heats the interior, but this released heat is not enough to prevent or remove the degeneracy unless a lot of gravitational energy is released, which requires the star (or core) to be quite massive. We shall return later to just what the required mass is. When a star becomes fully electron-degenerate, it can contract no further.

It is noteworthy that in a typical stellar environment even a degenerate electron gas is mostly empty space. The electrons are by no means packed into contact; their densities are (typically) a million to a thousand million times lower than the density of the atomic nucleus. The nuclei themselves still move about freely among the electrons, obey the usual perfect gas law, and exert the normal pressure of particles of their masses and temperature. The pressure exerted by the degenerate electrons, however, generally swamps that of the nuclei, so it is the electron pressure that dominates and controls the structure of that region of the star.

Other particles become degenerate as well—in particular, a *degenerate neutron gas* is important in astronomy. But neutrons are nearly 2000 times as massive as electrons and at the same temperature have very much greater momentum. Neutrons, therefore, must be crowded to enormously higher densities before filling their available momentum states and becoming degenerate.

32.2 THE HELIUM FLASH

Now back to our story of how a star with a mass similar to that of the sun becomes a red giant. (We will define what we mean by "similar" in Section

32.4.) After hydrogen in the core of the star has been exhausted, energy generation through the fusion of hydrogen to form helium must come to a halt. The center of the star is still the hottest part of it, however, and heat energy continues to leak out. But all of stellar evolution is governed by the battle between pressure and gravity. Once energy generation in the core of the star stops, and as heat leaks out, then the central temperature does not remain hot enough to keep the gas pressure high enough to resist the weight of those layers of the star that lie outside the core. Gravity gains the upper hand, at least for a while, and the overlying layers crush the core and force it to shrink in size. As the core contracts, it releases gravitational potential energy, which heats up both the core and the region immediately surrounding it. Eventually hydrogen begins to fuse to form helium in a shell around the core.

The core itself still has no source of thermonuclear energy and continues to shrink in diameter and grow still hotter. As the core shrinks, not only does it get very hot but also it becomes very dense, and matter in the innermost part becomes electron-degenerate. Meanwhile, the surrounding shell soon exhausts its hydrogen and also contracts until it becomes degenerate and joins the core. With its increased mass and consequent release of additional gravitational energy, the core becomes still smaller and the nondegenerate nuclei become even hotter. Thus the degenerate core continues to contract and heat. By the time the star reaches its maximum luminosity at the top of the red giant branch on the H–R diagram, its central temperature exceeds 100 million degrees Kelvin.

The triple-alpha process, that is, the fusion of three helium atoms to form a single carbon nucleus, is expected to begin abruptly in the central core of a red giant. As soon as the temperature becomes high enough to start the triple-alpha process, the extra energy released is transmitted quickly through the entire degenerate core, producing a rapid heating of all the helium there. Because the gas in the core is degenerate, the red giant lacks a safety valve that was available to the star when it was still on the main sequence. In a main-sequence star, if the core of the star becomes hotter for any reason whatsoever, then the core can expand slightly, the temperature will drop, nuclear reactions will slow, and the star will become stable again.

In a degenerate core, the pressure is determined primarily by degeneracy effects. Raising the temperature slightly does not significantly alter the pressure, but it will speed nuclear reactions that will in turn raise the temperature still further. Helium burning accelerates, and a runaway generation of energy results. Once helium burning begins, the entire core is reignited in a flash.

The new energy released in this *helium flash* is so great, in fact, that it removes the degeneracy, expands the core, and reverses the expansion of the outer parts of the red giant. The outer layers of the star shrink rapidly. Calculations indicate that the star's surface temperature increases and its luminosity decreases. The point representing the star on the H–R diagram takes on a new position to the left of and somewhat below its place as a red giant. At this time, the core of the star is nondegenerate and is stable, fusing helium to form carbon. Often a newly formed carbon nucleus is joined by another helium nucleus to produce a nucleus of oxygen. Surrounding this helium-burning core is a shell in which hydrogen is fusing to form helium. Such a star is called a horizontal branch star because of its location on the H–R diagram.

Figure 32.1 shows the post-main-sequence evolutionary track for a star that evolves in the manner we have just described. As soon as the helium is exhausted in the central region, however, energy production by the triple-alpha process is shut off, creating a situation analogous to that of a main-sequence star when its central hydrogen is used up and hydrogen burning ceases in its center. Now, however, there is a core of carbon and oxygen surrounded by a shell where helium is still burning; farther out in the star is another shell where hydrogen is burning. The star now moves back to the red giant domain on the H–R diagram. Calculations indicate that all of these evolutionary stages occur in tens or hundreds of millions of years or less—a brief time compared with the main-sequence lifetime.

In stars with masses similar to that of the sun, the formation of a carbon-oxygen core marks the end of the generation of nuclear energy at the center of the star. What happens to the star then?

32.3 WHITE DWARFS: ONE FINAL STAGE OF EVOLUTION

When a star with a mass similar to that of the sun exhausts its store of nuclear energy, it can only contract and release more of its potential energy. Eventually, the shrinking star will attain an enormous

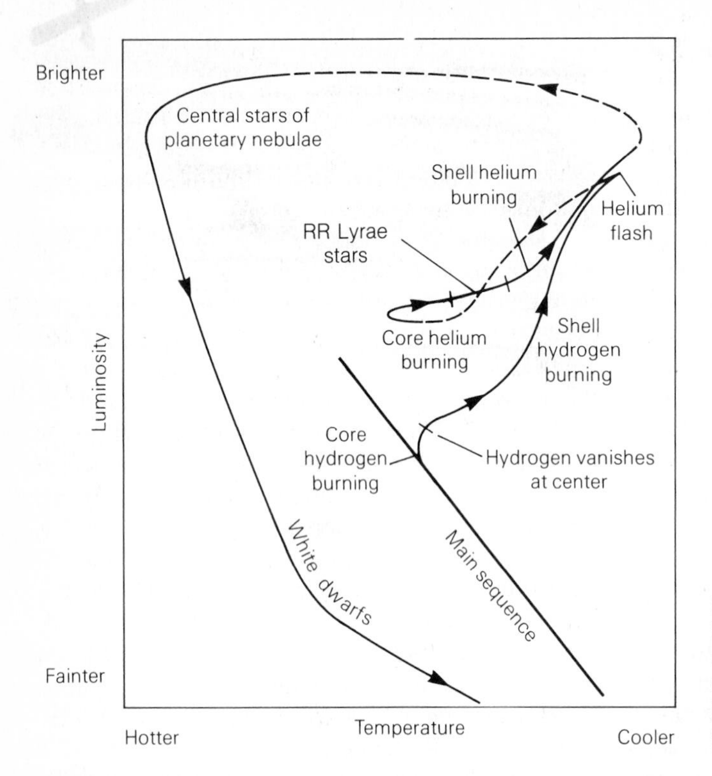

Figure 32.1 The changes in temperature and luminosity that occur to a star with a mass similar to that of the sun as it evolves from the main sequence and becomes a white dwarf. Dashed lines indicate rapid phases of evolution for which theoretical calculations are still very uncertain.

density. We observe such stars—the extremely compact white dwarf stars (Section 26.2d), whose mean densities range up to over 1 million times that of water. In a white dwarf the electrons are completely degenerate throughout the star, save for a very thin layer at the surface.

(a) Structure of White Dwarfs

The structure of a white dwarf was first studied by R. H. Fowler. White dwarfs are simpler than most stars because the pressure that supports a white dwarf in hydrostatic equilibrium is supplied almost entirely by the degenerate electrons, and therefore does not depend on the temperature, but only on the density. The volume to which a star can be compressed before the electrons become degenerate (that is, its density) depends on the amount of gravitational potential energy that can be released by the collapsing star, which in turn depends on its mass. The size of a white dwarf, therefore, depends on its mass—the more massive the white dwarf, the smaller its size. A white dwarf of one solar mass must have a radius of about 1 percent of the sun's—about the size of the Earth!

In the more massive white dwarfs some of the electrons have speeds that are an appreciable fraction of that of light, and a rigorous treatment must include the effects of special relativity (Chapter 11). The first such rigorous models of white dwarfs were constructed by the Nobel prize–winning astrophysicist S. Chandrasekhar. Chandrasekhar's analysis shows that white dwarfs of masses successively greater than the sun's are successively smaller than 1 percent of its radius, until a mass of 1.4 solar masses is reached, at which point hydrostatic equilibrium cannot be achieved (Figure 32.2). Thus, 1.4 solar masses is the upper limit to the mass of a white dwarf. Astronomers refer to this upper limit as the Chandrasekhar limit. Stars that are more massive than 1.4 solar masses at the time they run out of sources of nuclear energy must continue to collapse to a size that is far smaller than that of a white dwarf. Precisely what these stars become will be discussed in Section 32.8.

White dwarfs have hot interiors—tens of millions of Kelvins. At those temperatures and at the high densities of these stars, any remaining hydrogen would undergo violent fusion into helium, giving the stars luminosities many times higher than observed. Consequently, white dwarfs can have no hydrogen in their interiors. Their most probable internal composition is a mixture of carbon and oxygen, the principal products of helium burning.

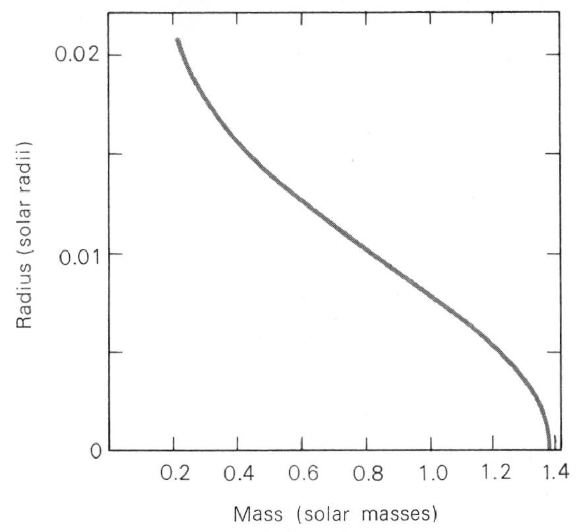

Figure 32.2 Theoretical relation between the masses and the radii of white dwarf stars.

Recent theoretical studies indicate that at least some white dwarfs probably have cores in which the matter has crystallized. Some also have very strong magnetic fields—up to hundreds of millions of gauss. Moreover, a few display light variations, with periods of several minutes.

(b) Evolution of White Dwarfs

A white dwarf is presumed to have exhausted its available nuclear-energy sources. It cannot contract and release gravitational potential energy because of the great pressure of the degenerate-electron gas. Thus its only source of energy is the thermal energy (that is, kinetic energy) of the nondegenerate nuclei of atoms, behaving as ordinary gas particles, scattered throughout the degenerate electrons. As these nuclei slow down (cool), the electron gas conducts their thermal energy to the surface. At the boundary of the star, the very thin skin-like layer of nondegenerate gas radiates this energy into space. Only the opacity of this outer layer keeps the nuclei in the interior of the star from cooling off at once.

Gradually, however, a white dwarf does cool off, much like a hot pan when it is removed from a stove. The cooling is relatively rapid at first, but as the star's internal temperature drops, so does its cooling rate. Calculations indicate that its luminosity should drop to about 1 percent of the sun's in the first few hundred million years of its existence as a white dwarf. Since the radius of a white dwarf is constant, its luminosity (by Stefan's law) is proportional to the fourth power of its effective temperature. Therefore, as a white dwarf cools, its track on the H–R diagram is along a diagonal line toward the lower right, that is, toward low temperature and low luminosity.

Eventually, a white dwarf will cease to shine at all. It will then be a *black dwarf*, a cold mass of degenerate gas floating through space. (Do not confuse *black dwarf* with *black hole*—see Chapter 33.) A long time may be required, however, for a star to cool off to the black dwarf stage. It may be that the Galaxy is not old enough for any star to have yet had time to become a black dwarf.

(c) Novae—A Last Burst of Glory

While isolated white dwarfs die in the unspectacular fashion we have just described, those white dwarfs that have a nearby stellar companion may call attention to their demise by becoming *novae*.

Nova literally means "new." Actually, a nova is an existing star that suddenly emits an outburst of light. In ancient times, when such an outburst brought a star's luminosity up to naked-eye visibility, it seemed like a new star. Novae remain bright for only a few days or weeks and then gradually fade. They seldom remain visible to the unaided eye for more than a few months. The Chinese, whose annals record novae from centuries before Christ, called them "guest stars." Only occasionally are novae visible to the naked eye, but, on the average, two or three are found telescopically each year. Many must escape detection; altogether there may be as many as two or three dozen nova outbursts per year in our galaxy. The light curve of a typical nova is shown in Figure 32.3.

Novae occur in close binary star systems, in each of which one member is a white dwarf and the other member is a star transferring mass to the white dwarf by the mechanism described at the end of this section. For a while (years to centuries) hydrogen-rich material from the outer layers of the nonwhite dwarf just piles up on the surface of the white dwarf. Gradually, however, the weight of this matter increases, and hence so does the temperature, until it approaches that of the degenerate interior of the dwarf. Then, explosively, hydrogen burning ignites through the CNO cycle, and like a nuclear bomb it blows off the outer layer of matter that had accumulated on the white dwarf.

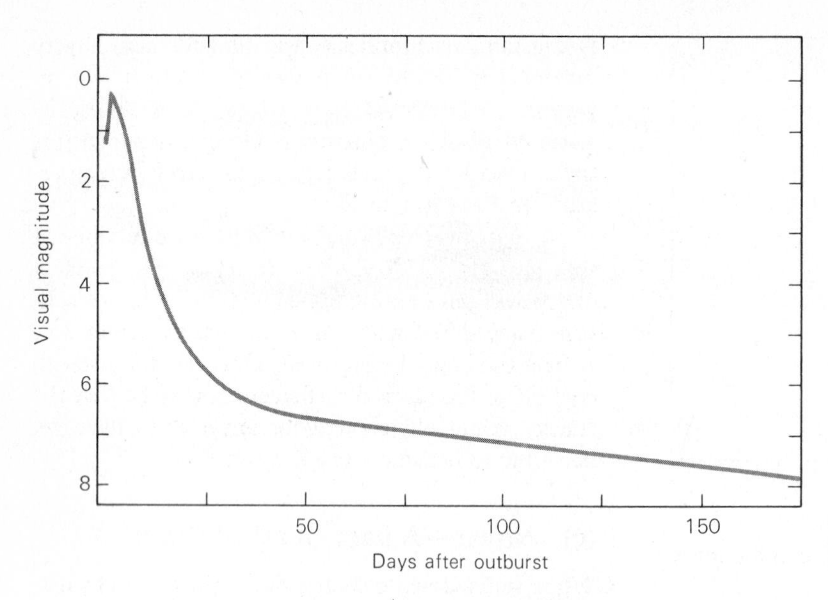

Figure 32.3 Light curve of Nova Puppis, 1942.

After the outburst, the white dwarf settles down, but since mass is continually flowing onto it from its companion, the process eventually repeats. Some novae have long been known to be recurrent. Generally, the more intense the outburst, the longer the period of quiescence between nova flareups. According to the model, *all* novae recur on some time scale or other; the most violent *classical* novae, which reach absolute visual magnitudes of -6 to -9 (about 10^5 times the luminosity of the sun), may wait hundreds or thousands of years or more between outbursts.

We learn something of the physics of novae from their spectra. Just after outburst, the dark lines show sudden shifts to the violet—indicating that the star's photosphere is rising and approaching us. Shortly after the maximum light output, however, the expanding ejected shell of matter thins out enough to become transparent. It then no longer emits a continuous spectrum but still absorbs some of the star's light and re-emits it in bright emission lines. These emission lines, however, originate from different parts of the shell, moving at different speeds along our line of sight, and consequently the lines display a wide range of Doppler shifts (Figure 32.5). Light re-emitted from parts *A* and *B* of the shell, moving away from the Earth, is shifted to longer wavelengths; light from parts *C*

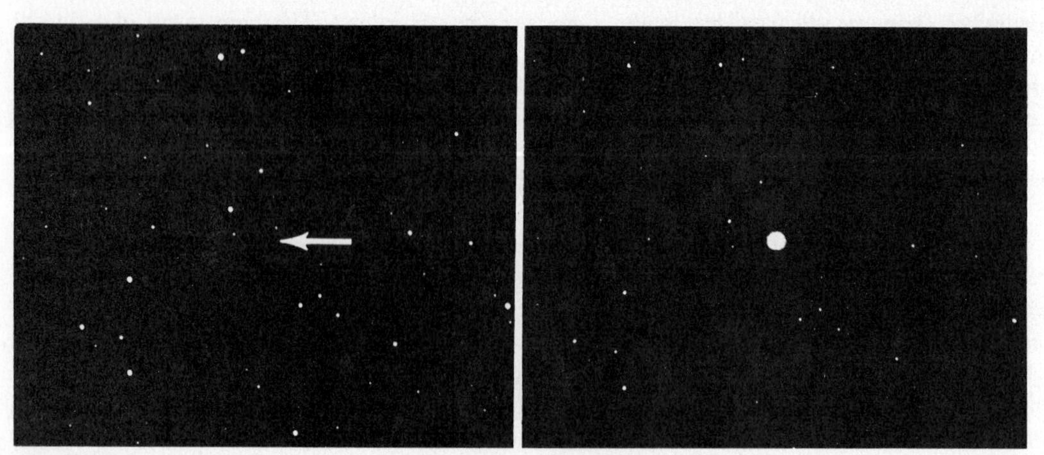

Figure 32.4 Nova Herculis, as it appeared before and after its outburst in 1934. (Yerkes Observatory)

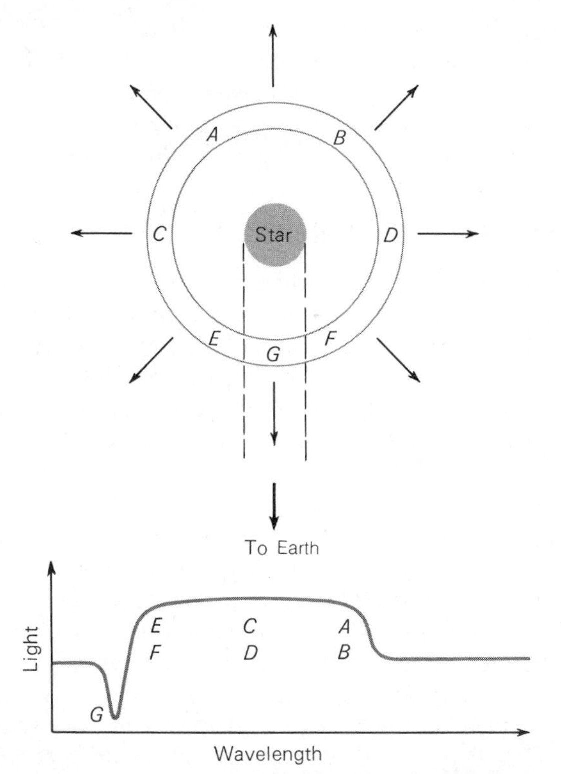

Figure 32.5 Diagram of an expanding nova shell. Below is a profile of an emission and absorption line formed by the shell.

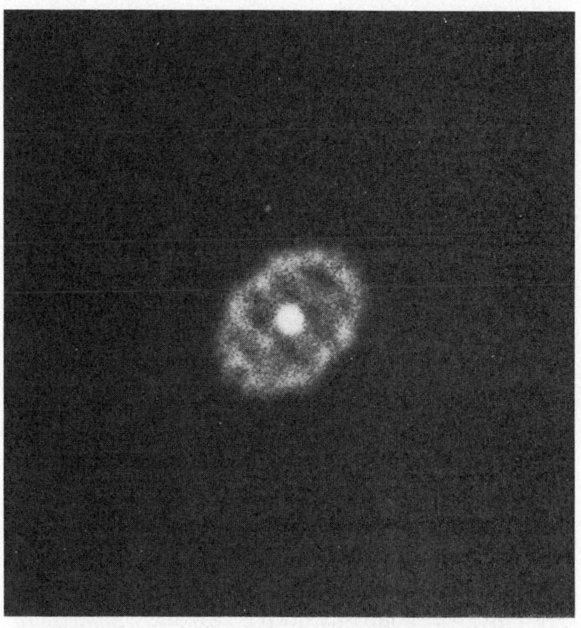

Figure 32.6 The shell of Nova Herculis photographed with the Lick Observatory's 3-m telescope in 1972. (Courtesy H. Ford, UCLA)

and D is not shifted; light from parts E and F is shifted to the violet. Only part G of the shell, directly between the Earth and the star itself, produces absorption lines in the observed spectrum of the star; these are shifted farthest of all to the violet. The emitted light from the shell, therefore, does not appear to *us* in the form of sharp bright lines but as broadened lines, or emission bands. At the violet end of each emission band is the sharp absorption line produced by the part of the shell directly between us and the star.

From the widths of the emission bands, or from the displacement of the absorption lines at their violet ends, we can calculate the velocity with which the shell is expanding. Velocities of ejection of up to 1000 km/s or more are found. From the total light emitted by the shell, we can calculate the amount of material it contains. A typical nova shell is found to contain from 10^{-5} to 10^{-4} solar mass. Some months or years after the outburst, the expanding envelope may become visible on telescopic photographs, as in the case of Nova Herculis (Figure 32.6).

Some post novae (or, presumably, novae between outbursts) are X-ray sources. X-rays are normally emitted by a very hot gas (millions of Kelvins). Gas can be heated to such temperatures by friction if it falls together at very high speeds. The matter falling on the white dwarf star from its companion accelerates to such high speeds as it approaches the dense compact star, so the X-rays are a natural consequence of mass transfer in a binary system when the recipient star is a small dense one.

(d) Mass Exchange in Binary Stars

Novae and some types of supernovae (Section 32.7) apparently occur when mass is transferred from a normal star to a white dwarf companion. This mass transfer is a consequence of gravity acting on the outer atmosphere of a star that is expanding as it evolves. Specifically, on the line between the centers of the two stars of a binary system there is a point where the gravitational attraction between a small body and one of the stars would be equal to that between the body and the other star. If the two stars were of equal mass, for example, that point would lie halfway between them. There are, in fact, entire surfaces that partially or completely surround either or both stars; a small body can move freely along one of these surfaces under the combined effects of the gravitational forces of the two stars and their mutual revolution.

Now suppose that one of the stars of a binary system, in becoming a red giant, increases in size, so that its outer surface extends through one of these regions between the stars. The stellar material along such a surface "doesn't know which star it belongs to" and is no longer gravitationally bound to the original star. An exchange of material from one star to the other may result. Also, some material may flow away from both stars and form a ring encircling both, or even a common envelope. Such a gaseous envelope surrounding a binary system can reveal itself by producing absorption lines superposed upon the spectra of both stars. These absorption lines do not show large variations in Doppler shift, since the material in the outer envelope does not move as rapidly as do the two stars in their orbits about the system's center of gravity.

32.4 MASS LOSS

What kinds of stars eventually end their lives as white dwarfs? The critical determining factor is mass. White dwarfs can be no more massive than about 1.4 times the mass of the sun. Yet somewhat surprisingly, observations indicate that stars that have masses larger than 1.4 times the mass of the sun *at the time they are on the main sequence* also complete their evolution by becoming white dwarfs. One argument that this must be so comes from the sheer number of white dwarfs that have been identified. There are far too many to be accounted for if only stars with main-sequence masses less than 1.4 solar masses become white dwarfs.

A second argument is even more compelling. White dwarfs have been found in young open clusters—clusters so young that only stars with masses in the range of six to eight times the mass of the sun have had time within the lifetime of the cluster to exhaust their supply of nuclear energy and complete their evolution to the white dwarf stage. In the Pleiades, for example, stars with masses five times the mass of the sun are still on the main sequence. This cluster also has at least one white dwarf, and its main sequence mass must, therefore, have exceeded five solar masses.

If stars that initially have masses in the range of six to eight times the mass of the sun are to become white dwarfs, then somehow they must get rid of enough matter so that their total mass at the time nuclear-energy generation ceases is less than 1.4 masses. In this section, we examine one of the ways in which such high-mass stars can lose mass.

(a) Planetary Nebulae

For stars of no more than a few solar masses—including red giants in globular clusters and in the galactic corona, nucleus, and disk of the Galaxy—one of the most important mass-ejection mechanisms may be the planetary nebula phenomenon. Planetary nebulae are shells of gas ejected from red giant stars. The red giants then quickly evolve to become low-luminosity, hot stars, which illuminate the expanding gas shells (see Figure 32.1). Planetary nebulae derive their name from the fact that a few bear a superficial telescopic resemblance to planets; actually they are thousands of times larger than the entire solar system, and have nothing whatever to do with planets.

Planetary nebulae are identified in two ways. Often they appear large enough to see or to photograph with a telescope. The most famous example is the ring nebula, in Lyra. It is typical of many planetaries in that, although actually a hollow shell of material emitting light, it appears as a ring. The explanation is that we are looking through the *thin* dimensions of the front and rear parts of the shell, while along its periphery our line of sight encounters a long path through the glowing material. Similarly, a soap bubble often appears to be a thin ring. The other way in which planetary nebulae are identified is by their spectra. Those that are very distant from us are unresolved and look like stars, but their spectra show emission lines that indicate the existence of luminous shells of gas surrounding stars, as do spectra of nova shells. Altogether, there are about 1000 planetary nebulae catalogued. Doubtless there are many distant ones that have escaped detection, so there must be some tens of thousands in the Galaxy. Nevertheless, among the tens of billions of stars in the system, planetary nebulae must be classed as rare objects. Some examples of planetary nebulae are shown in Color Plates 28b and 28c.

An appreciable amount of material is ejected in the shell of a planetary nebula. From the light emitted by the shells, we calculate that they must have masses of 10 to 20 percent that of the sun. The shells, typically, expand about their parent stars at speeds of 20 to 30 km/s. There is also evidence that a given star may eject several planetary nebula shells.

The linear diameters of planetary nebula shells can be calculated from their angular diameters and distances (although the latter are known only with considerable uncertainty). A typical planetary appears to have a diameter of about ½ LY to 1 LY. If it is assumed that the gas shell has always expanded at the speed with which it is now enlarging about its parent star, its age can be calculated. Most of the gas shells have been ejected within the past 50,000 years; an age of 20,000 years is more or less typical. After about 100,000 years, the shell is so enlarged that it is too thin and tenuous to be seen. The rarity of planetary nebulae, therefore, is due entirely to the fact that they cannot be seen for very long; they are temporary phenomena. When we take account of the relatively short time over which planetary nebulae exist, we find that they are actually very common, and that an appreciable fraction of all stars must sometime evolve through the planetary nebula phase.

The gas shells of planetary nebulae shine by the process of fluorescence. They absorb ultraviolet radiation from their central stars and re-emit this energy as visible light. The physical processes are the same as in emission nebulae (Section 27.3b). During the first few tens of thousands of years after the ejection of the shell, the gas in it is dense and thick enough to prevent the star's ultraviolet radiation from penetrating all the way through it. This energy is completely absorbed, therefore, within the inner part of the shell, and only that inner part is luminous. The outer portion of the gaseous shell is dark and cold; it is transparent to visible light, however, so we see all the reradiated, longer-wavelength energy emitted from within.

The mechanism of the fluorescence process enables us to calculate, or at least to estimate, the temperatures of the central stars of planetary nebulae. All the visible light emitted by the gas shell is converted from ultraviolet energy originally emitted by the star. Knowing the details of the atomic processes of absorption and emission of light that are involved in the fluorescence phenomenon, we can calculate the rate at which ultraviolet radiation must be leaving the star to account for the visible light coming from the gas shell; it turns out to be a far greater amount of energy than the star radiates in its observable, visible spectrum. Most of the sun's radiant energy, on the other hand, is in the form of visible light. The central star of a planetary nebula, therefore, must be many times hotter than the sun for so large a fraction of its luminosity to be in the ultraviolet (see Section 8.1). Nearly all these stars are hotter than 20,000 K, and some have temperatures well in excess of 100,000 K, which makes them among the hottest known stars.

Despite their high temperatures, the central stars of planetary nebulae do not have exceedingly high luminosities—some emit little more total energy than does the sun. They must, therefore, be stars of small size; some, in fact, appear to have the dimensions of white dwarfs. Thus a planetary nebula may be the last ejection of matter by a star before it collapses to a white dwarf.

The central stars of planetary nebulae constitute a class of very hot, and usually very small, dense stars. It does not follow, however, that the stars were small and hot when the gas shells were ejected. The velocity of escape from such a small dense star is extremely high—up to thousands of kilometers per second. It is very difficult to imagine an ejection mechanism whereby a shell of gas can be shot off at such speeds, and a few thousand years later be expanding at a leisurely 20 to 30 km/s. It is more likely that the nebulae were ejected from their parent stars when the latter were, at an earlier stage in their evolution, large red giants, from which the escape velocity would be under 100 km/s.

(b) Consequences of Mass Loss

As we have seen, many stars manage to qualify for white-dwarfhood by ejecting part of their matter into space. Very significantly, however, the material they shed is not the same as that from which they were formed, for the nuclear reactions by which they shine alter the chemical composition of their constituent gases.

We saw in Section 29.3b that stars convert hydrogen into helium and, moreover, at least some stars in some stages of their evolution are building up helium into carbon and heavier elements. Thus, inside stars, some of the lighter elements of the universe are gradually being converted into heavier ones. As these stars eject matter into the interstellar medium, that matter is richer in heavy elements than was the material from which the stars were formed. In other words, a gradual enrichment of the heavy-element abundance in interstellar matter is taking place. The heavy-element abundance in stars that are forming now is thus higher than in those that formed in the past. The fact that the oldest known stars (those in globular clusters) are the stars with the lowest known abundance of heavy elements provides evidence for this scenario.

We have reason to think (Chapter 37) that originally *all* stars in the Galaxy were formed of nearly pure hydrogen and helium, and that all the other elements were synthesized in the hot centers of stars at advanced stages of their evolution, and/or during explosive stages, such as in supernovae. Stars such as the sun, in whose outer layers heavy elements are observed spectroscopically, thus have to be of the second (or higher) "generation"; that is, they have been formed of matter that was once part of other stars. It is a grand concept that the planets (and we!) are composed of atoms that were synthesized in earlier generation stars—that we are literally made of "stardust."

32.5 SOLAR EVOLUTION

Figure 32.7 summarizes our current ideas on the evolution of a star of about 1.2 solar masses on the H–R diagram. In its early stages, the star contracts and moves to the left, reaching the main sequence with a size only slightly greater than that of the sun. In its subsequent evolution to the red giant stage, it grows to a radius of tens of millions of kilometers. The further evolution is uncertain. Perhaps the star goes through stages of variability, or emits material as a planetary nebula. Its final size as a white dwarf is only about that of the Earth.

The sun is a typical star, and much of the theory of stellar evolution that we have discussed applies to it. From theoretical calculations we can now form a fairly clear picture of the approximate past history of the sun, and we can make at least educated guesses about its future.

(a) Early Solar Evolution

Several calculations of the sun's early evolution, analogous to those described in Section 30.1, have been made. (See Section 12.4 for a discussion of the formation of the solar system.) The exact track of evolution of the sun on the H–R diagram depends on its assumed initial chemical composition and on rather uncertain data involving the opacity of the outer layers of the young sun. The time required for it to contract to the main sequence was probably a few tens of millions of years.

Since it reached the main sequence, the sun has increased somewhat in luminosity, probably by about 30 to 50 percent. During that interval of from 4 to 5 billion years, it has depleted much of the hydrogen at its very center, but a pure helium core has not yet had time to form. It is not certain how much more time the sun has before starting to evolve to the red giant stage, but a good guess is that it has lived out about half of its main-sequence life. We can probably look forward to at least another 5 billion years before the sun's structure undergoes large changes.

(b) The Future of the Solar System

All available evidence leads us to expect that sometime in the future the sun will leave the main sequence and evolve to a red giant. That time will occur in about 5 billion years, when the sun's photosphere will reach nearly to the orbit of Mars. The Earth, then well inside the sun, and exposed to temperatures of thousands of degrees, will gradually vaporize. The gases in the greatly distended outer layers of the sun will, of course, be very tenuous, but they should still offer enough resistance to the partially vaporized Earth to slow it in its orbital motion. Calculations suggest that the Earth will spiral inward toward the very hot interior of the sun, reaching its final end about ten thousand years after being swallowed by the sun.

If, in that remote time, humans could leave the doomed Earth but remain in the solar system, they would find that most of the naked-eye stars in our 20th-century sky would long since have exhausted their nuclear fuel and evolved to white dwarfs (or perhaps some other end). Main-sequence stars less massive than the sun would still be shining, but only those few of them passing temporarily through the solar neighborhood would be near enough to see with the unaided eye. It is doubtful, however, if by then *all* star formation in the Galaxy would have ceased. Luminous young stars might be shining in remote clouds of interstellar matter; a Milky Way might still stretch around the sky.

32.6 EVOLUTION OF MASSIVE STARS

Stars of masses up to six to eight times the mass of the sun probably end their lives as white dwarfs.

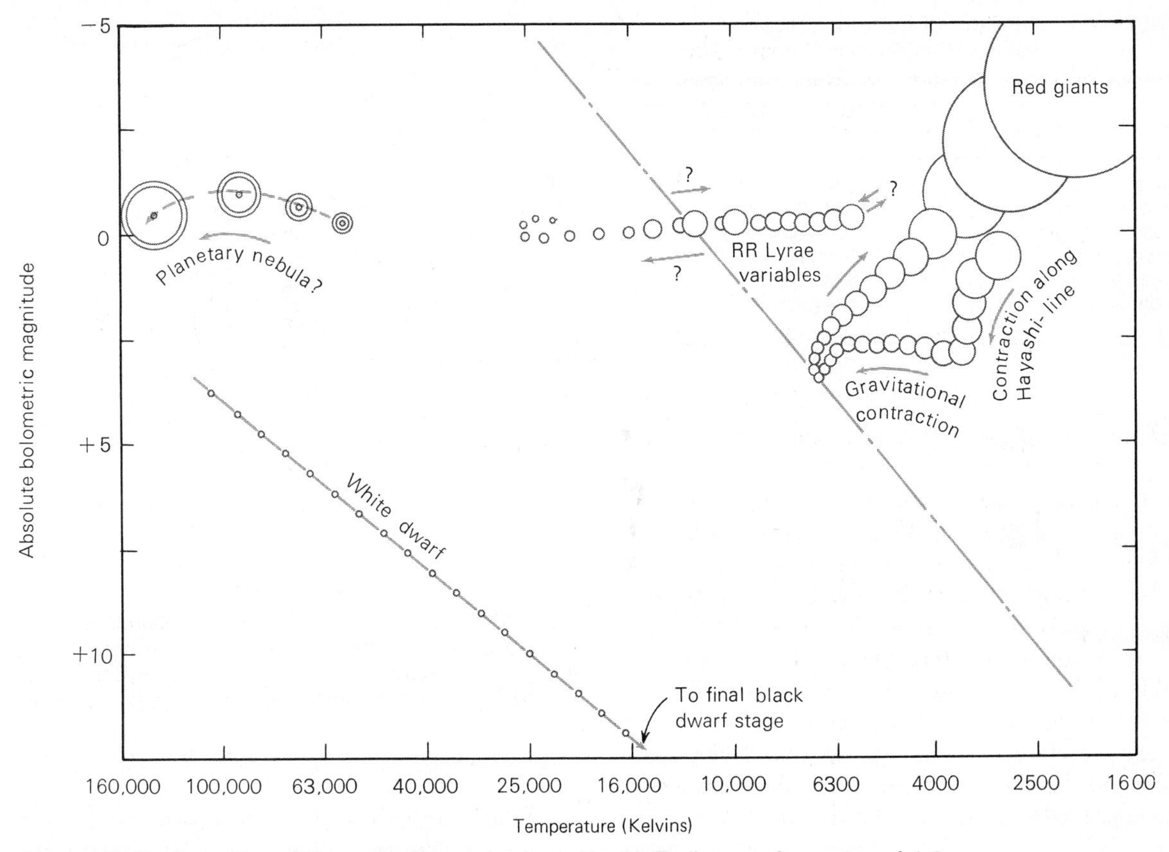

Figure 32.7 Summary of the evolutionary track on the H–R diagram for a star of 1.2 solar masses.

But stars can have masses as large as 100 times the mass of the sun. What is their fate?

The initial stages of evolution of a massive star are quite similar to what happens to stars with masses comparable to that of the sun. On the main sequence, massive stars are, of course, fusing four hydrogen atoms to form helium, converting mass to energy in the process. After the hydrogen in the core of the star is exhausted, the core contracts, since there is no longer enough pressure produced by the generation of heat to resist the force of gravity. The core contracts, releases gravitational potential energy, and so heats both itself and its environs. Hydrogen begins to burn in a shell surrounding the core, and helium fusion begins to produce carbon and then oxygen in the core.

There are two primary differences between what happens up to this point in a massive star and the scenario that we described for stars with masses similar to that of the sun. First, massive stars evolve much more rapidly. While a star like the sun may spend about 10 billion years in its main-sequence phase of evolution burning hydrogen in its core, a star with a mass of 60 to 100 times the mass of the sun can exhaust its supply of hydrogen in a few million years. Massive stars have much more available fuel, but they use it up at such a prodigious rate that their lifetimes are very short. The second difference between high- and low-mass stars is that helium ignition begins when the core is not degenerate, and so there is no helium flash.

It is only after helium is exhausted that the evolution of a massive star takes a very different course from that of solar-type stars. In a massive star, the weight of the outer layers is sufficient to force the core of carbon and oxygen to contract. Again, this contraction heats both the core and its surroundings. Helium-burning begins in a shell surrounding the carbon-oxygen core, and outside this shell there is a second shell in which hydrogen is fusing to

TABLE 32.1 Duration of Various Stages of Nuclear Fusion in a Massive Star

SOURCE OF ENERGY	TIME REQUIRED TO EXHAUST ENERGY SOURCE
Hydrogen burning	7 Million Years
Helium burning	500,000 Years
Carbon burning	600 Years
Neon burning	1 Year
Oxygen burning	6 Months
Silicon burning	1 Day

form helium. The carbon-oxygen core continues to shrink until it becomes hot enough to ignite carbon, which can then form neon, still more oxygen, and finally silicon. After each of the possible sources of nuclear fuel is exhausted, the core contracts until it reaches a temperature high enough to lead to the fusion of still-heavier nuclei. Each cycle lasts for a shorter period of time than the one that preceded it (Table 32.1 shows results of calculations by T. A. Weaver of the Lawrence Livermore National Laboratory), and each releases energy until the last. When the temperature in the core finally becomes hot enough to fuse silicon to form iron, then energy generation must cease. Up to this point, each fusion reaction has *released* energy. Iron nuclei are, however, so tightly bound, that it *requires* energy to fuse iron with any other atomic nucleus.

At this stage of its evolution, a massive star resembles an onion, with an iron core, and at progressively larger distances from the center, layers of decreasing temperature that are burning successively silicon, oxygen, neon, carbon, helium, and finally hydrogen.

What happens next to a star with an iron core? The computations become very complicated, but we can trace what happens in a schematic way. The core of the star, which has no source of energy, must contract. But this core has a mass that exceeds the Chandrasekhar limit of 1.4 times the mass of the sun, and so the collapse does not stop when the electrons in the core become degenerate. The collapse continues at a very high speed to ever-higher densities and temperatures. The iron nuclei break up. Electrons and protons are forced to combine to form neutrons and neutrinos. The neutrons become degenerate, and in a manner analogous to the degenerate electrons in the case of a white dwarf, the pressure produced by this degeneracy resists the further collapse.

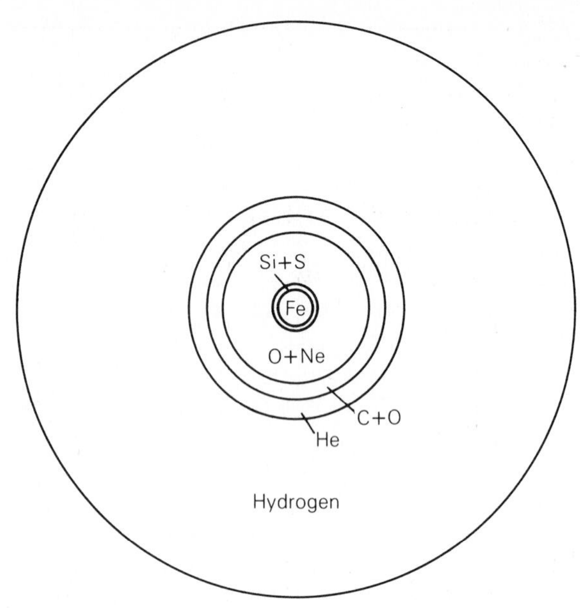

Figure 32.8 Just before its final gravitational collapse, a massive star resembles an onion. The iron core is surrounded by layers of silicon, sulfur, oxygen, neon, carbon mixed with some oxygen, helium, and finally hydrogen.

When the radius of the core is about 10 kilometers, its density is about that of an atomic nucleus. In effect, the matter in the core has merged to form a single gigantic nucleus. This material strongly resists further compression. At this point the collapse, which lasted only about one second, halts abruptly. The shock of this abrupt jolt generates waves throughout the outer layers of the star and causes the star to blow off those outer layers in a violent *supernova* explosion.

32.7 SUPERNOVAE

Supernovae were discovered long before astronomers realized that these spectacular cataclysms marked the death throes of stars. Five supernovae have been observed in our own Galaxy. The Chinese reported the temporary appearance of "guest stars" in the years 1006, 1054, and 1181 A.D. The first of these was reported to be nearly as bright as the half moon, yet there are no European records of the observation of this remarkable event. One cannot help but wonder if the ability to report objectively on celestial phenomena was strongly impaired by the prevailing European belief that the fixed stars were absolutely immutable.

The supernova in 1054 A.D. may have been observed by Indians in the American southwest. Several petroglyphs of about the right age in Arizona and New Mexico depict a bright star next to the crescent Moon, in approximately the orientation the supernova and Moon would have had when they rose together on July 5, 1054, the day after the outburst, and at about the time when it was most spectacular. The 1054 supernova remained visible to the unaided eye in broad daylight for several weeks and in the nighttime sky until April 1056.

The two remaining galactic supernovae occurred in 1572 and 1604 and were observed in considerable detail by Tycho Brahe and Johannes Kepler, respectively. From these historical records, from studies of the remnants of supernovae explosions in our own Galaxy, and from analyses of supernovae in other galaxies, we estimate that one supernova explosion occurs somewhere in the Milky Way Galaxy every 25 to 100 years. Unfortunately, however, no supernova explosion has been detected in our Galaxy since the invention of the telescope. Most of what we know about what happens during the explosion is derived from observations of supernovae in other galaxies.

The light curve of a supernova is similar to that of an ordinary nova, except for the far greater luminosity of the supernova and its greater duration of visibility. Supernovae rise to maximum light extremely quickly (sometimes in a few days or less). In contrast to an ordinary nova, which increases in luminosity a paltry few thousands or at most tens of thousands of times, a supernova is a star that flares up to hundreds of millions of times its former brightness. At maximum light, supernovae reach absolute magnitude -14 to -18, or possibly even -20, or about 10 billion times the luminosity of the sun. (Supernovae appear faint, however, because they are at large distances from the sun.) For a brief time, a supernova may outshine the entire galaxy in which it appears. Just after maximum, the gradual decline sets in, and the star fades in light until it disappears from telescopic visibility within a few months or years after its outburst. Bright emission lines in the spectra of supernovae indicate that they, like ordinary novae, eject material at the time of their outbursts. The velocities of ejection may be substantially greater than in ordinary novae, however, and speeds of up to 10,000 km/s have been observed. Moreover, a much larger amount of material is ejected; in fact, a large fraction of the original star may go off in the expanding envelope.

(a) Types of Supernovae

The available evidence suggests that there are two distinct types of supernovae. A Type I supernova is thought to occur in a binary system that initially contains a white dwarf and a nearby companion. The intense gravitational force exerted by the white dwarf attracts matter from the companion star (see Section 32.3d). The mass of the white dwarf, which initially must have been less than 1.4 solar masses, begins to build up, and eventually it may exceed the Chandrasekhar limit. At this point, the white dwarf must begin to collapse. As it does so, it heats up, new nuclear reactions begin, and the energy released is so great that it completely disrupts the star. Gases are blown out into space at velocities of several thousand kilometers per second, and the temperature of the gas is typically several million degrees. No central star remains behind. The explosion has completely destroyed the white dwarf.

Observations are consistent with this model, in that Type I supernovae occur primarily in types of galaxies (ellipticals) or regions of galaxies (away from spiral arms) where there are large numbers of old, low-mass stars. In contrast, Type II supernovae appear in regions where there are large numbers of young massive stars. Type II supernovae are found in spiral arms and in regions where there is a great deal of dust and gas. They are seldom seen in elliptical galaxies. Because of their association with regions of recent star formation, we think that it is the Type II supernovae that mark the deaths of massive stars. If this picture is correct, then we would expect that some Type II supernovae would leave behind neutron stars.

When it explodes, a supernova sends the debris of the disrupted star hurtling out into space (Color Plate 28a). This debris is so hot that it radiates most of its energy in the X-ray region of the electromagnetic spectrum. In both types of supernovae, this outrushing gas has a higher abundance of such heavy elements as silicon, sulfur, argon, and calcium than does the sun. In Type I, but not in Type II, supernovae, the abundances of nickel, iron, and cobalt are also abnormally high. The fact that supernovae remnants are enriched in these chemical elements offers brilliant confirmation of our model that heavy elements are manufactured in the interiors of stars. When the stars explode, these elements are returned to the interstellar medium and mixed with it, and so become available to be incorporated in succeeding generations of stars. Along with plan-

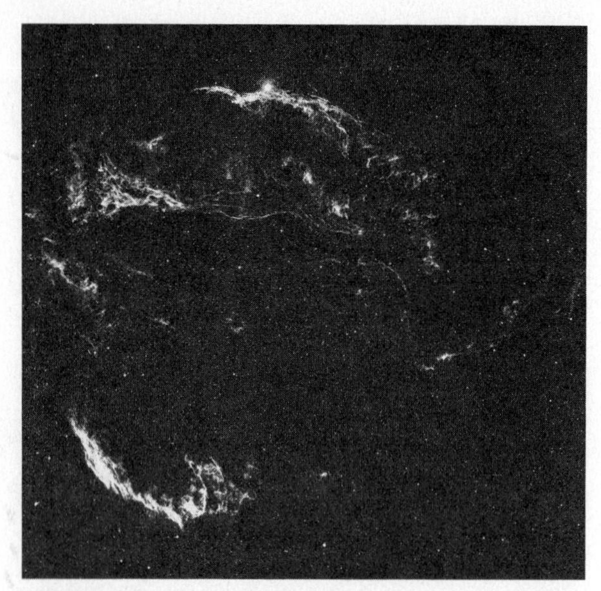

Figure 32.9 Filamentary nebula in Cygnus, photographed with the 1.2-m Schmidt telescope. The nebula is believed to be the remnant of an ancient supernova. (Palomar Observatory; California Institute of Technology)

etary nebulae, supernovae play a major role in building up the chemical elements.

The two types of supernovae can be distinguished observationally by their light curves. In the case of Type I supernovae, all of the light curves are closely similar. The rise to maximum brightness occurs over a period of a few weeks, and the star fades over a period of about six months. Type II supernovae light curves show considerable variety, but usually they fade more slowly than the light curves of Type I supernovae. Type II supernovae are also typically about five times fainter at maximum light than Type I supernovae.

Even though they were made without telescopes, the observations by Tycho Brahe of the supernova of 1572 and by Kepler of the supernova of 1604 are so accurate that we can construct light curves that show how the brightnesses of these two objects varied. Both were Type I supernovae. The supernova of 1006 was probably also a Type I supernova because it was extremely bright. The remnants of these explosions have been identified, and there is no evidence of a central star in any of them.

The light curves of the supernova of 1054 are imprecise but suggest that it was probably a Type II. The expanding gases blown off by this explosion are called the Crab nebula, and there is a star located at the center of this nebula. In fact, it is primarily through detailed studies of the Crab nebula that we have gained absolutely convincing observational confirmation that a massive star can end its life in a catastrophic explosion that leaves behind a neutron star.

(b) The Crab Nebula

The best studied example of the remnant of a supernova explosion is the Crab nebula in Taurus, a chaotic, expanding mass of gas, visible telescopically and appearing spectacular on telescopic photographs.

The outer parts of this gas cloud are observed to be moving away from the center at rates roughly proportional to their distances from it. Angular motions of up to 0.″222 per year have been measured. The Doppler shifts of light from the center of the nebula show the gases there to be moving toward us at speeds of up to 1450 km/s. If the nebula were to expand at the same rate in all directions, we could conclude that at the distance of the Crab, 1450 km/s produces an annual proper motion of 0.″222. In Section 22.3c we developed the formula

$$T = 4.74\ \mu r \text{ km/s},$$

which relates tangential velocity, T, proper motion, μ, and distance, r. Upon substituting 1450 km/s for T and 0.″222 for μ, we find for the distance to the Crab nebula 1380 pc. However, the nebula is elongated in shape (Figure 32.10), not spherical, and when all factors are considered, a distance for the nebula of about 2000 pc is thought to be most probable.

Now if we were to assume that the nebula has always expanded at the same rate, we could derive its age by calculating how long the expansion at its edge (0.″222/yr) would take for it to reach its present maximum radius of 180″. With their present speeds, the filaments of gas would have had to start their outward motion at about 1140 A.D. Radiation from the supernova blast should accelerate the gas, however, so that it must have started expanding at an earlier epoch. It turns out that both the location and computed time of formation of the Crab nebula are in good agreement with the occurrence of the supernova of 1054. The Crab nebula, therefore, must be the material ejected during that stellar explosion.

Figure 32.10 The Crab nebula, photographed in red light with the 5-m telescope. (Palomar Observatory; California Institute of Technology)

The Crab nebula is a strong source of radio waves, infrared radiation, X-rays, and gamma rays, as well as of light. As an X-ray source, it is known as Tau X-1. The radio spectrum (variation of radio energy with wavelength) has characteristics that led the Soviet astrophysicist I. S. Shklovsky, in 1953, to propose that the radiation is from the synchrotron process (Section 34.4a). The Crab nebula is the first astronomical object from which synchrotron radiation was recognized. When the nature of the radio radiation from the Crab nebula was discovered, it was suggested that some of its visible light might similarly originate from the synchrotron mechanism. Observations show that the red filaments (see Color Plate 28a) derive their light mostly from hydrogen ions recombining with electrons, but the white light, and the other radiation, from radio to gamma rays is synchrotron, showing the Crab nebula to possess strong magnetism, and a large source of relativistic electrons. We shall see also that at the center of the nebula is a neutron star, which also is referred to as a pulsar because it emits pulses or bursts of radiation.

32.8 PULSARS

In 1967, Jocelyn Bell, graduate research student at Cambridge University, was studying scintillations of radio sources with one of the Cambridge radio telescopes. (The apparatus she was using more nearly resembles an array of low wire clotheslines stretched out a few feet above the ground than a telescope; the Observatory even brings in sheep to graze the grass under the wires in summer, because there is not room for a person with a lawnmower. Yet, it *is* a radio telescope—Figure 32.11). In the course of Bell's investigation she made a remarkable discovery—one that won her advisor, Antony Hewish, the Nobel Prize in physics, because his analysis of the object (and other similar ones) revealed the first evidence for *neutron stars*.

(a) Discovery of Pulsars

What Bell had found, in the constellation of Vulpecula, was a source of rapid, sharp, intense, and extremely regular pulses of radio radiation, the

Figure 32.11 The Cambridge dipole array radio telescope used by Jocelyn Bell when she discovered the first pulsar. (Courtesy Jocelyn Bell Burnell)

pulses arriving exactly every 1.33728 s. For a time there was speculation that they might be signals from an intelligent civilization, and the radio astronomers half-jokingly dubbed the source, "LGM," for "little green men," and withheld announcement pending more careful study. Soon, however, three additional similar sources were discovered in widely separated directions in the sky, making it highly unlikely that they were signals from other civilizations. By 1981 hundreds of such sources had been discovered at radio observatories throughout the world. They are called *pulsars,* for *pulsating radio sources.*

The first pulsar to be discovered is typical. Detailed studies show each pulse to be complex and to consist of at least three much shorter pulses, each lasting only a few milliseconds, and the entire pulse sequence only a few tens of milliseconds. Although extremely regular in period, the pulses vary considerably from one to the next in intensity. The radiation in all pulses is polarized.

The pulses are emitted over a wide range of radio frequencies, presumably simultaneously, but they are received at the Earth first at high frequency, with successively lower frequencies being received with successively longer delay times. The delay in the arrival of the radiation is believed to be due to the retardation of radio waves by free electrons in interstellar space. The interstellar electrons, in effect, give interstellar space an index of refraction (at radio wavelengths) greater than unity, so that the speed of radio radiation through space is reduced below its value in a perfect vacuum; the reduction in speed is greater the greater the wavelength, accounting for the observed relation between frequency and delay time of the received pulse. We can estimate the density of free electrons in space and calculate the distances to the pulsars from the relation between frequency and the delay of reception time of the pulses; typical distances for the pulsars are a few hundred parsecs. A few pulsars have 21-cm radio absorption lines (Section 27.3k), which show them to lie at least as far as certain interstellar hydrogen clouds producing those lines. Some pulsars, we find, are more than 1000 pc away. At such distances, the radio energy emitted in each pulse must be enormous for us to observe it as strongly as we do. Moreover, because of the sharpness of the pulses, that radio energy must be coming from a region at most a few hundred kilometers in diameter; otherwise, the light-travel time across the emitting region would result in lengthening the pulse time. The pulse periods of different pulsars range from a little longer than 1/1000 s to nearly 10 s.

One pulsar is in the middle of the Crab nebula (described above). It has a pulse period of 0.033 s, and the period is observed to be very slowly increasing, showing that pulsars evolve, pulsing gradually more slowly as they age. This pulsar is also observed to emit optical (visible) light and X-ray pulses with that same 0.033-second period. About 10 to 15 percent of the X-ray radiation from the Crab nebula comes from the pulsar. It has been identified with what appears to be a star of about 16th magnitude (Figure 32.12).

In addition to these pulsating radio sources, dozens of X-ray sources pulse in short regular periods. Of these, only the Crab pulsar is seen also in visible light and radio waves, but the other X-ray pulsars are believed to be objects similar to radio pulsars, but more energetic in their emission. At least some are members of binary star systems, and for four of these, enough information is available to calculate masses by techniques described in Chapter 25. These four X-ray pulsars have masses in the range 1.4 to 1.8 times that of the sun.

(b) Theory of Pulsars—Neutron Stars

The energy emitted by pulsars is not small; that from the Crab pulsar is considerably more than the energy emitted by the sun. Thus pulsars are like

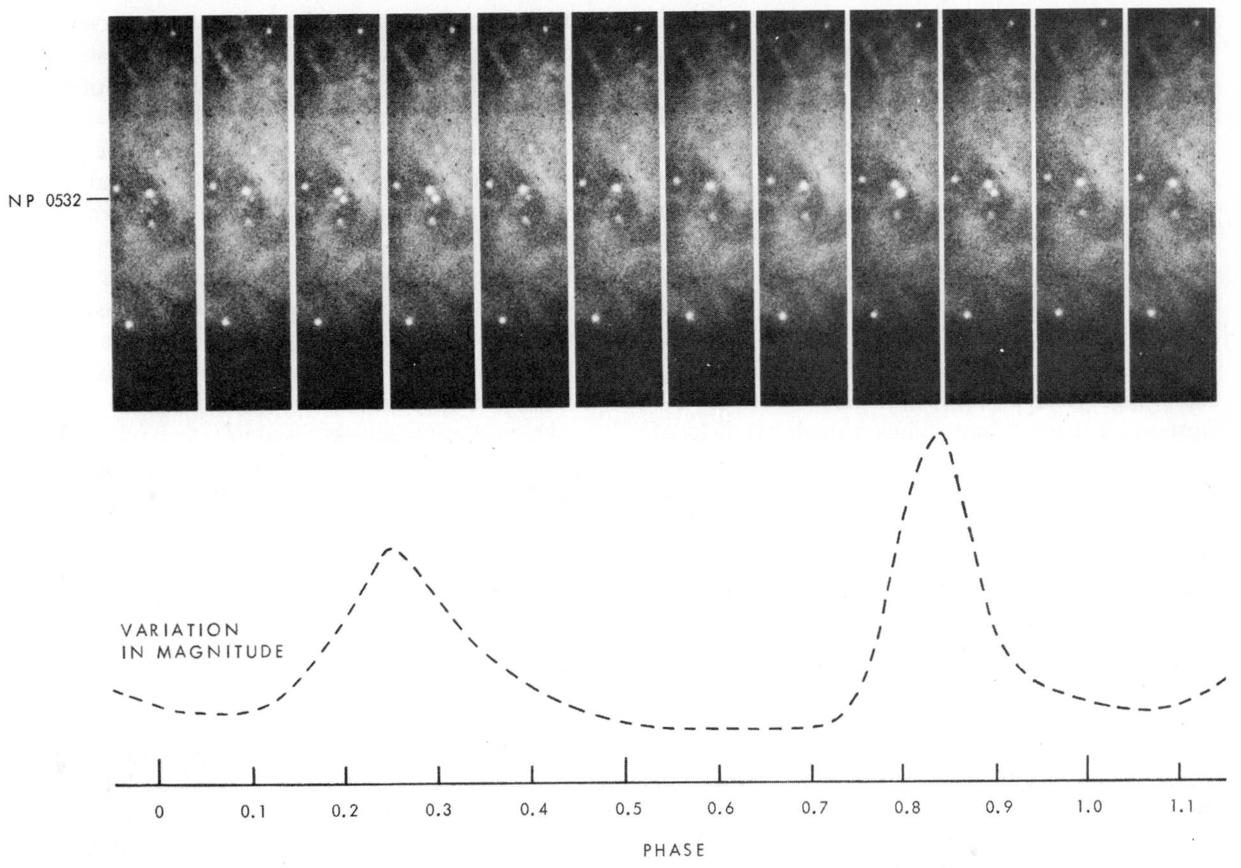

Figure 32.12 A series of photographs of the central part of the Crab nebula taken by S. P. Maran at Kitt Peak National Observatory. Note the star that seems to blink on and off; it is the pulsar, which has a period of 1/30 s. (Kitt Peak National Observatory and National Optical Astronomy Observatories)

stars in their output of radiation. Yet they emit this energy in pulses of up to 650 per second, as if they were pulsating at such high frequency. But what kind of object can pulsate so rapidly? Even a white dwarf would pulsate with a period of several seconds or more—hardly fast enough for the pulsars. Theoretical calculations carried out long before the discovery of pulsars proved to hold the answer to this question.

The solution to the problem of the pulsars suggested itself in 1932, when Chadwick discovered the neutron. Theoreticians then speculated that if the matter in a star could be subjected to such high pressure as to force the free electrons into the atomic nuclei, the star could become a body composed entirely of neutrons. A few years later, Mount Wilson astronomers Walter Baade and Fritz Zwicky suggested that supernova explosions might form neutron stars. At least some pulsars are associated with the remnants of supernovae. Could pulsars be neutron stars?

Neutrons, like electrons, obey the Pauli principle and can become degenerate if crowded into a sufficiently small volume for a given momentum range, so perhaps a star could collapse into degenerate neutrons if it somehow escaped becoming a white dwarf. The neutrons, in such a condition, cannot decay into protons and electrons, for by the time the star is that collapsed, the allowable states for electrons would be filled. The structure of a neutron star is analogous to that of a white dwarf, except that neutron stars are much smaller. A neutron star of one solar mass would have a radius of only about 10 km. Such a star would have a density of 10^{14} to 10^{15} g/cm^3—comparable to that of the atomic nucleus itself.

We note in Table 31.2 that a star of such dimensions would have a natural period of pulsation of less than one ten-thousandth of a second. Since a neutron star could rotate with any period much longer than this, we believe that pulsars must be rotating neutron stars.

There exists a mass-radius relation for neutron stars, and an upper mass limit as well; although the exact theory is not yet certain, the mass limit for a neutron star is believed to be from two to three solar masses.

Any magnetic field that existed in the original star is highly compressed if the core of the star collapses to a neutron star. Thus a moderate field of the order of 1 gauss in a star the size of the sun increases to the order of 10^{10} to 10^{12} gauss around the neutron star. At the very surface of the collapsed star neutrons decay into protons and electrons. Many of these charged particles should leave the stellar surface and move out in the vicinity of the magnetic poles into the circumstellar magnetic field. With such intense fields and high densities many of the particles, especially the electrons, move at speeds close to the speed of light and emit electromagnetic energy by the synchrotron mechanism (Section 34.4a). The radiation is very directional, however, and if the magnetic field of the star happens not to be coaxial with the star's rotation (which also increases greatly during the collapse, to conserve angular momentum), the rotation carries first one and then the other magnetic pole into our view. The consequence can be that the radiation from the rotating magnetic field can be directed toward us once each time the star turns on its axis.

This, at least, is one model for the pulsed radiation from a pulsar. Meanwhile, the atomic nuclei are thought also to be accelerated by the magnetic field to speeds close to that of light, producing cosmic rays. If this entire picture (or some modification of it) is correct, we can understand how supernovae produce heavy elements in the universe, nebulae like the Crab, neutron stars, pulsars, and cosmic rays.

(c) Evolution of Pulsars

The energy radiated by a pulsar, evidently from matter ejected from the star interacting with the stellar magnetic field, robs the star of rotational energy. Thus, theory predicts that the rotating neutron stars should gradually slow down, and that the pulsars should slowly increase their periods. Indeed, as we have already mentioned, the Crab pulsar is actually observed to be increasing the interval between its pulses; similarly, the Vela pulsar (in the center of the Gum nebula) is also found to be slowing down. According to present ideas, the Crab pulsar is rather young and of short period (we know it is only about 900 years old) while the other, older pulsars have already slowed to longer periods. Typical pulsars thousands of years old have lost too much energy to emit appreciably in the visible and X-ray wavelengths, and are observed only as radio pulsars; their periods are a second or more.

One other possible end state for a star is that of a *black hole*, which involves one of the more bizarre predictions of general relativity theory. We shall discuss black holes in Chapter 33.

32.9 SS 433

One of the most remarkable objects in the Galaxy—possibly the first of many such objects to be discovered—is SS 433.

SS 433 is in a catalogue of objects showing H-alpha (hydrogen) emission lines compiled by C. G. Stephenson and N. Sanduleak at Case Western Reserve University in 1977. More recently, it attracted attention when it was found to be a source of variable radio emission and X-rays. Moreover, it is near the center of a supernova remnant and may be related to a supernova outburst. But the really astonishing features of SS 433 became known when its optical spectrum was studied in a long series of observations—especially by B. Margon and his associates at UCLA.

The spectrum shows bright emission lines of hydrogen and helium arising from three different sources. One source appears to be the main object, and the Doppler shift of its lines indicates a relatively low speed, like that expected for an object in our Galaxy. The other two sources, however, have very high speeds with respect to the main body and are evidently streams of gas ejected from it in oppositely directed jets. Analysis shows the gas to be streaming out (in both directions) at 26 percent the speed of light. Moreover, the whole system is rotating in such a way that the directions of the jets describe a conical motion (suggesting precession) with a period of 164 days.

The object giving rise to the remarkable radiation is a member of an eclipsing binary system with

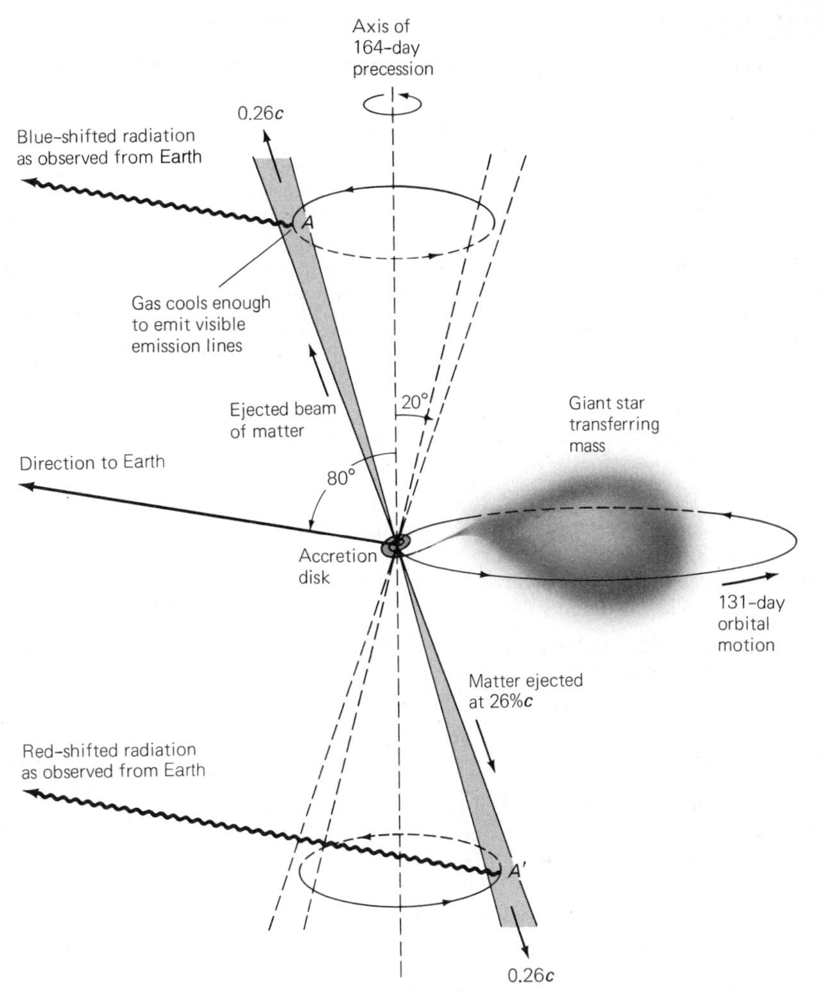

Figure 32.13 A model for SS 433, proposed by UCLA astronomers Jonathan Katz, Bruce Margon, and George Abell. A binary system consists of a giant star and a collapsed star (probably a neutron star or a black hole) in orbit about each other with a period of 13.1 days. Matter from the giant transfers to the compact star and accumulates into an accretion disk, from which jets of gas are ejected at 26 percent the speed of light. The disk precesses so that the jets describe a motion along the surface of an imaginary cone of half angle 20°. SS 433 is viewed from the Earth at an angle of 80° to the cone axis, so that at a typical time one jet has a component of motion toward the Earth, and its radiation appears blueshifted by the Doppler effect, whereas the other beam has a component of motion away from the Earth, and its radiation appears redshifted. The material in the jets probably does not cool enough to emit visible radiation until some distance from the disk—say, at *A* and *A*′.

a period of 13.1 days. One component of this system is a star with a mass in the range 13 to 24 times the mass of the sun. The other object is compact, and its mass is somewhere between 1.4 and 12 times the mass of the sun. Whether this compact object is a neutron star or a black hole remains an open question that cannot be resolved until its mass is determined more precisely, but in the best current model the compact star is assumed to be a neutron star.

The model that best accounts for the jets of high-velocity gas involves mass transfer from the companion star to the neutron star. Because of the orbital revolution, the transferring matter does not fall directly on the neutron star, but ends up in a flat rotating disk of matter surrounding the neutron star in its equatorial plane. (In this *accretion disk,* some of the matter may lose angular momentum to other matter and fall into the star.) According to the model, it is from the accretion disk that matter is ejected, perpendicularly to the disk, into the streams that give rise to the observed radiation. Under tidal forces produced by the non-neutron-star companion, the accretion disk is presumed to precess with the 164-day period, accounting for the variations in radial velocity of the radiation from the jets (Figure 32.13). This model, however, is not the only one possible and is therefore not universally accepted at this writing.

The importance of SS 433 is the study of the mechanisms for the mass ejection at so high a speed and the periodic change in direction. In some respects, SS 433 has properties that on a small scale remind us of those of radio galaxies and quasars (Chapter 35). Perhaps our eventual understanding of SS 433 will bring us the insight needed to fathom the powerhouses of those far grander objects associated with remote galaxies.

EXERCISES

1. Where on the H–R diagram does a star *end?*
2. Suppose a star spends 10×10^9 years on the main sequence and burns up 10 percent of its hydrogen. Then it quickly becomes a red giant with a luminosity 100 times as great as that it had while on the main sequence and remains a red giant until it burns up the rest of its hydrogen. How long a time would it be a red giant? Ignore helium-burning and other nuclear reactions, and assume that the star brightens from main sequence to red giant almost instantaneously.
3. Since supernovae occur so rarely in any one galaxy, how might a search for them be conducted? (It may help to glance at Chapter 35.)

★4. Explain why a rotating ring surrounding a shell star produces a sharp absorption line at the *center* of each emission line rather than at the violet edge, as in the case of an expanding shell of gas.

★5. Calculate the diameter of a Wolf-Rayet star, in terms of the sun's diameter, if the star has 10,000 times the sun's luminosity and seven times its temperature.

6. Compare and contrast nova shells, supernova shells, and planetary nebulae.
7. Suppose the luminous shell of a planetary nebula is easily resolved with a telescope. Now suppose the spectrum of the nebula is photographed, with the slit of the spectrograph extending completely along one diameter of the shell. Sketch the appearance of a typical emission line in the spectrogram, and explain your sketch. It may help to look over the description of the construction of a spectrograph in Chapter 9.
8. Suppose the central star of a planetary nebula is 16 times as luminous as the sun, and 20 times as hot (about 110,000 K). Find its radius, in terms of the sun's. Does this star have the dimensions of a white dwarf?
9. The gas shell of a particular planetary nebula is expanding at the rate of 20 km/s. Its diameter is 1 LY. Find its age. For this calculation, assume that there are 3×10^7 s/yr, and 10^{13} km/LY.
10. The angular radius of an expanding spherical shell of gas about a star is observed to increase at 0.″19/year. The radial velocity of expansion of the shell, with respect to the star, is observed to be 1200 km/s. What is the distance to the object? (See Section 22.3c).
    Answer: About 1330 pc
11. Assume that a pulsar is 100 pc away. Suppose that no star brighter than apparent magnitude 23 shows up in that position of the sky. What is the brighter limit to the absolute magnitude that a star associated with the pulsar could have?
12. Show that the surface temperature of a particular gradually cooling white dwarf is proportional to the fourth root of its luminosity. (See Chapter 8.)
13. By the time the sun becomes a white dwarf, the constellations familiar to us now will not be seen, even if their stars never change in luminosity. Why?
14. Calculate, very roughly, the density of a hypothetical neutron star of one solar mass.
15. Even when the jets of gas in SS 433 are moving *across* our line of sight, the wavelengths of the spectral lines are longer (as observed from Earth) than they would be if the jets were motionless, or if the source were in the laboratory. Can you suggest an explanation? (*Hint:* See Chapter 11. How fast is the material moving? What happens to its time, as we observe it? What does this have to do with the wavelengths [or frequencies] of the light it emits?)

Albert Einstein (1879–1955) received the Nobel Prize in 1921, not for his theory of relativity but for the photoelectric effect. At that time his ideas on relativity were still at the frontier. Einstein believed that such seemingly diverse areas of physics as mechanics and electromagnetic phenomena—and even gravitation—were guided in the same way by underlying principles.

33

GENERAL RELATIVITY: CURVED SPACETIME AND BLACK HOLES

In his special theory of relativity (Chapter 11), Einstein showed that different observers in uniform relative motion perceive space and time differently—that is, they disagree with each other on measurements such as length, time, momentum, and energy. The special theory also shows how those measurements by relatively moving observers compare with each other. But the rules of special relativity do not apply to the comparison of measurements by observers that are *accelerated* with respect to each other. Einstein's general theory of relativity, published in 1916, shows how we can extend the special theory to take account of observers in relative acceleration produced by gravitation. In doing so, Einstein discloses a new formulation of gravitation, because a gravitational force on a system is indistinguishable from an acceleration of that system, that is, gravitation and acceleration are equivalent.

33.1 PRINCIPLE OF EQUIVALENCE

Galileo noted that all bodies, despite their different masses, if dropped together fall to the ground at the same rate. According to Newton's law of gravitation, the Earth pulls on a more massive object with a greater force than it does on a less massive one. The two objects fall together, however, because according to Newton's second law, a proportionately greater force is required to impart the same acceleration to the heavier object. The mass that appears in Newton's law of gravitation, however, is *gravitational mass,* and that which appears in his formulation of the laws of motion is *inertial mass.* The fact that bodies of different mass fall together implies that inertial mass and gravitational mass are the same.

In 1909 this equality was verified to better than 5 parts in 10^9 in a famous experiment by the Hungarian physicist, Baron Roland von Eötvös. Today, similar experiments have pushed any difference to below one part in a million million (10^{12}). It is widely assumed that gravitational and inertial mass are, in fact, identical.

Einstein had the genius to turn that isolated fact* into a powerful principle of physics—the principle of equivalence. In doing so, he took the first

---

* As Julian Schwinger described it; we owe several of the examples in this chapter for describing the general theory of relativity to the unique talent of Professor Schwinger for peeling away the rind and getting to the heart of the idea that describes a fundamental physical concept.

Figure 33.1 In Skylab everything stays put or moves uniformly because there is no apparent gravitation acting inside the laboratory. (NASA)

giant step toward formulating his general theory of relativity.

Because two bodies falling together side by side approach the ground together, they are obviously *not* accelerated with respect to each other. Thus they are aware of no force acting between them. For example, suppose a brave boy and girl simultaneously jump into a bottomless chasm from opposite sides of its banks. If we ignore air friction, while they fall they accelerate downward together and feel no external force acting on them. They can throw a ball back and forth between them, aiming always in a straight line, as if there were no gravitation, and the ball, falling along with them, would move directly to its target.

It's very different on the surface of the Earth. Everyone knows that a ball, once thrown, falls to the ground. Thus in order to reach its target (the catcher) the ball must be aimed upward somewhat, so that it follows a parabolic arc—falling as it moves forward—until it is caught at the other end.

Because our freely falling boy, girl, and ball are all falling together, we could enclose them in a large box falling with them. Inside that box, no one can be aware of any gravitational force; nothing falls to the ground, or anywhere else, but moves in a straight line in the most simple natural way, obeying Newton's laws. By having our box fall with the boy and girl, we have removed the force of gravitation by selecting a coordinate system that is accelerating at just the right rate to compensate for gravitation. Here is the principle of equivalence—a force of gravitation is equivalent to an acceleration of the coordinate system of the observer, and such a force can therefore be completely compensated for by an appropriate choice of an accelerated coordinate system. Einstein himself pointed out how a rapidly descending elevator seems to reduce our weight and a rapidly ascending one increases it. In a *freely falling* elevator, with no air friction, we would lose our weight altogether.

This idea is not hypothetical. In 1973 and 1974, astronauts in the Skylab lived for months in just such an environment (Figure 33.1). The Skylab was, of course, falling freely around the Earth, as it continued to do until, unmanned, it finally suffered too much friction with the Earth's tenuous upper atmosphere and plunged to a fiery doom in the summer of 1979. But while in free fall the astronauts lived in a seemingly magical world where there were no outside forces. One could give a wrench a shove, and it would move at constant speed across the orbiting laboratory. One could lay a pencil in midair and it would remain there, as if no force acted on it.

Mind, there *was* a force; neither the Skylab nor the astronauts were *really* weightless, for they continually fell around the Earth, pulled by its gravity. But since all fell together—lab, astronauts, wrench, and pencil—*locally* all gravitational forces were absent.

Thus the Skylab provides an excellent example of the principle of equivalence—how local effects of gravitation can be removed by a suitable accelera-

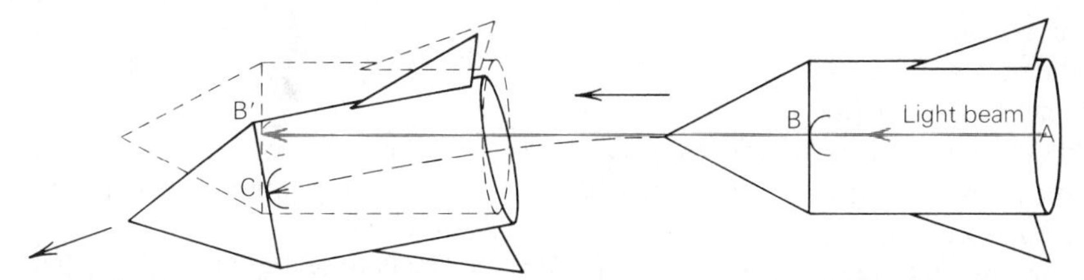

Figure 33.2 If in a spaceship moving to the left (in this figure) in its orbit about a planet, light is beamed from the rear, *A*, toward the front, *B*, we might expect the light to strike at *B'*, above the target in the ship, which has fallen out of its straight path in its orbit about the planet. Instead the light, bent by gravity, follows the curved path and strikes at *C*.

tion of the coordinate system. To the astronauts it was as if they were far off in space, remote from all gravitating objects. But what if astronauts *were* in remote space, and were to activate the engines of their ship, producing acceleration. The ship would then push up against their feet giving the impression of a gravitational tug. If one were to drop a small coin and a hammer, the floor of the ship would move up to meet both objects at the same time; to the astronauts, though, it would seem that the hammer and coin fell to the floor together. To them it would be exactly the same situation as that isolated fact made famous by Galileo—that heavy and light objects fall together. In other words, an acceleration of one's local environment produces exactly the same effect as a gravitational attraction; the two are indistinguishable—again, the principle of equivalence.

(a) Trajectories of Light and Matter

Einstein postulated that the principle of equivalence is a fundamental fact of nature. If so, however, there must be *no* way in which an astronaut, at least by experiments within his local environment, can distinguish between his weightlessness in remote interstellar space and his free fall in a gravitational field about a planet like the Earth.

But how about light? If the astronauts shone a beam of light along the length of their ship and if the ship were falling in a free-fall orbit about a planet, would the ship not then surely fall away from a straight line path, which the beam must follow, causing the light to strike above its target?

Not so, according to Einstein. If the principle of equivalence is correct, there must be no way of knowing whether one is accelerated (any more than he can detect his own absolute motion) and hence the experiment must fail. Thus the light beam *must fall with the ship* if that ship is in orbit about a gravitating body (Figure 33.2). The idea that light, as well as material bodies, must be affected by gravity led Einstein to the prediction that stars seen by light from them that passes near the sun must appear displaced because of bending of their light by the gravitational field of the sun. This prediction, when formulated precisely, was, as we shall see, eventually confirmed by observation during a solar eclipse.

(b) The Gravitational Redshift

Let us consider another possible experiment in a freely falling laboratory (the Einstein elevator). Suppose we shine a light beam—say, a laser beam of a precise frequency—upward from floor to ceiling. Now the laboratory accelerates downward, gaining speed, so by the time the light beam travels up to the ceiling, that ceiling is moving downward faster than the source on the floor was when the light left it. In other words, the receiver at the ceiling is *approaching* the source (where it was when the light left it). Therefore, wouldn't we expect to find the light at the ceiling blueshifted slightly because of the Doppler effect (Section 8.3b)? But this would violate the principle of equivalence, for the blueshift would reveal our downward acceleration and show us we could not be weightless in free space. Therefore, Einstein postulated, there must be a *redshift*, due to the light moving upward against gravity, that exactly compensates the Doppler shift that would otherwise be observed. If so, that gravitational redshift should be observed in radiation climbing upward in a gravitational field—in principle at least—even at the surface of the Earth. Is such an effect observed?

The Earth's gravitational field is too weak to show the effect on visible light because we know of no source for which the frequency can be so sharply defined that the extremely tiny redshift would be noticeable. Yet, it has been observed, not in visible wavelengths of electromagnetic radiation, but in gamma radiation. In the 1960s, at the Jefferson Physical Laboratory at Harvard University, gamma rays were sent from a source in the basement to a detector at the top of the building. The source of gamma rays was radioactive cobalt; the radiation was confined to a very sharp frequency interval by a technique invented by the Nobel Prize–winning physicist Rudolf Mössbauer. If a similar detecting layer of cobalt were placed directly above the emitting layer, the gamma rays would be absorbed by the former. However, the detector was placed at the top of the building, 20 m above the source. The gamma rays, traveling upward against the Earth's gravitation, suffered a gravitational redshift, and were not absorbed by the upper cobalt detector. In order to absorb them, the detector had to be moved slowly downward to produce a blueshift to compensate the Earth's gravitational redshift. The actual motion of the detecting cobalt needed to make it absorb the gamma rays from the emitting cobalt in the basement was so slow that it would have required a full year to close the 20-m gap between emitter and detector. That speed produced a Doppler shift that agreed with the value needed to compensate for Einstein's predicted redshift to within one percent.

We can calculate the amount of gravitational redshift with elementary algebra. Let the elevator, of height h, be falling freely in, say, the Earth's gravitational field, and for definiteness, near the Earth's surface, where the uniform gravitational acceleration is g. Light, at speed c, takes a time $t = h/c$ to travel from the elevator floor to the ceiling. During this time, the elevator, accelerating uniformly, has increased its downward speed by

$$v = gt = gh/c.$$

The value gh/c is, then, the speed of the detector (when light reaches it) relative to the source (when light left it). The Doppler shift would then be (Section 8.3b)

$$\frac{\Delta\lambda}{\lambda} = z = \frac{v}{c} = \frac{gh}{c^2}.$$

So long as the acceleration is uniform and $v << c$, so that the simple formula for the Doppler shift can be used, gh/c^2 is the gravitational redshift.

In the Jefferson Laboratory experiment, h = 20 m = 2 x 10^4 cm, so the gravitational redshift is

$$z = \frac{gh}{c^2} = \frac{980 \times 2 \times 10^4}{(3 \times 10^{10})^2} = 2.18 \times 10^{-14}$$

The speed required to produce a Doppler shift of that amount is

$$v = 2.18 \times 10^{-14}\, c = 6.53 \times 10^{-4} \text{ cm/s}.$$

At that speed, the time for the detector to travel 2 x 10^4cm is

$$t = \frac{2 \times 10^4}{6.53 \times 10^{-4}} = 3.06 \times 10^7 \text{ s} \approx 1 \text{ year}.$$

According to the principle of equivalence, one should be able to *produce* a gravitational redshift by merely accelerating a spaceship far away from all gravitating bodies. Clearly, such a redshift would be produced: if radiation is beamed from one passenger to another in the direction of the spaceship's acceleration, the receiver, where the radiation is absorbed, is moving away from the source, where that same radiation is emitted. Thus, since the source and receiver are separating, there is a redshift (due to the Doppler effect) that is indistinguishable from that produced by a gravitational field that produces the same acceleration.

Within a freely falling spaceship (like Skylab) or in the Jefferson Laboratory, the gravitational field is essentially uniform. Such is not the case, however, for the light we observe leaving a star, because that light has to pass from the strong field near the star's surface on out through the continually weakening one as it gets farther and farther from it. However, Einstein showed that we need only add up the tiny effects as the light passes through each small region within which gravity can be regarded as effectively constant to calculate the total gravitational redshift of light leaving the star. It works out that the wavelengths of light from the sun should be increased by about 2 parts in a million—a redshift too small to be distinguished from other effects.

The acceleration of gravity at a distance r from the center of a star of mass M is GM/r^2 (Section 4.2), where G is the gravitational constant. Over a small distance, dr, the small contribution, dz, to the gravitational redshift is

$$dz = \frac{1}{c^2}\frac{GM}{r^2}dr.$$

Readers familiar with calculus will recognize that a simple integration—adding up of contributions—from the surface, R, of the star to a point very far away gives for the total gravitational redshift of light from the star:

$$z = \frac{1}{c^2}\frac{GM}{R}.$$

White dwarf stars, however, being very dense, have a much stronger surface gravity than the sun, and Einstein suggested that the gravitational redshift of the light from white dwarfs might be detectable. It can only be observed, however, for white dwarfs whose radial velocities are known from independent methods so that it can be separated from the Doppler shift due to the stars' motions. Fortunately, several white dwarfs are members of binary star systems, and their radial velocities can be deduced from those of their non–white dwarf companions, for which the gravitational redshift is negligible. The first reliable confirmation of the effect was made in 1954 by UCLA astronomer D. M. Popper, who measured the gravitational redshift of the white dwarf companion of the star 40 Eridani.

The precision of Popper's observations was such as to verify Einstein's predictions to within about 20 percent. Far higher accuracy has been attained recently in the near-Earth environment with space-age technology. In the mid-1970s, a hydrogen maser carried by a rocket to an altitude of 10,000 km was used to detect the radiation from a similar maser on the ground. That radiation showed a gravitational redshift due to the Earth's field that confirmed the relativity predictions to within a few parts in ten thousand.

(c) Limitations to the Principle of Equivalence

We have seen that the force of gravitation can be compensated in a suitably accelerated coordinate system locally—that is, over dimensions small enough that within them the acceleration produced by gravitation can be regarded as constant. Thus within a freely falling spaceship in orbit about the Earth, gravity appears absent, and all bodies behave according to the rules of special relativity; everything either remains at rest or moves uniformly in a straight line—a straight line as defined by the path of a light beam. If there are two spaceships in orbit, but one, say, 100 km above the other, the principle of equivalence applies to each. However, the motions of objects within one ship would *not* appear unaccelerated as seen from the other, for the force of Earth's gravitation varies with distance from the Earth, and is appreciably different between the two ships.

Because the ships are in different gravitational fields, the gravitational redshifts within the two of them are different as well. Now all that we have said about frequency of light applies to the rate of all other physical processes as well; the rate of passing of light waves is just one of many ways to measure the passage of time. Time flows differently in different gravitational fields too. A clock in the spaceship of lower orbit (hence in a region of stronger gravity) runs more slowly than one in the other ship. Astronauts in the different ships would therefore disagree on the rate of time passage, as well as on the paths of unaccelerated bodies.

The above considerations suggest how an astronaut could tell that he was in orbit in a gravitational field—even without observing the planet beneath him. Suppose he fires a rocket probe straight ahead. For a while, the probe would continue in essentially the same orbit as the ship and would appear to hang motionless as viewed through a forward porthole. After a time, however, its motion away from the ship would carry the probe into a higher orbit where it would be accelerated downward less strongly, and it would slowly drift upward in the window, and eventually out of view. A differently accelerated reference frame now applies to the probe, just as it does to a separate orbiting spaceship.

Einstein's problem in formulating general relativity was to unify these separate descriptions of

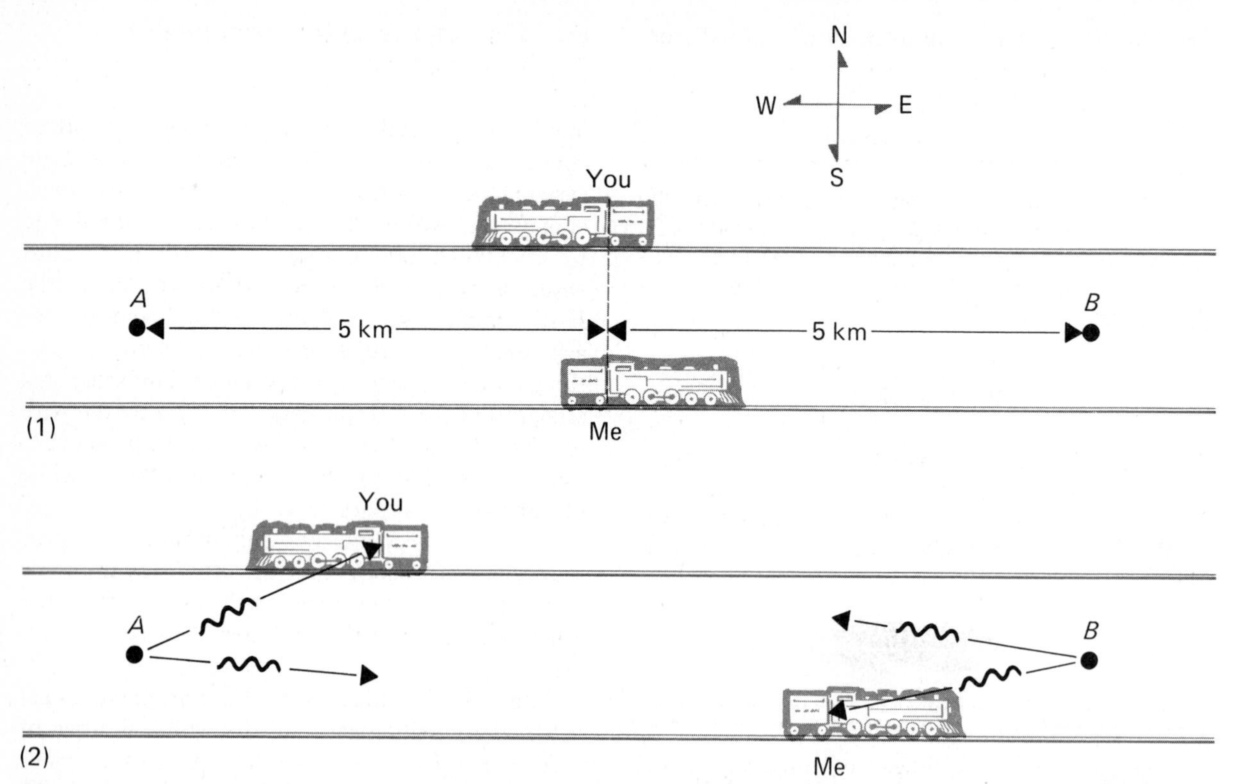

Figure 33.3 An experiment demonstrating the nonabsoluteness of simultaneity.

motion in different parts of a gravitational field into a connected whole, in order to find how to define a reference frame in which all objects, no matter where they are, are unaccelerated. To succeed, Einstein had to employ two ideas: spacetime and curvature.

33.2 SPACETIME

There is nothing mysterious about four-dimensional spacetime. Imagine yourself in the rear seats at an outdoor concert at the Hollywood Bowl. The sound from the orchestra in the shell, hundreds of feet away, takes a goodly fraction of a second to reach you, and the players seem to be behind the beat of the conductor. When a piece is finished, you first hear the applause from people near you, and slightly later from the front of the amphitheater. Because of the finite travel time of sound, all people do not hear the same note of music at the same time; nor do events that appear simultaneous *visually* seem to be audibly simultaneous.

Light also has a finite speed, so we never see an instantaneous snapshot of events around us (as we saw in Chapter 11). The speed of light is so great that within a single room we obtain *effectively* an instantaneous snapshot, but it is certainly not the case astronomically. We see the Moon as it was just over a second ago, and the sun as it was about 8 minutes ago. At the same time, we see the stars by light that left them years ago, and the other galaxies as they were millions of years in the past. We do not observe the world about us at an instant in time, but rather we see different things about us as different *events* in spacetime.

Relatively moving observers do not even agree on the order of events. As an example, suppose you and I approach each other along an east-west direction in rapidly moving trains (Figure 33.3). When we are abreast of each other, a person on the ground pulls a switch to set off two flares, one 5 km east of us and the other 5 km west of us, and he observes the flares to go off simultaneously. I am moving toward flare *B*, while you are moving toward flare *A*. By the time the light from flare *B*

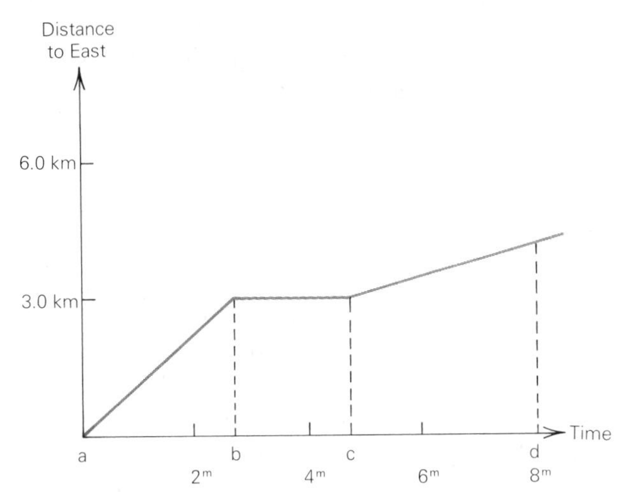

Figure 33.4 The progress of a motorist traveling east across town.

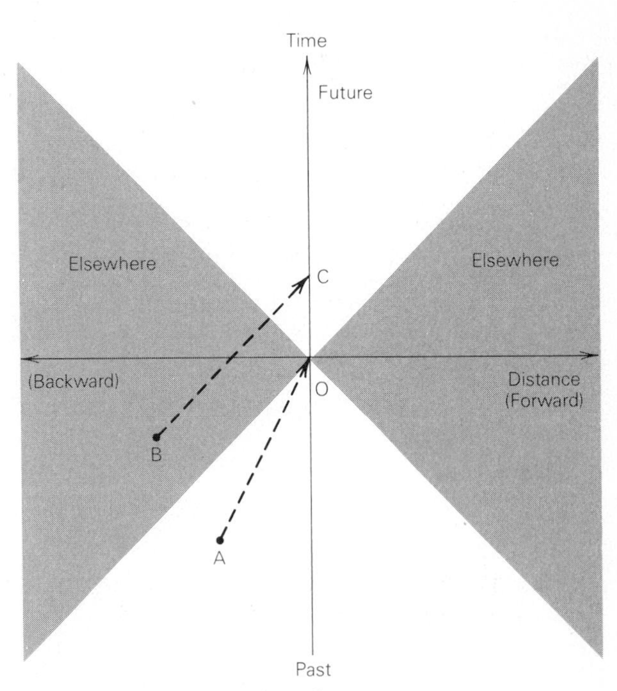

Figure 33.5 A spacetime diagram.

reaches me, I am as shown in Figure 33.3 (2). You, meanwhile, are the same distance in the opposite direction from where we passed and are just receiving the light from flare A. Sometime later, I receive the light from flare A and you from flare B. Thus two happenings that appear simultaneous to one are not simultaneous to the other. Space and time are inextricably connected. We need to describe the universe not just in terms of three-dimensional space, but in terms of four-dimensional spacetime.

We can easily represent the spatial positions of objects in two dimensions on a flat sheet of paper (for example, the plan of a city). To plot three dimensions on a page, the draftsman uses projections. Architectural drawings of a home generally show three projections: floor plan and two different elevations—say, the house as seen from the east and from the north—to give all necessary information. By the use of perspectives (which rely on our learned experience), we can also give an impression of a three-dimensional view. There is no easy way, though, to draw a four-dimensional perspective to include time.

There is no problem, however, in showing a two-dimensional projection of four-dimensional spacetime. Figure 33.4, for example, shows the progress of a motorist driving to the east across town. How much time has elapsed since he left home is shown on the horizontal axis, and how far he has traveled eastward is shown on the vertical axis. From a to b he drove at a uniform speed; from b to c he stopped for a traffic light and made no progress, and from c to d he drove more slowly because of increased traffic.

Figure 33.5 shows a rather conventional two-dimensional representation of spacetime. Time increases upward in the figure, and one of the three spatial dimensions is shown horizontally. If we measure time in years and distance in light years, light goes one unit of distance in one unit of time, so flows along diagonal lines as shown. "Here and now" is at the origin of the diagram. At this instant we can receive information of a past event along such a line as AO; in this case the messenger was going slower than light, so he covered less distance than light would in the same time. Because nothing can go faster than light, we cannot, right now, know of something happening at point B in spacetime, for the message along BO would have to travel faster than light. We will have to wait until we are at C in the future, before a light or radio beam can get us the word along path BC.

We can also show three dimensions of spacetime in a perspective drawing, as in Figure 33.6. Here time flows to the right, the height above the ground is upwards in the figure, and one of the two dimensions along the ground is shown obliquely as

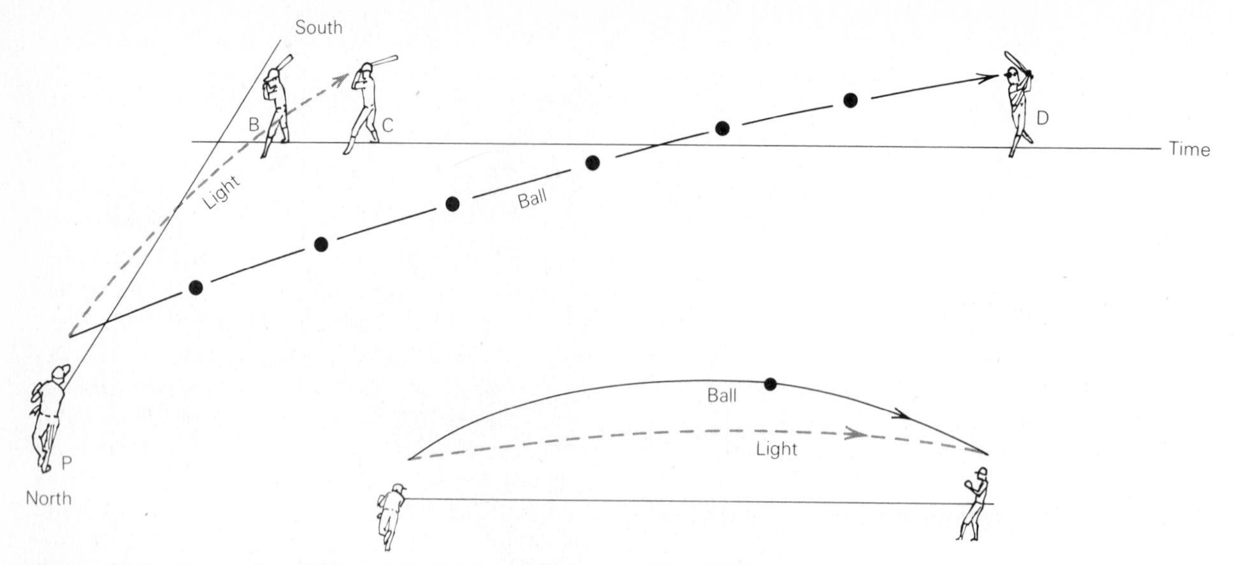

Figure 33.6 The paths of light and of a baseball in spacetime. Below are shown the corresponding paths in space alone.

the north-south line. Suppose a batter is at point B (homeplate) and a baseball pitcher is at P, north of homeplate. The pitcher throws the ball to the batter, but the ball, traveling at a finite speed, flows along path PD in spacetime, arriving at the batter at D. Light, however, travels much faster than the ball, and arrives at the batter at C. In the gravitational field of the Earth both the ball and light beam fall to the ground enroute to the plate at the same rate, but the light reaches the plate so quickly that it falls only a slight distance compared to the ball, which takes much longer. Thus in space (lower part of Figure 33.6) the path of the light is much more nearly straight than that of the ball. In spacetime, however, the path of the ball is very long compared with that of the light beam, and it is easy to see how in the uniform gravitational field (as on the baseball diamond) both ball and light fall along paths of the same shape (that is, the same *curvature*) in spacetime.

33.3 CURVATURE OF SPACETIME

Because of the principle of equivalence, Euclidean geometry applies in a freely falling reference frame in a uniform gravitational field. But freely falling systems at different places on Earth fall in different directions, because "down" is always toward the center of the Earth. Thus we cannot describe the behavior of objects in different widely separated freely falling systems with spacetime coordinates for which Euclidean geometry holds. Let's consider a familiar analogy. A simple Mercator-type map, with lines of constant latitude running horizontally and lines of constant longitude running vertically, is fine for showing a small area of the Earth—say a single city—without noticeable distortion. But such a map cannot show a large area of the curved Earth without distortion; everyone knows how distorted and enlarged lands of extreme latitude (those near the poles) appear on the usual flat world maps. We cannot map the Earth with Euclidean plane geometry.

Indeed, if we travel far enough in a straight line on the surface of the Earth, we end back at our starting point; our path is a *great circle* (the equator is such a great circle). More generally, if we take into account the slight polar flattening of the Earth, as well as effects of such irregularities as mountains, our "straight line" path is called a *geodesic*, which means "Earth divider."

Einstein showed how to find spacetime coordinates within which all objects move as they would if there were no forces. In a small local region, where a gravitational field is uniform, those coordinates are Euclidean, just as a city plan can be well described with plane geometry. But to describe paths of objects over a large region, where the gravitational field varies, spacetime must be curved, just as we must use curved geometry to describe a large area of the spherical Earth.

The distribution of matter determines the nature of a gravitational field, so it is the distribution of matter that determines the geometry of curved spacetime. Within this curved spacetime, everything moves in the simplest possible way as if no gravitation were there at all. In analogy with Earth geometry, the paths of light and material objects in spacetime are called *geodesics.*

People usually say that the geometry of spacetime is determined by matter, but we think this description makes the subject sound unduly mysterious. What meaning can there be to "curved space" if space is the absence of matter? It is not space in itself that is curved; rather it is the system of coordinates that we can conveniently use to describe the motions of objects and light. By selecting non-Euclidean geometry—curved coordinates—we can describe the paths of light and objects as "straight" in the same sense that great circles (or geodesics) are "straight" on the curved surface of the Earth.

The mathematics needed to handle the problem, on the other hand, was not available in Euclid's time. The geometry needed to describe curved spacetime was developed after the pioneering work of the great German mathematician, physicist, and astronomer, Karl Friedrich Gauss. Gauss became involved in the invention of new geometry when he was commissioned to survey the German State of Hanover by its king, George IV—also king of Great Britain. The new geometry received its full expression in the hands of Gauss' student, Bernhard Riemann.

Riemann, in applying for a university position at Göttingen, had submitted three possible topics for a lecture he would deliver. Traditionally, the judges selected one of the first two topics offered, however, so Riemann had not bothered to prepare a lecture on the third. That, however, was the very topic that Gauss had been pondering for decades. Consequently, in only a few weeks Riemann wrote out the lecture on the third topic, which was to be his masterpiece: "On the Hypotheses Which Lie at the Foundations of Geometry."

By the end of the 19th century the new Riemannian geometry was further facilitated with the invention of tensor calculus. By 1915, Einstein was able to use these new mathematical techniques to derive the *field equations* of general relativity, which describe the curvature of spacetime by matter, and the *geodesic* equations, which describe the unaccelerated paths of objects in spacetime.

33.4 TESTS OF GENERAL RELATIVITY

Is general relativity, then, essentially different from Newtonian gravitational theory, or merely a different but equivalent mathematical formulation? Relativity *is* different from Newtonian theory in that the signals that govern gravitational interactions are not instantaneous, but travel with the speed of light, and also, of course, in that matter and energy are equivalent, so that not only matter itself but also energy contributes to gravitation—that is, to the geometry of spacetime—and energy (light, for example) as well as mass is affected by that geometry. Naturally, where speeds are low compared with that of light, and where the gravitational field is relatively weak—both conditions of which are met throughout most of the solar system—the predictions of general relativity must agree with those of Newton's theory, which has served us so admirably in our technology and in guiding space probes to the other planets. In familiar territory, therefore, the differences between predictions of the two theories are subtle, and consequently very difficult to detect.

Einstein himself proposed three observational tests of general relativity. One is the gravitational redshift, already described, the second is the deflection of starlight that passes close to the sun, and the third is a subtle effect on the motion of the planet Mercury.

(a) Deflection of Starlight

The strength of the gravitational acceleration at the surface of the sun is 28 times its value at the surface of the Earth. Einstein calculated from general relativity theory that starlight just grazing the sun's surface should be deflected by an angle of 1''.75. Stars cannot be seen or photographed near the sun in bright daylight, but with difficulty they *can* be photographed close to the sun at times of total solar eclipses. Einstein suggested an eclipse observation to test the light deflection in a paper he published during World War I. A single copy of that paper, passed through neutral Holland, reached the British astronomer Arthur S. Eddington. The next suitable

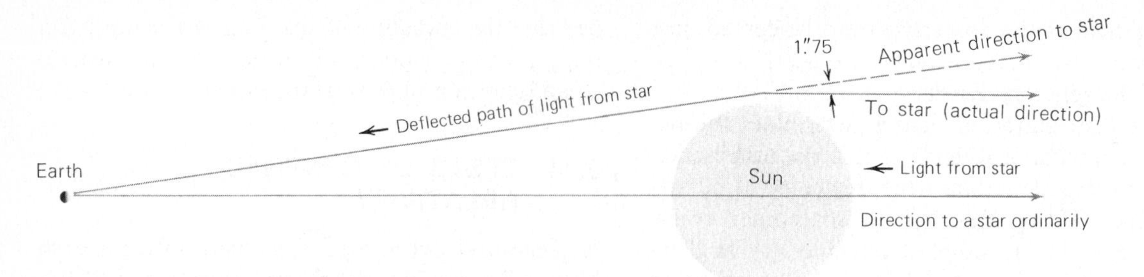

Figure 33.7 Deflection of starlight passing near the sun.

eclipse was on May 29, 1919. The British organized two expeditions to observe it, one on the island of Principe, off the coast of West Africa, and the other in Sobral, in North Brazil. Despite some problems with the weather, both expeditions obtained successful photographs of stars near the sun. Measures of their positions were then compared with measurements on photographs of the same stars taken at other times of the year when the sun was elsewhere in the sky. The stars seen near the sun were indeed displaced and, to the accuracy of the measurements, the shifts were consistent with the predictions of relativity. It was a triumph that made Einstein a world celebrity overnight.

Eclipse observations to test the relativity effect have continued over the years, but the measures are very difficult to make and the precision of the confirmation is not high. Far higher accuracy has been obtained recently at radio wavelengths. Simultaneous observations of the same source with two radio telescopes far apart can pinpoint the direction of the source very precisely. The United States National Radio Astronomy Observatory at Greenbank, West Virginia, with radio telescopes 35 km apart, observed several remote astronomical radio sources (quasars—see Chapter 35) when the sun was nearly in front of them. The apparent directions of the quasars showed shifts similar to those of stars seen near the sun. The accuracy of these observations is high enough to confirm the Einstein prediction to within 1 percent.

Notice (Figure 33.7) how the sun acts like a lens in deflecting light passing near it. A very remote object, acting as such a *gravitational lens,* can produce distorted and multiple images of another object far beyond, but nearly in line with it. For decades astronomers have suspected that such gravitational lens effects might be found in nature, and in the late 1970s the first such example was discovered: the double quasar 0975 + 561 is believed to be two images of the same object produced when its light passes near a foreground galaxy. It is described in Chapter 35.

In addition to sources of light or radio waves appearing displaced slightly when seen near the sun, the radiation from them is also delayed slightly in reaching the Earth. We have no way of measuring that delay in light from stars or quasars, but we can detect it in the radio pulses broadcast from space probes, because we know where they are and when the signals should arrive at Earth. The experiment has been performed with several planetary probes, but most precisely with the Viking landers on Mars. When Mars is on the far side of the sun, signals from the Vikings must pass through a region of spacetime that is relatively strongly curved by the sun (Figure 33.8) and are observed to be delayed by about 100 microseconds, as if Mars had jumped some 30 km out of its orbit. It is just the delay expected by relativity theory to within 1 part in a thousand.

(b) Advance of Perihelion of Mercury

According to relativity, the energy and momentum associated with the motion of a body, and even its gravitational energy, all contribute to its effective mass, and hence to the force of gravitation on it. Now, Mercury has a fairly eccentric orbit, so that it is only about two-thirds as far from the sun at perihelion as it is at aphelion. As required by Kepler's second law, Mercury moves fastest when nearest the sun, and all of the motional and gravitational effects add to the attraction between the two. Consequently, relativity predicts that a very tiny additional push on Mercury, over and above that predicted by Newtonian theory, should occur at each

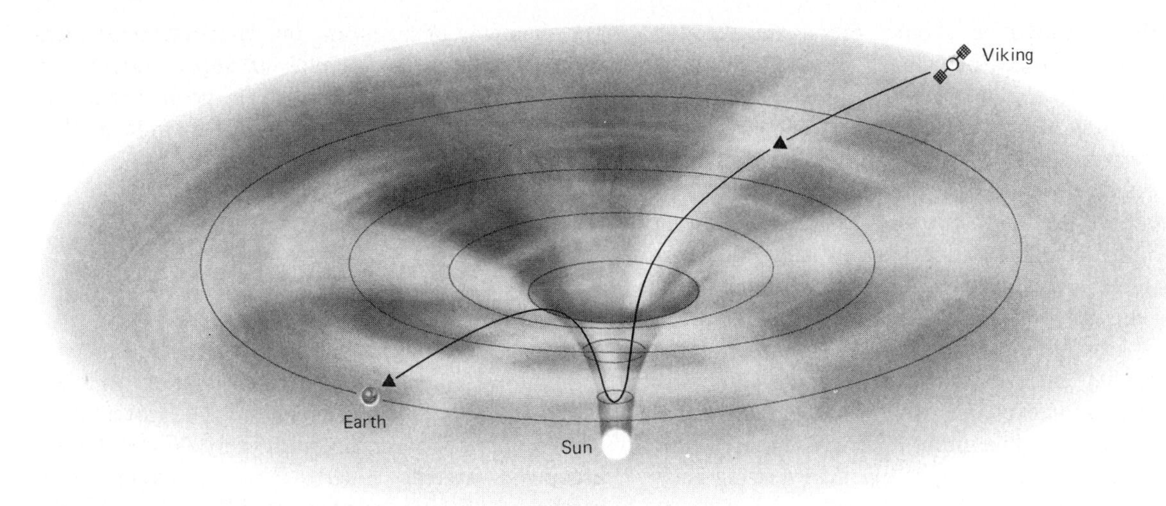

Figure 33.8 Radio signals from Viking are delayed because they have to pass near the sun, where spacetime is curved relatively strongly.

perihelion. The result of this effect is to make the *line of apsides,* which is the long dimension (major axis) of Mercury's orbit, slowly rotate in space, so that each successive perihelion occurs in a slightly different direction as seen from the sun (Figure 33.9). The prediction of relativity is that the direction of perihelion should change by only 43″ per century; it would thus take about 30,000 years for the line of apsides to make a complete rotation.

The gravitational effects (perturbations) of the other planets on Mercury also produce an advance of its perihelion, and to a far greater extent than the relativistic prediction. According to Newtonian theory, the perihelion of Mercury should advance by 531″ per century. Even in the last century, however, it was observed that the actual advance is 574″ per century. The discrepancy was first called to attention by Leverrier, co-discoverer of Neptune, and in analogy with Neptune, it was assumed that an intramercurial planet was responsible. The hypothetical planet was even named for the god Vulcan. Vulcan, of course, never materialized, but that 43″ anomaly was entirely explained by relativity. The relativistic advance of perihelion can also be observed in the orbits of several minor planets that come close to the sun.

Additional tests for relativity theory are in the planning or experimental stages—some at the frontier of modern technology. In a satellite experiment expected to fly in the early 1990s, the behavior of a gyroscope will be carefully monitored. Relativity predicts an angular change in orientation of the axis of the gyroscope due to its motion through the Earth's gravitational field of only 0.″05 per year, but it is believed that even that small change can be measured accurately enough to check the theory.

33.5 GRAVITATIONAL WAVES

Because the geometry of spacetime depends on the distribution of matter, any rearrangement of that matter must result in an alteration of spacetime—that is, it creates a disturbance. Relativity predicts

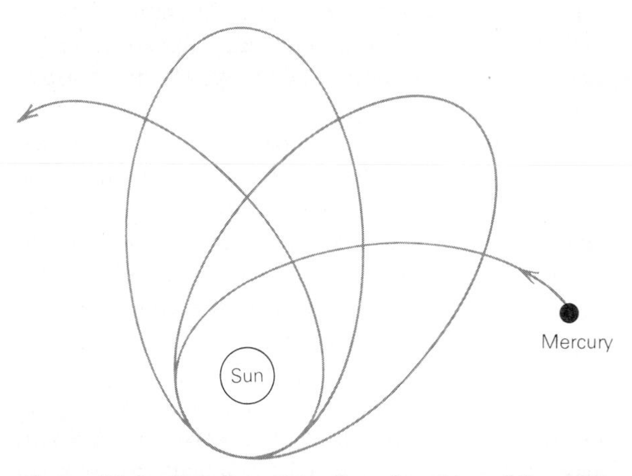

Figure 33.9 Rotation of the line of apsides of the orbit of a planet, such as Mercury, because of various perturbations.

that that disturbance should propagate through space with the speed of light. It is called a *gravitational wave*.

We think we should be able to detect gravitational waves, as we do other kinds of radiation. Whereas electromagnetic radiation can be detected, for example, in the way it sets charges oscillating back and forth, gravitational waves should set material objects vibrating. Detection of gravitational waves, however, is extremely difficult. One reason is that gravitation is an exceedingly weak force (Chapter 1), and its radiation is correspondingly weak. But moreover, the disturbance gravitational waves should produce in an object is far more subtle than simple oscillation; rather than moving back and forth as a unit, the body suffers very slight compressions and lengthenings.

Gravitational waves produced by mass motions on the Earth, or even of the Earth itself, would be far too weak to detect by technology we can currently imagine. We would need a motion of very large mass, and at a speed approaching that of light. One possible source could be the sudden collapse of a massive star, perhaps in a supernova explosion, that is, a stellar catastrophe.

The pioneering attempt to detect such gravitational waves from collapsing stars was by Joseph Weber, physicist at the University of Maryland. In the 1960s he suspended large metal cylinders equipped with very delicate sensors. Weber had hoped to detect gravitational waves by the vibrations they set up in the cylinders. To distinguish true gravitational waves from space from purely local disturbances, he placed one cylinder in Maryland and a second one at the Argonne National Laboratory outside Chicago; only signals recorded simultaneously at both stations would be of astronomical origin.

To date, Weber has not detected signals that can unequivocally be attributed to gravitational waves from space. Other laboratories, however, are developing far more sensitive detectors. At Stanford University, for example, an extremely delicate microphone is arranged to pick up the vibrations that a passing gravitational wave sets up in a metal bar. The bar is a practical detector because the two ends would be set vibrating differently, which sets up an oscillation in the bar itself. The gravitational waves passing the Earth produced by the collapse of a solar mass star at the center of the Galaxy would displace the ends of the bar by only one ten-millionth the diameter of an atom, but with advanced technology, utilizing principles of superconductivity, the Stanford experiment has a good chance of detecting such gravitational waves if, indeed, they exist.

(a) The Binary Pulsar PSR1913+16

It may be that nature has already provided us with indirect evidence of gravitational radiation. In 1974, R. A. Hulse and J. H. Taylor, of the University of Massachusetts, observing with the 1000-ft radio telescope at Arecibo, Puerto Rico, discovered a remarkable pulsar, now designated PSR1913+16 (the numbers give the coordinates of its location in the sky). The unique thing about PSR1913+16 is that the period of the pulses itself shows cyclic variations over a short time interval of 7^h45^m. These period changes are due to the Doppler effect caused by the pulsar's revolution about another object. When the pulsar is approaching us in its orbit, each successive pulse has less far to travel to reach us on Earth, and we receive the pulses slightly closer together than average. Conversely, when the pulsar is moving away from us, we receive the pulses slightly spread out in time. Thus we can analyze the orbital motion of the pulsar just as we do that of a spectroscopic binary star (Section 25.2d).

Such an analysis, combined with our knowledge of the expected properties of neutron stars, indicates that the pulsar is in mutual revolution in that 8-hr period about a mute companion of comparable mass that is probably either a white dwarf or another neutron star. Because the size of the orbit is only a little bigger than the diameter of the sun, it is exceedingly unlikely that the companion is a normal star, for then the pulses would be eclipsed by it during part of each revolution (contrary to observation) unless the orbit were almost face on, in which case we should not see appreciable variations in the pulse period.

Now there are two important points about this binary pulsar. First, at its orbital speed of approximately 0.1 percent that of light, it should radiate gravitational waves at a rate great enough to carry appreciable energy away from the system, causing the pulsar and its companion to spiral slowly closer together. Second, pulsars are superb clocks, remaining stable to 0.001 s over several years.

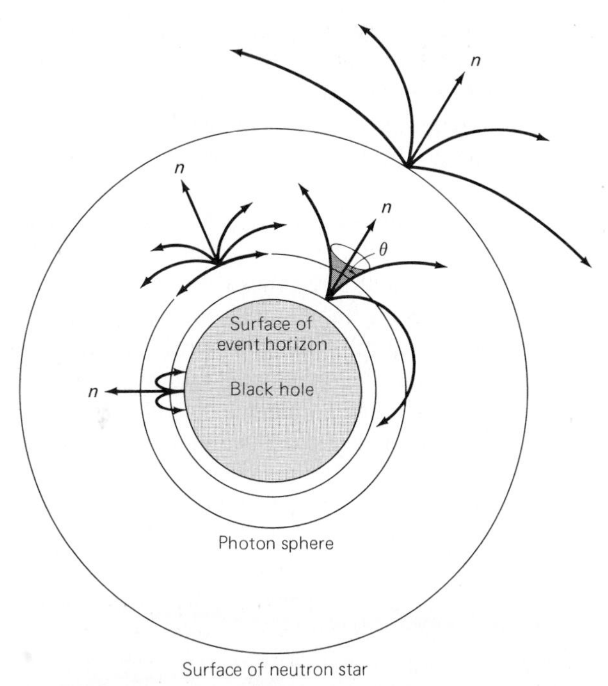

Figure 33.10 The deflection of light from a very dense star. At a radius smaller than that of the photon sphere, to escape, light must flow into a cone of half-angle θ with respect to the normal, *n,* to the surface. At the event horizon. θ = 0.

As the pulsar and its companion spiral together, their period of revolution shortens. The shortening is only about one ten-millionth of a second per orbit, but the effect accumulates like a clock that runs a little faster each day. During the first seven years this system was observed, the time of periastron (when the two objects are closest together) shifted by more than a full second of time relative to the time it would have been occurring if the period had remained constant. Pulsars are such good timekeepers that this shift is easily observed.

In fact, the shift in periastron time is just what general relativity predicts it should be due to the emission of gravitational radiation by two stars, each of about 1.5 solar masses, with an orbit like that of the binary pulsar. PSR1913+16 is now regarded as providing strong evidence that gravitational waves do exist, as Einstein predicted.

33.6 BLACK HOLES

Until now, the manifestations of general relativity theory we have discussed have been so subtly different from Newtonian theory that one might well wonder if the new ideas about space and time brought about by Einstein are really important enough to concern us. Indeed, many physicists, during the first half century after Einstein introduced his revolutionary theory, regarded the subject as almost academic.

But hold! In the last chapter we encountered pulsars, which provide strong evidence for the existence of neutron stars. Neutron stars have very great density and extremely strong surface gravity. We recall that light grazing the surface of the sun is deflected by about 1″.75. Light grazing the surface of a white dwarf would be deflected by about 1′, and that grazing a typical neutron star by about 30°.

In 1796, the French mathematician Pierre Simon, Marquis de Laplace speculated about the properties of an object that had so great a gravitational field that light could not escape at all, but would be bent right around and stay with the object. Laplace's "corps obscurs" were later reconsidered by modern physicists, armed with the new rigor of general relativity theory. John Wheeler, the Princeton physicist who has become intimately associated with general relativity, has dubbed such objects "black holes."

Consider the light radiated from the surface of a neutron star. That which emerges normal (perpendicular) to the surface flows out radially from the star. That emitted at an angle of, say, 30° to the normal leaves the star at an angle somewhat greater than 30° to the normal, because of the gravitational deflection. Now imagine a more massive star that shrinks to a smaller size and higher density than a neutron star. As the surface gravity increases, the deflection of light increases too. Eventually the star reaches a size at which a horizontal beam of light enters a circular orbit. A surface of that radius is called the *photon sphere.*

As the star shrinks to a size smaller than the photon sphere, to escape the star light must flow into a cone about the normal to the surface of half-angle θ (Figure 33.10), and light at a greater angle falls back on the star. The angle θ becomes smaller and smaller as the star collapses, until the radius of the star is two-thirds that of the photon sphere, near which θ becomes zero, and no light at all can escape. At this point the velocity of escape from the star equals the speed of light. As the star contracts still more, light and everything else is trapped inside, unable to escape through that surface where the escape velocity is the speed of light. That surface is called the *event horizon,* and its radius is the *Schwarzschild radius,* named for Karl Schwarzschild, who first described the situation a few years after Einstein introduced general relativity. This surface is the boundary of the black hole. All that is inside is hidden forever from us; as the star shrinks through the event horizon it literally disappears from the universe.

The size of the Schwarzschild radius is proportional to the mass of the star. For a star of one solar mass, the black hole is about 3 km in radius; thus the entire black hole, some 6 km in diameter, is about one-third the size of a one-solar-mass neutron star.

The event horizons of larger and smaller black holes—if they exist—have greater and lesser radii, respectively. For example, if a globular cluster of 100,000 stars could collapse to a black hole, it would be 300,000 km in radius, a little less than half the radius of the sun. If the entire Galaxy could collapse to a black hole, it would be only about 10^{12} km in radius—about 0.03 pc. On the other hand, for the Earth to become a black hole it would have to be compressed to a radius of only 1 cm— about the size of a golf ball. A typical minor planet, if crushed to a small enough size to be a black hole, would have the dimensions of an atomic nucleus!

It happens that the correct size of the event horizon can be calculated by pretending that Newton's laws apply. The formula for the velocity of escape is $\sqrt{2GM/R}$, where M and R are the mass and radius. By equating this to c, the speed light, and solving for R, we find that the radius of the event horizon of a black hole of mass M is

$$R = \frac{2GM}{c^2}.$$

(a) One Way to Make a Black Hole

Theory tells us what a black hole must be like. But do black holes exist? And if so, how were they created?

The most plausible way to manufacture a black hole is through the collapse of a star that has used up all of its store of nuclear energy. As we have seen, however, only fairly massive stars can end their lives as black holes. Stars with initial main-sequence masses of six to eight times the mass of the sun apparently complete their evolution by becoming white dwarfs. At the time that nuclear burning ceases, the mass of the core of these stars is less than 1.4 times the mass of the sun, and electron degeneracy can halt the collapse. Stars with masses of 8 to about 100 times the mass of the sun are thought to burn nuclear fuel until a core of iron and nickel is created. When this core forms, no more energy can be produced by the fusion of atomic nuclei, and the core begins to collapse. If the mass of this core is less than two to three times the mass of the sun, then neutron degeneracy is able to halt the collapse abruptly, and the resulting shock waves blow off the outer layers of the star and produce a Type II supernova explosion.

The critical question, and one which theory has not yet been able to answer, is whether or not some stars are so massive that even neutron degeneracy is not able to stop the collapse. If stars exist in which the core exceeds the upper limit on the mass of a neutron star (two to three times the mass of the sun), then nothing that we now know of can halt the collapse, and a black hole will be formed.

If theory cannot yet tell us what conditions produce a black hole rather than a neutron star, then do observations provide any evidence that black holes actually do exist? And how would we go about looking for an object that we cannot see? The answer is that we can look for the gravitational effects of a black hole on a nearby star. As stars collapse into black holes they leave behind their gravitational fields, and if a black hole should be a member of a double star system, then we may be able to detect the black hole by studying the orbital motion of its companion.

(b) Candidates for Black Holes

To find a black hole: (1) We must find a star whose motion (found from the Doppler shift of its spectral lines) shows it to be a member of a binary star system, and to have a companion of mass too high to be a white dwarf or a neutron star. (2) That companion star must not be visible, for a black hole, of course, gives off no light. But being invisible is not enough, for a relatively faint star might be unseen next to the light of a brilliant companion. Therefore, (3) we must have evidence that the unseen star, of mass too high to be a neutron star, is also a collapsed object—one of extremely small size—for then our theory predicts that it must be a black hole—or at least a star on the way to becoming one.

Modern space astronomy has come to the rescue in (3). One way to know we have a small object of high gravity (and possibly a black hole) is if matter falling toward or into it is accelerated to high speed. Near the event horizon of a black hole, matter is moving at near the speed of light. Internal friction can heat it to very high temperatures—up to 100 million Kelvins or more. Such hot matter emits radiation in the form of X-rays. Modern orbiting X-ray telescopes—especially the Einstein telescope, HEAO 2—can and do reveal such intense sources of X-radiation.

So we want X-ray sources associated with binary stars with invisible companions of high mass. We cannot prove that such a system contains a black hole, but at present we have no other theory for what the invisible massive companion can be if, indeed, the X-rays are coming from gas heated by falling toward it.

We can easily understand the origin of such infalling gas. We have already seen (Section 32.3d) how stars in close binary systems can exchange mass. Suppose one star in such a double star system has evolved to a black hole and that the second star has now evolved to a red giant so large that its outer layers pass through that point of no return between the stars and some of its matter falls to the black hole. The mutual revolution of the giant star and black hole cause the material from the former to flow not directly onto the black hole, but, because of conservation of angular momentum, to spiral around it, collecting in a flat disk of matter called the *accretion disk*. In the inner part of the accretion disk the matter is revolving about the black hole so fast that its internal friction heats it up to the temperature where it emits X-rays. In the course of this friction, some material in the accretion disk is given extra momentum, and escapes from the double star system, and other material loses momentum and falls into the black hole—lost forever to observation from the rest of the universe.

Another way to form an accretion disk in a binary star system is from material ejected from the companion of the black hole as a stellar wind; some of the ejected gas will flow close enough to the black hole to be captured by it into the disk. Such a case, we think, is the binary system containing the first X-ray source discovered in Cygnus—Cygnus X-1 (Figure 33.11). The visible star is a normal B-type star. The spectrographic observations, however, show it to have an unseen companion of mass near ten times that of the sun. That companion would be a black hole if it were a small, collapsed object. The X-rays from it strongly suggest that it is, for we have no other explanation for the source of those X-rays than gas heated by an infall toward a tiny massive object. Of course we cannot be certain that Cygnus X-1 is a black hole, but many astronomers think that it probably is.

(c) Properties of Black Holes

Much of the modern folklore about black holes is misleading. One idea is that black holes are monsters that go about sucking things up with their gravity. Actually, the gravitational attraction surrounding a black hole at a large distance is the same as that around any other star (or object) of the same mass. Even if another star, or a spaceship, were to pass one or two solar radii from a black hole, Newton's laws would give an excellent account of what would happen to it. It is only very near the surface of a black hole that its gravitation is so strong that Newton's laws break down; for a black hole of the mass of the sun, light would have to come within 4.5 km of its center to be trapped. A solar mass black hole, remember, is only 3 km in radius—a very tiny target. Even collisions between ordinary stars, hundreds of thousands of times bigger in diameter, are so rare as to be essentially nonexistent. A star would be far, far safer to us as an interloping black hole than it would have been in its former stellar dimensions.

(d) A Trip Into a Black Hole

Still, it is interesting to contemplate a trip into a black hole. Suppose that the invisible companion of the star associated with Cygnus X-1 is a black hole of ten solar masses. What would you see if a daring astronaut bravely flies into it in a spaceship?

At first he darts away from you as though he were approaching any massive star. However, when he nears the event horizon of Cygnus X-1—some 30 km in radius, and presumably near the center of the accretion disk—things change. The strong gravitational field around the black hole makes his clocks run more slowly as seen by you. Signals from

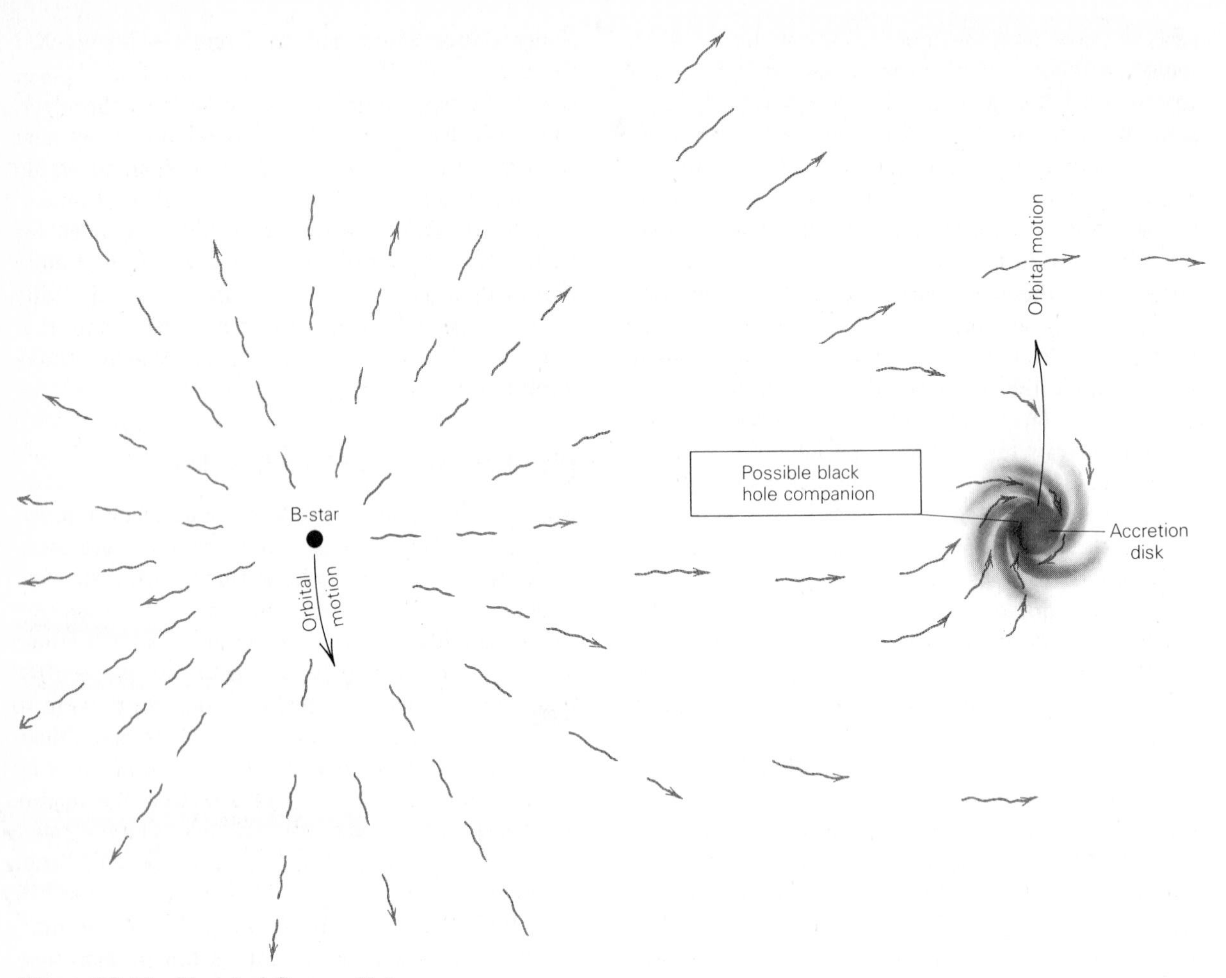

Figure 33.11 Model of Cygnus X-1.

him reach you at greatly increased wavelengths because of the gravitational redshift. As he approaches the event horizon, his time slows to a stop—as seen by you—and his signals are redshifted through radio waves to infinite wavelength; he fades from view as he seems to you to come to a stop, frozen at the event horizon. All matter falling into a black hole appears to an outside observer to stop and fade at the event horizon, frozen in place, and taking an infinite time to fall through it—including the matter of a star itself that is collapsing into a black hole. For this reason, black holes are sometimes called *frozen stars*.

This, however, is only as you, well outside the black hole, see things. To the astronaut, time goes at its normal rate and he crashes right on through the event horizon, noticing nothing special as he does so, except for enormously strong tidal forces that rip him apart. At least this is true of ten-solar-mass black holes. If there were a very massive black hole, say, thousands of millions of solar masses, its event horizon, although small in astronomical dimensions, would be large enough that its tidal forces are not severe, and an astronaut, in principle, could survive a trip into it.

But in no case is there an escape. Once inside, astronaut, light, and everything else is doomed to remain hidden forever from the universe outside. Moreover, current theory predicts that after entering the black hole, the astronaut races irreversibly to the center, to a point of zero volume and infinite density.

Mathematicians call such a point a *singularity*, but many physicists doubt that such singularities of infinite density can exist in the real universe and suspect that new physics will eventually emerge to help us understand what really happens to matter at extraordinary densities. In particular, mathematical solutions suggesting that black holes can collapse through *wormholes* to emerge as "white holes" in "other universes"—ideas often exploited in the popular literature—now appear to be abstractions without physical basis; recent theoretical studies, in fact, show that even general relativity theory does not predict that black holes emerge elsewhere as white holes.

On the other hand, massive stars collapsing through their own event horizons to become black holes do not have to reach densities beyond those understood by present-day physics. We can never know what goes on inside the event horizon, but we know of no reason why black holes themselves should not exist.

(e) Other Ways to Make Black Holes

While the collapse of a massive star seems to be the mechanism most likely to produce a black hole, calculations suggest that other processes may lead to the formation of black holes of all sizes, with masses ranging from 10^{-5} grams to 10^{8} times the mass of the sun.

For a long time, theoretical models suggested that stars more massive than about 100 times the mass of the sun were unstable. There is now (controversial) observational evidence that a few superstars with masses a few hundred to a few thousand times the mass of the sun may exist. When such objects finally collapse, it seems likely that they would form black holes, although there is as yet no observational evidence that black holes with such large masses really do occur.

As we shall see in Chapter 35, there is evidence that supermassive black holes (up to 10^{8} times the mass of the sun) can be found at the centers of many galaxies. Such massive black holes might be formed from the collapse of supermassive stars, through the collision and coalescence of stars in a very dense star cluster, through the coalescence of smaller black holes, or through the accretion and accumulation of matter into an originally much smaller black hole.

Most intriguing of all is the possibility that very small black holes—as small as one hundred-thousandth of a gram—may have formed during the first fraction of a second after the universe began. The density of material as it crosses the event horizon to form a black hole varies inversely as the mass of the black hole. An object with a mass of 10^{9} times the mass of the sun would have a density about equal to the density of water as it fell inside its event horizon. For an object with a mass equal to that of the sun, the corresponding density is a thousand times the density of an atomic nucleus. Objects much less massive than the sun would have to be compressed to incredible densities to form a black hole. We can think of no circumstances at the present time that seem likely to produce such an enormous compression. The necessary densities may, however, have been achieved during the first fraction of second after the universe was formed, when all of the matter of the universe was confined to a very small volume (see Chapter 37).

There is no observational evidence that such primordial black holes actually exist, but they are of great interest to theoreticians because their properties are determined by quantum mechanics as well as by relativity, and in a most amazing way. This possibility has been explored by the brilliant British theoretical astrophysicist Stephen Hawking.

We have seen (Chapter 29) that all fundamental particles have their antiparticles; for example, electrons and positrons, protons and antiprotons, and so on. Whenever a particle and its antiparticle come into contact, they annihilate each other, transforming completely into energy. Similarly, pure energy can be converted into pairs of particles—an electron and a positron, for example. This process is known as *pair production* and is observed regularly in the nuclear physics laboratory (Figure 33.12).

Now all this is possible because mass and energy are equivalent. But obviously mass cannot be created from nothing—we need energy to do it. Yet, according to quantum theory, it is possible for matter (or energy) to be created from nothing for an exceedingly brief period of time. This possibility comes about because of the innate uncertainty in nature, at the microscopic level, of the measures of physical quantities such as mass and energy. The principle does not violate conservation laws because any matter that comes into being almost immediately disappears again spontaneously, so on the average mass and energy (combined) are conserved.

But, Hawking points out, what if a positron and electron (say) come into existence momentarily in the vicinity of a black hole (Figure 33.13). There is a chance that one or the other will fall into the

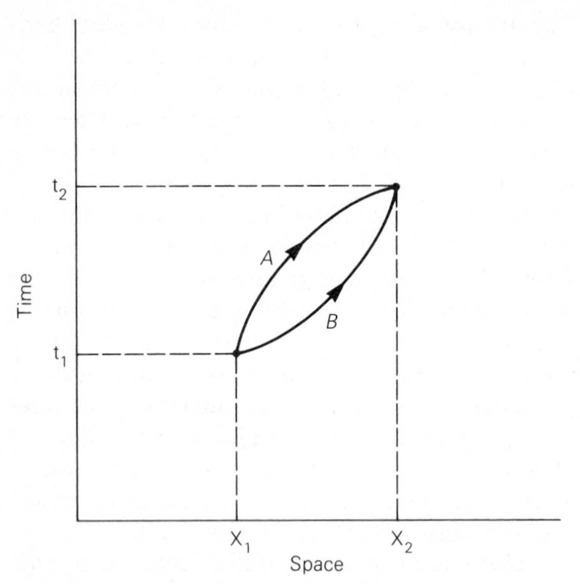

Figure 33.12 A spacetime diagram showing the creation of a particle and an antiparticle and their subsequent annihilation. At time t_1 and position x_1, the two particles are created. They may then move at different velocities and so separate in space. For example, the particle may move along path *A* in spacetime, and the antiparticle may move along trajectory *B*. Ultimately their mutual attraction will bring them back together to the same point in space at the same time (point t_2 and x_2), and they will destroy each other. Particles created in this way do not separate far enough for long enough to be observable.

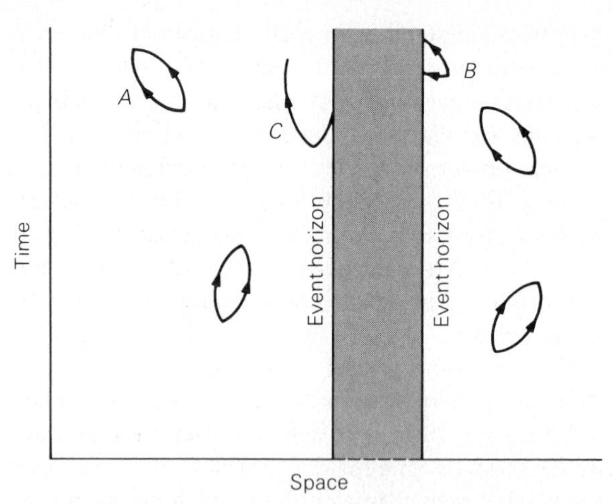

Figure 33.13 The shaded area is a schematic representation of a black hole in spacetime (schematic because we do not show the curvature of spacetime that is present near a black hole). Far from the black hole (at *A*, for example), particles and antiparticles appear and annihilate each other. If a particle and an antiparticle are created near a black hole, however, the possibility exists that both particles will fall into the black hole (as at *B*) or that only one particle will fall into the black hole (as at *C*), leaving its partner free to escape.

hole and hence not be able to annihilate with its antiparticle, returning the energy it "borrowed" from nature. Its antiparticle, therefore, can escape unscathed. However, many such positrons and electrons so created near black holes and escaping from them do annihilate each other, creating energy. That energy cannot come from nothing; according to Hawking's theory it must come from the black hole itself. Robbing the black hole of energy in this way robs it of mass (for they are equivalent), so the black hole must slowly evaporate through this process of pair production.

As esoteric as this idea may seem, it is generally thought to be possible by theoretical physicists. The process is only important, however, near very tiny black holes. Solar-mass black holes would evaporate in this way at an absolutely negligible rate. In fact, the only black holes that would have had time to so evaporate in the age of the universe would be those of original mass less than about 10^{15} g (like a minor planet). Smaller ones would already be gone; those of about 10^{15} g should be finishing off about now, if they were formed in the big bang in the first place. Because the evaporation rate increases as the mass of the black hole goes down, at the end one would go off explosively, emitting a final burst of gamma radiation.

Nobody knows whether such mini black holes were formed in the early universe, nor, if so, whether the evaporation process Hawking envisions would really occur. If so, we would expect to see bursts of gamma rays from exploding mini black holes from time to time. So far we have not, but the speculation remains an interesting possibility.

We have touched on many ideas in this chapter. Some, near the end, are highly speculative, but the main thrust of general relativity theory now appears to be rather firmly established, even though tests of some of its predictions are still going on. It is probably safe to say that general relativity has come of age and is an important part of modern astronomy.

And general relativity is almost singlehandedly the inspiration of one genius—Albert Einstein. For half a century it was an intellectually stimulating, but still largely academic subject. Today it is in the mainstream, and tomorrow it may well be absolutely fundamental to our understanding of the universe.

EXERCISES

1. Consider a bucket nearly full of water. A spring is attached to the middle of the inside of the bottom of the bucket, and at the other end of the spring is a cork. The cork, trying to float to the top of the water, stretches the spring somewhat. Now suppose the bucket, with its water, spring, and cork, is dropped from a high building so that it remains upright as it falls. What happens to the spring and cork? Explain your answer thoroughly.
2. A monkey hanging from a branch of a tree sees a hunter aiming a rifle directly at him. The monkey then sees a flash, telling him that the rifle has been fired. The monkey, reacting quickly, lets go of the branch and drops, so that the bullet will pass harmlessly over his head. Does this act save the monkey's life? Why?
3. Some of the Skylab astronauts exercised by running around the inside wall of their cylindrical vehicle. How could they stay against the wall while running, rather than float aimlessly inside the Skylab? What physical principles are involved?
4. ★ What is the radial velocity that would produce the same shift of spectral lines as the gravitational redshift of a star of one solar mass (2×10^{33} g) and of radius 4447 km? (The gravitational constant, G, is 6.67×10^{-8} dyne·cm^2/g^2.)
   Answer: 100 km/s
5. Draw a diagram showing the progress in spacetime of an automobile traveling northward. For the first hour, in city traffic, it goes only 30 km/hr. Then, in the country, for 3 hr it goes 90 km/hr. Finally, because the driver is late, he drives the car for 1 hr at 140 km/hr.
6. Make up a new example of a geodesic in spacetime and show it on a spacetime diagram.
7. The Earth moves in its nearly circular orbit of 1 AU radius at a speed of only 1/10,000 that of light. What is the radius of the circle that most nearly matches the path of a beam of light passing the sun at the Earth's distance from it?
8. As the binary pulsar loses energy through gravitational radiation, why do its members *speed up* and why does the period get *shorter?*
9. What would be the radius of a black hole with the mass of the planet Jupiter?
10. Why is the time dilation in a gravitational field equivalent to a gravitational redshift? (*Hint:* What is the definition of frequency? How is the frequency of radiation affected by a redshift?)
11. Why would we not expect X-rays from a disk of matter about an ordinary star, or even a white dwarf?
12. If the sun could suddenly collapse to a black hole, how would the period of the Earth's revolution about it differ from what it is now?
13. Could any of the dark globules, small round opaque objects seen superimposed against the starlight of the Milky Way, be stars that have become black holes? Why or why not?

Jan Hendrik Oort (b.1900), Dutch astronomer, has received many honors, including Knighthood in the Order of the Nederlandse Leeuw. He was a founding member of the International Astronomical Union and was the Union's president from 1958 to 1961. He originated the generally accepted theory of the origin of comets, but he is most famous for his pioneering work in the study of the rotation of the Galaxy. *(Sky and Telescope)*

34

THE GALAXY

In 1610 Galileo described his telescopic observations of the Milky Way, which showed it to be composed of a multitude of individual stars. In 1750 Thomas Wright published a speculative explanation, which turned out to be substantially correct—that the sun is located within a disk-shaped system of stars, and that the Milky Way is the light from the surrounding stars that lie more or less in the *plane* of the disk. The disk shape of the stellar system to which the sun belongs—the *Galaxy*—was demonstrated quantitatively in 1785 by Herschel's "star gauging." We have already described the results of Herschel's investigations (Section 26.3).

It was the second decade of the 20th century before astronomers had deduced, approximately, the true size of the Galaxy and the fact that the sun is located far from the center of the disk-shaped system. In 1925 the existence of other, similar, stellar systems was proved; the "pinwheel" shape of many of these other galaxies suggested that our own system might also contain spiral arms winding outward from a massive central nucleus.

Astronomers now picture the Galaxy as a thin round disk of luminous matter distributed across a region about 30,000 parsecs in diameter, with a thickness of a few hundred parsecs. This luminous portion of the Galaxy is apparently embedded in a halo of nonluminous or dark matter that extends to a distance of at least 50,000 parsecs from the galactic center. The sun orbits the center of the Galaxy at a speed of about 220 km/s at a distance of about 8500 parsecs. Within a few thousand parsecs of the galactic center, the stars are no longer confined to the disk but rather form a spheroidal *nuclear bulge* of old stars. Globular clusters, spherically shaped groups of stars that may contain up to a million old stars, are also distributed in a spherical halo around the galactic center and are found as much as 50,000 parsecs from it. Dust and gas, the raw material from which stars form, are fairly closely confined to the galactic disk. The center of the Galaxy may harbor a black hole. In this chapter, we will survey the properties of the various kinds of objects that populate our Galaxy.

34.1 ARCHITECTURE OF THE GALAXY

Until early in the 20th century, the Galaxy was generally believed to be centered approximately at the sun and to extend only a few thousand light years from it. The primary reason for this misconception was the (at that time unrecognized) presence of interstellar dust, which obscures the light from distant stars. With optical telescopes, we can observe

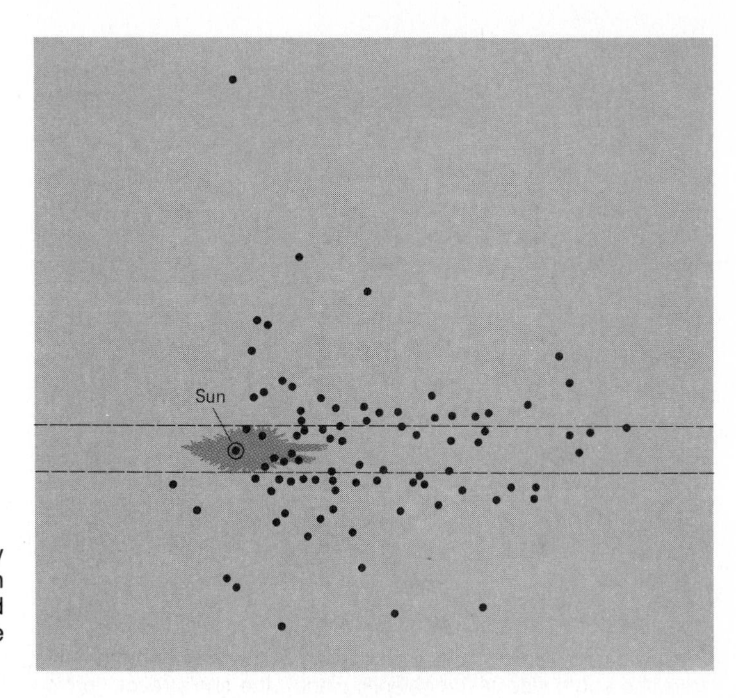

Figure 34.1 Copy of a diagram by Shapley showing the distribution of globular clusters in a plane perpendicular to the Milky Way, and containing the sun and the center of the Galaxy.

only a small portion of the Milky Way Galaxy. The shift from the "heliocentric" to the "galactocentric" view of our system, as well as the first knowledge of its true size, came about largely through the efforts of Harlow Shapley and his investigation of the distribution of the globular clusters. Because of their brilliance, and the fact that they are not confined to the central plane of the Galaxy where they would otherwise be largely obscured by interstellar dust, they can be observed (with telescopes) to very large distances.

(a) Distribution of the Globular Clusters

Most globular clusters contain at least a few RR Lyrae variable stars (or cluster-type variables—Section 31.2c), whose absolute magnitudes are known to lie between 0 and +1. The distance to an RR Lyrae star in a globular cluster, and hence to the cluster itself, can therefore be calculated from its observed apparent magnitude (Section 23.2b). Shapley was able to determine the distances to the closer globular clusters that contained RR Lyrae stars that could be observed individually. Distance estimates to globular clusters that do not contain RR Lyrae stars, or that are too far away to permit resolution of the variables, were obtained indirectly. Shapley measured angular diameters of globular clusters of known distance, thus obtaining their true diameters. Assuming a statistical average for the true diameter of the clusters, he was then able to obtain distance estimates for the remote ones from their observed angular diameters.

From their directions and derived distances, Shapley, in 1917, mapped out the three-dimensional distribution in space of the 93 globular clusters then known. He found that the clusters formed a roughly spheroidal system. The center of that spheroidal system was not at the sun, however, but at a point in the middle of the Milky Way in the direction of Sagittarius, and at a distance of some 25,000 to 30,000 LY (Figure 34.1). Shapley then made the bold—and correct—assumption that the system of globular clusters is centered upon the center of the Galaxy. Today this assumption has been verified by many pieces of evidence, including the observed distributions of globular clusters in other spiral galaxies. A schematic of the Galaxy is shown in Figure 34.2. Although the sun lies far from the galactic center, the main disk of the Galaxy probably extends a nearly equal distance beyond the sun and comprises a gigantic system, which is at least 100,000 LY across. Today, the center of the galactic nucleus is estimated to be 8500 pc from the sun.

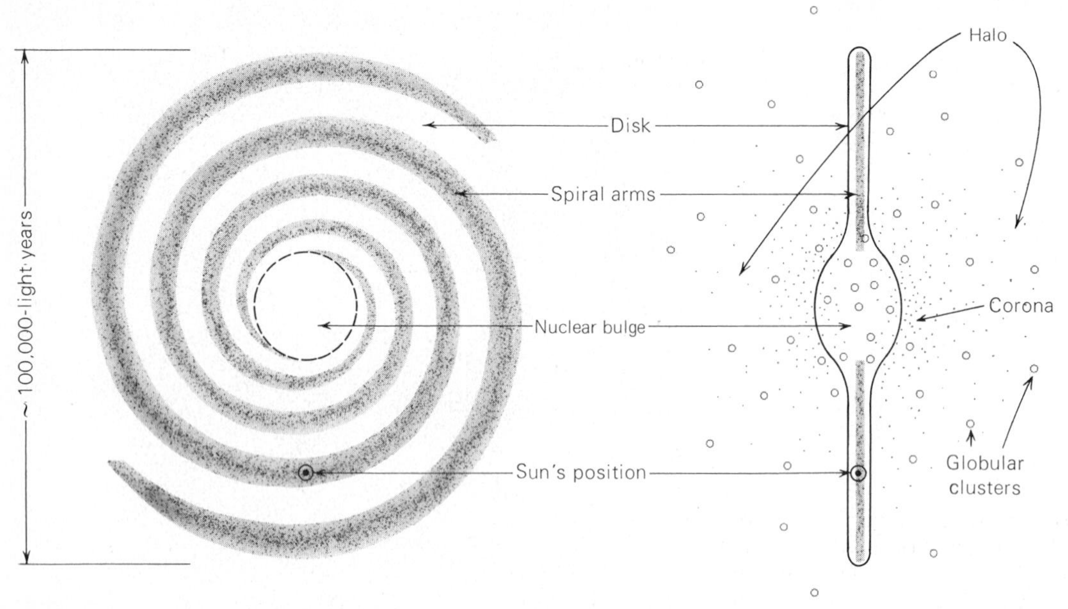

Figure 34.2 Schematic representation of the Galaxy. In the planar view on the left (seen from the south side of the galactic plane), the sun's revolution is counterclockwise.

(b) The Galactic Halo

The main body of the Galaxy is confined to a relatively flat disk, while the globular clusters define a more or less spheroidal system superimposed upon the disk. A sparse "haze" of individual stars—not members of clusters but far outnumbering the cluster stars—also exists in the region outlined by the globular clusters. This haze of stars and clusters forms the galactic halo, a region whose volume exceeds that of the main disk of the Galaxy by many times. The presence of stars in the halo was first suspected when RR Lyrae stars not belonging to clusters were found lying in such directions and at such distances as to place them far from the galactic plane. These variables, of course, represent only an extremely minute fraction of all halo stars, but they serve as tracers, indicating the distribution of stars in the halo, not only because their distances can be determined easily, but because they have fairly high luminosities and can be seen to relatively large distances. Some X-ray radiation appears to be emitted by gas in the halo. In order to emit X-rays, this gas must be very hot—of the order of 10^6 K—too hot to produce the emission lines seen in ordinary H II regions. The gas, probably heated by supernova shells and/or stellar winds (Section 27.4a), extends only into the inner part of the halo and defines a region now usually referred to as the *galactic corona*.

The spatial density of stars and clusters in the halo of the Galaxy increases toward the Milky Way plane, particularly toward the galactic nucleus. When we look to either side of the Milky Way (that is, slightly above or below its plane) in the directions of Scorpius, Ophiuchus, and Sagittarius, our line of sight skims near the nuclear bulge in the middle of the disk of the Galaxy. In those directions we find the greatest numbers of globular clusters and stars in the halo. The largest number of RR Lyrae stars seen in these directions have apparent magnitudes near 15, which means that they must be at distances of from 7000 to 10,000 pc.

Individual RR Lyrae stars have been found as far away as 30,000 to 50,000 LY on either side of the galactic plane, which shows that the halo must have an overall thickness of at least 100,000 LY. A few globular clusters discovered on the Palomar Sky Survey appear to have distances from the sun of more than ¼ million light years. If these systems are true members of the Galaxy rather than intergalactic objects, then, in the direction of the galactic plane the halo may have a diameter of two or three hundred thousand light years, extending far beyond the "rim" of the main disk of the Galaxy. Halos of some other galaxies have been traced to similar distances. We shall see in Section 34.3 that modern data on the rotation of our Galaxy give good reason to think that the halo contains a large fraction of the Galaxy's mass.

(c) Interstellar Matter

Interstellar matter, that is, gas and dust not now incorporated into stars, is confined fairly closely to the plane of the Milky Way Galaxy. Radio astronomy provides the techniques necessary to locate and study clouds of gas and dust. Hydrogen is the most abundant element in the universe, and the first interstellar clouds to be studied extensively were those that could be observed at 21 cm, the wavelength of a line produced by neutral atomic hydrogen. Approximately 80 percent of all the neutral hydrogen in the Milky Way Galaxy is to be found outside the sun's orbit around the galactic center.

Studies of the 21-cm line have been a key element in efforts to determine the nature of the spiral structure in our own Galaxy. Observations show clearly that we do not live in a Galaxy with only two well-defined spiral arms. While the interpretation of the measurements is somewhat controversial, it appears likely that the Galaxy has four spiral arms.

Our current picture is that the outer regions of the disk of the Galaxy are dominated by large-scale arcs (spiral arms) that extend over distances of 5000 to 25,000 parsecs. These structures are complex, with spurs extending from them in various directions. Young stars, as well as gas and dust, are concentrated in the arms. The formation of massive young stars does not seem to be occurring at distances further than 20,000 parsecs from the center of the Galaxy, but neutral hydrogen extends to at least 30,000 parsecs.

Our picture of the inner part of the Galaxy was revolutionized by the discovery of giant molecular clouds. Also discovered by radio astronomy, these clouds are the largest structures in the Galaxy and are the dominant component of the interstellar medium inside the sun's orbit around the Galaxy.

Typical giant molecular clouds have masses in the range 10^5 to 3×10^6 times the mass of the sun. Somewhat cigar-shaped, their longest dimension is typically about 40 parsecs. The temperature in their interiors is only about 10 K, and at these low temperatures simple molecules like H_2 and CO are abundant. These clouds are found in greatest numbers at distances between 4000 and 8000 parsecs from the galactic center, and there are probably about 4000 such clouds in the Galaxy. Most star formation takes place inside these clouds.

34.2 REVOLUTION OF THE SUN IN THE GALAXY

Like a gigantic "solar system," the entire Galaxy is rotating. The sun, partaking of the galactic rotation, moves with a speed of about 220 km/s in a nearly circular path about the nucleus. The method has been described (Chapter 22) by which we determine the motions of stars with respect to the sun (*space motions*), or with respect to the local standard of rest (*peculiar velocities*). Proper motions, however, and thus space motions and peculiar velocities, can be detected only for those stars that are relatively near the sun and that occupy a volume of space that is very small compared with the size of the Galaxy. These are stars moving along with us about the center of the Galaxy; their space motions, most of which are less than a few tens of kilometers per second, result from the slight differences between the inclinations, eccentricities, and sizes of their galactic orbits and the sun's. Only by observing far more distant objects can we determine our own true motion in the Galaxy.

(a) The Sun's Galactic Orbit

The motion of the sun in the Galaxy is deduced from the apparent motions of objects surrounding us that do not share in the general galactic rotation. The globular clusters are the most convenient such objects. These clusters are moving, to be sure, but the fact that they are found in a spheroidal distribution, rather than being confined to the flat plane of the Galaxy, is evidence that the system of globular clusters as a whole is not rotating as rapidly as the disk of the Galaxy. By analyzing the radial velocities of the globular clusters in various directions, we can determine the motion of the sun with respect to them very much in the same way that we determine the solar motion with respect to the local standard of rest (Section 22.4a). In one direction, the globular clusters, on the average, seem to be approaching us, while in the opposite direction they seem to recede from us. This procedure yields only a lower limit for the speed of the sun's revolution

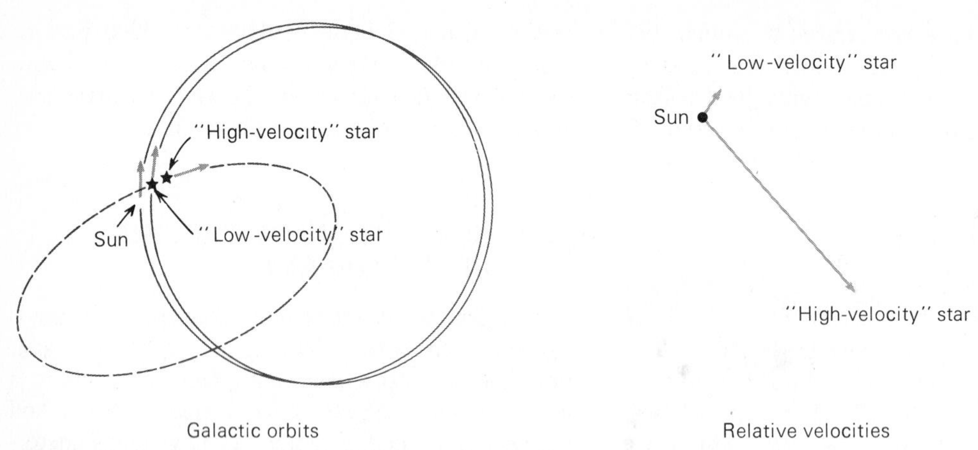

Figure 34.3 Galactic orbits of high- and low-velocity stars *(left),* and their relative velocities.

about the galactic center if there is any net rotation of the system of globular clusters.

The motion of the sun in the Galaxy can also be deduced from the radial velocities of the nearby external galaxies. The Large Magellanic Cloud, for example, has a radial velocity of recession of about 270 km/s, while the Andromeda galaxy has a velocity of approach of about 300 km/s (see Chapter 37). Studies of the radial velocities of other nearby galaxies show that if they are moving at random, their individual velocities with respect to each other and to the center of the Galaxy are relatively small, and that most of the high velocities observed for them are actually due to the orbital motion of the sun about the center of our own Galaxy.* We conclude, therefore, that the sun is moving in the general direction away from the Large Magellanic Cloud and toward the Andromeda galaxy.

When the data from various sources are combined, they indicate that the sun is moving in the direction of the constellation Cygnus, with a speed of about 220 km/s. This direction lies in the Milky Way and is about 90° from the direction of the galactic center, which shows that the sun's orbit is nearly circular and lies in the main plane of the Galaxy. As viewed from the north side of the galactic plane, the orbital motion of the sun is clockwise. The period of the sun's revolution about the nucleus, the *galactic year,* can be found by dividing the circumference of the sun's orbit by its speed; it comes out roughly 200 million (2×10^8) of our terrestrial years. We can observe, therefore, only a "snapshot" of the Galaxy in rotation; we do not actually see stars traverse appreciable portions of their orbits.

(b) High- and Low-Velocity Stars

As already mentioned, the majority of the stars near the sun move nearly parallel to the sun's path about the galactic nucleus, and their speeds with respect to the sun are generally less than 40 or 50 km/s. These are said to be *low-velocity stars.* The radial velocities of nearby gas clouds, as indicated by the Doppler displacements of the interstellar absorption lines and the bright lines of emission nebulae (Section 27.3), are also low and show that they, like the sun, move in roughly circular orbits about the galactic nucleus. In other words, the interstellar material in our part of the Galaxy also belongs to the class of low-velocity objects.

Some stars, on the other hand, have speeds relative to the sun in excess of 80 km/s and are called *high-velocity stars.* They move along orbits of rather high eccentricity that cross the sun's orbit in the plane of the Galaxy at rather large angles (Figure 34.3). Nearby stars moving on such orbits are passing through the solar neighborhood and are only temporarily near us. Stars in the galactic halo and globular clusters also have orbits very different from the sun's and are high-velocity objects.

---

* This statement applies only to the very nearest external galaxies—those that belong to the Local Group; see Chapers 35 and 36.

The term "high velocity" or "low velocity" refers to the speed of an object *with respect to the sun* and has nothing to do with its motion in the Galaxy. Most high-velocity stars lag behind the sun in its motion about the galactic center and hence are actually revolving about the Galaxy with speeds *less* than those of the low-velocity stars near the sun.

The component of velocity of an individual star in a direction perpendicular to the plane of the Galaxy (sometimes called the z component of its velocity) cannot, in general, be determined from its radial velocity alone. However, the average of the z-velocity components for stars of a given class or group can be found from a statistical analysis of their radial velocities. It is learned that low-velocity stars usually have lower velocity components perpendicular to the galactic plane than do high-velocity stars. Consequently, high-velocity stars tend to be less strongly concentrated to the plane of the Galaxy than low-velocity stars. Stars in the halo are extreme examples and usually have very high z-velocity components. It is, of course, the high velocities of coronal objects perpendicular to the galactic plane that carries them high into the galactic corona and accounts for their spheroidal distribution in the Galaxy. Globular clusters, in particular, are believed to revolve about the nucleus of the Galaxy in orbits of high eccentricity and inclination to the galactic plane, perhaps rather like the comets revolving about the sun in the solar system. A globular cluster must pass through the plane of the Galaxy twice during each revolution. The large distances between stars, both within the cluster and within the Galaxy itself, make stellar collisions during the cluster's penetration of the galactic disk exceedingly improbable.

The different kinds of motion of stars in the Galaxy may be related to the times and places of formation of those stars; the high-velocity objects are believed to be among the older members of the Galaxy.

34.3 SPIRAL STRUCTURE OF THE GALAXY

Observations have been made of many other spiral galaxies that contain interstellar matter and in which individual stars can be resolved (Chapter 35). In many of these systems both the optically conspicuous interstellar matter and the most luminous resolved stars are generally concentrated in the *spiral arms*. The association of the brightest stars and the gas and dust is not mysterious; highly luminous stars are relatively young objects and have recently formed from clouds of interstellar gas and dust (Chapter 30.) In mapping out the spiral structure of our own Galaxy, therefore, we can use the gaseous nebulae and the very luminous main-sequence O and B stars as "tracers" to identify spiral arms.

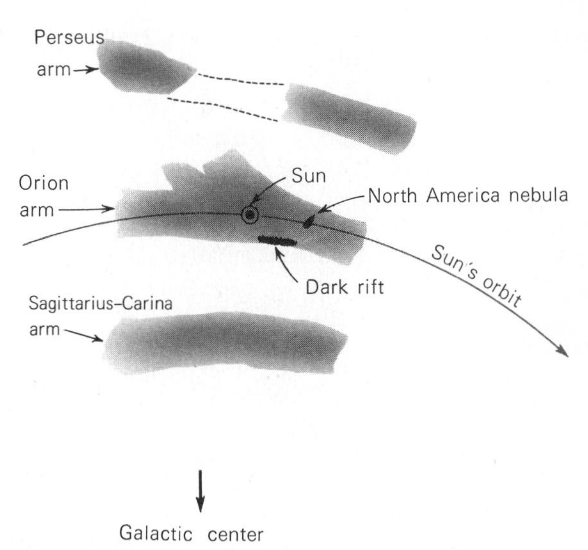

Figure 34.4 Spiral structure of the Galaxy as deduced from optical observations.

(a) The Spiral Structure from Optical Observations

Because of our position in the disk of the Galaxy, and because we are surrounded by interstellar dust, it is difficult for us to identify even the nearby spiral arms optically. Nevertheless, in 1951 short pieces of three nearby arms were detected by W. W. Morgan and his students at the Yerkes Observatory from the observed directions and distances of a large number of O and B stars and from the distribution of emission nebulae. The fragmentary data on the spiral structure of the Galaxy, as deduced from such optical observations, are summarized in Figure 34.4.

The sun appears to be near the inner edge of an arm called the *Orion arm,* or *spur*, which contains such conspicuous features as the North America nebula, the Coalsack (near the Southern Cross), the Cygnus Rift (great dark nebula in the summer Milky Way), and the Orion nebula. More distant,

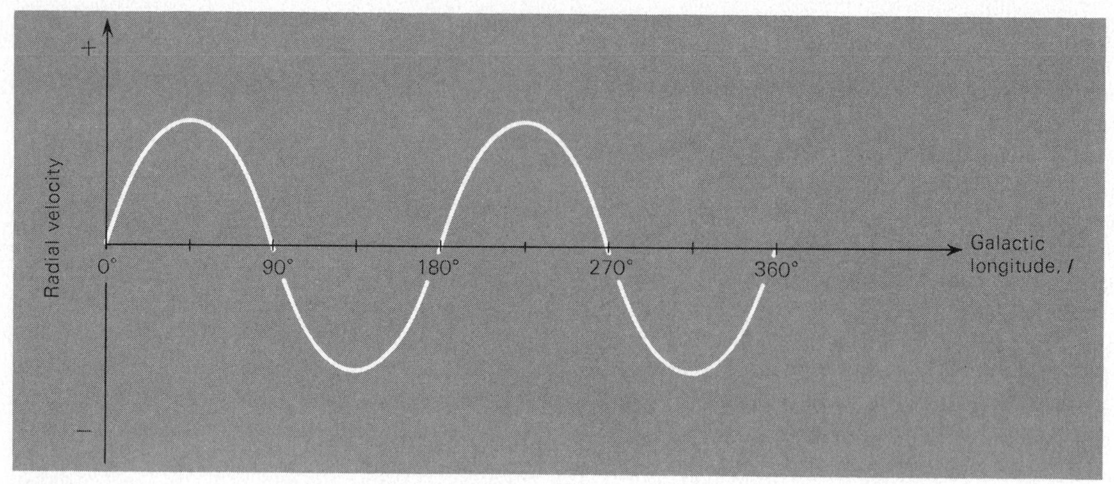

Figure 34.5 Plot of the observed radial velocities of stars in the plane of the galactic equator at a certain distance from the sun against their galactic longitudes.

and therefore less conspicuous, emission nebulae can be identified in the *Sagittarius-Carina* and *Perseus arms,* located, respectively, about 2000 pc inside and outside the sun's position with respect to the galactic nucleus.

(b) Differential Galactic Rotation

At the sun's distance from its center, the Galaxy does not rotate as a solid wheel. Stars in larger orbits do not keep abreast of those in smaller ones, but trail behind. This effect produces a shearing motion in the plane of the Galaxy, called *differential galactic rotation.* We can detect the differential galactic rotation from observations of proper motions and radial velocites of stars around us.

As an illustration, let us consider the effect of differential rotation on the radial velocities of stars in different directions in the plane of the Galaxy. Let us assume that all stars have circular orbits (only approximately true). If we look in the direction of the galactic center or in the opposite direction, the stars are moving across our line of sight and show on average no radial velocity. Neither do stars directly in front of us or behind us in our galactic orbit, for they are moving the same as we are. Stars in front of us, but at an angle of only 45° from the direction to the galactic center, on the other hand, are pulling away from us, for we, in our larger orbit, are not moving fast enough to keep up with them, so they show positive radial velocities. We, in turn, are pulling ahead of stars in the opposite direction (behind us, but 135° from the center), so they too have positive radial velocities. We are gaining, however, on stars ahead of us in larger orbits (say, 135° from the center), so they show negative velocities, as do stars opposite them that are gaining on us.

If we prepare a plot of radial velocities of stars of a given distance but in different directions in the plane of the Galaxy, we obtain a curve like that of Figure 34.5. The abscissas are galactic longitude, the angle measured eastward in the plane of the Galaxy from the direction to its center. Such a curve is called a *double sine* curve, because it goes through two cycles in 360°. The amplitude of the curve (the height of its waves) depends on the distances to the stars observed.

(c) The Rotation Curve of the Galaxy

A graph of the orbital speeds of stars in the Galaxy against orbital radius is called the *rotation curve.* The inner part of the curve has been known from the 1950s; it was initially obtained from observations of the Doppler shifts of the 21-cm radiation from neutral hydrogen in the Galaxy (Section 27.3k), now supplemented by radio studies of molecular lines, particularly of CO, because in most directions absorption by interstellar dust limits our range of optical and ultraviolet observations.

Suppose we point a radio telescope at galactic longitude *l* in the plane of the Galaxy (Figure 34.6). If *l* is between 0° and 90° (as in the figure), all gas clouds along the line of sight that are within the sun's orbit must be moving *away* from the sun, and the 21-cm line from each will be shifted to a slightly longer wavelength. We actually receive 21-cm radiation simulta-

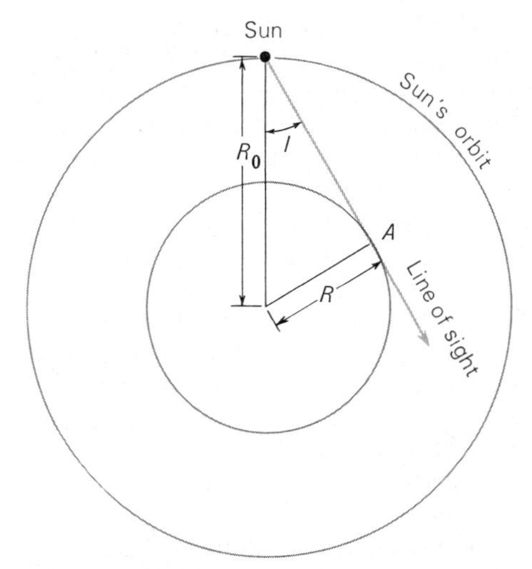

Figure 34.6 Determination of the orbital velocities of objects with smaller orbits than that of the sun.

neously from many clouds at various distances, moving away from us at various speeds. That radiation, therefore, is received as a *band* covering a small range of wavelength, each part of the band being 21-cm radiation from a different cloud that is along our line of sight, and is therefore shifted by a different amount. The long-wavelength edge of the band corresponds to radiation coming from the interstellar hydrogen moving away from us with the greatest radial velocity. It can be shown (if all galactic orbits are assumed to be circular) that at a given galactic longitude, the gas with the greatest radial velocity will be at position *A* (in Figure 34.6), where our line of sight passes closest to the galactic center. The distance, *R*, of *A* from the galactic center is a simple function of the radius of the sun's orbit R_o, and the galactic longitude, *l*.* Since our line of sight is tangent to the orbit of a cloud at *A*, the maximum observed radial velocity, after correction for the sun's motion, is the actual orbital speed of a cloud moving on a circular orbit of radius *R*. By directing the radio telescope at various other galactic longitudes, we find the circular orbital speeds at other distances from the galactic center.

Beyond the sun's orbit it is more difficult to detect the rotation curve because the distances to the emitting interstellar clouds are less well determined. Nevertheless, observations since 1978 have provided pretty good evidence for the rotation curve out to nearly 20,000 pc from the galactic center.

In its inner parts, the Galaxy rotates like a solid body (or wheel), but by a distance of about 4000 parsecs from the center the curve flattens out at a rotational velocity of about 250 km/s (Figure 34.7). Radial velocities of distant globular clusters indicate that the rotational velocity of the Galaxy remains near 250 km/s to much greater distances—possibly to as far as 50,000 pc. We shall see (chapter 35) that many other galaxies have similar flat rotation curves to very great distances from their centers.

(d) Spiral Structure as Found From 21-cm Observations

Once the rotation of the Galaxy at various distances from its center is known, we can use 21-cm observa-

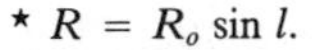

* $R = R_o \sin l$.

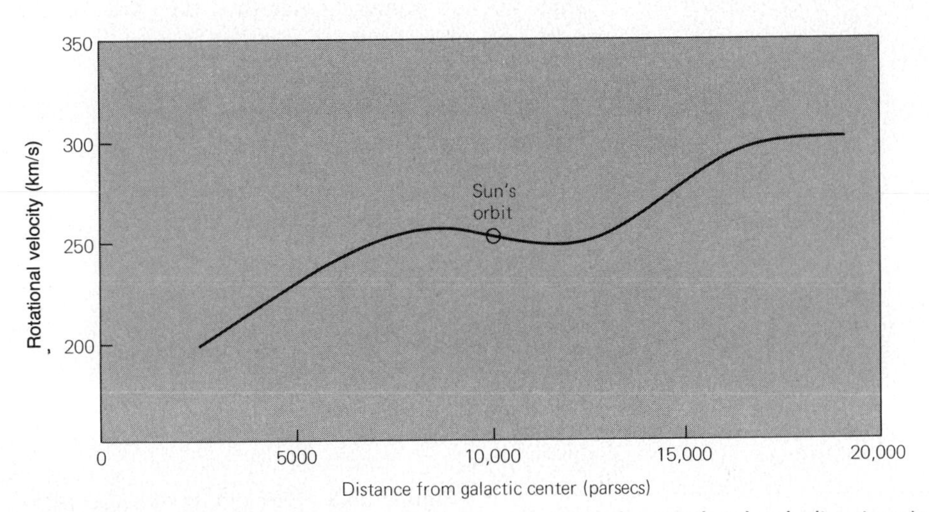

Figure 34.7 Rotation curve for our Galaxy, showing the rotational velocity at various distances from the galactic center.

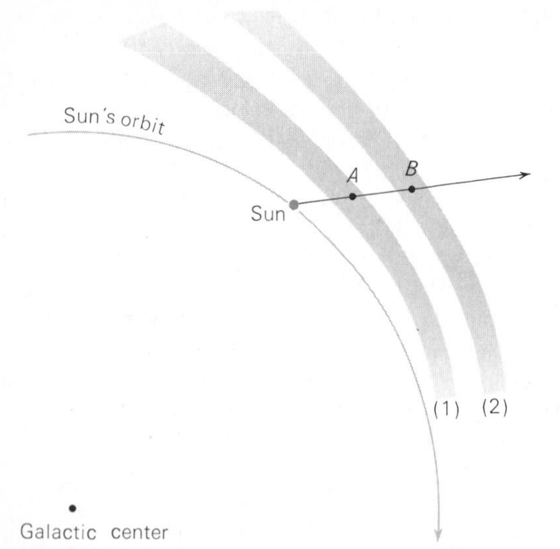

Figure 34.8 Observation of spiral arms at a wavelength of 21 cm.

tions to map out the distribution of gas in our stellar system. The problem is actually very complicated because the neutral hydrogen clouds are known not to revolve about the galactic center in perfectly circular orbits. In principle, however, the procedure is straightforward. Suppose a radio telescope is pointed in the direction shown in Figure 34.8; our line of sight (in this example) passes through two spiral arms, arm (1) at point *A* and arm (2) at point *B*. Most of the interstellar gas in that line of sight is concentrated at those points. Near 21 cm, the signal strength increases whenever radiation is received from neutral hydrogen. In Figure 34.8, the strongest radiation, coming from points *A* and *B*, is received at wavelengths λ_1 and λ_2, respectively; λ_1 and λ_2 are both slightly less than 21 cm because the sun is overtaking *A* and *B* in their galactic revolution, giving them *negative* radial velocities. A plot of observed signal strength versus wavelength would look something like Figure 34.9. The difference between λ_1 (or λ_2) and the nominal wavelength of the 21-cm line (actually 21.11 cm) gives the radial velocity of point *A* (or *B*); these data, plus a knowledge of how the Galaxy rotates, enable the distances to *A* and *B* to be determined and thus locate them in the galactic plane. Similar observations made at other galactic longitudes reveal the distances of gas clouds in other directions. Thus, a map of the Galaxy is built up. Such is shown in Figure 34.10.

In practice, uncertainties in the outer part of the rotation curve and particularly the noncircular motions of the gas clouds make the determinations of the locations of gas clouds very uncertain.

(e) Formation and Permanence of Spiral Structure

It is not surprising that much of the interstellar material is concentrated into elongated features that resemble spiral arms (Figure 34.11). No matter what the original distribution of the material might be, the differential rotation of the Galaxy would be expected to form it into spirals. Figure 34.12 shows the development of spiral arms from two irregular blobs of interstellar matter, as the portions of the blobs closest to the galactic center move fastest, while those farther away trail behind.

It is harder to understand, however, why the arms are not wound tighter than they are. At the

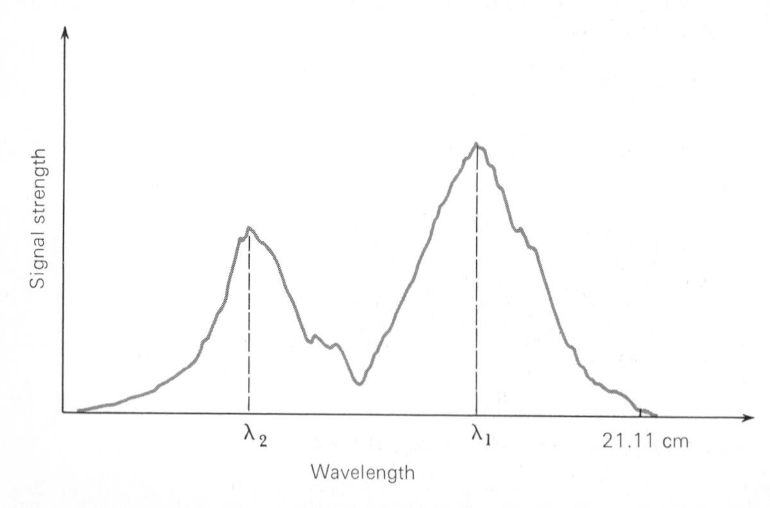

Figure 34.9 Strength of received radio signals versus wavelength for the observation illustrated in Figure 34.8.

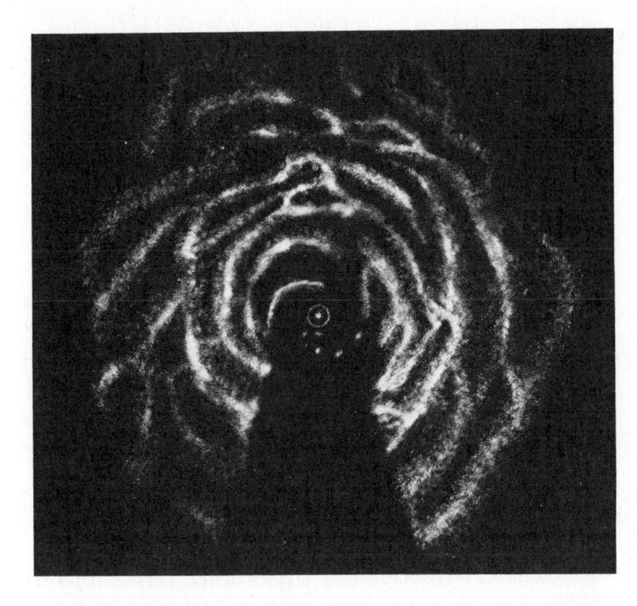

Figure 34.10 A drawing of the spiral structure of the Galaxy, deduced from 21-cm observations made at Leiden, The Netherlands, and at Sydney, Australia. The large and small circles show the location of the galactic center and the sun, respectively. (Courtesy of Gart Westerhout)

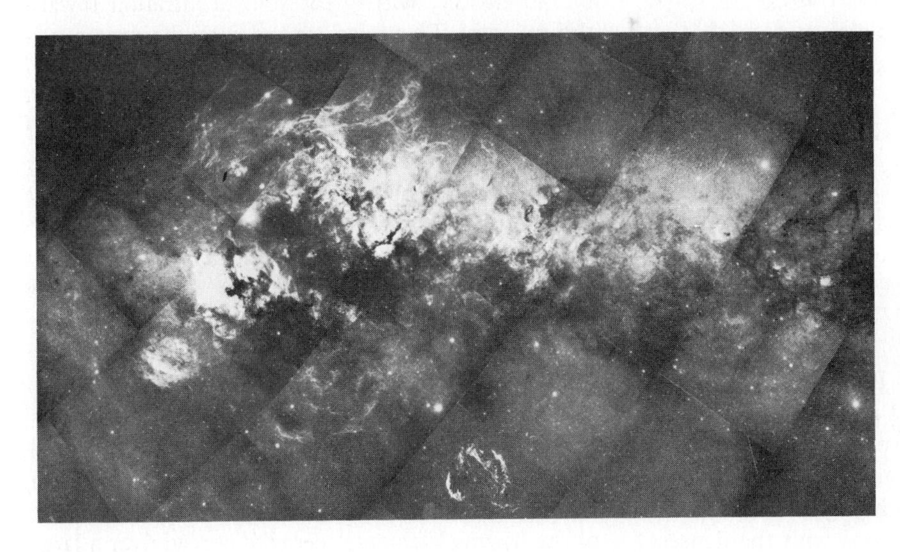

Figure 34.11 Mosaic of photographs of the Milky Way in Cygnus, photographed with the 124-cm Schmidt telescope. (National Geographic Society—Palomar Observatory Sky Survey, reproduced by permission of Caltech/Palomar Observatory)

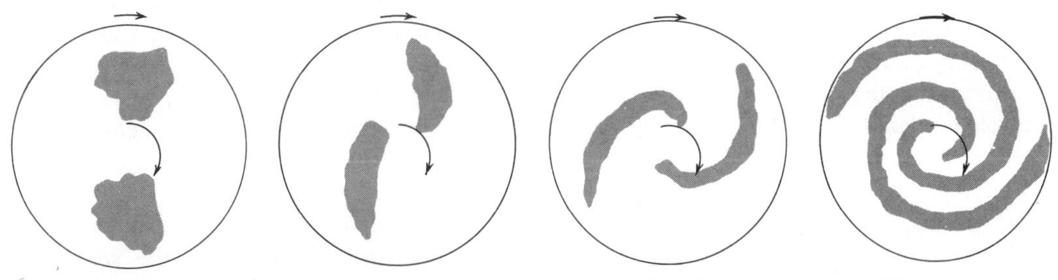

Figure 34.12 Hypothetical formation of the two spiral arms from irregular clouds of interstellar material.

sun's distance from the center of the Galaxy, the Galaxy rotates once in about 2×10^8 years. Its total age, however, is believed to be about 10^{10} years, in which case a point near the sun has made at least 50 revolutions. With so many turns, we would expect the spiral arms to be wound very tightly and to lie very much closer together than they do.

A step toward an understanding of how spiral structure can be maintained is provided by a theory developed by C. C. Lin at the Massachusetts Institute of Technology and his associates Frank Shu and Chi Yuan. Lin and his coworkers calculated the manner in which stars and gas clouds would move if they had circular paths about the galactic center and were influenced both by the gravitational fields produced by the Galaxy as a whole and by the matter forming the spiral arms themselves. They find that objects should slow down slightly in the regions of the spiral arms, and linger there longer than elsewhere in their orbits, thus building up a wave of higher than average density where the spiral arms are. This *density wave* model predicts that an equilibrium state is reached in which there are two trailing arms, inclined about 6° to the circular orbits of the stars and gas clouds, spaced about 3000 pc apart—in rough agreement with observations. The entire spiral pattern rotates more slowly than the actual material in the Galaxy, so that the stars, gas, and dust pass slowly through the spiral arms.

As gas and dust clouds approach the inner boundaries of an arm and encounter the higher density of slower moving matter, they collide with it, producing a shock boundary. It is here that the theory predicts star formation is most likely to occur, with protostars (like the solar nebula—Section 12.4) condensing in higher density regions. In some other galaxies we do see the highly luminous, and hence certainly young, stars along with the densest dust clouds near the inner boundaries of spiral arms, as well as radio radiation from the shocks. Density waves have also been observed directly in the rings of Saturn (Section 18.5).

Calculations indicate, however, that there may be problems in sustaining density wave patterns over the lifetime of a galaxy. There may also be ways, unrelated to spiral density waves, to produce elongated structures that look like spiral arms. For example, progressive star formation in molecular clouds may mimic spiral arms (Section 30.1). We have one fairly satisfactory theory—the spiral density wave theory—for explaining the structure of the Milky Way Galaxy, but we should be aware that there may be other ways to develop spiral structure, and we cannot yet be certain about how our Galaxy evolved.

(f) Different Stellar Populations in the Galaxy

Striking correlations are found between the characteristics of stars and other objects and their locations in the Galaxy. Some classes of objects, for example, are found only in regions of interstellar matter, that is, in the spiral arms of the Galaxy. Examples are bright supergiants, main-sequence stars of high luminosity (spectral classes O and B), Wolf-Rayet stars, type I cepheid variables, and young open star clusters.

The distributions of some other classes of objects show no correlation with the location of spiral arms. These objects are found throughout the disk of the Galaxy, with greatest concentration toward the nucleus. They also extend into the sparse galactic halo. Examples are planetary nebulae, novae, type II cepheids, RR Lyrae variables, and Mira variables of periods less than 250 days. The globular clusters, which also belong to this group of objects, are found almost entirely in the halo and nuclear bulge of the Galaxy.

Main-sequence stars of spectral types F through M exist in all parts of the Galaxy, as do red giants and, probably, white dwarfs (observations of white dwarfs, of course, are limited to our immediate neighborhood because of their low luminosities).

There are also differences in the chemical compositions of stars in different parts of the Galaxy. Nearly all stars appear to be composed mostly of hydrogen and helium, but the residual abundance of the heavier elements seems to spread over a large range for different stars. In the sun and in other stars associated with the interstellar matter of the spiral arms, the heavy elements (elements heavier than hydrogen and helium) account for about 1 to 4 percent of the total stellar mass. Stars in the outer galactic halo and in globular clusters, however, have much lower abundances of the heavy elements—often less than one-tenth, or even one-hundredth that of the sun. There is also evidence that the abundance of heavy elements of stars in the disk varies systematically with the time that the stars were born. Measurements show, for example, that stars formed 10 billion or more years ago have a lower

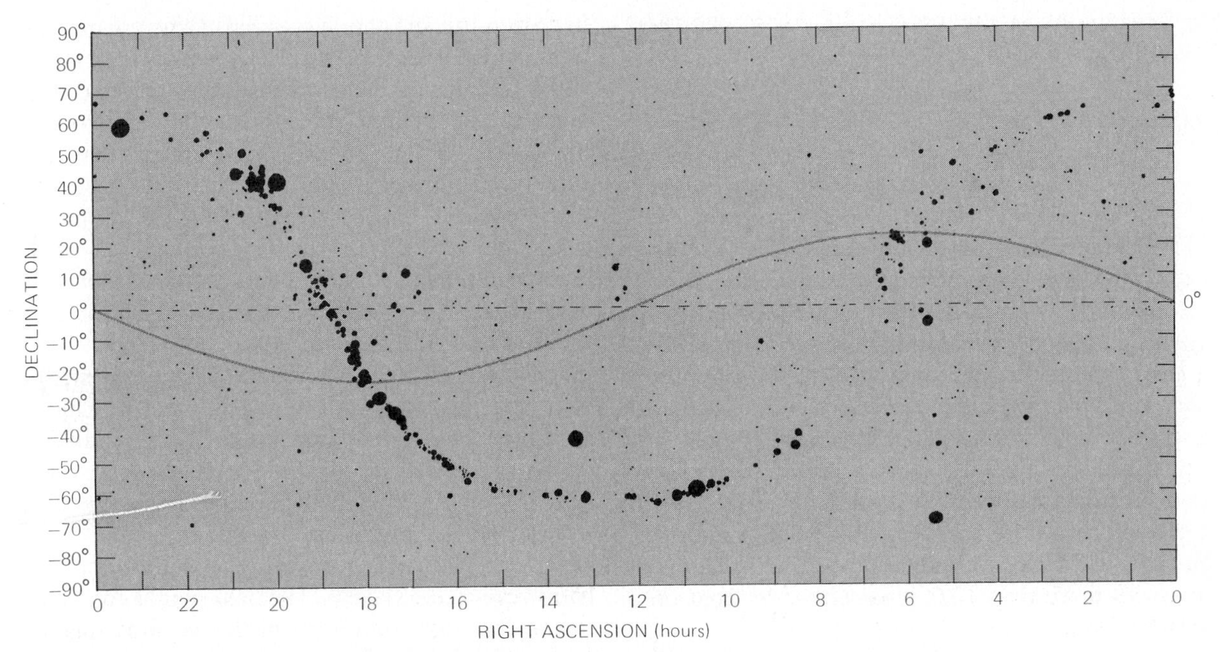

Figure 34.13 The distribution of discrete radio sources in the Galaxy. The symbols on this map of the sky show the positions of galactic radio sources emitting continuous radiation near 21 cm (not the 21-cm line). The larger symbols denote stronger sources. The dashed line across the middle is the celestial equator and the solid wavy line is the ecliptic. Note how most of the radio sources are concentrated along the plane of the Milky Way, which, although really a great circle in the sky, appears on this kind of map as a large open U. (Courtesy G. Verschuur)

abundance of heavy elements than does the sun, and this result suggests that the metal content of the Galaxy has increased over time.

The stars associated with the spiral arms are sometimes said to belong to *population I*, while those found elsewhere in the Galaxy are said to belong to *population II*. The terms "population I" and "population II" were first applied to different classes of stars by W. Baade, of the Mount Wilson and Palomar Observatories. During World War II, Baade was impressed by the similarity of the stars in the nuclear bulge of the Andromeda galaxy to those in the globular clusters in the halo of our own Galaxy; he concluded that the stars situated in spiral arms must, collectively, display different properties from those located elsewhere in the Galaxy.

Today we can interpret the phenomenon of different stellar populations in the light of stellar evolution. Thus, population I comprises stars of many different ages, including some that were recently formed or are still forming from gas and dust in the galactic disk. Population II, on the other hand, consists entirely of old stars, probably formed early in the history of the Galaxy.

It is clear today that two stellar populations are insufficient to account completely for the distribution of all the different kinds of stars in the Galaxy. Modern investigators now generally define several different stellar populations, ranging from "extreme population I" (spiral-arm objects), through "disk population" objects, to "extreme population II" (halo objects).

(g) The Mass of the Galaxy

We can make an estimate of the mass of the inner part of the Galaxy (lying inside the sun's orbit) with an application of Kepler's third law (as modified by Newton). Assume the sun's orbit to be circular and the Galaxy to be roughly spherical so we can treat it as though its mass internal to the sun were concentrated to a point at the galactic center. If the sun is 10,000 pc from the center, its orbit has a radius of 2×10^9 AU (there are 2×10^5 AU in one parsec). Since its period is 2×10^8 yr, we have

$$\text{Mass (Galaxy)} = \frac{(2 \times 10^9)^3}{(2 \times 10^8)^2} = 2 \times 10^{11} \text{ solar masses.}$$

More sophisticated calculations based on complicated models give a similar result.

It must be emphasized that this is only the mass contained in the volume inside the sun's orbit. It is a good estimate for the total mass of the Galaxy if and only if no more than a small fraction of its mass is to be found beyond the radius that marks the sun's distance from the galactic center. For many years, astronomers thought this was a reasonable assumption, since the number of bright stars and the amount of luminous matter drop dramatically at distances more than 10,000 parsecs or so from the galactic center.

Observations now show, however, that while there is relatively little luminous matter lying beyond 10,000 parsecs, there must be a lot of nonluminous, invisible, *dark matter* at large distances from the galactic center.

We can understand how astronomers reached this conclusion by rewriting Kepler's third law (Section 4.4) in the form

$$M = \frac{rv^2}{G},$$

where M is the mass of the Galaxy inside a circular orbit of radius r, and v, is the orbital speed of an object (star, interstellar cloud, or globular cluster) around the galactic center. (In order to obtain this equation from Kepler's law, we have simply replaced the period P by $P = 2\pi r/v$.)

What this form of Kepler's third law says is that if all the mass M of the Galaxy is concentrated inside a radius R, e.g., the sun's orbital radius, then as one moves beyond R further out in the Galaxy, that is, to the larger radius r, the rotational velocity v must decrease so that the quantity rv^2 remains constant and equal to the total mass of the Galaxy, M.

In fact, this is not at all what actually happens. As we look outwards to objects between 10,000 and 50,000 parsecs from the sun, we find that their orbital velocities remain constant at about 250 km/s. (The objects typically observed are clouds of atomic hydrogen gas and the few globular clusters that are found at these large distances.) Since r is increasing, and v remains constant, M must also increase. Detailed measurements indicate that the total mass of the Galaxy is four to five times greater than the value given above for the amount of matter inside the solar orbit. The total mass of the Galaxy out to a distance of 50,000 parsecs is about 10^{12} solar masses. Theoretical arguments suggest that this dark matter is distributed in a spherical halo around the Galaxy. But what is it?

Since this matter is invisible, it cannot be in the form of ordinary stars. It cannot be gas in any form. If it were neutral hydrogen, its 21-cm radiation would have been detected. If it were ionized hydrogen, it would also emit measurable radiation. If it were molecular hydrogen, absorption bands would have been observed in ultraviolet spectra of objects lying beyond the Galaxy. The halo cannot consist of interstellar dust, since dust in the required quantities would absorb and scatter so much radiation as to make galaxies unobservable. The possibilities that remain are low-mass objects (less than about 0.08 times the mass of the sun), which can never ignite nuclear reactions, or exotic subatomic particles. As we shall see, dark matter is a major constituent of other galaxies as well (Section 35.4).

34.4 MAGNETISM IN THE GALAXY

There is ample evidence that magnetism is an important phenomenon in the Galaxy. In the sun, we have seen (Chapter 28) that there are very strong magnetic fields associated with the active regions around sunspots, and a general solar field of about the strength of the Earth's. We have also seen that some stars have extremely strong magnetic fields and that objects like pulsars must have incredibly strong fields. But in addition to magnetism associated with particular objects in the Galaxy, there are general fields whose force lines usually run along the spiral arms, and possibly a weak field throughout the halo as well. These general interstellar magnetic fields of the Galaxy are of low intensity, but cover such vast regions of space that they contain a great deal of total energy. In a few favorable cases, magnetic fields have been measured directly in interstellar material through detection of Zeeman splitting of the spectral lines (Section 24.2). The

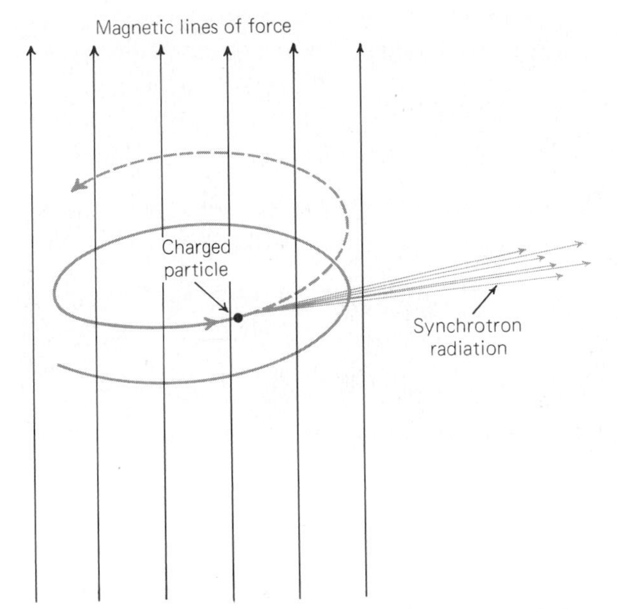

Figure 34.14 The emission of synchrotron radiation by a charged particle moving at nearly the speed of light in a magnetic field.

fields are on the order of 10^{-5} gauss or less. In most cases, we must rely on clues that allow us to detect indirectly the general magnetism of the Galaxy.

(a) Synchrotron Radiation

We have already described some of the mechanisms by which atoms can radiate electromagnetic energy (Chapter 8). In addition, charged particles, such as electrons and ions, radiate electromagnetic energy if they are accelerated by a magnetic field. When a charged particle enters a magnetic field, the field compels it to move in a circular or spiral path around the lines of force; the particle is thus accelerated and radiates energy. If the speed of the particle is nearly the speed of light, the particle is said to be *relativistic.* In this case, the energy it radiates is called *synchrotron radiation,* because particles so radiate when they are accelerated to relativistic speeds in a laboratory synchrotron. Any solid or liquid body or gas that is not at absolute zero temperature radiates electromagnetic energy (Section 8.1). To distinguish it from this normal *thermal* radiation from gases or bodies, synchrotron radiation is sometimes called *nonthermal* radiation.

Both the intensity and frequency of synchrotron radiation are greater, the greater the energy of the particle and the stronger the magnetic field. This nonthermal radiation has properties that make it easy to recognize: it has a distinctive distribution of intensity with wavelength, it is highly polarized, and the energy radiated by a particle is primarily in the direction of the particle's instantaneous motion (Figure 34.14). Both atomic nuclei (positive ions) and electrons radiate when accelerated, but the nuclei are thousands of times as massive as electrons, and consequently for the same energy have much lower speeds. Thus the nuclei do not generally move fast enough to emit significant synchrotron radiation.

We find many astronomical examples in which relativistic electrons are spiraling through magnetic fields and are emitting synchrotron radiation, although we do not yet, in all cases, know the origin of these energetic electrons, nor the mechanisms that give them their great speeds. We have already described the radiation belt around the planet Jupiter and the lesser one about the Earth, in which relativistic electrons in the magnetic fields of those planets emit synchrotron radiation at radio wavelengths. Such energy is emitted far more strongly from electrons in the magnetic fields associated with remnants of supernovae (Chapter 32). One such supernova remnant is the Crab nebula, in which many of the electrons have such high energy that much of the synchrotron radiation is in visible, ultraviolet, and even X-ray, as well as in radio wavelengths.

The Galaxy, in fact, abounds in sources of nonthermal radiation. Some are in its very nucleus (Section 34.5). In certain other galaxies (Chapter 35) we find synchrotron radiation emanating from their inner halos, and often from invisible regions far outside the parts of those galaxies that we can observe optically. Evidently those galaxies have very extended magnetic fields that sometimes reach far beyond their visible images.

The nonthermal radiation shows us that there are interstellar magnetic fields in our Galaxy. When polarized radiation passes through ionized gas in a magnetic field, the plane of the polarization rotates as the radiation moves forward—an effect called *Faraday rotation.* The amount of Faraday rotation, however, depends on the frequency. The highly polarized synchrotron radiation from nonthermal sources shines through the partially ionized gases of interstellar space to reach the Earth. We find that

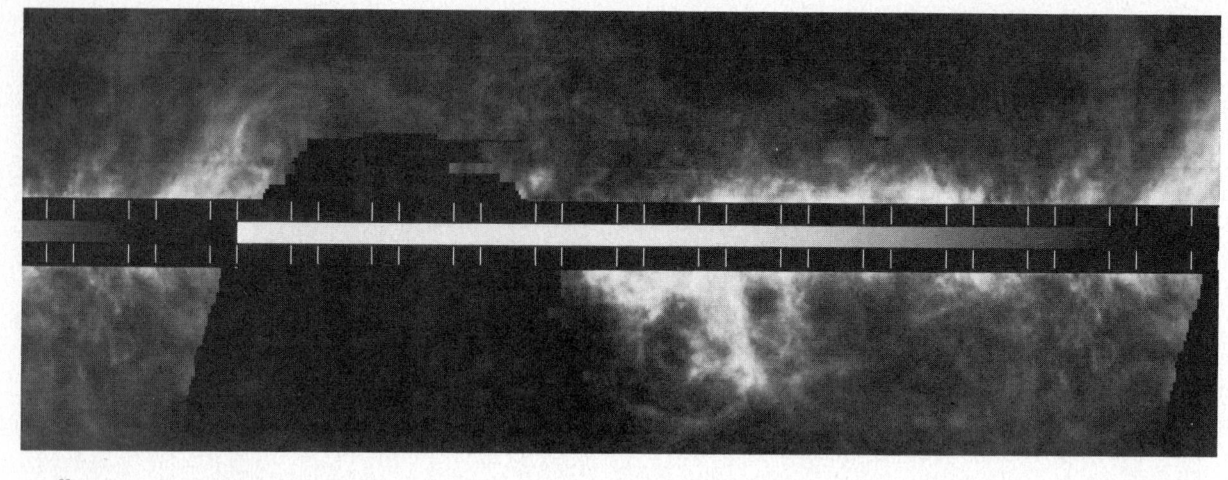

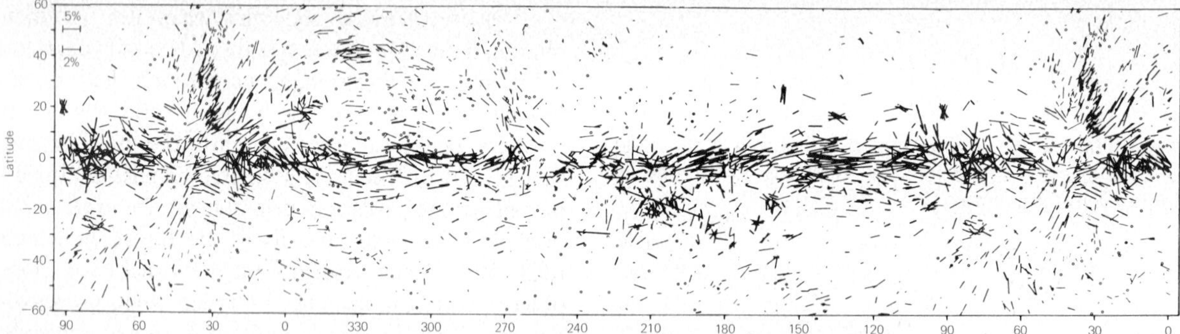

Figure 34.15 Two charts plotted in galactic coordinates, in which the plane of the Galaxy (Milky Way) runs horizontally across the middle. *(Above)* The lightness indicates the relative strength of 21-cm radiation, which is emitted by neutral hydrogen in the spiral arms. *(Below)* A map prepared by D. S. Mathewson and V. C. Ford (Mount Stromlo Observatory) showing the relative strength and direction of starlight polarization. Note how the polarization directions match many of the filaments of neutral hydrogen, showing the alignment of polarization with the spiral arms. (Courtesy C. Heiles, University of California, Berkeley, and E. B. Jenkins, Princeton University)

different wavelengths of this radiation have undergone different amounts of Faraday rotation, showing that even the nearly empty interstellar space has magnetic fields. We can estimate what fraction of the interstellar gas is ionized and measure the amount of Faraday rotation produced at different wavelengths. With these data we can calculate the intensity of the magnetic field in space. Because of their inherent uncertainties the calculations are not very precise, but they indicate an average field in the spiral arms of the Galaxy of from 10^{-6} to 10^{-5} gauss; by comparison, the field at the surface of the Earth is a little less than 1 gauss.

(b) Polarization of Starlight

We have seen (Section 27.2e) that interstellar dust polarizes starlight shining through it, which is evidence that some of the dust particles are elongated and are aligned in space. An interstellar magnetic field is the only mechanism suggested to account for this alignment of dust grains. Moreover, we find that the directions of the polarization for all of the stars we observe in the general direction toward or away from the galactic center are roughly parallel to each other. This is what we would expect if the grains were aligned by magnetic fields whose lines of force run along the spiral arms, for in those directions we are looking more or less broadside at spiral arms (Figure 34.15). On the other hand, the polarization of light from stars that lie in directions nearly at right angles to the galactic center, and hence along our own spiral arm, has random orientation. Again this is what we expect, for here we look along the lines of force of the interstellar field so that there is no preferential alignment of particles across our line of sight. Theory suggests that the magnetic field strength required to produce enough alignment of dust grains to account for the observed polarization effects is of the same order as that calculated from the Faraday rotation of synchrotron radiation.

(c) Containment of Cosmic Rays

Cosmic rays approach the Earth almost equally from all directions in space. Most of them cannot originate from the sun or another part of the solar system, for then they would come from certain directions, and, for example, show strong differences in influx between day and night. Some may come from beyond the Galaxy, but it is hard to believe that very many do, for we cannot imagine what sources could fill all of extragalactic space with cosmic rays to the observed density. Most investigators, therefore, consider it probable that most cosmic rays originate in our Galaxy. One likely source is believed to be supernovae (Chapter 32).

On the other hand, the atomic nuclei that compose most cosmic rays travel at very nearly the speed of light. If they travel in straight lines and originate in the Galaxy, they would escape it in at most a few tens of thousands of years. Now, the suggested sources of cosmic rays are all very unusual objects (such as supernovae), and there cannot be enough of them situated uniformly around us to provide a continuous, nearly isotropic flux of cosmic rays all traveling in straight lines. We conclude that cosmic rays must be produced by scattered sources over the entire Galaxy, and that they are trapped or stored in it for very long periods of time, moving about in interstellar space in all possible directions.

Because cosmic rays are charged particles (atomic nuclei), they can be captured into spiraling orbits around magnetic lines of force, and this, in fact, is the only way we know of that they can be kept in the Galaxy. Calculations show that interstellar fields of strength like that predicted from Faraday rotation and polarization can trap cosmic rays of energy up to 10^{17} or 10^{18} eV in the Galaxy for hundreds of millions of years. Cosmic rays of very much higher energy—say, 10^{20} eV—would escape the Galaxy in a relatively short time, but ones of such very high energy are observed extremely rarely at the Earth, and could originate from relatively few nearby sources, such as supernovae that occurred within the past few thousand years. Some very high-energy cosmic rays, alternatively, could have an extragalactic origin. In any case, we can understand the general cosmic ray phenomena only if the Galaxy has extensive magnetic fields.

34.5 THE NUCLEUS OF THE GALAXY

Near the center of the Galaxy is a large concentration of stars, generally called the *nuclear bulge*. At its center, lying behind the constellation Sagittarius, is the *nucleus* of the Galaxy. We cannot see the nucleus in visible light or in the ultraviolet, because those wavelengths are absorbed by the intervening interstellar dust. In the optical region of the spectrum, light from the central regions of the Galaxy is dimmed by factors of 10^6 to 10^{12} (15 to 30 magnitudes). High-energy X-rays and gamma rays, however, force their way through the interstellar medium and are recorded by instruments on rockets and satellites. Also, the infrared and radio radiation, whose wavelengths are long compared with the sizes of the interstellar grains, flow around them and reach us from the center of the Galaxy. The very bright radio source in that region is known as *Sagittarius A*, and was, in fact, the first cosmic radio source ever discovered.

(a) The Central 2 Parsecs

What then do all of these new observational techniques tell us about the structure of the nucleus of the Milky Way Galaxy? At a distance of 1 to 2 parsecs from the center of the Galaxy, there is a ring—not a sphere—of high-density molecular gas and dust that lies in same plane as do the spiral arms. The dust in this ring is hot and so must be heated by some source of energy that lies within it. Since the dust is radiating heat but is not cooling off, all of the energy radiated away must be replaced by energy from this central source. We can therefore estimate how much energy must be supplied every second to the dust ring from this central source, and we find that there must be a source of energy near the center of the Galaxy that has a luminosity that is 10 to 30 million times greater than the luminosity of the sun.

It is also possible to measure the gas in the ring spectroscopically and so derive its velocity from the Doppler effect. The ring appears to be rotating about the galactic center. In order to constrain the dust in the ring to follow a circular orbit around the nucleus of the Galaxy, mass must be concentrated in the nucleus in sufficient quantity to exert the required gravitational force. From the observed rotational velocity of the dust ring, we estimate that the total mass contained within it must be 2 to 5 million times greater than the mass of the sun.

What is the nature of this concentration of mass at the center of the Galaxy? Observations show that there is very little dust within the central 2 parsecs. There is also very little cold gas; the gas inside the ring is not in the form of either molecules or neutral atoms. There is some hot gas, which is ionized and distributed in clumps throughout the central region of the Galaxy. Energy is required to strip one or more electrons out of their orbits around an atomic nucleus, and from the observations of the ionized gas, we can estimate that the source of the ionizing radiation must have a temperature no higher than 35,000 K. The total mass of the hot gas in the central two parsecs of the Galaxy is about 70 times the mass of the sun.

From these measurements, it is obvious that most of the mass in the center of the Galaxy does *not* appear in the form of gas. In addition to the clumps of hot gas, there are numerous other individual sources of radiation. These sources appear to be predominantly cool K and M giant stars, and there are enough to account for all or nearly all of the several million solar masses of material that are required to keep the dust in the rotating ring in orbit.

(b) The Central Energy Source

Two models have been proposed to account for the source of energy that lies at the galactic center. The first suggests that there was a burst of star formation not more than a few million years ago, in which a large number of stars formed at very nearly the same time. Those stars that had very large masses and temperatures on the main sequence greater than 35,000 K have now all evolved to become cool supergiants or even possibly neutron stars or black holes. About 100 stars remain with temperatures near 35,000 K, and it is these stars that produce the energy that is responsible for ionizing the gas and heating the dust ring.

Unfortunately, this model does not account for all of the observations. There is evidence for a very compact source of nonthermal radiation in the galactic center. Perhaps the best evidence of this kind is the discovery of a line of gamma radiation produced when an electron and its antiparticle, a positron, collide and annihilate each other (Chapter 29.2b). Where does this antimatter come from? Since antimatter cannot exist for long periods of time in our Galaxy of normal matter, the positrons must be produced within the galactic center itself. We know from Einstein's equation

$$E = mc^2$$

that matter and energy are equivalent and that one can be converted entirely to the other. In particular, an electron and a positron can be produced by the collision of two extremely energetic photons (Chapter 37.3). Photons with the required energy are not radiated from the surfaces of normal stars.

A second argument against stars as the only source of energy in the center of the Galaxy comes from radio astronomy. At the center of the Galaxy, there is a source of radio emission. This source is less than 20 astronomical units in diameter. For comparison, the diameter of Saturn's orbit is also equal to about 20 AU. It seems very unlikely that 100 very hot, very massive stars would be crammed into so small a volume of space.

The radio and gamma ray observations can be explained only if there is a source in the galactic center that has a very small diameter, and this requirement leads us to suspect that the central source may be a *black hole.* A black hole, of course, can itself emit no electromagnetic radiation. Matter—gas, dust, and even perhaps stars—will be attracted by the gravitational force of the black hole. This material will spiral in towards the black hole and will form a disk of material around it. As the material spirals ever closer to the black hole, it accelerates and heats through compression to millions of degrees. This hot matter then would be the source of the energy radiated by the compact object at the galactic center.

Which of these two models is correct? Perhaps both. As so often happens in science, it now appears that neither model is entirely satisfactory, but that each is partly valid and both may be required to account for the observations of the galactic center. New observations suggest that there are *two* sources, slightly separated from one another, near the galactic center. One, which is a source of infrared radiation, is apparently a cluster of very luminous stars. Nearby is Sagittarius A, the compact source of radio energy, which at least right now can only be explained by the assumption that it is a black hole accreting matter.

From the proximity of these two sources, it is even possible to set a limit on the mass of the black hole by requiring that it be of low enough mass so that its gravitational forces will not disrupt the star cluster near it. It seems likely that the mass of the black hole can be no more than a few hundred times greater than the mass of the sun, and this limit is consistent with models that attempt to explain the gamma radiation produced by the annihilation of positrons and electrons.

A word of caution is in order. It is only in the last five years, with the availability of new observational techniques, that we have begun to amass observational evidence that there really is a black hole at the center of the Galaxy. Research on this subject is continuing very vigorously, and the interpretation of the observations may be changed as new measurements are made. For example, the mass of the black hole at the galactic center remains quite uncertain, and some astronomers believe it could be as massive as a few million times the mass of the sun rather than the few hundred solar masses suggested here. Such a massive object in our Galaxy would be most remarkable, since we think it could not have been produced by the evolution of normal stars. As we shall see, however, there is independent evidence that much more massive black holes lie at the heart of energetic, active galaxies and of quasars (Chapter 35). The circumstantial evidence that black holes do indeed exist in the nuclei of our own and other galaxies is becoming very strong.

EXERCISES

1. Sketch the distribution of globular clusters about the Galaxy, and show the sun's position. Show how they would appear on a Mercator map of the sky, with the central line of the Milky Way chosen as the "equator."
2. The globular clusters probably have highly eccentric orbits, and either oscillate through the plane of the Galaxy or revolve about its nucleus. Suppose the latter is the case; where would the clusters spend most of their time? (Think of Kepler's second law.) At any given time, would you expect most globular clusters to be moving at high or low speeds with respect to the center of the Galaxy? Why?
3. The period of the sun's revolution about the center of the Galaxy was calculated from its measured speed and distance from the center of the Galaxy. How would the period be changed if the sun's distance from the galactic center were 20 percent greater than the figure assumed?
4. If the galactic halo has an overall radius of 100,000 LY, what is the volume occupied by the Galaxy in cubic parsecs? If the mass of the Galaxy is 1×10^{12} suns, and if it were uniformly distributed throughout this volume, what would be the mean density?
   Answer: 1.5×10^{13} pc^3; 6.6×10^{-2} solar masses/pc^3
5. What would the mass of the Galaxy interior to the sun be if the sun's distance from the center were 10,000 pc but its period of revolution about the nucleus were only 100 million years?
6. Suppose we correctly knew the sun's distance from the center of the Galaxy but had derived a value for the speed of the sun in its orbit that is too high by 10 percent. How much would our calculated mass of the Galaxy interior to the sun be in error, and in which direction would the error be?
7. Suppose the mean mass of a star in the Galaxy were only ⅓ solar mass. Using the value for the mass of the Galaxy found in the text, find how many stars the system contains. What did you assume about the total mass of interstellar matter in finding your answer?
8. ★ If its orbital speed is 300 km/s, what is the period of a gas cloud moving in a circular orbit of radius 20,000 pc?
9. ★ Suppose 21-cm observations are made in the galactic plane in a direction 45° from that of the galactic center. What is the distance from the galactic center to the gas cloud with the maximum observed radial velocity?
   Answer: About 7000 pc
10. ★ Why are we not able to map out the spiral structure of the Galaxy in directions $l = 0°$ and 180° from 21-cm observations? Why do you suppose we *are* able to map out its spiral structure in directions $l = 90°$ and 270°?
11. Describe the details of an experiment in which you witness the formation of "spiral arms" of cream in a cup of coffee that you have stirred vigorously before putting in the cream.
12. Distinguish clearly between the orbital motion of the sun, toward galactic longitude 90°, and the *solar motion*, toward the *solar apex*, which was described in Chapter 22.

35

GALAXIES

Walter Baade (1893–1960), born in Westphalia, joined the staff of the Mount Wilson Observatory in 1930. He discovered the two populations of stars, expanded the distance scale and age of the universe, and was an early investigator of supernovae and radio sources. He is still recognized as one who used large telescopes to the very best advantage. (California Institute of Technology)

The "analogy [of the nebulae] with the system of stars in which we find ourselves . . . is in perfect agreement with the concept that these elliptical objects are just [island] universes—in other words, Milky Ways. . . ."

So wrote Immanuel Kant (1724–1804) in 1755* concerning the faint patches of light that telescopes revealed in large numbers. Unlike the true gaseous nebulae that populate the Milky Way (Chapter 27), the nebulous-appearing luminous objects referred to by Kant are found in all directions in the sky *except* where obscuring clouds of interstellar dust intervene. Despite Kant's (and others') speculation that these patches of light are actually systems like our own Milky Way Galaxy, the weight of astronomical opinion rejected the hypothesis, and their true nature remained a subject of controversy until 1924. The realization, less than a century ago, that our Galaxy is not unique and central in the universe ranks with the acceptance of the Copernican system as one of the great advances in cosmological thought.

* *Universal Natural History and Theory of the Heavens.*

35.1 GALACTIC OR EXTRAGALACTIC?

The discovery and cataloguing of nebulae had reached full swing by the close of the 18th century. A very significant contribution to our knowledge of these objects was provided by the work of William Herschel and his only son, John (1792–1871). William surveyed the northern sky by scanning it visually with the world's first large reflecting telescopes, instruments of his own design and manufacture (Figure 35.1). John took his father's telescopes to the southern hemisphere and extended the survey to the rest of the sky.

For a while, the elder Herschel himself, who had discovered thousands of "nebulae,"† regarded these objects as galaxies, like the Milky Way sys-

† *Nebula* (plural *nebulae*) literally means "cloud." Faint star clusters, glowing gas clouds, dust clouds reflecting starlight, and galaxies all appear as faint, unresolved luminous patches when viewed visually with telescopes of only moderate size. Since the true natures of these various objects were not known to the early observers, all of them were called "nebulae." Today, we usually reserve the word "nebula" for the true gas or dust clouds, but some astronomers still refer to galaxies as nebulae or *extragalactic nebulae*.

Figure 35.1 Herschel's 40-foot telescope. (Yerkes Observatory)

tem; he was known to remark once that he had discovered more than 1500 "universes." He found, however, that many of the nebulae appeared as individual (although rather indistinct) "stars," surrounded by hazy glows of light. The fact that it is difficult to reconcile the appearance of such an object with that of a remote stellar system led Herschel, in 1791, to abandon the island-universe hypothesis. Still, the concept of the possibility of other galaxies never quite disappeared from astronomical thought.

(a) Catalogues of Nebulae

One of the earliest catalogues of nebulous-appearing objects was prepared in 1781 by the French astronomer Charles Messier (1730–1817). Messier was a comet hunter, and as an aid to himself and others in his field he placed on record 103 objects that might be mistaken for comets. Because Messier's list contains some of the most conspicuous star clusters, nebulae, and galaxies in the sky, these objects are often referred to by their numbers in his catalog—for example, M31, the great galaxy in Andromeda.

In the years from 1786 to 1802, William Herschel presented to the Royal Society three catalogues, containing a total of 2500 nebulae. The *General Catalogue of Nebulae,* published by John Herschel in 1864, contains 5079 objects, of which 4630 had been discovered by him and his father. The *General Catalogue* was revised and enlarged into a list of 7840 nebulae and clusters by J. L. E. Dreyer in 1888. Today most bright galaxies are known by their numbers in Dreyer's *New General Catalogue*—for example, NGC 224 = M31. Two supplements to the *New General Catalogue,* known as the first and second *Index Catalogues* (abbreviated "IC"), were published in 1895 and in 1908.

By 1908 nearly 15,000 nebulae had been catalogued and described. Some had been correctly identified as star clusters and others as gaseous nebulae (such as the Orion nebula). The nature of most of them, however, still remained unexplained. If they were nearby, with distances comparable to those of observable stars, they would have to be luminous clouds, probably of gas, possibly intermixed with stars, within our Galaxy. If, on the other hand, they were very remote, far beyond the foreground stars of the Galaxy, they could be unresolved *systems* of thousands of millions of stars, galaxies in their own right, or as Kant had described them, "island universes." The resolution of the problem required the determination of the distances to at least some of the nebulae.

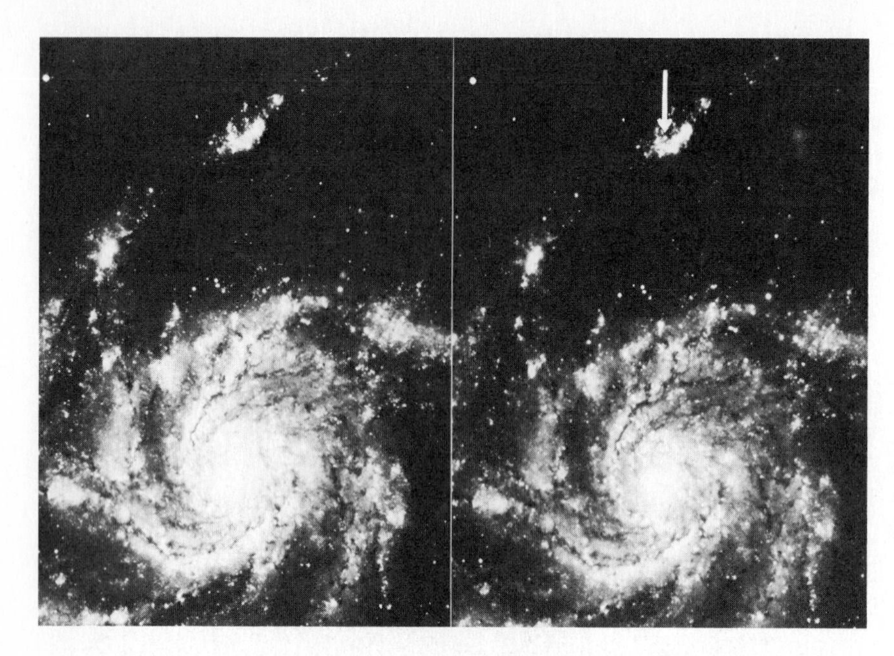

Figure 35.2 Two photographs of NGC 5457 made with the 5-m telescope. Bright nova appears on the photograph taken on February 7, 1951 *(right)*, but not on the one taken on June 9, 1950. (Palomar Observatory, California Institute of Technology)

(b) Arguments For and Against the Island-Universe Hypothesis

By the early 20th century the nebulae could be divided into two distinct groups: those largely irregular and amorphous, which are concentrated near the Milky Way, and those showing either elliptical or wheel-shaped symmetry, which are most numerous in parts of the sky far from the Milky Way. Those of the former type have bright-line spectra and were clearly recognized as gaseous nebulae. The few available spectra of the brighter nebulae in the latter category, on the other hand, showed absorption lines like the spectra of stars. Moreover, most of their radial velocities were found to be very high, and directed away from the sun (Chapter 37), and it seemed that such rapidly moving objects would escape the Galaxy. These data supported the hypothesis that the nebulae of the second type are extragalactic.

By 1917 several novae (Section 32.3c) had been discovered in the more conspicuous nebulae (Figure 35.2). If those novae were as luminous as the 26 novae then known to have occurred in our own Galaxy, they, and the nebulae in which they appeared, would have to be at distances of about 1 million LY—far beyond the limits of our Galaxy. Similar distances were found for the nebulae if the brightest resolved stars in them were assumed to have the same intrinsic brightness as the most luminous stars in the Galaxy.

On the other hand, not all astronomers agreed that real stars had actually been resolved in any of the nebulae. Moreover, two of the novae that had been observed in the nebulae were far brighter than the rest. S Andromedae, for example, which appeared in the nebula M31 in 1885, reached magnitude 7.2. Today we recognize those two novae as *supernovae* (Section 32.7), but supernovae were unknown in the early part of the century. If those two bright "novae" were assumed to be only as luminous as ordinary novae, the distances calculated for them turned out to be only a few thousand light-years, and the nebulae would not be extragalactic.

A remark should be made here concerning the Magellanic Clouds, two stellar systems that are not visible from as far north as the United States (Figure 35.3). The distances to the Clouds, determined from the cepheid variables in them, were thought at that time to be about 75,000 LY (they are actually at least twice that distance). In the years preceding 1924, however, the diameter of our own Galaxy had been overestimated at about 300,000 LY (inadequate corrections having been made for the absorption of light from the globular clusters by interstellar dust). The Magellanic Clouds, therefore, although now regarded as neighboring external galaxies, were then considered to be outlying sections of our own Galaxy.

Two of the major protagonists in the controversy over the nature of the nebulae were Harlow Shapley (1885–1972), of the Mount Wilson Observatory, and H. D. Curtis (1872-1932), of the Lick Observatory. Their opposing views culminated on April 26, 1920, in the famous Shapley-Curtis debate before the National Academy of Sciences. Curtis supported the island-universe theory, and Shapley opposed it. Of course, the controversy was not settled by the debate; according to A. R. Sandage, "Perhaps the fairest statement that can be made is that Shapley used many of the correct arguments but came to the wrong conclusion. Curtis, whose intuition was better in this case, gave rather weak and sometimes incorrect arguments from the facts, but reached the correct conclusion."*

Figure 35.3 The Large and Small Magellanic Clouds *(top* and *lower left). (Sky and Telescope)*

(c) The Resolution of the Controversy

The final resolution of the controversy was brought about by the discovery of variable stars in some of the nearer "nebulae" in 1923 and 1924 (Figure 35.4). Edwin Hubble, working with the 100-inch

* A. R. Sandage, *The Hubble Atlas of Galaxies.* Carnegie Institution, Washington, D.C., 1961, p.3.

Figure 35.4 A field of variable stars in the Andromeda galaxy, with two variables marked. Photographed with the 5-m telescope. (Palomar Observatory, California Institute of Technology)

(2.5-m) telescope at the Mount Wilson Observatory, analyzed the light curves of variables he had discovered in M31, M33, and NGC 6822 and found that they were cepheids (Section 31.2). Although cepheid variables are supergiant stars, the ones studied by Hubble appeared very faint—near magnitude 18. Those stars, therefore, and the systems in which they were found, must be very remote; the "nebulae" had been established as galaxies. Hubble's exciting results were presented to the American Astronomical Society at its 33rd meeting, which began on December 30, 1924.

35.2 THE EXTRAGALACTIC DISTANCE SCALE

One of the most important, difficult, and controversial problems in modern observational astronomy is that of the scale of distances to galaxies. Galaxies are far too remote, of course, to display parallaxes or proper motions. In order to determine distances to galaxies, we must therefore resort to a multistep process. The traditional approach to determining distances goes roughly as follows. First, we derive distances to individual nearby stars in our own Galaxy by measuring parallaxes and proper motions. With knowledge of the absolute magnitudes of these nearby stars, we can then determine distances to *clusters,* which contain stars similar to those with known absolute magnitudes. Once we measure the distance to a cluster, we know the absolute magnitude of every star within the cluster. Fortunately, clusters contain some stars, including cepheid variables, that are much more luminous than any of the nearby stars for which we can obtain parallaxes by direct measurement. These stars are so luminous, in fact, that ones just like them can be detected in other galaxies. Since we can measure the apparent magnitudes of these stars and already know their absolute magnitudes from studies of stars in clusters in our own Galaxy, we can use the inverse-square law for the propagation of light (Section 8.1b) to determine the distances to the galaxies to which they belong. These luminous stars thus serve as *standard candles* for measuring extragalactic distances.

Individual stars can be detected only in relatively nearby galaxies. At larger distances, we must use objects that are even brighter than the brightest stars as standard candles. Globular clusters, H II regions, and supernovae have all proven to be useful. At the greatest distances of all, we use entire galaxies as standard candles for determining distances to clusters of galaxies.

(a) The Distances to Galaxies

Some of the more important ways by which we estimate distances to galaxies are the following:

1. *Cepheids.* Cepheid variables gave Hubble the first clue to the remote nature of galaxies and are still our first important link to galaxian distances. That is why these relatively rare stars are so significant to us. If we can recognize a cepheid in a galaxy, we can find its luminosity or absolute magnitude from its period through the period-luminosity relation (Section 23.3e). Thus, the cepheids can serve as standard candles, for comparison of their known absolute magnitudes and observed apparent magnitudes enables us to find their distances and hence the distances to the galaxies in which they occur, with the inverse-square law of light (Section 8.1b). The most luminous cepheids are about 20,000 times more luminous than the sun, which makes them supergiant stars, but even so they can be detected in only about 30 of the nearest galaxies—even with the world's largest optical telescopes.
2. *Brightest stars.* The most luminous stars are even brighter than cepheids and can be seen to greater distances. Thus, once calibrated in those galaxies whose distances are known from observations of cepheids, the brighest stars can be used as standard candles. Young, high-mass supergiant stars range in absolute magnitude to as bright as about -9 or even -10, or up to one million times the luminosity of the sun, and so can extend the distance scale to more than six times the distance to which cepheids can be seen, but this is still our very immediate cosmic neighborhood!
3. *Novae.* Novae can be recognized in nearby galaxies, and from their light curves we know what their approximate luminosities are. Observations of novae with the Hubble Space Telescope are expected to play a key role in establishing the extragalactic distance scale. The brightest novae, however, even at maximum light, do not outshine the brightest stars, so

they corroborate the distances in our neighborhood but do not extend the scale. What is needed are brighter standard candles.

4. *Globular clusters.* Although globular clusters range considerably in total light (because they differ in their numbers of stars), we sometimes recognize many such clusters in one galaxy. If we assume that the brightest of them is like the brightest globular cluster in our own Galaxy (absolute magnitude about -10), that object becomes a standard candle. Distances determined this way are rather uncertain, and do not extend the distance scale beyond that determined by the brightest individual young stars, but the method is still useful because globular clusters are often seen in galaxies without a young population of supergiant stars.
5. *Supernovae.* Because of their high luminosities (up to 10^{10} times the luminosity of the sun), supernovae can be seen in very remote galaxies and would seem to be ideal standard candles, except that supernovae differ considerably among themselves in absolute magnitude at maximum light. Some progress is being made in calibrating different kinds of supernovae, and in the future they may well hold the key to the extragalactic distance scale. The difficulty at present is that supernovae occur rarely in any one galaxy, and the sample that have appeared in galaxies of well-known distances is rather small.
6. *21-cm line width.* A very promising technique of distance determination was developed in the late 1970s. A typical spiral galaxy contains a great deal of neutral hydrogen gas in revolution about its center. The 21-cm radiation from this hydrogen in different parts of the galaxy, moving at different speeds in our line of sight, therefore displays a range of Doppler shifts, so that the entire 21-cm line radiation from the galaxy is observed as a broad band (think of how the rotation of a star affects its spectral lines—Section 24.2d). After a simple correction for the tilt of the plane of the galaxy to our line of sight, the width of that 21-cm line gives a measure of the maximum rotational velocity in the galaxy. That rotational velocity is correlated with the galaxy's mass, and we might expect the luminosity to be correlated with the mass as well, and hence with the 21-cm line width. Indeed, it is; the correlation was first noted between the line width and the total visual light from a galaxy, but the relation is tighter if the galaxy's infrared luminosity is used instead, because infrared radiation is less affected by interstellar dust. This method of distance determination, then, is to observe the 21-cm line width, thereby learning the galaxy's absolute infrared magnitude, then to observe the galaxy's apparent infrared magnitude, and finally to calculate the distance from the inverse-square law.
7. *Total light of galaxies.* If all galaxies were identical, they would all emit the same total amount of light, and the magnitude of a galaxy as a whole would indicate its distance. Galaxies range enormously in total luminosity; nevertheless, some types of galaxies display a relatively small range of luminosities. For those galaxies, rough distances can be estimated from their total apparent magnitudes. Of course, we can only establish what those mean luminosities are for the types of galaxies that are well represented among those whose distances can be determined from other more direct methods. Unfortunately, many galaxies, probably the majority, do not have distinguishing characteristics that enable us to estimate their absolute magnitudes; we can only tell which ones are highly luminous and which ones are not if we see a collection of them of various brightnesses, side by side in a cluster.

These techniques are summarized in Table 35.1, which gives the type of galaxy for which the specific technique is useful (galaxy types are described in Section 35.3), the range of distances over which the technique can be applied, and the reliability of the distance estimates derived with each technique.

With so many ways of finding distances to galaxies, one might think that the distance scale is well settled. Unfortunately, it is not; the various standard candles are difficult to calibrate, and the measurements are difficult to make. Experts differ in their judgment of the proper interpretation of the observations, the calibration of the standard candles and yardsticks, and what standards are the most reliable. These differences in judgment translate to a difference in the distances of remote galaxies of a factor of two.

TABLE 35.1 Methods for Estimating the Distance to Galaxies

METHOD	RELIABILITY	GALAXY TYPE FOR WHICH METHOD IS USEFUL	APPROXIMATE DISTANCE RANGE OVER WHICH METHOD IS USEFUL (millions of pc)
Cepheids	Very	Spirals, irregulars	0–4
Brightest stars	Moderate	Spirals, irregulars	0–20
Novae	Very	Sc, irregulars	0–20
Globular clusters	Moderate	All	0–20
Supernovae	Moderate	All	0–200
21-cm line width	Very	Spirals, irregulars	0–25
Total light of galaxies	Low	Spirals, irregulars	0–100
Brightest galaxy in cluster*	Very	Ellipticals in clusters of galaxies	20–5000
Radial velocities*	Very	All	100–1000

* Relative distances only.

Perhaps an uncertainty of whether a particular galaxy is two billion or four billion light years away may seem to be pretty poor precision in an exact science like astronomy, where we know the distances between the planets in the solar system to better than one part in a million. But uncertainty is common at the frontier of research. At least we are far beyond the controversy of whether or not the "nebulae" are extragalactic.

(b) Relative Extragalactic Distances

While the absolute distances to galaxies in, for example, light years, remain uncertain by as much as a factor of two, there are ways in which we can estimate the *relative* distances of two galaxies with far more assurance than we know the actual distance to either.

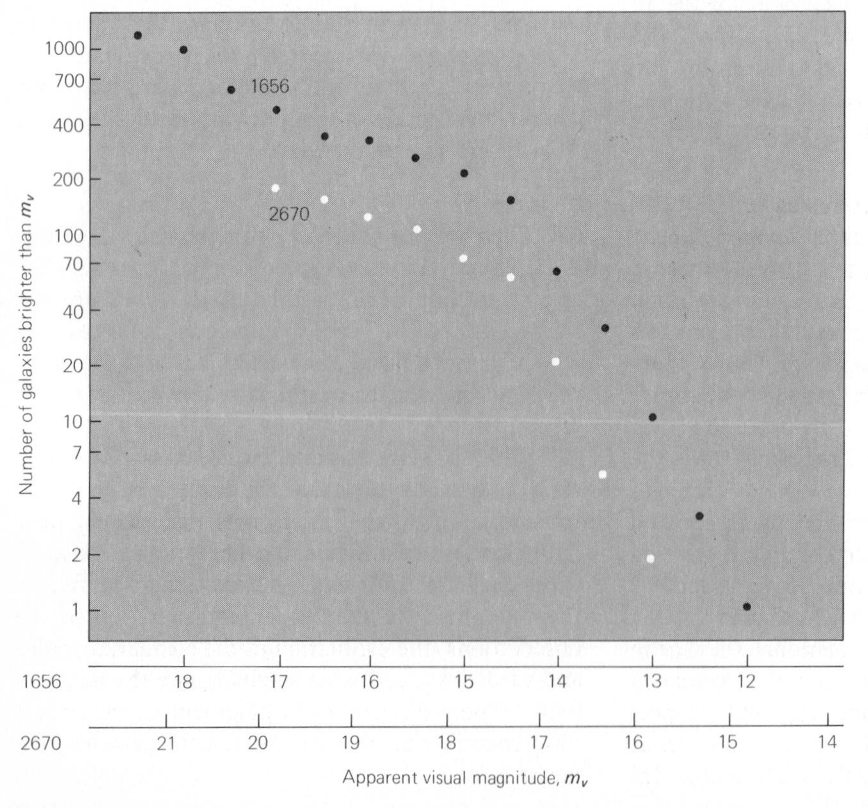

Figure 35.5 The luminosity functions of two clusters of galaxies, identified by their numbers in the Abell catalogue of rich clusters. The horizontal scale is apparent visual magnitude, with fainter magnitudes to the left. The vertical scale shows the number of galaxies in each cluster that are brighter than each corresponding magnitude. Cluster 2670 is more distant and fainter appearing than cluster 1656, but the points have been plotted in a shifted magnitude scale to show the similarity of its luminosity function to that of the nearer cluster. The two magnitude scales are shown below; corresponding points in the luminosity function of 2670 are 2.8 magnitudes fainter than in the luminosity function 1656. From that difference in magnitude we can easily calculate the relative distances of the two clusters.

Figure 35.6. The Andromeda galaxy, M31, photographed with the 1.2-m Schmidt telescope. (Palomar Observatory, California Institute of Technology)

One way involves clusters of galaxies. We shall see (Chapter 36) that most, if not all, galaxies are members of groups or clusters of anywhere from a few galaxies to a few thousand galaxies each. Although the galaxies in a cluster display a wide range of luminosity, if the galaxies in different clusters were formed under the same initial conditions and according to the same probabilistic laws, we would expect the brightest galaxy (or perhaps the second or tenth brightest galaxy) in each cluster to be intrinsically similar. If, for example, the brightest galaxy in one cluster were four times as bright-appearing as the brightest galaxy in another, we would expect the second cluster to be twice as distant as the first. Alternatively, we can compare the distributions of the numbers of galaxies of various apparent brightnesses (or *luminosity functions*) of the two clusters. If the distributions were similar except for a shift in the apparent magnitude scale, in the sense that galaxies at some point in the distribution of one cluster were four times as bright-appearing as those at the corresponding point in the other, we could conclude that the second cluster is twice as far away as the first (Figure 35.5).

Another way of finding relative galaxian distances is from the correlation between the distances of galaxies and the Doppler shifts of the lines in their spectra to longer wavelengths. This relation, known as the *Hubble law,* is a direct consequence of the uniform expansion of the universe, and we shall return in Chapter 37 to this extremely important relation and its significance. Let it suffice for now, however, that the more remote a galaxy, the greater is the shift (to longer wavelengths) of the light coming from it, and at least for the not terribly remote galaxies, the speeds with which they are moving away from us are proportional to their distances. In visible light, a shift to longer wavelengths is a shift toward the red end of the spectrum, so the Doppler shift to longer wavelengths of a distant galaxy is usually called its *redshift.*

Knowing that a particular galaxy, moving away from us at twice the speed of another, is twice as far away does not, in itself, tell us the distance of either, but only their relative distances. The velocity-distance relation (Hubble law) must be calibrated; we must know the actual distance of a galaxy of a particular velocity, and this requires knowing the absolute extragalactic distance scale.

35.3 TYPES OF GALAXIES

Galaxies differ a great deal among themselves, but the majority of optically bright galaxies fall into two general classes: spirals and ellipticals. A minority are classed as irregular.

(a) Spiral Galaxies

Our own Galaxy and the Andromeda galaxy (Figure 35.6), M31, which is believed to be much like it, are typical large spiral galaxies. Like our Galaxy (Chapter 34), a spiral consists of a nucleus, a disk, a halo, and spiral arms. Interstellar material is usually spread throughout the disks of spiral galaxies. Bright emission nebulae are present, and absorption of light by dust is also often apparent, especially in those systems turned almost edge-on to our line of sight (Figure 35.7). The spiral arms contain the young stars, which include luminous supergiants. These bright stars and the emission nebulae make the arms of spirals stand out like the arms of a

Figure 35.7 NGC 4565, a spiral galaxy in Coma Berenices, seen edge-on. Photographed in red light with the 5-m telescope. (Palomar Observatory, California Institute of Technology)

Figure 35.8 The Sc galaxy NGC 5194 (M51) and its irregular II companion, NGC 5195. Photographed with the 5-m telescope. (Palomar Observatory, California Institute of Technology)

fourth-of-July pinwheel. Open star clusters can be seen in the arms of nearer spirals, and globular clusters are often visible in their halos; in M31, for example, more than 200 globular clusters have been identified. Spiral galaxies contain both young and old stars.

Some famous spirals are illustrated in these pages (Color Plates 29c, 30a, 30b). Galaxies M51 and M33 (Figure 35.8 and 35.9, respectively) are seen nearly face-on; NGC 4565 (Figure 35.7) is nearly edge-on. Note the absorbing lane of interstellar dust in NGC 4565—a thin slab in the central plane of the disk—which is silhouetted against the nucleus. M81 (Figure 35.10), like M31 (Figure 35.6), is viewed obliquely.

A large minority (perhaps a third or more) of spiral galaxies display conspicuous "bars" running through their nuclei; the spiral arms of such a system usually begin from the ends of the bar, rather than winding out directly from the nucleus. These are called *barred spirals*. A famous example is NGC 1300 (Figure 35.11). Some astronomers believe that almost all spirals contain at least a weak bar. Studies of the rotations of some barred spirals show that their inner parts (out to the ends of the bars) are rotating approximately as solid bodies. In the absence of differential shearing rotation, the straight bar can persist, rather than winding up; the detailed structures and dynamics of barred spirals, however, are just beginning to be understood.

In both normal and barred spirals we observe a gradual transition of morphological types. At one extreme, the nuclear bulge is large and luminous, the arms are faint and tightly coiled, and bright emission nebulae and supergiant stars are inconspicuous. At the other extreme are spirals in which the nuclear bulges are small—almost lacking—and the arms are loosely wound, or even wide open. In these latter galaxies, there is a high degree of resolution of the arms into luminous stars, star clusters, and emission nebulae. Our galaxy and M31 are both

Figure 35.9 NGC 598 (M33), a spiral galaxy in Triangulum, photographed with the 5-m telescope. (Palomar Observatory, California Institute of Technology)

Figure 35.10 NGC 3031 (M81), a spiral galaxy in Ursa Major, photographed with the 5-m telescope. (Palomar Observatory, California Institute of Technology)

intermediate between these two extremes. Photographs of spiral galaxies, illustrating this transition of types, are shown in Figures 35.12 and 35.13. All spirals and barred spirals rotate in the sense that their arms trail, as does our own Galaxy.

Spiral galaxies range in diameter from about 20,000 to more than 100,000 LY, and the atomic hydrogen in the disks often extends to far greater diameters. From the limited observational data available, their masses are estimated to range from 10^9 to 10^{12} times the mass of the sun. The absolute magnitudes of most spirals fall in the range -16 to -22.5, corresponding to luminosities of about 10^8 to 10^{11} suns. Our galaxy and M31 are relatively large and massive, as spirals go.

(b) Elliptical Galaxies

More than two-thirds of the thousand most conspicuous galaxies in the sky are spirals. For this reason

Figure 35.11 NGC 1300, a barred spiral galaxy in Eridanus, photographed with the 5-m telescope. (Palomar Observatory, California Institute of Technology)

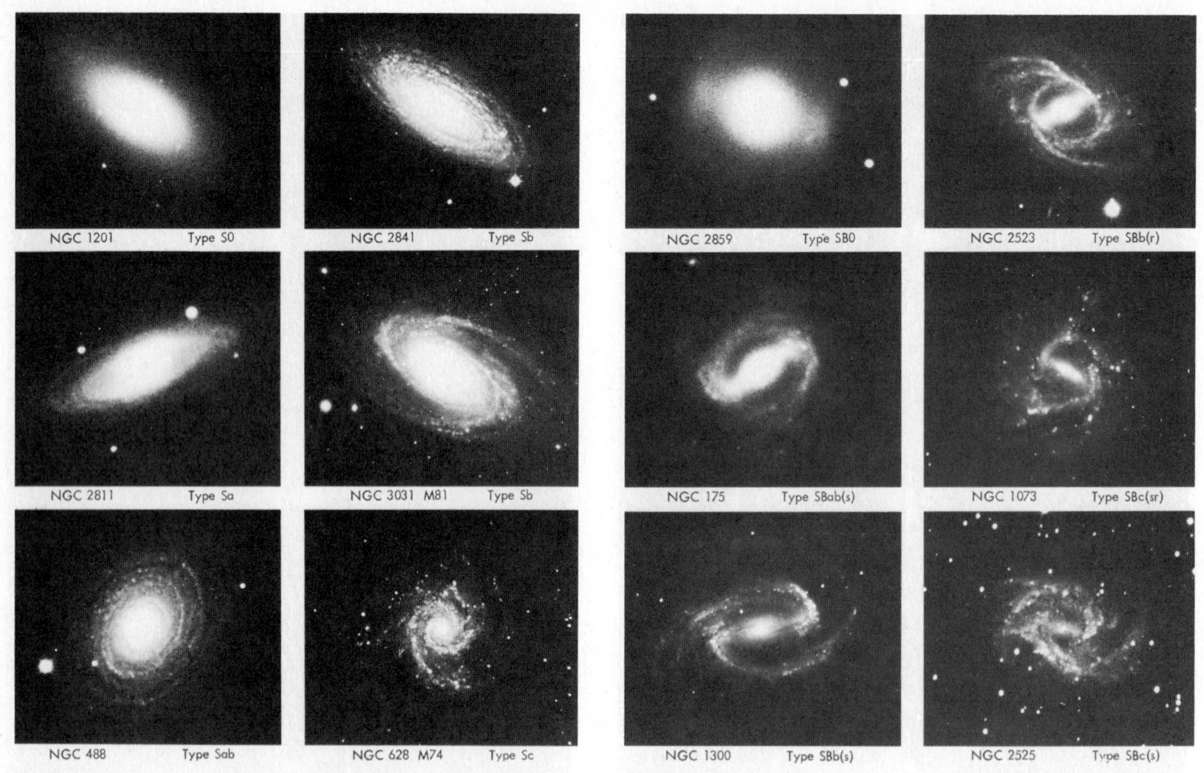

Figure 35.12 Types of spiral galaxies. (Palomar Observatory, California Institute of Technology)

Figure 35.13 Types of barred spirals. (Palomar Observatory, California Institute of Technology)

it is often said that most galaxies are spirals. Actually, however, the most numerous galaxies in any given volume of space are those of relatively low luminosity, which cannot be seen at large distances and which, therefore, are not among the brightest-appearing galaxies. (Similarly, the most numerous stars are faint main-sequence stars, very few of which can be seen with the unaided eye—see Section 26.1.) In nearby clusters half or more of these dwarf galaxies fall into the class of *elliptical* galaxies. Moreover, the rich clusters, which contain a good fraction of all galaxies, are composed mostly of ellipticals. Elliptical galaxies, therefore, are really far more numerous than spirals.

Elliptical galaxies are spherical or ellipsoidal systems that are thought to consist almost entirely of old stars; they contain no trace of spiral arms. Their light is dominated by red stars, and in this respect, ellipticals resemble the nucleus and halo components of spiral galaxies. Dust and emission nebulae are not conspicuous in elliptical galaxies, but ellipticals are not devoid of interstellar matter. Many do contain narrow lanes of absorbing dust, and X-ray data indicate that 1 to 2 percent of the total mass of ellipticals may be in the form of gas at a temperature that exceeds a million degrees. In the

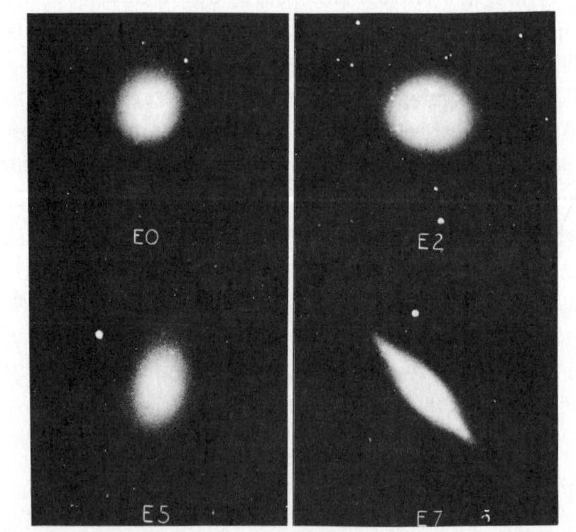

Figure 35.14 Types of elliptical galaxies. (Palomar Observatory, California Institute of Technology)

larger nearby ellipticals, many globular clusters can be identified. The elliptical galaxies show various degrees of flattening, ranging from systems that are approximately spherical to those that approach the flatness of spirals (Figure 35.14). The distribution of light in a typical luminous elliptical galaxy shows that, while it has many stars concentrated toward its center, a sparse scattering of stars extends for very great distances and merges imperceptibly into the void of intergalactic space. For this reason, it is nearly impossible to define the total size of an elliptical galaxy. Similarly, it is not obvious how far the halo of a spiral galaxy extends.

Elliptical galaxies have a much greater range in size, mass, and luminosity than do the spirals. The rare giant ellipticals (for example, M87—Figure 35.15) are more luminous than any known spiral. The brightest ellipticals in some rich clusters (for example, NGC 4886, in the Coma cluster of galaxies—see Section 36.2) have absolute magnitudes that are brighter than -23—more than 10^{11} times the luminosity of the sun, and about five times the luminosity of the Andromeda galaxy. The mass of the stars in giant ellipticals is typically at least 10^{12} times the mass of the sun. While, as stated above, the diameters of these large galaxies are difficult to define, they certainly extend over at least several hundred thousand light years, considerably larger than the largest spirals.

Figure 35.16 Leo II, a dwarf elliptical galaxy (negative print), photographed with the 5-m telescope. (Palomar Observatory, California Institute of Technology)

Figure 35.15 NGC 4486 (M87), giant elliptical galaxy in Virgo, photographed with the 5-m telescope. Note the many visible globular clusters in the galaxy. (Palomar Observatory, California Institute of Technology)

Elliptical galaxies range all the way from the giants, just described, to dwarfs, which we think are the most common kind of galaxy. An example of a dwarf elliptical is the Leo II system, shown in Figure 35.16. There are so few bright stars in this galaxy that even its central regions are transparent. The total number of stars, however, (most of which are too faint to show in Figure 35.16), is probably at least several million. The absolute magnitude of this typical dwarf is about -10; its luminosity is about one million times that of the sun and is about equal to the luminosity of the brighest known individual stars. It is so near to us (about 750,000 LY) that its diameter (about 5000 LY) is probably lim-

Figure 35.17 NGC 6822, a nearby irregular galaxy. (Palomar Observatory, California Institute of Technology

Figure 35.18 The Large Magellanic Cloud. (California Institute of Technology)

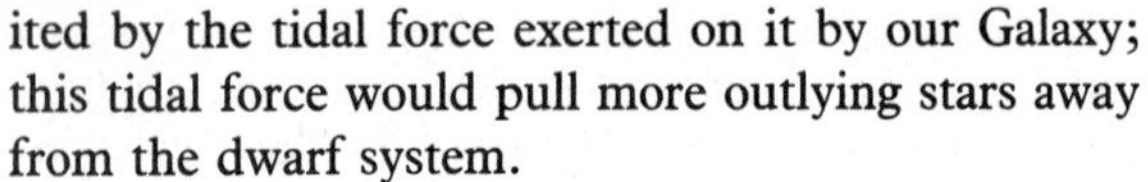

ited by the tidal force exerted on it by our Galaxy; this tidal force would pull more outlying stars away from the dwarf system.

Whether still smaller galaxies than dwarfs like Leo II exist depends on how galaxies are defined. Several globular clusters are known that are more than 200,000 LY from the nucleus of our Galaxy. It is not known whether objects like these are distributed through intergalactic space or whether they are outlying members of our Galaxy. If the former is the case, they must be galaxies in their own right. Perhaps even individual stars exist in intergalactic space.

Intermediate between the giant and dwarf elliptical galaxies are systems such as M32 and NGC 205, two near companions to M31. They can be seen in the photograph of M31 (Figure 35.6); NGC 205 is the one that is farther from M31.

(c) Irregular Galaxies

A few percent of the brightest-appearing galaxies in the northern sky are classed as irregular (Figure 35.17). They show no trace of circular symmetry but have an irregular or chaotic appearance. The irregular galaxies divide into two groups. The first group, denoted *Irr I* galaxies, consists of objects containing many O and B stars and emission nebulae. The best-known examples are the Large and Small Magellanic Clouds (Figure 35.3 and Color Plate 29a), our nearest galaxian neighbors (although some astronomers would classify the Large Magellanic Cloud—Figure 35.18—as a barred spiral). We find many star clusters in these galaxies, as well as variable stars, supergiants, and gaseous nebulae; they contain both old and young stars. Both of the Magellanic Clouds lack conspicuous dust clouds, and a deficiency of dust clouds is typical of this kind of irregular galaxy.

Galaxies of the second irregular type (*Irr II*) resemble the Irr I objects in their lack of symmetry. These objects, however, are completely amorphous in texture. The Irr II galaxies generally also show conspicuous dark lanes of absorbing interstellar dust. Examples are M82 (Figure 35.19) and the companion to the spiral galaxy M51 (Figure 35.8).

(d) Classification of Galaxies

Of the several classification schemes that have been suggested for galaxies, one of the earliest and simplest, and the one most used today, was invented by Hubble during his study of galaxies in the 1920s.

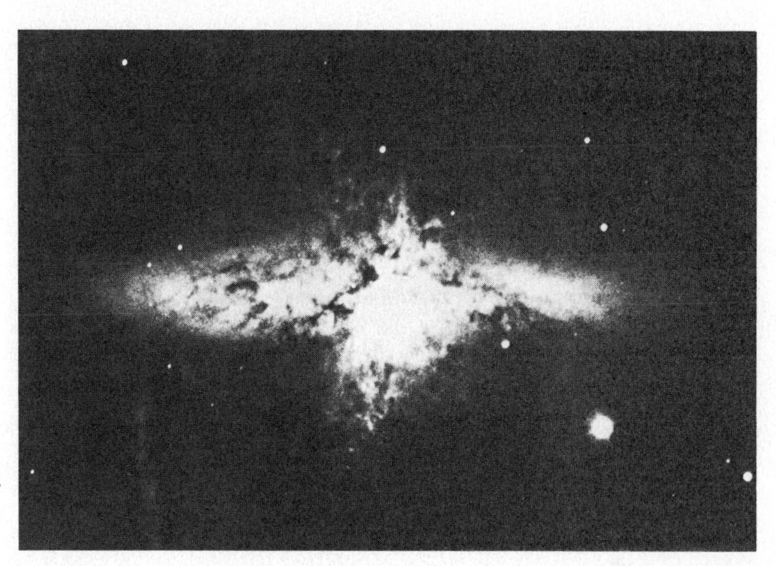

Figure 35.19 NGC 3034 (M82), an irregular II galaxy in Ursa Major. (Palomar Observatory, California Institute of Technology)

Hubble's scheme consists of three principal classification sequences: ellipticals, spirals, and barred spirals. The irregular galaxies (Irr I and Irr II) form a fourth class of objects in Hubble's classification.

The ellipticals are classified according to their degree of flattening or *ellipticity*. Hubble denoted the spherical galaxies by E0, and the most highly flattened by E7. The classes E1, E2, . . ., E6, are used for galaxies of intermediate ellipticity.* Hubble's classification of elliptical galaxies is based on the appearance of their *images,* not upon their true shapes. An E7 galaxy, for example, must really be a relatively flat elliptical galaxy seen nearly edge on, but an E0 galaxy could be one of any degree of ellipticity, seen face on. Analyses indicate that some elliptical galaxies are *oblate* (like a pumpkin), others are *prolate* (like a football), and still others are *triaxial,* that is, the three perpendicular axes through the center to the edge are unequal in length.

Hubble classed the normal spirals as S and the barred spirals as SB. Lowercase letters a, b, and c are added to denote the extent of the nucleus and the tightness with which the spiral arms are coiled. For example, Sa and SBa galaxies are spirals and barred spirals in which the nuclei are large and the arms tightly wound. Sc and SBc are spirals of the opposite extreme. Our Galaxy and M31 are classed as Sb.

In rich clusters, galaxies are observed that have the disk shape of spirals but no trace of spiral arms. Hubble regarded these as galaxies of type intermediate between spirals and ellipticals and classed them S0.

Hubble's classification scheme for all but irregular galaxies is illustrated in Figure 35.20, in which the morphological forms are sketched and labeled, and with the three principal sequences joined at S0. The diagram is based on one by Hubble himself.

Hubble's classification scheme has been modified and expanded since his time to give a more complete description, but such refinements need not concern us here.

(e) Some Unusual Classes of Galaxies

There are also some other classes of unusual galaxies that were not defined by Hubble:

cD Galaxies cD galaxies are supergiant elliptical galaxies, usually E0 or E1, that are frequently found in (or near) the centers of clusters of galaxies. They are the largest galaxies known and tend to outshine the next brightest cluster galaxies by as much as a factor of two. Often they are strong radio sources as well.

Compact Galaxies The class of compact galaxies consists of a large number of galaxies of relatively small size and high surface brightness. They are usually elliptical or irregular.

* Each of the numbers 0 through 7 that describe the flattening of a galaxy is defined in terms of the major and minor axes of the image of the galaxy, a and b, respectively, by $10(a - b)/a$.

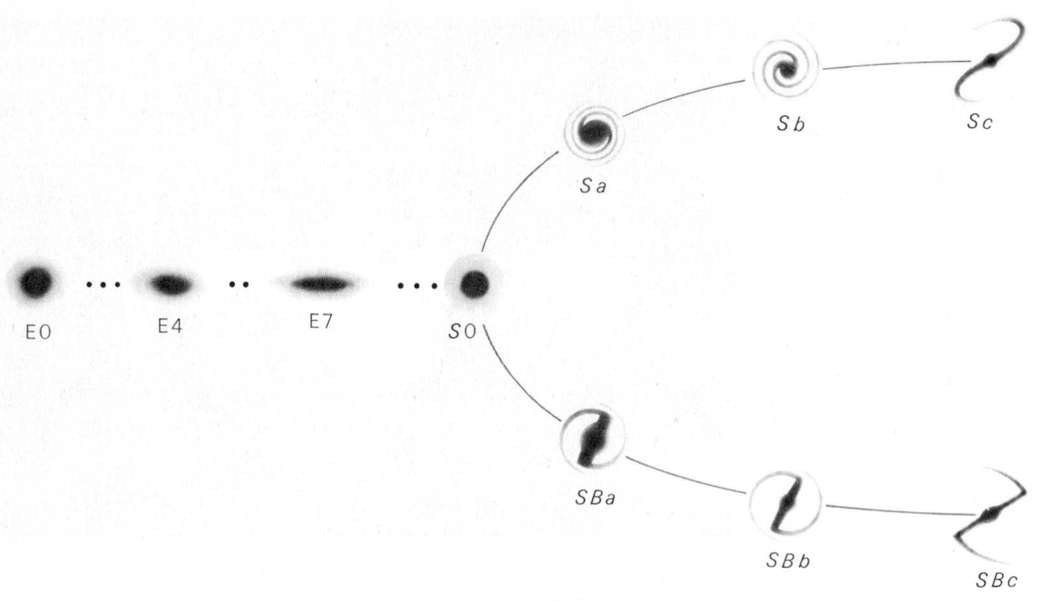

Figure 35.20 Hubble's classification scheme for galaxies.

N Galaxies An N galaxy is a galaxy with a very bright, nearly stellar-appearing nucleus. The rest of the galaxy appears as a sort of faint, extended haze. Today, N galaxies are regarded as belonging to a class of galaxies with active nuclei, which are described in Section 35.6d.

Seyfert Galaxies About a dozen galaxies of this class were first described by Karl Seyfert (1911–1960) of Vanderbilt University, from whom the class derives its name. A Seyfert galaxy is a spiral that has a small bright region in its nucleus, whose spectrum shows bright emission lines arising from hot gases there. Seyfert galaxies are sometimes strong infrared and radio emitters. They are also considered to be galaxies with active nuclei (Section 35.6d).

(f) Evolution of Galaxies

The continuity of the morphological forms of galaxies along classification sequences suggests that these different forms might represent stages of evolution for galaxies. There is much doubt, however, that galaxies evolve from one type to another at all. The fact that different kinds of galaxies are flattened by different amounts almost certainly results from their having different amounts of angular momentum—that is, from their different rotation rates. In other words, galaxies might always have had essentially their present forms (at least since their formation and in the absence of interactions with other galaxies), the form of a particular galaxy depending mostly on its mass and angular momentum per unit mass.

We do expect that stars within galaxies will evolve, as outlined in Chapters 30 and 32. Elliptical galaxies may always have been elliptical, but they may have had supergiant stars when they were young. Spirals may never become elliptical, but eventually their spiral arms may become less conspicuous when (and if) virtually all of their interstellar matter is converted into stars. As we noted earlier, there are S0 galaxies that have the disk shape and nuclear bulges that resemble those of spirals, but have stellar populations like those of ellipticals and lack spiral arms. The eventual fate of spirals must be to evolve into S0 galaxies. We will return to the issue of the evolution of galaxies in Chapter 36.

(g) Summary

The gross features of the different kinds of galaxies are summarized in Table 35.2. Many of the figures given, especially for mass, luminosity, and diameter, are very rough and are intended to illustrate only orders of magnitude.

TABLE 35.2 Gross Features of Galaxies of Different Types

	SPIRALS	ELLIPTICALS	IRREGULARS
Mass (solar masses)	10^9 to 10^{12}	10^6 to 10^{13}	10^8 to 10^{11}
Diameter (thousands of light-years)	20 to 300 or more	2 to 500 (?)	5 to 30
Luminosity (solar units)	10^8 to 10^{11}	10^6 to 10^{11}	10^7 to 2×10^9
Absolute visual magnitude	-15 to -22.5	-9 to -23	-13 to -20
Population content of stars	Old and young	Old	Old and young
Composite spectral type	A to K	G to K	A to F
Interstellar matter	Both gas and dust	Almost no dust; little gas	Much gas; some are deficient in dust, others contain large quantities of dust

35.4 DETERMINATION OF PROPERTIES OF GALAXIES

The linear size of that part of a galaxy that corresponds to an observed angular size can be calculated once the distance to the galaxy is known, just as we calculate the diameter of the sun or of a planet. Also if we know the distance to a galaxy, we can apply the inverse-square law of light and calculate its total luminosity, or the absolute visual magnitude, from the amount of light flux we receive from it. Thus our knowledge of the radii and luminosities of galaxies is dependent on the accuracy of our estimates of the distances to galaxies. It also follows that an error in the extragalactic distance scale will lead to systematic errors in the derived sizes and luminosities.

The determination of masses of galaxies, however, is more difficult and, in fact, is possible (by present techniques) for only a small fraction of them. There are several techniques for measuring galaxian masses.

(a) Mass of Galaxies from Internal Motions

We determine the masses of galaxies, like those of other astronomical bodies, by measuring their gravitational influences on other objects or on the stars within them. We must assume, of course, that Newton's law of gravitation is valid over extragalactic distances. Some theorists have challenged this assumption, but most astronomers agree that observations to date are consistent with it.

Internal motions in galaxies provide the most reliable methods of measuring their masses. The procedure for spiral galaxies is to observe the rotation of a galaxy from the Doppler shifts of either features in the optical spectrum or the 21-cm line of neutral hydrogen, and then to compute its mass with the help of Kepler's third law.

As an illustration we shall consider the rotation of M31, the Andromeda galaxy (Figure 35.6). The galaxy is inclined at an angle of only about 15° to our line of sight, so we see it highly foreshortened. There is evidence for rapid rotation of the central part of the nuclear bulge of the galaxy, but far from the nucleus we can represent the rotation curve by the simplified version of Figure 35.21, which, while not accurately reflecting the observed details, serves better for the illustration of how the mass of the galaxy can be determined.

The radial velocity of the brilliant nucleus shows the galaxy as a whole approaches us at nearly 300 km/s. The still more negative radial velocities of regions southwest of the nucleus indicate that that side of the galaxy is turning toward us and the northeast side away from us. We see that the maximum rotational speed in M31 is reached at about 50′ from its center and is almost constant to 150′, at about 230 km/s with respect to the center of the galaxy. At the distance of M31 (680 thousand parsecs—680 kpc), 150′ corresponds to about 30,000 pc, or 6×10^9 AU, so the circumference of the orbit of a star at that distance would be 3.8×10^{10} AU. A speed of 230 km/s is 48 AU/yr, so the period of the star would be about 7.9×10^8 yr. If we apply Kepler's third law to the mutual revolution of such a star and M31, we find that the mass of the galaxy contained within 3×10^4 pc is about

$$\text{mass} = \frac{(6 \times 10^9)^3}{(7.9 \times 10^8)^2} = 3.5 \times 10^{11} \text{ solar masses,}$$

which is about the same as the mass of our own

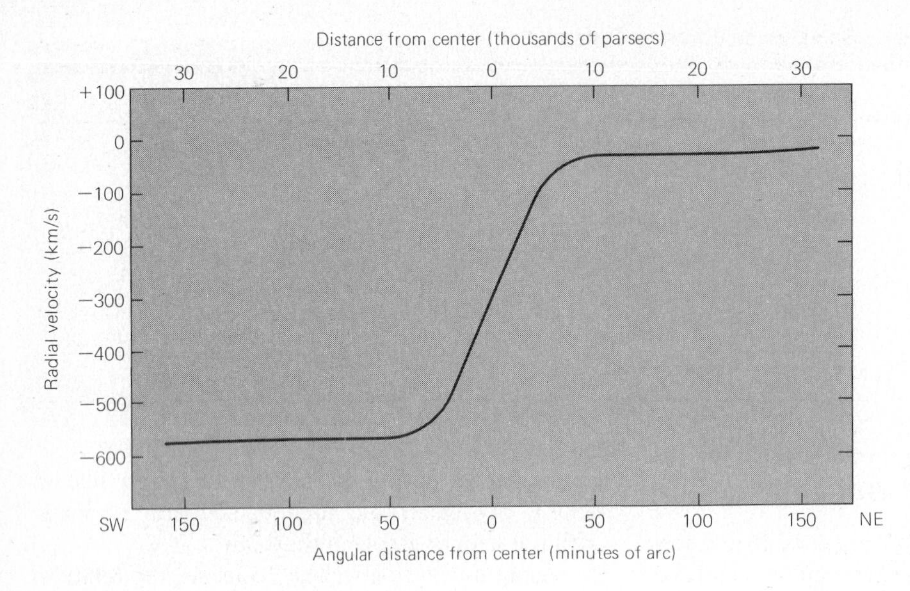

Figure 35.21 A simplified representation of the rotation curve of the Andromeda galaxy (M31).

Galaxy. The actual mass of M31 can be higher because we have not included the material that lies more than 3×10^4 pc from the center of the galaxy. Lower limits to the masses of many spiral galaxies have been found from their rotation curves by this procedure.

Elliptical galaxies are not highly flattened and are not in rapid rotation. Nevertheless, the velocities of the stars in such a galaxy depend on its gravitational attraction for them, and hence on its mass. The spectrum of a galaxy is a composite of the spectra of its many stars, whose different motions produce different Doppler shifts. The lines in the composite spectrum, therefore, are broadened, and the amount by which they are broadened indicates the range of speeds with which the stars are moving with respect to the center of mass of the galaxy. Application of the virial theorem, then, enables us to calculate the mass of the galaxy, just as we calculate the mass of a star cluster (Section 31.3a).

(b) Masses of Systems of Galaxies

Like stars, galaxies are often observed in close pairs. The periods of revolution, however, are typically hundreds of millions of years, so we do not "see" the motion of one galaxy about the other, as we can observe the mutual revolution of the members of a binary star system. Nevertheless, in analogy with studies of double stars, it has long been hoped that studies of binary galaxies could tell us something about galaxian masses. Unfortunately, there are a variety of problems with this approach. Two galaxies seen close together on the sky may be at quite different distances from us and may not be physically related. Even if two galaxies are gravitationally bound, their motions may be affected by other nearby galaxies. The observed radial velocities depend on the eccentricity of the galaxian orbits and on the inclination of the plane of the orbit to our line of sight. Accordingly, intrinsic uncertainties of factors of four or more remain in typical masses assigned to samples of binary galaxies.

The masses of *clusters* of galaxies can be calculated by the same technique used to "weigh" star clusters. The radial velocities of many galaxies in a cluster are first measured. The average of these velocities is that of the center of mass of the cluster, and the differences between the velocities of individual galaxies and this mean value tell us how fast they are moving within the cluster. With the help of the virial theorem (Section 31.3a), we can then calculate the gravitational potential energy of the cluster, and hence its mass.

(c) Mass-to-Light Ratio

An important datum is the ratio of the mass of a galaxy, in units of the solar mass, to its light output, in units of the solar luminosity. For the sun, of course, this ratio would be unity. Galaxies, however, are not composed of stars that are all identical to the sun. The overwhelming majority of stars are less luminous than the sun, and usually these stars

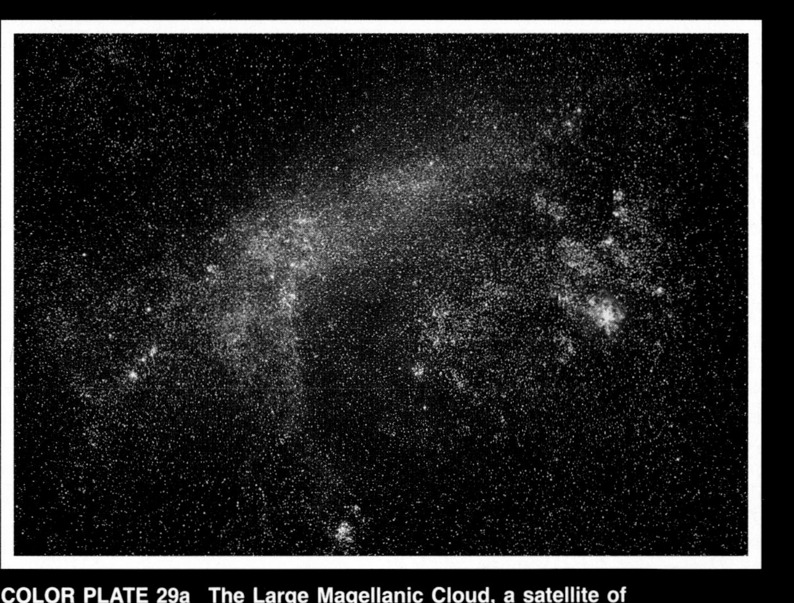

COLOR PLATE 29a The Large Magellanic Cloud, a satellite of our own Galaxy, is visible to the naked eye from the Southern Hemisphere. Because of its proximity, astronomers are able to study the Magellanic Clouds in detail, thus obtaining insights into processes within our own Galaxy. The large gaseous nebula is 30 Doradus, the Tarantula Nebula. The CCD image was made at the Cerro Tololo Inter-American Observatory/University of Michigan Schmidt telescope. (National Optical Astronomy Observatories)

COLOR PLATE 29b M 32. A small Local Group elliptical galaxy in the constellation Andromeda. (U.S. Naval Observatory)

COLOR PLATE 29c The spiral galaxy M 31 in Andromeda, photographed with the Palomar Schmidt telescope. (California Institute of Technology/Palomar Observatory)

COLOR PLATE 30a NGC 4535. A type S(B)c spiral galaxy with possible barred structure. (National Optical Astronomy Observatories)

COLOR PLATE 30b M 83. Type Sc spiral galaxy. The spiral has two principal arms and a third, fainter one. M 83 is 10 million LY away and has a diameter of 30,000 LY. (Cerro Tololo Inter-American Observatory photograph, courtesy National Optical Astronomy Observatories)

COLOR PLATE 30c Map of the observed velocity of galaxies plotted as a function of right ascension for a narrow range of declination (26.5° to 32.5°). Right ascension ranges from 17 hours at the left to 8 hours at the right; radial velocities range from zero at the apex of the pie-shaped wedge to 15,000 km/s at the rim. Note that galaxies appear to be distributed in elongated structures that surround empty regions where there are almost no galaxies. These empty regions have typical diameters of about 30 million parsecs. (Courtesy V. Lapparent, M. Geller, and J. Huchra, Harvard Smithsonian Center for Astrophysics)

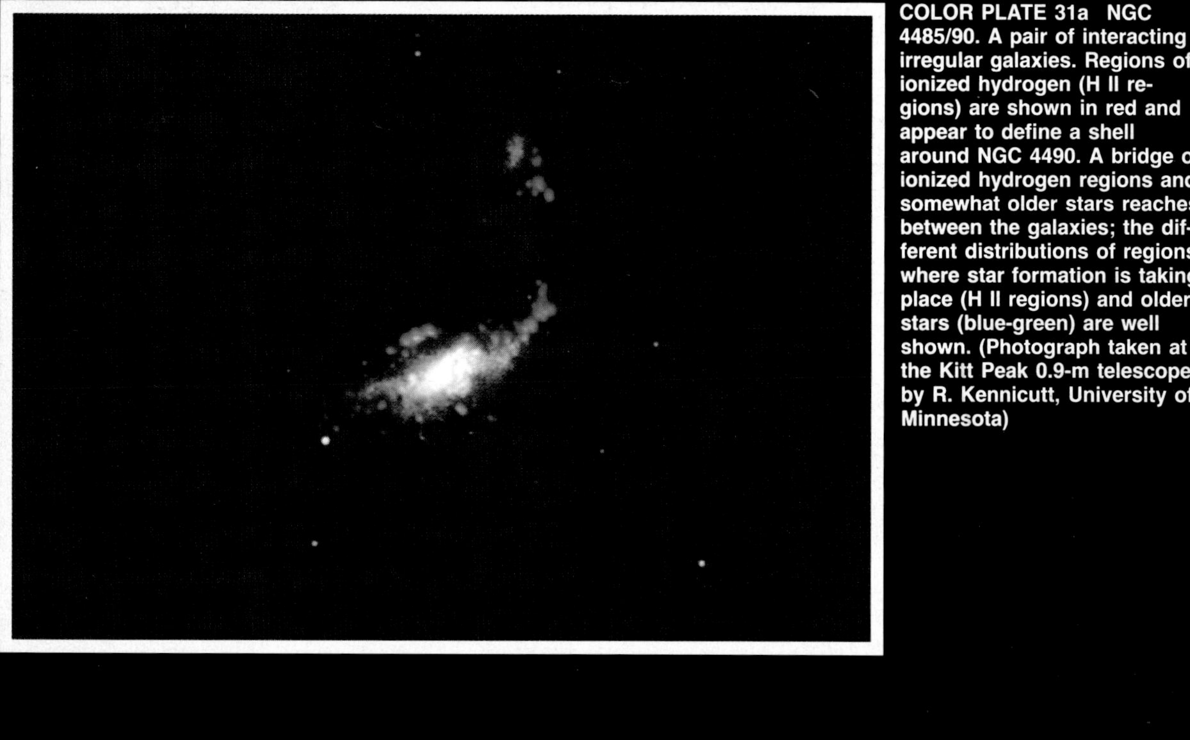

COLOR PLATE 31a NGC 4485/90. A pair of interacting irregular galaxies. Regions of ionized hydrogen (H II regions) are shown in red and appear to define a shell around NGC 4490. A bridge of ionized hydrogen regions and somewhat older stars reaches between the galaxies; the different distributions of regions where star formation is taking place (H II regions) and older stars (blue-green) are well shown. (Photograph taken at the Kitt Peak 0.9-m telescope by R. Kennicutt, University of Minnesota)

COLOR PLATE 31b NGC 274/5. A spiral and an elliptical in a close pair. (Hydrogen emission is shown in red, while starlight is blue-green.) The spiral has very active star formation, presumably induced by the stress of interaction. The elliptical, by contrast, is almost completely undisturbed, with very little evidence of hydrogen emission or recent star formation. This probably reflects the very small amount of gas available for star formation in the elliptical. (Images obtained at the 2.1-m telescope at Kitt Peak, by W. Keel and R. Kennicutt)

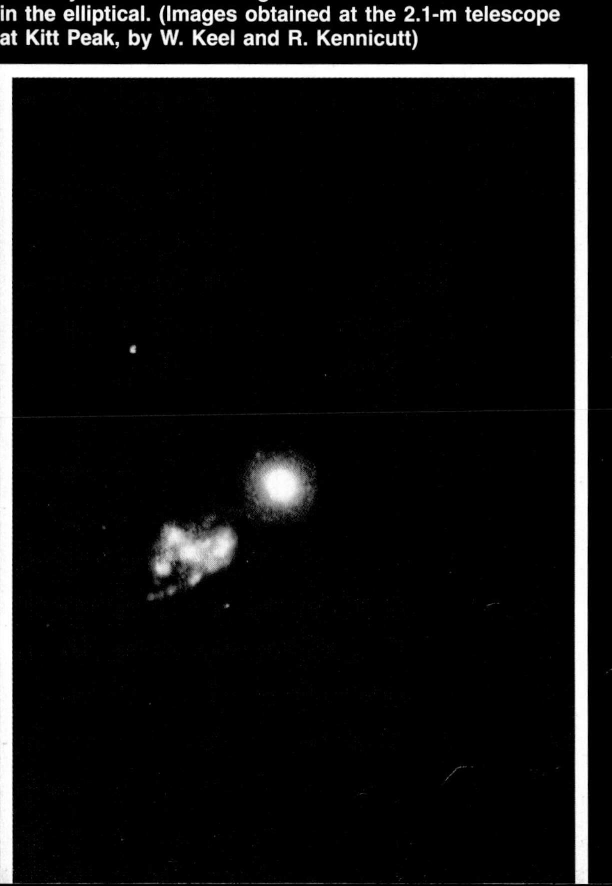

COLOR PLATE 31c NGC 4676AB (the Mice). A classic system of colliding galaxies which have produced narrow tidal tails as a consequence of the interaction. Star formation is present in the tails. This is a pseudocolor image, in which different colors correspond to different intensities. (Image taken with a CCD at the 2.1-m telescope at Kitt Peak by W. Keel and R. Kennicutt, courtesy National Optical Astronomy Observatories)

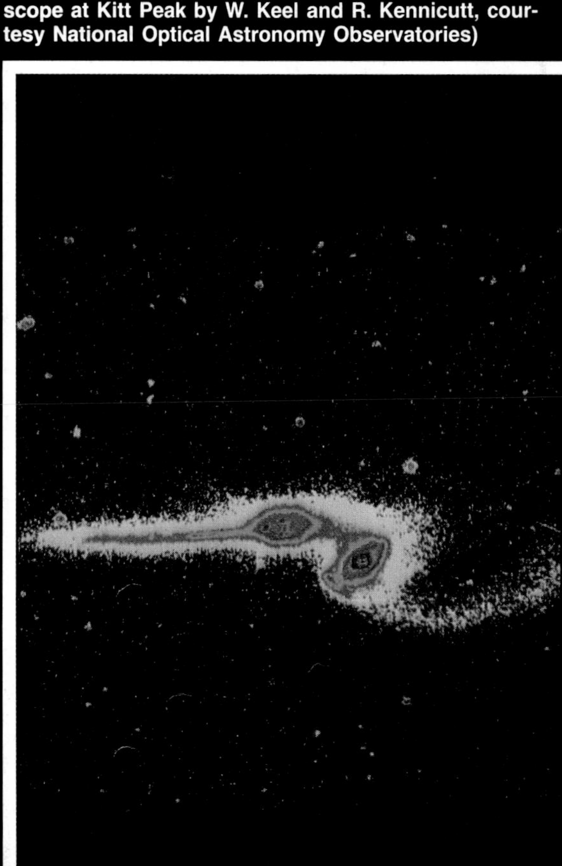

COLOR PLATE 32a ESO 2330–38. The blue streamer appears to be the remnant of a galaxy like the Small Magellanic Cloud, which has been disrupted by a passage close to the main galaxy. The encounter was nearly perpendicular to the plane of the spiral, and much of the material of the small galaxy will remain in nearly polar orbits, forming a "spindle" galaxy. (Image taken at Cerro Tololo Inter-American Observatory with a CCD by W. Keel)

COLOR PLATE 32b NGC 5544/5. A pair of galaxies that show no internal signs of interaction. Although they appear in nearly the same direction, they are probably at very different distances from the sun. The Sc galaxy, seen more edge-on, shows H II regions (red) in the arms, while its S0/a companion shows very smooth structure with little or no star-forming activity. (Image obtained by W. Keel and R. Kennicutt with a CCD at the 2.1-m telescope at Kitt Peak, Courtesy National Optical Astronomy Observatories)

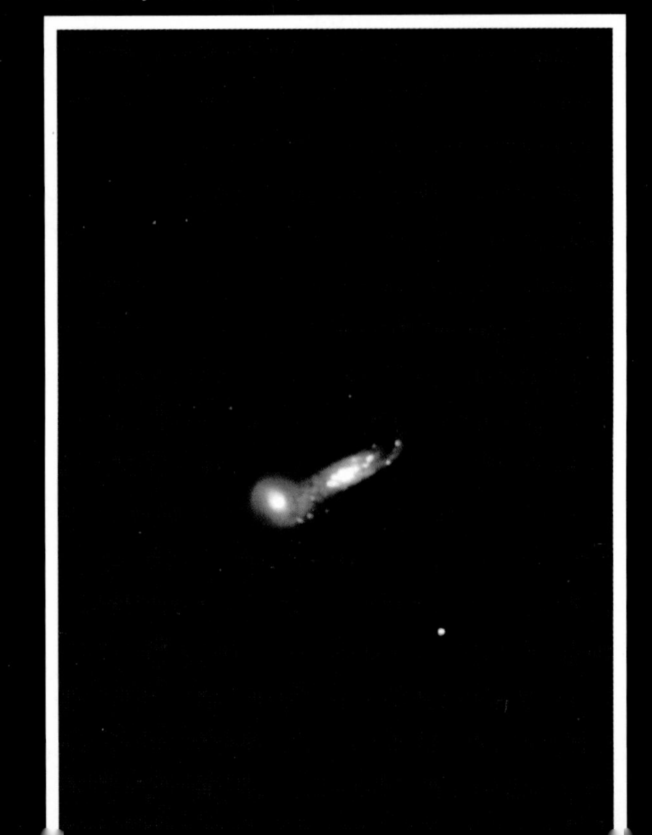

COLOR PLATE 32c A pseudocolor image of NGC 5754 and 5755, a peculiar interacting galaxy pair. The different colors correspond to different intensities of radiation. (National Optical Astronomy Observatories)

contribute most to the mass of a system, without accounting for very much light; thus, the mass-light ratio is generally greater than one. Only in systems, or regions within systems, where there are many young, highly luminous stars, can the ratio be less than one. Galaxies in which star formation is still occurring tend to have mass-light ratios in the range one to ten, while in galaxies consisting mostly of an older stellar population, the ratio is 10 to 20.

But the above comments refer only to the inner, more conspicuous parts of galaxies; there is growing evidence that galaxies are surrounded by massive halos, which have very high mass-light ratios. Rotation curves for spiral galaxies, for example, are usually more or less like that for M31 (Figure 35.21), which rises to a maximum velocity and then flattens. Furthermore, from observations of the 21-cm line of atomic hydrogen, we know that the rotation curves remain flat even beyond the point where the visible radiation begins to drop off. The only way that the rotational velocity can remain high is if the visible matter is supplemented by invisible matter, that is, by matter that does not emit detectable radiation. This dark matter still does, of course, exert gravitational force, and this force is what keeps the material in the outer portion of the galaxy, with its high rotational velocity and correspondingly large centripetal force, from flying off into space (see Section 34.3g for a more detailed explanation). Detailed measurements by Vera Rubin of the Carnegie Institution of Washington, Albert Bosma of the Marseilles Observatory, and other astronomers indicate that in the visible portions of spiral galaxies the total amount of matter, light plus dark, exceeds the luminous mass by about a factor of two.

Statistical analyses of galaxies in clusters lead to still larger mass-light ratios, providing further evidence for substantial amounts of dark matter. Masses derived from the virial theorem for clusters of galaxies, when compared with the light from the galaxies in the clusters, sometimes yield mass-light ratios as high as 200 or 300. In typical clusters, then, there is 5 to 15 times more dark matter than luminous matter. We shall see (Chapter 36) that clusters often contain hot intergalactic gas, which may contribute up to half of the cluster mass, but intracluster gas cannot account entirely for the high mass-light ratios.

The outer parts of at least some galaxies, therefore, and perhaps multiple systems of galaxies as units, evidently contain matter that has not yet been identified by its light or by means other than its gravitational influence. The anomaly is sometimes called the *missing mass* problem. In fact, the mass is there, it is the light that is missing.

What is this dark matter? Extensive searches have been made for both hot and cold gas, and neither is present in sufficient quantity to account for the dark matter. The most likely explanation seems to be that the dark matter in galaxies is composed of massive collapsed objects (e.g., black holes), stellar-like objects that are not massive enough to burn hydrogen (brown dwarfs), or massive neutrinos or exotic subnuclear particles (see Chapter 37).

Whatever the composition of the dark matter, it is startling to realize that probably 90 percent or more of all the material in the universe cannot be observed directly in any part of the electromagnetic spectrum. The light that we see from galaxies does not trace the bulk of material that is present in space, and so may give us a very misleading picture of the large-scale structure of the universe. An understanding of the properties and distribution of this invisible matter is crucial, however. Through the gravitational force that it exerts, dark matter probably plays a dominant role in the formation of galaxies. As we shall see in Chapter 37, it may also determine the ultimate fate of the universe.

35.5 GALAXY ENCOUNTERS AND COLLISIONS

In the existing catalogues of peculiar galaxies are many examples of strange-appearing pairs of galaxies interacting with each other. We can now understand many of these in terms of gravitational tidal effects. The effects of tides between pairs of galaxies that chance to pass close to each other at a low relative velocity have been studied by Alar and Juri Toomre. They point out three fundamental properties of tidal interactions: (1) The tidal force is proportional to the inverse cube of the separation of the galaxies (see Section 5.4). (2) Tidal forces on an object tend to elongate it; thus there are tidal bulges on both the near and far sides of each galaxy with respect to the other. (3) The perturbed galaxies are generally rotating before the tidal encounter, and the subsequent distributions of their material must therefore reflect the conservation of their angular momenta.

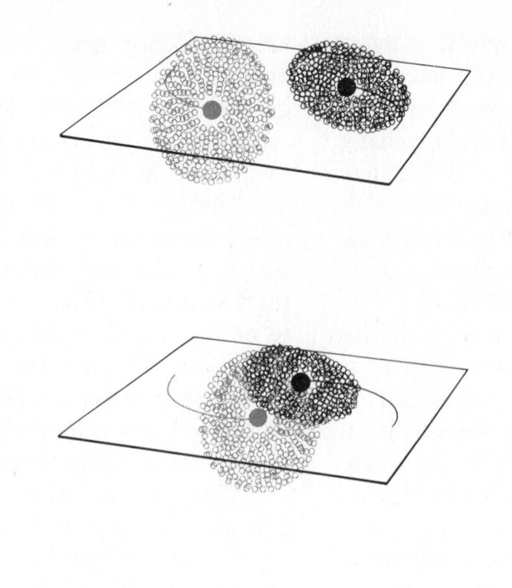

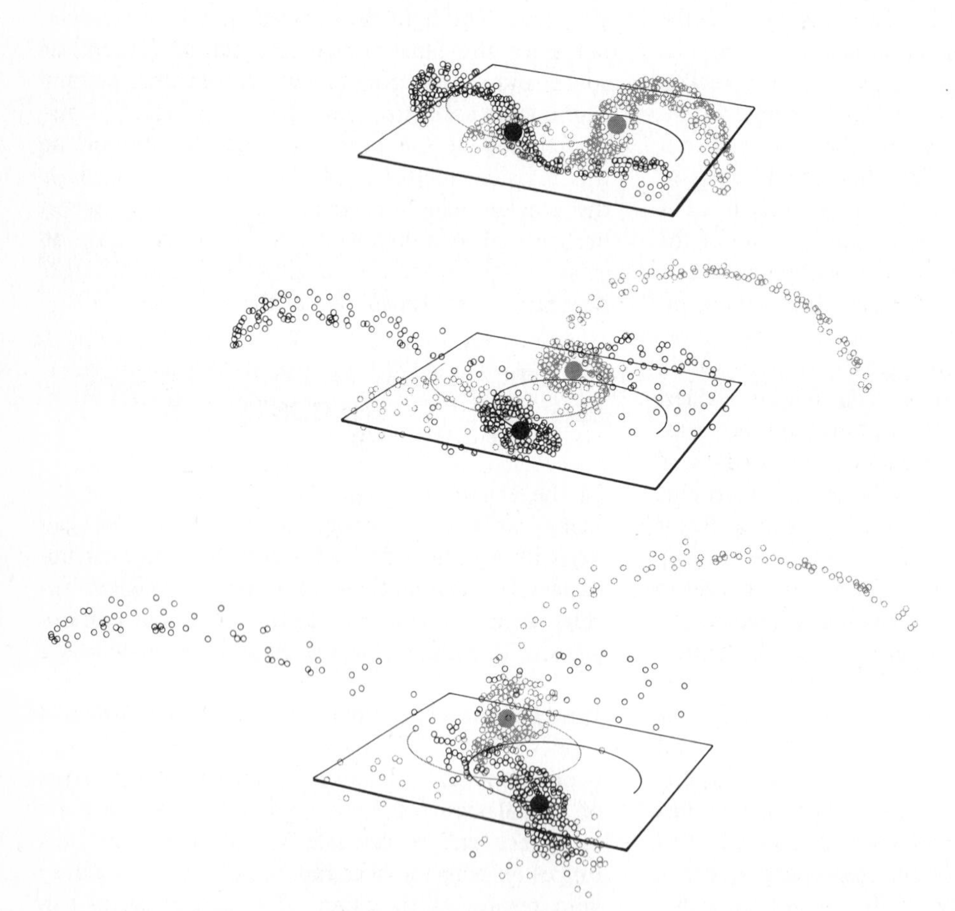

Figure 35.22 A sequence of five frames from a computer-produced motion picture that simulates the tidal distortion of two interacting galaxies. In this computer run, the initial conditions were chosen to see if the strange appearance of the pair of galaxies NGC 4038 and NGC 4039 could be accounted for in terms of tidal effects. Compare the last two frames with the photograph of the galaxies in Figure 35.23. (Courtesy Alar Toomre, MIT)

At first we might expect a tidal interaction between two galaxies to pull matter out of each toward the other. Such bridges of matter may form between the galaxies, but also there are "tails" of material that string out away from each galaxy in a direction opposite to that of the other. Because of the rotation of the galaxies, the tails and bridges can take on unusual shapes, especially when account is

Figure 35.23 The interacting pair of galaxies NGC 4038 and 4039. (Palomar Observatory, California Institute of Technology)

taken of the fact that the orbital motions of the galaxies can lie in a plane at any angle to our line of sight. The Toomre brothers have been able to calculate models of interacting galaxies that mimic the appearances of a number of strange-looking pairs actually seen in the sky (Figures 35.22 and 35.23). Color Plates 31a to 32c show examples of interacting galaxies.

(a) Galactic Mergers and Cannibalism

If galaxies collide with slow enough relative speed, they may avoid the usual tidal disruption. Calculations show that some parts of slowly colliding galaxies can be ejected, while the main masses become binary (or multiple) systems with small orbits about each other. Such a newly formed binary galaxy, surrounded by a mutual envelope of stars and possibly interstellar matter, may eventually coalesce into a single large galaxy. This process is especially likely in the collisions of the most massive members of a cluster of galaxies, which tend to have the lowest relative speeds and to be concentrated toward the center. Mergers may convert spirals to ellipticals.

Other processes within clusters can also affect the morphology and evolution of galaxies. While we use the term *merger* to refer to the interaction of two galaxies of comparable size, the swallowing of a small galaxy by one that is much larger is described as galactic cannibalism. Two mechanisms are relevant. The first is *tidal stripping*. If a small galaxy approaches a large one too closely, then its self-

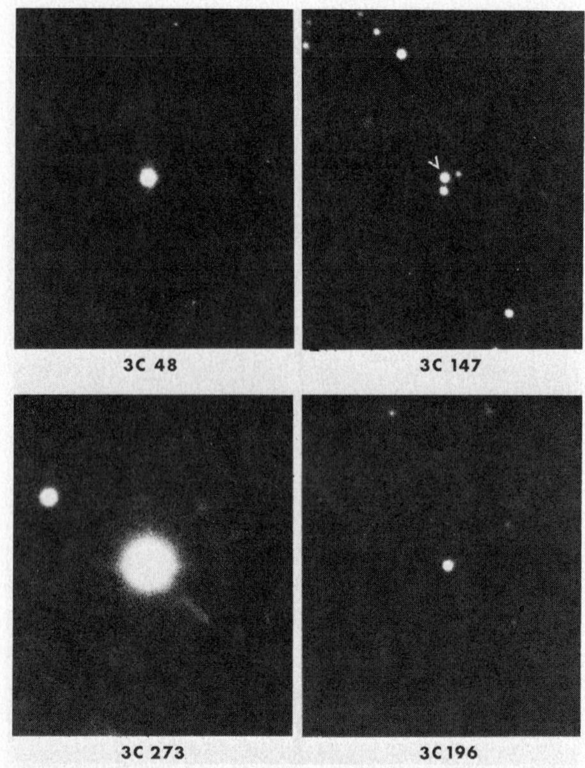

Figure 35.24 Quasi-stellar radio sources photographed with the 5-m telescope. (Palomar Observatory, California Institute of Technology)

gravity may be inadequate to retain the stars and gas in its outer regions. The tidal forces of the larger galaxy will dominate and will rip stars away from the lower mass galaxy. The physics is the same as that discussed in Section 5.4f, which considered what happens to a small satellite in the vicinity of a large planet.

A large galaxy can swallow or cannibalize the dense core of a smaller galaxy through a second mechanism, which is referred to as *dynamical friction*. The basic idea is that if the core of the smaller galaxy is moving rapidly through the envelope of stars of the larger galaxy, it will lose energy and decelerate while the stars in the larger galaxy will accelerate. This process causes the smaller galaxy to slow and spiral into the massive one. Rich clusters are often observed to have one or more supergiant galaxies (of type cD—see Section 35.3e) near its center. It is likely that these galaxies were formed by galactic cannibalism, that is, by the swallowing of cores of smaller galaxies, slowed down by dynamical friction, or of their tidally stripped envelopes.

35.6 QUASARS

If the sun were typical among stars as a radio emitter, we would not expect to observe strong radio emission from stars. It was with considerable surprise, therefore, that in 1960 two radio sources were identified with what appeared to be stars. There seemed to be no chance that the identifications were in error, because the precise positions of the radio sources were pinned down by noting the exact instants they were occulted by the Moon. By 1963 the number of such "radio stars" had increased to four (Figure 35.24). They were especially perplexing objects because their optical spectra showed emission lines that at first could not be identified with known chemical elements.

The breakthrough came in 1963 when M. Schmidt, at Caltech's Palomar Observatory, recognized the emission lines in one of the objects to be the Balmer lines of hydrogen (Section 8.3) shifted far to the red from their normal wavelengths. If the redshift is a Doppler shift, the object must be receding from us at about 15 percent the speed of light! With this hint, the emission lines in the other objects were re-examined to see if they too might be well-known lines with large redshifts. Such proved, indeed, to be the case, but the other objects were found to be receding from us at even greater speeds. Evidently, they could not be neighboring stars; their stellar appearance must be due to the fact that they are very distant. They are called, therefore, *quasi-stellar radio sources*, or simply *quasi-stellar objects* (abbreviated QSO). Later, similar objects were found, which are *not* sources of strong radio emission. Today they are all designated by the term *quasar*, and about 90 percent of all known quasars are *not* radio sources. Some astronomers think that radio-emitting quasars are a temporary phase in the evolution of quasars.

By 1980, hundreds of quasars had been catalogued, and systematic surveys indicate that there must be more than 20,000 brighter than the 18th magnitude. The number of still fainter—and presumably more distant—quasars is not known, but there are certainly very many. All have spectra that show large to very large redshifts. In a few cases, the relative shifts of wavelength $\Delta\lambda/\lambda$ exceed 3.5, and for the majority, $\Delta\lambda/\lambda$ is greater than 1.0. If we apply the exact formula for the Doppler shift (Section 8.3b), we find that $\Delta\lambda/\lambda = 4.0$, the largest redshift measured as of this writing (August 1986), corresponds to a velocity of more than 92 percent the speed of light.

Quasars thus have much higher speeds than any known galaxy and must, *if* they follow the same relationship between velocity and distance that characterizes normal galaxies (the Hubble law—Section 37.1), be even more distant. We will examine the arguments pro and con about whether or not the quasars conform to the Hubble law after we look in detail at the properties of quasars.

(a) Characteristics of Quasars

When quasars were first discovered, they all seemed to be unresolved optically—that is, they appear stellar, and most of them as very faint stars, at that. One of the brightest, 3C 273, is several hundred times too faint to see with the unaided eye. A few quasars, though, proved to be associated with tiny wisps or filaments of nebulous-appearing matter. Some are resolved at radio wavelengths, which indicates that the radio energy (at least for some) comes from regions outside the visible photographic images. The radio radiation is believed to be synchrotron.

Although they differ considerably from each other in luminosity, the quasars are nevertheless extremely luminous at all wavelengths. In visible light most are far more luminous than the brightest elliptical galaxies—their absolute magnitudes are as bright as -25 or -26, a luminous energy equal to more than 10^{12} suns. Despite their enormous redshifts, they are very blue in color. The reason is that quasars are much brighter in the ultraviolet region of the spectrum than are normal galaxies. Their redshifts are large enough to shift visible light to the infrared, and ultraviolet light into the visible part of the spectrum, where it can be observed with ground-based telescopes. The excess ultraviolet radiation that becomes observable in this way makes quasars look blue relative to normal galaxies.

Most surprising of all is that almost all quasars are variable, both in radio emission and in visible light. Their variation is irregular, evidently at random, by a few tenths of a magnitude or so, but sometimes flare-ups of more than a magnitude are observed in an interval of a few weeks. Since quasars are highly luminous, a change in brightness by a magnitude (a factor of 2.5 in light) means an extremely large amount of energy is released rather suddenly. Moreover, because the fluctuations occur in such short times, the part of a quasar responsible for the light (and radio) variations must be smaller than the distance light travels in a month or so; otherwise light emitted at one time from different parts of the object would reach Earth at different times (because of the range of distances light would have to travel to reach us), and the increase in brightness would seem to us to last for more than the few weeks that is typically observed.

(b) The Radiation from Quasars

The spectra of quasars have several components. First, there is a continuous spectrum, ranging all the way from radio waves to X-rays (and probably gamma rays as well). The relative intensities in various regions of the spectrum vary from quasar to quasar; some are too weak in radio radiation, for example, to be observed. Data from the Einstein X-ray telescope, however, show that all or nearly all quasars are X-ray emitters.

Some quasars emit radio radiation from their central regions, but in most cases the radio source is double, with the radiation coming from extended regions on opposite sides of the quasar. The radiation is synchrotron emission. So, evidently, is that in the visible and X-ray spectral regions. Some quasars, however, have excessive infrared radiation—more than would be expected from relativistic electrons accelerated in magnetic fields so as to produce the radiation observed in other wavelengths. The excessive infrared, when present, is thought to be caused by dust re-emitting radiation that is absorbed at shorter wavelengths.

Quasars also have emission lines (with which we measure their redshifts). These must originate from ionized gas at not so high a temperature that its atoms cannot capture electrons and emit light (Section 8.3). Moreover, these lines are all broad, suggesting that many emitting clouds have a large range of velocities in our line of sight. The best guess is that the emission is from many clouds ejected from, or simply moving at relatively high speed around, a central object.

Finally, quasars often have absorption lines. Sometimes there are two or more different sets of absorption lines displaying different redshifts. The

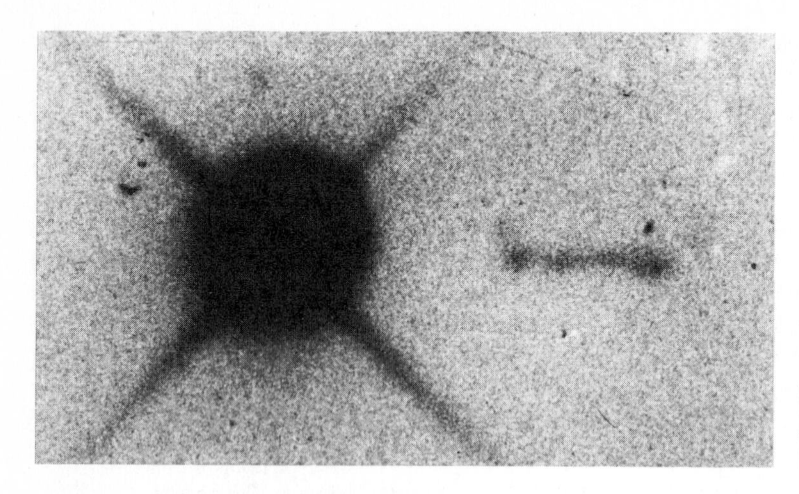

Figure 35.25 The jet associated with the quasar 3C 273. The quasar is the bright star-like image with diffraction spikes. The jet is the not quite straight horizontal feature to the right of the quasar. (Courtesy Gerard Lelièvre, Canada-France-Hawaii Telescope Corporation)

absorption-line redshifts, however, are always less than those of the emission lines. The currently accepted interpretation is that the absorption lines are formed in gas clouds that lie between us and the quasar and move at velocities of recession lower than that of the (sometimes much) more distant quasar. These clouds are believed to be material previously ejected by the quasar (cooled-off gas that formerly showed emission lines?), interstellar gas in intervening unseen galaxies, or possibly intergalactic gas that was never incorporated into a galaxy, through which the quasar's light must pass to reach us.

(c) An Energy Powerhouse

What is the source of energy that powers the quasars? It must be remarkable indeed, because:

1. It must provide power up to 10^{47} ergs/s—equivalent to nearly 10^{14} times the luminosity of the sun.
2. It must, in some cases, account for variations in the total radiated power by as much as a factor of two or more, and over time scales of years or months or, in some cases, only days.
3. In at least the objects that vary in luminosity, the powerhouse must be compact enough that light can travel across it in a time less than that of its variations, that is, in a few light days or light weeks. For comparison, this distance is much smaller than that between the sun and the nearest star.
4. It must be able to eject relativistic electrons in directed jets (Figure 35.25) and in sufficient numbers to provide synchrotron radiation as intense as the total visible energy emitted by a bright galaxy.
5. It must possess a powerful magnetic field, or there must be a very strong field in the material surrounding the central engine that powers the quasar, and the energy in these magnetic fields must be comparable to the total nuclear energy available in all of the stars in a large galaxy. It is the spiraling of electrons around magnetic lines of force that gives rise to the synchrotron emission.

It was the great difficulty of devising a physical model to explain all of these characteristics that led some astronomers to suggest that the redshifts of the quasars were not cosmological in origin. (Redshifts are described as cosmological if they conform to the Hubble law, which relates redshifts and distances and attributes the redshifts to the Doppler effect.) These astronomers argued that rather than being a consequence of the expansion of the universe, the redshift of the spectral lines in quasars was produced by some physical mechanism that Earth-bound scientists had not previously observed. If this hypothesis were correct, then the measured redshift could not be used to estimate the distances of quasars. Since there were at the time no alternative methods for estimating distances, the quasars could be assumed to be close enough to us so that their energy output was within the range of that associated with normal galaxies.

As support for this point of view, some astronomers, most notably Halton Arp of the Mount Wilson and Las Campañas Observatories, have sought

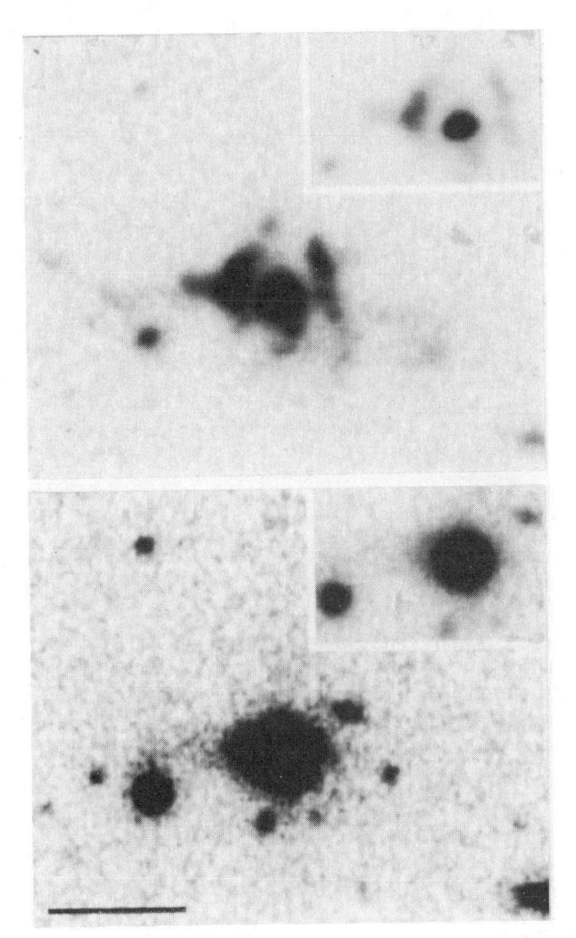

Figure 35.26 A quasar (4C 37.43) with a redshift of 0.371 embedded in an extended region that has emission lines characteristic of hot gas. The upper image was taken with a filter chosen to transmit only the redshifted forbidden emission line of O III at 5007 Ångstroms. The lower photograph was taken in a wavelength region where the quasar has no emission lines. There is a compact galaxy to the east *(left)* and slightly south of the quasar, and it has approximately the same redshift as the quasar. The compact galaxy is connected to the quasar and its host galaxy by a bridge, which can be seen in the lower photograph. The existence of this bridge shows that the compact galaxy and the host galaxy of the quasar are interacting. The smaller inset photographs show the central portion of the larger photographs but at higher contrast. Images were taken with the Canada-France-Hawaii Telescope. (Courtesy Alan Stockton, University of Hawaii)

evidence for physical associations between high-redshift quasars and low-redshift normal galaxies. If two objects are physically associated, they must be at the same distance. If they also have very different redshifts, then redshifts must—sometimes, at least—not be a reliable indicator of distance. Logic therefore demands that there must be some physical process other than the Doppler effect that can produce very large redshifts. There are indeed many cases in which quasars appear close to galaxies on the sky. It is always possible, however, that these are chance superpositions of two objects that are really at very different distances. There remain too few examples of an apparent association between a quasar and a galaxy with discordant redshifts to convince most astronomers that the redshifts of quasars are not cosmological in origin.

In fact, several astronomers have turned this argument around and have searched for clusters of galaxies in the near vicinity of quasars. This task is not easy observationally because normal galaxies are fainter than quasars and are therefore more difficult to detect. Nevertheless, studies to date show that quasars are often surrounded on the sky by small clusters of galaxies, and the cluster galaxies exhibit the same redshift as the quasar. It is highly improbable that the apparent velocities of quasar and galaxy would coincide unless the two objects were physically associated and at the same distance.

There have been several other developments, both observational and theoretical, that support the hypothesis that quasars are at the distance indicated by their redshifts and that their prodigious output of energy can indeed be accounted for in terms of physical processes that we are familiar with. One key observation is the discovery that many relatively nearby quasars ($\Delta\lambda/\lambda = 0.5$) are not true point sources but rather are embedded in a faint, fuzzy-looking patch of light. The colors of this fuzz are like those of spiral galaxies. In a few cases, spectra have been obtained and indicate that the light of the fuzz is derived from stars, again demonstrating that quasars are located in galaxies (Figures 35.26 and 35.27).

(d) Violent Activity in Galaxies

When quasars were first discovered, it was thought that they were much more luminous than galaxies. They are indeed more luminous than normal galaxies, but we have now found bona fide galaxies—albeit peculiar ones—that fill in the luminosity gap. These peculiar galaxies share many of the properties of the quasars, although to a less spectacular degree (Figure 35.28). For example, the nuclei of Seyfert galaxies exhibit strong, broad emission lines. Like

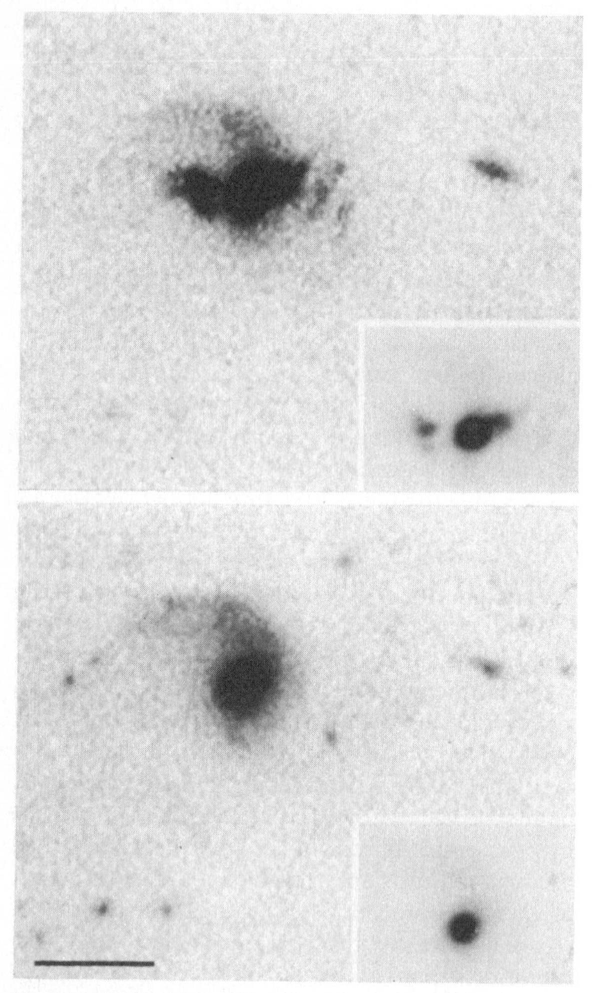

Figure 35.27 Photographs taken in the same way as those in Figure 35.26, but of Markarian 1014, a quasar with a redshift of 0.163. The galaxy associated with this quasar is one of the most luminous known. The emission from the spiral-like feature to the north is starlight. (The appearance of this feature in the upper image, which was taken with a filter designed to transmit emission from O III is due entirely to continuum radiation coming through the filter.) There are two conspicuous clumps of ionized gas to the east and west of the nucleus of the galaxy in which the quasar is embedded. Images were taken with the Canada-France-Hawaii Telescope. (Courtesy Alan Stockton, University of Hawaii)

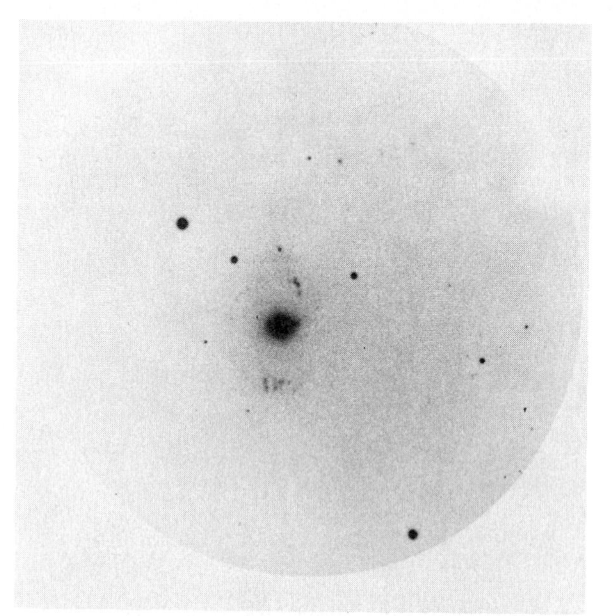

Figure 35.28 The Seyfert galaxy NGC 4151 (negative print). Although the exposure is so short that the spiral structure of the galaxy is barely visible, the brilliant nucleus is already burned out. Photographed with the Lick Observatory 3-m telescope. (Courtesy H. Ford, UCLA)

quasars, these galaxies apparently contain hot gas in a small central region, and the width of the lines indicates that the gas is moving at speeds up to thousands of kilometers per second. Some Seyferts are radio and/or X-ray sources, and all emit strongly in the infrared. Some show brightness variations over a period of a few months, and so, again like quasars, the region from which the radiation comes can be no more than a few light months across. The visual luminosities of Seyferts are about normal for spiral galaxies, but when account is taken of their infrared emission, their total luminosities are found to be about 100 or so times normal.

The crucial point about Seyferts and other active galaxies is that a significant fraction of their power output comes from a source other than individual stars.

The Seyfert properties can be recognized easily only in relatively nearby galaxies. It is quite possible that 1 or 2 percent of all spiral galaxies have these active nuclei. Alternatively, it is possible that all spiral galaxies (even our own?) have these properties 1 or 2 percent of the time.

M87 is another interesting galaxy that is a strong radio source. Short exposures of it show a luminous jet directed away from its nucleus, and a faint hint of a second radial jet in the opposite direction (Figure 35.29). Both the nucleus and the brighter jet emit synchrotron radiation indicating magnetic fields and a source of relativistic electrons. M87 and its jet are also strong sources of X-rays,

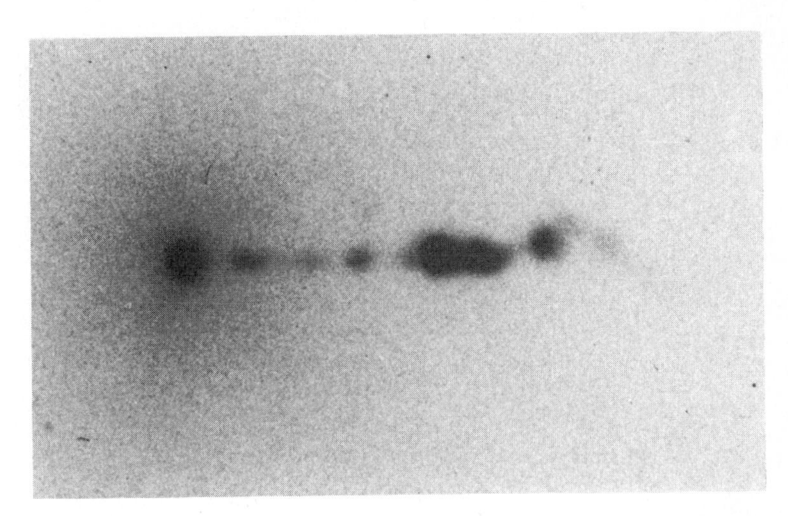

Figure 35.29 A very short exposure of the elliptical galaxy M87 showing the jet that emanates from it. The jet is not continuous but consists of several knots of hot gas. (Courtesy Gerard Lelièvre, Canada-France-Hawaii Telescope Corporation)

implying a hot gas throughout the entire galaxy and out into its halo. Finally, the optical spectrum of M87 shows very broad lines, indicating high velocities of the stars there, as though they were being accelerated by a very dense, massive core.

Intermediate between the quasars and such galaxies as M87 or the Seyfert galaxies is a class known as *N-type galaxies*. N galaxies have small nuclei that are very bright compared with the main parts of those galaxies; often they appear as stellar images superimposed on faint wispy or nebulous backgrounds. Their bright nuclei indicate that enormous amounts of energy are being emitted from those regions.

Objects of another class believed to be related to Seyfert galaxies and quasars are the *BL Lac* objects, named for the prototype, BL Lacertae. BL Lac is a stellar-appearing object that shows large irregular variations in luminosity, a fact that accounts for its variable-star designation. Like other BL Lac objects, it has no spectral lines, but it is a strong radio source, and its continuous radiation appears to be synchrotron.

In 1974, a spectrum was obtained at Palomar of the light passing through a ring-shaped aperture centered on BL Lac itself, but blocking out light from its central image. The source of faint light passing through the annular opening surrounding BL Lac proved to have a spectrum like that of a normal galaxy with a radial velocity of 21,000 km/s (or $\Delta\lambda/\lambda = 0.07$). BL Lac (and presumably other objects in its class) is evidently the brilliant nucleus of a distant galaxy.

Finally, it has been known since 1948 that many giant elliptical galaxies that appear comparatively normal in the optical region of the spectrum are powerful emitters of radio energy. Some, M87 being one example, emit thousands of times as much radio energy as is typical of bright galaxies. In some radio galaxies, the bulk of the radio emission comes from small regions within them, while in some others—the *core-halo* sources—there are bright sources in the nucleus of the galaxy surrounded by larger extended regions of radio emission. In about three-quarters of the radio galaxies, however, the radio source is double, with most of the radiation coming from extended regions on opposite sides of the galaxy. Typically, the two emitting regions are far larger than the galaxy itself and are centered a hundred thousand parsecs away from it.

In the radio galaxy 3C 276, so-called because of its listing in the Third Cambridge radio source catalog, the radio-emitting clouds extend to six million light years on either side of the galaxy. Often radio observations reveal two well-delineated jets of radio radiation pointing away from the galaxy to the large extended sources (Figure 35.30). These jets can be more than a million light years long.

Presumably, ionized gases are shot out along the jets into the radio "clouds" by an extremely intense source of energy in the nucleus of the galaxy. It is thought that the gas eventually collides with neutral gas in the space beyond (previously ejected material?) and slows to a stop, defining the sometimes rather sharp outer edges of the emitting re-

Figure 35.30 A schematic sketch of a double-source radio galaxy. The small spot in the middle represents the optical image of the galaxy, and the shaded regions on either side, the relative intensity of the radio radiation. The radiation is usually most intense at the outer limits of the radio lobes. Sometimes radio "jets" are observed, as shown in the sketch.

gions. Similar jets are seen in more than half of all quasars.

Another interesting structure displayed by some radio galaxies is that of the *head-tail* source. These are probably associated only with galaxies in clusters or at least in regions of space pervaded with intergalactic gas. In head-tail radio galaxies, the two opposing radio lobes fold back along themselves, so that the radio galaxy resembles a comet (Figure 35.31). We think this is a result of the drag of intergalactic gas through which the radio galaxy is moving.

(e) Black Holes—The Power Behind the Quasars?

The observations of quasars and of all the various types of galaxies that are unusually active emitters of optical, X-ray, and radio radiation suggest that what these objects have in common is a compact source of enormous energy, evidently buried in the nucleus of a galaxy. Many models have been offered to account for this energy source, including stellar collisions in dense galactic cores, supermassive stars, extraordinarily powerful supernovae, and others.

The most widely accepted model at the time that this is written is that quasars, and presumably other types of active galaxies as well, derive their energy output from an enormous *black hole* at the center of what would otherwise be a normal galaxy. The black hole must be very large—perhaps a billion solar masses. Given such a massive black hole, then relatively modest amounts of additional material—only about 10 solar masses per year—falling into the black hole would be adequate to produce as much energy as a thousand normal galaxies and could account for the total energy of a quasar. The idea is that the black hole would capture matter—stars, dust, and gas—swirling about in the dense nuclear regions of the galaxy. This material would spiral in towards the black hole and would form a disk of material around it. As the material spirals ever closer to the black hole, it accelerates and heats through compression to millions of degrees. This hot matter can radiate prodigious amounts of energy as it falls into the black hole.

A number of the phenomena that we observe can be explained naturally in terms of this model. First and foremost, it can produce the amount of energy that is actually observed to be emitted by quasars and active galactic nuclei. The black hole is also fairly compact in terms of its circumference and so the emission produced by infalling matter comes from a small volume of space. As we recall, this condition is required to explain the fact that quasars vary on a time scale of weeks to months.

The accretion disk around the black hole may be dense enough to prevent radiation from escaping in all but the two directions perpendicular to the

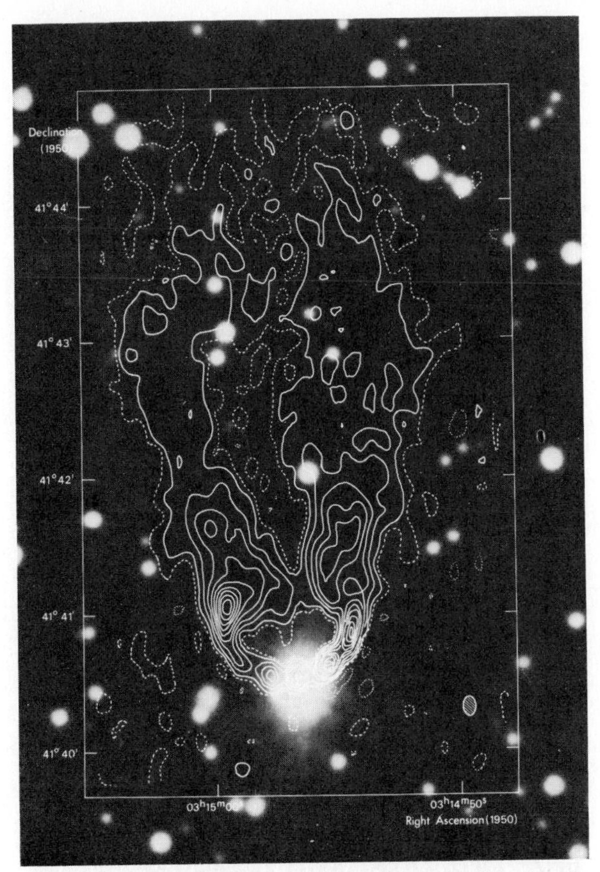

Figure 35.31 Contours showing the intensity of radio radiation at 5×10^9 Hertz of the head-tail radio source 3C83.1, in the Perseus cluster of galaxies. The map is superimposed on a print from the Palomar Sky Survey; the radio galaxy is the large image at the bottom. (Courtesy K. J. Wellington, G. K. Miley, and H. van der Laan; photograph copyright, National Geographic Society—Palomar Observatory Sky Survey).

disk (Figure 35.32), and so the disk may be responsible for the formation of the *jets* that are often associated with quasars and galaxies with active nuclei. The basic idea behind the formation of jets is that matter in the accretion disk will move inwards toward the black hole. Some of this matter will not actually fall into the black hole but will feed the jets. That is, some infalling matter will be accelerated by the intense radiation pressure in the vicinity of the black hole and will be blown out into space along the rotation axis of the black hole in a direction perpendicular to the plane of the accretion disk.

If matter in the accretion disk is continually being depleted by falling into the black hole or being blown out from the galaxy in the form of jets, then a quasar can continue to radiate only so long as there is gas available to replenish the accretion disk. Where does this matter come from? One possibility is that very dense star clusters form near the centers of galaxies and that these stars can supply the fuel, either through gas that is lost during the normal course of stellar evolution through stellar winds and supernovae explosions or because the tidal forces exerted by the black hole are strong enough to tear the stars apart. An alternate source of fuel may come from collisions of galaxies or from galactic cannibalism. That is, if two galaxies collide and merge, then gas and dust from one galaxy may come close enough to the black hole in the other to be devoured by it and so provide the necessary fuel. An example of an interaction between a galaxy and a quasar is shown in Figure 35.26.

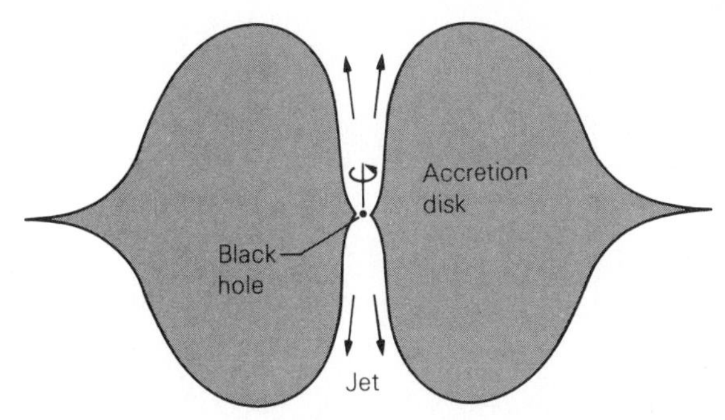

Figure 35.32 A schematic diagram of the accretion disk around a rotating black hole in the center of a galaxy. The material in the disk is responsible for channeling the outflow in the two directions perpendicular to the disk and thus producing jets.

Observations show that very bright quasars were much more common a few billion years ago than they are now. That is, there are many more quasars at great distances, where we are seeing the universe as it was several billion years back in time, than there are nearby. One possible explanation is that as quasars age they simply run out of fuel—that as time passes, all of the gas, dust, and stars available to fuel the black hole are consumed by it. Indeed, many of the relatively nearby, still-active quasars prove to be embedded in galaxies that have recently been involved in collisions with other galaxies. Gas and dust from this second galaxy has apparently been swept up by a dormant black hole and so has provided the new source of fuel required to rekindle it.

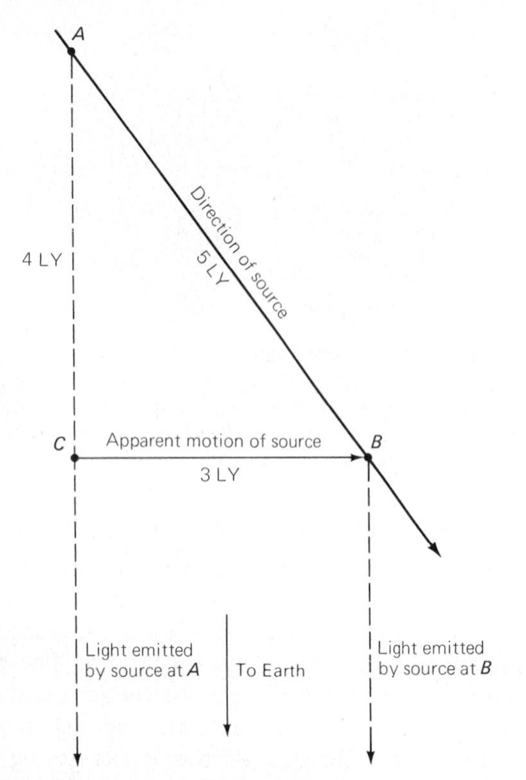

Figure 35.33 Explanation of the apparently superluminal velocities in quasars.

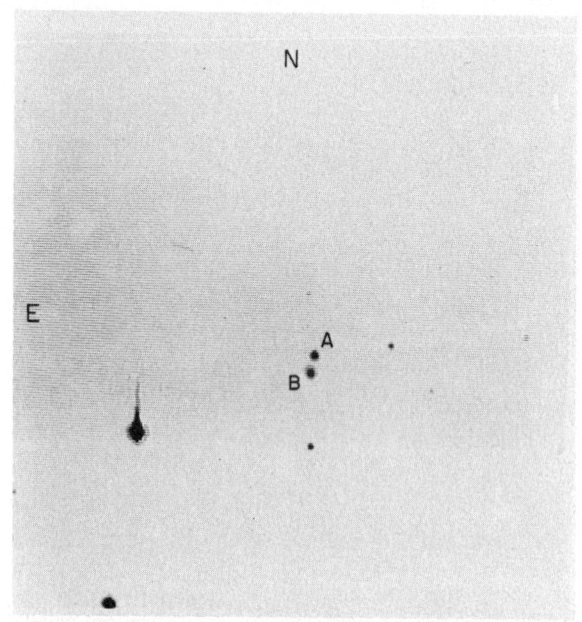

Figure 35.34 The double quasar 0957 + 561. The two components, believed to be different images of one quasar produced by a gravitational lens, are labeled *A* and *B*. (Courtesy Jerome Kristian, Mount Wilson and Las Campañas Observatories; Carnegie Institution of Washington)

This model suggests that there should be low-level quasar-like activity in some nearby galaxies. That is, black holes should exist in the nuclei of nearby galaxies but should be producing only low levels of activity either because they have relatively small masses or because they have already accreted nearly all of the matter available in their vicinity. Recent observations by A. Filippenko and W. Sargent at the California Institute of Technology have shown that emission lines like those seen in quasars and Seyferts but of much lower intensity are indeed seen in the nuclei of otherwise normal nearby galaxies. As we have already seen (Section 34.5), there is also some evidence that a black hole of at least modest size may lie hidden at the center of our own Milky Way Galaxy.

Of course, these ideas are still speculative. We know we have a small compact source of enormous energy at the heart of every quasar. It seems plausible to associate that source of energy with a giant black hole, but this subject is at the frontier of research in astrophysics. New observations and theories may modify the picture presented here substantially during the next few years.

35.7 QUASARS AND GENERAL RELATIVITY

(a) Superluminal Velocities in Quasars

Several quasars have been observed (about 50 by mid-1981) with very-long-baseline interferometry (VLBI—see Section 10.1), and some show structure. In at least six cases, small discrete sources in the quasars have been found that change position from one observation to the next. The motions are generally radially outward from the center of the quasar image. The velocities have been measured for these moving sources, and if the redshifts of the quasars are cosmological, some of the velocities of the moving sources are in the range of five to ten times the speed of light. Such velocities are, of course, impossible, and for a while this was taken as evidence that the quasars must be relatively nearby.

The high measured speeds, however, are illusory. If an object is moving almost toward the observer, it can appear to have a far greater speed than it really does, because of the finite speed of light. To see how this can be, imagine a quasar ejecting

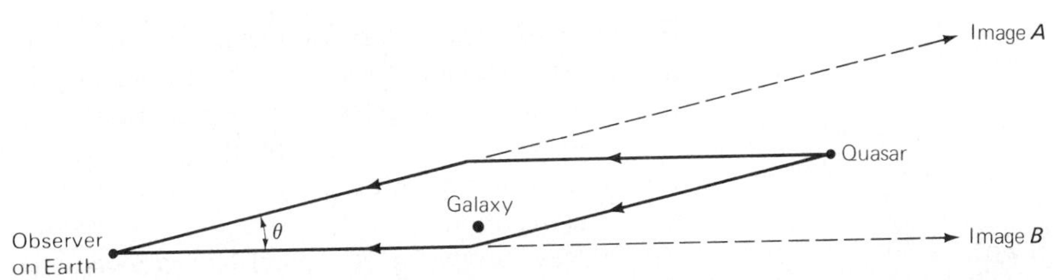

Figure 35.35 Gravitational lens associated with the double quasar 0957 + 561. Two light rays from the quasar are shown being bent in passing a foreground galaxy, and they arrive together at Earth. We thus see two images of the quasar, in directions *A* and *B*. This simple schematic does not reflect the subtle complexities produced by the finite size of the galaxy, by the cluster of which it is a member, nor of relativity theory itself, but it does illustrate how we can observe multiple images of one object. The angular separation, Θ, of the two images at Earth is greatly exaggerated, being only 6″ in reality.

small sources in various directions at just under the speed of light. One such source passes through point *A* in Figure 35.33. The direction toward Earth from the quasar is along the line $\overline{AC}$, but the source moves in a somewhat different direction, $\overline{AB}$. Let the source move 5 pc from *A* to *B*, and let *A* be 4 light years farther from us than *B* (that is, $\overline{AC}$ is 4 LY). Now the source, moving barely under the speed of light, takes barely over five years to reach *B*, while its radiation, emitted at *A*, reaches *C* in 4 years. That radiation continues on toward Earth, being one year ahead of the radiation emitted in our direction by the source when it is at *B*. Perhaps billions of years later the radiation reaches our radio telescopes, that from *A* still one year ahead of that from *B*. What we observe is the source apparently moving tangentially outward from the center of the quasar, from *C* to *B*, and in one year going a distance found from the theorem of Pythagoras to be $\sqrt{(5^2 - 4^2)} = 3$ LY. The source apparently had three times the speed of light, but only because of the projection effect and because the light took four years to go from *A* to *C* while the source went from *A* to *B* in five years. If the direction of motion were more nearly toward the Earth, the effect would be even more dramatic (see Exercise 17).

(b) The Double Quasar 0957 + 561: A Gravitational Lens?

In 1979, astronomers D. Walsh, R. F. Carswell, and R. J. Weymann of the University of Arizona noticed that a pair of quasars, separated by only 6″ and known collectively as 0957 + 561 (the numbers give their coordinates in the sky), are remarkably similar in appearance and spectra (Figure 35.34). They are both at about 17th magnitude, and both have a redshift ($\Delta\lambda/\lambda$) of 1.4. The astronomers suggested that the two quasars might actually be only one, and that we are seeing two images produced by an intervening object, acting as a gravitational lens (Section 33.4a).

We now know that there is an 18th-magnitude galaxy that lies in the same direction as one of the quasars. In fact, the galaxy turns out to be a member of a rich cluster of galaxies, which has a redshift of 0.39 and thus is much closer than the quasar. The geometry and estimated mass of the galaxy are correct to produce the gravitational lens effect, and there is now little question that 0957 + 561 is, indeed, a pair of images produced by a gravitational lens. A schematic of the lens is shown in Figure 35.35.

As this is written, there is fairly convincing evidence for seven gravitational lenses in all, and searches are under way to discover still more (Figure 35.36). The search is difficult. If theoretical calculations are correct, the light from only about one quasar in a thousand will pass close enough to a galaxy so that the galaxy can act as a gravitational lens and produce a double image of the background quasar.

Galaxies may not be the only gravitational lenses. As we have already seen (Section 35.4c), a large fraction of the material in the universe is not luminous. This dark matter may also act as a gravitational lens. If this dark matter is in the form of point-like objects (e.g., black holes), the separation of the two images of the background quasar will be too small to be detected with optical telescopes, al-

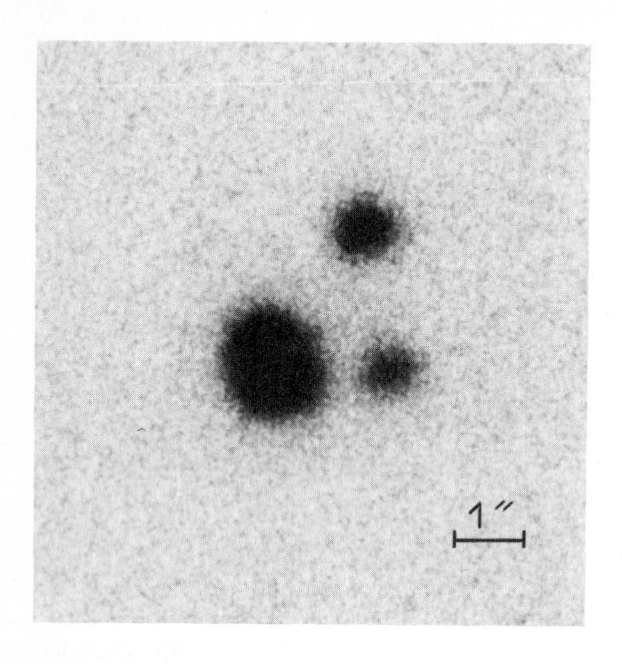

Figure 35.36 Gravitational lenses can produce very complex images. This photograph shows the triple quasar PG 1115 + 080. All three of the point-like objects have identical spectra and so are presumed to be different images of the same quasar produced by a gravitational lens. (Courtesy Gerard Lelièvre, Canada-France-Hawaii Telescope Corporation)

though if the black holes are massive (about a million times the mass of the Sun) the images may be resolvable with radio telescopes used as interferometers. Searches for such double images are under way and, if successful, may tell us what the dark matter is made of.

Images produced by point-like gravitational lenses can appear much brighter than the actual source would appear to be in the absence of lensing. Some of the brightest quasars and BL Lac objects may owe their apparently high luminosities to enhancement by gravitational lensing.

EXERCISES

1. Why is the term "island universe" a misnomer?
2. In a hypothetical galaxy, a cepheid variable is observed, which, at median light, is at magnitude +15. From its period, the absolute magnitude of the cepheid is determined to be −5. What is the distance to the galaxy?
   Answer: 10^5 pc; 3.26×10^5 LY
3. In a very remote galaxy, a supernova is observed which reaches magnitude +17. Assume that the absolute magnitude of the supernova was −18, and calculate the distance to the galaxy.
   Answer: 10^8 pc; 3.26×10^8 LY
4. How can the visual light of an emission nebula in another galaxy exceed that from the star whose energy produced the nebula?
5. Starting with the determination of the size of the Earth, outline all the steps one has to go through to obtain the distance to a remote cluster of galaxies.
6. Suppose a supernova explosion occurred in a galaxy at a distance of 10^8 pc, and that the supernova reached an absolute magnitude of −19 at its brightest. If we are only now detecting it, how long ago did the event actually occur? What is the *apparent* magnitude of the supernova at its brightest? (See Section 23.2b.)
7. Why do we use the *brightest* galaxies in a cluster as indicators of its distance, rather than average galaxies in it? (*Hint:* There are two reasons, one involving the definition of "average," the other involving the distances to typical clusters.)
8. The tenth brightest galaxy in cluster A is at apparent magnitude +10, while the tenth brightest galaxy in cluster B is at apparent magnitude +15. Which cluster is more distant, and by how many times?
9. How can we determine the inclination of M31 (the Andromeda galaxy) to our line of sight?
10. Suppose the rotation curve of a galaxy is flat beyond a certain radial distance, R, from the center of the galaxy, and suppose the galaxy to have perfect spherical symmetry. Because of the spherical symmetry, it works out that only the part of the galaxy interior to r contributes to the gravitational acceleration of a star or gas cloud at a distance r from the center.
    a) How does that part of the galaxy's mass act, gravitationally?
    b) If the mass of the galaxy interior to r, expressed in solar masses, is M, and if distance is expressed in AUs and time in years, what is the period of a star or gas cloud in a circular orbit at a distance r from the center?
    c) What is the velocity of that star, or gas cloud, in AUs/year?
    d) For r greater than R, how does M depend on r?
11. ★ Classify the following galaxies according to Hubble type: (a) a galaxy that is chaotic in appearance, with no symmetry and no resolved stars; (b) a galaxy with an elliptically shaped image whose major axis is twice its minor axis; (c) a galaxy with very tightly wound spiral arms and a large nucleus.
12. Where might the gas and dust (if any) in an elliptical galaxy come from?

13. If extragalactic globular clusters exist as "galaxies," as what kind of galaxies would you classify them? How about extragalactic stars?

*14. Assume that its redshift is a Doppler shift, and verify that a quasi-stellar source with a redshift of $\Delta\lambda/\lambda$ = 3.53 has a radial velocity of 91 percent the speed of light (See Section 8.3.)

15. Suppose a quasar recedes at $^{15}\!/_{17}$ (=0.882) times the speed of light. What is the observed relative shift of its spectral lines ($\Delta\lambda/\lambda$)?
Answer: $\Delta\lambda/\lambda = 3$

16. a) Suppose a quasar has an absolute visual magnitude of −25. What is its visual luminosity in terms of the sun's? (Assume that for the sun, $M_v = +5$.)
b) Suppose that the quasar has a distance of 5×10^9 pc. Suppose, further, that the fact that it appears stellar implies that its angular diameter is less than 1 arc-second. What is the upper limit to its linear diameter in parsecs?
c) If it is variable, can you suggest a way of fixing a smaller upper limit to its linear size?
Answers: (a) 10^{12} (b) 2.5×10^4 pc

17. A quasar emits two radiating clouds in our general direction at $^{13}\!/_{14}$ the speed of light. They are first seen by us when they appear to have been first produced at the central powerhouse, and are subsequently observed to move, apparently radially outward in opposite directions from the center of the quasar image. Fourteen years after the quasar actually emitted the clouds, they are, in reality, 12 LY closer to us than the quasar itself.
a) How many years, on Earth, elapse between our first observation of the clouds and when we observe them at their positions when they are 12 LY nearer to us?
b) At what speed do they appear to us to be separating?
Answer to (b): Five times the speed of light.

18. Let us imagine four classes of objects: those that are resolved in photographic images (say, whose angular sizes are more than about 1″) and those that are not (smaller than 1″), and for each of these cases, those that have radial velocities less than 1000 km/s, and those that have radial velocities greater than 1000 km/s. Thus we can classify astronomical objects into four groups as follows:

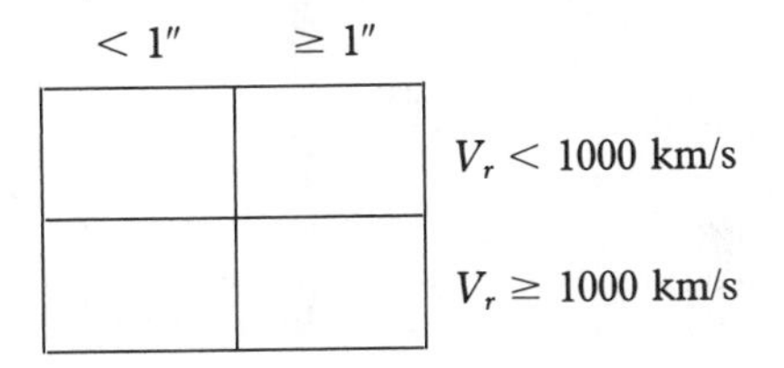

The Moon, for example, would go into the upper right-hand box. Now classify each of the following:
a) the sun
b) Jupiter
c) a typical Trojan minor planet
d) stars in our Galaxy
e) the Orion nebula
f) galaxies in the Coma cluster
g) Seyfert galaxies
h) quasars

Edwin Powell Hubble (1889–1953) left a career in law to study astronomy. In 1919 he joined the staff of the Mount Wilson Observatory, where he was the first to show that the "nebulae" are actually galaxies, like our own. He was also the first to show that the large-scale structure of the universe is homogeneous. In 1929 he gave the first evidence for the expansion of the universe. (California Institute of Technology)

36 THE STRUCTURE OF THE UNIVERSE

Celestial objects rarely travel through space alone. The Earth is but one of nine planets orbiting the sun. The sun itself seems somewhat unusual in that, so far as we know, it is a single star. Most stars are at least double and many are members of either galactic or globular clusters. All of the stars and clusters in the Milky Way Galaxy are gravitationally bound, orbit around a common center, and will complete their evolution in close proximity. Does this cosmic togetherness persist on still larger scales? Are most galaxies to be found in clusters of galaxies? What is the structure of the universe as a whole? How are galaxies distributed in space? Are there as many in one direction of the sky as in any other? And if we count fainter and fainter galaxies, presumably farther and farther away, do we find that their numbers increase in the way they should if galaxies are distributed uniformly in depth? Hubble, a man of no small perspective, began to try to answer these questions only a few years after he showed that galaxies—that is, systems of stars lying far beyond our own Milky Way—even existed.

36.1 HUBBLE'S FAINT GALAXY SURVEY

Hubble had at his disposal what were then the world's largest telescopes—the 100-inch (2.5-m) and 60-inch (1.5-m)—reflectors on Mount Wilson. But although those telescopes can probe to great depths, they can do so only in small fields of view. To photograph the entire sky with the 100-inch telescope would take not just a lifetime, but thousands of years. So instead, Hubble sampled the sky in many regions, much as Herschel did with his star gauging (Section 26.3a). In the 1930s Hubble photographed 1283 sample areas or fields with the telescopes, and on each photograph he carefully counted the numbers of galaxy images to various limits of brightness, the faintest corresponding to the greatest depth of space that could then be probed.

The results of Hubble's survey are shown in Figure 36.1, which is a map of the sky shown in what are called *galactic coordinates:* the Milky Way,

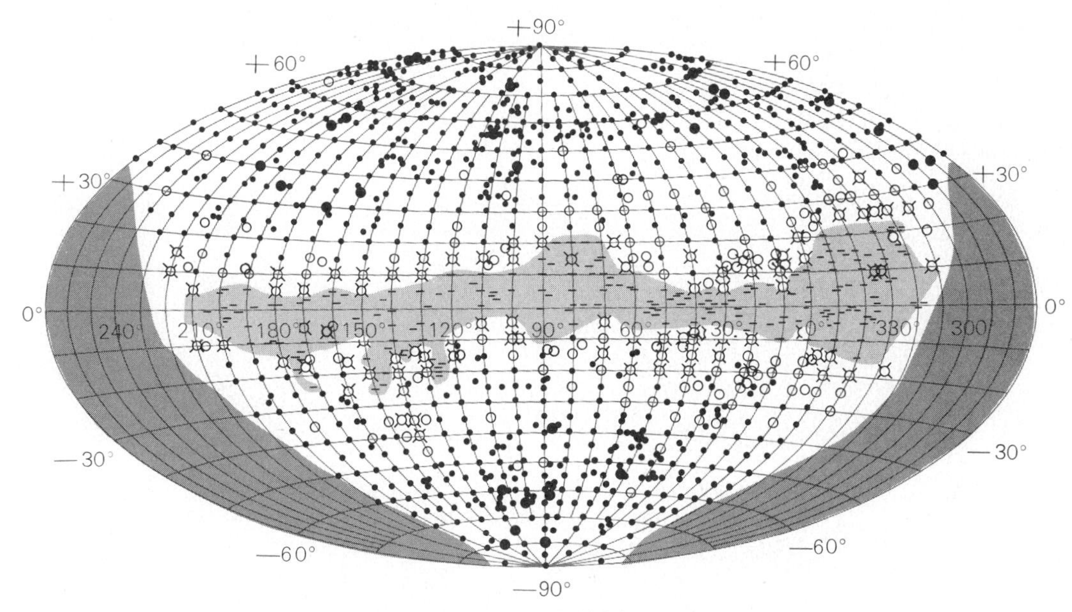

Figure 36.1 Distribution of galaxies according to Hubble's survey.

across the middle of the plot, defines the galactic equator, and the top and bottom of the map—the galactic poles—are 90° away from the Milky Way. The empty sectors at the lower right and left are the parts of the sky too far south to observe from Mount Wilson. Each symbol represents one of the regions of the sky surveyed by Hubble, and the size of the symbol indicates the relative number of galaxies he could observe in that area.

The first obvious thing to notice is that we do not see galaxies in the direction of the Milky Way; the obscuring clouds of dust in our Galaxy hide what lies beyond in those directions. Hubble called this part of the sky the *zone of avoidance*. Near the Milky Way in direction, the counts of galaxies are below average and are denoted by open circles. The farther we look from the Milky Way, the less obscuring foreground dust lies in our line of sight and the more galaxies we see. From the counts of galaxies in different directions, Hubble determined that light is dimmed by about 0.25 magnitudes (blue, or photographic magnitudes) in traversing a half-thickness of the Galaxy at the sun's position. Having derived how foreground dust dims the light from galaxies, Hubble could correct his counts to allow for the effect.

After such correction, Hubble found that on the large scale the distribution of galaxies is *isotropic*, which means that if we look at a large enough area of the sky, we find as many galaxies in one direction as in any other. From the 44,000-odd galaxies Hubble counted in his selected regions, he calculated that about 100 million were potentially photographable with the 100-inch telescope. (This means that with the larger telescopes and more sensitive detectors available today, we could, given enough time, record about one billion.) Moreover, Hubble found that the numbers of galaxies increase with faintness about as we would expect if they were distributed uniformly in depth.

These findings of Hubble were enormously important, for they indicated, at least to the precision of his data, that the universe is isotropic and homogeneous—the same in all directions and at all distances. In other words, his results indicate that the universe is not only about the same everywhere, but that the part we can see around us, aside from small-scale local differences, is representative of the whole. This idea of the uniformity of the universe is called the *cosmological principle* and is the starting assumption for nearly all theories of cosmology (next chapter).

36.2 CLUSTERING OF GALAXIES

The uniformity of the universe applies only to very large volumes of space. Galaxies, as we shall see, do tend to cluster, and clusters of galaxies do tend to

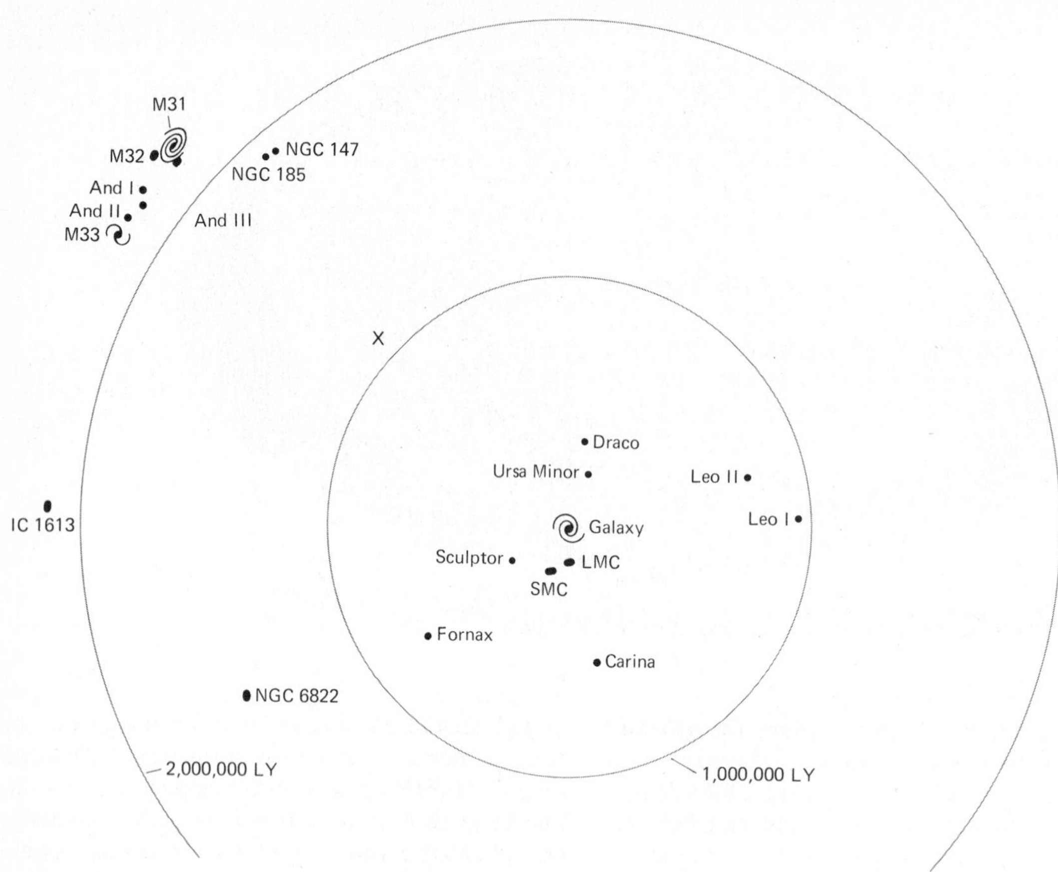

Figure 36.2 The Local Group. The "X" marks the approximate center of mass of the group.

form superclusters. Between the superclusters, there are great voids where few if any galaxies can be found. The universe is uniform only on scale sizes large enough to include a number of superclusters and voids. This idea can be understood through an analogy. The residents of your neighborhood may be fairly typical of the residents of the city as a whole in which you live. But the people within the particular house in which you live may not be typical in number, age, or other characteristics. Just as we must take a fairly large sample of population to obtain a representative group of people, so must we consider a fairly large volume of space in order to find within it the characteristics and kinds of objects that are typical of other large volumes of space.

The fact that galaxies often occur in clusters has been known for a long time. When Hubble began his work on galaxy counts, he was aware of several conspicuous clusters, and he made a point of avoiding them in his photographic survey so that they would not bias his results. However, when he investigated the statistics of the numbers of galaxy images on the photographs, he realized that the galaxies are not distributed in space with perfect randomness, but show a clumpy distribution, as if they are clustered. Indeed, so clumpy was the distribution that Hubble was led to think that probably *all*, or at least *most*, galaxies are in groups or clusters, rather than being single in space.

Observations now available to astronomers confirm that galaxies do indeed tend to group together on all scale lengths from one million parsecs (a small group) up to certainly scales of 20 million parsecs (a supercluster). There is growing evidence that superclusters themselves may form much larger-scale structures, perhaps as large as 100 to 300 million parsecs in length.

(a) The Local Group

One small group of galaxies is the one to which our own Galaxy belongs—the Local Group. It is spread over about a million parsecs and contains at least 21 known members (some investigators count as members a somewhat larger number of galaxies at a somewhat greater distance). There are three large spiral galaxies (our own, the Andromeda galaxy, and M33), at least 11 dwarf irregulars, four intermediate ellipticals, and ten known dwarf ellipticals. Appendix 17 gives the properties of the galaxies that are generally accepted to be members of the Local Group. Figure 36.2 is a plot of the Local Group; the galaxies have been projected onto an arbitrary plane centered on our Galaxy, then their distances from the center of the plot have been increased so that they are shown at the correct relative distances from us.

The radial velocities of some of the Local Group galaxies have been measured. If we assume that they are moving at random within the Local Group, we can find the motion of the sun with respect to the center of mass of the Group. The procedure is analogous to the determination of the motion of the sun compared with its neighboring stars in our part of our own Galaxy, that is, with respect to the local standard of rest (Section 22.4a). The Andromeda galaxy, for example, approaches us at about 266 km/s, while the Large Magellanic Cloud recedes from us at about 276 km/s. Most of the apparent radial velocity of each of these objects is due to the revolution of the sun about the center of the Galaxy and the motion of the Galaxy in the Local Group. From an analysis of the radial velocities of all the other galaxies, we find that the sun's orbital velocity is about 220 km/s and that our Galaxy is moving at a speed of 100 to 150 km/s in the Local Group. The average of the motions of all the galaxies in the Group indicates that its total mass is of the order 5×10^{12} solar masses. Although this mass estimate, which is based on the virial theorem, is rather uncertain, if correct, it implies that extensive amounts of dark matter must be present in the Local Group.

(b) The Neighboring Groups and Clusters

Well beyond the edge of the Local Group we find other similar systems. Among them is a small group of galaxies centered on M81, and another on M51 (the galaxies shown in Figures 35.10 and 35.8, respectively). Another example of a small group of galaxies is shown in Figure 36.4. At a distance of several tens of millions of light years, however, a group like the Local Group, seen in projection against the background of very many more distant galaxies, would not be noticed. At large distances, we recognize only very rich clusters (that is, clusters of many member galaxies) that stand out like "sore thumbs" against the background.

Figure 36.3 NGC 185, an intermediate elliptical galaxy in the Local Group. (Lick Observatory)

The nearest moderately rich cluster is the famous Virgo cluster, a system with thousands of members (Figure 36.5). It contains a concentration of mostly elliptical galaxies that includes M87—the radio galaxy and X-ray source shown in Figures 35.15 and 35.29. Within several degrees of M87 are many spirals as well as ellipticals and, associated with the brightest galaxies, very many dwarfs, like the dwarf ellipticals in the Local Group. All of the galaxies in the Virgo cluster are too remote for us to observe their cepheids, and distances estimated by some of the other techniques described in Section 35.2 are quite uncertain. The concentration of galaxies around M87 is probably between 30 and 60 million light years away (10 to 20 million parsecs). If its distance is 40 million LY, its linear diameter is about 4 million LY. The entire Virgo complex contains several other concentrations of galaxies and thus is much larger. Even the entire system, however, cannot be considered a *great* cluster—one that would be recognizable in remote parts of the universe.

Figure 36.4 Stephan's quintet, a small group of galaxies. (Lick Observatory)

A good example of a cluster that is much larger than the Virgo complex is the Coma cluster, which has a linear diameter of at least 10 million LY and thousands of observable galaxies (Figure 36.6). The cluster is centered on two giant ellipticals, whose absolute visual magnitudes are between -23 and -24, or about 4×10^{11} times more luminous than the sun. The Coma cluster contains more and more

Figure 36.5 The central region of the Virgo cluster of galaxies. (Kitt Peak National Observatory)

Figure 36.6 The central part of the Coma cluster of galaxies.

members at magnitudes that are successively fainter. There is every reason to expect dwarf elliptical galaxies to be present; if so, the total number of galaxies in the cluster might be tens of thousands. Rich clusters like Coma usually show marked spherical symmetry and high central concentration. They contain few if any spiral galaxies in the cluster core, but rather have a membership dominated by ellipticals and those galaxies that resemble spirals but without spiral arms or obvious evidence of interstellar matter (those classed as S0—Section 35.3d). Two other clusters of galaxies are shown in Figures 36.7 and 36.8.

Clusters of galaxies, particularly rich clusters like Coma, are usually sources of X-rays. The X-rays from a cluster are thermal radiation from intracluster gas at a temperature of 10^7 to 10^8 K, with the most intense radiation generally coming from the center of the cluster. The mass of gas required to produce the X-radiation is typically 10 to 20 percent, but may range up to half of the total cluster mass and so can account for a part of the missing mass in rich clusters (Section 35.4c).

There may be a significant relation between the presence of hot gas and of S0 galaxies in a cluster. X-ray emission lines of heavy elements such as iron have been observed at such intensity as to suggest that the heavy-element abundance in the hot gas is similar to that in the sun, rather than the matter being all or nearly all hydrogen and helium, as current theories predict for the primordial matter from which the clusters formed. This suggests that at least some of the X-ray emitting gas must have undergone nucleosynthesis in stellar interiors and that processed matter would then have been ejected into interstellar space within the cluster galaxies by such mechanisms as supernova outbursts. Finally, this material was swept from the galaxies by collisions between them and by their moving through the intracluster gas, or possibly by internal processes such as stellar winds and supernova explosions. Such sweeping of interstellar matter from galaxies stops star formation in them, and the spiral arms gradually disappear, leaving the galaxies as type S0. (Some S0 galaxies are found outside of dense clusters, so there may be other ways to form S0 galaxies.) The swept gas is hot because the galaxies collide with each other or pass through intracluster gas at speeds of up to thousands of kilometers per second.

(c) Evolution of Galaxies

There are many types of galaxies, including spirals, ellipticals, and irregulars. In Chapter 35, we said that while the stars within these galaxies formed,

Figure 36.7 Irregular cluster of galaxies in Hercules, photographed with the 5-m telescope. (Palomar Observatory, California Institute of Technology)

evolved, and died, the galaxies themselves did not change from one type to another—once an elliptical, always an elliptical. In the case of human beings there has been continuing controversy over the question of nature versus nurture. How many of our characteristics are determined by the genes we inherit and to what extent are we a product of

Figure 36.8 A distant cluster of galaxies in Hydra, with a radial velocity of about 20 percent that of light. (Palomar Observatory, California Institute of Technology)

our environment? Similar questions can be asked about galaxies, and observations of galaxies in clusters raise again the question of the possible impact of the surrounding environment on the form and structure of galaxies.

The crucial observation that requires a re-examination of this question is the discovery that, while all types of galaxies are found in both high- and low-density environments, about 80 to 90 percent of the galaxies in the high-density environments in the centers of clusters of galaxies are ellipticals and S0s. Conversely, isolated galaxies found in regions outside of clusters or groups of galaxies, where the density of material is low, are mostly spirals. Furthermore, this correlation between galaxy type and density may hold even outside of clusters. A small dense region may contain many elliptical galaxies even if that region is not part of a cluster.

Is it possible that ellipticals are made, not born? Is it possible that collisions and mergers of galaxies in the dense central regions of clusters can convert spirals to giant ellipticals? For a long time, astronomers believed that such collisions were extremely rare and only a few freak galaxies were altered by interactions with near neighbors. Two things have happened to alter this view. First, we now think that galaxies are about ten times more massive than we had guessed simply by looking at the luminous matter. The dark matter increases their gravitational pull and makes it more likely that they will draw other galaxies close enough to interact and possibly merge. Second, quite apart from the theoretical arguments we see evidence that mergers do occur (see Section 35.5). In fact, we see so many examples of galaxies currently undergoing collisions, that it seems likely that at least 5 percent of all galaxies have been so affected at some time in their lives. Mergers of two galaxies of comparable size occur only if the interaction takes place at low velocity. Low velocities are unlikely in the current cores of galaxy clusters, but conditions may have been very different, and conducive to mergers, when clusters were much younger.

Given that mergers do occur fairly frequently, can we then conclude that merged spirals make ellipticals? The answer in most cases is probably no. First of all, there seem to be simply too many ellipticals in the centers of dense clusters to be explained this way. Second, ellipticals of a given luminosity have many more globular clusters around them than do spirals, and mergers cannot produce this difference. Third, elliptical galaxies range in mass and radius continuously from dwarf ellipticals to giant ellipticals. The dwarfs certainly did not form from the merger of spirals, since they are much smaller than spirals, and scientists are always reluctant to appeal to two different mechanisms to explain the formation of what appears to be a single class of objects. The whole subject of how ellipticals form is a controversial one. Perhaps the safest conclusion is that there are many ways an elliptical *could* have been produced, and it remains likely that initial conditions and environment have both played a part.

36.3 THE LARGE-SCALE DISTRIBUTION OF MATTER: SUPERCLUSTERS

If stars form clusters of stars and galaxies are often found in clusters of galaxies, are there then clusters of clusters of galaxies? The answer is yes, and we will call very large-scale structures of matter containing one or more clusters of galaxies *superclusters*. Determining the nature of superclusters is currently one of the most active of all fields of observational astronomy, and the people who have contributed significant results are too numerous to name. It is worth noting, however, that one of the pioneers in this type of research was the senior author of this text.

The best studied of the superclusters is the one that includes the Milky Way Galaxy. The most prominent grouping of galaxies within the *Local Supercluster* is the Virgo cluster. The Milky Way Galaxy lies in the outskirts of the Local Supercluster. The diameter of the Local Supercluster is at least 20 million parsecs, and its mass probably lies in the range 10^{15} to 10^{16} times that of the sun.

Are galaxies distributed uniformly throughout the volume of space encompassed by the Local Supercluster? Observations indicate that the vast majority of bright galaxies lie in a small number of clumps or groups, and that most of these groupings are narrowly confined to a plane. In fact, 60 percent of all of the galaxies contained within the Local Supercluster are to be found within a disk-like region whose diameter is about six times its thickness. Both the Local Group and the Virgo cluster itself lie in the plane of this disk. The dominance of the Virgo cluster is indicated by the fact that one-third of all the clusters in the disk are members of the Virgo cluster.

Perhaps the most startling fact revealed by this detailed study of the Local Supercluster is that space is mostly empty. Most of the galaxies are concentrated into individual clusters and these clusters occupy only about 5 percent of the total volume of space contained within the boundaries of the local supercluster. A major problem for any theory of the formation of galaxies and of large-scale structure in the universe is to explain why galaxies are so closely clumped and why most of the universe is devoid of luminous matter.

Studies of more distant galaxies show that the Local Supercluster is not unique. Galaxies are found preferentially in large filamentary superclusters that extend over distances as large as 60 million parsecs (Figure 36.9 and Color Plate 30c). These filamentary structures are separated by voids, that is, by large holes where few or no galaxies can be found. This result has come as a surprise to astronomers. Most would probably have predicted that the regions between giant clusters of galaxies should be filled with many small groups of galaxies or even with isolated individual galaxies. Assiduous searches within these voids have so far only confirmed that they do indeed contain *very* few galaxies.

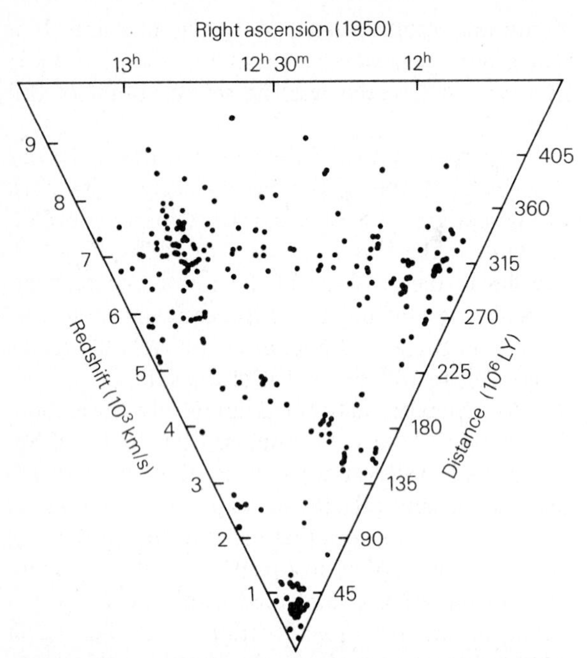

Figure 36.9 Distribution of galaxies in a particular direction in space. The left-hand side of the triangle indicates the redshift of each galaxy. The right-hand side of the triangle indicates the distance to each galaxy on the assumption that the Hubble constant is equal to 75 kilometers per second per million parsecs. The Coma cluster is at a right ascension of 13 hours with a recession velocity of about 7000 km/s. The Coma cluster is only a richly populated region in a continuous band of galaxies that stretches across the sky at a distance from our Galaxy of about 315 million light years. Note also that there are huge voids where there are no galaxies. (Based on data obtained by Stephen A. Gregory and Laird A. Thompson)

36.4 FORMATION OF GALAXIES

As we look off into space, we look back into time, for we see remote objects as (and where) they were far in the past, when light left them to begin its long journey across space to reach our telescopes. Remote objects, therefore, are in a sense historical documents in the universe, even though we may have difficulty in interpreting their message.

(a) The Age of Galaxies

Two great revolutions in physics have led to major advances in our understanding of astronomical objects. The development of quantum mechanics provided the basis for interpreting spectra of gases. This knowledge in turn allowed astronomers to determine the physical conditions in stellar atmospheres and in gaseous nebulae and to derive the compositions of both. Physicists then turned their attention to the nuclei of atoms, and determined both their structure and the nature of the protons and neutrons that combine to form the chemical elements that compose our world. This research led directly to a determination of the source of the prodigious amounts of energy radiated by the sun and other stars, and provided the basis for modeling stellar formation and evolution. We are on much less secure ground when we attempt to discuss the origin and evolution of galaxies, but the key to understanding the formation of large-scale structures in the universe may lie in particle physics—that is, in the attempts now being made to understand the subatomic particles that make up protons and neutrons and control their interactions.

It is only during the last decade that astrophysicists have begun seriously to tackle the question of how galaxies form. One starting point for all theories of galaxy formation is the fact that most galaxies are very old indeed, and there are several observations that lead inescapably to this conclusion. For example, there are stars in globular clusters in our own Galaxy that are 12 to 20 billion years old. Therefore, the Milky Way must be at least this old. Yet the universe itself is not significantly older than this amount. (The age of the universe is derived from the observed rate of expansion. All galaxies are moving farther and farther apart. If we project this expansion back in time, we find that all of the galaxies were very close together sometime between 10 and 20 billion years ago, with the uncertainty being due to uncertainties in our estimates of how far away galaxies are from us at the present time. These ideas are discussed in the next chapter.) It appears that the Milky Way must have formed during the first two to three billion years after the expansion of the universe began.

There is other evidence that galaxies formed very early, and have been around for billions of years. For example, if we look at galaxies that are several billions of light years away, we are seeing light that left those galaxies several billion years ago. We are, in effect, looking back in time, and seeing those galaxies as they were when they were much younger than our own Galaxy is now. The most distant normal galaxies that we have studied in this way emitted the light that we observe when the universe was about half its present age. Yet these young galaxies have about the same luminosity and colors, and hence about the same stellar content, as do nearby and much older galaxies. This similarity of galaxies that span half the age of the universe suggests that galaxies were fully formed and quite mature several billion years ago.

We can probe still farther back in time, and still closer to the beginning of the universe, by observing quasars, which are much brighter than normal galaxies and can be seen at much larger distances. When we do so, we find that the composition of the gas in quasars is very much like the composition of the gas in our own Galaxy. Specifically, the gas in quasars contains not only hydrogen and helium but also heavier elements such as carbon, nitrogen, and oxygen. As we shall see in the next chapter, we think that these heavy elements were not present when the universe began, but rather were manufactured in the first generation of stars that evolved within newly formed galaxies. The fact that quasars contain large amounts of heavy elements means that at least one generation of stars had already completed its evolution even before the light that we now see was emitted. Given the distance to quasars, this means that some galaxies must have formed when the age of the universe was no more than 20 percent as old as it is now.

(b) Which Came First—Galaxies or Superclusters?

Given that galaxies are billions of years old, we can next ask which came first—galaxies or clusters of galaxies? One can imagine two extreme possibilities. Individual galaxies may have formed first and then clumped together, attracted by gravity, to form clusters and superclusters. An alternative point of view has been espoused by the Soviet theoretician Y. Zeldovich, who suggests that pancake-shaped blobs of primordial gas with masses as large as whole superclusters of galaxies collapsed due to gravity and fragmented to form sheets and filaments of galaxies. Yet a third idea has been proposed by J. Ostriker of Princeton, who suggests that a series of cosmic explosions, not the simple working of gravity, determined the large-scale structure of the universe. Unfortunately, we cannot yet decide which of these three ideas most nearly describes what really happened.

There is observational evidence that in at least some cases clusters form, or at least add to their membership, by accumulating galaxies that have already formed. For example, the expansion velocities of some nearby groups of galaxies, including possibly our own, are apparently slowing down, and these groups may even begin to fall towards the giant cluster in Virgo (Section 37.3c) many billions of years from now. If we could return then and observe the Virgo cluster, it might contain many more galaxies than it does now.

There are problems, however, with trying to build clusters and superclusters from the bottom up by forming galaxies first and then letting them come together afterwards. As we have seen, the centers of dense clusters are dominated by elliptical galaxies, while spirals are found preferentially in sparsely populated regions outside the clusters. It is true that interactions between galaxies can change their character somewhat. Galactic mergers or cannibalism can form giant elliptical galaxies in dense cluster cores, but such interactions cannot account for the large number of moderate-sized ellipticals seen in a cluster like Virgo, in which the density of galaxies is only moderately high. Interactions between galaxies and intracluster gas can strip much of the gas from spirals but cannot convert a spiral into an elliptical galaxy. If we assume that galaxies form first and only afterwards come together to form clusters, then we must also explain how a galaxy knows at the time of its formation what its future environment will be. In other words, if galaxy formation and cluster formation are separate events, how do we explain the fact that the higher the density of the present environment in which we search for galaxies, the higher the ratio of ellipticals to spirals?

The alternative is to build the universe from the top down. Large-scale sheets and "pancakes" of gas—filamentary structures that constitute protoclusters and protosuperclusters—form first. Only afterwards, as this gas fragments and undergoes gravitational collapse, are galaxies formed. The major attraction of this approach is that it seems easier to explain the existence of superclusters of galaxies and of voids where there is no luminous matter if very large–scale structure forms first. Even so, there may be problems in understanding the size and emptiness of some voids. Furthermore, there is as yet no well-developed theory for how such a protocluster cloud fragments to form galaxies, and thus detailed comparisons of theory and observations for these top-down models are not yet possible.

Another problem with the top-down models of galaxy formation is that it may be difficult to form galaxies on the right time scale. If large-scale structures form first and galaxies form afterwards, then theory indicates that galaxy formation should still have been taking place when the universe was half as old as it is now. Observations, however, tell us that many galaxies were probably fully formed by that time.

A recent survey of the distribution of galaxies in space by V. de Lapparent, M. Geller, and J. Huchra, at the Harvard-Smithsonian Center for Astrophysics, provides some support for the idea that cosmic explosions may play a role in the formation of galaxies. These observers confirm that there are indeed great voids where the density of galaxies is very low. They also report that the galaxies themselves are to be found on bubble-like surfaces enveloping the voids, with a typical bubble having a diameter in the range 25 to 50 million parsecs. Such a structure might be formed if energetic explosions swept through the protogalactic gas, prior to the formation of this gas into galaxies, thereby producing dense, cool shells that could subsequently fragment to form galaxies. The process is somewhat similar to what happens as star formation proceeds through a molecular cloud (Section 30.1). Again, however, there are problems in reconciling theory and observation. In particular, it may be difficult to produce voids as large as those that have been observed.

We are then left with several possibilities, none entirely satisfactory, for explaining how groups, clusters, and superclusters of galaxies formed and why they are distributed so nonuniformly in space. As we shall see in the next section, there are comparable uncertainties in theories of the formation of individual galaxies.

(c) A Universe of (Mostly) Dark Matter?

As we have seen, many galaxies formed in the first few billion years after the expansion of the universe began. Forming galaxies that rapidly is a problem for many theories. What is required is a way to force ordinary matter—protons and neutrons and electrons—to clump together quickly into galaxy-sized masses. It is here that particle physics may provide a vital clue. Particle physics theorists predict that a whole zoo of particles, as yet undetected in Earth-bound laboratories, should have been produced during the first minute or so after the universe began. Among this zoo are particles that have small masses, move very slowly, and radiate no light at all. Because these particles move so slowly, but attract each other gravitationally, it is very easy for them to form massive clumps. Some astrophysicists suggest that these exotic particles account for at least some of the large amounts of dark matter associated with galaxies and clusters of galaxies. Remember that there is evidence that most of the matter in the universe is dark and does not emit light. These theorists also suggest that this dark matter plays a crucial role in the formation of galaxies. Because the dark matter clumps so easily, it can come together and form a kind of gravitational trap that will then capture ordinary matter—protons and neutrons and electrons—and concentrate it and thus speed the formation of galaxies.

This theory, bizarre though it may sound, does successfully predict a number of things when its implications are calculated quantitatively. For example, the theory predicts that galaxies will have approximately the masses that they actually do have. The theory also predicts that spirals will be more common than large ellipticals, as is observed to be the case, and that ellipticals will occur preferentially in denser regions in the cores of large clusters.

There are problems with this theory, however. For example, the dark matter theory predicts that galaxies should form everywhere throughout space. There should be no great voids. The dark matter, according to the calculations, does not tend to form structure on the scale of clusters or superclusters of galaxies. One solution to this problem is to assume that galaxies and dark matter are not distributed in the same way. Perhaps galaxy formation is biased in some way, so that only some matter turns into galaxies and other matter does not. If this idea is correct, then the voids would not be empty at all but rather would be filled with failed galaxies that for some reason never formed stars. Astronomers are now trying to develop convincing theories for why galaxy formation should be so strongly biased that it occurs in some places and not others.

Finally, we must stress that there is no experimental evidence that dark particles of the kind required to encourage galaxy formation actually exist. They are predicted by certain types of theories devised to explain the properties of subatomic matter. There are legitimate reasons for thinking they exist that have nothing to do with cosmology. The ideas discussed here are currently very fashionable. It remains to be seen whether they are also correct. What is required is some critical observational test that would prove one way or another that dark matter with the required properties actually exists.

It should be obvious from this discussion that we are on very uncertain ground when we attempt to describe the formation of galaxies. Both observational and theoretical efforts to understand this process are being pursued very vigorously, and it is quite likely that the ideas presented here will be greatly modified by future research.

EXERCISES

1. Assume for the sake of illustration that the dust in our Galaxy all lies in a relatively thin flat disk, with the sun in the central plane of that disk. Under these circumstances, show by a diagram that the obscuration of distant galaxies is less and less at greater and greater directions from the Milky Way (that is, at greater and greater galactic latitudes).
2. Suppose on one survey you count galaxies to a certain limiting faintness. On a second survey you count galaxies to a limit that is four times fainter.
   a) To how much greater distance does your second survey probe?
   b) How much greater is the volume of space you are reaching in your second survey?
   c) If galaxies are distributed homogeneously, how many times as many galaxies would you expect to count on your second survey?
3. Assume that all quasars have the same intrinsic luminosity and are distributed at random in space. Derive a formula for the number that should be seen that deliver to Earth a light flux greater than a given flux, ℓ. Ignore the expansion of the universe and evolutionary effects.
4. If galaxies are distributed at random in space, statistical theory enables us to predict how many fields should contain a certain number of galaxies each. Hubble found that a far larger number of fields had too few galaxies, and also that a far larger number of fields had too many galaxies, than would be expected for a random distribution. Explain why this result suggests that galaxies tend to be clustered.
5. Are there subunits or subcondensations in the Local Group? If so, discuss them.
6. Would an indefinite hierarchy of clustering (clusters upon clusters, without limit) be consistent or inconsistent with the cosmological principle? Why?
7. Observations show that the number of quasars is vastly greater at faint magnitudes and large redshifts than would be predicted for a uniform distribution in space. Explain clearly and completely why the cosmological principle is violated if quasars are *not* temporary phenomena at an earlier epoch of the history of the universe.

Abbé Georges Lemaître (1894–1966), Belgian cosmologist, studied theology at Mechelen and mathematics and physics at the University of Leuven. It was there that he studied the expansion of the universe and postulated its explosive beginning. He predicted the Hubble law two years before its verification, and he was the first to consider seriously the physical processes by which the universe began. (Yerkes Observatory)

37 THE BIG BANG: THAT VANISHED BRILLIANCE

When space turned around, the earth heated
When space turned over, the sky reversed
When the sun appeared standing in shadows
To cause light to make bright the moon,
When the Pleiades are small eyes in the night,

From the source in the slime was the earth formed
From the source in the dark was darkness formed
From the source in the night was night formed
From the depths of the darkness, darkness so deep
Darkness of day, darkness of night
Of night alone

Did night give birth
Born was Kumulipo in the night, a male
Born was Po'ele in the night, a female*

So begins the *Kumulipo*, the 2102-line Hawaiian chant of creation composed in the 18th century. The purpose of this chant was to trace the lineage of a newborn chief to his ultimate origins. The *Kumulipo* records much of what Hawaiians knew about natural history, and also indicates clearly that they believed that life evolved on the Earth, rather than being created by some divine force. Because of its emphasis on evolution, the *Kumulipo* received considerable attention in 19th-century Europe, which was wrestling with the controversial ideas of Darwin as set out in his book on the *Origin of the Species by Means of Natural Selection.*

Unique in its treatment of the subject matter, the *Kumulipo* is far from unique in its choice of subject. Throughout all of recorded history, in myths, in religious traditions, in literature, and in poetry, we find attempts to account for the origin of the universe and all that is contained within it. It is one of the triumphs of 20th-century science that we now think that we can trace the evolution of the universe back to about 10^{-45} seconds after it began. The study of the organization and evolution of the universe is called cosmology, and modern views of cosmology will be described in this chapter. It is important to recognize, however, that the ideas that we are discussing lie on the frontiers of science. It is quite likely that new observations and a better understanding of elementary particle physics will modify the picture that is presented here. Our description of the evolution of the universe, particu-

* Translation from *Kumulipo, The Hawaiian Hymn of Creation, Volume 1,* by Rubellite Kawena Johnson (Honolulu: Topgallant Publishing Co., Ltd.), 1981.

larly during the first second after its beginning, may seem as primitive to scientists of the 22nd century as the *Kumulipo* does to us now.

37.1 THE EXPANDING UNIVERSE

(a) The Evidence: Slipher's Observations

The universe is expanding. This fundamental observation underlies all of modern cosmological thought. Curiously, this fact was discovered as a by-product of the search for distant solar systems.

In 1894, Percival Lowell established an observatory in Flagstaff, Arizona, in order to study the planets and to search for life in the universe. Lowell thought that the spiral nebulae might be solar systems in the process of formation—like the solar nebula (Chapter 12). He therefore asked one of the Observatory's young astronomers, Vesto M. Slipher, to photograph the spectra of some of the spiral nebulae to see if their spectral lines suggest chemical compositions like those expected for newly forming planets.

The Lowell Observatory's major instrument was a 24-inch refracting telescope, which was not at all well suited to observations of faint spiral nebulae. With technology available in those days, exposure times were 20 to 40 hours long! Slipher's first spectrum, in 1912, was of the Andromeda galaxy (not then known to be a galaxy, of course); its spectrum did not suggest new planets but did display a Doppler shift of its absorption lines that indicated a motion of that object toward the Earth of about 300 km/s. We know today that most of that speed is caused by the revolution of the sun about the center of our own Galaxy, carrying us roughly in the direction of M31. Nevertheless, the observed radial velocity was enormous compared with those of most stars. Over the next 20 years, Slipher photographed spectra of more than 40 additional nebulae (now known to be galaxies). Only a few (now recognized as members of our Local Group) are approaching us; Slipher found the overwhelming majority to be receding at speeds as large as 1800 km/s.

(b) The Hubble Law

The profound implications of Slipher's work became apparent only during the 1920s, when Hubble succeeded not only in proving that the spiral nebulae are galaxies but also found ways of estimating their distances. Hubble carried out the key observations in collaboration with a remarkable man: Milton Humason (Figure 37.1). Humason began his astronomical career by driving a mule train up the trail on Mount Wilson to the observatory. In those early days, supplies had to be brought up that way; even astronomers hiked up to the mountaintop for their turns at the telescope. Humason became interested in the work of the astronomers and took a job as janitor at the observatory. After a time he became a night assistant, helping the astronomers run the telescope and take data. Eventually, he made such a mark that he became a full astronomer at the observatory.

Figure 37.1 Milton Humason. (Courtesy of the Archives, California Institute of Technology)

By the late 1920s Humason was collaborating with Hubble by photographing the spectra of faint galaxies with the 100-inch telescope. Humason concentrated on galaxies in clusters whose relative distances could be estimated with some confidence. In 1931 Hubble and Humason jointly published their classic paper in the *Astrophysical Journal,* which compared distances and velocities of remote galaxies moving away from us at speeds of up to nearly 20,000 km/s. Their *law of redshifts* (Figure 37.2), now known as the *Hubble law,* established that the velocities of recession of galaxies are proportional to their distances from us. These observations also

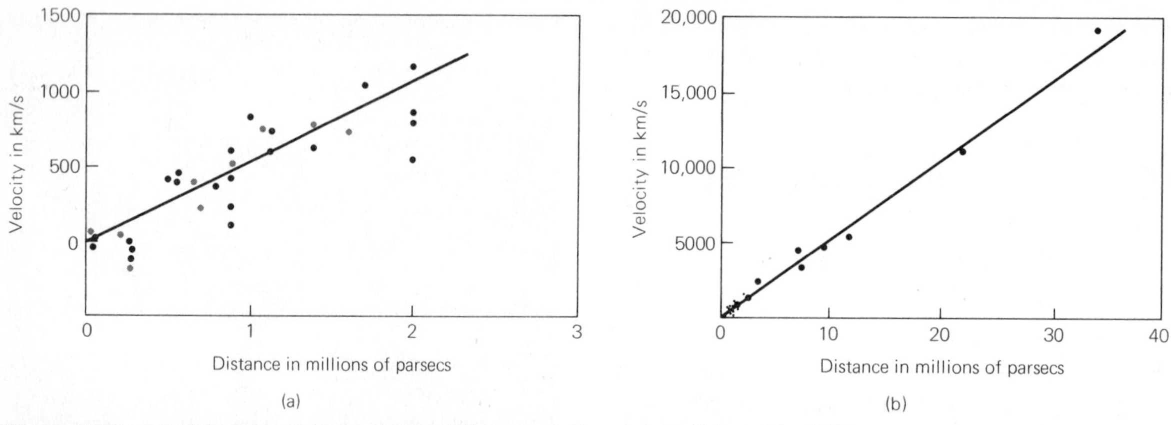

Figure 37.2 (a) Hubble's original velocity-distance relation, adapted from his 1929 paper in the *Proceedings of the National Academy of Sciences.* (b) Hubble and Humason's velocity-distance relation, adapted from their 1931 paper in the *Astrophysical Journal.* The small dots at the lower left are the points in the diagram in the 1929 paper (a).

show beyond doubt that the universe is expanding uniformly, even though Hubble and Humason themselves, at that time, were cautious about so interpreting their observations.

(c) Expansion Requires a Velocity-Distance Relation

At this point, let us be quite clear why a uniformly expanding universe requires that we and all other observers within it, no matter where they are located, must observe a proportionality between the velocities and distances of remote galaxies. Imagine a ruler made of flexible rubber, with the usual lines marked off at each centimeter. Now suppose someone with strong arms grabs each end of the ruler and slowly stretches it, so that, say, it doubles in length in one minute (Figure 37.3). Consider an intelligent ant sitting on the mark at 2 cm—intentionally *not* at either end or in the middle. He measures how fast other ants, sitting at the 4-, 7-, and 12-cm marks move away from him as the ruler stretches. The one at 4 cm, originally 2 cm away, has doubled its distance; it has moved 2 cm/min. Similarly, the ones at 7 cm and 12 cm originally 5 and 10 cm distant, have had to move away at 5 and 10 cm/min, respectively. All ants move at speeds proportional to their distance. Now repeat the analysis, but put the intelligent ant on some other mark, say on 7 or 12, and you'll find that in all cases, as long as the ruler stretches uniformly, this ant finds that every other ant moves away at a speed proportional to its distance.

For a three-dimensional analogy, look at the raisin bread in Figure 37.4. The cook has put too much yeast in the dough, and when he sets the bread out to rise, it doubles in size during the next hour and all the raisins move farther apart. Some representative distances from one of the raisins (chosen arbitrarily, but not at the center) to several others are shown in the figure. Since each distance doubles during the hour, each raisin must move away from the one selected as origin at a speed proportional to its distance. The same is true, of

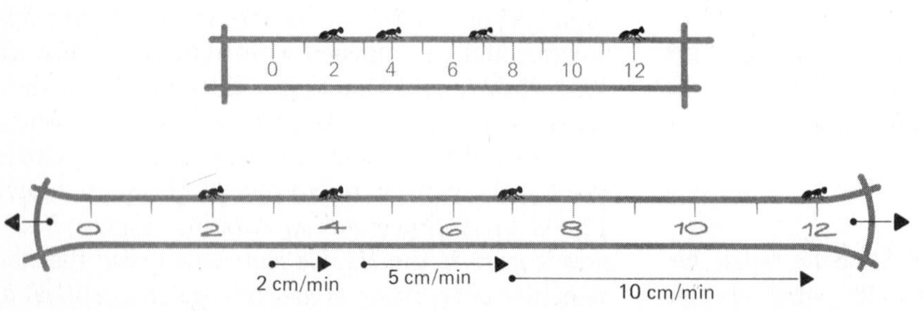

Figure 37.3 Stretching a ruler.

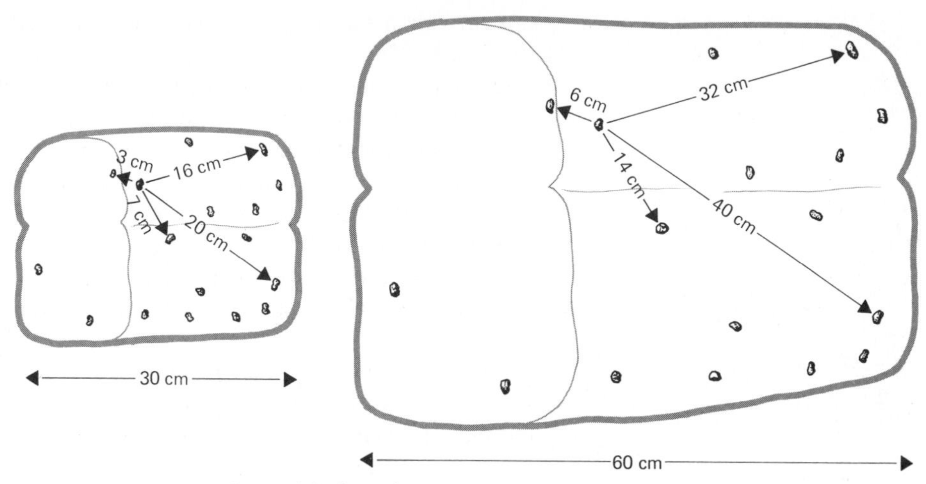

Figure 37.4 Expanding raisin bread.

course, no matter which raisin you start with. But the analogy must not be carried too far; in the bread it is the expanding dough that carries the raisins apart, but in the universe no pervading medium is presumed to separate the galaxies.

From the foregoing, it should be clear that if the universe is uniformly expanding, all observers everywhere, including us, must see all other objects moving away from them at speeds that are greater in proportion to their distances. As Hubble and Humason showed, that is precisely what observers on Earth do see.

As telescopes have grown larger and detectors have become more sensitive it has become possible to observe more and more remote galaxies with greater and greater speeds of recession (Figure 37.5). The cluster of galaxies shown in Figure 37.6 moves away from us with a speed of 108,000 km/s—36 percent of the speed of light. Even more remote galaxies have been found. The current (October 1985) record-holder is 3C 256, a radio galaxy for which H. Spinrad and S. Djorgovski at the University of California at Berkeley have measured a velocity of about 78 percent of the speed of light. There are quasars that have even higher velocities, provided that the redshift of their spectral lines is indeed produced by the cosmological Doppler effect.

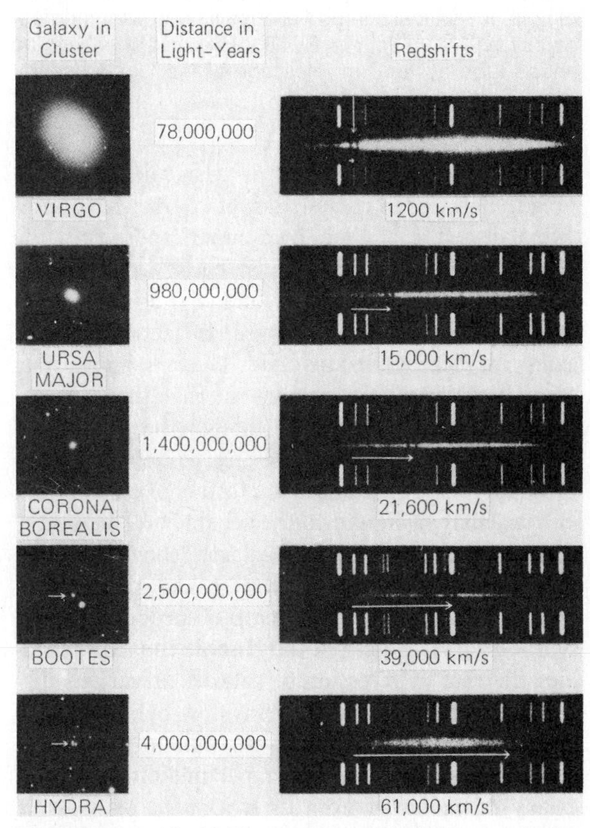

Figure 37.5 *(Left)* Photographs of individual galaxies in successively more distant clusters; *(right)* the spectra of those galaxies, showing the Doppler shift of two strong absorption lines due to ionized calcium. The distances are estimated from the faintness of galaxies in each cluster. (Palomar Observatory, California Institute of Technology)

The expansion of the universe does not imply that the galaxies and clusters of galaxies themselves are expanding. The raisins in our raisin bread analogy do not grow in size as the loaf expands. Similarly, their mutual gravitation holds galaxies and clusters together, and they simply separate as the

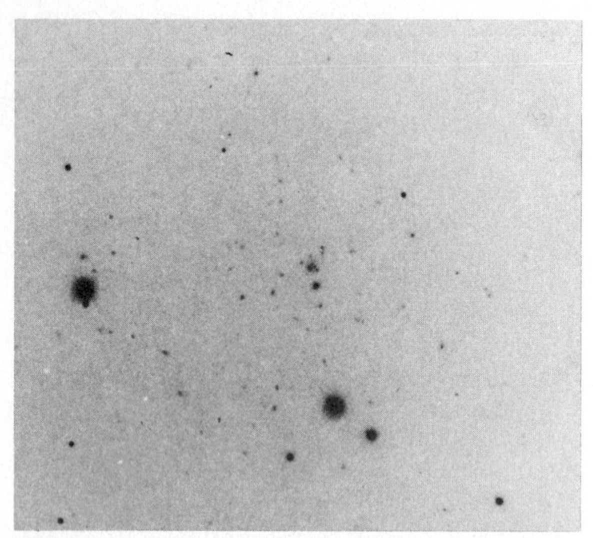

Figure 37.6 One of the remote clusters of galaxies for which there are measured redshifts (negative print). The cluster is receding from us at 36 percent the speed of light. It was discovered because its brightest member galaxy is a radio source, 3C295. Photographed with an image tube at the 3-m telescope of the Lick Observatory. (Courtesy H. Ford, UCLA)

universe expands, just as do the raisins in the bread. Galaxies in clusters do, of course, have individual motions of their own superimposed on the general expansion. Galaxies in pairs, for example, revolve about each other, and those in clusters move about within the clusters. In fact, a few galaxies in nearby groups and clusters move fast enough within those systems so that they are actually approaching us even though the clusters of which they are a part are moving, as units, away.

The relative distances to clusters of galaxies are known fairly well, and to the accuracy of the observations, remote clusters of galaxies show the same proportionality between velocity and distance as nearby galaxies. The constant of proportionality, symbolized H and called the Hubble constant, specifies the rate of recession of galaxies at various distances. The Hubble constant is now believed to lie in the range 50 to 100 km/s per million parsecs; in other words, if H is 75 km/s per million parsecs, a galaxy moves away from us at a speed of 75 km/s for every million parsecs of its distance. As an example, a galaxy that is 100 million parsecs away will be moving away from us at a speed of 7500 km/s.

Many people, concerned about such high speeds for galaxies, wonder if the large redshifts of their spectral lines might be due to some other cause than expansion of the universe. But such an explanation would require new physics, because no other mechanism is known that can produce the observed redshifts; gravitational redshifts, for example, would be accompanied by other effects not observed. But we hope the reader is convinced by now that the *easiest* way to interpret the redshifts is as cosmological Doppler shifts, indicating an expanding universe; otherwise, we would have to come up with a new physical principle that could produce redshifts.

If the redshifts were not Doppler shifts, we would also have to explain why the universe is *not* expanding. The general theory of relativity actually predicted the expansion of the universe before the confirming observations were made. Newton himself recognized that if the theory of gravitation applied to all of the objects in the universe, and if the universe was *finite* in size, then the universe must necessarily collapse as all of the objects were attracted to one another. Newton thought, however, that an *infinite* universe could perhaps be stable.

In 1917, Einstein applied the new gravitational theory embodied in his general theory of relativity (Chapter 11) to the universe as a whole; he could find no solutions to the field equations that gave a homogeneous static universe, *even if it is infinite.* Einstein consequently altered the field equations with the introduction of a new term, called the *cosmological constant,* which represents a repulsion that can balance gravitational attraction over large distances and permit a static universe. There is, however, no evidence for such a repulsion in nature, and it seemed to spoil the elegance of the theory to introduce the cosmological constant *ad hoc* so that the universe would conform to preconceived notions.

In the decade following Einstein's introduction of his static model of the universe, a number of other investigators found that the field equations also have solutions for nonstatic universes. A nonstatic universe could either contract or expand, but if it is now contracting, unless it is very young, it is hard to see why it has not already collapsed; expanding models were therefore more attractive. The most famous models for an expanding universe were those of the Dutch astronomer W. de Sitter, the Russian mathematician Aleksandr Friedmann, and the Belgian priest and cosmologist Abbé Georges Lemaître. Friedmann, whose solutions are most seriously considered today, showed that relativity theory is compatible with an infinite, homogeneous expanding universe for all values of the cosmological constant, including zero. Lemaître favored an expanding universe that still had a cosmological constant. All expanding solutions, however, predict that there should be a proportionality between the distances of remote objects and the speeds with which they recede from us. Lemaître, in particular, was aware of the significance of observations being made in the United States that might demonstrate the expansion of the universe.

American physicist George Gamow reports that Einstein said, upon learning that his relativity theory is compatible with an expanding universe, as observed, that the introduction of the cosmological constant was "the biggest blunder of my life."

37.2 Cosmological Models

Let us now see what theory predicts in detail for the past and future of the universe. Common to most cosmological theories is the cosmological principle—the assumption that the universe is homogeneous—and we will make that assumption here. We shall also adopt the best description of gravitation yet found, that of general relativity, and for simplicity and in common with nearly all cosmologists today, we assume that the cosmological constant is zero. Of course, the universe may not be, in reality, described by so simple a model as the best theory we know today. But it is the purpose of science to find descriptions of the way nature behaves in terms of models, and cosmology is no exception. The question is, can we find a model (theory, hypothesis) that *works?*

The models that we will consider here all presume that the universe began at a particular finite time in the past and that the universe is evolving today. Since the models are also consistent with general relativity, they are called *relativistic evolving models* of cosmology. These models make solid, testable predictions about such observable quantities as the age of the universe and the abundance of helium relative to hydrogen. As we shall see, these predictions appear to be consistent with what is actually observed.

(a) Maximum Age of the Universe

As the universe expands, galaxies and clusters of galaxies separate from each other. Thus if we extrapolate backward in time, we would find them coming together, until some time in the distant past when all matter was crowded to an extreme density—a condition that marks a unique beginning of the universe, or at least of that universe we can know about. At that beginning, the universe suddenly began its expansion with a phenomenon called the *big bang*. We return in the next section to the physical conditions of the universe at and just after the big bang.

Now the total amount of matter and energy of the universe—presumed to be conserved (constant in time)—creates gravitation, whereby all objects pull on all other objects (including light). This mutual attraction must *slow* the expansion, which means that in the past the expansion must have been at a greater rate than it is today. How much greater depends on the importance of gravitation in decelerating the expansion. At the extreme, if the total mass-energy density has always been low enough that gravitation is ineffective (an essentially "empty" universe), the deceleration would be zero, and only in that case would the universe always have been expanding at the present rate.

Clearly that extreme of an empty universe corresponds to the greatest age of the universe (since the big bang), because if the expansion were faster in the past, galaxies would have reached their present separations in a smaller time. Consequently, we can obtain an estimate of the upper limit to the age of the universe by asking how long it would take for distant galaxies, always moving away from us at their present rates, to have reached those distances. Call that maximum possible age T_o. Now the Hubble law states

$$V_r = Hr,$$

where V_r is the radial velocity of a galaxy of distance r. But velocity is just distance divided by time; that is, $V_r = r/T_o$. Hence,

$$\frac{r}{T_o} = Hr,$$

or

$$T_o = \frac{1}{H}.$$

We see, then, that the maximum age of the universe is just the reciprocal of the Hubble constant, which is believed to be no smaller than about 50 km/s per million parsecs. Since there are 3.086×10^{19} km in a million parsecs, H = 50 km/s per 3.086×10^{19} km, or 1.62×10^{-18} per second. The corresponding value of T_o is 6.172×10^{17} s or 1.95×10^{10} years. The actual age of the universe must be less than T_o, because the expansion has to be decelerating. If H is equal to 100 km/s, as some recent observations indicate, then the universe can be no more than about 10 billion years old.

The best alternative method of estimating the age of the universe is from calculations of the evolution of the oldest stars we know, namely, the members of globular clusters. The best current models yield ages in the range 12 to 20 billion years, and thus are in approximate agreement with the age derived from determination of H. If H does eventually prove to be as large as 100 km/s, however, then we would be faced with a problem, since the oldest stars would be older than the universe itself—and that, of course, cannot be! We would then have to reconsider both our models for stellar evolution and our basic cosmological model.

Radial velocities of galaxies can be observed unambiguously from the Doppler shifts of their spectral lines, so the accuracy of H depends on the accuracy of our measurements of the distances to galaxies. Thus, the age of the universe, and the extragalactic distance scale are inextricably connected.

(b) The Scale of the Universe

By assumption (the cosmological principle), the universe is always homogeneous on the large scale. Consequently, the expansion rate must be uniform (the same in all directions), so that the universe must undergo a uniform change in scale with time. It is customary to represent that scale by R. The actual value of the scale is arbitrary—we could think of it as being the distance between any two representative objects or points in space, since R changes in the same way everywhere. R plays the same role as the "scale of miles" on a terrestrial map; it tells us by how much the universe has expanded (or contracted) at any time. For a static universe, R would, of course, be constant, but in an expanding universe, R increases with time. The entire dynamical history of the universe is provided by the mathematical description of R with time.

The field equations of general relativity can be solved to learn how R varies with time if the amount by which the universe is decelerating is specified. It is difficult, however, to measure whether or not the rate of expansion of the universe is decreasing with time. In order to do so, one must either measure both the local Hubble constant (or distance scale) and the mean density of matter in space, or alternatively one can look back in time and try to detect differences in the expansion rates then and now.

By "mean density" we mean the mass of matter (including the equivalent mass of energy) that would be contained in each unit of volume (say, one cubic centimeter) if all of the stars, galaxies, and other objects were taken apart, atom by atom, and if all of those particles, along with the light and other energy, were distributed throughout all space with absolute uniformity.

The reason that H and the mean density are the two relevant parameters can be seen by drawing an analogy with the launching of a rocket ship from the surface of the Earth. If we launch a rocket at too low a velocity, the Earth's gravitational force will be great enough to overcome the rocket's upward motion and force it to return to the ground. If, on the other hand, the speed of the rocket exceeds a critical value (about 11 km/s), the rocket can escape from the Earth. In the case of the expansion of the universe, the expansion velocity H is analogous to the launch speed of the rocket. If H is high enough, then the mutual gravitation of all the matter in the universe, which can be estimated from the mean density, will not be strong enough to halt the expansion. If H is low enough or the mean density high enough, then the expansion may be slowed or even reversed, in which case the universe would begin to shrink.

As the universe expands, the Hubble constant and mean density both decrease, but the current values of these observational parameters can be combined to form a single quantity, denoted q_o, and called the *deceleration parameter*.* In the evolving relativistic cosmologies under discussion, q_o must have a positive value, indicating a slowing of the expansion. We shall return in Section 37.4 to the problem of determining q_o observationally. For now, we shall treat it as a free parameter to see what

---

* $q_o = \frac{8\pi G}{3}\frac{\rho}{H^2}$, where G is the constant of gravitation, and ρ and H are the instantaneous values of the mean density and the Hubble constant, respectively.

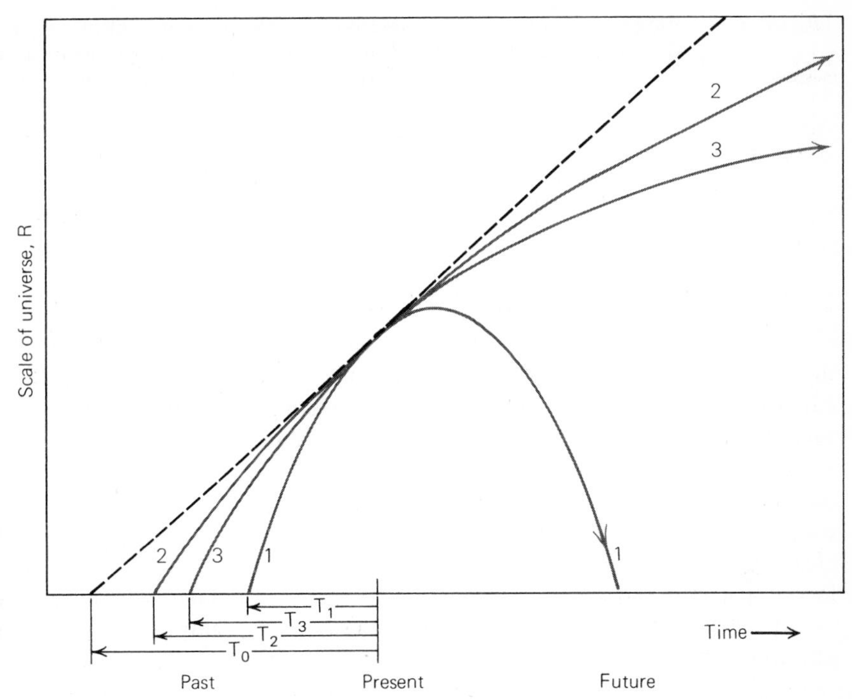

Figure 37.7 A plot of *R(t),* the scale of the universe, against time for various cosmological models. Curve 1 represents the class of solutions for closed universes, curve 2 represents the class for open universes, and curve 3 is the critical solution for the boundary between open and closed universes. The dashed line is the case for an empty universe.

are the possible solutions for R that are allowed by general relativity.

The solutions are shown in Figure 37.7. Time increases to the right and the scale, R, upward in the figure. Today, marked "present" along the time axis, R is increasing. The straight dashed line corresponds to the empty universe with no deceleration; it intercepts the time axis at a time T_o in the past. The other curves represent varying amounts of deceleration and start from the big bang at shorter times in the past.

If the deceleration is above a critical value, corresponding to q_o greater than one-half, R is given by curve 1; we see that in this case the universe stops expanding some time in the future and begins contracting. Eventually, the scale drops to zero, with what University of Texas physicist John A. Wheeler calls the "big crunch." In this case, the universe is said to be *closed,* for it cannot expand forever. It is tempting to speculate that another big bang might follow the "crunch," giving rise to a new expansion phase, and ensuing contraction, perhaps oscillating between successive big bangs indefinitely in the past and future. Such speculation is sometimes referred to as the *oscillating* theory of the universe, but it is not really a theory, for we know of no mechanism that can produce another big bang. General relativity (and other theories) predicts, instead, that at the crunch the universe will collapse into a universal black hole. (Of course, we have no complete theory for the first big bang, either!) In any case, the oscillating theory is a speculation on a possible variation of the closed model of the universe.

Alternatively, if q_o is *less* than one-half (curve 2 in Figure 37.7), gravitation is never important enough to stop the expansion, and the universe expands forever. In this case the universe is said to be *open.* At the critical value, $q_o = \frac{1}{2}$ (curve 3), the universe can just barely expand forever; it is the boundary between the families of open and closed universes. The mean density to make $q_o = \frac{1}{2}$ is referred to as the "critical density" to close the universe. The $q_o = \frac{1}{2}$ universe has an age of exactly $\frac{2}{3}T_o$—the age of the empty universe. Open universes have ages between $\frac{2}{3}T_o$ and T_o, and closed universes ages of less than $\frac{2}{3}T_o$.

(c) Curvature of Spacetime in the Universe

We saw in Section 33.3 that the presence of matter and energy produces a curvature of spacetime, and that in general relativity we describe objects as mov-

TABLE 37.1 Parameters of Evolving Relativistic Cosmological Models for Which the Cosmological Constant is Zero and for Which H = 75 km/s per Million Parsecs

MODEL (Curve in Figure 37.7)	KIND OF UNIVERSE	AGE, T, SINCE BIG BANG (Units of 10^9 yr)	DECELERATION PARAMETER, q_0	SIGN OF CURVATURE, k	MEAN DENSITY, ρ (g/cm^3)
1	Closed (oscillating?)	$T < 9$	$q_0 > 1/2$	$+1$	$\rho > 10^{-29}$
2	Open	$9 < T \leq 13$	$q_0 < 1/2$	-1	$\rho \leq 10^{-29}$
3	Flat	$T = 9$	$q_0 = 1/2$	0	$\rho = 10^{-29}$

ing unaccelerated along curved geodesics in spacetime. The mathematics required to calculate the curvature is very complex around separated bodies such as those in the solar system. For the universe as a whole, however, the assumption of the cosmological principle vastly simplifies the problem. Locally (near stars and galaxies) the curvature is complex, but on the large scale it must be uniform and the same everywhere.

If the universe is closed, in the sense that material bodies cannot separate forever, then neither can light nor other radiation. All geodesics are closed, and the curvature of spacetime is said to be positive. It does *not* mean that theoretically a light ray beamed from Earth will literally come back to its source, however; first, in passing near discrete stars and galaxies it will be deflected slightly (see the discussion of gravitational lenses in Section 33.4a), and second, as the universe expands and changes scale, so does the radius of curvature of the light beam. It does mean, though, that light, as well as material particles, starting at the big bang must return to the big bang (or crunch); there is no possibility of ever communicating with an infinite universe. The universe is closed, then, in the broadest meaning of that word. The closed universe is *unbounded*, for no edge can ever be observed, but it is still *finite*.

On the other hand, if the universe is open, the curvature of spacetime is said to be negative. Geodesics are curved, but never close on themselves, like the path of a tiny being walking always straight ahead on a surface shaped like a saddle. The open universe is curved but infinite.

Spacetime in the critical universe with $q_o = 1/2$, separating the open and closed ones, has zero curvature; in this case geodesics are the kinds of straight lines to which Euclid's geometry applies. This universe is said to be *flat* or *Euclidean*. The flat, $q_o = 1/2$ universe is also called the *Einstein–de Sitter* universe, for it is mathematically equivalent to a particular model considered by those scientists subsequent to Einstein's early ideas about a static universe.

(d) Summary of Relativistic Evolving Models of the Universe

The possibilities described in the previous paragraphs are summarized in Table 37.1. To be specific, a particular value of the Hubble constant, H, has been assumed: 75 km/s per million parsecs. If H is actually 50 km/s per million parsecs, the ages should be 1.5 times the values given in the table and the mean densities 4/9 as great as those given. The sign of the curvature of spacetime is conventionally indicated by the quantity k, which takes on the values -1, 0, and $+1$ for negative, zero, or positive curvature.

(e) Other Cosmologies

We emphasize that the foregoing discussion is based on particular assumptions—the cosmological principle, conservation of mass-energy, and general relativity theory with zero cosmological constant. Of course, any or all of these assumptions could be wrong, and if so, there is an infinity of other possibilities.

Another model of cosmology, which received wide attention in the 1950s, is the *steady-state* theory of Hermann Bondi, Thomas Gold, and Fred Hoyle. The steady state is based on a generalization of the cosmological principle called the *perfect cosmological principle*, in which it is assumed that on the large scale the universe is not only the same everywhere, but for all time. In the steady state, mass-energy is *not* conserved, for as the universe expands and would otherwise thin out, new matter is continuously being created to keep the mean density the same at all times. (Hence the steady state is also called the *continuous creation* theory.) The theory also predicts at what rate matter is created to assure that there will always be the same admixture of young and old stars and galaxies. The creation would occur so gradually (presumably as individual atoms coming into being here and there) that we would not notice it. The steady-state universe is infinite and eternal and has much philosophical appeal. But we have already seen that certain objects—especially quasars—seem to suggest a secular evolution of matter in the universe, which violates the perfect cosmological principle. Moreover, we shall see below that there is direct evidence that the universe has evolved from a hot dense state, strongly supporting the idea of a big bang.

The models of the universe, other than those summarized in Table 37.1, that have received the most interest are those based on general relativity, but with nonzero cosmological constant. Although there is no evidence for a cosmological constant, its introduction by Einstein was nevertheless allowed mathematically in his derivation of the field equations, and general relativity could be quite correct even if the cosmological constant is *not* zero. By assigning various values to the constant, a whole new range of possibilities becomes available, and in particular, the universe can be much older than $1/H$ (Figure 37.8). If H is as high as 100 km/s, as suggested by some recent observations, then the time since the big bang would have to be less than 10^{10} years, which is less than our current estimate for the age of the oldest globular star clusters. One way out of this difficulty is to adopt a value of the cosmological constant greater than zero. Most cosmologists, however, prefer to stick to the simplest case (zero cosmological constant) unless or until a more definitive determination of the value for the Hubble constant requires us to abandon those models.

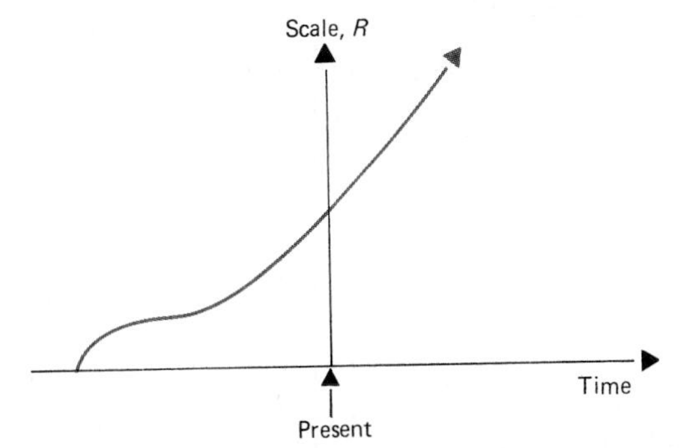

Figure 37.8 The change in scale with time for a universe with a positive cosmological constant.

37.3 THE BEGINNING

Georges Lemaître was probably the first to propose a specific model for the big bang itself. He envisioned all the matter of the universe starting in one great bulk he called the *primeval atom*. The primeval atom broke into tremendous numbers of pieces, each of them further fragmenting, and so on, until what was left were the present atoms of the universe, created in a vast nuclear fission. Lemaître thought the left-over radiation from that cosmic fireball was the cosmic rays. In a popular account of his theory he wrote, "The evolution of the world could be compared to a display of fireworks just ended—some few red wisps, ashes and smoke. Standing on a well-cooled cinder we see the slow fading of the suns and we try to recall the vanished brilliance of the origin of the worlds."

We know today that most cosmic rays are not from the big bang, but probably from supernovae in our Galaxy. We also know much more about nuclear physics, and that the primeval fission model cannot be correct. Yet Lemaître's vision inspired more modern work, and in some respects was quite prophetic.

In the 1940s the American physicist George Gamow suggested a universe with the opposite kind of beginning—nuclear fusion. He worked out the details with Ralph Alpher, and they published the results in 1948. (They added the name of physicist Hans Bethe to their paper, so that the coauthors would be Alpher, Bethe, and Gamow, a pun on the first three letters of the Greek alphabet: alpha, beta, and gamma.) Gamow's universe started with fundamental particles that built up the heavy elements by fusion in the big bang. His ideas were close to our modern view, except that the conditions in the primordial universe were not right for atoms to fuse to carbon and beyond, and only hydrogen and helium should have been formed in appreciable abundances. The heavier elements, we think, formed later in stars (Chapter 32).

(a) Standard Model of the Big Bang

The modern theory for the evolution of the early universe is called the *standard model* of the big bang. The details were worked out in a 1967 paper by

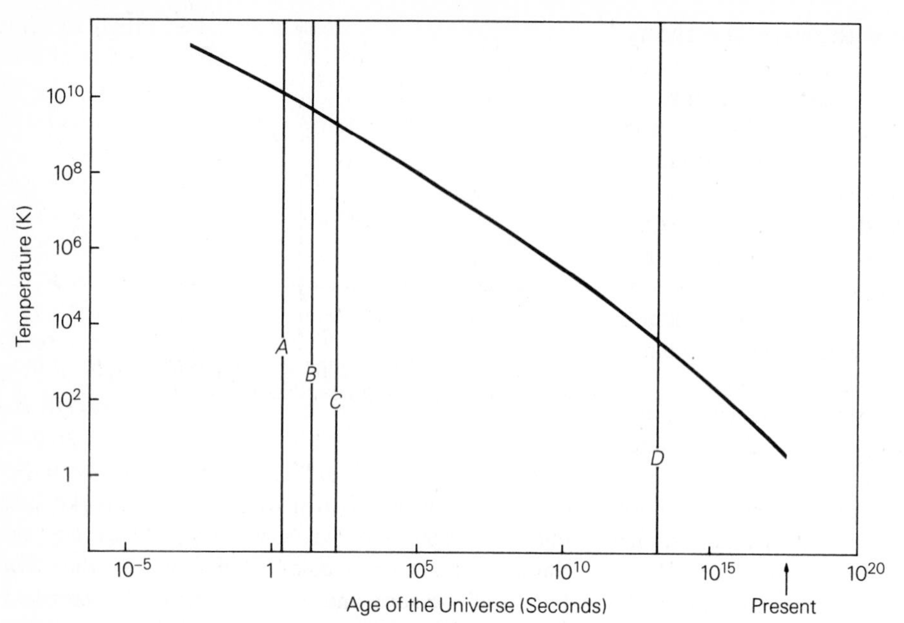

Figure 37.9 Standard model of the early universe showing how the temperature varies with time. The vertical line labeled *A* designates approximately the time at which neutrinos stop interacting with matter. *B* denotes the time when positrons and electrons annihilate. Helium synthesis occurs at time *C*, and the universe becomes transparent to radiation at time *D*.

Robert Wagoner, at Stanford University, and William Fowler and Fred Hoyle at Caltech. Three simple ideas hold the key to tracing the changes that occurred during the first few minutes after the universe began.

The first essential piece of information is that the universe cools as it expands. Figure 37.9 shows how the temperature changes with the passage of time. In the first fraction of a second, the universe was unimaginably hot. By the time 1⁄100 of a second had elapsed, the temperature had dropped to 100 billion (10^{11}) degrees. After about three minutes, the temperature had reached about one billion (10^9) degrees, still some 70 times hotter than the interior of the sun. After 700,000 years, the temperature was down to a mere 3000 degrees, and the universe has continued to cool since that time. All of these temperatures but the last are derived from theoretical calculations, since obviously no one was there to measure them directly. As we shall see, however, we have actually detected the feeble glow of radiation emitted at a time when the universe was about 700,000 years old. Indeed the fact that we have done so is one of the strongest arguments in favor of the validity of the big bang model.

The second step in understanding the evolution of the universe is to realize that at very early times it was so hot that collisions of photons—that is, collisions of discrete units of pure electromagnetic energy—could produce material particles. That is a startling concept that may seem to defy common sense. It is important to realize, however, that the conditions we are talking about are far beyond the realm of our daily experience. Only in high-energy accelerators have we succeeded in achieving the temperatures and densities that are needed to produce matter from pure energy.

Suppose that two photons do collide and disappear with all of their energy going into the production of two or more particles of matter. What kind of particles will be produced? Even at rest, a particle has an energy E associated with it, and the amount of that energy can be calculated from Einstein's equation

$$E = mc^2,$$

where m is the mass of the particle and c is the speed of light (see Chapter 11). If two photons are to collide and produce two particles of mass m, then each photon must have an energy equal to mc^2.

But what determines the energy of a photon? That question brings us to our third concept. The energy of a photon can be estimated well enough for our purposes by multiplying the temperature of the radiation by a constant k, which is known as Boltzmann's constant. In other words, the higher the temperature, the higher the energy of a typical photon, and the more massive the particles that can be produced by the collision of two such photons. To take a specific example, at a temperature of 6 billion (6×10^9) degrees, the collision of two typical photons can create a positron and an electron. The much more massive proton can be created only in an environment that has a temperature in excess of 10^{14} degrees.

Keeping these three ideas in mind, we will now trace the evolution of the universe from the time that it was about 0.01 second old and had a temperature of about 10^{11} degrees. Why not begin at the very beginning? The reason is that when the universe was less than about one one-hundredth of a second old and the temperature was hotter than 10^{12} degrees, the universe was filled with strongly interacting subatomic particles. The theory of these particles is difficult to deal with, and it is only very recently that theoretical physicists have begun to speculate about what the universe might have been like at this time. We will look at some of those speculations later in this chapter. There are no theories, however, that allow us to penetrate to a time when the universe was less than about 10^{-45} seconds old. When the universe was that young, the density was so high that the theory of general relativity is no longer applicable, and we have as yet no theory that can deal with such extreme conditions.

The universe, one one-hundredth of a second after the beginning, consists of a soup of matter and radiation. Each particle collides rapidly with other particles. The temperature is not high enough so that colliding photons can produce neutrons and protons, but it is high enough for the production of electrons and positrons. Neutrons and protons are present, however, leftovers from an even younger and hotter universe. There may also be a sea of exotic particles that will later play a role as "dark matter." The neutrons can interact with positrons to produce protons and antineutrinos (and vice versa), and electrons can combine with protons to produce neutrons and neutrinos (and again vice versa). The picture then is of a seething cauldron of a universe, with photons colliding and interchanging energy, often meeting in such a violent impact that the photons themselves are destroyed and leave in their wake an electron-positron pair, of neutrons being converted to protons and protons to neutrons through collisions of particles. It is much too hot for protons and neutrons to combine to form heavier atomic nuclei.

As it happens, the neutron is 1/7 of 1 percent heavier than the proton, and the free neutron—that is, one that is not contained within an atomic nucleus with other neutrons and protons—is unstable and decays spontaneously to a proton and an electron. Since the neutron is more massive than the proton, it takes slightly more energy to convert a proton to a neutron than vice versa. As the temperature of the universe drops, there comes a time when the typical electron no longer has enough energy to convert a proton to a neutron, but protons themselves can still be produced by collisions between neutrons and positrons or by the spontaneous decay of neutrons. This tiny difference in mass explains why there are many times more protons than neutrons—about 7 times more, in fact—in the universe.

By the time the universe is a little more than 1 second old, and the temperature has dropped to a mere 10 billion degrees, the density has dropped to the point where neutrinos no longer interact with matter, but simply travel freely through space. In fact, these neutrinos should now be all around us. Since they have been traveling through space unimpeded and hence unchanged since the universe was 1 second old, observations of them and measurement of their properties would offer one of the best tests of the big bang model. Unfortunately, the very characteristic that makes them so useful, the fact that they interact so weakly with matter that they have survived unaltered for all but the first second of time, also renders them undetectable, at least with present techniques. Perhaps someday someone will devise a way to capture these messengers from the past.

The universe continues to expand and cool. When it is nearly 14 seconds old, the temperature has reached 3 billion degrees. Typical photons have too little energy to produce electron-positron pairs through collisions, and so the electrons and positrons themselves begin to collide and annihilate each other (see Chapter 29 for a discussion of matter and antimatter). Fortunately for us there is a slight excess of matter over antimatter, and we owe our existence to that excess!

Protons and neutrons begin to be bound into stable atomic nuclei only when the universe is about 3 minutes and 46 seconds old and when the temperature is down to 900 million degrees. The first step in building atomic nuclei is the collision of a neutron and proton to form deuterium (heavy hydrogen), and essentially all of the neutrons are used up by this reaction. At higher temperatures, the deuterium is immediately blasted apart by interactions with a photon before it can interact with a third particle to form a stable nucleus. At the temperatures and densities reached between 3 and 4 minutes after the beginning, however, deuterium can survive long enough so that collisions can convert nearly all of it to helium, which has an atomic mass of 4 (2 protons and 2 neutrons). There is no stable particle of mass 5, however, and we believe that very few heavier elements are formed at this early stage of the universe. Instead, heavy elements are predominantly produced later deep in the interiors of stars (Chapter 32).

According to models of the big bang, there are approximately 2 neutrons for every 14 protons at the time that helium nucleosynthesis occurs. If 2 neutrons and 2 protons are required to form a helium nucleus, then 12 protons remain. One prediction of the big bang model is therefore that there should be 1 helium atom for every 12 hydrogen atoms. In units of mass, this means that about 25 per cent of the matter in the universe should be helium and about 75 percent should be hydrogen. Within the uncertainty of the observations, this ratio is precisely what is observed. Of course, a small enhancement of helium must have resulted from nucleosynthesis in stars, but we estimate that ten times as much helium was manufactured in the first 200 seconds of the universe as in all the generations of stars during the succeeding 10 to 20 billion years.

For the next few hundred thousand years, the universe was much like a stellar interior—hot and opaque, with radiation being scattered from one particle to another. By about 700,000 years after the big bang, the temperature had dropped to about 3000 K and the density of atomic nuclei to about 1000 per cubic centimeter. Under these conditions, the electrons and nuclei combined to form stable atoms of hydrogen and helium. With no free electrons to scatter photons, the universe became transparent, and matter and radiation no longer interacted; each evolved in its separate way.

One billion years after the big bang, stars and galaxies had probably begun to form, but we are not sure of the precise mechanisms. Certainly, however, deep in the interiors of stars matter was reheated, stars began to shine, nuclear reactions were ignited, and the gradual synthesis of the heavier elements began.

Now we must emphasize that the fireball must not be thought of as a localized explosion—like an exploding superstar. There were no boundaries and no site of the explosion. It was everywhere. The fireball still exists, in a sense. It has expanded greatly, but the original matter and radiation are still present and accounted for. The stuff of our bodies came from material in the fireball. We were and are still in the midst of it; it is all around us.

But what happened to that radiation released when the universe became transparent at the tender age of just under a million years?

(b) The Cosmic Background Radiation

The question was first considered by Alpher and Robert Herman, both associates of Gamow. They realized that just before the universe became transparent it must have been radiating like a black body at a temperature of 3000 K. If we could have seen that freed radiation just after neutral atoms formed, it would have resembled that from a reddish star. But that was at least ten thousand million years ago, and in the meantime the scale of the universe has increased a thousandfold. The light emitted by the once hot gas in our part of the universe is now thousands of millions of light years away.

To observe that glow of the early universe, we must look out in all directions in space to such great distance—10 to 20 thousand million light years—that we are looking back in time through those 10 to 20 thousand million years. Now those remote parts of the universe, because of its expansion, should be receding from us at a speed within two parts in a million of that of light. The radiation from them would be redshifted to wavelengths a thousand times those at which it was emitted.

When a black body approaches us, the Doppler shift shortens the wavelengths of its light and causes it to mimic a black body of higher temperature. When a black body recedes, it mimics a cooler black body. Alpher and Herman predicted that the glow from the fireball should now be at radio wavelengths and should resemble the radiation from a black body at a temperature of only 5 K—just a few degrees above absolute zero. But there was no way, in 1948, of observing such radiation from space, so

Figure 37.10 Arno A. Penzias *(right)* and Robert W. Wilson. (Bell Labs)

the prediction did not attract much attention and was forgotten.

In the mid-1960s, however, the idea occurred independently to Princeton physicist Robert H. Dicke, who realized that microwave radio telescopes could then be built that might detect that dying glow of the big bang. He and his Princeton colleagues confirmed that the theory was correct and began construction of a suitable microwave receiver on the roof of the Princeton biology building. They were not, however, the first to observe the radiation.

Unknown to them, a few miles away, in Holmdel, New Jersey, Arno Penzias and Robert Wilson (Figure 37.10) of the Bell Telephone Laboratories, were using the Laboratories' delicate microwave horn antenna to make careful measures of the absolute intensity of radio radiation coming from certain places in the Galaxy. But they were plagued with some unexpected background noise in the system that they could not get rid of. They checked everything, and eliminated the Galaxy as a source, also the sun, the sky, the ground, and even the equipment.

At one point they realized that a couple of pigeons had made their home in the antenna, and nested up near the throat of the horn where it was warmer. Penzias and Wilson could chase the birds away while they observed, but they found that the birds left, as Penzias puts it, a layer of white, sticky dielectric substance coating the inside of the horn. That substance would radiate, producing radio interference. They disassembled the horn and cleaned it, and the unwanted noise did go down somewhat, but it did not go away completely.

Finally, Penzias and Wilson decided that they had to be detecting radiation from space. Penzias mentioned it in a telephone conversation with another radio astronomer, B. Burke, who was aware of the Princeton work. Burke got Penzias and Wilson in touch with Dicke, and it was soon realized that the predicted glow from the primeval fireball had been observed. Since then the radiation has been very thoroughly checked throughout the entire radio spectrum, with observations from ground-based radio telescopes, with instruments carried aloft in balloons, with a receiver in a U2 reconnaissance airplane, and with detectors flown in spacecraft. The microwave background radiation closely matches that expected from a black body with a temperature of 2.8 K.

Penzias and Wilson received the Nobel Prize for their work in 1978. And perhaps almost equally fitting, just before his death in 1966, Lemaître learned about the discovery of his "vanished brilliance."

(c) Further Information from the Cosmic Background Radiation

The faint glow of radio radiation is now called the *cosmic background radiation* (sometimes, *CBR*). It has now been observed at many wavelengths, and all observations are compatible with the CBR being redshifted radiation emitted by a hot gas. Among other things, if correctly interpreted, the CBR shows that the universe has evolved from a hot, uniform state—as opposed, for example, to the steady-state theory.

At any given wavelength, the cosmic background radiation is extremely isotropic on the small scale. In directions that differ by only a few minutes of arc any fluctuations in its intensity must be less than one part in ten thousand. The uniformity of the radiation tells us that at an age of less than a million years the universe was extremely uniform in density. But at least *some* density variations had to be present to allow matter to gravitationally clump up to form stars and galaxies. The isotropy of the CBR, therefore, puts interesting constraints on theories of supercluster, cluster, and galaxy formation.

On the other hand, a large-scale anisotropy in the cosmic background radiation has now been established, in the sense that it is slightly brighter in one direction than in the opposite direction in the sky. This is because of our own motion through space. If you approach a black body, its radiation is all Doppler-shifted to shorter wavelengths and resembles that from a slightly hotter black body; if you move away from it, the radiation appears like that from a slightly cooler black body. Such an effect has been searched for and observed in the microwave background.

The measurements of this relative speed are very difficult because the difference in intensity is very tiny compared with the radiation from the Earth's own atmosphere. The measurements must therefore be made from high-flying balloons, aircraft, or spacecraft. The experiment has been carried out so far by three independent groups of researchers, whose results are in excellent agreement. The data imply that our Local Group of galaxies is moving at a speed of about 630 km/s with respect to the CBR, or with respect to the uniform expansion of the universe as a whole. It can be thought of as a peculiar motion of the Local Group, superimposed on the general expansion.

But what can cause the Local Group to have such a high peculiar velocity? One interpretation is that it is due to the gravitational attraction of the Local Supercluster. The Supercluster is expanding, but less rapidly than the universe as a whole. Thus we find that the Virgo cluster (which lies near the direction to the center of the Supercluster) is separating from us at about 1000 km/s, but if it were not for the gravitation of the Supercluster, we and the Virgo cluster would be flying apart at about 1400 km/s. (Some popular accounts have described the Local Group as "falling toward the Virgo cluster," but as we see, this is misleading.) If the interpretation is correct, we learn something of the dynamics of the Local Supercluster from the large-scale anisotropy of the cosmic background radiation.

This simple interpretation cannot, however, be entirely correct. The motion of the Local Group and the anisotropy of the cosmic background radiation do not point directly toward Virgo, but rather at an angle to this direction and more toward the Hydra-Centaurus cluster of galaxies. Whether the component of motion away from Virgo indicates that there are still larger-scale structures that influence the flow gravitationally, or whether there are simply random or turbulent motions of several hundred km/s superposed on the otherwise smooth expansion is a question that remains to be answered.

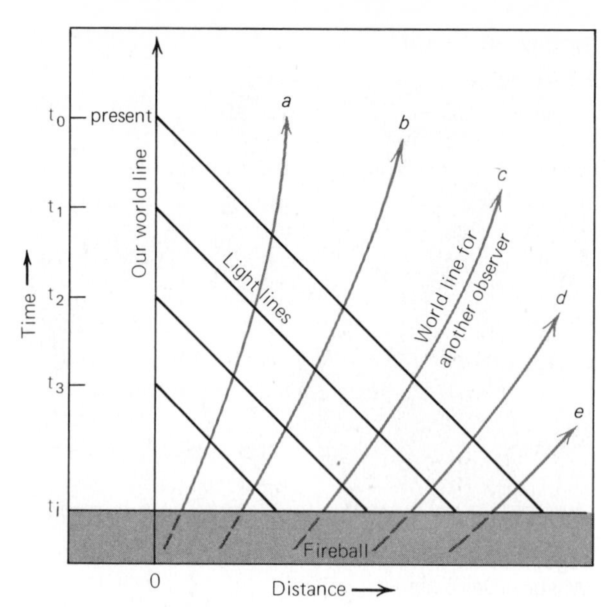

Figure 37.11 Schematic representation of world lines for us and for several other hypothetical observers on distant galaxies in the universe on a two-dimensional cross section of spacetime. The straight diagonal lines are light paths.

(d) A Look Back to the Big Bang

Since the cosmic background radiation comes from the time when the fireball first became transparent, it is at the farthest point in space to which we can presently observe. If we could see that radiation visually, it would be as if it were coming from an opaque wall, and no radiation from a more distant source could ever reach us—for that source would have to lie farther back into time where it would be behind that opaque wall.

Figure 37.11 is a schematic cross section of four-dimensional spacetime (see Chapter 33). The distance from us, in some arbitrary direction, is shown along the horizontal axis, and time increases along the vertical axis. Light paths are shown as 45° lines, as in Chapter 33. Our world line (geodesic) is upward along the time axis, and we are presently at time t_0. Times t_1, t_2, and t_3 are successively farther back into the past. The world lines of several other hypothetical observers on distant galaxies are shown and are labeled a, b, c, and so forth, in order of increasing distance. Note that they all recede from us with time as the universe expands. (Because of the peculiarity of spacetime diagrams of an expanding universe, the diagram is not strictly correct for uniform time and distance measures along the axes, but nevertheless serves to illustrate the situation.)

When we look out into space, we look back into time along a light geodesic (diagonal lines). Note that the light we see comes from successively more distant objects successively farther into the past at times when they were actually separated less far from us in spatial coordinates. There is a limit, however, to how far back we can look. The fireball is presumed to have become transparent at time t_i, and the shaded region at the bottom of the diagram corresponds to the opaque fireball into which we cannot see. Thus the limiting distance we could observe in a particular direction is at the point where the light path to us intercepts the top of the shaded region at time t_i.

As time goes on, we can see farther and farther away, and more galaxies would come into view (if we had a large enough telescope), for as the time from the big bang becomes greater we are looking farther into the past to see the fireball, and hence farther away in space. Note that at earlier times such as t_3, t_2, and t_1, we saw only relatively nearer objects, and there were fewer galaxies between us and the threshold provided by the fireball itself. Thus, not only does the universe expand with time, but the part of it accessible to observation becomes greater as well.

On Earth, the microwave radiation is very feeble compared with, say, sunlight. But far off in intergalactic space, that radio background is by far and away the most intense radiation around. The observed radiation comes equally from all directions and gives no direction to a "center" of the universe; the universe, its "center" and its origin are all around us. There is a boundary to the universe that is revealed to us by these photons, but it is a boundary in time, not in space.

The concept that there is a limiting distance beyond which we cannot see, combined with a finite age for the universe, provides an answer to an apparently simple yet surprisingly profound question: Why is the sky dark at night? Suppose we lived in an infinite, Euclidean, static universe that is uniformly filled with stars. Then in every direction in which we look, we should see, at some distance great or small, a star. Since the total number of stars is infinite, their total contribution to the night sky is potentially very great. In fact, it is possible to show mathematically that if the universe were infinite, the entire night sky should be as bright as the sun.

The universe, however, is no more than 20 billion years old. Light from stars more than 20 billion years away simply has not had time to reach the Earth. The universe may indeed be infinite, but we can observe only that portion that lies within a sphere centered on the Earth with a radius of 20 billion light years. Within this sphere, the density of stars is so low in most directions that our line of sight does not intercept a star.

37.4 THE FUTURE OF THE UNIVERSE

Is the universe open or closed? Will the expansion of the universe continue forever, or will it eventually reverse itself? There are several observational tests by which we hope to be able to distinguish between the evolving cosmological models. As of this writing, the observations are still not precise enough to reach a definite conclusion, but it is worth describing some examples of these tests.

(a) The Mean Density

One test requires the determination of the mean density of matter in space. We have noted that knowledge of this quantity is sufficient to determine whether or not the gravitational force of the matter in the universe is strong enough to overpower and halt the expansion. We can estimate the mean density from the number of galaxies and clusters we observe out to a given distance, and from a knowledge of the masses of these objects. There is considerable uncertainty in the masses of clusters of galaxies, and we do not know how much matter (if any) may exist in intergalactic space. Nevertheless, the best data available to date indicate that the total mass associated with the luminous parts of galaxies is only about 0.01 the critical density of matter required to halt the expansion of the universe. Even if we allow for the dark matter that is apparently associated with clusters of galaxies, the density is still no more than a few tenths of the critical density. Direct measurements of the total mass in the universe suggest, therefore, that the universe is open, but the estimates are too uncertain to be sure of this conclusion.

(b) The Deuterium/Hydrogen Ratio

The production of deuterium during the first few minutes after the big bang is extremely sensitive to the density of matter in the universe. If the density is high, then most of the deuterium will interact with other particles of matter to form helium. If the density is low, then a significant amount of deuterium will escape unscathed, and may still be observable today.

The proportion of deuterium in interstellar space is thought to be a measure of that formed in the fireball, for in stellar interiors it is rather quickly converted to helium. It is very difficult to detect deuterium in space, but careful measures show that the ratio of deuterium to hydrogen is probably in the range 10^{-4} to 10^{-5}. From this crude estimate the density of the fireball at the critical time can be inferred, and from that knowledge it is possible to predict what present-day density would result. The calculation suggests a present density of about 10^{-31} g/cm, again pointing to an open universe.

(c) The Hubble Law

Another less obvious test involves the speeds of receding galaxies at great distances. The evolutionary models we have described predict the expansion to slow down due to the gravitational forces between galaxies. Because of these changes in the rate of expansion of the universe, the radial velocities of remote galaxies should deviate from a relation exactly proportional to their observed distances; that is, a graph of radial velocity versus distance should not necessarily be a straight line for very distant galaxies (see Exercise 6). The exact form of the redshift-distance relation is different as predicted by each cosmological model.

The differences between the observable quantities predicted by various models are small until very large distances are reached—that is, until we look back through an appreciable interval of time. Unfortunately, precise distances of remote objects, as well as their luminosities and other characteristics, are very difficult to determine, and critically accurate observations are required.

One procedure for obtaining relative distances of clusters is from their brightest member galaxies. Figure 37.12 is a plot of the radial velocities (on a logarithmic scale) of nearly a hundred clusters of galaxies against the apparent magnitudes of their brightest members, adapted from a diagram by A. Sandage. The magnitudes of the brightest cluster galaxies are indicators of distance (Section 35.2a); they have been corrected for interstellar absorption in our Galaxy and for certain effects of the redshift. The different lines on the plot correspond to predictions of different cosmological models. We can see from the scatter of the points that more precise observations are needed to apply this test. Also, we do not yet know how to take account of changes in the luminosities of galaxies due to evolution of the stars in them.

In summary, observational tests of cosmological models appear to rule out the steady-state theory, but are not yet definitive enough to choose between the various evolving models predicted by general relativity. And of course the observations may well be compatible with many other cosmological theories not yet considered.

37.5 THE NEW INFLATIONARY UNIVERSE

While observations cannot yet tell us the fate of the universe, a remarkable new theory has been developed during the last decade that not only predicts the future but also resolves a number of problems with the big bang model that we have just described. The big bang model is successful in explaining the relationship between velocity and distance that is observed for galaxies; it accounts for the cosmic background radiation; and it explains why about 25 percent of the mass of the universe is in the form of helium.

There are, however, two important characteristics of the universe that the simple big bang model cannot explain. The first is the uniformity of the universe. The cosmic background radiation is the same, no matter which direction we look, to an accuracy of at least one part in ten thousand. There is, however, a maximum distance that light can have traveled since the time the universe began. This distance is called the horizon distance, because any two objects separated by this distance cannot ever have been in contact. No information, no

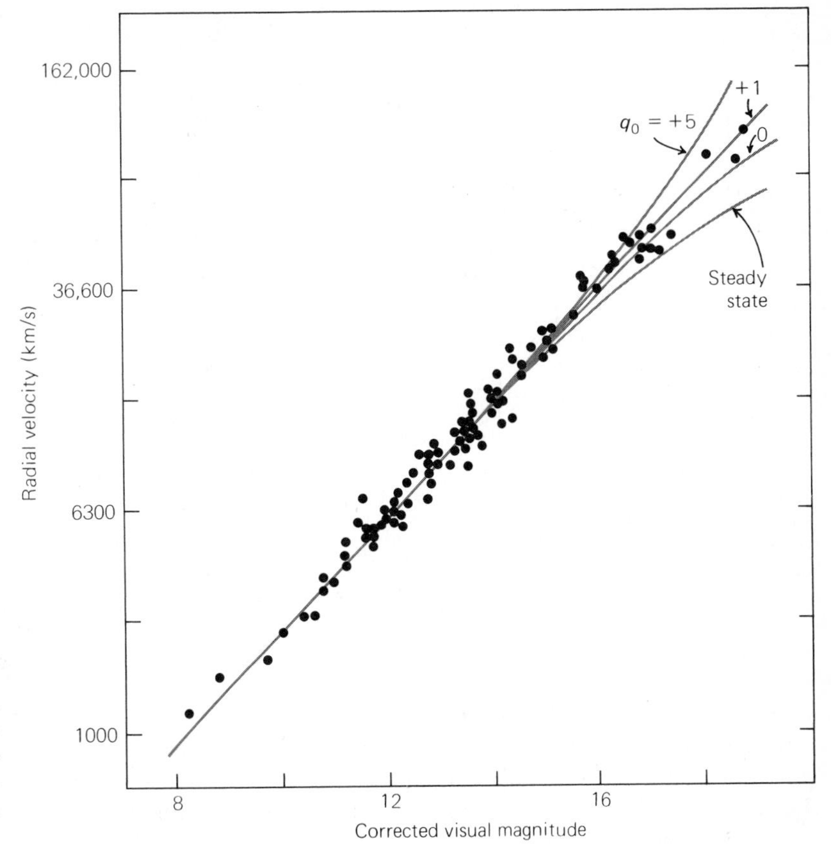

Figure 37.12 Hubble diagram for clusters of galaxies. Ordinates are radial velocities (on a logarithmic scale) and abscissas are the apparent magnitudes of the brightest cluster galaxies, corrected for interstellar extinction and certain effects of redshift. Points are observed values, and the solid curves are the predicted relations for various cosmological models. (Adapted from a diagram by A. Sandage)

physical process can propagate faster than the speed of light. One region of space separated by more than the horizon distance from another is truly beyond the horizon.

If we measure the CBR in two opposite directions in the sky, we are observing regions that are separated by more than 90 times the horizon distance. Why then are their temperatures so precisely the same? According to the standard big bang model, they have never been causally connected. The only explanation is simply that the universe started out being absolutely uniform. Scientists are always very uncomfortable, however, when they must appeal to a special set of initial conditions to account for the phenomena that they observe.

The second problem with the standard big bang model is the so-called flatness problem. As we have seen, observations are unable to tell us whether the expansion of the universe will continue forever or will ultimately slow and perhaps even come to a halt and reverse itself. The remarkable fact, however, is that the universe is so precisely balanced between these two possibilities that we cannot yet determine which is correct. There could have been, after all, so little matter that it would be obvious that the universe is open and that the expansion will continue forever. Alternatively, there could have been so much matter that the universe would be clearly and unambiguously closed. Instead, the amount of matter present is within a factor of ten of the value that corresponds to precise balance between these two situations. As we have seen, spacetime in the critical universe that separates the open and closed universes has zero curvature. It is termed a flat universe, and the question is why our universe is so very nearly flat.

The answers to the horizon and flatness problems may lie in the application of grand unified theories (GUTs) to cosmological problems. Theoretical physicists speculate (see Chapter 29) that the strong, weak, and electromagnetic forces are not three independent forces but rather different aspects of a single unified force. Theories based on this hypothesis make some predictions for energies

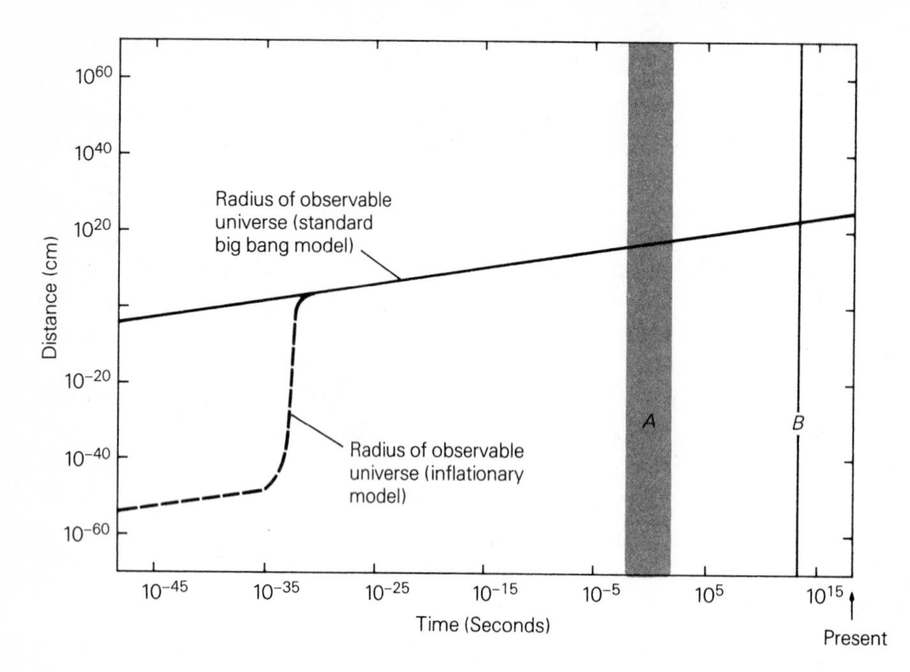

Figure 37.13 Radius of the observable universe as a function of time for the standard big bang model (solid line) and for the inflationary model (dashed line). The two models are the same for all times after 10^{-30} second. Electrons, positrons, and the lightest atomic nuclei are formed during the time interval labeled *A*. The universe becomes transparent to radiation at the time designated *B*.

that can be attained in high-energy accelerators. The most dramatic consequences occur, however, at energies ten times greater than any achieved on Earth. In fact, only the universe in its infancy offers a suitable laboratory for testing the validity of GUTs.

One model universe described by GUTs is called the new inflationary universe. The inflationary universe is identical to the big bang universe for all time after the first 10^{-30} second. Prior to that time there was a brief period of extraordinarily rapid expansion or inflation during which the scale of the universe increased by a factor of 10^{50} times more than was suggested by standard big bang models (Figure 37.13). Prior to the inflation, all of the universe that we can now see was causally connected. That is, the horizon distance was large enough to include all of the observable universe and there was adequate time for the universe to homogenize itself and come to thermal equilibrium.

The inflationary universe is also a universe that is now flat. This result can best be understood through an analogy. Consider a balloon that is being inflated. When it is partly filled, it has a small radius and is sharply curved. As it is inflated, the curvature becomes less and less, and it becomes more and more difficult to distinguish a portion of the surface of the balloon from a flat surface. So it is with the universe. The expansion has been so enormous that the universe is now very close to flat.

GUTs make some other predictions about the nature of the universe. For example, for reasonable choices for the parameters in the theory, it is possible to produce the excess of matter over antimatter that is crucial to our very existence. The theory also predicts that protons should not have an infinite lifetime but should decay on average after a period of time that is in excess of 10^{30} years. Fortunately, there are so many protons in our world, that it should be possible to detect a few protons that decay sooner than this average value, and experiments are being conducted to try to do so.

One solid prediction of the inflationary universe model is that the universe should be exactly flat, that is, that the mean density should precisely equal the critical density. (The only way to avoid this result is to assume that some special conditions occurred during the early expansion of the universe, and, of course, the desire to avoid an appeal to special initial conditions was one of the original motivations for devising the inflationary universe.) As we have seen, observations indicate that the amount of matter that we have so far found in galaxies and in clusters of galaxies is a factor of five to ten below the critical density. If the required additional matter is indeed present, then it must not emit light or we would have seen it. It cannot lie within galaxies, or even within the local supercluster, or we would have detected its gravitational effects. And it cannot be in the form of protons and neutrons or combinations thereof, including black holes, or there would be far less deuterium in the universe. What is this dark matter? As yet theory has produced no entirely satisfactory theoretical candidates, and of course we have not observed it.

This dark matter may, if it exists, be exactly what is required to explain the formation of galaxies (Section 36.4c). Because dark matter would solve so many problems in the theory of the origin and evolution of the universe and of large-scale structures within it, many physicists and astrophysicists have come to believe that it must exist. Before these ideas can be accepted as anything more than simply current fashion, however, they must stand the test of time and of experiments designed to prove or disprove the existence of certain types of particles that theory suggests may make up the dark matter.

37.6 CONCLUSION

This chapter may seem to conclude on an unsatisfactory note. Throughout this book we have traced the fascinating and often puzzling properties of the luminous matter in the universe. And now we find that visible matter may not even be the most important constituent of the universe. It may be that dark particles of a kind completely unknown in everyday experience have dominated the evolution of the universe. Do we even need this complication? Apart from the philosophical problems raised by the homogeneity and flatness of the universe, observations are still entirely consistent with an open universe with an age of 10 to 20 billion years. Is a model that requires that 90 percent or more of the matter in the universe be invisible really an improvement?

As is so often the case with new scientific theories, the model of the inflationary universe, while it answers many questions, raises a whole host of new ones. Whatever the ultimate judgment of the validity of this model, it is quite clear that the marriage of particle physics with cosmology has allowed us to probe much closer to the beginning of our universe than we would have thought possible only 20 years ago.

Thus we have concluded our exploration of the universe with descriptions about how it *might* be, but no assertion of how it *is*. Science does not, nor can it, provide definitive answers to all questions. There will always be a new frontier.

The following words, written by Edwin Hubble in 1936, are still appropriate:*

> Thus the explorations of space end on a note of uncertainty. And necessarily so. We are, by definition, in the very center of the observable region. We know our immediate neighborhood rather intimately. With increasing distance, our knowledge fades, and fades rapidly. Eventually, we reach the dim boundary—the utmost limits of our telescopes. There, we measure shadows, and we search among ghostly errors of measurement for landmarks that are scarcely more substantial.
>
> The search will continue. Not until the empirical resources are exhausted, need we pass on to the dreamy realms of speculation.

---

* *The Realm of the Nebulae*, Yale University Press, 1936. pp. 201–202. Quoted by permission of the publisher.

EXERCISES

1. A cluster of galaxies is observed to have a radial velocity of 60,000 km/s. Assuming that $H = 50$ km/s per million parsecs, find the distance to the cluster.
2. Plot the "velocity-distance relation" for the raisins in the bread analogy from the data given in Figure 37.4.
3. Repeat Exercise 2, but use some other raisin than *A* for a reference. Is your new plot the same as the last one?
4. Why can the redshifts in the spectra of galaxies *not* be explained by the absorption of their light by intergalactic dust?
5. Calculate the maximum possible age for the universe for the case where the Hubble constant, $H = 75$ km/s $\cdot$ 10^6 pc.
6. Assume that the radial velocities of galaxies have always been the same and are given *at this instant of time* by $V = Hr$, where $H = 50$ km/s per million parsecs. Note, however, that we do not observe the *present* distances of galaxies, but the distances they had when light left them on its journey to us. Now plot the relation between velocity and distance that would be obtained directly from *observations* (that is, corresponding to *measured* distances, not present distance). Consider several distances, out to 2×10^9 pc. Discuss the shape of the curve. How would this curve differ if the expansion rate were decreasing

(say, because of gravitational attraction between galaxies)? What if it were increasing?

7. Suppose we were to count all galaxies out to a certain distance in space. If the universe were not expanding, the total number counted should be proportional to the *cube* of the limiting distance of our counts. (Why?) Taking account of the finite time required for light to reach us, describe the relation that would be *observed* between the total count and the limiting distance for (a) the steady-state theory, and (b) the evolutionary theory of the universe.
8. Draw a diagram like that in Figure 37.7 for an evolving cosmological model with $q_0 < ½$ (open space). Indicate on the time axis about where the most remote quasars are, and also where the 2.7 K black-body radiation is coming from.
9. Refer to Figure 37.11. Describe how the background black-body radiation must have differed at times t_3, t_2, and t_1 from how it appears now.
10. Sketch and discuss a schematic spacetime diagram like that in Figure 37.11, showing the world lines of several galaxies for the case of a closed universe that will contract again sometime in the future.
11. The Hubble law, $V = Hr$, relates the velocity of recession and the distance to an object. But we do not actually observe velocity and distance. What are the *observable* quantities?
12. Can you see any problems in postulating a closed universe if the Hubble constant, H, were 100 km/s · 10^6 pc?
13. Suppose the universe will expand forever (that is, an open model). Describe what will become of the radiation from the primeval fireball.
14. Are you perplexed with the task of imagining an infinite universe? What if it is finite; then what lies beyond? Do you suppose the experts can visualize infinities and eternities any better than you can? If not, can they simply accept these concepts, because they are used to talking and thinking about them? Collect opinions of such experts if you can find them. Who are the experts on infinities and eternities?
15. Suppose $q_0 > ½$, so that the universe is closed. Can there be other similar "universes" beyond the closed light paths in our own universe, with the same or even different properties? What might they be like, and how do they interrelate with our universe? Should cosmologists be concerned with studying such possibilities? Why or why not?

Frank Drake (b. 1930), the American astronomer who is a pioneer in the search for extraterrestrial intelligence. His *Project Ozma* (about 1960) was the first organized attempt to detect radio signals from extraterrestrial civilizations. The project (as expected) did not succeed, but the continuing advocacy by Drake, Carl Sagan, and others has since led to much more realistic and powerful searches for evidence of other intelligent creatures.

38

OTHER PLANETS, OTHER LIFE: ARE WE ALONE?

It is a legitimate question whether intelligent life, or any kind of life, exists elsewhere than on the Earth. At the outset we must say that we cannot answer this question with certainty, but just because the problem is difficult does not mean that it must remain unsolved for long. There exist reasonable, and even promising, approaches to answering this question of the prevalence of intelligent life in the universe.

A closely related question, as long as we are restricting our attention to life as we know the term, concerns the existence of other planets, perhaps including some with conditions similar to those on Earth. The search for planets revolving about other stars is currently an area in astronomy in which great advances are possible, bringing the answer to this question within our grasp. We begin this chapter with a discussion of other planets, proceeding from there to the more difficult issues of life in the universe and the possibilities for communicating with other intelligences.

38.1 OTHER PLANETARY SYSTEMS

(a) The Problem of Planetary Detection

At this writing not one planet has been identified orbiting another star. At first we might think that this negative statement indicates that other planets are rare or even nonexistent, and that the processes that formed our solar system (see Chapter 12) are unusual or unique. But such sweeping conclusions would be unwarranted. Other planets would be extremely difficult to discover, and the fact that none are known is probably just a consequence of the inadequacy of present techniques for search and discovery.

There are two basic approaches to finding planets revolving about other stars. First is direct detection of radiation from the planet. Second is indirect detection, based on the planet's influence on the star it accompanies.

The difficulty with direct detection is that any planet will be extremely faint in comparison to its star. Suppose, for example, that you were at a great distance and wished to detect reflected light from the planet Jupiter. Jupiter intercepts and reflects just about one-billionth of the radiation from the sun, so its apparent brightness in visible light is only 10^{-9} that of the sun, or a difference of about 22 magnitudes. As seen from Alpha Centauri, the nearest star, the sun would have an apparent magnitude of about one and Jupiter would be magnitude 23.

Of course, an astronomer on a planet in the Alpha Centauri system, if equipped with telescopes as large as ours, could detect stars of 23rd magnitude. There would be an additional problem, however, because Jupiter would be separated from the sun by only about 4 arcseconds, and its light would be very difficult to detect in the presence of its brilliant companion. If we consider instead looking from a star ten times farther away, Jupiter would be magnitude 28 and it would be only 0.4 arcsec from the sun—clearly beyond the limits of detection with contemporary astronomical techniques.

From the above discussion we conclude that a Jupiter-size planet orbiting one of the nearest stars might just barely be detected directly in visible light. The Hubble Space Telescope will carry out such a search. But most astronomers feel that we would be extremely fortunate if this search yielded positive results.

What about planets larger (and therefore brighter) than Jupiter? Apparently, such planets are not possible. As noted in Chapter 17, Jupiter has very nearly the largest possible size for a substellar object: if its mass were less, it would be smaller (like Saturn), but if its mass were greater, it would also assume a smaller size, due to gravitational self-compression. Generally speaking, detection of planets in the visible is not within our capability, so we must consider other possible approaches.

One way around the difficulty posed by the great difference in brightness between a star and its planet is to search in a region of the spectrum where the star is less bright. In the infrared near 20 micrometers wavelength, for example, Jupiter emits considerable thermal radiation, while the sun is relatively weak. The sun is still about 10,000 times brighter than Jupiter, however, for a difference of 10 magnitudes. When large infrared telescopes are built in space, observations in this part of the spectrum may provide a practical way to search for large planets.

The second approach to the search for other planetary systems is more promising, since it requires observation of the star and not the planet. Consider a planet in orbit about a star. As described in Chapter 4, both the planet and the star actually revolve about the *barycenter,* and the sizes of their orbits are inversely proportional to their masses. Suppose the planet, like Jupiter, has a mass about one-thousandth that of the star; the size of the star's orbit is then 10^{-3} that of the planet's.

To return to our example of observations of our own system carried out from Alpha Centauri: the diameter of the apparent orbit of Jupiter is 10 arcsec, and that of the orbit of the sun is .010 arcsec, or 10 milliarcsec. If astronomers there could measure the apparent position of the sun to this precision, they would see it describe an orbit of diameter 10 milliarcsec with a period of 12 years, equal to that of Jupiter. From the observed motion and the period, they could then deduce the mass and distance of Jupiter using Kepler's laws.

Such an approach to the search for other planets is based on very high–precision *astrometry,* the technique of exact position measurement in astronomy. The same kinds of observations that are used to determine the aberration of starlight, the proper motions of stars, and their parallax can also yield evidence of invisible companions if the measurements are carried out with sufficient precision (Figure 38.1). To estimate the precision required, we can look again at our solar system example. To detect Jupiter from Alpha Centauri we would need to measure positions to about 1 milliarcsec. To detect the Earth, which has much less mass and a smaller orbit as well, the precision goes up to a few microarcsec. From a distance of 10 parsecs, the precision needed is ten times greater: 100 microarcsec for a Jovian planet and better than 1 microarcsec for a terrestrial planet. On the other hand, searches for planets around stars less massive than the sun require proportionately less precision, since the size of the apparent orbit about the barycenter is inversely proportional to the mass of the star.

Contemporary astrometry has a precision of a few milliarcsec, sufficient to detect Jovian-mass companions around the nearer low-mass stars. Such observations have been carried out for several decades, with varying degrees of success. At one time, observers reported with considerable confidence

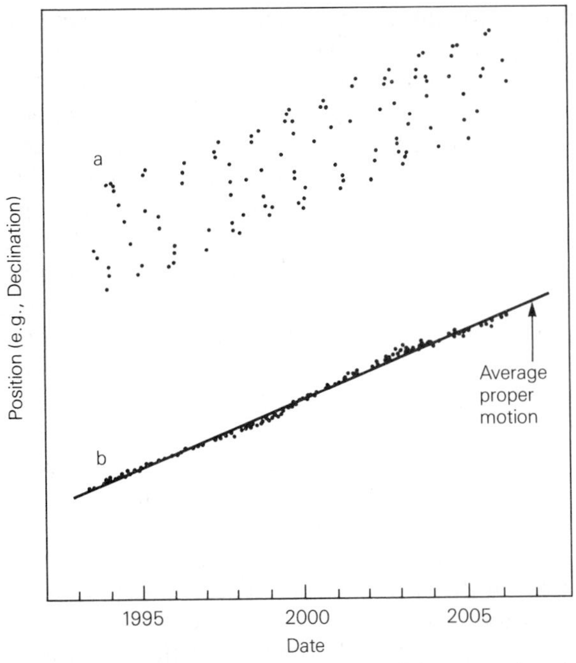

Figure 38.1 One of the most powerful approaches to searching for planets around other stars is called the astrometric technique. Although the companion planets cannot be detected directly, their presence is seen in tiny periodic shifts in the position of the star they orbit.

that they had found several cases of companions with masses similar to that of Jupiter, but subsequent analyses failed to confirm these claims. At this writing, there are no firm detections of invisible companions of planetary mass. The searches are only significant for stars of very low mass, since the precision now possible would not detect a Jupiter orbiting a solar-type star. Considerable advances in astrometric precision are being made, however, and within a few years it should be possible to search with precision of better than a milliarcsec from the ground.

An astrometric telescope in space, where there is no distortion from the Earth's atmosphere, would be a much more powerful tool for the detection of other planetary systems. Telescopes with capabilities of 10 microarcsec are under study. Such an instrument could carry out a definitive search for planetary companions to nearby stars, while it would also greatly advance astronomy in other areas, particularly the improvement in the astronomical distance scale that would result from extending stellar parallax measurements to a thousand parsecs or more.

(b) Brown Dwarfs

While astrometric observations have so far failed to detect other planets of Jovian mass (or smaller), they have probed into the very low end of the stellar mass distribution. A number of objects of substellar mass seem to be orbiting nearby stars. Clearly, the existence of these bodies is related to the issue of the prevalence of planetary systems, since if there are many such objects, we may be encouraged to hope that the smaller, true planets have formed also.

As noted in Chapter 29, a star is usually defined as a condensed object with sufficient mass to generate its own energy by the fusion of hydrogen to helium in its interior. This definition applies whether or not such hydrogen fusion reactions are taking place now; a star is still a star late in its evolution when it derives its luminosity primarily from other nuclear reactions (a red giant) or even after its nuclear fuel is exhausted (a white dwarf). The main issue is whether there is sufficient mass to trigger self-sustaining hydrogen fusion as the original protostar contracts.

Theoretical calculations have shown that the minimum mass for such reactions is approximately 0.08 times the mass of the sun—or equivalent to about 70 times the mass of Jupiter. One star is known—a small nearby dwarf called Ross 614—that is apparently very near this lower mass limit. Anything smaller than this limit is called a *brown dwarf,* which is an object intermediate between a star and a planet.

Since brown dwarfs do not have major internal energy sources from nuclear fusion, they should actually resemble the giant planets Jupiter and Saturn more closely than they do true stars. One way to study what a brown dwarf is probably like is to begin with theories for the interior of Jupiter and adapt the calculations for progressively more massive objects. Such theoretical models indicate that brown dwarfs, even though they may be much more massive than Jupiter, should all have nearly the same size, with diameters near 150,000 km. Their interiors should be electron-degenerate, and metallic hydrogen like that in the giant planets is expected to be a major constituent. Because of their larger masses, however, brown dwarfs produce substantially more gravitational heat than Jupiter or Saturn, resulting in surface temperatures near 1,000 K rather than about 100 K for the giant planets.

We discuss brown dwarfs in this chapter because they should be easier to detect than their cooler cousins the planets. Any search for other planetary systems must also consider brown dwarfs, which can be thought of as representing the high-mass part of a distribution of planetary sizes.

In 1985 the first brown dwarf was observed directly by D. McCarthy and colleagues using the 4-m telescope at Kitt Peak. They employed speckle interferometric techniques to distinguish the brown dwarf from its brighter companion, the main-sequence M dwarf called Van Biesbroeck 8 (VB8). Although the results are preliminary at this writing, the new object (designated VB8B) appears to have a surface temperature of about 1300 K, a diameter just slightly smaller than that of Jupiter, and a mass about 50 times greater than Jupiter's.

Calculations suggest that brown dwarfs form in much the same way as stars do, contracting and heating until their central temperatures are high enough for deuterium (heavy hydrogen) to react to form helium, resulting in a brief stage of nuclear energy release. Soon this source of energy is exhausted, however, and the brown dwarf begins a long period of slow cooling. Over billions of years, its diameter remains nearly constant but its surface temperature gradually declines from initial values near 2000 K down to below 1000 K. The object VB8B is apparently about midway in this evolution, probably having formed billions of years ago.

The relationship between brown dwarfs and the less massive objects we would call true planets is not clear. Perhaps brown dwarfs and planets coexist in the same systems, or perhaps they represent alternative evolutionary paths. It has also been suggested that substellar objects of all sorts might be common between the stars, where they could contribute significantly to the mass in the Galaxy without being readily detectable through either emission or absorption of light. In any case, it seems clear that if many brown dwarfs are in orbit around nearby stars, astronomers should be greatly encouraged in their search for true planets as well.

(c) Small Bodies Orbiting Stars

Brown dwarfs can be detected because they are relatively massive and have substantial luminosities, with surface temperatures of 1000 K or more. At the opposite extreme, it has also been possible to detect disks consisting of billions of small solid particles revolving about several stars. These disks provide another clue to the question of the existence of other planetary systems.

Disks can be detected because the total surface area of their myriad particles can be much greater than that of the star they surround. Imagine taking the mass now in the Earth and distributing it into particles only 1 meter in diameter. If these particles had the same density as the Earth, there would be more than 10^{24} of them, with a total surface area equal to more than one billion times that of the sun. Such a cloud of meter-sized particles would be relatively easy to detect in either infrared or visible radiation.

The initial evidence that disks of solid particles are associated with stars came in 1983, when the Infrared Astronomy Satellite (IRAS) carried out the first survey at wavelengths longer than 10 micrometers, in the spectral region most sensitive to emission from objects with temperatures of 100 K or less. Several stars, including the bright nearby stars Vega and Fomalhaut, were found to be more luminous at these wavelengths than could be understood from stellar emission alone. Subsequent observations with ground-based telescopes confirmed that the excess radiation was coming from an extended cloud of cool solid particles. Such clouds can only be dynamically stable if they are shaped like a disk or ring, like the rings of Saturn.

The best studied stellar disk is that of the fourth-magnitude A5 star Beta Pictoris. Because the disk is seen nearly edge-on from the Earth, it has been possible to image it in red light (Figure 38.2) and to detect its effect on the starlight passing through, as well as to measure its thermal emission at longer wavelengths. The disk extends out to at least 200 AU, with apparently an inner boundary at a few AU from the star. Its thickness is no more than a few AU and possibly much less. Unfortunately, there is no way to determine the sizes of the particles in the disk or their total mass, although probably they are small—perhaps considerably below a meter in diameter.

These disks are not planetary systems, of course, but their existence does demonstrate that solid matter can form in a disk-like configuration very similar to that postulated for the solar nebula out of which our solar system condensed (Chapter 12). It is possible that the disks we see are the outer parts of systems in which planets have already formed within a few AU of the star; in this case,

Figure 38.2 Although it is not a planetary system as we know the term, this photograph clearly shows a disk of solid material revolving around the star Beta Pictoris. Evidence of this material was first picked up by IRAS, which discovered excess infrared radiation from this star. (NASA/JPL)

the disks are perhaps much larger versions of our own asteroid belt and Oort cloud of comets. In any case, these stars certainly did not form entirely alone, even if their companions may never have pulled themselves together into true planets.

(d) Other Planets: Where Do We Stand?

The evidence available on the formation of other planetary systems is frustratingly incomplete. We know that most stars form as part of binary systems, a situation that probably makes the presence of planets impossible. But several tens of percent of the stars do not have stellar companions. Some of these are seen to have substellar companions—brown dwarfs—and others are surrounded by disks or rings of small solid particles.

The circumstantial evidence seems to favor the formation of planetary systems around some single stars. Since there are cases known where such stars are accompanied by a disk of small solid particles and others by objects much more massive than Jupiter, it seems reasonable to suppose that intermediate cases exist as well. It is simply our misfortune not to be able at present to detect companions in the mass range of greatest interest to us.

If the search for other planetary systems has a sufficiently high scientific priority, the technology exists to mount this effort with an Earth-orbiting astrometric telescope. Such an instrument could answer the question once and for all whether there are planets in the mass range from that of Uranus up to that of Jupiter in orbit around any of the hundred or so nearest single solar-type stars. Many people believe that this search would be almost sure to succeed, yielding knowledge of many other planetary systems to compare with the one case we already know of—our solar system. However, even a negative result would be important. If there are no other planetary systems in the sample searched, there is something drastically wrong with our theories of the origin of the solar system, since these theories predict that planetary systems should be rather common. If we are wrong about something so basic, perhaps we should find out about our error.

38.2 LIFE IN THE UNIVERSE

(a) Life in the Solar System

In the quest for life in the universe, it is appropriate to begin with the planets we know best, those in our own solar system. By examining the nature of life on Earth and its possible adaptability to other planets, we can gain a perspective on the environments elsewhere that might also support living things.

The life that we are familiar with developed with our planet, influencing the Earth as it evolved (Chapter 13). While we do not know the details of its origin, life clearly seems to have been the product of the chemical evolution of the early atmosphere and oceans, at a time when there was no free oxygen and conditions were at least mildly reducing (that is, dominated chemically by hydrogen rather than oxygen). Temperatures on Earth must have been roughly similar to those today, with at least part of the oceans consisting of liquid water at all seasons, and with plentiful sunlight to provide the energy needed for life to grow and diversify.

Laboratory experiments have reproduced many of the early chemical steps along the road to life. We know that naturally occurring reactions would

Figure 38.3 The surface of Mars as seen in the first Viking 1 photograph transmitted to Earth. Did this barren planet once support life? (NASA/JPL)

have resulted in plentiful organic chemicals, enough to make the early oceans a sort of organic "soup." (Similar reactions are taking place today in the atmosphere of Saturn's satellite Titan.) Among the prebiological compounds that must have been present are amino acids, sugars, and proteins. Within the first billion years of its formation, the Earth had developed self-replicating molecules that could use this organic soup to create copies of themselves, beginning the long evolutionary course leading to plants, pigs, and even people.

Could a similar sequence of events have taken place on other planets in the solar system? In the outer planets and their larger satellites, we see reducing atmospheres and nonbiological organic chemistry that mimics in many ways the conditions on the primitive Earth. Certainly the atmospheres of Jupiter, Saturn, and Titan contain many of what are believed to be the building blocks of life. Yet there is no evidence that life actually developed, perhaps because temperatures were too low for liquid water, or because the lack of solid surfaces on the giant planets denied them a stable enough environment for life to prosper. In any case, the evidence suggests that life is probably not present, since we do not see any unusual chemistry in the atmospheres of the outer planets that indicates the activity of large numbers of living organisms.

Among the inner planets, Mars has always seemed the most likely abode of life. It is interesting to note that early in the 20th century educated persons all over the world believed that the scientific evidence (as promoted by Percival Lowell and others) indicated the existence of not just life, but *intelligent life,* on Mars. But with increasing knowledge of the red planet, this hope has faded. Following the Viking landings in 1976, it seems clear that conditions on Mars are too harsh for the development of life such as we know it on Earth (Figure 38.3). The absence of liquid water, and particularly the lethal ultraviolet radiation that reaches the surface unimpeded by ozone or other absorbers in the thin Martian atmosphere, are not conducive to biological activity.

The issue is more complicated, however, for we know that life can evolve to adapt itself to harsh conditions. Virtually every ecological niche on our planet is occupied by living things, many of them highly specialized and exotic by ordinary standards. It is really not difficult to imagine Martian organisms that are adapted to conditions there (as, indeed, generations of science fiction writers have shown). But in order to evolve to tolerate present conditions, life would have to have started on Mars under much more favorable circumstances, and these warmer, wetter conditions would have had to persist for a long time—perhaps a billion years or more—for life to get a good start. We know that rivers once flowed on Mars and its atmosphere was much thicker than at present, but the duration of such a climate may have been brief (see Chapter 15). It is an open question whether life could have developed then, and if so, what its subsequent history might have been as Mars cooled and lost most of its atmosphere. Someday we may find evidence of past life on Mars, but until then we can only speculate.

We find, then, several planets in our solar system that seem to offer possibilities for the development of life, but only one—the Earth—where living things have proliferated and survived to the present. Perhaps not coincidentally, Earth is also the only planet with liquid water on its surface. If Mars were a little larger and a little closer to the sun, it might also have a climate like our own. Similarly, Venus could perhaps have developed differently if it were farther from the sun, escaping the runaway greenhouse effect that led to the loss of its water and the scorching temperatures it now experiences. But only on the Earth were conditions just right for life to prosper.

One of the fascinating properties of life is that its chemical building blocks are composed primarily of some of the most common elements in the universe: hydrogen, oxygen, carbon, and nitrogen. These elements should be present in any planetary system that forms anywhere in the Galaxy. Organic compounds are even prevalent in interstellar space, as we saw in Chapter 27. There is nothing special or unusual about the chemistry of life; if self-replicating organic molecules once form, the universe is full of the chemical food that they require for their growth.

But under what circumstances will such complex, self-replicating molecules form? That is a very difficult question. Part of the problem arises because we have only a single form of life to study, chemically speaking. All the tremendous diversity of living things on Earth represents the same genetic material. Only the packaging differs: from paramecium to parakeet, from rose to rhinoceros. Therefore we do not know what other kinds of self-replicating molecules, and hence what other chemical variants, are possible. If alternative forms once existed on Earth, they were destroyed by the dominant DNA-based life that we know today.

One reason that biologists were so interested in the search for life on Mars was the possibility of discovering a different, independently formed life. The faint hope of such discovery in the future dictates that we maintain planetary quarantine, so that no terrestrial lifeforms can accidentally contaminate Mars or any other planet where indigenous and independent life might exist. We owe this precaution to future generations. But meanwhile, the hope of finding life elsewhere in the solar system fades with every year, and most astronomers feel we must look elsewhere if we are to find evidence of extraterrestrial life.

(b) The Possibility of Intelligent Life in the Galaxy

We have seen that other planetary systems are difficult to detect, and questions of the origin and evolution of life are also hard to answer when the only life we know is that on our own planet. How much harder, then, is the problem of detecting *intelligent* life elsewhere! Yet there is hope, for intelligent life may be broadcasting its presence to us. If so, then finding evidence of intelligent life may be even easier than searching for planets around nearby stars.

There are probably 10^9 potentially observable galaxies in the universe, and probably many times this number of galaxies too faint to observe. Each may contain thousands of millions of stars, a large fraction of which might have planets. The possibilities of life throughout the universe, therefore, would seem to be enormously greater than in our own Galaxy. Yet because even light requires millions of years to travel the great distances between galaxies, any other societies that we have a chance of discovering are probably in our own Galaxy.

In the following paragraphs we will outline an approach to the search for intelligent life in our Galaxy. The first steps are theoretical and highly speculative: we will try to estimate the number of intelligent civilizations that might exist. Although the numerical results are extremely uncertain, the exercise is instructive in many ways, partly because it leads to a strategy for possible discovery and communication with other intelligent creatures. Besides, trying to deal with such a question brings in many of the ideas developed in this text, and most students consider it fun!

(c) The Number of Galactic Civilizations

University of California astronomer Frank Drake has pioneered the attempt to estimate the number of potentially communicative civilizations in the Galaxy. Drake's famous equation expresses the number, N, of currently extant civilizations in the Galaxy as the product of seven factors:

$$N = R_s f_p n_p f_b f_i f_c L_c,$$

where R_s is the rate of star formation in the Galaxy, f_p is the fraction of those stars with planetary systems, n_p is the mean number of planets suitable for life per planetary system, f_b is the fraction of those planets suitable for life on which life has actually developed, f_i is the fraction of those planets with life on which intelligent organisms have evolved, f_c is the fraction of those intelligent species that have developed communicative civilizations, and L_c is the mean lifetime of those civilizations. The first three factors are essentially astronomical in nature, the next two are biological, and the last two are sociological. We are able to make some educated estimates regarding the astronomical factors, we may be on shaky ground with the biological ones, and we are almost playing numbers games in trying to estimate values for the last two. Yet some interesting estimates can be made, and limits derived.

The mass of the Galaxy (Chapter 34) is believed to be from 2×10^{11} to 10^{12} solar masses. We do not know the form of much of the mass in the galactic halo, but it is a safe assumption that there are at least 4×10^{11} stars in the Galaxy. Recent estimates indicate that the number of new stars formed per year is about 10, and that this is a reasonable guess for the average star formation rate over the past few billion years. Thus we adopt $R_s = 10$ stars per year.

The sun originated from a cloud of gas and dust—the solar nebula—whose rotation caused it to flatten into a disk from which the planets formed. As discussed above, we expect similar formation of planetary systems elsewhere to be commonplace around single stars. However, taking into account other uncertainties, we will estimate that only one star in ten actually forms a planetary system, adopting $f_p = 0.1$.

In our own solar system, at least one and possibly two or three planets originally had suitable conditions for the development of life. Other systems may have none, but if we assume that on the average a star of the right sort with a planetary system has one suitable planet, we can estimate $n_p = 1.0$. Let us be more conservative, however, and adopt instead $n_p = 0.1$.

Many biologists are of the opinion that given the right kind of planet and enough time, the development of life is inevitable. Perhaps it is so, but in the spirit of a devil's advocate, we argue that the certainty of life forming has not been demonstrated as yet. Let us consider the liberal estimate that life, given the right conditions, is certain to develop, but also the devil's advocate guess that it happens only, say, 1 percent of the time. The corresponding values of f_b are thus 1.0 and 0.01.

Similarly, given the emergence of life, there is a widespread view that with enough time and natural selection a highly intelligent species will certainly evolve. Even were it inevitable that an intelligent species evolve on every planet with life, however, how long should it take? On Earth, it took 4.5×10^9 years. What if we happened to be quick about it, but that the average intelligent species takes, say, 20×10^9 years? Moreover, of the many parallel lines of evolution on Earth, only one (so far) has produced a being with enough intelligence to build a technology. Certainly, one could not rule out that the probability could be as low as 1 percent. Again, we take f_i to be either 1.0 or 0.01.

Not all intelligent societies would necessarily develop a technology capable of interstellar communication. We are on the threshold of that capability and possess a natural curiosity about the rest of the universe. Insects, while sometimes highly organized, do not appear to have any curiosity at all, however, and it is not certain if this human trait is fundamental to intelligence. Even if a society were curious, it might have good reason for wishing to have nothing to do with any other civilization. Some investigators suppose that most intelligent species will form communicative technological societies. As a conservative alternative, let us assume that only one-tenth of them do; that is, $f_c = 0.1$.

It is generally agreed that the final factor, L_c, is the most uncertain. It is useful, therefore, to leave L_c as an unknown and see what the rest of the equation yields with the numbers we have suggested. If we substitute the more or less optimistic estimates into the equation, we find

$$N = 10 \times 0.1 \times 0.1 \times 1 \times 1 \times 0.1 \times L_c = 0.01 \times L_c.$$

Our devil's advocate estimates, on the other hand, lead to

$$N = 10 \times 0.1 \times 0.1 \times 0.01 \times 0.01 \times 0.1 \times L_c = 10^{-6} \times L_c.$$

In both cases, we can obtain the actual number of

communicative civilizations present at any time in the Galaxy by replacing L_c with our estimate of the average lifetime in years.

Some have speculated that a technology might survive for an average of 10^9 years, a substantial fraction of the age of the Galaxy. The only known technology, of course, is our own, and we have only just reached the capability of interstellar communication. Below we shall suggest arguments that our technology might well end in a few decades. If so, and if we are typical, then L_c might be about 100 years. In rebuttal, some contend that if a communicative society can manage to survive for 100 years, it might well maintain itself for a billion years. We simply do not know.

If $L_c = 100$, then the optimistic estimate for the number of communicative civilizations in the Galaxy at any time is just one—at the present, ourselves! In the devil's advocate case, this average number drops to 0.0001, suggesting that there are long intervals in which no one is out there to communicate with. In either case, the search for other civilizations will be fruitless.

On the other hand, if $L_c = 1$ billion years, the two values for N are ten million and one thousand—a Galaxy teeming with civilizations with which to communicate. The corresponding distances to the nearest such civilization are less than 100 LY and somewhat more than 1000 LY, respectively. While we cannot choose among these very different estimates, we see the rationale by which estimates are made. We also come to one very important conclusion that will influence any strategy to search for other communicative civilizations.

38.3 SETI: THE SEARCH FOR EXTRATERRESTRIAL INTELLIGENCE

(a) The Basic Problem

What is the nature of the other intelligent beings with whom we might want to communicate? Clearly, we cannot know. Often we imagine some vaguely anthropomorphic creatures that would share our basic values, or at least our interests in science and mathematics. We also tend to think of these creatures as existing on planets something like the Earth and being made of organic chemicals like those of terrestrial life. But we should be open to other possibilities, many of which have been explored by the better science fiction writers. For one entirely different idea of what intelligent life might be like, read Fred Hoyle's novel, *The Black Cloud,* in which the protagonist is an interstellar cloud.

But we are dealing in this discussion with interstellar communication. Therefore, we should consider only what we have termed *technical* or *communicative* civilizations. It is the lifetime of these communicative civilizations that enters into the Drake equation discussed above.

The Drake equation tells us that if the lifetime of communicative civilizations is short, then basically we are alone in the Galaxy. This conclusion is valid whatever the particular choice you make of numbers for the various factors in the calculations. Putting it the other way around, a search for extraterrestrial civilizations will succeed only if the average such civilization has a lifetime much longer than our own—perhaps many millions of years longer.

The important conclusion that follows is that *any civilization we contact is likely to be very much older, and very much more advanced, than our own.* The chance of coming across another civilization like ourselves, just a few decades after its discovery of radio astronomy, is vanishingly small. Thus if we are going to search, it must be with the expectation of discovering a civilization far in advance of our own. It also follows that the best approach is to let them do the talking and to assume that they will be ahead of us in considering the problems of interstellar communication. *The proper strategy for SETI, the search for extraterrestrial intelligence, is to wait or, better, to listen.*

(b) Direct Contact

One means of detecting other Galactic civilizations might be by interstellar travel. We have seen, however, that the nearest neighboring civilization is expected to be at least a few hundred, and probably a thousand, or even tens of thousands of light years away. Because nothing can travel faster than light, a visit to another civilization would involve at least hundreds, and more likely thousands of years.

Now, to be sure, a space traveler's time slows down (with respect to ours) near enough the speed of light (Chapter 11). To make travel to other possible civilizations feasible in a human lifetime, however, the traveler's time would have to slow down at the very least by a factor of 5 (so that a 400-year round trip could be accomplished in 80 years of the crew's time). This much time dilation requires a speed of 98 percent that of light, and the energy requirements to reach that speed are absolutely enormous.

Consider, as an example, calculations presented by S. von Hoerner: Suppose we wish to send a moderate payload of 10 tons (three to five automobiles) into interstellar space and accelerate it to 0.98 c. In this relatively small payload we must provide an environment to provide life support for the crew for several decades. We add another 10 tons for engines and propulsion systems. The total energy required, no matter how it is obtained or how fast it is expended, is about 4×10^{29} ergs—roughly enough energy to supply the entire world's needs (at the present global expenditure rate of energy) for some 200 years. It would probably require the complete annihilation of matter, which we do not know how to accomplish at present. If we wanted the crew to reach its final speed of 0.98 c at an acceleration equal to the Earth's gravity, which would take 2.3 years, it would require the equivalent of 40 million annihilation plants of 15 million watts each, producing energy to be transmitted (with perfect efficiency) by 6×10^9 transmitting stations of 10^5 watts each, and all of this apparatus must be contained within a mass of 10 tons!

These enormous energy requirements apply only if we need to travel close to the speed of light to take advantage of the relativistic time dilation. Interstellar travel is possible if we are willing to take a long time to do it—which requires many generations for the crew. We may wonder about the morality of subjecting the crew's offspring to a life in a spaceship destined for an unknown fate generations in the future. But at least it is possible.

On the other hand, even now we can send, and have sent, material messages into interstellar space. Both the Pioneer and Voyager spacecraft, for example, have entered orbits on which they will eventually escape the solar system. It is unlikely that they will ever be seen or recovered by another intelligent species, but it is remotely possible. Partly for this reason, each Pioneer carries a plaque bearing line drawings of human beings, and cryptic messages describing the world from which it came, and the Voyagers contain phonograph recordings with messages from and descriptions of Earth. It is doubtful that the message on the plaque will ever be decoded or the recording heard, but publicity about them has called attention to the possibility of intelligent life in the Galaxy. More important by far is the message carried by each spacecraft itself; its discovery would convey a great deal of information about the species that launched it and the state of our technology.

A final possibility for direct communication is a visit to the Earth by extraterrestrial visitors. If the nearest civilizations are hundreds of light years away, they cannot have come to see us as a result of learning about us, for even radio waves that we have inadvertently been emitting into space—our radio and television programs—have only been on their way for a few decades. They could have reached only the very nearest stars, and it is highly unlikely that anyone there has received them and dispatched spaceships to look us over. If we have been visited, it would have to have been by random selection by interstellar travelers, and it is extraordinarily unlikely that among the billions of stars in the Galaxy, we should have been singled out for surveillance.

Yet the popular literature is full of accounts of sightings of UFO's, presumably operated by some intelligence, and even of alleged evidence for highly intelligent beings that have visited the Earth and taught people to build such magnificent structures as the pyramids, Easter Island statues, and other marvels. Not only do the latter accounts fail to acknowledge the great amount of work that has been done by competent professional archaeologists, but they are racist in their implication that earlier civilizations could not have had the talent to create great works of art and engineering.

Most scientists are highly skeptical of the extraterrestrial interpretation of reports of lights or erratically accelerating shiny objects in the sky, and of alien beings with unhuman countenances, yet with the human characteristics of two legs, two arms, a head, a mouth, a nose, two eyes, and other anthropomorphic features. Hard evidence of objects from space is lacking. Scientists, more than anyone, would delight in finding concrete evidence of alien life—there is so much we have to learn from it! But we still need evidence that can be analyzed by any competent scientist qualified to judge its extraterrestrial origin. Rumors, hearsay, secondhand reports, and eyewitness accounts by lay and inexperienced observers all must be given the benefit of the doubt, but still require positive verification before being taken as final evidence for life in the universe beyond the Earth.

(c) Contact by Radio

On the other hand, if there are other Galactic civilizations, there is a very real possibility that we may be able to communicate with them by radio. We do not necessarily mean two-way communication, for the radio waves would probably require hundreds of years at the very least for their round-trip travel between each question and answer. (Interstellar communication would be between civilizations, not individuals.) On the other hand, if there are communicative civilizations in the Galaxy, they may already be trying to communicate or at least to send one-way messages to other possible civilizations, just to inform them of their existence, and probably to convey much information in addition. With even our present technology we could send such messages ourselves to other stars in the Galaxy, but as argued above, it would be presumptuous for a young emerging civilization such as ours to do so. The first step is to try receiving messages.

There is a good chance that we would recognize an intelligent message—for example, a binary-coded broadcast of the number π repeated over and over. The discovery of extraterrestrial intelligence would be one of the most stupendous events in human history. If there were thousands or millions of communicating societies throughout the Galaxy, we could imagine a vast system of intercommunication, whereby many civilizations are sending messages to many other places in the Galaxy, not necessarily with the hope of receiving answers, for the messages themselves may well be received centuries after the sending civilizations had ceased to exist, but to pass on information about life in the Galaxy. It is a romantic and exciting idea, and many of us would like to be in on the network if, in fact, it exists.

(d) Searches for Extraterrestrial Broadcasts

With existing radio astronomy facilities, a number of limited searches for intelligently coded signals have already been made by radio astronomers in the United States and the Soviet Union. To date, there has been no success, but no success would have been expected from such meager efforts. Just what would it take to learn if messages are being beamed in our direction from other civilizations in the Galaxy?

The problem is difficult because we do not know in advance either the location or the nature of any possible broadcast. The entire electromagnetic spectrum is available for potential communication. Only the most general guidelines can be established without knowledge of the civilizations whose signals we hope to pick up.

Most of those who have considered this problem have concluded that radio waves—more specifically, microwaves—offer the most promising part of the electromagnetic spectrum. At these wavelengths, the absorption of energy by the interstellar medium and the emission of competing background radiation from natural sources are both near their minimum values. The presumption is that if a civilization wished to be found, it would probably choose to broadcast microwaves. Some have even suggested further that, given the importance of water as a basis for life, frequencies might be selected near the OH lines at about 18 cm wavelength—the so-called Galactic "water hole." This seems like a plausible suggestion, but it is also rather anthropomorphic, for who is to say what the thinking patterns of an alien, advanced race might be?

Artificial radio broadcasts tend to be at very specific wavelengths or frequencies. The only way that our own terrestrial civilization can accommodate the many demands for microwave communication channels is to restrict each transmitter to a very narrow frequency band. In addition, it is only within a single narrow band that a distant signal can be picked up in the presence of various sources of interference and noise. The more sensitive the receiver, the narrower the channel or band to which it is tuned must be.

All of the early searches for extraterrestrial radio signals were limited to one band or a very few frequency bands, and this is the primary reason that no success was expected. The key to an effective SETI program is the development of sensitive receivers that can listen at many bands simultaneously. This challenge, incidentally, is not too different from that addressed by optical astronomers when they try to design spectrographs that measure many wavelengths simultaneously. Since all the radiation is being collected by the telescope anyway, the best way to improve efficiency is to move to multichannel detectors.

The first modern SETI program, based on a thousand-channel receiver, began in the early 1980s. Financed in part by public contributions to the Planetary Society, this search uses an old 60-

Figure 38.4 The one telescope in the world now being used regularly to search for radio signals from extraterrestrial civilizations is this dish in Massachusetts. A part of this SETI effort is supported by donations made by private citizens through the Planetary Society. (D. Morrison)

foot radio telescope that Harvard University was planning to decommission (Figure 38.4). The strategy is to point the telescope sequentially at each part of the sky and to measure any radiation received in each of the thousand frequency channels, analyzing the results with sophisticated computer codes that can identify an artificial signal amid the natural babble of the Galaxy.

The next step in this search involves a much more powerful million-channel receiver, paid for by movie-maker Steven Spielberg and placed into operation by the Planetary Society in 1986. Eventually it is also hoped to make available a larger telescope to collect the signal. Of course, the greater sensitivity of a larger telescope implies a narrower beam, increasing the time required to survey the entire sky. Even more powerful would be an array of telescopes, each equipped with a million-channel receiver (Figure 38.5). Although such an effort would cost only a fraction of what we now spend on space exploration, there seems to be little chance that a SETI program of really large magnitude will be undertaken in the United States in the present century.

The consequences of the successful detection of a message from an advanced civilization are hard to imagine. In part, the significance would depend on our success in deciphering the message. We can hope that any civilization that wishes to be found by the relatively primitive techniques we know of will also have worked out a way to make its message intelligible to us. On the other hand, we might pick up a signal intended for other purposes, such as a galactic navigation beacon or a communication directed toward some other, equally advanced creatures. For one plausible scenario of the detection and decryption of an interstellar signal, read Carl Sagan's novel *Contact*.

38.4 EPILOGUE: CIVILIZATION ON EARTH

One of the interesting consequences of the Drake equation and the SETI activities that have recently begun is that they make us look at the issue of how long a technical civilization might last. Are there natural limits to the lifetime of a technical civilization such as our own? Are there any limits to our growth and progress? Let us conclude by examining these provocative questions.

Consider the sheet of paper on which the page you are reading is printed; it is roughly 0.1 mm thick. Imagine that you were to cut it into two sheets and stack them on top of one another. Then cut the stack of two sheets and combine the four sheets into a single stack. Next, make a third cut and stack up the eight sheets. If you were to continue the process until 100 cuts had been made, guess how thick the stack of paper would be.

The answer is 10^{10} LY—the distance to a remote quasar. When we double a number at each step, it is an example of a *geometric progression*. This progression increases slowly at first, but almost suddenly it explosively rises and rapidly approaches infinity as the number of doublings increases. If a quantity (say, inflation, population, use of energy, or the price of homes) increases at i percent per year, the time for that quantity to double is approximately 70 years divided by i. Thus, if (as has been the case in recent decades) our use of energy increases by 7 percent per year, in the next ten years

Figure 38.5 Artist's conception of the huge SETI antenna array proposed under the name of Project Cyclops in 1971. Such a telescope would have far greater capability to detect faint radio signals than anything now in operation or under serious consideration, but no government has shown interest in supporting a search of this magnitude for extraterrestrial civilizations. (NASA)

we will have used as much energy as has been used by all of mankind throughout history until now. Another example of a similar geometric progression is the world population, which at present is doubling about every 35 years (2 percent per year).

In New York City there are a little more than 100 square meters of land area for every person living in that city. Suppose we imagine New York to be more than 100 times as crowded as it is now, that there is only 1 m^2 per person. Moreover, let us suppose that the same crowding applies over the entire land area of the Earth. At the present rate of population growth, the land area of our planet would be filled to this density in about 550 years.

Perhaps one could argue about how crowded the Earth can be. An uncontested ultimate limit to the number of people, however, is set by the fact that we are all made of atoms of matter, and matter is conserved; that which constitutes our bodies has had to come from the Earth itself. Most of the matter in our bodies is in the form of water which, we have seen, outgassed from the Earth's crust. More and more of the water needed to build the bodies of people must come from the seas of the world as the population grows. At a doubling time of 35 years, the oceans would be entirely converted to people in about 1200 years (never mind about water to wash with, swim in, drink, and water our fields with, or even for fish). At the same doubling rate, in 1600 years the mass of people would be equal to the mass of the Earth, and in 2300 years to that of the solar system. If we could create matter from nothing (to make people), after about 5300 years a great sphere of humanity 150 LY in radius would be expanding at its surface with the speed of light!

The above examples are not meant as predictions but rather to illustrate how rapidly the numbers increase with a constant doubling time. In fact, during the past thousand years the doubling time for the population of the world has actually been decreasing with time. An excellent representation of the world population over the past two millennia is given by a formula graphed in Figure 38.6. A literal interpretation of this formula predicts that the world population will become infinite and the space available to each of us will go to zero in 2026, on Friday, November 13—a date dubbed "Doomsday."

Of course the world population cannot become infinite; various limitations will slow its present growth rate before 2026. But the problem is that most of the growth occurs so suddenly that the population can become unmanageable without warning; the problem may not be recognized until it is too late to prevent catastrophe.

Nor does expansion into space provide a solution. It has been seriously suggested that we can relieve overpopulation by emigrating to other planets. But even if we wanted to condemn our children to lives in airtight living enclosures on the Moon

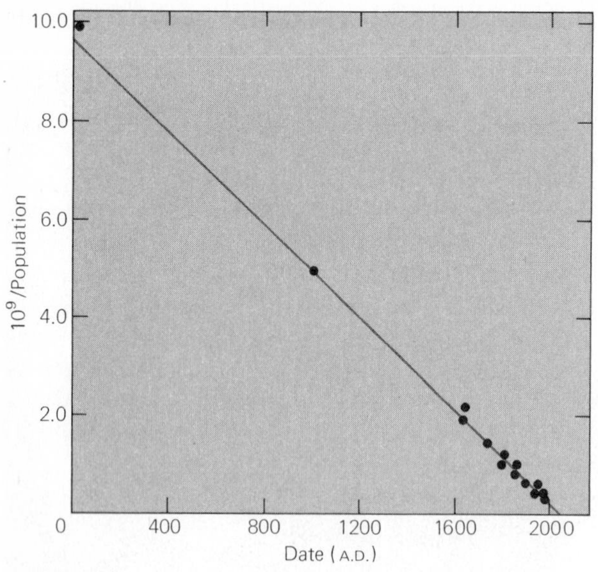

Figure 38.6 The increase in the world's population during the last 2000 years. Against the date is plotted one billion divided by the population, so the points are proportional to the reciprocal of the population. Equivalently, the vertical scale represents how much space there is on Earth for each person in the population. The straight line is a plot of an equation derived by von Förster, Mora, and Amiot, by which the population becomes infinite and the space per person drops to zero about the year 2026.

Figure 38.7 An ominous exhibit at the Bishop Museum in Honolulu, Hawaii. (Bishop Museum)

and Mars, without any hope of playing in the fields or hiking in the woods, we could only extend our time another 35 years, for the entire surface area of the Moon and Mars combined is only about the same as that of the land area of the Earth. Within about 500 years all possible planets about other stars within 150 LY would be occupied to the extent the Earth is now, and to find room for our ever-increasing number of people, we would have to transport them faster than the speed of light to ever more remote planets, which, of course, cannot be done.

In short, we are in danger of using up available space (and resources) so suddenly that we will have scarcely any warning to prepare for the onslaught of aggression and suffering that almost inevitably will result. The aggression prompted by population pressure (observed, for example, in experiments with colonies of rats that are allowed to multiply without check) is in itself a threat to survival. Nuclear bombs now stockpiled in the world are equivalent to about 10 tons of TNT per person—enough energy to raise a 10,000-ton apartment house 500 m into the sky to drop on every man, woman, and child now living. And the nuclear stockpile is increasing at a faster rate than the population.

In some developed countries, including the United States, the population, while not yet stable, is rising more slowly, but the worldwide growth is unabated. In most of Latin America the population doubles in 20 years. Compared with the United States, Mexico is a poor nation. Yet, to maintain their existing level of poverty, the Mexicans must double their homes, schools, hospitals, roads, factories, and in general, everything in their economy in just 20 years! Even in the less developed areas there are similar contrasts: China, the world's largest nation with a population of nearly a billion, has reduced its birth rate to a value lower than that of the United States, with prospects of a stable population in the 21st century, whereas India, the second-largest nation, continues to grow at the rate of tens of millions per year.

Pollution of the atmosphere and exhaustion of our fossil fuels may be altering our climate and potential for food production. If its population were to give up eating meat, the United States could probably feed its people for another population doubling or so, but this requires maintaining the present efficiency in farming and assumes no climatological degradation of the ability to grow food.

To the problems of overcrowding, the violence it breeds, and energy and food shortages, we should add the gradual increase in the temperature of the Earth due to the heat produced by the use of energy and by an increasing greenhouse effect. This factor alone could alter our climate enough to affect food production in less than a century. Another effect of the present trends of the evolution of our civilization is genetic deterioration (medicine increases the life span, but encourages survival of persons with genetically unfavorable mutations). Even if we could withstand all of these threats to our survival, there remains a possible crisis caused by the boredom and stagnation of a stable society trying to endure without substantial innovation on a completely filled planet for hundreds of centuries.

Possibly these threats to our survival can be circumvented by enlightened action. Or perhaps violent struggles for survival in a chaotic near future will result in a strong enlightened portion of society surviving to maintain a stable civilization. At present, thoughtful, rational international planning to preserve our planet as an abode for most of our race is missing. It may be overly pessimistic to predict that our technological society is doomed to an early end, but if we continue our present course, it is hard to imagine an alternative. Do other civilizations (if they exist) similarly destroy themselves, or have they learned, as we hope to learn, to preserve their longevity? Perhaps the discovery that another civilization has "made it" will one day be our salvation.

EXERCISES

1. Verify the statement in the text that Jupiter would be about 22 magnitudes fainter than the sun as seen in visible light from a distant star.
2. Determine the angular distance of the Earth from the sun as seen from distances of: (a) 1 pc; (b) 10 pc; (c) 100 pc.
3. Calculate the total surface area if the Earth were broken up into particles with diameters of: (a) 1 km; (b) 1 m; (c) 1 cm.
4. ★ Estimate the total brightness in visible light (relative to the brightness of the star) of a belt of asteroids circling a star at a distance of 10 AU and consisting of 10,000 objects with average diameter 10 km and average reflectivity 50 percent.
5. Compare and contrast the conditions in interstellar space and at the surface of a planet as they affect the formation of organic molecules.
6. Why might the presence of oxygen in a planetary atmosphere suggest that life exists there?
7. Suppose we could carry on a two-way radio communication with another civilization. How long would be the minimum time to receive an answer to a transmitted question if that civilization is: a) on the Moon? b) on Jupiter? c) on a planet in the Alpha Centauri system? d) on a planet in the Tau Ceti system (see Appendix 13)? and e) at the Galactic center?
8. All considered, what would you judge to be the extreme limits for the number of communicative civilizations in the Galaxy?
9. All considered, what would you judge to be the extreme limits for the number of intelligent civilizations of any kind in the Galaxy?
10. ★ Verify the statement in the text that a 0.1 mm sheet of paper cut and stacked 100 times would have a final thickness of 10^{10} LY.
11. A sample of bacteria in a large bottle doubles in number every 10 minutes. At 11:50 A.M. the bottle in only $\frac{1}{128}$ full. In the next few minutes, some bacteria scouts discover three additional identical bottles that are still empty, and reason that they now have room for four times the bacteria population that their original bottle can support. Thus they continue to reproduce at the same rate, judging that when their original bottle is filled, they or their offspring can move on to the other bottles. How much extra time does this buy them, before all four bottles are full?

Appendix 1

BIBLIOGRAPHY

(Technical references are marked with an asterisk.)

SOME OTHER GENERAL TEXTBOOKS IN ASTRONOMY

Abell, G. O., *Realm of the Universe,* 3rd ed. Philadelphia: Saunders College Publishing, 1984.

Berman, L., and Evans, J. C., *Exploring the Cosmos,* 5th ed. Boston: Little, Brown, 1986.

Goldsmith, D., *The Evolving Universe,* 2nd ed. Menlo Park, CA: Benjamin/Cummings, 1985.

Hartmann, W. K., *Astronomy: The Cosmic Journey,* 3rd ed. Belmont, CA: Wadsworth, 1985.

Jastrow, R., and Thompson, M. H., *Astronomy: Fundamentals and Frontiers,* 4th ed. New York: Wiley, 1984.

Kaufmann, W. J., *Universe.* New York: W. H. Freeman and Company, 1985.

Pasachoff, J. M., *Contemporary Astronomy,* 3rd ed. Philadelphia: Saunders College Publishing, 1985.

Pasachoff, J. M., *Astronomy: From the Earth to the Universe,* 3rd ed. Philadelphia: Saunders College Publishing, 1987.

Sagan, C., *Cosmos.* New York: Ballantine, 1980.

Shu, F. H., *The Physical Universe: An Introduction to Astronomy.* Mill Valley, CA: University Science Books, 1982.

Seeds, M., *Horizons: Exploring the Universe,* 2nd ed. New York: Wadsworth, 1985.

Snow, T. P., *The Dynamic Universe: An Introduction to Astronomy,* 2nd ed. St. Paul, MN: West Publishing Co., 1986.

Snow, T. P., *Essentials of the Dynamic Universe.* St. Paul, MN: West Publishing Co., 1984.

Zeilik, M., *Astronomy: The Evolving Universe,* 4th ed. New York: Harper and Row, 1985.

Zeilik, M., and Gaustad, J., *Astronomy, The Cosmic Perspective.* New York: Harper and Row, 1983.

HISTORIES OF ASTRONOMY

Abbott, D. (ed.), *Astronomers.* New York: Peter Bedrick Books, 1984. (Historical introduction and brief biographies of about 300 astronomers from past and present.)

Ashbrook, J., *The Astronomical Scrapbook.* Cambridge, MA: Sky Publishing Corp., 1984. (Excellent nontechnical discussions of fascinating facts and forgotten lore in the history of astronomy.)

Berendzen, R., Hart, R., and Seeley, D., *Man Discovers the Galaxies.* New York: Neale Watson Academic Publications, 1976. (Excellent history of the development of modern extragalactic astronomy.)

Boorstin, D. J., *The Discoverers.* New York: Random House, 1983. (Brilliant essays on the development of astronomy, timekeeping, and geography.)

Cooper, H. S. F., *The Search for Life on Mars.* New York: Harper and Row, 1980. (Eyewitness account of the Viking program, especially the Martian biology experiments.)

Cooper, H. S. F., *Imaging Saturn: The Voyager Flights to Saturn.* New York: Holt, Rinehart and Winston, 1982. (Eyewitness account of the Voyager exploration of Saturn.)

Ezell, E. C., and Ezell, L. N., *On Mars* (NASA SP-4212). Washington, DC: U.S. Government Printing Office, 1984. (Definitive history of the Mariner and Viking programs to explore Mars.)

Ferris, T., *The Red Limit,* 2nd ed. New York: Morrow, Quill, 1983. (Superbly written history of modern cosmology and the astronomers who played a role in it.)

*Gingerich, O. (ed.), *Astrophysics and Twentieth-Century Astronomy to 1950, Part A.* New York: Cambridge University Press, 1984. (Collection of original papers and commentary covering primarily developments in astronomical instrumentation and observing techniques, including the early days of radio astronomy and space exploration.)

Grosser, M., *The Discovery of Neptune*. New York: Dover, 1979. (An engrossing account of the discovery of new planets in the solar system.)

Harwit, M., *Cosmic Discovery*. New York: Basic Books, 1981. (More than a history of astronomy, this book explores the process of astronomical discovery. Controversial and somewhat technical, but a fascinating exploration of the way in which research is done.)

Hoskin, M., *Stellar Astronomy: Historical Studies*. Buckinghamshire, England: Science History Publications Ltd., 1982. (Account of the development of stellar astronomy by a leading historian of astronomy.)

Hoyt, W. G., *Lowell and Mars*. Tucson, AZ: University of Arizona Press, 1976. (Excellent account of the Martian canal controversy.)

Hoyt, W. G., *Planets X and Pluto*. Tucson, AZ: University of Arizona Press, 1980. (Excellent account of the discovery of Pluto.)

*Kellerman, K., and Sheets, B. (ed.). *Serendipitous Discoveries in Radio Astronomy*. Green Bank, WV: National Radio Astronomy Observatory, 1984. (Collection of important papers in radio astronomy.)

King, H. C., *Exploration of the Universe*. New York: New American Library, 1964. (A scholarly history; not to be confused with a famous text on astronomy of the same title.)

Koestler, A., *The Sleepwalkers*. New York: Macmillan, 1959. (A famous and very interesting history of the beginnings of modern science.)

Krupp, E. C., *Echoes of the Ancient Skies*. New York: Harper and Row, 1983. (Excellent discussion of archaeoastronomy for the scientifically literate layperson.)

*Lang, K. R., and Gingerich, O. (eds.), *A Source Book in Astronomy and Astrophysics, 1900–1975*. Cambridge, MA: Harvard University Press, 1979. (A valuable collection of important original papers with commentary by the editors.)

Newell, H. E., *Beyond the Atmosphere* (NASA SP-4211). Washington, DC: U.S. Government Printing Office, 1980. (The early history of the U.S. space science program, written by NASA's Associate Administrator for Space Science.)

Pannekoek, A., *A History of Astronomy*. New York: Interscience, 1961. (A respected history.)

*Shapley, H. (ed.), *A Source Book in Astronomy 1900–1950*. Cambridge, MA: Harvard University Press, 1960. (A collection of original papers.)

Struve, O., and Zebergs, V., *Astronomy of the Twentieth Century*. New York: Macmillan, 1962. (Excellent history with source material, now somewhat dated.)

*Sullivan, W. T., III (ed.), *Classics in Radio Astronomy*. Hingham, MA: Reidel, 1982. (Reprints of 37 classic papers from the development of radio astronomy, with commentary.)

Tucker, W. H., *The Star Splitters* (NASA SP-466). Washington, DC: U.S. Government Printing Office, 1984. (History of the beginnings of high-energy astronomy, with detailed accounts of the NASA High-Energy Observatories.)

Tucker, W. H., and Giacconi, R., *The X-ray Universe*. Cambridge, MA: Harvard University Press, 1985. (Insider's account of the history of X-ray astronomy, with emphasis on the highly successful Einstein observatory.)

Van Helden, A., *Measuring the Universe*. Chicago: University of Chicago Press, 1985. (A survey of human views of the scale of the universe from the time of Aristarchus to that of Halley.)

Washburn, M. L., *Mars at Last*. New York: Putnam, 1977. (Well-written journalist's account of the Viking program.)

Washburn, M. L., *Distant Encounters: The Exploration of Jupiter and Saturn*. New York: Harcourt Brace Jovanovich, 1983. (Well-written journalist's account of the Voyager program.)

CELESTIAL MECHANICS

Ahrendt, M. H., *The Mathematics of Space Exploration*. New York: Holt, Rinehart and Winston, 1965. (A relatively elementary survey of celestial mechanics.)

Ryabov, Y., *An Elementary Survey of Celestial Mechanics*. New York: Dover, 1961. (A semipopular survey.)

Van de Kamp, P., *Elements of Astromechanics*. San Francisco: Freeman, 1964. (For the lower-division student.)

TELESCOPES, INSTRUMENTATION, AND THE ELECTROMAGNETIC SPECTRUM

*Christianson, W. N., and Hogborn, J. A., *Radio Telescopes*. London: Cambridge University Press, 1969. (Fundamentals for the serious student.)

*Eccles, M. J., Sim, M. E., and Tritton, K. P., *Low Light Level Detectors in Astronomy*. New York: Cambridge University Press, 1983. (A book based on lectures given at the University of Edinburgh. Detectors, including photographic plates, image tubes, and CCD's, used to measure optical radiation are described and compared.)

Fichtel, C. E., and Trombka, J. I., *Gamma-Ray Astrophysics* (NASA SP-453). Washington, DC: U.S. Government Printing Office, 1981. (Introductory text for the science student on the new field of gamma-ray astronomy.)

Field, G. B., and Chaisson, E. J., *The Invisible Universe*.

Boston: Birkhauser, 1985. (Semipopular book on the new regions of the electromagnetic spectrum and the space observatories that astronomers would like to build to explore them.)

Henbest, N., and Marten, M., *The New Astronomy*. New York: Cambridge University Press, 1983. (A beautiful collection of X-ray, UV, IR, and radio images of astronomical objects.)

Longair, M. S., *High-Energy Astrophysics*. Cambridge, England: Cambridge University Press, 1981. (An introductory text, for upper division students of physics and astronomy, that focuses on the origin of cosmic rays in order to provide a broad introduction to contemporary astrophysics.)

Marx, S., and Pfau, W., *Observatories of the World*. New York: Van Nostrand Reinhold, 1982. (Incomplete and dated, this book is still of interest for its coverage of Eastern European and Soviet observatories.)

Rowan-Robinson, M., *Cosmic Landscape*. New York: Oxford University Press, 1979. (Beautifully written popular account of the various regions of the electromagnetic spectrum and what each tells us of the universe, from planets to the big bang.)

Tucker, W., and Tucker, K., *The Cosmic Inquirers: Modern Telescopes and their Makers*. Cambridge, MA: Harvard University Press, 1986. (An inside view of major new telescopes—the Very Large Array, Einstein, Gamma Ray Observatory, Infrared Astronomy Satellite, and Hubble Space Telescope—with emphasis on the persons who struggled to make them a reality.)

PLANETARY SYSTEM

Beatty, J. K., O'Leary, B., and Chaikin, A. (eds.), *The New Solar System,* 2nd ed. Cambridge, MA: Sky Publishing Corp., 1982. (Well-edited semipopular descriptions written by leading planetary scientists, illustrated in color.)

Brandt, J. C., and Chapman, R. D., *Introduction to Comets*. Cambridge, England: Cambridge University Press, 1981. (Introductory text on comets for the advanced undergraduate.)

*Burns, J., and Matthews, M. H. (eds.), *Satellites*. Tucson, AZ: University of Arizona Press, 1986. (The most comprehensive reference available, with chapters written by leading experts on planetary satellites.)

Carr, M. H., *The Surface of Mars*. New Haven, CT: Yale University Press, 1981. (The most comprehensive and authoritative book on Mars, beautifully illustrated with Viking photos, written at a semitechnical level.)

Carr, M. J., Saunders, R. S., Strom, R. G., and Wilhelms, D. E., *The Geology of the Terrestrial Planets* (NASA SP-469). Washington, DC: U.S. Government Printing Office, 1984. (Comprehensive reference on the geology of the terrestrial planets, very well illustrated with spacecraft photos.)

Chapman, C. R., *Planets of Rock and Ice*. New York: Scribners, 1982. (An excellent introduction to the terrestrial planets and satellites, well-written and authoritative, aimed at the scientifically literate layperson; highly recommended.)

Cortright, E. M., *Apollo Expeditions to the Moon* (NASA SP-350). Washington, DC: U.S. Government Printing Office, 1975. (Beautiful illustrations, well captioned.)

Elliot, J., and Kerr, R., *Rings*. Cambridge, MA: MIT Press, 1984. (Collaboration between a research scientist and a science writer that discusses the history of ring studies and current knowledge for the scientifically literate layperson.)

Frazier, K., *Solar System*. Alexandria, VA: Time-Life Books, 1985. (Very well-written and illustrated introduction for the layperson.)

French, B. M., *The Moon Book*. New York: Penguin Press, 1977. (Readable introduction to the Moon, written for the layperson.)

*Gehrels, T. (ed.), *Asteroids*. Tucson, AZ: University of Arizona Press, 1979. (The most comprehensive reference available, with chapters written by leading experts in the study of asteroids.)

*Greenberg, R., and Brahic, A. (eds.), *Planetary Rings*. Tucson, AZ: University of Arizona Press, 1984. (The most comprehensive reference available, with chapters written by leading experts on rings.)

Goldsmith, D., *Nemesis: The Death Star and Other Theories of Mass Extinction*. New York: Walker and Company, 1985. (Well-written semipopular account of the current debate on the astronomical causes of mass extinctions.)

Hartmann, W. K., *Moons and Planets* (2nd ed.). Belmont, CA: Wadsworth Publishing Company, 1983. (Text on planetary science for the advanced undergraduate.)

Hartmann, W., Miller, R., and Lee, P., *Out of the Cradle: Exploring the Frontiers Beyond Earth*. New York: Workman, 1984. (Beautiful collection of paintings and essays on the exploration of the planetary system.)

*Hunten, D. M., Colin, L., Donahue, T., and Moroz, V. I. (eds.), *Venus*. Tucson, AZ: University of Arizona Press, 1983. (The most comprehensive reference available, with chapters written by leading experts, including a number of Soviet contributors.)

Hutchison, R., *The Search for Our Beginning*. New York: Oxford University Press, 1983. (Highly readable introduction to meteoritics and the origin of the solar system.)

Masursky, H., Colton, G. W., and El-Baz, F., *Apollo Over the Moon: A View from Orbit* (NASA SP-362). Washington, DC: U.S. Government Printing Office, 1978. (Beautiful illustrations, well captioned.)

Morrison, D., *Voyage to Saturn* (NASA SP-451). Washington, DC: U.S. Government Printing Office, 1982. (The Voyager encounters with Saturn and their results; written for the layperson.)

*Morrison, D. (ed.), *Satellites of Jupiter*. Tucson, AZ: University of Arizona Press, 1983. (Comprehensive reference, with chapters written by leading experts.)

Morrison, D., and Owen, T., *The Planetary System*. Reading, MA: Addison–Wesley, 1987. (College text in planetary science.)

Morrison, D., and Samz, J., *Voyage to Jupiter* (NASA SP-439). Washington, DC: U.S. Government Printing Office, 1982. (The Voyager encounters with Jupiter and their results; written for the layperson.)

Murray, B., Malin, M. C., and Greeley, R., *Earthlike Planets*. San Francisco: W. H. Freeman, 1981. (Authoritative text on the geology of the terrestrial planets aimed at the advanced undergraduate.)

Raup, D. M., *The Nemesis Affair: A Story of the Death of Dinosaurs and the Ways of Science*. New York: W. W. Norton, 1986. (Outstanding account of the theory that periodic mass extinctions are due to collisions of comets with the Earth, written by a leading paleontologist and major participant in this exciting and controversial debate.)

Taylor, S. R., *Planetary Science: A Lunar Perspective*. Houston: Lunar and Planetary Institute, 1982. (Excellent treatment of the terrestrial planets by a lunar scientist, with emphasis on geochemistry.)

Schneider, S. H., and Londer, R., *The Coevolution of Climate and Life*. San Francisco: Sierra Club Books, 1984. (A modern discussion of weather, climate, human society, and planetary evolution written for the interested layperson.)

*Silver, L. T., and Schultz, P. H. (eds.), *Geological Implications of Impacts of Large Asteroids and Comets with the Earth*. Boulder, CO: Geological Society of America, 1983. (The most comprehensive reference available, with chapters written by experts from astronomy, geology, and paleontology.)

Sullivan, W. S., *Landprints*. New York: Times Books, 1985. (Highly readable account of geology for the layperson, written by a leading science journalist.)

Wasson, J. T., *Meteorites: Their Record of Early Solar-System History*. New York: Freeman, 1985. (Comprehensive introduction to meteoritics written for the upper-division undergraduate.)

Whipple, F. L., *The Mystery of Comets*. Washington, DC: Smithsonian Institution Press, 1985. (Up-to-date discussion of comets for the layperson, written by one of the leading cometary experts of the world.)

Wood, J. A., *Meteorites and the Origin of Planets*. New York: McGraw-Hill, 1968. (Short, readable book aimed at the scientifically literate layperson.)

SUN, STARS, AND INTERSTELLAR MEDIUM

Aller, L. G., *Atoms, Stars, and Nebulae*, 2nd ed. Cambridge, MA: Harvard University Press, 1971. (Excellent but somewhat dated, aimed at the scientifically literate layperson.)

Bok, B. J., and Bok, P. E., *The Milky Way*, 4th ed. Cambridge, MA: Harvard University Press, 1973. (The classic—although now somewhat dated—description of our Galaxy for the interested layperson.)

*Bowers, R., and Deeming, T., *Astrophysics* I: Stars; Astrophysics II: Interstellar Matter and Galaxies. Boston: Jones and Bartlett, 1984. (Comprehensive texts for the undergraduate science student.)

Clark, D. H., *Superstars*. New York: McGraw-Hill, 1984. (A layperson's introduction to supernovae, which shows the role these objects play in the formation of the atomic elements.)

Frazier, K., *Our Turbulent Sun*. Englewood Cliffs, NJ: Prentice-Hall, 1982. (Good popular-level account of the sun, with particular attention to solar-terrestrial relations and possible effects of the sun on Earth's climate and history.)

Gibson, E. G., *The Quiet Sun* (NASA SP-303). Washington, DC: U.S. Government Printing Office, 1973. (Excellent account of the sun, written for the science student by a scientist-astronaut.)

Greenstein, G., *Frozen Star*. New York: Scribner, 1983. (An eloquently written introduction to pulsars and black holes, which also portrays the scientists involved in this area of research.)

*Jordan, S. (ed.), *The Sun as a Star* (NASA SP-450). Washington, DC: U.S. Government Printing Office, 1981. (A highly technical discussion of solar atmospheric structures and processes, with emphasis on relating solar phenomena to observations of stars.)

Kippenhahn, R., *100 Billion Suns* (trans. J. Steinberg). New York: Basic Books, 1983. (Excellent popular account of stellar evolution.)

Malin, D., and Murdin, P., *Colours of the Stars*. New York: Cambridge University Press, 1984. (A beautifully produced coffee table–style book of astronomical images made with modern photographic techniques.)

Mitton, S., *Daytime Star: The Story of Our Sun*. New York: Scribners, 1983. (Well-written discussion of the sun for the layperson.)

Noyes, R. W., *The Sun, Our Star*. Cambridge, MA: Harvard University Press, 1982. (A well-presented picture of the sun, including chapters on climate and solar energy.)

Spitzer, L., *Searching Between the Stars*. New Haven, CT: Yale University Press, 1982. (An authoritative introduction to the modern study of the interstellar medium.)

Sullivan, W., *Black Holes*. Garden City, NJ: Doubleday, 1979. (Well-written account for the layperson by a leading science journalist.)

Washburn, M., *In the Light of the Sun*. New York: Harcourt Brace Jovanovich, 1981. (Well-written popular account of the sun and modern astrophysical ideas.)

RELATIVITY AND MODERN PHYSICS

Couderc, P., *The Expansion of the Universe*. London: Faber and Faber, 1952. (Although it does not contain recent observational and theoretical results, this is a very good development of relativistic cosmology with only minimal mathematics—through intermediate algebra.)

Davies, P., *Other Worlds*. New York: Simon and Schuster, 1980. (Popular discussion of quantum theory and cosmology.)

Davies, P. C. W., *The Forces of Nature*. Cambridge, England: Cambridge University Press, 1979. (An excellent semipopular account of modern physics.)

Einstein, A., *Relativity: The Special and General Theory*. New York: Crown Publishers, 1961. (A classic popular account, by Einstein himself, of his epoch-making new physics.)

Feynman, R., *The Character of Physical Law*. Cambridge, MA: MIT Press, 1965. (A series of popular lectures by one of the most articulate theoretical physicists of our time; an excellent insight into modern physics.)

Gardner, M., *The Relativity Explosion*. New York: McGraw-Hill, 1966. (A very readable description of the meaning of relativity for the layperson.)

Harrison, E. R., *Cosmology: The Science of the Universe*. New York: Cambridge University Press, 1981. (An introduction to past and present cosmological theory. Familiar objects of the universe, including stars and galaxies, as well as black holes, curved space, and cosmic horizons are discussed with grace and authority.)

Kaufmann, W. J., III., *The Cosmic Frontiers of General Relativity*. Boston: Little, Brown, 1977. (A layperson's account of relativity theory and some of the possible esoteric consequences of it.)

Polkinghorne, J. C., *The Particle Play*. San Francisco: W. H. Freeman, 1979. (A layperson's introduction to the exciting new world of the particles of nature.)

Russell, B., *The ABC of Relativity*. New York: Mentor Book, from the New American Library, 1969. (An interpretation of relativity theory for the nonscientist by the famous philosopher.)

Sciama, D. W., *The Physical Foundations of General Relativity*. Garden City, NY: Doubleday, 1969. (A description of the physical basis of Einstein's theory for the person with only a smattering of algebra.)

GALAXIES AND COSMOLOGY

Barrow, J., and Silk, J., *The Left Hand of Creation: Origin and Evolution of the Expanding Universe*. New York: Basic Books, 1983. (Authoritative account of the origin and evolution of the universe, with a good discussion of the relationship between subatomic physics and cosmology.)

Chaisson, E., *Cosmic Dawn*. Boston: Little, Brown, 1981. (An eloquent introduction to the development of "particles, galaxies, stars, planets, life, and culture."

Ferris, T., *Galaxies*. San Francisco: Sierra Club Books, 1981. (A magnificently illustrated coffee table–style book with informative text.)

Harrison, E. R., *Cosmology*. Cambridge, England: Cambridge University Press, 1981. (A college text that provides an outstanding introduction to cosmology. The book requires some knowledge of algebra.)

Hubble, E., *The Realm of the Nebulae*. New Haven, CT: Yale University Press, 1936; also New York: Dover, 1958. (A classic book for the educated layperson by one of the great astronomers of our time, describing his exploration of the extragalactic universe.)

Sandage, A. R., *The Hubble Atlas of Galaxies*. Washington, DC: Carnegie Institution, 1961. (A photographic atlas of galaxies of different types.)

*Sciama, D. W., *Modern Cosmology*. London: Cambridge University Press, 1971. (Although somewhat technical, this is an exceptionally readable text for upper-division science students.)

Sciama, D. W., *The Unity of the Universe*. New York: Doubleday, 1959. (An account of the interplay of modern physics and cosmology for the intelligent layperson.)

Shapley, H., *Galaxies*. Cambridge, MA: Harvard University Press, 1972. (One of the Harvard Series in Astronomy, this is an excellent although dated semipopular description of the extragalactic universe by one of the great astronomers who played a role in its investigation.)

Shipman, H. L., *Black Holes, Quasars, and the Universe*, 2nd ed. Boston: Houghton Mifflin, 1980. (Probably the best popular or semipopular account of black holes, relativity, and active galactic nuclei available.)

Silk, J., *The Big Bang*. San Francisco: Freeman, 1980. (An authoritative and readable book on cosmology for the intermediate undergraduate science student; highly recommended.)

Trefil, J. S., *The Moment of Creation*. New York: Scribners, 1983. (Excellent discussion of modern cosmological thinking for the interested layperson, incorporating elementary particle physics, grand unified theories, and the inflationary universe.)

Verschuur, G. L., *The Invisible Universe*. New York: Springer-Verlag, 1974. (An account of the role of radio astronomy in cosmology by a radio astronomer; written for the general public.)

Wagoner, R., and Goldsmith, D., *Cosmic Horizons: Understanding the Universe*. New York: W. H. Freeman, 1983. (Clearly written guide to the latest theories and observations about the origin, structure, and evolution of the cosmos.)

Weinberg, S., *The First Three Minutes*. New York: Basic Books, 1977. (Although somewhat dated, this remains one of the best popular accounts of modern cosmology, lucidly written by a Nobel Laureate physicist who has been at the forefront in the study of the basic forces of nature; highly recommended.)

LIFE IN THE UNIVERSE

Berendzen, R. (ed.), *Life Beyond the Earth and the Mind of Man*. Washington, DC: NASA, 1973. (A symposium of lectures by famous scientists and philosophers on the subject of extraterrestrial life.)

Bracewell, R. N., *The Galactic Club*. Stanford, CA: Stanford Alumni Association, 1974. (Speculations and scientific analyses of the kinds of civilizations that could exist in the Galaxy.)

*Billingham, J. (ed.), *Life in the Universe*. Cambridge, MA: MIT Press, 1981. (Comprehensive collection of papers from a conference held at NASA Ames Research Center in 1980.)

Drake, F., *Intelligent Life in Space*. New York: Macmillan, 1962. (A classic, although perhaps too optimistic, account of the possibilities of extraterrestrial communication.)

Goldsmith, D., and Owen, T., *The Search for Life in the Universe*. Menlo Park, CA: Benjamin/Cummings, 1980. (Comprehensive and well-written text, covering all of astronomy from the perspective of the search for life; a good successor to Shklovskii and Sagan.)

Hart, M. H., and Zuckerman, B. (eds.), *Extraterrestrials: Where Are They?* New York: Pergamon, 1982. (Papers from a conference discussing the provocative questions raised by the apparent absence of visitors to Earth from other planets.)

*Ponnamperuma, C., and Cameron, A. G. W. (eds.), *Interstellar Communications; Scientific Perspectives*. Boston: Houghton Mifflin, 1974. (Discussions of the scientific possibilities of communicating with extraterrestrial civilizations.)

Sagan, C., *The Cosmic Connection*. New York: Anchor Press-Doubleday, 1973. (Essays by America's best-known scientist that establish a perspective on our place in the universe.)

Shklovskii, I. S., and Sagan, C., *Intelligent Life in the Universe*. New York: Dell, 1966. (The classic work that investigates extraterrestrial life in a thoughtful and authoritative manner—dated but still fascinating.)

PSEUDOSCIENCE

Abell, G. O., and Singer, B. (eds.), *Science and the Paranormal*. New York: Scribners, 1981. (A collection of critical essays on allegedly paranormal topics by well-known scientists and science writers.)

Condon, E. U. (ed.), *Scientific Study of Unidentified Flying Objects*. New York: Bantam Books, 1969. (The famous report of a study commissioned by the U.S. Air Force to investigate UFOs.)

Culver, R. B., and Ianna, P. A., *The Gemini Syndrome: Star Wars of the Oldest Kind*. Tucson, AZ: Pachart, 1979. (An excellent account of astrology.)

Frazier, K. (ed.), *Paranormal Borderlands of Science*. Buffalo, NY: Prometheus Books, 1981. (A collection of articles from *The Skeptical Inquirer* that scrutinize a variety of claims of the paranormal.)

Gardner, M., *Fads and Fallacies in the Name of Science*. New York: Dover, 1957. (Fascinating look at pseudoscience.)

Gardner, M., *Science: Good, Bad, and Bogus*. Buffalo, NY: Prometheus Books, 1981. (A collection of essays from the famous philosopher, mathematician, and science writer.)

Klass, P. J., *UFOs Explained*. New York: Vintage Books, 1976. (A revealing expose of the UFO phenomenon by the foremost skeptical investigator of the subject.)

Radner, D., and Radner, M., *Science and Unreason*. Belmont, CA: Wadsworth, 1982. (A splendid little book on how to distinguish pseudoscience from science.)

Randi, J., *Flim-Flam*. New York: Lippincott and Crowell, 1980. (World-famous magician exposes frauds, some of which duped scientists as well as the public.)

Sheaffer, R., *The UFO Verdict: Examining the Evidence*. Buffalo, NY: Prometheus Books, 1981. (A skeptical survey of the evidence for extraterrestrial spaceships by a well-known UFO investigator.)

Standon, A., *Forget Your Sun Sign*. Baton Rouge, LA: Legacy Publishing Co., 1977. (Skeptical discussion of astrology.)

STAR ATLASES, SKY GUIDES, and SKY LORE

Allen, R. H., *Star Names*. New York: Dover, 1963. (An exhaustive reference on star names and their origins.)

Menzel, D. H., *A Field Guide to the Stars and Planets*. Boston: Houghton Mifflin, 1964. (A famous guide for the amateur.)

Minnaert, M., *The Nature of Light and Color in the Open Air*. New York: Dover, 1954. (The definitive book on the origin of optical phenomena in the sky, including rainbows, halos, shadow bands, and hundreds of others.)

Norton, W. W., *Sky Atlas*. Cambridge, Ma.: Sky Publishing Company, 1971. (One of the most popular sky atlases for knowledgeable amateurs.)

Olcott, W. T., *Olcott's Field Book of the Skies*. New York: Putnam, 1954. (One of the standard amateur guide books.)

Rükl, A., *The Amateur Astronomer*. London: Octopus Books, Ltd., 1979. (English version of original Czech). (An excellent handbook for the beginner, with easy-to-use star charts and charts and tables of lunar and planetary information through the year 2000.)

POPULAR JOURNALS ON ASTRONOMY

Astronomy, published monthly by Astromedia Corporation, 441 Mason Street, P.O. Box 92788, Milwaukee, WI 53202.

Mercury, published bimonthly by the Astronomical Society of the Pacific, 1290 24th Avenue, San Francisco, CA 94122.

The Planetary Report, published bimonthly by The Planetary Society, 65 N. Catalina Ave., Pasadena, CA 91106.

Sky and Telescope, published monthly by Sky Publishing Corporation, 49 Bay State Road, Cambridge, MA 02139.

Popular articles on astronomy also appear frequently in *Scientific American*, *Discover*, and *Science News*.

CAREER INFORMATION

Information about a career in astronomy is available from the Executive Officer of the American Astronomical Society, at Suite 300, 2000 Florida Avenue, N.W., Washington, DC 20009.

Appendix 2

GLOSSARY

aberration (of starlight). Apparent displacement in the direction of a star due to the Earth's orbital motion.

absolute magnitude. Apparent magnitude a star would have at a distance of 10 pc.

absolute zero. A temperature of −273°C (or 0 K), where all molecular motion stops.

absorption spectrum. Dark lines superimposed on a continuous spectrum.

accelerate. To change velocity; either to speed up, slow down, or change direction.

acceleration of gravity. Numerical value of the acceleration produced by the gravitational attraction on an object at the surface of a planet or star.

accretion. Gradual accumulation of mass, as by a planet forming by the building up of colliding particles in the solar nebula.

achromatic. Free of chromatic aberration.

active galactic nucleus. A galaxy is said to have an active nucleus if unusually violent events are taking place in its center, emitting in the process very large quantities of electromagnetic radiation. Seyfert galaxies and quasars are examples of galaxies with active nuclei.

active sun. The sun during times of unusual solar activity—spots, flares, and associated phenomena.

airglow. Fluorescence in the upper atmosphere.

albedo. The fraction of incident sunlight that a planet or minor planet reflects.

Allende meteorite. A primitive meteorite (carbonaceous chondrite) that fell in northern Mexico in 1969 and has yielded a wealth of information on the formation of the solar system.

almanac. A book or table listing astronomical events.

alpha particle. The nucleus of a helium atom, consisting of two protons and two neutrons.

altitude. Angular distance above or below the horizon, measured along a vertical circle, to a celestial object.

amplitude. The range in variability, as in the light from a variable star.

angstrom (Å). A unit of length equal to 10^{-8} cm.

angular diameter. Angle subtended by the diameter of an object.

angular momentum. A measure of the momentum associated with motion about an axis or fixed point.

annular eclipse. An eclipse of the sun in which the Moon is too distant to appear to cover the sun completely, so that a ring of sunlight shows around the Moon.

Antarctic Circle. Parallel of latitude 66½°S; at this latitude the noon altitude of the sun is 0° on the date of the summer solstice.

antimatter. Matter consisting of antiparticles: *antiprotons* (protons with negative rather than positive charge), *positrons* (positively-charged electrons), and *antineutrons*.

aperture. The diameter of an opening, or of the primary lens or mirror of a telescope.

aphelion. Point in its orbit where a planet is farthest from the sun.

apogee. Point in its orbit where an Earth satellite is farthest from the Earth.

Apollo asteroid. An Earth-crossing asteroid—with perihelion inside the Earth's orbit and aphelion outside the Earth's orbit.

apparent magnitude. A measure of the observed light flux received from a star or other object at the Earth.

apparent relative orbit. The projection onto a plane perpendicular to the line of sight of the relative orbit of the fainter of the two components of a visual binary star about the brighter.

apparent solar day. The interval between two successive transits of the sun's center across the meridian.

apparent solar time. The hour angle of the sun's center *plus* 12 hours.

apse (or **apsis**; pl. **apsides**). The point in a body's orbit where it is nearest or farthest from the object it revolves about.

Arctic Circle. Parallel of latitude 66½° N; at this latitude the noon altitude of the sun is 0° on the date of the winter solstice.

ascending node. The point along the orbit of a body

where it crosses from the south to the north of some reference plane, usually the plane of the celestial equator or of the ecliptic.

association. A loose cluster of stars whose spectral types, motions, or positions in the sky indicate that they have probably had a common origin.

asteroid. An object orbiting the sun that is smaller than a major planet, but that shows no evidence of an atmosphere or of other types of activity associated with comets. Also called a minor planet.

asteroid belt. The region of the solar system between the orbits of Mars and Jupiter in which most asteroids are located. The main belt, where the orbits are generally the most stable, extends from 2.2 to 3.3 AU from the sun.

astigmatism. A defect in an optical system whereby pairs of light rays in different planes do not focus at the same place.

astrology. The pseudoscience that treats with supposed influences of the configurations and locations in the sky of the sun, Moon, and planets on human destiny; a primitive religion having its origin in ancient Babylonia.

astrometric binary. A binary star in which one component is not observed, but its presence is deduced from the orbital motion of the visible component.

astrometry. That branch of astronomy that deals with the determination of precise positions and motions of celestial bodies.

astronomical unit (AU). Originally meant to be the semimajor axis of the orbit of the Earth; now defined as the semimajor axis of the orbit of a hypothetical body with the mass and period that Gauss assumed for the Earth. The semimajor axis of the orbit of the Earth is 1.000 000 230 AU.

astronomy. The branch of science that treats of the physics and morphology of that part of the universe that lies beyond the Earth's atmosphere.

astrophysics. The part of astronomy that deals principally with the physics of stars, stellar systems, and interstellar material. Astrophysics also deals, however, with the structures and atmospheres of the sun and planets.

atmospheric refraction. The bending, or refraction, of light rays from celestial objects by the Earth's atmosphere.

atom. The smallest particle of an element that retains the properties that characterize that element.

atomic clock. A time-keeping device regulated by the natural frequency of the emission or absorption of radiation of a particular kind of atom.

atomic mass unit. *Chemical:* one-sixteenth of the mean mass of an oxygen atom. *Physical:* one-twelfth of the mass of an atom of the most common isotope of carbon. The atomic mass unit is approximately the mass of a hydrogen atom, 1.67×10^{-24} g.

atomic number. The number of protons in each atom of a particular element.

atomic time. The time kept by a cesium (atomic) clock, based on the **atomic second**—the time required for 9,192,631,770 cycles of the radiation emitted or absorbed in a particular transition of an atom of cesium-133.

atomic transition. A change in the state of energy of an atom; the atom may gain or lose energy by collision with another particle or by the emission or absorption of a photon.

atomic weight. The mean mass of an atom of a particular element in atomic mass units.

aurora. Light radiated by atoms and ions in the ionosphere, mostly in the magnetic polar regions.

autumnal equinox. The intersection of the ecliptic and celestial equator where the sun crosses the equator from north to south.

azimuth. The angle along the celestial horizon, measured eastward from the north point, to the intersection of the horizon with the vertical circle passing through an object.

Baily's beads. Small "beads" of sunlight seen passing through valleys along the limb of the Moon in the instant preceding and following totality in a solar eclipse.

Balmer lines. Emission or absorption lines in the spectrum of hydrogen that arise from transitions between the second (or first excited) and higher energy states of the hydrogen atom.

bands (in spectra). Emission or absorption lines, usually in the spectra of chemical compounds or radicals, so numerous and closely spaced that they coalesce into broad emission or absorption bands.

barred spiral galaxy. Spiral galaxy in which the spiral arms begin from the ends of a "bar" running through the nucleus rather than from the nucleus itself.

barycenter. The center of mass of two mutually revolving bodies.

baryons (and **antibaryons**). The heavy atomic nuclear particles, such as protons and neutrons.

basalt. Igneous rock, composed primarily of silicon, oxygen, iron, aluminum, and magnesium, produced by the cooling of lava. Basalts make up most of Earth's oceanic crust and are also found on other planets that have experienced extensive volcanic activity.

base line. That side of a triangle used in triangulation or surveying whose length is known (or can be measured), and which is included between two angles that are known (or can be measured).

basin (impact). A large circular depression (generally greater than 200 km in diameter), usually rimmed by mountains, produced by the explosive impact of an asteroid or similar-sized projectile on a planetary surface.

big bang theory. A theory of cosmology in which the

expansion of the universe is presumed to have begun with a primeval explosion.

billion. In the United States and France and in this text, one thousand million (10^9); in Great Britain and Germany, originally, one million million (10^{12}). Usage in Great Britain is changing to conform to usage in the United States.

binary star. A double star; two stars revolving about each other.

binding energy. The energy required to completely separate the constituent parts of an atomic nucleus.

black body. A hypothetical perfect radiator, which absorbs and re-emits all radiation incident upon it.

black dwarf. A presumed final state of evolution for a star, in which all of its energy sources are exhausted and it no longer emits radiation.

black hole. A hypothetical body whose velocity of escape is equal to or greater than the speed of light; thus no radiation can escape from it.

Bode's Law. See *Titius-Bode Law*.

Bohr atom. A particular model of an atom, invented by Niels Bohr, in which the electrons are described as revolving about the nucleus in circular orbits.

bolide. A very bright fireball or meteor; sometimes defined as a fireball accompanied by sound.

bolometric correction. The difference between the visual and bolometric magnitudes of a star.

bolometric magnitude. A measure of the flux of radiation from a star or other object received just outside the Earth's atmosphere, as it would be detected by a device sensitive to *all* forms of electromagnetic energy.

breccia. Any rock made up of recemented fragments of material, usually the result of extensive impact cratering.

bremsstrahlung. Radiation from free-free transitions in which electrons gain or lose energy while being accelerated in the field of an atomic nucleus or ion.

brown dwarf. An object intermediate in size between a planet and a star. The approximate mass range is from about twice the mass of Jupiter up to the lower mass limit for self-sustaining nuclear reactions, which is 0.08 solar masses.

bubble chamber. A chamber in which bubbles form along the electrically charged path of a high-energy charged particle, rendering the track of that particle visible.

calculus. A branch of mathematics that permits computations involving rates of change (*differential* calculus) or of the contribution of an infinite number of infinitesimal quantities (*integral* calculus).

caldera. A volcanic crater, often resulting from the partial collapse of the summit of a shield volcano.

canals (on Mars). Supposed long, narrow, straight dark lines first reported on Mars by Schiaparelli and believed by Lowell and others to be the work of intelligent Martians. The Martian canals (not to be confused with Martian channels) were later shown to have been an illusion.

carbon cycle. A series of nuclear reactions in the interiors of stars involving carbon as a catalyst, by which hydrogen is transformed to helium.

carbonaceous asteroid. An asteroid made of dark, chemically primitive material, in some cases apparently similar to the carbonaceous meteorites but often even darker (typical reflectivity = 3 to 4 percent).

carbonaceous meteorite. A primitive meteorite made primarily of silicates but often including chemically bound water, free carbon, and complex organic compounds. Also called carbonaceous chondrites.

Cassegrain focus. An optical arrangement in a reflecting telescope in which light is reflected by a second mirror to a point behind the primary mirror.

CBR. See *cosmic background radiation*.

CCD. See *charge-coupled device*.

cD galaxy. A supergiant elliptical galaxy frequently found at the center of a cluster of galaxies.

celestial equator. A great circle on the celestial sphere 90° from the celestial poles; the circle of intersection of the celestial sphere with the plane of the Earth's equator.

celestial mechanics. The branch of astronomy that deals with the motions and gravitational influences of the members of the solar system.

celestial navigation. The art of navigation at sea or in the air from sightings of the sun, Moon, planets, and stars.

celestial poles. Points about which the celestial sphere appears to rotate, intersections of the celestial sphere with the Earth's polar axis.

celestial sphere. Apparent sphere of the sky; a sphere of large radius centered on the observer. Directions of objects in the sky can be denoted by the position of those objects on the celestial sphere.

center of gravity. Center of mass.

center of mass. The mean position of the various mass elements of a body or system, weighted according to their distances from that center of mass; that point in an isolated system that moves with constant velocity, according to Newton's first law of motion.

centripetal force (or **acceleration**). The force required to divert a body from a straight path into a curved path (or the acceleration experienced by the body); it is directed toward the center of curvature.

cepheid variable. A star that belongs to one of two classes (type I and type II) of yellow supergiant pulsating stars. These stars vary periodically in brightness, and the relationship between their periods and luminosities is useful in deriving distances to them.

Chandrasekhar limit. The upper limit to the mass of a white dwarf (equals 1.4 times the mass of the sun).

charged-coupled device (CCD). An array of electronic detectors of electromagnetic radiation, used at the focus of a telescope (or camera lens). A CCD acts like a photographic plate of very high sensitivity.

chemical condensation sequence. The calculated chemical compounds and minerals that would form in a cooling gas of cosmic composition, presented as a function of the temperature in the gas; used to infer the composition of grains that formed in the solar nebula at different distances from the protosun.

chondrite. A primitive stony meteorite that contains small spherical particles called *chondrules*.

chromatic aberration. A defect of optical systems whereby light of different colors is focused at different places.

chromosphere. That part of the solar atmosphere that lies immediately above the photospheric layers.

chronograph. A device for recording and measuring the times of events.

chronometer. An accurate clock.

circular velocity. The critical speed with which a revolving body can have a circular orbit.

circumpolar regions. Portions of the celestial sphere near the celestial poles that are either always above or always below the horizon.

circumstellar dust. Solid particles surrounding a star, which either condensed from material ejected from the star or are part of the interstellar material from which the star is forming.

climate. The weather conditions (temperature, precipitation, seasonal variations) averaged over a long enough span of time to eliminate most random variations—usually representing an average over a 10- to 20-year period.

cloud chamber. A chamber in which droplets of liquid condense along the electrically charged path of a high-energy charged particle, rendering the track of that particle visible.

cluster of galaxies. A system of galaxies containing from several to thousands of member galaxies.

cluster variable (RR Lyrae variable). A member of a certain large class of pulsating variable stars, all with periods less than one day. These stars are often present in globular star clusters.

coherent radiation. Electromagnetic radiation of a particular wavelength traveling in the same direction and in which the waves are all in phase with each other.

cold cloud (cold molecular cloud). Region of the interstellar medium where molecules, excited by collisions, radiate microwave and infrared radiation away, thereby cooling the region. Subsequent collapse of a cold cloud is believed to be a first step toward star formation.

color excess. The amount by which the color index of a star is increased when its light is reddened in passing through interstellar absorbing material.

color index. Difference between the magnitudes of a star or other object measured in light of two different spectral regions, for example, blue *minus* photovisual magnitudes.

color-magnitude diagram. Plot of the magnitudes (apparent or absolute) of the stars in a cluster against their color indices.

coma. A defect in an optical system in which off-axis rays of light striking different parts of the objective do not focus in the same place.

coma (of comet). The diffuse gaseous component of the head of a comet.

comet. A small body of icy and dusty matter that revolves about the sun. When a comet comes near the sun, some of its material vaporizes, forming a large *head* of tenuous gas, and often a *tail*.

commensurability. State of a quantity (such as the period of a minor planet) that can be reduced to another quantity (such as the period of another minor planet) by multiplication by a ratio of two small whole numbers.

compact galaxy. A galaxy of small size and high surface brightness.

comparative planetology. A way of looking at planetary science that emphasizes the similarities and differences among the planets and attempts to account for their current conditions in terms of common processes that act throughout the planetary system.

comparison spectrum. The spectrum of a vaporized element (such as iron) photographed beside the image of a stellar spectrum, and with the same camera, for purposes of comparison of wavelengths.

compound. A substance composed of two or more chemical elements.

compound nucleus. An excited nucleus, usually temporary, formed by nuclei of two or more simpler atoms.

conduction. The transfer of energy by the direct passing of energy or electrons from atom to atom.

conic section. The curve of intersection between a circular cone and a plane; these curves can be ellipses, circles, parabolas, or hyperbolas.

conjunction. The configuration of a planet when it has the same celestial longitude as the sun, or the configuration when any two celestial bodies have the same celestial longitude or right ascension.

conservation of angular momentum. The law that angular momentum is conserved in the absence of any force not directed toward or away from the point or axis about which the angular momentum is referred—that is, in the absence of a torque.

constellation. A configuration of stars named for a particular object, person, or animal; or the area of the sky assigned to a particular configuration.

contacts (of eclipses). The instants when certain stages of an eclipse begin.

continental drift. A gradual drift of the continents over the surface of the Earth due to *plate tectonics*.

continuous spectrum. A spectrum of light comprised of radiation of a continuous range of wavelengths or colors rather than only certain discrete wavelengths.

convection. The transfer of energy by moving currents of a fluid containing that energy.

core (of a planet). The central part of a planet, consisting of higher-density material, often metallic.

coriolis effect. The deflection (with respect to the ground) of projectiles moving across the surface of the rotating Earth.

corona. Outer atmosphere of the sun.

corona of Galaxy. The extension of the nuclear bulge of the Galaxy on either side of the plane of the Milky Way; a region containing hot gas that emits X-rays.

coronagraph. An instrument for observing the chromosphere and corona of the sun outside of eclipse.

coronal hole. A region in the sun's outer atmosphere where visible coronal streamers are absent.

cosine. One of the trigonometric functions of an angle. In a right triangle, the ratio of the lengths of the shorter side adjacent to the angle and the hypotenuse.

cosmic background radiation (CBR). The microwave radiation coming from all directions that is believed to be the redshifted glow of the big bang.

cosmic composition. The relative amounts of the elements that are thought to be present in the interstellar medium in the part of the Galaxy where the sun is located—hence approximately the elemental composition of the sun and the solar nebula at the time the solar system formed.

cosmic rays. Atomic nuclei (mostly protons) that are observed to strike the Earth's atmosphere with exceedingly high energies.

cosmogony. The study of the origin of the world or universe.

cosmological constant. A term that arises in the development of the field equations of general relativity, which represents a repulsive force in the universe. The cosmological constant is often assumed to be zero.

cosmological model. A specific model, or theory, of the organization and evolution of the universe.

cosmological principle. The assumption that, on the large scale, the universe at any given time is the same everywhere.

cosmology. The study of the organization and evolution of the universe.

coudé focus. An optical arrangement in a reflecting telescope whereby light is reflected by two or more secondary mirrors down the polar axis of the telescope to a focus at a place separate from the moving parts of the telescope.

crater. A circular depression (from the Greek word for cup), generally of impact origin. The rarer volcanic craters are usually identified as such (see *caldera*); crater by itself is used in this text to refer to an impact crater.

crescent moon. One of the phases of the Moon when its elongation is less than 90° from the sun and it appears less than half full.

crust (of Earth). The outer layer of the Earth.

cyclonic motion. A counterclockwise circular circulation of winds (in the northern hemisphere) that results from the coriolis effect.

dark matter. Nonluminous mass, whose presence can be inferred only because of its gravitational influence on luminous matter. Dark matter makes up at least half of the matter contained within the luminous parts of spiral galaxies and may constitute as much as 99 percent of all the mass in the universe. The composition of the dark matter is not known.

dark nebula. A cloud of interstellar dust that obscures the light of more distant stars and appears as an opaque curtain.

daughter. In referring to the process of radioactive decay, the name given to the isotope that results from the decay reaction.

daylight saving time. A time one hour more advanced than standard time, usually adopted in spring and summer to take advantage of long evening twilights.

deceleration parameter (q_0). A quantity that characterizes the future evolution of the various models of the universe based on general relativity.

declination. Angular distance north or south of the celestial equator to some object, measured along an hour circle passing through that object.

deferent. A stationary circle in the Ptolemaic system along which moves the center of another circle (epicycle), along which moves an object or another epicycle.

degenerate gas. A gas in which the allowable states for the electrons have been filled; it behaves according to different laws from those that apply to "perfect" gases.

density. The ratio of the mass of an object to its volume.

density-wave theory. A theory proposed by C. C. Lin and his associates for the spiral structure of the Galaxy.

descending node. The point along the orbit of a body where it crosses from the north to the south of some reference plane, usually the plane of the celestial equator or of the ecliptic.

deuterium. A "heavy" form of hydrogen, in which the nucleus of each atom consists of one proton and one neutron.

differential galactic rotation. The rotation of the galaxy, not as a solid wheel, but so that parts adjacent to each other do not always stay close together.

differential gravitational force. The difference between the respective gravitational forces exerted on two bodies near each other by a third, more distant body.

differentiation (geological). A separation or segregation of different kinds of material in different layers in the interior of a planet.

diffraction. The spreading out of light in passing the edge of an opaque body.

diffraction grating. A system of closely spaced equidistant slits or reflecting strips that, by diffraction and interference, produce a spectrum.

diffraction pattern. A pattern of bright and dark fringes produced by the interference of light rays, diffracted by different amounts, with each other.

diffuse nebula. A reflection or emission nebula produced by interstellar matter (not a planetary nebula).

disk (of planet or other object). The apparent circular shape that a planet (or the sun, or Moon, or a star) displays when seen in the sky or viewed telescopically.

disk of Galaxy. The central disk or "wheel" of our Galaxy, superimposed on the spiral structure.

dispersion. Separation, from white light, of different wavelengths being refracted by different amounts.

distance modulus. Difference between the apparent and absolute magnitudes of an object, which provides a measure of its distance through the inverse-square law of light.

diurnal. Daily.

diurnal circle. Apparent path of a star in the sky during a complete day due to the Earth's rotation.

diurnal motion. Motion during one day.

diurnal parallax. Apparent change in direction of an object caused by a displacement of the observer due to the Earth's rotation.

Doppler shift. Apparent change in wavelength of the radiation from a source due to its relative motion in the line of sight.

dust tail (of comet). A cometary tail, usually broad, somewhat curved, and yellow-white in color, made up of dust grains released from the nucleus of the comet.

dwarf (star). A main-sequence star (as opposed to a giant or supergiant).

dynamical parallax. A distance (or parallax) for a binary star derived from the period of mutual revolution, the mass-luminosity relation, and the laws of mechanics.

dyne. The metric unit of force; the force required to accelerate a mass of 1 gram in the amount 1 centimeter per second per second.

Earth-approaching asteroid. An asteroid with an orbit that crosses the Earth's orbit or that will at some time cross the Earth's orbit as it evolves under the influence of the planets' gravity—subdivided into the Apollo asteroids, which presently cross the Earth's orbit, and the Amor and Aten asteroids, which do not.

east point. The point on the horizon 90° from the north point (measured clockwise as seen from the zenith).

eccentric. The off-center position of the Earth in the presumed circular orbits of the sun, Moon, and planets in the Ptolemaic system.

eccentricity (of ellipse). Ratio of the distance between the foci to the major axis.

eclipse. The cutting off of all or part of the light of one body by another; in planetary science, the passing of one body into the shadow of another. (See *occultation, transit.*)

eclipse path. The track along the Earth's surface swept out by the tip of the shadow of the Moon (or the extension of its shadow) during a total (or annular) solar eclipse.

eclipsing binary star. A binary star in which the plane of revolution of the two stars is nearly edge-on to our line of sight, so that the light of one star is periodically diminished by the other passing in front of it.

ecliptic. The apparent annual path of the sun on the celestial sphere.

ejecta. Material excavated from an impact crater; includes the ejecta blankets surrounding lunar craters, crater rays, and (in the case of the Earth) dust released into the atmosphere by an impact.

ejecta blanket. Rough, hilly region surrounding an impact crater made up of ejecta that has fallen back to the surface, usually extending one to three crater radii from the rim.

electromagnetic force. One of the four fundamental forces of nature; the force that acts between charges and binds atoms and molecules together.

electromagnetic radiation. Radiation consisting of waves propagated through the building up and breaking down of electric and magnetic fields; these include radio, infrared, light, ultraviolet, X-rays, and gamma rays.

electromagnetic spectrum. The whole array or family of electromagnetic waves.

electron. A negatively charged subatomic particle that normally moves about the nucleus of an atom.

electron volt. The kinetic energy acquired by an electron that is accelerated through an electric potential of 1 volt; 1 electron volt is 1.60207×10^{-12} erg.

element. A substance that cannot be decomposed, by chemical means, into simpler substances.

elements (of orbit). Any of several quantities that describe the size, shape, and orientation of the orbit of a body.

ellipse. A conic section: the curve of intersection of a circular cone and a plane cutting completely through the cone.

elliptical galaxy. A galaxy whose apparent photometric contours are ellipses, and which contains no conspicuous interstellar material.

ellipticity. The ratio (in an ellipse) of the major axis *minus* the minor axis to the major axis.

elongation. The difference between the celestial longitudes of a planet and the sun.

emission line. A discrete bright spectral line.

emission nebula. A gaseous nebula that derives its visible light from the fluorescence of ultraviolet light from a star in or near the nebula.

emission spectrum. A spectrum consisting of emission lines.

encounter (gravitational). A near passing, on hyperbolic orbits, of two objects that influence each other gravitationally.

endogenic. In geology, referring to processes that derive their energy from the interior of a planet, such as earthquakes and volcanic eruptions.

energy. The ability to do work.

energy level (in an atom or ion). A particular level, or amount, of energy possessed by an atom or ion above the energy it possesses in its least energetic state.

energy equation. An equation that expresses the conservation of energy for two mutually revolving bodies; it relates their relative speed to their separation and the semimajor axis of the relative orbit.

energy spectrum. A table or plot showing the relative numbers of particles (in cosmic rays, for example) of various energies.

ephemeris A table that gives the positions of a celestial body at various times, or other astronomical data.

ephemeris time. A kind of time that passes at a strictly uniform rate; used to compute the instants of various astronomical events.

epicycle. A circular orbit of a body in the Ptolemaic system, the center of which revolves about another circle (the deferent).

equant. A stationary point in the Ptolemaic system not at the center of a circular orbit about which a body (or the center of an epicycle) revolves with uniform angular velocity.

equation of state. An equation relating the pressure, temperature, and density of a substance (usually a gas).

equation of time. The difference between apparent and mean solar time.

equator. A great circle on the Earth, 90° from its poles.

equatorial mount. A mounting for a telescope, one axis of which is parallel to the Earth's axis, so that a motion of the telescope about the axis can compensate for the Earth's rotation.

equinox. One of the intersections of the ecliptic and celestial equator.

equivalent width. A measure of the strength of a spectral line; the width of an absorption line of rectangular profile and zero intensity at its center equivalent in strength to the actual line.

erg. The metric unit of energy; the work done by a force of one dyne acting through a distance of one centimeter.

eruptive variable. A variable star whose changes in light are erratic or explosive.

ether. A hypothetical medium once supposed to exist in space and to transport electromagnetic radiation.

Euclidean. Pertaining to Euclidean geometry, or *flat space*.

eucrite meteorite. One of a class of basaltic meteorites believed to have originated on the asteroid 4 Vesta.

event. A point in four-dimensional spacetime.

event horizon. The surface through which a collapsing star is hypothesized to pass when its velocity of escape is equal to the speed of light, that is, when the star becomes a black hole.

evolutionary cosmology. A theory of cosmology that assumes that all parts of the universe have a common age and evolve together.

excitation. The process of imparting to an atom or an ion an amount of energy greater than that it has in its normal or least-energy state.

exclusion principle. see *Pauli exclusion principle*.

exogenic. In geology, referring to processes that are caused by external forces, such as impact cratering.

extinction. Attenuation of light from a celestial body produced by the Earth's atmosphere, or by interstellar absorption.

extragalactic. Beyond our own Milky Way Galaxy.

eyepiece. A magnifying lens used to view the image produced by the objective of a telescope.

faculus (pl. **faculae**). Bright region near the limb of the sun.

family (of asteroids). A group of asteroids with similar orbital elements, indicating a probable common origin in a collision sometime in the past.

Faraday rotation. The rotation of the plane of polarization of polarized radiation passing through ionized gas.

fault. In geology, a crack or break in the crust of a planet along which slippage or movement can take place, accompanied by seismic activity.

fermions. Fundamental particles, such as protons, electrons, and neutrons, all of spin ½, that obey certain laws formulated by Enrico Fermi.

field equations. A set of equations in general relativity that describe the curvature of spacetime in the presence of matter.

filtergram. A photograph of the sun (or part of it) taken through a special narrow-bandpass filter.

fireball. A spectacular meteor.

fission. The breakup of a heavy atomic nucleus into two or more lighter ones.

flare. A sudden and temporary outburst of light from an extended region of the solar surface.

flash spectrum. The spectrum of the very limb of the sun obtained in the instant before or after totality in a solar eclipse.

fluorescence. The absorption of light of one wavelength and re-emission of it at another wavelength; especially the conversion of ultraviolet into visible light.

flux. The rate at which energy or matter crosses a unit area of a surface.

focal length. The distance from a lens or mirror to the point where light converged by it comes to a focus.

focal ratio (speed). Ratio of the focal length of a lens or mirror to its aperture.

focus. Point where the rays of light converged by a mirror or lens meet.

focus of a conic section. Mathematical point associated with a conic section, whose distance to any point on the conic bears a constant ratio to the distance from that point to a straight line known as the *directrix*.

forbidden lines. Spectral lines that are not usually observed under laboratory conditions because they result from atomic transitions that are highly improbable.

force. That which can change the momentum of a body; numerically, the rate at which the body's momentum changes.

Foucault pendulum. A pendulum that seems to change direction of oscillation as the Earth turns. The experiment was first conducted in 1851 by Jean Foucault to demonstrate the rotation of the Earth.

Fraunhofer line. An absorption line in the spectrum of the sun or of a star.

Fraunhofer spectrum. The array of absorption lines in the spectrum of the sun or a star.

free-free transition. An atomic transition in which the energy associated with an atom or ion and passing electron changes during the encounter, but without capture of the electron by the atom or ion.

frequency. Number of vibrations per unit time; number of waves that cross a given point per unit time (in radiation).

fringes (interference). Successive dark and light lines, caused by interference of light waves with each other before they strike a screen or detecting device and are observed.

full moon. That phase of the Moon when it is at opposition (180° from the sun) and its full daylight hemisphere is visible from the Earth.

fusion. The building up of heavier atomic nuclei from lighter ones.

galactic cluster. An "open" cluster of stars located in the spiral arms or disk of the Galaxy.

galactic equator. Intersection of the principal plane of the Milky Way with the celestial sphere.

galactic latitude. Angular distance north or south of the galactic equator to an object, measured along a great circle passing through that object and the galactic poles.

galactic longitude. Angular distance, measured eastward along the galactic equator from the galactic center, to the intersection of the galactic equator with a great circle passing through the galactic poles and an object.

galactic poles. The poles of the galactic equator; the intersections with the celestial sphere of a line through the observer that is perpendicular to the plane of the galactic equator.

galactic rotation. Rotation of the Galaxy.

galaxy. A large assemblage of stars; a typical galaxy contains millions to hundreds of thousands of millions of stars.

Galaxy. The galaxy to which the sun and our neighboring stars belong; the Milky Way is light from remote stars in the Galaxy.

Galilean satellite. Any of the four largest of Jupiter's satellites, discovered by Galileo.

gamma rays. Photons (of electromagnetic radiation) of energy higher than those of X-rays; the most energetic form of electromagnetic radiation.

gauss. A unit of magnetic flux density.

Geiger counter. A device for counting high-energy charged particles and hence for measuring the intensity of corpuscular radiation.

geodesic. The path of a body in spacetime.

geodesic equations. A set of equations in general relativity by which the paths of objects in spacetime can be calculated.

geology. The study of the Earth's crust and surface and of the processes that influence them; by extension, similar studies of any solid planetary object.

geomagnetic. Referring to the Earth's magnetic field.

geomagnetic poles. The poles of a hypothetical bar magnet whose magnetic field most nearly matches that of the Earth.

giant (star). A star of large luminosity and radius.

gibbous moon. One of the phases of the Moon in which more than half, but not all, of the Moon's daylight hemisphere is visible from the Earth.

globular cluster. One of about 120 large star clusters that form a system of clusters centered on the center of the Galaxy.

globule. A small, dense, dark nebula; believed to be a possible site of star formation.

gluon. The hypothetical particle that mediates the strong nuclear force between quarks.

granulation. The "rice-grain" like structure of the solar photosphere.

gravitation. The mutual attraction of material bodies or particles.

gravitational constant, *G*. The constant of proportionality in Newton's law of gravitation; in metric units

G has the value 6.672×10^{-8} dyne · cm^2/gm^2.

gravitational energy. Energy that can be released by the gravitational collapse, or partial collapse, of a system.

gravitational lens. A configuration of celestial objects, one of which provides one or more images of the other by gravitationally deflecting its light.

gravitational redshift. The redshift caused by a gravitational field. The slowing of clocks in a gravitational field.

gravitational waves. Oscillations in spacetime, propagated by changes in the distribution of matter.

great circle. Circle on the surface of a sphere that is the curve of intersection of the sphere with a plane passing through its center.

greatest elongation (east or west). The largest separation in celestial longitude (to the east or west) that an inferior planet can have from the sun.

greenhouse effect. The blanketing of infrared radiation near the surface of a planet by, for example, carbon dioxide in its atmosphere.

Greenwich meridian. The meridian of longitude passing through the site of the old Royal Greenwich Observatory, near London; origin of longitude on the Earth.

Gregorian calendar. A calendar (now in common use) introduced by Pope Gregory XIII in 1582.

H. See *Hubble constant.*

H I region. Region of neutral hydrogen in interstellar space.

H II region. Region of ionized hydrogen in interstellar space.

hadron. A subnuclear particle; one of hundreds now known to exist, of mass from somewhat less to considerably more than that of the proton.

half-life. The time required for half of the radioactive atoms in a sample to disintegrate.

halo (around sun or Moon). A ring of light around the sun or Moon caused by refraction by the ice crystals of cirrus clouds.

halo (of galaxy). The outermost extent of our Galaxy or another, containing a sparse distribution of stars and globular clusters in a more or less spherical distribution.

harmonic law. Kepler's third law of planetary motion: the cubes of the semimajor axes of the planetary orbits are in proportion to the squares of the sidereal periods of the planets' revolutions about the sun.

harvest moon. The full moon nearest the time of the autumnal equinox.

Hayashi line. Track of evolution on the Hertzsprung-Russell diagram of a completely convective star.

head (of comet). The main part of a comet, consisting of its nucleus and coma.

heavy elements. In astronomy, usually those elements of greater atomic number than helium.

Heisenberg uncertainty principle. A principle of quantum mechanics that places a limit on the precision with which the simultaneous position and momentum of a body or particle can be specified.

helio-. Prefix referring to the sun.

heliocentric. Centered on the sun.

helium flash. The nearly explosive ignition of helium in the triple-alpha process in the dense core of a red giant star.

Helmholtz-Kelvin contraction. The gradual gravitational contraction of a cloud or a star, with the release of gravitational potential energy.

Hertz. A unit of frequency: one cycle per second. Named for Heinrich Hertz, who first produced radio radiation.

Hertzsprung gap. A V-shaped gap in the upper part of the Hertzsprung-Russell diagram where few stable stars are found.

Herzsprung-Russell (H–R) diagram. A plot of absolute magnitude against temperature (or spectral class or color index) for a group of stars.

highlands (lunar). The older, heavily cratered crust of the Moon, covering 83 percent of its surface and composed in large part of anorthositic breccias.

high-velocity star (or object). A star (or object) with high space motion relative to the sun; generally an object that does not share the orbital velocity of the sun about the galactic nucleus.

homogeneous star (or **stellar model**). A star (or theoretical model of a star) whose chemical composition is the same throughout its interior.

horizon (astronomical). A great circle on the celestial sphere 90° from the zenith.

horizon system. A system of celestial coordinates (altitude and azimuth) based on the astronomical horizon and the north point.

horizontal branch. A sequence of stars on the Hertzsprung-Russell diagram of a typical globular cluster of approximately constant absolute magnitude (near $M_v = 0$).

horizontal parallax. The angle by which an object appears displaced (after correction for atmospheric refraction) when viewed on the horizon from a place on the Earth's equator, compared with its direction if it were viewed from the center of the Earth.

horoscope. A chart showing the positions along the zodiac and in the sky of the sun, Moon, and planets at some given instant and as seen from a particular place on Earth—usually corresponding to the time and place of a person's birth.

hour angle. The angle measured westward along the celestial equator from the local meridian to the hour circle passing through an object.

hour circle. A great circle on the celestial sphere passing through the celestial poles.

house. A division or segment of the sky numbered according to its position with respect to the horizon and used by astrology in preparing a horoscope.

Hubble constant. Constant of proportionality between the velocities of remote galaxies and their distances. The Hubble constant is thought to lie in the range of 50 to 100 km/s per million parsecs.

Hubble law. The law of the redshifts. The radial velocities of remote galaxies are proportional to their distances from us.

hydrostatic equilibrium. A balance between the weights of various layers, as in a star or the Earth's atmosphere, and the pressures that support them.

hyperbola. A conic section of eccentricity greater than 1.0; the curve of intersection between a circular cone and a plane that is at too small an angle with the axis of the cone to cut all of the way through it, and is not parallel to a line in the face of the cone.

hyperfine transition. A change in the energy state of an atom that involves a change in the spin of its nucleus.

hypothesis. A tentative theory or supposition, advanced to explain certain facts or phenomena, which is subject to further tests and verification.

igneous rock. Any rock produced by cooling from a molten state.

image. The optical representation of an object produced by light rays from the object being refracted or reflected, as by a lens or mirror.

impact glass. Tiny beads of glass produced by the rapid cooling of hot droplets of silicate ejecta from a cratering impact. Impact glass resulting from micrometeorites is largely responsible for the low reflectivity of the lunar surface.

inclination (of an orbit). The angle between the orbital plane of a revolving body and some fundamental plane—usually the plane of the celestial equator or of the ecliptic.

Index Catalogue, IC. The supplement to Dreyer's *New General Catalogue* of star clusters and nebulae.

index of refraction. A measure of the refracting power of a transparent substance; specifically, the ratio of the speed of light in a vacuum to its speed in the substance.

inertia. The property of matter that requires a force to act on it to change its state of motion; momentum is a measure of inertia.

inertial system. A system of coordinates that is not itself accelerated, but that is either at rest or is moving with constant velocity.

inferior conjunction. The configuration of an inferior planet when it has the same longitude as the sun, and is between the sun and Earth.

inferior planet. A planet whose distance from the sun is less than the Earth's.

inflationary universe. A theory of cosmology in which the universe is assumed to have undergone a phase of very rapid expansion during the first 10^{-30} seconds. After this period of rapid expansion, the big bang and inflationary models are identical.

infrared radiation. Electromagnetic radiation of wavelength longer than the longest (red) wavelengths that can be perceived by the eye, but shorter than radio wavelengths.

interference. A phenomenon of waves that mix together such that their crests and troughs can alternately reinforce and cancel each other.

interferometer (stellar). An optical device, making use of the principle of interference of light waves, with which small angles can be measured.

International Date Line. An arbitrary line on the surface of the Earth near longitude 180° across which the date changes by one day.

interplanetary medium. The sparse distribution of gas and solid particles in the interplanetary space.

interstellar dust. Tiny solid grains in interstellar space, thought to consist of a core of rock-like material (silicates) or graphite surrounded by a mantle of ices. Water, methane, and ammonia are probably the most abundant ices.

interstellar gas. Sparse gas in interstellar space.

interstellar lines. Absorption lines superimposed on stellar spectra, produced by the interstellar gas.

interstellar matter. Interstellar gas and dust.

ion. An atom that has become electrically charged by the addition or loss of one or more electrons.

ionization. The process by which an atom gains or loses electrons.

ionization potential. The energy required to remove an electron from an atom.

ionosphere. The upper region of the Earth's atmosphere in which many of the atoms are ionized.

ion tail (of comet). See *plasma tail.*

irregular galaxy. A galaxy without rotational symmetry; neither a spiral nor elliptical galaxy.

irregular satellite. A planetary satellite with an orbit that is retrograde, or of high inclination or eccentricity.

irregular variable. A variable star whose light variations do not repeat with a regular period.

irreversible process. A process (for example, a chemical change) that cannot be reversed, for instance, because of the loss of one or more of the chemical products. The dissociation of water vapor in a planetary atmosphere is usually irreversible because the hydrogen escapes and thus cannot recombine with oxygen to form water again.

island universe. Historical synonym for galaxy.

isotope. Any of two or more forms of the same element, whose atoms all have the same number of protons but different numbers of neutrons.

isotropic. The same in all directions.
iteration. The "closing in" on the solution to a mathematical problem by repetitive calculations.
Jovian planet. Any of the planets Jupiter, Saturn, Uranus, and Neptune.
Julian calendar. A calendar introduced by Julius Caesar in 45 B.C.
Kepler's laws. Three laws, discovered by J. Kepler, that describe the motions of the planets.
kiloparsec (kpc). 1000 parsecs, or about 3260 LY.
kinetic energy. Energy associated with motion; the kinetic energy of a body is one-half the product of its mass and the square of its velocity.
kinetic theory (of gases). The science that treats the motions of the molecules that compose gases.
Kirkwood gaps. Gaps in the spacing of the minor planets that arise from perturbations produced by the major planets.
K/T event. The major break in the history of life on Earth that occurred 65 million years ago, between the Cretaceous and Tertiary periods, apparently due to the impact of an asteroidal object.
Lagrangian points. Five points in the plane of revolution of two bodies, revolving mutually about each other in circular orbits, where a third body of negligible mass can remain in equilibrium with respect to the other two bodies.
Lagrangian satellite (or asteroid). A small object that lies near the stable L-4 or L-5 Lagrangian points of an orbit of one object around another, forming an equilateral triangle with the two larger objects. Also called a Trojan, for the Trojan asteroids that occupy the Lagrangian points of the sun-Jupiter system.
Lagrangian surface. The surface in space between two objects revolving about each other on which a small particle is attracted equally to each object.
laser. An acronym for *light amplification by stimulated emission of radiation;* a device for amplifying a light signal at a particular wavelength into a coherent beam.
latitude. A north-south coordinate on the surface of the Earth; the angular distance north or south of the equator measured along a meridian passing through a place.
law. A statement of order or relation between phenomena that, under given conditions, is presumed to be invariable.
law of areas. Kepler's second law: the radius vector from the sun to any planet sweeps out equal areas in the planet's orbital plane in equal intervals of time.
law of the redshifts. The relation between the radial velocity and distance of a remote galaxy: the radial velocities are proportional to the distances of galaxies.
leap year. A calendar year with 366 days, inserted approximately every four years to make the average length of the calendar year as nearly equal as possible to the tropical year.
lepton. A subatomic particle of small mass, such as an electron or positron.
libration. Any of several phenomena by which an observer on Earth, over a period of time, can see more than one hemisphere of the Moon.
light. Electromagnetic radiation that is visible to the eye.
light curve. A graph that displays the time variation in light or magnitude of a variable or eclipsing binary star.
light year. The distance light travels in a vacuum in one year; 1 LY $= 9.46 \times 10^{17}$ cm, or about 6×10^{12} mi.
limb (of sun or Moon). Apparent edge of the sun or Moon as seen in the sky.
limb darkening. The phenomenon whereby the sun (or a star) is less bright near its limb than near the center of its disk.
limiting magnitude. The faintest magnitude that can be observed with a given instrument or under given conditions.
line broadening. The phenomenon by which spectral lines are not precisely sharp but have finite widths.
line profile. A plot of the intensity of light versus wavelength across a spectral line.
linear diameter. Actual diameter in units of length.
lithosphere. The upper layer of the Earth, to a depth of 50 to 100 km, involved in plate tectonics.
Local Group. The cluster of galaxies to which our Galaxy belongs.
local standard of rest. A coordinate system that shares the average motion of the sun and its neighboring stars about the galactic center.
Local Supercluster. The supercluster of galaxies to which the Local Group belongs.
longitude. An east-west coordinate on the Earth's surface; the angular distance, measured east or west along the equator, from the Greenwich meridian to the meridian passing through a place.
low-velocity star (or object). A star (or object) that has low space velocity relative to the sun; generally an object that shares the sun's high orbital speed about the galactic center.
luminosity. The rate of radiation of electromagnetic energy into space by a star or other object.
luminosity class. A classification of a star according to its luminosity for a given spectral class.
luminosity function. The relative numbers of stars (or other objects) of various luminosities or absolute magnitudes.
luminous energy. Light.
lunar. Referring to the Moon.
lunar eclipse. An eclipse of the Moon.

Lyman lines. A series of absorption or emission lines in the spectrum of hydrogen that arise from transitions to and from the lowest energy states of the hydrogen atoms.

Magellanic Clouds. Two neighboring galaxies visible to the naked eye from southern latitudes.

magnetic field. The region of space near a magnetized body within which magnetic forces can be detected.

magnetic pole. One of two points on a magnet (or the Earth) at which the greatest density of lines of force emerge. A compass needle aligns itself along the local lines of force on the Earth and points more or less toward the magnetic poles of the Earth.

magnetometer. A device for measuring the strength of magnetic fields.

magnetosphere. The region around a planet in which its intrinsic magnetic field dominates over the interplanetary field carried by the solar wind; hence, the region within which charged particles can be trapped by the planetary magnetic field.

magnifying power. The number of times larger (in angular diameter) an object appears through a telescope than with the naked eye.

magnitude. A measure of the amount of light flux received from a star or other luminous object.

main sequence. A sequence of stars on the Hertzsprung-Russell diagram, containing the majority of stars, that runs diagonally from the upper left to the lower right.

major axis (of ellipse). The maximum diameter of an ellipse.

major planet. A Jovian planet.

mantle (of Earth). The greatest part of the Earth's interior, lying between the crust and the core.

many-body problem. The problem of determining the positions and motions of more than two bodies in a system in which the bodies interact under the influence of their mutual gravitation. Also called the *n*-body problem.

mare. Latin for "sea"; name applied to the dark, relatively smooth features that cover 17 percent of the Moon.

maser. An acronym for *microwave amplification of stimulated emission radiation;* a device for amplifying a microwave (radio) signal at a particular wavelength into a coherent beam.

mass. A measure of the total amount of material in a body; defined either by the inertial properties of the body or by its gravitational influence on other bodies.

mass defect. The amount by which the mass of an atomic nucleus is less than the sum of the masses of the individual nucleons that compose it.

mass extinction. The sudden disappearance in the fossil record of a large number of species of life, to be replaced by new species in subsequent layers. Mass extinctions are indications of catastrophic changes in the environment, such as might be produced by a large impact on the Earth.

mass function. A numerical relation between the masses of the components of a binary star and the inclination of their plane of mutual revolution to the plane of the sky; the mass function is determined from an analysis of the radial-velocity curve of a spectroscopic binary when the spectral lines of only one of the stars are visible and gives a lower limit to the mass of the star whose spectrum is *not* observed.

mass-luminosity relation. An empirical relation between the masses and luminosities of many (principally main-sequence) stars.

mass-radius relation (for white dwarfs). A theoretical relation between the masses and radii of white dwarf stars.

Maunder minimum. The interval from 1645 to 1715 when solar activity was very low.

Maxwell's equations. A set of four equations that describe the fields around magnetic and electric charges, and how changes in those fields produce forces and electromagnetic radiation.

mean density of matter in the universe. The average density of the universe if all of its matter and energy could be smoothed out to absolute uniformity.

mean solar day. Average length of the apparent solar day.

mechanics. The branch of physics that deals with the behavior of material bodies under the influence of, or in the absence of, forces.

megaparsec (Mpc). One million (10^6) pc.

meridian (celestial). The great circle on the celestial sphere that passes through an observer's zenith and the north (or south) celestial pole.

meridian (terrestrial). The great circle on the surface of the Earth that passes through a particular place and the north and south poles of the Earth.

meson. A subatomic particle of mass intermediate between that of a proton and that of an electron.

mesosphere. The layer of the ionosphere immediately above the stratosphere.

Messier catalogue. A catalogue of nonstellar objects compiled by Charles Messier in 1787.

metamorphic rock. Any rock produced by the physical and chemical alteration (without melting) of another rock that has been subjected to high temperature and pressure.

metastable level. An energy level in an atom from which there is a low probability of an atomic transition accompanied by the radiation of a photon.

meteor. The luminous phenomenon observed when a meteoroid enters the Earth's atmosphere and burns up; popularly called a "shooting star."

meteor shower. Many meteors appearing to radiate from a common point in the sky caused by the collision of the Earth with a swarm of meteoritic particles.

meteorite. A portion of a meteroid that survives passage through the atmosphere and strikes the ground.

meteorite fall. The occurrence of a meteorite striking the ground.

meteoritics. The study of meteorites and of the formation of the solar system.

meteoroid. A meteoritic particle in space before any encounter with the Earth.

meteorology. The study of planetary atmospheres, weather, and climate.

Michelson-Morley experiment. The classic (1887) experiment by A. A. Michelson and E. W. Morley, in which they tried to measure the speed of the Earth in space by timing the speed of light in different directions. The failure of their experiment was later explained by special relativity.

micrometeorite. A meteoroid so small that, on entering the atmosphere of the Earth, it is slowed quickly enough that it does not burn up or ablate but filters through the air to the ground.

micron. Old term for micrometer (10^{-6} meter).

microwave. Short-wave radio wavelengths.

Milky Way. The band of light encircling the sky, which is due to the many stars and diffuse nebulae lying near the plane of the Galaxy.

minerals. The solid compounds (often primarily silicon and oxygen) that form rocks.

minor axis (of ellipse). The smallest or least diameter of an ellipse.

minor planet. See *asteroid.*

Mira–type variable star. Any of a large class of red-giant long-period or irregular pulsating variable stars, of which the star Mira is a prototype.

model atmosphere (or **photosphere**). The result of a theoretical calculation of the run of temperature, pressure, density, and so on, through the outer layers of the sun or a star.

molecule. A combination of two or more atoms bound together; the smallest particle of a chemical compound or substance that exhibits the chemical properties of that substance.

momentum. A measure of the inertia or state of motion of a body; the momentum of a body is the product of its mass and velocity. In the absence of a force, momentum is conserved.

monochromatic. Of one wavelength or color.

muon. A *mu meson,* a subatomic particle that behaves like an electron but that has about 200 times the electron's mass.

N galaxy. A galaxy with a stellar-appearing nucleus with the remainder of the galaxy appearing as a surrounding faint haze. Most or all N galaxies are probably either Seyfert galaxies or quasars.

nadir. The point on the celestial sphere 180° from the zenith.

nanosecond. One thousand-millionth (10^{-9}) second.

nautical mile. The mean length of one minute of arc on the Earth's surface along a meridian.

navigation. The art of finding one's position and course at sea or in the air.

nebula. Cloud of interstellar gas or dust.

nebular hypothesis. The basic idea that the sun and planets formed from the same cloud of gas and dust in interstellar space.

neutrino. A fundamental particle that has little or no rest mass and no charge but that does have spin and energy.

neutron. A subatomic particle with no charge and with mass approximately equal to that of the proton.

neutron star. A star of extremely high density composed almost entirely of neutrons.

New General Catalogue (NGC). A catalogue of star clusters, nebulae, and galaxies compiled by J. L. E. Dreyer in 1888.

new moon. Phase of the Moon when its longitude is the same as that of the sun.

Newtonian focus. An optical arrangement in a reflecting telescope, in which a flat mirror intercepts the light from the primary before it reaches the focus and reflects it to a focus at the side of the telescope tube.

Newton's laws. The laws of mechanics and gravitation formulated by Isaac Newton.

night sky light. The faint illumination of the night sky: the main source is usually fluorescence by atoms high in the atmosphere.

node. The intersection of the orbit of a body with a fundamental plane—usually the plane of the celestial equator or of the ecliptic.

nongravitational force. The force that acts on comets to change their orbits, due not to the gravitational influence of the other members of the planetary system, but rather to the rocket effect of gases escaping from the cometary nucleus.

nonthermal radiation. See *synchrotron radiation.*

north point. That intersection of the celestial meridian and astronomical horizon lying nearest the north celestial pole.

nova. A star that experiences a sudden outburst of radiant energy, temporarily increasing its luminosity by hundreds to thousands of times.

nuclear. Referring to the nucleus of the atom.

nuclear bulge. Central part of our Galaxy.

nuclear transformation. Transformation of one atomic nucleus into another.

nuclear winter. A term coined in the early 1980s for the period of global environmental disturbance, including blockage of sunlight and rapid decline in surface temperature, that would occur following a major nuclear weapons exchange and the widespread fires that would result.

nucleon. Any one of the subatomic particles that compose a nucleus.

nucleosynthesis. The building up of heavy elements from lighter ones by nuclear fusion.

nucleus (of atom). The heavy part of an atom, composed mostly of protons and neutrons, and about which the electrons revolve.

nucleus (of comet). The solid chunk of ice and dust in the head of a comet.

nucleus (of galaxy). Central concentration of matter at the center of a galaxy.

null geodesic. The path of a light ray in four-dimensional spacetime.

O-association. A stellar association in which the stars are predominantly of types O and B.

objective. The principal image-forming component of a telescope or other optical instrument.

objective prism. A prismatic lens that can be placed in front of a telescope objective to transform each star image into an image of its spectrum.

oblate spheroid. A solid formed by rotating an ellipse about its minor axis.

oblateness. A measure of the "flattening" of an oblate spheroid; numerically, the ratio of the difference between the major and minor diameters (or axes) to the major diameter (or axis).

obliquity. The angle between the equator of a planet and the plane of its orbit; hence the tilt of the rotational axis of a planet.

obliquity of the ecliptic. Angle between the planes of the celestial equator and the ecliptic; about 23½°.

obscuration (interstellar). Absorption of starlight by interstellar dust.

occultation. The passage of an object of large angular size in front of a smaller object, such as the Moon in front of a distant star or the rings of Saturn in front of the Voyager spacecraft.

Oort comet cloud. The spherical region around the sun from which most "new" comets come, representing objects with aphelia at about 50,000 AU, or extending about a third of the way to the nearest other stars.

opacity. Absorbing power; capacity to impede the passage of light.

open cluster. A comparatively loose or "open" cluster of stars, containing from a few dozen to a few thousand members, located in the spiral arms or disk of the Galaxy; galactic cluster.

opposition. Configuration of a planet when its elongation is 180°.

optical binary. Two stars at different distances nearly lined up in projection so that they appear close together, but which are not really dynamically associated.

optics. The branch of physics that deals with light and its properties.

orbit. The path of a body that is in revolution about another body or point.

outgassing. The process by which the gases of a planetary atmosphere work their way out from the crust of the planet.

oxidizing. In chemistry, referring to conditions in which oxygen dominates over hydrogen, so that most other elements form compounds with oxygen. In very oxidizing conditions, such as are found in the atmosphere of the Earth, free oxygen gas (O_2) or even atomic oxygen (O) are present.

Pangaea. Name given to the hypothetical continent from which the present continents of the Earth separated.

parabola. A conic section of eccentricity 1.0; the curve of the intersection between a circular cone and a plane parallel to a straight line in the surface of the cone.

paraboloid. A parabola of revolution; a curved surface of parabolic cross section. Especially applied to the surface of the primary mirror in a standard reflecting telescope.

parallactic ellipse. A small ellipse that a comparatively nearby star appears to trace out in the sky, which results from the orbital motion of the Earth about the sun.

parallax. An apparent displacement of an object due to a motion of the observer.

parallax (stellar). An apparent displacement of a nearby star that results from the motion of the Earth around the sun; numerically, the angle subtended by 1 AU at the distance of a particular star.

parallelogram of forces. A geometrical construction that permits the determination of the resultant of two different forces.

parent. In referring to the process of radioactive decay, the name given to the radioactive isotope that is destroyed in the decay process to produce a new, or daughter, isotope.

parent body. In planetary science, any larger original object that is the source of other objects, usually through breakup or ejection by impact cratering; for example, the asteroid 4 Vesta is thought to be the parent body of the eucrite meteorites.

parsec. The distance of an object that would have a stellar parallax of one second of arc; 1 parsec = 3.26 light years.

partial eclipse. An eclipse of the sun or Moon in which the eclipsed body does not appear completely obscured.

Pauli exclusion principle. Quantum-mechanical principle by which no two particles of the same kind can have the same position and momentum.

peculiar velocity. The velocity of a star with respect to the local standard of rest; that is, its space motion, corrected for the motion of the sun with respect to our neighboring stars.

penumbra. The portion of a shadow from which only part of the light source is occulted by an opaque body.

penumbral eclipse. A lunar eclipse in which the Moon passes through the penumbra, but not the umbra, of the Earth's shadow.

perfect cosmological principle. The assumption that, on the large scale, the universe appears the same from every place and at all times.

perfect gas. An "ideal" gas that obeys the perfect gas laws.

perfect gas laws. Certain laws that describe the behavior of an ideal gas: Charles' law, Boyle's law, and the equation of state for a perfect gas.

perfect radiator. Black body; a body that absorbs and subsequently re-emits all radiation incident upon it.

periastron. The place in the orbit of a star in a binary star system where it is closest to its companion star.

perigee. The place in the orbit of an Earth satellite where it is closest to the center of the Earth.

perihelion. The place in the orbit of an object revolving about the sun where it is closest to the sun's center.

period. A time interval; for example, the time required for one complete revolution.

period-density relation. Proportionality between the period and the inverse square root of the mean density for a pulsating star.

period-luminosity relation. An empirical relation between the periods and luminosities of cepheid variable stars.

periodic comet. A comet whose orbit has been determined to have an eccentricity of less than 1.0.

permafrost. A region in the crust of a planet where the temperature is always below the freezing point of water; also, the frozen water mixed with soil that is usually found in such cold regions.

perturbation. The disturbing effect, when small, on the motion of a body as predicted by a simple theory, produced by a third body or other external agent.

photocathode. The photoemissive surface in a television camera or photomultiplier from which electrons are dislodged when illuminated with an optical image.

photochemistry. Chemical reactions that are caused or promoted by the action of light, usually ultraviolet light that excites or dissociates some compounds and leads to the formation of new compounds.

photoconductive detector. A device for detecting radiation that contains a light sensitive surface. This surface absorbs photons, freeing electrons in the process. The electrons are stored in the device for subsequent readout. Charge-coupled devices (CCDs), which use silicon as a photosensitive surface, are one example.

photoelectric effect. The emission of an electron by the absorption of a photon by a substance.

photoemissive detector. A device for measuring light. Photons falling on a light-sensitive surface drive electrons away from it. The resulting electric current is amplified until it can be measured accurately. A photomultiplier is an example of a photoemissive device.

photometry. The measurement of light intensities.

photomultiplier. A device for detecting light in which the electric current generated when photons strike a photocathode is amplified at several stages within the tube. (See *photoemissive detector.*)

photon. A discrete unit of electromagnetic energy.

photon sphere. A surface surrounding a black hole, of radius about 1.4 times that of the event horizon, where a photon can have a closed circular orbit.

photosphere. The region of the solar (or a stellar) atmosphere from which continuous radiation escapes into space.

photosynthesis. The formation of carbohydrates in the chlorophyll-containing tissues of plants exposed to sunlight. In the process, oxygen is released to the atmosphere.

pion. A particular kind of *meson* or subatomic particle of mass intermediate between that of a proton and electron.

pixel. An individual picture element in a detector; for example, a particular silicon diode in a CCD or a grain in a photographic emulsion.

plage. A bright region of the solar surface observed in the monochromatic light of some spectral line.

Planck's constant. The constant of proportionality relating the energy of a photon to its frequency.

Planck's radiation law. A formula from which can be calculated the intensity of radiation at various wavelengths emitted by a black body.

planet. Any of the nine largest bodies revolving about the sun, or any similar non-self-luminous bodies that may orbit other stars.

planetarium. An optical device for projecting on a screen or domed ceiling the stars and planets and their apparent motions in the sky.

planetary nebula. A shell of gas ejected from, and enlarging about, a certain kind of extremely hot star.

planetary science. The study of the planetary system, an interdisciplinary field that includes elements of astronomy, physics, chemistry, geology, geophysics, meteorology, and space plasma physics.

planetary system. A term used in this text to refer to all of the solar system except the sun: the planets, their satellites, rings, comets, asteroids, meteoroids, dust, and the solar wind.

planetesimals. The hypothetical objects, from tens to hundreds of kilometers in diameter, that formed in the solar nebula as an intermediate step between tiny grains and the larger planetary objects we see today. The comets and some asteroids may be leftover planetesimals.

plasma. A hot ionized gas.

plasma tail (of a comet). A cometary tail, usually narrow and bluish in color, extending straight away from the sun and consisting of plasma streaming away from the head under the influence of the solar wind. Also called an ion tail.

plasma torus. In the Jupiter system, the donut-shaped region surrounding the planet near the orbit of Io that contains the most energetic part of the Jovian magnetosphere.

plate tectonics. The motion of segments or plates of the outer layer of the Earth over the underlying mantle.

polar axis. The axis of rotation of the Earth; also, an axis in the mounting of a telescope that is parallel to the Earth's axis.

polarization. A condition in which the planes of vibration (or the E vectors) of the various rays in a light beam are at least partially aligned.

polarized light. Light in which polarization is present.

Polaroid. Trade name for a transparent substance that produces polarization in light.

Population I and II. Two classes of stars (and systems of stars), classified according to their spectral characteristics, chemical compositions, radial velocities, ages, and locations in the Galaxy.

position angle. Direction in the sky of one celestial object from another; for example, the angle, measured to the east from the north, of the fainter component of a visual binary star in relation to the brighter component.

positron. An electron with a positive rather than negative charge; an antielectron.

postulate. An essential prerequisite to a hypothesis or theory.

potential energy. Stored energy that can be converted into other forms; especially gravitational energy.

Poynting-Robertson effect. An effect of the pressure of radiation from the sun on small particles that causes them to spiral slowly into the sun.

precession (of Earth). A slow, conical motion of the Earth's axis of rotation, caused principally by the gravitational torque of the Moon and sun on the Earth's equatorial bulge. *Lunisolar precession,* precession caused by the Moon and sun only; *planetary precession,* a slow change in the orientation of the plane of the Earth's orbit caused by planetary perturbations; *general precession,* the combination of these two effects on the motion of the Earth's axis with respect to the stars.

precession of the equinoxes. Slow westward motion of the equinoxes along the ecliptic that results from precession.

primary cosmic rays. The cosmic-ray particles that arrive at the Earth from beyond its atmosphere, as opposed to the secondary particles that are produced by collisions between primary cosmic rays and air molecules.

primary minimum (in the light curve of an eclipsing binary). The middle of the eclipse during which the most light is lost.

prime focus. The point in a telescope where the objective focuses the light.

prime meridian. The terrestrial meridian passing through the site of the old Royal Greenwich Observatory; longitude 0°.

primeval atom. A single mass whose explosion (in some cosmological theories) has been postulated to have resulted in all the matter now present in the universe.

primeval fireball. The extremely hot opaque gas that is presumed to have comprised the entire mass of the universe at the time of or immediately following the big bang; the exploding primeval atom.

primitive. In planetary science and meteoritics, an object or rock that is little changed, chemically, since its formation, and hence representative of the conditions in the solar nebula at the time of formation of the solar system. Also used to refer to the chemical composition of an atmosphere that has not undergone extensive chemical evolution.

primitive meteorite. A meteorite that has not been greatly altered chemically since its condensation from the solar nebula; called in meteoritics a chondrite (either ordinary chondrite or carbonaceous chondrite).

primitive rock. Any rock that has not experienced great heat or pressure and therefore remains representative of the original condensates from the solar nebula—never found on any object large enough to have undergone melting and differentiation.

Principia. Contraction of *Philosophiae Naturalis Principia Mathematica,* the great book by Newton in which he set forth his laws of motion and gravitation in 1687.

principle of equivalence. Principle that a gravitational force and a suitable acceleration are indistinguishable within a sufficiently local environment.

principle of relativity. Principle that all observers in uniform relative motion are equivalent; the laws of nature are the same for all, and no experiment can reveal an absolute motion or state of rest.

prism. A wedge-shaped piece of glass that is used to disperse white light into a spectrum.

prolate spheroid. The solid produced by the rotation of an ellipse about its major axis.

prominence. A phenomenon in the solar corona that commonly appears like a flame above the limb of the sun.

proper motion. The angular change in direction of a star per year as seen from the sun.

proton. A heavy subatomic particle that carries a positive charge; one of the two principal constituents of the atomic nucleus.

proton-proton chain. A chain of thermonuclear reactions by which nuclei of hydrogen are built up into nuclei of helium.

protoplanet (or **-star** or **-galaxy**). The original material from which a planet (or a star or galaxy) condensed.

pulsar. A variable radio source of small angular size that emits radio pulses in very regular periods that range from 0.03 to 5 seconds.

pulsating variable. A variable star that pulsates in size and luminosity.

q_0. See *deceleration parameter*.

quantum mechanics. The branch of physics that deals with the structure of atoms and their interactions with each other and with radiation.

quark. A hypothetical subatomic particle. Quarks of from 1 to 6 different kinds, in various combinations, are presumed to make up all other particles in the atomic nucleus.

quarter moon. Either of the two phases of the Moon when its longitude differs by 90° from that of the sun; the Moon appears half full at these phases.

quasar. A stellar-appearing object of very high redshift, presumed to be extragalactic and highly luminous; an active galactic nucleus.

RR Lyrae variable. One of a class of giant pulsating stars with periods less than one day; a cluster variable.

radar (imaging). The combination and computer processing of radar signals from a transmitter and receiver moving with respect to a surface (for example, mounted on an airplane or orbiting spacecraft) in order to produce a topographic image. Also called synthetic aperture radar.

radar (ranging). Radar (transmitted and received radiowave signals) used to measure the distance to, and motion of, a target object.

radial velocity. The component of relative velocity that lies in the line of sight.

radial velocity curve. A plot of the variation of radial velocity with time for a binary or variable star.

radiant (of meteor shower). The point in the sky from which the meteors belonging to a shower seem to radiate.

radiation. A mode of energy transport whereby energy is transmitted through a vacuum; also the transmitted energy itself.

radiation pressure. The transfer of momentum carried by electromagnetic radiation to a body that the radiation impinges upon.

radical. A bond of two or more atoms that does not, in itself, constitute a molecule, but that has characteristics of its own and enters into chemical reactions as if it were a single atom.

radio astronomy. The technique of making astronomical observations in radio wavelengths.

radio galaxy. A galaxy that emits greater amounts of radio radiation than average.

radio telescope. A telescope designed to make observations in radio wavelengths.

radioactive dating. The technique of determining the ages of rocks or other specimens by the amount of radioactive decay of certain radioactive elements contained therein.

radioactivity (radioactive decay). The process by which certain kinds of atomic nuclei naturally decompose with the spontaneous emission of subatomic particles and gamma rays.

ray (lunar). Any of a system of bright, elongated streaks on the Moon, produced by high-speed ejecta from relatively recent craters.

Rayleigh scattering. Scattering of light (photons) by molecules of a gas.

recurrent nova. A nova that has been known to erupt more than once.

red giant. A large, cool star of high luminosity; a star occupying the upper right portion of the Hertzsprung-Russell diagram.

reddening (interstellar). The reddening of starlight passing through interstellar dust, caused by the dust scattering blue light more effectively than red.

redshift. A shift to longer wavelengths of the light from remote galaxies; presumed to be produced by a Doppler shift.

reducing. In chemistry, referring to conditions in which hydrogen dominates over oxygen, so that most other elements form compounds with hydrogen. In very reducing conditions free hydrogen (H_2) is present and free oxygen (O_2) cannot exist.

reflecting telescope. A telescope in which the principal optical component (objective) is a concave mirror.

reflection. The return of light rays by an optical surface.

reflection nebula. A relatively dense dust cloud in interstellar space that is illuminated by starlight.

refracting telescope. A telescope in which the principal optical component (objective) is a lens or system of lenses.

refraction. The bending of light rays passing from one transparent medium (or a vacuum) to another.

regular satellites. Planetary satellites that have orbits of low or moderate eccentricity in approximately the plane of the planet's equator.

relative orbit. The orbit of one of two mutually revolving bodies referred to the other body as origin.

relativistic particle (or **electron**). A particle (electron) moving at nearly the speed of light.

relativity. A theory formulated by Einstein that describes the relations between measurements of physical phenomena by two different observers who are in relative motion at constant velocity (the *special theory of relativity*), or that describes how a gravitational field can be replaced by a curvature of spacetime (the *general theory of relativity*).

resolution. The degree to which fine details in an image are separated or resolved.

resolving power. A measure of the ability of an optical system to resolve or separate fine details in the image it produces; in astronomy, the angle in the sky that can be resolved by a telescope.

resonance. An orbital condition in which one object is subject to periodic gravitational perturbations by another, most commonly arising when two objects orbiting a third have periods of revolution that are simple multiples or fractions of each other.

rest mass. The mass of an object or particle as measured when it is at rest in the laboratory.

retrograde (rotation or revolution). Backwards with respect to the common direction of motion in the solar system; counterclockwise as viewed from the north, and going from east to west rather than from west to east.

retrograde motion. An apparent westward motion of a planet on the celestial sphere or with respect to the stars.

revolution. The motion of one body around another.

rift zone. In geology, a place where the crust is being torn apart by internal forces, generally associated with the injection of new material from the mantle and with the slow separation of tectonic plates.

right ascension. A coordinate for measuring the east-west positions of celestial bodies; the angle measured eastward along the celestial equator from the vernal equinox to the hour circle passing through a body.

Roche limit. See *tidal stability limit.*

rotation. Turning of a body about an axis running through it.

runaway greenhouse effect. A process whereby the heating of a planet leads to an increase in its atmospheric greenhouse effect and thus to further heating, thereby quickly altering the composition of its atmosphere and the temperature of its surface.

Russell-Vogt theorem. The theorem that the mass and chemical composition of a star determine its entire structure if it derives its energy entirely from thermonuclear reactions.

satellite. A body that revolves about a planet.

scale (of telescope). The linear distance in the image corresponding to a particular angular distance in the sky; say, so many centimeters per degree.

Schmidt telescope. A type of reflecting telescope invented by B. Schmidt, in which certain aberrations produced by a spherical concave mirror are compensated for by a thin objective correcting lens.

Schwarzschild radius. See *event horizon.*

science. The attempt to find order in nature or to find laws that describe natural phenomena.

scientific method. A specific procedure in science: (1) the observation of phenomena or the results of experiments; (2) the formulation of hypotheses that describe these phenomena, and that are consistent with the body of knowledge available; (3) the testing of these hypotheses by noting whether or not they adequately predict and describe new phenomena or the results of new experiments.

scintillation counter. Device for recording primary or secondary cosmic-ray particles from flashes of light produced when these particles strike fluorescent materials.

secondary cosmic rays. Secondary particles produced by interactions between primary cosmic rays from space and the atomic nuclei in molecules of the Earth's atmosphere.

secondary minimum (in an eclipsing binary light curve). The middle of the eclipse of the cooler star by the hotter, in which the light of the system diminishes less than during the eclipse of the hotter star by the cooler.

secular. Not periodic.

secular parallax. A mean parallax for a selection of stars, derived from the components of their proper motions that reflect the motion of the sun.

sedimentary rock. Any rock formed by the deposition and cementing of fine grains of material. On Earth, sedimentary rocks are usually the result of erosion and weathering, followed by deposition in lakes or oceans; however, breccias formed on the Moon by impact processes are also considered sedimentary rocks.

seeing. The unsteadiness of the Earth's atmosphere, which blurs telescopic images.

seismic waves. Vibrations traveling through the Earth's interior that result from earthquakes.

seismograph. An instrument used to record and measure seismic waves.

seismology (geology). The study of earthquakes and the conditions that produce them and of the internal structure of the Earth as deduced from analyses of seismic waves.

seismology (solar). The study of small changes in the radial velocity of the sun as a whole or of small regions on the surface of the sun. Analyses of these velocity changes can be used to infer the internal structure of the sun.

seismometer. An instrument used to measure and record seismic waves.

semimajor axis. Half the major axis of a conic section.

semiregular variable. A variable star, usually a red giant or supergiant, whose period of pulsation is far from constant.

separation (in a visual binary). The angular separation of the two components of a visual binary star.

SETI. The search for extraterrestrial intelligence, usually applied to searches for radio signals from other civilizations.

Seyfert galaxy. A galaxy belonging to the class of those with active galactic nuclei; one whose nucleus shows bright emission lines; one of a class of galaxies first described by C. Seyfert.

shadow cone. The umbra of the shadow of a spherical body (such as the Earth) in sunlight.

shell star. A type of star, usually of spectral-type B to F, surrounded by a gaseous ring or shell.

shepherd satellite. Informal term for a satellite that is thought to maintain the structure of a planetary ring through its close gravitational influence—specifically, the two Saturn satellites, Prometheus and Pandora, that orbit just inside and outside of the F ring.

shield volcano. A broad volcano built up through the repeated nonexplosive eruption of fluid basalts to form a low dome or shield shape, typically with slopes of only 4 to 6°, often with a large caldera at the summit. Examples include the Hawaiian volcanoes on Earth and the Tharsis volcanoes on Mars.

shock wave. A surface of highly compressed gas moving through the medium with the speed of sound; it is produced by an object moving through the medium at a speed greater than that of sound.

shower (of cosmic rays). A large "rain" of secondary cosmic-ray particles produced by a very energetic primary particle impinging on the Earth's atmosphere.

shower (meteor). Many meteors, all seeming to radiate from a common point in the sky, caused by the encounter by the Earth of a swarm of meteoroids moving together through space.

sidereal astrology. Astrology in which the horoscope is based on the positions of the planets with respect to the fixed stars rather than with respect to signs that, as a consequence of precession, slide through the zodiac.

sidereal day. The interval between two successive meridian passages of the vernal equinox.

sidereal month. The period of the Moon's revolution about the Earth with respect to the stars.

sidereal period. The period of revolution of one body about another with respect to the stars.

sidereal time. The local hour angle of the vernal equinox.

sidereal year. Period of the Earth's revolution about the sun with respect to the stars.

sign (of zodiac). Astrological term for any of twelve equal sections along the ecliptic, each of length 30°. Starting at the vernal equinox, and moving eastward, the signs are Aries, Taurus, Gemini, Cancer, Leo, Virgo, Libra, Scorpio, Sagittarius, Capricorn, Aquarius, and Pisces.

simultaneity. The occurrence of two events at the same time. In relativity, absolute simultaneity is seen not to have meaning, except for two simultaneous events occurring at the same place.

sine (of angle). One of the trigonometric functions; the sine of an angle (in a right triangle) is the ratio of the length of the side opposite the angle to that of the hypotenuse.

sine curve. A graph of the sine of an angle plotted against the angle.

small circle. Any circle on the surface of a sphere that is not a great circle.

SNC meteorite. One of a class of basaltic meteorites now believed by many planetary scientists to be impact-ejected fragments from Mars.

solar activity. Phenomena of the solar atmosphere: sunspots, plages, and related phenomena.

solar antapex. Direction away from which the sun is moving with respect to the local standard of rest.

solar apex. The direction toward which the sun is moving with respect to the local standard of rest.

solar constant. Mean amount of solar radiation received per unit time, by a unit area, just outside the Earth's atmosphere, and perpendicular to the direction of the sun; the numerical value is 1.37×10^6 $ergs/cm^2 \cdot s$.

solar motion. Motion of the sun, or the velocity of the sun, with respect to the local standard of rest.

solar nebula. The cloud of gas and dust from which the solar system formed.

solar parallax. Angle subtended by the equatorial radius of the Earth at a distance of 1 AU.

solar system. The system of the sun and the planets, their satellites, the minor planets, comets, meteoroids, and other objects revolving around the sun.

solar time. A time based on the sun; usually the hour angle of the sun *plus* 12 hours.

solar wind. A radial flow of plasma leaving the sun.

solidification age. The most common age determined by radioactive dating techniques; the time since the rock or mineral grain being tested solidified from the molten state, thus isolating itself from further chemical changes.

solstice. Either of two points on the celestial sphere where the sun reaches its maximum distances north and south of the celestial equator.

south point. Intersection of the celestial meridian and astronomical horizon 180° from the north point.

space motion. The velocity of a star with respect to the sun.

space plasma physics. The study of planetary magnetospheres, the solar corona, the solar wind, and their interactions.

spacetime. A system of one time and three spatial coordinates, with respect to which the time and place of an *event* can be specified; also called *spacetime continuum.*

speckle interferometry. A technique for circumventing the limitations on the resolution of large telescopes imposed by atmospheric turbulence. The technique relies on combining the information from a series of exposures of very short duration.

spectral class (or **type**). A classification of a star according to the characteristics of its spectrum.

spectral sequence. The sequence of spectral classes of stars arranged in order of decreasing temperatures of stars of those classes.

spectrogram. A photograph of a spectrum

spectrograph. An instrument for obtaining a spectrum (especially a spectrum recorded on a photographic plate); in astronomy, usually attached to a telescope to record the spectrum of a star, galaxy, or other astronomical object. (Often used interchangeably with spectrometer.)

spectroheliogram. A picture of the sun obtained with a spectroheliograph.

spectroheliograph. An instrument for imaging the sun, or part of the sun, in the monochromatic light of a particular spectral line.

spectrophotometry. The measurement of the intensity of light from a star or other source at different wavelengths.

spectroscope. An instrument for directly viewing the spectrum of a light source.

spectroscopic binary star. A binary star in which the components are not resolved optically, but whose binary nature is indicated by periodic variations in radial velocity, indicating orbital motion.

spectroscopic parallax. A parallax (or distance) of a star that is derived by comparing the apparent magnitude of the star with its absolute magnitude as deduced from its spectral characteristics.

spectroscopy. The study of spectra.

spectrum. The array of colors or wavelengths obtained when light from a source is dispersed, as in passing it through a prism or grating.

spectrum analysis. The study and analysis of spectra, especially stellar spectra.

spectrum binary. A binary star whose binary nature is revealed by spectral characteristics that can only result from the composite of the spectra of two different stars.

speed. The rate at which an object moves without regard to its direction of motion; the numerical or absolute value of velocity.

spherical aberration. A defect of optical systems whereby on-axis rays of light striking different parts of the objective do not focus at the same place.

spherical harmonics. A series of terms by which the shape of a body can be expressed mathematically to any desired degree of accuracy (by using enough terms in the series).

spicule. A jet of rising material in the solar chromosphere.

spiral arms. Arms of interstellar material and young stars that wind out in a plane from the central nucleus of a spiral galaxy.

spiral galaxy. A flattened, rotating galaxy with pinwheel-like arms of interstellar material and young stars winding out from its nucleus.

sporadic meteor. A meteor that does not belong to a shower.

spring tide. The highest tidal range of the month, produced when the Moon is near either the full or new phase.

sputtering. The process by which energetic atomic particles striking a solid alter its chemistry and eject additional atoms or molecular fragments from the surface.

standard time. The local mean solar time of a standard meridian, adopted over a large region to avoid the inconvenience of continuous time changes around the Earth.

star. A self-luminous sphere of gas.

star cluster. An assemblage of stars held together by their mutual gravitation.

statistical equilibrium. A condition in a system of particles in which, statistically, energy is divided equally among particles of all types.

statistical parallax. The mean parallax for a selection of stars, derived from the radial velocities of the stars and the components of their proper motions that cannot be affected by the solar motion.

steady state (theory of cosmology). A theory of cosmology embracing the perfect cosmological principle, and involving the continuous creation of matter.

Stefan's law or **Stefan-Boltzmann Law.** A formula from which the rate at which a black body radiates energy can be computed; the total rate of energy emission from a unit area of a black body is proportional to the fourth power of its absolute temperature.

stellar evolution. The changes that take place in the sizes, luminosities, structures, and so on, of stars as they age.

stellar model. The result of a theoretical calculation of the run of physical conditions in a stellar interior.

stellar parallax. The angle subtended by 1 AU at the distance of a star; usually measured in seconds of arc.

stimulated emission or **stimulated radiation.** Photons emitted by excited atoms undergoing downward transitions as a consequence of being stimulated by other photons of the same wavelength.

stony meteorite. A meteorite composed mostly of stony material.

stratosphere. The layer of the Earth's atmosphere above the troposphere (where most weather takes place) and below the ionosphere.

Strömgren sphere. A region of ionized gas in interstellar space surrounding a hot star; an H II region.

strong nuclear force. The force that binds together the parts of the atomic nucleus.

subduction zone. In terrestrial geology, a region where one crustal plate is forced under another, generally associated with earthquakes, volcanic activity, and the formation of deep ocean trenches.

subdwarf. A star of luminosity lower than that of main-sequence stars of the same spectral type.

subgiant. A star of luminosity intermediate between those of main-sequence stars and normal giants of the same spectral type.

subtend. To have or include a given angular size.

summer solstice. The point on the celestial sphere where the sun reaches its greatest distance north of the celestial equator.

sun. The star about which the Earth and other planets revolve.

sundial. A device for keeping time by the shadow a marker (gnomon) casts in sunlight.

sunspot. A temporary cool region in the solar photosphere that appears dark by contrast against the surrounding hotter photosphere.

sunspot cycle. The semiregular 11-year period with which the frequency of sunspots fluctuates.

supercluster. A large region of space (50 to 100 million parsecs across) where matter is concentrated into galaxies, groups of galaxies, and clusters of galaxies; a cluster of clusters of galaxies.

supergiant. A star of very high luminosity.

supergranulation. Large-scale convective patterns in the solar photosphere (up to 30,000 km in diameter).

superior conjunction. The configuration of a planet in which it and the sun have the same longitude, with the planet being more distant than the sun.

superior planet. A planet more distant from the sun than is the Earth.

supernova. An explosion that marks the final stage of evolution of a star. A Type I supernova is thought to occur when a white dwarf accretes enough matter to exceed the Chandrasekhar limit, collapses, and explodes. A Type II supernova is thought to mark the final collapse of a massive star.

surface gravity. The weight of a unit mass at the surface of a body.

surveying. The technique of measuring distances and relative positions of places over the surface of the Earth (or elsewhere); generally accomplished by triangulation.

synchrotron radiation. The radiation emitted by charged particles being accelerated in magnetic fields and moving at speeds near that of light.

synodic month. The period of revolution of the Moon with respect to the sun, or its cycle of phases.

synodic period. The interval between successive occurrences of the same configuration of a planet; for example, between successive oppositions or successive superior conjunctions.

T Tauri stars. Variable stars associated with interstellar matter that show rapid and erratic changes in light.

tachyon. A hypothetical particle that always moves with a speed greater than that of light. (There is no evidence that tachyons exist.)

tail (of a comet). See *dust tail* and *plasma tail.*

tangent (of angle). One of the trigonometric functions; the tangent of an angle (in a right triangle) is the ratio of the length of the side opposite the angle to that of the shorter of the adjacent sides.

tangential (transverse) velocity. The component of a star's space velocity that lies in the plane of the sky.

T-association. A stellar association containing T Tauri stars.

tau component (of proper motion). The component of a star's proper motion that lies perpendicular to a great circle passing through the star and the solar apex.

tectonics. See *plate tectonics.*

tektites. Small rounded glassy bodies found on the Earth, apparently ejecta from major impact craters.

telescope. An optical instrument used to aid in viewing or measuring distant objects.

temperature (absolute). Temperature measured in centigrade (Celsius) degrees from absolute zero.

temperature (Celsius; formerly **centigrade).** Temperature measured on scale where water freezes at 0° and boils at 100°.

temperature (color). The temperature of a star as estimated from the intensity of the stellar radiation at two or more colors or wavelengths.

temperature (effective). The temperature of a black body that would radiate the same total amount of energy that a particular body does.

temperature (excitation). The temperature of a star as estimated from the relative strengths of lines in its spectrum that originate from atoms in different stages of excitation.

temperature (Fahrenheit). Temperature measured on a scale where water freezes at 32° and boils at 212°.

temperature (ionization). The temperature of a star as estimated from the relative strengths of lines in its spectrum that originate from atoms in different stages of ionization.

temperature (Kelvin). Absolute temperature measured in centigrade degrees.

temperature (kinetic). A measure of the mean energy of the molecules in a substance.

temperature (radiation). The temperature of a black body that radiates the same amount of energy in a given spectral region as does a particular body.

tensor. A generalization of the concept of the vector, consisting of an array of numbers or quantities that transform according to specific rules.

terrestrial planet. Any of the planets Mercury, Venus, Earth, Mars, and sometimes Pluto.

***Tetrabiblos*.** A standard and widely used treatise on astrology by Ptolemy.

theory. A set of hypotheses and laws that have been well demonstrated as applying to a wide range of phenomena associated with a particular subject.

thermal energy. Energy associated with the motions of the molecules in a substance.

thermal equilibrium. A balance between the input and outflow of heat in a system.

thermal radiation. The radiation emitted by any body or gas that is not at absolute zero.

thermodynamics. The branch of physics that deals with heat and heat transfer among bodies.

thermonuclear energy. Energy associated with thermonuclear reactions or that can be released through thermonuclear reactions.

thermonuclear reaction. A nuclear reaction or transformation that results from encounters between nuclear particles that are given high velocities (by heating them).

thermosphere. The region of the Earth's atmosphere lying between the mesosphere and the exosphere.

tidal force. A differential gravitational force that tends to deform a body.

tidal heating. Generation of heat in a planetary object through repeated tidal stresses from a larger nearby object, probably important for the Moon early in its history and responsible today for the high level of volcanic activity on Io.

tidal stability limit. The distance—approximately 2.5 planetary radii from the center—within which differential gravitational forces (or tides) are stronger than the mutual gravitational attraction between two adjacent orbiting objects. Within this limit, fragments are not likely to accrete or assemble themselves into a larger object. Also called the Roche limit.

tide. Deformation of a body by the differential gravitational force exerted on it by another body; in the Earth, the deformation of the ocean surface by the differential gravitational forces exerted by the Moon and sun.

Titius-Bode law. A scheme by which a sequence of numbers can be obtained that give the approximate distances of the planets from the sun in astronomical units. Also called Bode's law.

ton (metric). One million grams (2204.6 lb).

topography. The configuration or relief of the surface of the Earth, Moon, or a planet.

total eclipse. An eclipse of the sun in which the sun's photosphere is entirely hidden by the Moon, or an eclipse of the Moon in which it passes completely into the umbra of the Earth's shadow.

transit. An instrument for timing the exact instant a star or other object crosses the local meridian. Also, the passage of a celestial body across the meridian; or the passage of a small body (say, a planet) across the disk of a large one (say, the sun).

triangulation. The operation of measuring some of the elements of a triangle so that other ones can be calculated by the methods of trigonometry, thus determining distances to remote places without having to span them directly.

triaxial ellipsoid. A solid figure whose cross sections along three planes at right angles to each other, and all passing through its center, are ellipses of different sizes and eccentricities.

trigonometry. The branch of mathematics that deals with the analytical solutions of triangles.

triple-alpha process. A series of two nuclear reactions by which three helium nuclei are built up into one carbon nucleus.

Trojan asteroid. One of a large number of asteroids that share Jupiter's orbit about the sun, but either preceding or following Jupiter by 60°.

Tropic of Cancer. Parallel of latitude 23½° N.

Tropic of Capricorn. Parallel of latitude 23½° S.

tropical astrology. The conventional practice of astrology, in which the horoscope is based on the signs that move through the zodiac with precession.

tropical year. Period of revolution of the Earth about the sun with respect to the vernal equinox.

troposphere. Lowest level of the Earth's atmosphere, where most weather takes place.

turbulence. Random motions of gas masses, as in the atmosphere of a star.

21-cm line. A spectral line of neutral hydrogen at the radio wavelength of 21 cm.

***U, B, V* system.** A system of stellar magnitudes consisting of measures in the ultraviolet, blue, and green-yellow spectral regions.

ultraviolet radiation. Electromagnetic radiation of wavelengths shorter than the shortest (violet) wavelengths to which the eye is sensitive; radiation of wavelengths in the approximate range 100 – 4000 Å.

umbra. The central, completely dark part of a shadow.

uncertainty principle. See *Heisenberg uncertainty principle*.

uncompressed density. The density that a planetary

object would have if it were not subject to self-compression from its own gravity, hence the density that is characteristic of its bulk material independent of the size of the object.

universal time. The local mean time of the prime meridian.

universe. The totality of all matter and radiation and the space occupied by same.

upsilon component (of proper motion). The component of a star's proper motion that lies along a great circle passing through the star and the solar apex.

variable star. A star that varies in luminosity.

vector. A quantity that has both magnitude and direction.

velocity. A vector that denotes both the speed and direction a body is moving.

velocity of escape. The speed with which an object must move in order to enter a parabolic orbit about another body (such as the Earth), and hence move permanently away from the vicinity of that body.

vernal equinox. The point on the celestial sphere where the sun crosses the celestial equator passing from south to north.

vertical circle. Any great circle passing through the zenith.

very-long-baseline interferometry (VLBI). A technique of radio astronomy whereby signals from telescopes thousands of kilometers apart are combined to obtain very high resolution with interferometry.

virial theorem. A relation between the potential and kinetic energies of a system of mutually gravitating bodies in statistical equilibrium.

visual binary star. A binary star in which the two components are telescopically resolved.

volatile materials. Materials that are gaseous at fairly low temperatures. This is a *relative* term, usually applied to the gases in planetary atmospheres and to common ices (H_2O, CO_2, etc.), but also sometimes used for elements such as cadmium, zinc, lead, and rubidium that form gases at temperatures up to 1000 K. (These are called *volatile elements*, as opposed to refractory elements.)

volume. A measure of the total space occupied by a body.

W Virginis star (type II cepheid). A variable star belonging to the relatively rare class of population II cepheids.

wandering of the poles. A semiperiodic shift of the body of the Earth relative to its axis of rotation; responsible for variation of latitude.

watt. A unit of power; 10 million ergs expended per second.

wavelength. The spacing of the crests or troughs in a wave train.

weak nuclear force. The nuclear force involved in radioactive decay. The weak force is characterized by the slow rate of certain nuclear reactions—such as the decay of the neutron, which occurs with a half-life of 11 min.

weather. The state of a planetary atmosphere—its composition, temperature, pressure, motion, etc.—at a particular place and time.

weight. A measure of the force due to gravitational attraction.

west point. The point on the horizon 270° around the horizon from the north point, measured in a clockwise direction as seen from the zenith.

white dwarf. A star that has exhausted most or all of its nuclear fuel and has collapsed to a very small size; believed to be near its final stage of evolution.

white hole. The hypothetical time reversal of a black hole, in which matter and radiation gush up. White holes are believed *not* to exist.

Wien's law. Formula that relates the temperature of a black body to the wavelength at which it emits the greatest intensity of radiation.

winter solstice. Point on the celestial sphere where the sun reaches its greatest distance south of the celestial equator.

Wolf-Rayet star. One of a class of very hot stars that eject shells of gas at very high velocity.

X-rays. Photons of wavelengths intermediate between those of ultraviolet radiation and gamma rays.

X-ray stars. Stars (other than the sun) that emit observable amounts of radiation at X-ray frequencies.

year. The period of revolution of the Earth around the sun.

Zeeman effect. A splitting or broadening of spectral lines due to magnetic fields.

zenith. The point on the celestial sphere opposite to the direction of gravity; or the direction opposite to that indicated by a plumb bob.

zenith distance. Arc distance of a point on the celestial sphere from the zenith; 90° minus the altitude of the object.

zero-age main sequence. Main sequence for a system of stars that have completed their contraction from interstellar matter and are now deriving all their energy from nuclear reactions, but whose chemical composition has not yet been altered by nuclear reactions.

zodiac. A belt around the sky 18° wide centered on the ecliptic.

zodiacal light. A faint illumination along the zodiac, believed to be sunlight reflected and scattered by interplanetary dust.

zone of avoidance. A region near the Milky Way where obscuration by interstellar dust is so heavy that few or no exterior galaxies can be seen.

zone time. The time, kept in a zone 15° wide in longitude, that is the local mean time of the central meridian of that zone. Zone time is used at sea, but over land the boundaries are irregular to conform to political boundaries, and it is called *standard time*.

Appendix 3

SOME MATHEMATICAL NOTES

1. Powers-of-ten Notation

It is often necessary to deal with very large or very small numbers. For example, the Earth is 150,000,000 kilometers from the sun and the mass of the hydrogen atom is 0.000 000 000 000 000 000 000 001 67 g. Instead of writing and carrying so many zeros, the numbers are usually written as figures between 1 and 10 multiplied by the appropriate power of 10. For example, 150,000,000 is 1.5 $\times$ 100,000,000, or 1.5×10^8. Similarly, 0.000 000 000 000 000 000 000 001 67 is 1.67/1,000 000 000 000 000 000 000 000 or $1.67/10^{24} = 1.67 \times 10^{-24}$. The rule in reading numbers written in this notation is that the exponent of 10 is the number of places the decimal point is to be moved to the right (if the exponent is positive) or to the left (if the exponent is negative).

Multiplication, division, and exponentiation of numbers are facilitated in powers-of-10 notation. Examples:

$$6{,}000{,}000 \times 400 = 6 \times 10^6 \times 4 \times 10^2 = (6 \times 4) \times (10^6 \times 10^2) = 24 \times 10^8 = 2.4 \times 10^9.$$

$$\frac{6 \times 10^{-26}}{9.3 \times 10^7} = \frac{6}{9.3} \times \frac{10^{-26}}{10^7} = \frac{6}{9.3} \times 10^{-26-7} = 0.645 \times 10^{-33} = 6.45 \times 10^{-34}$$

$$(4000)^3 = (4 \times 10^3)^3 = 4^3 \times (10^3)^3 = 64 \times 10^9 = 6.4 \times 10^{10}.$$

$$(64{,}000{,}000)^{1/2} = (64 \times 10^6)^{1/2} = 64^{1/2} \times (10^6)^{1/2} = 8 \times 10^3.$$

2. Angular Measure

The most common units of angular measure used in astronomy are the following:

1. Arc measure:
   one circle contains 360 degrees = 360°;
   1° contains 60 minutes of arc = 60′;
   1′ contains 60 seconds of arc = 60″.
2. Time measure:
   one circle contains 24 hours = 24^h;
   1^h contains 60 minutes of time = 60^m;
   1^m contains 60 seconds of time = 60^s.
3. Radian measure:
   one circle contains 2π radians.

A radian is the angle at the center of a circle subtended by a length along the circumference of the circle equal to its radius. Since the circumference of a circle is 2π times its radius, there are 2π radians in a circle.

Relations between these different units of angular measure are given in the following table.

ARC MEASURE	TIME MEASURE	RADIANS	SECONDS OF ARC
57°.2958	$3^h.820$	1.0	206,264.806
15°	1^h	0.2618	54,000
1°	4^m	1.745×10^{-2}	3,600
15′	1^m	4.363×10^{-3}	900
1′	4^s	2.090×10^{-4}	60
15″	1^s	7.27×10^{-5}	15
1″	$0^s.0667$	4.85×10^{-6}	1

3. Properties of Circles and Spheres

The ratio of the circumference of a circle to its diameter is always the same, regardless of the size of the circle. This ratio is universally symbolized by the Greek letter pi (π). Because it cannot be expressed as a simple ratio of two integers, π is a type of number called an *irrational number*, and so its value can never be specified exactly. It can, however, be approximated to any desired degree of accuracy by methods of mathematical analysis. Even the Greeks had determined the value of π by geometrical means to considerable accuracy. For many purposes, a sufficient approximation to π is

$$\pi = 3\frac{1}{7} = \frac{22}{7}.$$

The value of π has been evaluated to hundreds of decimal places; however, it is seldom needed to greater accuracy than

$$\pi = 3.14159265.$$

In a circle of circumference C, diameter D, and radius R ($R = \frac{1}{2} D$), we have from the definition of π,

$$\pi = \frac{C}{D}, \text{ or } C = \pi D = 2\pi R.$$

The circumference of a sphere of radius R is the circumference of any circle on the surface of the sphere whose center coincides with the center of the sphere (a *great circle*), and hence which also has a radius R. Thus, also on a sphere

$$C = 2\pi R.$$

For example, the radius of the Earth is about 6400 km. Its circumference is thus

$$C = 2\pi(6400) = 40{,}200 \text{ km}.$$

The *area* of a circle of radius R and diameter D is

$$A = \pi R^2 \text{ or } A = \frac{\pi D^2}{4}.$$

The surface area of a sphere of radius R is

$$A = 4\pi R^2.$$

This is the total area over the outside surface and must not be confused with the volume. For example, the surface area of the Earth is approximately:

$$A = 4\pi(6400)^2 = 5 \times 10^8 \text{ km}^2.$$

The *volume* of a sphere of radius R is

$$V = \frac{4}{3}\pi R^3.$$

Appendix 4

METRIC AND ENGLISH UNITS

In the English system of measure the fundamental units of length, mass, and time are the yard, pound*, and second, respectively. There are also, of course, larger and smaller units, which include the ton (2000 lb), the mile (1760 yd), the rod (16½ ft), the inch (1/36 yd), the ounce (1/16 lb), and so on. Such units are inconvenient for conversion and arithmetic computation.

In science, therefore, it is more usual to use the metric system, which has been adopted universally in nearly all countries. The fundamental units of the metric system are:

length: 1 meter (m)
mass: 1 kilogram (kg)
time: 1 second (sec)

A meter was originally intended to be 1 ten-millionth of the distance from the equator to the North Pole along the surface of the Earth. It is about 1.1 yd. A kilogram is about 2.2 lb. The second is the same in metric and English units. The most commonly used quantities of length and mass of the metric system are the following:

Length

1 km	= 1 kilometer	= 1000 meters	= 0.6214 mile	
1 m	= 1 meter	= 1.094 yards	= 39.37 inches	
1 cm	= 1 centimeter	= 0.01 meter	= 0.3937 inch	
1 mm	= 1 millimeter	= 0.001 meter	= 0.1 cm	= 0.03937 inch
1μm	= 1 micrometer	0.000 001 meter	= 0.0001 cm	= 3.937×10^{-5} inch

also: 1 mile = 1.6093 km
1 inch = 2.5400 cm

Mass

1 metric ton	= 10^6 grams	= 1000 kg	= 2.2046×10^3 lb
1 kg	= 1000 grams	= 2.2046 lb	
1 g	= 1 gram	= 0.0022046 lb	= 0.0353 oz
1 mg	= 1 milligram	= 0.001 g	= 2.2046×10^{-6} lb

also: 1 lb = 453.6 g
1 oz = 28.3495 g

---

* A pound is also used as a unit of force—especially by engineers. Then the corresponding unit of mass is the *slug*, which is 14.6 kg. In this book we use the more familiar system whereby a pound is a mass unit (in this system, the corresponding force unit is the *poundal*).

Appendix 5

TEMPERATURE SCALES

Three temperature scales are in general use:

1. Fahrenheit (F): water freezes at 32°F and boils at 212°F.
2. Celsius or centigrade* (C): water freezes at 0°C and boils at 100°C.
3. Kelvin or absolute (K): water freezes at 273 K and boils at 373 K.

All molecular motion ceases at −459°F = −273°C = 0 K. Thus Kelvin temperature is measured from this lowest possible temperature, called ***absolute zero***. It is the temperature scale most often used in astronomy. Kelvins are degrees that have the same value as centigrade or Celsius degrees, since the difference between the freezing and boiling points of water is 100 degrees in each.

On the Fahrenheit scale, water boils at 212 degrees and freezes at 32 degrees; the difference is 180 degrees. Thus to convert Celsius degrees or Kelvins to Fahrenheit it is necessary to multiply by 180/100 = 9/5. To convert from Fahrenheit to Celsius degrees or Kelvins, it is necessary to multiply by 100/180 = 5/9.

Example 1: What is 68°F in Celsius and in Kelvins?

$$68°\text{F} - 32°\text{F} = 36°\text{F above freezing.}$$

$$\frac{5}{9} \times 36° = 20°;$$

thus,

$$68°\text{F} = 20°\text{C} = 293\ \text{K}.$$

Example 2: What is 37°C in Fahrenheit and in Kelvins?

$$37°\text{C} = 273° + 37° = 310\ \text{K};$$

$$\frac{9}{5} \times 37° = 66.6\ \text{Fahrenheit degrees;}$$

thus,

37°C is 66.6°F above freezing
or

$$37°\text{C} = 32° + 66.6° = 98.6°\text{F}.$$

* Celsius is now the name used for centigrade temperature; it has a more modern standardization, but differs from the old centigrade scale by less than 0.1°.

Appendix 6

SOME USEFUL CONSTANTS

MATHEMATICAL CONSTANTS

$\pi = 3.1415926536$

1 radian $= 57^\circ.2957795$

$= 3437'.74677$

$= 206264''.806$

Number of square degrees on a sphere = 41 252.96124

PHYSICAL CONSTANTS

velocity of light	$c = 2.99792458 \times 10^{10}$ cm/s
constant of gravitation	$G = 6.672 \times 10^{-8}$ dyne · cm^2/g^2
Planck's constant	$h = 6.626 \times 10^{-27}$ erg · s
Boltzmann's constant	$k = 1.381 \times 10^{-16}$ erg/deg
mass of hydrogen atom	$m_H = 1.673 \times 10^{-24}$ g
mass of electron	$m_e = 9.1095 \times 10^{-28}$ g
charge on electron	$\epsilon = 4.803 \times 10^{-10}$ electrostatic units
Stefan-Boltzmann constant	$\sigma = 5.670 \times 10^{-5}$ erg/cm^2 · deg^4 · s
constant in Wien's law	$\lambda_{max}T = 0.28979$ cm/deg
Rydberg's constant	R $= 1.09737 \times 10^5$ per cm
1 electron volt	eV $= 1.6022 \times 10^{-12}$ erg
1 angstrom	Å $= 10^{-8}$ cm
1 ton TNT	$= 4.2 \times 10^{16}$ erg

ASTRONOMICAL CONSTANTS

astronomical unit	AU $= 1.495978707 \times 10^{13}$ cm
parsec	pc = 206265 AU
	= 3.262 LY
	$= 3.086 \times 10^{18}$ cm
light year	LY $= 9.4605 \times 10^{17}$ cm
	$= 6.324 \times 10^4$ AU
tropical year	= 365.242199 ephemeris days
sidereal year	= 365.256366 ephemeris days
	$= 3.155815 \times 10^7$ s
mass of Earth	$M_\oplus = 5.977 \times 10^{27}$ g
mass of sun	$M_\odot = 1.989 \times 10^{33}$ g

equatorial radius of Earth	$R_{\oplus} = 6378$ km
radius of sun	$R_{\odot} = 6.960 \times 10^{10}$ cm
luminosity of sun	$L_{\odot} = 3.83 \times 10^{33}$ erg/s
solar constant	$S = 1.37 \times 10^{6}$ erg/cm$^2 \cdot$ s
obliquity of ecliptic (1900)	$\epsilon = 23°27'8''.26$
direction of galactic center (1950)	$\alpha = 17^{h}42^{m}4^{s}$
	$\delta = -28°55'$
direction of north galactic pole (1950)	$\alpha = 12^{h}49^{m}$
	$\delta = +27°.4$

Appendix 7

ASTRONOMICAL COORDINATE SYSTEMS

Several astronomical coordinate systems are in common use. In each of these systems the position of an object in the sky, or on the celestial sphere, is denoted by two angles. These angles are referred to a *reference plane,* which contains the observer, and a *reference direction,* which is a direction from the observer to some arbitrary point lying in the reference plane. The intersection of the reference plane and the celestial sphere is a great circle, which defines the "equator" of the coordinate system. At two points, each 90° from this equator, are the "poles" of the coordinate system. Great circles passing through these poles intersect the equator of the system at right angles.

One of the two angular coordinates of each coordinate system is measured from the equator of the system to the object along the great circle passing through it and the poles. Angles on one side of the equator (or reference plane) are reckoned as positive; those on the opposite are negative. The other angular coordinate is measured along the equator from the reference direction to the intersection of the equator with the great circle passing through the object and the poles.

The system of terrestrial latitude and longitude provides an excellent analogue. Here the plane of the terrestrial equator is the fundamental plane, and the Earth's equator is the equator of the system; the North and South terrestrial Poles are the poles of the system. One coordinate, the *latitude* of a place, is reckoned north (positive) or south (negative) of the equator along a meridian passing through the place. The other coordinate, *longitude,* is measured along the equator to the intersection of the equator and the meridian of the place from the intersection of the equator and the Greenwich meridian. The direction (from the center of the Earth) to this latter intersection is the reference direction. Terrestrial longitude is either east or west (whichever is less), but the corresponding coordinate in celestial systems is generally reckoned in one direction from 0 to 360° (or, equivalently, from 0 to 24^h).

The following table lists the more important astronomical coordinate systems and defines how each of the angular coordinates is defined.

Astronomical Coordinate Systems

SYSTEM	REFERENCE PLANE	REFERENCE DIRECTION	"LATITUDE" COORDINATE	RANGE	"LONGITUDE" COORDINATE	RANGE
Horizon	Horizon plane	North point (formerly the south point was used by astronomers)	Altitude, *h*; toward the zenith (+) toward the nadir (−)	±90°	Azimuth, *A*; measured to the east along the horizon from the north point	0 to 360°
Equator	Plane of the celestial equator	Vernal equinox	Declination, δ; toward the north celestial pole (+) toward the south celestial pole (−)	±90°	Right ascension, α or R.A.; measured to the east along the celestial equator from the vernal equinox	0 to 24^h
Ecliptic	Plane of the Earth's orbit (ecliptic)	Vernal equinox	Celestial latitude, β; toward the north ecliptic pole (+) toward the south ecliptic pole (−)	±90°	Celestial longitude, λ; measured to the east along the ecliptic from the vernal equinox	0 to 360°
Galactic	Mean plane of the Milky Way	Direction to the galactic center	Galactic latitude, *b*; toward the north galactic pole (+) toward the south galactic pole (−)	±90°	Galactic longitude, *l*; measured along the galactic equator to the east from the galactic center	0 to 360°

Appendix 8

SOME NUCLEAR REACTIONS OF IMPORTANCE IN ASTRONOMY

Given here are the series of thermonuclear reactions that are most important in stellar interiors. The subscript to the left of a nuclear symbol is the atomic number; the superscript to the left is the atomic mass number. The symbols for the positive electron (positron) and electron are e^+ and e^-, respectively, for the neutrino is ν, and for a photon (generally of gamma-ray energy) is γ.

1. The Proton-Proton Chains

(Important below 15×10^6 K)

There are three ways the proton-proton chain can be completed. The first (a_1, b_1, c_1) is the most important, but depending on the physical conditions in the stellar interior, some energy is released by one or both of the following alternatives: a_1, b_1, c_2, d_2, e_2, and a_1, b_1, c_2, d_3, e_3, f_3.

(a_1) ${}^1_1\text{H} + {}^1_1\text{H} \rightarrow {}^2_1\text{H} + e^+ + \nu$

(b_1) ${}^2_1\text{H} + {}^1_1\text{H} \rightarrow {}^3_2\text{He} + \gamma$

(c_1) ${}^3_2\text{He} + {}^3_2\text{He} \rightarrow {}^4_2\text{He} + 2{}^1_1\text{H}$

or (c_2) ${}^3_2\text{He} + {}^4_2\text{He} \rightarrow {}^7_4\text{Be} + \gamma$

(d_2) ${}^7_4\text{Be} + e^- \rightarrow {}^7_3\text{Li} + \nu$

$(e)_2$ ${}^7_3\text{Li} + {}^1_1\text{H} \rightarrow 2{}^4_2\text{He}$

or (d_3) ${}^7_4\text{Be} + {}^1_1\text{H} \rightarrow {}^8_5\text{B} + \gamma$

(e_3) ${}^8_5\text{B} \rightarrow {}^8_4\text{Be} + e^+ + \nu$

(f_3) ${}^8_4\text{Be} \rightarrow 2{}^4_2\text{He}$

2. The Carbon-Nitrogen Cycle

(Important above 15×10^6 K)

(a) ${}^{12}_6\text{C} + {}^1_1\text{H} \rightarrow {}^{13}_7\text{N} + \gamma$

(b) ${}^{13}_7\text{N} \rightarrow {}^{13}_6\text{C} + e^+ + \nu$

(c) ${}^{13}_6\text{C} + {}^1_1\text{H} \rightarrow {}^{14}_7\text{N} + \gamma$

(d) ${}^{14}_7\text{N} + {}^1_1\text{H} \rightarrow {}^{15}_8\text{O} + \gamma$

(e) ${}^{15}_8\text{O} \rightarrow {}^{15}_7\text{N} + e^+ + \nu$

(f) ${}^{15}_7\text{N} + {}^1_1\text{H} \rightarrow {}^{12}_6\text{C} + {}^4_2\text{He}$

3. The Triple-Alpha Process

(Important above 10^8 K)

(a) ${}^4_2\text{He} + {}^4_2\text{He} \rightarrow {}^8_4\text{Be} + \gamma$

(b) ${}^4_2\text{He} + {}^8_4\text{Be} \rightarrow {}^{12}_6\text{C} + \gamma$

Appendix 9

ORBITAL DATA FOR THE PLANETS

PLANET	SYMBOL	SEMIMAJOR AXIS		SIDEREAL PERIOD		SYNODIC PERIOD (Days)	MEAN ORBITAL SPEED (km/s)	ORBITAL ECCEN-TRICITY	INCLINATION OF ORBIT TO ECLIPTIC
		AU	10^6 km	Tropical Years	Days				
Mercury	☿	0.3871	57.9	0.24085	87.97	115.88	47.9	0.206	7°004
Venus	♀	0.7233	108.2	0.61521	224.70	583.96	35.0	0.007	3.394
Earth	⊕	1.0000	149.6	1.000039	365.26	—	29.8	0.017	0.0
Mars	♂	1.5237	227.9	1.88089	686.98	779.87	24.1	0.093	1.850
(Ceres)	①	2.7671	414	4.603		466.6	17.9	0.077	10.6
Jupiter	♃	5.2028	778	11.86		399	13.1	0.048	1.308
Saturn	♄	9.538	1427	29.46		378	9.6	0.056	2.488
Uranus	⛢ or ♅	19.191	2871	84.07		370	6.8	0.046	0.774
Neptune	♆	30.061	4497	164.82		367	5.4	0.010	1.774
Pluto	♇	39.529	5913	248.6		367	4.7	0.248	17.15

Adapted from *The Astronomical Almanac* (U. S. Naval Observatory), 1981.

Appendix 10

PHYSICAL DATA FOR THE PLANETS

PLANET	DIAMETER km	DIAMETER Earth = 1	MASS (Earth = 1)	MEAN DENSITY (g/cm³)	ROTATION PERIOD (Days)	INCLINATION OF EQUATOR TO ORBIT	OBLATENESS	SURFACE GRAVITY (Earth = 1)	ALBEDO	VISUAL MAGNITUDE AT MAXIMUM LIGHT*	VELOCITY OF ESCAPE (km/s)
Mercury	4878	0.38	0.055	5.43	58.6	0°0	0	0.38	0.106	−1.9	4.3
Venus	12,104	0.95	0.82	5.24	−243.0	177.4	0	0.91	0.65	−4.4	10.4
Earth	12,756	1.00	1.00	5.52	0.9973	23.4	1/298.2	1.00	0.37	—	11.2
Mars	6,794	0.53	0.107	3.9	1.026	25.2	1/164	0.38	0.15	−2.0	5.0
Jupiter	142,796	11.2	317.8	1.3	0.41	3.1	1/16	2.53	0.52	−2.7	60
Saturn	120,000	9.41	94.3	0.7	0.43	26.7	1/9.2	1.07	0.47	+0.7	36
Uranus	52,400	4.11	14.6	1.3	−0.65	97.9	1/30	0.92	0.50	+5.5	21
Neptune	50,450	3.81	17.2	1.5	0.72	29	1/40	1.18	0.5	+7.8	24
Pluto†	2,500	0.20	0.0025	1.6	6.387	118	?	0.08	0.6	+15.1	1

* At mean opposition for superior planets.
Adapted from *The Astronomical Almanac* (U. S. Naval Observatory), 1981.
† Pluto data from D. Tholen, University of Hawaii.

Appendix 11

SATELLITES OF THE PLANETS

PLANET	SATELLITE NAME	DISCOVERY	SEMI-MAJOR AXIS (km × 1000)	PERIOD (Days)	DIAMETER (km)	MASS (10^{23} g)	DENSITY (g/cm³)	ALBEDO	OPPOSITION MAGNITUDE
Earth	Moon	—	384	27.32	3476	735	3.3	0.12	−12.5
Mars	Phobos	Hall (1877)	9.4	0.32	23	1×10^{-4}	2.2	0.05	12
	Diemos	Hall (1877)	23.5	1.26	13	2×10^{-5}	1.7	0.06	13
Jupiter	Metis	Voyager (1979)	128	0.29	20	—	—	0.05	17
	Adrastea	Voyager (1979)	129	0.30	40	—	—	0.05	17
	Amalthea	Barnard (1892)	181	0.50	200	—	—	0.06	14
	Thebe	Voyager (1979)	222	0.67	90	—	—	0.05	16
	Io	Galileo (1610)	422	1.77	3630	894	3.6	0.6	5
	Europa	Galileo (1610)	671	3.55	3138	480	3.0	0.6	5
	Ganymede	Galileo (1610)	1070	7.16	5262	1482	1.9	0.4	5
	Callisto	Galileo (1610)	1883	16.69	4800	1077	1.9	0.2	6
	Leda	Kowal (1974)	11090	239	15	—	—	—	20
	Himalia	Perrine (1904)	11480	251	180	—	—	0.03	15
	Lysithea	Nicholson (1938)	11720	259	40	—	—	0.05	18
	Elara	Perrine (1905)	11740	260	80	—	—	0.03	17
	Ananke	Nicholson (1951)	21200	631 (R)	30	—	—	—	19
	Carme	Nicholson (1938)	22600	692 (R)	40	—	—	—	18
	Pasiphae	Melotte (1908)	23500	735 (R)	40	—	—	—	17
	Sinope	Nicholson (1914)	23700	758 (R)	40	—	—	—	18
Saturn	Unnamed	Voyager (1985)	118.2	0.48	15?	3×10^{-5}	—	—	—
	Unnamed	Voyager (1985)	133.6	0.58	15?	3×10^{-5}	—	—	—
	Atlas	Voyager (1980)	137.7	0.60	40	—	—	0.5	18
	Prometheus	Voyager (1980)	139.4	0.61	80	—	—	0.5	16
	Pandora	Voyager (1980)	141.7	0.63	100	—	—	0.5	17
	Janus	Dollfus (1966)	151.4	0.69	190	—	—	0.5	15
	Epimetheus	Fountain, Larson (1980)	151.4	0.69	120	—	—	0.5	16
	Mimas	Herschel (1789)	186	0.94	394	0.4	1.2	0.8	13
	Enceladus	Herschel (1789)	238	1.37	502	0.8	1.2	1.0	12
	Tethys	Cassini (1684)	295	1.89	1048	7.5	1.3	0.8	10
	Telesto	Reitsema et al. (1980)	295	1.89	25	—	—	0.6	19
	Calypso	Pascu et al. (1980)	295	1.89	25	—	—	0.9	19
	Dione	Cassini (1684)	377	2.74	1120	11	1.4	0.6	10
	Helene	Lecacheux, Laques (1980)	377	2.74	30	—	—	0.6	19
	Rhea	Cassini (1672)	527	4.52	1530	25	1.3	0.6	10
	Titan	Huygens (1655)	1222	15.95	5150	1346	1.9	0.2	8
	Hyperion	Bond, Lassell (1848)	1481	21.3	270	—	—	0.3	14
	Iapetus	Cassini (1671)	3561	79.3	1435	19	1.2	0.5	10
	Phoebe	Pickering (1898)	12950	550 (R)	220	—	—	0.06	17

PLANET	SATELLITE NAME	DISCOVERY	SEMI-MAJOR AXIS (Km × 1000)	PERIOD (Days)	DIAMETER (km)	MASS (10^{23} g)	DENSITY (g/cm^3)	ALBEDO	OPPOSITION MAGNITUDE
Uranus	1986U7	Voyager (1986)	49.7	0.34	40?	—	—	—	24
	1986U8	Voyager (1986)	53.8	0.38	50?	—	—	—	23
	1986U9	Voyager (1986)	59.2	0.44	50?	—	—	—	23
	1986U3	Voyager (1986)	61.8	0.46	60?	—	—	—	23
	1986U6	Voyager (1986)	62.7	0.48	60?	—	—	—	23
	1986U2	Voyager (1986)	64.6	0.50	80?	—	—	—	22
	1986U1	Voyager (1986)	66.1	0.51	80?	—	—	—	22
	1986U4	Voyager (1986)	69.9	0.56	60?	—	—	—	23
	1986U5	Voyager (1986)	75.3	0.63	60?	—	—	—	23
	1985U1	Voyager (1985)	86.0	0.76	170	—	—	0.07	20
	Miranda	Kuiper (1948)	130	1.41	485	0.8	1.3	0.3	17
	Ariel	Lassell (1851)	191	2.52	1160	13	1.6	0.4	14
	Umbriel	Lassell (1851)	266	4.14	1190	13	1.4	0.2	15
	Titania	Herschel (1787)	436	8.71	1610	35	1.6	0.3	14
	Oberon	Herschel (1787)	583	13.5	1550	29	1.5	0.2	14
Neptune	Triton	Lassell (1846)	354	5.88 (R)	3500	—	—	0.4	14
	Nereid	Kuiper (1949)	552	360	500?	—	—	—	19
Pluto	Charon	Christy (1978)	19.7	6.39	1200	—	—	0.4	17

Appendix 12

TOTAL SOLAR ECLIPSES FROM 1972 THROUGH 2030

DATE	DURATION OF TOTALITY (min)	WHERE VISIBLE
1972 July 10	2.7	Alaska, Northern Canada
1973 June 30	7.2	Atlantic Ocean, Africa
1974 June 20	5.3	Indian Ocean, Australia
1976 Oct. 23	4.9	Africa, Indian Ocean, Australia
1977 Oct. 12	2.8	Northern South America
1979 Feb. 26	2.7	Northwest U.S., Canada
1980 Feb. 16	4.3	Central Africa, India
1981 July 31	2.2	Siberia
1983 June 11	5.4	Indonesia
1984 Nov. 22	2.1	Indonesia, South America
1987 March 29	0.3	Central Africa
1988 March 18	4.0	Philippines, Indonesia
1990 July 22	2.6	Finland, Arctic Regions
1991 July 11	7.1	Hawaii, Central America, Brazil
1992 June 30	5.4	South Atlantic
1994 Nov. 3	4.6	South America
1995 Oct. 24	2.4	South Asia
1997 March 9	2.8	Siberia, Arctic
1998 Feb. 26	4.4	Central America
1999 Aug. 11	2.6	Central Europe, Central Asia
2001 June 21	4.9	Southern Africa
2002 Dec. 4	2.1	South Africa, Australia
2003 Nov. 23	2.0	Antarctica
2005 April 8	0.7	South Pacific Ocean
2006 March 29	4.1	Africa, Asia Minor, U.S.S.R.
2008 Aug. 1	2.4	Arctic Ocean, Siberia, China
2009 July 22	6.6	India, China, South Pacific
2010 July 11	5.3	South Pacific Ocean
2012 Nov. 13	4.0	Northern Australia, South Pacific
2013 Nov. 3	1.7	Atlantic Ocean, Central Africa
2015 March 20	4.1	North Atlantic, Arctic Ocean
2016 March 9	4.5	Indonesia, Pacific Ocean
2017 Aug. 21	2.7	Pacific Ocean, U.S.A., Atlantic Ocean
2019 July 2	4.5	South Pacific, South America
2020 Dec. 14	2.2	South Pacific, South America, South Atlantic Ocean
2021 Dec. 4	1.9	Antarctica
2023 April 20	1.3	Indian Ocean, Indonesia
2024 April 8	4.5	South Pacific, Mexico, East U.S.A.
2026 Aug. 12	2.3	Arctic, Greenland, North Atlantic, Spain
2027 Aug. 2	6.4	North Africa, Arabia, Indian Ocean
2028 July 22	5.1	Indian Ocean, Australia, New Zealand
2030 Nov. 25	3.7	South Africa, Indian Ocean, Australia

Appendix 13

THE NEAREST STARS

STAR	RIGHT ASCENSION (1950) h	m	DECLINATION (1950) °	′	DISTANCE (pc)	PROPER MOTION ″	RADIAL VELOCITY (km/s)	SPECTRA OF COMPONENTS A	B	C	VISUAL MAGNITUDES OF COMPONENTS A	B	C	ABSOLUTE VISUAL MAGNITUDES OF COMPONENTS A	B	C
Sun								G2V			−26.8			+ 4.8		
Proxima Centauri*	14	26.3	−62	28	1.31	3.86	−16	M5V			+11.05			+15.4		
α Centauri	14	36.2	−60	38	1.35	3.68	−22	G2V	K0V		− 0.01	+ 1.33		+ 4.4	+ 5.7	
Barnard's Star	17	55.4	+4	33	1.81	10.34	−108	M5V			+ 9.54			+13.2		
Wolf 359	10	54.1	+7	19	2.35	4.70	+13	M8V			+13.53			+16.7		
Lalande 21185	11	00.6	+36	18	2.52	4.78	−84	M2V			+ 7.50			+10.5		
Luyten 726-8	1	36.4	−18	13	2.60	3.36	+30	M5.5V	M5.5V		+12.45	+12.95		+15.3	+15.8	
Sirius	6	42.9	−16	39	2.65	1.33	− 8	A1V	wd		− 1.46	+ 8.68		+1.4	+11.6	
Ross 154	18	46.7	−23	53	2.90	0.72	− 4	M4.5V			+10.6			+13.3		
Ross 248	23	39.4	+43	55	3.13	1.58	−81	M6V			+12.29			+14.8		
ϵ Eridani	3	30.6	−9	38	3.28	0.98	+16	K2V			+ 3.73			+ 6.1		
Ross 128	11	45.1	+1	6	3.31	1.37	−13	M5V			+11.10			+13.5		
Luyten 789-6	22	35.7	−15	36	3.31	3.26	−60	M6V			+12.18			+14.6		
61 Cygni	21	4.7	+38	30	3.42	5.22	−64	K5V	K7V		+ 5.22	+ 6.03		+ 7.6	+ 8.4	
ϵ Indi	21	59.6	−57	00	3.44	4.69	−40	K5V			+ 4.68			+ 7.0		
τ Ceti	1	41.7	−16	12	3.46	1.92	−16	G8V			+ 3.50			+ 5.7		
Procyon	7	36.7	+5	21	3.51	1.25	− 3	F5IV-V	wd		+ 0.37	+10.7		+ 2.6	+13.0	
BD + 59°1915	18	42.2	+59	33	3.52	2.28	+ 5	M4V	M5V		+ 8.90	+ 9.69		+11.2	+11.9	
BD + 43°44	0	15.5	+43	44	3.55	2.89	+17	M1V	M6V		+ 8.07	+11.04		+10.3	+13.3	
CD − 36°15693	23	2.6	−36	9	3.58	6.90	+10	M2V			+ 7.36			+ 9.6		
G51-15	8	26.9	+26	57	3.66	1.26		MV			+14.8			+17.0		
Luyten 725-32	1	10.1	−17	16	3.79	1.22		M5V			+11.5			+13.6		
BD + 5°1668	7	24.7	+5	23	3.79	3.73	+26	M5V			+ 9.82			+12.0		
CD − 39°14192	21	14.3	−39	4	3.85	3.46	+21	M0V			+ 6.67			+ 8.8		
Kapteyn's Star	5	09.7	−45	00	3.91	8.89	+245	M0V			+ 8.81			+10.8		
Kruger 60	22	26.3	+57	27	3.94	0.86	−26	M3V	M4.5V		+ 9.85	+11.3		+11.9	+13.3	
Ross 614	6	26.8	−2	46	4.12	0.99	+24	M7V	?		+11.07	+14.8		+13.1	+16.8	
BD − 12°4523	16	27.5	−12	32	4.20	1.18	−13	M5V			+10.12			+12.1		
Wolf 424	12	30.9	+9	18	4.27	1.75	− 5	M5.5V	M6V		+13.16	+13.4		+15.0	+15.2	
v. Maanen's Star	0	46.5	+5	9	4.31	2.95	+54	wd			+12.37			+14.3		
CD − 37°15492	0	2.5	−37	36	4.44	6.08	+23	M3V			+ 8.63			+10.4		
Luyten 1159 − 16	1	57.4	+12	51	4.52	2.08		M8V			+12.27			+13.9		
BD + 50°1725	10	8.3	+49	52	4.61	1.45	−26	K7V			+ 6.59			+ 8.3		
CD − 46°11540	17	24.9	−46	51	4.63	1.13		M4V			+ 9.36			+11.0		
CD − 49°13515	21	30.2	−49	13	4.67	0.81	+ 8	M3V			+ 8.67			+10.3		
CD − 44°11909	17	33.5	−44	17	4.69	1.16		M5V			+11.2			+12.8		

STAR	RIGHT ASCENSION (1950) h	m	DECLINATION (1950 °	′	DISTANCE (pc)	PROPER MOTION ″	RADIAL VELOCITY (Km/s)	SPECTRA OF COMPONENTS A	B	C	VISUAL MAGNITUDES OF COMPONENTS A	B	C	ABSOLUTE VISUAL MAGNITUDES OF COMPONENTS A	B	C
BD + 68°946	17	36.7	+68	23	4.69	1.33	−22	M3.5V			+ 9.15			+10.8		
G158 − 27	0	4.2	−7	48	4.72	2.06		MV			+13.7			+15.3		
G208-44/45	19	53.3	+44	17	4.76	0.75		MV	MV		+13.4	+14.0		+15.0	+15.6	
Ross 780	22	50.6	−14	31	4.78	1.16	+ 9	M5V			+10.7			+11.8		
40 Eridani	4	13.0	−7	44	4.83	4.08	−43	K0V	wd	M4.5V	+ 4.43	+ 9.53	+11.17	+ 6.0	+11.1	+12.7
Luyten 145-141	11	43.0	−64	33	4.85	2.68		wd			+11.44			+13.0		
BD + 20°2465	10	16.9	+20	7	4.93	0.49	+11	M4.5V			+ 9.43			+11.0		
70 Ophiuchi	18	2.9	+2	31	4.93	1.13	− 7	K1V	K5V		+ 4.2	+ 6.0		+ 5.7	+ 7.5	
BD + 43°4305	22	44.7	+44	5	5.00	0.83	− 2	M4.5V			+10.2			+11.7		

* Proxima Centauri is sometimes considered an outlying member of the α Centauri system.
Adapted from data supplied by the U.S. Naval Observatory.

Appendix 14

THE TWENTY BRIGHTEST STARS

STAR	RIGHT ASCENSION (1950) h	m	DECLINATION (1950) °	′	DISTANCE* (pc)	PROPER MOTION ″	SPECTRA OF COMPONENTS A	B	C	VISUAL MAGNITUDES OF COMPONENTS A	B	C	ABSOLUTE VISUAL MAGNITUDES OF COMPONENTS A	B	C
Sirius	6	42.9	−16	39	2.7	1.33	A1V	wd		−1.46	+8.7		+1.4	+11.6	
Canopus	6	22.8	−52	40	30	0.02	F01b-II			−0.72			−3.1		
α Centauri	14	36.2	−60	38	1.3	3.68	G2V	K0V		−0.01	+1.3		+4.4	+5.7	
Arcturus	14	13.4	+19	27	11	2.28	K2IIIp			−0.06			−0.3		
Vega	18	35.2	+38	44	8.0	0.34	A0V			+0.04			+0.5		
Capella	5	13.0	+45	57	14	0.44	GIII	M1V	M5V	+0.05	+10.2	+13.7	−0.7	+9.5	+13
Rigel	5	12.1	−8	15	250	0.00	B8 Ia	B9		+0.14	+6.6		−6.8	−0.4	
Procyon	7	36.7	+5	21	3.5	1.25	F5IV-V	wd		+0.37	+10.7		+2.6	+13.0	
Betelgeuse	5	52.5	+7	24	150	0.03	M2Iab			+0.41v			−5.5		
Achernar	1	35.9	−57	29	20	0.10	B5V			+0.51			−1.0		
β Centauri	14	00.3	−60	08	90	0.04	B1III	?		+0.63	+4		−4.1	−0.8	
Altair	19	48.3	+8	44	5.1	0.66	A7IV-V			+0.77			+2.2		
α Crucis	12	23.8	−62	49	120	0.04	B1IV	B3		+1.39	+1.9		−4.0	−3.5	
Aldebaran	4	33.0	+16	25	16	0.20	K5III	M2V		+0.86	+13		−0.2	+12	
Spica	13	22.6	−10	54	80	0.05	B1V			+0.91v			−3.6		
Antares	16	26.3	−26	19	120	0.03	MIIb	B4eV		+0.92v	+5.1		−4.5	−0.3	
Pollux	7	42.3	+28	09	12	0.62	K0III			+1.16			+0.8		
Fomalhaut	22	54.9	−29	53	7.0	0.37	A3V	K4V		+1.19	+6.5		+2.0	+7.3	
Deneb	20	39.7	+45	06	430	0.00	A2Ia			+1.26			−6.9		
β Crucis	12	44.8	−59	24	150	0.05	B0.5IV			+1.28v			−4.6		

* Distances of the more remote stars have been estimated from their spectral types and apparent magnitudes, and are only approximate.

Note: Several of the components listed are themselves spectroscopic binaries. A "v" after a magnitude denotes that the star is variable, in which case the magnitude at median light is given. A "p" after a spectral type indicates that the spectrum is peculiar. An "e" after a spectral type indicates that emission lines are present. When the luminosity classification is rather uncertain, a range is given.

Appendix 15

PULSATING VARIABLE STARS

TYPE OF VARIABLE	SPECTRA	PERIOD (days)	MEDIAN ABSOLUTE MAGNITUDE	AMPLITUDE (Magnitudes)	DESCRIPTION	EXAMPLE	NUMBER KNOWN IN GALAXY*
Cepheids (type I)	F to G supergiants	3 to 50	−1.5 to −5	0.1 to 2	Regular pulsation; period-luminosity relation exists	δ Cep	706
Cepheids (type II)	F to G supergiants	5 to 30	0 to −3.5	0.1 to 2	Regular pulsation; period-luminosity relation exists	W Vir	(About 50; included with type I)
RV Tauri	G to K yellow and red bright giants	30 to 150	−2 to −3	Up to 3	Alternate large and small maxima	RV Tau	104
Long-period (Mira-type)	M red giants	80 to 600	+2 to −2	>2.5	Brighten more or less periodically	ο Cet (Mira)	4566
Semiregular	M giants and supergiants	30 to 2000	0 to −3	1 to 2	Periodicity not dependable; often interrupted	α Ori	2221
Irregular	All types	Irregular	<0	Up to several magnitudes	No known periodicity; many may be semiregular, but too little data exist to classify them as such	π Gru	1687
RR Lyrae or cluster-type	A to F blue giants	<1	0 to +1	<1 to 2	Very regular pulsations	RR Lyr	4433
β Cephei or β Canis Majoris	B blue giants	0.1 to 0.3	−2 to −4	0.1	Maximum light occurs at time of highest compression	β Cep	23
δ Scuti	F subgiants	<1	0 to +2	<0.25	Similar to, and possibly related to, RR Lyrae variables	δ Sct	17
Spectrum variables	A main sequence	1 to 25	0 to +1	0.1	Anomalously intense lines of Si, Sr, and Cr vary in intensity with same period as light; most have strong variable magnetic fields	α^2C Vn	28

* According to the 1968 edition of the Soviet *General Catalogue of Variable Stars.*

Appendix 16

ERUPTIVE VARIABLE STARS

TYPE OF VARIABLE	SPECTRA	DURATION OF INCREASED BRIGHTNESS	NORMAL ABSOLUTE MAGNITUDE	AMPLITUDE (Magnitudes)	DESCRIPTION	EXAMPLE	NUMBER KNOWN IN GALAXY*
Novae	O to A hot subdwarfs	Months to years	>0	7 to 16	Rapid rise to maximum; slow decline; ejection of gas shell	GK Per	166
Nova-like variables or P Cygni stars	Hot B stars	Erratic	-3 to -6	Several magnitudes	Slow, erratic, and nova-like variations in light; may be unrelated to novae. Gas shell ejected	P Cyg	39
Supernovae	?	Months to years	?	15 or more	Sudden, violent flareup, followed by decline and ejection of gas shell	CM Tau (Crab nebula)	7
R Coronae Borealis	F to K supergiant	10 to several hundred days	-5	1 to 9	Sudden and irregular drops in brightness. Low in hydrogen abundance, but high in carbon abundance	R CrB	32
T Tauri or RW Aurigae	B to M main sequence and subgiants	Rapid and erratic	0 to $+8$	Up to a few magnitudes	Rapid and irregular light variations. Generally associated with interstellar material. Subtypes from G to M are called T Tauri variables	RW Aur T Tau	1109
U Geminorum or SS Cygni or "dwarf novae"	A to F hot subdwarfs	Few days to few weeks	>0	2 to 6	Nova-like outbursts at mean intervals that range from 20 to 600 days. Those with longer intervals between outbursts tend to have greater amplitudes. Many, if not all, are members of binary-star systems	SS Cyg, U Gem	215
Flare stars	M main sequence	Few minutes	>8	Up to 6	Sudden flareups in light; probably localized flares on surface of star	UV Cet	28
Z Camelopardalis variables	A to F hot subdwarfs	Few days	>0	2 to 5	Similar to U Geminorum, except that variations are sometimes interrupted by constant light for several cycles. Intervals between outbursts normally range from 10 to 40 days.	Z Cam	20

* According to the 1968 edition of the Soviet *General Catalogue of Variable Stars.*

Appendix 17

THE LOCAL GROUP OF GALAXIES

GALAXY	TYPE	RIGHT ASCENSION (1980) h	m	DECLINATION (1980) °		VISUAL MAGNITUDE (m)	DISTANCE (kpc)	DISTANCE (1000 LY)	DIAMETER (kpc)	DIAMETER (1000 LY)	ABSOLUTE MAGNITUDE (M_v)	RADIAL VELOCITY (km/s)	MASS (Solar Masses)
Our Galaxy	Sb	—	—	—		—	—	—	30	100	(−21)	—	2×10^{11}
Large Magellanic Cloud	Irr I	5	26	−69		0.9	48	160	10	30	−17.7	+276	2.5×10^{10}
Small Magellanic Cloud	Irr I	0	51	−73		2.5	56	180	8	25	−16.5	+168	
Ursa Minor system	E4 (dwarf)	15	8.6	+67	11		70	220	1	3	(−9)		
Sculptor system	E3 (dwarf)	0	58.9	−33	48	8.0	83	270	2.2	7	−11.8		(2 to 4×10^{6})
Draco system	E2 (dwarf)	17	19.9	+57	56		100	330	1.4	4.5	(−10)		
Carina system	E3 (dwarf)	6	41.1	−50	57		(170)	(550)	1.5	4.8	(−10)		
Fornax system	E3 (dwarf)	2	38.9	−34	36	8.4	250	800	4.5	15	−13.6	+39	(1.2 to 2×10^{7})
Leo II system	E0 (dwarf)	11	12.4	+22	16		230	750	1.6	5.2	−10.0	+39	(1.1×10^{6})
Leo I system	E4 (dwarf)	10	7.4	+12	24	12.0	280	900	1.5	5	−10.4		
NGC 6822	Irr I	19	43.8	−14	50	8.9	460	1500	2.7	9	−14.8	−32	
NGC 147	E6	0	32.0	+48	23	9.73	570	1900	3	10	−14.5		
NGC 185	E2	0	37.8	+48	14	9.43	570	1900	2.3	8	−14.8	−305	
NGC 205	E5	0	39.2	+41	35	8.17	680	2200	5	16	−16.5	−239	
NGC 221 (M32)	E3	0	41.6	+40	46	8.16	680	2200	2.4	8	−16.5	−214	
IC 1613	Irr I	1	2.1	+1	51	9.61	680	2200	5	16	−14.7	−238	
Andromeda galaxy (NGC 224; M31)	Sb	0	41.6	+41	10	3.47	680	2200	40	130	−21.2	−266	3×10^{11}
And I	E0 (dwarf)	0	44.6	+37	54	(14)	(680)	(2200)	0.5	1.6	(−11)		
And II	E0 (dwarf)	1	15.2	+33	18	(14)	(680)	(2200)	0.7	2.3	(−11)		
And III	E3 (dwarf)	0	34.2	+36	24	(14)	(680)	(2200)	0.9	0.9	(−11)		
NGC 598 (M33)	Sc	1	32.7	+30	33	5.79	720	2300	17	60	−18.9	−189	8×10^{9}

Appendix 18

THE MESSIER CATALOGUE OF NEBULAE AND STAR CLUSTERS

M	NGC or (IC)	RIGHT ASCENSION (1980) h	m	DECLI-NATION (1980) °	′	APPARENT VISUAL MAGNITUDE	DESCRIPTION
1	1952	5	33.3	+22	01	8.4	"Crab" nebula in Taurus; remains of SN 1054
2	7089	21	32.4	−0	54	6.4	Globular cluster in Aquarius
3	5272	13	41.2	+28	29	6.3	Globular cluster in Canes Venatici
4	6121	16	22.4	−26	28	6.5	Globular cluster in Scorpio
5	5904	15	17.5	+2	10	6.1	Globular cluster in Serpens
6	6405	17	38.8	−32	11	5.5	Open cluster in Scorpio
7	6475	17	52.7	−34	48	3.3	Open cluster in Scorpio
8	6523	18	02.4	−24	23	5.1	"Lagoon" nebula in Sagittarius
9	6333	17	18.1	−18	30	8.0	Globular cluster in Ophiuchus
10	6254	16	56.1	−4	05	6.7	Globular cluster in Ophiuchus
11	6705	18	50.0	−6	18	6.8	Open cluster in Scutum Sobieskii
12	6218	16	46.3	−1	55	6.6	Globular cluster in Ophiuchus
13	6205	16	41.0	+36	30	5.9	Globular cluster in Hercules
14	6402	17	36.6	−3	14	8.0	Globular cluster in Ophiuchus
15	7078	21	28.9	+12	05	6.4	Globular cluster in Pegasus
16	6611	18	17.8	−13	47	6.6	Open cluster with nebulosity in Serpens
17	6618	18	19.6	−16	11	7.5	"Swan" or "Omega" nebula in Sagittarius
18	6613	18	18.7	−17	08	7.2	Open cluster in Sagittarius
19	6273	17	01.4	−26	14	6.9	Globular cluster in Ophiuchus
20	6514	18	01.2	−23	02	8.5	"Trifid" nebula in Sagittarius
21	6531	18	03.4	−22	30	6.5	Open cluster in Sagittarius
22	6656	18	35.2	−23	56	5.6	Globular cluster in Sagittarius
23	6494	17	55.8	−19	00	5.9	Open cluster in Sagittarius
24	6603	18	17.3	−18	26	4.6	Open cluster in Sagittarius
25	(4725)	18	30.5	−19	16	6.2	Open cluster in Sagittarius
26	6694	18	44.1	−9	25	9.3	Open cluster in Scutum Sobieskii
27	6853	19	58.8	+22	40	8.2	"Dumbbell" planetary nebula in Vulpecula
28	6626	18	23.2	−24	52	7.6	Globular cluster in Sagittarius
29	6913	20	23.3	+38	27	8.0	Open cluster in Cygnus
30	7099	21	39.2	−23	16	7.7	Globular cluster in Capricornus
31	224	0	41.6	+41	10	3.5	Andromeda galaxy
32	221	0	41.6	+40	46	8.2	Elliptical galaxy; companion to M31
33	598	1	32.7	+30	33	5.8	Spiral galaxy in Triangulum
34	1039	2	40.7	+42	43	5.8	Open cluster in Perseus
35	2168	6	07.5	+24	21	5.6	Open cluster in Gemini
36	1960	5	35.0	+34	05	6.5	Open cluster in Auriga
37	2099	5	51.1	+32	33	6.2	Open cluster in Auriga
38	1912	5	27.3	+35	48	7.0	Open cluster in Auriga
39	7092	21	31.5	+48	21	5.3	Open cluster in Cygnus
40		12	21	+59			Close double star in Ursa Major
41	2287	6	46.2	−20	43	5.0	Loose open cluster in Canis Major
42	1976	5	34.4	−5	24	4	Orion nebula
43	1982	5	34.6	−5	18	9	Northeast portion of Orion nebula
44	2632	8	39	+20	04	3.9	Praesepe; open cluster in Cancer
45		3	46.3	+24	03	1.6	The Pleiades; open cluster in Taurus
46	2437	7	40.9	−14	46	6.6	Open cluster in Puppis
47	2422	7	35.7	−14	26	5	Loose group of stars in Puppis
48	2548	8	12.8	−5	44	6	"Cluster of very small stars"
49	4472	12	28.8	+8	06	8.5	Elliptical galaxy in Virgo
50	2323	7	02.0	−8	19	6.3	Loose open cluster in Monoceros
51	5194	13	29.1	+47	18	8.4	"Whirlpool" spiral galaxy in Canes Venatici
52	7654	23	23.3	+61	30	8.2	Loose open cluster in Cassiopeia
53	5024	13	12.0	+18	16	7.8	Globular cluster in Coma Berenices

M	NGC or (IC)	RIGHT ASCENSION (1980) h	m	DECLI-NATION (1980) °	′	APPARENT VISUAL MAGNITUDE	DESCRIPTION
54	6715	18	53.8	−30	30	7.8	Globular cluster in Sagittarius
55	6809	19	38.7	−30	59	6.2	Globular cluster in Sagittarius
56	6779	19	15.8	+30	08	8.7	Globular cluster in Lyra
57	6720	18	52.8	+33	00	9.0	"Ring" nebula; planetary nebula in Lyra
58	4579	12	36.7	+11	55	9.9	Spiral galaxy in Virgo
59	4621	12	41.0	+11	46	10.0	Spiral galaxy in Virgo
60	4649	12	42.6	+11	40	9.0	Elliptical galaxy in Virgo
61	4303	12	20.8	+4	35	9.6	Spiral galaxy in Virgo
62	6266	16	59.9	−30	05	6.6	Globular cluster in Scorpio
63	5055	13	14.8	+42	07	8.9	Spiral galaxy in Canes Venatici
64	4826	12	55.7	+21	39	8.5	Spiral galaxy in Coma Berenices
65	3623	11	17.9	+13	12	9.4	Spiral galaxy in Leo
66	3627	11	19.2	+13	06	9.0	Spiral galaxy in Leo; companion to M65
67	2682	8	50.0	+11	53	6.1	Open cluster in Cancer
68	4590	12	38.4	−26	39	8.2	Globular cluster in Hydra
69	6637	18	30.1	−32	23	8.0	Globular cluster in Sagittarius
70	6681	18	42.0	−32	18	8.1	Globular cluster in Sagittarius
71	6838	19	52.8	+18	44	7.6	Globular cluster in Sagitta
72	6981	20	52.3	−12	38	9.3	Globular cluster in Aquarius
73	6994	20	57.8	−12	43	9.1	Open cluster in Aquarius
74	628	1	35.6	+15	41	9.3	Spiral galaxy in Pisces
75	6864	20	04.9	−21	59	8.6	Globular cluster in Sagittarius
76	650	1	41.0	+51	28	11.4	Planetary nebula in Perseus
77	1068	2	41.6	−0	04	8.9	Spiral galaxy in Cetus
78	2068	5	45.7	0	03	8.3	Small emission nebula in Orion
79	1904	5	23.3	−24	32	7.5	Globular cluster in Lepus
80	6093	16	15.8	−22	56	7.5	Globular cluster in Scorpio
81	3031	9	54.2	+69	09	7.0	Spiral galaxy in Ursa Major
82	3034	9	54.4	+69	47	8.4	Irregular galaxy in Ursa Major
83	5236	13	35.4	−29	31	7.6	Spiral galaxy in Hydra
84	4374	12	24.1	+13	00	9.4	Elliptical galaxy in Virgo
85	4382	12	24.3	+18	18	9.3	Elliptical galaxy in Coma Berenices
86	4406	12	25.1	+13	03	9.2	Elliptical galaxy in Virgo
87	4486	12	29.7	+12	30	8.7	Elliptical galaxy in Virgo
88	4501	12	30.9	+14	32	9.5	Spiral galaxy in Coma Berenices
89	4552	12	34.6	+12	40	10.3	Elliptical galaxy in Virgo
90	4569	12	35.8	+13	16	9.6	Spiral galaxy in Virgo
91	omitted						
92	6341	17	16.5	+43	10	6.4	Globular cluster in Hercules
93	2447	7	43.7	−23	49	6.5	Open cluster in Puppis
94	4736	12	50.0	+41	14	8.3	Spiral galaxy in Canes Venatici
95	3351	10	42.9	+11	49	9.8	Barred spiral galaxy in Leo
96	3368	10	45.7	+11	56	9.3	Spiral galaxy in Leo
97	3587	11	13.7	+55	07	11.1	"Owl" nebula; planetary nebula in Ursa Major
98	4192	12	12.7	+15	01	10.2	Spiral galaxy in Coma Berenices
99	4254	12	17.8	+14	32	9.9	Spiral galaxy in Coma Berenices
100	4321	12	21.9	+15	56	9.4	Spiral galaxy in Coma Berenices
101	5457	14	02.5	+54	27	7.9	Spiral galaxy in Ursa Major
102	5866(?)	15	05.9	+55	50	10.5	Spiral galaxy (identification as M102; in doubt)
103	581	1	31.9	+60	35	6.9	Open cluster in Cassiopeia
104*	4594	12	39.0	−11	31	8.3	Spiral galaxy in Virgo
105*	3379	10	46.8	+12	51	9.7	Elliptical galaxy in Leo
106*	4258	12	18.0	+47	25	8.4	Spiral galaxy in Canes Venatici
107*	6171	16	31.4	−13	01	9.2	Globular cluster in Ophiuchus
108*	3556	11	10.5	+55	47	10.5	Spiral galaxy in Ursa Major
109*	3992	11	56.6	+53	29	10.0	Spiral galaxy in Ursa Major
110*	205	0	39.2	+41	35	9.4	Elliptical galaxy (companion to M31)

* Not in Messier's original (1781) list; added later by others.

Appendix 19

THE CHEMICAL ELEMENTS

ELEMENT	SYMBOL	ATOMIC NUMBER	ATOMIC WEIGHT* (Chemical Scale)	NUMBER OF ATOMS PER 10^{12} HYDROGEN ATOMS†
Hydrogen	H	1	1.0080	1×10^{12}
Helium	He	2	4.003	8×10^{10}
Lithium	Li	3	6.940	<10
Beryllium	Be	4	9.013	1.4×10^{1}
Boron	B	5	10.82	$<13 \times 10^{2}$
Carbon	C	6	12.011	4.2×10^{8}
Nitrogen	N	7	14.008	8.7×10^{7}
Oxygen	O	8	16.0000	6.9×10^{8}
Fluorine	F	9	19.00	3.6×10^{4}
Neon	Ne	10	20.183	1.3×10^{8}
Sodium	Na	11	22.991	1.9×10^{6}
Magnesium	Mg	12	24.32	3.2×10^{7}
Aluminum	Al	13	26.98	3.3×10^{6}
Silicon	Si	14	28.09	4.5×10^{7}
Phosphorus	P	15	30.975	3.2×10^{5}
Sulfur	S	16	32.066	1.6×10^{7}
Chlorine	Cl	17	35.457	3.2×10^{5}
Argon	Ar(A)	18	39.944	1.0×10^{5}
Potassium	K	19	39.100	1.4×10^{5}
Calcium	Ca	20	40.08	2.2×10^{6}
Scandium	Sc	21	44.96	1.1×10^{3}
Titanium	Ti	22	47.90	1.1×10^{5}
Vanadium	V	23	50.95	1.0×10^{4}
Chromium	Cr	24	52.01	5.1×10^{5}
Manganese	Mn	25	54.94	2.6×10^{5}
Iron	Fe	26	55.85	3.2×10^{7}
Cobalt	Co	27	58.94	3.2×10^{4}
Nickel	Ni	28	58.71	1.9×10^{6}
Copper	Cu	29	63.54	1.1×10^{4}
Zinc	Zn	30	65.38	2.8×10^{4}
Gallium	Ga	31	69.72	6.3×10^{2}
Germanium	Ge	32	72.60	3.2×10^{3}
Arsenic	As	33	74.91	2.5×10^{2}
Selenium	Se	34	78.96	2.5×10^{3}
Bromine	Br	35	79.916	5.0×10^{2}
Krypton	Kr	36	83.80	2.0×10^{3}
Rubidium	Rb	37	85.48	4.0×10^{2}
Strontium	Sr	38	87.63	7.9×10^{2}
Yttrium	Y	39	88.92	1.3×10^{2}
Zirconium	Zr	40	91.22	5.6×10^{2}
Niobium (Columbium)	Nb(Cb)	41	92.91	7.9×10^{1}
Molybdenum	Mo	42	95.95	1.4×10^{2}
Technetium	Tc(Ma)	43	(99)	—
Ruthenium	Ru	44	101.1	68
Rhodium	Rh	45	102.91	32
Palladium	Pd	46	106.4	32
Silver	Ag	47	107.880	4
Cadmium	Cd	48	112.41	71
Indium	In	49	114.82	45
Tin	Sn	50	118.70	100
Antimony	Sb	51	121.76	10

ELEMENT	SYMBOL	ATOMIC NUMBER	ATOMIC WEIGHT* (Chemical Scale)	NUMBER OF ATOMS PER 10^{12} HYDROGEN ATOMS†
Tellurium	Te	52	127.61	2.5×10^2
Iodine	I(J)	53	126.91	40
Xenon	Xe(X)	54	131.30	2.1×10^2
Cesium	Cs	55	132.91	<80
Barium	Ba	56	137.36	1.2×10^2
Lanthanum	La	57	138.92	13
Cerium	Ce	58	140.13	35
Praseodymium	Pr	59	140.92	4
Neodymium	Nd	60	144.27	18
Promethium	Pm	61	(147)	—
Samarium	Sm(Sa)	62	150.35	5
Europium	Eu	63	152.0	5
Gadolinium	Gd	64	157.26	13
Terbium	Tb	65	158.93	2
Dysprosium	Dy(Ds)	66	162.51	11
Holmium	Ho	67	164.94	3
Erbium	Er	68	167.27	7
Thulium	Tm(Tu)	69	168.94	2
Ytterbium	Yb	70	173.04	8
Lutecium	Lu(Cp)	71	174.99	6
Hafnium	Hf	72	178.50	6
Tantalum	Ta	73	180.95	1
Tungsten	W	74	183.86	50
Rhenium	Re	75	186.22	<2
Osmium	Os	76	190.2	5
Iridium	Ir	77	192.2	28
Platinum	Pt	78	195.09	56
Gold	Au	79	197.0	6
Mercury	Hg	80	200.61	$<10^2$
Thallium	Tl	81	204.39	8
Lead	Pb	82	207.21	85
Bismuth	Bi	83	209.00	<80
Polonium	Po	84	(209)	—
Astatine	At	85	(210)	—
Radon	Rn	86	(222)	—
Francium	Fr(Fa)	87	(223)	—
Radium	Ra	88	226.05	—
Actinium	Ac	89	(227)	—
Thorium	Th	90	232.12	2
Protactinium	Pa	91	(231)	—
Uranium	U(Ur)	92	238.07	<4
Neptunium	Np	93	(237)	—
Plutonium	Pu	94	(244)	—
Americium	Am	95	(243)	—
Curium	Cm	96	(248)	—
Berkelium	Bk	97	(247)	—
Californium	Cf	98	(251)	—
Einsteinium	E	99	(254)	—
Fermium	Fm	100	(253)	—
Mendeleevium	Mv	101	(256)	—
Nobelium	No	102	(253)	—

* Where mean atomic weights have not been well determined, the atomic mass numbers of the most stable isotopes are given in parentheses.

† Provided by L. H. Aller.

Appendix 20

THE CONSTELLATIONS

CONSTELLATION (Latin name)	GENITIVE CASE ENDING	ENGLISH NAME OR DESCRIPTION	ABBRE-VIATION	APPROXIMATE POSITION α	
				α	δ
				h	°
Andromeda	Andromedae	Princess of Ethiopia	And	1	+40
Antlia	Antliae	Air pump	Ant	10	−35
Apus	Apodis	Bird of Paradise	Aps	16	−75
Aquarius	Aquarii	Water bearer	Aqr	23	−15
Aquila	Aquilae	Eagle	Aql	20	+5
Ara	Arae	Altar	Ara	17	−55
Aries	Arietis	Ram	Ari	3	+20
Auriga	Aurigae	Charioteer	Aur	6	+40
Boötes	Boötis	Herdsman	Boo	15	+30
Caelum	Caeli	Graving tool	Cae	5	−40
Camelopardus	Camelopardis	Giraffe	Cam	6	+70
Cancer	Cancri	Crab	Cnc	9	+20
Canes Venatici	Canum Venaticorum	Hunting dogs	CVn	13	+40
Canis Major	Canis Majoris	Big dog	CMa	7	−20
Canis Minor	Canis Minoris	Little dog	CMi	8	+5
Capricornus	Capricorni	Sea goat	Cap	21	−20
Carina*	Carinae	Keel of Argonauts' ship	Car	9	−60
Cassiopeia	Cassiopeiae	Queen of Ethiopia	Cas	1	+60
Centaurus	Centauri	Centaur	Cen	13	−50
Cepheus	Cephei	King of Ethiopia	Cep	22	+70
Cetus	Ceti	Sea monster (whale)	Cet	2	−10
Chamaeleon	Chamaeleontis	Chameleon	Cha	11	−80
Circinus	Circini	Compasses	Cir	15	−60
Columba	Columbae	Dove	Col	6	−35
Coma Berenices	Comae Berenices	Berenice's hair	Com	13	+20
Corona Australis	Coronae Australis	Southern crown	CrA	19	−40
Corona Borealis	Coronae Borealis	Northern crown	CrB	16	+30
Corvus	Corvi	Crow	Crv	12	−20
Crater	Crateris	Cup	Crt	11	−15
Crux	Crucis	Cross (southern)	Cru	12	−60
Cygnus	Cygni	Swan	Cyg	21	+40
Delphinus	Delphini	Porpoise	Del	21	+10
Dorado	Doradus	Swordfish	Dor	5	−65
Draco	Draconis	Dragon	Dra	17	+65
Equuleus	Equulei	Little horse	Equ	21	+10
Eridanus	Eridani	River	Eri	3	−20
Fornax	Fornacis	Furnace	For	3	−30
Gemini	Geminorum	Twins	Gem	7	+20
Grus	Gruis	Crane	Gru	22	−45
Hercules	Herculis	Hercules, son of Zeus	Her	17	+30
Horologium	Horologii	Clock	Hor	3	−60
Hydra	Hydrae	Sea serpent	Hya	10	−20
Hydrus	Hydri	Water snake	Hyi	2	−75
Indus	Indi	Indian	Ind	21	−55
Lacerta	Lacertae	Lizard	Lac	22	+45

CONSTELLATION (Latin name)	GENITIVE CASE ENDING	ENGLISH NAME OR DESCRIPTION	ABBRE-VIATION	APPROXI-MATE POSITION α	δ
				h	°
Leo	Leonis	Lion	Leo	11	+15
Leo Minor	Leonis Minoris	Little lion	LMi	10	+35
Lepus	Leporis	Hare	Lep	6	−20
Libra	Librae	Balance	Lib	15	−15
Lupus	Lupi	Wolf	Lup	15	−45
Lynx	Lyncis	Lynx	Lyn	8	+45
Lyra	Lyrae	Lyre or harp	Lyr	19	+40
Mensa	Mensae	Table Mountain	Men	5	−80
Microscopium	Microscopii	Microscope	Mic	21	−35
Monoceros	Monocerotis	Unicorn	Mon	7	−5
Musca	Muscae	Fly	Mus	12	−70
Norma	Normae	Carpenter's level	Nor	16	−50
Octans	Octantis	Octant	Oct	22	−85
Ophiuchus	Ophiuchi	Holder of serpent	Oph	17	0
Orion	Orionis	Orion, the hunter	Ori	5	+5
Pavo	Pavonis	Peacock	Pav	20	−65
Pegasus	Pegasi	Pegasus, the winged horse	Peg	22	+20
Perseus	Persei	Perseus, hero who saved Andromeda	Per	3	+45
Phoenix	Phoenicis	Phoenix	Phe	1	−50
Pictor	Pictoris	Easel	Pic	6	−55
Pisces	Piscium	Fishes	Psc	1	+15
Piscis Austrinus	Piscis Austrini	Southern fish	PsA	22	−30
Puppis*	Puppis	Stern of the Argonauts' ship	Pup	8	−40
Pyxis* (= Malus)	Pyxidus	Compass on the Argonauts' ship	Pyx	9	−30
Reticulum	Reticuli	Net	Ret	4	−60
Sagitta	Sagittae	Arrow	Sge	20	+10
Sagittarius	Sagittarii	Archer	Sgr	19	−25
Scorpius	Scorpii	Scorpion	Sco	17	−40
Sculptor	Sculptoris	Sculptor's tools	Scl	0	−30
Scutum	Scuti	Shield	Sct	19	−10
Serpens	Serpentis	Serpent	Ser	17	0
Sextans	Sextantis	Sextant	Sex	10	0
Taurus	Tauri	Bull	Tau	4	+15
Telescopium	Telescopii	Telescope	Tel	19	−50
Triangulum	Trianguli	Triangle	Tri	2	+30
Triangulum Australe	Trianguli Australis	Southern triangle	TrA	16	−65
Tucana	Tucanae	Toucan	Tuc	0	−65
Ursa Major	Ursae Majoris	Big bear	UMa	11	+50
Ursa Minor	Ursae Minoris	Little bear	VMi	15	+70
Vela*	Velorum	Sail of the Argonauts' ship	Vel	9	−50
Virgo	Virginis	Virgin	Vir	13	0
Volans	Volantis	Flying fish	Vol	8	−70
Vulpecula	Vulpeculae	Fox	Vul	20	+25

* The four constellations Carina, Puppis, Pyxis, and Vela originally formed the single constellation, Argo Navis.

Appendix 21

STAR MAPS

The star maps, one for each month, are printed on six removable sheets at the back of this book. To learn the stars and constellations, the sheet containing the map for the current month should be taken outdoors and compared directly with the sky. The maps were designed for a latitude of about 35° N but are useful anywhere in the continental United States. Each map shows the appearance of the sky at about 9:00 P.M. (Standard Time) near the middle of the month for which it is intended; near the beginning and end of the month, it shows the sky as it appears about 10:00 P.M. and 8:00 P.M., respectively. To use a map, hold the sheet vertically, and turn it so that the direction you are facing is shown at the bottom. The middle of the map corresponds to your zenith (the point overhead), so the stars and constellations in the lower part of the chart (with the printing upright) will match, approximately, the sky in front of you.

These star maps were originally prepared by C. H. Cleminshaw for the *Griffith Observer* (published by the Griffith Observatory, P.O. Box 27787, Los Angeles, California, 90027), and are reproduced here by the very kind permission of the Griffith Observatory, Edwin C. Krupp, Director.

INDEX

Note: Page numbers in **boldface** refer to principal discussions. Index does not include Appendices, which are easily consulted through the Table of Contents.

A

B

C

H

I

J

K

L

Q

R

S

T

U

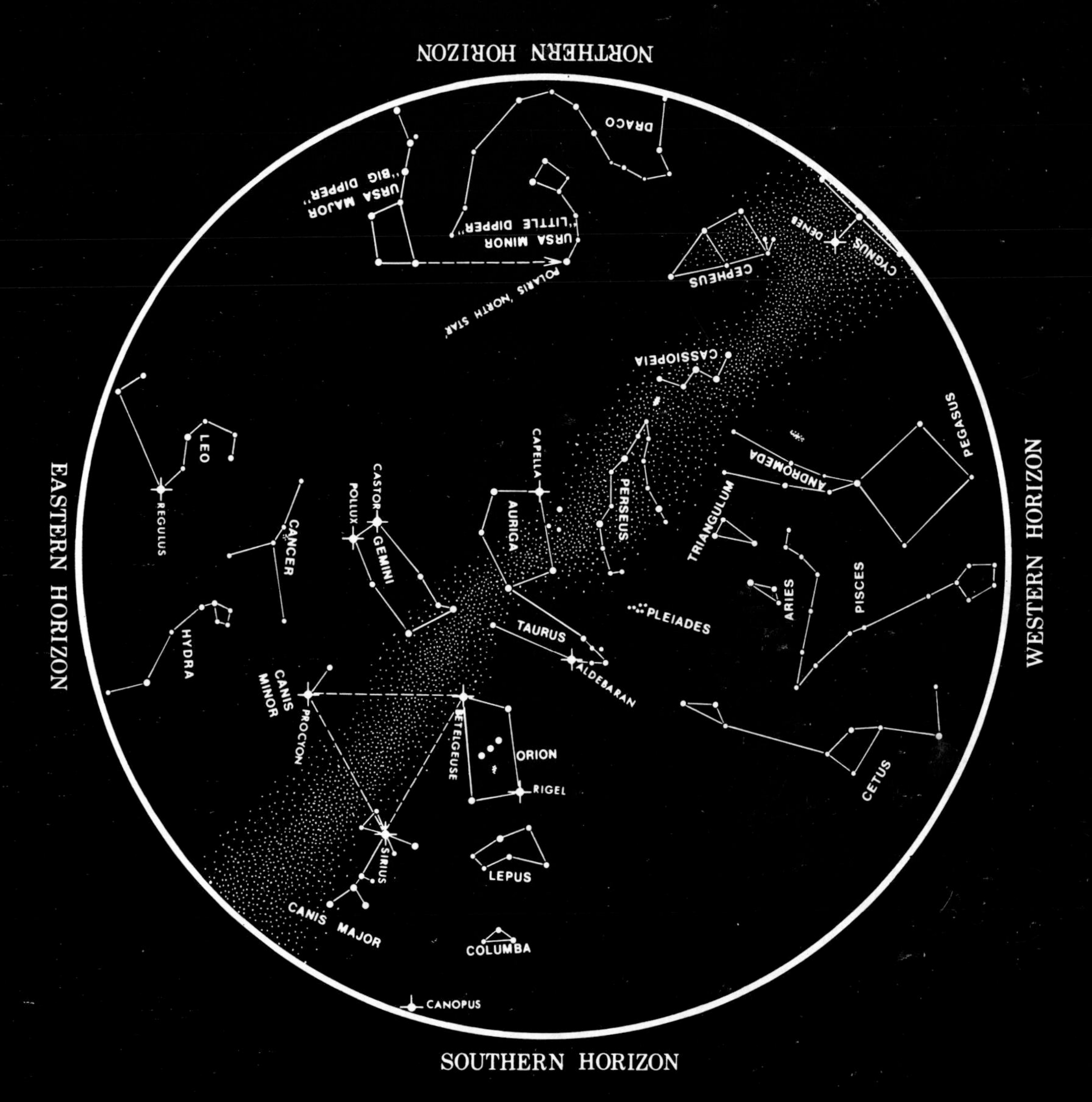

THE NIGHT SKY IN JANUARY

Latitude of chart is 34°N, but it is practical throughout the continental United States.

To use: Hold chart vertically and turn it so the direction you are facing shows at the bottom.

Chart time (Local Standard):
10 p.m. First of month
9 p.m. Middle of month
8 p.m. Last of month

Star Chart from *GRIFFITH OBSERVER*, Griffith Observatory, Los Angeles

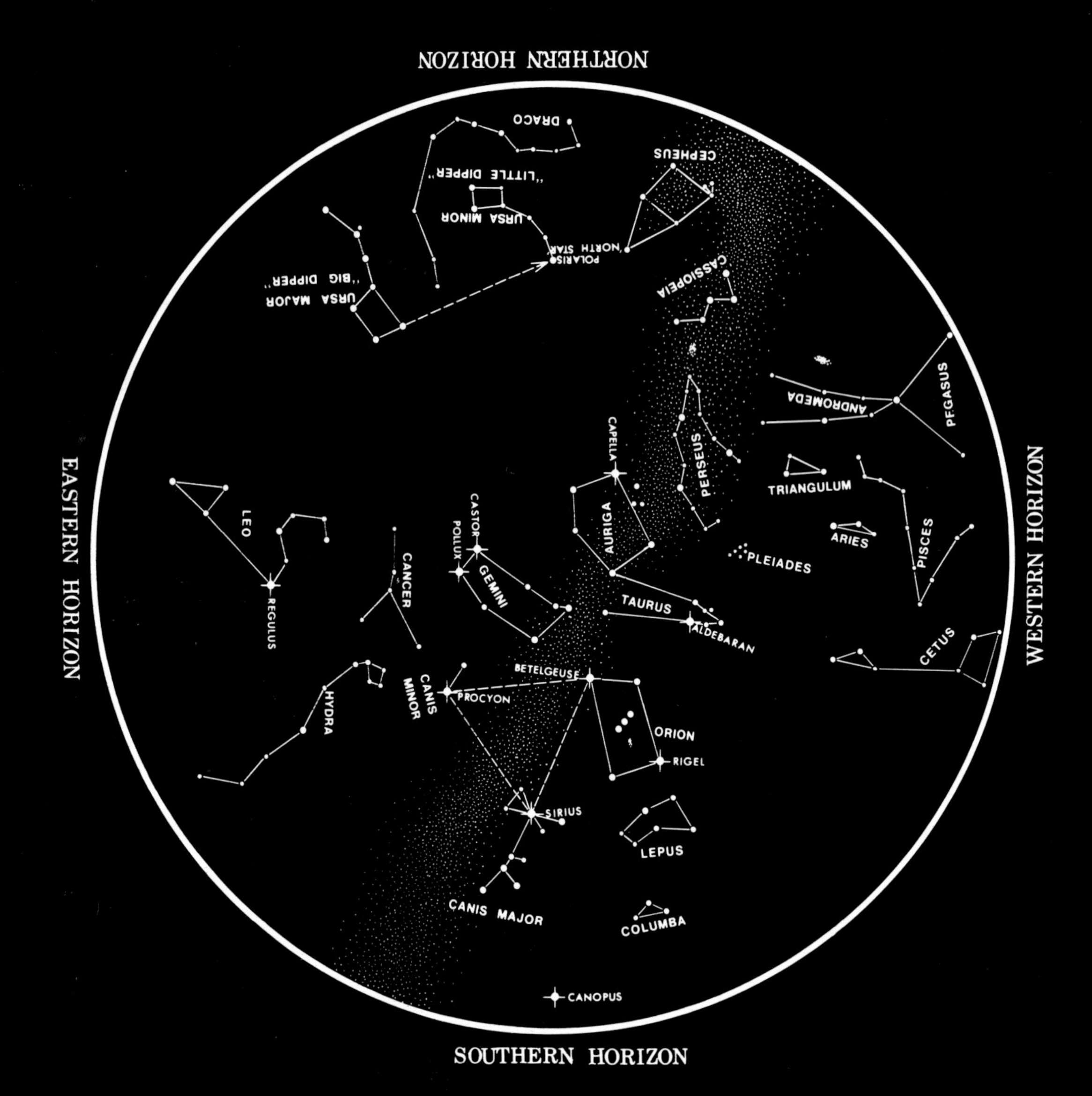

THE NIGHT SKY IN FEBRUARY

Latitude of chart is 34°N, but it is practical throughout the continental United States.

To use: Hold chart vertically and turn it so the direction you are facing shows at the bottom.

Chart time (Local Standard):
10 p.m. First of month
9 p.m. Middle of month
8 p.m. Last of month

Star Chart from *GRIFFITH OBSERVER*, Griffith Observatory, Los Angeles

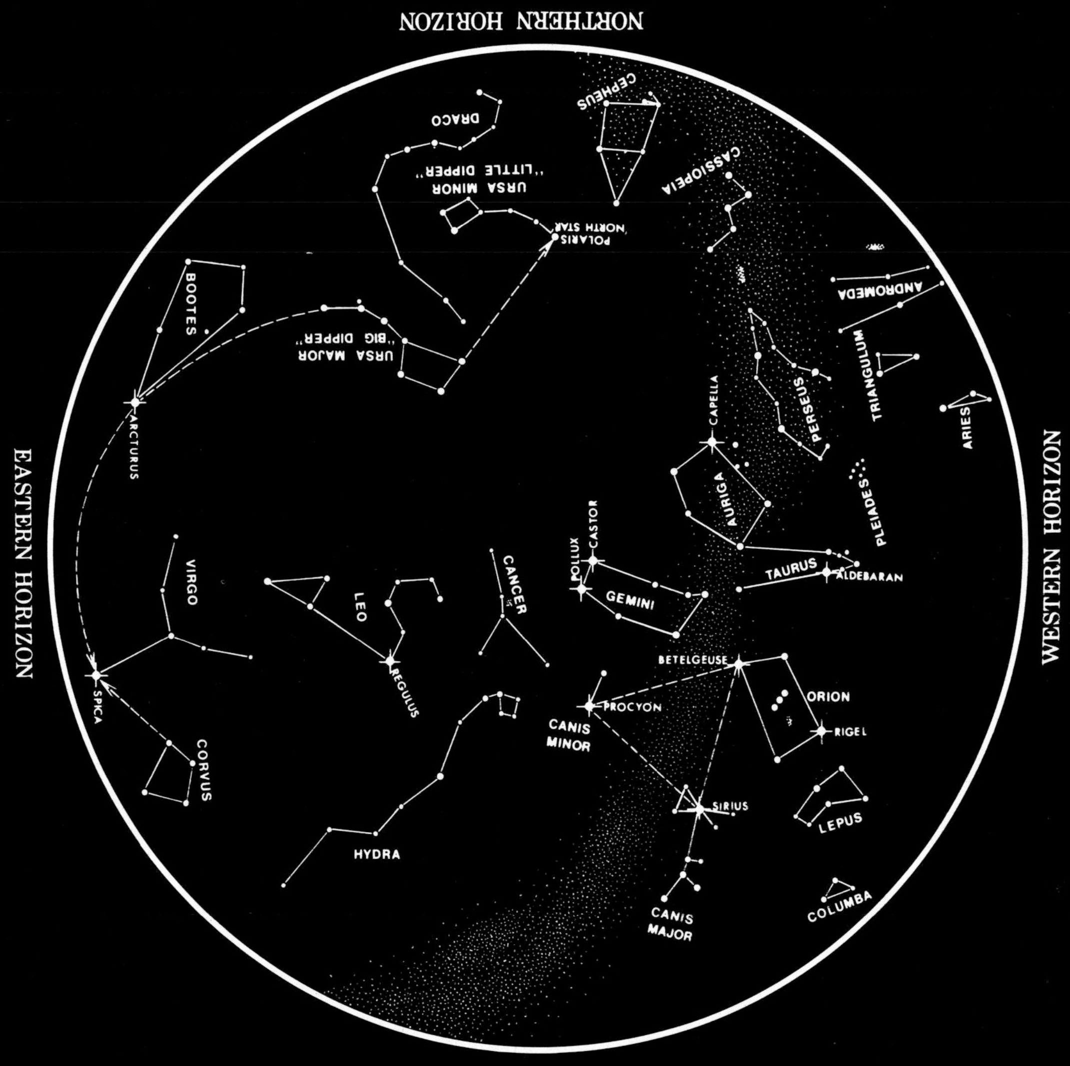

THE NIGHT SKY IN MARCH

Latitude of chart is 34°N, but it is practical throughout the continental United States.

To use: Hold chart vertically and turn it so the direction you are facing shows at the bottom.

Chart time (Local Standard):
10 p.m. First of month
9 p.m. Middle of month
8 p.m. Last of month

Star Chart from *GRIFFITH OBSERVER*, Griffith Observatory, Los Angeles

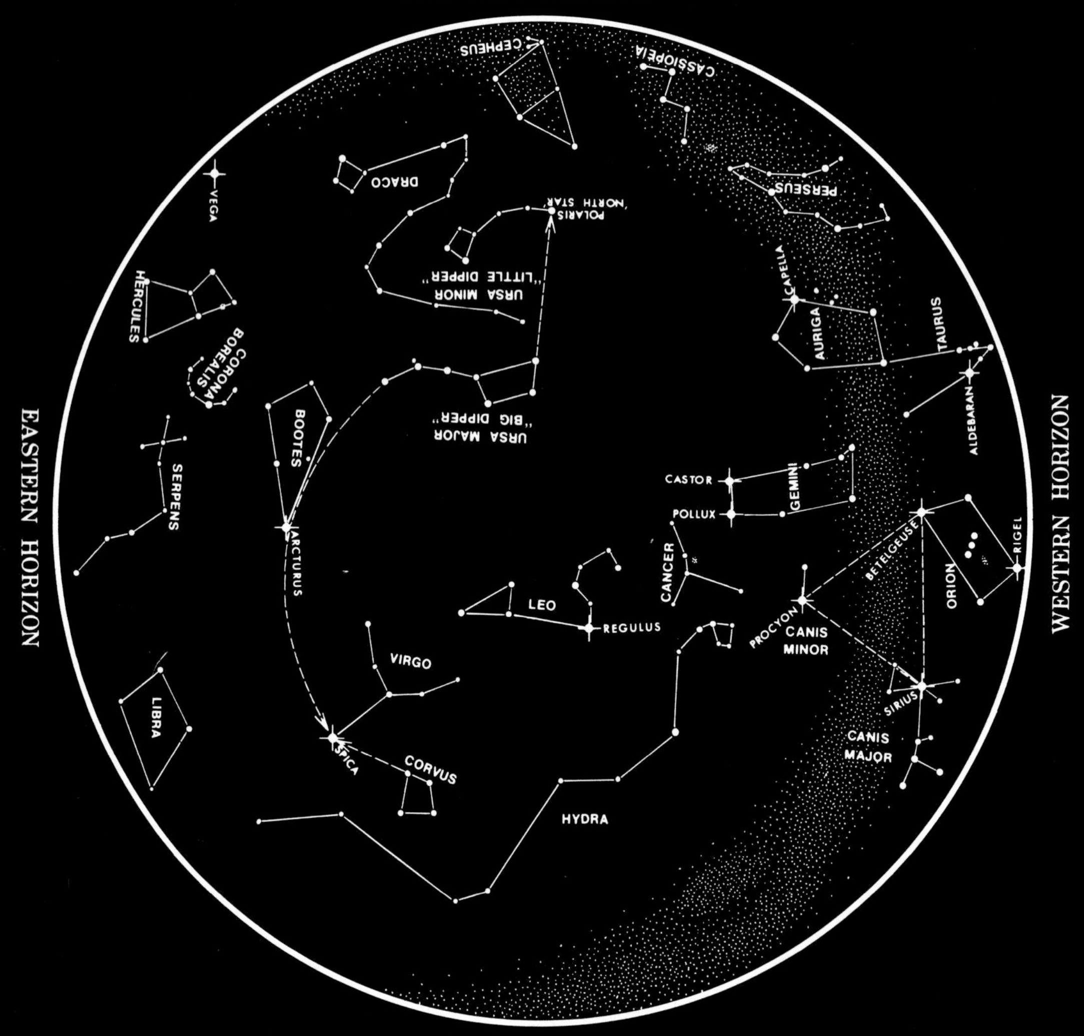

THE NIGHT SKY IN APRIL

Latitude of chart is 34°N, but it is practical throughout the continental United States.

To use: Hold chart vertically and turn it so the direction you are facing shows at the bottom.

Chart time (Local Standard):
10 p.m. First of month
9 p.m. Middle of month
8 p.m. Last of month

Star Chart from *GRIFFITH OBSERVER*, Griffith Observatory, Los Angeles

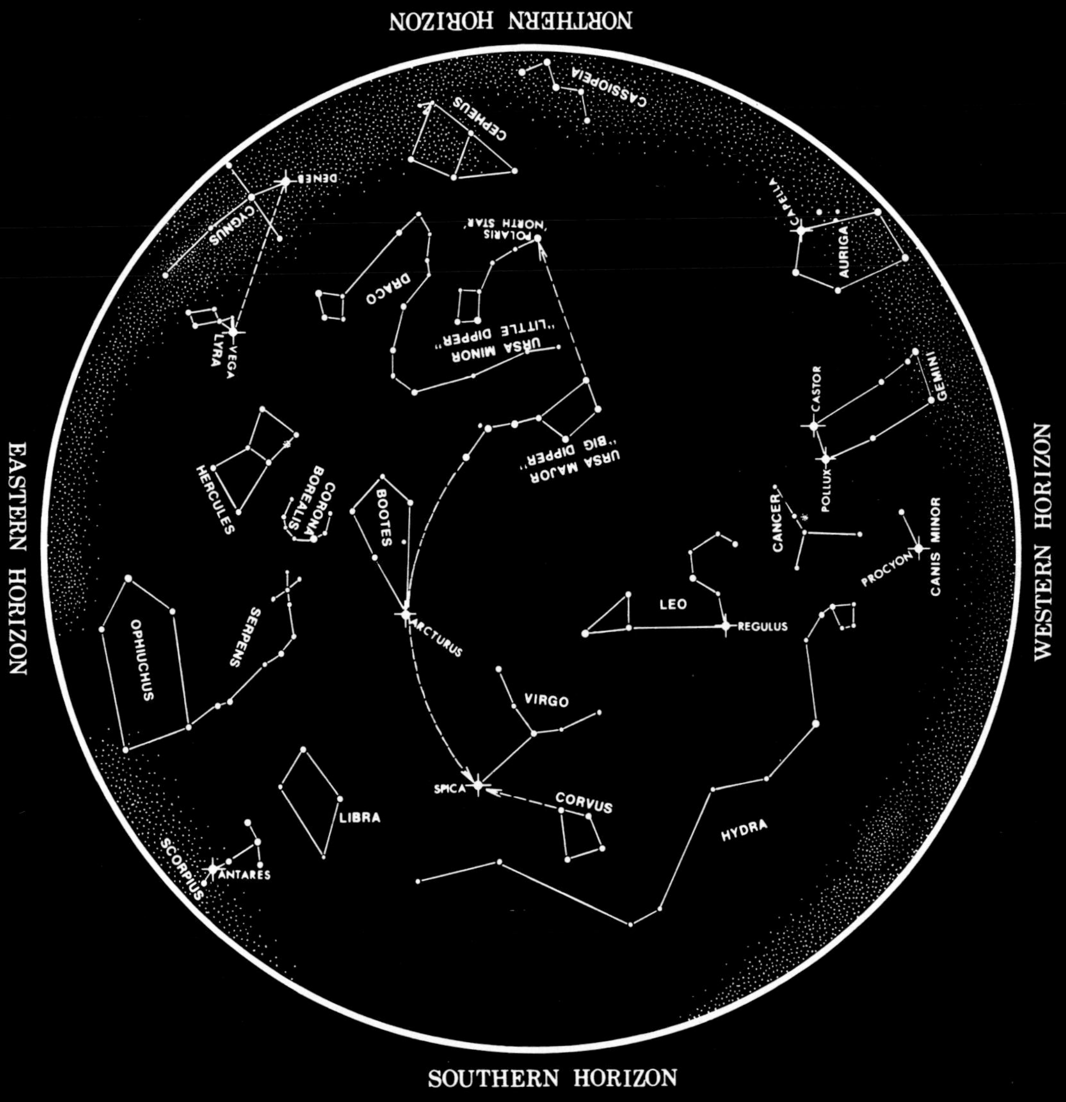

NORTHERN HORIZON

EASTERN HORIZON

WESTERN HORIZON

SOUTHERN HORIZON

THE NIGHT SKY IN MAY

Latitude of chart is 34°N, but it is practical throughout the continental United States.

To use: Hold chart vertically and turn it so the direction you are facing shows at the bottom.

Chart time (Local Standard):
10 p.m. First of month
9 p.m. Middle of month
8 p.m. Last of month

Star Chart from *GRIFFITH OBSERVER*, Griffith Observatory, Los Angeles

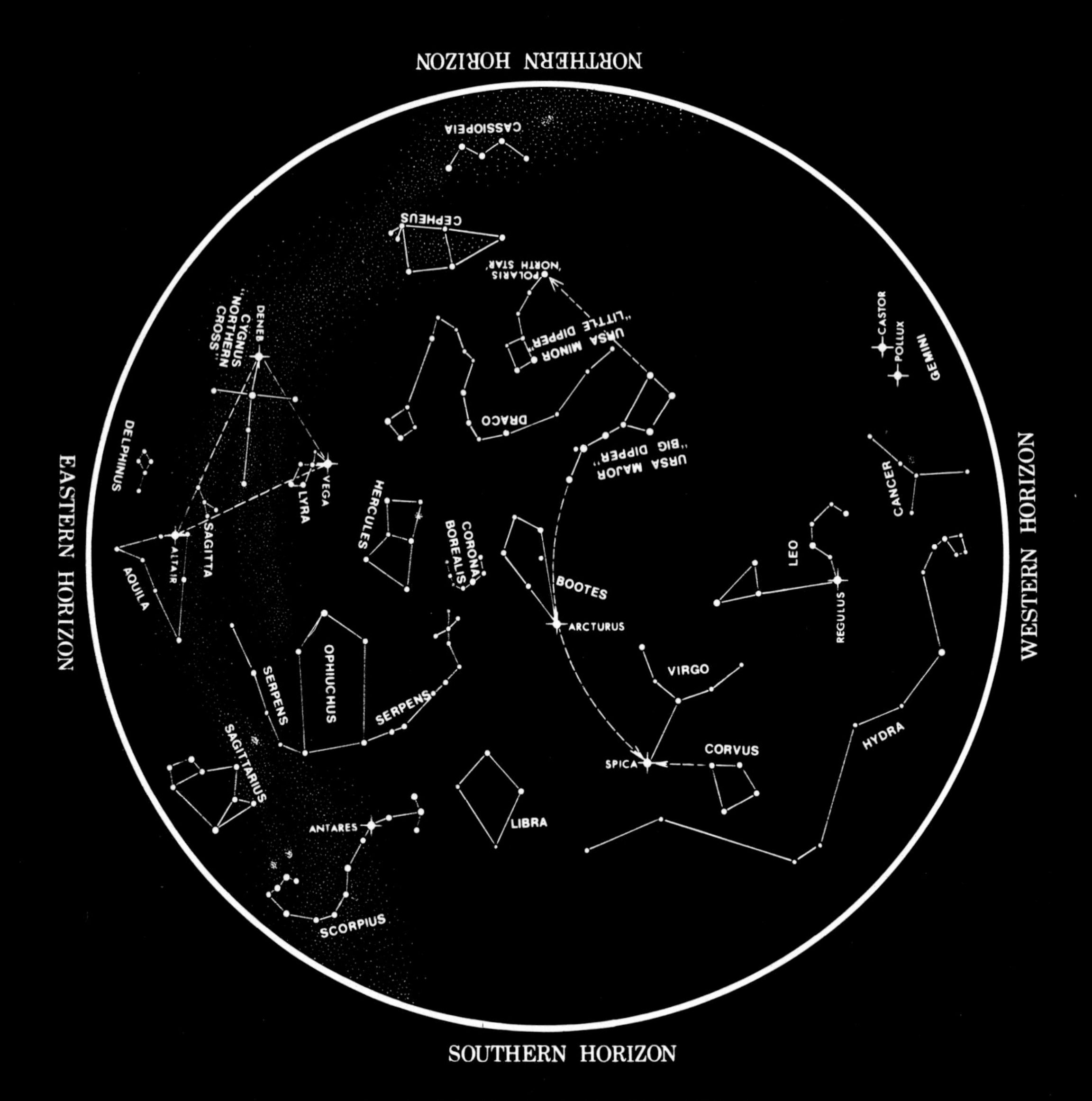

THE NIGHT SKY IN JUNE

Latitude of chart is 34°N, but it is practical throughout the continental United States.

To use: Hold chart vertically and turn it so the direction you are facing shows at the bottom.

Chart time (Local Standard):
10 p.m. First of month
9 p.m. Middle of month
8 p.m. Last of month

Star Chart from *GRIFFITH OBSERVER*, Griffith Observatory, Los Angeles

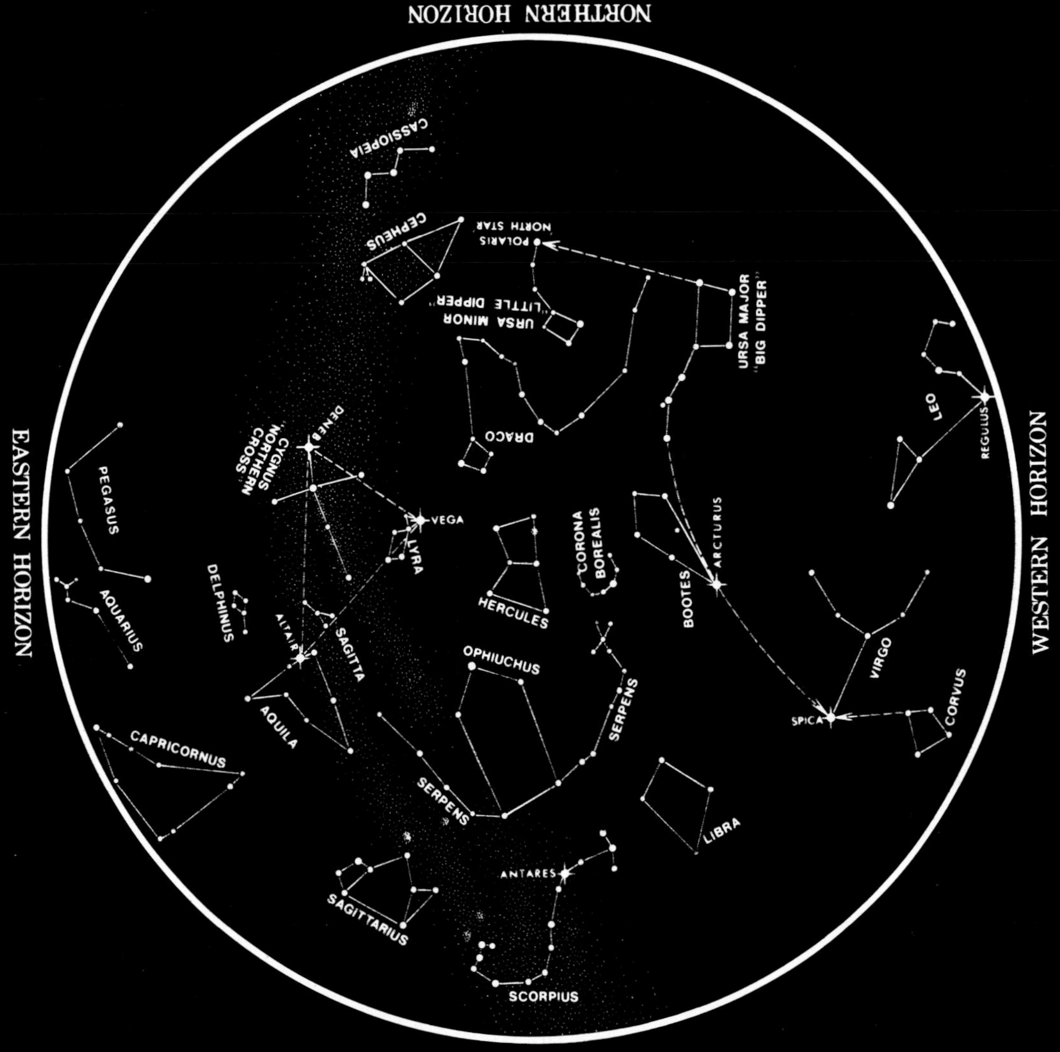

THE NIGHT SKY IN JULY

Latitude of chart is 34°N, but it is practical throughout the continental United States.

To use: Hold chart vertically and turn it so the direction you are facing shows at the bottom.

Chart time (Local Standard):

10 p.m. First of month
9 p.m. Middle of month
8 p.m. Last of month

Star Chart from *GRIFFITH OBSERVER*, Griffith Observatory, Los Angeles

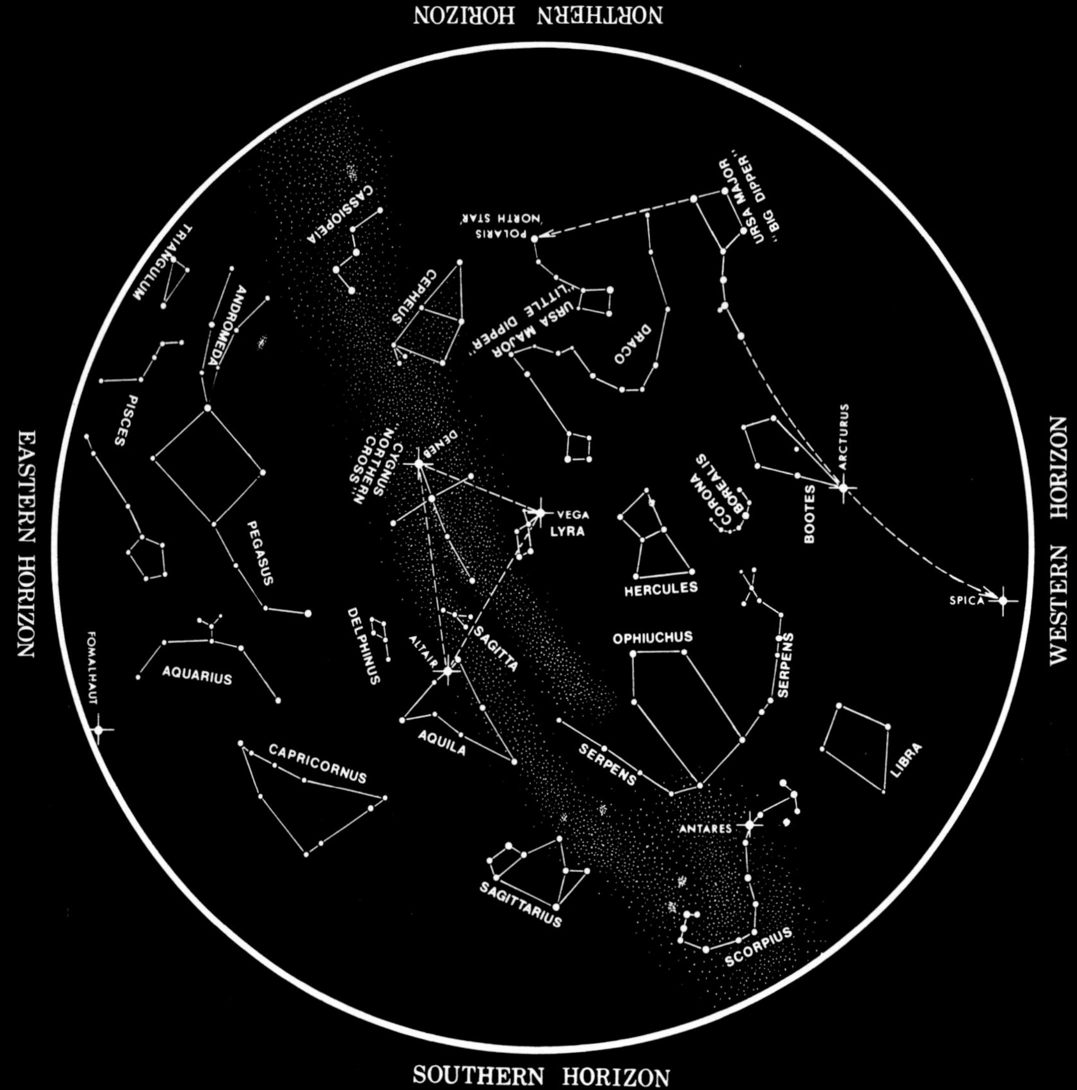

THE NIGHT SKY IN AUGUST

Latitude of chart is 34°N, but it is practical throughout the continental United States.

To use: Hold chart vertically and turn it so the direction you are facing shows at the bottom.

Chart time (Local Standard):
10 p.m. First of month
9 p.m. Middle of month
8 p.m. Last of month

Star Chart from *GRIFFITH OBSERVER*, Griffith Observatory, Los Angeles

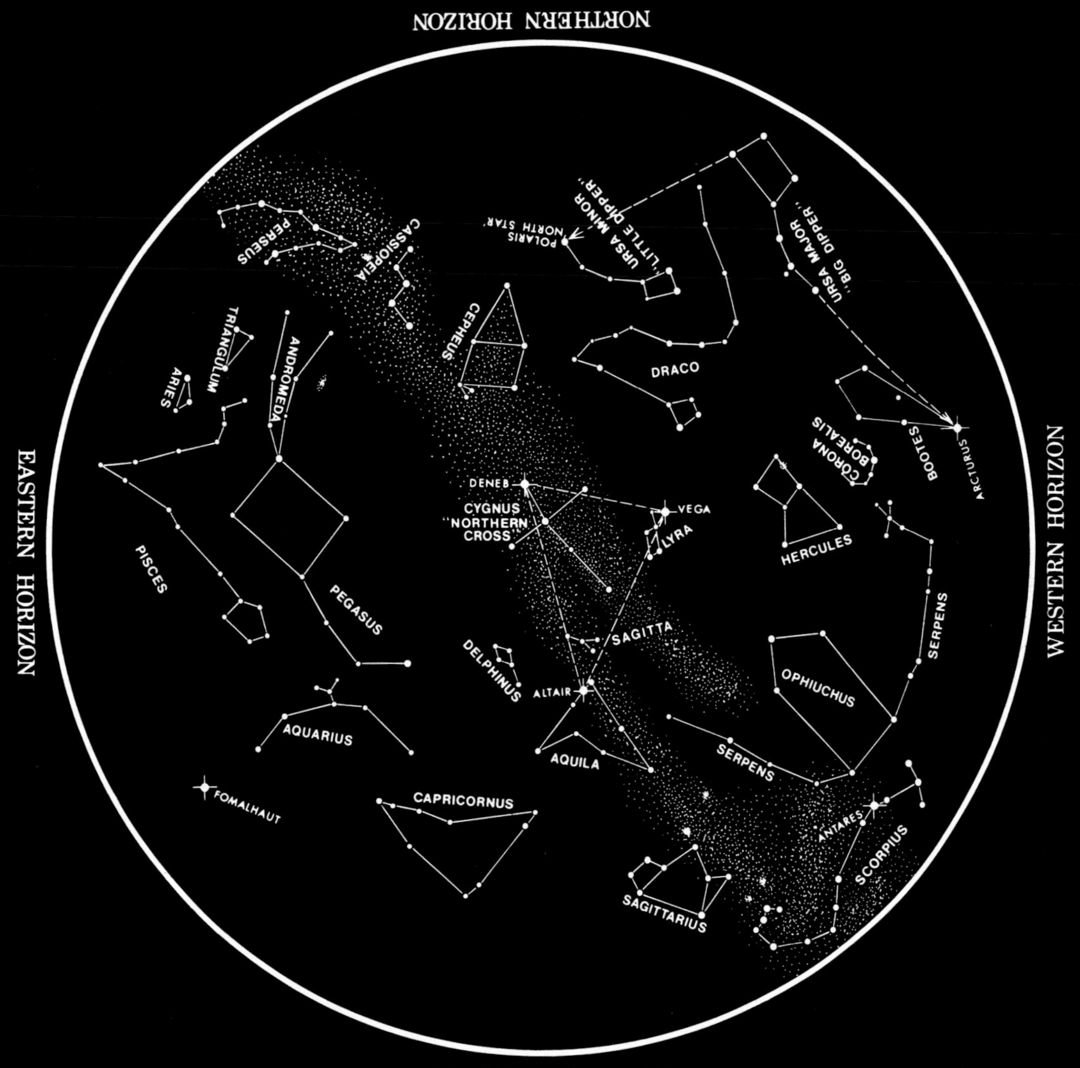

THE NIGHT SKY IN SEPTEMBER

Latitude of chart is 34°N, but it is practical throughout the continental United States.

To use: Hold chart vertically and turn it so the direction you are facing shows at the bottom.

Chart time (Local Standard):

10 p.m. First of month
9 p.m. Middle of month
8 p.m. Last of month

Star Chart from *GRIFFITH OBSERVER*, Griffith Observatory, Los Angeles

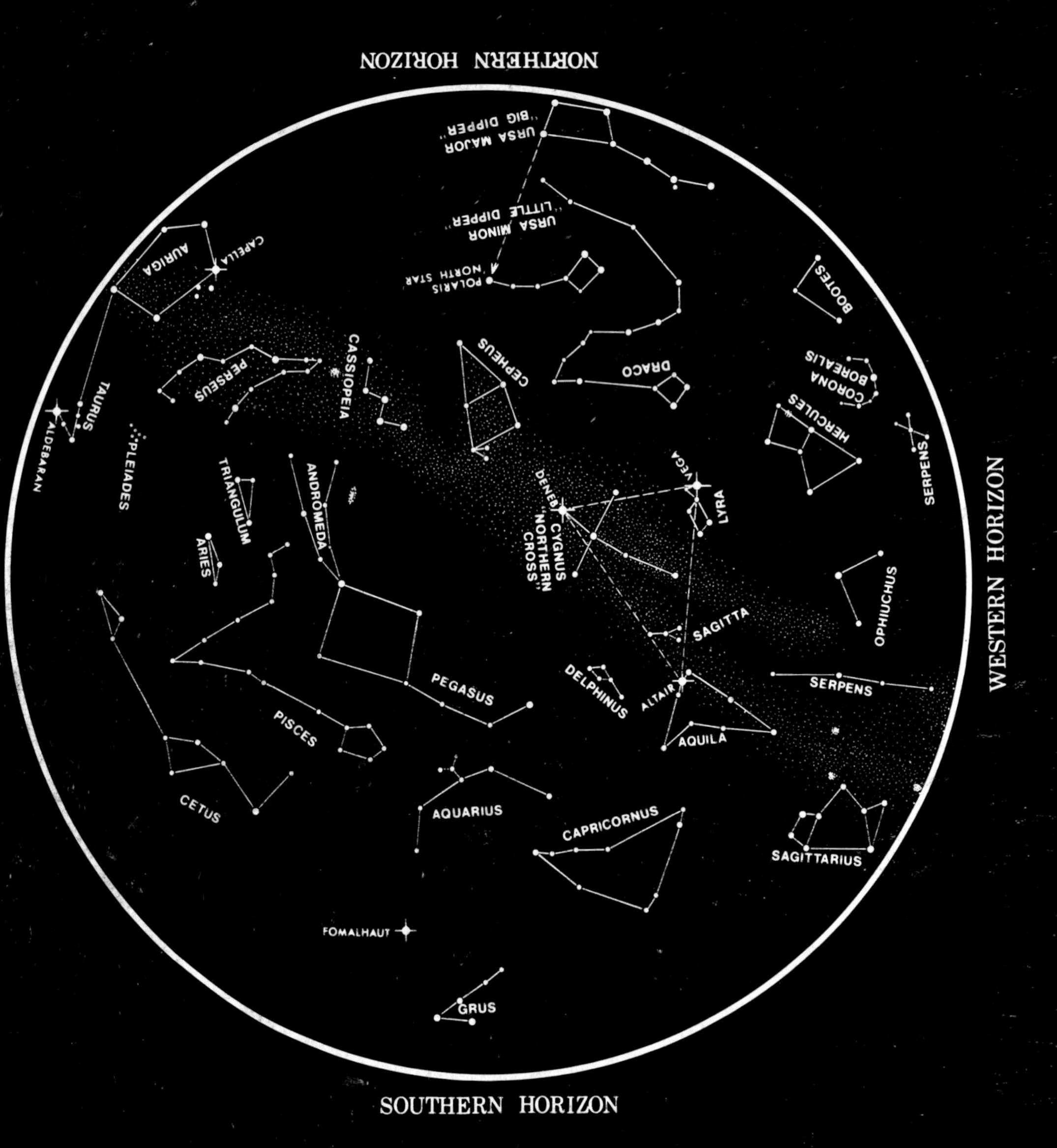

THE NIGHT SKY IN OCTOBER

Latitude of chart is 34° N, but it is practical throughout the continental United States.

To use: Hold chart vertically and turn it so the direction you are facing shows at the bottom.

Chart time (Local Standard):
10 p.m. First of month
9 p.m. Middle of month
8 p.m. Last of month

Star Chart from *GRIFFITH OBSERVER*, Griffith Observatory, Los Angeles